Mathematics: Theory & Applications

Series Editor
Nolan Wallach

Gabriel Daniel Villa Salvador

Topics in the Theory of Algebraic Function Fields

Birkhäuser
Boston • Basel • Berlin

Gabriel Daniel Villa Salvador
Centro de Investigación y de Estudios Avanzados del I.P.N.
Departamento de Control Automático
Col. Zacatenco, C.P. 07340
México, D.F.
México

Mathematics Subject Classification (2000): 11R58, 11R60, 14H05, 11G09, 11R32, 12F05, 12F10, 12F15, 11S20, 14H55, 11R37, 11R29, 14G10, 14G15, 14G50, 11S31, 11S20, 14H25, 12G05

Library of Congress Control Number: 2006927769

ISBN-10 0-8176-4480-6 e-IBSN 0-8176-4515-2
ISBN-13 978-0-8176-4480-2

Printed on acid-free paper.

Birkhäuser

Based on the original Spanish edition, *Introducción a la Teoría de las Functiones Algebraicas*, Fondo de Cultura Económica, México, 2003

Printed in the United States of America. (TXQ/MP)

9 8 7 6 5 4 3 2 1

www.birkhauser.com

To Martha, Sofía, and my father

Give a man a fish and you feed him for a day. Teach him how to fish and you feed him for a lifetime.
Lao Tse

He who is continually thinking things easy is sure to find them difficult.
Lao Tse

La educación es un seguro para la vida y un pasaporte para la eternidad.
(Education is an insurance for life and a passport for eternity.)
Aparisi y Guijarro

Preface

What are function fields, and what are they useful for? Let us consider a compact Riemann surface, that is, a surface in which every point has a neighborhood that is isomorphic to an open set in the complex field $\mathbb{C}$. Now assume the surface under consideration to be the Riemann sphere S^2; then the meromorphic functions defined in S^2, by which we mean functions from S^2 to $\mathbb{C} \cup \{\infty\}$ whose only singularities are poles, are precisely the rational functions $\frac{f(z)}{g(z)}$, where $f(z)$ and $g(z)$ are polynomials with coefficients in $\mathbb{C}$. These functions form a field $\mathbb{C}(z)$ called the field of rational functions in one variable over $\mathbb{C}$. In general, if R is a compact Riemann surface, let us consider the meromorphic functions defined on R. The set of such functions forms a field, which is called the field of meromorphic functions of R; it turns out that this field is a finite extension of $\mathbb{C}(z)$, or, in other words, *a field of algebraic functions of one variable over* $\mathbb{C}$.

Now, two Riemann surfaces are isomorphic as Riemann surfaces if and only if their respective fields of meromorphic functions are $\mathbb{C}$-isomorphic fields. This tells us that such Riemann surfaces are completely characterized by their fields of meromorphic functions.

In algebraic geometry, let us consider an arbitrary field k, and let C be a nonsingular projective curve defined on k. It turns out that the set of regular functions over C is a finite extension of the field $k(x)$ of rational functions over k. This field of regular functions on C is a *field of algebraic functions of one variable over* k.

The correspondence between curves and function fields is as follows. Assume k to be algebraically closed. If C is a nonsingular projective curve, consider the field $k(C)$ consisting of all regular functions in C. Conversely, for a given function field K/k (see Chapter 1), there exists a nonsingular projective curve C (which is unique up to isomorphism), such that $k(C)$ is k-isomorphic to K. On the other hand, the places (see Chapter 2) are in one-to-one correspondence with the points of C: to each point P of C we associate the maximal ideal m_P of the valuation ring ϑ_P.

There exists a third area of study in which function fields show up. This is number theory. Here a field of functions of one variable will play a role similar to that of a finite extension of the field $\mathbb{Q}$ of rational numbers. This is the point of view that we will be adopting in the course of this book.

The reader who is familiar with elementary number theory may consider that the field $k(x)$ of rational functions over k is the analogue of the rational field $\mathbb{Q}$, the polynomial ring $k[x]$ is the analogue of the ring of rational integers $\mathbb{Z}$, and finally that a field of functions of one variable is the analogue of a finite extension of $\mathbb{Q}$. It turns out that the analogy is much stronger when the field k is finite.

The mentioned analogy works in both directions. Oftentimes a problem that gets posed in number fields or, in other words, in finite extensions of $\mathbb{Q}$, admits an analogous problem in function fields, and the other way around. For example, if we consider the classical Riemann zeta function $\zeta(s)$, it is still unknown whether Riemann's conjecture on nontrivial zeros of $\zeta(s)$ holds (although a proof of its validity has been announced, this has not been confirmed yet). The analogue of this problem in function fields was solved by Weil in the middle of the last century (Chapter 7).

In a similar way, the classical theorem of Kronecker–Weber on abelian extensions of $\mathbb{Q}$ has its analogue in function fields. The Kronecker–Weber theorem establishes that any abelian extension of $\mathbb{Q}$ is contained in a cyclotomic extension. In other words, the maximal abelian extension of $\mathbb{Q}$ is the union of all its cyclotomic extensions. The analogue to this result is the theory of Carlitz–Hayes, which establishes, first of all, the analogues in function fields of the usual cyclotomic fields. The mere fact of adding roots of unity, as in the classical case, does not get us very far, since it would provide us only with what we shall call extensions of constants, which is far away from giving us all abelian extensions of a rational function field $k(T)$, where k is a finite field. The theory of Carlitz–Hayes (Chapter 12) provides us with the authentic analogue of cyclotomic fields, which leads us to the equivalent to the Kronecker–Weber theorem in function fields. This same theory may be generalized by considering not only $k(T)$ but also finite extensions. The study of this generalization gives as a result the so-called Drinfeld modules, or elliptic modules, as Drinfeld called them. A brief introduction to Drinfeld modules will be presented in Chapter 13.

In the other direction we have Iwasawa's theory in number fields. The origins of this theory are similar (in number fields) to considering a curve over a finite field and extending the field of constants k to its algebraic closure; in order to do this one must adjoin all roots of unity. In the number field case, adjoining all roots of unity gives a field too big, and for this reason one must consider only roots of unity whose order is a power of a given prime number. In this way, Iwasawa obtained the $\mathbb{Z}_p$-cyclotomic extensions of number fields, where $\mathbb{Z}_p$ is the ring of p-adic integers.

In the study of function fields, one may put the emphasis on the algebraic–arithmetic aspects or on the geometric–analytic ones. As Claude Chevalley rightly points out in his book [22], it is absolutely necessary to study both aspects of the theory, since each one has its own strengths in a natural way. However, even though both viewpoints may be treated in a textbook, one of them must be selected as the main focus of the book, since keeping both at the same time would be like superposing two photographs of the same object taken from different angles; the result would be a blurred and dull image of the object.

Our point of view in all the book will be the algebraic–arithmetic approach, and our principal interest will be the study of function fields as part of the algebraic theory of numbers. This by no means should be interpreted in the sense that we consider unimportant the analytic and the geometric approaches.

As we mentioned before, when the base field k of a function field is a finite field, the analogy between these fields and number fields is much closer. In this situation it is possible to define zeta functions, L-series, class numbers, etc. However, it must me stressed that there are fundamental differences between these two families of fields: the number fields have archimedean absolute values and the function fields do not (see Chapter 2); the ring of rational integers $\mathbb{Z}$ and the rational field $\mathbb{Q}$ are essentially unique, as opposed to polynomial rings $k[x]$ and rational function fields $k(x)$, which are respectively isomorphic to many rings and fields. Consequently, the situation of $\mathbb{Z}$ being contained in $\mathbb{Q}$ admits not only one analogue in function fields, but an infinity of them. Therefore, it is very important to keep in mind both aspects: the similarities between both families of fields as well as their fundamental differences.

This book may be used for a first-year graduate course on number theory. We tried to make it self-contained whenever possible, the only prerequisites being the following: a basic course in field theory; a first course in complex analysis; some basic knowledge of commutative algebra, say at the level of the Atiyah–Macdonald book [4]; and the mathematical maturity required to learn new concepts and relate them to known ones.

The first four chapters can be used for an introductory undergraduate course for mathematics majors, and Chapters 5, 6, 7, and 9 for a second course, avoiding the most technical parts, for instance the proofs of the Riemann hypothesis, Čebotarev's density theorem, the computation of the different, and Tate's genus formula.

The introductory chapter was written mainly to motivate the study of transcendental extensions, absolute values of $\mathbb{Q}$, and compact Riemann surfaces. However, in order to avoid making it long and tedious, we will establish the results needed for each topic at the moment they are required. The reason for this selection is as follows. A function field K over k is really just a finitely generated transcendental extension of k, with transcendence degree one. On the other hand, the study of such fields leads us to the study of their absolute values, whose analogues are, up to a certain point, the absolute values in $\mathbb{Q}$. Finally, compact Riemann surfaces constitute a splendid geometric representation of function fields. In the case of Riemann surfaces we shall not provide proofs of the presented results, since our interest is only that the reader know the fundamental results on compact Riemann surfaces, and use them as a motivation to study more general situations.

Chapter 2 is the introduction to our main objective. There, we define general concepts that will be necessary in the course of this volume, such as fields of constants, valuations, places, valuation rings, absolute values, etc. Once these concepts are mastered, we shall study the completions of a field with respect to an absolute value. The usefulness of the study of completions with respect to a metric is well known in the area of analysis. In our case, we shall use these completions as a basic tool for the study of the arithmetic properties of places in field extensions (Chapter 5). For this chapter it is convenient, but not necessary, that the reader be familiar with the completion of a metric space or at least with the standard completion of $\mathbb{Q}$ with respect to the usual absolute value obtaining the field of real numbers $\mathbb{R}$. We finish the chapter with Artin's approximation theorem, which can be considered as the generalization of the Chinese remainder theorem and which establishes the following: Given a finite number of absolute values and an equal number of elements of the field, we can find an element of the field that approximates the given elements in each absolute value as much as we want. Theorem 2.5.20 is the characterization of the completion of a function field.

Chapter 3 is dedicated to the famous Riemann–Roch theorem (Theorem 3.5.4 and corollaries) which is, without any doubt, the most important result of our book. The Riemann–Roch Theorem states the equality between dimensions of vector spaces, degree of a field extension and a very important field invariant: the genus. In order to establish the Riemann-Roch Theorem one requires various preliminary concepts, which will be defined in this chapter and will play a central role in the rest of the book: divisors, adeles or repartitions, Weil differentials, class groups, etc. The whole theory of function fields depends heavily on the Riemann–Roch theorem.

An important part of the work of any mathematician at any level is to develop and know examples concerning the topic on which he or she is working. Chapter 4 is dedicated to giving examples of the results found in Chapter 2 and 3. In the first two sections we present examples and characterize the function fields of genus 0 and 1 respectively, and in the last section we calculate the genus of a quadratic extension of a rational function field. Even though the genus can be found much more easily using the Riemann–Hurwitz genus formula (Theorem 9.4.2), the methods we use in this chapter are valuable by themselves.

Chapter 5 deals with Galois theory of function fields. After Chapter 3, this chapter can be considered as the second in importance. It is dedicated to the arithmetic of function fields (decomposition of places in the extensions, ramification, inertia, etc.). Here we study the relationship between the decomposition of places in an extension of function fields and the decomposition in the corresponding completions. Section 5.6 contains many technical details necessary to understand the notion of a different in an extension and the different in an extension of Dedekind domains, which is the way we study the arithmetic of number fields (Theorem 5.7.12). The last section of the chapter concerns the study of the different by means of the local differents (Theorem 5.7.21). The proof can be omitted without any loss of continuity. We end this chapter with an introduction to ramification groups.

Chapter 6 deals with congruence function fields, that is, function fields whose constant field is finite. As we said previously, the analogy between this kind of function fields and number fields is much closer. In this chapter we study zeta functions and L-series, as well as their functional equations.

Chapter 7 is dedicated to the Riemann hypothesis in function fields (Theorem 7.2.9). The proof that we present here is essentially due to Bombieri [7]. The reader can omit the details of the proof without any loss of continuity. As an application of the Riemann hypothesis we present an estimation on the number of prime divisors in a congruence function field, as well as the determination of the fields of class number 1.

Chapter 8 studies constant extensions in general, a particular case of which was seen in Chapter 6, namely the case that the constant field is finite. We have preferred to present first this special case for the readers that are interested in the most usual cases, that is, when the constant field is a perfect field, in order to avoid all the technical details of the general case. In this chapter we study the concepts of separability and of a separably generated field extension. We also study the genus change in this kind of extension and will see that the genus of the field decreases.

Chapter 9 concerns the Riemann–Hurwitz genus formula for geometric and separable extensions, which is probably the best technique for calculating the genus of an arbitrary function field. For inseparable extensions, Tate [152] used a substitute for the ordinary trace and found a genus formula for this type of extension. That substitute is the one used in the Riemann–Hurwitz formula. In Section 9.5, we present Tate's results. In the last section of the chapter, we revisit function fields of genus 0 and 1 and present the automorphism group of elliptic function fields. We conclude with hyperelliptic function fields, which will be used in Chapter 10 for cryptosystems.

In Chapter 10 we apply the theory of function fields, especially Chapter 6 and 7, to cryptography. We begin with a brief general introduction to cryptography: symmetric and asymmetric systems, public-key cryptosystems, the discrete logarithm problem, etc. Once these concepts are introduced we apply the theory of elliptic and hyperelliptic function fields to cryptosystems. In this way, we shall see that some groups that are determined by elliptic function fields, as well as some Jacobians, may be used both for public-key cryptosystems and for digital signatures and authentication.

Chapter 11 is a brief introduction to class field theory. We study Čebotarev's density theorem and briefly introduce profinite groups. Finally we present, without proofs, basic results of global as well as local class field theory. These results will be used in Chapter 12 to prove Hayes's theorem, which is analogous to the Kronecker–Weber theorem on the maximal abelian extension of a congruent function field, that is, a function field whose constant field is finite.

Chapter 12 is dedicated to the theory of cyclotomic function fields due to L. Carlitz and D. Hayes [15, 61]. We shall see that these fields are the analogue of the usual cyclotomic fields.

In Chapter 13 we give a brief introduction to Drinfeld, or elliptic, modules. The original objective of Drinfeld's module theory was to generalize the analogue of the Kronecker–Weber theorem to a function field over a general finite field, as well as complex multiplication and elliptic curves. We begin by presenting the Carlitz module, which is studied in Chapter 12 and is the simplest Drinfeld module. Using the analytic theory of exponential functions and lattices, we shall see that Drinfeld modules are ubiquitous. On the other hand, these modules provide us with an explicit class theory for general function fields over a finite field. We end the chapter with the application of Drinfeld modules to cryptography.

The last chapter is a study of the automorphism group of a function field. First we give a notion of differentiation due to H. Hasse and F. Schmidt [58] and then we use it to study the Wronskian determinant and Weierstrass points in characteristic p. We will see that the behavior in characteristic p is different from that in characteristic 0. We will use Weierstrass points to prove the classical result about the finiteness of the automorphism group of a function field K/k of genus larger than 1, where k is an algebraically closed field.

The appendix, which deals with group cohomology, is independent from the rest of the book. The reason why we decided to include it is that anyone interested in a further study of the arithmetic properties of function and local fields needs as a fundamental tool the cohomology of groups, particularly Theorem A.3.6.

Sometimes the way we present the topics is not the shortest possible, but since our main purpose was to write a textbook for graduate students, we chose to present particular cases first and later on give the general result. For instance, in Chapter 4 we state a formula for the genus of a quadratic extension of a rational function field and in Chapter 9 we present the Riemann–Hurwitz genus formula that generalizes what was done in Chapter 4. The same happens with the study of constant extensions.

It is important to specify that many of our results are a lot more general than what is presented here. For example, in Chapter 5 we study Galois theory of function fields, but most results hold for field extensions in general. Our motivation for emphasizing the particular case of function fields is to stress the beauty of this theory, independently of the fact that some of its particularities are really not particular but apply to the general case.

In order to limit the size of the book, we had to leave aside various topics such as the inverse Galois problem, topics in class field theory, the algebraic study of Riemann surfaces, holomorphic differentials, the Hasse–Witt theory, Jacobians, $\mathbb{Z}_p$-extensions, the Deuring–Šafarevič formula, etc.

The taste of this book is classical. We tried to preserve most of the original presentations. Our exposition owes a great deal to Deuring's monograph [28] and Chevalley's book [22].

There are many people to thank, but I will mention just a few of them. First of all, I am grateful to Professor Manohar Madan for teaching me this beautiful theory. I would like to thank Professors Martha Rzedowski Calderón and Fernando Barrera Mora for the time they spent doing a very careful reading of previous versions of this work, giving invaluable suggestions and correcting many errors. I also want to thank Ms. Anabel Lagos Cordoba and Ms. Norma Acosta Rocha for typing part of this book. I gratefully acknowledge Professor Simone Hazan for correcting the English version. I also thank Ms. Ann Kostant, executive editor of Birkhäuser Boston, and Mr. Craig Kavanaugh, assistant editor, for their support and interest in publishing this book. Thanks to David Kramer and to Elizabeth Loew for their great work on the final presentation of this book. Finally, many thanks to the Department of Automatic Control of CINVESTAV del Instituto Politécnico Nacional, for providing the necessary facilities for the making of this book. Part of the material was written during my sabbatical leave in the Mathematics Department of the Universidad Autónoma Metropolitana Iztapalapa. Part of this work was supported by CONACyT, project 36552-E.

México City,
November 2005

Gabriel D. Villa Salvador

Contents

Topics in the Theory of Algebraic Function Fields

1

Algebraic and Numerical Antecedents

In this introductory chapter we present three topics. The first one is the basic theory of transcendental fields, which is needed due to the fact that any function field is a finitely generated transcendental extension of a given field.

The second section is on distinct absolute values in the field of rational numbers $\mathbb{Q}$. In the development of number theory, it happens in a similar way as with continuous functions, that the "local" study of a field provides information on its "global" properties, and vice versa. The local structure of function fields and of number fields is closely related to that of the absolute values defined in them. We shall explore the existing parallelisms and differences between absolute values in $\mathbb{Q}$ and in rational function fields respectively.

The third topic of the chapter is Riemann surfaces, which serve as an infinite source of inspiration for a similar study, namely when the base field is completely arbitrary instead of being the complex field $\mathbb{C}$. Several concepts of a totally analytic nature such as those of differentials, distances, and meromorphic functions may be studied from an algebraic viewpoint and are consequently likely to be translated into arbitrary fields, including fields of positive characteristic.

We will not present here all prerequisites that will be needed in the rest of the book. Instead, these will be presented only at the moment they are necessary.

1.1 Algebraic and Transcendental Extensions

Definition 1.1.1. Let L/K be any field extension. A subset S of L is called *algebraically dependent (a. d.)* over K if there exist a natural number n, a nonzero polynomial $f(x_1, x_2, \ldots, x_n) \in K[x_1, x_2, \ldots, x_n]$ and n distinct elements $s_1, s_2, \ldots, s_n$ of S such that $f(s_1, s_2, \ldots, s_n) = 0$. If S is not algebraically dependent over K, it is called *algebraically independent (a. i.)* over K.

Example 1.1.2. Let $K[X, Y]$ be a polynomial ring of two variables over an arbitrary field K and let $f(X, Y) = X^2 - Y - 1$. Consider the field $L := K/(f(X, Y))$. Then $S := \{x\}$, where $x := X \bmod f(X, Y)$ is algebraically independent over K

and $T := \{x, y\}$, where $y := \mod f(X, Y)$ is algebraically dependent over K since $f(x, y) = 0$.

It is easy to see that if $S = \{s_1, s_2, \ldots, s_n\}$ is an algebraically independent set over K, then $K(s_1, s_2, \ldots, s_n)$ is isomorphic to the field $K(x_1, x_2, \ldots, x_n)$ of rational functions with n variables.

The algebraically independent sets can be ordered by inclusion, and applying Zorn's lemma, we can prove easily prove the existence of maximal algebraically independent sets.

Definition 1.1.3. Let L/K be a field extension. A *transcendental basis* of L over K is a maximal subset of L algebraically independent over K.

If S is a transcendental basis, it follows from the definition that L/K is algebraic if and only if S is the empty set.

Example 1.1.4. In Example 1.1.2 we have that $\{x\}$ and $\{y\}$ are transcendental basis of L over K.

Proposition 1.1.5. *Let L/K be a field extension, S an algebraically independent set over K, and $x \in L \setminus K(S)$. Then $S \cup \{x\}$ is algebraically independent over K if and only if x is transcendental over $K(S)$.*

Proof. Assume that $S \cup \{x\}$ is algebraically independent over K but x is not transcendental over $K(S)$. Then there exists a nonzero relation

$$f_n(s_1, \ldots, s_n) x^n + f_{n-1}(s_1, \ldots, s_n) x^{n-1} + \cdots + f_1(s_1, \ldots, s_n) x + f_0(s_1, \ldots, s_n) = 0$$

with $f_i(s_1, s_2, \ldots, s_n) \in K[s_1, s_2, \ldots, s_n]$. But this contradicts the fact that $S \cup \{x\}$ is algebraically independent

The proof of the converse is similar. □

Corollary 1.1.6. *Let L/K be a field extension and $S \subseteq L$ be an algebraically independent set. Then S is a transcendental basis over K if and only if $L/K(S)$ is an algebraic extension.* □

Corollary 1.1.7. *If $L/K(S)$ is an algebraic extension, then S contains a transcendental basis.* □

Theorem 1.1.8. *Any two transcendental bases have the same cardinality.*

Proof. Let S be a transcendental basis. First we assume that S is finite, say $S = \{s_1, s_2, \ldots, s_n\}$ with $|S| = n$. If T is any algebraically independent set, we will show that $|T| \leq n$. Let $\{x_1, x_2, \ldots, x_m\} \subseteq T$ be any finite subset of T and assume that $m \geq n$. By hypothesis, there exists a nonzero polynomial g_1 with $n + 1$ variables such that

$$g_1(x_1, s_1, s_2, \ldots, s_n) = 0.$$

Since $\{x_1\}$ and $\{s_1, s_2, \ldots, s_n\}$ are algebraically independent, it follows that x_1 and some s_i (say s_1) appear in g_1, so that s_1 is algebraic over $K(x_1, s_2, \ldots, s_n)$.

Repeating this process r times, $r < m$, and permuting the indices $s_2, \ldots, s_n$ if necessary, by induction on r we obtain that the field L is algebraic over $K(x_1, x_2, \ldots, x_r, s_{r+1}, \ldots, s_n)$. Therefore, there exists a nonzero polynomial g_2 with $n+1$ variables such that

$$g_2(x_{r+1}, x_1, \ldots, x_r, s_{r+1}, \ldots, s_n) = 0$$

and such that x_{r+1} appears in g_2. Since the x_i are algebraically independent, some s_j with $r+1 \leq j \leq n$ also appears in g_2. By permuting the indices if necessary, we may assume that s_{r+1} is the one that appears in g_2, that is, s_{r+1} is algebraic over

$$K\big(x_1, \ldots, x_r, x_{r+1}, s_{r+2}, \ldots, s_n\big),$$

so that L is algebraic over $K\big(x_1, \ldots, x_r, x_{r+1}, s_{r+2}, \ldots, s_n\big)$. Since the process can be repeated, it follows that we can replace the s's by x's and hence L is algebraic over $K(x_1, \ldots, x_n)$. This proves that $m = n$.

In short, if a given transcendental basis is finite, any other basis is also finite and has the same cardinality.

Now we assume that a transcendental basis S is infinite. The previous argument shows that any other basis is infinite. Let T be any other transcendental basis. For $s \in S$, there exists a finite set $T_s \subseteq T$ such that s is algebraic over $K(T_s)$. Since L is algebraic over $K(S)$ and S is algebraic over $K\left(\bigcup_{s\in S} T_s\right)$, it follows that L is algebraic over $K\left(\bigcup_{s\in S} T_s\right)$. Finally, since $\bigcup_{s\in S} T_s \subseteq T$, we have $\bigcup_{s\in S} T_s = T$, where T_s is a finite set.

Therefore $|T| \leq \sum_{s\in S} |T_s| \leq \aleph_0 |S| = |S|$. By symmetry we conclude that $|T| = |S|$. □

Definition 1.1.9. A field extension L/K is called *purely transcendental* if $L = K(S)$, where S is a transcendental basis of L over K. In this case, $K(S)$ is called a field of rational functions in $|S|$ variables over K.

Definition 1.1.10. Let L/K be a field extension. The cardinality of any transcendental basis of L over K is called the *transcendental degree of L over K* and is denoted by $\operatorname{tr} L/K$.

Example 1.1.11. In Examples 1.1.2 and 1.1.4 we have that the transcendental degree of L/K is 1 since $K(x)/K$ is purely transcendental and $L/K(x)$ is algebraic ($y^2 = x - 1$).

Proposition 1.1.12. *If $K \subseteq L \subseteq M$ is a tower of fields, then* $\operatorname{tr} M/K = \operatorname{tr} M/L + \operatorname{tr} L/K$. □

1.2 Absolute Values over $\mathbb{Q}$

Definition 1.2.1. Let k be any field. An *absolute value* over k is a function $\varphi : k \longrightarrow \mathbb{R}$, $\varphi(a) = |a|$, satisfying:

(i) $|a| \geq 0$ for all $a \in k$, and $|a| = 0$ if and only if $a = 0$,
(ii) $|ab| = |a||b|$ for all a and $b \in k$,
(iii) $|a + b| \leq |a| + |b|$ for all a and $b \in k$.

Note that if $|\ |$ is an absolute value then $|1| = 1$ and $|-x| = |x|$ for all $x \in K$ (Exercise 1.4.10).

The usual absolute value in $\mathbb{Q}$ is the most immediate example of the previous definition. Also, for any field k, the *trivial absolute value* is defined by $|a| = 1$ for $a \neq 0$ and $|0| = 0$.

Example 1.2.2. Let $p \in \mathbb{Z}$ be a prime number. For each nonzero $x \in \mathbb{Q}$, we write $x = p^n \frac{a}{b}$ with $p \nmid ab$ and $n \in \mathbb{Z}$. Let $|x|_p = p^{-n}$ and $|0| = 0$. We leave to the reader to verify that this defines an absolute value over $\mathbb{Q}$. It is called the *p-adic absolute value*, and it satisfies

$$|x + y|_p \leq \max\left\{|x|_p, |y|_p\right\}$$

for all $x, y \in \mathbb{Q}$. An absolute value with this last property is called *nonarchimedean.* We note that $\lim_{n\to\infty} |p^n|_p = 0$.

Definition 1.2.3. An absolute value $|\ | : k \longrightarrow \mathbb{R}$, is called *nonarchimedean* if $|a + b| \leq \max\{|a|, |b|\}$ for all $a, b \in k$. Otherwise, $|\ |$ is called *archimedean.*

Definition 1.2.4. Two nontrivial absolute values $|\ |_1$ and $|\ |_2$ over a field k are called *equivalent* if $|a|_1 < 1$ implies $|a|_2 < 1$ for all $a \in k$.

The relation given in Definition 1.2.4 is obviously reflexive and transitive. We also have the following result:

Proposition 1.2.5. *For any two nontrivial equivalent absolute values $|\ |_1$ and $|\ |_2$, we have $|a|_2 < 1$ whenever $|a|_1 < 1$, that is, the relation is symmetric. Therefore the relation defined above is an equivalence relation.*

Proof. Let $|a|_2 < 1$. If $|a|_1 > 1$, we have $\left|a^{-1}\right|_1 = |a|_1^{-1} < 1$. Therefore $\left|a^{-1}\right|_2 = |a|_2^{-1} < 1$, which is impossible. Hence $|a|_1 \leq 1$. If $|a|_1 = 1$, let $b \in k$ be such that $0 < |b|_1 < 1$. Such a b exists since $|\ |_1$ is nontrivial. Now $\left|ba^{-n}\right|_1 = |b|_1|a|_1^{-n} = |b|_1 < 1$. Thus $\left|ba^{-n}\right|_2 = |b|_2|a|_2^{-n} < 1$. Therefore $|b|_2^{1/n} < |a|_2$, which implies that

$$1 = \lim_{n\to\infty} |b|_2^{1/n} \leq |a|_2 < 1,$$

a contradiction that proves $|a|_1 < 1$. □

Remark 1.2.6. If $|\ |_1$ and $|\ |_2$ are two absolute values and $|a|_1 < 1$ implies $|a|_2 < 1$, then if $|\ |_1$ is nontrivial, $|\ |_2$ is nontrivial. Indeed, if $b \in k$ is such that $0 < |b|_1 < 1$, then we have $0 < |b|_2 < 1$.

From this point on all absolute values under consideration will be nontrivial.

Theorem 1.2.7. *Let $|\ |_1$ and $|\ |_2$ be two equivalent absolute values. Then there exists a positive real number c such that $|a|_1 = |a|_2^c$ for all $a \in k$.*

Proof. Let $0 < |b|_1 < 1$, so that $0 < |b|_2 < 1$. Put

$$c = \frac{\ln |b|_1}{\ln |b|_2}.$$

We have $|b|_1 = |b|_2^c$ with $c > 0$ and $c \in \mathbb{R}$. Now let $a \in k, a \neq 0$ and let $|a|_1 = |b|_1^r$ for some $r \in \mathbb{R}$. Let $\alpha_n, \beta_n \in \mathbb{Z}$, $\beta_n > 0$, be such that

$$\frac{\alpha_n}{\beta_n} \leq r \quad \text{and} \quad \lim_{n\to\infty} \frac{\alpha_n}{\beta_n} = r.$$

Then, since $|b|_1 < 1$, we have

$$|a|_1 = |b|_1^r \leq |b|_1^{\alpha_n/\beta_n},$$

that is,

$$\left|a^{\beta_n} b^{-\alpha_n}\right|_1 \leq 1,$$

so that $\left|a^{\beta_n} b^{-\alpha_n}\right|_2 \leq 1$, which implies that

$$|a|_2 \leq |b|_2^{\alpha_n/\beta_n}.$$

Therefore we have $|a|_2 \leq |b|_2^r$.

Now taking $\frac{\alpha_n}{\beta_n} \geq r$, it can be shown in a similar fashion that $|a|_2 \geq |b|_2^r$. Therefore $|a|_1 = |b|_1^r = |b|_2^{cr} = |a|_2^c$. □

Corollary 1.2.8. *If $|\ |_1$ and $|\ |_2$ are two equivalent absolute values in a field k, they define the same topology in k.* □

Proposition 1.2.9. *Let k be a field, and M the subring of k generated by 1, that is, $M = \{n \times 1 \mid n \in \mathbb{Z}\}$. Let $|\ |$ be an absolute value in k. Then $|\ |$ is nonarchimedean if and only if $|\ |$ is bounded in M.*

Proof. If $|\ |$ is nonarchimedean, we have for $n \in \mathbb{Z}$, $n > 0$,

$$|n \times 1| = |1 + \cdots + 1| \leq \max\{|1|, \ldots, |1|\} = |1| = 1,$$

and for $n \in \mathbb{Z}$, $n < 0$,

$$|n \times 1| = |-n \times 1| \leq |1| = 1,$$

so $|\ |$ is bounded in M.

Now assume that $|\ |$ is bounded in M, say $|m \times 1| < s$ for all $m \in \mathbb{Z}$. If $a, b \in k$ and $n \in \mathbb{N}$, we have

$$|a+b|^n = \left|\sum_{i=0}^{n}\binom{n}{i}a^i b^{n-i}\right| \leq \sum_{i=0}^{n}\left|\binom{n}{i}\right| |a|^i\,|b|^{n-i}$$
$$\leq s\sum_{i=0}^{n}|a|^i|b|^{n-i} \leq s(n+1)|a|^n,$$

where it is assumed that $|a| = \max\{|a|, |b|\}$.

Hence

$$|a+b| \leq s^{1/n}\sqrt[n]{n+1}\;|a| \xrightarrow[n\to\infty]{} |a| = \max\{|a|, |b|\},$$

and $|\;|$ is nonarchimedean. □

Corollary 1.2.10. *Every absolute value in a field of positive characteristic is nonarchimedean.* □

We finish this section characterizing the absolute values over the field of rational numbers.

Theorem 1.2.11 (Ostrowski). *Let φ be an absolute value in $\mathbb{Q}$. Then φ is trivial or it is equivalent to the usual absolute value or it is equivalent to some p-adic absolute value.*

Proof. Let φ be a nontrivial absolute value. Let us assume that there exists $n \in \mathbb{N}$, $n > 1$, such that $\varphi(n) \leq 1$. For $m \in \mathbb{N}$, we write

$$m = a_0 + a_1 n + \cdots + a_r n^r$$

with $0 \leq a_i \leq n-1$, $a_r \neq 0$. Now

$$\varphi\,(a_i) = \varphi(1+\cdots+1) \leq \varphi(1)+\cdots+\varphi(1) = a_i < n,$$

so

$$\varphi(m) \leq \sum_{i=0}^{r}\varphi\left(a_i n^i\right) = \sum_{i=0}^{r}\varphi\,(a_i)\,\varphi\,(n)^i < n\sum_{i=0}^{r}1 = n(1+r).$$

Since $m \geq n^r$, we have

$$r \leq \frac{\ln m}{\ln n} \quad \text{and} \quad \varphi(m) < \left(1+\frac{\ln m}{\ln n}\right)n.$$

Applying the above to m^s, $s \in \mathbb{N}$, we have

$$\varphi(m)^s = \varphi\left(m^s\right) < \left(1+\frac{\ln m^s}{\ln n}\right)n = \left(1+s\frac{\ln m}{\ln n}\right)n,$$

which implies

$$\varphi(m) < \left(1 + s\frac{\ln m}{\ln n}\right)^{1/s} n^{1/s} \xrightarrow[s\to\infty]{} 1.$$

We have shown that $\varphi(m) \leq 1$, so φ is bounded in $\mathbb{Z}$ and φ is nonarchimedean.

Let $\mathfrak{A} = \{m \in \mathbb{Z} \mid \varphi(m) < 1\}$. It can be verified that $\mathfrak{A}$ is an ideal. Now if $ab \in \mathfrak{A}$, then $\varphi(ab) = \varphi(a)\varphi(b) < 1$, so $\varphi(a) < 1$ or $\varphi(b) < 1$. Therefore $\mathfrak{A}$ is a prime ideal. Let $\mathfrak{A} = (p)$, where p is prime and $\varphi(p) < 1$. Let $c \in \mathbb{R}$, $c > 0$ be such that $\varphi(p) = p^{-c}$. If $m \notin \mathfrak{A}$, we have $p \nmid m$ and $\varphi(m) = 1$. Therefore, for

$$x \in \mathbb{Q} \quad \text{such that} \quad x = p^n \frac{a}{b}$$

with $p \nmid ab$, we have

$$\varphi(x) = \varphi(p)^n \frac{\varphi(a)}{\varphi(b)} = \varphi(p)^n = p^{-cn} = |x|_p^c,$$

so φ is equivalent to $|\ |_p$.

Now we assume that $\varphi(n) > 1$ for $n \in \mathbb{N}$, $n > 1$. Let $m, n \in \mathbb{Z}$, $m, n > 1$, and put

$$m^t = a_0 + a_1 n + \cdots + a_r n^r, \quad \text{where} \quad 0 \leq a_i \leq n-1, \quad a_r \neq 0.$$

We have $r \leq \frac{\ln m^t}{\ln n}$. Now we have

$$\begin{aligned}\varphi\left(m^t\right) = \varphi(m)^t &\leq \sum_{i=0}^{r} \varphi\left(a_i\right) \varphi\left(n\right)^i < \sum_{i=0}^{r} n\varphi\left(n\right)^r = n(1+r)\varphi(n)^r \\ &\leq n\left(1 + \frac{\ln m^t}{\ln n}\right)\varphi(n)^{(\ln m^t)/(\ln n)}.\end{aligned}$$

Therefore,

$$\begin{aligned}\varphi(m) &\leq n^{1/t}\left(1 + t\frac{\ln m}{\ln n}\right)^{1/t} \varphi(n)^{(1/t)((\ln m^t)/(\ln n))} \\ &= n^{1/t}\left(1 + t\frac{\ln m}{\ln n}\right)^{1/t} \varphi(n)^{(\ln m)/(\ln n)} \xrightarrow[t\to\infty]{} \varphi(n)^{(\ln m)/(\ln n)}.\end{aligned}$$

That is, $\varphi(m) \leq \varphi(n)^{(\ln m)/(\ln n)}$ or, equivalently,

$$\varphi(m)^{1/(\ln m)} \leq \varphi(n)^{1/(\ln n)}.$$

By symmetry we obtain $\varphi(m)^{1/(\ln m)} = \varphi(n)^{1/(\ln n)}$. Let $c \in \mathbb{R}$, $c > 0$, be such that $\varphi(m)^{1/(\ln m)} = e^c$ for all $m \in \mathbb{Z}$ such that $m > 1$.

We have $\varphi(m) = e^{c \ln m} = e^{\ln m^c} = m^c = |m|^c$ for all $m > 1$, $m \in \mathbb{Z}$.

For $m = 1$, $\varphi(1) = 1 = 1^c$.

For $m = 0$, $\varphi(0) = 0 = |0|^c$.

For $m < 0, m \in \mathbb{Z}$, $\varphi(m) = \varphi(-m) = |-m|^c = |m|^c$.

Finally, let $x \in \mathbb{Q}$ such that $x = \frac{a}{b}$. We have

$$\varphi(x) = \frac{\varphi(a)}{\varphi(b)} = \frac{|a|^c}{|b|^c} = |x|^c.$$

Therefore $\varphi(x) = |x|^c$ for all $x \in \mathbb{Q}$. This shows that φ is equivalent to $|\ |$, the usual absolute value of $\mathbb{Q}$. □

1.3 Riemann Surfaces

First we recall the definition of a Riemann surface.

Definition 1.3.1. Let R be a connected Hausdorff topological space. Then R is called a *Riemann surface* if there exists a collection $\{U_i, \Phi_i\}_{i\in I}$, such that:

(i) $\{U_i\}_{i\in I}$ is an open cover of R and $\Phi_i : U_i \longrightarrow \mathbb{C}$ is a homeomorphism over an open set of the complex plane $\mathbb{C}$ for each $i \in I$.

(ii) For every pair (i, j) such that $U_i \cap U_j \neq \emptyset$, $\Phi_j \Phi_i^{-1}$ is a conformal transformation of $\Phi_i\left(U_i \cap U_j\right)$ onto $\Phi_j\left(U_i \cap U_j\right)$.

In other words, a Riemann surface is a manifold that is obtained by gluing in a biholomorphic way neighborhoods that are homeomorphic to open sets of $\mathbb{C}$.

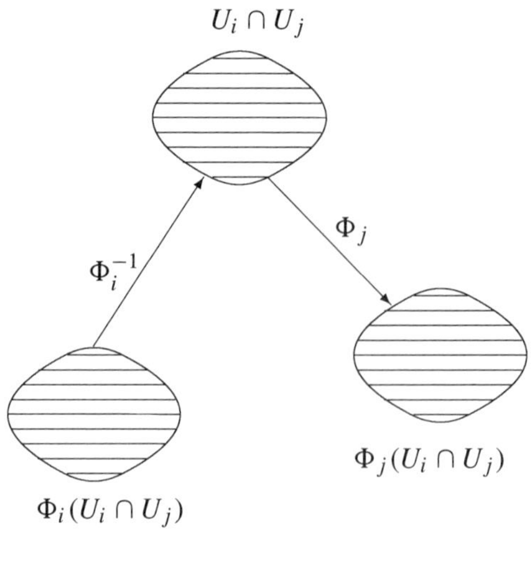

Definition 1.3.2. An *algebraic function* $w(z)$ of a complex variable z is a function satisfying a functional equation of the type

$$a_0(z)w^n + a_1(z)w^{n-1} + \cdots + a_n(z) = 0,$$

where $a_0(z) \neq 0$ and $a_i(z) \in \mathbb{C}[z]$ for $0 \leq i \leq n$.

Definition 1.3.3. A *Riemann surface R of an algebraic function* $w(z)$ is a connected complex manifold (that is, "locally" the same as $\mathbb{C}$) where $w(z)$ can be defined as an analytic function ($w\colon R \to \mathbb{C} \cup \{\infty\}$) and $w(z)$ is single-valued. (If $A \subseteq B$ are two Riemann surfaces of $w(z)$, A is open and closed in B, so $A = B$.)

If R and R' are two such connected complex manifolds, then R and R' are conformally equivalent. That is, R is essentially unique, and therefore we will say that R is *the* Riemann surface of $w(z)$.

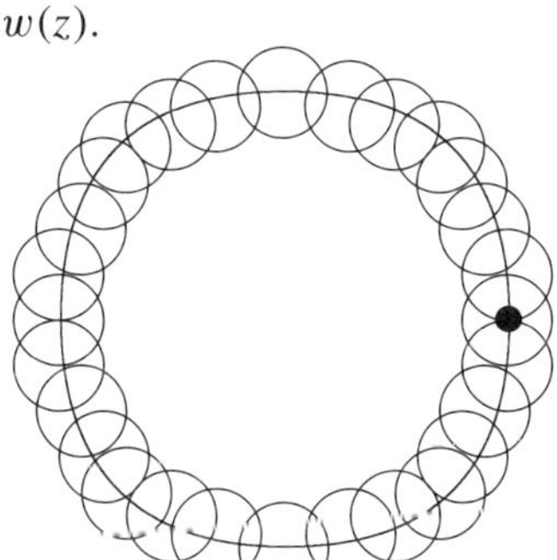

In order to clarify the previous definition, we consider the "function" defined by $w(z) = \sqrt{z}$ (that is, $w(z)^2 - z = 0$). When we begin to evaluate $w(1)$ we have two possible choices, $w(1) = 1$ or $w(1) = -1$. Say that we choose $w(1) = 1$. If we take the analytic continuation of $w(z)$ around the curve of equation $\varrho(t) = e^{it}, 0 \leq t \leq 2\pi$, we obtain, when we come back to the point $z = 1$, the value $w(1) = -1$ (and vice versa). If we go around for a second time with the analytic continuation, we obtain $w(1) = 1$. This procedure tells us that in order to obtain a solution to this problem, the point 1 is to be "divided" into two points, or, more precisely, all real values between 0 and ∞ included are to be divided into two parts. In other words, when we consider the Riemann surface S^2, we must remove the positive real curve starting at 0 and ending at ∞. When we separate this cut, the set obtained may be assumed to be the same as a half Riemann sphere with the ray of positive real numbers as the border and such that it appears twice. When we continue $w(z)$ through the curve $\varrho(t) = e^{it}$ and we come back to the point 1, we take the point 1 in

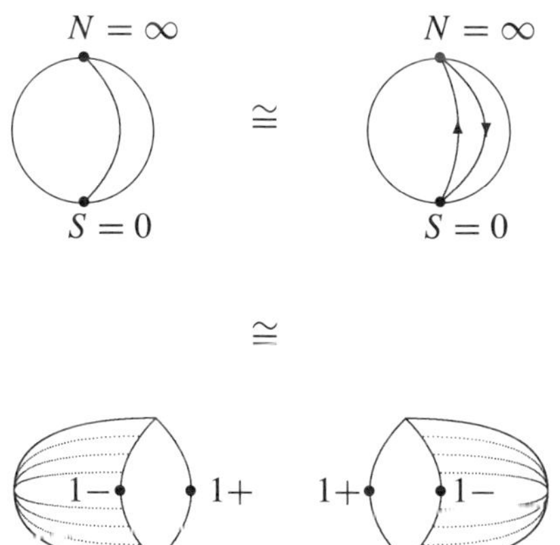

the second hemisphere instead of the first one. If we identify the respective borders we will obtain again the Riemann sphere, but with the previous process, $w(z)$ will be single-valued.

This is fundamentally the Riemann approach to make single-valued functions from multivalued ones.

We point out that this problem is defined not only for algebraic functions but also for many other multivalued functions, for instance the logarithmic function. Although in this case the problem can be solved in a similar fashion, the Riemann surface obtained will be different from the Riemann surfaces obtained from algebraic functions, which are compact.

Now we state some basic results of the theory of Riemann surfaces that will be generalized later to other situations. For the moment, they will serve us as a motivation and a basis of our general theory of algebraic functions.

Theorem 1.3.4. *The Riemann surface of an algebraic function is a compact Riemann surface (according to Definition* 1.3.1*).*

Proof. [72, Theorem 4.2, p. 156], [34, Corollary, p. 248]. □

The converse also holds.

Theorem 1.3.5. *If a Riemann surface is compact, then it is conformally equivalent to a Riemann surface of an algebraic function.*

Proof. [72, Theorem 4.3, p. 161], [34, Corollary IV.11.8, p. 249]. □

Theorem 1.3.6. *Every compact Riemann surface R is homeomorphic to a Riemann surface with g handles, where g is a nonnegative integer called the* genus *of R. Therefore two Riemann surfaces are topologically equivalent if and only if they have the same genus.*

Proof. [72, Theorems 4.8 and 4.9, p. 172], [164, Teorema 5.92, p. 261]. □

Theorem 1.3.7. *Every compact Riemann surface R of genus g is conformally equivalent to a cover of $(g+1)$ sheets of the Riemann sphere.* □

The previous results characterize all compact Riemann surfaces: on the one hand, the compact Riemann surfaces are exactly the Riemann surfaces of algebraic functions; on the other hand, they are topologically equivalent to a bidimensional sphere with g handles and conformally equivalent to a cover of a Riemann sphere.

We observe that the genus g characterizes the compact Riemann surfaces topologically but not analytically. For instance, there are infinitely many Riemann surfaces of genus 1 that are conformally inequivalent pairwise. This topic will be studied later and in a much more general setting.

Let $P \in R$ and $P \in U$ where U is an open set of R. Let $\varphi : U \longrightarrow \varphi(U) =$

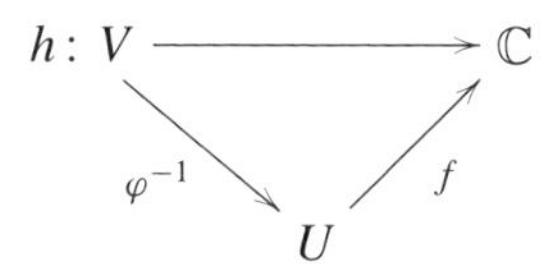

$V \subseteq \mathbb{C}$ be a homeomorphism given in Definition 1.3.1. For a given $f : U \longrightarrow \mathbb{C}$, let $h = f \circ \varphi^{-1}$. We say that f is holomorphic (meromorphic) in U if h is holomorphic (meromorphic) in V. The same definitions are given for a global function $f : R \longrightarrow \mathbb{C}$.

Theorem 1.3.8. *Let R be a Riemann surface and let $X(R) = \{f : R \to \mathbb{C} \mid f$ is meromorphic$\}$. Then $X(R)$ is a finitely generated field over $\mathbb{C}$ with transcendence degree 1; that is, $X(R) \cong \mathbb{C}(x, y)$ where x and y are two indeterminates over $\mathbb{C}$ satisfying a nonzero relation $F(x, y) = 0$, for F a polynomial in two variables.*

Proof. [72, Theorem 3.4, p. 95 and Theorem 4.3, p. 161], [34, Corollary, p. 250]. □

Finally we have the following theorem.

Theorem 1.3.9. *Let R_1, R_2 be two compact Riemann surfaces. Then R_1 and R_2 are conformally equivalent (that is, isomorphic as Riemann surfaces) if and only if $X(R_1)$ and $X(R_2)$ are $\mathbb{C}$-isomorphic as fields (that is, there exists a field isomorphism $\varphi : X(R_1) \to X(R_2)$ such that $\varphi(\alpha) = \alpha$ for all $\alpha \in \mathbb{C}$).*

Proof. [72, Theorems 4.5 and 4.6, p. 164]. □

Thus, we see that the study of compact Riemann surfaces can be done by means of their fields of meromorphic functions. This allows us to view algebraic function fields as Riemann surfaces over an arbitrary field (in place of $\mathbb{C}$). Of course we do not have all the analytic machinery available as in the field of complex numbers, but we can algebrize the properties of the Riemann surfaces and in this way find results of the same kind over an arbitrary field of constants.

By this method we will obtain the Riemann–Roch theorem, the Riemann–Hurwitz genus formula, the concept of a holomorphic differential or abelian differential of the first type, differentials, etc. On the other hand, when k is an arbitrary field, in particular not necessarily algebraically closed or of characteristic 0, k may have proper algebraic extensions or inseparable extensions. This necessarily implies that the theory will differ substantially from the analytical case.

1.4 Exercises

Exercise 1.4.1. Verify that the function $| \ |_p$ defined in Example 1.2.2 is an absolute value.

Exercise 1.4.2. Prove that the p-adic absolute value $| \ |_p$ is nonarchimedean.

Exercise 1.4.3. Prove Proposition 1.1.12.

Exercise 1.4.4. What is the topology on $\mathbb{Q}$ given by the trivial absolute value?

Exercise 1.4.5. Prove that if p and q are two different rational prime numbers, then the p-adic and the q-adic topologies in $\mathbb{Q}$ are different.

Exercise 1.4.6. Find $\operatorname{tr} \mathbb{C}/\mathbb{Q}$, $\operatorname{tr} \mathbb{R}/\mathbb{Q}$, and $\operatorname{tr} \mathbb{C}/\mathbb{R}$.

Exercise 1.4.7. Show that $\operatorname{Aut} \mathbb{C} := \{f : \mathbb{C} \to \mathbb{C} \mid f \text{ is a field automorphism}\}$ is an infinite set.

Exercise 1.4.8. Prove that if $S = \{s_1, \ldots, s_n\}$ is an algebraically independent set over a field K, then $K(s_1, \ldots, s_n)$ is isomorphic to the field $K(x_1, \ldots, x_n)$ of rational functions in n variables.

Exercise 1.4.9. Prove that an extension L/K is algebraic if and only if any transcendental basis of L/K is the empty set.

Exercise 1.4.10. If $| \ |$ is an absolute value on a field K, prove that $|1| = 1$ and $|-x| = |x|$ for all $x \in K$.

2

Algebraic Function Fields of One Variable

This chapter will serve as an introduction to our theory of function fields. Using as a source of inspiration compact Riemann surfaces, and especially their fields of meromorphic functions, we first generalize the concept of a function field. In this way we will obtain the general definition of a function field, and establish its most immediate properties.

Our second goal in this chapter will be to study absolute values in function fields, following the philosophy according to which the local study of an object provides information on its global properties, and vice versa. We will use the fact that the concept of an absolute value is equivalent to other concepts of a more algebraic nature: valuation rings, valuations, places, etc. This equivalence will be studied in Section 2.2, together with its basic properties. The places (Definition 2.2.10) correspond to points on a projective, nonsingular algebraic curve (at least over an algebraically closed field).

Next, we shall recall the definition of the completion of a field with respect to an absolute value, which is a particular case of a metric space. Such completions constitute the mentioned local study of function fields, which will be used for the global study of these fields.

In Section 2.4 we characterize all valuations of a field of rational functions that are trivial on the field of constants. Together with Chevalley's lemma, which states that places extend to overfields, this characterization will allow us to study valuations over an arbitrary function field.

In the last section we will present Artin's approximation theorem, which states the following: Given a finite number of distinct absolute values and the same number of arbitrary elements of a function field, we can find an element of the field that approximates the given elements as much as we want, each one in the corresponding absolute value.

We conclude the chapter with a characterization of the completion of a function field with respect to a given place. As we shall see, such completions are simply Laurent series, which makes their study easier than that of number fields; indeed, although the latter admit series representations, the series involved are not Laurent series, due to the difference in characteristics.

2.1 The Field of Constants

Definition 2.1.1. Let k be an arbitrary field. A *field of algebraic functions K over k* is a finitely generated field extension of k with transcendence degree $r \geq 1$. K is called a *field of algebraic functions of r variables.*

Example 2.1.2. Let k be any field and let $K = k[X, Y]/(f(x, y))$, where $k[X, Y]$ is the polynomial ring of two variables, k is any field, and $f(X, Y) = X^3 - Y^2 + 1$. Then if $x := X \bmod (f(X, Y))$ and $y := Y \bmod (f(X, Y))$, we have $K = k(x, y)$ with $x^3 = y^2 - 1$. Therefore $K = k(x, y)$ is a field of algebraic functions of one variable.

From this point on we will study only the case $r = 1$, that is, K will be a field of functions of one variable. We will call such a field a *function field* and it will be denoted by K/k.

We observe that if $x \in K$ is transcendental over k, then $K/k(x)$ is a finite extension (since it is algebraic and finitely generated).

Now, if z is any other element of K that is transcendental over k, then since K/k has transcendence degree 1, $\{x, z\}$ cannot be algebraically independent. Therefore there exists a nonzero polynomial $p(T_1, T_2) \in k[T_1, T_2]$ such that $p(x, z) = 0$. Since x and z are transcendental over k, x and z must appear in the expression of $p(x, z)$. Therefore, it follows immediately that x is algebraic over $k(z)$ (and z is algebraic over $k(x)$). Thus

$$[K : k(z)] = [K : k(x, z)][k(x, z) : k(z)] \leq [K : k(x)][k(x, z) : k(z)] < \infty,$$

as we mentioned before. This shows that any two elements x, z of K that are transcendental over k satisfy similar conditions, that is, $K/k(x)$ and $K/k(z)$ are finite. However, in general $[K : k(z)]$ and $[K : k(x)]$ are distinct. This is one of the principal differences with number fields, since a number field E has as base subfield its prime field, namely $\mathbb{Q}$, and $[E : \mathbb{Q}]$ is well and uniquely defined. In the case of algebraic functions K, we take as base field $k(x)$ with $x \in K$ transcendental over k, but $k(x)$ is not uniquely determined. On the other hand, if $x, z \in K$ are transcendental over k, we have $k(x) \cong k(z)$.

As a simple example of the previous remarks, we consider $K = \mathbb{Q}(x, z)$, where x, z are variables over $\mathbb{Q}$ that satisfy $x^2 + z^4 = 1$. We have $[K : \mathbb{Q}(x)] = 4$, $[K : \mathbb{Q}(z)] = 2$, $\left[K : \mathbb{Q}\left(x^2\right)\right] = 8$, etc.

Definition 2.1.3. Let K/k be a function field. The algebraic closure of k in K, that is, the field $k' = \{\alpha \in K \mid \alpha \text{ is algebraic over } k\}$, is called the *field of constants of K*.

Example 2.1.4. Let $K = \mathbb{R}(x, y)$ with x, y two variables over $\mathbb{R}$ satisfying

$$x^6 + 2x^3y^2 + y^4 = -1.$$

Since $x^3 + y^2 = i = \sqrt{-1}$, it follows that the field of constants of K is $\mathbb{C}$.

Example 2.1.5. If $k=\mathbb{R}$, $K=k(x,y)$ with $x^2=-y^2-1$, then $i\notin K$ since otherwise $x=i\sqrt{y^2+1}\in K$ and $\sqrt{y^2+1}\in K$ and it would follow that $K=k(x,y)=k(i\sqrt{y^2+1},y)$. However, it is easy to see that $i=p(i\sqrt{y^2+1},y)$ has no solution for any $p(X,Y)\in\mathbb{R}[X,Y]$. Therefore in this case the field of constants is $k=\mathbb{R}$.

Note that $k\subseteq k'$ and since K/k' cannot be algebraic, we have

$$1\leq \operatorname{tr} K/k'\leq \operatorname{tr} K/k=1.$$

Thus K/k' is also a function field, now over k', with the additional property that every element $x\in K\setminus k'$ is transcendental.

Proposition 2.1.6. *If $x\in K\setminus k'$, we have $[k':k]=[k'(x):k(x)]$. More generally, if x is a transcendental element over k and k', then $[k':k]=[k'(x):k(x)]$.*

Proof. Let $[k':k]=n$ with n finite or infinite. We will see later that n must be finite.

Let $\{\alpha_i\}_{i\in I}$ be a basis of the vector space k' over k, $|I|=n$. Let $p(x)\in k'(x)$, say $p(x)=\frac{a(x)}{b(x)}$, with $a(x),b(x)\in k'[x]$. We write $a(x)=\sum_{i=0}^m a_ix^i$, with $a_i\in k'$.

$$\begin{array}{ccc} k' & \text{———} & k'(x) \\ | & & | \\ k & \text{———} & k(x) \end{array}$$

We have

$$a_i=\sum_{j=1}^{r_i} a_{ij}\alpha_j,\quad a_{ij}\in k,\quad 0\leq i\leq m.$$

Let $t=\max\{r_i\mid i=0,\dots,m\}$ and $a_{ij}=0$ for $r_i<j\leq t$. We may write $a_i=\sum_{j=1}^t a_{ij}\alpha_j$. Thus

$$a(x)=\sum_{i=0}^m a_ix^i=\sum_{i=0}^m\left(\sum_{j=1}^t a_{ij}\alpha_j\right)x^i=\sum_{j=1}^t\alpha_j\left(\sum_{i=0}^m a_{ij}x^i\right)=\sum_{j=1}^t p_j(x)\alpha_j,$$

with $p_j(x)=\sum_{i=0}^m a_{ij}x^i\in k[x]$.

Therefore $a(x)$ is algebraic over $k(x)$.

If we apply the above argument to $b(x)\in k'[x]$, we obtain as a particular case that there exists a relation

$$\sum_{\ell=0}^r t_\ell(x)b(x)^\ell=0\quad\text{with}\quad t_\ell(x)\in k[x],\quad t_0(x)\neq 0,\quad\text{and}\quad t_r(x)\neq 0.$$

In particular,

$$b(x)\left\{\sum_{\ell=1}^r t_\ell(x)b(x)^{\ell-1}\right\}\left\{-t_0(x)^{-1}\right\}=1,$$

that is,

$$b(x)^{-1}=-\sum_{\ell=1}^r\frac{t_\ell(x)}{t_0(x)}b(x)^{\ell-1}.$$

Hence, $p(x) = a(x)b(x)^{-1} = \sum_{i=0}^{s} c_i(x)\alpha_i$ for $c_i(x) \in k(x)$. Therefore $\{\alpha_i\}_{i\in I}$ generates $k'(x)$ over $k(x)$.

Assume that there exists a relation $\sum_{i=0}^{s} q_i(x)\alpha_i = 0$, with $q_i(x) \in k(x)$ and such that some $q_j(x)$ is nonzero. Clearing denominators, we may assume that $q_i(x) \in k[x]$. Now, in case $x \mid q_i(x)$ for all i, we take $q_i(x) = xq_i'(x)$ and we obtain $x\sum_{i=0}^{s} q_i'(x)\alpha_i = 0$, so that $\sum_{i=0}^{s} q_i'(x)\alpha_i = 0$. Therefore, we may assume that $x \nmid q_j(x)$ for some j, or equivalently, $q_j(0) \neq 0$. Now, $\sum_{i=0}^{s} q_i(x)\alpha_i = 0$ implies $\sum_{i=0}^{s} q_i(0)\alpha_i = 0$, but then $q_i(0) \in k$ and $q_j(0) \neq 0$ imply that $\{\alpha_i\}_{i\in I}$ is not linearly independent over k.

Hence, $\{\alpha_i\}_{i\in I}$ is also a basis of $k'(x)/k(x)$ and therefore $[k'(x) : k(x)] = [k' : k]$. □

Coming back to the function field K/k, we have

$$[K : k(x)] = [K : k'(x)][k'(x) : k(x)] = [K : k'(x)][k' : k] < \infty,$$

so $n = [k' : k]$ is finite in Proposition 2.1.6.

From now on, unless otherwise stated, we will always assume that $k' = k$, that is, when mentioning a function field K/k, we will be assuming that the field of constants of K is k or, equivalently, that k is algebraically closed in K.

2.2 Valuations, Places, and Valuation Rings

Definition 2.2.1. An *ordered group* G is an abelian group $(G, +)$ with a relation $<$ satisfying, for $\alpha, \beta, \gamma \in G$:

(i) $\alpha < \beta$ or $\beta < \alpha$ or $\alpha = \beta$ (trichotomy),
(ii) If $\alpha < \beta$ and $\beta < \gamma$ then $\alpha < \gamma$ (transitivity),
(iii) If $\alpha < \beta$ then $\alpha + \gamma < \beta + \gamma$ (preservation of the group operation).

As usual, $\alpha \leq \beta$ will denote $\alpha < \beta$ or $\alpha = \beta$.

For an ordered group G, we define $G_0 = \{\alpha \in G \mid \alpha < 0\}$, where 0 denotes the identity of G. Then we have the disjoint union $G = G_0 \cup \{0\} \cup \{-G_0\}$. Furthermore, for all $\alpha, \beta \in G$ we have $\alpha < \beta$ if and only if $\alpha - \beta \in G_0$.

Conversely, if $(G, +)$ is an abelian group with identity 0 such that there exists a semigroup $H \subseteq G$ satisfying that $G = H \cup \{0\} \cup \{-H\}$ is a disjoint union, we can define for $\alpha, \beta \in G, \alpha < \beta \iff \alpha - \beta \in H$. It is easy to see that $<$ satisfies the conditions of Definition 2.2.1 and G is an ordered group whose set of "negative elements" is H.

We observe that if G is a nontrivial finite group, then G cannot be ordered since if $\alpha \in G$ and $\alpha \neq 0$, say $\alpha > 0$, then for any $n \in \mathbb{N}$,

$$n\alpha = \alpha + \cdots + \alpha > 0 + \cdots + 0 = 0,$$

that is, $n\alpha \neq 0$. In particular, if G is an ordered group then every nonzero element of G is of infinite order, that is, G is torsion free.

The most obvious examples of ordered groups are $\mathbb{Z}$, $\mathbb{Q}$, and $\mathbb{R}$ with the sum and the usual order.

Definition 2.2.2. Let K be an arbitrary field. A *valuation* v over K is a surjective function $v : K^* \longrightarrow G$, where G is an ordered group called the *value group* or *valuation group, satisfying*

(i) *For* $a, b \in K^*$, $v(ab) = v(a) + v(b)$, *that is,* v *is a group epimorphism,*
(ii) *For* $a, b \in K^*$ *such that* $a + b \neq 0$, $v(a+b) \geq \min\{v(a), v(b)\}$.

We define $v(0) = \infty$, where ∞ is a symbol such that $\infty \notin G, \alpha < \infty$ for all $\alpha \in G$ and $\infty + \infty = \alpha + \infty = \infty + \alpha = \infty$ for all $\alpha \in G$.

The purpose of including the symbol ∞ is simply to be able to define $v(0)$ in such a way that conditions (i) and (ii) of the definition are also satisfied.

As an example of valuation we have $K = \mathbb{Q}, G = \mathbb{Z}$, and $v = v_p$ the p-adic valuation, for $p \in \mathbb{Z}$ a rational prime. That is, for $x \in \mathbb{Q}^*$ we write

$$x = p^n \frac{a}{b}, \quad n \in \mathbb{Z}, \quad p \nmid ab \quad \text{and} \quad v_p(x) = n.$$

We leave it to the reader to verify that this is in fact a valuation. Also, observe the similarity of v_p with the p-adic absolute value (Example 1.2.2).

A fancier example, which is a simple generalization of the previous one, is the following. Consider a number field K, that is, $[K : \mathbb{Q}] < \infty$, and let ϑ_K be the integral closure of $\mathbb{Z}$ in K, that is,

$$\vartheta_K = \{\alpha \in K \mid \operatorname{Irr}(\alpha, x, K) \in \mathbb{Z}[x]\},$$

where $\operatorname{Irr}(\alpha, x, K)$ denotes the irreducible polynomial of α in $\mathbb{Q}[x]$.

Let $\mathcal{P}$ be a nonzero prime ideal of ϑ_K. It is known that ϑ_K is a Dedekind domain (see Definition 5.7.1), so that if $x \in K^*$, the principal fractional ideal (x) can be written as $\mathcal{P}^n \frac{\mathfrak{A}}{\mathfrak{B}}$ with $n \in \mathbb{Z}$, where $\mathfrak{A}, \mathfrak{B}$ are ideals of ϑ_K that are relatively prime to $\mathcal{P}$. Then we define $v_{\mathcal{P}}(x) = n$. As in the case of $\mathbb{Q}$, $v_{\mathcal{P}}$ is a valuation that is an extension of the p-adic valuation v_p of $\mathbb{Q}$, where $(p) = \mathcal{P} \cap \mathbb{Z}$.

In general we have the following result:

Proposition 2.2.3. *Let K be any field and let v be a valuation over K. Then*

(i) $v(1) = 0$,
(ii) $v\left(a^{-1}\right) = -v(a)$ *for all* $a \neq 0$,
(iii) $v(a) = v(-a)$,
(iv) *if* $v(a) \neq v(b)$, *then* $v(a+b) = \min\{v(a), v(b)\}$,
(v) $v\left(\sum_{i=1}^n a_i\right) \geq \min_{1 \leq i \leq n}\{v(a_i)\}$ *and equality holds if* $v(a_i) \neq v\left(a_j\right)$ *for all* $i \neq j$,
(vi) *if* $\sum_{i=1}^n a_i = 0, n \geq 2$, *then there exist* $i \neq j$ *such that* $v(a_i) = v\left(a_j\right)$.

Proof.

(i) We have $v(1) = v(1 \times 1) = v(1) + v(1)$, so, by the cancellation law property of abelian groups, it follows that $v(1) = 0$.
(ii) We have $0 = v(1) = v\left(aa^{-1}\right) = v(a) + v\left(a^{-1}\right)$. Therefore $v\left(a^{-1}\right) = -v(a)$.

(iii) We have

$$v(1) = 0 = v((-1)(-1)) = v(-1) + v(-1),$$

that is, $2v(-1) = 0$. Since the unique torsion element of an ordered abelian group is 0, we have $v(-1) = 0$. Therefore we obtain that

$$v(-a) = v((-1)a) = v(-1) + v(a) = 0 + v(a) = v(a).$$

(iv) We have $v(a+b) \geq \min\{v(a), v(b)\}$. Now if $v(a) \neq v(b)$, say $v(a) > v(b)$, then

$$\begin{aligned} v(b) &= v(b + a - a) \geq \min\{v(a+b), v(-a)\} \\ &= \min\{v(a+b), v(a)\} \geq v(b). \end{aligned}$$

Then from $v(b) = \min\{v(a+b), v(a)\}$ and $v(b) < v(a)$ we conclude that

$$v(a+b) = v(b) = \min\{v(a), v(b)\}.$$

(v) The case $n = 2$ is given in (iv). For $n > 2$, by induction on n we obtain

$$v\left(\sum_{i=1}^{n} a_i\right) = v\left(\sum_{i=1}^{n-1} a_i + a_n\right) \geq \min_{1 \leq i \leq n} \{v(a_i)\},$$

and if $v(a_i) \neq v(a_j)$ for all $i \neq j$, then

$$v\left(\sum_{i=1}^{n-1} a_i\right) = \min_{1 \leq i \leq n-1} \{v(a_i)\} \neq v(a_n).$$

Therefore

$$\begin{aligned} v\left(\sum_{i=1}^{n} a_i\right) &= \min\left\{v\left(\sum_{i=1}^{n-1} a_i\right), v(a_n)\right\} \\ &= \min\left\{\min_{1 \leq i \leq n-1} \{v(a_i)\}, v(a_n)\right\} = \min_{1 \leq i \leq n} \{v(a_i)\}. \end{aligned}$$

(vi) For $n \geq 2$, if $\sum_{i=1}^{n} a_i = 0$, then $v\left(\sum_{i=1}^{n} a_i\right) = v(0) = \infty$.
If $\min_{1 \leq i \leq n} \{v(a_i)\} = \infty$, then $v(a_i) = \infty$, that is, $a_i = 0$ for all i.
If $\min_{1 \leq i \leq n} \{v(a_i)\} < \infty$, then $v\left(\sum_{i=1}^{n} a_i\right) \neq \min_{1 \leq i \leq n} \{v(a_i)\}$. Hence, from (v), we have $v(a_i) = v(a_j)$ for two different indices $i \neq j$.

□

Now we consider an arbitrary field K and a valuation of K with values in an ordered group G. Let $\vartheta_v = \{x \in K \mid v(x) \geq 0\}$. Then, since

$$v(x) = v(-x), \quad v(xy) = v(x) + v(y),$$

and since G is an ordered group, it follows that ϑ_v is a ring. Furthermore, for $x \in K$, then if $x \notin \vartheta_v$, we have $v(x) < 0$. Thus $v\left(x^{-1}\right) = -v(x) > 0$, that is, $x^{-1} \in \vartheta_v$. Hence, given $x \in K$, we have $x \in \vartheta_v$ or $x^{-1} \in \vartheta_v$. Furthermore, for $x \in K$, if $x \in \vartheta_v$ then $x = \frac{x}{1} \in \vartheta_v$, and if $x \notin \vartheta_v$, then $x^{-1} \in \vartheta_v$ and therefore $x = \frac{1}{x^{-1}} \in \operatorname{quot} \vartheta_v$, where $\operatorname{quot} \vartheta_v$ denotes the field of quotients of ϑ_v, which proves that $K = \operatorname{quot} \vartheta_v$.

Now, $x \in \vartheta_v$ is a unit if and only if $x^{-1} \in \vartheta_v$, that is, $v(x) \geq 0$ and $v\left(x^{-1}\right) = -v(x) \geq 0$. Therefore

$$\vartheta_v^* = \{x \in K \mid v(x) = 0\}.$$

Let $\mathcal{P}_v = \{x \in K \mid v(x) > 0\}$ consist of the nonunits of ϑ_v. We will see that in fact $\mathcal{P}_v$ is an ideal. If $x \in \mathcal{P}_v$ and $y \in \vartheta_v$, we have

$$v(xy) = v(x) + v(y) \geq v(x) > 0,$$

so $xy \in \mathcal{P}_v$. On the other hand, if $x, y \in \mathcal{P}_v$, then

$$v(x+y) \geq \min\{v(x), v(y)\} > 0.$$

Therefore ϑ_v is a local ring with maximal ideal $\mathcal{P}_v$. Finally, $v\colon (K^*, \cdot) \longrightarrow (G, +)$ is a group epimorphism with $\ker v = \vartheta_v^*$. Thus

$$(G, +) \cong \left(K^*/\vartheta_v^*, \cdot\right).$$

The above discussion can be summed up as follows.

Proposition 2.2.4. *If K is a field and v a valuation over K, then $\vartheta_v = \{x \in K \mid v(x) \geq 0\}$ is a subring of K such that for all $x \in K$, $x \in \vartheta_v$ or $x^{-1} \in \vartheta_v$. In particular, ϑ_v is a local ring with maximal ideal*

$$\mathcal{P}_v = \{x \in K \mid v(x) > 0\} = \vartheta_v \setminus \vartheta_v^*, \quad \vartheta_v^* = \{x \in K \mid v(x) = 0\}.$$

Furthermore, we have $\operatorname{quot} \vartheta_v = K$ *and the value group of v is isomorphic to K^*/ϑ_v^*.* □

Definition 2.2.5. Every integral domain A that is not a field and such that each $x \in \operatorname{quot} A$ satisfies $x \in A$ or $x^{-1} \in A$ is called a *valuation ring.*

Proposition 2.2.6. *If A is a valuation ring and $K =$* quot *A, then K^*/A^* is an ordered group and the natural projection is a valuation with valuation ring A and value group K^*/A^**

Proof. We know that K^*/A^* is an abelian group. If $x, y \in K^*$, define

$$x \bmod A^* \leq y \bmod A^* \quad \text{if} \quad yx^{-1} \in A$$
$$(x \bmod A^* < y \bmod A^* \iff yx^{-1} \in A \setminus A^*),$$

Observe that if $x \bmod A^* = x_1 \bmod A^*$ and $y \bmod A^* = y_1 \bmod A^*$, then $x = ax_1$, $y = by_1$ with $a, b \in A^*$. Therefore $yx^{-1} = by_1(ax_1)^{-1} = ba^{-1}y_1x_1^{-1}$. Thus $yx^{-1} \in A \iff y_1x_1^{-1} \in A$, which proves that the order relation does not depend on the representatives.

Given three elements $\alpha, \beta, \gamma \in K^*/A^*$, we take $x, y, z \in K^*$ such that $\alpha = x \bmod A^*$, $\beta = y \bmod A^*$, $\gamma = z \bmod A^*$. Since A is a valuation ring, we have $xy^{-1} \in A$ or $\left(xy^{-1}\right)^{-1} = yx^{-1} \in A$, so that $\alpha \le \beta$ or $\beta \le \alpha$. Therefore, the relation is trichotomic.

Now if $\alpha \le \beta$ and $\beta \le \gamma$, then $yx^{-1} \in A$, $zy^{-1} \in A$ and $yx^{-1}zy^{-1} = zx^{-1} \in A$, which shows that $\alpha \le \gamma$. If $\alpha < \beta$ and $\beta < \gamma$, it is easy to see that $\alpha < \gamma$.

Finally, if $\alpha \le \beta$, then $yx^{-1} \in A$ so $yzz^{-1}x^{-1} = yz(zx)^{-1} \in A$, that is, $\alpha\gamma \le \beta\gamma$.

Therefore K^*/A^* is an ordered group; now consider the natural projection

$$v : K^* \longrightarrow K^*/A^*.$$

We have

$$v(xy) = xy \bmod A^* = \left(x \bmod A^*\right)\left(y \bmod A^*\right)$$

for any $x, y \in K^*$. If $x + y \neq 0$ then $v(x+y) = (x+y) \bmod A^*$. Let us assume that $x \bmod A^* \le y \bmod A^*$, that is, $yx^{-1} \in A$. We have

$$(x+y)x^{-1} = 1 + yx^{-1} \in A,$$

that is,

$$v(x+y) = (x+y) \bmod A^* \ge x \bmod A^* = \min\left\{x \bmod A^*, y \bmod A^*\right\}.$$

This proves that v is a valuation.

Finally, the valuation ring of v is given by

$$\vartheta_v = \left\{x \in K^* \mid v(x) \ge \bar{1}\right\} \cup \{0\} = \left\{x \in K^* \mid x1^{-1} = x \in A\right\} \cup \{0\} = A. \qquad \square$$

Propositions 2.2.4 and 2.2.6 show that the concepts of valuation rings and valuations are essentially the same.

Definition 2.2.7. Let $v_1 : K^* \longrightarrow (G_1, +)$ and $v_2 : K^* \longrightarrow (G_2, +)$ be two valuations of a field K. We say that v_1 and v_2 are *equivalent* if $v_1(\alpha) > 0 \iff v_2(\alpha) > 0$ for all $\alpha \in K^*$.

Observe that if $\alpha \in K^*$, then $v_1(\alpha) < 0 \iff v_1\left(\alpha^{-1}\right) > 0 \iff v_2\left(\alpha^{-1}\right) > 0 \iff v_2(\alpha) < 0$ and by complementation, we obtain $v_1(\alpha) = 0 \iff v_2(\alpha) = 0$. Therefore we have shown that if v_1 and v_2 are equivalent, then $\vartheta_{v_1} = \vartheta_{v_2}$; in particular, the value groups are isomorphic since both are isomorphic to $K^*/\vartheta_{v_1}^*$.

Now let v_1 and v_2 be two equivalent valuations with value groups G_1 and G_2 respectively. For $\alpha \in G_1$, let $a \in K^*$ be such that $v_1(a) = \alpha$ and define

$$\sigma : G_1 \longrightarrow G_2 \quad \text{such that} \quad \sigma(\alpha) = v_2(a).$$

Clearly, σ is defined by means of the formula $\sigma v_1 = v_2$. The first fact we have to verify is that σ is well defined, i.e., if $v_1(a) = v_1(b)$, then $v_2(a) = v_2(b)$. Let $a, b \in G$. We have

$$\begin{aligned} v_1(a) = v_1(b) &\Longrightarrow v_1\left(ab^{-1}\right) = v_1(a) - v_1(b) = 0 \\ &\Longrightarrow v_2\left(ab^{-1}\right) = v_2(a) - v_2(b) = 0 \Longrightarrow v_2(a) = v_2(b). \end{aligned}$$

Now, if $v_1(a) = \alpha$ and $v_1(b) = \beta$, then $v_1(ab) = v_1(a) + v_1(b) = \alpha + \beta$, so

$$\sigma(\alpha + \beta) = v_2(ab) = v_2(a) + v_2(b) = \sigma(\alpha) + \sigma(\beta),$$

and hence σ is a group homomorphism. Now given $\gamma \in G_2$, let $v_2(a) = \gamma$. If $v_1(a) = \alpha$, we have $\sigma(\alpha) = \gamma$. Therefore σ is an epimorphism. Also, if $\sigma(\alpha) = \sigma(\beta)$, then $v_2(a) = v_2(b)$ with a, b satisfying $v_1(a) = \alpha$, $v_1(b) = \beta$. Now

$$\begin{aligned} v_2(a) = v_2(b) \Longrightarrow v_2\left(ab^{-1}\right) = 0 &\Longrightarrow v_1\left(ab^{-1}\right) = 0 \\ &\Longrightarrow \alpha = v_1(a) = v_1(b) = \beta, \end{aligned}$$

that is, σ is injective. We have shown that σ is a group isomorphism.

Finally, if $\alpha < \beta$ with $\alpha, \beta \in G_1$, that is, $\beta - \alpha > 0$, we have $v_1\left(ab^{-1}\right) > 0$, where $v_1(a) = \alpha$, $v_1(b) = \beta$. Then $v_2\left(ab^{-1}\right) > 0$, so $\sigma(\alpha) < \sigma(\beta)$, which means that σ is order-preserving.

Conversely, let v_1, v_2 be two valuations over a field K with value groups G_1, G_2 respectively such that there exists an order-preserving isomorphism $\varphi : G_1 \to G_2$ such that $\varphi v_1 = v_2$. If $v_1(a) > 0$ we have $(\varphi v_1)(a) = v_2(a) > 0$, which tells us that v_1 and v_2 are equivalent.

We collect all the above discussion in the following proposition:

Proposition 2.2.8. *Two valuations v_1, v_2 over a field K with value groups G_1, G_2 respectively are equivalent if and only if there exists an order-preserving group isomorphism $\varphi : G_1 \longrightarrow G_2$ such that $\varphi v_1 = v_2$.* □

On the other hand, if $\vartheta_{v_1} = \vartheta_{v_2}$, then $\mathcal{P}_{v_1} = \mathcal{P}_{v_2}$ is the unique maximal ideal of $\vartheta_{v_1} = \vartheta_{v_2}$. We have $v_1(\alpha) > 0 \Longleftrightarrow \alpha \in \vartheta_{v_1} \setminus \mathcal{P}_{v_1} = \vartheta_{v_2} \setminus \mathcal{P}_{v_2} \Longleftrightarrow v_2(\alpha) > 0$. We have proved the following result:

Proposition 2.2.9. *Two valuations over a field are equivalent if and only if they have the same valuation ring.* □

Next, we will define the concept of a *place*.

Let E be an arbitrary field, and let ∞ be a symbol such that $\infty \notin E$. We define the set $E_1 = E \cup \{\infty\}$ and partially extend the field operations to E_1 in the following way:

$$x + \infty = \infty + x = \infty \quad \text{for all} \quad x \in E,$$
$$x \cdot \infty = \infty \cdot x \quad \text{for all} \quad x \in E^*,$$

and

$$\infty \cdot \infty = \infty.$$

Note that $\infty + \infty$, $0 \cdot \infty$, and $\infty \cdot 0$ are *not* defined.

Definition 2.2.10. A *place* on a field K is a function $\varphi : K \longrightarrow E \cup \{\infty\}$ (E a field) satisfying:

(i) $\varphi(a + b) = \varphi(a) + \varphi(b)$ for all $a, b \in K$;
(ii) $\varphi(ab) = \varphi(a)\varphi(b)$ for all $a, b \in K$;
(iii) There exists an element $a \in K$ such that $\varphi(a) = \infty$;
(iv) There exists an element $b \in K$ such that $\varphi(b) \neq \infty$ and $\varphi(b) \neq 0$.

Conditions (iii) and (iv) are given in order to keep φ from being trivial. Observe that $\varphi(0) = 0$ and $\varphi(1) = 1$ (Exercise 2.6.3). Given a place φ we define

$$\vartheta_\varphi = \{a \in K \mid \varphi(a) \neq \infty\} = \varphi^{-1}(E).$$

Proposition 2.2.11. *ϑ_φ is an integral subdomain of K, $\vartheta_\varphi \neq K$, and $\vartheta_\varphi \neq 0$.*

Proof. If $a, b \in \vartheta_\varphi$ we have $\varphi(a + b) = \varphi(a) + \varphi(b) \in E$, that is, $a + b \in \vartheta_\varphi$. If $a \in \vartheta_\varphi$, then $\varphi(a) \neq \infty$ and since

$$0 = \varphi(0) = \varphi(a - a) = \varphi(a) + \varphi(-a), \quad \text{we have} \quad \varphi(-a) = -\varphi(a) \in E.$$

It follows that $-a \in \vartheta_\varphi$.

Now for $a, b \in \vartheta_\varphi$ we have $\varphi(ab) = \varphi(a)\varphi(b) \in E$. Therefore ϑ_φ is an integral domain.

Since there exist $a, b \in K$ such that $\varphi(a) = \infty$, $\vartheta_\varphi \neq K$, $\varphi(b) \neq 0$, and $\varphi(b) \neq \infty$, we have $\varphi(b) \in E$ and $b \neq 0$, so $\vartheta_\varphi \neq 0$. □

Observe that $\varphi : \vartheta_\varphi \to E$ is a homomorphism such that $\ker\varphi = \{a \in \vartheta_\varphi \mid \varphi(a) = 0\} = \mathcal{P}_\varphi = \varphi^{-1}(0)$. Then $\vartheta_\varphi/\mathcal{P}_\varphi \cong \varphi(\vartheta_\varphi) \subseteq E$, and $\ker\varphi = \mathcal{P}_\varphi$ is a prime ideal of ϑ_φ.

If $b \in K \setminus \vartheta_\varphi$, $\varphi(b) = \infty$ and

$$1 = \varphi(1) = \varphi\left(b\frac{1}{b}\right) = \varphi(b)\varphi\left(\frac{1}{b}\right),$$

we have $\varphi\left(\frac{1}{b}\right) \neq \infty$ since $\infty\infty = \infty$. Thus $\varphi\left(\frac{1}{b}\right) \in E$. If $\varphi\left(\frac{1}{b}\right) \neq 0$, then $1 = \infty\varphi\left(\frac{1}{b}\right) = \infty$, which is absurd. Hence $\varphi\left(\frac{1}{b}\right) = 0$ and in particular $\frac{1}{b} \in \vartheta_\varphi$. This proves that for any $x \in K$ we have $x \in \vartheta_\varphi$ o $x^{-1} \in \vartheta_\varphi$, i.e., ϑ_φ is a valuation ring.

The maximal ideal $\mathcal{P}$ of ϑ_φ is the nonunit set of ϑ_φ, that is, $x \in \mathcal{P}$ if $x = 0$ or $x \neq 0$ and $x^{-1} \notin \vartheta_\varphi$. Therefore $\varphi\left(x^{-1}\right) = \infty$ or $x = 0$. The relations

$$1 = \varphi(1) = \varphi\left(xx^{-1}\right) = \varphi(x)\,\varphi\left(x^{-1}\right) \quad \text{and} \quad \varphi\left(x^{-1}\right) = \infty$$

imply $\varphi(x) = 0$, i.e., $x \in \ker\varphi$, and conversely, so $\mathcal{P} = \mathcal{P}_\varphi$.

We saw above how to obtain a valuation ring from a place. Conversely, consider a valuation ring ϑ, $\mathcal{P}$ its maximal ideal and $K = \operatorname{quot}\vartheta$. Let E be the field $\vartheta/\mathcal{P}$ and $E_1 = E \cup \{\infty\}$. Let $\varphi : K \longrightarrow E_1$ be given by

$$\varphi(x) = \begin{cases} x \bmod \mathcal{P} & \text{if } x \in \vartheta \\ \infty & \text{if } x \notin \vartheta. \end{cases}$$

We leave it as an exercise to verify that φ is a place. We have by definition

$$\vartheta_\varphi = \{a \in K \mid \varphi(a) \neq \infty\} = \vartheta.$$

Therefore we have shown that the concepts of place and valuation ring are the same.

Definition 2.2.12. Two places $\varphi_1 : K \longrightarrow E_1 \cup \{\infty\}$ and $\varphi_2 : K \longrightarrow E_2 \cup \{\infty\}$ are called *equivalent* if there exists a field isomorphism $\lambda : T_1 \longrightarrow T_2$, where $T_1 = \varphi_1\left(\vartheta_{\varphi_1}\right)$ and $T_2 = \varphi_2\left(\vartheta_{\varphi_2}\right)$, such that $\varphi_2 = \lambda\varphi_1$ (with the convention that $\lambda(\infty) = \infty$).

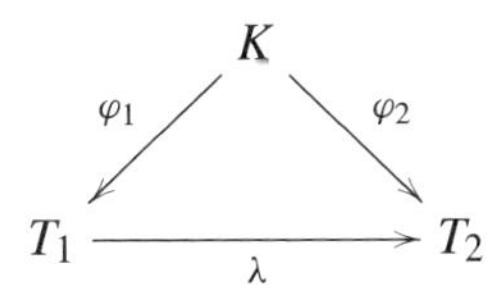

If φ_1 and φ_2 are equivalent, then

$$\vartheta_{\varphi_1} = \varphi_1^{-1}(E_1) = \varphi_1^{-1}(T_1) = \varphi_2^{-1}(\lambda(T_1)) = \varphi_2^{-1}(T_2) = \varphi_2^{-1}(E_2) = \vartheta_{\varphi_2}.$$

Conversely, if $\vartheta_{\varphi_1} = \vartheta_{\varphi_2}$, we have $\mathcal{P}_{\varphi_1} = \mathcal{P}_{\varphi_2}$ and it follows that $T_1 \cong \vartheta_{\varphi_1}/\mathcal{P}_{\varphi_1} = \vartheta_{\varphi_2}/\mathcal{P}_{\varphi_2} \cong T_2$.

In short, we have the following:

Proposition 2.2.13. *Two places φ_1 and φ_2 over a field K are equivalent if and only if $\vartheta_{\varphi_1} = \vartheta_{\varphi_2}$.* □

Let K be a field and let v be a valuation over K. If the value group G of v is contained in $(\mathbb{R}, +)$, then the valuation defines a function $|\ | : K \longrightarrow \mathbb{R}$ given by $|x|_v = e^{-v(x)}$, where $v(0) = \infty$, and $e^{-\infty} = 0$ by definition.

Proposition 2.2.14. *The function $|x|_v$ defined by the valuation v over K is a nonarchimedean absolute value that is nontrivial over K.*

Proof. For all $x, y \in K$ we have:

(i) $|x|_v = e^{-v(x)} \geq 0$ and $|x|_v = e^{-v(x)} = 0 \iff v(x) = \infty \iff x = 0$.
(ii) $|xy|_v = e^{-v(xy)} = e^{(-v(x)-v(y))} = e^{-v(x)}e^{-v(y)} = |x|_v|y|_v$.

(iii) $|x+y|_v = e^{-v(x+y)}$. Now, $v(x+y) \geq \min\{v(x), v(y)\}$, so that

$$-v(x+y) \leq -\min\{v(x), v(y)\} = \max\{-v(x), -v(y)\}.$$

Since the exponential function is increasing, we have

$$\begin{aligned} |x+y|_v &= e^{-v(x+y)} \leq e^{\max\{-v(x),-v(y)\}} \\ &= \max\left\{e^{-v(x)}, e^{-v(y)}\right\} = \max\left\{|x|_v, |y|_v\right\}. \end{aligned}$$

Finally, from the fact that v is nontrivial, it follows that $| \ |_v$ is nontrivial. □

The converse of Proposition 2.2.14 also holds. The proof is straightforward.

Proposition 2.2.15. *Let* $| \ | : K \longrightarrow \mathbb{R}$ *be a nonarchimedean absolute value over* K*. Then the function* $v_{| \ |}$ *defined by* $v_{| \ |} = -\ln|x|$*, where by definition* $-\ln|0| = +\infty$*, is a valuation with value group contained in* $(\mathbb{R}, +)$*.* □

Proposition 2.2.16. *Let* $| \ |_1$ *and* $| \ |_2$ *be two absolute values over a field* K *and let* v_1, v_2 *be the valuations associated with* $| \ |_1$ *and* $| \ |_2$ *respectively. Then* $| \ |_1$ *and* $| \ |_2$ *are equivalent if and only if* v_1 *and* v_2 *are equivalent.*

Proof. We have $v_i = -\ln|x|_i$, $|x|_i = e^{-v_i(x)}$, $i = 1, 2$. Assume that $| \ |_1$ and $| \ |_2$ are equivalent valuations, that is, $|x|_1 < 1 \iff |x|_2 < 1$. Then

$$v_1(x) > 0 \iff |x|_1 = e^{-v_1(x)} < 1 \iff |x|_2 = e^{-v_2(x)} < 1 \iff v_2(x) > 0.$$

So v_1 and v_2 are equivalent.

The converse is analogous. □

The above discussion proves that the concepts of nonarchimedean absolute value, valuation with value group contained in $\mathbb{R}$, valuation ring, and place are essentially the same concept and they correspond to their respective equivalence classes. This correspondence can be summarized as follows:

Absolute value nonarchimedean		**Valuation with value group contained in $\mathbb{R}$**
$\lvert x\rvert$	$\longrightarrow$	$v(x) = -\ln\lvert x\rvert$
$e^{-v(x)}$	$\longleftarrow$	$v(x)$
Valuation		**Valuation ring**
$v(x)$	$\longrightarrow$	$\vartheta = \{x \in K \mid v(x) \geq 0\}$
$v : K^* \longrightarrow K^*/\vartheta^*$, $v(x) = x \bmod \vartheta^*$	$\longleftarrow$	ϑ

Valuation ring		**Place**
ϑ with maximal ideal $\mathcal{P}$	$\longrightarrow$	$\varphi(x) = \begin{cases} x \bmod \mathcal{P} & \text{if } x \in \vartheta \\ \infty & \text{if } x \notin \vartheta \end{cases}$
$\vartheta = \{x \in K \mid \varphi(x) \neq \infty\}$	$\longleftarrow$	φ

In the number field case, there exist archimedean absolute values. In our case, the function field case, all absolute values are nonarchimedean, and in fact, they are discrete, that is, the value group is isomorphic to the ring $\mathbb{Z}$ of rational integers. So even though this section is of a general nature, the reader may consider only, if he or she wishes to, discrete valuations.

Proposition 2.2.17. *Let v_1, v_2 be two valuations over a field K with value group contained in $\mathbb{R}$. Then v_1 and v_2 are equivalent if and only if there exists $\alpha \in \mathbb{R}, \alpha > 0$, such that $v_1 = \alpha v_2$.*

Proof. If $|x|_i = e^{v_i(x)}$ are the associated absolute values, then $v_1 \sim v_2 \iff |\ |_1 \sim |\ |_2 \iff$ there exists $c > 0$ such that $|\ |_1 = |\ |_2^c$, $v_1 = -\ln|\ |_1 = -\ln|\ |_2^c = c\,(-\ln|\ |_2) = cv_2$. □

Definition 2.2.18. Let K be a field. A *prime divisor*, or simply a *prime*, of K is an equivalence class of the set of nontrivial absolute values of K. If the absolute values in the class are archimedean, the prime is called *infinite*; it is called *finite* otherwise.

Hence, in the nonarchimedean case, a prime divisor can be considered a place or the maximal ideal of the valuation ring associated with the absolute value. When we study function fields, the prime divisors will be identified with the maximal ideal of the valuation ring.

Note 2.2.19. Given a nonarchimedean absolute value $|\ |$ over a field K, the ring $\{x \in K \mid |x| \leq 1\}$ is a valuation ring whose maximal ideal is $\{x \in K \mid |x| < 1\}$. This is an immediate consequence of the fact that $v(x) = -\ln|x|$ defines a valuation with valuation ring

$$\begin{aligned}\vartheta_v &= \{x \in K \mid v(x) \geq 0\} = \{x \in K \mid -\ln|x| \geq 0\} \\ &= \{x \in K \mid \ln|x| \leq 0\} = \left\{x \in K \mid |x| \leq e^0 = 1\right\}\end{aligned}$$

and maximal ideal of ϑ_v

$$\mathcal{P}_v = \{x \in K \mid v(x) > 0\} = \{x \in K \mid |x| < 1\}.$$

In Exercise 2.6.14, the reader is asked to give an independent proof of these facts using only the properties of a nonarchimedean absolute value and not the valuation v.

We will end this section with the study of *discrete valuation rings*. Let K be a field and $v : K^* \longrightarrow \mathbb{Z}$ a valuation with valuation ring ϑ and maximal ideal $\mathcal{P}$. Let

$\pi \in \mathcal{P}$ be such that $v(\pi) = 1$ (π is called a *prime element* or *uniformizing element* of the valuation). Then given $x \in K^*$ such that $v(x) = n$, we have $v\left(\pi^{-n}x\right) = 0$, that is, $\pi^{-n}x \in \vartheta^*$, so that x can be written $x = a\pi^n$ with $a \in \vartheta^*$ and $n \in \mathbb{Z}$. This representation is unique since if $x = b\pi^m$ with $b \in \vartheta^*$ and $m \in \mathbb{Z}$, we have $v(x) = v(b\pi^m) = m = n$. Thus $a = b$. In particular, if $x \in \mathcal{P}$ then $x = a\pi^n$ with $n \geq 1$, and $a \in \vartheta^*$ so $\mathcal{P} = (\pi)$. Therefore $\mathcal{P}$ is principal.

Let $\mathfrak{A}$ be any ideal of ϑ such that $\mathfrak{A} \neq 0$ and $\mathfrak{A} \subseteq \mathcal{P}$. Let $n = \min\{v(x) \mid x \in \mathfrak{A}\}$. Then $n \geq 1$. Then there exists $x \in \mathfrak{A}$ such that $v(x) = n$, that is, $x = a\pi^n \in \mathfrak{A}$ with $a \in \vartheta^*$. This implies that $\pi^n \in \mathfrak{A}$ and $(\pi^n) \subseteq \mathfrak{A}$. If y is an arbitrary nonzero element of $\mathfrak{A}$, we have $v(y) \geq n$. Then $y = b\pi^m$ with $m \geq n$ and $b \in \vartheta^*$. Hence $y = \left(b\pi^{m-n}\right)\pi^n$, $b\pi^{m-n} \in \vartheta$. Therefore $y \in (\pi^n)$, that is, $\mathfrak{A} = (\pi^n) = \mathcal{P}^n$. Hence, every nonzero ideal of ϑ is a power of $\mathcal{P}$. We have the following theorem:

Theorem 2.2.20. *If v is a discrete valuation over a field K, the valuation ring ϑ (which is called a* discrete valuation ring*) satisfies that its maximal ideal $\mathcal{P}$ is principal and is generated by any prime element. Every nonzero ideal of ϑ is a power of $\mathcal{P}$ and the groups K^* and $\vartheta^* \times \mathbb{Z}$ are isomorphic.*

Proof. The first part of the statement was proved in the above discussion. To prove the last part, let $x \in K^*$. We can write $x = a\pi^n$ in a unique way, and therefore the function $\varphi : K^* \longrightarrow \vartheta^* \times \mathbb{Z}$, defined by $\varphi(x) = (a, n)$, is the isomorphism needed. □

2.3 Absolute Values and Completions

In this section, we will use the notation $|\ |$ for the usual absolute value in the field of real numbers $\mathbb{R}$. Let K be a field with absolute value $\|\ \|$.

Definition 2.3.1. Let K be any field. A sequence $\{a_n\}_{n=0}^{\infty} \subseteq K$ is called *Cauchy* if $\lim_{n,m\to\infty} \|a_n - a_m\| = 0$. We say that a_n *converges* to an element a if $\lim_{n\to\infty} \|a_n - a\| = 0$, or in other words, if a_n converges to a with respect to the topology given by the absolute value.

Definition 2.3.2. A field K is called *complete* if every Cauchy sequence in K converges to some element of K.

Example 2.3.3. Let $\mathbb{Q}$ with $|\ |_p$ the p-adic absolute value, that is, $|x|_p = e^{-v(x)}$, $v_p(x) = n$, where $x = p^n \frac{a}{b}$, $p \nmid ab$.

We consider the sequence $a_n = 1+p+p^4+\cdots+p^{n^2}$. If $n \leq m$ we have $a_m - a_n = p^{(n+1)^2} + \cdots + p^{m^2}$ and $|a_m - a_n|_p = e^{-(n+1)^2} \xrightarrow[n\to\infty]{} 0$. That is, a_n is a Cauchy sequence in $\left(\mathbb{Q}, |\ |_p\right)$; however, it can be proved that $\{a_n\}_{n=0}^{\infty}$ does not converge in $\mathbb{Q}$ (see Exercise 2.6.2), and so $\mathbb{Q}$ is not complete with respect to the absolute value $|\ |_p$. It is well known that $\mathbb{Q}$ is not complete with respect to the archimedean absolute value either, and since there are no other absolute values in $\mathbb{Q}$, it follows that $\mathbb{Q}$ is not complete with respect to any absolute value.

The completion of $\mathbb{Q}$ with respect to its usual absolute value is done using the same procedure as with a metric space. The completion obtained is the set $\mathbb{R}$ of real numbers.

We say that two Cauchy sequences $\{b_n\}_{n=1}^{\infty}$ and $\{a_n\}_{n=1}^{\infty}$ are *equivalent*, and we write $\{a_n\} \sim \{b_n\}$, if $\lim_{n\to\infty} ||a_n - b_n|| = 0$. It is easy to see that this defines an equivalence relation. Let $\bar{K}$ be the collection of all these equivalence classes and let $[\{a_n\}] \in \bar{K}$. We define $||[\{a_n\}]|| = \lim_{n\longrightarrow\infty} ||a_n||$. The latter is well defined since $\{||a_n||\} \subseteq \mathbb{R}$. Now if $\{a_n'\}$ defines the same element in $\bar{K}$, we will have

$$\left|||a_n|| - ||a_n'||\right| \le \left||a_n - a_n'\right|| \quad \text{so that} \quad \lim_{n\longrightarrow\infty} ||a_n|| = \lim_{n\longrightarrow\infty} \left||a_n'\right||.$$

Thus, $\|\ \|$ is well defined in $\bar{K}$.

Let $\alpha, \beta, \gamma \in \bar{K}$ where $\alpha = [\{a_n\}]\,, \beta = [\{b_n\}]\,, \gamma = [\{c_n\}]$. We define $\alpha + \beta = [\{a_n + b_n\}]$ and $\alpha\beta = [\{a_n b_n\}]$. We leave it as an exercise to verify that $\{a_n + b_n\}$ and $\{a_n b_n\}$ are Cauchy sequences and that the definitions of $\alpha + \beta$ and $\alpha\beta$ do not depend on the representatives.

With this structure, $\bar{K}$ is a commutative ring with unit, 0 and 1 being the representatives of the constant sequences 0 and 1 respectively.

If $\alpha \neq 0$, $\{a_n\}$ is not equivalent to the constant sequence 0. So

$$\lim_{n\to\infty} ||a_n - 0|| = \lim_{n\longrightarrow\infty} ||a_n|| \neq 0.$$

That is, there exists n_0 such that for $n \ge n_0$, $a_n \neq 0$. Therefore $\{a_n^{-1}\}_{n=n_0}^{\infty}$ is defined and it is a Cauchy sequence. Now, since $a_n a_n^{-1} = 1$ for $n \ge n_0$,

$$\alpha^{-1} = \left[\left\{a_n^{-1}\right\}_{n=n_0}^{\infty}\right]$$

is defined and $\alpha\alpha^{-1} = 1$. Thus $\bar{K}$ is a field.

Now the function $\varphi : K \longrightarrow \bar{K}$, defined by $\varphi(a) = \bar{a}$, where $\bar{a}$ is a representative of the sequence $\{a_n\}$ and $a_n = a$ for all n, is a field monomorphism. We note that

$$||\varphi(a)|| = \lim_{n\to\infty} ||a_n|| = \lim_{n\to\infty} ||a|| = ||a||.$$

Therefore the function $\|\ \|$ in $\bar{K}$ is an extension of the absolute value defined in K. It is easy to see that $\|\ \|$ defined in $\bar{K}$ is an absolute value.

Now, $\|\ \|$ is a nonarchimedean absolute value in K if and only if $\|\ \|$ is a nonarchimedean absolute value in $\bar{K}$. We will see that K is dense in $\bar{K}$. Given the monomorphism φ, we can assume without loss of generality that K is contained in $\bar{K}$. Let $\alpha \in K$, $\varepsilon > 0$, and

$$B(\alpha, \varepsilon) = \left\{\beta \in \bar{K} \mid ||\beta - \alpha|| < \varepsilon\right\}.$$

We will see that $B(\alpha, \varepsilon) \cap K \neq \emptyset$. There exists n_0 such that for $n \ge n_0$, $\left||a_n - a_{n_0}\right|| < \varepsilon$. We take the constant sequence $\overline{a_{n_0}} = \left\{a_{n_0}\right\} \in K$. Then $\overline{a_{n_0}} \in B(\alpha, \varepsilon) \cap K$. This proves that K is dense in $\bar{K}$.

Finally, let us see that $\bar{K}$ is complete, that is, every Cauchy sequence in $\bar{K}$ converges in $\bar{K}$. Let $\alpha_m = \left[\{a_{m,n}\}_{n=1}^{\infty}\right]$, with $\{\alpha_m\}_{m=1}^{\infty}$ a Cauchy sequence in $\bar{K}$. Since K is dense in $\bar{K}$, for each $m \in \mathbb{N}$ there exists a constant sequence (i.e., an element of K)

$$x_m = \left[\left\{x_n^{(m)}\right\}_{n=1}^{\infty}\right] \in K, \quad \text{such that} \quad ||x_m - \alpha_m|| < \frac{1}{m}$$

for all $m \in \mathbb{N}$. We will see that $\{x_m\}_{m=1}^{\infty}$ is a Cauchy sequence.

We have

$$||x_m - x_n|| \le ||x_m - \alpha_m|| + ||\alpha_m - \alpha_n|| + ||\alpha_n - x_n|| .$$

Now, since $\{\alpha_m\}_{m=1}^{\infty}$ is a Cauchy sequence, given $\varepsilon > 0$ there exists N such that if $n, m \ge N$,

$$\frac{1}{n} < \frac{\varepsilon}{3}, \ \frac{1}{m} < \frac{\varepsilon}{3}, \ \text{and} \ ||\alpha_m - \alpha_n|| < \frac{\varepsilon}{3}.$$

Therefore $||x_m - x_n|| < \varepsilon$ for $n, m \ge N$. Hence $\{x_m\}_{m=1}^{\infty}$ is Cauchy sequence.

Let $x_n^{(m)} = x^{(m)} \in K$ for all n. We have

$$||x_m - x_n|| = \lim_{t\to\infty} \left|\left|x_t^{(m)} - x_t^{(n)}\right|\right| = \left|\left|x^{(m)} - x^{(n)}\right|\right| < \varepsilon$$

for $n, m \ge N$, whence $\left\{x^{(m)}\right\}_{m=1}^{\infty} \subseteq K$ is a Cauchy sequence and it defines an element $\alpha \in \bar{K}$, $\alpha = \left[\{x^{(m)}\}\right]_{m=1}^{\infty}$.

We have

$$||\alpha_n - \alpha|| \le ||\alpha_n - x_n|| + ||x_n - \alpha|| < \frac{1}{n} + \lim_{p\to\infty} \left|\left|x^{(n)} - x^{(p)}\right|\right| .$$

Since $\left\{x^{(r)}\right\}_{r=1}^{\infty}$ is a Cauchy sequence, given $\varepsilon > 0$ there exists $N \in \mathbb{N}$ such that for $n, p \ge N$, $\left|\left|x^{(n)} - x^{(p)}\right|\right| < \frac{\varepsilon}{2}$ and $\frac{1}{n} < \frac{\varepsilon}{2}$. Thus, for $n \ge N$ we have $||\alpha_n - \alpha|| < \varepsilon$. Therefore $\{\alpha_n\}_{n=0}^{\infty}$ converges to $\alpha \in \bar{K}$, so $\bar{K}$ is complete.

Let Y be any other complete metric space such that there exists a metric space isometry $\lambda : K \longrightarrow Y$ (that is, λ is a distance-preserving map) and such that $\lambda(K)$ is dense in Y. We will see that there exists a bijective isometry $\psi : Y \longrightarrow \bar{K}$. Consider the diagram. If $y \in Y$, where $y = \lim_{n\to\infty} \lambda(y_n)$ and $y_n \in K$, then $\{\varphi(y_n)\}$ is a Cauchy sequence in $\bar{K}$ and we can define $z = \lim_{n\to\infty} \varphi(y_n)$. Let $\psi(y) = z$. It can be verified that $\psi(y)$ does not depend on the sequence $\{y_n\}_{n=0}^{\infty}$ and that ψ is an isometry. Since the process can be inverted, we obtain a function $\phi : \bar{K} \longrightarrow Y$, with $\phi\psi = \mathrm{Id}_Y$ and $\psi\phi = \mathrm{Id}_{\bar{K}}$. It is easy to see that ϕ and ψ are inverse isometries. We sum up the previous development in the following theorem:

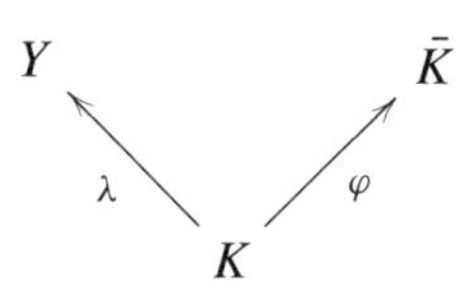

Theorem 2.3.4. *Let K be any field and let $|\ |$ be an absolute value in K. There exists a unique field $\bar{K}$ (up to isometry) such that (i) $K \subseteq \bar{K}$ and (ii) there is a unique way of extending $|\ |$ to $\bar{K}$ such that $\left(\bar{K}, |\ |\right)$ is a complete field and K is dense in $\bar{K}$.* □

Definition 2.3.5. The field obtained in Theorem 2.3.4 is called the *completion of K with respect to* $|\ |$.

Example 2.3.6. Given $\mathbb{Q}$ with the usual absolute value, the completion of $\mathbb{Q}$ is the field of real numbers $\mathbb{R}$.

Example 2.3.7. Given $\mathbb{Q}$ with the p-adic absolute value, the completion is denoted by $\mathbb{Q}_p$ and it can be represented as

$$\mathbb{Q}_p = \left\{ \sum_{n=m}^{\infty} a_n p^n \mid m \in \mathbb{Z}, a_n \in \{0, 1, \ldots, p-1\} \right\}.$$

$\mathbb{Q}_p$ is called the field of *p-adic numbers*. For instance, -1 is represented as follows:

$$-1 = \frac{p-1}{1-p} = (p-1)\sum_{n=0}^{\infty} p^n = \sum_{n=0}^{\infty} (p-1)p^n.$$

In fact,

$$S_m = \sum_{n=0}^{m} (p-1)p^n = \sum_{n=0}^{m} (p^{n+1} - p^n) = p^{m+1} - 1 \xrightarrow[m\to\infty]{v_p} 0 - 1 = -1.$$

The closure of $\mathbb{Z}$ in $\mathbb{Q}_p$ is called *the ring of p-adic integers* and denoted by $\mathbb{Z}_p$. We can represent it as

$$\mathbb{Z}_p = \left\{ \sum_{n=0}^{\infty} a_n p^n \mid a_n \in \{0, 1, \ldots, p-1\} \right\}.$$

Notation 2.3.8. Given a field K with a nonarchimedean absolute value $|\ |$, let $v_{|\ |}$ be the valuation associated to $|\ |$. Then the completion of K with respect to $|\ |$ will be denoted by $K_{\mathcal{P}}$, where $\mathcal{P}$ is the maximal ideal of the valuation ring associated to the valuation.

Definition 2.3.9. Let $|\ |$ be a nonarchimedean absolute value over a field K, $\vartheta = \{x \in K \mid |x| \leq 1\}$, and let $\mathcal{P} = \{x \in K \mid |x| < 1\}$ be the maximal ideal of ϑ. The field $\vartheta/\mathcal{P}$ is called the *residue field* of K with respect to $\mathcal{P}$.

Assume that ϑ is a discrete valuation ring. Let $K_{\mathcal{P}}$ be the completion of K with respect to $|\ |$. For $x \in K_{\mathcal{P}}$, we can write x as the limit of a sequence $\{x_n\}_{n=0}^{\infty} \subseteq K$. We have $|x| = \lim_{n\to\infty} |x_n|$, so the absolute value is nonarchimedean in $K_{\mathcal{P}}$. On the other hand, the valuation v can be extended to $K_{\mathcal{P}}$ by setting

$$v(x) = \lim_{n\to\infty} v(x_n).$$

Indeed, we have

$$|x| = e^{-v(x)} = \lim_{n\to\infty} |x_n| = \lim_{n\to\infty} e^{-v(x_n)} = e^{\lim_{n\to\infty} v(x_n)}.$$

In particular $v\left(K_{\mathcal{P}}^*\right) = \mathbb{Z} = v\left(K^*\right)$ since $\{v\left(x_n\right)\}_{n=0}^{\infty}$ is constant starting from some index n_0.

Hence $\hat{\vartheta} = \{x \in K_{\mathcal{P}} \mid |x| \leq 1\}$ is a discrete valuation ring and

$$\hat{\mathcal{P}} = \{x \in K_{\mathcal{P}} \mid |x| < 1\}$$

is its maximal ideal. It follows form the definitions that $\hat{\vartheta}$ and $\hat{\mathcal{P}}$ are the closures of ϑ and $\mathcal{P}$ in $K_{\mathcal{P}}$ respectively. Furthermore, if $\mathcal{P} = (\pi) = \pi\vartheta$ then $v(\pi) = 1$. Thus π also generates $\hat{\mathcal{P}}$ in $\hat{\vartheta}$, that is, $\hat{\mathcal{P}} = \pi\hat{\vartheta}$.

Proposition 2.3.10. *For any $n \in \mathbb{N}$, we have $\vartheta/\mathcal{P}^n \cong \hat{\vartheta}/\hat{\mathcal{P}}^n$.*

Proof. Let $\varphi : \vartheta \to \hat{\vartheta}/\hat{\mathcal{P}}^n$ be the natural homomorphism, that is, $\varphi\left(x\right) = x \bmod \hat{\mathcal{P}}^n$. First we will see that φ is an epimorphism. If $x \in \hat{\vartheta}$, there exists $\{x_m\}_{m=1}^{\infty} \subseteq \vartheta$ such that $x = \lim_{n\to\infty} x_m$. In particular, there exists $N \in \mathbb{N}$ such that for $m \geq N$,

$$x - x_m \in \hat{\mathcal{P}}^n = \{y \in K_{\mathcal{P}} \mid v(y) \geq n\} = \left\{y \in K_{\mathcal{P}} \mid |y| \leq e^{-n}\right\}.$$

Then $x \bmod \hat{\mathcal{P}}^n = x_m \bmod \hat{\mathcal{P}}^n = \varphi\left(x_m\right)$.

Finally,

$$\ker\varphi = \left\{x \in \vartheta \mid x \in \hat{\mathcal{P}}^n\right\} = \{x \in \vartheta \mid v(x) \geq n\} = \mathcal{P}^n,$$

from which we obtain the result. □

Corollary 2.3.11. *The residue fields of K and $K_{\mathcal{P}}$ are isomorphic.*

Proof. This is just the case $n = 1$ of Proposition 2.3.10. □

Notation 2.3.12. When we consider a convergent sequence $s_n = \sum_{i=m}^{n} a_i$, the limit is written as the series $\sum_{i=m}^{\infty} a_i = \lim_{n\to\infty} s_n$.

Proposition 2.3.13. *Each element $\alpha \neq 0$ in $K_{\mathcal{P}}$ has a unique series representation of the form*

$$\alpha = \pi^m \sum_{i=0}^{\infty} s_i \pi^i,$$

with $v\left(\alpha\right) = m \in \mathbb{Z}$, $s_i \in S \subseteq \vartheta$, $s_0 \neq 0$, S any set of representatives of $\vartheta/\mathcal{P} \cong \hat{\vartheta}/\hat{\mathcal{P}}$, and $0 \in S$.

Proof. First we note that for any $m \in \mathbb{Z}$ and $\{s_n\}_{n=0}^{\infty} \subseteq S$, $\pi^m \sum_{i=0}^{\infty} s_i \pi^i$ is an element of $K_{\mathcal{P}}$. This follows from the fact that the sequence $a_n = \pi^m \sum_{i=0}^{n} s_i \pi^i$ is Cauchy and that $K_{\mathcal{P}}$ is complete.

Now let us see that the representation is unique. If

$$\alpha = \pi^m \sum_{i=0}^{\infty} s_i \pi^i = \pi^{m_1} \sum_{i=0}^{\infty} s_i' \pi^i, \quad \text{with} \quad s_0 \neq 0 \quad \text{and} \quad s_0' \neq 0,$$

then $v(\alpha) = m = m_1$. Hence, $\sum_{i=0}^{\infty} s_i \pi^i = \sum_{i=0}^{\infty} s_i' \pi^i$, that is,

$$s_0 + s_1 \pi + \cdots = s_0' + s_1' \pi + \cdots .$$

Therefore

$$\left(s_0 - s_0'\right) + s_1 \pi + \cdots = s_1' \pi + \cdots .$$

The right side has valuation greater than or equal to 1, so $s_0 - s_0' = 0$. By induction on i it is easy to conclude that $s_i = s_i'$ for all i.

Finally, let us see that every element of $K_{\mathcal{P}}$ admits this kind of representation. Let $\alpha \in K_{\mathcal{P}}$ with $\alpha \neq 0$ and $v(\alpha) = m$. Then $v\left(\pi^{-m}\alpha\right) = 0$, that is, $\alpha = \pi^m \alpha_0$ with $\alpha_0 \in \hat{\vartheta}^*$. We have

$$\alpha_0 \equiv s_0 \bmod \hat{\mathcal{P}}, \quad s_0 \in S, \quad \text{and} \quad s_0 \neq 0.$$

Since $\alpha_0 - s_0 \in \hat{\mathcal{P}}$ it follows that $v(\alpha_0 - s_0) \geq 1$. Therefore $\alpha_0 = s_0 + \pi\alpha_1, \alpha_1 \in \hat{\vartheta}$. Repeating the process we obtain, for each n,

$$\alpha_0 = s_0 + s_1 \pi + \cdots + s_n \pi^n + \alpha_{n+1} \pi^{n+1}, \quad \text{with} \quad s_i \in S \quad \text{and} \quad \alpha_{n+1} \in \hat{\vartheta}.$$

The sequence $r_n = \sum_{i=0}^{n} s_i \pi^i$ satisfies $\alpha_0 - r_n = \alpha_{n+1}\pi^{n+1}$, that is, $v(\alpha_0 - r_n) \geq n + 1$. Thus r_n converges to α_0 and $\alpha = \pi^m \sum_{i=0}^{\infty} s_i \pi^i$. □

In the particular case of the p-adic valuation v in $\mathbb{Q}$, we have

$$\vartheta = \left\{x \in \mathbb{Q} \mid v_p(x) \geq 0\right\} = \left\{\frac{a}{b} \mid p \nmid b\right\} = \mathbb{Z}_{(p)},$$

which is the localization of $\mathbb{Z}$ at (p). The maximal ideal is $(p)\mathbb{Z}_{(p)}$ and the residue field is

$$\mathbb{Z}_{(p)}/(p)\mathbb{Z}_{(p)} \cong \mathbb{Z}/(p)\mathbb{Z} \cong \mathbb{F}_p,$$

the finite field of p elements.

A set of representatives is $\{0, 1, \ldots, p-1\} = S$. Therefore,

$$\mathbb{Q}_p = \left\{p^m \sum_{n=0}^{\infty} s_n p^n \mid m \in \mathbb{Z}, s_n \in S\right\}.$$

Furthermore,

$$\begin{aligned} \hat{\vartheta} &= \text{closure of } \mathbb{Z}_{(p)} \text{ in } \mathbb{Q}_p = \left\{x \in \mathbb{Q}_p \mid v(x) \geq 0\right\} \\ &= \left\{\sum_{n=0}^{\infty} s_n p^n \mid s_n \in S\right\} = \mathbb{Z}_p, \end{aligned}$$

the ring of p-adic integers.

An interesting observation is that there is no analogue to the uniqueness in the archimedean case. For instance in $\mathbb{R}$,

$$0.0999\cdots = \sum_{n=2}^{\infty} 9 \times \left(10^{-1}\right)^n = 0.1 = 1 \times \left(10^{-1}\right).$$

Theorem 2.3.14 (Hensel's lemma). *Let K be a complete field with respect to a nonarchimedean absolute value. Let $\bar{K}$ be the residue field, ϑ the valuation ring, which we assume to be a discrete valuation ring, and let $\mathcal{P}$ be the maximal ideal, i.e., $\bar{K} \cong \vartheta/\mathcal{P}$. We assume that $f(x) \in \vartheta[x]$ is a monic polynomial. Let $\bar{f}(x) = f(x) \bmod \mathcal{P} \in \bar{K}[x]$ and suppose that $\bar{f}(x) = h(x)g(x)$ with $h(x), g(x) \in \bar{K}[x]$ and $h(x), g(x)$ relatively prime. Then there exist $H(x), G(x) \in K[x]$ such that*

$$f(x) = H(x)G(x), \quad \bar{H}(x) = h(x), \quad \bar{G}(x) = g(x),$$

and

$$\deg H(x) = \deg h(x), \quad \deg G(x) = \deg g(x).$$

Proof. Since $f(x)$ is a monic polynomial, it follows that $\deg f(x) = \deg \bar{f}(x) = n$. Now let $h(x), g(x)$ be of degrees r and $n - r$ respectively. Let $H_1(x), G_1(x) \in \vartheta[x]$ be such that

$$\bar{H}_1(x) = h(x), \quad \bar{G}_1(x) = g(x), \quad \deg H_1 = \deg h, \quad \deg G_1 = \deg g.$$

Then

$$f(x) - G_1(x)H_1(x) \in \mathcal{P}[x].$$

Assume that for $k \geq 1$ we have constructed $G_k(x), H_k(x) \in \vartheta[x]$ such that

$$f(x) - G_k(x)H_k(x) \in \mathcal{P}^k[x], \quad \deg G_k(x) \leq \deg g(x), \ \deg H_k(x) \leq \deg h(x),$$
$$\bar{G}_k(x) = g(x), \quad \text{and} \quad \bar{H}_k(x) = h(x).$$

Now define

$$G_{k+1}(x) = G_k(x) + \pi^k m(x) \quad \text{and} \quad H_{k+1}(x) = H_k(x) + \pi^k n(x),$$

with $m(x), n(x)$ to be determined and π a prime element for $\mathcal{P}$. We have

$$\begin{aligned} &f(x) - G_{k+1}(x)H_{k+1}(x) \\ &\quad = f(x) - G_k(x)H_k(x) - \pi^k(m(x)H_k(x) + n(x)G_k(x)) - \pi^{2k}m(x)n(x). \end{aligned}$$

Now $\mathcal{P} = (\pi)$, $\mathcal{P}^k = \left(\pi^k\right)$, and $f(x) - G_k(x)H_k(x) \in \mathcal{P}^k[x]$. Therefore

$$u(x) = \frac{f(x) - G_k(x)H_k(x)}{\pi^k} \in \vartheta[x]$$

and

$$f(x) - G_{k+1}(x)H_{k+1}(x) \in \mathcal{P}^{k+1}[x] \Longleftrightarrow$$
$$\pi^k \left(u(x) - m(x)H_k(x) - n(x)G_k(x)\right) - \pi^{2k} m(x)n(x) \in \mathcal{P}^{k+1}[x].$$

Since $2k \geq k+1$ we need to find $m(x), n(x) \in \vartheta[x]$ such that

$$u(x) - m(x)H_k(x) - n(x)G_k(x) \subset \mathcal{P}[x].$$

Given that $\bar{H}_k(x) = h(x)$ and $\bar{G}_k(x) = g(x)$ are relatively prime, we choose $m(x)$ and $n(x)$ such that

$$\bar{u}(x) = \bar{m}(x)\bar{H}_k(x) + \bar{n}(x)\bar{G}_k(x).$$

Furthermore $m(x)$ and $n(x)$ can be chosen such that

$$\deg m(x) \leq n - r \quad \text{and} \quad \deg n(x) \leq r.$$

Then

$$\deg G_{k+1}(x) \leq \deg G_k(x) \leq n - r \quad \text{and} \quad \deg H_{k+1}(x) \leq \deg H_k(x) \leq r,$$

and therefore

$$v\left(G_{k+1} - G_k\right) \geq k \quad \text{and} \quad v\left(H_{k+1} - H_k\right) \geq k.$$

It follows that

$$\{G_k(x)\}_{k=1}^{\infty} \quad \text{and} \quad \{H_k(x)\}_{k=1}^{\infty} \subseteq \vartheta[x]$$

are Cauchy. Since K is complete, these sequences converge to polynomials $G(x)$, $H(x)$. Further, since

$$\bar{G}_k(x) = g(x) \quad \text{and} \quad \bar{H}_k(x) = h(x),$$

we have

$$\begin{aligned} &\bar{G}(x) = g(x), && \bar{H}(x) = h(x), \\ &\deg G(x) \leq \deg g(x) = n - r, && \deg H(x) \leq \deg h(x) = r. \end{aligned}$$

Finally, since

$$f(x) - G_k(x)H_k(x) \in \mathcal{P}^k[x],$$

we have

$$\lim_{k\to\infty} \left(f(x) - G_k(x)H_k(x)\right) = 0,$$

that is, $f(x) = H(x)G(x)$ with all the required properties. □

Example 2.3.15. As an application of Hensel's lemma we will prove that the p-adic field $\mathbb{Q}_p$, $p > 2$, contains the $(p-1)$th roots of unity. We consider the monic polynomial

$$f(x) = x^{p-1} - 1 \in \mathbb{Z}_p[x] \subseteq \mathbb{Q}_p[x].$$

The residue field of $\mathbb{Q}_p$ is $\mathbb{F}_p$, $(\vartheta = \mathbb{Z}_p, \mathcal{P} = p\mathbb{Z}_p)$. Hence, $\bar{f}(x) = x^{p-1} - 1 \in \mathbb{F}_p[x]$. We know that in $\mathbb{F}_p[x]$ we have

$$x^{p-1} - 1 = \prod_{\alpha \in \mathbb{F}_p^*} (x - \alpha),$$

and if $\alpha, \beta \in \mathbb{F}_p^*$ with $\alpha \neq \beta$, then $x - \alpha$ and $x - \beta$ are relatively prime. Therefore, by Hensel's lemma, $f(x)$ splits into linear factors of $\mathbb{Q}_p[x]$, that is, the $(p-1)$th roots of unity belong to $\mathbb{Q}_p$.

Proposition 2.3.13 tells us that every complete field under a nonarchimedean valuation can be represented as a "Laurent series" with "indeterminate" a prime element and coefficients in a set of representatives of the residue field. Here we note that the algebraic structure of the field does not always correspond to the structure of the field of Laurent series in an indeterminate with coefficients in a field. More precisely, let k be an arbitrary field and let t be a transcendental element over k. We define the *ring of formal series* as

$$k[[t]] = \left\{ \sum_{i=0}^{\infty} a_i t^i \mid a_i \in k \right\}$$

with the usual operations, that is,

$$\sum_{i=0}^{\infty} a_i t^i + \sum_{i=0}^{\infty} b_i t^i = \sum_{i=0}^{\infty} (a_i + b_i) t^i;$$

$$\sum_{i=0}^{\infty} a_i t^i \sum_{i=0}^{\infty} b_i t^i = \sum_{i=0}^{\infty} \left(\sum_{k=0}^{i} a_k b_{i-k} \right) t^i.$$

It is easy to see that $k[[t]]$ is an integral domain with field of quotients equal to

$$k((t)) = \left\{ \sum_{i=m}^{\infty} a_i t^i \mid m \in \mathbb{Z}, a_i \in k \right\} = \bigcup_{n=1}^{\infty} \frac{1}{t^n} k[[t]].$$

The latter field is called the field of *Laurent series*.

In $k((t))$ we define the natural valuation $v\colon k((t)) \to \mathbb{Z}$ as follows. If $f(t) \in k((t))$, $f(t) \neq 0$, we write $f(t) = t^n g(t)$ with $n \in \mathbb{Z}$, $g(t) \in k[[t]]$ and $g(0) \neq 0$. Then $v(f(t)) := n$. The valuation ring of v is $\vartheta_v = k[[t]]$, the maximal ideal is (t), the residue field is $k \cong k[[t]]/(t)$, and the absolute value is given by $\|f(t)\| = e^{-n}$.

Coming back to the case of a complete field, let us consider $\mathbb{Q}_p$ as an example. Each element $\mathbb{Q}_p$ is represented as $\sum_{i=m}^{\infty} a_i p^i$ or $\sum_{i=m}^{\infty} a_i \pi^i$, where π is a prime element and $a_i \in \{0, 1, \dots, p-1\}$. However, $\mathbb{Q}_p$ is not isomorphic to $\mathbb{F}_p((\pi))$. Indeed, on the one hand, $\mathbb{Q} \subseteq \mathbb{Q}_p$ implies that $\mathbb{Q}_p$ is of characteristic 0 and on the other hand, since $\mathbb{F}_p \subseteq \mathbb{F}_p((\pi))$, $\mathbb{F}_p((\pi))$ is of characteristic $p > 0$. Later on we will prove that in a function field, the completions are in fact fields of Laurent series.

Now a natural question is what happens with complete fields with respect to an archimedean valuation. The answer is very simple: the only complete archimedean fields are $\mathbb{R}$ and $\mathbb{C}$. We finish this section with a proof of this fact.

Proposition 2.3.16. *Let F be a field containing $\mathbb{C}$. Suppose that F is complete under an archimedean absolute value $\| \ \|$ defined such that $\|\alpha\| = |\alpha|$ for $\alpha \in \mathbb{C}$, where $| \ |$ is the usual absolute value of $\mathbb{C}$. Then for $x \in F$, $\sigma(x) = \{\lambda \in \mathbb{C} \mid x - \lambda 1 = 0\}$ is nonempty. Therefore $F = \mathbb{C}$.*

Proof. We can consider F as a vector space over $\mathbb{C}$. Furthermore, F is a normed space (with norm its absolute value), so that in particular, F is a Banach space. Let $x \in F$ and $\lambda_0 \notin \sigma(x)$, so that $(x - \lambda_0 1)^{-1} \neq 0$. From the Hahn–Banach theorem [130, Theorem 5.16], we know that there exists a bounded linear functional

$$\Phi \colon F \to \mathbb{C} \quad \text{such that} \quad \Phi\left[(x - \lambda_0 1)^{-1}\right] \neq 0.$$

Let

$$f \colon \mathbb{C} \setminus \sigma(x) \to \mathbb{C} \quad \text{be defined by} \quad f(\lambda) = \Phi\left((x - \lambda 1)^{-1}\right).$$

Then $f(\lambda_0) \neq 0$, and f is a differentiable function since

$$\begin{aligned}
\frac{f(\lambda) - f(\mu)}{\lambda - \mu} &= \frac{\Phi\left((x-\lambda 1)^{-1}\right) - \Phi\left((x-\mu 1)^{-1}\right)}{\lambda - \mu} \\
&= \frac{\Phi\left(\frac{1}{x-\lambda 1} - \frac{1}{x - \mu 1}\right)}{\lambda - \mu} = \frac{\Phi\left(\frac{\lambda - \mu}{(x-\lambda 1)(x-\mu 1)}\right)}{\lambda - \mu} \\
&= \Phi\left(\frac{1}{(x - \lambda \cdot 1)(x - \mu 1)}\right) \xrightarrow[\mu \to \lambda]{} \Phi\left((x - \lambda \cdot 1)^{-2}\right).
\end{aligned}$$

Therefore, if $\sigma(x) = \emptyset$, then f is an entire function. Now we have

$$\lambda f(\lambda) - \Phi\left[\lambda (x - \lambda 1)^{-1}\right] = \Phi\left[\left(\frac{x}{\lambda} - 1\right)^{-1}\right] \xrightarrow[\lambda \to \infty]{} \Phi(-1),$$

that is, $\lim_{\lambda \to \infty} f(\lambda) = \lim_{\lambda \to \infty} \frac{\Phi(-1)}{\lambda} = 0$, which tells us that f is bounded at the infinite point, and by Liouville's theorem [130, Theorem 10.23], f is constant and equal to 0. Therefore $f(\lambda_0) = 0$, which contradicts our choice.

For $x \in F$, there exists $\lambda \in \mathbb{C}$ such that $x - \lambda 1 = 0$, that is, $x = \lambda 1 = \lambda \in \mathbb{C}$. Therefore, $F \subset \mathbb{C}$. ◻

Theorem 2.3.17. *Let F be any field and assume that F is complete under an archimedean absolute value. Then $F = \mathbb{R}$ or $F = \mathbb{C}$.*

Proof. Since F has an archimedean absolute value, F is of characteristic 0. Therefore $\mathbb{Q} \subseteq F$. When we restrict the absolute value of F to $\mathbb{Q}$, we obtain the unique archimedean absolute value of $\mathbb{Q}$, which is the usual one. Since F is complete, F contains the completion of $\mathbb{Q}$ with respect to this absolute value, that is, $\mathbb{R} \subseteq F$. Now if $i = \sqrt{-1}$ we have $\mathbb{R}(i) = \mathbb{C} \subseteq F(i)$, so $[F(i) : F]$ is equal to 1 or 2. If $F(i) = F$, then using Proposition 2.3.16 we set $F = F(i) = \mathbb{C}$. If $F(i) \neq F$, then $F(i) = \{a + bi \mid a, b \in F\}$. The absolute value of F can be extended to $F(i)$ by putting

$$||a + bi|| = \sqrt{|a|^2 + |b|^2},$$

and it is easy to see that $F(i)$ is complete. Then from Proposition 2.3.16, we conclude that $F(i) = \mathbb{C}$ and since $\mathbb{R} \subseteq F$, it follows that $F = \mathbb{R}$. □

Remark 2.3.18. Proposition 2.3.16 is essentially the Gelfand–Mazur theorem [130, Theorem 18.7], and Theorem 2.3.17 is called the theorem of Ostrowski. For the proof of Proposition 2.3.16 we have used the theorem of Hahn–Banach, which is a standard result in the theory of functional analysis that can be found in any book in that area, for instance [130, Theorem 5.16]. The other ingredient, Liouville's theorem, should be well known from any basic course in complex analysis [130, Theorem 10.23].

Corollary 2.3.19. *The only archimedean fields are the subfields of $\mathbb{C}$ with the usual absolute value.* □

2.4 Valuations in Rational Function Fields

The purpose of this section is to find the analogue of Theorem 1.2.11, that is, to characterize all valuations in $k(x)$, for k an arbitrary field, such that the valuation is trivial on k.

First we study all valuations defined in a similar way as the p-adic valuations in $\mathbb{Q}$. Let $f(x) \in k[x]$ be an irreducible monic polynomial. For $\alpha(x) \in k(x)$, we write

$$\alpha(x) = \frac{h(x)}{g(x)} = f(x)^s \frac{u(x)}{v(x)}$$

with $u(x)$ and $v(x) \in k[x]$ relatively prime to $f(x)$, and $s \in \mathbb{Z}$. Let

$$v_f : k(x)^* \to \mathbb{Z} \quad \text{be defined by} \quad v_f(\alpha(x)) = s.$$

Then v_f is a valuation. We have

$$\vartheta_{v_f} = \vartheta_f = \left\{ \frac{a(x)}{b(x)} \in k(x) \mid (b(x), f(x)) = 1 \right\} = k[x]_{(f)},$$

and

$$\mathcal{P}_{v_f} = \mathcal{P}_f = \left\{ \frac{a(x)}{b(x)} \in k(x) \mid f(x)|a(x), (b(x), f(x)) = 1 \right\},$$

where $k[x]_{(f)}$ denotes the localization of $k[x]$ at $S = \left\{ f(x)^n \right\}_{n=0}^{\infty}$. Now, $\vartheta_f \neq k(x)$ since $\frac{1}{f} \notin \vartheta_f$. If $f \neq g$ with $f, g \in k[x]$ monic and irreducible polynomials, we have $v_f(f) = 1 > 0$ and $v_g(f) = 0$. Therefore v_f and v_g are inequivalent. Furthermore, if $\alpha \in k^*$ then $v_f(\alpha) = 0$, that is, v_f is trivial over k.

The residue field is

$$\vartheta_f / \mathcal{P}_f = k[x]_{(f)}/(f)k[x]_{(f)} \cong k[x]/(f)$$

and $k[x]/(f)$ is a finite extension of k of degree equal to the degree of f.

Now, if $y = \frac{1}{x}$ we have $k(y) = k(x)$. Each monic polynomial that is irreducible in $k[y]$ has an associated valuation; in particular, for $y \in k[y]$ we have a valuation that we denote by $v_y = v_\infty$. Now we study v_∞. Let $\alpha(x) \in k(x)^*$. Then $\alpha(x) = \frac{a(x)}{b(x)}$ and we have

$$\alpha(x) = \frac{a\left(\frac{1}{y}\right)}{b\left(\frac{1}{y}\right)} = \frac{y^{-\deg a} a_1(y)}{y^{-\deg b} b_1(y)} = y^{-(\deg a - \deg b)} \frac{a_1(y)}{b_1(y)},$$

with $a_1(y), b_1(y)$ relatively prime to y. Therefore

$$v_\infty(\alpha(x)) = v_y\left(y^{-(\deg a - \deg b)} \frac{a_1(y)}{b_1(y)} \right) = -(\deg a - \deg b) = -\deg \alpha(x),$$

where we define $\deg \frac{a(x)}{b(x)} = \deg a(x) - \deg b(x)$.

Now if $f(x) \in k[x]$ is a monic and irreducible polynomial, we have $v_f(f) = 1$ and $v_\infty(f) = -\deg f < 0$. Therefore v_f and v_∞ are inequivalent.

Finally, we have

$$\vartheta_{v_\infty} = \vartheta_\infty = k[y]_{(y)} = \left\{ \frac{f(x)}{g(x)} \mid \deg f - \deg g \le 0 \right\},$$

$$\mathcal{P}_{v_\infty} = \mathcal{P}_\infty = \left\{ \frac{f(x)}{g(x)} \mid \deg f - \deg g < 0 \right\},$$

and the residue field is

$$\vartheta_\infty / \mathcal{P}_\infty \cong k[y]_{(y)} / y k[y]_{(y)} \cong k[y]/(y) \cong k.$$

The result we are looking for is given in the following theorem:

Theorem 2.4.1. *The set of valuations v over $k(x)$ such that $v(a) = 0$ for $a \in k^*$ is exactly*

$$\left\{v_f \mid f(x) \in k[x] \text{ is a monic and irreducible polynomial}\right\} \cup \{v_\infty\}.$$

Furthermore, all of them are pairwise inequivalent and the residue field is a finite extension of k. In case the valuation is v_f, the degree of the residue field is equal to the degree of the polynomail f and in case the valuation is v_∞, the degree of the residue fiels is equal to one. Finally, all these valuations are discrete.

Proof. It remains only to verify that given any nontrivial valuation $v : k(x)^* \to G$ such that G is an ordered group and $v(a) = 0$ for all $a \in k^*$, then v is equivalent to v_∞ or to some v_f, where $f(x) \in k[x]$ is monic and irreducible.

Let ϑ be the valuation ring of v and let $\mathcal{P}$ be its maximal ideal. Now, if $x \in \vartheta$, then $k[x] \subseteq \vartheta$. Let $\wp = \mathcal{P} \cap k[x]$. We have $\wp$ is a prime ideal of $k[x]$, $k \cap \wp = \{0\}$, and $1 \notin \wp$. It follows that $\wp = (f)$, where f is a monic and irreducible polynomial or $f = 0$.

If $f = 0$, then $v\,(k[x]^*) = \{0\}$, so $v\,(k(x)^*) = 0$. But then we have $k(x) = \vartheta$, which contradicts the hypothesis that v is nontrivial. Therefore $\wp = (f)$ with $f \neq 0$. Let $g, h \in k[x]$ with $f \nmid h$ and $h \notin \wp$, that is, h is a unit in ϑ. We have $v\left(\frac{g}{h}\right) \geq 0$, which implies $\vartheta_f \subseteq \vartheta$.

Now assume that $u(x) \in k(x) \setminus \vartheta_f$. Then $u = \frac{g}{h}$ with $(g, h) = 1$ and $f \mid h$. If $u \in \vartheta$, since $g \in \vartheta_f^* \subseteq \vartheta$, it follows that $h^{-1} = g^{-1}u \in \vartheta$. However, we have $h \in \mathcal{P} \subseteq \vartheta$, and this implies that h is a nonunit, which is absurd. Hence $u \notin \vartheta$ and we have $\vartheta \subseteq \vartheta_f$, so $\vartheta = \vartheta_f$. Therefore v and v_f are equivalent.

If $x \notin \vartheta$, then $y = \frac{1}{x} = x^{-1} \in \vartheta$. From the above discussion, we conclude that $\mathcal{P} \cap k[y] = (\ell(y))$, where $\ell(y)$ is a monic and irreducible polynomial. Now $x = y^{-1} \notin \vartheta$, which is equivalent to saying that y is not a unit. Thus $y \in \mathcal{P} \cap k[y] = (\ell(y))$ and $\ell(y) \mid y$, which proves that $\ell(y) = y$. Therefore $\vartheta = \vartheta_\infty$ and v is equivalent a v_∞. □

Note 2.4.2. From this point on, a valuation in a function field K with $K/k(x)$ finite will mean a nontrivial valuation v such that $v(a) = 0$ for all $a \in k^*$.

Now we will study the case of a function field K with field of constants k. If $x \in K \setminus k$, $K/k(x)$ is a finite extension. If v is a valuation in K, $v|_{k(x)}$ is a valuation in $k(x)$. Therefore we need to study extensions of valuations, or equivalently, extensions of places.

Let $K \subseteq L$ be a field extension and let $\varphi_K : K \longrightarrow E \cup \{\infty\}$ be a place over K. We want to show that there exists a place over L, $\varphi_L : L \longrightarrow E_1 \cup \{\infty\}$, such that $E \subseteq E_1$ and $\varphi_L|_K = \varphi_K$. For this purpose, we will prove the following result:

Theorem 2.4.3 (Chevalley's lemma). *Let K be a field, ϑ a subring of K, and let $\varphi : \vartheta \longrightarrow F$ be a ring homomorphism, where F is an algebraically closed field. Let $x \in K^*$. Then φ can be extended to at least one of the rings $\vartheta\,[x]$ and $\vartheta\left[\frac{1}{x}\right]$.*

Proof. We may assume that $\varphi \neq 0$, since otherwise the result is trivial. Let $\mathcal{P} = \ker \varphi$. Then $\mathcal{P}$ is a prime ideal of ϑ. Let $\vartheta_{\mathcal{P}} = \left\{\frac{a}{b} \mid a, b \in \vartheta, b \notin \mathcal{P}\right\} \supseteq \vartheta$. The map φ can be extended to $\tilde{\varphi} : \vartheta_{\mathcal{P}} \longrightarrow F$ by putting

$$\tilde{\varphi}\left(\frac{a}{b}\right) = \frac{\varphi(a)}{\varphi(b)}.$$

We have $\tilde{\varphi}(\vartheta_{\mathcal{P}}) = \operatorname{quot}\varphi(\vartheta) = E$. Set $\tilde{\varphi}(a) = \bar{a}$. Let $T, \bar{T}$ be two indeterminates over $\vartheta_{\mathcal{P}}$ and E respectively. Then $\tilde{\varphi}$ can be extended in a unique way to $\bar{\varphi} : \vartheta_{\mathcal{P}}[T] \longrightarrow E\left[\bar{T}\right]$ such that

$$\bar{\varphi}\left(\sum_{i=0}^{n} a_i T^i\right) = \sum_{i=0}^{n} \bar{a}_i \bar{T}^i.$$

Define

$$\mathfrak{A} = \{p(T) \in \vartheta_{\mathcal{P}}[T] \mid p(x) = 0\} \quad \text{and} \quad \bar{\mathfrak{A}} = \bar{\varphi}(\mathfrak{A}).$$

Then $\mathfrak{A}$ is an ideal of $\vartheta_{\mathcal{P}}[T]$ and $\bar{\mathfrak{A}}$ is an ideal of $E\left[\bar{T}\right]$.

If $\bar{\mathfrak{A}} = (0)$, we define $\Phi : \vartheta_{\mathcal{P}}[x] \longrightarrow F$ by $\Phi(x) = \alpha \in F$ for some arbitrary α. Then

$$\Phi\left(\sum_{i=0}^{n} a_i x^i\right) = \sum_{i=0}^{n} \bar{a}_i \alpha^i.$$

If $\sum_{i=0}^{n} a_i x^i = 0$, then $\sum_{i=0}^{n} a_i T^i \in \mathfrak{A}$, so that

$$\sum_{i=0}^{n} \bar{a}_i \bar{T}^i = \varphi\left(\sum_{i=0}^{n} a_i T^i\right) \in \bar{\mathfrak{A}} = (0) \quad \text{and} \quad \sum_{i=0}^{n} \bar{a}_i \alpha^i = 0.$$

Thus Φ is the required extension.

If $\bar{\mathfrak{A}} \neq 0$ and $\bar{\mathfrak{A}} \neq E\left[\bar{T}\right]$, we have $\bar{\mathfrak{A}} = \left(f\left(\bar{T}\right)\right)$, where f is a nonconstant polynomial over E. Let α be a root of $f\left(\bar{T}\right)$ in F. Such a root exists since F is algebraically closed. Let $\bar{\varphi}(x) = \alpha$. Then $\bar{\varphi}$ can be extended in a unique way to a homomorphism of $\vartheta_{\mathcal{P}}[x]$ since the image $\bar{\varphi}$ of any polynomial that vanishes at x is of the form $g\left(\bar{T}\right) f\left(\bar{T}\right)$ and therefore vanishes at $\bar{T} = \alpha$.

Finally, if $\bar{\mathfrak{A}} = E\left[\bar{T}\right]$, then $\bar{\varphi}$ cannot be extended to $\vartheta_{\mathcal{P}}[x]$. Indeed, for $\bar{P}\left(\bar{T}\right) \in E[\bar{T}] \setminus \{0\}$, let $P(T) \in \mathfrak{A}$ be such that $\varphi(P(T)) = \bar{P}(\bar{T})$; if φ could be extended to $\bar{\varphi}\colon \vartheta_{\mathcal{P}}[x] \to F$, then we would have

$$0 = \bar{\varphi}(0) = \bar{\varphi}(P(x)) = \bar{P}\left(\bar{\varphi}(x)\right),$$

which is impossible since $\bar{\varphi}(x)$ would be a root of any polynomial.

Now we assume that $\bar{\varphi}$ cannot be extended to $\vartheta_{\mathcal{P}}\left[\frac{1}{x}\right]$ either. Let

$$\mathfrak{B} = \left\{p(T) \in \vartheta_{\mathcal{P}}[T] \mid p\left(\frac{1}{x}\right) = 0\right\} \quad \text{with} \quad \bar{\mathfrak{B}} = \bar{\varphi}(\mathfrak{B}).$$

Then we must have $\bar{\mathfrak{B}} = E\left[\bar{T}\right]$. Thus there exist $f(T), g(T) \in \vartheta_{\mathcal{P}}[T]$ with

$$f(T) = a_n T^n + \cdots + a_1 T + a_0 \quad \text{and} \quad g(T) = b_m T^m + \cdots + b_1 T + b_0$$

such that $\bar{\varphi}(f) = 1 = \bar{\varphi}(g)$ and $f(x) = g\left(\frac{1}{x}\right) = 0$.

We choose n, m to be minimal with this property. Without lost of generality, we may assume $m \leq n$. Therefore

$$\bar{a}_0 = \bar{b}_0 = 1 \quad \text{and} \quad \bar{a}_i = \bar{b}_j = 0 \quad \text{for} \quad i, j > 0.$$

Let $g_0(T) = b_0 T^m + \cdots + b_{m-1} T + b_m$. Using the division algorithm, we obtain

$$b_0^n f(T) = g_0(T) Q(T) + R(T)$$

with $Q(T), R(T) \in \vartheta_{\mathcal{P}}[T]$, and $\deg R < m = \deg g_0(T)$.

Now

$$g_0(x) = x^m \left(b_0 + \cdots + b_{m-1} \left(\frac{1}{x}\right)^{m-1} + b_m \left(\frac{1}{x}\right)^m \right) = x^m g\left(\frac{1}{x}\right) = 0$$

and so

$$b_0^n f(x) = 0 = g_0(x) Q(x) + R(x) = 0 + R(x) = R(x),$$

that is, $R(x) = 0$.

On the other hand, we have

$$1 = \bar{b}_0^n \bar{f}\left(\bar{T}\right) = \bar{g}_0\left(\bar{T}\right) \bar{Q}\left(\bar{T}\right) + \bar{R}\left(\bar{T}\right) = \bar{Q}\left(\bar{T}\right) \bar{T}^m + \bar{R}\left(\bar{T}\right).$$

Therefore $\bar{Q}\left(\bar{T}\right) = 0$, $\bar{R}\left(\bar{T}\right) = 1$, which contradicts the minimality of $n = \deg f$, since $R(T)$ satisfies

$$\bar{R}\left(\bar{T}\right) = 1, \quad R(x) = 0, \quad \text{and} \quad \deg R < m \leq n.$$

Hence φ can be extended to $\vartheta_{\mathcal{P}}\left[\frac{1}{x}\right]$. □

As a consequence of Chevalley's lemma, we will obtain the existence of extensions of places:

Theorem 2.4.4. *Let K be a field, and let $\vartheta \subseteq K$ be a subring. Let $\varphi : \vartheta \longrightarrow F$ be a ring homomorphism, where F is an algebraically closed field. Then φ can be extended to a monomorphism of K to F or to a place of K to $F \cup \{\infty\}$.*

Proof. We may assume that $\varphi \neq 0$. Let

$$\begin{aligned} \mathfrak{X} = \Big\{ & (\varphi_\alpha, \vartheta_\alpha) \mid \vartheta \subseteq \vartheta_\alpha \subseteq K, \vartheta_\alpha \text{ subring of } K, \\ & \varphi_\alpha : \vartheta_\alpha \longrightarrow F \text{ a homomorphism such that } \varphi_\alpha|_\vartheta = \varphi \Big\}. \end{aligned}$$

We define an order in $\mathfrak{X}$ by $(\varphi_\alpha, \vartheta_\alpha) \leq \left(\varphi_\beta, \vartheta_\beta\right)$ if and only if $\vartheta_\alpha \subseteq \vartheta_\beta$ and $\varphi_\beta|_{\vartheta_\alpha} = \varphi_\alpha$.

We have $(\varphi, \vartheta) \in \mathfrak{X}$, so $\mathfrak{X} \neq \emptyset$. Now if $\{(\varphi_\alpha, \vartheta_\alpha)\}_{\alpha \in I} \subseteq \mathfrak{X}$ is a chain, let $\vartheta_I = \bigcup_{\alpha \in I} \vartheta_\alpha$ and $\varphi_I : \vartheta_I \longrightarrow F$ be defined by $\varphi_I(x) = \varphi_\alpha(x)$ for all $x \in \vartheta_\alpha$. Then (φ_I, ϑ_I) is an upper bound of the chain.

By Zorn's lemma, $\mathfrak{X}$ has a maximal element (Φ, ϑ'). First we will see that ϑ' is a valuation ring or $\vartheta' = K$. If not, there exists $x \in K$ such that $x \notin \vartheta'$ and $x^{-1} \notin \vartheta'$. By Chevalley's lemma, Φ can be extended to a homomorphism of $\vartheta'[x]$ or a homomorphism of $\vartheta'\left[\frac{1}{x}\right] = \vartheta'\left[x^{-1}\right]$ in F, but in any case this contradicts the maximality of (Φ, ϑ').

Now if $\vartheta' = K$, Φ is a monomorphism. If $\vartheta' \neq K$, then for $x \in \vartheta'$ with $x \notin (\vartheta')^*$, we must have $\Phi(x) = 0$, since otherwise the formula $\Phi\left(x^{-1}\right) = \Phi(x)^{-1}$ would define an extension of Φ to $\vartheta'\left[\frac{1}{x}\right]$, a contradiction to the maximality of (Φ, ϑ').

Hence, we have $\Phi(x) = 0$ for $x \in \vartheta' \setminus (\vartheta')^*$, the latter being the maximal ideal $\mathcal{P}$ of ϑ'. Finally, Φ can be extended to a place of K by defining $\Phi(y) = \infty$ for $y \in K \setminus \vartheta'$. □

Corollary 2.4.5. *If $K \subseteq L$ is a field extension and $\varphi : K \longrightarrow E \cup \{\infty\}$ is a place of L, then φ can be extended to a place of L.*

Proof. Let F be an algebraic closure of E and consider the ring

$$\vartheta_\varphi = \{x \in K \mid \varphi(x) \neq \infty\}.$$

It follows from the remark after Proposition 2.2.11 that ϑ_φ is a valuation ring. Since ϑ_φ is a subring of L, by Theorem 2.4.4, φ can be extended either to a monomorphism of L or to a place of L. However, since there exists $x \in K$ such that $\varphi(x) = \infty$, the extension is necessarily a place of L. □

Corollary 2.4.6. *If v is a valuation in a field K and L is an extension of K, then v can be extended to a valuation of L.*

Proof. The statement follows from the correspondence between valuations and places and from Corollary 2.4.5. □

Corollary 2.4.7. *If K/k is a function field and $x \in K$ is a transcendental element over k, then there exists at least a valuation v over K such that $v(x) > 0$.*

Proof. In $k(x)$ we have $v_x(x) = 1 > 0$. If v is any extension of v_x in K, then $v(x) > 0$. □

Now we will show that every valuation in a function field is discrete, which will allow us to assume that the value group of the valuation is $\mathbb{Z}$. We will need two lemmas.

Lemma 2.4.8. *Let W be an ordered group that contains $\mathbb{Z}$ and such that $[W : \mathbb{Z}] < \infty$. Then $W \cong \mathbb{Z}$.*

Proof. Since $W/\mathbb{Z}$ is finite, there exists $n \in \mathbb{N}$ such that $0 \neq nW \subseteq \mathbb{Z}$. Therefore $nW \cong \mathbb{Z}$ and since W is torsion free, we have $nW \cong W$. □

Lemma 2.4.9. *Let L/K be a finite field extension with $[L : K] = n$. Let v be a valuation over L with value group V. If $W = v(K^*) \subseteq V$, then $[V : W] = m \leq n$.*

Proof. See Exercise 2.6.5. □

As an immediate consequence we obtain the following theorem:

Theorem 2.4.10. *Every valuation on a function field K/k is discrete.*

Proof. If v is a valuation over K, let $x \in K$ be transcendental over K. It follows from Theorem 2.4.1 that $v|_{k(x)}$ is discrete. The fact that v is discrete is a consequence of Lemmas 2.4.8 and 2.4.9. □

Next we will define the degree of a place in a function field. Let K/k be a function field, where k is the exact field of constants. If φ is a place over K, let ϑ and $\mathcal{P}$ be the valuation ring and the maximal ideal associated to φ respectively, that is,

$$\varphi : K \to k(\mathcal{P}) \cup \{\infty\},$$

where $k(\mathcal{P}) = \varphi(\vartheta) \cong \vartheta/\mathcal{P}$ (recall that $\vartheta = \big\{x \in K \mid \varphi(x) \neq \infty\big\}$, $\mathcal{P} = \big\{x \in K \mid \varphi(x) = 0\big\}$, $k \subseteq \vartheta$, and $k \cap \mathcal{P} = (0)$).

Notation 2.4.11. If v is a valuation in K, $\mathcal{P}$ the associated ideal, and $\vartheta_{\mathcal{P}}$ the valuation ring, we will write $k(\mathcal{P}) \cong \vartheta_{\mathcal{P}}/\mathcal{P}$ for the residue field associated to $\mathcal{P}$.

Resuming the above development, we have $k \subseteq \vartheta$, $k \cap \mathcal{P} = (0)$, so $\varphi : k \longrightarrow k(\mathcal{P})$ is a monomorphism. Therefore $k(\mathcal{P})$ is an extension of k. The importance of this extension is that it is finite.

Theorem 2.4.12. *Let K/k be a function field and let $\mathcal{P}$ be a maximal ideal associated to a place of K. Then $f_{\mathcal{P}} = d_K(\mathcal{P}) = [k(\mathcal{P}) : k] < \infty$.*

Proof. Let φ be the place associated to $\mathcal{P}$, i.e.,

$$\varphi : K \longrightarrow k(\mathcal{P}) \cup \{\infty\}, \quad \text{with} \quad \varphi(\mathcal{P}) = 0.$$

For $x \in \mathcal{P} \setminus \{0\}$, we have $\varphi(x) = 0$. Since $k \subseteq \vartheta$ where ϑ is the associated valuation ring, $\varphi|_k : k \longrightarrow k(\mathcal{P})$ is a field monomorphism. Therefore $\varphi(x) = 0$ implies that $x = 0$ or x is transcendental. Since we chose $x \neq 0$, x is necessarily transcendental. We have $[K : k(x)] = n < \infty$. It will be shown that in fact $[k(\mathcal{P}) : k] \leq n$.

Let $\alpha_1, \alpha_2, \ldots, \alpha_{n+1} \in k(\mathcal{P})$ be all distinct (if this is not possible, that is, if $|k(\mathcal{P})| \leq n$, the result is immediate). Let $a_i \in \vartheta$ be such that $\varphi(a_i) = \alpha_i$. Since $[K : k(x)] = n$, there exist polynomials $\{f_i(x)\}_{i=1}^{n+1} \subseteq k[x]$ such that

$$\sum_{i=1}^{n+1} a_i f_i(x) = 0 \tag{2.1}$$

with some $f_j(x) = b_m x^m + \cdots + b_1 x + b_0, b_0 \neq 0$. Let $f_i(x) = c_i + x g_i(x)$ with $c_i \in k$, and, of course, $c_j = b_0 \neq 0$. Then from (2.1), we obtain the relation

$$\sum_{i=1}^{n+1} a_i c_i = -x \sum_{i=1}^{n+1} a_i g_i(x). \tag{2.2}$$

Applying φ to each side of (2.2) we obtain

$$\sum_{i=1}^{n+1} c_i \alpha_i = -\varphi(x) \sum_{i=1}^{n+1} \alpha_i g_i(\varphi(x)) = 0$$

with some $c_j \neq 0$, which implies that the set $\{\alpha_i\}_{i=1}^{n+1}$ is linearly dependent over k. Hence $[k(\mathcal{P}) : k] \leq n$. □

Definition 2.4.13. The number $f_{\mathcal{P}} = d_K(\mathcal{P}) = [k(\mathcal{P}) : k]$ is called the *degree of the place $\mathcal{P}$ or the inertia degree of $\mathcal{P}$.*

Example 2.4.14. If $K = k(x)$ and $\mathcal{P}$ corresponds to the valuation given by the monic polynomial $f(x) \in k[x]$, then $k(\mathcal{P}) \cong k[x]/(f(x))$. Hence $[k(\mathcal{P}) : k] = \deg f$. Also, if $\mathcal{P}$ corresponds to the valuation given by $1/x$, we have $[k(\mathcal{P}) : k] = 1$.

Corollary 2.4.15. *For any place $\mathcal{P}$, $f_{\mathcal{P}}$ satisfies $1 \leq f_{\mathcal{P}} \leq n$, where $n = [K : k(x)]$, x is any element such that $v_{\mathcal{P}}(x) \neq 0$, and $v_{\mathcal{P}}$ is the associated valuation.*

Proof. If φ is the place associated to $v_{\mathcal{P}}$, $v_{\mathcal{P}}(x) \neq 0$ is equivalent to $\varphi(x) = 0$ or $\varphi(x) = \infty$. The case $\varphi(x) = \infty$ can be reduced to $\varphi\left(x^{-1}\right) = 0$, $k(x) = k\left(x^{-1}\right)$. □

Corollary 2.4.16. *If the field of constants of k is algebraically closed, then $f_{\mathcal{P}} = 1$ for every place $\mathcal{P}$.*

Proof. Since k is algebraically closed and $k(\mathcal{P})$ is a finite extension of k, in particular algebraic, then $k(\mathcal{P}) = k$. Therefore $f_{\mathcal{P}} = [k(\mathcal{P}) : k] = 1$. □

2.5 Artin's Approximation Theorem

The theorem that we will prove in this section, as indicated by the title, is due to Emil Artin. This result essentially establishes that given a finite number of pairwise inequivalent absolute values over a field K, and given the same number of elements of K, we can approximate simultaneously all those elements by a single element of K, each approximation being given in the respective absolute value. Here the phrase "a finite number of absolute values" is necessary in the sense that there does not exist an approximation theorem for an infinite number of absolute values. The approximation theorem can be considered as a generalization of the Chinese remainder theorem.

For instance, given $\varepsilon_1 = 10^{-25}$, $\varepsilon_2 = 10^{-30}$, $|\cdot|_1, |\cdot|_2$ the 5-adic and the 17-adic absolute values respectively, there exists $x \in \mathbb{Z}$ such that $|x-2|_1 < \varepsilon_1$ and $|x-7|_2 < \varepsilon_2$. We use the Chinese remainder theorem to find x satisfying $x \equiv 2 \bmod 5^n$ and $x \equiv 7 \bmod 17^n$ for some n to be given later. Thus we may write $x = 2 + 5^n t$ and $x = 7 + 17^n s$, so

$$|x-2|_1 = |5^n t|_1 \leq |5^n|_1 = 5^{-n} \quad \text{and} \quad |x-7|_2 = |17^n s|_2 \leq |17^n|_2 = 17^{-n}.$$

Therefore if we choose $n > 25 \log_5 10$ and $n > 30 \log_7 10$, x satisfies $|x-2|_1 < \varepsilon_1$ and $|x-7|_2 < \varepsilon_2$.

Recall that two nontrivial absolute values $|\ |_1, |\ |_2$ over a field K are called equivalent if $|x|_1 < 1 \iff |x|_2 < 1$, or, which is the same, if they define the same topology on K.

Proposition 2.5.1. *Let K be an arbitrary field, and let $|\ |_1, \ldots, |\ |_n$ be n nontrivial pairwise inequivalent absolute values over K. Then there exists an element a of K such that $|a|_1 > 1$ and $|a|_2 < 1, \ldots, |a|_n < 1$.*

Proof. We will proceed by induction on n. If $n = 2$, then there exist elements $b, c \in K^*$ such that

$$|b|_1 > 1, \quad |b|_2 \leq 1 \quad \text{and} \quad |c|_2 > 1, \quad |c|_1 \leq 1.$$

Let $a = \frac{b}{c}$. We have

$$|a|_1 = \frac{|b|_1}{|c|_1} \geq |b|_1 > 1 \quad \text{and} \quad |a|_2 = \frac{|b|_2}{|c|_2} < |b|_2 \leq 1.$$

Therefore a is the element we are looking for.

Let's assume that the result holds for $n-1 \geq 1$. For n, we begin by choosing $b \in K$ such that

$$|b|_1 > 1 \quad \text{and} \quad |b|_2 < 1, \ldots, |b|_{n-1} < 1,$$

and $c \in K$ such that

$$|c|_1 > 1, \quad |c|_n < 1.$$

Now if $|b|_n \leq 1$, then for $m \in \mathbb{N}$, $a = b^m c$ satisfies

$$\begin{aligned} |a|_1 &= |b|_1^m |c|_1 > 1, \\ |a|_i &= |b|_i^m |c|_i \xrightarrow[m \to \infty]{} 0, \quad 2 \leq i \leq n-1, \\ |a|_n &= |b|_n^m |c|_n < 1. \end{aligned}$$

Hence, taking m to be large enough, $a = b^m c$ satisfies $|a|_1 > 1$ and $|a|_i < 1, i = 2, \ldots, n$.

Now assume that $|b|_n > 1$. Then

$$\frac{b^m}{1+b^m} = \frac{1}{\frac{1}{b^m}+1} \xrightarrow[m\to\infty]{|\ |_n} \frac{1}{0+1} = 1$$

since

$$\left(\frac{1}{b}\right)^m \xrightarrow[m\to\infty]{|\ |_n} 0 \quad \text{and} \quad \frac{b^m}{1+b^m} \xrightarrow[m\to\infty]{|\ |_i} 0, \quad i = 2, \ldots, n-1.$$

Thus

$$\begin{aligned}
\left|\frac{b^m c}{1+b^m}\right|_n &\xrightarrow[m\to\infty]{} |c|_n < 1;\\
\left|\frac{b^m c}{1+b^m}\right|_i &\xrightarrow[m\to\infty]{} 0, \quad i = 2, \ldots, n-1;\\
\left|\frac{b^m c}{1+b^m}\right|_1 &\xrightarrow[m\to\infty]{} |c|_1 > 1.
\end{aligned}$$

Therefore $a = \frac{b^m c}{1+b^m}$, for a large enough natural number m, is the required element. □

Proposition 2.5.2. *Let $|\ |_1, \ldots, |\ |_n$ be nontrivial pairwise inequivalent absolute values over a field K. Given $\varepsilon > 0$, $\varepsilon \in \mathbb{R}$, there exists $x \in K$ such that $|1-x|_1 < \varepsilon$ and $|x|_i < \varepsilon$ for $i = 2, \ldots, n$.*

Proof. Let $a \in K$ be such that

$$|a|_1 > 1, \quad \text{and} \quad |a|_i < 1, \quad i = 2, \ldots, n.$$

Let

$$x = \frac{a^m}{1+a^m} \xrightarrow[m\to\infty]{} \begin{cases} 1 \text{ for } |\ |_1, \\ 0 \text{ for } |\ |_i, \quad 2 \le i \le n. \end{cases}$$

For m large enough, x satisfies the conditions of the proposition. □

Theorem 2.5.3 (Approximation Theorem). *Let $|\ |_1, \ldots, |\ |_n$ be nontrivial pairwise inequivalent absolute values over a field K. Given $\varepsilon > 0$, $\varepsilon \in \mathbb{R}$, and $a_1, a_2, \ldots, a_n \in K$, there exists $y \in K$ such that $|y - a_i|_i < \varepsilon$ for $1 \le i \le n$.*

Proof. Let $M = \max\left\{|a_i|_j \mid 1 \le i, j \le n\right\}$. If $M = 0$, the result is immediate. Let $M \neq 0$. It follows from Proposition 2.5.2 that there exist $b_1, b_2, \ldots, b_n$ such that

$$|1 - b_i|_i < \frac{\varepsilon}{Mn} \quad \text{for} \quad i = 1, \ldots, n \quad \text{and} \quad \left|b_j\right|_i < \frac{\varepsilon}{Mn} \quad \text{for} \quad 1 \le i \neq j \le n.$$

Let $y = a_1 b_1 + \cdots + a_n b_n$. Then we have for $1 \le i \le n$, $y - a_i = \sum_{\substack{j=1\\ j\neq i}}^{n} a_j b_j + a_i(b_i - 1)$, so that

$$|y-a_i|_i \le \sum_{\substack{j=1\\ j\neq i}}^{n} \left|a_j b_j\right|_i + |a_i|_i\, |b_i - 1|_i \le M \sum_{\substack{j=1\\ j\neq i}}^{n} \left|b_j\right|_i + M\, |b_i - 1|_i <$$

$$< M(n-1)\frac{\varepsilon}{Mn} + M\frac{\varepsilon}{Mn} = \left(\frac{n-1}{n} + \frac{1}{n}\right)\varepsilon = \varepsilon.$$

Hence y satisfies the conditions of the theorem. □

The next results are applications of some versions of the approximation theorem. In particular, Example 2.5.7 will be very useful.

Corollary 2.5.4. *Let $|\ |_1, \dots, |\ |_n$ be pairwise nontrivial inequivalent absolute values over a field K. Denote by K_i the topological space whose underlying set is K and the topology is generated by $|\ |_i$. Let $K_1 \times \cdots \times K_n$ be given with the product topology and let* $\begin{array}{ccc} K & \to & K_1 \times \cdots \times K_n \\ x & \mapsto & (x, \dots, x) \end{array}$ *be the diagonal map. Then K is dense in $K_1 \times \cdots \times K_n$.*

□

Corollary 2.5.5. *Let $v_1, \dots, v_n$ be n nontrivial pairwise inequivalent absolute values over a field K whose value groups are contained in $\mathbb{R}$. Then given $a_1, a_2, \dots, a_n \in K$ and $M \in \mathbb{R}$, there exists $x \in K$ such that $v_i(x - a_i) \ge M$ for $i = 1, \dots, n$.*

Proof. Let $|x|_i = e^{-v_i(x)}$. Then $|\ |_1, \dots, |\ |_n$ are nontrivial pairwise inequivalent absolute values satisfying $v_i(x) = -\ln |x|_i$. The required solution is

$$v_i(x-a_i)_i = -\ln |x-a_i|_i \ge M \iff |x-a_i|_i \le e^{-M}.$$

□

Corollary 2.5.6. *Let $v_1, \dots, v_n$ be nontrivial inequivalent pairwise absolute values over a field K with respective value groups $G_1, \dots, G_n$ satisfying $G_i \subseteq \mathbb{R}$. Then given $g_1 \in G_1, \dots, g_n \in G_n$ and $a_1, \dots, a_n \in K$, there exists $z \in K$ such that $v_i(z - a_i) = g_i$ for $i = 1, \dots, n$.*

Proof. Let x be such that $v_i(x - a_i) > g_i$ for $i = 1, \dots, n$. Such x exists by Corollary 2.5.5. Let $c_i \in K$ be such that $v_i(c_i) = g_i$ and let $y \in K$ be such that $v_i(y - c_i) > g_i$. Then

$$v_i(y) = v_i(y - c_i + c_i) = \min\{v_i(y-c_i), v_i(c_i)\} = g_i, \quad i = 1, \dots, n.$$

Let $z = x + y$. Then

$$v_i(z-a_i) = v_i(y + x - a_i) = \min\{v_i(y), v_i(x-a_i)\} = g_i, i = 1, \dots, n.$$ □

Example 2.5.7. Let K be a number field or a function field. Let $\mathcal{P}_1, \dots, \mathcal{P}_n$ be n distinct places of K. Let $a_1, a_2, \dots, a_n \in K$ be arbitrary elements and let $m_1, m_2, \dots, m_n$ be arbitrary natural numbers. Then the system of congruences $x \equiv a_i \bmod \mathcal{P}_i^{m_i}$ has a solution in K. The notation $x \equiv a \bmod \mathcal{P}^s$ means that $x - a \in \mathcal{P}^s$, where $\mathcal{P}$ is the ideal of the valuation. The existence of the solution follows from the fact that $x - a \in \mathcal{P}_i^{m_i} \iff v_i(x - a_i) \ge m_i$, which in turn follows from Corollary 2.5.6.

Remark 2.5.8. Corollaries 2.5.5 and 2.5.6 can be proved assuming only that the value groups are archimedean. The proof is similar to that of Theorem 2.5.3. We say that an ordered group G is *archimedean* if for any $a, b \in G$ such that $a > 0$, there exists $n \in \mathbb{N}$ such that $na > b$.

In Proposition 2.3.13 we proved that given a field K with discrete valuation v and prime ideal $\mathfrak{p}$, if $K_{\mathfrak{p}}$ is the completion of K with respect to v and π is a prime element, then every element x of $K_{\mathfrak{p}}$ can be written in a unique way as

$$\sum_{i=m}^{n} \alpha_i \pi^i, \quad m \in \mathbb{Z}, \quad \alpha_i \in S,$$

where S is a set of representatives of the residue field and $0 \in S$.

In the case of a number field, the residue field is of characteristic $p > 0$ and the completion is of characteristic 0, so that $K_{\mathfrak{p}}$ cannot be isomorphic to the field of Laurent series $k(\mathfrak{p})((\pi))$ since these two fields are of different characteristic.

In the case of a function field K/k, the residue field $k(\mathfrak{p})$ is a finite extension of the field of constants k. Therefore $k(\mathfrak{p})$, k, K, and $K_{\mathfrak{p}}$ all have the same characteristic. We will prove that in this case, $K_{\mathfrak{p}}$ and $k(\mathfrak{p})((x))$ are isomorphic, where $k(\mathfrak{p})((x))$ is the field of Laurent series in the indeterminate x.

Definition 2.5.9. Let K/k be a function field, $\mathfrak{p}$ be a place of K, and $K_{\mathfrak{p}}$ the completion of K with respect to $\mathfrak{p}$. Let ϑ and $\hat{\vartheta}$ be the rings of integers of K and $K_{\mathfrak{p}}$ respectively. Let $k(\mathfrak{p}) := \bar{k} \cong \vartheta/\mathfrak{p} \cong \hat{\vartheta}/\hat{\mathfrak{p}}$ be the residue field. A field $S \subseteq \hat{\vartheta}$ that can be mapped isomorphically onto $\bar{k}$ is called a *coefficient field* in $\hat{\vartheta}$.

Proposition 2.5.10. *If S is a coefficient field, then $K_{\mathfrak{p}}$ is isomorphic to $S((x))$ algebraically and topologically. Here the topology of $M((x))$ is the one corresponding to the valuation*

$$v\Big(\sum_{n=m}^{\infty} a_n x^n\Big) = m,$$

where $a_m \neq 0$.

Proof. If $\varphi\colon S \to \bar{k}$ is the isomorphism defined by $\varphi(s) = s \bmod \mathfrak{p}$, it follows from Proposition 2.3.13 that the map

$$\psi : S((x)) \to K_{\mathfrak{p}}$$
$$\sum_{n=m}^{\infty} a_n x^n \mapsto \sum_{n=m}^{\infty} \varphi(a_n)\pi^n$$

is an algebraic and topological isomorphism since $\psi(x) = \pi$. □

The next result proves that $\hat{\vartheta}$ always contains a coefficient field. The hard case is that in which k is not perfect.

Definition 2.5.11. Let $\bar{k}$ be of characteristic $p > 0$. A set $S = \{\theta_i\}_{i \in I} \subseteq \bar{k}$ is called a *p-basis of $\bar{k}$* if

$$\bar{k} = \bar{k}^p[S] \quad \text{and} \quad \left[\bar{k}^p[\theta_1, \dots, \theta_n] : \bar{k}^p\right] = p^n$$

for any distinct elements $\theta_1, \dots, \theta_n \in S$.

It is easy to see that the empty set is a p-basis if and only if $\bar{k}$ is perfect.

Proposition 2.5.12. *Let $\bar{k}$ be an imperfect field. Then there exist p-bases for $\bar{k}$.*

Proof. Let $\mathcal{A} = \left\{S \subseteq \bar{k} \mid \text{ for any distinct } \theta_1, \dots, \theta_n \in S, \left[\bar{k}^p[\theta_1, \dots, \theta_n] : \bar{k}^p\right] = p^n\right\}$. Then $\emptyset \in \mathcal{A}$ and $\mathcal{A} \neq \emptyset$. We define a partial order in $\mathcal{A}$ as follows:

$$S_1 \leq S_2 \iff S_1 \subseteq S_2.$$

Clearly every chain $\left\{S_\alpha\right\}_{\alpha \in I}$ has an upper bound $S := \bigcup_{\alpha \in I} S_\alpha \in \mathcal{A}$, so by Zorn's lemma, $\mathcal{A}$ contains a maximal element S. We have $\bar{k} = \bar{k}^p[S]$, since otherwise we may choose $a \in \bar{k} \setminus \bar{k}^p[S]$, and if $\theta_1, \dots, \theta_n$, are n distinct elements of S, we have $a \notin \bar{k}^p[\theta_1, \dots, \theta_n]$ and $a^p \in \bar{k}^p$, so that

$$\begin{aligned}
&\left[\bar{k}^p[\theta_1, \dots, \theta_n, a] : \bar{k}^p\right] \\
&\quad = \left[\bar{k}^p[\theta_1, \dots, \theta_n, a] : \bar{k}^p[\theta_1, \dots, \theta_n]\right]\left[\bar{k}^p[\theta_1, \dots, \theta_n] : \bar{k}^p\right] = pp^n = p^{n+1}.
\end{aligned}$$

Thus $S \cup \{a\} \in \mathcal{A}$ and $S \subsetneqq S \cup \{a\}$. The result follows. □

Definition 2.5.13. Assume that $\operatorname{char} \bar{k} = p > 0$. Let $a \in \bar{k}$. An element $\alpha \in \hat{\vartheta}$ is called a *multiplicative representative* or *Teichmüller representative* of a if $\bar{\alpha} = \alpha \bmod \mathfrak{p} = a$ and $\alpha \in \bigcap_{m=0}^{\infty} K_{\mathfrak{p}}^{p^m}$.

Proposition 2.5.14. *Let $\alpha, \beta \in \hat{\vartheta}$ and $v_{\mathfrak{p}}(\alpha - \beta) \geq m$ with $m \in \mathbb{N}$. Then $v_{\mathfrak{p}}\left(\alpha^{p^n} - \beta^{p^n}\right) \geq n + m$.*

Proof. We have $\alpha - \beta \in \hat{\mathfrak{p}}^m$. If π is a prime element for $\mathfrak{p}$, let $\alpha = \beta + \pi^m \delta$ with $\delta \in \hat{\vartheta}$. Then

$$\alpha^p = \sum_{j=1}^{p} \binom{p}{j} \beta^{p-j} \left(\pi^m \delta\right)^j + \beta^p. \tag{2.3}$$

We have $p \bmod \mathfrak{p} = \bar{p} = 0$ in $\bar{k}$. Thus $v_{\mathfrak{p}}(p) \geq 1$. For $1 \leq j \leq p - 1$, p divides $\binom{p}{j}$; hence $v_{\mathfrak{p}}\left(\binom{p}{j}\right) \geq 1$ and

$$v_{\mathfrak{p}}\left(\binom{p}{j} \beta^{p-j} \left(\pi^m \delta\right)^j\right) \geq 1 + 0 + mj \geq m + 1$$

for $j = 1, \dots, p - 1$. For $j = p$, we have

$$v_{\mathfrak{p}}\left(\binom{p}{p}\beta^{p-p}(\pi^m\delta)^p\right) = pm \geq m+1.$$

Thus by (2.3) we have $v_{\mathfrak{p}}(\alpha^p - \beta^p) \geq m+1$. The result follows by induction. □

Proposition 2.5.15. *An element $a \in \bar{k}$ has a multiplicative representative if and only if $a \in \bigcap_{m=0}^{\infty} \bar{k}^{p^m}$. In this case the multiplicative representative is unique. Furthermore, if α and β are the multiplicative representatives of a and b respectively, then $\alpha\beta$ is the multiplicative representative of ab.*

Proof. First let $a \in \bigcap_{m=0}^{\infty} \bar{k}^{p^m}$. Since k is of characteristic p, for each m there exists a unique $a_m \in \bar{k}$ such that $a_m^{p^m} = a$. Choose $\beta_m \in \hat{\vartheta}$ such that $\bar{\beta}_m = a_m$. We have

$$\overline{\beta_{m+1}^p} = a_{m+1}^p = a_m = \bar{\beta}_m.$$

Hence, $v_{\mathfrak{p}}(\beta_{m+1}^p - \beta_m) \geq 1$. From Proposition 2.5.14 we obtain

$$v_{\mathfrak{p}}\left(\beta_{m+1}^{p^{n+1}} - \beta_m^{p^n}\right) \geq n+1 \quad \text{for all} \quad n \geq 1.$$

In particular, the sequence $\{\beta_{i+n}^{p^n}\}_{n=0}^{\infty}$ is Cauchy.

Let $\alpha_i = \lim_{n\to\infty} \beta_{i+n}^{p^n} \in \hat{\vartheta}$. Then

$$\alpha_i^{p^i} = \lim_{n\to\infty} \beta_{i+n}^{p^{i+n}} = \lim_{n\to\infty} \beta_n^{p^n} = \alpha_0 \in K_{\mathfrak{p}}^{p^i}$$

for $i \geq 0$. Since $a_0 = \overline{\beta_n^{p^n}} = a$ for all n, we have $\bar{\alpha}_0 = a_0 = a$, that is, α_0 is a multiplicative representative of a.

Conversely, if $a \in \bar{k}$ has a multiplicative representative α, then $\alpha \in \bigcap_{m\geq 0} K_{\mathfrak{p}}^{p^m}$ so that $a = \bar{\alpha} \in \bigcap_{m\geq 0} \bar{k}^{p^m}$.

To show the uniqueness, let α and β be two multiplicative representatives of $a \in \bar{k}$. Then, writing $\alpha = \alpha_m^{p^m}$ and $\beta = \beta_m^{p^m}$ with $\alpha_m, \beta_m \in \hat{\vartheta}$, we get $\bar{\alpha}_m^{p^m} = \bar{\beta}_m^{p^m}$. It follows that $\bar{\alpha}_m = \bar{\beta}_m$ since $\operatorname{char} \bar{k} = p$. Hence $v_{\mathfrak{p}}(\alpha_m - \beta_m) \geq 1$. By Proposition 2.5.14 we have

$$v_{\mathfrak{p}}(\alpha - \beta) = v_{\mathfrak{p}}\left(\alpha_m^{p^m} - \beta_m^{p^m}\right) \geq m+1$$

for all m. Thus $\alpha = \beta$.

Finally, if α and β are the multiplicative representatives of a and b respectively, then $\overline{\alpha\beta} = \bar{\alpha}\bar{\beta} = ab$ and $\alpha\beta \in \bigcap_{m\geq 0} K_{\mathfrak{p}}^{p^m}$. Therefore, $\alpha\beta$ is the multiplicative representative of ab. □

Corollary 2.5.16. *Let $\mathfrak{R}$ be the set of multiplicative representatives of $\bar{k}$ in $\hat{\vartheta}$. If $\bar{k}$ is a perfect field, then every element of $\bar{k}$ has its multiplicative representative in $\mathfrak{R}$. The map $r : \bar{k} \to \mathfrak{R}$, $a \mapsto \alpha$, induces an isomorphism $\bar{k}^* \xrightarrow{\sim} \mathfrak{R} \setminus \{0\}$.*

Proof. Since $\bar{k}$ is perfect, we have $\bar{k}^{p^m} = \bar{k}$ for all $m \geq 0$. □

Definition 2.5.17. The correspondence $r : \bar{k} \to \mathfrak{R}$ defined in Corollary 2.5.16 is called the *Teichmüller map.*

If $\bar{k}$ is finite then $\mathfrak{R} \setminus \{0\}$ is a cyclic group of order $|\bar{k}| - 1$.

Corollary 2.5.18. *If α and β are the multiplicative representatives of a and $b \in \bar{k}$ respectively, then $\alpha + \beta$ is the multiplicative representative of $a + b$.*

Proof. Let $\alpha = \alpha_m^{p^m}$ and $\beta = \beta_m^{p^m}$ with $m \geq 0$. Then

$$\alpha + \beta = \alpha_m^{p^m} + \beta_m^{p^m} = (\alpha_m + \beta_m)^{p^m}.$$

Hence $\alpha + \beta \in \bigcap_{m \geq 0} K_{\mathfrak{p}}^{p^m}$ and $\overline{\alpha + \beta} = a + b$. □.

Proposition 2.5.19. *Let $\Theta = \{\theta_i\}_{i \in I}$ be a p-basis of $\bar{k}$. For each $i \in I$, let $\alpha_i \in \hat{\vartheta}$ be such that $\bar{\alpha}_i = \theta_i$. Then there exists an extension L of $K_{\mathfrak{p}}$, where L is a complete field, such that*

$$\bar{L} = \bigcup_{m=0}^{\infty} \bar{k}^{p^{-m}}.$$

Here $\bar{L}$ is the residue field of L and for each $i \in I$, α_i is the multiplicative representative of θ_i in L, and $\bar{k}^{p^{-m}}$ is the field of the roots of the polynomials $T^{p^m} - y$, $y \in \bar{k}$.

Proof. For each $m \in \mathbb{N}$, let $L_m = L_{m-1}(\{\alpha_{i,m}\}_{i \in I})$ where for all $i \in I$ $\alpha_{i,m}^p = \alpha_{i,m-1}$, $L_0 = K_{\mathfrak{p}}$ and $\alpha_{i,0} = \alpha_i$. If L is the completion of $L^* = \bigcup_{m \geq 0} L_m$, then L satisfies the conditions of the proposition. Since $\alpha_i \in \bigcap_{m=0}^{\infty} L^{p^m}$, it follows that α_i is the multiplicative representative of θ_i. □

Now we are ready to prove our main result.

Theorem 2.5.20. *Let K/k be a function field, $\mathfrak{p}$ a place of K, $K_{\mathfrak{p}}$ the completion of K with respect to $\mathfrak{p}$, and π a prime element of $\mathfrak{p}$. Then $K_{\mathfrak{p}}$ is isomorphic to $k(\mathfrak{p})((\pi))$, where $k(\mathfrak{p})$ is the residue field of $\mathfrak{p}$. More precisely, $\hat{\vartheta}$ contains a coefficient field S. If $k(\mathfrak{p})/k$ is separable we may choose $k \subseteq S$, and S is unique satisfying this property. If $k(\mathfrak{p})/k$ is not separable, then S is not necessarily unique.*

Proof.

I.-Separable Case: We have $k(\mathfrak{p}) \cong \vartheta/\mathfrak{p} \cong \hat{\vartheta}/\hat{\mathfrak{p}}$. Let $k(\mathfrak{p})/k$ be separable. We write $k(\mathfrak{p}) = k(\alpha)$, with $\alpha \in k(\mathfrak{p})$. Let $f(x)$ be the irreducible polynomial of α over k. Since α is separable, we have

$$f(x) = (x - \alpha)\, g(x) \quad \text{with} \quad g(x) \in k(\mathfrak{p})[x], \quad \text{and} \quad g(\alpha) \neq 0.$$

Therefore, $x - \alpha$ and $g(x)$ are relatively prime. Now consider $f(x)$ as a polynomial with coefficients in $\hat{\vartheta}$. If we apply Hensel's lemma to f, we can see that f admits a factor of degree one, $ax + b \in \hat{\vartheta}[x]$, and such that the residue is $a \equiv 1 \bmod \hat{\mathfrak{p}}$, $-b \equiv \alpha \bmod \hat{\mathfrak{p}}$.

Let $\alpha_1 = -\frac{b}{a} \in \hat{\vartheta}$ be such that $\alpha_1 \bmod \hat{\mathfrak{p}} = \alpha$. Now, α_1 is algebraic over k since $f(\alpha_1) = 0$. Let $n = \deg f$. The elements $1, \alpha, \ldots, \alpha^{n-1}$ are linearly independent over k, so that $k(\alpha_1)$ is a set of representatives of $k(\mathfrak{p})$ and $k(\alpha_1)$ is a field with $k \subseteq K_{\mathfrak{p}}, \alpha_1 \in K_{\mathfrak{p}}$, and so $k(\alpha_1) \subseteq \hat{\vartheta}$.

In order to prove the uniqueness of the field $k(\alpha_1)$, consider a subfield $E \subseteq \hat{\vartheta}$ such that E is a set of representatives of $k(\mathfrak{p})$. Let $\alpha_2 \in E$ be such that $\alpha_2 \bmod \hat{\mathfrak{p}} = \alpha$. We have $k(\mathfrak{p}) = k[\alpha]$, so that $E = k[\alpha_2] = k(\alpha_2)$. Now $f(\alpha_2) \bmod \mathfrak{p} \equiv f(\alpha) \equiv 0$, and hence $f(\alpha_2) = 0$. Recall that

$$f(x) = (ax + b)g(x), \quad f(\alpha_2) = (a\alpha_2 + b)\, g(\alpha_2),$$

but

$$g(\alpha_2) \bmod \mathfrak{p} \equiv g(\alpha) \neq 0.$$

Thus $a\alpha_2 + b = 0$, that is, $\alpha_2 = -\frac{b}{a} = \alpha_1$ and $E = k(\alpha_1)$.

Now if π is a prime element, then $S = k(\alpha_1)$ is the set of representatives of $k(\mathfrak{p})$. We have $S \subseteq K_{\mathfrak{p}}$ and $\pi \in K_{\mathfrak{p}}$. Any element $\sum_{n=m}^{\infty} a_n \pi^n$ of $S((\pi))$ is the limit of the Cauchy sequence $\left\{\sum_{i=m}^{n} a_i \pi^i\right\}_{n=m}^{\infty} \subseteq K_{\mathfrak{p}}$ and therefore converges in $K_{\mathfrak{p}}$. Conversely, every element of $K_{\mathfrak{p}}$ can be represented as a series. By all the above, the theorem follows.

II.-Inseparable Case: In this case, we have $\operatorname{char} k = p > 0$. By Proposition 2.5.10, it suffices to show that there exists a coefficient field in $\hat{\vartheta}$. Let L be as in Proposition 2.5.19. We have $\bar{L}^p = \bar{L}$, so $\bar{L}$ is a perfect field and by the first case there is a unique coefficient field N of $\bar{L}$ in ϑ_L. Let S be the subfield of N corresponding to $\bar{k} = k(\mathfrak{p})$. If $\gamma \in S$, then $\bar{\gamma} \in \bar{k}^{p^m}[\Theta]$ for some m, where $\Theta = \{\theta_i\}_{i \in I}$ is a p basis of $\bar{k}$. With the notation of Proposition 2.5.19 there exists an element

$$\beta_m \in \hat{\vartheta}\left[\{\alpha_{i,m}\}_{i \in I}\right] \quad \text{such that} \quad \bar{\beta}_m = \bar{\gamma}^{p^{-m}}.$$

It follows that

$$\beta_m \equiv \gamma^{p^{-m}} \bmod \mathfrak{p}_L,$$

where $\mathfrak{p}_L$ is the maximal ideal of the valuation ring ϑ_L. From Proposition 2.5.14 we obtain that $\beta_m^{p^m} \equiv \gamma \bmod \mathfrak{p}_L^{m+1}$. Since

$$\beta_m^{p^m} \in \hat{\vartheta}^{p^m}\left[\{\alpha_i\}_{i \in I}\right] \subseteq \hat{\vartheta},$$

it follows that

$$\gamma = \lim_{m\to\infty} \beta_m^{p^m} \in \hat{\vartheta}.$$

Therefore $S \subseteq \hat{\vartheta}$ and S is a coefficient field of $\bar{k}$ in $\hat{\vartheta}$. □

Remark 2.5.21. When $k(\mathfrak{p})/k$ is inseparable, there exist infinitely many coefficient fields. This follows from the proof of Theorem 2.5.20. That is, if we apply the given construction to another set of elements $\alpha_i' \in \hat{\vartheta}$ with $\bar{\alpha}_i = \bar{\alpha}_i'$ (see Proposition 2.5.19), then we obtain a coefficient field S' containing α_i'. Since $\hat{\mathfrak{p}} \cap S = \hat{\mathfrak{p}} \cap S' = (0)$, we have $S \neq S'$.

Remark 2.5.22. When $k(\mathfrak{p})/k$ is not separable, then it is not always possible to choose the coefficient field S so that $k \subseteq S$.

Example 2.5.23. Let k be a nonperfect field of characteristic p, that is, $k^p \neq k$. Let $a \in k \setminus k^p$ and

$$K = k(x),\ f(x) = x^p - a \in k[x].$$

Then f is irreducible and defines a place $\mathfrak{p}$ with $v_{\mathfrak{p}}(x^p - a) = 1$. We have $k(\mathfrak{p}) \cong k(b)$ with $b^p = a$. Let us see that a is not a p-power in $K_{\mathfrak{p}}$. We have that $x^p - a$ is a prime element for $\bar{\mathfrak{p}}$ (see after Definition 2.3.9). Assume that there exists $y \in K_{\mathfrak{p}}$ such that $y^p = a$. Then

$$(y - x)^p = y^p - x^p = a - x^p = -f(x),$$

whence

$$1 = v_{\mathfrak{p}}(f(x)) = v_{\mathfrak{p}}\left((y - x)^p\right) = p v_{\mathfrak{p}}(y - x),$$

which is impossible. Hence, if S is any field contained in $\hat{\vartheta}$ that is a system of representatives, then $a \notin S$. Indeed, if $a \in S \cong k(\mathfrak{p})$, then there exists $b_1 \in S \subseteq K_{\mathfrak{p}}$ such that $b_1^p = a$. Therefore $k \not\subseteq S$.

2.6 Exercises

Exercise 2.6.1. Let $K = k(x)$ and $y = 1/x$. Let $g(y) \in k[y]$ be a monic irreducible polynomial in y and v_g be the valuation associated to $g(y)$. Which valuation in the set $\{v_f, v_\infty \mid f(x) \in k[x] \text{ irreducible}\}$ does v_g correspond to?

Exercise 2.6.2. Let $x = \sum_{n=m}^{\infty} a_n p^n \in \mathbb{Q}_p$, where $a_n \in \{0, 1, 2, \ldots, p-1\}$. Prove that $x \in \mathbb{Q}$ if and only if there exists $n_0 \in \mathbb{Z}$, $n_0 \geq m$, and $k \in \mathbb{N}$ such that $a_{n+k} = a_n$ for all $n \geq n_0$, that, is x is periodic after a certain index.

Exercise 2.6.3. Let φ be a place of K. Show that $\varphi(0) = 0$ and $\varphi(1) = 1$.

Exercise 2.6.4. Let $p \in \mathbb{Z}$ be a prime number. Let $v_p \colon \mathbb{Q} \to \mathbb{Z}$ be the p-adic valuation, that is, if $x = p^\alpha \frac{a}{b} \in \mathbb{Q}^*$, $a, b \in \mathbb{Z}$, $p \nmid a$, $p \nmid b$, then $v_p(x) = \alpha$.

Let v be any valuation of $\mathbb{Q}$. We have $v(n) \geq v(1) \geq 0$ for all $n \in \mathbb{N}$. Prove that there exists $p \in \mathbb{N}$ minimum such that $v(p) > 0$.

Show that p is a prime number and that v is equivalent to v_p.

Exercise 2.6.5. Let L/E be a field extension and $\omega \colon L \to G \cup \{\infty\}$ be a valuation such that $\omega(L^*) = G$. Let $H = \omega(E^*) < G$.

Show that if $x_1, \ldots, x_n \in L$ are such that $\omega(x_1), \ldots, \omega(x_n)$ are distinct classes of G modulo H, then $x_1, x_2, \ldots, x_n$ are linearly independent over E. In particular, $[G : H] \leq [L : E]$.

Exercise 2.6.6. Let K/k be a function field. Show that all valuations of K that are trivial on k^* are discrete.

Exercise 2.6.7. Let $f(x) \in k[x]$ be a monic and irreducible polynomial. Let v_f be the valuation associated with the valuation ring ϑ_f and maximal ideal $\wp_f$. Prove that $\vartheta_f/\wp_f \cong k[x]/(f(x))$.

Exercise 2.6.8. Let k be an arbitrary field and $K = k(x)$ be the rational function field. Let $y = \frac{f(x)}{g(x)} \in k(x)$ with $(f(x), g(x)) = 1$ and $y \notin k$.

Prove that $[k(x) : k(y)] = \max\{\deg f(x), \deg g(x)\}$.

Let $\varphi \colon K \to K$ be such that

$$\varphi \in \operatorname{Aut}_k K = \\ \left\{\varphi \colon K \to K \mid \varphi \text{ is automorphism of } K \text{ and } \varphi(\alpha) = \alpha \ \forall\, \alpha \in k\right\}.$$

Prove that $\varphi(x) = \frac{ax+b}{cx+d}$ with $a, b, c, d \in k$, and $ad - bc \neq 0$.

Exercise 2.6.9. Let k be any field, $K = k(x)$ be a rational function field over k, and $z = \frac{ax+b}{cx+d}$ with $a, b, c, d \in k$ and $ad - bc \neq 0$. Let $f(z) \in k[z]$ be a monic and irreducible polynomial. Then there exists a unique place $\mathfrak{p}$ of K such that $v_{\mathfrak{p}}(f(z)) = 1$. Describe $\mathfrak{p}$ in terms of x.

Exercise 2.6.10. Find $|\operatorname{Aut}_k k(x)|$ when $k = \mathbb{F}_q$ is the finite field containing q elements.

Exercise 2.6.11. Let K be a number field, that is, $[K : \mathbb{Q}] < \infty$. Let $\wp_1, \ldots, \wp_s$ be different places of K (in this case we may consider place = ideal of ϑ_K), $m_1, \ldots, m_s \in \mathbb{N}$, and $a_1, \ldots, a_s \in K$ arbitrary. Show that there exists $x \in K$ such that $x \equiv a_i \bmod \wp_i^{m_i}$, $1 \leq i \leq s$, where $\wp_i^{m_i}$ denotes the m_ith power of the prime ideal $\wp_i$.

Exercise 2.6.12. Let $E \subseteq F$ be two arbitrary fields. Let x be any element in some field containing F such that x is transcendental over F. Prove that $[F : E] = [F(x) : E(x)]$ (finite or infinite).

Exercise 2.6.13. Let ϑ be a valuation ring, $\mathcal{P}$ its maximal ideal, let $K = \operatorname{quot}\vartheta$ and $E = \vartheta/\mathcal{P}$. Let $E_1 = E \cup \{\infty\}$ and consider $\varphi\colon K \to E_1$ given by

$$\varphi(x) = \begin{cases} x \bmod \mathcal{P} & \text{if } x \in \vartheta, \\ \infty & \text{if } x \notin \vartheta. \end{cases}$$

Prove that φ is a place and $\vartheta_\varphi = \vartheta$.

Exercise 2.6.14. Given a nonarchimedean absolute value $|\ |$ over a field K, prove using only the properties of a nonarchimedean absolute value that $\{x \in K \mid |x| \leq 1\}$ is a valuation ring with maximal ideal $\{x \in K \mid |x| < 1\}$.

Exercise 2.6.15. Prove Corollaries 2.5.5 and 2.5.6 assuming only that the values are archimedean instead of being contained in $\mathbb{R}$.

Exercise 2.6.16. Let ϑ be a discrete valuation ring and let $K = \operatorname{quot}\vartheta$. Prove that if $\vartheta \subseteq R \subsetneqq K$ for a ring R, then $\vartheta = R$.

3

The Riemann–Roch Theorem

The Riemann–Roch theorem relates various numbers and invariants of a function field, by means of an equality that plays a central role in our whole theory: It allows us to obtain elements that satisfy given properties, to construct automorphisms or homomorphisms with given characteristics, etc. On the other hand, this equality introduces an arithmetic invariant that is intrinsic to any function field, namely its genus.

We begin by defining divisors, which codify a finite number of places and provide us with relevant information on elements of the field that satisfy given conditions. We study basic properties of divisors as well as some vector spaces associated to them. Thanks to these vector spaces, which are subsets of the function field, we are able to introduce in a natural way the genus of the field and obtain Riemann's theorem.

Riemann's theorem is just an inequality that relates the dimension of the vector space associated to a divisor, the degree of the divisor, and the genus of the field. The missing quantity that would allow us to have equality corresponds to the Riemann–Roch theorem, and in order to find out what the inequality is, we will need the concept of a differential.

We will motivate the definition of a differential by means of the line complex integral. Using the residue theorem, we shall make these analytic concepts algebraic, obtaining in this way the general definition of a Weil differential and the missing term in Riemann's theorem.

From this point on, by K/k we will mean a function field with field of constants k.

3.1 Divisors

Notation 3.1.1. For a function field K, let $\mathbb{P}_K$ (or simply $\mathbb{P}$ when there is no confusion possible), be the set of all places of K, that is,

$$\mathbb{P}_K = \{\mathcal{P} \mid \mathcal{P} \text{ is a place of } K\}.$$

Definition 3.1.2. Given a function field K, the free abelian group generated by all the elements of $\mathbb{P}_K$ is called *the divisor group of* K and will be denoted by D_K. The places are also called *prime divisors*. The divisor group will be written multiplicatively.

Hence, an arbitrary divisor $\mathfrak{A}$ can be written uniquely as $\prod_{\mathcal{P}\in\mathbb{P}_K} \mathcal{P}^{v_{\mathcal{P}}(\mathfrak{A})}$, where $v_{\mathcal{P}}(\mathfrak{A}) \in \mathbb{Z}$ and $v_{\mathcal{P}}(\mathfrak{A}) = 0$ for almost all $\mathcal{P}$ (almost all means all but a finite number). The *unit divisor*, that is, the divisor $\prod_{\mathcal{P}\in\mathbb{P}_K} \mathcal{P}^0$, is denoted by $\mathfrak{N}$. The divisor $\mathfrak{N}$ is the only one satisfying $v_{\mathcal{P}}(\mathfrak{N}) = 0$ for every place $\mathcal{P}$.

Definition 3.1.3. A divisor $\mathfrak{A}$ is called *integral* if $v_{\mathcal{P}}(\mathfrak{A}) \geq 0$ for every place $\mathcal{P}$. We say that a divisor $\mathfrak{A}$ *divides* another divisor $\mathfrak{B}$ if there exists an integral divisor $\mathfrak{C}$ such that $\mathfrak{B} = \mathfrak{A}\mathfrak{C}$. This is equivalent to saying that $v_{\mathcal{P}}(\mathfrak{B}) \geq v_{\mathcal{P}}(\mathfrak{A})$ for all $\mathcal{P}$. When $\mathfrak{A}$ divides $\mathfrak{B}$ we will write $\mathfrak{A} \mid \mathfrak{B}$.

Definition 3.1.4. We say that two divisors $\mathfrak{A}$, $\mathfrak{B}$ are *relatively prime* or *coprime* if $v_{\mathcal{P}}(\mathfrak{A}) \neq 0 \Longrightarrow v_{\mathcal{P}}(\mathfrak{B}) = 0$, that is, $\mathfrak{A}$ and $\mathfrak{B}$ have no common prime divisors.

Note that all the above are just generalizations of definitions and notation that are used in the usual arithmetic.

Recall that given a place $\mathcal{P}$, $f_{\mathcal{P}} = [k(\mathcal{P}) : k]$ denotes the degree of $\mathcal{P}$ (Definition 2.4.13), where $k(\mathcal{P})$ is the residue field. We extend this definition to any divisor.

Definition 3.1.5. Let $\mathfrak{A}$ be a divisor. We define the *degree of* $\mathfrak{A}$, which will be denoted by $d_K(\mathfrak{A})$, or $d(\mathfrak{A})$ in case there is no possible confusion, by

$$d_K(\mathfrak{A}) = \sum_{\mathcal{P}\in\mathbb{P}_K} f_{\mathcal{P}} v_{\mathcal{P}}(\mathfrak{A}), \quad \text{where} \quad \mathfrak{A} = \prod_{\mathcal{P}\in\mathbb{P}_K} \mathcal{P}^{v_{\mathcal{P}}(\mathfrak{A})}.$$

Definition 3.1.6. Let S be a set of prime divisors of K and let $\mathfrak{A}$ be a divisor. We define $\Gamma(\mathfrak{A}|S) = \{x \in K \mid v_{\mathcal{P}}(x) \geq v_{\mathcal{P}}(\mathfrak{A}) \text{ for all } \mathcal{P} \in S\}$.

Note that $x \in \Gamma(\mathfrak{A}|S)$ if and only if $|x|_{\mathcal{P}} = e^{-v_{\mathcal{P}}(x)} \leq e^{-v_{\mathcal{P}}(\mathfrak{A})}$ for all $\mathcal{P}$ in S, that is, $\Gamma(\mathfrak{A}|S)$ measures how many elements in K have their absolute values $|\ |_{\mathcal{P}}$ less than or equal to the values $e^{-v_{\mathcal{P}}(\mathfrak{A})}$ for every prime divisor $\mathcal{P}$ in S.

For instance, if $K = k(x)$, $\mathfrak{A} = \mathcal{P}_1^3\mathcal{P}_2^{-2}\mathcal{P}_7^{-4}$, where $\mathcal{P}_i$ corresponds to the polynomial $x - i$ and $S = \{\mathcal{P}_1, \mathcal{P}_2\}$, then $\Gamma(\mathfrak{A}|S) = \{(x-1)^n(x-2)^m h(x) \mid n \geq 3, m \geq -2, h(x) \in k(x), v_{\mathcal{P}_1}(h(x)) = v_{\mathcal{P}_2}(h(x)) = 0\}$.

Proposition 3.1.7. *$\Gamma(\mathfrak{A}|S)$ is a vector space over the field k of constants of K.*

Proof. Exercise 3.6.3. □

The proof of the next proposition is left to the reader.

Proposition 3.1.8.

(i) *If $\mathfrak{A} \mid \mathfrak{B}$, then $\Gamma(\mathfrak{B}|S) \subseteq \Gamma(\mathfrak{A}|S)$.*
(ii) *If $S \subseteq S_1$ then $\Gamma(\mathfrak{A}|S_1) \subseteq \Gamma(\mathfrak{A}|S)$.*

(iii) *If* $\mathfrak{C} := \mathfrak{A}\mathfrak{B}^{-1} = \prod_{\mathcal{P}\in\mathbb{P}_K} \mathcal{P}^{v_{\mathcal{P}}(\mathfrak{C})}$ *satisfies* $v_{\mathcal{P}}(\mathfrak{C}) = 0$ *for all* $\mathcal{P} \in S$, *then* $\Gamma(\mathfrak{A}|S) = \Gamma(\mathfrak{B}|S)$. □

From Proposition 3.1.8, we obtain that given S and $\mathfrak{A}$, we can define $\mathfrak{A}_0 = \prod_{\mathcal{P}\in S} \mathcal{P}^{v_{\mathcal{P}}(\mathfrak{A})}$ (that is, $\mathfrak{A}_0$ has support in S and its components are equal to those of $\mathfrak{A}$). Then $\Gamma(\mathfrak{A}_0|S) = \Gamma(\mathfrak{A}|S)$.

The next theorem, which is very important, allows us to measure the relative dimension of the vector spaces $\Gamma(\mathfrak{A}|S)$.

Theorem 3.1.9. *Let S be finite and $\mathfrak{A}|\mathfrak{B}$. Then*

$$\dim_k \frac{\Gamma(\mathfrak{A}|S)}{\Gamma(\mathfrak{B}|S)} = d(\mathfrak{B}_0) - d(\mathfrak{A}_0) = d\left(\mathfrak{B}_0\mathfrak{A}_0^{-1}\right).$$

Proof. By Proposition 3.1.8 (iii) we may assume $\mathfrak{B} = \mathfrak{B}_0$ and $\mathfrak{A} = \mathfrak{A}_0$. Since $\mathfrak{A}|\mathfrak{B}$, we have $\mathfrak{B} = \mathfrak{A}\mathcal{P}_1 \cdots \mathcal{P}_n$ with $\mathcal{P}_i \in S$ (not necessarily distinct). We have $\Gamma(\mathfrak{A}|S) \supseteq \Gamma(\mathfrak{A}\mathcal{P}_1|S) \supseteq \Gamma(\mathfrak{A}\mathcal{P}_1\mathcal{P}_2|S) \supseteq \cdots \supseteq \Gamma(\mathfrak{A}\mathcal{P}_1\cdots\mathcal{P}_n|S) = \Gamma(\mathfrak{B}|S)$. Therefore

$$\begin{aligned}\dim_k \frac{\Gamma(\mathfrak{A}|S)}{\Gamma(\mathfrak{B}|S)} &= \dim_k \frac{\Gamma(\mathfrak{A}|S)}{\Gamma(\mathfrak{A}\mathcal{P}_1|S)} + \dim_k \frac{\Gamma(\mathfrak{A}\mathcal{P}_1|S)}{\Gamma(\mathfrak{A}\mathcal{P}_1\mathcal{P}_2|S)} + \cdots \\ &\qquad \cdots + \dim_k \frac{\Gamma(\mathfrak{A}\mathcal{P}_1\cdots\mathcal{P}_{n-1}|S)}{\Gamma(\mathfrak{B}|S)}\end{aligned} \tag{3.1}$$

If we prove $\dim_k \frac{\Gamma(\mathfrak{C}|S)}{\Gamma(\mathfrak{C}\mathcal{P}|S)} = d(\mathcal{P})$ for $\mathcal{P} \in S$, then by (3.1) it will follow that

$$\dim_k \frac{\Gamma(\mathfrak{A}|S)}{\Gamma(\mathfrak{B}|S)} = d(\mathcal{P}_1) + \cdots + d(\mathcal{P}_n) = d\left(\mathfrak{B}\mathfrak{A}^{-1}\right).$$

Hence it suffices to consider the case $\mathfrak{B} = \mathfrak{A}\mathcal{P}$, $\mathcal{P} \in S$, which means we must prove the equality

$$\dim_k \frac{\Gamma(\mathfrak{A}|S)}{\Gamma(\mathfrak{A}\mathcal{P}|S)} = d(\mathcal{P}) = f_{\mathcal{P}} = [k(\mathcal{P}) : k] = f.$$

First, from the approximation theorem (Corollary 2.5.6) there exists $u \in K$ such that $v_{\mathfrak{S}}(u) = v_{\mathfrak{S}}(\mathfrak{A})$ for all $\mathfrak{S} \in S$. In particular, $u \in \Gamma(\mathfrak{A}|S)$.

If $x_1, x_2, \ldots, x_f, x_{f+1}$ are any $f+1$ elements in $\Gamma(\mathfrak{A}|S)$, then

$$v_{\mathcal{P}}\left(x_i u^{-1}\right) = v_{\mathcal{P}}(x_i) - v_{\mathcal{P}}(u) = v_{\mathcal{P}}(x_i) - v_{\mathcal{P}}(\mathfrak{A}) \geq 0.$$

Thus, for $i = 1, \ldots, f+1$, $x_i u^{-1} \in \vartheta_{\mathcal{P}}$, where $\vartheta_{\mathcal{P}}$ is the valuation ring of $\mathcal{P}$. Since $k(\mathcal{P}) = \vartheta_{\mathcal{P}}/\mathcal{P}$ is of degree f over k, there exist $a_1, a_2, \ldots, a_f, a_{f+1} \in k$, not all zero, such that $\sum_{i=1}^{f+1} a_i x_i u^{-1} \in \mathcal{P}$. Equivalently, $\sum_{i=1}^{f+1} a_i x_i \in \mathcal{P}u$. Therefore $\sum_{i=1}^{f+1} a_i x_i \in \Gamma(\mathfrak{A}\mathcal{P}|S)$. This shows that

$$\dim_k \frac{\Gamma(\mathfrak{A}|S)}{\Gamma(\mathfrak{A}\mathcal{P}|S)} \leq f.$$

Conversely, let $y_1, y_2, \ldots, y_f \in \vartheta_{\mathcal{P}}$ be such that their classes $y_i \bmod \mathcal{P} = \bar{y}_i \in k(\mathcal{P})$ are linearly independent over k. Again by the approximation theorem (Corollary 2.5.5), there exist $y_i' \in K$ such that

$$v_{\mathcal{P}}\left(y_i' - y_i\right) > 0 \quad \text{and} \quad v_{\mathfrak{S}}\left(y_i'\right) \geq 0 \quad \text{whenever} \quad \mathfrak{S} \in S \quad \text{and} \quad \mathfrak{S} \neq \mathcal{P}.$$

Then $y_i' \equiv y_i \bmod \mathcal{P}$, and $\bar{y}_i' = \bar{y}_i \in k(\mathcal{P})$. Now if u is as before, we will have $v_{\mathfrak{S}}\left(uy_i'\right) = v_{\mathfrak{S}}(u) + v_{\mathfrak{S}}\left(y_i'\right) \geq v_{\mathfrak{S}}(u) + 0 = v_{\mathfrak{S}}(u) = v_{\mathfrak{S}}(\mathfrak{A})$ for all $\mathfrak{S} \in S$ such that $\mathfrak{S} \neq \mathcal{P}$.

On the other hand,

$$v_{\mathcal{P}}\left(uy_i'\right) = v_{\mathcal{P}}(u) + v_{\mathcal{P}}\left(y_i'\right) = v_{\mathcal{P}}(\mathfrak{A}) + 0 = v_{\mathcal{P}}(\mathfrak{A})$$

since $\bar{y}_i = \bar{y}_i' \in k(\mathcal{P})$ and $\bar{y}_i \neq 0$. Hence $y_i', y_i \in \vartheta_{\mathcal{P}} \setminus \mathcal{P}$, that is, $v_{\mathcal{P}}\left(y_i'\right) = v_{\mathcal{P}}(y_i) = 0$.

Therefore, $\left\{uy_i'\right\}_{i=1}^f \subseteq \Gamma(\mathfrak{A}|S)$. Now we will see that these elements are linearly independent modulo $\Gamma(\mathfrak{A}\mathcal{P}|S)$. Let $\sum_{i=1}^f a_i u y_i' \in \Gamma(\mathfrak{A}\mathcal{P}|S)$ with $a_i \in k$. Then for all $\mathfrak{S} \in S$ we have

$$\begin{aligned} v_{\mathfrak{S}}\left(\sum_{i=1}^f a_i u y_i'\right) &= v_{\mathfrak{S}}(u) + v_{\mathfrak{S}}\left(\sum_{i=1}^f a_i y_i'\right) = v_{\mathfrak{S}}(\mathfrak{A}) + v_{\mathfrak{S}}\left(\sum_{i=1}^f a_i y_i'\right) \\ &\geq v_{\mathfrak{S}}(\mathfrak{A}\mathcal{P}) = v_{\mathfrak{S}}(\mathfrak{A}) + v_{\mathfrak{S}}(\mathcal{P}). \end{aligned}$$

Thus

$$v_{\mathfrak{S}}\left(\sum_{i=1}^f a_i y_i'\right) \geq v_{\mathfrak{S}}(\mathcal{P}) \quad \text{for all} \quad \mathfrak{S} \in S.$$

In particular, taking $\mathfrak{S} = \mathcal{P}$, we obtain

$$v_{\mathcal{P}}\left(\sum_{i=1}^f a_i y_i'\right) \geq v_{\mathcal{P}}(\mathcal{P}) = 1,$$

that is, $\sum_{i=1}^f a_i y_i' \in \mathcal{P}$, whence $\sum_{i=1}^f a_i \bar{y}_i' = 0 \in k(\mathcal{P})$. Since $\left\{\bar{y}_i'\right\}_{i=1}^f$ is linearly independent over k, it follows that $a_i = 0$, $i = 1, \ldots, f$. Therefore

$$\dim_k \frac{\Gamma(\mathfrak{A}|S)}{\Gamma(\mathfrak{A}\mathcal{P}|S)} \geq f. \qquad \square$$

Definition 3.1.10. Let $\mathfrak{A}$ be any divisor of K. We denote by $L_K(\mathfrak{A})$ or $L(\mathfrak{A})$ the k-vector spaces $\Gamma(\mathfrak{A} \mid \mathbb{P}_K)$. That is,

$$L(\mathfrak{A}) = \{x \in K \mid v_{\mathcal{P}}(x) \geq v_{\mathcal{P}}(\mathfrak{A}) \text{ for all } \mathcal{P} \in \mathbb{P}_K\}.$$

For instance, if $K = k(x)$, $\mathfrak{A} = \mathcal{P}_1^3 \mathcal{P}_2^{-2} \mathcal{P}_7^{-4}$, where $\mathcal{P}_i$ corresponds to the polynomial $x - i$, we have

$$L(\mathfrak{A}) = \{(x-1)^3(x-2)^{-2}(x-7)^{-4}h(x) \mid h(x) \in k[x], \deg h(x) \le 3\}.$$

Note that $L(\mathfrak{A})$ measures how many elements of K have all their absolute values less than or equal to the values $e^{-v_{\mathcal{P}}(\mathfrak{A})}$ for every prime divisor $\mathcal{P}$ of the field.

We have that $L(\mathfrak{A})$ is a k-vector space and if $\mathfrak{A} \mid \mathfrak{B}$, then $L(\mathfrak{A}) \supseteq L(\mathfrak{B})$. These vector spaces play a central roll in the Riemann–Roch theorem.

Theorem 3.1.11. *For any divisor $\mathfrak{A}$, we have $\ell(\mathfrak{A}) := \dim_k L(\mathfrak{A}) < \infty$. If $\mathfrak{A} \mid \mathfrak{B}$, then*

$$\ell(\mathfrak{A}) + d(\mathfrak{A}) \le \ell(\mathfrak{B}) + d(\mathfrak{B}).$$

Proof. Let S be the set of prime divisors $\mathcal{P}$ such that $v_{\mathcal{P}}(\mathfrak{A}) \ne 0$ or $v_{\mathcal{P}}(\mathfrak{B}) \ne 0$. Then S is finite.

We have

$$L(\mathfrak{A}) \cap \Gamma(\mathfrak{B}|S) = L(\mathfrak{B}). \tag{3.2}$$

On the other hand, $L(\mathfrak{A}) + \Gamma(\mathfrak{B}|S) \subsetneqq \Gamma(\mathfrak{A}|S)$, so applying the isomorphism theorems we obtain that there exists a monomorphism $\frac{L(\mathfrak{A})}{L(\mathfrak{B})} \to \frac{\Gamma(\mathfrak{A}|S)}{\Gamma(\mathfrak{B}|S)}$, which shows that $\dim_k \frac{L(\mathfrak{A})}{L(\mathfrak{B})} \le \dim_k \frac{\Gamma(\mathfrak{A}|S)}{\Gamma(\mathfrak{B}|S)} = d(\mathfrak{B}) - d(\mathfrak{A}) < \infty$ (see Exercise 3.6.24)

Let $\mathfrak{B}$ be an integral divisor with $\mathfrak{B} \ne \mathfrak{N}$, where $\mathfrak{N}$ is the unit divisor. For $x \in L(\mathfrak{B}) \setminus \{0\}$, we have $x \notin k$. Indeed, since $\mathfrak{B}$ is an integral divisor that is different from $\mathfrak{N}$, there exists a prime divisor $\mathcal{P}$ such that $v_{\mathcal{P}}(x) \ge v_{\mathcal{P}}(\mathfrak{B}) > 0$, that is, $v_{\mathcal{P}}(x) > 0$, and therefore x is transcendental. Furthermore, $v_{\mathfrak{S}}(x) \ge v_{\mathfrak{S}}(\mathfrak{B}) \ge 0$ for all $\mathfrak{S}$. This is impossible since the valuation v_∞ in $k(x)$ is such that $v_\infty(x) = -1 < 0$. If we extend v_∞ to K, then if v is such an extension we have $v(x) < 0$.

Hence, $L(\mathfrak{B}) = \{0\}$ for an integral divisor $\mathfrak{B} \ne \mathfrak{N}$. Given $\mathfrak{A}$ arbitrary, we will prove that there exists an integral divisor $\mathfrak{B} \ne \mathfrak{N}$ such that $\mathfrak{A} \mid \mathfrak{B}$. Let

$$\mathfrak{B} = \mathfrak{S} \prod_{\substack{\mathcal{P} \in \mathbb{P}_K \\ v_{\mathcal{P}}(\mathfrak{A}) \ne 0}} \mathcal{P}^{|v_{\mathcal{P}}(\mathfrak{A})|+1} \quad \text{with} \quad \mathfrak{S} \in \mathbb{P}_K \quad \text{such that} \quad v_{\mathfrak{S}}(\mathfrak{A}) = 0.$$

Then there exists an integral divisor $\mathfrak{B}$ such that $v_{\mathfrak{S}}(\mathfrak{B}) = 1 > 0$, $\mathfrak{B} \ne \mathfrak{N}$, and

$$\mathfrak{C} = \mathfrak{B}\mathfrak{A}^{-1} = \mathfrak{S} \prod_{\substack{\mathcal{P} \in \mathbb{P}_K \\ v_{\mathcal{P}}(\mathfrak{A}) \ne 0}} \mathcal{P}^{|v_{\mathcal{P}}(\mathfrak{A})| - v_{\mathcal{P}}(\mathfrak{A})+1}$$

is an integral divisor. Therefore, $\mathfrak{A} \mid \mathfrak{B}$ and we have

$$\frac{L(\mathfrak{A})}{L(\mathfrak{B})} = \frac{L(\mathfrak{A})}{\{0\}} = L(\mathfrak{A}) \quad \text{and} \quad \ell(\mathfrak{A}) = \dim_k \frac{L(\mathfrak{A})}{L(\mathfrak{B})} \le d(\mathfrak{B}) - d(\mathfrak{A}) < \infty.$$

This shows that $\ell(\mathfrak{A}) < \infty$ for any divisor $\mathfrak{A}$. The second part follows immediately since $\ell(\mathfrak{A}) - \ell(\mathfrak{B}) = \dim_k \frac{L(\mathfrak{A})}{L(\mathfrak{B})} \le d(\mathfrak{B}) - d(\mathfrak{A})$. □

In the process of proving the above theorem, we have obtained the following corollary:

Corollary 3.1.12. *If $\mathfrak{B}$ is an integral divisor and $\mathfrak{B} \neq \mathfrak{N}$, then $L(\mathfrak{B}) = 0$.* □

For the next proposition, and only for it, we consider the possibility that the field k' of constants of a function field over k properly contains k. In any case, we have $[k' : k] < \infty$ (Proposition 2.1.6).

Proposition 3.1.13. *Let K be a function field over k. Let k' be the field of constants of K. Then if $\mathfrak{N}$ is the principal divisor of K, we have $L(\mathfrak{N}) = k'$.*

Proof. If x is transcendental over k, the valuation v_∞ in $k'(x)$ satisfies $v_\infty(x) = -1$. When we extend v_∞ to a valuation v in K, we obtain $v(x) < 0$. On the other hand, we have

$$L(\mathfrak{N}) = \{z \in K \mid v_{\mathcal{P}}(z) \geq v_{\mathcal{P}}(\mathfrak{N}) = 0 \text{ for all } \mathcal{P}\}.$$

Therefore $L(\mathfrak{N}) \subseteq k'$.

Now if $\alpha \in k'$ and $\alpha \neq 0$, then α is algebraic over k. Hence there exist $a_0, \ldots, a_{n-1} \in k$ such that

$$\alpha^n + a_{n-1}\alpha^{n-1} + \cdots + a_1\alpha + a_0 = 0, \quad \text{that is,} \quad \alpha^n = -\sum_{i=0}^{n-1} a_i\alpha^i \neq 0.$$

Assume that $v_{\mathcal{P}}(\alpha) \neq 0$ for some prime divisor $\mathcal{P}$. Then for $a_i \neq 0$,

$$v_{\mathcal{P}}\left(a_i\alpha^i\right) = v_{\mathcal{P}}(a_i) + i v_{\mathcal{P}}(\alpha) = i v_{\mathcal{P}}(\alpha) \neq j v_{\mathcal{P}}(\alpha) \quad \text{for} \quad i \neq j.$$

That is,

$$v_{\mathcal{P}}\left(-\sum_{i=0}^{n-1} a_i\alpha^i\right) = \min_{a_i \neq 0}\{i v_{\mathcal{P}}(\alpha)\} \neq n v_{\mathcal{P}}(\alpha),$$

which is absurd.

Hence, we have obtained that $v_{\mathcal{P}}(\alpha) = 0$ for all $\alpha \in k'$ such that $\alpha \neq 0$, so $k' \subseteq L(\mathfrak{N})$, proving the equality. □

Corollary 3.1.14. *If $\alpha \in k'$ is nonzero, then $v_{\mathcal{P}}(\alpha) = 0$ for any prime divisor $\mathcal{P}$.* □

Coming back to our usual notation, namely when k denotes the exact field of constants of K, we have the following corollary:

Corollary 3.1.15. $L(\mathfrak{N}) = k$ *and* $\dim_k L(\mathfrak{N}) = 1$. □

3.2 Principal Divisors and Class Groups

The first part of this section will be dedicated to proving two important results, which are:

(i) If $x \in K$ is nonzero there exist only a finite number of places $\mathcal{P}$ such that $v_{\mathcal{P}}(x) \neq 0$.
As a consequence of (i), for any $x \in K^*$ we can define the divisor of x by $(x)_K = \prod_{\mathcal{P} \in \mathbb{P}_K} \mathcal{P}^{v_{\mathcal{P}}(x)}$. This will allow us to prove:
(ii) $d\left((x)_K\right) = 0$ for all $x \in K^*$.

In other words, $(x)_K$ codifies all the absolute values or valuations of x in a single divisor, which will be of degree 0.

Theorem 3.2.1. *If $x \in K^*$, there exists only a finite number of places $\mathcal{P}$ such that $v_{\mathcal{P}}(x) \neq 0$.*

Proof. If $x \in k^*$, then $v_{\mathcal{P}}(x) = 0$ for all $\mathcal{P}$ and there is nothing to prove. Now assume that $x \in K \setminus k$, that is, x is transcendental. Let $[K : k(x)] = N < \infty$. Let $\mathcal{P}_1, \ldots, \mathcal{P}_n$ be n distinct places such that $v_{\mathcal{P}_i}(x) > 0$ for $i = 1, \ldots, n$. We will see that $n \leq N$. Let $\mathfrak{B} = \prod_{i=1}^{n} \mathcal{P}^{v_{\mathcal{P}_i}(x)}$. Clearly $\mathfrak{B}$ is an integral divisor. Let $S = \{\mathcal{P}_1, \ldots, \mathcal{P}_n\}$. From Theorem 3.1.9 we obtain

$$\dim_k \frac{\Gamma\left(\mathfrak{N}|S\right)}{\Gamma\left(\mathfrak{B}|S\right)} = d\left(\mathfrak{B}\right) - d\left(\mathfrak{N}\right) = d\left(\mathfrak{B}\right) = \sum_{i=0}^{n} f_{\mathcal{P}_i} v_{\mathcal{P}_i}(x). \tag{3.3}$$

Let $y_1, y_2, \ldots, y_N, y_{N+1}$, be $N+1$ distinct elements of $\Gamma\left(\mathfrak{N}|S\right)$. That is,

$$v_{\mathcal{P}}\left(y_j\right) \geq v_{\mathcal{P}}\left(\mathfrak{N}\right) = 0, \quad \text{with} \quad \mathcal{P} \in S \quad \text{and} \quad j = 1, 2, \ldots, N+1.$$

Since $[K : k(x)] = N$, there exist polynomials $f_j \in k[x]$ of which at least one has a nonzero constant term such that $\sum_{j=1}^{N+1} f_j(x) y_j = 0$. We write $f_j(x) = a_j + x g_j(x)$ with $a_j \in k$. Then $\sum_{j=1}^{N+1} a_j y_j = -x \sum_{j=1}^{N+1} g_j(x) y_j$, where some a_j is nonzero. Since $v_{\mathcal{P}_i}(x) > 0$, we have $v_{\mathcal{P}_i}\left(g_j(x)\right) \geq 0$. Therefore

$$\begin{aligned} v_{\mathcal{P}_i}\left(\sum_{j=1}^{N+1} a_j y_j\right) &= v_{\mathcal{P}_i}(x) + v_{\mathcal{P}_i}\left(\sum_{j=1}^{N+1} g_j(x) y_j\right) \\ &\geq v_{\mathcal{P}_i}(x) = v_{\mathcal{P}_i}\left(\mathfrak{B}\right), \quad i = 1, \ldots, n, \end{aligned}$$

that is, $\sum_{j=1}^{N+1} a_j y_j \in \Gamma\left(\mathfrak{B}|S\right)$. Hence

$$\dim_k \frac{\Gamma\left(\mathfrak{N}|S\right)}{\Gamma\left(\mathfrak{B}|S\right)} = \sum_{i=1}^{n} f_{\mathcal{P}_i} v_{\mathcal{P}_i}(x)) \leq N.$$

In particular, $n \leq N$.

We have proved that there are at most N distinct places $\mathcal{P}$ such that $v_{\mathcal{P}}(x) > 0$. Taking $y = \frac{1}{x}$, we show the existence of at most N different places $\mathfrak{S}$ such that $v_{\mathfrak{S}}(y) = -v_{\mathfrak{S}}(x) > 0$, or $v_{\mathfrak{S}}(x) < 0$. Therefore there are at most $2N$ different places $\mathcal{P}$ such that $v_{\mathcal{P}}(x) \neq 0$. □

Definition 3.2.2. Given $x \in K^*$, we define the *principal divisor of x in K* as $(x)_K = \prod_{\mathcal{P} \in \mathbb{P}_K} \mathcal{P}^{v_{\mathcal{P}}(x)}$. If there is no possible confusion, we will write (x) instead of $(x)_K$.

Definition 3.2.3. Given $x \in K^*$, we define the *divisor of zeros of x* by

$$\mathfrak{Z}_x = \prod_{\substack{\mathcal{P} \in \mathbb{P}_K \\ v_{\mathcal{P}}(x) > 0}} \mathcal{P}^{v_{\mathcal{P}}(x)}$$

and the *pole divisor of x* by

$$\mathfrak{N}_x = \prod_{\substack{\mathcal{P} \in \mathbb{P}_K \\ v_{\mathcal{P}}(x) < 0}} \mathcal{P}^{-v_{\mathcal{P}}(x)}.$$

We observe that both $\mathfrak{Z}_x$ and $\mathfrak{N}_x$ are integral divisors and that

$$(x)_K = \mathfrak{Z}_x \mathfrak{N}_x^{-1} = \frac{\mathfrak{Z}_x}{\mathfrak{N}_x}.$$

Proposition 3.2.4. *The set of all principal divisors $\{(x)_K \mid x \in K^*\}$ is a subgroup of D_K.*

Proof. From the properties of valuations it follows that $(xy)_K = (x)_K(y)_K$ and that $\left(x^{-1}\right)_K = (x)_K^{-1}$. □

Definition 3.2.5. The subgroup of principal divisors is denoted by P_K and it is called the *principal divisor subgroup of K*. The quotient $C_K = D_K/P_K$ is called the *complete group of divisor classes of K* or *class group of K*.

Remark 3.2.6. Theorem 3.2.1 proves that for $x \in K \setminus k$, we have $d(\mathfrak{Z}_x) \leq N$ and $d(\mathfrak{N}_x) \leq N$, where $[K : k(x)] = N$. The next theorem proves that equality holds.

Theorem 3.2.7. *For $x \in K \setminus k$, $d(\mathfrak{Z}_x) = d(\mathfrak{N}_x) = N = [K : k(x)]$.*

Proof. Let $y \in K$ be an integral element over $k[x]$. Then y satisfies an equation of the form

$$y^m + f_{m-1}(x)y^{m-1} + \cdots + f_1(x)y + f_0(x) = 0 \tag{3.4}$$

with $f_i(x) \in k[x]$.

If $\mathcal{P} \nmid \mathfrak{N}_x$ (that is, $\mathcal{P}$ is not a pole of x), then $v_{\mathcal{P}}(x) \geq 0$ and

$$\begin{aligned} v_{\mathcal{P}}\left(y^m\right) &= m v_{\mathcal{P}}(y) = v_{\mathcal{P}}\left(-\sum_{i=0}^{m-1} f_i(x) y^i\right) \\ &\geq \min_{0\leq i\leq m-1} \{v_{\mathcal{P}}(f_i(x)) + i v_{\mathcal{P}}(y)\} \\ &\geq \min_{0\leq i\leq m-1} \{i v_{\mathcal{P}}(y)\} = t v_{\mathcal{P}}(y), \quad t \in \{0, 1, \ldots, m-1\}. \end{aligned}$$

It follows that $(m-t)v_{\mathcal{P}}(y) \geq 0$ with $m - t > 0$, so that $v_{\mathcal{P}}(y) \geq 0$. Therefore $\mathcal{P} \nmid \mathfrak{N}_y$.

Now let y be an arbitrary element of K^*. Since y is algebraic over $k(x)$, it satisfies an equation of the form

$$g_r(x)y^r + g_{r-1}(x)y^{r-1} + \cdots + g_1(x)y + g_0(x) = 0 \tag{3.5}$$

with $g_i(x) \in k[x]$ and $g_r(x) \neq 0$. Multiplying the equation (3.5) by $g_r(x)^{r-1}$, we obtain

$$\begin{aligned} &(g_r(x)y)^r + g_{r-1}(x)\left(g_r(x)y\right)^{r-1} + \cdots \\ &\qquad + g_r(x)^{r-2} g_1(x)\left(g_r(x)y\right) + g_r(x)^{r-1} g_0(x) = 0, \end{aligned}$$

that is, $z = g_r(x)y$ is an integral element over $k[x]$.

Let $[K : k(x)] = N$ and let $y_1, y_2, \ldots, y_N$ be a basis of $K/k(x)$. From the above remarks, we may assume that $y_1, y_2, \ldots, y_N$ are integral elements over $k[x]$. For any $r \geq 0$, the set

$$\left\{x^i y_j\right\}_{j=1,\ldots,N}^{i=0,\ldots,r}$$

is linearly independent over k. Now, from the previous observations we obtain that if $\mathcal{P} \mid \mathfrak{N}_{y_j}$ then $\mathcal{P} \mid \mathfrak{N}_x$, say $\mathfrak{N}_x = \mathcal{P}^a \mathfrak{A}$ and $\mathfrak{N}_{y_j} = \mathcal{P}^b \mathfrak{B}$, where $\mathfrak{A}$ and $\mathfrak{B}$ are integral divisors that are relatively prime to $\mathcal{P}$ and $a, b > 0$.

Let $\alpha_j \geq \frac{b}{a}$, with α_j an integer. Then

$$\mathfrak{N}_x^{\alpha_j}(y_j) = \frac{\mathfrak{N}_x^{\alpha_j} \mathfrak{Z}_{y_j}}{\mathcal{P}^b \mathfrak{B}} = \frac{\mathcal{P}^{\alpha_j a - b} \mathfrak{Z}_{y_j} \mathfrak{A}^{\alpha_j}}{\mathfrak{B}} \quad \text{with} \quad \alpha_j a - b \geq 0,$$

so $v_{\mathcal{P}}\left(\mathfrak{N}_x^{\alpha_j}(y_j)\right) \geq 0$. This shows that there exists a natural number s such that $\mathfrak{N}_x^s(y_j)$ is integral for all j.

Also, we have that $\mathfrak{N}_x^{r+s}(x^i)(y_j)$ are integral for $i = 0, \ldots, r$ and $j = 1, \ldots, N$. In particular, $x^i y_j \in L\left(\mathfrak{N}_x^{-r-s}\right)$ for $i = 0, \ldots, r$ and $j = 1, \ldots, N$ and these $(r+1)N$ elements are linearly independent over k. Since $\mathfrak{N}_x^{-r-s} | \mathfrak{N}_x$, by Theorem 3.1.11 we have

$$\ell\left(\mathfrak{N}_x^{-r-s}\right) + d\left(\mathfrak{N}_x^{-r-s}\right) \leq \ell\left(\mathfrak{N}_x\right) + d\left(\mathfrak{N}_x\right).$$

On the other hand, since x is transcendental, then $\mathfrak{N}_x$ is different from $\mathfrak{N}$ and $\mathfrak{N}_x$ is an integral divisor, so by Corollary 3.1.12, $\ell\left(\mathfrak{N}_x\right) = 0$.

We obtain

$$\begin{aligned}(r+1)N &\le \ell\left(\mathfrak{N}_x^{-r-s}\right) \le \ell\left(\mathfrak{N}_x\right) + d\left(\mathfrak{N}_x\right) - d\left(\mathfrak{N}_x^{-r-s}\right) \\ &= 0 + d\left(\mathfrak{N}_x\right) + (r+s)d\left(\mathfrak{N}_x\right) = (r+s+1)d\left(\mathfrak{N}_x\right) \quad \text{for all} \quad r \ge 0.\end{aligned}$$

Thus we have $d\left(\mathfrak{N}_x\right) \ge \frac{N(r+1)}{r+s+1} \underset{r\to\infty}{\longrightarrow} N$, and $d\left(\mathfrak{N}_x\right) \ge N$. Since we obtained $d\left(\mathfrak{N}_x\right) \le N$ in the proof of Theorem 3.2.1, we have the equality $d\left(\mathfrak{N}_x\right) = N$.

Finally, we have $\mathfrak{Z}_x = \mathfrak{N}_{1/x}$. Since $k(x) = k\left(\frac{1}{x}\right)$ we apply the above argument to $\frac{1}{x}$ with $\left[K : k\left(\frac{1}{x}\right)\right] = N$. Hence, we obtain

$$d\left(\mathfrak{Z}_x\right) = d\left(\mathfrak{N}_{1/x}\right) = N. \qquad \square$$

Remark 3.2.8. Observe that for $x \in K^*$, $(x)_K = \mathfrak{N}$ if and only if $x \in k^*$.

Corollary 3.2.9. *For* $x \in K^*$, $d\left((x)_K\right) = 0$.

Proof. If $x \in k^*$ then $(x)_K = \mathfrak{N}$ with $d\left((x)_K\right) = d\left(\mathfrak{N}\right) = 0$. If $x \in K \setminus k$, then

$$[K : k(x)] = N \quad \text{and} \quad d\left((x)_K\right) = d\left(\mathfrak{Z}_x\right) - d\left(\mathfrak{N}_x\right) = N - N = 0. \qquad \square$$

Definition 3.2.10. We say that an element x of K^* is *divisible by a divisor* $\mathfrak{A}$, and we write $\mathfrak{A} \mid x$, if $\mathfrak{A} \mid (x)_K$. If $x, y \in K^*$, we write $x \equiv y \bmod \mathfrak{A}$ whenever $x = y$ or $\mathfrak{A} \mid x - y$.

Note 3.2.11. With the previous notation we have $L\left(\mathfrak{A}\right) = \{x \in K \mid \mathfrak{A} \mid x\}$. Also note that for $x \in K^*$, $x \in L(\mathfrak{A})$ if and only if $(x)_K = \mathfrak{A}\mathfrak{C}$ for an integral divisor $\mathfrak{C}$.

Now let $d : D_K \longrightarrow \mathbb{Z}$ be the degree function. By definition d is a group homomorphism and the image of d is a nonzero subgroup of $\mathbb{Z}$, that is, $d\left(D_K\right) = m\mathbb{Z}$ with $m \in \mathbb{N}$. Therefore $d\left(D_K\right)$ and $\mathbb{Z}$ are isomorphic as groups. Let

$$\ker d = D_{K,0} = \{\mathfrak{A} \in D_K \mid d\left(\mathfrak{A}\right) = 0\}$$

be the subgroup of divisors of degree 0. We have

$$P_K \subseteq D_{K,0} \quad \text{and} \quad D_K/D_{K,0} = D_K/\ker d \cong d\left(D_K\right) \cong \mathbb{Z}.$$

We have the exact sequence

$$1 \longrightarrow D_{K,0} \longrightarrow D_K \stackrel{d}{\longrightarrow} m\mathbb{Z} \longrightarrow 0.$$

It follows that $D_K \cong D_{K,0} \oplus \mathbb{Z}$ (Exercise 3.6.2).

On the other hand, consider the function $i : K^* \longrightarrow P_K$ defined by $i(x) = (x)_K$. Clearly, i is a group epimorphism and $\ker i = k^*$ (Exercise 3.6.2). Therefore we obtain the exact sequence

$$1 \longrightarrow k^* \longrightarrow K^* \longrightarrow P_K \longrightarrow 1 \quad \text{and} \quad P_K \cong K^*/k^*.$$

Since $P_K \subseteq D_{K,0}$, d induces an epimorphism $\tilde{d} : C_K = D_K/P_K \longrightarrow m\mathbb{Z}$, and $\ker \tilde{d} = C_{K,0} = \{\mathfrak{A} \bmod P_K \mid d(\mathfrak{A}) = 0\} \cong D_{K,0}/P_K$.

That is, the degree function can be defined in a class $C \in C_K$ as $d(C) = d(\mathfrak{A})$ where $\mathfrak{A} \in C$. This definition does not depend on the representative $\mathfrak{A}$ since if $\mathfrak{A}$ and $\mathfrak{B}$ determine the same class C of C_K, then there exists $x \in K^*$ such that

$$\mathfrak{A} = \mathfrak{B}(x)_K \quad \text{and} \quad d(\mathfrak{A}) = d(\mathfrak{B}) + d((x)_K) = d(\mathfrak{B}) + 0 = d(\mathfrak{B}).$$

Definition 3.2.12. The *degree* of a class $C \in C_K$ is defined by $d(C) = d(\mathfrak{A})$, where $\mathfrak{A}$ is any divisor belonging to C.

Definition 3.2.13. The group $C_{K,0}$ is called *the group of classes of divisors of degree* 0.

We observe that since

$$1 \longrightarrow C_{K,0} \longrightarrow C_K \xrightarrow{d} m\mathbb{Z} \longrightarrow 0$$

is exact it follows that $C_K \cong C_{K,0} \oplus \mathbb{Z}$ (see Exercise 3.6.2). In particular C_K is never a finite group.

Definition 3.2.14. If $C_{K,0}$ is finite, the number $h_K = |C_{K,0}|$ is called the *class number of the field K*.

We collect the above discussion into the following theorem:

Theorem 3.2.15. *Let K/k be a function field. The degree function $d : D_K \to \mathbb{Z}$ defines an exact sequence*

$$1 \longrightarrow D_{K,0} \longrightarrow D_K \xrightarrow{d} m\mathbb{Z} \longrightarrow 0,$$

where $m \in \mathbb{N}$, $m\mathbb{Z} \cong \mathbb{Z}$, $D_K \cong D_{K,0} \oplus \mathbb{Z}$, $D_{K,0} = \ker d$ is the subgroup consisting of all divisors of degree 0 of K, and $P_K \subseteq D_{K,0}$. This sequence induces the exact sequence

$$1 \longrightarrow C_{K,0} \longrightarrow C_K \xrightarrow{d} m\mathbb{Z} \longrightarrow 0,$$

which implies

$$C_K \cong C_{K,0} \oplus \mathbb{Z}.$$

Finally, we have the exact sequence

$$1 \longrightarrow k^* \longrightarrow K^* \xrightarrow{i} P_K \longrightarrow 1,$$

where $i(x) = (x)_K$, and as a consequence the sequence

$$1 \longrightarrow k^* \longrightarrow K^* \xrightarrow{i} D_K \xrightarrow{\pi} C_K \longrightarrow 1$$

is exact, where π is the natural projection. □

For further reference, we list all the exact sequences obtained:

$$1 \longrightarrow D_{K,0} \longrightarrow D_K \xrightarrow{\frac{1}{m}d} \mathbb{Z} \longrightarrow 0, \quad D_K \cong D_{K,0} \oplus \mathbb{Z}, \tag{3.6}$$

$$1 \longrightarrow C_{K,0} \longrightarrow C_K \xrightarrow{\frac{1}{m}d} \mathbb{Z} \longrightarrow 0, \quad C_K \cong C_{K,0} \oplus \mathbb{Z}, \tag{3.7}$$

$$1 \longrightarrow k^* \longrightarrow K^* \xrightarrow{i} P_K \longrightarrow 1, \tag{3.8}$$

$$1 \longrightarrow k^* \longrightarrow K^* \xrightarrow{i} D_K \xrightarrow{\pi} C_K \longrightarrow 1, \tag{3.9}$$

$$1 \longrightarrow P_K \longrightarrow D_K \longrightarrow C_K \longrightarrow 1. \tag{3.10}$$

Example 3.2.16. Let $K = k(x)$ be a rational function field. Let $\mathfrak{A}$ be a divisor of degree 0, that is, $\mathfrak{A} \in D_{K,0}$. We write $\mathfrak{A} = \prod_{i=1}^r \mathcal{P}_i^{\alpha_i}$, where each $\mathcal{P}_i$ $(1 \le i \le r)$ is a prime divisor of K. We have $d\,(\mathfrak{A}) = \sum_{i=0}^r \alpha_i d\,(\mathcal{P}_i) = 0$.

Now choose $\mathcal{P}_r$ to be $\mathcal{P}_\infty$, i.e., the place corresponding to the valuation v_∞. Each $\mathcal{P}_i$ $(1 \le i \le r-1)$ is associated to some irreducible polynomial $f_i(x)$ of $k[x]$. We have $d\,(\mathcal{P}_\infty) = 1$.

Therefore $\alpha_r = -\sum_{i=0}^{r-1} \alpha_i \deg f_i$. Now, for any valuation $v \neq v_{f_i}, v_\infty$, we have $v\,(f_i) = 0$, $v_{f_i}\,(f_i) = 1$, and $v_\infty\,(f_i) = -\deg f_i$. Hence the divisor of f_i is $(f_i)_K = \frac{\mathcal{P}_i}{\mathcal{P}_\infty^{\deg f_i}}$, where $\mathcal{P}_i$ is the divisor corresponding to v_{f_i} and $\mathcal{P}_r = \mathcal{P}_\infty$ is the prime divisor corresponding to v_∞. Therefore

$$\begin{aligned} \left(\prod_{i=1}^{r-1} f_i(x)^{\alpha_i}\right)_K &= \prod_{i=1}^{r-1} (f_i(x))_K^{\alpha_i} = \prod_{i=1}^{r-1} \frac{\mathcal{P}_i^{\alpha_i}}{\mathcal{P}_\infty^{\alpha_i \deg f_i}} = \\ &= \left(\prod_{i=1}^{r-1} \mathcal{P}_i^{\alpha_i}\right) \mathcal{P}_\infty^{-\sum_{i=1}^{r-1} \alpha_i \deg f_i} = \prod_{i=1}^{r} \mathcal{P}_i^{\alpha_i} = \mathfrak{A}, \end{aligned}$$

that is, $\mathfrak{A}$ is principal since $\mathfrak{A} = (\alpha(x))_K$, where $\alpha(x) = \prod_{i=1}^{r-1} f_i(x)^{\alpha_i} \in k(x)^*$. We observe that if $r = 0$, then $\mathfrak{A} = \mathfrak{N} = (1)_K$, $1 \in k^*$.

This shows that $D_{K,0} = P_K$. Thus $D_{K,0}/P_K = C_{K,0} = \{1\}$ and $h_K = 1$.

In short, we have proved that any rational function field has class number 1.

Finally, since $d\,(\mathcal{P}_\infty) = 1$, the degree function d is surjective: $d\,(D_K) = \mathbb{Z}$ and $C_K \cong \mathbb{Z}$.

Note 3.2.17. If $d\,(D_K) = m\mathbb{Z}$ with $m \in \mathbb{N}$, we have

$$m = \min\left\{n \in \mathbb{N} \mid \text{ there exists a divisor } \mathfrak{A} \text{ such that } d\,(\mathfrak{A}) = n\right\}.$$

When $K = k(x)$ we have $m = 1$. If k is algebraically closed every prime divisor is of degree 1 so that $m = 1$. This is not true in general. Later on we will see an example where $m = 2$ (Proposition 4.1.9). An important result is that when k is a finite field, $m = 1$. This will be proved in Chapter 6 (Theorem 6.3.8).

We end this section with a generalization of Corollary 3.1.12.

Proposition 3.2.18. *If $\mathfrak{B}$ is a divisor such that $d(\mathfrak{B}) > 0$ or $d(\mathfrak{B}) = 0$ and $\mathfrak{B}$ is not principal, then $L(\mathfrak{B}) = \{0\}$. In particular, if $\mathfrak{B}$ is integral and $\mathfrak{B} \neq \mathfrak{N}$ then $L(\mathfrak{B}) = \{0\}$. If $\mathfrak{B} = (x)_K$ is principal, we have $L(\mathfrak{B}) = \{\alpha x \mid \alpha \in k\}$ and $\ell(\mathfrak{B}) = 1$.*

Proof. If $d(\mathfrak{B}) > 0$ and $x \in L(\mathfrak{B}) \setminus \{0\}$, then $(x)_K = \mathfrak{B}\mathfrak{C}$, where $\mathfrak{C}$ is an integral divisor. Thus $0 = d((x)_K) = d(\mathfrak{B}) + d(\mathfrak{C}) \geq d(\mathfrak{B}) > 0$, which is absurd. Hence, we have $L(\mathfrak{B}) = \{0\}$.

Now if $d(\mathfrak{B}) = 0$ and $\mathfrak{B}$ is not principal, assume that there exists $x \in L(\mathfrak{B}) \setminus \{0\}$. Then $(x)_K = \mathfrak{B}\mathfrak{C}$ for some integral divisor $\mathfrak{C}$. Therefore $0 = d((x)_K) = d(\mathfrak{B}) + d(\mathfrak{C}) = d(\mathfrak{C})$, that is, $\mathfrak{C}$ is integral and of degree 0, so $\mathfrak{C} = \mathfrak{N}$ and $\mathfrak{B} = (x)_K$, which contradicts the hypothesis.

In particular, if $\mathfrak{B}$ is an integral divisor, we have $\mathfrak{B} \neq \mathfrak{N}$ with $d(\mathfrak{B}) > 0$ and $L(\mathfrak{B}) = \{0\}$.

Finally, if $\mathfrak{B} = (x)_K$ is principal, then if $y \in L((x)_K) \setminus \{0\}$, we have $(y)_K = (x)_K$. Hence $y = \alpha x$ for some $\alpha \in k^*$ and $L((x)_K) = \{\alpha x \mid \alpha \in k\}$. □

3.3 Repartitions or Adeles

We start this section by proving Riemann's theorem, which constitutes half of the Riemann–Roch theorem, the most important result of this book. For this purpose we need the following proposition:

Proposition 3.3.1. *Let $x \in K$ be a transcendental element. Then there exists an integer $a \in \mathbb{Z}$ depending only on x such that $\ell\left(\mathfrak{N}_x^{-m}\right) + d\left(\mathfrak{N}_x^{-m}\right) \geq a$ for all $m \in \mathbb{Z}$.*

Proof. In the proof of Theorem 3.2.7 we obtained that there exists $s \in \mathbb{N}$ such that for all $r \geq 0$ we have

$$\ell\left(\mathfrak{N}_x^{-s-r}\right) \geq N(r+1) = d(\mathfrak{N}_x)(r+1), \quad \text{and} \quad N = d(\mathfrak{N}_x) = [K : k(x)].$$

For $m = r + s \geq s$ we have

$$\begin{aligned}\ell\left(\mathfrak{N}_x^{-m}\right) + d\left(\mathfrak{N}_x^{-m}\right) &\geq (r+1)d(\mathfrak{N}_x) - md(\mathfrak{N}_x) = (r+1-m)d(\mathfrak{N}_x)\\ &= (-s+1)d(\mathfrak{N}_x) = a,\end{aligned}$$

where we define a to be $(-s+1)d(\mathfrak{N}_x)$.

Now, for $m < s$, we have $\mathfrak{N}_x^{-s} \mid \mathfrak{N}_x^{-m}$, so from Theorem 3.1.11 we obtain

$$\ell\left(\mathfrak{N}_x^{-m}\right) + d\left(\mathfrak{N}_x^{-m}\right) \geq \ell\left(\mathfrak{N}_x^{-s}\right) + d\left(\mathfrak{N}_x^{-s}\right) \geq a.$$ □

Theorem 3.3.2 (Riemann). *Let x be a transcendental element and let*

$$1 - g = \sup\left\{a \mid \ell\left(\mathfrak{N}_x^{-m}\right) + d\left(\mathfrak{N}_x^{-m}\right) \geq a \text{ for all } m \in \mathbb{Z}\right\},$$

that is, $1 - g$ is the greatest lower bound of the set

$$\left\{\ell\left(\mathfrak{N}_x^{-m}\right) + d\left(\mathfrak{N}_x^{-m}\right) \mid m \in \mathbb{Z}\right\} \subseteq \mathbb{Z}.$$

Then for any divisor $\mathfrak{A} \in D_K$ we have $\ell(\mathfrak{A}) + d(\mathfrak{A}) \geq 1 - g$.

Proof. If $\mathfrak{A}, \mathfrak{B}$ are integral divisors and $\mathfrak{C} = \frac{\mathfrak{A}}{\mathfrak{B}} = \mathfrak{A}\mathfrak{B}^{-1}$, then $\mathfrak{B}^{-1}|\mathfrak{C}$ and by Theorem 3.1.11, we have

$$\ell(\mathfrak{C}) + d(\mathfrak{C}) \geq \ell\left(\mathfrak{B}^{-1}\right) + d\left(\mathfrak{B}^{-1}\right).$$

This shows that the theorem holds in general if it holds for divisors of the type $\mathfrak{B}^{-1}$, where $\mathfrak{B}$ is an integral divisor. Now let $z \in K^*$, and let $\mathfrak{A} \in D_K$ be arbitrary.

Let

$$\varphi : L(\mathfrak{A}) \longrightarrow K \quad \text{be defined by} \quad \varphi(y) = zy.$$

Since $z \neq 0$, φ is k-linear and injective. Its image is contained in $L((z)\mathfrak{A})$. On the other hand, consider the function

$$\psi : L((z)\mathfrak{A}) \longrightarrow K \quad \text{defined by} \quad \psi(y) = z^{-1}y.$$

Clearly ψ is injective and its image is contained in $L(\mathfrak{A})$. Therefore $L((z)\mathfrak{A}) \cong \operatorname{im}\varphi$ and

$$L((z)\mathfrak{A}) \cong L(\mathfrak{A}) \tag{3.11}$$

as k-vector spaces. In particular,

$$\ell((z)\mathfrak{A}) = \ell(\mathfrak{A}) \tag{3.12}$$

for all $z \in K^*$ and $\mathfrak{A} \in D_K$. On the other hand, we have $d((z)\mathfrak{A}) = d((z)) + d(\mathfrak{A}) = d(\mathfrak{A})$, that is,

$$\ell(\mathfrak{A}) + d(\mathfrak{A}) = \ell((z)\mathfrak{A}) + d((z)\mathfrak{A}). \tag{3.13}$$

Let $\mathfrak{B}$ be an arbitrary integral divisor and $m \geq 0$. By Theorem 3.1.11, we have

$$\ell\left(\mathfrak{N}_x^{-m}\mathfrak{B}\right) + d\left(\mathfrak{N}_x^{-m}\mathfrak{B}\right) \geq \ell\left(\mathfrak{N}_x^{-m}\right) + d\left(\mathfrak{N}_x^{-m}\right) \geq 1 - g.$$

Now since x is transcendental, then $d(\mathfrak{N}_x) > 0$, so

$$\ell\left(\mathfrak{N}_x^{-m}\mathfrak{B}\right) \geq -d\left(\mathfrak{N}_x^{-m}\mathfrak{B}\right) + 1 - g = md(\mathfrak{N}_x) - d(\mathfrak{B}) + 1 - g \xrightarrow[m\to\infty]{} \infty.$$

Pick an m large enough so that $\ell\left(\mathfrak{N}_x^{-m}\mathfrak{B}\right) > 0$. In particular there exists $y \in L\left(\mathfrak{N}_x^{-m}\mathfrak{B}\right)$, so we obtain the following implications

$$\begin{aligned}
\mathfrak{N}_x^{-m}\mathfrak{B} \mid (y) &\Longrightarrow (y)\mathfrak{N}_x^{m}\mathfrak{B}^{-1} \text{ is integral } \Longrightarrow \mathfrak{N}_x^{-m} \mid (y)\mathfrak{B}^{-1} \\
&\Longrightarrow \ell\left(\mathfrak{B}^{-1}\right) + d\left(\mathfrak{B}^{-1}\right) = \ell\left((y)\mathfrak{B}^{-1}\right) + d\left((y)\mathfrak{B}^{-1}\right) \\
&\qquad \geq \ell\left(\mathfrak{N}_x^{-m}\right) + d\left(\mathfrak{N}_x^{-m}\right) \geq 1 - g,
\end{aligned}$$

which is what we wanted to prove. □

Corollary 3.3.3. *The number* $1-g$ *is the greatest lower bound of the set*

$$\{\ell(\mathfrak{A}) + d(\mathfrak{A}) \mid \mathfrak{A} \in D_K\}$$

and also the greatest lower bound of the set

$$\left\{\ell\left(\mathfrak{N}_z^{-m}\right) + d\left(\mathfrak{N}_z^{-m}\right) \mid m \in \mathbb{Z}\right\}$$

for any $z \in K \setminus k$. *In particular,* $1-g$ *is independent of* z. □

Definition 3.3.4. The number $g = g_K$ is called the *genus* of the field K.

Example 3.3.5. Let $K = k(x)$. Then it will be proved that $\ell(\mathfrak{p}_\infty^{-t}) = t+1$ (Proposition 4.1.3), where $\mathfrak{p}_\infty$ is the pole divisor of x. Therefore

$$\ell(\mathfrak{p}_\infty^{-t}) + d(\mathfrak{p}_\infty^{-t}) = t+1-t = 1.$$

It follows that $g_{k(x)} = 0$.

Example 3.3.6. Let $K = k(x, y)$ where $y^2 = f(x) \in k[x]$ is a polynomial of even degree m, $f(x)$ is square-free and $\operatorname{char} k \neq 2$. We will see (Corollary 4.3.6) that $\ell(\mathfrak{N}_x^{-t}) = 2t + 2 - \frac{m}{2}$ and $(\mathfrak{N}_x^{-t}) = -td(\mathfrak{N}_x) = 2t$. Thus

$$\ell(\mathfrak{N}_x^{-t}) + d(\mathfrak{N}_x^{-t}) = 2t + 2 - \frac{m}{2} - 2t = 2 - \frac{m}{2}.$$

Therefore $g_K = \frac{m}{2} - 1$.

Proposition 3.3.7. *We have* $g \geq 0$.

Proof. The statement follows from $\ell(\mathfrak{N}) + d(\mathfrak{N}) = 1 + 0 \geq 1 - g$. □

Definition 3.3.8. Let $\mathfrak{A} \in D_K$. The nonnegative integer

$$\delta(\mathfrak{A}) = \ell\left(\mathfrak{A}^{-1}\right) + d\left(\mathfrak{A}^{-1}\right) + g - 1 = \ell\left(\mathfrak{A}^{-1}\right) - d(\mathfrak{A}) + g - 1$$

is called the *specialty degree of* $\mathfrak{A}$.

If $\delta(\mathfrak{A}) = 0$, $\mathfrak{A}$ is called *nonspecial.*

If $\delta(\mathfrak{A}) > 0$, $\mathfrak{A}$ is called *special.*

Remark 3.3.9. From the proof of Riemann's theorem, we have obtained that for all $x \in K^*$ and for all $\mathfrak{A} \in D_K, \ell((x)\mathfrak{A}) = \ell(\mathfrak{A})$, that is, if $C \in C_K$ and $\mathfrak{A} \in C$, $\ell\left(\mathfrak{A}^{-1}\right)$ does not depend on $\mathfrak{A}$ but only on C. In other words, if $\mathfrak{A}, \mathfrak{B} \subset C$, we have

$$\mathfrak{A} = \mathfrak{B}(x), \quad \mathfrak{A}^{-1} = \mathfrak{B}^{-1}\left(x^{-1}\right), \quad \text{and} \quad \ell\left(\mathfrak{A}^{-1}\right) = \ell\left(\mathfrak{B}^{-1}\left(x^{-1}\right)\right) = \ell\left(\mathfrak{B}^{-1}\right).$$

Definition 3.3.10. Let $C \in C_K$. We define the *dimension* $N(C)$ of the class C by $N(C) = \ell\left(\mathfrak{A}^{-1}\right)$ for an arbitrary $\mathfrak{A} \in C$. Equivalently, $N(C) - \ell(\mathfrak{A})$ for any $\mathfrak{A}^{-1} \in C$.

For each place $\mathcal{P}$ of K, let $\xi_{\mathcal{P}} \in K_{\mathcal{P}}$, where $K_{\mathcal{P}}$ is the completion of K with respect to $\mathcal{P}$. The approximation theorem establishes that given a finite set $\mathcal{P}_1, \mathcal{P}_2, \ldots, \mathcal{P}_n$ of distinct places of K, there exists $x \in K$ such that $v_{\mathcal{P}_i}\left(x - \xi_{\mathcal{P}_i}\right) > 0$ for all $i = 1, \ldots, n$. In fact, the approximation theorem shows this for $\xi_{\mathcal{P}_i} \in K$, but if $\xi_{\mathcal{P}_i} \in K_{\mathcal{P}_i}$, we choose $\xi'_{\mathcal{P}_i} \in K$ such that $v_{\mathcal{P}_i}\left(\xi_{\mathcal{P}_i} - \xi'_{\mathcal{P}_i}\right) > m$ for m sufficiently large.

A natural question is whether the approximation theorem is also true for an infinite number of places, even with a weaker condition: given $\xi_{\mathcal{P}} \in K_{\mathcal{P}}$ for each place $\mathcal{P}$ of K, does there exist $x \in K$ such that $v_{\mathcal{P}}(x - \xi_{\mathcal{P}}) \geq 0$ for all $\mathcal{P}$?

A necessary condition for the answer to the above question to be positive is that $v_{\mathcal{P}}(\xi_{\mathcal{P}}) \geq 0$ for almost all $\mathcal{P}$, since if $\mathcal{P}$ is such that $v_{\mathcal{P}}(\xi_{\mathcal{P}}) < 0$, the condition $v_{\mathcal{P}}(x - \xi_{\mathcal{P}}) \geq 0$ implies

$$v_{\mathcal{P}}(x) = v_{\mathcal{P}}(x - \xi_{\mathcal{P}} + \xi_{\mathcal{P}}) = \min\{v_{\mathcal{P}}(x - \xi_{\mathcal{P}}), v_{\mathcal{P}}(\xi_{\mathcal{P}})\} = v_{\mathcal{P}}(\xi_{\mathcal{P}}) < 0,$$

and this is possible only for a finite number of places $\mathcal{P}$.

The above condition motivates the following definition.

Definition 3.3.11. A *repartition* or *adele* is a function $\varphi : \mathbb{P}_K \longrightarrow \bigcup_{\mathcal{P} \in \mathbb{P}_K} K_{\mathcal{P}}$ such that $\varphi(\mathcal{P}) \in K_{\mathcal{P}}$ for all $\mathcal{P}$ and $v_{\mathcal{P}}(\varphi(\mathcal{P})) \geq 0$ for almost all $\mathcal{P}$.

Equivalently, a repartition ξ is a sequence $\xi = \{\xi_{\mathcal{P}}\}_{\mathcal{P} \in \mathbb{P}_K} \in \prod_{\mathcal{P} \in \mathbb{P}_K} K_{\mathcal{P}}$ such that $\xi_{\mathcal{P}} \in \vartheta_{\mathcal{P}}$ for almost all $\mathcal{P}$, where $\vartheta_{\mathcal{P}}$ denotes the valuation ring of $K_{\mathcal{P}}$. For a repartition θ, $\theta_{\mathcal{P}}$ denotes its component at $\mathcal{P}$.

The space of all repartitions of K will be denoted by $\mathfrak{X}_K = \Lambda_K$, or $\Lambda = \mathfrak{X}$ in case that the underlying field is clearly K.

We leave the proof of the next proposition to the reader.

Proposition 3.3.12. *The set $\mathfrak{X}_K$ is a k-algebra, that is, $\mathfrak{X}_K$ is a k-vector space and it is also a ring with its operations defined componentwise. In other words, for $a \in k, \xi, \theta \in \mathfrak{X}$ we define $(a\xi)_{\mathcal{P}} = a\xi_{\mathcal{P}}$; $(\xi + \theta)_{\mathcal{P}} = \xi_{\mathcal{P}} + \theta_{\mathcal{P}}$; $(\theta\xi)_{\mathcal{P}} = \xi_{\mathcal{P}}\theta_{\mathcal{P}}$.* □

The function $K \xrightarrow{\phi} \mathfrak{X}$, defined by $\phi(x) = \xi_x$, where $(\xi_x)_{\mathcal{P}} = x$ for all $\mathcal{P}$, is a monomorphism. Thus, under this injection we will assume that $K \subseteq \mathfrak{X}$ by identifying each $x \in K$ with the constant repartition equal to x for every component.

Proposition 3.3.13. *For a place $\mathcal{P}$, the valuation $v_{\mathcal{P}}$ can be extended to $\mathfrak{X}$ by defining $v_{\mathcal{P}}(\xi) = v_{\mathcal{P}}(\xi_{\mathcal{P}})$ for all $\xi \in \mathfrak{X}$. This extension satisfies the same properties as the original valuation on K, that is:*

(i) $v_{\mathcal{P}}(\xi + \theta) \geq \min\{v_{\mathcal{P}}(\xi), v_{\mathcal{P}}(\theta)\}$ *for all* $\xi, \theta \in \mathfrak{X}$,
(ii) $v_{\mathcal{P}}(\xi\theta) = v_{\mathcal{P}}(\xi) + v_{\mathcal{P}}(\theta)$ *for all* $\xi, \theta \in \mathfrak{X}$,
(iii) $v_{\mathcal{P}}(\xi_x) = v_{\mathcal{P}}(x)$ *for all* $x \in K$.

Proof. The result follows immediately from the definition. □

Definition 3.3.14. Let $\mathfrak{A} \in D_K$ and $\xi \in \mathfrak{X}_K$. We say that $\mathfrak{A}$ *divides* ξ or that ξ *is divisible by* $\mathfrak{A}$ and we write $\mathfrak{A} \mid \xi$ if $v_{\mathcal{P}}(\xi) \geq v_{\mathcal{P}}(\mathfrak{A})$ for all $\mathcal{P} \in \mathbb{P}_K$. We say that two repartitions ξ, θ *are congruent modulo* $\mathfrak{A}$, and we write $\xi \equiv \theta \bmod \mathfrak{A}$, if $\xi - \theta$ is divisible by $\mathfrak{A}$.

Notation 3.3.15. Let $\mathfrak{A} \in D_K$. We denote by

$$\mathfrak{X}(\mathfrak{A}) = \{\xi \in \mathfrak{X} \mid \mathfrak{A} \mid \xi\} = \{\xi \in \mathfrak{X} \mid v_{\mathcal{P}}(\xi) \geq v_{\mathcal{P}}(\mathfrak{A}) \text{ for all } \mathcal{P} \in \mathbb{P}_K\}$$

the set of repartitions that are divisible by $\mathfrak{A}$. Clearly $\mathfrak{X}(\mathfrak{A})$ is a k-vector space. We will also write $\Lambda_K(\mathfrak{A}) = \mathfrak{X}_K(\mathfrak{A})$.

The set $\mathfrak{X}(\mathfrak{A})$ is similar to $L(\mathfrak{A})$ with repartitions instead of elements. Since we may consider that the set of repartitions contains K^* this allows us a greater degree of flexibility in the study of valuations.

The question previous to Definition 3.3.11 can be reformulated and generalized in the following way: given $\xi \in \mathfrak{X}$, does there exist $x \in K$ such that $x \equiv \xi \bmod \mathfrak{A}$? This will be true if and only if $\mathfrak{A} \mid \xi_x - \xi$, which is equivalent to $v_{\mathcal{P}}(x - \xi_{\mathcal{P}}) \geq v_{\mathcal{P}}(\mathfrak{A})$ for all $\mathcal{P} \in \mathbb{P}_K$. The original question corresponds to the case $\mathfrak{A} = \mathfrak{N}$.

Theorem 3.3.16. *Let $\mathfrak{A}, \mathfrak{B} \in D_K$ be such that $\mathfrak{A} \mid \mathfrak{B}$. Let $S = \{\mathcal{P} \in \mathbb{P}_K \mid v_{\mathcal{P}}(\mathfrak{A}) \neq 0$ or $v_{\mathcal{P}}(\mathfrak{B}) \neq 0\}$. Then S is finite and*

$$\frac{\Gamma(\mathfrak{A}|S)}{\Gamma(\mathfrak{B}|S)} \cong \frac{\mathfrak{X}(\mathfrak{A})}{\mathfrak{X}(\mathfrak{B})} \tag{3.14}$$

as k-vector spaces. In particular,

$$\dim_k \frac{\mathfrak{X}(\mathfrak{A})}{\mathfrak{X}(\mathfrak{B})} = d(\mathfrak{B}) - d(\mathfrak{A}) < \infty. \tag{3.15}$$

Proof. For $x \in \Gamma(\mathfrak{A}|S)$, we define the repartition μ_x by

$$(\mu_x)_{\mathcal{P}} = \begin{cases} x & \text{if } \mathcal{P} \in S \\ 0 & \text{if } \mathcal{P} \notin S. \end{cases}$$

Observe that $v_{\mathcal{P}}(\mu_x) = v_{\mathcal{P}}((\mu_x)_{\mathcal{P}}) \geq v_{\mathcal{P}}(x) \geq v_{\mathcal{P}}(\mathfrak{A})$ for all $\mathcal{P} \in \mathbb{P}_K$, that is, $\mu_x \in \mathfrak{X}(\mathfrak{A})$. Define $\varphi : \Gamma(\mathfrak{A}|S) \longrightarrow \mathfrak{X}(\mathfrak{A})$ by $\varphi(x) = \mu_x$. It is easy to verify that φ is k-linear

For $x \in \Gamma(\mathfrak{A}|S)$ we have

$$\varphi(x) = \mu_x \in \mathfrak{X}(\mathfrak{B}) \iff v_{\mathcal{P}}(x) \geq v_{\mathcal{P}}(\mathfrak{B}) \text{ for all } \mathcal{P} \in S \iff x \in \Gamma(\mathfrak{B}|S),$$

which means that the function $\tilde{\varphi} : \frac{\Gamma(\mathfrak{A}|S)}{\Gamma(\mathfrak{B}|S)} \longrightarrow \frac{\mathfrak{X}(\mathfrak{A})}{\mathfrak{X}(\mathfrak{B})}$ induced by φ is a k-monomorphism.

We will see that $\tilde{\varphi}$ is also surjective. Let $\xi \in \mathfrak{X}(\mathfrak{A})$. By the approximation theorem, there exists $x \in K$ such that

$$v_{\mathcal{P}}(x-\xi) \geq v_{\mathcal{P}}(\mathfrak{B}) \text{ for all } \mathcal{P} \in S.$$

Since $\xi \in \mathfrak{X}(\mathfrak{A})$, we have $\mu_x \in \mathfrak{X}(\mathfrak{A})$. Indeed, if $\mathcal{P} \notin S$, then

$$v_{\mathcal{P}}(\mu_x) = v_{\mathcal{P}}(0) = \infty,$$

and if $\mathcal{P} \in S$, then

$$\begin{aligned} v_{\mathcal{P}}(\mu_x) = v_{\mathcal{P}}(x) = v_{\mathcal{P}}(x - \xi_{\mathcal{P}} + \xi_{\mathcal{P}}) &\geq \min\{v_{\mathcal{P}}(x-\xi_{\mathcal{P}}), v_{\mathcal{P}}(\xi_{\mathcal{P}})\} \\ &\geq \min\{v_{\mathcal{P}}(\mathfrak{B}), v_{\mathcal{P}}(\mathfrak{A})\} = v_{\mathcal{P}}(\mathfrak{A}). \end{aligned}$$

Furthermore,

$$v_{\mathcal{P}}(\mu_x - \xi) \geq v_{\mathcal{P}}(\mathfrak{B}) \text{ for all } \mathcal{P} \in \mathbb{P}_K,$$

so $\mu_x \equiv \xi \bmod \mathfrak{X}(\mathfrak{B})$. Thus, for $\mathcal{P} \in S$, we have $v_{\mathcal{P}}(x) \geq v_{\mathcal{P}}(\mathfrak{A})$. Therefore $x \in \Gamma(\mathfrak{A}|S)$ and we have

$$\tilde{\varphi}(x) = \mu_x + \mathfrak{X}(\mathfrak{B}) = \xi + \mathfrak{X}(\mathfrak{B}),$$

that is, $\tilde{\varphi}$ is surjective and we have proved the first part of the theorem. The second part is an immediate consequence of Theorem 3.1.9. □

3.4 Differentials

Our main goal in this section is to define the concept of differential in a general function field. The original concept of differential is, naturally, analytic. Our first objective is, starting from its analytic nature, to extract an algebraic representation of a differential in the complex plane in order to be able to give a general definition. It would have been possible to give the definition directly without any previous motivation, but the reason why we call this object a differential would be obscure as well as its similarity with the differentials that everyone knows. The differentials defined here are the Weil differentials. In Chapter 9 we will study the Hasse differentials, and in Chapter 14 we will study successive differentials, namely the Hasse–Schmidt differentials.

First, let $K = \mathbb{C}(x)$ be the rational function field over the field of complex numbers. Let $u \in K$. The object $u\ dx$ can be viewed as a "linear integral element" in the following way: if γ is any path in $\mathbb{C}$ not containing any pole of u, then the linear integral $\int_\gamma u\ dx$ is well defined. For $a \in \mathbb{C}$, let $\mathcal{P}_a$ be the zero divisor of $x-a$, that is, $(x-a)_K = \frac{\mathcal{P}_a}{\mathcal{P}_\infty}$, and write

$$v_{\mathcal{P}_a}(u) =: \text{ order of } u\ dx \text{ in } \mathcal{P}_a.$$

In the Riemann sphere, for $a = \infty \in \mathbb{C}_\infty = S^2$, we have

$$y = \frac{1}{x}, \quad dy = -\frac{1}{x^2}dx$$

and

$$\int_\gamma u dx = -\int_\gamma u y^{-2} dy = -\int_\gamma \frac{u}{y^2} dy.$$

Since $\mathbb{C}(x) = \mathbb{C}(y)$ and $\mathfrak{N}_x = \mathfrak{Z}_y$, it is reasonable to define the order of $u\ dx$ in $\mathcal{P}_\infty$ to be $v_{\mathcal{P}_\infty}(u) - 2$.

In short, we define the

$$\text{order of } u\ dx \text{ in } a \in S^2 \text{ as } \begin{cases} v_{\mathcal{P}_a}(u) & \text{if } a \in \mathbb{C}, \\ v_{\mathcal{P}_\infty}(u) - 2 & \text{if } a = \infty. \end{cases}$$

If $a \in \mathbb{C}$ and γ is a simple positively oriented closed path such that a is in the interior of γ and γ does not contain any other pole of u in its interior, we have

$$\frac{1}{2\pi i}\oint_\gamma u\ dx = \text{ Residue of } u \text{ in } (x = a) = \operatorname*{Res}_{x=a} u := \text{ Residue of } u\ dx \text{ in } \mathcal{P}_a.$$

For $a = \infty$, we choose γ to be a simple positively oriented closed path containing every pole of u in the interior of γ in the finite plane $\mathbb{C}$. In other words, ∞ is not contained in the interior of the path when this path is considered in $\mathbb{C}$. We have

$$\operatorname*{Res}_{x=\infty} u = -\frac{1}{2\pi i}\oint_\gamma u\ dx := \text{ Residue of } u\ dx \text{ in } \mathcal{P}_\infty.$$

Hence, by definition we have

$$\operatorname*{Res}_{\mathcal{P}_a} u\ dx = \operatorname*{Res}_{x=a} u, \ a \in \mathbb{C}_\infty.$$

Now, if $a_1, a_2, \ldots, a_h \in \mathbb{C}$ are all the poles of u in $\mathbb{C}$ and Γ is a simple positively oriented closed path containing $a_1, a_2, \ldots, a_h$ in its interior, then by the residue theorem, we have

$$\frac{1}{2\pi i}\oint_\gamma u\ dx = \sum_{i=1}^{h} \operatorname*{Res}_{x=a_i} u = \quad \operatorname*{Res}_{x=\infty} u,$$

that is,

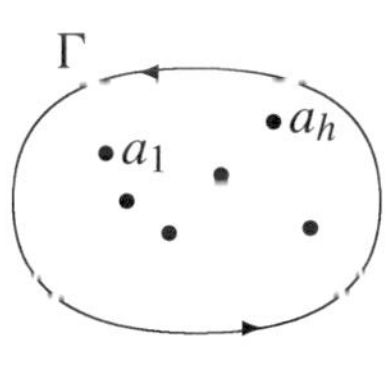

$$\textstyle\sum_{a\in\mathbb{C}_\infty} \operatorname{Res}_{\mathcal{P}_a} u\ dx = 0.$$

If $\mathcal{P}$ is any place of K and α is an element of the completion of $K_\mathcal{P}$ of K with respect to $\mathcal{P}$, then we can define $\operatorname{Res}_\mathcal{P} \alpha\ dx$ in an analogous way to the case $\alpha \in K$.

To that end, first we write $\mathcal{P} = \mathcal{P}_a$ with $a \in \mathbb{C}_\infty$. By means of a change of variable $x \longrightarrow x - a$ or $x \longrightarrow \frac{1}{x}$, we may assume that $\mathcal{P}$ is the divisor of zeros $\mathcal{P}_0$ of x. Then $\alpha \in \mathbb{C}((x))$ (Theorem 2.5.20). If $\alpha = \sum_{n=m}^{\infty} a_n x^n$, then

$$\alpha\, dx = \sum_{n=m}^{\infty} a_n x^n\, dx \quad \text{and} \quad \operatorname*{Res}_{\mathcal{P}} \alpha\, dx = a_{-1}.$$

By this observation we may take $\xi \in \mathfrak{X}_K = \mathfrak{X}$, and $u\, dx$ is as before and $K = \mathbb{C}(x)$. Then if $\mathcal{P}$ is any place, we define

$$\omega^{\mathcal{P}}(\xi) = \text{ residue in } \mathcal{P} \text{ of } \xi_{\mathcal{P}} u\, dx.$$

We note that since $v_{\mathcal{P}}(\xi_{\mathcal{P}}) \geq 0$ and $v_{\mathcal{P}}(u) \geq 0$ for almost all $\mathcal{P}$, then $\omega^{\mathcal{P}}(\xi) = 0$ for all but a finite number of places $\mathcal{P}$. Then the function

$$\omega : \mathfrak{X} \longrightarrow \mathbb{C} \quad \text{given by} \quad \omega(\xi) = \sum_{\mathcal{P} \in \mathbb{P}_K} \omega^{\mathcal{P}}(\xi)$$

is well defined and clearly $\mathbb{C}$-linear. Our objective now is to study $\ker \omega$.

If $t \in K$, then $\xi_t \in \mathfrak{X}$ satisfies

$$\omega(\xi_t) = \sum_{\mathcal{P} \in \mathbb{P}_K} \omega^{\mathcal{P}}((\xi_t)_{\mathcal{P}}) = \sum_{\mathcal{P} \in \mathbb{P}_K} \omega^{\mathcal{P}}(t) = \sum_{a \in \mathbb{C}_\infty} \operatorname*{Res}_{a} tu\, dx = 0,$$

that is, $K \subseteq \ker \omega$.

If $\mathfrak{A} = \prod_{\mathcal{P} \in \mathbb{P}_K} \mathcal{P}^{v_{\mathcal{P}}(\mathfrak{A})}$ is any divisor, we say that $u\, dx \equiv 0 \bmod \mathfrak{A}$ if the order of $u\, dx$ in $\mathcal{P}$ is greater than or equal to $v_{\mathcal{P}}(\mathfrak{A})$ for all $\mathcal{P} \in \mathbb{P}_K$.

Let $\mathfrak{A}$ be a divisor such that $u\, dx \equiv 0 \bmod \mathfrak{A}$. Set

$$\begin{aligned}\mathfrak{X}\left(\mathfrak{A}^{-1}\right) &= \left\{\xi \in \mathfrak{X} \mid \mathfrak{A}^{-1} \mid \xi\right\} = \left\{\xi \in \mathfrak{X} \mid \xi \equiv 0 \bmod \mathfrak{A}^{-1}\right\} \\ &= \{\xi \in \mathfrak{X} \mid v_{\mathcal{P}}(\xi) \geq -v_{\mathcal{P}}(\mathfrak{A}), \mathcal{P} \in \mathbb{P}_K\}.\end{aligned}$$

If $\xi \in \mathfrak{X}\left(\mathfrak{A}^{-1}\right)$, then $v_{\mathcal{P}}(\xi) \geq -v_{\mathcal{P}}(\mathfrak{A})$ for all $\mathcal{P} \in \mathbb{P}_K$. Therefore

$$\begin{aligned}\text{order } \xi_{\mathcal{P}} u dx &= \begin{cases} v_{\mathcal{P}}(\xi_{\mathcal{P}}) + v_{\mathcal{P}}(u) & \text{if } \mathcal{P} \neq \mathcal{P}_\infty \\ \\ v_{\mathcal{P}}(\xi_{\mathcal{P}}) + v_{\mathcal{P}}(u) - 2 & \text{if } \mathcal{P} = \mathcal{P}_\infty \end{cases} \\ &\geq -v_{\mathcal{P}}(\mathfrak{A}) + v_{\mathcal{P}}(\mathfrak{A}) = 0,\end{aligned}$$

so $\omega^{\mathcal{P}}(\xi) = 0$ for all $\mathcal{P} \in \mathbb{P}_K$. In particular, we have $\omega(\xi) = 0$, that is, $\mathfrak{X}\left(\mathfrak{A}^{-1}\right) \subseteq \ker \omega$.

Therefore ω vanishes on $K + \mathfrak{X}\left(\mathfrak{A}^{-1}\right)$, where $\mathfrak{A}$ is any divisor such that $u\, dx \equiv 0 \bmod \mathfrak{A}$.

All the previous discussion motivates the general definition of differential in an arbitrary function field.

From this point on, K/k will denote an arbitrary function field.

Definition 3.4.1. Let K/k be an arbitrary function field. A *differential* (Weil differential) in K is a k-linear function $\omega : \mathfrak{X}_K \longrightarrow k$ such that there exists a divisor $\mathfrak{A} \in D_K$ with the property that $\ker \omega \supseteq K + \mathfrak{X}\left(\mathfrak{A}^{-1}\right)$. In this case we say that $\mathfrak{A}$ *divides* ω and we write $\mathfrak{A} \mid \omega$.

Definition 3.4.2. A differential ω in a function field K is said to be of the *first kind* or a *holomorphic differential* if $\mathfrak{N} \mid \omega$, that is, if $K + \mathfrak{X}(\mathfrak{N}) \subseteq \ker \omega$.

Proposition 3.4.3. *If $\mathfrak{A}$ and $\mathfrak{B}$ are divisors such that $\mathfrak{B} \mid \mathfrak{A}$, then if $\mathfrak{A} \mid \omega$, we have $\mathfrak{B} \mid \omega$.*

Proof. Since $\mathfrak{B} \mid \mathfrak{A}$, we have $\mathfrak{A}^{-1} | \mathfrak{B}^{-1}$. Therefore $\mathfrak{X}\left(\mathfrak{B}^{-1}\right) \subseteq \mathfrak{X}\left(\mathfrak{A}^{-1}\right)$ and $\mathfrak{X}\left(\mathfrak{B}^{-1}\right) + K \subseteq \mathfrak{X}\left(\mathfrak{A}^{-1}\right) + K \subseteq \ker \omega$, so $\mathfrak{B} \mid \omega$. □

Theorem 3.4.4. *If $\mathfrak{A}$ and $\mathfrak{B}$ are divisors in a function field K/k such that $\mathfrak{A} \mid \mathfrak{B}$, then we have the following exact sequence of k-vector spaces:*

$$0 \longrightarrow \frac{L(\mathfrak{A})}{L(\mathfrak{B})} \longrightarrow \frac{\mathfrak{X}(\mathfrak{A})}{\mathfrak{X}(\mathfrak{B})} \longrightarrow \frac{\mathfrak{X}(\mathfrak{A}) + K}{\mathfrak{X}(\mathfrak{B}) + K} \longrightarrow 0. \tag{3.16}$$

In particular,

$$\dim_k \frac{\mathfrak{X}(\mathfrak{A}) + K}{\mathfrak{X}(\mathfrak{B}) + K} = (\ell(\mathfrak{B}) + d(\mathfrak{B})) - (\ell(\mathfrak{A}) + d(\mathfrak{A})).$$

Furthermore,

$$\dim_k \frac{\mathfrak{X}}{\mathfrak{X}(\mathfrak{B}) + K} = \delta\left(\mathfrak{B}^{-1}\right) = \ell(\mathfrak{B}) + d(\mathfrak{B}) + g - 1$$

for any divisor $\mathfrak{B}$.

Proof. The natural injection $i : \mathfrak{X}(\mathfrak{A}) \longrightarrow \mathfrak{X}(\mathfrak{A}) + K$, composed with the natural projection $\pi : \mathfrak{X}(\mathfrak{A}) + K \longrightarrow \frac{\mathfrak{X}(\mathfrak{A})+K}{\mathfrak{X}(\mathfrak{B})+K}$, gives an epimorphism

$$f = \pi \circ i : \mathfrak{X}(\mathfrak{A}) \longrightarrow \frac{\mathfrak{X}(\mathfrak{A}) + K}{\mathfrak{X}(\mathfrak{B}) + K}.$$

Clearly $\mathfrak{X}(\mathfrak{B}) \subseteq \ker f$, so f induces an epimorphism

$$\tilde{f} : \frac{\mathfrak{X}(\mathfrak{A})}{\mathfrak{X}(\mathfrak{B})} \longrightarrow \frac{\mathfrak{X}(\mathfrak{A}) + K}{\mathfrak{X}(\mathfrak{B}) + K}.$$

To finish we use two equalities (see Exercise 3.6.10):

(1) $\mathfrak{X}(\mathfrak{A}) \cap (\mathfrak{X}(\mathfrak{B}) + K) = L(\mathfrak{A}) + \mathfrak{X}(\mathfrak{B})$,
(2) $L(\mathfrak{A}) \cap \mathfrak{X}(\mathfrak{B}) = L(\mathfrak{B})$.

Applying (1) and (2) we have

$$\ker \tilde{f} = \frac{\mathfrak{X}(\mathfrak{A}) \bigcap (\mathfrak{X}(\mathfrak{B}) + K)}{\mathfrak{X}(\mathfrak{B})} = \frac{L(\mathfrak{A}) + \mathfrak{X}(\mathfrak{B})}{\mathfrak{X}(\mathfrak{B})} \cong \frac{L(\mathfrak{A})}{L(\mathfrak{A}) \bigcap \mathfrak{X}(\mathfrak{B})} = \frac{L(\mathfrak{A})}{L(\mathfrak{B})}.$$

This proves (3.16).
Then we have

$$\dim_k \frac{\mathfrak{X}(\mathfrak{A}) + K}{\mathfrak{X}(\mathfrak{B}) + K} = \dim_k \frac{\mathfrak{X}(\mathfrak{A})}{\mathfrak{X}(\mathfrak{B})} - \dim_k \frac{L(\mathfrak{A})}{L(\mathfrak{B})}.$$

From Theorem 3.3.16, we obtain that $\dim_k \frac{\mathfrak{X}(\mathfrak{A})}{\mathfrak{X}(\mathfrak{B})} = d(\mathfrak{B}) - d(\mathfrak{A})$. Therefore,

$$\begin{aligned} \dim_k \frac{\mathfrak{X}(\mathfrak{A}) + K}{\mathfrak{X}(\mathfrak{B}) + K} &= d(\mathfrak{B}) - d(\mathfrak{A}) - (\ell(\mathfrak{A}) - \ell(\mathfrak{B})) \\ &= (\ell(\mathfrak{B}) + d(\mathfrak{B})) - (\ell(\mathfrak{A}) + d(\mathfrak{A})). \end{aligned}$$

In order to prove the last equality, consider $\mathfrak{C}$ to be any divisor such that $\ell(\mathfrak{C}) + d(\mathfrak{C}) = 1 - g_K$, where g_K is the genus of K. For each $\mathcal{P} \in \mathbb{P}_K$, let $u_{\mathcal{P}} = \min\{v_{\mathcal{P}}(\mathfrak{B}), v_{\mathcal{P}}(\mathfrak{C})\}$ and let $\mathfrak{A}_1 = \prod_{\mathcal{P} \in \mathbb{P}_K} \mathcal{P}^{u_{\mathcal{P}}}$. Then $\mathfrak{A}_1 \mid \mathfrak{B}$ and $\mathfrak{A}_1 \mid \mathfrak{C}$. From Theorems 3.1.11 and 3.3.2 we obtain

$$1 - g \le \ell(\mathfrak{A}_1) + d(\mathfrak{A}_1) \le \ell(\mathfrak{C}) + d(\mathfrak{C}) = 1 - g,$$

that is, $\ell(\mathfrak{A}_1) + d(\mathfrak{A}_1) = 1 - g$. Therefore

$$\begin{aligned} \dim_k \frac{\mathfrak{X}}{\mathfrak{X}(\mathfrak{B}) + K} &\ge \dim_k \frac{\mathfrak{X}(\mathfrak{A}_1) + K}{\mathfrak{X}(\mathfrak{B}) + K} = (\ell(\mathfrak{B}) + d(\mathfrak{B})) - (\ell(\mathfrak{A}_1) + d(\mathfrak{A}_1)) \\ &= (\ell(\mathfrak{B}) + d(\mathfrak{B})) - (1 - g) = \ell(\mathfrak{B}) + d(\mathfrak{B}) - 1 + g \\ &= \delta\left(\mathfrak{B}^{-1}\right). \end{aligned}$$

For the other equality, consider $\tau_1, \ldots, \tau_m$ to be m elements of $\mathfrak{X}$ that are linearly independent over the k-module $\mathfrak{X}(\mathfrak{B}) + K$. Set

$$w_{\mathcal{P}} = \min_{1 \le i \le m} \{v_{\mathcal{P}}(\tau_i), v_{\mathcal{P}}(\mathfrak{B})\}.$$

Let $\mathfrak{A}_2 = \prod_{\mathcal{P} \in \mathbb{P}_K} \mathcal{P}^{w_{\mathcal{P}}}$. Then $\mathfrak{A}_2 \mid \tau_i$ for all $1 \le i \le m$, that is, $\tau_i \in \mathfrak{X}(\mathfrak{A}_2)$. Thus

$$\begin{aligned} m \le \dim_k \frac{\mathfrak{X}(\mathfrak{A}_2) + K}{\mathfrak{X}(\mathfrak{B}) + K} &= (\ell(\mathfrak{B}) + d(\mathfrak{B})) - (\ell(\mathfrak{A}_2) + d(\mathfrak{A}_2)) \le \\ &\le \ell(\mathfrak{B}) + d(\mathfrak{B}) - (1 - g) = \delta\left(\mathfrak{B}^{-1}\right). \end{aligned}$$

Therefore $\dim_k \frac{\mathfrak{X}}{\mathfrak{X}(\mathfrak{B}) + K} = \delta\left(\mathfrak{B}^{-1}\right)$. □

Proposition 3.4.5. *Let $\mathfrak{A}$ be a divisor in K. We define*

$$D(\mathfrak{A}) = \{\omega \mid \omega \text{ is a differential such that } \mathfrak{A} \mid \omega\}.$$

Then $D(\mathfrak{A})$ is isomorphic to the dual of the k-vector space $\frac{\mathfrak{X}}{\mathfrak{X}(\mathfrak{A}^{-1})+K}$. In particular,

$$\dim_k D(\mathfrak{A}) = \dim_k \frac{\mathfrak{X}}{\mathfrak{X}\left(\mathfrak{A}^{-1}\right)+K} = \delta(\mathfrak{A}) = \ell\left(\mathfrak{A}^{-1}\right) + d\left(\mathfrak{A}^{-1}\right) - 1 + g.$$

Proof. Recall that given a vector space V over k, the dual V^* of V is the vector space of all linear functionals from V to k. Furthermore, if $\dim_k V < \infty$ we have $\dim_k V = \dim_k V^*$. Here, taking $V = \frac{\mathfrak{X}}{\mathfrak{X}(\mathfrak{A}^{-1})+K}$, we have

$$V^* = \left\{ f : \frac{\mathfrak{X}}{\mathfrak{X}\left(\mathfrak{A}^{-1}\right)+K} \longrightarrow k \mid f \text{ is } k\text{-linear} \right\}.$$

Now,

$$\begin{aligned} D(\mathfrak{A}) &= \{\omega \mid \omega \text{ is a differential such that } \mathfrak{A} \mid \omega\} \\ &= \left\{ \omega : \mathfrak{X} \longrightarrow k \mid \ker \omega \supseteq \mathfrak{X}\left(\mathfrak{A}^{-1}\right) + K \right\}. \end{aligned}$$

Therefore $\omega \in D(\mathfrak{A})$ induces in a unique way

$$\tilde{\omega} : \frac{\mathfrak{X}}{\mathfrak{X}\left(\mathfrak{A}^{-1}\right)+K} \longrightarrow k, \ \tilde{\omega} \in V^*, \ \tilde{\omega}\left(\xi \bmod \left(\mathfrak{X}\left(\mathfrak{A}^{-1}\right)+K\right)\right) = \omega(\xi).$$

Conversely, given $f \in V^*$, let $\omega = f \circ \pi$, where π is the natural projection of $\mathfrak{X}$ in $\frac{\mathfrak{X}}{\mathfrak{X}(\mathfrak{A}^{-1})+K}$. The functions

$$D(\mathfrak{A}) \xrightarrow{\phi} V^*, \quad V^* \xrightarrow{\psi} D(\mathfrak{A})$$

defined by $\phi(\omega) = \tilde{\omega}$ and $\psi(f) = f \circ \pi$ respectively, are clearly k-linear, and we have

$$\begin{aligned} (\phi \circ \psi)(f) &= \phi(\psi(f)) = \phi(f \circ \pi) = \widetilde{(f \circ \pi)} = f; \\ (\psi \circ \phi)(\omega) &= \psi(\tilde{\omega}) = \tilde{\omega} \circ \pi = \omega. \end{aligned}$$

In other words, ϕ and ψ are inverse isomorphisms, which proves the proposition. □

Corollary 3.4.6. *We have $\dim_k D(\mathfrak{N}) = g$, where g is the genus of K. That is, the dimension of the vector space of holomorphic differentials is g.*

Proof. By Proposition 3.4.5, we have

$$\begin{aligned} \dim_k D(\mathfrak{N}) &= \dim_k \frac{\mathfrak{X}}{\mathfrak{X}(\mathfrak{N}) + K} = \delta(\mathfrak{N}) \\ &= \ell(\mathfrak{N}) + d(\mathfrak{N}) - 1 + g = 1 + 0 - 1 + g = g. \end{aligned}$$

□

Proposition 3.4.7. *Let ω_1, ω_2 be two differentials, and $\mathfrak{A}_1 \mid \omega_1$, $\mathfrak{A}_2 \mid \omega_2$, $\mathfrak{A}_1$, $\mathfrak{A}_2 \in D_K$. Then if $\mathfrak{A}$ is the greatest common divisor of $\mathfrak{A}_1$ and $\mathfrak{A}_2$, that is, $\mathfrak{A} = \prod_{\mathcal{P}\in\mathbb{P}_K} \mathcal{P}^{u_\mathcal{P}}$, $u_\mathcal{P} = \min\{v_\mathcal{P}(\mathfrak{A}_1), v_\mathcal{P}(\mathfrak{A}_2)\}$, we have $\mathfrak{A} \mid \omega$, where $\omega = \omega_1 + \omega_2$.*

Proof. Exercise 3.6.11. □

Proposition 3.4.8. *The set Dif_K of all differentials over K is a K-vector space with the operations $(X\omega)(\xi) = \omega(X\xi)$, $X \in K$, $\omega \in \mathrm{Dif}_K$, $\xi \in \mathfrak{X}$. Furthermore, if $\mathfrak{A} \mid \omega$, and $X \neq 0$, then $(X)_K \mathfrak{A} \mid X\omega$.*

Proof. First let us see that $X\omega$ is k-linear. If $\xi, \theta \in \mathfrak{X}$ and $\alpha, \beta \in k$, we have

$$\begin{aligned}(X\omega)(\alpha\xi + \beta\theta) &= \omega(X(\alpha\xi + \beta\theta)) = \omega(\alpha X\xi + \beta X\theta) \\ &= \alpha\omega(X\xi) + \beta\omega(X\theta) = \alpha(X\omega)(\xi) + \beta(X\omega)(\theta).\end{aligned}$$

Now, if $\mathfrak{A} \mid \omega$ and $\xi \in \mathfrak{X}\left((X)_K^{-1}\mathfrak{A}^{-1}\right)$, we have

$$v_\mathcal{P}(\xi) \geq v_\mathcal{P}\left((X)_K^{-1}\mathfrak{A}^{-1}\right) = -v_\mathcal{P}(X) - v_\mathcal{P}(\mathfrak{A}),$$

so $v_\mathcal{P}(X\xi) = v_\mathcal{P}(X) + v_\mathcal{P}(\xi) \geq -v_\mathcal{P}(\mathfrak{A})$, i.e., $X\xi \in \mathfrak{X}\left(\mathfrak{A}^{-1}\right)$. Therefore $(X\omega)(\xi) = \omega(X\xi) = 0$. This proves that $X\omega$ is a differential and that $(X)_K \mathfrak{A} \mid X\omega$.

The equalities

$$\begin{aligned}(XY)\omega &= X(Y\omega), \\ (X+Y)\omega &= X\omega + Y\omega, \\ X\left(\omega + \omega'\right) &= X\omega + X\omega',\end{aligned}$$

for $X, Y \in K$, and $\omega, \omega' \in \mathrm{Dif}_K$ are immediate and show that Dif_K is a K-vector space. □

The next result proves that the differentials are of dimension 1 over K. In particular, it says that the differentials $u\,dx$ that we considered at the beginning of this section are all such differentials existing in $\mathbb{C}(x)$.

Theorem 3.4.9. *Let $\omega_0 \in \mathrm{Dif}_K$ with $\omega_0 \neq 0$. Then every differential ω can be expressed in a unique way as $\omega = X\omega_0$ for some $X \in K$. In particular, $\dim_K \mathrm{Dif}_K = 1$.*

Proof. If $\omega = 0$, it suffices to take $X = 0$. Let $\omega \neq 0$. Let $\mathfrak{B}_0 \mid \omega_0$, $\mathfrak{B} \mid \omega$. Let $\mathfrak{A}$ be an integral divisor different from $\mathfrak{N}$. We consider

$$\varphi : L\left(\mathfrak{A}^{-1}\mathfrak{B}_0^{-1}\right) \longrightarrow D\left(\mathfrak{A}^{-1}\right), \quad \text{defined by} \quad \varphi(X) = X\omega_0,$$

and

$$\psi : L\left(\mathfrak{A}^{-1}\mathfrak{B}^{-1}\right) \longrightarrow D\left(\mathfrak{A}^{-1}\right), \quad \text{defined by} \quad \psi(X) = X\omega.$$

Then ϕ and ψ are k-monomorphisms (see Exercise 3.6.12).

By Theorem 3.3.2, we have

$$\left(\ell\left(\mathfrak{A}^{-1}\mathfrak{B}_0^{-1}\right)+\ell\left(\mathfrak{A}^{-1}\mathfrak{B}^{-1}\right)\right)+\left(d\left(\mathfrak{A}^{-1}\mathfrak{B}_0^{-1}\right)+d\left(\mathfrak{A}^{-1}\mathfrak{B}^{-1}\right)\right)$$
$$\geq (1-g)+(1-g)=2-2g.$$

Therefore $\ell\left(\mathfrak{A}^{-1}\mathfrak{B}_0^{-1}\right)+\ell\left(\mathfrak{A}^{-1}\mathfrak{B}^{-1}\right)\geq 2d\left(\mathfrak{A}\right)+d\left(\mathfrak{B}_0\right)+d\left(\mathfrak{B}\right)+2-2g$.

We have

$$\dim_k D\left(\mathfrak{A}^{-1}\right)=\delta\left(\mathfrak{A}^{-1}\right)=\ell\left(\mathfrak{A}\right)+d\left(\mathfrak{A}\right)+g-1=d\left(\mathfrak{A}\right)+g-1.$$

Thus, if we choose $d\left(\mathfrak{A}\right)$ such that

$$2d\left(\mathfrak{A}\right)+d\left(\mathfrak{B}_0\right)+d\left(\mathfrak{B}\right)+2-2g>d\left(\mathfrak{A}\right)+g-1,$$

or equivalently,

$$d\left(\mathfrak{A}\right)>-d\left(\mathfrak{B}_0\right)-d\left(\mathfrak{B}\right)+3g-3,$$

we will have $\dim_k \operatorname{im}\varphi+\dim_k \operatorname{im}\psi>\dim_k D\left(\mathfrak{A}^{-1}\right)$, which implies that $\operatorname{im}\varphi\cap\operatorname{im}\psi\neq\{0\}$. Therefore there exist $A, B\in K^*$ such that $A\omega_0=B\omega$. Equivalently, $\omega=\frac{A}{B}\omega_0$. The uniqueness follows from the fact that Dif_K is a K-vector space. □

The next step is to assign to each differential $\omega\neq 0$ a unique divisor.

Proposition 3.4.10. *Assume that $\omega\in\operatorname{Dif}_K$ and $\mathfrak{A}, \mathfrak{B}\in D_K$ are such that $\mathfrak{A}\mid\omega$ and $\mathfrak{B}\mid\omega$, and that $\mathfrak{C}$ is the least common multiple of $\mathfrak{A}$ and $\mathfrak{B}$, that is, $\mathfrak{C}=\prod_{\mathcal{P}\in\mathbb{P}_K}\mathcal{P}^{u_{\mathcal{P}}}$, where $u_{\mathcal{P}}=\max\left\{v_{\mathcal{P}}\left(\mathfrak{A}\right), v_{\mathcal{P}}\left(\mathfrak{B}\right)\right\}$. Then $\mathfrak{C}\mid\omega$.*

Proof. Let $\xi\in\mathfrak{X}\left(\mathfrak{C}^{-1}\right)$, that is, $v_{\mathcal{P}}\left(\xi\right)\geq -v_{\mathcal{P}}\left(\mathfrak{C}\right)=-\max\left\{v_{\mathcal{P}}\left(\mathfrak{A}\right), v_{\mathcal{P}}\left(\mathfrak{B}\right)\right\}$. We define $\xi', \xi''\in\mathfrak{X}$ with the property

$$\xi'_{\mathcal{P}}=\xi_{\mathcal{P}} \text{ and } \xi''_{\mathcal{P}}=0 \text{ for } \mathcal{P} \text{ such that } v_{\mathcal{P}}\left(\mathfrak{A}\right)\geq v_{\mathcal{P}}\left(\mathfrak{B}\right);$$
$$\xi'_{\mathcal{P}}=0 \text{ and } \xi''_{\mathcal{P}}=\xi_{\mathcal{P}} \text{ for } \mathcal{P} \text{ such that } v_{\mathcal{P}}\left(\mathfrak{A}\right)<v_{\mathcal{P}}\left(\mathfrak{B}\right).$$

Then $\xi=\xi'+\xi''$. We also observe that if $v_{\mathcal{P}}\left(\mathfrak{A}\right)\geq v_{\mathcal{P}}\left(\mathfrak{B}\right)$, then $v_{\mathcal{P}}\left(\mathfrak{C}\right)=v_{\mathcal{P}}\left(\mathfrak{A}\right)$, so

$$v_{\mathcal{P}}\left(\xi'\right)=v_{\mathcal{P}}\left(\xi'_{\mathcal{P}}\right)=v_{\mathcal{P}}\left(\xi_{\mathcal{P}}\right)\geq -v_{\mathcal{P}}\left(\mathfrak{C}\right)=-v_{\mathcal{P}}\left(\mathfrak{A}\right).$$

On the other hand, if $v_{\mathcal{P}}\left(\mathfrak{A}\right)<v_{\mathcal{P}}\left(\mathfrak{B}\right)$, then

$$v_{\mathcal{P}}\left(\mathfrak{C}\right)=v_{\mathcal{P}}\left(\mathfrak{B}\right) \quad\text{and}\quad v_{\mathcal{P}}\left(\xi'\right)=v_{\mathcal{P}}\left(0\right)=\infty>-v_{\mathcal{P}}\left(\mathfrak{A}\right),$$

that is, $\mathfrak{A}^{-1}\mid\xi'$. Similarly, we obtain $\mathfrak{B}^{-1}\mid\xi''$. Thus

$$\omega\left(\xi\right)=\omega\left(\xi'+\xi''\right)=\omega\left(\xi'\right)+\omega\left(\xi''\right)=0+0=0,$$

which shows that $\mathfrak{X}\left(\mathfrak{C}^{-1}\right)+K\subseteq\ker\omega$. Therefore, $\mathfrak{C}\mid\omega$. □

Theorem 3.4.11. *For each differential $\omega \neq 0$, there exists a unique divisor, which will denoted by $(\omega)_K$, such that $\mathfrak{A} \mid \omega \iff \mathfrak{A} \mid (\omega)_K$. The divisor $(\omega)_K$ is the divisor associated to the differential ω.*

Proof. First let us see that the degrees of all possible divisors $\mathfrak{A}$ such that $\mathfrak{A} \mid \omega$ have an upper bound.

Let $\mathfrak{A} \mid \omega$. Consider $\varphi : L\left(\mathfrak{A}^{-1}\right) \longrightarrow D(\mathfrak{N})$, defined by $\varphi(X) = X\omega$. Then φ is well defined since $\mathfrak{A} \mid \omega$ and $\mathfrak{A}^{-1} \mid (X)_K$ imply $\mathfrak{N} \mid X\omega$.

Furthermore, φ is a k-monomorphism, so $\ell\left(\mathfrak{A}^{-1}\right) \leq \dim_k D(\mathfrak{N}) = g$. On the other hand,

$$\ell\left(\mathfrak{A}^{-1}\right) + d\left(\mathfrak{A}^{-1}\right) \geq 1 - g,$$

so

$$d\left(\mathfrak{A}^{-1}\right) = -d(\mathfrak{A}) \geq 1 - g - \ell\left(\mathfrak{A}^{-1}\right) \geq 1 - g - g = 1 - 2g.$$

Thus, we have

$$d(\mathfrak{A}) \leq 2g - 1.$$

We define $(\omega)_K$ to be a divisor of maximum degree such that $(\omega)_K \mid \omega$. We will see that $(\omega)_K$ is unique.

If $\mathfrak{A}$, $\mathfrak{B}$ are two divisors of maximum degree such that $\mathfrak{A} \mid \omega$ and $\mathfrak{B} \mid \omega$, then if $\mathfrak{C}$ is the least common multiple of $\mathfrak{A}$ and $\mathfrak{B}$, then $\mathfrak{C} \mid \omega$ and $d(\mathfrak{C}) \leq d(\mathfrak{A})$. Now, since $\mathfrak{A} \mid \mathfrak{C}$ and $\mathfrak{B} \mid \mathfrak{C}$, we have $d(\mathfrak{C}) \geq d(\mathfrak{A})$. Hence, $d(\mathfrak{C}) = d(\mathfrak{A})$, which implies that $\mathfrak{C} = \mathfrak{A} = \mathfrak{B}$. Therefore $(\omega)_K$ is unique.

Let $\mathfrak{A} \mid \omega$. Let $\mathfrak{B}$ be the least common multiple of $\mathfrak{A}$ and $(\omega)_K$. Then $\mathfrak{B} \mid \omega$ and $d(\mathfrak{B}) \geq d((\omega)_K)$, which implies that $d(\mathfrak{B}) = d((\omega)_K)$. Therefore $\mathfrak{B} = (\omega)_K$ and $\mathfrak{A} \mid (\omega)_K$. Conversely, if $\mathfrak{A} \mid (\omega)_K$, then ω vanishes on

$$\mathfrak{X}\left((\omega)_K^{-1}\right) + K \supseteq \mathfrak{X}\left(\mathfrak{A}^{-1}\right) + K,$$

that is $\mathfrak{A} \mid \omega$. □

Corollary 3.4.12. *If $X \in K^*$ and $\omega \in \mathrm{Dif}_K$ with $\omega \neq 0$ then $(X\omega)_K = (X)_K (\omega)_K$.*

Proof. If $\mathfrak{A} \mid \omega$, by Proposition 3.4.8 we obtain $(X)_K \mathfrak{A} \mid X\omega$. Therefore $(X)_K \mathfrak{A} \mid (X\omega)_K$ and since $(\omega)_K \mid \omega$, we have $(X)_K (\omega)_K \mid (X\omega)_K$.

Conversely, $\omega = X^{-1}X\omega = X^{-1}(X\omega)$, so that from the above argument we obtain $\left(X^{-1}\right)_K (X\omega)_K = (X)_K^{-1} (X\omega)_K \mid (\omega)_K$, which is equivalent to $(X\omega)_K \mid (X)_K (\omega)_K$. It follows that $(X\omega)_K = (X)_K (\omega)_K$. □

An important consequence of Corollary 3.4.12 is that the set consisting of the divisors of the nonzero differentials form exactly a class in C_K. More precisely, let $\omega \in \mathrm{Dif}_K$ with $\omega \neq 0$. Let $(\omega)_K \in C$ and $C \in C_K = D_K/P_K$. If ω' is another nonzero differential, then $\omega' = X\omega$, $X \in K^*$. Therefore $\left(\omega'\right)_K = (X)_K (\omega)_K$, that is, $\left(\omega'\right)_K$ and $(\omega)_K$ differ just by a principal divisor, and $\left(\omega'\right)_K \in C$. Conversely, if $\mathfrak{A} \in C$ and $\mathfrak{A} \in D_K$, then $\mathfrak{A}$, $(\omega)_K \in C$, so

$$\mathfrak{A} \equiv (\omega)_K \bmod P_K, \quad \text{that is,} \quad \mathfrak{A} = (X)_K (\omega)_K = (X\omega)_K$$

for some $X \in K^*$. Therefore $\mathfrak{A}$ is the divisor of the nonzero differential $X\omega$. We have

$$C = \{(\omega)_K \mid \omega \in \mathrm{Dif}_K, \omega \neq 0\}.$$

Definition 3.4.13. The class C consisting of all divisors of the nonzero differentials of a function field is called the *canonical class* and is denoted by $W = W_K$.

3.5 The Riemann–Roch Theorem and Its Applications

In Sections 3.3 and 3.4 of this chapter, we have developed the concepts of repartitions or adeles and that of differentials. On the other hand, Riemann's theorem (Theorem 3.3.2) essentially establishes that for each divisor $\mathfrak{A} \in D_K$ we have the formula

$$\delta\left(\mathfrak{A}^{-1}\right) = \ell(\mathfrak{A}) + d(\mathfrak{A}) + g_K - 1,$$

where g_K is the genus of the field.

Furthermore, Proposition 3.4.5 establishes that

$$\delta\left(\mathfrak{A}^{-1}\right) = \dim_k D\left(\mathfrak{A}^{-1}\right) = \dim_k \frac{\mathfrak{X}}{\mathfrak{X}(\mathfrak{A}) + K},$$

that is, $\delta\left(\mathfrak{A}^{-1}\right)$ is the dimension of the k-vector space of all differentials vanishing on $\mathfrak{X}(\mathfrak{A}) + K$, or equivalently all differentials such that $\mathfrak{A}^{-1} \mid \omega$.

What remains to do in order to obtain the Riemann–Roch theorem is to interpret $\delta\left(\mathfrak{A}^{-1}\right)$ as the dimension of a certain space $L(\mathfrak{B})$, and on the other hand, to determine the dimension of a class $C \in C_K$ by means of the divisors $\mathfrak{A} \in C$. We proceed to do this immediately.

Definition 3.5.1. Let $C \in C_K$ be an arbitrary class and let $\mathfrak{A}$ be any divisor in C. If $\mathfrak{A}_1, \ldots, \mathfrak{A}_n \in C$, we have $\frac{\mathfrak{A}_i}{\mathfrak{A}} = (x_i)_K$ for each $x_i \in K^*$. We say that the divisors $\mathfrak{A}_1, \ldots, \mathfrak{A}_n$ are *linearly independent* if $x_1, \ldots, x_n$ are linearly independent over k.

An apparent problem with this definition is that it seems to depend on the divisor $\mathfrak{A}$. The next result proves that this in not the case.

Proposition 3.5.2. *Definition* 3.5.1 *does not depend on* $\mathfrak{A}$ *or on the elements* x_i, $1 \leq i \leq n$.

Proof. Let $\mathfrak{A}$ and $x_1, x_2, \dots, x_n$ be as in the definition. We need to prove that if $(x_i)_K = \left(x_i'\right)_K$, then $\left\{x_1', \dots, x_n'\right\}$ is also linearly independent over k. To this end we observe that if $u, v \in K^*$ are such that $(u)_K = (v)_K$, then $\left(u^{-1}v\right)_K = \mathfrak{N}$, so that $v = \alpha u, \alpha \in k^*$. Therefore $x_i' = \alpha_i x_i$, with $\alpha_i \in k^*, i = 1, \dots, n$. Thus $\left\{x_1', \dots, x_n'\right\}$ are linearly independent over k.

Finally, if $\mathfrak{B} \in C$ is arbitrary and $\frac{\mathfrak{A}_i}{\mathfrak{B}} = (y_i)_K$ for $i = 1, \dots, n$, we must prove that $\{y_1, y_2, \dots, y_n\}$ is a linearly independent set over k.

Observe that $\mathfrak{A}, \mathfrak{B} \in C$, and hence $\frac{\mathfrak{A}}{\mathfrak{B}} = (z)_K$ with $z \in K^*$. Therefore $(y_i)_K = \frac{\mathfrak{A}_i}{\mathfrak{B}} = \frac{\mathfrak{A}_i}{\mathfrak{A}}\frac{\mathfrak{A}}{\mathfrak{B}} = (x_i)_K\,(z)_K = (x_i z)_K$. That is, $y_i = \alpha_i z x_i$ with $\alpha_i \in k^*, z \in K^*$, $i = 1, \dots, n$. From this, it follows immediately that $\{y_1, y_2, \dots, y_n\}$ is a linearly independent set over k. □

In Definition 3.3.10 we defined the dimension of a class $C \in C_K$ as $N(C) = \ell\left(\mathfrak{A}^{-1}\right)$, for an arbitrary $\mathfrak{A} \in C$. The following proposition relates the dimension to the maximum size of a subset of C consisting of linearly independent integral divisors.

Proposition 3.5.3. *Let $C \in C_K$ be any class. Then $N(C)$ is equal to the maximum number of linearly independent integral divisors belonging to C. In particular, this number is finite.*

Proof. Let n be the maximum size of a linearly independent subset of C consisting of integral divisors and let $\mathfrak{A}_1, \dots, \mathfrak{A}_n$ be such a subset. Let $\mathfrak{A} \in C$. Put $\frac{\mathfrak{A}_i}{\mathfrak{A}} = (x_i)_K$ for $i = 1, \dots, n$. Then $x_1, x_2, \dots, x_n$ are linearly independent over k. Therefore we have

$$(x_i)_K = \mathfrak{A}^{-1}\mathfrak{A}_i \Longrightarrow x_i \in L\left(\mathfrak{A}^{-1}\right), \quad \text{so} \quad n \le \ell\left(\mathfrak{A}^{-1}\right) = N(C).$$

On the other hand, if $y_1, y_2, \dots, y_{N(C)}$ is a basis of $L\left(\mathfrak{A}^{-1}\right)$, then $(y_i)_K = \mathfrak{A}^{-1}\mathfrak{C}_i$, where the $\mathfrak{C}_i$'s are integral divisors and $\mathfrak{C}_i \in C, 1 \le i \le N(C)$, with $\left\{y_1, y_2, \dots, y_{N(C)}\right\}$ linearly independent. Thus $N(C) \le n$, proving the result. □

We are ready to state and prove the Riemann–Roch theorem.

Theorem 3.5.4 (Riemann–Roch). *Let K/k be a function field and $C \in C_K$ any class. Let W be the canonical class and g the genus of K. Then*

$$N(C) = d(C) - g + 1 + N\left(WC^{-1}\right).$$

Equivalently, if $\mathfrak{A}$ is any divisor and ω is any nonzero differential, we have

$$\ell\left(\mathfrak{A}^{-1}\right) = d\left(\mathfrak{A}\right) - g + 1 + \ell\left((\omega)_K^{-1}\,\mathfrak{A}\right).$$

In other words,

$$\delta\left(\mathfrak{A}\right) = \ell\left(\mathfrak{A}^{-1}\right) + d\left(\mathfrak{A}^{-1}\right) + g - 1 = \ell\left((\omega)_K^{-1}\,\mathfrak{A}\right) = N\left(WC^{-1}\right)$$

for all $\mathfrak{A} \in C$.

Proof. Let C be an arbitrary class and let $\mathfrak{A} \in C$. We have

$$N(C) = \ell\left(\mathfrak{A}^{-1}\right) = d(\mathfrak{A}) - g + 1 + \delta(\mathfrak{A}) = d(C) - g + 1 + \delta(\mathfrak{A}).$$

Furthermore, $\delta(\mathfrak{A}) = \dim_k D(\mathfrak{A})$, where $D(\mathfrak{A}) = \{\omega \in \mathrm{Dif}_K \mid \mathfrak{A} \mid \omega\}$. By Theorem 3.4.11, we have

$$\begin{aligned} D(\mathfrak{A}) &= \{\omega \in \mathrm{Dif}_K \setminus\{0\} \mid \mathfrak{A} \mid (\omega)_K\} \cup \{0\} \\ &= \left\{\omega \in \mathrm{Dif}_K \setminus\{0\} \mid \mathfrak{A}^{-1}(\omega)_K \text{ is integral divisor}\right\} \cup \{0\}. \end{aligned}$$

Therefore

$$\begin{aligned} \delta(\mathfrak{A}) = \max \big\{ & n \mid \omega_1, \omega_2, \ldots, \omega_n \in \mathrm{Dif}_K \setminus\{0\} \text{ linearly independent over} \\ & k \text{ such that } (\omega_1)^K \mathfrak{A}^{-1}, \ldots, (\omega_n)^K \mathfrak{A}^{-1} \text{ are integral divisors} \big\} \\ = N & \left(WC^{-1}\right) = \ell\left((\omega)_K^{-1} \mathfrak{A}\right), \end{aligned}$$

which proves the theorem. □

Corollary 3.5.5. *Let W be the canonical class. Then $N(W) = g$ and $d(W) = 2g-2$. In particular, the dimension of the holomorphic differentials is g (see Corollary 3.4.6).*

Proof. Clearly, $\ell\left(\mathfrak{A}^{-1}\right) = d(\mathfrak{A}) - g + 1 + \ell\left((\omega)_K^{-1}\mathfrak{A}\right)$. Therefore, taking $\mathfrak{A} = \mathfrak{N}$, we have

$$\ell(\mathfrak{N}) = 1 = d(\mathfrak{N}) - g + 1 + \ell\left((\omega)_K^{-1}\right) = 0 - g + 1 + \ell\left((\omega)_K^{-1}\right).$$

Thus $N(W) = \ell\left((\omega)_K^{-1}\right) = 1+g-1 = g$ (this has already been obtained in Corollary 3.4.6).

Now if $\mathfrak{A} = (\omega)_K^{-1}$, we have

$$N(W) = g = d(W) - g + 1 + N\left(WW^{-1}\right) = d(W) - g + 1 + N(P_K).$$

On the other hand,

$$N(P_K) = \ell(\mathfrak{N}) = 1,$$

so

$$d(W) = g + g - 1 - 1 = 2g - 2.$$ □

Corollary 3.5.6. *If $\mathfrak{A}$ is a divisor such that $d(\mathfrak{A}) > 2g-2$ or $d(\mathfrak{A}) = 2g-2$ and $\mathfrak{A} \notin W$, then $\ell\left(\mathfrak{A}^{-1}\right) = d(\mathfrak{A}) - g + 1$. In particular, $\ell\left(\mathfrak{A}^{-1}\right) \geq g-1$.*

Proof. If $d(\mathfrak{A}) > 2g-2$ we have $d\left((\omega)_K^{-1}\mathfrak{A}\right) > (2g-2)-(2g-2)=0$, so $\ell\left((\omega)_K^{-1}\mathfrak{A}\right)=0$ (Proposition 3.2.18).

If $d(\mathfrak{A}) = 2g-2$ and $\mathfrak{A} \notin W$, then $(\omega)_K^{-1}\mathfrak{A}$ is a nonprincipal divisor of degree 0. Hence, from Proposition 3.2.18 we obtain $\ell\left((\omega)_K^{-1}\mathfrak{A}\right)=0$.

In any case, we obtain $\ell\left(\mathfrak{A}^{-1}\right) = d(\mathfrak{A}) - g + 1 + \ell\left((\omega)_K^{-1}\mathfrak{A}\right) = d(\mathfrak{A}) - g + 1$ and, in particular, $\ell\left(\mathfrak{A}^{-1}\right) = d(\mathfrak{A}) - g + 1 \geq 2g-2-g+1 = g-1$. □

Corollary 3.5.7. *If $W' \in C_K$ and $g' \in \mathbb{Z}$ are such that $N(C) = d(C) - g' + 1 + N\left(W'C^{-1}\right)$ for all classes C, then $W' = W$ and $g' = g$. In other words, W and g are uniquely determined by the Riemann–Roch theorem.*

Proof. Taking $C = W'$, we have

$$\begin{aligned} N(W') &= d(W') - g' + 1 + N\left(W'(W')^{-1}\right) \\ &= d(W') - g' + 1 + N(P_K) \\ &= d(W') - g' + 1 + 1 \\ &= d(W') - g' + 2. \end{aligned}$$

If $C = P_K$, then

$$N(P_K) = 1 = d(P_K) - g' + 1 + N\left(W'P_K^{-1}\right) = 0 - g' + 1 + N(W'),$$

whence

$$N(W') = 1 + g' - 1 = g' \quad \text{and} \quad d(W') = N(W') + g' - 2 = 2g' - 2.$$

If C is now any class such that $d(C) > 2g'-2$, then $N\left(W'C^{-1}\right) = 0$ by Proposition 3.2.18. Therefore $N(C) = d(C) - g' + 1$.

Hence, applying Corollary 3.5.6 and the above, we obtain that for any class C such that $d(C) > \max\left\{2g-2, 2g'-2\right\}$, we have $N(C) = d(C) - g' + 1 = d(C) - g + 1$, which implies that $g = g'$.

In particular, $N(W') = g' = g$, $d(W') = 2g' - 2 = 2g - 2$, whence, $W'W^{-1}$ is of degree zero and

$$\begin{aligned} g = N(W') &= d(W') - g + 1 + N\left(W'W^{-1}\right) \\ &= 2g - 2 - g + 1 + N\left(W'W^{-1}\right), \end{aligned}$$

which implies that $N\left(W'W^{-1}\right) = g - 2g + 2 + g - 1 = 1$. It follows that $W'W^{-1} = P_K$, since P_K is the only class of degree 0 and positive dimension. Therefore, $W' = W$. □

The following corollary states that there always exist elements with a unique given pole (or zero).

Corollary 3.5.8. *Let $\mathcal{P}$ be a prime divisor and let $n > 2g-1$ ($n > 0$ if $g = 0$). Then there exists an element x in K such that $\mathfrak{N}_x = \mathcal{P}^n$, that is, there exists an integral divisor $\mathfrak{B}$ such that $\mathfrak{B}$ is relatively prime to $\mathcal{P}$ and $(x)_K = \frac{\mathfrak{B}}{\mathcal{P}^n}$.*

Proof. Exercise 3.6.13. □

Definition 3.5.9. We say that a divisor $\mathfrak{A}$ *divides* a class C, and we write $\mathfrak{A} \mid C$, if $\mathfrak{A}$ divides $\mathfrak{B}$ for every integral divisor $\mathfrak{B}$ of C.

For the next result we use the notation $C\,\mathfrak{A}$ to denote the class $C\,C'$, where $\mathfrak{A} \in C'$.

Theorem 3.5.10. *Let $C \in C_K$ and $\mathfrak{A} \in D_K$, with $\mathfrak{A}$ an integral divisor. Then*

$$N(C) \le N(C\,\mathfrak{A}) \le N(C) + d(\mathfrak{A}).$$

Furthermore,

$$N(C) = N(C\,\mathfrak{A}) \iff \mathfrak{A} \mid C\,\mathfrak{A} \text{ and } N(C\,\mathfrak{A}) = N(C) + d(\mathfrak{A}) \iff \mathfrak{A} \mid W\,C^{-1}.$$

Proof. Let $\mathfrak{B} \in C$ so $N(C) = \ell\left(\mathfrak{B}^{-1}\right)$. Let $x \in L\left(\mathfrak{B}^{-1}\right)$. Then $(x)_K$ has the form $\frac{\mathfrak{C}}{\mathfrak{B}}$, where $\mathfrak{C}$ is an integral divisor. Since $\mathfrak{A}$ is an integral divisor we have $(x)_K = \frac{\mathfrak{C}\mathfrak{A}}{\mathfrak{B}\mathfrak{A}}$. Hence $x \in L\left(\mathfrak{B}^{-1}\mathfrak{A}^{-1}\right)$, so $L\left(\mathfrak{B}^{-1}\right) \subseteq L\left(\mathfrak{B}^{-1}\mathfrak{A}^{-1}\right)$ and $N(C) = \ell\left(\mathfrak{B}^{-1}\right) \le \ell\left(\mathfrak{B}^{-1}\mathfrak{A}^{-1}\right) = N(C\mathfrak{A})$.

Now, if $N(C) = N(C\mathfrak{A})$, then $L\left(\mathfrak{B}^{-1}\right) = L\left(\mathfrak{B}^{-1}\mathfrak{A}^{-1}\right)$ for all $\mathfrak{B} \in C$. Let $T \in C\mathfrak{A}$, where T is an integral divisor and $T = \mathfrak{B}\mathfrak{A}$, $\mathfrak{B} \in C$. In this case, $N(C) = N(C\mathfrak{A}) > 0$. Therefore there exists an integral divisor $\mathfrak{B}_0 \in C$. Thus T and $\mathfrak{B}_0\mathfrak{A} \in C\,\mathfrak{A}$. Then $\frac{T}{\mathfrak{B}_0\mathfrak{A}} = (x)_K$ is a principal divisor and $x \in L\left(\mathfrak{B}_0^{-1}\mathfrak{A}^{-1}\right) = L\left(\mathfrak{B}_0^{-1}\right)$. Therefore, $(x)_K = \frac{\mathfrak{B}\mathfrak{A}}{\mathfrak{B}_0\mathfrak{A}} = \frac{\mathfrak{B}}{\mathfrak{B}_0}$ and $\mathfrak{B}$ is an integral divisor. We have $T = \mathfrak{B}\mathfrak{A}$, which means that $\mathfrak{A} \mid T$.

Conversely, if $\mathfrak{A} \mid C\mathfrak{A}$ let $x \in L\left(\mathfrak{B}^{-1}\mathfrak{A}^{-1}\right)$ with $\mathfrak{B} \in C$. Then $(x)_K = \frac{\mathfrak{C}}{\mathfrak{B}\mathfrak{A}}$, where $\mathfrak{C}$ is an integral divisor. Since $\frac{\mathfrak{C}}{\mathfrak{B}\mathfrak{A}}$ is principal, we have $\mathfrak{C} \in C\,\mathfrak{A}$ and $\mathfrak{A} \mid C\mathfrak{A}$. Hence $(x)_K = \frac{\mathfrak{C}}{\mathfrak{B}\mathfrak{A}} = \frac{\mathfrak{C}_1}{\mathfrak{B}}$, that is, $x \in L\left(\mathfrak{B}^{-1}\right)$. Therefore

$$L\left(\mathfrak{B}^{-1}\mathfrak{A}^{-1}\right) \subseteq L\left(\mathfrak{B}^{-1}\right) \subseteq L\left(\mathfrak{B}^{-1}\mathfrak{A}^{-1}\right),$$

which implies that $N(C\mathfrak{A}) = N(C)$.

For the remaining part of the proof we apply the Riemann–Roch theorem, and we obtain

$$\begin{aligned} N(C\mathfrak{A}) &= d(C\mathfrak{A}) - g + 1 + N\left(WC^{-1}\mathfrak{A}^{-1}\right) \\ &= d(C) + d(\mathfrak{A}) - g + 1 + N\left(WC^{-1}\mathfrak{A}^{-1}\right). \end{aligned}$$

Using the first part, we obtain

$$N\left(WC^{-1}\mathfrak{A}^{-1}\right) \leq N\left(WC^{-1}\mathfrak{A}^{-1}\mathfrak{A}\right) = N\left(WC^{-1}\right),$$

and by applying again the Riemann–Roch theorem, we get

$$\begin{aligned} N\left(C\mathfrak{A}\right) &= d(C) + d\left(\mathfrak{A}\right) - g + 1 + N\left(WC^{-1}\mathfrak{A}^{-1}\right) \\ &\leq d(C) + d\left(\mathfrak{A}\right) - g + 1 + N\left(WC^{-1}\right) = N(C) + d\left(\mathfrak{A}\right). \end{aligned}$$

Finally, again by the first part we have

$$\begin{aligned} N\left(C\mathfrak{A}\right) = d(C) + d\left(\mathfrak{A}\right) &\iff N\left(WC^{-1}\mathfrak{A}^{-1}\right) = N\left(WC^{-1}\right) \\ &\iff \mathfrak{A} \mid WC^{-1}\mathfrak{A}^{-1}\mathfrak{A} = WC^{-1}. \end{aligned}$$

□

Corollary 3.5.11. *For any class C, we have $N(C) \leq \max\{0, 1 + d(C)\}$.*

Proof. If $N(C) = 0$, there is nothing to prove. If $N(C) > 0$, there exists an integral divisor $\mathfrak{A} \in C$ such that

$$N(C) = N\left(P_K\mathfrak{A}\right) \leq N\left(P_K\right) + d\left(\mathfrak{A}\right) = 1 + d\left(\mathfrak{A}\right) = 1 + d(C).$$

□

The next result will make clearer the reason why we use the term *special* for a divisor $\mathfrak{A} \in D_K$.

Proposition 3.5.12. *Let $\mathfrak{A} \in D_K$.*

(i) *$\mathfrak{A}$ is nonspecial if and only if $\ell_K\left(\mathfrak{A}^{-1}\right) = d_K\left(\mathfrak{A}\right) + 1 - g_K$.*
(ii) *If $d_K\left(\mathfrak{A}\right) > 2g_K - 2$, then $\mathfrak{A}$ is nonspecial.*
(iii) *The property of a divisor $\mathfrak{A}$ being special or nonspecial depends only on the class $\mathfrak{A} \in C \in C_K$ of $\mathfrak{A}$ in the divisor class group.*
(iv) *If $\mathfrak{A} \in W_K$, then $\mathfrak{A}$ is special.*
(v) *If $\mathfrak{A}$ satisfies $\ell_K\left(\mathfrak{A}^{-1}\right) > 0$ and $d_K\left(\mathfrak{A}\right) < g_K$, then $\mathfrak{A}$ is special.*
(vi) *If $\mathfrak{A}$ is nonspecial and $\mathfrak{A} \mid \mathfrak{B}$, then $\mathfrak{B}$ is nonspecial.*

Proof.

(i) This follows from Definition 3.3.8.
(ii) This follows from Corollary 3.5.6 and (i).
(iii) This follows from Remark 3.3.9.
(iv) This $\mathfrak{A} \in W_K$, then $\delta_K\left(\mathfrak{A}\right) = \ell_K\left((\omega)_K^{-1}\mathfrak{A}\right) = \ell_K\left((\omega)_K^{-1}(\omega)_K\right) = \ell_K\left(\mathfrak{N}\right) = 1 \neq 0$.
(v) We have $1 \leq \ell_K\left(\mathfrak{A}^{-1}\right) = d_K\left(\mathfrak{A}\right) + 1 - g_K + \delta_K\left(\mathfrak{A}\right)$. Thus, $\delta_K\left(\mathfrak{A}\right) \geq g_K - d_K\left(\mathfrak{A}\right) > 0$, and $\mathfrak{A}$ is special.
(vi) If $\mathfrak{A} \mid \mathfrak{B}$, then $\mathfrak{B}^{-1} \mid \mathfrak{A}^{-1}$, and by Theorem 3.1.11 we have

$$\begin{aligned} \delta(\mathfrak{B}) &= \ell_K\left(\mathfrak{B}^{-1}\right) + d_K\left(\mathfrak{B}^{-1}\right) + g_K - 1 \\ &\leq \ell_K\left(\mathfrak{A}^{-1}\right) + d_K\left(\mathfrak{A}^{-1}\right) + g_K - 1 = \delta(\mathfrak{A}). \end{aligned}$$

Thus $0 \leq \delta(\mathfrak{B}) \leq \delta(\mathfrak{A}) = 0$. It follows that $\delta(\mathfrak{B}) = 0$ and $\mathfrak{B}$ is nonspecial. □

Lemma 3.5.13. *Let K/k be any function field of genus $g > 0$. Let T be any set consisting of prime divisors of K of degree 1. If $|T| \geq g$, then given any integral divisor $\mathfrak{A}$ such that $\ell_K(\mathfrak{A}^{-1}) = 1$ and $d_K(\mathfrak{A}) \leq g - 1$, there exists $\mathfrak{p} \in T$ such that $\ell_K(\mathfrak{A}^{-1}\mathfrak{p}^{-1}) = 1$.*

Proof. Let $\mathfrak{p}_1, \ldots, \mathfrak{p}_g \in T$ be any set of g distinct elements of T such that $d_K(\mathfrak{p}_i) = 1$ for $i = 1, \ldots, g$. Assume that for all $i = 1, \ldots, g$,

$$\ell_K(\mathfrak{A}^{-1}\mathfrak{p}_i^{-1}) > 1.$$

There exist elements $x_i \in L_K(\mathfrak{A}^{-1}\mathfrak{p}_i^{-1}) \setminus L_K(\mathfrak{A}^{-1})$ for $i = 1, \ldots, g$. We have

$$v_{\mathfrak{p}_i}(x_i) = -v_{\mathfrak{p}_i}(\mathfrak{A}) - 1 \quad \text{and} \quad v_{\mathfrak{p}_j}(x_i) \geq -v_{\mathfrak{p}_j}(\mathfrak{A}) \quad \text{for} \quad i \neq j.$$

It follows from Proposition 2.2.3 (v) that $\{1, x_1, \ldots, x_g\}$ is a linearly independent set over k. Let $\mathfrak{C}$ be any divisor such that

$$\mathfrak{A}\mathfrak{p}_1 \cdots \mathfrak{p}_g \mid \mathfrak{C}$$

with $d_K(\mathfrak{C}) = 2g - 1$. Such $\mathfrak{C}$ clearly exists since we have $d_K(\mathfrak{A}\mathfrak{p}_1 \cdots \mathfrak{p}_g) = d_K(\mathfrak{A}) + g \leq 2g - 1$. Then

$$1, x_1, \ldots, x_g \in L_K(\mathfrak{C}^{-1}).$$

Thus, $\ell_K(\mathfrak{C}^{-1}) \geq 1 + g$. On the other hand, from Corollary 3.5.6 we obtain that

$$\ell_K(\mathfrak{C}^{-1}) = d_K(\mathfrak{C}) - g + 1 = g.$$

This contradiction proves the lemma. □

Proposition 3.5.14. *With the conditions of Lemma 3.5.13, there exists a nonspecial integral divisor $\mathfrak{A}$ with $\deg_K \mathfrak{A} = g$ and if $\mathfrak{p}$ is a prime divisor such that $\mathfrak{p} \mid \mathfrak{A}$, then $\mathfrak{p} \in T$.*

Proof. Let $\mathfrak{p}_1, \ldots, \mathfrak{p}_g$ be any set of g distinct prime divisors in T. Using Lemma 3.5.13 we can find divisors

$$\mathfrak{p}_{i_1} \mid \mathfrak{p}_{i_1}\mathfrak{p}_{i_2} \mid \cdots \mid \mathfrak{p}_{i_1}\mathfrak{p}_{i_2} \cdots \mathfrak{p}_{i_g} =: \mathfrak{A},$$

with $1 \leq i_j \leq g$ for all j, such that

$$\ell_K(\mathfrak{p}_{i_1}^{-1} \cdots \mathfrak{p}_{i_j}^{-1}) = 1$$

for $j = 1, \ldots, g$. In particular, $\ell_K(\mathfrak{A}^{-1}) = 1$. We have

$$d_K(\mathfrak{A}) + 1 - g = g + 1 - g = 1 = \ell_K(\mathfrak{A}^{-1}).$$

From Proposition 3.5.12 (i) we conclude that $\mathfrak{A}$ is nonspecial. □

Definition 3.5.15. Let K/k be an arbitrary function field of genus $g > 0$. A set of g different prime divisors $\mathfrak{p}_1, \dots, \mathfrak{p}_g$ of degree 1 is called a *nonspecial system* if

$$\ell_K\big((\mathfrak{p}_1 \cdots \mathfrak{p}_g)^{-1}\big) = 1,$$

or equivalently if $\delta_K(\mathfrak{p}_1 \cdots \mathfrak{p}_g) = 0$.

Proposition 3.5.16. *Let k be an algebraically closed field. Let K/k be a function field of genus $g > 0$. Then there exists a nonspecial system $\mathfrak{p}_1, \dots, \mathfrak{p}_g$ in K. Furthermore, $\mathfrak{p}_1$ may be chosen arbitrarily and $\mathfrak{p}_2, \dots, \mathfrak{p}_g$ may be chosen arbitrarily with finitely many exceptions.*

Proof. Let $\mathfrak{p}_1 \in \mathbb{P}_K$ be arbitrary. Then, since $g > 0$, K is not a rational function field and $L_K\big(\mathfrak{p}_1^{-1}\big) = k$. Thus, $\ell_K\big(\mathfrak{p}_1^{-1}\big) = 1$. It follows that

$$\delta_K(\mathfrak{p}_1) = \ell_K\big(\mathfrak{p}_1^{-1}\big) + d_K\big(\mathfrak{p}_1^{-1}\big) + g - 1 = g - 1.$$

From the proof of Lemma 3.5.13, we see that there are at most $g-1$ prime divisors $\mathfrak{p}$ such that $\mathfrak{p} \neq \mathfrak{p}_1$ and

$$\ell_K\big(\mathfrak{p}_1^{-1}\mathfrak{p}^{-1}\big) \neq 1.$$

For any $\mathfrak{p}_2$ not in this set and such that $\mathfrak{p}_2 \neq \mathfrak{p}_1$ we have

$$\ell_K\big(\mathfrak{p}_1^{-1}\mathfrak{p}_2^{-1}\big) = 1 \quad \text{and} \quad \delta_K(\mathfrak{p}_1\mathfrak{p}_2) = g - 2.$$

The result follows immediately by induction. □

Remark 3.5.17. Proposition 3.5.16 provides an explanation of the terminology of a nonspecial divisor. That is, $\mathfrak{A} = \mathfrak{p}_1 \cdots \mathfrak{p}_g$ is nonspecial for all but finitely many $\mathfrak{p}_1, \dots, \mathfrak{p}_g$.

Corollary 3.5.18. *If k is not an algebraically closed field, and K/k is a function field of genus $g > 0$, then there exists a finite constant extension k' such that we can find a nonspecial system in $K' = Kk'$.*

Proof. Let $\bar{k}$ be a separable closure of k. Then in $\bar{K} = K\bar{k}$ there exist nonspecial systems. Let $\mathfrak{p}_1, \dots, \mathfrak{p}_g$ be one of them. Then $\mathfrak{p}_1, \dots, \mathfrak{p}_g$ are of degree 1 in some finite constant extension of K. □

3.6 Exercises

Exercise 3.6.1. Let K be a function field over k. Let k' be the exact field of constants, $k' \supseteq k$. Show that if $\alpha \in (k')^*$ then $v_\wp(\alpha) = 0$ for all places $\wp$. Conclude that $k' = \big\{x \in K \mid v_\wp(x) = 0 \text{ for every place } \wp\big\} \cup \{0\}$.

Exercise 3.6.2. Let K be a function field with constant field k. Let D_K be the divisor group of K, P_K the principal divisor group, and $C_K = D_K/P_K$ (analogously, let $D_{K,0}$, $P_{K,0}$, $C_{K,0}$ be the respective groups of degree 0). Show that:

(i) $D_K \cong D_{K,0} \oplus \mathbb{Z}$.
(ii) $C_K \cong C_{K,0} \oplus \mathbb{Z}$.
(iii) $P_K \cong K^*/k^*$.

Exercise 3.6.3. Let K be a function field, $\mathfrak{A}$ a divisor, and $S \subseteq \mathbb{P}_K$ the set of prime divisors of K. Prove that $\Gamma(\mathfrak{A}|S)$ is a vector space over the field of constants k.

Exercise 3.6.4. Consider $K = k(x)$ in the context of the previous exercise. Under what conditions does it hold that $\dim_k \Gamma(\mathfrak{A}|S) < \infty$?

Exercise 3.6.5. Let $K = k(x)$ and let $\wp$ be the place corresponding to the irreducible polynomial $p(x) \in k[x]$. Prove that $f_\wp = \deg p(x)$.

Exercise 3.6.6. Under what conditions does it hold that $\dim_k \Lambda(\mathfrak{A}) < \infty$, where

$$\Lambda(\mathfrak{A}) = \mathfrak{X}(\mathfrak{A}) = \{\xi \in \mathfrak{X}_K \mid \mathfrak{A} \mid \xi\}?$$

Exercise 3.6.7. Give an example of a function field K and a repartition $\xi \in \mathfrak{X}_K$ such that there does not exist $x \in K$ with $v_\wp(x - \xi_\wp) \geq 0 \ \forall \wp \in \mathbb{P}_K$. (It is not necessary to give explicitly the example, just to show that such an example in fact exists. You may assume that there exist function fields with genus $g > 0$.)

Exercise 3.6.8. If $\mathfrak{B} \mid \mathfrak{A}$, show that $\Lambda(\mathfrak{B}^{-1}) + K \subseteq \Lambda(\mathfrak{A}^{-1}) + K$.

Exercise 3.6.9. Let $K = k(x)$. Describe the divisors of the form $(x - a)_K$, with $a \in k$. More generally, describe $(\alpha)_K$ with $\alpha \in K^*$.

Exercise 3.6.10. Let $\mathfrak{A}$, $\mathfrak{B}$ be divisors such that $\mathfrak{A} \mid \mathfrak{B}$. Prove that

$$(\mathfrak{X}(\mathfrak{B}) + K) \cap \mathfrak{X}(\mathfrak{A}) = \mathfrak{X}(\mathfrak{B}) + L(\mathfrak{A}) \quad \text{and} \quad L(\mathfrak{A}) \cap \mathfrak{X}(\mathfrak{B}) = L(\mathfrak{B}).$$

Exercise 3.6.11. If $\mathfrak{A}$, $\mathfrak{B}$ are divisors and ω, δ are two differentials such that $\mathfrak{A} \mid \omega$ and $\mathfrak{B} \mid \delta$, prove that $\mathfrak{C} \mid \Omega$, where $\Omega = \omega + \delta$ and $\mathfrak{C} = (\mathfrak{A}, \mathfrak{B})$ is the greatest common divisor of $\mathfrak{A}$ and $\mathfrak{B}$.

Exercise 3.6.12. Let $\mathfrak{A}$ be an integral divisor, $\omega \neq 0$ a nonzero differential, and let $\mathfrak{B}$ be a divisor such that $\mathfrak{B} \mid \omega$. Let $\varphi\colon L(\mathfrak{A}^{-1}\mathfrak{B}^{-1}) \to D(\mathfrak{A}^{-1})$ be defined by $\varphi(x) = x\omega$.

Prove that in fact $x \in L(\mathfrak{A}^{-1}\mathfrak{B}^{-1}) \Rightarrow \varphi(x) \in D(\mathfrak{A}^{-1})$ and that φ is a k-monomorphism.

Exercise 3.6.13. Let $\wp$ be a prime divisor and let $n > \max\{2g-1, 0\}$. Prove that there exists $x \in K$ with a unique pole $\wp$ of order n, that is, $\mathfrak{N}_x = \wp^n$.

Exercise 3.6.14. Let K/k be a function field.

(i) Prove that if $\mathfrak{A}$ is not principal and $d(\mathfrak{A}) = 0$, then $\ell(\mathfrak{A}^{-1}) = 0$.
(ii) If $\mathfrak{A} \notin \mathcal{W}$, $d(\mathfrak{A}) = 2g - 2$, show that $\ell(\mathfrak{A}^{-1}) = g - 1$.

Exercise 3.6.15. Let k be a finite field such that $k \cong \mathbb{F}_q$. Let $\mathcal{C}$ be a class and let $N(\mathcal{C})$ be its dimension. Prove that the number of integral divisors in $\mathcal{C}$ is $\frac{q^{N(\mathcal{C})}-1}{q-1}$.

Exercise 3.6.16. For a function field K/k we could have defined a repartition in K as a function

$$\varphi \colon \mathbb{P}_K \to K$$

such that $v_{\mathfrak{p}}(\varphi(\mathfrak{p})) \geq 0$ for almost all $\mathfrak{p}$. Prove that all the results of this chapter hold with this definition of repartition.

Exercise 3.6.17. Let K/k be a function field of genus $g > 0$. Let $\mathfrak{p}$ be any place of K. Prove that there exists a holomorphic differential ω in K such that $v_{\mathfrak{p}}(\omega) = 0$, that is, $\mathfrak{p}$ is not a zero of ω.

Exercise 3.6.18. Let K/k be a function field. Let $\mathfrak{A}$ be an integral divisor. If $\ell\big(\mathfrak{A}^{-1}\big) = d(\mathfrak{A}) + 1$ with $d(\mathfrak{A}) > 0$, prove that K is of genus 0.

Exercise 3.6.19. Let K/k be a function field of genus $g_K \geq 1$ and let W be its canonical class. If $\mathfrak{A} \in W^{-1}$, prove that if $\mathfrak{A} \mid \mathfrak{B}$ with $\mathfrak{A} \neq \mathfrak{B}$, then $\ell(\mathfrak{A}) \neq \ell(\mathfrak{B})$, that is, $\ell(\mathfrak{B}) < \ell(\mathfrak{A})$.

Exercise 3.6.20. With the notation of Exercise 3.6.19, let $\mathfrak{p}$ be a prime divisor of degree 1. Prove that $\ell(\mathfrak{A}) = \ell\big(\mathfrak{A}\mathfrak{p}^{-1}\big)$.

Exercise 3.6.21. With the notation of Exercise 3.6.20, show that if $\deg \mathfrak{p} > 1$ then $\ell(\mathfrak{A}) \neq \ell\big(\mathfrak{A}\mathfrak{p}^{-1}\big)$.

Exercise 3.6.22. Let K/k be any function field. Let $\mathfrak{A}$ be any divisor such that $L_K\big(\mathfrak{A}^{-1}\big) \neq \{0\}$. Prove that there exists an integral divisor $\mathfrak{B}$ in the divisor class of $\mathfrak{A}$.

Exercise 3.6.23. If W' is any class in the function field K/k such that $d_K(W') = 2g_K - 2$ and $\ell_K(W_K^{-1}) = g_K$, prove that $W' = W$ is the canonical class of K.

Exercise 3.6.24. Let $\mathfrak{a} \mid \mathfrak{b}$ and let $S = \big\{\mathfrak{p} \in \mathbb{P}_K \mid v_{\mathfrak{p}}(\mathfrak{a}) \neq 0 \text{ or } v_{\mathfrak{p}}(\mathfrak{b}) \neq 0\big\}$. Show that there exists a natural monomorphism $\frac{L(\mathfrak{a})}{L(\mathfrak{b})} \xrightarrow{\varphi} \frac{\Gamma(\mathfrak{a}|S)}{\Gamma(\mathfrak{b}|S)}$. In particular, $\ell_K(\mathfrak{a}) - \ell_K(\mathfrak{b}) \leq d_K(\mathfrak{b}) - d_K(\mathfrak{a})$.

Exercise 3.6.25. Prove Proposition 3.3.12.

Exercise 3.6.26. Let k be a finite field such that $|k| = q$ and let K/k be a function field.

(i) Prove that the number of integral divisors of degree $m \in \mathbb{N}$ is finite.
(ii) If $m \geq g_K$, prove that each class of degree m contains an integral divisor. Therefore the set C_m consisting of the classes of degree m is finite.
(iii) If $\mathfrak{M}$ is a divisor of degree $m \geq g_K$, then

$$\varphi \colon C_{K,0} \to C_m, \quad \text{defined by} \quad \varphi(\overline{\mathfrak{A}}) = \overline{\mathfrak{A}\mathfrak{M}},$$

is a bijection. Therefore

$$\left|C_{k,0}\right| = |C_m| < \infty.$$

4

Examples

In this chapter we present examples that illustrate how one can apply our results of Chapters 2 and 3. We shall first recall a few facts about rational function fields and characterize fields of genus 0.

Our second goal is to examine function fields of genus 1, among which are found elliptic function fields that correspond to the most important and widely investigated elliptic curves of algebraic geometry.

Finally, we present quadratic extensions of $k(x)$ in characteristic different from 2, and we compute the genus of such extensions. Among these fields are found hyperelliptic function fields, which, up to an abuse of the formal definition, contain elliptic function fields. We shall study those fields in detail in Section 9.6.4.

The reader will encounter hyperelliptic and elliptic function fields in Chapter 10 again, where they will be used in their applications to cryptography.

It should be mentioned that the computation of the genus could be done in a faster and more efficient way using the Riemann–Hurwitz genus formula, which will be studied in Chapter 9. However, the methods presented in this chapter, aside from their mathematical beauty, allow us to investigate the fields involved in detail and to get acquainted in a deeper way with their structure.

4.1 Fields of Rational Functions and Function Fields of Genus 0

First we consider the field $K = k(x)$ of rational functions where k is an arbitrary field and x a transcendental element over k. We recall some results about $k(x)$ that we have already obtained.

In Section 2.4 we characterized the set of all valuations on K, namely

$$\left\{v_f \mid f(x) \in k[x] \text{ is monic and irreducible}\right\} \cup \{v_\infty\}$$

(Theorem 2.4.1).

Example 3.2.16 shows that every divisor of degree 0 is principal; in particular, $C_{K,0} = 1$ and the class number h_K is equal to 1.

Proposition 4.1.1. *Let K be a purely transcendental extension over a field F. Then F is algebraically closed in K. In particular, the field of constants of a rational function field $k(x)$ is k.*

Proof. Let $\{x_i\}_{i\in I}$ be a transcendence base of K over F, that is, $K = F\left(\{x_i\}_{i\in I}\right)$. Let $\alpha \in K$ be algebraic over F. We must prove that $\alpha \in F$. Since $\alpha \in K$, α is a polynomial in a finite number of variables, that is, there exists a finite subset J of I such that $\alpha \in F\left(\{x_i\}_{i\in J}\right)$. This shows that we may assume, without loss of generality, that I is finite, or, which is the same, that $K = F(x_1, x_2, \dots, x_n)$.

We will prove the result by induction on n. For $n = 0$, K is equal to F and there is nothing to prove. If $n = 1$, then $\alpha \in F(x)$. If $\alpha \in F(x) \setminus F$, we have

$$\alpha = \frac{f(x)}{g(x)}, \qquad f(x), g(x) \in F[x],$$

and f, g relatively prime. Then x satisfies the equation

$$h(T) = f(T) - \alpha g(T) \in F(\alpha)[T].$$

Therefore

$$[F(x) : F(\alpha)] \leq \deg h(T) = \max\{\deg f, \deg g\} < \infty,$$

so $[F(\alpha) : F] = \infty$. Thus α is transcendental over F.

We assume that the result holds for $n - 1$ with $n \geq 2$. In order to prove it for n, let $\alpha \in F(x_1, \dots, x_n)$ be algebraic over F. In particular, α is an algebraic element over $F(x_1, \dots, x_{n-1})$ and it follows from the case $n = 1$ that $\alpha \in F(x_1, \dots, x_{n-1})$. By the induction hypothesis, we conclude that $\alpha \in F$. □

Corollary 4.1.2. *Let $\alpha \in k(x) \setminus k$ be of the form $\alpha = \frac{f(x)}{g(x)}$, where $f(x), g(x) \in k[x]$ are relatively prime. Then $[k(x) : k(\alpha)] = \max\{\deg f, \deg g\}$ (see Exercise 2.6.8).*

Proof. Since $\alpha \in k(x) \setminus k$, α is transcendental. The divisor of α is

$$(\alpha)_K = \frac{\mathfrak{A}_f}{\mathfrak{A}_g}\mathcal{P}_\infty^{(\deg g - \deg f)},$$

where

$$\mathfrak{A}_f = \mathcal{P}_{p_1}^{\alpha_1} \dots \mathcal{P}_{p_r}^{\alpha_r}, \qquad f(x) = p_1(x)^{\alpha_1} \dots p_r(x)^{\alpha_r},$$

$p_i(x)$ are distinct irreducible polynomials, and similarly for $\mathfrak{A}_g$. Now, since $k(\alpha) = k\left(\frac{1}{\alpha}\right)$, we may assume $\deg g \geq \deg f$. By applying Theorem 3.2.7 to $k(x)/k(\alpha)$, we obtain

$$[k(x) : k(\alpha)] = d(\mathfrak{N}_\alpha) = d\left(\mathfrak{A}_g\right) = \deg g = \max\{\deg f, \deg g\}. \qquad \square$$

Proposition 4.1.3. *The genus of $k(x)$, $g_{k(x)}$, is zero.*

Proof. If $f(x) \in k(x)$ is a rational function, we write $f(x) = p_1(x)^{\alpha_1} \dots p_s(x)^{\alpha_s}$, where $p_1(x), \dots, p_s(x)$ are distinct irreducible polynomials in $k[x]$ and $\alpha_i \in \mathbb{Z}$. Then

$$(f(x))_{k(x)} = \left(\prod_{i=1}^{s} \mathcal{P}_{p_i}^{\alpha_i}\right) \mathcal{P}_\infty^{-\deg f}.$$

Let $t \geq 0$ be arbitrary. Then

$$L\left(\mathcal{P}_\infty^{-t}\right) = \left\{ f(x) \in k(x) \mid (f(x))_{k(x)} = \frac{\mathfrak{A}}{\mathcal{P}_\infty^t}, \quad \mathfrak{A} \text{ is an integral divisor} \right\},$$

and this is the set of polynomials of degree at most t.

Therefore $\ell\left(\mathcal{P}_\infty^{-t}\right) = t + 1$.

Let g be the genus of $k(x)$ and let $t > 2g - 2$ be such that $d\left(\mathcal{P}_\infty^t\right) = td\left(\mathcal{P}_\infty\right) = t > 2g - 2$. By Corollary 3.5.6, we have

$$t + 1 = \ell\left(\mathcal{P}_\infty^{-t}\right) = d\left(\mathcal{P}_\infty^t\right) - g + 1 = t - g + 1,$$

whence, $g = 0$. □

Now, if W is the canonical class, we have

$$d(W) = 2g - 2 = 0 - 2 = -2.$$

On the other hand, since $C_{K,0} = 1$, for each $n \in \mathbb{Z}$ there exists a unique class C_n of degree n, which implies that

$$W = C_{-2} = \mathcal{P}_\infty^{-2} P_{k(x)}.$$

Since $\mathcal{P}_\infty^{-2}$ belongs to C_{-2}, there exists a differential ω such that $(\omega)_{k(x)} = \mathcal{P}_\infty^{-2}$ and every differential is of the form $f(x)\,\omega$, with $f(x) \in k(x)$. We will now describe this differential ω.

Let $\xi \in \mathfrak{X}$ be given by

$$\xi_{\mathcal{P}_\infty} = \frac{1}{x}, \quad \text{and} \quad \xi_{\mathcal{P}} = 0 \quad \text{for all} \quad \mathcal{P} \neq \mathcal{P}_\infty.$$

From Theorem 3.3.16, we obtain

$$\dim_k \frac{\mathfrak{X}\left(\mathcal{P}_\infty\right)}{\mathfrak{X}\left(\mathcal{P}_\infty^2\right)} = d\left(\mathcal{P}_\infty^2\right) - d\left(\mathcal{P}_\infty\right) = 2 - 1 = 1.$$

Since $v_\infty(\xi) = v_\infty\left(\xi_{\mathcal{P}_\infty}\right) = 1$, we have $\xi \in \mathfrak{X}\left(\mathcal{P}_\infty\right) \setminus \mathfrak{X}\left(\mathcal{P}_\infty^2\right)$ and furthermore, $\xi \in \left(\mathfrak{X}\left(\mathcal{P}_\infty\right) + K\right) \setminus \left(\mathfrak{X}\left(\mathcal{P}_\infty^2\right) + K\right)$. On the other hand, we have

$$\delta\left(\mathcal{P}_\infty^{-1}\right) = \dim_k \frac{\mathfrak{X}}{\mathfrak{X}\left(\mathcal{P}_\infty\right) + K} = d\left(\mathcal{P}_\infty\right) + \ell\left(\mathcal{P}_\infty\right) + g - 1 = 1 + 0 + 0 - 1 = 0.$$

Therefore, $\mathfrak{X} = \mathfrak{X}(\mathcal{P}_\infty) + K$ and $L(\mathcal{P}_\infty) = \{0\}$. From Theorem 3.4.4, we get

$$\frac{\mathfrak{X}(\mathcal{P}_\infty)}{\mathfrak{X}(\mathcal{P}_\infty^2)} \cong \frac{\mathfrak{X}(\mathcal{P}_\infty) + K}{\mathfrak{X}(\mathcal{P}_\infty^2) + K} = \frac{\mathfrak{X}}{\mathfrak{X}(\mathcal{P}_\infty^2) + K},$$

the latter being of dimension 1. Therefore every repartition θ can be written as

$$\theta = a\xi + \xi_1, \quad \text{with} \quad \xi_1 \in \mathfrak{X}\left(\mathcal{P}_\infty^2\right) + K.$$

Let

$$\omega : \mathfrak{X} \to k \quad \text{be such that} \quad (\omega) = \mathcal{P}_\infty^{-2}, \quad \text{that is,} \quad \mathfrak{X}\left(\mathcal{P}_\infty^2\right) + K \subseteq \ker \omega.$$

Then

$$\omega(\theta) = a\omega(\xi) + \omega(\xi_1) = a\omega(\xi).$$

We define $\omega(\xi) = -1$. This is approximately something like the following:

$$\operatorname*{Res}_{\mathcal{P}_a} \omega = \begin{cases} 0 \ , \ a \neq \infty, \\ -1 \ , \ a = \infty. \end{cases}$$

Then $(\omega)_{k(x)} = \mathcal{P}_\infty^{-2}$ and ω is uniquely determined by the conditions

$$\omega\left(\mathfrak{X}\left(\mathcal{P}_\infty^2\right) + K\right) = 0 \quad \text{and} \quad \omega(\xi) = -1.$$

Indeed, if ω' is any other differential with the same conditions, then for any repartition

$$\theta = a\xi + \xi_1, \quad \text{with} \quad \xi_1 \in \mathfrak{X}\left(\mathcal{P}_\infty^2\right) + K \quad \text{and} \quad a \in k,$$

we have

$$\begin{aligned}\left(\omega - \omega'\right)(\theta) &= a\left(\omega(\xi) - \omega'(\xi)\right) + \left(\omega(\xi_1) - \omega'(\xi_1)\right) \\ &= a(-1 - (-1)) + (0 - 0) = 0.\end{aligned}$$

Thus $\omega = \omega'$.

Definition 4.1.4. The differential ω of $k(x)$, defined by

$$\omega\left(\mathfrak{X}\left(\mathcal{P}_\infty^2\right) + K\right) = 0, \quad \omega(\xi) = -1,$$

where

$$\xi_{\mathcal{P}_\infty} = \frac{1}{x} \quad \text{and} \quad \xi_{\mathcal{P}} = 0 \quad \text{for all} \quad \mathcal{P} \neq \mathcal{P}_\infty,$$

will be denoted by $\omega = dx$.

Every differential is of the form $f(x)\,dx$, with $f(x) \in k(x)$. We have

$$(dx)_{k(x)} = \frac{1}{\mathcal{P}_\infty^2}.$$

Proposition 4.1.3 shows that a rational function field $k(x)$ is of genus zero. A natural question is the following: Is every function field K/k of genus zero a rational function field? The answer is, as we will see immediately, no. It is necessary to have an extra condition, namely that there exist a prime divisor of degree one. This situation holds if k is algebraically closed or if k is finite but may not hold in other cases (in the case that k is finite it will be necessary to use the Riemann hypothesis, Chapter 7).

Independently from the above discussion, what we have in any case the following result:

Proposition 4.1.5. *If K/k is any field of functions such that $g_K = 0$, then $C_{K,0} = \{1\}$ and consequently, $h_K = 1$.*

Proof. Let C be a class of degree 0. We wish to prove that $C = P_K$. Since $d(C) = 0 > -2 = 2g_K - 2$, it follows by Corollary 3.5.6 that

$$N(C) = d(C) - g_K + 1 = 0 - 0 + 1 = 1,$$

whence, there exists an integral divisor $\mathfrak{A}$ in C with degree 0. The only integral divisor of degree 0 is $\mathfrak{N}$, so $\mathfrak{N} \in C$. Therefore $C = P_K$. □

Proposition 4.1.6. *If K/k is a field of functions of genus 0, then K contains integral divisors of degree 2, and in particular it contains prime divisors of degree 1 or 2. Moreover, there exists $x \in K \setminus k$ such that $[K : k(x)] \leq 2$.*

Proof. Let W be the canonical class of K, $d(W) = 2g_K - 2 = -2$. We have $d\left(W^{-1}\right) = 2 > -2 = 2g_K - 2$. By Corollary 3.5.6,

$$N\left(W^{-1}\right) = d\left(W^{-1}\right) - g_K + 1 = 2 - 0 + 1 = 3,$$

that is, there exist at least three integral divisors in W^{-1}, and all of them are of degree 2. Since every divisor is a product of prime divisors, it follows that there exist prime divisors of degree 1 or 2. Indeed, if $\mathfrak{A}$ is an integral of degree 2, then

$$\mathfrak{A} = \mathcal{P}, \quad \mathcal{P}_1\mathcal{P}_2, \quad \text{or} \quad \mathcal{P}^2$$

for some prime divisors $\mathcal{P}, \mathcal{P}_1, \mathcal{P}_2$.

Since $N\left(W^{-1}\right) = 3$, there exist two integral divisors $\mathfrak{A}_1, \mathfrak{A}_2$ of degree 2 with $\mathfrak{A}_1 \neq \mathfrak{A}_2$. Since $\mathfrak{A}_1, \mathfrak{A}_2 \in W^{-1}$, $\frac{\mathfrak{A}_1}{\mathfrak{A}_2} = (x)_K$ is principal and $x \notin k$. By eliminating all common prime factors in $\mathfrak{A}_1$ and $\mathfrak{A}_2$, we obtain $(x)_K = \frac{\mathfrak{B}_1}{\mathfrak{B}_2}$, where $\mathfrak{B}_1$ and $\mathfrak{B}_2$ are relatively prime integral divisors of degree 1 or 2 and $\mathfrak{B}_1 \neq \mathfrak{B}_2$. By Theorem 3.2.7 we have $[K : k(x)] = d\left(\mathfrak{N}_x\right) = d\left(\mathfrak{B}_2\right) \leq 2$. □

We observe that if $K = k(x)$, then K contains prime divisors of degree 1, for instance $\mathcal{P}_\infty$; furthermore, for each $a \in k$ with $(x-a)_K = \frac{\mathcal{P}_a}{\mathcal{P}_\infty}$, $\mathcal{P}_a$ is of degree 1 and in fact $\{\mathcal{P}_a, \mathcal{P}_\infty \mid a \in k\}$ is the set of all prime divisors of degree 1.

Theorem 4.1.7. *Let K/k be a function field. If $K = k(x)$ then $g_K = 0$. Conversely, if $g_K = 0$, then K is a rational function field or a quadratic extension of $k(x)$. Furthermore, K contains prime divisors of degree 1 or 2. Finally, $K = k(x)$ if and only if there exists at least one prime divisor of degree 1.*

Proof. It remains to prove that if $g_K = 0$ and K contains a prime divisor of degree 1, then K is a rational function field.

Let $\mathcal{P}$ be a place of degree 1. We have $d(\mathcal{P}) = 1 > -2 = 2g_K - 2$. By Corollary 3.5.6,

$$\ell\left(\mathcal{P}^{-1}\right) = d(\mathcal{P}) - g_K + 1 = 1 - 0 + 1 = 2.$$

Therefore, there exist elements x_1, x_2 in $L\left(\mathcal{P}^{-1}\right)$ that are linearly independent over k, which implies $\frac{x_1}{x_2} \in K \setminus k$. On the other hand, we have

$$(x_1)_K = \frac{\mathfrak{A}}{\mathcal{P}} \quad \text{and} \quad (x_2)_K = \frac{\mathfrak{B}}{\mathcal{P}},$$

where $\mathfrak{A}$, $\mathfrak{B}$ are integral divisors and $d(\mathfrak{A}) = d(\mathfrak{B}) = 1$. Hence, if $x = \frac{x_1}{x_2}$, then $(x)_K = \frac{\mathfrak{A}}{\mathfrak{B}}$ and $x \notin k$. Thus, by Theorem 3.2.7, $[K : k(x)] = d(\mathfrak{N}_x) = d(\mathfrak{B}) = 1$, so $K = k(x)$. □

Corollary 4.1.8. *If K/k is a function field of genus 0 and k is algebraically closed, then $K = k(x)$ is a rational function field.*

Proof. If $\mathcal{P}$ is a place of K, then $k(\mathcal{P})$ is an algebraic extension of k. Therefore $k(\mathcal{P}) = k$ and $f_{\mathcal{P}} = [k(\mathcal{P}) : k] = 1$, that is, every place is of degree 1. □

We finish this section with an example of a field of genus 0 that is not a rational function field.

Let $\mathbb{R}$ be the field of real numbers and let $K = \mathbb{R}(x, y)$, where x, y are transcendental elements over $\mathbb{R}$ satisfying the equation

$$x^2 + y^2 + 1 = 0.$$

Let $K_0 = \mathbb{R}(x)$. Then since $y^2 = -x^2 - 1$, we have $y \notin K_0$, so $[K : K_0] = 2$.

The field of constants of K is a finite extension of $\mathbb{R}$. Therefore it is $\mathbb{R}$ or $\mathbb{C}$. Let us see that it is in fact $\mathbb{R}$. For the sake of contradiction, let us assume that $\mathbb{C}$ is the field of constants of K, that is, $i = \sqrt{-1} \in K$. Since $i \notin K_0$, it follows that $[K_0(i) : K_0] = 2$. Therefore $K_0(i) = K$. On the other hand, $K_0(i) = \mathbb{R}(x)(i) = \mathbb{C}(x)$ implies $y \in \mathbb{C}(x)$. However, since $y^2 = -x^2 - 1$, we have $y = \pm i\sqrt{x^2 + 1}$, which is not a rational function of x. Therefore the field of constants of K is $\mathbb{R}$.

Now we will see that K is not a rational function field. If this were the case, we would have $K = \mathbb{R}(z)$ with $z \in K \setminus \mathbb{R}$. Now, by the remark we made before Theorem 4.1.7, there would exist infinitely many places of degree 1. To prove that this is not the case, we will show that there can only be finitely many degree-1 places.

Let $\mathcal{P}$ be a place of K such that $v_{\mathcal{P}}(x) \geq 0$. Observe that all but finitely many places satisfy this condition, that is, there are only finitely many places $\mathfrak{S}$ such that $v_{\mathfrak{S}}(x) < 0$ (Theorem 3.2.1). Let

$$\varphi : K \longrightarrow (\vartheta_{\mathcal{P}}/\mathcal{P}) \cup \{\infty\}$$

be the corresponding place (see Section 2.2, particularly 2.2.10–2.2.13). We have $[\vartheta_{\mathcal{P}}/\mathcal{P} : \mathbb{R}] < \infty$ (Theorem 2.4.12). Hence $\vartheta_{\mathcal{P}}/\mathcal{P}$ is isomorphic to $\mathbb{R}$ or $\mathbb{C}$, so $\mathcal{P}$ is of degree 1 (in the case $\vartheta_{\mathcal{P}}/\mathcal{P} \cong \mathbb{R}$) or 2 (in the case $\vartheta_{\mathcal{P}}/\mathcal{P} \cong \mathbb{C}$).

We will prove that $\vartheta_{\mathcal{P}}/\mathcal{P} \cong \mathbb{C}$. The condition $v_{\mathcal{P}}(x) \geq 0$ is equivalent to $\varphi(x) \neq \infty$ (see Definition 2.2.10). If $\varphi(x) \in \mathbb{C} \setminus \mathbb{R}$ there is nothing to prove. If $\varphi(x) \in \mathbb{R}$, the equation $x^2+y^2+1 = 0$ implies $\varphi(x)^2+\varphi(y)^2+1 = 0$, so that $\varphi(y)^2 = -\varphi(x)^2-1 \in \mathbb{R}$. Since the latter is negative, we have $\varphi(y) = \pm i\sqrt{\varphi(x)^2 + 1} \in \mathbb{C} \setminus \mathbb{R}$. In any case, we get $\varphi(K) \not\subset \mathbb{R} \cup \{\infty\}$. Therefore $\vartheta_{\mathcal{P}}/\mathcal{P} \cong \mathbb{C}$ and $d\,(\mathcal{P}) = 2$.

By the above, K contains at most finitely many places of degree 1, which implies that K is not a rational function field over $\mathbb{R}$.

We will prove that in fact K has no degree-1 places. The case that remains to analyze is $v_{\mathcal{P}}(x) < 0$. If this is the case, let $x' = \frac{1}{x}$ and $y' = \frac{y}{x}$ and observe that $(x')^2 + (y')^2 + 1 = 0$. If φ is the corresponding place, then $\varphi\,(x) = \infty$, which implies $\varphi\,(x') = 0 \neq \infty$. Hence, as before, we obtain $\vartheta_{\mathcal{P}}/\mathcal{P} \cong \mathbb{C}$ and $d\,(\mathcal{P}) = 2$. This shows that every place in K is of degree 2.

Finally, we will prove that the genus of K is 0. In $\mathbb{R}(x)$ we write $(x)_{\mathbb{R}(x)} = \frac{\mathcal{P}_0}{\mathcal{P}_\infty}$ and in K, $(x)_K = \frac{\mathfrak{B}_0}{\mathfrak{B}_\infty}$. We observe that since $[K : \mathbb{R}(x)] = d\,(\mathfrak{B}_0) = d\,(\mathfrak{B}_\infty) = 2$ and every prime divisor of K is of degree 2, both $\mathfrak{B}_0$ and $\mathfrak{B}_\infty$ are prime divisors. Now, if v_∞ is the valuation corresponding to $\mathfrak{B}_\infty$, which is the extension of $\mathcal{P}_\infty$ to K, we have $v_\infty(x) = -1$. Thus

$$\begin{aligned} v_\infty\left(-x^2 - 1\right) &= \min\left\{v_\infty\left(x^2\right), v_\infty(-1)\right\} = \min\left\{2v_\infty\left(x\right), v_\infty(-1)\right\} \\ &= \min\{-2,\ 0\} = -2. \end{aligned}$$

In particular, we have

$$2\,v_\infty(y) = v_\infty\left(y^2\right) = v_\infty\left(-x^2 - 1\right) = -2,$$

which implies that $v_\infty(y) = -1$.

For $m \geq 1$, we have

$$L\left(\mathfrak{N}_x^{-m}\right) \supseteq \left\{a(x) + yb(x) \mid a(x), b(x) \in \mathbb{R}[x], v_{\mathfrak{B}_\infty}\left(a(x) + yb(x)\right) \geq -m\right\}.$$

We have

$$v_{\mathfrak{B}_\infty}\left(a(x)\right) = \begin{cases} \infty & \text{if } a(x) = 0 \\ -\deg a(x) & \text{if } a(x) \neq 0. \end{cases}$$

If $\deg a(x) \neq \deg b(x) + 1$, then since $v_\infty(y) = -1$, we have $v_\infty(a(x)) \neq v_\infty(yb(x))$, in which case

$$v_\infty(a(x)+yb(x)) = \min\{v_\infty(a(x), v_\infty(yb(x))\} = \min\{-\deg a, -1-\deg b\},$$

and

$$a(x)+yb(x) \in L\left(\mathfrak{N}_x^{-m}\right) \quad \text{if and only if} \quad \deg a \le m, \deg b \le m-1.$$

If $\deg a(x) = \deg b(x) + 1$, we write

$$a(x) = rx^n + a_1(x) \quad \text{and} \quad b(x) = sx^{n-1} + b_1(x),$$

with $\deg a_1(x) \le n-1$, $\deg b_1(x) \le n-2$, and $rs \neq 0$. Therefore

$$a(x) + yb(x) = x^{n-1}(rx+sy) + a_1(x) + yb_1(x).$$

Now we have

$$\begin{aligned} v_\infty\left(a_1(x)+yb_1(x)\right) &\ge \min\left\{v_\infty\left(a_1(x)\right), v_\infty\left(yb_1(x)\right)\right\} \\ &= \min\left\{-\deg a_1(x), -1-\deg b_1(x)\right\} \\ &\ge \min\left\{1-n, -1+2-n\right\} = 1-n. \end{aligned}$$

Since $K = K_0(y) = K_0(rx+sy)$, we have

$$[K_0(rx+sy) : K_0] = 2 = d\left(\mathfrak{N}_{rx+sy}\right),$$

and for every place $\mathfrak{B} \neq \mathfrak{B}_\infty$, $v_{\mathfrak{B}}(rx+sy) \ge 0$. It follows that $(rx+sy)_K = \frac{\mathfrak{A}}{\mathfrak{B}_\infty}$, that is, $v_\infty(rx+sy) = -1$.

Since $v_\infty\left(x^{n-1}\right) = 1-n$, we have

$$v_\infty\left(x^{n-1}(rx+sy)\right) = 1-n-1 = -n < 1-n.$$

Using Proposition 2.2.3 (iv), we conclude that

$$v_\infty\left(a(x)+yb(x)\right) = -n.$$

Therefore, the following also holds in this case:

$$a(x)+yb(x) \in L\left(\mathfrak{N}_x^{-m}\right) \text{ if and only if } -n = -\deg a = -\deg b - 1 \ge -m,$$

or equivalently,

$$\deg a(x) \le m \quad \text{and} \quad \deg b(x) \le m-1.$$

In short,

$$\begin{aligned} L\left(\mathfrak{N}_x^{-m}\right) \supseteq \big\{ & a(x)+yb(x) \mid a(x), b(x) \in \mathbb{R}[x], \\ & \deg a(x) \le m, \deg b(x) \le m-1 \big\}. \end{aligned}$$

It follows that

$$\ell\left(\mathfrak{N}_x^{-m}\right) = \dim_{\mathbb{R}} L\left(\mathfrak{N}_x^{-m}\right) \geq (m+1) + m = 2m+1.$$

On the other hand, we have

$$d\left(\mathfrak{N}_x^{m}\right) = md\left(\mathfrak{N}_x\right) = md\left(\mathfrak{P}_\infty\right) = m(2) = 2m.$$

By the Riemann–Roch Theorem (Corollary 3.5.6), when m is large enough, we have

$$2m+1 \leq \ell\left(\mathfrak{N}_x^{-m}\right) = d\left(\mathfrak{N}_x^{m}\right) - g_K + 1 = 2m+1-g_K.$$

Therefore $g_K \leq 0$. Hence $g_K = 0$.

We sum up the previous discussion into the following proposition:

Proposition 4.1.9. *Let $K = \mathbb{R}(x, y)$, where x and y are transcendental elements over $\mathbb{R}$ satisfying $x^2 + y^2 + 1 = 0$. Then the field of constants of K is $\mathbb{R}$, and K has genus 0 and is not a rational function field. Finally, every place of K is of degree 2.* □

Remark 4.1.10. Proposition 4.1.9 provides an example in which the degree function $d : D_K \longrightarrow \mathbb{Z}$ is not surjective, since every prime divisor is of degree 2. It follows that $d\left(D_K\right) = 2\mathbb{Z} \neq \mathbb{Z}$.

4.2 Elliptic Function Fields and Function Fields of Genus 1

In the previous section we studied function fields of genus 0 and we saw that they are "almost" fields of rational functions. Now we will study the function fields of genus 1 that "almost" are fields of elliptic functions.

Definition 4.2.1. Let K/k be a function field of genus $g_K = 1$. Then K is called an *elliptic function field* if K contains a prime divisor of degree 1.

Example 4.2.2. Let $K = \mathbb{R}(x, y)$ where x, y are transcendental elements over $\mathbb{R}$ satisfying the equation

$$x^2 + y^4 + 1 = 0.$$

Then K is of genus 1 (see Section 4.3, in particular Corollary 4.3.9) but every prime divisor of K is of degree 2. The proof is exactly the same as in Proposition 4.1.9.

In this section we characterize the elliptic function fields of characteristic different from 2. The case char $k = 2$ will be studied in Section 9.6.2.

Let $\mathcal{P}$ be a prime divisor of degree 1 in the elliptic function field K/k with $g = g_K = 1$. If W denotes the canonical class of K, we have $d(W) = 2g-2 = 2-2 = 0$, and on the other hand, $N(W) - g = 1$. Thus W is a class of degree 0 and positive dimension, which implies, by Proposition 3.2.18, that $W = P = P_K$. Therefore the canonical class and the principal class are the same.

Now we have $d(\mathcal{P}) = 1 > 0 = 2g - 2$, so by Corollary 3.5.6,

$$\ell\left(\mathcal{P}^{-n}\right) = d\left(\mathcal{P}^{n}\right) - g + 1 = nd(\mathcal{P}) - 1 + 1 = n, \quad \text{for} \quad n \geq 1.$$

In particular we have $\ell\left(\mathcal{P}^{-1}\right) = 1$ and $\ell\left(\mathcal{P}^{-2}\right) = 2$. Let $\{1, x\}$ be a basis of $L\left(\mathcal{P}^{-2}\right)$. Then $(x)_K \mathcal{P}^2$ is an integral divisor, that is, $\mathfrak{N}_x \mid \mathcal{P}^2$. On the other hand, since $K \neq k(x)$, we have $[K : k(x)] = d(\mathfrak{N}_x) \leq 2$, which implies that

$$\mathfrak{N}_x = \mathcal{P}^2 \quad \text{and} \quad [K : k(x)] = 2 = d(\mathfrak{N}_x).$$

We have $L\left(\mathcal{P}^{-2}\right) \subseteq L\left(\mathcal{P}^{-3}\right)$ and $\ell\left(\mathcal{P}^{-3}\right) = 3$, so there exists $y \in K$ such that $y \notin L\left(\mathcal{P}^{-2}\right)$ and $\{1, x, y\}$ is a basis of $L\left(\mathcal{P}^{-3}\right)$. Since $y \notin L\left(\mathcal{P}^{-2}\right)$ it follows that $\mathfrak{N}_y = \mathcal{P}^3$. Now the denominators of the divisors of $1, x, y, x^2, xy, x^3$, and y^2 are, respectively,

$$\mathfrak{N}, \ \mathcal{P}^2, \ \mathcal{P}^3, \ \mathcal{P}^4, \ \mathcal{P}^5, \ \mathcal{P}^6, \quad \text{and} \quad \mathcal{P}^6.$$

Since the first six elements listed have distinct denominators, they are linearly independent over k and all of them belong to $L\left(\mathcal{P}^{-6}\right)$, which is of dimension $\ell\left(\mathcal{P}^{-6}\right) = 6$. Thus, they form a basis and there exist $\gamma, \delta, \alpha_i \in k, i = 0, 1, 2, 3$, such that the relation

$$y^2 + \gamma xy + \delta y = \alpha_3 x^3 + \alpha_2 x^2 + \alpha_1 x + \alpha_0 \tag{4.1}$$

holds. We will see that $y \notin k(x)$. Let us assume that $y = \frac{f(x)}{h(x)} \in k(x)$ with $f(x), h(x)$ relatively prime.

We have from (4.1)

$$\frac{f^2 + \gamma xfh + \delta fh}{h^2} = \alpha_3 x^3 + \alpha_2 x^2 + \alpha_1 x + \alpha_0.$$

Then $h \mid f^2$, which implies that $h = 1$. That is, we have

$$f^2 + \gamma xf + \delta f = \alpha_3 x^3 + \alpha_2 x^2 + \alpha_1 x + \alpha_0. \tag{4.2}$$

From (4.2) it follows that f is a polynomial of degree at most 1. Now we have

$$y = f(x), \quad \mathfrak{N}_y = \mathcal{P}^3, \quad -3 = v_{\mathcal{P}}(y) = v_{\mathcal{P}}(f(x)) = v_{\mathcal{P}}(ax + b) = v_{\mathcal{P}}(x) = -2,$$

which is absurd. Hence, we have $y \notin k(x)$. Therefore

$$[k(x, y) : k(x)] \geq 2 = [K : k(x)],$$

which implies that $K = k(x, y)$.

Let char $K \neq 2$. We have

$$\begin{aligned} y^2+\gamma xy+\delta y &= y^2+y\left(\gamma x+\delta\right) \\ &= y^2+y\left(\gamma x+\delta\right)+\left(\frac{\gamma x+\delta}{2}\right)^2-\left(\frac{\gamma x+\delta}{2}\right)^2 \\ &= \left(y+\left(\frac{\gamma x+\delta}{2}\right)\right)^2-\left(\frac{\gamma x+\delta}{2}\right)^2. \end{aligned}$$

Therefore, if $z=y+\left(\frac{\gamma x+\delta}{2}\right)$, then

$$K=k(x,z) \quad \text{with} \quad z^2=f(x) \quad \text{and} \quad \deg f(x)\leq 3.$$

If $f(x)$ has degree 1, then $z=\sqrt{\alpha_2x+\alpha_3}$ and $K=k\left(\sqrt{\alpha_2x+\alpha_3}\right)$ is a rational function field, and hence of genus 0. If $\deg f(x)=2$, then K is of genus 0 (see Corollary 4.3.10 below). Thus $\deg f(x)=3$ and $\alpha_3\neq 0$. By multiplying (4.1) by α_3^2 and making a change of variables $y_1=\alpha_3 y$, $x_1=\alpha_3 x$, we may assume that $\alpha_3=1$. On the other hand, $f(x)$ has no repeated irreducible factors since if $z^2=f(x)=h(x)^2g(x)$ with $\deg h(x)=\deg g(x)=1$, then

$$z_1^2=\left(\frac{z}{h(x)}\right)^2=g(x)$$

and

$$K=k\left(x,z_1\right)=k\left(x,\sqrt{ax+b}\right)=k\left(\sqrt{ax+b}\right).$$

Therefore K is a rational function field and thus is of genus 0.

In short, we have the following result:

Proposition 4.2.3. *Let K/k be an elliptic function field. Then $K=k(x,y)$, where x and y are transcendental over k and satisfy a relation $g(y)=f(x)$ for some monic separable polynomials $f(x)\in k[x]$ and $g(y)\in k[y]$ of respective degrees 3 and 2. Furthermore, if* char $K\neq 2$, *then $f(x)$ and $g(y)$ can be chosen such that $f(x)$ is square-free and $g(y)=y^2$.* □

The converse also holds when char $K\neq 2$.

Theorem 4.2.4. *Let K/k be a function field such that* char $K\neq 2$. *Then K/k is an elliptic function field if and only if $K=k(x,y)$ where x and y are transcendental elements over k, $y^2=f(x)$ and $f(x)$ is a square free polynomial of degree 3.*

Proof.

($\Longrightarrow$) This is just Proposition 4.2.3.

($\Longleftarrow$) By Corollary 4.3.11 below, K is of genus 1. Now it suffices to see that K contains a place of degree 1.

Since $y^2=f(x)$ with $f(x)$ of degree 3, $\mathcal{P}\mid\mathfrak{N}_x$ implies $\mathcal{P}^3\mid\mathfrak{N}_{f(x)}$. Therefore $\mathfrak{N}_{f(x)}=\mathfrak{N}_x^3$ and $\mathfrak{N}_y^2\mid\mathfrak{N}_x^3$. On the other hand,

$$[K : k(x)] = [k(x, y) : k(x)] = 2 \quad \text{and} \quad [K : k(y)] = [k(x, y) : k(y)] = 3.$$

Thus, we obtain

$$d\left(\mathfrak{N}_x^3\right) = 3d\left(\mathfrak{N}_x\right) = 3[K : k(x)] = 6 = 2[K : k(y)] = 2d\left(\mathfrak{N}_y\right) = d\left(\mathfrak{N}_y^2\right).$$

Hence $\mathfrak{N}_y^2 = \mathfrak{N}_x^3$. Since $d\left(\mathfrak{N}_x\right) = 2$, we have

$$\mathfrak{N}_x = \mathcal{P}_1, \quad \mathfrak{N}_x = \mathcal{P}_2\mathcal{P}_3 \quad \text{or} \quad \mathfrak{N}_x = \mathcal{P}_4^2$$

with $\mathcal{P}_i$ prime divisors, $d\left(\mathcal{P}_1\right) = 2$ and $d\left(\mathcal{P}_i\right) = 1$, $i = 2, 3, 4$.

Now $\mathfrak{N}_x^3$ is $\mathcal{P}_1^3$ or $\mathcal{P}_2^3\mathcal{P}_3^3$ or $\mathcal{P}_4^6$, but $\mathfrak{N}_y^2 = \mathfrak{N}_x^3$ implies that the exponents of $\mathfrak{N}_x^3$ must be divisible by 2, whence it follows that $\mathfrak{N}_x = \mathcal{P}^2$, $d\left(\mathcal{P}\right) = 1$ and $\mathfrak{N}_y = \mathcal{P}^3$. In particular, K contains a prime divisor of degree 1. □

Now assume that $\operatorname{char} k \neq 2, 3$. By (4.1) and the case $\operatorname{char} k \neq 2$, we have $K = k(x, y)$ with

$$y^2 = x^3 + \alpha_2 x^2 + \alpha_3 x + \alpha_4. \tag{4.3}$$

Let $x' := x - \frac{\alpha_2}{3}$. Then

$$\begin{aligned} x^3 + \alpha_2 x^2 + \alpha_3 x + \alpha_4 &= \left(x' - \frac{\alpha_2}{3}\right)^3 + \alpha_2\left(x' - \frac{\alpha_2}{3}\right)^2 + \alpha_3\left(x' - \frac{\alpha_2}{3}\right) + \alpha_4 \\ &= (x')^3 + ax' + b. \end{aligned}$$

Thus

$$(2y)^2 = 4(x')^3 + 4ax' + 4b.$$

In short, when $\operatorname{char} k \neq 2, 3$, there exist $x, y \in K$ such that

$$y^2 = 4x^3 - g_2 x - g_3, \quad \text{with} \quad g_2, g_3 \in k. \tag{4.4}$$

Definition 4.2.5. The equation (4.4) is called the *Weierstrass form.*

Finally, we consider a function field K/k of any characteristic and $g_K = 1$. If K contains a divisor of degree 1, then there exists an integral divisor of degree 1 (Exercise 3.6.22). Thus there exists a prime divisor of degree 1 and K/k is an elliptic function field.

We sum up the above discussion into the following theorem.

Theorem 4.2.6. *Let K/k be a function field of genus 1. Then K/k is an elliptic function field if and only if there exists a divisor of degree 1.*

If $\operatorname{char} k \neq 2$, *$K/k$ is an elliptic function field if and only if $K = k(x, y)$ with*

$$y^2 = f(x), \tag{4.5}$$

where $f(x)$ is a monic separable polynomial of degree 3.

Furthermore, if $\operatorname{char} k \neq 2, 3$, *then $K = k(x, y)$ with*

$$y^2 = 4x^3 - g_2 x - g_3 \quad \text{and} \quad g_2, g_3 \in k. \tag{4.6}$$

□

4.3 Quadratic Extensions of $k(x)$ and Computation of the Genus

In Sections 4.1 and 4.2, the study of the fields of genus 0 and 1 led us to encounter several function fields K such that $[K : k(x)] = 2$. When $g_K \geq 2$ these fields are a special type of *hyperelliptic function field*, which will be examined in Section 9.6.4. In this section we fill in the gaps remaining from Section 4.2, namely the computation of the genus (Example 4.2.2 and Theorem 4.2.4). We could have proceeded differently and started with this section and then applied directly the results obtained here. However, we consider that the way we chose provides the reader with a motivation consisting in seeing the examples first and calculating the genus in quadratic extensions of $k(x)$. It is also important to clarify that later on, when we develop ramification theory and the Riemann–Hurwitz genus formula, we will have at our disposal a much more general method for calculating the genus of a function field.

In this section we consider a function field K/k such that there exists $x \in K$ with $[K : k(x)] = 2$ and char $K \neq 2$.

Lemma 4.3.1. *We have* $K = k(x, y)$, *where* $y^2 = f(x)$ *and* $f(x) \in k[x]$ *is square-free.*

Proof. Let $y \in K \setminus k(x)$. Then

$$k(x) \subseteq k(x, y) \subseteq K \quad \text{and} \quad [k(x, y) : k(x)] \geq 2 = [K : k(x)],$$

which implies that $K = k(x, y)$. Now, since y is of degree 2 over $k(x)$, the irreducible polynomial of y is of the form $y^2 + ay + b = 0$ with $a, b \in k(x)$. Since char $K \neq 2$, by completing squares we obtain

$$y^2 + ay + \frac{a^2}{4} = \frac{a^2}{4} - b, \quad \text{or} \quad \left(y + \frac{a}{2}\right)^2 = \frac{a^2}{4} - b.$$

Now let $z = y + \frac{a}{2}$, $K = k(x, z)$, and $z^2 = c$ with $c \in k(x)$. We can write $c = \frac{h(x)}{g(x)}$ for some relatively prime elements $h(x)$, $g(x)$ of $k[x]$. Then

$$(g(x)z)^2 = h(x)g(x).$$

Put $u = g(x)z$ and $t(x) = h(x)g(x)$. We then have $K = k(x, u)$ and $u^2 = t(x)$. Finally, we can write $t(x) = r(x)^2 f(x)$ with $f(x)$ square-free. Then if $v = \frac{u}{r(x)}$, then $K = k(x, v)$ and $v^2 = f(x)$, where $f(x)$ is square-free. □

From this point on, K will denote a field of the form

$$k(x, y), \quad \text{where} \quad y^2 = f(x)$$

for some square-free polynomial $f(x)$ of degree m. Since $[K : k(x)] = 2$ and char $K \neq 2$, $K/k(x)$ is a Galois extension. Let $\operatorname{Gal}(K/k(x)) = \{1, \sigma\}$ with

$$K = k(x, y), \quad y^2 = f(x) \quad \text{and} \quad \sigma(y) = -y.$$

Let $\mathcal{P}$ be an arbitrary place with valuation ring ϑ and associated valuation $v_{\mathcal{P}}$. We define $v_{\mathcal{P}^\sigma}$ by $v_{\mathcal{P}^\sigma}(z) := v_{\mathcal{P}}\left(\sigma^{-1}(z)\right) = v_{\mathcal{P}}(\sigma(z))$.

Lemma 4.3.2. *$v_{\mathcal{P}^\sigma}$ is a valuation with maximal ideal $\mathcal{P}^\sigma = \{\sigma(\alpha) \mid \alpha \in \mathcal{P}\}$ and valuation ring ϑ^σ.*

Proof. It is straghtforward. □

Now, σ can be extended to D_K in a natural way; that is, if

$$\mathfrak{A} = \prod_{i=1}^{r} \mathcal{P}_i^{\alpha_i} \in D_K \quad \text{we define} \quad \mathfrak{A}^\sigma := \prod_{i=1}^{r} \left(\mathcal{P}_i^\sigma\right)^{\alpha_i} \in D_K.$$

For $z \in K$, we have

$$v_{\mathcal{P}^\sigma}(\sigma(z)) = v_{\mathcal{P}}\left(\sigma^{-1}(\sigma(z))\right) = v_{\mathcal{P}}(z).$$

Therefore we obtain the following lemma:

Lemma 4.3.3. *If $z \in K^*$, then $(z)_K^\sigma = (z^\sigma)_K$.*

Proof. If $(z)_K = \frac{\mathfrak{Z}_z}{\mathfrak{N}_z}$, then $v_{\mathcal{P}^\sigma}(\sigma(z)) = v_{\mathcal{P}}(z)$, that is, $\mathfrak{Z}_z^\sigma = \mathfrak{Z}_{\sigma(z)}$ and $\mathfrak{N}_z^\sigma = \mathfrak{N}_{\sigma(z)}$. Therefore

$$(z)_K^\sigma = \frac{\mathfrak{Z}_z^\sigma}{\mathfrak{N}_z^\sigma} = \frac{\mathfrak{Z}_{\sigma(z)}}{\mathfrak{N}_{\sigma(z)}} = (\sigma(z))_K = \left(z^\sigma\right)_K.$$ □

Proposition 4.3.4. *Let $t \in \mathbb{N}$ and let $\mathfrak{N}_x$ be the pole divisor of x. If $z \in L\left(\mathfrak{N}_x^{-t}\right)$, then $\sigma(z) \in L\left(\mathfrak{N}_x^{-t}\right)$. In particular, if $z = a(x) + yb(x)$ with $a(x), b(x) \in k(x)$ and $z \in L\left(\mathfrak{N}_x^{-t}\right)$, then $\sigma(z) = a(x) - yb(x) \in L\left(\mathfrak{N}_x^{-t}\right)$.*

Proof. Let $z \in L\left(\mathfrak{N}_x^{-t}\right)$ be nonzero. Then $(z)_K = \frac{\mathfrak{A}}{\mathfrak{N}_x^t}$, for some integral divisor $\mathfrak{A}$. Therefore $\mathfrak{A}^\sigma$ is an integral divisor and

$$(z)_K^\sigma = (\sigma(z))_K = \frac{\mathfrak{A}^\sigma}{\left(\mathfrak{N}_x^t\right)^\sigma} = \frac{\mathfrak{A}^\sigma}{\mathfrak{N}_{\sigma(x)}^t} = \frac{\mathfrak{A}^\sigma}{\mathfrak{N}_x^t},$$

which implies $\sigma(z) \in L\left(\mathfrak{N}_x^{-t}\right)$. □

Proposition 4.3.5. *For $t \in \mathbb{N}$ we have*

$$L\left(\mathfrak{N}_x^{-t}\right) = \left\{a(x) + yb(x) \mid a(x), b(x) \in k[x], \deg a \le t \text{ and } \deg b \le t - \frac{m}{2}\right\}.$$

Proof. Let $z \in L\left(\mathfrak{N}_x^{-t}\right)$ be of the form

$$z = a(x) + yb(x) \quad \text{with} \quad a(x),\ b(x) \in k(x).$$

We have

$$\sigma(z) = a(x) - yb(x) \in L\left(\mathfrak{N}_x^{-t}\right),$$

and hence

$$z + \sigma(z) = 2a(x) \in L\left(\mathfrak{N}_x^{-t}\right).$$

Therefore $a(x) \in L\left(\mathfrak{N}_x^{-t}\right)$ since char $K \neq 2$. Now, if

$$a(x) = \frac{s(x)}{r(x)}, \quad \text{where} \quad s(x),\ r(x) \in k[x]$$

are relatively prime and $r(x)$ is a nonconstant polynomial, there exists an irreducible polynomial $g(x)$ in $k[x]$ such that $g(x) \mid r(x)$, that is,

$$v_g(a(x)) < 0 \quad \text{in} \quad k(x) \quad \text{and} \quad v_g \neq v_\infty.$$

Now if v is an extension of v_g to K, we have $v(a(x)) < 0$, where $v \neq v'_\infty$ and v'_∞ is any extension of v_∞ to K. However, since $a(x) \in L\left(\mathfrak{N}_x^{-t}\right)$, $t \geq 1$ implies that $v(a(x)) \geq 0$. This contradiction proves that $a(x) \in k[x]$.

Now we write

$$a(x) = a_n x^n + \cdots + a_1 x + a_0, \quad \text{with} \quad a_n \neq 0.$$

If $\mathcal{P} \mid \mathfrak{N}_x$, then

$$v_{\mathcal{P}}\left(a_i x^i\right) = \begin{cases} \infty & \text{if } a_i = 0, \\ i v_{\mathcal{P}}(x) & \text{if } a_i \neq 0. \end{cases}$$

Therefore since $v_{\mathcal{P}}(x) < 0$ we get

$$v_{\mathcal{P}}(a(x)) = \min\{i v_{\mathcal{P}}(x) \mid 0 \leq i \leq n, a_i \neq 0\} = n v_{\mathcal{P}}(x).$$

In particular, we have $\mathfrak{N}_{a(x)} = \mathfrak{N}_x^n$. Since $a(x) \in L\left(\mathfrak{N}_x^{-t}\right)$, it follows that $n \leq t$. In short, $a(x)$ is a polynomial of degree at most t.

On the other hand, $y^2 = f(x)$ implies

$$\begin{aligned} zz^\sigma &= (a(x) + yb(x))(a(x) - yb(x)) = a(x)^2 - y^2 b(x)^2 \\ &= a(x)^2 - f(x)b(x)^2 \in L\left(\mathfrak{N}_x^{-2t}\right). \end{aligned}$$

Indeed, from

$$(z)_K = \frac{\mathfrak{A}}{\mathfrak{N}_x^t}, \quad \text{we get} \quad \left(z^\sigma\right)_K = \frac{\mathfrak{A}^\sigma}{\mathfrak{N}_x^t}, \quad \text{so} \quad \left(zz^\sigma\right)_K = \frac{\mathfrak{A}\mathfrak{A}^\sigma}{\mathfrak{N}_x^{2t}}.$$

It follows from the previous discussion that $a(x)^2 - f(x)b(x)^2$ is a polynomial of degree at most $2t$, which implies that $f(x)b(x)^2$ is a polynomial of degree at most $2t$. Since f is square-free, it follows that $b(x)$ must be a polynomial and since $\deg f = m$, we have $\deg b \leq \frac{2t-m}{2} = t - \frac{m}{2}$.

Conversely, let $a(x) \in k[x]$ be of degree at most t and let $b(x) \in k[x]$ be of degree at most $t - \frac{m}{2}$. Observe that for any valuation v such that $v(x) \geq 0$, we have $v(y) \geq 0$ since $y^2 = f(x)$ and $v(f(x)) \geq 0$. Then

$$(y)_K^2 = \left(y^2\right)_K = (f(x))_K = \frac{\mathfrak{A}}{\mathfrak{N}_x^{\deg f}} = \frac{\mathfrak{A}}{\mathfrak{N}_x^m},$$

so $(y)_K = \frac{\mathfrak{A}_1}{\mathfrak{N}_x^{m/2}}$ for some integral divisor $\mathfrak{A}_1$.

Let $z = a(x) + yb(x)$. Then $z^\sigma = a(x) - yb(x)$ and $z \in L\left(\mathfrak{N}_x^{-n}\right)$ for some n. Now, if $\mathcal{P} \mid \mathfrak{N}_x$, we have

$$\begin{aligned} v_{\mathcal{P}}\left(z + z^\sigma\right) &= v_{\mathcal{P}}\left(2a(x)\right) = v_{\mathcal{P}}(a(x)) = \deg a(x) v_{\mathcal{P}}(x) \\ &\geq t v_{\mathcal{P}}(x) = v_{\mathcal{P}}\left(x^t\right), \end{aligned}$$

and

$$\begin{aligned} v_{\mathcal{P}}\left(z - z^\sigma\right) &= v_{\mathcal{P}}\left(2yb(x)\right) = v_{\mathcal{P}}(y) + v_{\mathcal{P}}(b(x)) \\ &= \frac{m}{2} v_{\mathcal{P}}(x) + \deg b(x) v_{\mathcal{P}}(x) = \left(\frac{m}{2} + \deg b(x)\right) v_{\mathcal{P}}(x) \\ &\geq t v_{\mathcal{P}}(x) = v_{\mathcal{P}}\left(x^t\right). \end{aligned}$$

Therefore $z + z^\sigma$ and $z - z^\sigma$ belong to $L\left(\mathfrak{N}_x^{-t}\right)$, which implies that

$$2z = \left(z + z^\sigma\right) + \left(z - z^\sigma\right) \in L\left(\mathfrak{N}_x^{-t}\right),$$

whence $z \in L\left(\mathfrak{N}_x^{-t}\right)$. □

Corollary 4.3.6. *We have*

$$\ell\left(\mathfrak{N}_x^{-t}\right) = \begin{cases} 0 & \text{if } t < 0, \\ t + 1 & \text{if } 0 \leq t \leq \left[\frac{m+1}{2}\right] - 1, \\ 2t + 2 - \left[\frac{m+1}{2}\right] & \text{if } t \geq \left[\frac{m+1}{2}\right]. \end{cases}$$

Proof. If $t < 0$, then $\mathfrak{N}_x^{-t}$ is an integral divisor, so $L\left(\mathfrak{N}_x^{-t}\right) = \{0\}$ and $\ell\left(\mathfrak{N}_x^{-t}\right) = 0$.

Let $t \geq 0$. We have

$$L\left(\mathfrak{N}_x^{-t}\right) = \left\{a(x) + yb(x) \mid \deg a \leq t, \deg b \leq t - \frac{m}{2}\right\}.$$

If

$$t \leq \left[\frac{m+1}{2}\right] - 1 = \left[\frac{m+1-2}{2}\right] = \left[\frac{m-1}{2}\right] < \frac{m}{2},$$

then

$$t - \frac{m}{2} < 0, \quad \text{so} \quad b(x) = 0.$$

Therefore

$$L\left(\mathfrak{N}_x^{-t}\right) = \{a(x) \mid \deg a \leq t\} \quad \text{and} \quad \ell\left(\mathfrak{N}_x^{-t}\right) = t + 1.$$

Finally, if $t \geq \left[\frac{m+1}{2}\right] \geq \frac{m}{2}$, we have

$$\deg b \leq \left[t - \frac{m}{2}\right] = \begin{cases} t - \frac{m}{2} & \text{if } m \text{ is even,} \\ t - 1 - \frac{m-1}{2} = t - \frac{m+1}{2} & \text{if } m \text{ is odd} \end{cases}$$

Therefore,

$$\ell\left(\mathfrak{N}_x^{-t}\right) = t + 1 + \left[t - \frac{m}{2}\right] + 1 = \begin{cases} t + 1 + t - \frac{m}{2} + 1 & \text{if } m \text{ is even} \\ t + 1 + t - \frac{m+1}{2} + 1 & \text{if } m \text{ is odd} \end{cases}$$

$$= \begin{cases} 2t + 2 - \frac{m}{2} & \text{if } m \text{ is even} \\ 2t + 2 - \frac{m+1}{2} & \text{if } m \text{ is odd} \end{cases} = 2t + 2 - \left[\frac{m+1}{2}\right]. \qquad \square$$

Corollary 4.3.7. *We have*

$$g = g_K = \left[\frac{m+1}{2}\right] - 1 = \begin{cases} \frac{m}{2} - 1 & \text{if } m \text{ is even,} \\ \frac{m-1}{2} & \text{if } m \text{ is odd.} \end{cases}$$

Proof. We have $[K : k(x)] = 2 = d(\mathfrak{N}_x)$. If $t > g$, then $t \in \mathbb{N}$ and $d\left(\mathfrak{N}_x^t\right) = t d(\mathfrak{N}_x) = 2t > 2g - 2$. By Corollary 3.5.6, $\ell\left(\mathfrak{N}_x^{-t}\right) = d\left(\mathfrak{N}_x^t\right) - g + 1$. Therefore for $t > \max\left\{0, g, \left[\frac{m+1}{2}\right]\right\}$, we have

$$\ell\left(\mathfrak{N}_x^{-t}\right) = 2t + 2 - \left[\frac{m+1}{2}\right] = d\left(\mathfrak{N}_x^t\right) - g + 1 = 2t - g + 1.$$

Hence $g = 2t + 1 - (2t + 2) + \left[\frac{m+1}{2}\right] = \left[\frac{m+1}{2}\right] - 1.$ $\square$

Now all cases pending from Section 4.2 are an immediate consequence of Corollary 4.3.7.

Corollary 4.3.8 (see Proposition 4.1.9). *If $K = \mathbb{R}(x, y)$ with $x^2 + y^2 + 1 = 0$, then $g_K = 0$.*

Proof. Since $y^2 = -\left(x^2 + 1\right)$, we have $m = 2$ and $g = \left[\frac{m+1}{2}\right] - 1 = \left[\frac{3}{2}\right] - 1 = 1 - 1 = 0.$ $\square$

Corollary 4.3.9 (see Example 4.2.2). *If $K = \mathbb{R}(x, y)$ with $x^2 + y^4 + 1 = 0$, then $g = 1$.*

Proof. We have $x^2 = -\left(y^4+1\right)$, so $K = k(y)(x)$ with $m = 4$. Then $g = \left[\frac{m+1}{2}\right] - 1 = \left[\frac{4+1}{2}\right] - 1 = \left[\frac{5}{2}\right] - 1 = 2 - 1 = 1$. □

Corollary 4.3.10. *If $K = \mathbb{R}(x, y)$ where $y^2 = f(x)$ and f is square-free and of degree 2, then $g = 0$.*

Proof. Put $m = 2$ and $g = \left[\frac{2+1}{2}\right] - 1 = 1 - 1 = 0$. □

Corollary 4.3.11 (see Theorem 4.2.4). *If $K = \mathbb{R}(x, y)$ is such that $y^2 = f(x)$, with $f(x)$ square-free and* $\deg f(x) = 3$, *then $g = 1$.*

Proof. $g = \left[\frac{3+1}{2}\right] - 1 = 2 - 1 = 1$. □

Remark 4.3.12. In Proposition 4.1.9, we obtained that if

$$K = \mathbb{R}(x, y) \quad \text{and} \quad x^2 + y^2 + 1 = 0,$$

then K is not a rational function field. Now, if

$$K = \mathbb{C}(x, y) \quad \text{with} \quad x^2 + y^2 + 1 = 0, \quad \text{and} \quad g = 0,$$

then since $\mathbb{C}$ is algebraically closed, K surely is a rational function field. It is natural to ask what the difference is between this and the real case. To answer this question, observe that

$$y^2 = -\left(x^2+1\right) = -(x+i)(x-i) = -(x+i)^2 \frac{x-i}{x+i}.$$

Then

$$y = i(x+i)\sqrt{\frac{x-i}{x+i}},$$

so

$$K = \mathbb{C}(x, y) = \mathbb{C}(x, z),$$

where

$$z = \sqrt{\frac{x-i}{x+i}} \quad \text{or} \quad z^2 = \frac{x-i}{x+i},$$

whence $x = -i\frac{z^2+1}{z^2-1}$, that is, $x \in \mathbb{C}(z)$. Thus $K = \mathbb{C}(z)$. The previous argument would not have been possible with $\mathbb{R}$ in place of $\mathbb{C}$.

4.4 Exercises

Exercise 4.4.1. Let K/k be a function field of genus 0 that is not a rational function field. Prove that there exists a constant extension k'/k of degree 2 such that Kk' is a rational function field.

Exercise 4.4.2. Let $K = \mathbb{R}(x, y)$ with $x^4 + y^2 + 1 = 0$. Prove that every place of K is of degree 2.

Exercise 4.4.3. Let K/k be a function field. Let

$$\varrho := \min\{n \in \mathbb{N} \mid \text{ there exists } \mathfrak{p} \in \mathbb{P}_K, d_K(\mathfrak{p}) = n\},$$

and

$$d := \min\{n \in \mathbb{N} \mid \text{ there exists } \mathfrak{A} \in D_K, d_K(\mathfrak{A}) = n\}.$$

Prove that d divides ϱ and if $g_K = 1$, then $d = \varrho$.

Exercise 4.4.4. Let $K = \mathbb{R}(x, y)$ with $x^n + y^2 + 1 = 0$. Characterize the set of positive integers $n \in \mathbb{N}$ such that every place of K is of degree 2.

Exercise 4.4.5. Let $\operatorname{char} k = 2$ and consider $K = k(x, y)$ given by $x^3 + y^2 + 1 = 0$. Show that $g_K = 0$ and conclude that Corollary 4.3.7 does not hold for characteristic 2.

Exercise 4.4.6. Let $\operatorname{char} k = 2$ and let $f(x) \in k[x]$ be a separable polynomial of degree 3. Let $K = k(x, y)$ be given by $y^2 - y = f(x)$. Show that K contains a prime divisor of degree 1 and $g_K \leq 1$.

Exercise 4.4.7. With the conditions of Exercise 4.4.6, if $f(x)$ is separable of degree 4, what can we say about g_K?

5

Extensions and Galois Theory

This chapter is about the Galois theory of function fields. Many of the results presented here are of a general nature, but our interest and emphasis will be focused on function fields.

Most of our main results are based on the situation in which the constant field k is perfect. When the field of constants is not perfect, strange things may happen, and we shall mention a few of them in Chapter 9.

In Section 5.4 we study the completions of a field extension; as we shall see, the knowledge of extensions of such completions, or in other words the local case, is useful for the study of the global case.

Section 5.5 is dedicated to entire bases, which will be indispensable when we study Tate's genus formula for inseparable extensions in Chapter 9.

We shall consider ramification in cyclic extensions, both Kummer extensions and Artin–Schreier extensions. Moreover, we shall obtain Kummer's theorem on the decomposition type of a prime in an extension.

We end the chapter with ramification groups, which are useful for the study of extensions with wild ramification.

After Chapter 3, which treats the Riemann–Roch theorem, this chapter may be considered as the second in importance of our book, due to the fact that it contains basic concepts and results of the theory such as ramification, decomposition of places, norm, and different.

5.1 Extensions of Function Fields

Definition 5.1.1. Let K/k and L/ℓ be two function fields. We say that L is an *extension* of K if $K \subseteq L$ and $\ell \cap K = k$.

Proposition 5.1.2. *Let L/ℓ be an extension of K/k, and let $x \in K$ be transcendental over k. Then x is transcendental over ℓ.*

Proof. We have $x \in K \setminus k$, so $x \notin K \cap \ell = k$. Thus $x \notin \ell$, that is, $x \in L \setminus \ell$. Therefore x is transcendental over ℓ. $\square$

Definition 5.1.3. Let L be an extension of K. A place $\mathcal{P}$ of L is called *variable* or *trivial* over K if $v_{\mathcal{P}}(x) = 0$ for all $x \in K^*$. This is equivalent to saying that $K \subseteq \vartheta_{\mathcal{P}}$.

If $\mathcal{P}$ is nontrivial over K, then $v_{\mathcal{P}|_K}$ defines a nontrivial valuation in K. In other words, there exists a prime divisor $\wp$ of K such that $v_{\mathcal{P}|_K} \cong v_{\wp}$ (here the symbol $\cong$ is used to mean that the two valuations are equivalent).

Definition 5.1.4. When $\mathcal{P}$ is nontrivial over K, and hence $v_{\mathcal{P}|_K} \cong v_{\wp}$, we say that $\mathcal{P}$ *is over* $\wp$ or that $\mathcal{P}$ *is above* $\wp$ or that $\mathcal{P}$ *divides* $\wp$, and this is denoted by $\mathcal{P} \mid \wp$ or $\mathcal{P}|_K = \wp$.

Consider an extension L of K, $\mathcal{P}$ a nontrivial place of L over K and $\mathcal{P}|_K = \wp$. Since the valuations are discrete and normalized, it follows that $v_{\mathcal{P}} : L^* \longrightarrow \mathbb{Z}$ and $v_{\wp} : K^* \longrightarrow \mathbb{Z}$ are surjective. On the other hand, $v_{\mathcal{P}|_K}$ is not surjective in general, so $v_{\mathcal{P}}(K^*) = e\mathbb{Z}$ for some $e \geq 1$. Thus we have $v_{\mathcal{P}}(x) = e v_{\wp}(x)$ for all $x \in K$.

Definition 5.1.5. The number e obtained above is called *the ramification index of* $\mathcal{P}$ *over* $\wp$ and it is denoted by $e = e(\mathcal{P}|\wp) = e_{L/K}(\mathcal{P}|\wp)$.

Example 5.1.6. Let $K = k(x, y)$ be defined by $y^2 = x$. Let $\mathfrak{P}_0$ be the zero divisor of y. Then $v_{\mathfrak{P}_0}(x) = v_{\mathfrak{P}_0}(y^2) = 2$. Therefore if $\mathfrak{p}_0$ is the zero divisor of x, $\mathfrak{P}_0|_{k(x)} = \mathfrak{p}_0$ and $e(\mathfrak{P}_0|\mathfrak{p}_0) = 2$.

Proposition 5.1.7. *If L/ℓ is any extension of K/k, and $\mathcal{P}$ is a place of L over a place $\wp$ of K, then $k(\wp) = \vartheta_{\wp}/\wp$ can be embedded in a natural way in $\ell(\mathcal{P}) = \vartheta_{\mathcal{P}}/\mathcal{P}$.*

Proof. Since $\mathcal{P}|_K = \wp$, we have $\vartheta_{\mathcal{P}} \cap K = \vartheta_{\wp}$ and $\mathcal{P} \cap K = \wp$. Hence the natural map from $\vartheta_{\wp}/\wp$ to $\vartheta_{\mathcal{P}}/\mathcal{P}$ is a monomorphism of fields. □

Proposition 5.1.8. *Let L/ℓ be an extension of K/k. The following conditions are equivalent:*

(1) $[\ell : k] < \infty$.
(2) $[L : K] < \infty$.
(3) *If $\mathcal{P}$ is any place of L over a place $\wp$ of K, then* $[\ell(\mathcal{P}) : k(\wp)] < \infty$.

Proof. By Theorem 2.4.12 we have $[k(\wp) : k] < \infty$ and $[\ell(\mathcal{P}) : \ell] < \infty$. From

$$[\ell(\mathcal{P}) : k] = [\ell(\mathcal{P}) : k(\wp)]\,[k(\wp) : k] = [\ell(\mathcal{P}) : \ell]\,[\ell : k],$$

it follows that

$$[\ell(\mathcal{P}) : k(\wp)] < \infty \iff [\ell : k] < \infty,$$

which proves the equivalence of (1) and (3).

Now let $x \in K \setminus k$. Then $x \in L \setminus \ell$. By definition we have $[K : k(x)] < \infty$ and $[L : \ell(x)] < \infty$, so

$$[L : k(x)] = [L : K][K : k(x)] = [L : \ell(x)][\ell(x) : k(x)] .$$

Therefore,

$$[L : K] < \infty \iff [\ell(x) : k(x)] < \infty.$$

By Proposition 2.1.6, we have $[\ell(x) : k(x)] = [\ell : k]$, which implies that (1) and (2) are equivalent. □

Similarly, we obtain the following proposition:

Proposition 5.1.9. *Let L/ℓ be an extension of K/k. The following conditions are equivalent:*

(1) ℓ is algebraic over k,
(2) L is algebraic over K,
(3) If $\mathcal{P}$ is a prime divisor of L over the prime divisor $\wp$ of K, then $\ell(\mathcal{P})$ is algebraic over $k(\wp)$.

Proof. Exercise 5.10.8. □

Definition 5.1.10. Let L/K be an extension of function fields, and let $\mathcal{P}$ be a place of L over a place $\wp$ of K. We define the *relative degree of $\mathcal{P}$ over $\wp$* by $d_{L/K}(\mathcal{P}|\wp) = \left[\ell(\mathcal{P}) : k(\wp)\right]$ (which can be finite or infinite).

Proposition 5.1.11. *If $d_L(\mathcal{P}) = [\ell(\mathcal{P}) : \ell]$ and $d_K(\wp) = \left[k(\wp) : k\right]$, then*

$$d_L(\mathcal{P})[\ell : k] = d_{L/K}(\mathcal{P}|\wp)\, d_K(\wp) .$$

Proof. The result follows from the following diagram, which allows us to calculate $[\ell(\mathcal{P}) : k]$ in two different ways.

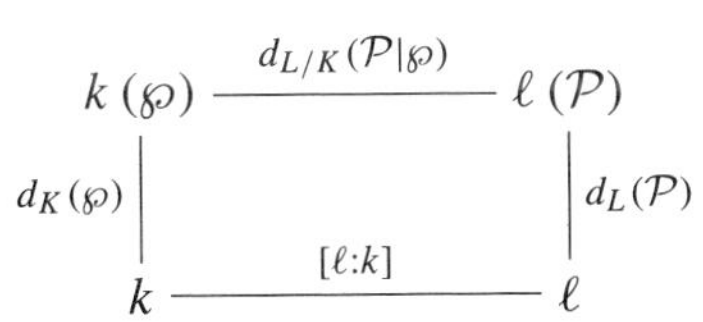

□

Proposition 5.1.12. *If L/ℓ is an algebraic extension of K/k, then no place of L is variable over K.*

Proof. Assume that there exists a valuation v of L that is trivial over K. For each $\alpha \in L$, consider

$$f(x) = x^n + a_{n-1}x^{n-1} + \cdots + a_0 \in K[x],$$

where $f(x)$ is the irreducible polynomial of α. Then

$$\alpha^n + a_{n-1}\alpha^{n-1} + \cdots + a_0 = 0 \quad \text{with} \quad a_i \in K, \quad i = 0, \ldots, n, \quad \text{and} \quad a_0 \neq 0.$$

We have

$$0 = v(a_0) = v\left(-\alpha\left(\alpha^{n-1} + a_{n-1}\alpha^{n-2} + \cdots + a_1\right)\right)$$
$$= v(\alpha) + v\left(\alpha^{n-1} + a_{n-1}\alpha^{n-2} + \cdots + a_1\right).$$

Therefore, if we choose $\alpha \in L$ such that $v(\alpha) > 0$, we obtain

$$v\left(\alpha^{n-1} + a_{n-1}\alpha^{n-2} + \cdots + a_1\right) \geq \min\{(n-1)v(\alpha), \ldots, 0\} = 0.$$

Thus

$$0 = v(\alpha) + v\left(\alpha^{n-1} + a_{n-1}\alpha^{n-2} + \cdots + a_1\right) \geq v(\alpha) > 0,$$

which is impossible. □

Theorem 5.1.13. *Let L/ℓ be an algebraic extension of K/k. Given a place $\wp$ of K, the number of places of L over $\wp$ is finite and nonzero.*

Proof. Let $g = g_K$ be the genus of K and let $C \in C_K$ be the class of the divisor $\wp^{g+1}$. Then

$$d(C) = d_K\left(\wp^{g+1}\right) = (g+1)d_K(\wp) \geq g+1,$$

so

$$N(C) \geq d(C) - g + 1 \geq 2.$$

Hence there exist another integral divisor $\mathfrak{S} \in C$ and $x \in K \setminus k$ such that $\frac{\wp^{g+1}}{\mathfrak{S}} = (x)_K$. Then x is transcendental over k, $v_\wp(x) > 0$, and $v_{\wp'}(x) > 0$ if and only if $\wp' = \wp$. It follows from the definition of extension of function fields that $x \notin \ell$. Now the divisor of x in L is

$$(x)_L = \frac{\mathcal{P}_1^{a_1} \ldots \mathcal{P}_h^{a_h}}{(\mathfrak{N}_x)_L}, \quad \text{with} \quad h \geq 1 \quad \text{and} \quad a_i > 0.$$

We will see that $\mathcal{P}_1, \ldots, \mathcal{P}_h$ are precisely the places of L over $\wp$. If $\mathcal{P}$ is any place of L over $\wp$, we have $v_\mathcal{P}(x) = e(\mathcal{P}|\wp)\, v_\wp(x) > 0$. Therefore $\mathcal{P} \mid (\mathfrak{Z}_x)_L = \mathcal{P}_1^{a_1} \ldots \mathcal{P}_h^{a_h}$, that is, $\mathcal{P} \in \{\mathcal{P}_1, \ldots, \mathcal{P}_h\}$ and conversely. □

The most important arithmetical result in algebraic extensions of function fields is the following formula:

Theorem 5.1.14. *Let L/ℓ be an extension of K/k (finite or infinite). Let $\wp$ be a place of K and let $\mathcal{P}_1, \ldots, \mathcal{P}_h$ be the places of L over $\wp$. Then*

$$[L : K] = \sum_{i=1}^{h} d_{L/K}(\mathcal{P}_i|\wp)\, e_{L/K}(\mathcal{P}_i|\wp).$$

Proof. If $h = \infty$, the result follows immediately. Assume that h is finite. By Proposition 5.1.8, we have

$$[L : K] = \infty \iff d_{L/K}(\mathcal{P}_i|\wp) = \infty \quad \text{for} \quad i = 1, \dots, h.$$

Therefore the formula holds trivially in this case.

Now suppose that $[L : K] < \infty$, and let $x \in K \setminus k$ be such that $(x)_K = \frac{\wp^{g+1}}{\mathfrak{S}}$ for some integral divisor $\mathfrak{S} \neq \wp^{g+1}$. Let

$$A = \sum_{i=1}^{h} d_{L/K}(\mathcal{P}_i|\wp)\, e_{L/K}(\mathcal{P}_i|\wp).$$

We have

$$(x)_L = \frac{(\mathfrak{Z}_x)_L}{(\mathfrak{N}_x)_L} = \frac{\mathcal{P}_1^{a_1} \dots \mathcal{P}_h^{a_h}}{(\mathfrak{N}_x)_L} = \frac{\mathcal{P}_1^{v_{\mathcal{P}_1}(x)} \dots \mathcal{P}_h^{v_{\mathcal{P}_h}(x)}}{(\mathfrak{N}_x)_L}.$$

It follows by Theorem 3.2.7 that

$$\begin{aligned}
[L : \ell(x)] = d_L((\mathfrak{Z}_x)_L) &= \sum_{i=1}^{h} v_{\mathcal{P}_i}(x) d_L(\mathcal{P}_i) \\
&= \sum_{i=1}^{h} v_\wp(x) e_{L/K}(\mathcal{P}_i|\wp)\, d_L(\mathcal{P}_i) \\
&= \frac{v_\wp(x) d_K(\wp)}{[\ell : k]} \sum_{i=1}^{h} d_{L/K}(\mathcal{P}_i|\wp)\, e_{L/K}(\mathcal{P}_i|\wp) \qquad \text{(Proposition 5.1.11)} \\
&= \frac{d_K\left(\wp^{v_\wp(x)}\right)}{[\ell : k]} A = \frac{d_K((\mathfrak{Z}_x)_K)}{[\ell : k]} A = \frac{[K : k(x)]}{[\ell : k]} A \qquad \text{(Theorem 3.2.7).}
\end{aligned}$$

On the other hand, we have

$$[L : \ell(x)] = \frac{[L : K][K : k(x)]}{[\ell(x) : k(x)]} = \frac{[K : k(x)]}{[\ell : k]} [L : K].$$

Hence we obtain $A = [L : K]$. □

Corollary 5.1.15. *With the above notation, we have*

$$h \le [L : K], \quad d_{L/K}(\mathcal{P}_i|\wp) \le [L : K] \quad \text{and} \quad e_{L/K}(\mathcal{P}_i|\wp) \le [L : K]$$

for $i = 1, \dots, h$. □

Proposition 5.1.16. *Consider any tower of function fields of the form $K/k \subseteq L/\ell \subseteq M/m$. For any prime divisor $\mathfrak{P}$ of M that is nontrivial over K, let $\mathcal{P} = \mathfrak{P}\mid_L$ and $\wp = \mathfrak{P}\mid_K = \mathcal{P}\mid_K$. Then*

$$e_{M/K}(\mathfrak{P}|\wp) = e_{M/L}(\mathfrak{P}|\mathcal{P})\, e_{L/K}(\mathcal{P}|\wp)$$

and

$$d_{M/K}(\mathfrak{P}|\wp) = d_{M/L}(\mathfrak{P}|\mathcal{P})\, d_{L/K}(\mathcal{P}|\wp).$$

Proof. If $x \in K^*$, we have $v_{\mathfrak{P}}(x) = e_{M/K}(\mathfrak{P}|\wp)\, v_{\wp}(x)$, and on the other hand,

$$v_{\mathfrak{P}}(x) = e_{M/L}(\mathfrak{P}|\mathcal{P})\, v_{\mathcal{P}}(x) = e_{M/L}(\mathfrak{P}|\mathcal{P})\, e_{L/K}(\mathcal{P}|\wp)\, v_{\wp}(x).$$

Picking $x \in K^*$ such that $v_{\wp}(x) \neq 0$, we obtain the first equality.

Furthermore, we have

$$\begin{aligned} d_{M/K}(\mathfrak{P}|\wp) &= \left[m(\mathfrak{P}) : k(\wp)\right] = \left[m(\mathfrak{P}) : \ell(\mathcal{P})\right]\left[\ell(\mathcal{P}) : k(\wp)\right] \\ &= d_{M/L}(\mathfrak{P}|\mathcal{P})\, d_{L/K}(\mathcal{P}|\wp). \end{aligned}$$

□

5.2 Galois Extensions of Function Fields

We first recall some general results of field theory. Let L/K be an algebraic extension of fields and let $L_s = \{x \in L \mid x \text{ is separable over } K\}$; L_s is called the *separable closure of* K *in* L, L/L_s is purely inseparable, and L_s/K is separable. Furthermore,

$$[L:K]_s = [L_s : K] \qquad \text{separability degree of } L/K,$$

$$[L:K]_i = [L : L_s] \qquad \text{inseparability degree of } L/K,$$

and

$$[L:K] = [L:K]_s\,[L:K]_i\,.$$

Now let

$$L_i = \{x \in L \mid x \text{ is purely inseparable over } K\}.$$

Then L_i is a subfield of L and clearly L_i/K is purely inseparable. However, if L/K is not normal, then L/L_i is not necessarily separable.

Example 5.2.1. If X, T are two variables over $k = \mathbb{F}_2$, consider the fields $K = k\left(T, X^4 + TX^2 + 1\right)$ and $L = k(T, X)$. We leave it to the reader to verify the following assertions: $L_s = k\left(T, X^2\right)$, and $L_i = K$ (see Exercise 5.10.3).

Hence, in this case we have $L_s L_i = L_s \neq L$. In fact, in general we have $L_s L_i = L$ if and only if L/L_i is separable.

Definition 5.2.2. Let $K \subseteq L$ be an arbitrary field extension. We define the group of K-automorphisms of L by

$$\operatorname{Aut}(L/K) = \operatorname{Aut}_K(L) := \{\sigma : L \to L \mid \sigma \text{ is an automorphism}, \sigma|_K = \operatorname{Id}_K\}.$$

If L/K is any Galois extension, we have $\operatorname{Gal}(L/K) = \operatorname{Aut}(L/K)$. If H is any group of automorphisms of a field L, the fixed field of L under H is

$$L^H = \{a \in L \mid \sigma(a) = a \text{ for all } \sigma \in H\}.$$

If $\operatorname{Gal}(L/K)$ is finite, by Artin's theorem, L/L^H is a Galois extension such that $\operatorname{Gal}\left(L/L^H\right) \cong H$.

Now if L/K is any finite normal extension and $G = \operatorname{Aut}(L/K)$, then L/L^G is a Galois extension that is separable, and L^G/K is purely inseparable. In this case, we have $L_i = L^G$ and $L_i L_s = L^G L_s = L$ (compare with Example 5.2.1).

Hence, in the normal case we obtain

$$\left[L : L^G\right] = [L : K]_s \quad \text{and} \quad \left[L^G : K\right] = [L : K]_i.$$

Definition 5.2.3. Assume that L/ℓ is a finite extension of K/k, where L/ℓ and K/k are function fields. If $\mathcal{P}$ is a place of L and $\wp = \mathcal{P}|_K$, we define

$$d_{L/K}(\mathcal{P}|\wp)_i = \left[\ell(\mathcal{P}) : k(\wp)\right]_i$$

and

$$d_{L/K}(\mathcal{P}|\wp)_s = \left[\ell(\mathcal{P}) : k(\wp)\right]_s.$$

A prime divisor $\mathcal{P}$ is called *separable* if $d_{L/K}(\mathcal{P}|\wp)_i = 1$, *inseparable* if $d_{L/K}(\mathcal{P}|\wp)_i > 1$, and *purely inseparable* if $d_{L/K}(\mathcal{P}|\wp) = d_{L/K}(\mathcal{P}|\wp)_i$.

Definition 5.2.4. Let L/ℓ and M/m be two extensions of K/k and let $\sigma : L \longrightarrow M$ be a field isomorphism such that $\sigma(\ell) = m$ and $\sigma|_K = \operatorname{Id}_K$. Then for a place $\mathcal{P}$ of L we define the place $\sigma(\mathcal{P})$ of M by means of the valuation $v_{\sigma\mathcal{P}}$, defined by $v_{\sigma\mathcal{P}}(x) = v_{\mathcal{P}}\left(\sigma^{-1}x\right)$ for all $x \in M$.

Proposition 5.2.5. *If we interpret $\mathcal{P}$ as the maximal ideal of the valuation ring $\vartheta_{\mathcal{P}}$ corresponding to $v_{\mathcal{P}}$, then $\sigma(\mathcal{P})$ is simply the image of $\mathcal{P}$ under σ, that is, $\sigma(\mathcal{P}) = \{\sigma(\alpha) \mid \alpha \in \mathcal{P}\}$.*

Proof. This is clear. □

Proposition 5.2.6. *The map that associates $\sigma(\mathcal{P})$ to each place $\mathcal{P}$ is a permutation of the prime divisors of L and M. Furthermore, we have $\ell(\mathcal{P}) \overset{\sigma}{\cong} m(\sigma\mathcal{P})$ and $\vartheta_{\mathcal{P}} \overset{\sigma}{\cong} \vartheta_{\sigma\mathcal{P}}$. Finally, if $\mathcal{P}$ is over the place $\wp$, then $\sigma(\mathcal{P})$ is over $\wp$ and the isomorphism $\bar{\sigma} : \ell(\mathcal{P}) \xrightarrow{\cong} m(\sigma\mathcal{P})$ is such that $\bar{\sigma}|_{k(\wp)} = \operatorname{Id}_{k(\wp)}$. In particular, we have*

$$d_{L/K}(\mathcal{P}|\wp) = d_{M/K}(\sigma\mathcal{P}|\wp) \quad \text{and} \quad e_{L/K}(\mathcal{P}|\wp) = e_{M/K}(\sigma\mathcal{P}|\wp).$$

Proof. All assertions follow immediately from the definitions. □

Theorem 5.2.7. *Let L/ℓ be a normal finite extension of K/k. Let $\mathcal{P}$ be a place of L over the place $\wp$ of K. Let $\mathcal{P}'$ be any other place of L over $\wp$. Then there exists $\sigma \in G = \operatorname{Aut}(L/K)$ such that $\sigma\mathcal{P} = \mathcal{P}'$. In other words, G acts transitively on the places of L that divide a given place of K.*

Proof. Exercise 5.10.9. □

Definition 5.2.8. Let L/ℓ be a finite normal extension of K/k. If $\mathcal{P}$ is a place of L over $\wp$ of K, we define the *decomposition group* of $\mathcal{P}$ by

$$D(\mathcal{P}|\wp) = D_{L/K}(\mathcal{P}|\wp) = \{\sigma \in \operatorname{Aut}(L/K) \mid \sigma(\mathcal{P}) = \mathcal{P}\}.$$

By Theorem 5.2.7, $G = \operatorname{Aut}(L/K)$ acts transitively on

$$A = \{\mathcal{P} \mid \mathcal{P} \text{ is a prime of } L \text{ such that } \mathcal{P}|_K = \wp\}.$$

Thus

$$|A| = \frac{|G|}{|D(\mathcal{P}|\wp)|}, \text{ which is the number of prime divisors of } L \text{ over } \wp.$$

Proposition 5.2.9. *Let L/ℓ be a finite normal extension of K/k. Let $\sigma \in \operatorname{Aut}(L/K)$. Then $D(\sigma\mathcal{P}|\wp) = \sigma D(\mathcal{P}|\wp)\sigma^{-1}$.*

Proof. We have

$$\begin{aligned}\theta \in D(\sigma\mathcal{P}|\wp) &\iff \theta\sigma\mathcal{P} = \sigma\mathcal{P} \iff \left(\sigma^{-1}\theta\sigma\right)(\mathcal{P}) = \mathcal{P}\\ &\iff \sigma^{-1}\theta\sigma \in D(\mathcal{P}|\wp) \iff \theta \in \sigma D(\mathcal{P}|\wp)\sigma^{-1}.\end{aligned}$$ □

Theorem 5.2.10. *Let L/ℓ be a finite normal extension of K/k. Let $\mathcal{P}$ be a place of L over the place $\wp$ of K. Then $\ell(\mathcal{P})$ is a normal extension of $k(\wp)$. Furthermore, there exists a natural epimorphism from $D(\mathcal{P}|\wp)$ to $\operatorname{Aut}(\ell(\mathcal{P})/k(\wp))$.*

Proof. Let $\mathcal{P} = \mathcal{P}_1, \dots, \mathcal{P}_h$ be all prime divisors of L over $\wp$. Let $\bar{y} \in \ell(\mathcal{P}) = \vartheta_{\mathcal{P}}/\mathcal{P}$, with $y \in \vartheta_{\mathcal{P}}$. Let $y' \in L$ be such that $v_{\mathcal{P}_1}(y - y') > 0$ and $v_{\mathcal{P}_j}(y') > 0$ for all $j = 2, \dots, h$. By the approximation theorem (Corollary 2.5.6), such y' exists. Then $y - y' \in \mathcal{P}$. In particular, we have $y' \in \bar{y}$. Hence, replacing y by y', we may assume that $v_{\mathcal{P}_1}(y) \geq 0$ and $v_{\mathcal{P}_j}(y) > 0$ for $j = 2, \dots, h$.

Let $G = \operatorname{Aut}(L/K)$. We have

$$f(x) = \Big\{\prod_{\sigma \in G}(x - \sigma y)\Big\}^{[L:K]_i} \in \vartheta_{\mathcal{P}}[x] \subseteq K[x].$$

For $\sigma \notin D(\mathcal{P}|\wp)$, we have $\sigma^{-1}\mathcal{P} \neq \mathcal{P}$, so $v_{\mathcal{P}}(\sigma y) = v_{\sigma^{-1}\mathcal{P}}(y) > 0$. Therefore, if we set

$$\overline{f(x)} = f(x) \bmod \wp, \quad \text{then} \quad \sigma \notin D(\mathcal{P}|\wp) \quad \text{implies} \quad \overline{\sigma y} = 0.$$

Thus, we have

$$\overline{f(x)} = \left\{ \prod_{\sigma \in D(\mathcal{P}|\wp)} (x - \overline{\sigma y}) \right\}^{[L:K]_i} x^s, \text{ with } s \in \mathbb{N} \cup \{0\}, \text{ and } \overline{f(x)} \in k(\wp)[x].$$

This implies that $\overline{f(x)}$ has all its roots in $\ell(\mathcal{P})$, and since $\bar{y}$ a root of $\overline{f(x)}$, it follows that $\ell(\mathcal{P})$ is a normal extension over $k(\wp)$.

If $\sigma \in D(\mathcal{P}|\wp)$, we have $\sigma(\mathcal{P}) = \mathcal{P}$ and $\sigma(\vartheta_{\mathcal{P}}) = \vartheta_{\mathcal{P}}$, so $\bar{\sigma}$ is an automorphism of $\ell(\mathcal{P}) = \vartheta_{\mathcal{P}}/\mathcal{P}$. Since $\sigma|_K = \mathrm{Id}$, we have that $\sigma|_{k(\wp)} = \mathrm{Id}_{k(\wp)}$. Thus $\bar{\sigma} \in \mathrm{Aut}(\ell(\mathcal{P})/k(\wp))$.

It is clear that the function

$$D(\mathcal{P}|\wp) \xrightarrow{\varphi} \mathrm{Aut}(\ell(\mathcal{P})/k(\wp)) = H$$

is a group homomorphism. Notice that $\ell(\mathcal{P})$ is a Galois extension over $k_1 = \ell(\mathcal{P})^H \supseteq k(\wp)$. Let $\ell(\mathcal{P}) = k_1(\bar{y})$ with $\bar{y} \in \ell(\mathcal{P})$ and $y \in \vartheta_{\mathcal{P}}$. Clearly, every element of H is uniquely determined by its action on $\bar{y}$. The conjugate elements of $\bar{y}$ are of the form $\bar{\sigma}(\bar{y})$ for some $\sigma \in D(\mathcal{P}|\wp)$ (this follows from the above arguments). That is, every $\theta \in H$ is of the form $\theta = \bar{\sigma}, \sigma \in D(\mathcal{P}|\wp)$. Therefore φ is an epimorphism. □

Definition 5.2.11. The kernel of the natural epimorphism

$$D(\mathcal{P}|\wp) \to \mathrm{Aut}(\ell(\mathcal{P})/k(\wp))$$

is called the *inertia group* of $\mathcal{P}$ over $\wp$, and it is denoted by

$$I(\mathcal{P}|\wp) = I_{L/K}(\mathcal{P}|\wp).$$

We will assume that L/K is a (finite) normal extension for Corollary 5.2.12 up to Corollary 5.2.19.

We have

$$\begin{aligned} I(\mathcal{P}|\wp) &= \left\{\sigma \in D(\mathcal{P}|\wp) \mid \bar{\sigma} = \mathrm{Id}_{\ell(\mathcal{P})}\right\} \\ &= \{\sigma \in D(\mathcal{P}|\wp) \mid \sigma x \equiv x \bmod \mathcal{P} \text{ for all } x \in \vartheta_{\mathcal{P}}\} \\ &= \{\sigma \in \mathrm{Aut}(L/K) \mid \sigma x \equiv x \bmod \mathcal{P} \text{ for all } x \in \vartheta_{\mathcal{P}}\}. \end{aligned}$$

Corollary 5.2.12. $\mathrm{Aut}(\ell(\mathcal{P})/k(\wp))$ *is isomorphic to* $D(\mathcal{P}|\wp)/I(\mathcal{P}|\wp)$. □

Corollary 5.2.13. *If h is the number of places in L over the place $\wp$ of K, we have* $|\mathrm{Aut}(L/K)| = h\,|D(\mathcal{P}|\wp)|$.

Proof. If $G = \mathrm{Aut}(L/K)$, we have $\mathrm{Aut}(L/K) = \mathrm{Gal}\left(L/L^G\right)$. Therefore

$$|G| = \left[L : L^G\right] = [L : K]_s = \frac{|G|}{|D(\mathcal{P}|\wp)|}\,|D(\mathcal{P}|\wp)| = h\,|D(\mathcal{P}|\wp)|. \quad \square$$

Corollary 5.2.14. $\left[D\left(\mathcal{P}|\wp\right) : I\left(\mathcal{P}|\wp\right)\right] = d_{L/K}\left(\mathcal{P}|\wp\right)_s$.

Proof. We have

$$\left[D\left(\mathcal{P}|\wp\right) : I\left(\mathcal{P}|\wp\right)\right] = \left|\operatorname{Aut}\left(\ell\left(\mathcal{P}\right)/k\left(\wp\right)\right)\right| = \left[\ell\left(\mathcal{P}\right) : k\left(\wp\right)\right]_s = d_{L/K}\left(\mathcal{P}|\wp\right)_s. \quad \square$$

Proposition 5.2.15. *With the same conditions as in Theorem* 5.2.10, *we have* $I\left(\sigma\mathcal{P}|\wp\right) = \sigma I\left(\mathcal{P}|\wp\right)\sigma^{-1}$.

Proof. We have

$$\begin{aligned} I\left(\sigma\mathcal{P}|\wp\right) &= \ker\left(D\left(\sigma\mathcal{P}|\wp\right) \longrightarrow \operatorname{Aut}\left(\ell\left(\sigma\mathcal{P}\right)/k\left(\wp\right)\right)\right) \\ &= \ker\left(\sigma D\left(\mathcal{P}|\wp\right)\sigma^{-1} \longrightarrow \operatorname{Aut}\left(\ell\left(\sigma\mathcal{P}\right)/k\left(\wp\right)\right)\right) \\ &= \sigma\left(\ker\left(D\left(\mathcal{P}|\wp\right) \longrightarrow \operatorname{Aut}\left(\ell\left(\mathcal{P}\right)/k\left(\wp\right)\right)\right)\right)\sigma^{-1} = \sigma I\left(\mathcal{P}|\wp\right)\sigma^{-1}. \quad \square \end{aligned}$$

Proposition 5.2.16. *For all* $i = 1, \ldots, h$, *we have*

$$d = d_{L/K}\left(\mathcal{P}|\wp\right) = d_{L/K}\left(\mathcal{P}_i|\wp\right) \quad \text{and} \quad e = e_{L/K}\left(\mathcal{P}|\wp\right) = e_{L/K}\left(\mathcal{P}_i|\wp\right).$$

Proof. Let $\mathcal{P}_i = \sigma\left(\mathcal{P}\right)$. Then

$$\ell\left(\mathcal{P}_i\right) = \ell\left(\sigma\left(\mathcal{P}\right)\right) = \vartheta_{\sigma(\mathcal{P})}/\sigma\mathcal{P} = \bar{\sigma}\left(\vartheta_{\mathcal{P}}/\mathcal{P}\right) \cong \vartheta_{\mathcal{P}}/\mathcal{P} = \ell\left(\mathcal{P}\right).$$

Hence

$$d_{L/K}\left(\mathcal{P}_i|\wp\right) = \left[\ell\left(\mathcal{P}_i\right) : k\left(\wp\right)\right] = \left[\ell\left(\mathcal{P}\right) : k\left(\wp\right)\right] = d_{L/K}\left(\mathcal{P}|\wp\right).$$

If $x \in K^*$ satisfies $v_{\wp}(x) \neq 0$, we have

$$v_{\mathcal{P}_i}(x) = e_{L/K}\left(\mathcal{P}_i|\wp\right)v_{\wp}(x)$$

and

$$v_{\mathcal{P}_i}(x) = v_{\sigma\mathcal{P}}(x) = v_{\mathcal{P}}\left(\sigma^{-1}x\right) = v_{\mathcal{P}}(x) = e_{L/K}\left(\mathcal{P}|\wp\right)v_{\wp}(x).$$

Therefore $e_{L/K}\left(\mathcal{P}_i|\wp\right) = e_{L/K}\left(\mathcal{P}|\wp\right)$. $\square$

Corollary 5.2.17. *We have*

$$[L : K] = edh, \quad \text{where} \quad e = e_{L/K}\left(\mathcal{P}|\wp\right) \quad \text{and} \quad d = d_{L/K}\left(\mathcal{P}|\wp\right).$$

Proof. This is an immediate consequence of Theorem 5.1.14 and Proposition 5.2.16. $\square$

Corollary 5.2.18. *With the notation of the previous corollary, we have*

$$ed = [L : K]_i\left|D\left(\mathcal{P}|\wp\right)\right|.$$

Proof. We have

$$ed = \frac{[L:K]}{h} = \frac{[L:K]}{|\mathrm{Aut}(L/K)|}\,|D(\mathcal{P}|\wp)| \qquad \text{(Corollary 5.2.13)}$$
$$= \frac{[L:K]}{[L:K]_s}\,|D(\mathcal{P}|\wp)| = [L:K]_i\,|D(\mathcal{P}|\wp)|\,. \qquad \square$$

Whenever there is no confusion possible, we will denote $e_{L/K}(\mathcal{P}|\wp)$ by e, $d_{L/K}(\mathcal{P}|\wp)$ by d, $d_{L/K}(\mathcal{P}|\wp)_i$ by d_i, etc.

Corollary 5.2.19. $|I| = \frac{ed_i}{[L:K]_i}$.

Proof.

$$|I| = \frac{|D|}{[D:I]} = \frac{ed}{[L:K]_i d_s} = \frac{ed_i}{[L:K]_i} \qquad \text{(Corollaries 5.2.14 and 5.2.18).} \quad \square$$

Proposition 5.2.20. *If L/K is a separable algebraic extension, then ℓ/k is also a separable extension.*

Proof. If ℓ/k is infinite and not separable, there exists an element α of ℓ that is not separable over k. Thus $k(\alpha)/k$ is inseparable and $[k(\alpha):k] < \infty$. Hence we may assume that ℓ/k is finite.

Next, we may assume that ℓ/k is normal since if $\tilde{\ell}/k$ is the normal closure, then $K\tilde{\ell} \subseteq \tilde{L}$, where $\tilde{L}$ is the Galois closure of L/K. In the case that $\tilde{\ell}/k$ is not separable, we have $\tilde{\ell}_i \neq k$. Therefore there exists $x \in \tilde{\ell}_i \setminus k$ such that $x^{p^t} \in k$ and $p = \operatorname{char} k$. We have $x \in \tilde{L} \setminus K$ and $x^{p^t} \in K$, which is impossible since $\tilde{L}/K$ is separable.

Hence we may assume that ℓ/k is normal. If ℓ/k is not separable, there exists $\alpha \in \ell \setminus k$ such that $\alpha^{p^t} \in k$ for some $t \geq 1$. We have $\alpha \in L$ and since $K \cap \ell = k$, $\alpha \notin K$. This together with $\alpha^{p^t} \in K$ contradicts the separability of L/K. $\square$

Theorem 5.2.21. *If L/K is an algebraic separable extension and the field ℓ of constants of L is a perfect field, then for every place $\mathcal{P}$ of L and $\wp = \mathcal{P}|_K$, $\ell(\mathcal{P})/k(\wp)$ is a separable extension.*

Proof. Since L/K is a separable extension, it follows that $E = K\ell$ is a separable extension of K. Now $\ell \subseteq E \subseteq L$, so the field of constants of E is ℓ. Let $\mathfrak{P} = \mathcal{P}|_E$. Then $k(\wp) \subseteq \ell(\mathfrak{P}) \subseteq \ell(\mathcal{P})$. Since ℓ is a perfect field and $\ell(\mathfrak{P})$ is a finite extension of ℓ (Theorem 2.4.12), $\ell(\mathfrak{P})$ is a perfect field too. Therefore $\ell(\mathcal{P})/\ell(\mathfrak{P})$ is a separable extension, and we may assume that $L = K\ell$.

Let us assume that $\mathcal{P}$ is an inseparable place. Thus $\ell(\mathcal{P})/k(\wp)$ is not separable. Let $y \in L$ be such that $\bar{y} \in \ell(\mathcal{P})$ is an inseparable element over $k(\wp)$. Since $y \in K\ell$ is a finite linear combination of elements of K and ℓ, we have $y \in L_1 = K(\alpha_1, \ldots, \alpha_n)$, where $\alpha_i \in \ell$ and L_1/K is a finite extension. Taking every conjugate of each of the α_i, we may assume that L_1/K is a normal extension. That is, it is a finite Galois extension.

Since $\bar{y}$ is inseparable in $\ell(\mathcal{P})/k(\wp)$, if $\mathcal{P}_1$ is a place of L_1 over $\wp$, then $\bar{y} \in \ell_1(\mathcal{P}_1)$ is inseparable over $k(\wp)$, where ℓ_1 is the field of constants of L_1. We have (Corollary 5.2.19)

$$|I(\mathcal{P}_1|\wp)| = e_{L_1/K}(\mathcal{P}_1|\wp)\, d_{L_1/K}(\mathcal{P}_1|\wp)_i \geq d_{L_1/K}(\mathcal{P}_1|\wp)_i > 1.$$

Thus there exists $\sigma \in I = I(\mathcal{P}_1|\wp)$, with $\sigma \neq \mathrm{Id}$. Since L_1/K is normal, it follows that ℓ_1/k is normal (see Exercise 5.10.20).

Now we have $L_1 = K(\alpha_1, \ldots, \alpha_n)$, with $\alpha_i \in \ell_1 \subseteq \vartheta_{\mathcal{P}_1}$, and

$$\sigma(\alpha_i) \equiv \alpha_i \bmod \mathcal{P}_1 \quad \text{for} \quad i = 1, \ldots, n.$$

Equivalently,

$$v_{\mathcal{P}_1}(\alpha_i - \sigma(\alpha_i)) > 0 \quad \text{for} \quad i = 1, \ldots, n.$$

Since all α_i and $\sigma(\alpha_i)$ are constants, it follows that $v_{\mathcal{P}_1}(\alpha_i - \sigma(\alpha_i)) > 0$ implies $\alpha_i = \sigma(\alpha_i)$. Therefore $\sigma = \mathrm{Id}$. □

Remark 5.2.22. If ℓ is not a perfect field in Theorem 5.2.21, then there may exist inseparable places (see Exercise 5.10.18 and Theorem 5.2.33).

Corollary 5.2.23. *Let L/K be a finite separable normal extension, i.e., a Galois extension. Assume that the field ℓ of constants of L is a perfect field. If $\mathcal{P}$ is a place of L, put $\wp = \mathcal{P}|_K$, $e = e_{L/K}(\mathcal{P}|\wp)$ and $d = d_{L/K}(\mathcal{P}|\wp)$; let h be the number of places of L over $\wp$, $I = I_{L/K}(\mathcal{P}|\wp)$, and $D = D_{L/K}(\mathcal{P}|\wp)$. Then*

$$[L:K] = edh, \quad |D| = ed, \quad |I| = e, \quad \text{and} \quad [D:I] = d.$$

Proof. By Proposition 5.2.20 and Theorem 5.2.21, all inseparability degrees are equal to 1. The result follows using Corollaries 5.2.14, 5.2.17, 5.2.18, and 5.2.19. □

For the purely inseparable case we have the following theorem:

Theorem 5.2.24. *Let L/ℓ be a finite purely inseparable field extension of K/k. Then for each place $\wp$ of K, there exists a unique place $\mathcal{P}$ of L such that $\mathcal{P}|_K = \wp$. Furthermore, if $p = \operatorname{char} k$, then $e_{L/K}(\mathcal{P}|\wp) = p^t$ for some $t \geq 0$. Finally, $\ell(\mathcal{P})/k(\wp)$ is purely inseparable.*

Proof. Let $y \in L$. There exists $n \in \mathbb{N}$ such that $y_0 = y^{p^n} \in K$. Let $\mathcal{P}$ be any place of L over $\wp$, so that

$$p^n v_{\mathcal{P}}(y) = v_{\mathcal{P}}\left(y^{p^n}\right) = v_{\mathcal{P}}(y_0) = e_{L/K}(\mathcal{P}|\wp)\, v_{\wp}(y_0).$$

Therefore, if $\mathcal{P}_1$ and $\mathcal{P}_2$ are two places of L over $\wp$, and if we choose y such that $v_{\mathcal{P}_1}(y) \neq 0$, then $v_{\mathcal{P}_2}(y) \neq 0$, $v_{\wp}(y_0) \neq 0$, and

$$v_{\mathcal{P}_1}(y) = v_{\mathcal{P}_2}(y) = \frac{e_{L/K}(\mathcal{P}|\wp)\, v_\wp(y_0)}{p^n}.$$

Thus $v_{\mathcal{P}_1} = v_{\mathcal{P}_2}$, which means that $\mathcal{P}_1 = \mathcal{P}_2$.

Now if $y \in L$ is such that $v_{\mathcal{P}}(y) = 1$, then $p^n = e_{L/K}(\mathcal{P}|\wp)\, v_\wp(y_0)$. Hence $e = e_{L/K}(\mathcal{P}|\wp) \mid p^n$, which implies that $e = p^t$ for some $t \geq 0$.

Finally, if $\alpha \in \ell(\mathcal{P})$, let $y \in \vartheta_{\mathcal{P}}$ be such $y \bmod \mathcal{P} = \alpha$. Then $y^{p^t} \in K$, so $\alpha^{p^t} \in k(\wp)$. Thus $\ell(\mathcal{P})/k(\wp)$ is purely inseparable. □

Example 5.2.25. Let k be an algebraically closed field of characteristic $p > 0$ and let x be a transcendental element over k. Set

$$y = x^p, \quad K = k\left(x^p\right) = k(y), \quad \text{and} \quad L = k(x).$$

Let $\wp$ be a place of K. If $\wp$ is the infinite place, then

$$(y)_K = \frac{\wp_0}{\wp} \quad \text{and} \quad (y)_L = \left(x^p\right)_L = (x)_L^p = \frac{\mathfrak{P}_0^p}{\mathfrak{P}_\infty^p}.$$

Thus $\wp$ is ramified.

If $\wp$ is not the infinite place, there exists $a \in k$ such that

$$(y-a)_K = \frac{\wp}{\wp_\infty}.$$

We have

$$(y-a)_L = \left(x^p - \left(a^{1/p}\right)^p\right)_L = \left(\left(x - a^{1/p}\right)\right)_L^p = \frac{\mathfrak{P}^p}{\mathfrak{P}_\infty^p}.$$

Therefore $\wp$ is ramified. Hence every place of K is ramified in L/K.

We will see that the phenomenon of the previous example can occur only in inseparable extensions.

In fact, we have the following corollary:

Corollary 5.2.26. *Let L/ℓ be a purely inseparable finite extension of K/k. If k is a perfect field, then every place $\wp$ of K is fully ramified in L.*

Proof. Let $\mathcal{P}$ be any place of L that divides $\wp$. We have

$$[L:K] = he(\mathcal{P}|\wp) f(\mathcal{P}|\wp).$$

Now since $\ell(\mathcal{P})/k$ is separable and $\ell(\mathcal{P})/k(\wp)$ is purely inseparable, it follows that $\ell(\mathcal{P}) = k(\wp)$ and

$$h = 1 \quad \text{and} \quad f(\mathcal{P}|\wp) = [\ell(\mathcal{P}) : k(\wp)] = 1.$$

Thus $e(\mathcal{P}|\wp) = [L:K]$. □

Definition 5.2.27. In any extension L/ℓ of K/k, a place $\mathcal{P}$ of L is called *ramified* if $e = e_{L/K}(\mathcal{P}|\wp) > 1$, where $\wp = \mathcal{P}|_K$. Also, we say that $\wp$ *is ramified* in L/K.

When L/K is an infinite extension, by $e > 1$ we will mean that $e > 1$ in some finite subextension.

Proposition 5.2.28. *Let $K \subseteq L \subseteq E$ be a tower of function fields with $[E : K] < \infty$, and let $\mathfrak{P}$ be a place of E. Let $\mathcal{P} := \mathfrak{P}|_L$ and $\wp := \mathfrak{P}|_K$. Then $\mathfrak{P}$ is ramified in E/K if and only if $\mathcal{P}$ is ramified in E/L or $\wp$ is ramified in L/K.*

Proof. The statement follows from Proposition 5.1.16. □

Definition 5.2.29. Let L/ℓ be an extension of K/k. We say that L/K is a *constant extension* if $L = K\ell$, and that L/K is a *geometric extension* if $\ell = k$.

Remark 5.2.30. Given a function field K/k and an extension ℓ of k such that $\ell \cap K = k$, the field of constants of $L = K\ell$ may contain ℓ properly.

Example 5.2.31. Let k_0 be a field of characteristic $p > 0$, and u, v be two elements that are algebraically independent over k_0. Let $k = k_0(u, v)$ and x be a variable over k. Let

$$K = k(x, y) \quad \text{be such that} \quad y^p = ux^p + v.$$

Let k' be the field of constants of K. Then $[K : k(x)]$ is equal to 1 or p. We will see that $k' = k$. If $k' \neq k$, then $[k' : k] = [k'(x) : k(x)] \mid [K : k(x)]$, that is, $[k' : k] = p$ and $K = k'(x)$. Therefore $y = u^{1/p}x + v^{1/p} \in k'(x)$, so $u^{1/p}, v^{1/p} \in k'$ and

$$\begin{aligned} p = [k' : k] &\geq \left[k\left(u^{1/p}, v^{1/p}\right) : k\right] \\ &= \left[k\left(u^{1/p}, v^{1/p}\right) : k\left(u^{1/p}\right)\right]\left[k\left(u^{1/p}\right) : k\right] = pp = p^2, \end{aligned}$$

which is absurd. Whence, we have $k' = k$.

Let $\ell_0 = k\left(v^{1/p}\right)$ and $L = K\ell_0$. Then

$$\ell_0 \cap K = k \quad \text{and} \quad u^{1/p} = \frac{y - v^{1/p}}{x} \in K\ell_0 = L.$$

Therefore the field ℓ of constants of L contains ℓ_0 properly since

$$\ell \supseteq k\left(u^{1/p}, v^{1/p}\right) \supsetneqq \ell_0.$$

In Chapter 8 we will study the general constant extension $L = K\ell$.

Theorem 5.2.32. *Let L/ℓ be an algebraic separable extension of K/k and assume that $L = K\ell$. That is, L is an extension of constants of K. Then no place of L is ramified or inseparable over K.*

Proof. For the sake of contradiction, let $\mathcal{P}$ be a ramified or inseparable place of L and let $\wp := \mathcal{P}|_K$. If $\mathcal{P}$ is ramified, choose $y \in L$ such that $v_{\mathcal{P}}(y) = 1$. Since y is of the form $\frac{\sum_{i=1}^{n} \alpha_i x_i}{\sum_{j=1}^{m} \beta_j z_j}$ with $\alpha_i, \beta_j \in \ell$ and $x_i, z_j \in K$, y must lie in a finite extension $K(\gamma_1, \ldots, \gamma_r)$ of K with $\gamma_i \in \ell$. By adding the conjugates of the elements γ_i, $1 \leq i \leq r$, we may assume that $L_1 = K(\gamma_1, \ldots, \gamma_r)$ is a finite normal separable extension of K. Let $\mathfrak{P} := \mathcal{P}|_{L_1}$. Since $y \in L_1$, we have $v_{\mathfrak{P}}(y) = v_{\mathcal{P}}(y)$, so $\mathcal{P}|\mathfrak{P}$ is unramified and it follows that $\mathfrak{P}$ is ramified over K. If $\mathcal{P}$ is inseparable, pick $\bar{y} \in \ell(\mathcal{P})$ inseparable over $k(\wp)$. Since $\bar{y} \in \ell_1(\mathfrak{P})$, we have $\operatorname{Irr}(\bar{y}, T, \ell(\mathcal{P})) = \operatorname{Irr}(\bar{y}, T, \ell_1(\mathfrak{P}))$, so $\bar{y}$ is inseparable over $k(\wp)$.

Thus we may assume that $L = K\ell$ is a finite Galois extension over K. Therefore $|I| = |I(\mathcal{P}|\wp)| = e_{L/K}(\mathcal{P}|\wp) d_{L/K}(\mathcal{P}|\wp)_i > 1$.

Let $\sigma \in I$ with $\sigma \neq \mathrm{Id}$. Since $\sigma(\gamma_i) \equiv \gamma_i \bmod \mathcal{P}$ for all $1 \leq i \leq r$, we have $v_{\mathcal{P}}(\sigma\gamma_i - \gamma_i) > 0$. Finally, $\gamma_i \in \ell$, so we obtain that $\sigma\gamma_i = \sigma_i$ for all $1 \leq i \leq r$. Hence $\sigma = \mathrm{Id}$. □.

Theorem 5.2.33. *Let L/ℓ be an algebraic separable extension of K/k. Then there are at most finitely many prime divisors of L that are ramified or inseparable.*

Proof. First assume that L/K is a finite Galois extension. We have $L = K(z) = K\left(\frac{1}{z}\right)$ for some $z \in L$. Let $\mathcal{P}$ be a place of L. Then z or $\frac{1}{z}$ belongs to the valuation ring of $\mathcal{P}$. Therefore

$$\begin{aligned}
&\mathcal{P} \text{ ramified or inseparable}\\
&\quad \iff |I| = |I(\mathcal{P}|\wp)| = e_{L/K}(\mathcal{P}|\wp)\, d_{L/K}(\mathcal{P}|\wp)_i > 1\\
&\quad \iff \text{there exists } \sigma \in I, \sigma \neq \mathrm{Id} \iff\\
&\quad \iff v_{\mathcal{P}}(\sigma(z) - z) > 0 \text{ when } z \in \vartheta_{\mathcal{P}} \text{ or } v_{\mathcal{P}}\left(\frac{1}{\sigma(z)} - \frac{1}{z}\right) > 0 \text{ when } \frac{1}{z} \in \vartheta_{\mathcal{P}}.
\end{aligned}$$

Now since $|\operatorname{Gal}(L/K)| < \infty$, there are only finitely many places satisfying $v_{\mathcal{P}}(\sigma(z) - z) > 0$ or $v_{\mathcal{P}}\left(\frac{1}{\sigma(z)} - \frac{1}{z}\right) > 0$, namely, only the divisors appearing in the support of $(\sigma(z) - z)_L$ or in the support of $\left(\frac{1}{\sigma(z)} - \frac{1}{z}\right)_L$, where $\sigma \in G, \sigma \neq \mathrm{Id}$.

When L/K is a finite separable extension, we take the Galois closure $\tilde{L}$. Since the theorem holds for $\tilde{L}/K$, it also holds for L/K.

Now let L/K be an arbitrary algebraic separable extension. Let $x \in K \setminus k$. Then $x \notin \ell$, so $L/\ell(x)$ is a finite extension. Since $K\ell \supseteq \ell(x)$, it follows that $L/K\ell$ is a finite extension. Therefore the theorem holds for $L/K\ell$. Finally, by Theorem 5.2.32 there are no places in $K\ell/K$ that are ramified or inseparable, so the theorem holds for L/K. □

Definition 5.2.34. A field k is called *separably closed* if any algebraic extension k'/k is purely inseparable. Any separably closed field is infinite.

Corollary 5.2.35. *If k is a separably closed field and K/k is separably generated, that is, there exists $x \in K \setminus k$ such that $K/k(x)$ is separable, then K contains infinitely many divisors of degree 1 and there exist nonspecial systems in K.*

Proof. Let $x \in K \setminus k$ be such that $K/k(x)$ is a finite separable extension. Since k is separably closed, k is infinite. Thus $k(x)$ contains infinitely many prime divisors of degree 1 (for any $a \in k$, $(x-a)_{k(x)} = \frac{\mathcal{P}_a}{\mathcal{P}_\infty}$, where $\mathcal{P}_a$ is a prime divisor of degree 1). By Theorem 5.2.33 there exist finitely many inseparable prime divisors in K over $k(x)$. If $\wp$ is a separable prime divisor of K, then $k(\wp)/k$ is separable and thus $k(\wp) = k$. Therefore if $\wp$ is above a prime divisor of degree 1 in $k(x)$, $\wp$ is of degree 1.

Finally, the existence of nonspecial systems in K follows immediately from the proof of Lemma 3.5.13 (see also Proposition 3.5.16). □

5.3 Divisors in an Extension

Given a finite extension L/ℓ of K/k we want to define a group monomorphism

$$\varphi : D_K \longrightarrow D_L \quad \text{such that} \quad \varphi(P_K) \subseteq P_L,$$

that is, $\varphi((x)_K) = (x)_L$.

If $(x)_K = \prod_{i=1}^{m} \wp_i^{v_{\wp_i}(x)}$, we have

$$(x)_L = \prod_{i=1}^{m} \prod_{j=1}^{h_i} \mathcal{P}_{ij}^{e_{ij} v_{\wp_i}(x)} = \prod_{i=1}^{m} \prod_{j=1}^{h_i} \mathcal{P}_{ij}^{v_{\mathcal{P}_{ij}}(x)},$$

where $e_{ij} = e_{L/K}\left(\mathcal{P}_{ij}|\wp_i\right)$ and for $i = 1, \ldots, m$, the $\mathcal{P}_{ij}$'s ($1 \leq j \leq h_i$) are all the places of L over $\wp_i$. This justifies the following definition:

Definition 5.3.1. Let $\varphi : D_K \longrightarrow D_L$ be defined on the set of generators of D_K by $\varphi(\wp) = \prod_{i=1}^{h} \mathcal{P}_i^{e_i}$, where $e_i = e_{L/K}(\mathcal{P}_i|\wp)$, $\wp$ is a place of K, and $\mathcal{P}_1, \ldots, \mathcal{P}_h$ are all the places of L that are above $\wp$. Then φ extends in a natural way to D_K.

More precisely, if $\mathfrak{A} = \prod_{i=1}^{m} \wp_i^{v_{\wp_i}(\mathfrak{A})}$, then $\varphi(\mathfrak{A}) = \prod_{i=1}^{m} \prod_{j=1}^{h_i} \mathcal{P}_{ij}^{e_{ij} v_{\wp_i}(\mathfrak{A})} = \prod_{i=1}^{m} \prod_{j=1}^{h_i} \mathcal{P}_{ij}^{v_{\mathcal{P}_{ij}}(\mathfrak{A})}$.

The function φ is called the *conorm* of K to L, and it is denoted by $\operatorname{con}_{K/L}$.

From the definition we have the following result:

Proposition 5.3.2. *The map $\operatorname{con}_{K/L}$ is a monomorphism from D_K to D_L such that $\operatorname{con}_{K/L}(P_K) \subseteq P_L$ and such that if $x \in K^*$, then $\operatorname{con}_{K/L}((x)_K) = (x)_L$. Finally, $\operatorname{con}_{K/L}$ induces a group homomorphism $\overline{\operatorname{con}_{K/L}} : C_K \longrightarrow C_L$.* □

We will see later that in fact, $\operatorname{con}_{K/L}\left(D_{K,0}\right) \subseteq D_{L,0}$.

Remark 5.3.3. Observe that $\overline{\mathrm{con}_{K/L}}$ is not necessarily injective (see Exercise 5.10.21). Also, since $\mathrm{con}_{K/L}$ is injective, we will assume that $D_K \subseteq D_L$.

Theorem 5.3.4. *Let L/K be an arbitrary extension of function fields. There exists $\lambda_{L/K} \in \mathbb{Q}$ such that $\lambda_{L/K} > 0$, $\lambda_{L/K}$ depends only on L and K, and for all $\mathfrak{A} \in D_K$,*

$$d_L(\mathfrak{A}) = \frac{d_K(\mathfrak{A})}{\lambda_{L/K}}.$$

In particular, $d_L(\mathfrak{A}) = 0$ if and only if $d_K(\mathfrak{A}) = 0$. Therefore $\mathrm{con}_{K/L}$ induces a group homomorphism

$$\overline{\mathrm{con}_{K/L}} : C_{K,0} \longrightarrow C_{L,0}.$$

Finally, if $[L : K] < \infty$, then $\lambda_{L/K} = \frac{[\ell:k]}{[L:K]}$.

Proof. Since d_L and d_K are group homomorphisms, it suffices to prove our assertions for a place $\wp$ of K.

First, assume that $[L : K] < \infty$ and let $\wp = \mathcal{P}_1^{e_1} \cdots \mathcal{P}_h^{e_h}$. Since $\mathrm{con}_{K/L}(\wp) = \mathcal{P}_1^{e_1} \cdots \mathcal{P}_h^{e_h}$ with $e_i = e_{L/K}(\mathcal{P}_i|\wp)$, we have

$$\begin{aligned} d_L(\wp) &= \sum_{i=1}^{h} e_i d_L(\mathcal{P}_i) \\ &= \sum_{i=1}^{h} e_i d_{L/K}(\mathcal{P}_i|\wp) \frac{d_K(\wp)}{[\ell : k]} && \text{(Proposition 5.1.11)} \\ &= \frac{d_K(\wp)}{[\ell : k]} \sum_{i=1}^{h} e_i d_{L/K}(\mathcal{P}_i|\wp) = \frac{d_K(\wp)}{[\ell : k]}[L : K] && \text{(Theorem 5.1.14).} \end{aligned}$$

Therefore $\lambda_{L/K} = \frac{[\ell:k]}{[L:K]}$.

Now let L/K be an arbitrary extension. To finish the proof of the theorem, it suffices to show that for any prime divisors $\mathfrak{A}, \mathfrak{B} \in D_K$ (of degree different from 0),

$$\frac{d_L(\mathfrak{A})}{d_K(\mathfrak{A})} = \frac{d_L(\mathfrak{B})}{d_K(\mathfrak{B})} > 0.$$

Indeed, $\lambda_{L/K}$ can then be defined as $\dfrac{d_K(\mathfrak{A})}{d_L(\mathfrak{A})}$ for any prime divisor $\mathfrak{A}$.

Assume that there are two places $\mathfrak{A}, \mathfrak{B}$ of K such that

$$\frac{d_L(\mathfrak{A})}{d_K(\mathfrak{A})} < \frac{d_L(\mathfrak{B})}{d_K(\mathfrak{B})}, \quad \text{that is,} \quad \frac{d_L(\mathfrak{A})}{d_L(\mathfrak{B})} < \frac{d_K(\mathfrak{A})}{d_K(\mathfrak{B})}.$$

Let $\frac{n}{m} \in \mathbb{Q}$ be such that

$$\frac{d_L(\mathfrak{A})}{d_L(\mathfrak{B})} < \frac{n}{m} < \frac{d_K(\mathfrak{A})}{d_K(\mathfrak{B})}. \tag{5.1}$$

Then from (5.1) we obtain, for $t \in \mathbb{N}$ large enough,

$$d_K\left(\mathfrak{A}^{mt}\mathfrak{B}^{-nt}\right) = t\left(md_K(\mathfrak{A}) - nd_K(\mathfrak{B})\right) > 2g_K - 1$$

and

$$d_L\left(\mathfrak{A}^{mt}\mathfrak{B}^{-nt}\right) = t\left(md_L(\mathfrak{A}) - nd_L(\mathfrak{B})\right) < 0.$$

By the Riemann–Roch theorem (Corollary 3.5.6) there exists $x \in K$ such that $x \in L_K\left(\mathfrak{A}^{-mt}\mathfrak{B}^{nt}\right)$. We have $(x)_K = \mathfrak{C}\mathfrak{A}^{-mt}\mathfrak{B}^{nt}$, where $\mathfrak{C}$ is an integral divisor. Therefore

$$d_L\left((x)_L\right) = d_L(\mathfrak{C}) - t\left(md_L(\mathfrak{A}) - nd_L(\mathfrak{B})\right) > 0$$

for t large enough. This contradicts Corollary 3.2.9 and proves the theorem. □

Let L/ℓ be a finite extension of K/k and let L_1 be the normal closure of L/K. Let ℓ_1 be the algebraic closure of ℓ in L_1, so ℓ_1 is the field of constants of L_1. Put $G = \operatorname{Aut}(L_1/K)$ and $H = \operatorname{Aut}(L_1/L) \subseteq G$. Consider the set G/H of left cosets of H in G.

Definition 5.3.5. We define the *norm* of $y \in L$ over K as

$$N_{L/K}(y) = \Big\{\prod_{\bar{\sigma}\in G/H} \sigma y\Big\}^{[L:K]_i} = \Big\{\prod_{\sigma\in T} \sigma y\Big\}^{[L:K]_i},$$

where $T = \{\sigma : L \longrightarrow L_1 \text{ monomorphism with } \sigma|_K = \mathrm{Id}\}$.

We have $|T| = [L:K]_s = \frac{[L_1:K]_s}{[L_1:L]_s} = \frac{|G|}{|H|}$.

Clearly, $\left\{\prod_{\sigma\in T}\sigma y\right\} \in L_1^G$, and this implies $\left\{\prod_{\sigma\in T}\sigma y\right\}^{[L:K]_i} \in K$. Therefore $N_{L/K}(y) \in K$.

Definition 5.3.6. We define the *norm* of D_L in D_K to be the function $N_{L/K} : D_L \longrightarrow D_K$ defined by $N_{L/K}(\mathfrak{A}) = \left\{\prod_{\bar{\sigma}\in G/H}\sigma\mathfrak{A}\right\}^{[L:K]_i}$.

In the above definition, what is meant by $\sigma\mathfrak{A}$ is $\{\sigma a \mid a \in \mathfrak{A}\} \subseteq L_1$. We will see that in fact, $N_{L/K}(\mathfrak{A}) \in D_K$, or, more precisely, $N_{L/K}(\mathfrak{A}) = \operatorname{con}_{K/L_1}(\mathfrak{B})$ for some $\mathfrak{B} \in D_K$.

Theorem 5.3.7. *The norm N defined above is multiplicative and satisfies:*

(1) *For all $\mathfrak{A} \in D_L$, $N_{L/K}(\mathfrak{A}) \in D_K$; more precisely, there exists $\mathfrak{B} \in D_K$ such that $N_{L/K}(\mathfrak{A}) = \operatorname{con}_{K/L_1}(\mathfrak{B})$.*
(2) *If $\mathcal{P}$ is a prime divisor of L over the prime divisor $\wp$ of K, we have $N_{L/K}(\mathcal{P}) = \wp^d$, where $d = d_{L/K}(\mathcal{P}|\wp)$.*
(3) *For all $y \in L$, $N_{L/K}((y)_L) = \left(N_{L/K}(y)\right)_K$.*
(4) *If $\mathfrak{A} \in D_K$, then $N_{L/K}(\mathfrak{A}) = \mathfrak{A}^{[L:K]}$, or, more precisely,*

$$N_{L/K}\left(\operatorname{con}_{K/L}(\mathfrak{A})\right) = \mathfrak{A}^{[L:K]}.$$

(5) *If $M \supseteq L \supseteq K$ is a tower of fields, we have $N_{M/K} = N_{L/K} \circ N_{M/L}$.*

Proof. It is clear that N is multiplicative.

(2) Let L_1 be the normal closure of L/K, $G = \operatorname{Aut}(L_1/K)$, $H = \operatorname{Aut}(L_1/L) \subseteq G$, and let $\mathfrak{S}$ be a prime divisor of L_1 over $\mathcal{P}$. Let $Z_K = D_{L_1/K}(\mathfrak{S}|\wp) \subseteq G$ and $Z_L = D_{L_1/L}(\mathfrak{S}|\mathcal{P}) \subseteq H$. Since L_1/L is a normal extension, it follows by Theorem 5.2.7 or Proposition 5.2.16 that

$$(\mathcal{P})_{L_1} = \operatorname{con}_{L/L_1}(\mathcal{P}) = \Big\{\prod_{\bar{\sigma}\in H/Z_L} \sigma\mathfrak{S}\Big\}^e \in D_{L_1}, \quad e = e_{L_1/L}(\mathfrak{S}|\mathcal{P}).$$

Then

$$\begin{aligned}
N_{L/K}(\mathcal{P})^{|Z_L|} &= \prod_{\bar{\theta}\in G/H} \theta\Big\{\Big(\prod_{\bar{\sigma}\in H/Z_L} \sigma\mathfrak{S}\Big)^{e|Z_L|[L:K]_i}\Big\} \\
&= \prod_{\bar{\theta}\in G/H} \theta\Big\{\prod_{\bar{\sigma}\in H/Z_L} (\sigma\mathfrak{S})^{|Z_L|}\Big\}^{e[L:K]_i} \\
&= \prod_{\bar{\theta}\in G/H} \theta\Big\{\Big(\prod_{\sigma\in H} \sigma\mathfrak{S}\Big)^{e[L:K]_i}\Big\} = \Big(\prod_{\bar{\theta}\in G/H}\prod_{\sigma\in H} \theta\sigma\mathfrak{S}\Big)^{e[L:K]_i} \\
&= \Big(\prod_{\delta\in G} \delta\mathfrak{S}\Big)^{e[L:K]_i} = \Big(\prod_{\bar{\delta}\in G/Z_K} (\delta\mathfrak{S})^{|Z_K|}\Big)^{e[L:K]_i} \\
&= \Big\{\Big(\prod_{\bar{\delta}\in G/Z_K} \delta\mathfrak{S}\Big)^{e_{L_1/K}(\mathfrak{S}|\mathcal{P})}\Big\}^{\frac{[L:K]_i|Z_K|}{e_{L/K}(\mathcal{P}|\wp)}} = \big(\operatorname{con}_{K/L_1}\wp\big)^r,
\end{aligned}$$

with $r = \frac{[L:K]_i|Z_K|}{e_{L/K}(\mathcal{P}|\wp)}$.
We have

$$\begin{aligned}
\frac{[L:K]_i\,|Z_K|}{e_{L/K}(\mathcal{P}|\wp)} &= \frac{[L_1:K]_i}{[L_1:L]_i}\frac{\big|D_{L_1/K}(\mathfrak{S}|\wp)\big|}{e_{L_1/K}(\mathfrak{S}|\wp)} e_{L_1/L}(\mathfrak{S}|\mathcal{P}) = \\
&\qquad\qquad \text{(Proposition 5.1.16)} \\
&= \frac{[L_1:K]_i\,\big|D_{L_1/K}(\mathfrak{S}|\wp)\big|}{e_{L_1/K}(\mathfrak{S}|\wp)}\frac{e_{L_1/L}(\mathfrak{S}|\mathcal{P})}{[L_1:L]_i} \\
&= \frac{d_{L_1/K}(\mathfrak{S}|\wp)}{d_{L_1/L}(\mathfrak{S}|\mathcal{P})}\big|D_{L_1/L}(\mathfrak{S}|\mathcal{P})\big| \qquad \text{(Corollary 5.2.18)} \\
&= d_{L/K}(\mathcal{P}\mid\wp)\,|Z_L| \qquad \text{(Proposition 5.1.16).}
\end{aligned}$$

Therefore, if $d = d_{L/K}(\mathcal{P}|\wp)$ we have obtained $N_{L/K}(\mathcal{P})^{|Z_L|} = \wp^{d|Z_L|}$, which implies that $N_{L/K}(\mathcal{P}) = \wp^d$.

(1) This is an immediate consequence of (2).

(4) Since $\sigma(\mathfrak{A}) = \mathfrak{A}$ for all $\sigma \in G$, it follows that $N_{L/K}(\mathfrak{A})$ $= \left(\prod_{\bar{\sigma}\in G/H} \sigma\mathfrak{A}\right)^{[L:K]_i} = \mathfrak{A}^{[L:K]_s [L:K]_i)} = \mathfrak{A}^{[L:K]}$.
(3) We have

$$N_{L/K}((y)_L) = \Big(\prod_{\bar{\sigma}\in G/H} \sigma((y)_L)\Big)^{[L:K]_i} = \Big(\prod_{\bar{\sigma}\in G/H} (\sigma(y))_L\Big)^{[L:K]_i}$$
$$= \Big(\Big(\prod_{\bar{\sigma}\in G/H} (\sigma(y))\Big)^{[L:K]_i}\Big)_K = \left(N_{L/K}(y)\right)_K .$$

(5) It suffices to prove the statement for a prime divisor $\mathfrak{P}$ of M. The result follows immediately from (2) and from Proposition 5.1.16. □

Corollary 5.3.8. *For $\mathfrak{A} \in D_L$, we have $d_K\left(N_{L/K}\mathfrak{A}\right) = [\ell : k]\, d_L(\mathfrak{A})$.*

Proof. Since the degree and the norm maps are multiplicative, it suffices to prove the statement for a prime divisor $\mathfrak{A}$. In this case $\mathfrak{A} = \mathcal{P}$ is a prime divisor of L, $\wp = \mathcal{P}|_K$, and we have

$$\begin{aligned} d_K\left(N_{L/K}\mathcal{P}\right) &= d_K\left(\wp^{d_{L/K}(\mathcal{P}|\wp)}\right) \qquad \text{(Proposition 5.1.11)} \\ &= d_{L/K}(\mathcal{P}|\wp)\, d_K(\wp) = [\ell : k]\, d_L(\mathcal{P}) . \end{aligned}$$
□

Corollary 5.3.9. *The norm map $N_{L/K}$ induces in a natural way maps*

$$N_{L/K} : C_L \to C_K \quad \textit{and} \quad N_{L/K} : C_{L,0} \to C_{K,0}.$$

Furthermore, we have

$$N_{L/K} \circ \overline{\operatorname{con}}_{K/L}(C) = C^n, \quad \text{where} \quad n = [L : K], \quad \text{and} \quad C \in C_K .$$

Proof. By Theorem 5.3.7, we have $N_{L/K}(P_L) \subseteq P_K$. Hence $N_{L/K}$ induces in a natural way the homomorphism $N_{L/K} : C_L = D_L/P_L \longrightarrow D_K/P_K = C_K$, and by Corollary 5.3.8 we obtain $N_{L/K}\left(D_{L,0}\right) \subseteq D_{K,0}$.

Finally, $N_{L/K} \circ \overline{\operatorname{con}}_{K/L}(C) = C^n$ follows by Theorem 5.3.7 (4). □

5.4 Completions and Galois Theory

Consider a finite extension L/K of function fields. For a place $\wp$ of K, let $\mathcal{P}_1, \ldots, \mathcal{P}_h$ be all the places of L over $\wp$. We will denote by $K_\wp$ the completion of K with respect to the valuation $v_\wp$ and by $L_{\mathcal{P}_i}$, $1 \le i \le h$, the completion of L with respect to the valuation $v_{\mathcal{P}_i}$.

For $1 \le i \le h$, let L_i be the topological field with underlying set L and the topology given by $v_{\mathcal{P}_i}$, $1 \le i \le h$. Observe that in spite of having the same underlying set, for $i \ne j$ the identity map is not a homeomorphism from L_i to L_j since $v_{\mathcal{P}_i}$

and $v_{\mathcal{P}_j}$ are inequivalent valuations (L_i and L_j might be, in some cases, topologically isomorphic, under an isomorphism different from the identity). On the other hand, K is considered with the topology given by $v_\wp$. Thus $K \subseteq L_i$ in both the algebraic and the topological sense.

Since $L_{\mathcal{P}_i}$ is the completion of L_i and $K \subseteq L_i$, it follows immediately that $K_\wp \subseteq L_{\mathcal{P}_i}$. The inclusion $K_\wp \subseteq L_{\mathcal{P}_i}$ means that when we obtain L_i by means of Cauchy sequences, we obtain a natural injection $L_i \xrightarrow{\lambda} L_{\mathcal{P}_i}$ defined by $\lambda(\alpha) = \left[\{\alpha_n\}_{n=1}^{\infty}\right]$, where $\alpha_n = \alpha$ for all n. Thus $\lambda(K)$ is a subfield of $L_{\mathcal{P}_i}$ and the closure $\overline{\lambda(K)}$ in $L_{\mathcal{P}_i}$ is a complete field containing $\lambda(K)$. Clearly the latter is a minimal complete field containing $\lambda(K)$. Therefore $\overline{\lambda(K)}$ is the completion of $\lambda(K) \cong K$, and $\overline{\lambda(K)} \cong K_\wp$ (algebraically and topologically). This is the meaning of the inclusion $K_\wp \subseteq L_{\mathcal{P}_i}$.

As we remarked above, the L_{p_i}'s are not necessarily topologically isomorphic. Furthermore, in some cases, they are not even algebraically isomorphic, and what is more they may satisfy $\left[L_{\mathcal{P}_i} : K_\wp\right] \neq \left[L_{\mathcal{P}_j} : K_\wp\right]$ for some pair of indices $i \neq j$. The reason that this phenomenon can happen is that in fact, we may have more than one minimal extension containing both L and $K_\wp$. Of course this would not occur if $K_\wp$ and L were both contained in a larger field, in which case $L_{\mathcal{P}_i}$ would be the subfield generated by $K_\wp$ and L.

In order to clarify why the fields $L_{\mathcal{P}_i}$ can be quite different, we present briefly the theory of composition of fields.

Definition 5.4.1. Let K be an arbitrary field and let E/K and L/K be two extensions of K. By a *composition* of the fields E and L we mean a triple (M, φ, σ), where M is a field containing K, and $\varphi : E \longrightarrow M$ and $\sigma : L \longrightarrow M$ are field monomorphisms such that $\sigma|_K = \varphi|_K = \mathrm{Id}_K$ and M is generated by $\varphi(E)$ and $\sigma(L)$.

Remark 5.4.2. When E and L are contained in a field Ω, unless otherwise stated, we will understand the composite $EL \subseteq \Omega$ as the minimum subfield of Ω containing E and L.

Definition 5.4.3. Two compositions (M, φ, σ), $\left(M', \varphi', \sigma'\right)$ of E/K and L/K are called *equivalent* if there exists an isomorphism $\lambda : M \longrightarrow M'$ such that

$$\lambda \circ \varphi = \varphi' \quad and \quad \lambda \circ \sigma = \sigma'$$

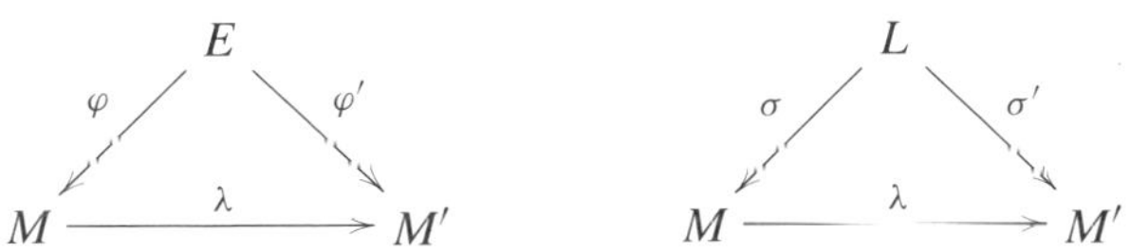

The above relation defines an equivalence relation. The problem now consists in determining all its equivalence classes. Even though Definitions 5.4.1 and 5.4.3 apply to the general case, for our purposes we will study only the case of a finite extension.

Consider L/K such that $[L : K] = n < \infty$ and let E/K be an arbitrary extension. Let (M, φ, σ) be a composition of E and L. Put $E' = \varphi(E)$, $L' = \sigma(L)$, and let $E'L' = \left\{ \sum_{i=1}^{r} e_i \ell_i \mid e_i \in E', \ell_i \in L', r \in \mathbb{N} \right\}$.

Clearly $E'L'$ is a subalgebra of M/K. Since $E'L' \subseteq M$, and M is a field, $E'L'$ is an integral domain. On the other hand, if $\{\alpha_1, \ldots, \alpha_n\}$ is a basis of L/K, the set $\{\sigma(\alpha_1), \ldots, \sigma(\alpha_n)\}$ generates $E'L'/E'$. Since E' is a field, it follows that $E'L'$ is a field. Therefore $E'L' = M$. Now, let

$$\theta : E \otimes_K L \to M \quad \text{be defined by} \quad \theta(e \otimes_K \ell) = \varphi(e)\,\sigma(\ell).$$

Clearly θ is a K-epimorphism and M is isomorphic to $(E \otimes_K L)/\ker\theta$. Since M is a field, $\mathfrak{M} = \ker\theta$ is a maximal ideal. Furthermore, K is isomorphic to $K \otimes_K K$ and $\theta|_K = \mathrm{Id}_K$, so $\mathfrak{M} \cap K = (0)$. Observe that the homomorphisms

$$E \xrightarrow{i} E \otimes_K L, \quad i(e) = e \otimes_K 1,$$

$$L \xrightarrow{j} E \otimes_K L, \quad j(\ell) = 1 \otimes_K \ell,$$

are injective since $\theta \circ i$ and $\theta \circ j$ are injective homomorphisms and $(\theta \circ i)(E) = \varphi(E)$, $(\theta \circ j)(L) = \sigma(L)$. Hence $\mathfrak{M} \cap E = \mathfrak{M} \cap L = (0)$. Furthermore, since $\mathfrak{M}$ is a maximal ideal, $\mathfrak{M}$ has no units. This implies the following theorem:

Theorem 5.4.4. *Let K be an arbitrary field and let E/K, L/K be two extensions of K. Then the equivalence classes of compositions of E with L over K are in a bijective correspondence with the maximal ideals of the K-algebra $E \otimes_K L$. In particular, the composition of fields always exists.*

Proof. We already have seen that to each composition corresponds a maximal ideal. Conversely, let $\mathfrak{M}$ be a maximal ideal of $E \otimes_K L$ and let M be the field $(E \otimes_K L)/\mathfrak{M}$. Define

$$E \xrightarrow{i} (E \otimes_K L)/\mathfrak{M} \text{ by } i(e) = (e \otimes_K 1) + \mathfrak{M}$$

and

$$L \xrightarrow{j} (E \otimes_K L)/\mathfrak{M} \text{ by } j(\ell) = (1 \otimes_K \ell) + \mathfrak{M}.$$

Since $\mathfrak{M}$ is a maximal ideal, it does not contain units, so i and j are injective, that is, i and j are monomorphisms and clearly M is generated by $i(E)$ and $j(L)$. Furthermore, $i|_K = j|_K = \mathrm{Id}_K$. Therefore (M, i, j) is a composition of E and L.

Now let (M, φ, σ) and $\left(M', \varphi', \sigma'\right)$ be two compositions with

$$M \cong (E \otimes_K L)/\mathfrak{M} \quad \text{and} \quad M' \cong (E \otimes_K L)/\mathfrak{M}'.$$

If M and M' are equivalent, then there exists an isomorphism

$$\lambda : M \to M' \quad \text{such that} \quad \lambda \circ \varphi = \varphi' \quad \text{and} \quad \lambda \circ \sigma = \sigma'.$$

Let $\sum_{i=1}^{r} e_i \otimes_K \ell_i \in \mathfrak{M}$. We have the implications

$$\sum_{i=1}^{r} \varphi(e_i)\,\sigma(\ell_i) = 0 \quad \text{in} \quad M$$

$$\begin{aligned}
&\Longrightarrow \lambda\left(\sum_{i=1}^{r} \varphi(e_i)\,\sigma(\ell_i)\right) = \sum_{i=1}^{r} (\lambda\varphi)(e_i)\,(\lambda\sigma)(\ell_i) \\
&\qquad = \sum_{i=1}^{r} \varphi'(e_i)\,\sigma'(\ell_i) = 0 \quad \text{in} \quad M' \\
&\Longrightarrow \sum_{i=1}^{r} e_i \otimes_K \ell_i \in \mathfrak{M}'.
\end{aligned}$$

Hence $\mathfrak{M} \subseteq \mathfrak{M}'$. Since both ideals are maximal, it follows that $\mathfrak{M} = \mathfrak{M}'$.

Conversely, let $\mathfrak{M} = \mathfrak{M}'$. Then if θ and θ' are the isomorphisms from $(E \otimes_K L)/\mathfrak{M}$ to M and M' respectively,

$$\begin{array}{ccc}
(E \otimes_K L)/\mathfrak{M} & \longrightarrow & (E \otimes_K L)/\mathfrak{M}' \\
\downarrow{\scriptstyle\theta} & & \downarrow{\scriptstyle\theta'} \\
M & \underset{\theta'\theta^{-1}}{\longrightarrow} & M'
\end{array}$$

then $\lambda = \theta'\theta^{-1}$ is the isomorphism from M to M' and

$$\begin{aligned}
\varphi = \theta \circ i \quad \varphi' &= \theta' \circ i = \theta' \circ \theta^{-1} \circ \theta \circ i = \lambda \circ \varphi, \\
\sigma = \theta \circ j\text{'} \quad \sigma' &= \theta' \circ j = \theta' \circ \theta^{-1} \circ \theta \circ j = \lambda \circ \sigma,
\end{aligned}$$

which gives equivalent extensions. □

The next result states that the number of maximal ideals in $E \otimes_K L$ is finite.

Theorem 5.4.5. *Let T be a field, and A an algebra over T such that A has finite dimension and an identity element. Then A contains a finite number of maximal ideals.*

Proof. Let $\dim_T A = n < \infty$ and $\mathfrak{M}_1, \dots, \mathfrak{M}_r$ be distinct maximal ideals of A. Let $\mathfrak{N} = \bigcap_{i=1}^{r} \mathfrak{M}_i$. By the Chinese remainder's theorem, we have

$$A/\mathfrak{N} \cong \bigoplus_{i=1}^{r} A/\mathfrak{M}_i.$$

Observe that $A/\mathfrak{N}$ and $A/\mathfrak{M}_i$ are T algebras. Furthermore,

$$n = \dim_T A \geq \dim_T A/\mathfrak{N} = \sum_{i=1}^{r} \dim_T A/\mathfrak{M}_i \geq r.$$

Thus $r \leq n$. □

Corollary 5.4.6. *If K is any field, and E/K and L/K are extensions of K such that $[L : K] = n < \infty$, then the number of composition classes of E and L over K is finite. In fact, the number of such composition classes is less than or equal to n.*

Proof. It is clear that if $\{\alpha_1, \ldots, \alpha_n\}$ is a basis of L/K, then $1 \otimes_K \alpha_1, \ldots, 1 \otimes_K \alpha_n$ generate $E \otimes_K L$ over E. That is, $\dim_E (E \otimes_K L) \leq \dim_K L = n$. The result follows by Theorem 5.4.5. □

We consider the case that $L = K(\theta)$ is a finite simple extension of K. Then $L \cong K[x]/(f(x))$, where $f(x) = \operatorname{Irr}(\theta, x, K)$. Let

$$f(x) = p_1(x)^{e_1} \cdots p_r(x)^{e_r} \in E[x]$$

be the composition of $f(x)$ as a product of irreducible factors in $E[x]$. We have

$$\begin{aligned} E \otimes_K L &\cong (E \otimes_K (K[x]/(f(x)))) \cong (E \otimes_K K[x])/(f(x)) \\ &\cong E[x]/(f(x)) \cong \bigoplus_{i=1}^{r} \left(E[x]/\left(p_i(x)^{e_i}\right)\right). \end{aligned}$$

The compositions of E with L over K are given by $E[x]/(p_i(x))$, since the maximal ideals of $E[x]/(f(x))$ are precisely $(p_i(x))/(f(x))$, $1 \leq i \leq r$.

We have the equalities

$$\dim_E (E \otimes_K L) = \deg f(x) = \sum_{i=1}^{r} e_i \deg p_i(x) = [L : K].$$

Let θ_i be a root of $p_i(x)$ for $i = 1, \ldots, r$. When L/K is a separable extension, we have $e_i = 1$, and

$$E \otimes_K L \cong \bigoplus_{i=1}^{r} (E[x]/(p_i(x))) \cong \bigoplus_{i=1}^{r} E(\theta_i)$$

is the direct sum of all the compositions of E with L over K.

Now we return to our main concern.

Theorem 5.4.7. *Let K be a complete field with respect to a valuation v and let L/K be a finite extension of fields. Then there exists a unique extension w of v to L. Furthermore, L is complete.*

Proof. The existence of w follows from Corollary 2.4.6. Let $|\ |_K$ and $|\ |_L$ be the corresponding absolute values. Let $\alpha \in L^*$ and $\beta = \alpha^n / N(\alpha)$, where $n = [L : K]$ and N denotes the norm of L in K. Then $N(\beta) = \frac{N(\alpha^n)}{N(\alpha)^n} = \frac{N(\alpha)^n}{N(\alpha)^n} = 1$.

We claim that if $\gamma \in L$ is such that $|\gamma|_L < 1$, then $|N(\gamma)|_K < 1$. Indeed, let $|\gamma|_L < 1$ and set

$$\gamma^t = x_1^{(t)} \omega_1 + \cdots + x_n^{(t)} \omega_n,$$

where $\{\omega_1, \dots, \omega_n\}$ is a basis of L/K. Since $|\gamma|_L < 1$, it follows that

$$\gamma^t \xrightarrow[t\to\infty]{} 0, \quad \text{so that} \quad x_i^{(t)} \xrightarrow[t\to\infty]{} 0 \quad \text{for} \quad 1 \le i \le n.$$

Now $N\left(\gamma^t\right)) = (N(\gamma))^t$ is a homogeneous polynomial in $x_1^{(t)}, \dots, x_n^{(t)}$, so

$$N\left(\gamma^t\right) \xrightarrow[t\to\infty]{} 0, \quad \text{which implies that} \quad |N(\gamma)|_K < 1.$$

Similarly, $|\gamma|_L > 1$ implies $|N(\gamma)|_K > 1$.

This shows that $|\beta|_L = 1$ whenever $N(\beta) = 1$. Hence,

$$1 = |\beta|_L = \frac{|\alpha|_L^n}{|N(\alpha)|_L} \Longrightarrow |\alpha|_L = \sqrt[n]{|N(\alpha)|_L} = \sqrt[n]{|N(\alpha)|_K}.$$

We have proved that the extension of the absolute value is unique. □

We now consider function fields L/ℓ and K/k such that $[L : K] = n$. We wish to show that if $\wp$ is a place of K and $\mathcal{P}_1, \dots, \mathcal{P}_h$ are all the places of L over $\wp$, then the result obtained from the discussion after Corollary 5.4.6 holds in the case that L/K is not simple.

Theorem 5.4.8. *Let $e_i = e_{L/K}(\mathcal{P}_i \mid \wp)$ and $f_i = d_{L/K}(\mathcal{P}_i \mid \wp)$. Then*

$$\left[L_{\mathcal{P}_i} : K_{\wp}\right] = e_i f_i.$$

Proof. Let π_K be a prime element of K and let π_L be a prime element of L. Then $v_L(\pi_K) = e = e_i$. By Theorem 2.5.20 and Proposition 2.3.10, we have $k' = k(\wp) = \vartheta_{\wp}/\wp \cong \hat{\vartheta}_{\wp}/\hat{\wp}$ and $\ell' = \ell(\mathcal{P}_i) = \vartheta_{\mathcal{P}_i}/\mathcal{P}_i \cong \hat{\vartheta}_{\mathcal{P}_i}/\hat{\mathcal{P}}_i$. Thus $K_{\wp} = S((\pi_K))$ and $L_{\mathcal{P}_i} = T((\pi_L))$, where S and T are fields such that $S \cong k'$ and $T \cong \ell'$.

Since $v_L\left(\pi_L^s\right) = s < e$ for $s = 1, \dots, e-1$, we have $\left[K_{\wp}(\pi_L) : K_{\wp}\right] \ge e$. Now assume $f = f_i = \left[\ell' : k'\right]$. Since $L_{\mathcal{P}_i} = K_{\wp}(\pi_L)\,T$, it follows that $\left[L_{\mathcal{P}_i} : K_{\wp}\right] \ge ef$.

On the other hand, L is dense in $L_{\mathcal{P}_i}$ and $L_{\mathcal{P}_i}$ is a complete field that is a finite extension of $K_{\wp}$. It follows that $L_{\mathcal{P}_i}$ must be the composition of the fields L and $K_{\wp}$ over K. By the proof of Theorem 5.4.5 (and also by Corollary 5.4.6 and Theorem 5.1.14), we have

$$[L : K] = n \ge \dim_{K_{\wp}}\left(L \otimes_K K_{\wp}\right) \ge \sum_{i=1}^h \dim_{K_{\wp}} L_{\mathcal{P}_i} \ge \sum_{i=1}^h e_i f_i = n.$$

Therefore these inequalities must be in fact equalities. In particular, $\left[L_{\mathcal{P}_i} : K_{\wp}\right] = e_i f_i$. □

Corollary 5.4.9. *With the notation above, we have $\left(L \otimes_K K_{\wp}\right) \cong \bigoplus_{i=1}^h L_{\mathcal{P}_i}$.*

Proof. For each $1 \leq i \leq h$, there exists a maximal ideal $\mathfrak{M}_i$ such that $L_{\mathcal{P}_i}$ is isomorphic to $(L \otimes_K K_\wp)/\mathfrak{M}_i$. Therefore, if $\mathfrak{N} = \bigcap_{i=1}^h \mathfrak{M}_i$,

$$(L \otimes_K K_\wp)/\mathfrak{N} \cong \bigoplus_{i=1}^h (L \otimes_K K_\wp)/\mathfrak{M}_i \cong \bigoplus_{i=1}^h L_{\mathcal{P}_i}.$$

On the other hand, since $\dim_{K_\wp} \left(L \otimes_K K_\wp\right) = n = \sum_{i=1}^h \left[L_{\mathcal{P}_i} : K_\wp\right]$, it follows that $\mathfrak{N} = (0)$. □

As a consequence of the fields $L_{\mathcal{P}_i}$ being exactly the distinct compositions of L with $K_\wp$ over K, we have the following:

Theorem 5.4.10. *Let L/ℓ be a finite extension of K/k. Let $\wp$ be a place of K and let $\mathcal{P}_1, \ldots, \mathcal{P}_h$ be the places of L over $\wp$. If L/K is separable, then $L_{\mathcal{P}_i}/K_\wp$ is separable for all $i = 1, \ldots, h$. If L/K is normal, then $L_{\mathcal{P}_i}/K_\wp$ is normal for $i = 1, \ldots, h$. Finally, if L/K is Galois then $L_{\mathcal{P}_i}/K_\wp$ is Galois and* $\operatorname{Gal}\left(L_{\mathcal{P}_i}/K_\wp\right) \cong D_{L/K}\left(\mathcal{P}_i|\wp\right)$.

Proof. If L/K is separable (normal) ((Galois)), then $L_{\mathcal{P}_i} = K_\wp L$ is separable (normal) ((Galois)) over $K_\wp$.

If L/K is Galois, then clearly $D_{L/K}\left(\mathcal{P}_i|\wp\right) \subseteq \operatorname{Gal}\left(L_{\mathcal{P}_i}/K_\wp\right)$. By Corollary 5.2.18 and Theorem 5.4.8, we have

$$\left|D_{L/K}\left(\mathcal{P}_i|\wp\right)\right| = e_i f_i = \left[L_{\mathcal{P}_i} : K_\wp\right] = \left|\operatorname{Gal}\left(L_{\mathcal{P}_i}/K_\wp\right)\right|.$$

It follows that $D_{L/K}\left(\mathcal{P}|\wp\right) \cong \operatorname{Gal}\left(L_{\mathcal{P}_i}/K_\wp\right)$. □

5.5 Integral Bases

We will use the results of this section in the study of the Tate genus formula for inseparable extensions in Chapter 9.

Let K/k be any function field.

Proposition 5.5.1. *Let $x \in K \backslash k$ and let R be the ring of elements of K that do not have any pole outside the set of zeros of x. Then there exists a finite subset $\{\omega_1, \ldots, \omega_m\}$ of R that contains a basis of K over $k(x)$ and such that every element of R is a linear combination of $\omega_1, \ldots, \omega_m$ with coefficients in $k[x^{-1}]$, that is, $R = \sum_{i=1}^m k[x^{-1}]\omega_i$.*

Proof. By definition we have

$$R = \{y \in K \mid v_{\mathfrak{q}}(y) \geq 0 \ \forall\, \mathfrak{q} \in \mathbb{P}_K, \mathfrak{q} \nmid \mathfrak{z}_x\} = \bigcup_{s=0}^{\infty} L_K(\mathfrak{z}_x^{-s}).$$

Set $n = [K : k(x)] = d(\mathfrak{z}_x)$. Let $\{u_1, \ldots, u_n\}$ be any basis of $K/k(x)$. Then for any $1 \leq i \leq n$, there exists a relation

$$u_i^n = \sum_{j=0}^{n-1} c_{ij} u_i^j, \quad \text{with} \quad c_{ij} \in k(x), \quad i = 1, \dots, n, \quad j = 0, \dots, n-1.$$

Define $c_{ij} = \frac{a_{ij}}{b_i}$ with $a_{ij}, b_i \in k[x^{-1}]$. Let $\omega_i := b_i u_i$. Then

$$\omega_i^n = b_i^n u_i^n = \sum_{j=0}^{n-1} b_i^n c_{ij} u_i^j = \sum_{j=0}^{n-1} a_{ij} b_i^{n-j-1} \omega_i^j$$

with $a_{ij} b_i^{n-j-1} \in k[x^{-1}]$.

Therefore ω_i is integral over $k[x^{-1}] \subseteq k(x)$ (see the proof of Theorem 3.2.7) and since $\omega_i = b_i u_i$, $b_i \in k[x^{-1}]$, and $\{u_1, \dots, u_n\}$ is a basis of $K/k(x)$, it follows that $\{\omega_1, \dots, \omega_n\}$ is a basis of K over $k(x)$.

Let $\mathfrak{P}_1, \dots, \mathfrak{P}_h$ be the prime divisors dividing $\mathfrak{Z}_x$. Let $r \in \mathbb{Z}$ be such that $r \geq 0$ and

$$v_{\mathfrak{P}_j}(\omega_i) \geq -r \quad \text{for all} \quad 1 \leq j \leq h \quad \text{and} \quad 1 \leq i \leq n.$$

Choose an integer M such that $M > r$ and consider $x^{-t}\omega_i$ for $0 \leq t \leq M - r$ and $1 \leq i \leq n$. Then

$$\begin{aligned} v_{\mathfrak{P}_j}(x^{-t}\omega_i) &= -t v_{\mathfrak{P}_j}(x) + v_{\mathfrak{P}_j}(\omega_i) \geq -t v_{\mathfrak{P}_j}(x) - r \\ &\geq -t v_{\mathfrak{P}_j}(x) - r v_{\mathfrak{P}_j}(x) \geq -M v_{\mathfrak{P}_j}(x). \end{aligned}$$

Thus

$$x^{-t}\omega_i \in L_K(\mathfrak{Z}_x^{-M}) =: \mathcal{L}_M.$$

Let $\mathcal{L}'_M$ be the k-vector space generated by $\left\{x^{-t}\omega_i\right\}_{0 \leq t \leq M-r}^{1 \leq i \leq n}$. As in the proof of Theorem 3.2.7, we have $\dim_k \mathcal{L}'_M = (M - r + 1)n$ and

$$\ell_K(\mathfrak{Z}_x^{-M}) \leq \ell_K(\mathfrak{Z}_x) + d(\mathfrak{Z}_x) - d(\mathfrak{Z}_x^{-M}) = (M+1)d(\mathfrak{Z}_x) = (M+1)n.$$

Therefore

$$\dim_k \mathcal{L}_M - \dim_k \mathcal{L}'_M \leq rn$$

for all $M \in \mathbb{Z}$. Put

$$a = \max_{M \in \mathbb{Z}} \{\dim_k \mathcal{L}_M - \dim_k \mathcal{L}'_M\}.$$

Let $z_1, \dots, z_b \in R$ be such that their residue classes modulo $\sum_{i=1}^n k[x^{-1}]\omega_i$ are linearly independent over k. Let $M > 0$ be such that $z_1, \dots, z_b \in \mathcal{L}_M$. Then any nontrivial k-linear combination of $\{z_1, \dots, z_b\}$ does not belong to $\mathcal{L}'_M$ since $\mathcal{L}'_M \subseteq \sum_{i=1}^n k[x^{-1}]\omega_i$. It follows that $b \le a$.

Thus, there exist elements $\omega_{n+1}, \dots, \omega_m$ (with $m-n \le a$) such that every element of R belongs to

$$\sum_{i=1}^n k[x^{-1}]\omega_i + \sum_{j=n+1}^m k\omega_i.$$

This proves the proposition. □

Now we consider a finite extension L/ℓ of K/k. Let $\mathfrak{p}$ be a prime divisor of K and let $\{\mathfrak{P}_1, \dots, \mathfrak{P}_n\}$ be the places of L above $\mathfrak{p}$. Let $\{y_1, \dots, y_n\}$ be a basis of L over K.

Proposition 5.5.2. *Let R be the ring of elements of L which do not have any pole outside $\{\mathfrak{P}_1, \cdots, \mathfrak{P}_n\}$. There exists a nonzero element u of K depending only on $\mathfrak{p}$ and on the basis $\{y_1, \dots, y_n\}$ such that if*

$$y = \sum_{i=1}^n x_i y_i \quad \text{with} \quad x_i \in K, \quad 1 \le i \le n, \quad \text{and} \quad y \in R,$$

then

$$ux_i \in \bigcup_{s=0}^{\infty} L_K(\mathfrak{p}^{-s}) = \Gamma \ \text{ for all } i = 1, \dots, n.$$

Proof. Let $x \in K \setminus k$ be such that $\mathfrak{p}$ is the only zero of x (Corollary 3.5.8). Let $\{\omega_1, \dots, \omega_N\} \subseteq K$ be such that $R = \sum_{i=1}^N \ell[x^{-1}]\omega_i$.

Since $[L : K] < \infty$, it follows that $[\ell(x) : k(x)] = [\ell : k] < \infty$. Let $\{v_1, \dots, v_m\}$ be a basis of ℓ over k. Then every element of $\ell[x^{-1}]$ can be written as a linear combination of $v_1, \dots, v_m$ with coefficients in $k[x^{-1}]$. Thus

$$R = \sum_{j=1}^m \sum_{i=1}^N k[x^{-1}]v_j\omega_i. \tag{5.2}$$

Let

$$v_j\omega_i = \sum_{t=1}^n a_{jit} y_t \quad \text{with} \quad a_{jit} \in K \quad \text{for all} \quad j, i, t, \tag{5.3}$$

and let $\mathfrak{q}_1, \dots, \mathfrak{q}_r$ be the places of K such that the poles of $\{a_{jit}\}$ are contained in $\{\mathfrak{q}_1, \dots, \mathfrak{q}_r\}$. Put

$$M_\nu = \min_{j,i,t} v_{\mathfrak{q}_\nu}(a_{jit})$$

and

$$d_K(\mathfrak{p}^s \prod_{\nu=1}^r \mathfrak{q}_\nu^{M_\nu}) \xrightarrow[s\to\infty]{} \infty.$$

By the Riemann–Roch theorem, there exists $u \in K \setminus k$ such that $u \in L_K(\mathfrak{p}^{-s} \prod_{\nu=1}^r \mathfrak{q}_\nu^{-M_\nu})$ for $s \gg 0$.

We have $v_{\mathfrak{q}_\nu}(ua_{jit}) = v_{\mathfrak{q}_\nu}(u) + v_{\mathfrak{q}_\nu}(a_{jit}) \geq -M_\nu + v_{\mathfrak{q}_\nu}(a_{jit}) \geq 0$, so the pole divisor of ua_{jit} is $\mathfrak{p}^{s_0}$ for some $s_0 \geq 0$ and $ua_{jit} \in \Gamma$.

Now, $k[x^{-1}] \subseteq \Gamma$, so if $y = \sum_{t=1}^n x_t y_t \in R$, then by (5.2) and (5.3) we have $x_t = \sum_{j=1}^m \sum_{i=1}^N z_{ij} a_{jit}$ with $z_{ij} \in k[x^{-1}] \subseteq \Gamma$ and $ua_{jit} \in \Gamma$. Hence $ux_t \in \Gamma$. □

Similarly we will prove the following result:

Proposition 5.5.3. *Let $\mathfrak{p}$ and $\{y_1, \ldots, y_n\}$ be as in Proposition 5.5.2. There exists a nonzero element v of K, depending only on $\mathfrak{p}$ and on the basis $\{y_1, \ldots, y_n\}$, such that if $y = \sum_{i=1}^n x_i y_i$ with $x_i \in K$, $1 \leq i \leq n$, and y satisfies*

$$y \in \vartheta := \bigcap_{j=1}^h \vartheta_{\mathfrak{P}_j} = \{\xi \in L \mid v_{\mathfrak{P}_j}(\xi) \geq 0, 1 \leq j \leq h\},$$

then $vx_i \in \vartheta_{\mathfrak{p}}$ for all $1 \leq i \leq n$.

Proof. Let $\mathfrak{p}'$ be a place of K such that $\mathfrak{p}' \neq \mathfrak{p}$. Put $\Gamma' := \bigcup_{s=0}^\infty L_K((\mathfrak{p}')^{-s})$ and $R' := \bigcup_{i=1}^{h'} \bigcup_{n=0}^\infty L_L((\mathfrak{P}'_i)^{-n})$, where $\mathfrak{P}'_1, \ldots \mathfrak{P}'_{h'}$ are the places of L above $\mathfrak{p}'$.

Let $\{\mathfrak{B}_1, \ldots, \mathfrak{B}_r\}$ be the set of poles $\mathfrak{B}_i$ of y such that $\mathfrak{B}_i$ lies above some place of K distinct from $\mathfrak{p}'$. Let $\mathfrak{q}_j := \mathfrak{B}_j|_K$, $1 \leq j \leq r$. For each j, we take $m_j \geq 0$ such that if $\xi \in K$, then $v_{\mathfrak{q}_j}(\xi) \geq m_j \Rightarrow v_{\mathfrak{B}_j}(\xi y) \geq 0$. For any $M > 0$,

$$\xi \in L_K\left((\mathfrak{p}')^{-M} \prod_{j=1}^r \mathfrak{q}_j^{m_j}\right) \tag{5.4}$$

implies $\xi y \in R'$.

Choose M to be large enough such that if $\mathfrak{A}_M := (\mathfrak{p}')^M \prod_{j=1}^r \mathfrak{q}_j^{-m_j}$, then $d_K(\mathfrak{A}_M) \geq 2g_K - 2 + d_K(\mathfrak{p})$.

By Corollary 3.5.6,

$$\ell_K(\mathfrak{A}_M^{-1}) = d_K(\mathfrak{A}_M) - g_K + 1 \quad \text{and} \quad \ell_K(\mathfrak{p}\mathfrak{A}_M^{-1}) = \ell_K(\mathfrak{A}_M^{-1}) - d_K(\mathfrak{p}).$$

Let $\xi \in L_K(\mathfrak{A}_M^{-1}) \setminus L_K(\mathfrak{p}\mathfrak{A}_M^{-1})$ and notice that $\xi y \in R'$ by (5.4). On the other hand, since $y \in \vartheta$, y is integral with respect to $\mathfrak{B}_1, \ldots, \mathfrak{B}_h$, so $\mathfrak{q}_j \neq \mathfrak{p}$ for all $1 \leq j \leq r$. Since $\xi \notin L_K(\mathfrak{p}\mathfrak{A}_M^{-1})$, we have $v_{\mathfrak{p}}(\xi) = 0$ and hence ξ is a unit of $\vartheta_{\mathfrak{p}}$.

Let u' be the element of Proposition 5.5.2 corresponding to $\mathfrak{p}'$ and the basis $\{y_1, \ldots, y_n\}$. Since $y = \sum_{i=1}^n x_i y_i \in \vartheta$ with $x_i \in K$, it follows that

$$\xi y = \sum_{i=1}^n (\xi x_i) y_i \in \vartheta, \quad \text{and} \quad u'\xi x_i \in \Gamma'.$$

In particular, $u'\xi x_i \in \vartheta_{\mathfrak{p}}$. Since ξ is a unit of $\vartheta_{\mathfrak{p}}$, it follows that $v = u'$ satisfies the conditions of Proposition 5.5.3. □

Remark 5.5.4. If we fix the basis $\{y_1, \ldots, y_n\}$ and the place $\mathfrak{p}'$ of Proposition 5.5.3, then the element v found in Proposition 5.5.3 works for every place $\mathfrak{p} \neq \mathfrak{p}'$. There are only finitely many places $\mathfrak{p}$ that fail to satisfy all the following conditions:

(1) $\mathfrak{p} \neq \mathfrak{p}'$.
(2) $v_{\mathfrak{B}}(y_i) \geq 0$ for all $1 \leq i \leq n$, $\mathfrak{B} \in \mathbb{P}_L$ and $\mathfrak{B} \mid \mathfrak{p}$.
(3) $v_{\mathfrak{p}}(v) = 0$.

Definition 5.5.5. With the same notation, we call $\{y_1, \ldots, y_n\}$ an *integral basis* at $\mathfrak{p}$ if

(a) $y_i \in \bigcap_{\mathfrak{B}|\mathfrak{p}} \vartheta_{\mathfrak{B}} = \vartheta$.
(b) If $y = \sum_{i=1}^{n} x_i y_i \in \vartheta$ with $x_i \in K$, $1 \leq i \leq n$, then $x_i \in \vartheta_{\mathfrak{p}}$ for all $1 \leq i \leq n$.

From Remark 5.5.4 we obtain the following:

Theorem 5.5.6. *Any field basis of L with respect to K is an integral basis at almost all places of K.* □

Now we consider a finite extension L/ℓ of K/k and a field basis $\{y_1, \ldots, y_n\}$ of L over K. Let $\mathfrak{p}$ be a place of K such that $\{y_1, \ldots, y_n\}$ is an integral basis at $\mathfrak{p}$. Let $\mathfrak{B} \mid \mathfrak{p}$ and consider the valuation ring $\vartheta_{\mathfrak{B}}$ at $\mathfrak{B}$.

We have $\{y_1, \ldots, y_n\} \subseteq \vartheta_{\mathfrak{B}}$ and

$$\vartheta := \bigcap_{\mathfrak{B}|\mathfrak{p}} \vartheta_{\mathfrak{B}} = \vartheta_{\mathfrak{p}} y_1 + \cdots + \vartheta_{\mathfrak{p}} y_n.$$

Let $\hat{\vartheta}_{\mathfrak{B}}$ be the completion of $\vartheta_{\mathfrak{B}}$. If $z \in \hat{\vartheta}_{\mathfrak{B}}$, by the approximation theorem (Corollary 2.5.5) there exist $y_m \in L$ with $m \in \mathbb{N}$ such that

$$v_{\mathfrak{B}}(z - y_m) > m \quad \text{for all} \quad m \in \mathbb{N}$$

and

$$v_{\mathfrak{B}'}(y_m) \geq 0 \quad \text{for all} \quad \mathfrak{B}' \neq \mathfrak{B} \quad \text{such that} \quad \mathfrak{B}'|\mathfrak{p}.$$

Therefore $\lim_{m\to\infty} y_m = z$ in $\hat{\vartheta}_{\mathfrak{B}}$ and $y_m \in \vartheta$. We have

$$y_m = \sum_{i=1}^{n} x_{im} y_i \quad \text{with all} \quad x_{im} \in \vartheta_{\mathfrak{p}}.$$

It is easy to see that $\{x_{im}\}_{m=1}^{\infty}$ is a Cauchy sequence in $\vartheta_{\mathfrak{p}}$ (see Theorem 5.4.7), so $\{x_{im}\}_{m=1}^{\infty}$ converges. Let $\hat{x}_i := \lim_{m\to\infty} x_{im} \in \hat{\vartheta}_{\mathfrak{p}}$. We have

$$z = \sum_{i=1}^{n} \hat{x}_i y_i \in \hat{\vartheta}_{\mathfrak{p}} y_1 + \cdots + \hat{\vartheta}_{\mathfrak{p}} y_n.$$

Furthermore, from Corollary 5.4.9 we obtain that

$$\hat{\vartheta}_{\mathfrak{p}} \otimes_{\vartheta_{\mathfrak{p}}} \vartheta \cong \bigoplus_{\mathfrak{B}|\mathfrak{p}} \hat{\vartheta}_{\mathfrak{B}}$$

and $\{y_1, \ldots, y_n\}$ is basis of $\bigoplus_{\mathfrak{B}|\mathfrak{p}} \hat{\vartheta}_{\mathfrak{B}}$ over $\hat{\vartheta}_{\mathfrak{p}}$.

We have proved the following theorem:

Theorem 5.5.7. *Let L/ℓ be any finite extension of the function field K/k and let $\{y_1, \ldots, y_n\}$ be any field basis of L over K. Then for almost all places $\mathfrak{p}$ of K, $y_1, \ldots, y_n$ generate the completion $\hat{\vartheta}_{\mathfrak{B}}$ of $\vartheta_{\mathfrak{B}}$ over $\hat{\vartheta}_{\mathfrak{p}}$, where $\mathfrak{B} \mid \mathfrak{p}$.* □

We state two corollaries of Theorem 5.5.7 that we will use in Chapter 9.

Corollary 5.5.8. *Let L/ℓ be any finite extension of K/k and let $\{y_1, \ldots, y_n\}$ be any field basis of L/K. Then*

$$\mathfrak{X}_L = \mathfrak{X}_K y_1 + \cdots + \mathfrak{X}_K y_n,$$

where $\mathfrak{X}_L$ and $\mathfrak{X}_K$ are the rings of repartitions of L and K respectively. Here $\mathfrak{X}_K$ may be considered as a subset of $\mathfrak{X}_L$; indeed, we can define $\phi \colon \mathfrak{X}_K \to \mathfrak{X}_L$ by $\phi(\xi) = \lambda$ for all $\xi \in \mathfrak{X}_K$, where $\lambda_{\mathfrak{B}} = \xi_{\mathfrak{p}}$ for any $\mathfrak{B} \mid \mathfrak{p}$.

Proof. Clearly, $\mathfrak{X}_K y_1 + \cdots + \mathfrak{X}_K y_n$ is a subset of $\mathfrak{X}_L$. Let $\lambda \in \mathfrak{X}_L$, let $\mathfrak{p}$ be a place of K, and let $\mathfrak{B}_1, \ldots, \mathfrak{B}_h$ be places of L above $\mathfrak{p}$. Since

$$\bigoplus_{\mathfrak{B}|\mathfrak{p}} L_{\mathfrak{B}} \overset{\theta}{\cong} L \otimes_K K_{\mathfrak{p}} = \Big(\sum_{i=1}^{n} K y_i\Big) \otimes_K K_{\mathfrak{p}},$$

we have $(\lambda_{\mathfrak{B}})_{\mathfrak{B}|\mathfrak{p}} = \theta\Big(\big(\sum_{i=1}^{n} x_i y_i\big) \otimes \sum_{j=1}^{m} z_j\Big)$ with $x_i \in K$, $z_j \in K_{\mathfrak{p}}$. Thus $\lambda \in V y_1 + \cdots + V y_n$, where

$$V := \prod_{\mathfrak{p} \in \mathbb{P}_K} K_{\mathfrak{p}}.$$

We need to prove that the "coefficients" of y_i belong to $\mathfrak{X}_K$, that is, that the components are integers for almost all $\mathfrak{p} \in \mathbb{P}_K$.

For almost all $\mathfrak{B}$, we have

$$\lambda_{\mathfrak{B}} \in \hat{\vartheta}_{\mathfrak{B}}, \quad \text{that is,} \quad v_{\mathfrak{B}}(\lambda_{\mathfrak{B}}) \geq 0.$$

If $\lambda_{\mathfrak{B}} = \sum_{i=1}^{n} x_i y_i$, then by Theorem 5.5.7 we have $x_i \in \hat{\vartheta}_{\mathfrak{p}}$ for almost all $\mathfrak{p}$. Hence

$$\mathfrak{X}_L = \mathfrak{X}_K y_1 + \cdots + \mathfrak{X}_K y_n.$$ □

The next corollary will be used when we consider the genus change in purely inseparable extensions.

Corollary 5.5.9. *Let L/ℓ be a purely inseparable extension of K/k of degree p. Let $L = K(\alpha)$, where $\alpha^p = a \in K$. Then, for almost all $\mathfrak{p} \in \mathbb{P}_K$, $\hat{\vartheta}_{\mathfrak{B}} = \bigoplus_{i=0}^{p-1} \hat{\vartheta}_{\mathfrak{p}}\alpha^i$, where $\mathfrak{B}$ is the only place of L above K.*

Proof. The statement follows from the facts that there is only one place above $\mathfrak{p}$ (Theorem 5.2.24), that $\hat{\vartheta}_{\mathfrak{B}}$ is a free $\hat{\vartheta}_{\mathfrak{p}}$-module of rank p (see, for example, Theorem 2.5.20), and from Theorem 5.5.7. □

Remark 5.5.10. Corollary 5.5.9 states that for almost all $\mathfrak{p}$, if

$$y = \sum_{i=0}^{p-1} x_i \alpha^i \in \hat{\vartheta}_{\mathfrak{B}}$$

then $x_i \in \hat{\vartheta}_{\mathfrak{p}}$, for all $0 \le i \le p-1$.

Now that we have studied the structure of the extensions $L_{\mathcal{P}_i}/K_{\wp}$, it is necessary to mention the role played by the places. When we start with a place $\wp$ of K, $\wp$ can be seen as the maximal ideal of the corresponding valuation ring ϑ, and similarly for $\mathcal{P}_i$. The place $\hat{\wp}$ of $K_{\wp}$ is the same ideal $\wp$ but considered in the valuation ring $\hat{\vartheta}$ that is the completion of ϑ with respect to the topology given by the valuation. More precisely, $\hat{\wp} = \wp\hat{\vartheta}$, where $\hat{\wp}$ is the completion of $\wp$. Furthermore, since $\hat{\vartheta}/\hat{\wp} \cong \vartheta/\wp$ (Proposition 2.3.10), we can consider that $\wp$ and $\hat{\wp}$ are one and the same place. Since $L_{\mathcal{P}_i}$ has only a unique extension of $\hat{\wp}$ (Theorem 5.4.7), namely $\hat{\mathcal{P}}_i$, the advantage of working with $L_{\mathcal{P}_i}/K_{\wp}$ is that there is only one place "above" and only one place "below," which does not hold in L/K, where there are infinitely many places. Furthermore, by the above argument, we have $e_{L_{\mathcal{P}_i}/K_{\wp}}\left(\hat{\mathcal{P}}_i|\hat{\wp}\right) = e_{L/K}\left(\mathcal{P}_i|\wp\right)$ and $d_{L_{\mathcal{P}_i}/K_{\wp}}\left(\hat{\mathcal{P}}_i|\hat{\wp}\right) = d_{L/K}\left(\mathcal{P}_i|\wp\right)$.

Finally, we prove the following results on bases:

Proposition 5.5.11. *Let $\alpha_1, \ldots, \alpha_f$ be elements of $\vartheta_{\mathcal{P}_i}$ such that $\left\{\bar{\alpha}_j\right\}_{j=1}^{f}$ is a basis of $\ell\left(\mathcal{P}_i\right)/k\left(\wp\right)$, and let π_i be a prime element of L with respect to $v_{\mathcal{P}_i}$. Then the elements $\left\{\alpha_j \pi_i^s\right\}_{s=0,\ldots,e-1}^{j=1,\ldots,f}$ form a basis of $L_{\mathcal{P}_i}/K_{\wp}$.*

Proof. This follows from the facts that $L_{\mathcal{P}_i} = \ell\left(\mathcal{P}_i\right) K_{\wp}\left(\pi_i\right)$, $K_{\wp} = k\left(\wp\right)\left(\left(\pi\right)\right)$, where π is a prime element of K, and $\left[L_{\mathcal{P}_i} : K_{\wp}\right] = ef$. □

Proposition 5.5.12. *Let k be an algebraically closed field of characteristic zero. Let L and K be function fields over k with $K \subseteq L$ and such that L/K is of finite degree. Assume that e is the ramification index of a place $\mathfrak{P}$ of L over $\mathfrak{p}$. Then if Π is a prime element of $L_{\mathfrak{P}}$, there exists a prime element π of $K_{\mathfrak{p}}$ such that*

$$\pi = \Pi^e.$$

Proof. Let Π_1 be any prime element of $L_{\mathfrak{P}}$ for $\mathfrak{P}$. For any prime element π of $K_{\mathfrak{p}}$, we have

$$v_{\mathfrak{P}}(\pi) = ev_{\mathfrak{p}}(\pi) = e.$$

Thus, π has an expansion in $L_{\mathfrak{P}} \cong k((\Pi_1))$ defined as follows:

$$\pi = a_e \Pi_1^e + a_{e+1}\Pi_1^{e+1} + \cdots, \quad a_i \in k, a_e \neq 0.$$

Let

$$\Pi = b_1 \Pi_1 + b_2 \Pi_1^2 + \cdots, \quad b_i \in k, b_1 \neq 0,$$

be another prime element. Then

$$\Pi^e = c_1 \Pi_1^e + c_2 \Pi_1^{e+1} + \cdots + c_n \Pi_1^{n+e-1} + \cdots,$$

where c_n is a polynomial of degree $n + e - 1$ in $b_1, \ldots, b_n$. Furthermore,

$$c_n = \sum_{\substack{(i_1,\ldots,i_e), i_j \geq 1 \\ i_1+i_2+\cdots+i_e = n+e-1}} b_{i_1} \cdots b_{i_e}.$$

We have $c_n = p^{(n)}(b_1, \ldots, b_{n-1}) + eb_1^{e-1} b_n$, where $p^{(n)}(b_1, \ldots, b_{n-1})$ is a polynomial in $b_1, \ldots, b_{n-1}$ with rational integer coefficients. Thus there exist $b_1 \neq 0$ and $b_2, \ldots, b_n, \ldots \in k$ satisfying $c_n = a_n$ for all $n \geq 1$. It follows that $\Pi^e = \pi$. □

Definition 5.5.13. Let E/F be an extension of fields and let $\alpha \in E$. The function $T_\alpha : E \longrightarrow E$, defined by $T_\alpha(z) = \alpha z$, is an F-linear transformation. The characteristic polynomial of T_α, namely $f_\alpha(x) = \det(xI - T_\alpha)$, is called the *characteristic polynomial* of α.

Let A be the matrix associated to T_α with respect to a basis of E/F.

Proposition 5.5.14. *We have*

$$\begin{aligned} N_{E/F}(\alpha) &= \text{norm of } \alpha = \det A = \det T_\alpha = (-1)^n f_\alpha(0) = (-1)^n b_0, \\ \operatorname{Tr}_{E/F}(\alpha) &= \text{trace of } \alpha = \text{trace of } A = \text{trace of } T_\alpha = -b_{n-1}, \end{aligned}$$

where $f_\alpha(x) = x^n + b_{n-1}x^{n-1} + \cdots + b_1 x + b_0$. □

Let L/ℓ be a finite extension of K/k, $\wp$ a place of K, and $\mathcal{P}_1, \ldots, \mathcal{P}_h$ the places of L above $\wp$.

Theorem 5.5.15. *Let $\alpha \in L$. If $f_\alpha(x)$ is the characteristic polynomial of α over K and $f_\alpha^{(i)}$ is the characteristic polynomial of $\alpha \in L_{\mathcal{P}_i}$ over $K_\wp$, then $f_\alpha(x) = \prod_{i=1}^h f_\alpha^{(i)}(x)$*

Proof. The K-linear transformation $T_\alpha : L \longrightarrow L$, corresponds to the $K_\wp$-linear transformation

$$T_\alpha \otimes 1 : L \otimes_K K_\wp \longrightarrow L \otimes_K K_\wp.$$

Furthermore, we have $L \otimes_K K_\wp \cong \bigoplus_{i=1}^h L_{\mathcal{P}_i}$. Thus

$$(T_\alpha \otimes 1)(x_1, \dots, x_n) = (\alpha x_1, \dots, \alpha x_n) \text{ with } x_i \in L_{\mathcal{P}_i}.$$

By Proposition 5.5.11, we can choose a basis of $L_{\mathcal{P}_i}/K_\wp$ whose members belong to L, and the result follows. □

Now we state several corollaries.

Corollary 5.5.16. *Let* $e = e_{L/K}(\mathcal{P}_i|\wp)$ *and* $f = d_{L/K}(\mathcal{P}_i|\wp)$. *Let* $y \in L_{\mathcal{P}_i}$ *be nonzero. Then* $v_\wp\left(N_{L_{\mathcal{P}_i}/K_\wp} y\right) = f v_{\mathcal{P}_i}(y)$.

Proof. Let $m = v_{\mathcal{P}_i}(y)$. Then $y = \omega\pi_i^m$, where π_i is a prime element for $v_{\mathcal{P}_i}$ and ω is a unit. The norm of ω is a unit, so $N_{L_{\mathcal{P}_i}/K_\wp} y = Ny = (N\omega)(N\pi_i)^m$. Now, $N\pi_i = \omega_1\pi_i^{ef}$ for a unity ω_1, and $\pi_i^e = \omega_2\pi_K$ for a unit ω_2 in $L_{\mathcal{P}_i}$ where π_K is a prime element for $v_\wp$. Therefore $Ny = \omega_3\pi_K^{fm}$, where ω_3 is a unit. Thus we obtain that $v_\wp(Ny) = fm = f v_{\mathcal{P}_i}(y)$. □

Corollary 5.5.17. *For* $i = 1, \dots, h$, *define*

$$N_{\mathcal{P}_i} = N_{L_{\mathcal{P}_i}/K_\wp}, \quad \mathrm{Tr}_{\mathcal{P}_i} = \mathrm{Tr}_{L_{\mathcal{P}_i}/K_\wp}, \quad \text{and} \quad N = N_{L/K}, \quad \mathrm{Tr} = \mathrm{Tr}_{L/K}.$$

Then for $\alpha \in L$ *we have*

$$\mathrm{Tr}\,\alpha = \sum_{i=1}^h \mathrm{Tr}_{\mathcal{P}_i}\,\alpha \quad \text{and} \quad N\alpha = \prod_{i=1}^h N_{\mathcal{P}_i}\alpha.$$

Proof. The statement follows from Theorem 5.5.15 and Proposition 5.5.14. □

Corollary 5.5.18. *Let* $\alpha \in L$. *Then* $v_\wp\left(N_{L/K}\alpha\right) = \sum_{i=1}^h f_i v_{\mathcal{P}_i}(\alpha)$, *where* $f_i = d_{L/K}(\mathcal{P}_i|\wp)$.

Proof. By Corollaries 5.5.16 and 5.5.17, we have

$$v_\wp\left(N_{L/K}\alpha\right) = v_\wp\left(\prod_{i=1}^h N_{\mathcal{P}_i}\alpha\right) = \sum_{i=1}^h v_\wp\left(N_{\mathcal{P}_i}\alpha\right) = \sum_{i=1}^h f_i v_{\mathcal{P}_i}(\alpha). \qquad \square$$

5.6 Different and Discriminant

Let L/K be a finite separable extension of function fields, $\mathcal{P}$ a place of L, and $\wp = \mathcal{P}|_K$. Denote by e and f the ramification index and relative degree of $\mathcal{P}$ over $\wp$ respectively. By Theorem 5.4.10, $L_{\mathcal{P}}$ is separable over $K_\wp$ and $[L_{\mathcal{P}} : K_\wp] = ef$. Let π_L and π_K be prime elements of $v_{\mathcal{P}}$ and $v_\wp$ respectively, with $v_{\mathcal{P}}(\pi_K) = e \geq 1$.

Now consider the Galois closure $\tilde{L}$ of L/K, and assume that $\mathfrak{S}$ is a place in $\tilde{L}$ over $\mathcal{P}$. Let $\tilde{\ell}$ be the field of constants of $\tilde{L}$. We have the following diagram:

$$\begin{array}{ccccccc}
\mathfrak{S} & \text{——} & \tilde{\ell}(\mathfrak{S}) & \text{——} & \tilde{L} & \text{——} & \tilde{L}_{\mathfrak{S}} \\
| & & | & & | & & | \\
\mathcal{P} & \text{——} & \ell(\mathcal{P}) & \text{——} & L & \text{——} & L_{\mathcal{P}} \\
| & & | & & | & & | \\
\wp & \text{——} & k(\wp) & \text{——} & K & \text{——} & K_\wp
\end{array}$$

Let $D = D(\mathfrak{S}|\wp) = \operatorname{Gal}\left(\tilde{L}_{\mathfrak{S}}/K_\wp\right)$, $D_1 = D(\mathfrak{S}|\mathcal{P}) = \operatorname{Gal}\left(\tilde{L}_{\mathfrak{S}}/L_{\mathcal{P}}\right)$, $I = I(\mathfrak{S}|\wp)$, and $I_1 = I(\mathfrak{S}|\mathcal{P})$.

The set of classes $\operatorname{Aut}\left(\tilde{\ell}(\mathfrak{S})/k(\wp)\right)/\operatorname{Aut}\left(\tilde{\ell}(\mathfrak{S})/\ell(\mathcal{P})\right)$ is in bijective correspondence with $\operatorname{Aut}(\ell(\mathcal{P})/k(\wp))$. Furthermore,

$$\operatorname{Aut}\left(\tilde{\ell}(\mathfrak{S})/k(\wp)\right) \cong D/I$$

and

$$\operatorname{Aut}\left(\tilde{\ell}(\mathfrak{S})/\ell(\mathcal{P})\right) \cong D_1/I_1 = D_1/(D_1 \cap I) \cong D_1 I/I.$$

Therefore the elements of $\operatorname{Aut}(\ell(\mathcal{P})/k(\wp))$ are in correspondence with the cosets of

$$(D/I)/(D_1 I/I) \sim D/D_1 I.$$

For $z \in \vartheta$, we denote by $\bar{z}$ its equivalence class modulo the ideal. That is, $\bar{z}$ is in the residue field. We have

$$\begin{aligned}
\overline{\left(\operatorname{Tr}_{L_{\mathcal{P}}/K_\wp}(z)\right)} &= \overline{\left(\sum_{\sigma \in D/D_1} \sigma z\right)} = \sum_{\sigma \in D_1 I/D_1} \sigma \sum_{\theta \in D/D_1 I} \overline{\theta z} \\
&= |D_1 I/D_1| \left(\operatorname{Tr}_{\ell(\mathcal{P})/k(\wp)}(\bar{z})\right).
\end{aligned}$$

Now, $|D_1 I/D_1| = |I/(I \cap D_1)| = |I/I_1| = \frac{|I|}{|I_1|} = e(\mathcal{P}|\wp) = e$. It follows that

$$\overline{\left(\operatorname{Tr}_{L_{\mathcal{P}}/K_\wp}(z)\right)} = e\left(\operatorname{Tr}_{\ell(\mathcal{P})/k(\wp)}(\bar{z})\right). \tag{5.5}$$

We write $\operatorname{Tr} = \operatorname{Tr}_{L_{\mathcal{P}}/K_\wp}$.

Theorem 5.6.1. *There exists $m \geq 0$ such that if $x \in L_{\mathcal{P}}$ satisfies $v_{\mathcal{P}}(x) \geq -m$, then $v_{\wp}(\operatorname{Tr} x) \geq 0$. Also, there exists x_0 with $v_{\mathcal{P}}(x_0) < -m$ and $v_{\wp}(\operatorname{Tr} x_0) < 0$.*

Proof. If $v_{\mathcal{P}}(x) \geq 0$, then $x \in \vartheta_{\mathcal{P}}$. Therefore $\operatorname{Tr} x \in \vartheta_{\wp}$ and $v_{\wp}(\operatorname{Tr} x) \geq 0$ (see Corollary 5.7.6). On the other hand, let $y \in \vartheta_{\mathcal{P}}$ be such that $\operatorname{Tr} y \neq 0$. This element exists since $L_{\mathcal{P}}/K_{\wp}$ is separable. If $x \in K$ is such that $v_{\wp}(x) < -v_{\wp}(\operatorname{Tr} y)$, we have

$$v_{\wp}(\operatorname{Tr} xy) = v_{\wp}(x \operatorname{Tr} y) = v_{\wp}(x) + v_{\wp}(\operatorname{Tr} y) < 0.$$

Let

$$A = \left\{ n \in \mathbb{Z} \mid v_{\mathcal{P}}(x) \geq n \Longrightarrow v_{\wp}(\operatorname{Tr} x) \geq 0 \right\}.$$

Notice that $0 \in A$, $\mathbb{N} \subseteq A$, but there exists $n \in \mathbb{Z}$ such that $n < 0$ and $n \notin A$ (for example, pick $n = n_0 = v_{\mathcal{P}}(xy)$ above). Furthermore, if $n_0 \notin A$, we have $n_0 - 1 \notin A$ since $v_{\mathcal{P}}(x) \geq n_0 \Longrightarrow v_{\mathcal{P}}(x) \geq n_0 - 1$.

Let $t = \inf A$. We have $t \in \mathbb{Z}$ and $t \leq 0$. Let $m = -t$. Then $m \geq 0$ and if $x \in L_{\mathcal{P}}$ is such that

$$v_{\mathcal{P}}(x) \geq -m = t \in A \quad \text{then} \quad v_{\wp}(\operatorname{Tr} x) \geq 0.$$

On the other hand, since $t - 1 \notin A$ there exists $x \in L_{\mathcal{P}}$ with

$$v_{\mathcal{P}}(x) \geq t - 1 = -m - 1 \quad \text{and} \quad v_{\wp}(\operatorname{Tr} x) < 0.$$

If $v_{\mathcal{P}}(x) > t - 1$, then $v_{\mathcal{P}}(x) \geq t = -m$, which contradicts the fact that $t \in A$. Thus, $v_{\mathcal{P}}(x) = t - 1 = -m - 1 < -m$. □

Definition 5.6.2. The maximum nonnegative integer satisfying Theorem 5.6.1 is denoted by $m(\mathcal{P})$ and called *the differential exponent of $\mathcal{P}$ with respect to K*.

The importance of this exponent is that it shows up only in the presence of ramification or inseparable residue field extensions. This is stated more precisely in the following theorem:

Theorem 5.6.3. *We have $m(\mathcal{P}) \geq e - 1$. Furthermore, $m(\mathcal{P}) > e - 1$ if and only if at least one of the following two conditions holds:*

(1) $p = \operatorname{char} k$ *divides* e.
(2) $\ell(\mathcal{P})/k(\wp)$ *is inseparable.*

Proof. If $y \in L_{\mathcal{P}}$ satisfies $v_{\mathcal{P}}(y) \geq -(e-1)$, then since $v_{\mathcal{P}}(\pi_K) = e$, we have $v_{\mathcal{P}}(\pi_K y) \geq 1$. Therefore $\pi_K y \in \mathcal{P}$.

It follows that $\overline{\operatorname{Tr}(\pi_K y)} = e \operatorname{Tr}(\overline{\pi_K y}) = 0$ and $\operatorname{Tr}(\pi_K y) \in \wp$. Hence $v_{\wp}(\operatorname{Tr}(\pi_K y)) = v_{\wp}(\pi_K \operatorname{Tr} y) = 1 + v_{\wp}(\operatorname{Tr} y) \geq 1$. That is, $\operatorname{Tr} y \geq 0$. We have obtained that $m(\mathcal{P}) \geq e - 1$.

Now if $\ell(\mathcal{P})/k(\wp)$ is not separable, let y be such that $v_{\mathcal{P}}(y) \geq -e$. We have $\pi_K y \in \vartheta_{\mathcal{P}}$. Since $\operatorname{Tr}_{\ell(\mathcal{P})/k(\wp)} \equiv 0$, we have $\overline{\operatorname{Tr} \pi_K y} = 0$. Thus

$$v_\wp (\operatorname{Tr} \pi_K y) = 1 + v_\wp (\operatorname{Tr} y) \geq 1 \quad \text{or, equivalently,} \quad v_\wp (\operatorname{Tr} y) \geq 0.$$

Hence $m(\mathcal{P}) \geq e$.

If $p \mid e$, again if $v_{\mathcal{P}}(y) \geq -e$ then $\pi_K y \in \vartheta_{\mathcal{P}}$ and $\overline{\operatorname{Tr}(\pi_K y)} = e \operatorname{Tr}(\overline{\pi_K y}) = 0$. Therefore $v_\wp (\operatorname{Tr} y) \geq 0$ and $m(\mathcal{P}) \geq e$.

Conversely, assume that $\ell(\mathcal{P})/k(\wp)$ is a separable extension and that $p \nmid e$. Since $\ell(\mathcal{P})/k(\wp)$ is separable, there exists $y \in \vartheta_{\mathcal{P}}$ such that $\operatorname{Tr}_{\ell(\mathcal{P})/k(\wp)}(\bar{y}) \neq 0$.

We have

$$\overline{\operatorname{Tr} y} = e \operatorname{Tr}_{\ell(\mathcal{P})/k(\wp)}(\bar{y}) \neq 0 \Longrightarrow v_\wp(\operatorname{Tr} y) = 0 \quad \text{and} \quad v_\wp\left(\operatorname{Tr}\left(\pi_K^{-1} y\right)\right) = -1.$$

On the other hand,

$$v_{\mathcal{P}}\left(\pi_K^{-1} y\right) = -e \quad \text{and} \quad v_\wp\left(\operatorname{Tr}\left(\pi_K^{-1} y\right)\right) = -1 < 0 \quad \text{implies that} \quad m(\mathcal{P}) < e.$$

Since $e - 1 \leq m(\mathcal{P}) < e$, it follows that $m(\mathcal{P}) = e - 1$. □

Now, since we are considering the case in which L/K is separable, we have the following corollary:

Corollary 5.6.4. *We have $m(\mathcal{P}) = 0$ for all but a finite number of places $\mathcal{P}$.*

Proof. If $\mathcal{P}$ is a separable nonramified place, then $m(\mathcal{P}) = e_{L/K}(\mathcal{P}|\wp) - 1 = 1 - 1 = 0$. By Theorem 5.2.33, the number of places $\mathcal{P}$ that are ramified or inseparable is finite. □

Definition 5.6.5. The divisor $\mathfrak{D}_{L/K} = \prod_{\mathcal{P} \in \mathbb{P}_L} \mathcal{P}^{m(\mathcal{P})}$ is called the *different of the extension*.

A similar definition can be made using completions exclusively.

Definition 5.6.6. For the completions $L_{\mathcal{P}}/K_\wp$ we define the *local different* as $\mathfrak{D}_{\mathcal{P}} = \hat{\mathcal{P}}^{\alpha(\mathcal{P})}$, where $\alpha(\mathcal{P})$ is the maximum integer such that $v_\wp(\operatorname{Tr} y) \geq 0$ whenever $y \in L_{\mathcal{P}}$ satisfies $v_{\mathcal{P}}(y) \geq -\alpha(\mathcal{P})$.

It is easy to see that $\alpha(\mathcal{P})$ is the same integer $m(\mathcal{P})$ defined before. Therefore we have the following result:

Proposition 5.6.7. *Identifying the place $\mathcal{P}$ of L with its completion $\hat{\mathcal{P}}$ in $L_{\mathcal{P}}$, we have $\mathfrak{D}_{L/K} = \prod_{\mathcal{P} \in \mathbb{P}_L} \mathfrak{D}_{\mathcal{P}}$. Furthermore, the equality $\mathfrak{D}_{\mathcal{P}} = (1)$ holds except when $\mathcal{P}$ is either ramified or inseparable.* □

Definition 5.6.8. We define the *discriminant* $\mathfrak{d}_{L/K}$ of the extension L/K as $N_{L/K}\mathfrak{D}_{L/K} = \mathfrak{d}_{L/K}$. The discriminant $\mathfrak{d}_{L/K}$ is a divisor of K.

Proposition 5.6.9. *A place $\wp$ divides $\mathfrak{d}_{L/K}$ if and only if $\wp$ is ramified or $\wp$ is inseparable, that is, if there exists a place $\mathcal{P}$ in L such that $\mathcal{P}|_K = \wp$ and $\mathcal{P}$ is ramified or $\ell(\mathcal{P})/k(\wp)$ is inseparable.*

Proof. The statement follows immediately from Definition 5.6.8. □

5.7 Dedekind Domains

Now we study the differents and discriminants in Dedekind domains in order to relate them later on to our definition. By an integral domain, we understand a commutative ring with unity and without nonzero zero divisors.

Definition 5.7.1. Let A be an integral domain that is not a field and let K be the field of quotients of A. We call A a *Dedekind domain* if it satisfies:

(i) Every nonzero prime ideal $\mathcal{P}$ is maximal.
(ii) A is Noetherian.
(iii) A is integrally closed. That is, if $x \in K$ satisfies a relation $x^n + a_{n-1}x^{n-1} + \cdots + a_1 x + a_0 = 0$ with $a_i \in A$, then $x \in A$.

Example 5.7.2. If k is a field, then the ring $k[x]$ of polynomials in one variable is a Dedekind domain. If K is any finite number field and ϑ_K is its ring of integers, then ϑ_K is a Dedekind domain. Indeed, we have $[K : \mathbb{Q}] < \infty$ and $\vartheta_K = \{\alpha \in K \mid \operatorname{Irr}(\alpha, x, \mathbb{Q}) \in \mathbb{Z}[x]\}$.

Definition 5.7.3. Let A be a Dedekind domain and let K be the quotient field corresponding to A. An A-module $M \subseteq K$ is called a *fractional ideal* if $M \neq 0$ and M is finitely generated. Equivalently, there exists $a \in A$ such that $a \neq 0$ and $aM \subseteq A$. A fractional ideal is called *invertible* if there exists another fractional ideal M' such that $MM' = A$.

Theorem 5.7.4. *If A is a Dedekind domain, every nonzero ideal $\mathfrak{A}$ of A can be written in a unique way as a product of prime ideals.*

Proof. Let $\mathcal{P}$ be a nonzero prime ideal. Let

$$\mathcal{P}^{-1} := \{x \in K \mid x\mathcal{P} \subseteq A\}.$$

Then $\mathcal{P}^{-1}$ is an A-module. If $a \in \mathcal{P}$ is nonzero we have $a\mathcal{P}^{-1} \subseteq \mathcal{P}\mathcal{P}^{-1} \subseteq A$, so $\mathcal{P}^{-1}$ is a fractional ideal. Since $\mathcal{P}\mathcal{P}^{-1} \subseteq A$, $\mathcal{P}\mathcal{P}^{-1} = \mathfrak{A}$ is an ideal of A. Clearly we have $A \subseteq \mathcal{P}^{-1}$, and hence $\mathcal{P}\mathcal{P}^{-1} \supseteq \mathcal{P}A = \mathcal{P}$. Now $\mathcal{P}$ is a maximal ideal, so we must have $\mathcal{P}\mathcal{P}^{-1} = \mathcal{P}$ or $\mathcal{P}\mathcal{P}^{-1} = A$.

First we will see that $A \subsetneqq \mathcal{P}^{-1}$. For this purpose we will prove that every nonzero ideal I of A contains a product of prime ideals $\mathcal{P}_1 \cdots \mathcal{P}_r$ such that $\mathcal{P}_i \supseteq I$, $1 \leq i \leq r$. For the sake of contradiction, assume that there exists some ideal I not satisfying the above property. Since A is Noetherian, we can choose I' to be maximal among those ideals not satisfying the property. Clearly I' is not a prime ideal. Therefore there exist $a, b \in A \setminus I'$ such that $ab \in I'$. Put $\mathfrak{A} = I' + (a)$ and $\mathfrak{B} = I' + (b)$. We have

$$I' \subsetneqq \mathfrak{A}, \quad I' \subsetneqq \mathfrak{B} \quad \text{and} \quad \mathfrak{A}\mathfrak{B} \subseteq I'.$$

Since I' is maximal, it follows that both $\mathfrak{A}$ and $\mathfrak{B}$ contain a product of prime ideals, which in turn contain $\mathfrak{A}$ and $\mathfrak{B}$. Therefore they contain I'. This contradicts our choice of I'.

Now we will show that $A \subsetneqq \mathcal{P}^{-1}$. Let $c \in \mathcal{P}$ be such that $c \neq 0$ and $(c) \neq \mathcal{P}$. Notice that if $(c) = \mathcal{P}$, then $(c^2) \subsetneqq (c)$ since c is not a unit. The ideal generated by c contains a product of r prime ideals $\mathcal{P}_1, \ldots, \mathcal{P}_r$ such that $\mathcal{P}_i \supseteq (c)$. Choose r to be the least integer satisfying the above property. Then

$$\mathcal{P}_1 \cdots \mathcal{P}_r \subseteq (c) \subsetneqq \mathcal{P}.$$

Since $\mathcal{P}$ is a prime ideal, $\mathcal{P}$ must contain some $\mathcal{P}_i$, say $\mathcal{P}_1$ (otherwise if $\mathcal{P}_i \nsubseteq \mathcal{P}$ for all $1 \leq i \leq r$, let $a_i \in \mathcal{P}_i \setminus \mathcal{P}$, $a = a_1 \cdots a_r \in \mathcal{P}_1 \cdots \mathcal{P}_r$, $a \notin \mathcal{P}$).

Since $\mathcal{P}_1 \subseteq \mathcal{P}$ and $\mathcal{P}_1$ is maximal, we have $\mathcal{P} = \mathcal{P}_1$. Observe that $r > 1$ since otherwise $r = 1$ and $\mathcal{P} \supseteq (c) \supseteq \mathcal{P}$, which would imply that $(c) = \mathcal{P}$. Since r is minimum, we have $(c) \nsupseteq \mathcal{P}_2 \ldots \mathcal{P}_r$. Let $a \in \mathcal{P}_2 \ldots \mathcal{P}_r$ be such that $a \notin (c)$. Then $\frac{a}{c} \notin A$ and

$$\left(\frac{a}{c}\right)\mathcal{P} \subseteq \left(\frac{1}{c}\right)\mathcal{P}(a) \subseteq \left(\frac{1}{c}\right)\mathcal{P}_1\mathcal{P}_2 \cdots \mathcal{P}_r = \left(\frac{1}{c}\right)(c) \subseteq A.$$

Therefore $\frac{a}{c} \in \mathcal{P}^{-1} \setminus A$.

Therefore, $A \subsetneqq \mathcal{P}^{-1}$. Now, if $\mathcal{P}\mathcal{P}^{-1} = \mathcal{P}$, then $\mathcal{P}\mathcal{P}^{-2} = \mathcal{P}\mathcal{P}^{-1} = \mathcal{P}$. It follows that in general, $\mathcal{P}\mathcal{P}^{-n} = \mathcal{P}$ for all $n \geq 1$. Hence, if $a \in \mathcal{P}$ and $b \in \mathcal{P}^{-1}$ are such that $a \neq 0$ and $b \notin A$, we have $ab^n \in \mathcal{P}$ for all $n \geq 0$. Put $J = \langle ab^n \mid n \geq 0 \rangle$ and $J_n = \langle a, ab, ab^2, \ldots, ab^n \rangle$. We have $I \subseteq \mathcal{P}$ and $J_m \subseteq J_{m+1}$ for all m. Since A is Noetherian, there exists n such that $J_n = J_{n-1}$. In other words, there exist $c_0, \ldots, c_{n-1} \in A$ such that $ab^n = \sum_{i=0}^{n-1} c_i ab^i$. Equivalently, $b^n = \sum_{i=0}^{n-1} c_i b^i$ with all $c_i \in A$, which implies that $b \in A$, a contradiction. Therefore $\mathcal{P}\mathcal{P}^{-1} = A$.

Now we will see that every nonzero ideal $\mathfrak{A}$ of A can be written in a unique way as a product of prime ideals. First we will show the existence.

If $\mathfrak{A} = A$, then $\mathfrak{A} = \mathcal{P}^0$, where $\mathcal{P}$ is a prime ideal. Assume that $\mathfrak{A} \neq A$ and let $\mathcal{P}_1 \cdots \mathcal{P}_r \subseteq \mathfrak{A}$ with $\mathcal{P}_i \supseteq \mathfrak{A}$, $i = 1, \ldots, r$, and assume that r is the minimum integer satisfying this condition. We will demonstrate the existence by induction on r. If $r = 1$, then $\mathcal{P}_1 \supseteq \mathfrak{A} \supseteq \mathcal{P}_1$ and therefore $\mathcal{P}_1 = \mathfrak{A}$. Now suppose $r > 1$. Let $\mathcal{P}$ be maximal such that $\mathcal{P}_1 \cdots \mathcal{P}_r \subseteq \mathfrak{A} \subseteq \mathcal{P}$, so that $\mathcal{P}$ contains some $\mathcal{P}_i$, say $\mathcal{P}_1$. Thus

$$\mathcal{P}_1 = \mathcal{P} \quad \text{and} \quad \mathcal{P}\mathcal{P}_2 \cdots \mathcal{P}_r \subseteq \mathfrak{A} \subseteq \mathcal{P}.$$

Multiplying by $\mathcal{P}^{-1}$, we have $\mathcal{P}_2 \cdots \mathcal{P}_r \subseteq \mathcal{P}^{-1}\mathfrak{A} \subseteq A$. Therefore $\mathcal{P}^{-1}\mathfrak{A} = \mathfrak{S}_1 \cdots \mathfrak{S}_s$ is a product of prime ideals, and $\mathfrak{A} = \mathcal{P}\mathfrak{S}_1 \ldots \mathfrak{S}_s$.

Now we will see the uniqueness. Assume

$$\mathfrak{A} = \mathcal{P}_1 \cdots \mathcal{P}_r = \mathcal{P}'_1 \cdots \mathcal{P}'_s.$$

If $r = 1$ or $s = 1$, say $r = 1$, we have $\mathfrak{A} = \mathcal{P}_1 = \mathcal{P}'_1 \cdots \mathcal{P}'_s$. Therefore there exists some index i such that $\mathcal{P}'_i \subseteq \mathcal{P}_1 = \mathcal{P}$, say $\mathcal{P}'_1 \subseteq \mathcal{P}$, which implies $\mathcal{P}'_1 = \mathcal{P}$. Therefore, $\mathfrak{A} = \mathcal{P} = \mathcal{P}\mathcal{P}'_2 \cdots \mathcal{P}'_s$. Multiplying by $\mathcal{P}^{-1}$, we obtain $A = \mathcal{P}'_2 \cdots \mathcal{P}'_s$, so $s - 1 = 0$. Indeed, otherwise $\mathcal{P}'_2 \cdots \mathcal{P}'_s$ would be a proper ideal. Now assume that $r > 1$ and $s > 1$. We have $\mathcal{P}_1 \supseteq \mathcal{P}_1 \cdots \mathcal{P}_r = \mathcal{P}'_1 \cdots \mathcal{P}'_s$ and, as before, $\mathcal{P}'_1 = \mathcal{P}_1$. Multiplying by $\mathcal{P}_1^{-1}$ we obtain

$$\mathcal{P}_2 \cdots \mathcal{P}_r = \mathcal{P}_2' \cdots \mathcal{P}_s'.$$

By the induction hypothesis we have $r = s$ and $\mathcal{P}_i' = \mathcal{P}_i$, for $i = 2, \ldots, r = s$. □

Now consider M to be an arbitrary fractional ideal. Let $a \in A$ be such that $a \neq 0$ and $aM = \mathfrak{A} \subseteq A$. By Theorem 5.7.4, $aM = \mathcal{P}_1 \cdots \mathcal{P}_r$. Setting $(a) = \mathfrak{S}_1 \ldots \mathfrak{S}_s$, we obtain for M an expression $M = \mathcal{P}_1 \cdots \mathcal{P}_r \mathfrak{S}_1^{-1} \cdots \mathfrak{S}_s^{-1}$. In other words, every fractional ideal M is expressed as a product $\mathcal{P}_1^{\alpha_1} \cdots \mathcal{P}_r^{\alpha_r}$ of prime ideals, where each $\mathcal{P}_i$ is a prime ideal of A and $\alpha_i \in \mathbb{Z}$.

Now we assume that there exist two different expressions:

$$\mathcal{P}_1^{\alpha_1} \cdots \mathcal{P}_r^{\alpha_r} = \mathcal{P}_1^{'\beta_1} \cdots \mathcal{P}_r^{'\beta_r}.$$

Writing positive and negative powers separately, we have

$$M = \mathfrak{A}\mathfrak{B}^{-1} = \mathfrak{C}\mathfrak{D}^{-1}, \quad \text{where} \quad \mathfrak{A},\ \mathfrak{B},\ \mathfrak{C},\ \mathfrak{D} \quad \text{are ideals of } A.$$

Therefore $\mathfrak{A}\mathfrak{D} = \mathfrak{B}\mathfrak{C}$. By the uniqueness of the ideals of A and since neither $\mathfrak{A}$ and $\mathfrak{B}$ nor $\mathfrak{C}$ and $\mathfrak{D}$ have any common factors, it follows that $\mathfrak{A} = \mathfrak{C}$ and $\mathfrak{B} = \mathfrak{D}$. The uniqueness is proved, and we have the following theorem:

Theorem 5.7.5. *Every fractional ideal of A can be written in a unique way as a product of prime ideals of A with powers in $\mathbb{Z}$.* □

Corollary 5.7.6. *The set of fractional ideals of A form a free abelian group whose generators are the nonzero prime ideals of A.* □

Theorem 5.7.7. *Let A be a Dedekind domain and let K be the field of quotients of A. Let L/K be a finite extension with $[L : K] = n$. Put*

$$B = \{\alpha \in L \mid \operatorname{Irr}(\alpha, x, K) \in A[x]\}.$$

Then B is a Dedekind domain called the integral closure of A in L.

Proof. We present the proof when L/K is separable. The proof of the general case can be found in [78, Chapter I, Theorem 6.1]. Let $\operatorname{Tr} : L \to K$ be the trace map. Since L/K is a separable extension, it follows that Tr is surjective. If $x \in B$, the conjugate elements of x have the same irreducible polynomials as x. Therefore $\operatorname{Tr} x \in A$. Let $\{e_1, \ldots, e_n\}$ be a basis of L/K with $e_i \in B$ (it is easy to see that if $\alpha \in L$, there exists $a \in A$ such that $a \neq 0$ and $a\alpha \in B$). Let C be the A-free module generated by $\{e_1, \ldots, e_n\}$, that is, $C = \bigoplus_{i=1}^n Ae_i$. For any A-submodule $M \subseteq L$, let $M^* = \{x \in L \mid \operatorname{Tr}(xy) \in A \text{ for all } y \in M\}$.

We have $C \subseteq B \subseteq B^* \subseteq C^*$. Since C^* is the A-free module generated by the dual basis of $\{e_1, \ldots, e_n\}$ with respect to the nondegenerate bilinear form $\operatorname{Tr}(xy)$, it follows that C^* is Noetherian. Therefore B is finitely generated as an A-module. In particular, B is Noetherian.

Now, if $\alpha \in L$ satisfies

$$\alpha^n + b_{n-1}\alpha^{n-1} + \cdots + b_1\alpha + b_0 = 0 \quad \text{with each} \quad b_i \in B,$$

then the A-module $A[\alpha]$ is finitely generated since B is. Set

$$A[\alpha] = \langle x_1, x_2, \ldots, x_m \rangle .$$

Then $\alpha x_i = \sum_{j=1}^m a_{ij}x_j$ for $1 \le i \le m$. Therefore

$$\sum_{j=1}^{m} \left(\delta_{ij}\alpha - a_{ij}\right) x_j = 0, \quad \text{where} \quad \delta_{ij} = \begin{cases} 1 \text{ if } i = j, \\ 0 \text{ if } i \neq j. \end{cases}$$

In terms of matrices we have

$$\begin{bmatrix} \alpha - a_{11} & -a_{12} & \cdots & -a_{1m} \\ \cdot & \cdot & \cdots & \cdot \\ \cdot & \cdot & \cdots & \cdot \\ \cdot & \cdot & \cdots & \cdot \\ -a_{m1} & -a_{m2} & \cdots & \alpha - a_{mm} \end{bmatrix} \begin{bmatrix} x_1 \\ \cdot \\ \cdot \\ \cdot \\ x_n \end{bmatrix} = \begin{bmatrix} 0 \\ \cdot \\ \cdot \\ \cdot \\ 0 \end{bmatrix}.$$

If $M = \left[\delta_{ij}\alpha - a_{ij}\right]_{1\le i,j\le m}$, let N be the adjoint matrix of M. Then $NM = (\det M)I_n$ and $(\det M)x_i = 0$ for $1 \le i \le m$. But $1 \in A \subseteq A[\alpha]$ implies that $(\det M)1 = \det M = 0$. On the other hand,

$$\det M = \alpha^n + c_{n-1}\alpha^{n-1} + \cdots + c_1\alpha + c_0 = 0$$

with $c_i \in A$, so $\alpha \in B$. Therefore B is integrally closed.

Finally, let $\mathcal{P}$ be a nonzero prime ideal of B. Assume for the sake of contradiction that $\mathcal{P}$ is not maximal, and let $\mathfrak{S}$ be a maximal ideal such that $\mathcal{P} \subsetneqq \mathfrak{S} \subsetneqq B$. Now $\mathcal{P} \cap A$ is a nonzero prime ideal of A, and so is $\mathfrak{S} \cap A$. Since A is a Dedekind domain and $\mathcal{P} \cap A$ is a prime ideal of A we have $\mathcal{P} \cap A = \mathfrak{S} \cap A$. Let $x \in \mathfrak{S} \setminus \mathcal{P}$. Then $x \in B$ and x satisfies a relation

$$x^n + a_{n-1}x^{n-1} + \cdots + a_1x + a_0 = 0, \quad \text{with} \quad a_i \in A \quad \text{and} \quad a_0 \neq 0.$$

We have $a_0 \in A \cap \mathfrak{S} = A \cap \mathcal{P}$, which means that

$$a_0 = -x\left(x^{n-1} + a_{n-1}x^{n-2} + \cdots + a_2x + a_1\right) \in \mathcal{P}.$$

Since $x \notin \mathcal{P}$, we have

$$x^{n-1} + a_{n-1}x^{n-2} + \cdots + a_2x + a_1 \in \mathcal{P}.$$

Therefore $a_1 \in \mathfrak{S} \cap A = \mathcal{P} \cap A$, which implies

$$x\left(x^{n-2} + a_{n-1}x^{n-3} + \cdots + a_3x + a_2\right) \in \mathcal{P},$$

and so on. It follows that $a_0, \ldots, a_{n-1} \in \mathcal{P}$. Thus, we obtain that $x + a_{n-1} \in \mathcal{P}$, and consequently $x \in \mathcal{P}$, which is absurd. This proves that $\mathcal{P}$ is in fact maximal, and B is a Dedekind domain. $\square$

5.7.1 Different and Discriminant in Dedekind Domains

The module B^* defined in the proof of Theorem 5.7.7 is a finitely generated A-module. Since $B \subseteq B^* \subseteq L$, the B-module B^* is finitely generated, and hence B^* is a fractional ideal. The inverse of this module is the different. More precisely:

Definition 5.7.8. Let A be a Dedekind domain and put $K = \operatorname{quot} A$. Let L/K be a separable finite extension and B the integral closure of A in L. Define

$$\mathfrak{D}_{B/A}^{-1} := \{x \in L \mid \operatorname{Tr}(xy) \in A \text{ for all } y \in B\}.$$

It is easy to see that $\mathfrak{D}_{B/A}^{-1}$ is a fractional B-module whose inverse $\mathfrak{D}_{B/A}$ is an ideal of B, called the *different* of B over A.

The norm $N_{L/K}\mathfrak{D}_{B/A}$ is an ideal of A called the *discriminant* of B over A.

We will now study the case of function fields in order to relate the two definitions of different.

Let K/k be a function field and let $x \in K \setminus k$. Then $K/k(x)$ is a finite extension.

Clearly, $k[x]$ is a Dedekind domain. Note that there exists a one-to-one correspondence between the prime ideals of $k[x]$ (considered as a ring) and the places of $k(x)$ distinct from the infinite place $\wp_\infty$, that is, from the place given by $(x)_{k(x)} = \frac{\wp_0}{\wp_\infty}$. More precisely, if $\wp$ is a place of $k(x)$ and $\wp \neq \wp_\infty$, the ring $\vartheta_\wp$ is the localization of $k[x]$ at a prime ideal $(f(x))$ of $k[x]$ (see Section 2.4) and the prime ideal $\wp = (f(x))$ corresponds to the ideal $\wp\vartheta_\wp$. Let $\mathcal{P}_1, \dots, \mathcal{P}_r$ be the places of K over $\wp_\infty$.

Theorem 5.7.9. *The integral closure of $k[x]$ in K is $\bigcap \vartheta_{\mathcal{P}}$, where $\mathcal{P}$ runs through all the places of K distinct from $\mathcal{P}_1, \dots, \mathcal{P}_r$.*

Proof. Let ϑ be the integral closure of $k[x]$ in K. If $\alpha \in \vartheta$, we have

$$\alpha^n + p_{n-1}(x)\alpha^{n-1} + \cdots + p_1(x)\alpha + p_0(x) = 0 \quad \text{with} \quad p_i(x) \in k[x].$$

It follows that if $\mathcal{P} \notin \{\mathcal{P}_1, \dots, \mathcal{P}_r\}$, then $v_{\mathcal{P}}(p_i(x)) \geq 0$ for each i. Therefore $v_{\mathcal{P}}(\alpha) \geq 0$ and $\alpha \in \vartheta_{\mathcal{P}}$ whenever $\mathcal{P}$ is distinct from all the $\mathcal{P}_i$'s. Thus $\vartheta \subseteq \bigcap_{\mathcal{P} \notin \{\mathcal{P}_1, \dots, \mathcal{P}_r\}} \vartheta_{\mathcal{P}}$.

Conversely, let $\alpha \in \bigcap_{\mathcal{P} \notin \{\mathcal{P}_1, \dots, \mathcal{P}_r\}} \vartheta_{\mathcal{P}}$ and

$$f(T) = \operatorname{Irr}(\alpha, T, k(x)) = T^n + a_{n-1}T^{n-1} + \cdots + a_1 T + a_0 \quad \text{with} \quad a_i \in k(x).$$

Let $\tilde{K}$ be the normal closure of $K/k(x)$ and let $\alpha^{(1)} = \alpha, \alpha^{(2)}, \dots, \alpha^{(n)}$ be the distinct conjugates of α. Then each a_i is a symmetric function of $\alpha^{(1)}, \alpha^{(2)}, \dots, \alpha^{(n)}$, and for any irreducible polynomial $f(x) \in k[x]$, $v_f(a_i(x)) \geq 0$. Indeed, all extensions $\mathcal{P}$ that are not extensions of $\wp_\infty$ satisfy $v_{\mathcal{P}}(a_i) \geq 0$. This proves that $a_i(x) \in k[x]$. Therefore α is integral over $k[x]$. □

Theorem 5.7.10. *Let K/k be any function field and let $\mathcal{P}_1, \dots, \mathcal{P}_r$, $r \geq 1$, be any finite set of distinct prime divisors. Then there exists an element x of K whose poles are precisely $\mathcal{P}_1, \dots, \mathcal{P}_r$, i.e., $v_{\mathcal{P}_i}(x) < 0$ for $1 \leq i \leq r$ and $v_{\mathcal{P}}(x) \geq 0$ for all $\mathcal{P} \notin \{\mathcal{P}_1, \dots, \mathcal{P}_r\}$.*

Proof. By the Riemann–Roch theorem, there exist $x_i \in K \setminus k$ and $n_i > 0$ such that $\mathfrak{N}_{x_i} = \mathcal{P}_i^{n_i}$ (Corollary 3.5.8). Clearly, $x = x_1 + \cdots + x_r$ is the element satisfying the required property. □

Corollary 5.7.11. *Let $\mathcal{P}_1, \ldots, \mathcal{P}_r$, $r \geq 1$, be any set of places of K. If $\vartheta = \bigcap_{\mathcal{P} \notin \{\mathcal{P}_1, \ldots, \mathcal{P}_r\}} \vartheta_{\mathcal{P}}$, there exists $x \in K \setminus k$ such that ϑ is the integral closure of $k[x]$ in K. In particular, ϑ is a Dedekind domain.*

Proof. Let x be given by the previous theorem. Then $\wp_\infty = \mathcal{P}_1^{n_1} \cdots \mathcal{P}_r^{n_r}$, and hence $\mathcal{P}_1, \ldots, \mathcal{P}_r$ are precisely the prime divisors of K above the infinite prime $\wp_\infty$ of $k(x)$. □

It follows from the above that given a finite collection of prime divisors $\mathcal{P}_1, \ldots, \mathcal{P}_r$ of K, $A = \bigcap_{\mathcal{P} \notin \{\mathcal{P}_1, \ldots, \mathcal{P}_r\}} \vartheta_{\mathcal{P}}$ is a Dedekind domain whose prime ideals are in bijective correspondence with the prime divisors of K distinct from $\mathcal{P}_1, \ldots, \mathcal{P}_r$; indeed, if $\mathcal{P}$ is a prime ideal of A, then $A_{\mathcal{P}}$ is a valuation ring of K. Therefore $A_{\mathcal{P}} = \vartheta_{\mathcal{P}'}$ for some $\mathcal{P}'$ and $\mathcal{P} A_{\mathcal{P}} = \mathcal{P}' \vartheta_{\mathcal{P}'}$ and conversely. In view of this correspondence we may assume that the prime ideals of A are the places of K distinct from $\mathcal{P}_1, \ldots, \mathcal{P}_r$.

In what follows, the set of prime divisors $T = \{\wp_1, \ldots, \wp_r\}$ of K will be fixed.

Let L/K be a finite separable extension of K and $T^* = \{\mathcal{P} \mid \mathcal{P}$ is a place of $L, \mathcal{P} \mid \wp_i$ for some $1 \leq i \leq r\}$. Put $\vartheta_K = \bigcap_{\wp \notin T} \vartheta_\wp$ and let ϑ_L be the integral closure of ϑ_K in L. It is easy to see that $\vartheta_L = \bigcap_{\mathcal{P} \notin T^*} \vartheta_{\mathcal{P}}$ (Exercise 5.10.25). Let $\mathfrak{D}_{L/K}$ be the different as defined in Definition 5.6.5 and let $\mathfrak{D}'_{L/K}$ be the different according to Definition 5.7.8, with $\mathfrak{D}'_{L/K} = \mathfrak{D}_{\vartheta_L/\vartheta_K}$.

Theorem 5.7.12. $\mathfrak{D}_{L/K} = \mathfrak{D}'_{L/K} \prod_{\mathcal{P} \in T^*} \mathcal{P}^{\alpha_{\mathcal{P}}}$ *for some* $\alpha_{\mathcal{P}} \geq 0$.

Proof. First note that if S is a multiplicative set of a Dedekind domain, then $S^{-1}A$ is a Dedekind domain (Exercise 5.10.24).

By Exercise 5.10.26, if A is a Dedekind domain and $K = \operatorname{quot} A$, L/K is a finite separable extension, and B is the integral closure of A in L, then $S^{-1}B$ is the integral closure of $S^{-1}A$ in L. We have $S^{-1}\mathfrak{D}_{B/A} = \mathfrak{D}_{S^{-1}B/S^{-1}A}$ since if $x \in \mathfrak{D}_{B/A}^{-1}$,

$$\operatorname{Tr}(xB) \subseteq A \Longrightarrow \operatorname{Tr}\left(S^{-1}xB\right) = S^{-1}\operatorname{Tr}(xB) \subseteq S^{-1}A$$

and conversely.

Applying the above argument to an arbitrary prime $\wp$ of A, we consider $S = A \setminus \wp$ and we set $S^{-1}\mathfrak{D}'_{L/K} = S^{-1}\mathfrak{D}_{\vartheta_L/\vartheta_K} = \mathfrak{D}_{(\vartheta_L)_\wp/(\vartheta_K)_\wp}$.

Now since A is a Dedekind domain, $A_\wp$ is a discrete valuation ring. In fact, if $\pi \in \wp \setminus \wp^2$ we have $(\pi) = \pi A = \wp\mathfrak{A}$ with $(\mathfrak{A}, \wp) = (1)$, so that $(A \setminus \wp) \cap \mathfrak{A} \neq \emptyset$. Therefore $\mathfrak{A} A_\wp = A_\wp$. Consequently $\pi A_\wp = \wp A_\wp \mathfrak{A} A_\wp = \wp A_\wp$. This shows that the maximal ideal $\wp A_\wp$ is principal. Next, if $\mathfrak{B} A_\wp$ is any nontrivial ideal of A, then $\mathfrak{B} A = \wp^n \mathfrak{C}$ with $(\mathfrak{C}, \wp) = (1)$, $n > 0$, so

$$\mathfrak{B} A_\wp = \wp^n A_\wp \mathfrak{C} A_\wp = \wp^n A_\wp = \left(\pi^n\right) A_\wp.$$

Hence $A_\wp$ is a valuation ring. Furthermore, $A_\wp = \vartheta_\wp$. If $\mathcal{P}$ is any prime ideal of B over $\wp$, we have $B_\mathcal{P} = \vartheta_\mathcal{P}$, from which we obtain that the completions of $\hat{B}_\mathcal{P}$ and $\hat{\vartheta}_\mathcal{P}$ are the same, whence we have $v_\mathcal{P}\left(\mathfrak{D}_{L/K}\right) = v_\mathcal{P}\left(\mathfrak{D}_\mathcal{P}\right) = v_\mathcal{P}\left(\mathfrak{D}_{\hat{B}_\mathcal{P}/\hat{A}_\wp}\right)$ by definition.

We will demonstrate the following: If $A_\wp$ is a discrete valuation ring and $\mathcal{P}$ is an ideal above the maximal ideal $\wp$ of $A_\wp$, let $\hat{A}_\wp$ and $\hat{B}_\mathcal{P}$ be the corresponding completions. Then $\mathfrak{D}_{B_\wp/A_\wp}\hat{B}_\mathcal{P} = \mathfrak{D}_{\hat{B}_\mathcal{P}/\hat{A}_\wp}$.

To prove the latter statement, it suffices to show that $v_\mathcal{P}\left(\mathfrak{D}_{B_\wp/A_\wp}\right) = v_\mathcal{P}\left(\mathfrak{D}_{\hat{B}_\mathcal{P}/\hat{A}_\wp}\right)$.

Let Tr be the trace of L to K, and let $\mathrm{Tr}_\mathcal{P}$ be the trace of $L_\mathcal{P}$ to $K_\wp$. By Corollary 5.5.17 we have $\mathrm{Tr} = \sum_{i=0}^h \mathrm{Tr}_{\mathcal{P}_i}$, where $\mathcal{P}_1, \ldots, \mathcal{P}_h$ are all the primes of B dividing $\wp$. We write $\mathcal{P} = \mathcal{P}_1$. Let $x \in L_\mathcal{P}$ and assume that $\mathrm{Tr}_\mathcal{P}\left(x\hat{B}_\mathcal{P}\right) \subseteq \hat{A}_\wp$, that is, $x \in \mathfrak{D}^{-1}_{\hat{B}_\mathcal{P}/\hat{A}_\wp}$. It follows from the approximation theorem (Theorem 2.5.3) that there exists $\xi \in L$ such that

$$|\xi - x|_\mathcal{P} < \varepsilon \quad \text{and} \quad |\xi|_{\mathcal{P}_i} < \varepsilon, \quad 2 \leq i \leq h, \quad \text{for some small enough } \varepsilon.$$

For $y \in B_\wp$, there exists ε' small enough such that $\left|\mathrm{Tr}_{\mathcal{P}_i}(\xi y)\right|_\wp < \varepsilon'$ for $2 \leq i \leq h$, and $\mathrm{Tr}_\mathcal{P}(\xi y) \in A$ (because the local trace is a continuous function). Therefore $\xi \in \mathfrak{D}^{-1}_{B_\wp/A_\wp}$, and we obtain $\overline{\mathfrak{D}^{-1}_{B_\wp/A_\wp}} \supseteq \mathfrak{D}^{-1}_{\hat{B}_\mathcal{P}/\hat{A}_\wp}$, where the bar denotes closure in $L_\mathcal{P}$.

Conversely, let $x \in \mathfrak{D}^{-1}_{B_\wp/A_\wp}$ and $y \in \hat{B}_\mathcal{P}$. Write $\mathfrak{D}_{B_\mathcal{P}/A_\wp} = \mathcal{P}_1^{n_1}\cdots\mathcal{P}_h^{n_h}$ for $n_i \geq 0$. Then $x \in \mathfrak{D}^{-1}_{B_\wp/A_\wp}$ if and only if $v_{\mathcal{P}_i}(x) \geq -n_i$ for $1 \leq i \leq h$.

Let $\xi \in L$ be such that $v_{\mathcal{P}_1}(\xi - x) = m_1 > v_{\mathcal{P}_1}(x)$ and $v_{\mathcal{P}_i}(\xi - x) = m_i \gg 0$. Notice that in particular, $\xi \in \mathfrak{D}^{-1}_{B_\wp/A_\wp}$.

Now let $\eta \in B_\wp$ be such that η is very close to y with respect to $\mathcal{P}_1$ and very close to 0 with respect to $\mathcal{P}_2, \ldots, \mathcal{P}_h$. Since $\xi \in \mathfrak{D}^{-1}_{B_\wp/A_\wp}$ and $\eta \in B_\wp$, we have $\mathrm{Tr}(\xi\eta) \in A_\wp$. On the other hand, for $1 \leq i \leq r$, $\mathrm{Tr}_{\mathcal{P}_i}(\xi\eta) \in \hat{A}_\wp$, since $\mathrm{Tr}_{\mathcal{P}_i}$ is continuous and ξ and η are very close to 0. Hence $\left|\mathrm{Tr}_{\mathcal{P}_i}(\xi\eta)\right|_\wp < 1$ for $2 \leq i \leq r$.

Since $\mathrm{Tr}(\xi\eta) = \mathrm{Tr}_{\mathcal{P}_1}(\xi\eta) + \sum_{i=2}^h \mathrm{Tr}_{\mathcal{P}_i}(\xi\eta) \in \hat{A}_\wp$, we have $\mathrm{Tr}_{\mathcal{P}_1}(\xi\eta) \in \hat{A}_\wp$. On the other hand, $|\xi\eta - xy|_{\mathcal{P}_1} < \varepsilon$ implies $\mathrm{Tr}_{\mathcal{P}_1}(xy) \in \hat{A}_\wp$ and $x \in \mathfrak{D}^{-1}_{\hat{B}_\mathcal{P}/\hat{A}_\wp}$. Thus we have $\mathfrak{D}^{-1}_{B_\wp/A_\wp}\hat{B}_\mathcal{P} \subseteq \mathfrak{D}^{-1}_{\hat{B}_\mathcal{P}/\hat{A}_\wp}$.

Therefore, $\mathfrak{D}_{B_\mathcal{P}/A_\wp}$ is dense in $\mathfrak{D}_{\hat{B}_\mathcal{P}/\hat{A}_\wp}$, from which we obtain the result.

Finally, we have

$$v_\mathcal{P}\left(\mathfrak{D}'_{L/K}\right) = v_\mathcal{P}\left(\mathfrak{D}'_{B/A}\right) = v_\mathcal{P}\left(\mathfrak{D}_{B_\wp/A_\wp}\right) = v_\mathcal{P}\left(\mathfrak{D}_{\hat{B}_\mathcal{P}/\hat{A}_\wp}\right) = v_\mathcal{P}\left(\mathfrak{D}_{L/K}\right),$$

which is what we wanted to prove. □

Remark 5.7.13. Theorem 5.7.12 can be used to obtain the different of L/K by means of the differents of certain Dedekind domains. For instance, if we take $A_1 = \bigcap_{\wp \neq \wp_1} \vartheta_\wp$, $A_2 = \bigcap_{\wp \neq \wp_2} \vartheta_\wp$, then

$$\mathfrak{D}_{L/K} = \mathfrak{D}_{B_1/A_1} \prod_{\mathcal{P}|\wp_1} \mathcal{P}^{\alpha_{\mathcal{P}}}, \quad \mathfrak{D}_{L/K} = \mathfrak{D}_{B_2/A_2} \prod_{\mathcal{P}|\wp_2} \mathcal{P}^{\beta_{\mathcal{P}}},$$

so $\mathfrak{D}_{L/K}$ is the least common multiple of $\mathfrak{D}_{B_1/A_1}$ and $\mathfrak{D}_{B_2/A_2}$.

By proving Theorem 5.7.12 we have also obtained the following:

Proposition 5.7.14. *Assume that A is a discrete valuation ring with maximal ideal $\mathfrak{p}$. Let $K :=$ quot A, L/K be a finite separable extension, B the integral closure of A in L, and $\mathfrak{P}$ is any ideal of B above $\mathfrak{p}$. Denote by $\hat{A}$ and $\hat{B}$ the completions of A and B at $\mathfrak{p}$ and $\mathfrak{P}$ respectively. Then $\mathfrak{D}_{B/A}\hat{B} = \mathfrak{D}_{\hat{B}/\hat{A}}$.* □.

Theorem 5.7.15. *Let $K \subseteq L \subseteq M$ be a tower of finite separable extensions of function fields. Then*

$$\mathfrak{D}_{M/K} = \mathfrak{D}_{M/L} \operatorname{con}_{L/M} \mathfrak{D}_{L/K}. \qquad \square$$

Proof. Since the number of ramified or inseparable places is finite (Theorem 5.2.33), we may take $A = \cap\, \vartheta_{\wp}$, where $\wp$ runs through any set containing all inseparable and ramified prime divisors. Then by Theorem 5.7.12, it suffices to demonstrate that $\mathfrak{D}_{C/A} = \mathfrak{D}_{C/B} \operatorname{con}_{B/C} \mathfrak{D}_{B/A}$, where B is the integral closure of A in L and C is the integral closure of A in M.

Assume that R is any Dedekind domain, $F =$ quot R, E/F is a finite separable extension, and S is the integral closure of R in E. For a fractional ideal $\mathfrak{B}$ of S, we have

$$\begin{aligned} \operatorname{Tr}\mathfrak{B} \subseteq \mathfrak{A} &\iff \mathfrak{A}^{-1}\operatorname{Tr}\mathfrak{B} \subseteq R \iff \operatorname{Tr}\left(\mathfrak{A}^{-1}\mathfrak{B}\right) \subseteq R \iff \mathfrak{A}^{-1}\mathfrak{B} \subseteq \mathfrak{D}_{S/R}^{-1} \\ &\iff \mathfrak{B} \subseteq \mathfrak{A}\,\mathfrak{D}_{S/R}^{-1}. \end{aligned}$$

Now, coming back to our case, we have

$$\begin{aligned} \mathfrak{C} \subseteq \mathfrak{D}_{C/B}^{-1} &\iff \operatorname{Tr}_{M/L}(\mathfrak{C}) \subseteq B \iff \mathfrak{D}_{B/A}^{-1}\operatorname{Tr}_{M/L}(\mathfrak{C}) \subseteq \mathfrak{D}_{B/A}^{-1} \\ &\iff \operatorname{Tr}_{L/K}\left(\mathfrak{D}_{B/A}^{-1}\operatorname{Tr}_{M/L}(\mathfrak{C})\right) \subseteq A. \end{aligned}$$

Notice that $\mathfrak{D}_{B/A}^{-1} \subseteq L$ and that $\mathfrak{D}_{B/A}^{-1}$ can be considered as a fractional ideal of C. Thus

$$\operatorname{Tr}_{M/L}\left(\operatorname{con}_{B/C}\mathfrak{D}_{B/A}^{-1}\mathfrak{C}\right) = \mathfrak{D}_{B/A}^{-1}\operatorname{Tr}_{M/L}(\mathfrak{C}),$$

or equivalently,

$$\operatorname{Tr}_{M/L}\left(\mathfrak{D}_{B/A}^{-1}\mathfrak{C}\right) = \mathfrak{D}_{B/A}^{-1}\operatorname{Tr}_{M/L}(\mathfrak{C}).$$

Hence

$$\begin{aligned}\operatorname{Tr}_{L/K}\left(\mathfrak{D}_{B/A}^{-1}\operatorname{Tr}_{M/L}(\mathfrak{C})\right)\subseteq A &\Longleftrightarrow \operatorname{Tr}_{L/K}\operatorname{Tr}_{M/L}\left(\operatorname{con}_{B/C}\mathfrak{D}_{B/A}^{-1}\mathfrak{C}\right)\\ =\operatorname{Tr}_{M/K}\left(\operatorname{con}_{B/C}\mathfrak{D}_{B/A}^{-1}\mathfrak{C}\right)\subseteq A &\Longleftrightarrow \operatorname{con}_{B/C}\mathfrak{D}_{B/A}^{-1}\mathfrak{C}\subseteq\mathfrak{D}_{C/A}^{-1}\\ &\Longleftrightarrow \mathfrak{C}\subseteq\operatorname{con}_{B/C}\mathfrak{D}_{B/A}\mathfrak{D}_{C/A}^{-1}.\end{aligned}$$

Therefore, $\mathfrak{D}_{C/B}^{-1}=\operatorname{con}_{B/C}\mathfrak{D}_{B/A}\mathfrak{D}_{C/A}^{-1}$. □

Corollary 5.7.16. *With the hypothesis of Theorem* 5.7.15, *we have*

$$\partial_{M/K}=\partial_{L/K}^{n}N_{L/K}\left(\partial_{M/L}\right),\quad n=[M:L].$$ □

5.7.2 Discrete Valuation Rings and Computation of the Different

Throughout this subsection we will assume that the residue field extensions are separable.

Theorem 5.7.17. *Let A be a Dedekind domain and $K=\operatorname{quot}A$. Let $L=K(\alpha)$ be a finite separable field extension of degree n and let B be the integral closure of A in L. If $B=A[\alpha]$, then $\mathfrak{D}_{B/A}=\left(f'(\alpha)\right)$, where $f(x)=\operatorname{Irr}(\alpha,x,K)$.*

Proof. Considering B as an A-module, we have the basis $\left\{1,\alpha,\dots,\alpha^{n-1}\right\}$. On the other hand, $T=\mathfrak{D}_{B/A}^{-1}$ is the fractional ideal $\{x\in L\mid \operatorname{Tr}(xB)\subseteq A\}$.

Since L/K is separable, the trace is surjective. It follows that $\phi(x,y)=\operatorname{Tr}(xy)$ is a nondegenerate bilinear form. Now assume that $\{\alpha_1,\dots,\alpha_n\}$ is any basis of the A-module B and $\{\beta_1,\dots,\beta_n\}$ is the dual basis. We have

$$\operatorname{Tr}\left(\beta_i\alpha_j\right)=\delta_{ij}=\begin{cases}1 \text{ if } i=j,\\ 0 \text{ if } i\neq j.\end{cases},$$

and hence $\{\beta_1,\dots,\beta_n\}\subseteq T$.

Conversely, if $x\in T$, let $a_i=\operatorname{Tr}(x\alpha_i)\in A$ and $y=x-\sum_{i=1}^{n}a_i\beta_i$. Then

$$\operatorname{Tr}\left(y\alpha_j\right)=\operatorname{Tr}\left(x\alpha_j\right)-\sum_{i=1}^{n}a_i\operatorname{Tr}\left(\beta_i\alpha_j\right)=a_j-a_j=0,\quad j=1,\dots,n,$$

which implies that $y=0$. Therefore

$$x=a_1\beta_1+\cdots+a_n\beta_n,\quad \text{so}\quad T\cong A\beta_1\oplus\cdots\oplus A\beta_n.$$

Put

$$g(x)=\frac{f(x)}{x-\alpha}=b_0+b_1x+\cdots+b_{n-1}x^{n-1}.$$

We will see that

$$\left\{\frac{b_i}{f'(\alpha)}\right\}_{i=0}^{n-1} \quad \text{is the dual basis of} \quad \left\{1, \alpha, \ldots, \alpha^{n-1}\right\}.$$

Let $\alpha = \alpha_1, \ldots, \alpha_n$ be the n distinct roots of $f(x) = \prod_{i=1}^{n}(x - \alpha_i)$. For $0 \leq r \leq n - 1$, consider the polynomial

$$h(x) = x^r - \sum_{i=1}^{n} \frac{f(x)\alpha_i^r}{(x - \alpha_i)\, f'(\alpha_i)}.$$

Since $\left.\frac{f(x)}{(x-\alpha_j)}\right|_{x\,=\,\alpha_j} = f'\left(\alpha_j\right)$, we have

$$h\left(\alpha_j\right) = \alpha_j^r - \sum_{i \neq j} \frac{f(\alpha_j)\alpha_i^r}{\left(\alpha_j - \alpha_i\right) f'(\alpha_i)} + \left(\left.\frac{f(x)}{\left(x - \alpha_j\right)}\right|_{x\,=\,\alpha_j}\right) \frac{\alpha_j^r}{f'\left(\alpha_j\right)} = 0.$$

The degree of $h(x)$ is at most $n - 1$; on the other hand, $h(x)$ has n roots, so $h(x) \equiv 0$. It follows that for $r = 0, 1, \ldots, n - 1$,

$$x^r = \sum_{i=1}^{n} \frac{f(x)\alpha_i^r}{(x - \alpha_i)\, f'(\alpha_i)}.$$

Now

$$\begin{aligned}
\operatorname{Tr} x^r = nx^r &= \sum_{i=1}^{n} \operatorname{Tr}\left(\frac{f(x)\alpha_i^r}{(x - \alpha_i)\, f'(\alpha_i)}\right) = n \operatorname{Tr}\left(\frac{f(x)\alpha^r}{(x - \alpha)\, f'(\alpha)}\right) \Longrightarrow \\
x^r = \operatorname{Tr}\left(\frac{f(x)\alpha^r}{(x - \alpha)\, f'(\alpha)}\right) &= \operatorname{Tr}\left(\frac{1}{f'(\alpha)}\alpha^r\left(b_0 + b_1 x + \cdots + b_{n-1}x^{n-1}\right)\right) \\
&= \sum_{i=0}^{n-1} \operatorname{Tr}\left(\frac{\alpha^r}{f'(\alpha)} b_i\right) x^i \Longrightarrow \operatorname{Tr}\left(\frac{\alpha^i}{f'(\alpha)} b_j\right) = \delta_{ij}.
\end{aligned}$$

Therefore the dual basis of $\{1, \alpha, \ldots, \alpha^n\}$ is $\left\{\frac{b_0}{f'(\alpha)}, \frac{b_1}{f'(\alpha)}, \ldots, \frac{b_{n-1}}{f'(\alpha)}\right\}$. We will see that $A[\alpha] = B = A\left[b_0, \ldots, b_{n-1}\right]$. Indeed,

$$\begin{aligned}
f(x) &= (x - \alpha)\left(b_0 + b_1 x + \cdots + b_{n-1}x^{n-1}\right) \\
&= \sum_{i=0}^{n-1} b_i x^{i+1} - \sum_{i=0}^{n-1} \alpha b_i x^i = b_{n-1}x^n + \sum_{i=1}^{n-1} (b_{i-1} - \alpha b_i)\, x^i - \alpha b_0.
\end{aligned}$$

Hence, if $f(x) = a_0 + a_1 x + \cdots + a_{n-1}x^{n-1} + x^n$ with $a_i \subset A$, we have

$$b_{n-1} = 1 \quad \text{and} \quad b_{i-1} - \alpha b_i = a_i \quad \text{for} \quad 1 \leq i \leq n - 1, \quad \text{and} \quad -\alpha b_0 = a_0.$$

In particular, $A[\alpha] = B = A\left[b_0, \ldots, b_{n-1}\right]$.
Therefore, we have

$$\mathfrak{D}_{L/K}^{-1} = T = \bigoplus_{i=0}^{n-1} A\left[\frac{b_i}{f'(\alpha)}\right] = \frac{A\left[b_0, \ldots, b_{n-1}\right]}{(f'(\alpha))} = \frac{B}{f'(\alpha)} = \left(f'(\alpha)\right)^{-1},$$

so $\mathfrak{D}_{B/A} = \left(f'(\alpha)\right)$. □

Unfortunately, the case $B = A[\alpha]$ is very rare, one instance of this case being our former case when we completed at each prime (Corollary 5.7.20 below). This is the reason why the way to calculate the different is reduced to the complete case.

We begin with the following theorem:

Theorem 5.7.18. *Assume that A is a discrete valuation ring with maximal ideal $\wp$ and that B has only one prime ideal $\mathcal{P}$ over $\wp$. Further, assume that $B/\mathcal{P}$ is a separable extension of $A/\wp$. Then $B = A[\alpha]$ for some $\alpha \in B$.*

Proof. Let $\beta \in B$ be such that $(A/\wp)\left[\bar{\beta}\right] = B/\mathcal{P}$, where $\bar{\beta} = \beta \bmod \mathcal{P}$. Let $f(x) \in A[x]$ be a monic polynomial such that $f(x) \bmod \wp = \operatorname{Irr}\left(\bar{\beta}, x, A/\wp\right)$. Let $\pi \in \mathcal{P}\backslash\mathcal{P}^2$. Then $v_{\mathcal{P}}(\pi) = 1$ and we have

$$f(x) = f(\beta) + f'(\beta)(x-\beta) + \cdots + \frac{f^{n-1}(\beta)}{(n-1)!}(x-\beta)^{n-1} + (x-\beta)^n .$$

Therefore $f(\beta+\pi) \equiv \left(f(\beta) + f'(\beta)\pi\right) \bmod \pi^2$.

Since $B/\mathcal{P}$ is separable over $A/\wp$ we have $f'(\beta) \not\equiv 0 \bmod \pi$.

On the other hand, $\bar{f}\left(\bar{\beta}\right) = 0$ implies $v_{\mathcal{P}}(f(\beta)) \geq 1$. If $v_{\mathcal{P}}(f(\beta)) = 1$, then $f(\beta)$ is a prime element of B. Assume $v_{\mathcal{P}}(f(\beta)) > 1$. Since

$$f(\beta+\pi) - f(\beta) = \pi f'(\beta) \bmod \pi^2,$$

we have

$$v_{\mathcal{P}}(f(\beta+\pi) - f(\beta)) = v_{\mathcal{P}}(\pi) + v_{\mathcal{P}}\left(f'(\beta)\right) = 1,$$

so $v_{\mathcal{P}}(f(\beta+\pi)) = 1$.

In any case, the ring $A[\beta]$ or $A[\beta+\pi]$ contains a prime element of $\mathcal{P}$. Let $\alpha = \beta$ or $\beta+\pi$ be such that $A[\alpha]$ contains a prime element π' of $\mathcal{P}$. Then $A\left[\alpha, \pi'\right] = A[\alpha]$. Furthermore, $\wp B = \mathcal{P}^e$ with $e \geq 1$. Let $C = A[\alpha]$. We will see that $C + \wp B = B$.

Since α generates the residue field, we have $B \subseteq C + \mathcal{P}B$. Now for all $r \geq 0$, $\mathcal{P}^r/\mathcal{P}^{r+1}$ is isomorphic to $B/\mathcal{P}$ under the isomorphism

$$\begin{aligned} B &\to \mathcal{P}^r/\mathcal{P}^{r+1}, \\ x &\mapsto x(\pi')^r, \end{aligned} \quad \text{with} \quad \pi' \in \mathcal{P} \setminus \mathcal{P}^2.$$

In other words, $\left\{\alpha^i(\pi')^j\right\}_{j=0,1,\ldots,e-1}^{i\geq 0}$ generates $B/\mathcal{P}^e B = B/\wp B$ over $A/\wp$. Therefore if $x \in B$, we have $x \equiv \sum_{i,j} d_{ij}\alpha^i(\pi')^j \bmod \wp B$ for some $d_{ij} \in A$, which proves that $C + \wp B = B$.

Now $B/C = (C + \wp B)/C$ implies $\wp(B/C) = (C + \wp B)/C = B/C$, so if $M = B/C$ then $\wp M = M$. Let $\{x_1, x_2, \ldots, x_n\}$ be a set of generators of M over A. We have

$$\left.\begin{array}{l} x_1 = p_{11}x_1 + \cdots + p_{1n}x_n \\ \quad \vdots \\ x_n = p_{n1}x_1 + \cdots + p_{nn}x_n \end{array}\right\} \text{ with } p_{ij} \in \wp.$$

In terms of matrices, this translates to

$$\left[\delta_{ij} - p_{ij}\right]_{i,j} [x_i]_{1 \le i \le n} = [0], \quad \text{with} \quad \delta_{ij} = \begin{cases} 1 \text{ if } i = j, \\ 0 \text{ if } i \neq j. \end{cases}$$

Multiplying by the adjoint matrix of

$$\left[\delta_{ij} - p_{ij}\right]_{i,j} = N, \quad \text{we obtain that} \quad (\det N)x_i = 0 \quad \text{for all} \quad i,$$

that is, $(\det N)M = 0$. Now, $\det N = 1 + x$ for some $x \in \wp$, so $\det N \in A \setminus \wp$. Therefore $\det N$ is a unit, and this implies that $M = 0$. We obtain $B = C = A[\alpha]$. □

Remark 5.7.19. The last part of the proof of Theorem 5.7.18 is known as *Nakayama's lemma*. More precisely, Nakayama's lemma establishes that if A is a ring, $\mathfrak{a}$ is an ideal contained in every maximal ideal of A and M is a finitely generated A-module such that $\mathfrak{a}M = M$, then $M = 0$.

We apply Theorem 5.7.18 to the complete fields case, in which the rings $\tilde{\vartheta}_\wp$ are discrete valuation rings and there exists a unique prime ideal over $\wp$.

Corollary 5.7.20. *Let L/K be a separable extension of function fields such that the field of constants ℓ of L is a perfect field. Let $\mathfrak{P} \in \mathbb{P}_L$ and $\mathfrak{p} := \mathfrak{P}|_K$. Then $\vartheta_{\mathfrak{P}} = \vartheta_{\mathfrak{p}}[\alpha]$ for some $\alpha \in \vartheta_{\mathfrak{P}}$.*

Proof. $\vartheta_{\mathfrak{P}}$ and $\vartheta_{\mathfrak{p}}$ are discrete valuation rings, and $\vartheta_{\mathfrak{P}}/\mathfrak{P} = \ell(\mathfrak{P})$ is a separable extension of $\vartheta_{\mathfrak{p}}/\mathfrak{p} = k(\mathfrak{p})$ (Theorem 5.2.21). □

Theorem 5.7.21. *Let A be a Dedekind domain, $K = \operatorname{quot} A$, let L/K be a finite separable extension, and let B be the integral closure of A in L. Then $\mathfrak{D}_{B/A}$ is the greatest common divisor of the set*

$$\left\{f'(\alpha) \mid \alpha \in B, L = K(\alpha), f(x) = \operatorname{Irr}(\alpha, x, K)\right\}$$
$$= \left\langle f'(\alpha) \mid \alpha \in B, L = K(\alpha), f(x) = \operatorname{Irr}(\alpha, x, K)\right\rangle.$$

Proof. Let $\alpha \in B$, so $A[\alpha] \subseteq B$. By Theorem 5.7.17 we have

$$\mathfrak{D}_{B/A}^{-1} = \{x \in L \mid \operatorname{Tr}(xB) \subseteq A\} \subseteq \{x \in L \mid \operatorname{Tr}(xA[\alpha]) \subseteq A\} = \left(f'(\alpha)\right)^{-1}.$$

Therefore $\left(f'(\alpha)\right) \subseteq \mathfrak{D}_{B/A}$, or, equivalently, $\mathfrak{D}_{B/A} \mid \left(f'(\alpha)\right)$.

To prove the converse, notice that since $\mathfrak{D}_{B/A} = \prod_{\mathcal{P}\in\mathbb{P}_L} \mathfrak{D}_{\mathcal{P}}$, the equality

$$\mathfrak{D}_{B/A} = \left\langle f'(\alpha) \mid \alpha \in B, L = K(\alpha), \quad f(x) = \operatorname{Irr}(\alpha, x, K)\right\rangle$$

holds if for each $\mathcal{P}$ of B, we can find $\alpha \in B$ such that $v_{\mathcal{P}}\left(\mathfrak{D}_{B/A}\right) = v_{\mathcal{P}}\left(f'(\alpha)\right)$ (note that we always have $v_{\mathcal{P}}\left(\mathfrak{D}_{B/A}\right) \leq v_{\mathcal{P}}\left(f'(\alpha)\right)$).

Let $\wp = \mathcal{P}|_A = \mathcal{P} \cap A$. Let $T = \left\{\sigma : L \longrightarrow \bar{K}_{\wp} \mid \sigma|_K = \operatorname{Id}_K\right\}$, where $\bar{K}_{\wp}$ denotes an algebraic closure of $K_{\wp}$. Clearly, $\sigma(L) K_{\wp}$ is a complete field that contains $K_{\wp}$. Thus $\sigma(L) K_{\wp} = L_{\mathcal{P}_i}$ for some i, where $\wp B = \mathcal{P}_1^{e_1} \cdots \mathcal{P}_g^{e_g}$. Hence, $\sigma(L) K_{\wp}$ is one of the completions of $K_{\wp}$.

We define an equivalence relation in T by: $\sigma \sim \tau$ if $\sigma(L) K_{\wp} = \tau(L) K_{\wp}$, or equivalently there exists a $\bar{K}_{\wp}$-automorphism λ such that $\lambda|_{K_{\mathcal{P}}} = \operatorname{Id}_{K_{\mathcal{P}}}$ and $\lambda\sigma|_L = \tau|_L$. Observe that if $[\sigma]$ denotes the equivalence class of $L_{\mathcal{P}_i}$, the distinct classes of $\sim$ are the $K_{\wp}$-monomorphisms of $L_{\mathcal{P}_i}$ in $\bar{K}_{\wp}$. Thus there are $\left[L_{\mathcal{P}_i} : K_{\wp}\right]$ elements in this class. On the other hand we have $|T| = [L : K]$, which coincides with our formula $[L : K] = \sum_{i=1}^{g}\left[L_{\mathcal{P}_i} : K_{\wp}\right]$.

Let $\sigma_1 \in T$ be in the class determined by $L_{\mathcal{P}}$ and $\mathcal{P} = \mathcal{P}_1$. Let

$$\alpha \in L, \quad L = K(\alpha), \quad f(x) = \operatorname{Irr}(\alpha, x, K) = \prod_{\sigma\in T}(x - \sigma\alpha),$$

$$f'(\alpha) = \prod_{\substack{\sigma\in T\\ \sigma\neq \operatorname{Id}}}(\alpha - \sigma\alpha), \quad \text{and} \quad \sigma_1\left(f'(\alpha)\right) = \prod_{\substack{\sigma\in T\\ \sigma\neq\sigma_1}}(\sigma_1\alpha - \sigma\alpha) = f'(\sigma_1\alpha).$$

By Theorem 5.7.18, there exists $\beta \in B_{\mathcal{P}}$ such that $B_{\mathcal{P}} = A_{\wp}[\beta]$. Since A and B are Dedekind domains, their localizations are discrete valuation rings (see the proof of Theorem 5.7.12). Observe that if $\beta' \in B_{\mathcal{P}}$, is such that $\left|\beta - \beta'\right| < \varepsilon$ for ε small enough, then $A_{\wp}\left[\beta'\right] = B_{\mathcal{P}}$. Indeed, put $B_{\mathcal{P}} = \bigoplus_{i=0}^{r-1} A_{\wp}\beta^i$ and let $\Pi \in \mathcal{P} \setminus \mathcal{P}^2$, $\wp B_{\mathcal{P}} = \mathcal{P}^e$, choose $\varepsilon \leq \frac{1}{r^2}|\Pi|^e$. Let $\left|\beta - \beta'\right| < \varepsilon$. We have that $x \in B_{\mathcal{P}}$ satisfies $|x|_{\mathcal{P}} \leq 1$, so

$$\begin{aligned}
&\left|\sum_{i=0}^{r-1} a_i\beta^i - \sum_{i=0}^{r-1} a_i(\beta')^i\right|_{\mathcal{P}} \\
&\qquad \leq \sum_{i=1}^{r-1} |a_i|_{\mathcal{P}} \left|\beta - \beta'\right|_{\mathcal{P}} \left(\left|\beta^{i-1} + \beta^{i-2}\beta' + \cdots + (\beta')^{i-1}\right|_{\mathcal{P}}\right) \\
&\qquad\qquad < \varepsilon r r = \varepsilon r^2 \leq |\Pi|^e,
\end{aligned}$$

so $A_{\wp}\left[\beta'\right] + \wp B_{\mathcal{P}} = B_{\mathcal{P}} = A_{\wp}[\beta]$. By the same argument as that given in the proof of Theorem 5.7.18 (Nakayama's lemma), we obtain $A_{\wp}\left[\beta'\right] = A_{\wp}[\beta] = B_{\mathcal{P}}$.

For $\lambda \in T$, we denote by $L_{\mathcal{P}_\lambda}$ the completion given by $\lambda(L) K_{\wp} \subseteq \bar{K}_{\wp}$.

Now, if $\{\lambda\}$ varies in a finite set of $K_{\wp}$-automorphisms of $\bar{K}_{\wp}$, then the elements $\lambda\beta$ have residue classes conjugated over $A_{\wp}/\wp$ since $\lambda|_{K_{\wp}} = \operatorname{Id}$. Therefore, if these classes are zero, then $|\lambda\beta|_{\mathcal{P}_\lambda} < 1$, and hence $|\lambda\beta - 1|_{\mathcal{P}_\lambda} = 1$. If these classes are nonzero, then $|\lambda\beta|_{\mathcal{P}_\lambda} = |\lambda\beta - 0|_{\mathcal{P}_\lambda} = 1$.

In any case, $|\lambda\beta - a|_{\mathcal{P}_\lambda} = 1$ for a equal to 0 or 1. Let $\sigma_1, \dots, \sigma_g \in T$ be representatives of the distinct classes corresponding to completions $L_{\mathcal{P}_1}, \dots, L_{\mathcal{P}_g}$. By Artin's approximation (Theorem 2.5.3), there exists $\alpha \in L$ such that $|\sigma_1\alpha - \beta|_{\mathcal{P}}$ and for $2 \le i \le g$, $|\sigma_i\alpha - a|_{\mathcal{P}_i}$ are very small. We may assume that $\alpha \in B$ (see Exercise 5.10.27).

If $L \neq K(\alpha)$, we write $\alpha_1 = \alpha + \pi^t\gamma$, where $L = K(\gamma)$, γ is an integral element, and $v_\wp(\pi) = 1$. Then α_1 is integral. We will see that for t large enough, $L = K(\alpha_1)$. Let $E = K(\alpha_1) \subseteq L$. If for each completion the equality $L_{\mathcal{P}'} = E_{\mathcal{P}'}$ holds, then

$$[E : K] = \sum_{\mathcal{P}'|\wp} \left[E_{\mathcal{P}'} : K_\wp\right] = \sum_{\mathcal{P}'|\wp} \left[L_{\mathcal{P}'} : K_\wp\right] = [L : K],$$

so $E = L$. Therefore, it suffices to see that $L_{\mathcal{P}'} = E_{\mathcal{P}'}$.

Assume that K is a complete field and $L = K(\gamma) = K(\alpha_1 - \alpha) = K(\alpha_1, \alpha)$. Let t be such that

$$|\alpha_1 - \alpha| = |\pi|^t\,|\gamma| < |\sigma\alpha - \alpha|$$

for any isomorphism σ of $K(\alpha)$ satisfying $\alpha \neq \mathrm{Id}$. Whenever τ is a $K(\alpha_1)$-monomorphism of $K(\alpha_1, \alpha)$ into an algebraic closure over $K(\alpha_1)$, we have $\tau(\alpha_1 - \alpha) = \alpha_1 - \tau\alpha$. Recall that the unique extension of the absolute value of a complete field is given by $|\xi| = |N\xi|^{1/n}$ (Theorem 5.4.7). Since $\tau(\alpha_1 - \alpha)$ and $\alpha_1 - \alpha$ have the same norm over $K(\alpha_1)$, we have

$$|\alpha_1 - \tau\alpha| = |\alpha_1 - \alpha| < |\sigma\alpha - \alpha| \quad \text{for} \quad \sigma \neq \mathrm{Id}.$$

Thus

$$\begin{aligned} |\tau\alpha - \alpha| = |\tau\alpha - \alpha_1 + \alpha_1 - \alpha| &\le \max\{|\tau\alpha - \alpha_1|, |\alpha_1 - \alpha|\} \\ &< |\alpha - \sigma\alpha| \quad \text{for} \quad \sigma \neq \mathrm{Id}. \end{aligned}$$

Therefore $\tau = \mathrm{Id}$, and in particular, $K(\alpha_1, \alpha) = K(\alpha_1)$.

Returning to our case, we may assume that $L = K(\alpha_1)$ by setting $\alpha_1 = \alpha + \pi^t\gamma$. Again, we denote α_1 by α.

It follows from that fact that $|\sigma_1\alpha - \beta|_{\mathcal{P}_1}$ is small that $B_{\mathcal{P}} = A_\wp[\sigma_1\alpha]$ (see the proof above). Now $\mathfrak{D}_{\mathcal{P}} = \mathcal{P}^s$ for some $s \ge 0$. Since this different is given by

$$\left(g'(\alpha)\right) = \Bigg(\prod_{\substack{\sigma\sim\sigma_1 \\ \sigma\neq\sigma_1}} (\sigma_1\alpha - \sigma\alpha)\Bigg),$$

we have $s = \sum_{\substack{\sigma\sim\sigma_1 \\ \sigma\neq\sigma_1}} v_{\mathcal{P}}(\sigma_1\alpha - \sigma\alpha) = v_{\mathcal{P}}\left(\mathfrak{D}_{B/A}\right)$.

Finally, it remains to prove that

$$v_{\mathcal{P}}\left(\prod_{\substack{\sigma\not\sim\sigma_1 \\ \sigma\in T}} (\sigma_1\alpha - \sigma\alpha)\right) = \sum_{\substack{\sigma\not\sim\sigma_1 \\ \sigma\in T}} v_{\mathcal{P}}(\sigma_1\alpha - \sigma\alpha) = 0,$$

or equivalently, that

$$|\sigma_1\alpha - \sigma\alpha|_{\mathcal{P}} = 1 \quad \text{whenever} \quad \sigma \not\sim \sigma_1.$$

Suppose that

$$\sigma \not\sim \sigma_1, \quad \text{where} \quad \sigma = \lambda\sigma_i \quad \text{for some} \quad 2 \leq i \leq g,$$

and

$$\begin{aligned} |\sigma_1\alpha - \sigma\alpha|_{\mathcal{P}} = |\sigma_1\alpha - \lambda\sigma_i\alpha|_{\mathcal{P}} &= \left|\lambda^{-1}\sigma_1\alpha - \sigma_i\alpha\right|_{\mathcal{P}_{\lambda^{-1}}} \\ &= \left|\lambda^{-1}\sigma_1\alpha - a + a - \sigma_i\alpha\right|_{\mathcal{P}_{\lambda^{-1}}} \end{aligned}$$

since $|a - \sigma_i\alpha|_{\mathcal{P}_{\lambda^{-1}}}$ was chosen to be small enough. We have

$$|\sigma_1\alpha - \sigma\alpha|_{\mathcal{P}} = \left|\lambda^{-1}\sigma_1\alpha - a\right|_{\mathcal{P}_{\lambda^{-1}}} = \left|\lambda^{-1}\sigma_1\alpha - \lambda^{-1}\beta + \lambda^{-1}\beta - a\right|_{\mathcal{P}_{\lambda^{-1}}}.$$

Also, $\left|\lambda^{-1}\sigma_1\alpha - \lambda^{-1}\beta\right|_{\mathcal{P}_{\lambda^{-1}}} = |\sigma_1\alpha - \beta|_{\mathcal{P}}$ is small enough, so we obtain $|\sigma_1\alpha - \sigma\alpha|_{\mathcal{P}} = \left|\lambda^{-1}\beta - a\right|_{\mathcal{P}_{\lambda^{-1}}} = 1$, which proves the theorem. □

For an application of Theorem 5.7.21 see Examples 5.8.8 and 5.8.9 below.

Remark 5.7.22. The argument used to prove $K(\alpha, \alpha_1) = K(\alpha_1)$ is known as *Krasner's lemma*:

Theorem 5.7.23 (Krasner's Lemma). *Let K be a field that is complete under a valuation. Let α, β belong to an algebraic closure of K and assume that α is separable over $K(\beta)$. If for any monomorphism $\sigma \neq \mathrm{Id}$ of $K(\alpha)$ into an algebraic closure of K over K we have*

$$|\beta - \alpha| < |\sigma\alpha - \sigma|,$$

then $K(\alpha) \subseteq K(\beta)$.

Proof. Exercise 5.10.28. □

5.8 Ramification in Artin–Schreier and Kummer Extensions

We begin this section with a theorem due to Kummer that establishes the decomposition of a prime ideal in Dedekind domains. First, we present a particular case that is much easier to prove, and next we give the general function field case.

Theorem 5.8.1 (Kummer's Theorem). *Let A be a Dedekind domain, $K = \operatorname{quot} A$, and let L/K be a finite separable extension of K. Let B be the integral closure of A in L. Assume that $B = A[\alpha]$ for some α. Put $f(x) = \operatorname{Irr}(\alpha, x, K)$ and let $\wp$ be a nonzero prime ideal of A. Let $\bar{f}$ be the reduction modulo $\wp$, i.e., $\bar{f}(x) \in A/\wp[x]$. Let $\bar{f}(x) = \bar{p}_1(x)^{e_1} \cdots \bar{p}_g(x)^{e_g}$ be the decomposition as a product of irreducible polynomials in $A/\wp[x]$. Then*

$$\wp B = \mathcal{P}_1^{e_1} \cdots \mathcal{P}_g^{e_g}$$

where

$$\mathcal{P}_i = \wp B + p_i(\alpha) B \quad \text{for} \quad 1 \le i \le g,$$

with $p_i(x)$ a monic polynomial in $A[x]$ whose reduction modulo $\wp$ is $\bar{p}_i(x)$.

Proof. Let $\bar{p}$ be any irreducible factor of $\bar{f}$, $\bar{\alpha}$ a root of $\bar{p}$, and $\mathfrak{S}$ the prime ideal of B that is the kernel of the natural epimorphism

$$B = A[\alpha] \longrightarrow \bar{A}[\bar{\alpha}], \quad \bar{A} = A/\wp.$$

Then $\wp B + p(\alpha) B \subseteq \mathfrak{S}$. Conversely, if $g(\alpha) \in \mathfrak{S}$ with $g(x) \in A[x]$, we have $\bar{g}(\bar{\alpha}) = 0$, which implies that $\bar{g} = \bar{p}\bar{h}$ with $\bar{h} \in \bar{A}[x]$. Hence $g - ph \in \wp[x]$ and $g(\alpha) \in \wp B + p(\alpha) B$, from which we obtain $\wp B + p(\alpha) B = \mathfrak{S}$.

Since $[B/\mathfrak{S} : A/\wp] = [\bar{A}[\bar{\alpha}] : \bar{A}] = \deg \bar{p}_i$, the inertia degree of $\mathfrak{S}$ is precisely the degree of $\bar{p}_i$, whence for each i such that $1 \le i \le g$,

$$\mathcal{P}_i = \wp B + p_i(\alpha) B$$

is a prime ideal that lies above $\wp$. Furthermore, if $i \neq j$ then $\mathcal{P}_i \neq \mathcal{P}_j$, since otherwise $p_i(\alpha) = p_j(\alpha) a + tb$, for $t \in \wp$ and $a, b \in B$. Therefore $\bar{p}_i(x) - a\bar{p}_j(x) = 0$, which is impossible since $p_i(x)$ and $p_j(x)$ are distinct irreducible polynomials of $A/\wp[x]$.

Let $\mathfrak{S}_i = \wp B + p_i(\alpha)^{e_i} B$. It is clear that $\mathfrak{S}_i = \mathcal{P}_i^{e_i'}$ for some e_i'. Now, we have

$$\prod_{i=1}^{g} \mathfrak{S}_i \subseteq \wp B + p_1(\alpha)^{e_1} \cdots p_g(\alpha)^{e_g} B \subseteq \wp B.$$

Therefore $\mathcal{P}_1, \ldots, \mathcal{P}_g$ are all the ideals over $\wp$. Furthermore, for $1 \le i \le g$, $\wp \subseteq \mathfrak{S}_i$, so

$$\wp B \subseteq \bigcap_{i=1}^{g} \mathfrak{S}_i = \prod_{i=1}^{g} \mathfrak{S}_i = \mathcal{P}_1^{e_1'} \cdots \mathcal{P}_g^{e_g'} \subseteq \wp B.$$

It follows that $\wp B = \mathcal{P}_1^{e_1'} \cdots \mathcal{P}_g^{e_g'}$. Moreover,

$$\mathcal{P}_1^{e_1} \cdots \mathcal{P}_g^{e_g} \subseteq \mathfrak{S}_1 \cdots \mathfrak{S}_g = \mathcal{P}_1^{e_1'} \cdots \mathcal{P}_g^{e_g'},$$

which implies that $e_i \geq e'_i$ for $1 \leq i \leq g$.

Finally, we have the analogue of Theorem 5.1.14, namely

$$[L : K] = \sum_{i=1}^{g} e'_i \deg \mathcal{P}_i = \sum_{i=1}^{g} e'_i \deg \bar{p}_i \leq \sum_{i=1}^{g} e_i \deg p_i = \deg f(x) = [L : K],$$

and hence $e_i = e'_i$ for $1 \leq i \leq g$. □

Theorem 5.8.2 (Kummer's Theorem). *Let K/k be a function field and let $\mathfrak{p}$ be a place of K. Assume that $L = K(\alpha)$, where α is integral over $\vartheta_{\mathfrak{p}}$. Let $p(T) = \operatorname{Irr}(\alpha, T, K) \in \vartheta_{\mathfrak{p}}[T]$ be the minimal polynomial of α over K, and let*

$$\bar{p}(T) := p(T) \bmod \mathfrak{p} = \prod_{i=1}^{r} \bar{p}_i(T)^{a_i}$$

be the decomposition of $\bar{p}(T)$ in $k(\mathfrak{p})[T]$. Let $p_i(T) \in \vartheta_{\mathfrak{p}}[T]$ be such that $\deg p_i(T) = \deg \bar{p}_i(T)$ and $p_i(T) \bmod \mathfrak{p} = \bar{p}_i(T)$ for $1 \leq i \leq r$.

Then there exist r different places $\mathfrak{P}_i$ of L above $\mathfrak{p}$ such that $p_i(\alpha) \in \mathfrak{P}_i$ and $d_{L/K}(\mathfrak{P}_i|\mathfrak{p}) \geq \deg \bar{p}_i(T)$.

Assume furthermore that $a_i = 1$ for $1 \leq i \leq r$ or $\{1, \alpha, \ldots, \alpha^{n-1}\}$ is an integral basis for $\mathfrak{p}$, where $n = [L : K]$. Then $\mathfrak{P}_1, \ldots, \mathfrak{P}_r$ are all the places of L above $\mathfrak{p}$,

$$\operatorname{con}_{K/L} \mathfrak{p} = \prod_{i=1}^{r} \mathfrak{P}_i^{a_i}, \quad \vartheta_{\mathfrak{P}_i}/\mathfrak{P}_i \cong \frac{k(\mathfrak{p})[T]}{(\bar{p}_i(T))},$$

and hence $d_{L/K}(\mathfrak{P}_i|\mathfrak{p}) = \deg \bar{p}_i(T)$.

Proof. Let $k(\mathfrak{p})_i := \frac{k(\mathfrak{p})[T]}{(\bar{p}_i(T))}$ for $1 \leq i \leq r$. Then $[k(\mathfrak{p})_i : k(\mathfrak{p})] = \deg \bar{p}_i(T)$. Consider the natural ring epimorphism

$$\pi : \vartheta_{\mathfrak{p}}[T] \to \vartheta_{\mathfrak{p}}[\alpha] \qquad \text{and} \qquad \pi_i : \vartheta_{\mathfrak{p}}[T] \to k(\mathfrak{p})_i,$$

defined by

$$\pi(f(T)) = f(\alpha) \qquad \text{and} \qquad \pi_i(f(T)) = \bar{f}(T) \bmod \bar{p}_i(T).$$

Then $\ker \pi = (p(T))$ and $\pi_i(p(T)) = 0$. Therefore $\ker \pi \subseteq \ker \pi_i$ for $1 \leq i \leq r$, and π_i induces a ring epimorphism

$$\varrho_i : \vartheta_{\mathfrak{p}}[\alpha] \to k(\mathfrak{p})_i$$

such that $\varrho_i \circ \pi = \pi_i$, i.e.,

$$\varrho_i(h(\alpha)) = \bar{h}(T) \bmod \bar{p}_i(T).$$

Notice that $\mathfrak{p}\vartheta_{\mathfrak{p}}[\alpha] \subseteq \ker \varrho_i$ and $p_i(\alpha)\vartheta_{\mathfrak{p}}[\alpha] \subseteq \ker \varrho_i$. It follows that

$$\mathfrak{p}\vartheta_{\mathfrak{p}}[\alpha] + p_i(\alpha)\vartheta_{\mathfrak{p}}[\alpha] \subseteq \ker \varrho_i.$$

Conversely, let $h(\alpha) = \sum_{j=0}^{n-1} b_j\alpha^j \in \ker \varrho_i$, with $h(T) \in \vartheta_{\mathfrak{p}}[T]$. We have

$$\bar{h}(T) = \bar{p}_i(T)\bar{g}(T) \quad \text{with} \quad g(T) \in \vartheta_{\mathfrak{p}}[T].$$

Thus

$$h(T) - p_i(T)g(T) \in \mathfrak{p}\vartheta_{\mathfrak{p}}[T] \quad \text{and} \quad h(\alpha) - p_i(\alpha)g(\alpha) \in \mathfrak{p}\vartheta_{\mathfrak{p}}[\alpha].$$

Therefore $h(\alpha) \in \mathfrak{p}\vartheta_{\mathfrak{p}}[\alpha] + p_i(\alpha)\vartheta_{\mathfrak{p}}[\alpha]$, and we have

$$\ker \varrho_i = \mathfrak{p}\vartheta_{\mathfrak{p}}[\alpha] + p_i(\alpha)\vartheta_{\mathfrak{p}}[\alpha]. \tag{5.6}$$

By Theorem 2.4.4 there exists a place $\mathfrak{P}_i$ of L extending ϱ_i (note that $\ker \varrho_i \neq 0$). Therefore $\vartheta_{\mathfrak{p}}[\alpha] \subseteq \vartheta_{\mathfrak{P}_i}$, so $\mathfrak{P}_i \mid \mathfrak{p}$ and $p_i(\alpha) \in \mathfrak{P}_i$. Furthermore,

$$k(\mathfrak{p}) \subseteq k(\mathfrak{p})_i \cong \vartheta_{\mathfrak{p}}[\alpha]/\ker \varrho_i \subseteq \vartheta_{\mathfrak{P}_i}/\mathfrak{P}_i.$$

Thus

$$d_{L/K}(\mathfrak{P}_i|\mathfrak{p}) = \left[\vartheta_{\mathfrak{P}_i}/\mathfrak{P}_i : \vartheta_{\mathfrak{p}}/\mathfrak{p}\right] = \left[k(\mathfrak{P}_i) : k(\mathfrak{p})\right] \geq \left[k(\mathfrak{p})_i : k(\mathfrak{p})\right] = \deg \bar{p}_i(T).$$

For $i \neq j$, $\bar{p}_i(T)$ and $\bar{p}_j(T)$ are distinct irreducible polynomials in $\left(\vartheta_{\mathfrak{p}}/\mathfrak{p}\right)[T] = k(\mathfrak{p})[T]$. Hence there exist $\bar{A}(T), \bar{B}(T) \in k(\mathfrak{p})[T]$ such that

$$1 = \bar{A}(T)\bar{p}_i(T) + \bar{B}(T)\bar{p}_j(T).$$

It follows that $\bar{A}(\alpha)\bar{p}_i(\alpha) + \bar{B}(\alpha)\bar{p}_j(\alpha) - 1 \in \mathfrak{p}\vartheta_{\mathfrak{p}}[\alpha]$. Thus $1 \in \ker \varrho_i + \ker \varrho_j$ and $\mathfrak{P}_i \neq \mathfrak{P}_j$ since $\ker \varrho_i \subseteq \mathfrak{P}_i$ and $\ker \varrho_j \subseteq \mathfrak{P}_j$. This proves the first part of the theorem.

Now assume that $a_i = 1$ for all $1 \leq i \leq r$. We have

$$p(T) = \prod_{i=1}^{r} \bar{p}_i(T).$$

From Theorem 5.1.14, we obtain

$$\begin{aligned}[L : K] = \deg p(T) &= \sum_{i=1}^{r} \deg \bar{p}_i(T) \leq \sum_{i=1}^{r} d_{L/K}(\mathfrak{P}_i|\mathfrak{p}) \\ &\leq \sum_{i=1}^{r} d_{L/K}(\mathfrak{P}_i|\mathfrak{p}) e_{L/K}(\mathfrak{P}_i|\mathfrak{p}) \leq [L : K].\end{aligned}$$

It follows that $e_{L/K}(\mathfrak{P}_i|\mathfrak{p}) = 1$, $d_{L/K}(\mathfrak{P}_i|\mathfrak{p}) = \deg \bar{p}_i(T)$, and $\mathfrak{P}_1, \ldots, \mathfrak{P}_r$ are all the prime divisors in L dividing $\mathfrak{p}$.

Now assume that $\{1, \alpha, \ldots, \alpha^{n-1}\}$ is an integral basis for $\mathfrak{p}$. If ϑ is the integral closure of $\vartheta_{\mathfrak{p}}$ in L, then $\vartheta = \vartheta_{\mathfrak{p}}[\alpha]$.

Let $\mathfrak{P}$ be any place of L above $\mathfrak{p}$. We have

$$0 = p(\alpha) \equiv \prod_{i=1}^{r} p_i(\alpha)^{a_i} \bmod \mathfrak{p},$$

so $p(\alpha) \in \mathfrak{P}$. Therefore $p_i(\alpha) \in \mathfrak{P}$ for some i such that $1 \leq i \leq r$. We have

$$\ker \varrho_i \subseteq \mathfrak{P} \cap \vartheta_{\mathfrak{p}}[\alpha]. \tag{5.7}$$

It follows from the maximality of the ideal $\ker \varrho_i$ that

$$\ker \varrho_i = \mathfrak{P} \cap \vartheta_{\mathfrak{p}}[\alpha] = \mathfrak{P}_i \cap \vartheta_{\mathfrak{p}}[\alpha]. \tag{5.8}$$

Since $\{1, \alpha, \ldots, \alpha^{n-1}\}$ is an integral basis for $\mathfrak{p}$, we have $\vartheta_{\mathfrak{p}}[\alpha] = \bigcap_{\mathfrak{P}|\mathfrak{p}} \vartheta_{\mathfrak{P}}$. By Artin's approximation theorem (Corollary 2.5.6), there exists $y \in L$ such that $v_{\mathfrak{P}}(y) > 0$ and $v_{\mathfrak{B}}(y) = 0$ for all $\mathfrak{B} \neq \mathfrak{P}$ such that $\mathfrak{B} \mid \mathfrak{p}$. It follows that $y \in \bigcap_{\mathfrak{B}|\mathfrak{p}} \vartheta_{\mathfrak{B}}$ and $y \in \mathfrak{P}$. Using (5.8) we obtain that $y \in \mathfrak{P}_i$ and $v_{\mathfrak{P}_i}(y) > 0$. Hence $\mathfrak{P} = \mathfrak{P}_i$ for some i, that is, $\mathfrak{P}_1, \ldots, \mathfrak{P}_r$ are all the prime divisors above $\mathfrak{p}$.

Next we will prove that $d_{L/K}(\mathfrak{P}_i|\mathfrak{p}) = \deg p_i(T)$. Again, using Artin's approximation theorem we obtain $\beta_i \in L$ such that $v_{\mathfrak{P}_i}(\beta_i) = 1$ and $v_{\mathfrak{P}_j}(\beta_i) = 0$ for $j \neq i$ and $1 \leq i \leq r$. Let π be a prime element of $\mathfrak{p}$, that is, $v_{\mathfrak{p}}(\pi) = 1$. Then by (5.6) and (5.8),

$$\beta_i \in \vartheta_{\mathfrak{p}}[\alpha] \cap \mathfrak{P}_i = p_i(\alpha)\vartheta_{\mathfrak{p}}[\alpha] + \mathfrak{p}\vartheta_{\mathfrak{p}}[\alpha] = p_i(\alpha)\vartheta_{\mathfrak{p}}[\alpha] + \pi\vartheta_{\mathfrak{p}}[\alpha].$$

We write $\beta_i = p_i(\alpha)s_i(\alpha) + \pi t_i(\alpha)$ with $s_i(\alpha), t_i(\alpha) \in \vartheta_{\mathfrak{p}}[\alpha]$. Then

$$\prod_{i=1}^{r} \beta_i^{a_i} = s(\alpha) \prod_{i=1}^{r} p_i(\alpha)^{a_i} + \pi t(\alpha)$$

for some $s(\alpha), t(\alpha) \in \vartheta_{\mathfrak{p}}[\alpha]$.

Since $p(\alpha) \equiv \prod_{i=1}^{r} p_i(\alpha)^{a_i} \bmod \pi\vartheta\mathfrak{p}[\alpha]$ and $p(\alpha) = 0$, we have

$$\prod_{i=1}^{r} \beta_i^{a_i} = \pi u(\alpha) \quad \text{with} \quad u(\alpha) \in \vartheta_{\mathfrak{p}}[\alpha].$$

In particular, $a_j = v_{\mathfrak{P}_j}\left(\prod_{i=1}^{r} \beta_i^{a_i}\right) \geq v_{\mathfrak{P}_j}(\pi) = e(\mathfrak{P}_j|\mathfrak{p})$.

Now, by (5.8) we have

$$k(\mathfrak{p})_i \cong \vartheta_{\mathfrak{p}}[\alpha]/\ker \varrho_i = \vartheta_{\mathfrak{p}}[\alpha]/(\mathfrak{P}_i \cap \vartheta_{\mathfrak{p}}[\alpha]).$$

Let $\varphi \colon \vartheta_{\mathfrak{p}}[\alpha] \to \vartheta_{\mathfrak{P}_i}/\mathfrak{P}_i$ be defined by $\varphi(h(\alpha)) = h(\alpha) \bmod \mathfrak{P}_i$. Clearly, φ is a ring homomorphism and $\ker \varphi = \mathfrak{P}_i \cap \vartheta_{\mathfrak{p}}[\alpha] = \ker \varrho_i$. If $y \in \vartheta_{\mathfrak{P}_i}$, by Artin's approximation theorem there exists $z \in L$ such that $v_{\mathfrak{P}_i}(y - z) > 0$ and $v_{\mathfrak{P}_j}(z) \geq 0$ for all $j = 1, \ldots, r$ such that $j \neq i$. Thus $z \in \bigcap_{j=1}^{r} \vartheta_{\mathfrak{P}_j} = \vartheta_{\mathfrak{p}}[\alpha]$ and $y \equiv z \bmod \mathfrak{P}_i$, so $\varphi(z) = y \bmod \mathfrak{P}_i$. Hence φ is an epimorphism and

$$k(\mathfrak{p})_i \cong \vartheta_{\mathfrak{p}}[\alpha]/\ker \varrho_i = \vartheta_{\mathfrak{p}}[\alpha]/(\mathfrak{P}_i \cap \vartheta_{\mathfrak{p}}[\alpha]) = \vartheta_{\mathfrak{p}}[\alpha]/\ker \varphi \cong \vartheta_{\mathfrak{P}_i}/\mathfrak{P}_i.$$

It follows that

$$d_{L/K}(\mathfrak{P}_i|\mathfrak{p}) = \left[\vartheta_{\mathfrak{P}_i}/\mathfrak{P}_i : k(\mathfrak{p})\right] = [k(\mathfrak{p})_i : k(\mathfrak{p})] = \deg p_i(T).$$

Using Theorem 5.1.14, we obtain that

$$\begin{aligned}[L:K] &= \sum_{i=1}^{r} e_{L/K}(\mathfrak{P}_i|\mathfrak{p}) d_{L/K}(\mathfrak{P}_i|\mathfrak{p}) \\ &\leq \sum_{i=1}^{r} a_i \deg p_i(T) = \deg p(T) = [L:K].\end{aligned}$$

In particular, we get $a_i = e_{L/K}(\mathfrak{P}_i \mid \mathfrak{p})$ and $\operatorname{con}_{K/L} \mathfrak{p} = \prod_{i=1}^{r} \mathfrak{P}_i^{a_i}$. □

Now we recall the basic facts about Kummer and Artin–Schreier extensions. Let K/k be any function field.

Theorem 5.8.3. *Let L/K be a cyclic extension of degree n. Let $G = \operatorname{Gal}(L/K) = \langle\sigma\rangle$. Consider $\alpha \in L$. Then*

(i) $\operatorname{Tr}_{L/K} \alpha = 0$ *if and only if there exists $\beta \in L$ such that $\alpha = \beta - \sigma\beta$.*
(ii) $N_{L/K}\alpha = 1$ *if and only if there exists $\beta \in L$ such that $\alpha = \beta/\sigma\beta$.*

Proof.

(i) (⇐) If $\alpha = \beta - \sigma\beta$, then

$$\operatorname{Tr}_{L/K} \alpha = \operatorname{Tr}_{L/K} \beta - \operatorname{Tr}_{L/K}(\sigma\beta) = \operatorname{Tr}_{L/K} \beta - \operatorname{Tr}_{L/K} \beta = 0.$$

(⇒) Since L/K is a separable extension, there exists $\gamma \in L$ such that $\operatorname{Tr}_{L/K} \gamma = a \neq 0$, with $a \in K$. Then $\operatorname{Tr}_{L/K}(a^{-1}\gamma) = a^{-1} \operatorname{Tr}_{L/K} \gamma = 1$. Assume that $\operatorname{Tr}_{L/K} \alpha = 0$. We have $\sigma^0\alpha = -\sum_{j=1}^{n-1} \sigma^j \alpha$.
Let $\beta = \sum_{i=0}^{n-2} \left(\sum_{j=0}^{i} \sigma^j \alpha\right)\sigma^i \gamma$. Then $\beta - \sigma\beta = \alpha$.

(ii) This is just Hilbert's Theorem 90 (Theorem A.2.16), for a cyclic group G. □

Theorem 5.8.4 (Artin–Schreier Extensions). *Let* $\operatorname{char} k = p > 0$. *Then L/K is a cyclic extension of degree p if and only if there exists $z \in L$ such that $L = K(z)$ with* $\operatorname{Irr}(z, T, K) = T^p - T - a \in K[T]$.

Proof. (⇒) Let $G = \operatorname{Gal}(L/K) = \langle\sigma\rangle$, with $o(\sigma) = p$. Then $\operatorname{Tr}_{L/K} 1 = p1 = 0$. By Theorem 5.8.3, there exists $z \in L$ such that $\sigma z - z = 1$ or $\sigma z = z + 1$. Hence $\sigma^i z = z + i$ and $\sigma^i z - z$ if and only if $p \mid i$. Therefore

$$\operatorname{Irr}(z, T, K) = \prod_{i=0}^{p-1} (T - (z + i))$$

is of degree p.

Notice that

$$\sigma(z^p - z) = (\sigma z)^p - \sigma z = (z+1)^p - (z+1) = z^p - z.$$

Hence

$$z^p - z = a \in K \quad \text{and} \quad z^p - z - a = 0.$$

It follows that $\operatorname{Irr}(z, T, K) = T^p - T - a$ (and $T^p - T - a = \prod_{i=0}^{p-1}(T - (z+i))$, $a = z^p - z$).

($\Leftarrow$) If $L = K(z)$ and $\operatorname{Irr}(z, T, K) = T^p - T - a$, then for any $i \in \mathbb{Z}$,

$$i^p \equiv i \bmod p \quad \text{and} \quad (z+i)^p - (z+i) = z^p + i^p - z - i = z^p - z = a.$$

Therefore $z, z+1, \ldots, z+(p-1)$ are the roots of $\operatorname{Irr}(z, T, K)$. In particular, z and $z+1$ are conjugates over K and $L = K(z)$ is a Galois extension over K. Let $G = \operatorname{Gal}(L/K)$. There exists $\sigma \in G$ such that $\sigma z = z + 1$. Then $\sigma^i z = z + i$ and $o(\sigma) = p$. Thus $G = \langle \sigma \rangle$ is a cyclic extension of degree p. □

Theorem 5.8.5 (Kummer Extensions). *Let* $\operatorname{char} k = p \geq 0$ *and let* $n \in \mathbb{N}$ *be such that* $p \nmid n$ *(*n *can be chosen arbitrarily in the case* $p = 0$*). Suppose that* k *contains a primitive root of unity* ζ_n*. Then* L/K *is a cyclic extension of degree* n *if and only if there exists* $z \in L$ *such that* $L = K(z)$ *and*

$$\operatorname{Irr}(z, T, K) = T^n - a \in K[T].$$

Proof. ($\Rightarrow$) Let $G = \operatorname{Gal}(L/K) = \langle \sigma \rangle$ and $o(\sigma) = n$. We have $N_{L/K}\zeta_n = \zeta_n^n = 1$. Thus, by Theorem 5.8.3 there exists $z \in L$ such that $\sigma z = \zeta_n z$. Since $\sigma^i z = \zeta_n^i z$ and $\sigma^i z = z$ if and only if $n \mid i$, it follows that $z, \zeta_n z, \ldots, \zeta_n^{n-1} z$ are distinct conjugates of z. Thus

$$\operatorname{Irr}(z, T, K) = \prod_{i=0}^{n-1}(T - \zeta_n^i z).$$

On the other hand, $\sigma(z^n) = (\sigma z)^n = (\zeta_n z)^n = z^n$. Hence $z^n = a \in K$ and $z, \zeta_n z, \ldots, \zeta_n^{n-1} z$ are the roots of $T^n - a \in K[T]$. Therefore

$$\operatorname{Irr}(z, T, K) = T^n - a \quad \text{and} \quad z^n = a \in K.$$

($\Leftarrow$) For $a \neq 0$, $T^n - a$ is a separable polynomial with distinct roots $z, \zeta_n z, \ldots, \zeta_n^{n-1} z$, where z is any element of the algebraic closure $\bar{K}$ of K such that $z^n = a$. Therefore $L = K(z)$ is a normal and separable extension of K, and L/K is a Galois extension. Now, since $T^n - a$ is assumed to be irreducible, z and $\zeta_n z$ are conjugates over K. Thus, there exists $\sigma \in G = \operatorname{Gal}(L/K)$ such that $\sigma z = \zeta_n z$. It follows that $o(\sigma) = n = o(G) = [L : K]$ and L/K is a cyclic extension of degree n. □

Next, we turn our attention to the case that two cyclic extensions L_1/K and L_2/K of the type considered in Theorems 5.8.4 and 5.8.5 are the same.

Proposition 5.8.6. *Let* char $k = p > 0$ *and let* $L_i = K(z_i)/K$, $i = 1, 2$, *be two cyclic extensions of degree* p *given by* $z_i^p - z_i = a_i \in K$, $i = 1, 2$. *The following are equivalent:*

(i) $L_1 = L_2$.
(ii) $z_1 = jz_2 + b$ *for* $1 \le j \le p-1$ *and* $b \in K$.
(iii) $a_1 = ja_2 + (b^p - b)$ *for* $1 \le j \le p-1$ *and* $b \in K$.

Proof. If $z_1 = jz_2 + b$, then $z_2 = j'z_1 - j'b$ with $jj' \equiv 1 \bmod p$. Thus $L_1 = L_2$. Conversely, if $L_1 = L_2$, then if $G = \mathrm{Gal}(L_1/K) = \mathrm{Gal}(L_2/K) = \langle\sigma\rangle$, we may choose σ such that $\sigma z_1 = z_1 + 1$. Now, since σz_2 is a conjugate of z_2 over K, we have $\sigma z_2 = z_2 + j'$ with $1 \le j' \le p-1$. Let $1 \le j \le p-1$ be such that $jj' \equiv 1 \bmod p$. Then

$$\sigma(jz_2) = j\sigma z_2 = jz_2 + jj' = jz_2 + 1.$$

Therefore $\sigma(z_1 - jz_2) = z_1 - jz_2$. It follows that $z_1 - jz_2 = b \in K$.

Next, if $z_1 = jz_2 + b$, then

$$\begin{aligned} z_1^p - z_1 = a_1 &= (jz_2 + b)^p - (jz_2 + b) = j(z_2^p - z_2) + (b^p - b) \\ &= ja_2 + (b^p - b). \end{aligned}$$

Conversely, if $a_1 = ja_2 + (b^p - b)$ we have $z_1^p - z_1 = (jz_2 + b)^p - (jz_2 + b)$, i.e.,

$$(z_1 - (jz_2 + b))^p - (z_1 - (jz_2 + b)) = 0.$$

It follows that $\omega = z_1 - jz_2 - b$ is a root of $\omega^p - \omega = 0$. Thus $\omega \in \mathbb{F}_p$. □

Proposition 5.8.7. *Let* char $k = p \ge 0$ *and let* K *contain a primitive* n*th root* ζ_n *of* 1 *with* $(n, p) = 1$. *Let* $L_i = K(z_i)$ $(i = 1, 2)$ *be two cyclic extensions of* K *of degree* n, *given by* $z_i^n = a_i$. *The following are equivalent:*

(i) $L_1 = L_2$.
(ii) $z_1 = z_2^j c$ *for all* $1 \le j \le n-1$ *such that* $(j, n) = 1$ *and* $c \in K$.
(iii) $a_1 = a_2^j c^n$ *for all* $1 \le j \le n-1$ *such that* $(j, n) = 1$ *and* $c \in K$.

Proof The equivalence of (ii) and (iii) is clear.

Assume $L_1 = L_2$. If $G = \mathrm{Gal}(L_1/K) = \mathrm{Gal}(L_2/K) = \langle\sigma\rangle$, choose σ such that $\sigma z_1 = \zeta_n z_1$. Now, σz_2 is a conjugate of z_2 over K, so

$$\sigma z_2 = \zeta_n^{j'} z_2 \quad \text{with} \quad 1 \le j' \le n-1.$$

Let $d = (j', n)$. Then $\sigma^{n/d} z_2 = \zeta_n^{j'n/d} z_2 = z_2$, and hence $\sigma^{n/d} = \mathrm{Id}$. Since $o(\sigma) = n$, we have $d = (j', n) = 1$. Choose j such that $jj' \equiv 1 \bmod n$. Thus $\sigma(z_2^j) = \zeta_n^{jj'} z_2^j = \zeta_n z_2^j$, and

$$\sigma(z_1 z_2^{-j}) = z_1 z_2^{-j}, \quad \text{so} \quad z_1 z_2^{-j} = c \in K.$$

Conversely, if $z_1 = z_2^j c \in L_2$, $(j, n) = 1$, and $c \in K$, then $L_1 \subseteq L_2$, and if $jj' \equiv 1 \bmod n$,

$$z_1^j = z_2^{jj'} c^j = z_2^{1+\ell n} c^j = z_2 a_2^\ell c^j, \quad \text{so} \quad z_2 = z_1^j a_2^{-\ell} c^{-j} \in L_1.$$

Therefore $L_1 = L_2$. □

In order to study ramification in Artin–Schreier and Kummer extensions we provide the following two examples due to Hasse [52]. These two examples are for the case of rational function fields. The general case will be given later on.

Example 5.8.8. Let $K = k(x)$ be a rational function field where k is a perfect field of characteristic $p > 0$. Let $L = K(y)$ be a cyclic extension of degree p. Then, since L/K is an Artin–Schreier extension, y satisfies an equation of the form

$$y^p - y = r(x), \ \text{where}\ r(x) \in k(x) \ \text{and}\ r(x) \notin \left\{ g(x)^p - g(x) \mid g(x) \in k(x) \right\}.$$

It is easy to see that $\alpha(T) = \operatorname{Irr}(y, T, k(x)) = T^p - T - r(x)$. The roots of the latter polynomial are all $y + i$ such that $i \in \mathbb{F}_p$. Observe that $L = K(z)$, where

$$z^p - z = h(x) \in k(x) \iff z = jy + m(x), \ \text{with}\ m(x) \in k(x) \ \text{and}\ j \in \mathbb{F}_p^*.$$

Note that by substituting y by $jy + m(x)$, with $m(x) \in k(x)$ and $j \in \mathbb{F}_p^*$, the resulting expression for $r(x)$ becomes $jr(x) + m(x)^p - m(x)$.

We will see that we can substitute y in such a way that $r(x)$ takes the form

$$(r(x))_K = \frac{\mathfrak{C}}{\wp_1^{\lambda_1} \cdots \wp_s^{\lambda_s}},$$

where $\mathfrak{C}$ is integral divisor relatively prime to $\wp_i$, $\lambda_i > 0$, and $\lambda_i \not\equiv 0 \bmod p$ for $i = 1, \ldots, s$.

First, write

$$r(x) = \frac{g(x)}{f(x)}, \quad \text{where} \quad f(x) = \prod_{i=1}^{n} p_i(x)^{\alpha_i},$$

$(f(x), g(x)) = 1$, and $p_1(x), \ldots, p_n(x)$ are distinct irreducible polynomials. Using partial fractions we obtain that the expression for $r(x)$ is

$$\frac{g(x)}{f(x)} = s(x) + \sum_{i=1}^{n} \sum_{k=0}^{\alpha_i - 1} \frac{t_k^{(i)}(x)}{p_i(x)^{\alpha_i - k}}$$

with

$$\deg t_k^{(i)}(x) < \deg p_i(x) \quad \text{for} \quad k = 0, 1, \ldots, \alpha_i - 1.$$

Let $v_{\wp_i}$ be the valuation over $k(x)$ corresponding to $p_i(x)$. We have

$$v_{\wp_i}(r(x)) = -\alpha_i \quad \text{and} \quad v_{\wp}(r(x)) \geq 0 \quad \text{for any} \quad \wp \neq \wp_1, \dots, \wp_n, \wp_\infty.$$

Then $v_\wp(y^p - y) \geq 0$, and since $v_\wp(y^p - y) \geq \min\{pv_\wp(y), v_\wp(y)\}$, it follows that $v_\wp(y) \geq 0$. Thus y is integral with respect to $A_\wp$, or in other words, $y \in \vartheta_{\mathcal{P}}$ for a place $\mathcal{P}$ above $\wp$.

Now,

$$\alpha(T) = \prod_{i=0}^{p-1}(T - y - i) \quad \text{and} \quad \alpha'(T) = \sum_{i=0}^{p-1}\prod_{j\neq i}(T - y - j),$$

so

$$\alpha'(y) = \prod_{j=1}^{p-1}(y - y - j) = \prod_{j=1}^{p-1}(-j),$$

and $\left(\alpha'(y)\right)_L$ is the unit divisor $\mathfrak{N}$. Therefore $\wp$ is unramified (Theorem 5.6.3).

It follows that the only ramified places can be $\wp_1, \dots, \wp_n, \wp_\infty$.

Returning to our decomposition, if p divides α_i, we write $\alpha_i = \lambda_i p$. Then

$$r(x) = \frac{t_0^{(i)}(x)}{p_i(x)^{\lambda_i p}} + t_1(x) \quad \text{with} \quad v_{\wp_i}(t_1(x)) > -\lambda_i p.$$

Since $[k[x]/(p_i(x)) : k] < \infty$ and k is a perfect field, $M = k[x]/(p_i(x))$ is perfect, that is, $M^p = M$. Thus there exists $m(x) \in k[x]$ such that

$$m(x)^p \equiv t_0^{(i)}(x) \bmod p_i(x).$$

Let $n(x) = -\frac{m(x)}{p_i(x)^{\lambda_i}}$. If $u = y + n(x)$, then $L = K(u) = K(y)$, and we have

$$\begin{aligned} u^p - u &= y^p - y + n(x)^p - n(x) = r(x) + n(x)^p - n(x) \\ &= \frac{t_0^{(i)}(x)}{p_i(x)^{\lambda_i p}} + t_1(x) - \frac{m(x)^p}{p_i(x)^{\lambda_i p}} + \frac{m(x)}{p_i(x)^{\lambda_i}} = h(x). \end{aligned}$$

Finally,

$$v_{\wp_i}(h(x)) \geq \min\left\{v_{\wp_i}\left(\frac{t_0^{(i)}(x) - m(x)^p}{p_i(x)^{\lambda_i p}}\right), v_{\wp_i}(t_1(x)), v_{\wp_i}\left(\frac{m(x)}{p_i(x)^{\lambda_i}}\right)\right\},$$

$$v_{\wp_i}\left(\frac{t_0^{(i)}(x) - m(x)^p}{p_i(x)^{\lambda_i p}}\right) \geq 1 - \lambda_i p > -\lambda_i p;$$

$$v_{\wp_i}(t_1(x)) > -\lambda_i p,$$

$$v_{\wp_i}\left(\frac{m(x)}{p_i(x)^{\lambda_i}}\right) \geq 0 - \lambda_i > -\lambda_i p.$$

Therefore $(h(x))_{k(x)} = \frac{\mathfrak{A}}{\wp_i^\beta}$, where $\beta < \lambda_i p$ and $\mathfrak{A}$ is relatively prime to $A_{\wp_i}$. Observe that for $j \neq i$, we have

$$v_{\wp_j}\left(\frac{t_0^{(i)}(x)}{p_i(x)^{\lambda_i p}}\right) \geq 0;$$
$$v_{\wp_j}\left(\frac{m(x)^p}{p_i(x)^{\lambda_i p}}\right) \geq 0;$$
$$v_{\wp_j}\left(\frac{m(x)}{p_i(x)^{\lambda_i}}\right) \geq 0;$$
$$v_{\wp_j}(t_1(x)) = v_{\wp_j}(r(x)) = -\alpha_j < 0.$$

Thus $v_{\wp_j}(h(x)) = v_{\wp_j}(t_1(x)) = v_{\wp_j}(r(x)) = -\alpha_j$. This means that in the previous argument, the values $v_{\wp_j}$ do not change for $j \neq i$. We also have $v_\wp(h(x)) \geq 0$ for $\wp \neq \wp_1, \ldots, \wp_n, \wp_\infty$.

Continuing with this process, we eventually transform our expression $L = K(\omega)$ into

$$\omega^p - \omega = \alpha(x) \in k(x) \quad \text{and} \quad (\alpha(x))_{k(x)} = \frac{\mathfrak{C}}{\wp_1^{\lambda_1} \cdots \wp_m^{\lambda_m}} \wp_\infty^s,$$

where $\mathfrak{C}$ is an integral divisor that is relatively prime to $\wp_1, \ldots, \wp_m, \wp_\infty$, and $\lambda_i > 0$, $(\lambda_i, p) = 1, i = 1, \ldots, m$.

Now working with $\wp_\infty$, if $s \geq 0$ or $s < 0$ and $(p, s) = 1$, $\wp_\infty$ is also of the required form. Finally, assume $s < 0$ and $p \mid s$, say $s = -pt$, with $t > 0$. Let

$$\alpha(x) = \frac{f_1(x)}{g_1(x)}, \quad s = v_\infty(\alpha(x)) = -\deg \alpha(x) = -\deg f_1(x) + \deg g_1(x) < 0.$$

Then $\deg g_1(x) < \deg f_1(x)$, and by the division algorithm,

$$f_1(x) = g_1(x)q_1(x) + r_1(x) \quad \text{with} \quad r_1(x) = 0 \quad \text{or} \quad \deg r_1(x) < \deg g_1(x),$$

so

$$\alpha(x) = \frac{f_1(x)}{g_1(x)} = q_1(x) + \frac{r_1(x)}{g_1(x)}.$$

We have $v_\infty\left(\frac{r_1(x)}{g_1(x)}\right) > 0$. Therefore

$$v_\infty(\alpha(x)) = s = -pt = -\deg q_1(x).$$

We can write $q_1(x)$ as the sum of ax^{pt} with terms of lower degree. Since $q_1(x) \in k[x]$ and k is a perfect field, there exists $b \in k$ such that $b^p = a$. Let $\omega_1 = \omega - bx^t$. Then

$$\omega_1^p - \omega_1 = q_2(x) + \frac{r_1(x)}{g_1(x)} \quad \text{with} \quad \deg q_2(x) \le pt - 1 < -s.$$

It is easy to see that any place $\wp$ satisfies the following: if $v_\wp(\alpha(x)) \ge 0$ then $v_\wp\left(q_2(x) + \frac{r_1(x)}{g_1(x)}\right) \ge 0$, and if $v_\wp(\alpha(x)) < 0$, then $v_\wp\left(q_2(x) + \frac{r_1(x)}{g_1(x)}\right) = v_\wp(\alpha(x))$.

By iterating this process we obtain an equation of the type

$$y^p - y = \alpha(x), \quad \text{where} \quad (\alpha(x))_K = \frac{\mathfrak{C}}{\wp_1^{\lambda_1} \cdots \wp_m^{\lambda_m}},$$

$\mathfrak{C}$ is an integral divisor relatively prime to $\wp_1, \ldots, \wp_m$, $\lambda_i > 0$, and $(\lambda_i, p) = 1$, $i = 1, \ldots, m$. We have already noted that if $\wp \neq \wp_1, \ldots, \wp_m$, then $\wp$ is unramified.

Now we will see that $\wp_1, \ldots, \wp_m$ are exactly the ramified prime divisors. If $\mathcal{P}$ is a place over some $\wp_i$, then

$$e = e(\mathcal{P}|\wp_i) \quad \text{and} \quad v_{\mathcal{P}}(\alpha(x)) = e v_{\wp_i}(\alpha(x)) = -e\lambda_i.$$

On the other hand,

$$v_{\mathcal{P}}(\alpha(x)) = v_{\mathcal{P}}\left(y^p - y\right) < 0, \quad \text{so} \quad v_{\mathcal{P}}(y) < 0.$$

Therefore $v_{\mathcal{P}}(y^p - y) = p v_{\mathcal{P}}(y)$. Thus p divides $e\lambda_i$ and since $(p, \lambda_i) = 1$, p divides e. Consequently $e \ge p$. But since $[L : K] = p \ge e$, we must have $e = p$, and furthermore, each $\wp_1, \ldots, \wp_m$ is ramified. Let $p_i(x) \in k[x]$ (or $p_i(x) = \frac{1}{x}$ in the case $\wp_i = \wp_\infty$), with $v_{\wp_i}(p_i(x)) = 1$. Let $\wp_i = \mathcal{P}_i^p$ in D_L. Set $\mathcal{P} = \mathcal{P}_i$.

We have $v_{\mathcal{P}_i}(y) = -\lambda_i$. We wish to compute $\mathfrak{D}_{\mathcal{P}}$, the different at $\mathcal{P}$.

Let π be a prime element for $\mathcal{P}$, that is, $v_{\mathcal{P}}(\pi) = 1$. Then $\vartheta_{\hat{\mathcal{P}}} = \vartheta_{\hat{\wp}_i}[\pi]$ (because $\wp_i$ is ramified) and we have $\mathfrak{D}_{\hat{\mathcal{P}}} = \hat{\mathcal{P}}^s$ with $s = v_{\mathcal{P}}\left(g'(\pi)\right)$ and $g(T) = \operatorname{Irr}(\pi, T, K)$ (Theorem 5.7.17).

Now, since $(\lambda_i, p) = 1$, there exist u, v such that $-u\lambda_i + vp = 1$. We have

$$v_{\mathcal{P}}\left(y^u p_i(x)^v\right) = u v_{\mathcal{P}}(y) + v v_{\mathcal{P}}(p_i(x)) = -u\lambda_i + vp = 1.$$

Therefore we may pick $\pi = y^u p_i(x)^v$. The conjugates of π are the elements $(y + j)^u p_i(x)^v$, so that $g(T) = \prod_{j=0}^{p-1}(T - (y + j)^u p_i(x)^v)$ and

$$g'(\pi) = \prod_{j=1}^{p-1}\left((y + j)^u - y^u\right) p_i(x)^v = \prod_{j=1}^{p-1}\left(ujy^{u-1} + s_j(y)\right) p_i(x)^v,$$

where $s_j(y) = \sum_{\ell=0}^{u-2} \binom{u}{\ell} y^\ell j^{u-\ell}$, and

$$v_{\mathcal{P}}\left(\binom{u}{\ell} y^\ell j^{u-\ell}\right) = \ell v_{\mathcal{P}}(y) > (u-1) v_{\mathcal{P}}(y).$$

It follows that

$$v_{\mathcal{P}}\left(g'(\pi)\right) = v_{\mathcal{P}}\left(\prod_{j=1}^{p-1}\left(juy^{u-1}p_i(x)^v\right)\right) = v_{\mathcal{P}}\left(\left(\frac{y^u p_i(x)^v}{y}\right)^{p-1}\right)$$
$$= \left(-(u-1)\lambda_i + vp\right)(p-1) = (\lambda_i+1)(p-1).$$

Therefore $\mathfrak{D}_{\mathcal{P}} = \mathcal{P}^{(\lambda_i+1)(p-1)}$.

In short, assume $L = K(y)$, where

$$y^p - y = \alpha(x) \quad \text{and} \quad (\alpha(x))_{k(x)} = \frac{\mathfrak{C}}{\wp_1^{\lambda_1}\cdots\wp_n^{\lambda_n}},$$

$\mathfrak{C}$ an integral divisor relatively prime to $\wp_1, \ldots, \wp_n$, $\lambda_i > 0$, and $(\lambda_i, p) = 1$, $1 \le i \le n$. Then $\wp_1, \ldots, \wp_n$ are the ramified primes in L/K and if $\wp_i = \mathcal{P}_i^p$ in D_L, we have

$$\mathfrak{D}_{L/K} = \prod_{i=1}^{n} \mathcal{P}_i^{(\lambda_i+1)(p-1)} \quad \text{and} \quad \partial_{L/K} = N_{L/K}\mathfrak{D}_{L/K} = \prod_{i=1}^{n} \wp_i^{(\lambda_i+1)(p-1)}.$$

Example 5.8.9. Let $K = k(x)$ and $L = K(y)$, where L/K is a cyclic extension of degree n, $p \nmid n$, and $p = \operatorname{char} k$ (or $\operatorname{char} k = 0$). Assume that k contains the nth roots of unity. Then, since L/K is a Kummer extension, we may assume $y^n = f(x)$ with $f(x) \in k[x]$ and $f(x)$ nondivisible nth-powers.

Let

$$(f(x))_{k(x)} = \frac{\wp_1^{\lambda_1}\cdots\wp_r^{\lambda_r}}{\wp_\infty^t}, \quad \text{where} \quad t = \deg f(x) \quad \text{and} \quad 0 < \lambda_i < n.$$

As in Example 5.8.8, $\wp_1, \ldots, \wp_r$ are the ramified prime divisors, and possibly $\wp_\infty$ too. For $\wp_\infty$, let $t = nq + r$, $1 \le r \le n$. Substituting y by $z = \frac{y}{x^{q+1}}$ we obtain

$$z^n = \frac{y^n}{x^{(q+1)n}} = \frac{f(x)}{x^{nq+n}}$$

and

$$v_\infty\left(\frac{f(x)}{x^{nq+n}}\right) = v_\infty\left(f(x)\right) - v_\infty\left(x^{(q+1)n}\right) = -t + qn + n = n - r$$

with $0 \le n - r \le n-1$. As before, $\wp_\infty$ is ramified $\iff n - r \neq 0 \iff n \neq r \iff n \nmid t = \deg f(x)$.

Let $\wp_i$ be one of the ramified prime divisors. Since $(p, n) = 1$, p does not divide the ramification index e of $\wp_i$. We have in D_L: $\wp_i = \left(\mathcal{P}_1^{(i)}\cdots\mathcal{P}_{g_i}^{(i)}\right)^e$. Let $\mathcal{P}$ be any prime above $\wp_i$. We have $v_{\mathcal{P}}\left(y^n\right) = n v_{\mathcal{P}}(y) = \lambda_i e$. Therefore $v_{\mathcal{P}}(y) = \frac{\lambda_i e}{n}$. Let $d_i = (\lambda_i, n)$. We have $\frac{n}{d_i} v_{\mathcal{P}}(y) = \frac{\lambda_i}{d_i} e$, and since $\left(\frac{n}{d_i}, \frac{\lambda_i}{d_i}\right) = 1$,

$$\frac{n}{d_i} \,\Big|\, e \quad \text{and} \quad \frac{\lambda_i}{d_i} \,\Big|\, v_{\mathcal{P}}(y).$$

Now if $z = y^{n/d_i}$ then $z^{d_i} = y^n$ and $d_i \mid \lambda_i$. Hence

$$\left(\frac{z}{p_i(x)^{\lambda_i/d_i}}\right)^{d_i} = z_1^{d_i} = h(x) \in k[x] \quad \text{and} \quad v_{\wp_i}(h(x)) = 0.$$

Therefore $\wp_i$ is unramified from K to $K(z)$. Since $[L : K(z)] = \frac{n}{d_i}$, we have $e \leq \frac{n}{d_i}$, which shows that $e = \frac{n}{d_i}$ and $v_{\mathcal{P}}(y) = \frac{\lambda_i}{d_i}$.

Since $p \nmid e$, it follows by Theorem 5.6.3 that $\mathfrak{D}_{\mathcal{P}} = \mathcal{P}^{e-1} = \mathcal{P}^{(n/d_i)-1}$. If $\wp_i = \left(\mathcal{P}_1^{(i)} \cdots \mathcal{P}_{g_i}^{(i)}\right)^{n/d_i}$, we have $\frac{n}{d_i} f_i g_i = n$, where each f_i is the relative degree of $\mathcal{P} = \mathcal{P}_j^{(i)}$ over $\wp_i$. Finally, note that if $\wp_\infty$ is ramified, then $n \nmid t = \deg f(x)$ and the ramification index is

$$e_\infty = \frac{n}{(n-r, n)} = \frac{n}{(r, n)} = \frac{n}{(t, n)}.$$

Therefore the discriminant at $\wp_i$ is given by

$$\mathfrak{d}_{\wp_i} = \wp_i^{(n/d_i - 1) f_i g_i} = \wp_i^{(n/d_i - 1) d_i} = \wp_i^{d_i(e_i - 1)}.$$

In the general case, it is not always possible to write all prime divisors at a time under the form prescribed in Examples 5.8.8 and 5.8.9. However, the following result shows that we can do so for any fixed prime divisor in the case of a perfect field of constants.

Theorem 5.8.10. *Let k be a perfect field of characteristic $p > 0$. Let $\mathfrak{p}$ be a fixed place in K. If L/K is a cyclic extension of degree p, then $L = K(y)$ with $y^p - y = a$ and*

$$v_{\mathfrak{p}}(a) \geq 0 \quad \text{or} \quad v_{\mathfrak{p}}(a) = \lambda < 0, \quad \text{and} \quad (\lambda, p) = 1.$$

Proof. Let $L = K(z)$ with $z^p - z = B$. If $v_{\mathfrak{p}}(B) \geq 0$, we set $a = B$ and we are done. Assume that $v_{\mathfrak{p}}(B) = \mu < 0$. If $(\mu, p) = 0$ there is nothing to prove. Otherwise, let $\mu = -p\lambda$, $\lambda > 0$. By Theorem 2.5.20, we have

$$B = \frac{b_{-p\lambda}}{\pi^{p\lambda}} + \frac{b_{-p\lambda+1}}{\pi^{p\lambda-1}} + \cdots + \frac{b_{-1}}{\pi} + b_0 + b_1\pi + \cdots, \tag{5.9}$$

where $b_i \in k(\mathfrak{p})$, $b_{-p\lambda} \neq 0$, and π is a prime element for $\mathfrak{p}$.

Since $k(\mathfrak{p})$ is a perfect field, we may choose $c \in k(\mathfrak{p})$ such that $c^p = b_{-p\lambda}$. Let $C \in \vartheta_{\mathfrak{p}}$ be such that $C \bmod \mathfrak{p} = c \in k(\mathfrak{p}) = \vartheta_{\mathfrak{p}}/\mathfrak{p}$. Set $y := z - C\pi^{-\lambda}$, $L = K(y)$, and

$$y^p - y = z^p - C^p\pi^{-p\lambda} - z + C\pi^{-\lambda} = B - C^p\pi^{-p\lambda} + C\pi^{-\lambda}.$$

Since $v_{\mathfrak{p}}(C) = 0$, it follows by (5.9) that

$$v_{\mathfrak{p}}(a) \geq -p\lambda + 1 \quad \text{with} \quad a = B - C^p\pi^{-p\lambda} + C\pi^{-\lambda}.$$

If $v_{\mathfrak{p}}(a) \geq 0$ or $v_{\mathfrak{p}}(a) < 0$ and $(v_{\mathfrak{p}}(a), p) = 1$ we are done. Otherwise, we repeat the process. We obtain the result in a finite number of steps. □

Theorem 5.8.11. *In the situation of Theorem 5.8.10, if $v_{\mathfrak{p}}(a) \geq 0$, then $\mathfrak{p}$ is unramified (in this case the hypothesis that k is a perfect field is not necessary), and if $v_{\mathfrak{p}}(a) < 0$ and $(v_{\mathfrak{p}}(a), p) = 1$ then $\mathfrak{p}$ is ramified and the local different is given by*

$$\mathfrak{D}_{\mathfrak{P}} = \mathfrak{P}^{(\lambda+1)(p-1)},$$

where $\mathfrak{p} = \mathfrak{P}^p$ and $\lambda = -v_{\mathfrak{p}}(a)$.

Proof. Let $f(T) = T^p - T - a = \operatorname{Irr}(y, T, K)$. First, assume that $v_{\mathfrak{p}}(a) \geq 0$. Since $y^p - y = a$, if $\mathfrak{P}$ is any place in L above $\mathfrak{p}$, we have $v_{\mathfrak{P}}(y) \geq 0$. Thus y is integral with respect to $\mathfrak{P}$. Now $f'(y) = -1$, and by Theorem 5.7.21 it follows that $\mathfrak{P}$ is unramified. Note that for this case we do not need the hypothesis that k is a perfect field.

Next, assume that $v_{\mathfrak{p}}(a) = -\lambda < 0$ and $(\lambda, p) = 1$. Let $\mathfrak{P}$ be a prime divisor in L dividing $\mathfrak{p}$. Then $v_{\mathfrak{P}}(y^p - y) = v_{\mathfrak{P}}(a) < 0$. Therefore $v_{\mathfrak{P}}(y) < 0$ and

$$v_{\mathfrak{P}}(y^p - y) = p v_{\mathfrak{P}}(y) = v_{\mathfrak{P}}(a) = e(\mathfrak{P}|\mathfrak{p}) v_{\mathfrak{p}}(a) = -\lambda e, \quad \text{where} \quad e = e(\mathfrak{P} \mid \mathfrak{p}).$$

The conditions that $(p, \lambda) = 1$, $p \mid e$, $e = p$, and $\mathfrak{p}$ is ramified in L/K imply $\mathfrak{p} = \mathfrak{P}^p$. We also have $v_{\mathfrak{P}}(y) = -\lambda$.

Let $u, v \in \mathbb{Z}$ be such that $-\lambda u + pv = 1$. Then if π is a prime element for $\mathfrak{p}$, we have

$$v_{\mathfrak{P}}(y^u \pi^v) = u v_{\mathfrak{P}}(y) + v v_{\mathfrak{P}}(\pi) = -\lambda u + pv = 1.$$

Therefore $\Pi = y^u \pi^v$ is a prime element for $\mathfrak{P}$. By Proposition 5.5.11, $\vartheta_{\hat{\mathfrak{P}}} = \vartheta_{\hat{\mathfrak{p}}}[\pi]$, where $\vartheta_{\hat{\mathfrak{P}}}$ and $\vartheta_{\hat{\mathfrak{p}}}$ denote the completions of $\vartheta_{\mathfrak{P}}$ and $\vartheta_{\mathfrak{p}}$ respectively. By Theorem 5.7.17, we have $v_{\hat{\mathfrak{P}}}(\hat{\mathfrak{D}}_{\hat{\mathfrak{P}}}) = v_{\mathfrak{p}}(\mathfrak{D}_{L/K}) = v_{\mathfrak{p}}(g'(\Pi))$, with $g(T) = \operatorname{Irr}(\Pi, T, K)$. The set of conjugates of Π is

$$\{\sigma^j \Pi = (y + j)^u \pi^v, \quad j = 0, 1, \ldots, p-1\} = \{\Pi_0, \Pi_1, \ldots, \Pi_{p-1}\}.$$

Thus

$$g(T) = \prod_{i=0}^{p-1} (T - \Pi_i) \quad \text{and} \quad g'(T) = \sum_{j=0}^{p-1} \prod_{i \neq j} (T - \Pi_i).$$

We have

$$\begin{aligned} g'(\Pi) &= \prod_{i=1}^{p-1}(y^u\pi^v - (y+i)^u\pi^v) \\ &= \pi^{v(p-1)}\prod_{i=1}^{p-1}\left(y^u - \sum_{\ell=0}^{u}\binom{u}{\ell}y^\ell i^{u-\ell}\right) \\ &= (-1)^{p-1}\pi^{v(p-1)}\prod_{i=1}^{p-1}(uiy^{u-1}+s_i(y)) \end{aligned}$$

with $s_i(y) = \sum_{\ell=0}^{u-2}\binom{u}{\ell}y^\ell i^{u-\ell}$. Thus $v_{\mathfrak{P}}(s_i(y)) > v_{\mathfrak{P}}(uiy^{u-1}) = (u-1)(-\lambda)$. Therefore

$$\begin{aligned} v_{\mathfrak{P}}(g'(\Pi)) &= vp(p-1) + \sum_{i=1}^{p-1}(u-1)(-\lambda) \\ &= vp(p-1) - \lambda(p-1)(u-1) \\ &= (p-1)(vp-\lambda u+\lambda) = (p-1)(1+\lambda). \end{aligned}$$

□

We obtain analogous results for Kummer extensions.

Theorem 5.8.12. *Let k be any field of characteristic $p \geq 0$. Let L/K be a cyclic extension of degree n with $(n,p)=1$. Assume that k contains a primitive nth root of unity ζ_n. Let $\mathfrak{p}$ be a fixed place of K. Then $L = K(y)$ with $y^n = a$ and $0 \leq v_{\mathfrak{p}}(a) \leq n-1$; $\mathfrak{p}$ is unramified in L/K if and only if $v_{\mathfrak{p}}(a) = 0$.*

If $v_{\mathfrak{p}}(a) = m > 0$ and $\mathfrak{P}$ is a prime divisor of L above $\mathfrak{p}$, we have

$$e(\mathfrak{P}|\mathfrak{p}) = \frac{n}{(n,m)} \quad \text{and} \quad v_{\mathfrak{P}}(\mathfrak{D}_{\mathfrak{P}}) = \frac{n}{(n,m)} - 1.$$

Proof. Let $L = K(z)$ with $z^n = b$, $v_{\mathfrak{p}}(b) = tn + r$ and $0 \leq r \leq n-1$. If π is a prime element for $\mathfrak{p}$, then $\left(\frac{z}{\pi^t}\right)^n = \frac{b}{\pi^{nt}}$ and $v_{\mathfrak{p}}\left(\frac{b}{\pi^{nt}}\right) = r$. The rest of the proof is the same as in Example 5.8.9. □

Definition 5.8.13. We say that the equation given in Theorem 5.8.11 or Theorem 5.8.12 is in *normal form* or *standard form at the prime* $\mathfrak{p}$.

Remark 5.8.14. The hypothesis that k is perfect is not necessary in Theorem 5.8.12. However, if k is not a perfect field, in general we cannot write an equation like the one in Theorem 5.8.11 in a normal form for a given prime divisor. For instance, assume that k is not a perfect field and let $a \in k \setminus k^p$. If $K = k(x)$ and $L = K(y)$ with

$$y^p - y = ax^p \tag{5.10}$$

then (5.10) cannot be modified in order to have the infinite prime of K written in normal form (see Exercise 5.10.18, Exercise 5.10.29, Example 14.3.12, and Exercise 14.5.16).

5.9 Ramification Groups

Theorem 5.6.3 shows a clear difference between ramification types, depending on the divisibility of the ramification index by the characteristic p. A more detailed study of this difference originates in the definition of the ramification groups, which we will study now.

Consider any Galois extension L/K of function fields with Galois group $G = \mathrm{Gal}(L/K)$. If $\mathcal{P}$ is a prime divisor of L and $\wp = \mathcal{P}|_K$, then the decomposition group satisfies $D_{L/K}(\mathcal{P}|\wp) = D = \mathrm{Gal}\left(L_{\mathcal{P}}/K_{\wp}\right)$ (Theorem 5.4.10). We will assume that the residue field extension $\ell(\mathcal{P})/k(\wp)$ is separable. To study the ramification, it suffices to consider the ramification in $L_{\mathcal{P}}/K_{\wp}$. Therefore we will assume that L/K is a Galois extension of complete fields. We also assume that the residue field extension is separable. Within this situation, K is complete with respect to the valuation $v_{\wp}$, the valuation ring is $\vartheta_{\wp}$, and the valuation has a unique extension $v_{\mathcal{P}}$ to L. We have $\vartheta_{\mathcal{P}} = \vartheta_{\wp}[\beta]$ for some β (Theorem 5.7.18). If $f(x) = \mathrm{Irr}(\beta, x, K)$, then $\mathfrak{D}_{\mathcal{P}} = \left(f'(\beta)\right)$ (Theorem 5.7.17) and the discriminant satisfies $\partial_{\wp} = \left(N_{L/K}f'(\beta)\right)$.

Proposition 5.9.1. *Let $\beta_1 = \beta, \beta_2, \ldots, \beta_n$ be the conjugates of β. Then*

$$N_{L/K}\left(f'(\beta)\right) = (-1)^{n(n-1)/2}\prod_{i<j}^{n}\left(\beta_i - \beta_j\right)^2 = \prod_{i\neq j}^{n}\left(\beta_i - \beta_j\right).$$

Proof. We leave the proof to the reader (Exercise 5.10.31). □

Definition 5.9.2. Let $e = e_{L/K}(\mathcal{P}|\wp)$ and let $\bar{K}$ be the residue field $\vartheta_{\wp}/\wp$. Let $p = \mathrm{char}\,\bar{K}$. If $p \mid e$, $\wp$ is called *wildly ramified,* and if $p \nmid e$, $\wp$ is called *tamely ramified.*

We write $A_L = \vartheta_{\mathcal{P}}$ and $A_K = \vartheta_{\wp}$. Let $x \in A_L$ be such that $A_L = A_K[x]$, and let $\pi \in A_L$ be such that $v_{\mathcal{P}}(\pi) = 1$. Let $G = \mathrm{Gal}(L/K)$ (which corresponds to the decomposition group before taking completions).

Proposition 5.9.3. *Let $\sigma \in G$, and $i \in \mathbb{Z}$ be such that $i \geq -1$. The following three conditions are equivalent:*

(a) *σ acts trivially on $A_L/\mathcal{P}^{i+1}$;*
(b) *$v_{\mathcal{P}}(\sigma(a) - a) \geq i + 1$ for all $a \in A_L$;*
(c) *$v_{\mathcal{P}}(\sigma(x) - x) \geq i + 1$.*

Proof. We leave the proof to the reader (Exercise 5.10.32). □

Theorem 5.9.4. *For each $i \geq -1$ put $G_i = \{\sigma \in G \mid v_{\mathcal{P}}(\sigma(x) - x) \geq i + 1\}$. Then $G_i \supseteq G_{i+1}$, each G_i is a normal subgroup of G, $G_{-1} = G$, and G_0 is the inertia group. Furthermore, for i large enough, $G_i = \mathrm{Id}$.*

Proof. Since $\mathcal{P}^\sigma = \mathcal{P}$, we have

$$v_\mathcal{P}(\sigma(x) - x) = v_{\mathcal{P}^{\sigma^{-1}}}\left(x - \sigma^{-1}(x)\right) = v_\mathcal{P}\left(\sigma^{-1}(x) - x\right),$$

so $\sigma \in G_i$ implies $\sigma^{-1} \in G_i$.

If $\sigma, \theta \in G_i$, we have

$$\begin{aligned} v_\mathcal{P}(\sigma\theta(x) - x) &= v_\mathcal{P}((\sigma\theta)(x) - \sigma(x) + \sigma(x) - x) \\ &\geq \min\{v_\mathcal{P}((\sigma\theta)(x) - \sigma(x)), v_\mathcal{P}(\sigma(x) - x)\} \\ &= \min\left\{v_{\mathcal{P}^{\sigma^{-1}}}(\theta(x) - x), v_\mathcal{P}(\sigma(x) - x)\right\} \geq i + 1. \end{aligned}$$

Therefore, G_i is a subgroup of G.

Now let $\sigma \in G_i$ and $\phi \in G$. We have

$$v_\mathcal{P}\left(\left(\phi^{-1}\sigma\phi\right)(x) - x\right) = v_{\mathcal{P}^\phi}((\sigma\phi)(x) - \phi x) = v_\mathcal{P}\left(\sigma x' - x'\right), \quad x' = \phi(x).$$

Since

$$A_L = \phi(A_L) = \phi(A_K[x]) = A_K[\phi(x)] = A_K[x'],$$

it follows, by Proposition 5.9.3, that $v_\mathcal{P}\left(\sigma(x') - x'\right) \geq i + 1$. Thus G_i is a normal subgroup of G.

Clearly, $G_i \supseteq G_{i+1}$. Furthermore,

$$G_0 = \{\sigma \in G \mid v_\mathcal{P}(\sigma x - x) \geq 1\} = \{\sigma \in G \mid \sigma y \equiv y \bmod \mathcal{P}\ \forall\, y \in A_L\},$$

which is the definition of the inertia group.

Finally, for $\sigma \neq \mathrm{Id}$, there exists x such that $\sigma x \neq x$, so $v_\mathcal{P}(\sigma x - x) = i_\sigma \neq \infty$. Let $r = \max\{i_\sigma \mid \sigma \neq \mathrm{Id}\}$. Then

$$\sigma \in G_r \iff v_\mathcal{P}(\sigma x - x) \geq r + 1 > i_\sigma \iff \sigma = \mathrm{Id}.$$

Thus $G_r = \mathrm{Id}$. □

Definition 5.9.5. For $i \geq -1$, the group G_i is called the *ith ramification group of G or ith ramification group of L/K.*

Definition 5.9.6. We define the function $i_G : G \longrightarrow \mathbb{Z} \cup \{\infty\}$ by $i_G(\sigma) = v_\mathcal{P}(\sigma x - x)$.

As a consequence of what we have already proved, we obtain the following result:

Proposition 5.9.7.

(1) $i_G(\sigma) = \infty$ *if and only if* $\sigma = \mathrm{Id}$,

(2) $i_G(\sigma) \geq i + 1$ *if and only if* $\sigma \in G_i$,

(3) $i_G\left(g\sigma g^{-1}\right) = i_G(\sigma)$ *for all* $\sigma, g \in G$. □

Proposition 5.9.8. $\sum_{\sigma\neq \mathrm{Id}} i_G(\sigma) = \sum_{i=0}^{\infty}(|G_i|-1)$.

Proof. Let $r_i = |G_i|-1$. If $\sigma \in G_{i-1}\backslash G_i$, then $v_{\mathcal{P}}(\sigma x - x) = i$, so $i_G(G_{i-1}\setminus G_i) = i$ and $|G_{i-1}\setminus G_i| = r_{i-1} - r_i$.

Therefore

$$\sum_{\sigma\neq \mathrm{Id}} i_G(\sigma) = \sum_{i=0}^{\infty}\sum_{\sigma\in G_{i-1}\backslash G_i} i_G(\sigma) = \sum_{i=0}^{\infty} i\left(r_{i-1}-r_i\right).$$

Let t be such that $G_t = \mathrm{Id}$. Then $r_{t+1} = 0$ and

$$\begin{aligned}\sum_{i=0}^{\infty} i\left(r_{i-1}-r_i\right) &= \sum_{i=0}^{t+1} i\left(r_{i-1}-r_i\right) = \sum_{i=0}^{t+1} i r_{i-1} - \sum_{i=0}^{t+1} i r_i\\ &= \sum_{i=0}^{t}(i+1)r_i - \sum_{i=0}^{t+1} i r_i = \sum_{i=0}^{t} r_i - (t+1)r_{t+1}\\ &= \sum_{i=0}^{t} r_i = \sum_{i=0}^{\infty} r_i = \sum_{i=0}^{\infty}\left(|G_i|-1\right).\end{aligned}$$ □

Theorem 5.9.9. *We have* $\mathfrak{D}_{\mathcal{P}} = \mathcal{P}^s$, *where* $s = \sum_{\sigma\neq\mathrm{Id}} i_G(\sigma) = \sum_{i=0}^{\infty}(|G_i|-1)$.

Proof. Let

$$A_L = A_K[x], \quad [L:K] = ef = n, \quad \text{and} \quad f(T) = \mathrm{Irr}(x,T,K) = \prod_{\sigma\in G}(T-\sigma x).$$

Then

$$f'(T) = \sum_{\sigma\in G}\prod_{\theta\neq\sigma}(T-\theta x) \quad \text{and} \quad f'(x) = \prod_{\sigma\neq\mathrm{Id}}(x-\sigma x).$$

By Theorem 5.7.17, we have

$$s = v_{\mathcal{P}}(f'(x)) = \sum_{\sigma\neq\mathrm{Id}} v_{\mathcal{P}}(\sigma x - x) = \sum_{\sigma\neq\mathrm{Id}} i_G(\sigma) = \sum_{i=0}^{\infty}\left(|G_i|-1\right).$$ □

Corollary 5.9.10. $\wp$ *is wildly ramified if and only if* $G_1 \neq \{\mathrm{Id}\}$.

Proof. We have $|G_0| = e$ (Corollary 5.2.23) and $|G_0|-1 = e-1$. By Theorem 5.6.3, $s > e-1 \iff p \mid e \iff \wp$ is wildly ramified.

On the other hand, $s > e-1$ if and only if $|G_1|-1>0$. □

Corollary 5.9.11. *If* char $\bar{K} = 0$, *then* $G_1 = \{\mathrm{Id}\}$. □

Example 5.9.12. Let $K = k(x, y)$ be the function field defined by

$$y^q - y = x^m \quad \text{where} \quad q = p^u, p = \operatorname{char} k, m > 1 \quad \text{and} \quad m \mid q + 1.$$

Set $q + 1 = mn$. Then we will prove that $g_K = \frac{(m-1)(q-1)}{2}$.

First we consider a root α of $T^q - T - x^m$. Then for any $\mu \in \mathbb{F}_{p^u} = \mathbb{F}_q$,

$$(\alpha + \mu)^q - (\alpha + \mu) = \alpha^q + \mu^q - \alpha - \mu = \alpha^q + \mu - \alpha - \mu = \alpha^q - \alpha = x^m.$$

Therefore $\{y + \mu \mid \mu \in \mathbb{F}_q\}$ is the set of roots of $T^q - T - x^m$.

In particular, $K/k(x)$ is a Galois extension. Let $\mathfrak{B}$ be a prime divisor in K dividing the infinite prime $\wp_\infty$ of $k(x)$. We have

$$v_{\mathfrak{B}}(y^q - y) = m v_{\mathfrak{B}}(x) = m e(\mathfrak{B}|\wp_\infty) v_{\wp_\infty}(x) = -m e(\mathfrak{B}|\wp_\infty) < 0.$$

Therefore $v_{\mathfrak{B}}(y) < 0$, since otherwise we would have $v_{\mathfrak{B}}(y^q - y) \geq 0$. Thus

$$v_{\mathfrak{B}}(y^q - y) = \min\{v_{\mathfrak{B}}(y^q), v_{\mathfrak{B}}(y)\} = q v_{\mathfrak{B}}(y).$$

It follows that $q v_{\mathfrak{B}}(y) = -m e(\mathfrak{B}|\wp_\infty)$. Since $(q, m) = 1$, q divides $e(\mathfrak{B}|\wp_\infty)$. Therefore $\wp_\infty$ is fully ramified, $e(\mathfrak{B}|\wp_\infty) = q$, and $[K : k(x)] = q$. We also have $v_{\mathfrak{B}}(y) = -m$.

For any $\mu \in \mathbb{F}_q$, let $\sigma_\mu \in \operatorname{Gal}(K/k(x))$ be defined by

$$\sigma_\mu(y) = y + \mu.$$

Then $\theta\colon (\mathbb{F}_q, +) \to \operatorname{Gal}(K/k(x))$ is a group isomorphism.

Now, for any prime divisor $\mathfrak{P}$ distinct from $\mathfrak{B}$, we have $v_{\mathfrak{P}}(y) \geq 0$ since $v_{\mathfrak{P}}(x^m) \geq 0$. Thus $y \in \vartheta_{\mathfrak{P}}$. We have

$$\alpha(T) = T^q - T - x^m = \prod_{\mu \in \mathbb{F}_q} (T - y - \mu).$$

Hence $\alpha'(T) = \sum\limits_{\beta \in \mathbb{F}_q} \prod\limits_{\mu \neq \beta} (T - y - \mu)$, so $\alpha'(y) = \prod_{\mu \neq 0}(y - y - \mu) = (-1)^{q-1} \prod_{\mu \in \mathbb{F}_q} \mu$.

Therefore $(\alpha'(y))_K$ is the unit divisor $\mathfrak{N}$. It follows by Theorem 5.7.21 that $\mathfrak{P}$ is unramified in $K/k(x)$. Hence $\mathfrak{D}_{K/k(x)} = \mathfrak{B}^s$ for some s. Next we determine the ramification groups G_i for $\mathfrak{B}$.

Since $v_{\mathfrak{B}}(y) = -m$ and $v_{\mathfrak{B}}(x) = -q$, we have $v_{\mathfrak{B}}(y^{-n}x) = nm - q = 1$. Thus $y^{-n}x$ is a prime element for $\mathfrak{B}$. Now, $G_{-1} = G_0 = G$ and for $\mu \in \mathbb{F}_q^*$,

$$\begin{aligned}
\sigma_\mu\left(y^{-n}x\right) - y^{-n}x &= (y + \mu)^{-n} x - y^{-n} x \\
&= x\left(\frac{y^n - (y + \mu)^n}{(y + \mu)^n y^n}\right) = \frac{-\mu y^{n-1} + \cdots}{(y^2 + \mu y)^n} x.
\end{aligned}$$

Thus

$$\begin{aligned} v_{\mathfrak{B}}(\sigma_\mu(y^{-n}x) - y^{-n}x) &= (n-1)v_{\mathfrak{B}}(y) + v_{\mathfrak{B}}(x) - 2nv_{\mathfrak{B}}(y) \\ &= (n+1)m - q = q + 1 + m - q = m + 1. \end{aligned}$$

It follows by Theorem 5.9.4 that $\sigma_\mu \in G_m$ and $\sigma_\mu \notin G_{m+1}$. Therefore

$$G = G_{-1} = G_0 = \cdots = G_m, \quad G_{m+1} = \{1\}.$$

Using Theorem 5.9.9 we obtain that

$$s = \sum_{i=0}^{\infty}(|G_i| - 1) = \sum_{i=0}^{m}(q-1) = (m+1)(q-1).$$

Applying the Riemann–Hurwitz genus formula we get

$$g_K = 1 + q(0-1) + \frac{1}{2}(m+1)(q-1) = \frac{(m-1)(q-1)}{2}.$$

Definition 5.9.13. Let $U_L = U_L^{(0)}$ be the set of units of A_L, i.e., $U_L = \{y \in A_L \mid v_{\mathcal{P}}(y) = 0\}$. For $i \geq 1$, let $U_L^{(i)} = 1 + \mathcal{P}^i$.

Proposition 5.9.14.

(1) $U_L^{(0)}/U_L^{(1)} \cong \ell(\mathcal{P})^*$.
(2) *For* $i \geq 1$, $U_L^{(i)}/U_L^{(i+1)} \cong \mathcal{P}^i/\mathcal{P}^{i+1} \cong \ell(\mathcal{P})$.

Proof.

(1) Let $\varphi : U_L^{(0)} \longrightarrow \ell(\mathcal{P})^* = (A_L/\mathcal{P})^*$ be the natural map. Clearly, φ is surjective and we have

$$\ker\varphi = \left\{x \in U_L^{(0)} \mid x + \mathcal{P} = 1 + \mathcal{P}\right\} = 1 + \mathcal{P} = U_L^{(1)}.$$

(2) Let $i \geq 1$ and let $\varphi : \mathcal{P}^i \to 1 + \mathcal{P}^i = U_L^{(i)}$ be defined by $\varphi(y) = 1 + y$. Then φ is a bijective function that is not a homomorphism. The function

$$\tilde{\varphi} : \mathcal{P}^i \longrightarrow U_L^{(i)}/U_L^{(i+1)}$$

is surjective. We will see that $\tilde{\varphi}$ is a homomorphism.
We have $\tilde{\varphi}(y+z) = 1 + (y+z) \bmod U_L^{(i+1)}$. On the other hand,

$$\begin{aligned} \tilde{\varphi}(y)\tilde{\varphi}(z) &= (1+y)(1+z) \bmod U_L^{(i+1)} \\ &= 1 + (y+z) + yz \bmod U_L^{(i+1)}. \end{aligned}$$

Since $y, z \in \mathcal{P}^i$, we have $yz \in \mathcal{P}^{2i} \subseteq \mathcal{P}^{i+1}$. Thus $1 + yz \equiv 1 \bmod U_L^{(i+1)}$, from which we obtain

$$\tilde{\varphi}(y)\tilde{\varphi}(z) = 1 + (y+z) \bmod U_L^{(i+1)} = \tilde{\varphi}(y+z).$$

This proves that $\tilde{\varphi}$ is an epimorphism.
Furthermore, it is clear that $\ker \tilde{\varphi} = \mathcal{P}^{i+1}$. Therefore $\left(\mathcal{P}^i/\mathcal{P}^{i+1}, +\right) \cong \left(U_L^{(i)}/U_L^{(i+1)}, \cdot\right)$.
Finally, the $A_L/\mathcal{P}$-modules $\mathcal{P}^i/\mathcal{P}^{i+1}$ and $A_L/\mathcal{P}$ are isomorphic and hence they are isomorphic as $\bar{L}$-vector spaces. It follows that $\mathcal{P}^i/\mathcal{P}^{i+1}$ has dimension 1. Indeed

$$\begin{aligned} A_L &\xrightarrow{\psi} \mathcal{P}^i/\mathcal{P}^{i+1} \\ x &\mapsto \pi^i x + \mathcal{P}^{i+1} \end{aligned}$$

is an epimorphism and $\ker \psi = \mathcal{P}$. □

Proposition 5.9.15. *$\sigma \in G_i$ if and only if $\sigma(\pi)/\pi \in U_L^{(i)}$.*

Proof. By substituting G by G_0 and K by K^{G_0} if necessary, we may assume that L/K is totally ramified. In this case $A_K[\pi] = A_L$ (Proposition 5.5.11).
By Proposition 5.9.3, it follows that

$$\begin{aligned} \sigma \in G_i &\iff v_{\mathcal{P}}\left(\sigma(\pi) - \pi\right) = 1 + v_{\mathcal{P}}\left(\frac{\sigma(\pi)}{\pi} - 1\right) \geq i+1 \\ &\iff v_{\mathcal{P}}\left(\frac{\sigma(\pi)}{\pi} - 1\right) \geq i \\ &\iff \frac{\sigma(\pi)}{\pi} = 1 + t, \quad t \in \mathcal{P}^i \iff \frac{\sigma(\pi)}{\pi} \in 1 + \mathcal{P}^i = U_L^{(i)}. \end{aligned}$$ □

Theorem 5.9.16. *The function that to each $\sigma \in G_i$ assigns $\frac{\sigma(\pi)}{\pi}$ induces, by taking quotients, a monomorphism of G_i/G_{i+1} into a subgroup of $U_L^{(i)}/U_L^{(i+1)}$. Furthermore, this monomorphism is independent of the prime element π chosen.*

Proof. If π' is any other prime element, then $\pi' = \pi u$ with $u \in U_L$. Therefore $\frac{\sigma(\pi')}{\pi'} = \frac{\sigma(\pi)}{\pi}\frac{\sigma(u)}{u}$. If $\sigma \in G_i$ we have $\sigma(u) \equiv u \bmod \mathcal{P}^{i+1}$. Thus

$$\frac{\sigma(u)}{u} \equiv 1 \bmod U_L^{(i+1)}, \quad \text{so} \quad \frac{\sigma(\pi')}{\pi'} \equiv \frac{\sigma(\pi)}{\pi} \bmod U_L^{(i+1)}.$$

Hence the function $\theta : G_i \longrightarrow U_L^{(i)}/U_L^{(i+1)}$, defined by $\theta(\sigma) = \frac{\sigma(\pi)}{\pi} \bmod U_L^{(i+1)}$, does not depend on the prime element
If $\sigma, \phi \in G_i$ we have

$$\frac{(\sigma\phi)(\pi)}{\pi} = \frac{(\sigma\phi)(\pi)}{\phi\pi}\frac{\phi\pi}{\pi} = \frac{\sigma(\pi)}{\pi}\frac{\phi(\pi)}{\pi}\frac{\sigma(v)}{v}$$

with $v = \frac{\phi(\pi)}{\pi}$. Since $\phi(\pi) \equiv \pi \bmod \mathcal{P}^{i+1}$, it follows that $v \in U_L$, $\frac{\sigma v}{v} \equiv 1 \bmod U_L^{(i+1)}$. Therefore θ is a homomorphism, and clearly

$$\ker \theta = \left\{ \sigma \;\middle|\; \frac{\sigma(\pi)}{\pi} \equiv 1 \bmod U_L^{(i+1)} \right\} = G_{i+1}. \qquad \square$$

Corollary 5.9.17. *G_0/G_1 is a cyclic group whose order is relatively prime to the characteristic of $\ell(\mathcal{P})$.*

Proof. We have $G_0/G_1 \subseteq U_L^{(0)}/U_L^{(1)} \cong \ell(\mathcal{P})^*$. Thus G_0/G_1 is a finite subgroup of the group of units of $\bar{L}^*$. Therefore it is a cyclic group whose order is relatively prime to the characteristic of $\ell(\mathcal{P})$. $\square$

Corollary 5.9.18. *If $\wp$ is tamely ramified, then G_0 is a cyclic group.*

Proof. In this case G_1 is trivial. $\square$

Corollary 5.9.19. *If the characteristic of $\ell(\mathcal{P})$ is $p > 0$, then the quotients G_i/G_{i+1} $(i \geq 1)$ are elementary abelian p-groups, i.e., $G_i/G_{i+1} \cong (\mathbb{Z}/p\mathbb{Z})^\alpha$ for some α. Also, G_1 is a p-group.*

Proof. For $i \geq 1$, $U_L^{(i)}/U_L^{(i+1)}$ is isomorphic to $\ell(\mathcal{P})$. Therefore it is an abelian group such that $p\left(U_L^{(i)}/U_L^{(i+1)}\right) = 0$. It follows that G_i/G_{i+1} is an elementary abelian p-group.

Since $|G_1| = \prod_{i=1}^{\infty} |G_i/G_{i+1}|$, each G_i/G_{i+1} is of order p^{r_i} for some $r_i \geq 0$. Furthermore, for i large enough we have $|G_i/G_{i+1}| = 1$. Hence G_1 is a p-group. $\square$

Corollary 5.9.20. *G_0 is a solvable group.*

Proof. This follows from the facts that G_0/G_1 is a cyclic group, in particular solvable, and that G_1 is a p-group. $\square$

5.10 Exercises

Exercise 5.10.1. Let K/k be a function field and let $x, y \in K \setminus k$ be such that $[K : k(x)]$ and $[K : k(y)]$ are relatively prime. Prove that $K = k(x, y)$.

Exercise 5.10.2. Give an explicit example of an extension of function fields L/K and a valuation v on L such that $v|_K : K^* \to \mathbb{Z}$ is not surjective.

Exercise 5.10.3. Let X, T be two variables over the field of two elements $k = \mathbb{F}_2$. Let $K = k\left(T, X^4 + TX^2 + 1\right)$ and $L = k\,(T, X)$. Prove that $L_s = k\left(T, X^2\right)$, and that $L_i = K$. In particular, we have $L_i L_s \neq L$.

Exercise 5.10.4. Let L/K be a finite extension of fields. Prove that $L = L_s L_i$ if and only if L/L_i is a separable extension.

Exercise 5.10.5. With the notation of Exercise 5.10.4, prove that if L/K is a normal extension, then $L = L_s L_i$.

Exercise 5.10.6. Prove or give a counterexample: Let L/ℓ be an arbitrary extension of K/k. Then no place of L is variable over K. (See Proposition 5.1.12).

Exercise 5.10.7. Give an example of a function field extension L/K, and places $\mathfrak{P}$ of L and $\mathfrak{p}$ of K, such that:

(i) $e_{L/K}(\mathfrak{P}|\mathfrak{p}) > 1$.
(ii) $d_{L/K}(\mathfrak{P}|\mathfrak{p}) > 1$.

Exercise 5.10.8. Let L/ℓ be an extension of K/k. Show that the following conditions are equivalent:

(i) ℓ is an algebraic extension of k.
(ii) L is an algebraic extension of K.
(iii) If $\mathfrak{P}$ is a prime divisor of L above the place $\mathfrak{p}$ of K, then $\ell(\mathfrak{P})$ is an algebraic extension of $k(\mathfrak{p})$.

Exercise 5.10.9. Let L/E be a finite normal field extension and let $G := \operatorname{Aut}(L/E)$.

Let v be a valuation of E. If w is an extension of v to L and $\sigma \in G$, we define $(\sigma w)(x) = w(\sigma^{-1}x)$ for $x \in L$. Equivalently, $(\sigma w)(\sigma y) = w(y)$.

Assume that there exist two extensions w and w' of v such that $\sigma w \neq w'$ for all $\sigma \in G$.

Then by the approximation theorem there exists $x \in L$ such that $w'(x) > 0$, $(\sigma^{-1}w)(x) = 0$, and $\sigma^{-1}w'(x) \geq 0$ for all $\sigma \in G$.

Consider $y = N_{L/E}x$.

Prove that the above implies $v(y) > 0$ and $v(y) = 0$.

This contradiction shows that given two arbitrary extensions w, w' of v, there exists $\sigma \in G$ such that $\sigma w = w'$. That is, G acts transitively over the extensions w of v.

Exercise 5.10.10. Let k be an algebraically closed field and let K/k be a function field. Let L/K be a finite Galois extension, $\mathfrak{p}$ be a prime divisor of K, and $\mathfrak{P}$ a prime divisor of L such that $\mathfrak{P} \mid \mathfrak{p}$. We have $\mathfrak{P}|_K = \mathfrak{p}$. Prove that $D(\mathfrak{P}|\mathfrak{p}) = I(\mathfrak{P}|\mathfrak{p})$.

Exercise 5.10.11. With the hypotheses of Exercise 5.10.10, assume that k is a finite field. Prove that $\frac{D(\mathfrak{P}|\mathfrak{p})}{I(\mathfrak{P}|\mathfrak{p})}$ is a cyclic group.

Exercise 5.10.12. Let L/K be a finite Galois extension of function fields and let F/K be an arbitrary extension such that $L \cap F = K$. Let $E = LF$. The function $\varphi\colon \operatorname{Gal}(E/F) \to \operatorname{Gal}(L/K)$ defined by $\varphi(\sigma) = \sigma|_L$ is an isomorphism.

Let $\mathfrak{P}$ be a prime divisor of F and $\mathfrak{Q}$ be a prime divisor of E over $\mathfrak{P}$. Put $\mathfrak{p} = \mathfrak{P}|_K$ and let $\wp = \mathfrak{Q}|_L$. Prove that:

(i) $D(\mathfrak{Q} \mid \mathfrak{P})|_L \subseteq D(\wp \mid \mathfrak{p})$;
(ii) $I(\mathfrak{Q} \mid \mathfrak{P})|_L \subseteq I(\wp \mid \mathfrak{p})$;

Deduce that if $\mathfrak{P}$ is ramified in E/F and the field of constants ℓ of L is perfect, then $\mathfrak{p}$ is ramified in L/K.

Note that if ℓ is not perfect, then $\mathfrak{p}$ may be unramified. In this case $\mathfrak{p}$ is inseparable. See Exercise 5.10.18.

Exercise 5.10.13. Let L/K be a finite separable extension and let $\tilde{L}$ be the Galois closure of L/K. We have

$$\tilde{L} = \prod_{\sigma \in H} L^\sigma, \quad H = \left\{\sigma : L \to \bar{K} \mid \sigma|_K = \mathrm{Id}\right\},$$

where $\bar{K}$ denotes the algebraic closure of K. Assume that the field of constants k of K is a perfect field.

Let $\mathfrak{p}$ be a prime divisor of K such that $\mathfrak{p}$ is nonramified in L. Prove that $\mathfrak{p}$ is nonramified in $\tilde{L}/K$.

Hint: Let $I(\mathfrak{P}|\mathfrak{p})$ be the inertia group of $\mathfrak{P} \mid \mathfrak{p}$ in $\tilde{L}/K$. Let $F = L^{I(\mathfrak{P}|\mathfrak{p})}$ be the fixed field. Prove that $L^\sigma \subseteq F$ for all $\sigma \in H$.

Exercise 5.10.14. Let k be an algebraically closed field and $K = k(x)$. Let $\mathfrak{p}_1, \mathfrak{p}_2, \mathfrak{p}_3$ be three distinct prime divisors of K and let $\sigma \in \operatorname{Aut}_k k(x)$ be such that $\mathfrak{p}_i^\sigma = \mathfrak{p}_i$ for $i = 1, 2, 3$.

Prove that $\sigma = \mathrm{Id}_K$.

Is the same result true in the case that k is not algebraically closed?

Exercise 5.10.15. Let k be a finite field and let K be a function field over k. Suppose L and E are two distinct Galois extensions of K of degree p, where p is a prime number, such that $L \cap E = K$.

Let $\mathfrak{P}_K$ be a prime divisor of K. Let $\mathfrak{P}_L$ and $\mathfrak{P}_E$ be places of L and E respectively such that $\mathfrak{P}_K = \mathfrak{P}_L^p$, $\mathfrak{P}_K = \mathfrak{P}_E^p$ in L/K and E/K respectively. In other words, we are assuming that $\mathfrak{P}_K$ is ramified in L/K as well as in E/K.

Set $F = LE$ and let $\mathfrak{P}_F$ be a place of F such that $\mathfrak{P}_F \mid \mathfrak{P}_K$. If p is different from the characteristic of k, the inertia group $I(\mathfrak{P}_F|\mathfrak{P}_K)$ is a cyclic group.

Using this fact, prove that there exists a unique field M satisfying $K \subsetneqq M \subsetneqq F$ (that is, $[M : K] = p$) such that $\mathfrak{P}_K$ is *not* ramified in M/K.

Exercise 5.10.16. Prove that $\ell = k\left(u^{1/p}, v^{1/p}\right)$ in Example 5.2.31.

Exercise 5.10.17. Let $K \subseteq M \subseteq L$ be any tower of function fields. Prove that $\lambda_{L/K} = \lambda_{L/M}\lambda_{M/K}$ (see Theorem 5.3.4).

Exercise 5.10.18. Let $L = k(x, y)$ be given by $y^p - y = ax^p$, where k is an imperfect field of characteristic p and $a \in k \setminus k^p$. Then $L/k(x)$ is a separable extension and the field of constants of L is k. Show that if $\mathfrak{p}_\infty$ is the infinite prime in $k(x)$ and $\mathfrak{q}$ is a place in L above $\mathfrak{p}_\infty$, then $\mathfrak{q} \mid \mathfrak{p}_\infty$ is purely inseparable. In particular, Theorem 5.2.21 is no longer true if k is not a perfect field.

Exercise 5.10.19. Let k be any field of characteristic p and let K/k be a function field over k. If L/K is a cyclic extension of degree p such that $L = K(y)$ with $y^p - y = \alpha$ and $v_{\mathfrak{p}}(\alpha) \geq 0$ for a place $\mathfrak{p}$ of K, prove that $\mathfrak{p}$ is unramified.

Exercise 5.10.20. Prove that if L/K is a normal extension of function fields then ℓ/k is a normal extension, where ℓ and k are the fields of constants of L and K respectively.

Exercise 5.10.21. Give an example in which $\overline{\text{con}}\colon C_K \to C_L$ and $\overline{\text{con}}\colon C_{0,K} \to C_{0,L}$ are not injective (see Exercise 8.7.20).

Exercise 5.10.22. Let L/ℓ be a finite extension of K/k. Is it true that $[\ell : k] \leq [L : K]$?

Exercise 5.10.23. Let A be a Dedekind domain with only a finite number of prime ideals. Prove that A is a principal ideal domain.

Exercise 5.10.24. Let A be a Dedekind domain and let S be a multiplicative subset of A. Prove that $S^{-1}A$ is a Dedekind domain.

Exercise 5.10.25. Let K be a function field and let $T = \{\mathfrak{p}_1, \ldots, \mathfrak{p}_r\}$, $r \geq 1$, be a finite set of prime divisors of K. Let L/K be a finite separable extension and let $T^* = \{\mathfrak{P} \mid \mathfrak{P}$ is a place of L, $\mathfrak{P} \mid \mathfrak{p}_i$ for some $1 \leq i \leq r\}$. Let $\vartheta_K := \bigcap_{\mathfrak{p}\in T} \vartheta_{\mathfrak{p}}$ and $\vartheta_L := \bigcap_{\mathfrak{P}\in T^*} \vartheta_{\mathfrak{P}}$. Prove that ϑ_K is the integral closure of ϑ_K in L.

Exercise 5.10.26. If A is a Dedekind domain, $K :=$ quot A, L/K is a finite separable extension, and B is the integral closure of A in L, prove that $S^{-1}B$ is the integral closure of $S^{-1}A$ in L, where $S \subseteq K$ is a multiplicative subset of K.

Exercise 5.10.27. Prove the claim that we may assume that the element α found in the proof of Theorem 5.7.21 belongs to B.

Exercise 5.10.28. Prove Theorem 5.7.23.

Exercise 5.10.29. Let L/K be the extension given in Exercise 5.10.18, and suppose $a \in k \setminus k^p$. Prove that L/K is an unramified extension, i.e., every place of $k(x)$ is unramified in L.

Exercise 5.10.30. Let L/K be a cyclic extension of function fields of degree p^n, where p is a prime number and $n \geq 1$. Assume that the field of constants of K is perfect. Let $\mathfrak{p}$ be a prime divisor of K. Let $K_0 = K \subseteq K_1 \subseteq \cdots \subseteq K_n = L$ be such that $[K_i : K_0] = p^i$.

Assume that $\mathfrak{p}$ is unramified in K_i/K_0 but ramified in K_{i+1}/K_0. Prove that any prime divisor $\mathfrak{P}$ of K_i that lies above $\mathfrak{p}$ is fully ramified in L/K_i. Deduce that $e(\mathfrak{B}|\mathfrak{p}) = e(\mathfrak{B}|\mathfrak{P}) = p^{n-i}$, where $\mathfrak{B}$ is a prime divisor in L above $\mathfrak{p}$. In other words, if a prime divisor in this kind of extension starts ramifying at some point, it keeps ramifying all the way.

Exercise 5.10.31. Prove Proposition 5.9.1.

Exercise 5.10.32. Prove Proposition 5.9.3.

Exercise 5.10.33. Give an example of a constant function field extension such that there exist ramified prime divisors and unramified prime divisors. That is, if the field of constants is not perfect, then Corollary 5.2.26 and Theorem 5.2.32 are no longer true.

Exercise 5.10.34. Let A be a Dedekind domain, and let $\mathfrak{a}$, $\mathfrak{b}$ be nonzero integral ideals such that $\mathfrak{a} \subseteq \mathfrak{b}$. Show that there exists $d \in A \setminus \{0\}$ such that $(\mathfrak{a}, (d)) = \mathfrak{a} + (d) = \mathfrak{b}$.

Exercise 5.10.35. Let A be a Dedekind domain and $\mathfrak{a}$, $\mathfrak{b}$, $\mathfrak{c}$ nonzero integral ideals such that $\mathfrak{a} \subseteq \mathfrak{b}$. Show that the A-modules $\frac{\mathfrak{b}\mathfrak{c}}{\mathfrak{a}\mathfrak{c}}$ and $\frac{\mathfrak{b}}{\mathfrak{a}}$ are isomorphic.

Exercise 5.10.36. Let A be a Dedekind domain, and $\mathfrak{a}$, $\mathfrak{b}$ nonzero integral ideals of A. Prove that there exists an integral ideal $\mathfrak{c}$ such that $\mathfrak{a}\mathfrak{c}$ is principal and $\mathfrak{b} + \mathfrak{c} = (\mathfrak{b}, \mathfrak{c}) = (1) = A$.

6

Congruence Function Fields

In this chapter we shall consider congruent function fields, which are fields whose field of constants is finite. This case is the one for which the analogy between number fields and function fields is deepest.

In this situation, the class number is finite and may even be computed by means of the numerator of the zeta function (Theorem 6.3.5). On the other hand, the zeta function and the L-series may be defined exclusively in this kind of function field.

It should be mentioned that the situation studied in this chapter is a particular case of the general case of extensions of constants. The latter will be presented in Chapter 8. In the family of extensions of constants presented in this chapter, we do not encounter the technical difficulties that are found in the general case, and for this reason the presentation here will be simpler and clearer.

Later on, in Chapter 10, on the applications of elliptic and hyperelliptic fields to cryptography, we shall apply the class number by means of the zeta function.

In this chapter k and ℓ will always mean finite fields such that $k \subseteq \ell$, say $k = \mathbb{F}_q$, $\ell = \mathbb{F}_{q^f}$, where $q = p^u$, p is a rational prime, and $u \geq 1$. We observe that $k = \left\{x \in \bar{\mathbb{F}}_p \mid x^q = 1\right\}$, $\ell = \left\{x \in \bar{\mathbb{F}}_p \mid x^{q^f} = 1\right\}$, and $[\ell : k] = f$, where $\bar{\mathbb{F}}_p$ is an algebraic closure of $\mathbb{F}_p$.

6.1 Constant Extensions

Definition 6.1.1. A function field K/k is called *congruent* or a *congruence function field* if k is a finite field.

We consider a congruence function field K/k and a finite extension ℓ of k. Let $L = K\ell$ be the constant extension, with $[\ell : k] = f$. In Example 5.2.31, we saw that in general, it is possible for the field of constants of $L = K\ell$ to contain ℓ properly. The next theorem shows that this is not the case for congruence function fields, nor in fact for much more general families of function fields. However, for the time being we will limit ourselves to congruence function fields. The general constant extension will be studied in Chapter 8.

Theorem 6.1.2. *Let K/k be a congruence function field, ℓ a finite extension of k, and $L = K\ell$ the constant extension. Then the field of constants of L is ℓ.*

Proof. Set $\ell = k(\xi)$ with $[\ell : k] = f$. Then $\operatorname{Irr}(\xi, T, k) \mid T^{q^f} - T = \prod_{\alpha\in\ell}(T-\alpha)$. We have $L = K\ell = Kk(\xi) = K(\xi)$ with $\operatorname{Irr}(\xi, T, K) \mid \operatorname{Irr}(\xi, T, k)$, so $\operatorname{Irr}(\xi, T, K) \in K[T] \cap \ell[T] = k[T]$. That is, $\operatorname{Irr}(\xi, T, K) = \operatorname{Irr}(\xi, T, k)$. It follows that $[L : K] = \deg \operatorname{Irr}(\xi, T, K) = \deg \operatorname{Irr}(\xi, T, k) = [\ell : k] = f$. If ℓ' denotes the field of constants of L, with $\ell \subseteq \ell'$, then $L = K\ell'$. By all the above, we have $\left[\ell' : k\right] = [L : K] = [\ell : k] = f$, which implies that $\ell' = \ell$. □

Theorem 6.1.3. *Let $L = K\ell$. Then $[L : K] = [\ell : k]$. For $\mathfrak{A} \in D_K$, we have $d_L(\mathfrak{A}) = d_K(\mathfrak{A})$. More precisely, $d_L\left(\operatorname{con}_{K/L}\mathfrak{A}\right) = d_K(\mathfrak{A})$. Finally, $g_L = g_K$.*

Proof. In Theorem 5.3.4, we proved the existence of $\lambda_{L/K} \in \mathbb{Q}^+$ such that $d_L(\mathfrak{A}) = \frac{d_K(\mathfrak{A})}{\lambda_{L/K}}$ for $\mathfrak{A} \in D_K$, and we obtained that $\lambda_{L/K} = \frac{[\ell:k]}{[L:K]}$. By Theorem 6.1.2, it follows immediately that $\lambda_{L/K} = 1$. Therefore $d_L(\mathfrak{A}) = d_K(\mathfrak{A})$ for $\mathfrak{A} \in D_K$.

Now let $\mathfrak{A} \in D_K$. We have

$$L_K(\mathfrak{A}) = \left\{x \in K \mid v_{\wp}(x) \geq v_{\wp}(\mathfrak{A}) \text{ for all } \wp \in \mathbb{P}_K\right\}$$

and

$$L_L(\mathfrak{A}) = \{y \in L \mid v_{\mathcal{P}}(x) \geq v_{\mathcal{P}}(\mathfrak{A}) \text{ for all } \mathcal{P} \in \mathbb{P}_L\}.$$

If $y = \sum_{i=1}^n a_i x_i$ with $a_i \in \ell$ and $x_i \in L_K(\mathfrak{A})$, then

$$\begin{aligned} v_{\mathcal{P}}(y) &\geq \min_{1\leq i\leq n} v_{\mathcal{P}}(a_i x_i) \geq \min_{1\leq i\leq n} v_{\mathcal{P}}(x_i) \\ &= \min_{1\leq i\leq n} v_{\wp}(x_i) \geq v_{\wp}(\mathfrak{A}) = v_{\mathcal{P}}\left(\operatorname{con}_{K/L}(\mathfrak{A})\right), \end{aligned}$$

where $\wp = \mathcal{P}|_K$ (recall that by Theorem 5.2.32, L/K has no ramified divisors). This shows that $\ell L_K(\mathfrak{A}) \subseteq L_L(\mathfrak{A})$.

Conversely, let $y \in L_L(\mathfrak{A})$; then $v_{\mathcal{P}}(y) \geq v_{\mathcal{P}}(\mathfrak{A})$ for all $\mathcal{P} \in \mathbb{P}_L$. Let $y = \sum_{i=1}^n a_i x_i$ with $a_i \in \ell$ and $x_i \in K$. In fact, if $\ell = k(\xi)$ with $[\ell : k] = f$, we may write $y = a_0 + a_1\xi + \cdots + a_{f-1}\xi^{f-1}$ with $a_i \in K$. Let $y^{(1)} = y, y^{(2)}, \ldots, y^{(f)}$ be the conjugates of y. We have $y^{(i)} = a_0 + a_1\xi^{(i)} + \cdots + a_{f-1}\left(\xi^{(i)}\right)^{f-1}$. Solving each of these equations for a_i, we obtain

$$a_i = \frac{\det\begin{bmatrix} 1 & \xi^{(1)} & \cdots & y^{(1)} & \cdots & \left(\xi^{(1)}\right)^{f-1} \\ \cdot & \cdot & \cdots & \cdot & \cdots & \cdot \\ \cdot & \cdot & \cdots & \cdot & \cdots & \cdot \\ \cdot & \cdot & \cdots & \cdot & \cdots & \cdot \\ 1 & \xi^{(f)} & \cdots & y^{(f)} & \cdots & \left(\xi^{(f)}\right)^{f-1} \end{bmatrix}}{\det\begin{bmatrix} 1 & \xi^{(1)} & \cdots & \left(\xi^{(1)}\right)^{i} & \cdots & \left(\xi^{(1)}\right)^{f-1} \\ \cdot & \cdot & \cdots & \cdot & \cdots & \cdot \\ \cdot & \cdot & \cdots & \cdot & \cdots & \cdot \\ \cdot & \cdot & \cdots & \cdot & \cdots & \cdot \\ 1 & \xi^{(f)} & \cdots & \left(\xi^{(f)}\right)^{i} & \cdots & \left(\xi^{(f)}\right)^{f-1} \end{bmatrix}} = \frac{b_i}{c_i}$$

with $c_i \in k^*$, $b_i = \sum_{j=1}^{f} t_j y^{(j)}$, and $t_j \in \ell$.

Since $y^{(j)}$ is a conjugate of y and $\mathfrak{A} \in D_K$, $y \in L_L(\mathfrak{A})$ implies that $y^{(j)} \in L_L(\mathfrak{A})$. Therefore

$$v_{\wp}(a_i) = v_{\wp}\left(\sum_{j=1}^{f} \frac{t_j}{c_j} y^{(j)}\right) \geq \min_{1\leq j\leq f} v_{\mathcal{P}}\left(y^{(j)}\right) \geq v_{\mathcal{P}}(\mathfrak{A}) = v_{\wp}(\mathfrak{A}),$$

so $a_i \in L_K(\mathfrak{A})$. This shows that $\ell L_K(\mathfrak{A}) = L_L(\mathfrak{A})$.

In particular, we have $\dim_{\ell} L_L(\mathfrak{A}) = \ell_{\ell}(\mathfrak{A}) = \dim_k L_K(\mathfrak{A}) = \ell_k(\mathfrak{A})$.

Let $\mathfrak{A} \in D_K$ be a divisor with $d_K(\mathfrak{A}) = d_L(\mathfrak{A}) > \max\{2g_K - 2, 2g_L - 2\}$. By the Riemann–Roch theorem (Corollary 3.5.6), we have

$$\ell_{\ell}\left(\mathfrak{A}^{-1}\right) = d_L(\mathfrak{A}) - g_L + 1$$

and

$$\ell_k\left(\mathfrak{A}^{-1}\right) = d_K(\mathfrak{A}) - g_K + 1.$$

Since $\ell_{\ell}\left(\mathfrak{A}^{-1}\right) = \ell_k\left(\mathfrak{A}^{-1}\right)$ and $d_L(\mathfrak{A}) = d_K(\mathfrak{A})$, it follows that $g_L = g_K$. □

Theorem 6.1.4. *Let $\mathcal{P}$ be a place of $L = K\ell$, and $\wp = \mathcal{P}|_K$. Then $\ell(\mathcal{P}) = k(\wp)\ell$.*

Proof. We have $k(\wp)\ell \subseteq \ell(\mathcal{P})$. Let $z \in \vartheta_{\mathcal{P}}$. We can write z as $\sum_{i=0}^{f-1} a_i \xi^i$, $a_i \in K$, where $\ell = k(\xi)$. It remains to prove that $a_i \in \vartheta_{\wp}$. The latter follows by the same argument given in the proof of Theorem 6.1.3, since if $z^{(i)}$ is a conjugate of z, then $z^{(i)} = z^{\sigma} \in \vartheta_{\mathcal{P}^{\sigma}}$ for some $\sigma \in \operatorname{Gal}(\ell/k)$ and $a_i = \sum_{i=0}^{f} t_i z^{(i)}$ with $t_i \in \ell$ and $z^{(i)} \in \vartheta_{\mathcal{P}^{(i)}}$. Therefore $a_i \in \vartheta_{\wp}$. □

6.2 Prime Divisors in Constant Extensions

Let K/k be a congruence function field and let $L = K\ell$ be the extension of constants. Let $\wp$ be a place in K and let $\mathcal{P}_1, \mathcal{P}_2, \ldots, \mathcal{P}_h$ be the places of L over $\wp$. We know that $\wp$ is unramified and $\operatorname{con}_{K/L}\wp = \mathcal{P}_1 \cdots \mathcal{P}_h$. Since ℓ/k is always a Galois extension (in fact cyclic), L/K is a Galois extension and $d_{L/K}(\mathcal{P}_i|\wp) = d$ for $1 \leq i \leq h$. By Theorem 5.1.14 we have $dh = f = [\ell : k]$. Now, the equality $d_K(\wp) = [k(\wp) : k] = r$ implies that $k(\wp) = \mathbb{F}_{q^r}$. Similarly, if $d_L(\mathcal{P}_i) = [\ell(\mathcal{P}_i) : \ell] = s$, then $\ell(\mathcal{P}_i) = \mathbb{F}_{q^{fs}}$.

On the other hand, $k(\wp)\ell = \mathbb{F}_{q^r}\mathbb{F}_{q^f} = \mathbb{F}_{q^{[r,f]}} = \mathbb{F}_{q^{fs}} = \ell(\mathcal{P}_i)$. Therefore $fs = [r, f] = \frac{rf}{(r,f)}$ and $s = \frac{r}{(r,f)}$.

It follows that $d_L(\mathcal{P}_i) = s = \frac{d_K(\wp)}{(d_K(\wp), f)}$. By Proposition 5.1.11 we have the equality $d_L(\mathcal{P}_i)[\ell : k] = d_{L/K}(\mathcal{P}_i|\wp)\, d_K(\wp)$. Hence, $sf = dr$, that is, $d = \frac{sf}{r} = \frac{r}{(r,f)}\frac{f}{r} = \frac{f}{(r,f)}$ and $h = \frac{f}{d} = (r, f)$. This is summarized in the following theorem:

Theorem 6.2.1. *Let $\wp$ be a place in K, $\mathcal{P}_1 \cdots \mathcal{P}_h$ the places of $L = K\ell$ over $\wp$, and $[\ell : k] = f$. Then:*

(1) $d_{L/K}(\mathcal{P}_i|\wp) = \frac{f}{(d_K(\wp), f)}$,
(2) $h = (d_K(\wp), f)$,
(3) $d_L(\mathcal{P}_i) = \frac{d_K(\wp)}{(d_K(\wp), f)}$. □

Now we will study the class group $C_{K,0} = D_{K,0}/P_K$. Recall that C_K is isomorphic to $C_{K,0} \oplus \mathbb{Z}$, with $C_K = D_K/P_K$.

Theorem 6.2.2. *Let K/k be a congruence function field. Then $|C_{K,0}| < \infty$.*

Proof. See Exercise 3.6.26. □

Notation 6.2.3. $h_K = h = |C_{K,0}|$ will denote the number of classes of the field K.

Proposition 6.2.4. *The number of integral divisors in a class C of K is equal to $\frac{q^{d(C)}-1}{q-1}$, $q = |k|$.*

Proof. See Exercise 3.6.15. □

Definition 6.2.5. Let $\varrho_K = \varrho := \min\{n \in \mathbb{N} \mid \exists\, \mathfrak{A} \in D_K \text{ such that } d(\mathfrak{A}) = n\}$.

We will prove later on (Theorem 6.3.8) that $\varrho = 1$.

We know that there exists an integral divisor of degree n if and only if $\varrho \mid n$. Since the canonical class is of degree $2g - 2$, we have $\varrho \mid 2g - 2$, with g denoting the genus of K.

Theorem 6.2.6. *Let A_n be the number of integral divisors of degree n, and $\varrho \mid n$. Then if $n > 2g - 2$, we have $A_n = h\left(\frac{q^{n-g+1}-1}{q-1}\right)$.*

Proof. We have $A_n = \sum_{\substack{C \in C_K \\ d(C)=n}} (\#\text{ of integral divisors in } C) = \sum_{\substack{C \in C_K \\ d(C)=n}} \frac{q^{N(C)}-1}{q-1}$.

By the Riemann–Roch theorem (Corollary 3.5.6),

$$N(C) = d(C) - g + 1 = n - g + 1.$$

Since there are h classes C of degree n, we have

$$A_n = h\left(\frac{q^{n-g+1} - 1}{q - 1}\right).$$ □

6.3 Zeta Functions and L-Series

Definition 6.3.1. For a prime divisor $\wp$ of K, the cardinality of $k(\wp)$ is called the *norm* of $\wp$ and it will be denoted by $N(\wp)$.

Observe that if $f_\wp = [k(\wp) : k] = d_K(\wp)$ and $|k| = q$, then $N(\wp) = |k(\wp)| = q^{d_K(\wp)}$.

Definition 6.3.1 can be extended to arbitrary integral divisors

$$\mathfrak{A} = \prod_{\wp \in \mathbb{P}_K} \wp^{v_\wp(\mathfrak{A})}$$

as follows:

$$N(\mathfrak{A}) = \prod_{\wp \in \mathbb{P}_K} N(\wp)^{v_\wp(\mathfrak{A})} = \prod_{\wp \in \mathbb{P}_K} q^{d_K(\wp) v_\wp(\mathfrak{A})} = q^{\sum_\wp d_K(\wp) v_\wp(\mathfrak{A})} = q^{d_K(\mathfrak{A})}.$$

Clearly we have $N(\mathfrak{A}\mathfrak{B}) = N(\mathfrak{A}) N(\mathfrak{B})$ for $\mathfrak{A}, \mathfrak{B} \in D_K$.

Definition 6.3.2. We define the *zeta function* of K as

$$\zeta_K(s) = \sum_{\mathfrak{A} \text{ integral}} \frac{1}{(N(\mathfrak{A}))^s} = \sum_{\mathfrak{A} \text{ integral}} q^{-d_K(\mathfrak{A})s}.$$

Theorem 6.3.3. *The series $\zeta_K(s)$ converges absolutely and uniformly in compact subsets of $\{s \in \mathbb{C} \mid \mathbb{R}\mathrm{e}\, s > 1\}$.*

Proof. Let $t = \frac{2g-2}{\varrho}$. We have

$$\begin{aligned}
\zeta_K(s) &= \sum_{\mathfrak{A} \text{ integral}} \frac{1}{(N(\mathfrak{A}))^s} = \sum_{\mathfrak{A} \text{ integral}} \frac{1}{q^{d_K(\mathfrak{A})s}} = \sum_{n=0}^{\infty} A_{\varrho n} q^{-n\varrho s} \\
&= \sum_{n=0}^{t} A_{\varrho n} q^{-n\varrho s} + \sum_{n=t+1}^{\infty} A_{\varrho n} q^{-n\varrho s}.
\end{aligned}$$

By Theorem 6.2.6,

$$\sum_{n=t+1}^{\infty} A_{\varrho n} q^{-n\varrho s} = \frac{h}{q-1} \sum_{n=t+1}^{\infty} \left(q^{n\varrho - g + 1} - 1\right) q^{-n\varrho s}.$$

Now

$$\begin{aligned}
\sum_{n=t+1}^{\infty} \left| \left(q^{n\varrho - g + 1} - 1\right) q^{-n\varrho s} \right| &= \sum_{n=t+1}^{\infty} \left(q^{n\varrho - g + 1} - 1\right) q^{-n\varrho\, \mathbb{R}\mathrm{e}\, s} \\
&= \frac{1}{q^{g-1}} \sum_{n=t+1}^{\infty} \left(q^{1 - \mathbb{R}\mathrm{e}\, s}\right)^{n\varrho} - \sum_{n=t+1}^{\infty} \left(q^{-\mathbb{R}\mathrm{e}\, s}\right)^{n\varrho},
\end{aligned}$$

from which the result follows. □

We make the substitution $u = q^{-\varrho s}$, $B_n = A_{\varrho n}$. Then $Z_K(u) = \zeta_K(s) = \sum_{n=0}^{\infty} B_n u^n$.

The canonical class W is of degree $2g-2$; there are $(h-1)$ classes C of degree $2g-2$ that are different from the class W, and we have

$$N(W) = g, \quad \text{and} \quad N(C) = (2g-2) - g + 1 = g - 1$$

for $C \neq W$, and $d(C) = 2g-2$ (Corollaries 3.5.5 and 3.5.6).

Therefore $A_{2g-2} = \frac{q^g-1}{q-1} + (h-1)\frac{q^{g-1}-1}{q-1}$ and $A_{\varrho n} = h\left(\frac{q^{\varrho n-g+1}-1}{q-1}\right)$ for $n > \frac{2g-2}{\varrho}$.

Proposition 6.3.4. *Let* $t = \frac{2g-2}{\varrho}$. *Then*

$$B_j - \left(q^{\varrho}+1\right) B_{j-1} + q^{\varrho} B_{j-2} = 0 \text{ for } j > t+2$$

and

$$B_{t+2} - \left(q^{\varrho}+1\right) B_{t+1} + q^{\varrho} B_t = q^{g+\varrho-1}.$$

Proof. For $j > t+2$, we have

$$\begin{aligned} j\varrho &> (t+2)\varrho = t\varrho + 2\varrho = 2g - 2 + 2\varrho \geq 2g, \\ (j-1)\varrho &> (t+1)\varrho = t\varrho + \varrho = 2g - 2 + \varrho \geq 2g - 1, \\ (j-2)\varrho &> t\varrho = 2g - 2. \end{aligned}$$

It follows that $B_j - (q^{\varrho}+1) B_{j-1} + q^{\varrho} B_{j-2} = 0$.

On the other hand, $B_{t+2} - (q^{\varrho}+1) B_{t+1} + q^{\varrho} B_t = q^{\varrho+g-1}$. □

Now we consider

$$\begin{aligned} &(1-u)\left(1-q^{\varrho}u\right) Z_K(u) \\ &= \left(1 - \left(1+q^{\varrho}\right) u + q^{\varrho}u^2\right) Z_K(u) \\ &= \sum_{n=0}^{\infty} B_n u^n - \sum_{n=0}^{\infty} \left(1+q^{\varrho}\right) B_n u^{n+1} + \sum_{n=0}^{\infty} q^{\varrho} B_n u^{n+2} \\ &= \sum_{n=0}^{\infty} \left(B_n - \left(1+q^{\varrho}\right) B_{n-1} + q^{\varrho} B_{n-2}\right) u^n \qquad \text{(with } B_{-1} = B_{-2} = 0\text{)} \\ &= \sum_{n=0}^{t+2} \left(B_n - \left(1+q^{\varrho}\right) B_{n-1} + q^{\varrho} B_{n-2}\right) u^n \qquad \text{(Proposition 6.3.4, with } A_0 = B_0 = 1\text{)} \\ &= 1 + \left(B_1 - \left(q^{\varrho}+1\right)\right) u + \sum_{n=2}^{t+2} \left(B_n - \left(1+q^{\varrho}\right) B_{n-1} + q^{\varrho} B_{n-2}\right) u^n. \end{aligned}$$

Thus the element $(1-u)(1-q^{\varrho}u) Z_K(u) = P_K(u)$ of $\mathbb{Z}[u]$ is a polynomial.

Let $P_K(u) = a_0 + a_1 u + a_2 u^2 + \cdots + a_{t+2} u^{t+2}$, $a_0 = 1$, $a_1 = B_1 - (q^{\varrho}+1)$, and $a_{t+2} = q^{g+\varrho-1}$ (Proposition 6.3.4).

Theorem 6.3.5. *The function $Z_K(u)$ is a rational function and satisfies*

$$Z_K(u) = \frac{P_K(u)}{(1-u)\,(1-q^{\varrho}u)},$$

where $P_K(u) \in \mathbb{Z}[u]$ is a polynomial of degree $t+2 = \frac{2g-2}{\varrho} + 2$.

Furthermore, $P_K(1) = h\frac{q^{\varrho}-1}{q-1} = \lim_{u\to 1}(1-u)\,(1-q^{\varrho}u)\,Z_K(u)$.

Proof. Setting $B_{-1} = B_{-2} = 0$, we have

$$\begin{aligned}
P_K(1) &= \sum_{n=0}^{t+2} \left(B_n - \left(1+q^{\varrho}\right) B_{n-1} + q^{\varrho} B_{n-2}\right) \\
&= \sum_{n=0}^{t+2} \left(B_n - B_{n-1} - q^{\varrho} B_{n-1} + q^{\varrho} B_{n-2}\right) \\
&= B_{t+2} - B_{-1} - q^{\varrho}\left(B_{t+1} - B_{-2}\right) \\
&= A_{t\varrho+2\varrho} - q^{\varrho} A_{t\varrho+\varrho} \\
&= A_{2g-2+2\varrho} - q^{\varrho} A_{2g-2+\varrho} \\
&= h\frac{q^{2g-2+2\varrho-g+1}-1}{q-1} - q^{\varrho} h \frac{q^{2g-2+\varrho-g+1}-1}{q-1} \\
&= \frac{h}{q-1}\left(q^{g+2\varrho-1} - 1 - q^{g+2\varrho-1} + q^{\varrho}\right) \\
&= \frac{q^{\varrho}-1}{q-1} h.
\end{aligned}$$

□

Corollary 6.3.6. *$Z_K(u)$ has a simple pole for $u = 1$.* □

In order to prove the equality $\varrho = 1$, we need another expression for $\zeta_K(s)$.

Theorem 6.3.7 (Product Formula).

$$\zeta_K(s) = \prod_{\wp\in\mathbb{P}_K} \left(1 - N(\wp)^{-s}\right)^{-1} \quad \text{with} \quad \mathbb{R}\text{e}\, s > 1.$$

Proof. Let $\wp$ be a prime divisor, and $d(\wp) = n$. Then

$$a_{\wp} = \left(1 - N(\wp)^{-s}\right)^{-1} - 1 = \frac{1}{1-q^{-ns}} - 1 = \frac{q^{-ns}}{1-q^{-ns}} = \frac{1}{q^{ns}-1}.$$

We have $|q^{ns}-1| \geq |q^{ns}| - 1 = q^{n\alpha} - 1$, with $\alpha = \mathbb{R}\text{e}\, s > 1$. Therefore $\left|a_{\wp}\right| \leq \frac{1}{q^{n\alpha}-1} \leq \frac{2}{q^{n\alpha}}$ for n sufficiently large.

Now,

$$|\{\wp \mid d(\wp) = n\}| \leq A_n = h\left(\frac{q^{n-g+1}-1}{q-1}\right), \quad \text{with} \quad n > 2g-2.$$

Therefore we have

$$\sum_{n\gg 0} |a_{\wp}| \leq \frac{h}{q-1} q^{-g+1} \sum_{n=0}^{\infty} \frac{2}{q^{n(\alpha-1)}} - \frac{h}{q-1} \sum_{n=0}^{\infty} \frac{2}{q^{n\alpha}} < \infty,$$

and hence $\prod_{\wp\in\mathbb{P}_K} \left(1 - N(\wp)^{-s}\right)^{-1}$ is absolutely convergent. Rearranging the terms of the product, we obtain

$$\begin{aligned}\prod_{\wp\in\mathbb{P}_K} \left(1 - N(\wp)^{-s}\right)^{-1} &= \prod_{\wp\in\mathbb{P}_K} \left(\frac{1}{1 - N(\wp)^{-s}}\right)\\ &= \prod_{\wp\in\mathbb{P}_K} \left(\sum_{n_{\wp}=0}^{\infty} \left(N(\wp)^{-n_{\wp}s}\right)\right) = \sum N\left(\wp_1^{\alpha_1}\cdots\wp_r^{\alpha_r}\right)^{-s},\end{aligned}$$

where the sum is taken over all powers of the divisors $\wp_1, \ldots, \wp_r$ and $\alpha_i \geq 0$ for $i = 1, \ldots, r$. Therefore

$$\begin{aligned}\prod_{\wp\in\mathbb{P}_K} \left(1 - N(\wp)^{-s}\right)^{-1} &= \sum_{\substack{\wp_1,\ldots,\wp_r\in\mathbb{P}_K\\ \alpha_i\geq 0}} N\left(\wp_1^{\alpha_1}\cdots\wp_r^{\alpha_r}\right)^{-s}\\ &= \sum_{\mathfrak{A}\in D_K \text{ integral}} \frac{1}{(N(\mathfrak{A}))^s} = \zeta_K(s). \qquad \square\end{aligned}$$

Let $|k| = q$, $\ell = \mathbb{F}_{q^f}$, and let $L = K\ell$ be the extension of constants. We wish to compare $\zeta_L(s)$ with $\zeta_K(s)$ when $f = \varrho$. For a place $\wp$ of K, ϱ divides $d_K(\wp)$, and if $\mathcal{P}_1, \ldots, \mathcal{P}_r$ are the prime divisors of L over $\wp$, by Theorem 6.2.1 we have $r = (d_K(\wp), \varrho) = \varrho$, whence there always exist ϱ factors in L over any given prime divisor of K. Furthermore, we have

$$d_L(\mathcal{P}_i) = \frac{d_K(\wp)}{(d_K(\wp), \varrho)} = \frac{d_K(\wp)}{\varrho}.$$

On the other hand, $N(\mathcal{P}_i) = (q^{\varrho})^{d_L(\mathcal{P}_i)} = q^{\varrho d_K(\wp)/\varrho} = q^{d_K(\wp)} = N(\wp)$. Therefore

$$\begin{aligned}\zeta_L(s) &= \prod_{\mathcal{P}\in\mathbb{P}_L} \left(1 - \frac{1}{N(\mathcal{P})^s}\right)^{-1} = \prod_{\wp\in\mathbb{P}_K} \left(\prod_{\mathcal{P}|\wp} \left(1 - \frac{1}{N(\mathcal{P})^s}\right)^{-1}\right)\\ &= \prod_{\wp\in\mathbb{P}_K} \left(1 - \frac{1}{N(\wp)^s}\right)^{-\varrho} = \left(\prod_{\wp\in\mathbb{P}_K} \left(1 - \frac{1}{N(\wp)^s}\right)\right)^{-\varrho} = \zeta_K(s)^{\varrho}.\end{aligned}$$

Thus $\zeta_L(s) = \zeta_K(s)^{\varrho}$. On the one hand, by Corollary 6.3.6, both $\zeta_L(s)$ and $\zeta_K(s)$ have a pole of order 1 at $s = 0$ (or at $u = 1$ with the change of variables $u = q^{-\varrho s}$). On the other hand, $\zeta_K(s)^{\varrho}$ has a pole of order ϱ at $s = 0$. It follows that $\varrho = 1$.

We have obtained the following theorem:

Theorem 6.3.8 (F.K. Schmidt). *Let K/k be any congruence function field and set*

$$\varrho = \min\{n \in \mathbb{N} \mid \text{ there exists } \mathfrak{A} \in D_K, d(\mathfrak{A}) = n\}.$$

Then $\varrho = 1$. □

Corollary 6.3.9. *$Z_K(u) = \frac{P_K(u)}{(1-u)(1-qu)}$, where $u = q^{-s}$, $P_K(u) \in \mathbb{Z}[u]$ is of degree $2g$, $P_K(u) = 1 + (A_1 - (q+1))u + \cdots + q^g u^{2g}$, and $P_K(1) = h$ is the class number of K.*

Proof. In Proposition 6.3.4 we wrote $t = \frac{2g-2}{\varrho}$, so that $B_n = A_{\varrho n} = A_n$, $t = 2g-2$, $\frac{q^{\varrho}-1}{q-1} = 1$, etc. Substituting these expressions in Theorem 6.3.5 we obtain the result. □

Corollary 6.3.10. *If K is a congruence function field of genus 0, then $Z_K(u) = \frac{1}{(1-u)(1-qu)}$.* □

Now we will study the L-series of a congruence function field.

Definition 6.3.11. A *character χ of finite order* of the group of classes C_K is a homomorphism $\chi : C_K \to \mathbb{C}^*$ defined so that there exists $n \in \mathbb{N}$ satisfying $\chi^n = 1$. In other words, $\chi(C_K) \subseteq \{\xi \in \mathbb{C} \mid \xi^n = 1 \text{ for some } n \in \mathbb{N}\}$.

A character χ can be extended to the group of divisors $\chi : D_K \to \mathbb{C}^*$, by setting $\chi(\mathfrak{A}) = \chi(\mathfrak{A}P_K)$, where P_K is the principal class. Note that $|\chi(\mathfrak{A})| = 1$.

Definition 6.3.12. Given a character χ of finite order over D_K, we define *the L-series associated to χ* by

$$L(s, \chi, K) = \sum_{\mathfrak{A} \text{ integral}} \chi(\mathfrak{A}) \frac{1}{(N(\mathfrak{A}))^s}, \quad \text{where} \quad s \in \mathbb{C} \quad \text{and} \quad \operatorname{Re} s > 1.$$

Theorem 6.3.13. *The series $\sum_{\mathfrak{A} \text{ integral}} \chi(\mathfrak{A}) \frac{1}{(N(\mathfrak{A}))^s}$ converges absolutely and uniformly in compact subsets of $\{s \in \mathbb{C} \mid \operatorname{Re} s > 1\}$.*

Proof. This follows from Theorem 6.3.3 and from the fact that $|\chi(\mathfrak{A})| = 1$ for all $\mathfrak{A} \in D_K$. □

We have the following product formula, which is an immediate consequence of Theorem 6.3.7:

Theorem 6.3.14. $L(s, \chi, K) = \prod_{\wp \in \mathbb{P}_K} \left(1 - \frac{\chi(\wp)}{N(\wp)^s}\right)^{-1}$ *for all s such that $\operatorname{Re} s > 1$.* □

6.4 Functional Equations

In this section, we consider the case $g = g_K = 0$, which implies that

$$Z_K(u) = \frac{1}{(1-u)(1-qu)} \quad \text{or} \quad \zeta_K(s) = \frac{1}{\left(1-q^{-s}\right)\left(1-q^{1-s}\right)}.$$

We have

$$\begin{aligned} q^{-s}\zeta_K(s) &= \frac{1}{\left(1-q^{-s}\right)(q^s-q)} = \frac{1}{q^{-s}(q^s-1)\,q\left(q^{s-1}-1\right)} \\ &= q^{s-1}\frac{1}{(1-q^s)\left(1-q^{s-1}\right)} = q^{s-1}\zeta_K(1-s). \end{aligned}$$

Therefore, $q^{-s}\zeta_K(s) = q^{s-1}\zeta_K(1-s)$ and $g = 0$.

For $g > 0$, consider $u = q^{-s}$ and $Z_K(u) = \zeta_K(s)$. Then

$$\begin{aligned} Z_K(u) &= \frac{P_K(u)}{(1-u)(1-qu)}, \\ P_K(u) &= a_0 + a_1u + \cdots + a_{2g}u^{2g}, \quad a_0 = 1, \quad \text{and} \quad a_{2g} = q^g. \end{aligned}$$

Theorem 6.4.1. *For $0 \le i \le 2g$, we have $a_{2g-i} = a_i q^{g-i}$.*

Proof. For $i = 0$, we have $a_{2g} = a_{2g-0} = q^g = a_0q^{g-0}$. In general, $a_i = A_i - (q+1)A_{i-1} + qA_{i-2}$, where A_i is the number of integral divisors of degree i (see the argument preceding Theorem 6.3.5). We obtain $A_i = \sum_{\substack{C\in C_K\\ d(C)=i}} \frac{q^{N(C)}-1}{q-1}$.

By the Riemann–Roch theorem, we have

$$N(C) = d(C) - g + 1 + N\left(WC^{-1}\right) = i - g + 1 + N\left(WC^{-1}\right).$$

Now, $d\left(WC^{-1}\right) = 2g-2-i$, and when C runs through all classes of degree i, WC^{-1} runs trough all classes of degree $2g-2-i$.

Since there are h classes of each degree, where h is the class number of K, we have

$$(q-1)A_i = \sum_{d(C)=i} q^{N(C)} - \sum_{d(C)=i} 1 = \sum_{d(C)=i} q^{N(C)} - h.$$

Hence,

$$\begin{aligned} (q-1)A_i + h &= \sum_{d(C)=i} q^{N(C)} = \sum_{d(C)=i} q^{i-g+1+N(WC^{-1})} \\ &= q^{i-g+1}\sum_{d(C)=i} q^{N(WC^{-1})} = q^{i-g+1}\sum_{d(C)=2g-2-i} q^{N(C)} \\ &= q^{i-g+1}\left((q-1)A_{2g-2-i} + h\right). \end{aligned}$$

Therefore

$$\begin{aligned}(q-1)A_{2g-2-i} &= \frac{(q-1)A_i+h}{q^{i-g+1}}-h,\\ (q-1)A_{2g-1-i} &= (q-1)A_{2g-2-(i-1)} = \frac{(q-1)A_{i-1}+h}{q^{i-g}}-h,\\ (q-1)A_{2g-i} &= (q-1)A_{2g-2-(i-2)} = \frac{(q-1)A_{i-2}+h}{q^{i-g-1}}-h.\end{aligned}$$

It follows that $a_{2g-i}=q^{g-i}a_i$. □

Corollary 6.4.2. *We have*

$$P_K\left(\frac{1}{qu}\right)=q^{-g}u^{-2g}P_K(u) \quad \text{and} \quad u^{1-g}Z_K(u)=(qu)^{g-1}Z_K\left(\frac{1}{qu}\right).$$

Proof. Notice that

$$\begin{aligned}P_K\left(\frac{1}{qu}\right) &= a_0+a_1\left(\frac{1}{qu}\right)+\cdots+a_{2g}\left(\frac{1}{qu}\right)^{2g}=\frac{1}{(qu)^{2g}}\sum_{i=0}^{2g}a_i(qu)^{2g-i}\\ &= q^{-g}u^{-2g}\sum_{i=0}^{2g}a_iq^{g-i}u^{2g-i}=q^{-g}u^{-2g}\sum_{i=0}^{2g}a_{2g-i}u^{2g-i}\\ &= q^{-g}u^{-2g}P_K(u).\end{aligned}$$

Also,

$$\begin{aligned}Z_K\left(\frac{1}{qu}\right) &= \frac{P_K\left(\frac{1}{qu}\right)}{\left(1-\frac{1}{qu}\right)\left(1-\frac{q}{qu}\right)}=\frac{q^{-g}u^{-2g}P_K(u)}{(qu-1)(u-1)}qu^2\\ &= q^{1-g}u^{2(1-g)}\frac{P_K(u)}{(1-u)(1-qu)}=q^{1-g}u^{2(1-g)}Z_K(u).\end{aligned}$$

□

Corollary 6.4.2 is the functional equation of the zeta function in terms of the variable $u=q^{-s}$. Since $\zeta_K(s)=Z_K\left(q^{-s}\right)$, we obtain, in terms of the variable s, the following result:

Theorem 6.4.3 (Functional Equation for the Zeta Function). *We have*

$$q^{s(g-1)}\zeta_K(s)=q^{(1-s)(g-1)}\zeta_K(1-s) \quad \text{for all} \quad s\in\mathbb{C}.$$

In particular, $\zeta_K(s)$ is a meromorphic function in the whole complex plane $\mathbb{C}$ with simple poles in

$$\left\{s \;\middle|\; q^{-s}=u\in\left\{1,\frac{1}{q}\right\}\right\}=\left\{a+\frac{2k\pi\iota}{\ln q} \;\middle|\; k\in\mathbb{Z}, a=0,1\right\}.$$

Proof. Setting $u = q^{-s}$, we obtain

$$\begin{aligned} q^{s(g-1)}\zeta_K(s) &= u^{1-g}Z_K(u) = (qu)^{g-1}Z_K\left(\frac{1}{qu}\right) \\ &= q^{(1-s)(g-1)}Z_K\left(q^{s-1}\right) = q^{(1-s)(g-1)}\zeta_K(1-s). \end{aligned}$$

In the expression $Z_K(u) = \frac{P_K(u)}{(1-u)(1-qu)}$, the denominator is equal to zero for $u = 1$ and $u = \frac{1}{q}$. On the other hand, $P_K(1) = h \neq 0$ (Corollary 6.3.9) and $P_K\left(\frac{1}{q}\right) = q^{-g}P_K(1) = q^{-g}h \neq 0$ (Corollary 6.4.2). Therefore $u = 1$ and $u = q^{-1}$ are the only poles of $Z_K(u)$ and they are simple.

In terms of the variable s we have the following equivalences:

$$u = q^{-s} = 1 \Leftrightarrow q^s = e^{s\ln q} = 1 \Leftrightarrow s\ln q = 2\pi ji,\ j \in \mathbb{Z} \Leftrightarrow s = \frac{2\pi ji}{\ln q},\ j \in \mathbb{Z},$$

$$u = q^{-s} = q^{-1} \Leftrightarrow q^s = q \Leftrightarrow q^{s-1} = 1 \Leftrightarrow s = 1 + \frac{2\pi ji}{\ln q},\ j \in \mathbb{Z}. \qquad \square$$

Coming back to the series L, let χ be a character of finite order.

Proposition 6.4.4. *If* $\chi\left(C_{K,0}\right) = 1$, *then*

$$L(s,\chi,K) = \zeta_K\left(s - \frac{2\pi i\alpha}{\ln q}\right),$$

where $\chi(C_0) = e^{2\pi i\alpha}$ *and* C_0 *is a class of degree* 1. *Equivalently,*

$$L(s,\chi,K) = Z_K\left(e^{2\pi i\alpha}u\right).$$

Proof. C_K is isomorphic to $C_{K,0} \oplus \langle C_0\rangle$ under the following identification: if C is an arbitrary class of degree n, $C = CC_0^{-n}C_0^n$. Then

$$\chi(C) = \chi\left(CC_0^{-n}\right)\chi\left(C_0^n\right) = \chi(C_0)^n = e^{2\pi i\alpha n}.$$

We have

$$\begin{aligned} L(s,\chi,K) &= \sum_{\mathfrak{A}\text{ integral}} \frac{\chi(\mathfrak{A})}{(N(\mathfrak{A}))^s} = \sum_{C'\in C_{K,0}} \sum_{\substack{\mathfrak{A}\in C'C_0^n \\ \mathfrak{A}\text{ integral}}} \sum_{n=0}^{\infty} \frac{\chi(\mathfrak{A})}{(N(\mathfrak{A}))^s} \\ &= \sum_{C'\in C_{K,0}} \sum_{\substack{\mathfrak{A}\in C'C_0^n \\ \mathfrak{A}\text{ integral}}} \sum_{n=0}^{\infty} e^{2\pi i\alpha n}q^{-ns} \\ &= \sum_{C'\in C_{K,0}} \sum_{\substack{\mathfrak{A}\in C'C_0^n \\ \mathfrak{A}\text{ integral}}} \sum_{n=0}^{\infty} q^{(\frac{2\pi i\alpha}{\ln q}-s)n} \\ &= \sum_{\mathfrak{A}\text{ integral}} (N(\mathfrak{A}))^{-(s-\frac{2\pi i\alpha}{\ln q})} = \zeta_K\left(s - \frac{2\pi i\alpha}{\ln q}\right). \end{aligned}$$

Also, $\zeta_K\left(s - \frac{2\pi i\alpha}{\ln q}\right) = Z_K\left(q^{-s} q^{2\pi i\alpha/\ln q}\right) = Z_K\left(e^{2\pi i\alpha} u\right)$. □

Corollary 6.4.5. *If $\chi\left(C_{K,0}\right) = 1$, then the series L satisfies the functional equation*

$$q^{s(g-1)} L(s, \chi, K) = \chi(W) q^{(1-s)(g-1)} L(1-s, \bar{\chi}, K),$$

where W is the canonical class and $\bar{\chi}$ is the conjugate of χ, i.e., $\bar{\chi}(\mathfrak{A}) := \overline{\chi(\mathfrak{A})} = \chi(\mathfrak{A}^{-1})$.

Proof. Using the functional equation of Corollary 6.4.2 and setting $u' = e^{2\pi i\alpha} u$, we obtain

$$\begin{aligned} q^{s(g-1)} L(s, \chi, K) &= q^{s(g-1)} Z_K\left(u'\right) = q^{s(g-1)} (qu')^{g-1} (u')^{g-1} Z_K\left(\frac{1}{qu'}\right) \\ &= q^{(s+1)(g-1)} q^{-s(2g-2)} \left(e^{2\pi i\alpha}\right)^{2g-2} Z_K\left(\frac{1}{qu} e^{-2\pi i\alpha}\right) \\ &= q^{(1-s)(g-1)} \left(e^{2\pi i\alpha}\right)^{2g-2} Z_K\left(\frac{e^{-2\pi i\alpha}}{qu}\right). \end{aligned}$$

Since $d(W) = 2g-2$, $\chi(W) = \left(e^{2\pi i\alpha}\right)^{(2g-2)}$, and $\bar{\chi}(C_0) = e^{-2\pi i\alpha}$, it follows that

$$\begin{aligned} q^{s(g-1)} L(s, \chi, K) &= q^{(1-s)(g-1)} \chi(W) Z_K\left(q^{s-1-\frac{2\pi i\alpha}{\ln q}}\right) \\ &= q^{(1-s)(g-1)} \chi(W) \zeta_K\left(1 - s + \frac{2\pi i\alpha}{\ln q}\right) \end{aligned}$$

and

$$L(1-s, \bar{\chi}, K) = \zeta_K\left(1 - s - \left(-\frac{2\pi i\alpha}{\ln q}\right)\right) = \zeta_K\left(1 - s + \frac{2\pi i\alpha}{\ln q}\right).$$

Therefore

$$q^{s(g-1)} L(s, \chi, K) = q^{(1-s)(g-1)} \chi(W) L(1-s, \bar{\chi}, K).$$ □

The functional equation given by Corollary 6.4.5 is satisfied for any character of finite order. However, we need to provide a different proof from the one given in the case $\chi\left(C_{K,0}\right) = 1$.

Let χ be such that $\chi\left(C_{K,0}\right) \neq 1$. Then $C_{K,0} \neq 1$ and $g > 0$ (Proposition 4.1.5). Let C_0' be a class of degree 0 such that $\chi\left(C_0'\right) \neq 1$. We have

$$\chi\left(C_0'\right) \sum_{C_0 \in C_{K,0}} \chi(C_0) = \sum_{C_0 \in C_{K,0}} \chi\left(C_0' C_0\right) = \sum_{C_0 \in C_{K,0}} \chi(C_0),$$

that is,

$$\left(\chi\left(C_0'\right)-1\right)\left(\sum_{C_0\in C_{K,0}}\chi\left(C_0\right)\right)=0.$$

Since $\chi\left(C_0'\right)\neq 1$, it follows that $\sum_{C_0\in C_{K,0}}\chi\left(C_0\right)=0$.

Let C_1 be a class of degree 1. We have

$$\begin{aligned}
(q-1)L\left(s,\chi,K\right) &= (q-1)\sum_{\mathfrak{A}\text{ integral}}\chi\left(\mathfrak{A}\right)\frac{1}{\left(N\left(\mathfrak{A}\right)\right)^s}\\
&=\sum_{d(C)=0}^{\infty}(q-1)\left\{\frac{q^{N(C)}-1}{q-1}\chi\left(C\right)q^{-d(C)s}\right\}\\
&=\sum_{n=0}^{\infty}\sum_{C_0\in C_{K,0}}\chi\left(C_0C_1^n\right)q^{-ns}\left(q^{N(C_0C_1^n)}-1\right)\\
&=\sum_{C_0\in C_{K,0}}\chi\left(C_0\right)\sum_{n=0}^{\infty}\chi\left(C_1\right)^n\left(q^{N(C_0C_1^n)}-1\right)q^{-ns}\\
&=\sum_{C_0\in C_{K,0}}\chi\left(C_0\right)\sum_{n=0}^{2g-2}\chi\left(C_1\right)^n\left(q^{N(C_0C_1^n)}-1\right)q^{-ns}\\
&\quad+\sum_{C_0\in C_{K,0}}\chi\left(C_0\right)\sum_{n=2g-1}^{\infty}\chi\left(C_1\right)^n\left(q^{n-g+1}-1\right)q^{-ns}.
\end{aligned}$$

The second sum is equal to 0 since $\sum_{C_0\in C_{K,0}}\chi\left(C_0\right)=0$. Therefore

$$\begin{aligned}
(q-1)L\left(s,\chi,K\right)=&\sum_{C_0\in C_{K,0}}\chi\left(C_0\right)\sum_{n=0}^{2g-2}\chi\left(C_1\right)^n q^{N(C_0C_1^n)}q^{-ns}\\
&-\sum_{C_0\in C_{K,0}}\chi\left(C_0\right)\sum_{n=0}^{2g-2}\chi\left(C_1\right)^n q^{-ns}.
\end{aligned}$$

Again using the fact that $\sum_{C_0\in C_{K,0}}\chi\left(C_0\right)=0$, we obtain

$$(q-1)L\left(s,\chi,K\right)=\sum_{C_0\in C_{K,0}}\chi\left(C_0\right)\sum_{n=0}^{2g-2}\chi\left(C_1\right)^n q^{N(C_0C_1^n)}q^{-ns}.$$

Writing $u=q^{-s}$, we have $(q-1)L\left(s,\chi,K\right)=\sum_{d(C)=0}^{2g-2}\chi\left(C\right)q^{N(C)}u^{d(C)}$, which is a polynomial in u of degree at most $2g-2$.

The coefficient of u^{2g-2} is

$$a=\sum_{d(C)=2g-2}\chi\left(C\right)q^{N(C)}=\sum_{C_0\in C_{K,0}}\chi\left(WC_0\right)q^{N(WC_0)}.$$

From the Riemann–Roch theorem we obtain

$$N(WC_0) = d(WC_0) - g + 1 + N\left(C_0^{-1}\right),$$
$$N\left(C_0^{-1}\right) = 0 \quad \text{if} \quad C_0 \neq P_K, \quad \text{and} \quad N(P_K) = 1.$$

Thus

$$N(WC_0) = 2g - 2 - g + 1 = g - 1 \text{ if } C_0 \neq P_K \text{ and } N(W) = g,$$

and

$$\begin{aligned} a &= \sum_{\substack{C_0 \in C_{K,0} \\ C_0 \neq P_K}} \chi(W)\chi(C_0)q^{g-1} + \chi(W)q^g \\ &= \chi(W) \sum_{C_0 \in C_{K,0}} \chi(C_0)q^{g-1} + \chi(W)\left(q^g - q^{g-1}\right) \\ &= \left(q^{g-1}\right)(q-1)\chi(W) \neq 0. \end{aligned}$$

Therefore $(q-1)L(s,\chi,K)$ is a polynomial of degree $2g-2$ and its coefficient of highest degree is $(q-1)\chi(W)q^{g-1}$.

Applying again the Riemann–Roch theorem we obtain

$$(q-1)L(s,\chi,K) = \sum_{d(C)=0}^{2g-2} \chi(C)q^{N(C)}u^{d(C)}$$

$$\begin{aligned} &= \sum_{d(C)=0}^{2g-2} \chi(C)q^{d(C)-g+1+N(WC^{-1})}u^{d(C)} \\ &= q^{g-1}u^{2g-2}\chi(W) \sum_{d(C)=0}^{2g-2} \chi\left(CW^{-1}\right)q^{-2g+2+d(C)+N(WC^{-1})}u^{d(C)-2g+2} \\ &= q^{g-1}u^{2g-2}\chi(W) \sum_{d(C)=0}^{2g-2} \overline{\chi\left(C^{-1}W\right)}q^{d(W^{-1}C)+N(WC^{-1})}u^{d(W^{-1}C)} \\ &= q^{g-1}u^{2g-2}\chi(W) \sum_{d(C)=0)}^{2g-2} \overline{\chi\left(WC^{-1}\right)}q^{N(WC^{-1})}\left(\frac{1}{qu}\right)^{d(WC^{-1})} \\ &= q^{g-1}u^{2g-2}\chi(W) \sum_{d(C)=0}^{2g-2} \bar{\chi}(C)q^{N(C)}\left(\frac{1}{qu}\right)^{d(C)} \\ &= q^{g-1}u^{2g-2}\chi(W)(q-1)L(1-s,\bar{\chi},K) \\ &= q^{g-1}q^{-2s(g-1)}(q-1)\chi(W)L(1-s,\bar{\chi},K). \end{aligned}$$

Therefore $L(s,\chi,K) = q^{(g-1)(1-2s)}\chi(W)L(1-s,\bar{\chi},K)$, or in other words,

$$q^{s(g-1)} L(s, \chi, K) = q^{(1-s)(g-1)} \chi(W) L(1-s, \bar{\chi}, K).$$

In short, we have the following theorem:

Theorem 6.4.6 (Functional Equation for L-Series). *Let K/k be a congruence function field with $|k| = q$, let W be the canonical class, and let χ be a character of finite order. Then*

$$q^{s(g-1)} L(s, \chi, K) = \chi(W) q^{(1-s)(g-1)} L(1-s, \bar{\chi}, K). \qquad \square$$

We end this chapter with a result that relates L series to the zeta function of an extension of constants.

Let K/k be a congruence function field with $k = \mathbb{F}_q$, $\ell = \mathbb{F}_{q^r}$, and $L = K\ell$. Let χ_j be the character of K that satisfies $\chi_j(C) = e^{\frac{2\pi i j}{r}}$ in every class of degree 1. Then $\chi_j\left(C_{K,0}\right) = 1$ for $j = 1, \ldots, r$, and we have the following result:

Theorem 6.4.7.

$$\zeta_L(s) = \prod_{j=1}^{r} L\left(s, \chi_j, K\right).$$

Proof. First, notice that if $a, b \in \mathbb{N}$,

$$\prod_{n=1}^{a} \left(1 - e^{\frac{2\pi i n}{a} b} z\right) = \left(1 - z^{\frac{a}{(a,b)}}\right)^{(a,b)}.$$

Now

$$\begin{aligned}
\zeta_L(s) &= \prod_{\mathcal{P} \in \mathbb{P}_L} \left(1 - \frac{1}{N(\mathcal{P})^s}\right)^{-1} = \prod_{\wp \in \mathbb{P}_K} \prod_{\mathcal{P} | \wp} \left(1 - q^{-s r d_L(\mathcal{P})}\right)^{-1} \\
&= \prod_{\wp \in \mathbb{P}_K} \prod_{\mathcal{P} | \wp} \left(1 - q^{-s \frac{r d_K(\wp)}{(d_{L/K}(\mathcal{P}|\wp), r)}}\right)^{-1} \qquad \text{(Theorem 6.2.1)}.
\end{aligned}$$

There are $(r, d_K(\wp))$ factors of the form $\mathcal{P} \mid \wp$. Therefore

$$\begin{aligned}
\zeta_L(s) &= \prod_{\wp \in \mathbb{P}_K} \left\{1 - \left(N_K(\wp)\right)^{-s \frac{r}{(r, d_K(\wp))}}\right\}^{-(r, d_K(\wp))} \\
&= \prod_{\wp \in \mathbb{P}_K} \prod_{n=1}^{r} \left(1 - \frac{1}{N(\wp)^s} e^{\frac{2\pi i n}{r} d_K(\wp)}\right)^{-1} \qquad (a = r, b = d_K(\wp)) \\
&= \prod_{n=1}^{r} \zeta_K\left(s - \frac{2\pi i n}{r \ln q}\right) \\
&= \prod_{n=1}^{r} L(s, \chi_n, K) \qquad \text{(Proposition 6.4.4)}. \quad \square
\end{aligned}$$

6.5 Exercises

Exercise 6.5.1. Let k be a finite field with $|k| = q$. Let K be an elliptic function field over k with class number h. Find $\zeta_K(s)$ explicitly.

Exercise 6.5.2. Let k be a finite field with $|k| = q$, and $K = k(x, y)$ with $y^m = x$, $m \in \mathbb{N}$. Find $\zeta_K(s)$ explicitly.

Exercise 6.5.3. Let $K/\mathbb{F}_q$ be a hyperelliptic function field of genus 2. Find $\zeta_K(s)$ explicitly (see Exercise 10.9.4).

7

The Riemann Hypothesis

In Chapter 6 we defined the zeta function of a congruence function field. This definition arises from the natural extension of the usual Riemann zeta function $\zeta(s) = \sum_{n=1}^{\infty} \frac{1}{n^s}$. It is known that $\zeta(s)$ has a meromorphic extension to the complex plane, with a unique pole at $s = 1$. This pole is simple with residue 1. Furthermore, $\zeta(s)$ has zeros at $s = -2n$ $(n \in \mathbb{N})$ and these are called the trivial zeros of $\zeta(s)$. On the other hand, $\zeta(s)$ has no zeros different from the trivial ones in $\mathbb{C} \setminus \{s \mid 0 \leq \mathbb{R}\mathrm{e}\, s \leq 1\}$. Finally, the Riemann hypothesis states that the zeros of $\zeta(s)$ other than the trivial ones lie on the line of equation $\mathbb{R}\mathrm{e}\, s = \frac{1}{2}$.

The latter is still an open problem. However, for function fields the answer is known and is positive. This was proved by André Weil in 1940–1941 [158, 159] and the main goal of this chapter is to give a proof of the Riemann hypothesis as well as some applications.

In particular, when considering extensions of constants whose degree is a power of a prime number, we find that the analogue of Iwasawa's invariant μ for number fields is 0 in our case. We end the chapter with the presentation of the analogue of the Brauer–Siegel theorem on number fields.

7.1 The Number of Prime Divisors of Degree 1

Let $k = \mathbb{F}_q$ be a finite field and $k_r = \mathbb{F}_{q^r}$ the extension of degree $r \geq 1$ of k. One of our goals is to estimate the number of prime divisors of degree n in K/k. For this purpose we will frequently use the Möbius function μ and the Newton identities. We now state the definitions and then will prove their main properties.

Definition 7.1.1. An *arithmetic function* in $\mathbb{Q}$ is any function $f : \mathbb{N} \longrightarrow \mathbb{Q}$. The *Möbius function* is the function $\mu : \mathbb{N} \longrightarrow \mathbb{Q}$ defined as follows. If $n \in \mathbb{N}$ and $\prod_{i=1}^{r} p_i^{a_i}$ is its decomposition into prime divisors, then

$$\mu(n) = \begin{cases} 1 & \text{if } n = 1, \\ (-1)^r & \text{if } a_1 = \cdots = a_r = 1, \\ 0 & \text{in any other case.} \end{cases}$$

Lemma 7.1.2. *We have*

$$\sum_{d|n} \mu(d) = \varepsilon(n) = \begin{cases} 1 \text{ if } n = 1, \\ 0 \text{ if } n > 1. \end{cases}$$

Proof. We leave the proof to the reader (Exercise 7.7.1). □

Theorem 7.1.3 (Inversion Formula of Möbius). *If f, g are two arithmetic functions such that*

$$g(n) = \sum_{d|n} f(d)$$

for all $n \in \mathbb{N}$, then

$$f(n) = \sum_{d|n} g(d)\mu\left(\frac{n}{d}\right) = \sum_{d|n} g\left(\frac{n}{d}\right)\mu(d).$$

Proof. For any two arithmetic functions h and k we define the product $h * k$ by

$$(h * k)(n) = \sum_{d|n} h\left(\frac{n}{d}\right) k(d) = \sum_{d|n} h(d)k\left(\frac{n}{d}\right).$$

This product is called the *convolution product*. The set of arithmetic functions together with $*$ is a commutative ring with unit element ε, where

$$\varepsilon(n) = \begin{cases} 1 \text{ if } n = 1, \\ 0 \text{ if } n > 1. \end{cases}$$

Furthermore, if we denote by 1 the function with constant value 1, then by Lemma 7.1.2, $\mu * 1 = \varepsilon$. Thus $\mu = 1^{-1}$.

Now, we have $g(n) = \sum_{d|n} f(d)$, that is, $g = f*1$. Therefore $f = g*1^{-1} = g*\mu$, and hence $f(n) = \sum_{d|n} g(d)\mu\left(\frac{n}{d}\right)$. □

Now let k be any field, $K = k(X_1, X_2, \ldots, X_n)$ the field of rational functions in n variables, and let $f(T) = \prod_{i=1}^{n}(T - X_i) \in K[T]$. Then

$$f(T) = T^n - \sigma_1 T^{n-1} + \sigma_2 T^{n-2} - \cdots + (-1)^s \sigma_s T^{n-s} + \cdots + (-1)^n \sigma_n,$$

where σ_s is the s-symmetric elementary function in $X_1, X_2, \ldots, X_n$, i.e.,

$$\begin{aligned} \sigma_0 &= 1, \\ \sigma_1 &= \sum_{i=1}^{n} X_i, \\ \sigma_2 &= \sum_{i<j} X_i X_j, \\ \cdots & \quad \cdots \\ \sigma_s &= \sum_{i_1 < \cdots < i_s} X_{i_1} \cdots X_{i_s}, \\ \cdots & \quad \cdots \\ \sigma_n &= X_1 \cdots X_n. \end{aligned}$$

Let $\varrho_m = X_1^m + \cdots + X_n^m$, $m \geq 1$ and $\varrho_0 = n$.

Theorem 7.1.4 (Newton identities). *We have*

$$\varrho_m - \varrho_{m-1}\sigma_1 + \cdots + (-1)^{m-1}\varrho_1\sigma_{m-1} + (-1)^m\sigma_m m = 0 \text{ for } 1 \leq m \leq n-1,$$

and

$$\varrho_m - \varrho_{m-1}\sigma_1 + \cdots + (-1)^n\varrho_{n-m}\sigma_n = 0 \quad \text{for} \quad m \geq n.$$

Proof. Consider the series $T^{1-n}f'(T)$ in the field of Laurent series $K((T))$ (K any field of characteristic 0). We have

$$\begin{aligned} T^{1-n}f'(T) &= T^{1-n}f(T)\frac{f'(T)}{f(T)} = T^{1-n}\Big(\sum_{i=0}^{n}(-1)^i\sigma_i T^{n-i}\Big)\Big(\sum_{i=1}^{n}\frac{1}{T-X_i}\Big) \\ &= T\Big(\sum_{i=0}^{n}(-1)^i\sigma_i T^{-i}\Big)\Big(\sum_{i=1}^{n}\sum_{m=0}^{\infty} X_i^m T^{-m-1}\Big) \\ &= \Big(\sum_{i=0}^{n}(-1)^i\sigma_i T^{-i}\Big)\Big(\sum_{m=0}^{\infty}\varrho_m T^{-m}\Big) \qquad (7.1) \\ &= \sum_{m=0}^{\infty}\Big(\sum_{s=0}^{m}(-1)^s\sigma_s\varrho_{m-s}\Big)T^{-m}, \end{aligned}$$

where $\sigma_j = 0$ for $j > n$.

On the other hand,

$$\begin{aligned} T^{1-n}f'(T) &= T^{1-n}\left(\sum_{m=0}^{n-1}(n-m)(-1)^m\sigma_m T^{n-m-1}\right) \\ &= \sum_{m=0}^{n-1}(n-m)(-1)^m\sigma_m T^{-m}. \qquad (7.2) \end{aligned}$$

Equating coefficients in (7.1) and (7.2) we obtain the Newton identities. □

Proposition 7.1.5. *Let $\psi(d)$ be the number of monic irreducible polynomials of degree d in $\mathbb{F}_q[T]$. Then $\psi(n) = \frac{1}{n}\sum_{d|n}\mu\left(\frac{n}{d}\right)q^d$.*

Proof. Exercise 7.7.2. □

If K is a function field with field of constants k, K_r will denote the extension of constants $Kk_r = K_r$; the field of constants of K_r is k_r (Theorem 6.1.2). Let $Z_K(u) = \zeta_K(s)$ be the zeta function of K, where $u = q^{-s}$, and let $Z_r(v) = \zeta_{K_r}(s)$ be the zeta function of K_r, where $v = (q^r)^{-s} = q^{-rs} = u^r$. Then $Z_r(v) = Z_r(u^r)$.

Theorem 6.4.7 demonstrates that $\zeta_{K_r}(s) = \prod_{j=1}^{r} L(s, \chi_j, K)$, where χ_j is the character satisfying $\chi_j(C) = \xi_r^j$ in every class of degree 1, and $\xi_r = e^{\frac{2\pi i}{r}}$ for $j = 1, \ldots, r$.

By Proposition 6.4.4, $L\left(s, \chi_j, K\right) = \zeta_K\left(s - \frac{2\pi i j}{r \ln q}\right)$.
We have

$$\zeta_K\left(s - \frac{2\pi i j}{r \ln q}\right) = Z_K\left(q^{-s} q^{\frac{2\pi i j}{r \ln q}}\right).$$

Since $q^{\frac{2\pi i j}{r \ln q}} = e^{\ln q^{\frac{2\pi i j}{r \ln q}}} = e^{\frac{2\pi i j}{r \ln q} \ln q} = \xi_r^j$, it follows that

$$Z_r\left(u^r\right) = Z_r(v) = \zeta_{K_r}(s) = \prod_{j=1}^{r} L\left(s, \chi_j, K\right) = \prod_{j=1}^{r} Z_K\left(\xi_r^j u\right).$$

Thus Theorem 6.4.7 yields the following:

Theorem 7.1.6. *If K_r is the extension of constants of degree r of the field K, we have $Z_{K_r}(u^r) = \prod_{j=1}^{r} Z_K\left(\xi_r^j u\right)$, where $u = q^{-s}$ and $\xi_r = e^{\frac{2\pi i}{r}}$.* □

If $K_0 = \mathbb{F}_q(x)$, we have $Z_{K_0}(u) = Z_0(u) = \frac{1}{(1-u)(1-qu)}$ by Corollary 6.3.10 and $Z_K(u) = \frac{P_K(u)}{(1-u)(1-qu)} = Z_0(u) P_K(u)$, $P_K(u) = \frac{Z_K(u)}{Z_0(u)}$.

Now $P_K(u) = \sum_{i=0}^{2g} a_i u^i$ with $a_{2g-i} = a_i q^{g-i}$ for $0 \leq i \leq 2g$ (Theorem 6.4.1). We have $a_0 = 1$, $a_{2g} = q^g$, and $a_1 = A_1 - (q+1)$, where A_1 is the number of integral divisors of degree 1 and is equal to the number of places of degree 1 (Corollary 6.3.9). We have $\deg(P_K(u)) = 2g$ and if $\omega_1^{-1}, \ldots, \omega_{2g}^{-1}$ are the roots of $P_K(u)$, then $P_K(u) = \prod_{i=1}^{2g}(1 - \omega_i u)$.

Proposition 7.1.7. *We have $q^g = \prod_{i=1}^{2g} \omega_i$ and $N - (q+1) = -\sum_{i=1}^{2g} \omega_i$, where N is the number of prime divisors of degree* 1. *Furthermore, $P_K\left(\omega_i^{-1}\right) = 0$ if and only if $P_K\left(\frac{\omega_i}{q}\right) = 0$.*

Proof. From $P_K(u) = \sum_{i=0}^{2g} a_i u^i = \prod_{i=1}^{2g}(1 - \omega_i u)$, it follows that

$$a_{2g} = q^g = \prod_{i=1}^{2g}(-\omega_i) = \prod_{i=1}^{2g} \omega_i; \quad a_1 = N - (q+1) = -\sum_{i=1}^{2g} \omega_i.$$

On the other hand, the functional equation of $P_K(u)$ (Corollary 6.4.2) establishes that $P_K\left(\frac{1}{qu}\right) = q^{-g} u^{-2g} P_K(u)$. Therefore $P_K\left(\omega_i^{-1}\right) = 0$ if and only if $P_K\left(\frac{1}{q\omega_i^{-1}}\right) = P_K\left(\frac{\omega_i}{q}\right) = 0$. □

We have

$$\frac{1}{\omega_i} = \frac{\omega_i}{q} \iff \omega_i^2 = q \iff \omega_i = \pm\sqrt{q}.$$

Therefore we may rearrange the inverses of the roots of $P_K(u)$ to obtain the sequence

$$\omega_1, \omega_1', \ldots, \omega_f, \omega_f', \sqrt{q}, \ldots, \sqrt{q}, -\sqrt{q}, \ldots, -\sqrt{q}$$

with

$$f \le g, \quad \omega_i \ne \omega_i', \quad \text{and} \quad \omega_i\omega_i' = q, \quad i = 1, \ldots, f.$$

Let t be the number of times that $\sqrt{q}$ appears and let s be the number of times that $-\sqrt{q}$ appears. Thus $2f + t + s = 2g$. Since $q^g = \prod_{i=1}^{2g} \omega_i$, we have $q^g = q^f q^{t/2}(-1)^s q^{s/2}$. It follows that s is even and so is t. In particular, we may take $f = g$, that is, $\omega_1, \omega_1', \ldots, \omega_g, \omega_g', \omega_i\omega_i' = q$ for all $1 \le i \le g$.

Thus we obtain $P_K(u) = \prod_{i=1}^{g} (1 - \omega_i u)\left(1 - \omega_i' u\right)$.

Theorem 7.1.8. *The following conditions are equivalent:*

(i) *The zeros of the zeta function $\zeta_K(s)$ lie on the line of equation $\mathbb{R}\mathrm{e}\, s = \frac{1}{2}$,*
(ii) *The zeros of the function $Z_K(u)$ lie on the circle of equation $|u| = q^{-1/2}$,*
(iii) *If $\omega_1, \ldots, \omega_{2g}$ are the inverses of the roots of $P_K(u)$, then $|\omega_i| = \sqrt{q}$ for $i = 1, \ldots, 2g$.*

Proof.
(i) $\Longleftrightarrow$ (ii): This equivalence follows from the facts that $u = q^{-s}$, $|u| = q^{-\mathbb{R}\mathrm{e}\, s}$, and $Z_K(u) = \zeta_K(s)$. Therefore $Z_K(u) = Z_K\left(q^{-s}\right) = \zeta_K(s)$.

(ii) $\Longleftrightarrow$ (iii): This follows from $Z_K(u) = \frac{P_K(u)}{(1-u)(1-qu)}$, $P_K(1) = h_K \ne 0$, and $P_K\left(\frac{1}{q}\right) = q^{-g} P_K(1) \ne 0$. Therefore the roots of $Z_K(u)$ are the roots of $P_K(u)$, which are the ω_i^{-1}'s. Hence (ii) is equivalent to $\left|\omega_i^{-1}\right| = |\omega_i|^{-1} = q^{-1/2}$, that is, $|\omega_i| = \sqrt{q}$. □

Our goal is to prove the following analogue of the classical Riemann hypothesis:

Riemann hypothesis: The conditions in Theorem 7.1.8 hold for any congruence function field.

The proof will be done in several steps.

Proposition 7.1.9. *Let N be the number of prime divisors of degree 1 in K. If the Riemann hypothesis holds, then $|N - (q+1)| \le 2g\sqrt{q}$.*

Proof. We have

$$N - (q+1) = -\sum_{i=1}^{2g} \omega_i, \quad \text{so} \quad |N - (q+1)| \le \sum_{i=1}^{2g} |\omega_i| = 2g\sqrt{q}. \qquad \square$$

Proposition 7.1.10. *The Riemann hypothesis holds for the field K if and only if it holds for the field K_r.*

Proof. By Theorem 7.1.6, we have (with the natural notation)

$$P_{K_r}\left(u^r\right) = \frac{Z_{K_r}\left(u^r\right)}{Z_{0,r}\left(u^r\right)} = \prod_{j=1}^{r} \frac{Z_K\left(\xi_r^j u\right)}{Z_0\left(\xi_r^j u\right)} = \prod_{j=1}^{r} P_K\left(\xi_r^j u\right)$$

$$= \prod_{j=1}^{r}\prod_{i=1}^{2g}\left(1 - \omega_i \xi_r^j u\right) = \prod_{i=1}^{2g}\left(1 - \omega_i^r u^r\right).$$

Hence, $P_{K_r}\left(u^r\right) = \prod_{i=1}^{2g}\left(1 - \omega_i^r u^r\right)$. Therefore $\omega_1^r, \ldots, \omega_{2g}^r$ are the inverses of the zeros of P_{K_r}, whence $|\omega_i| = \sqrt{q}$ if and only if $\left|\omega_i^r\right| = \sqrt{q^r}$, $q^r = \left|\mathbb{F}_{q^r}\right|$, and $\mathbb{F}_{q^r}$ is the field of constants of K_r. □

Let N_r be the number of prime divisors of degree 1 in K_r.

Proposition 7.1.11. *If there exists $c > 0$ such that $|N_r - (q^r + 1)| \leq cq^{r/2}$ for all r, then the Riemann hypothesis holds for K.*

Proof. Applying the operator $D = -u\frac{d}{du}\ln$ to both sides of the equality $P_K(u) = \prod_{i=1}^{2g}(1 - \omega_i u)$, we obtain

$$D\left(P_K(u)\right) = -u\frac{d}{du}\ln\left(\prod_{i=1}^{2g}(1 - \omega_i u)\right) = -u\left(\sum_{i=1}^{2g}\frac{d}{du}\ln(1 - \omega_i u)\right)$$

$$= \sum_{i=1}^{2g}\frac{\omega_i u}{1 - \omega_i u} = \sum_{i=1}^{2g}\sum_{n=1}^{\infty}\omega_i^n u^n = \sum_{n=1}^{\infty}\left(\sum_{i=1}^{2g}\omega_i^n\right)u^n.$$

We have $-\sum_{i=1}^{2g}\omega_i^n = N_n - (q^n + 1)$. Our hypothesis implies that

$$\left|N_n - \left(q^n + 1\right)\right| = \left|\sum_{i=1}^{2g}\omega_i^n\right| \leq cq^{n/2}.$$

Therefore, if R is the radius of convergence of the series, we have

$$R = \limsup_{n\to\infty}\left(\left|\sum_{i=1}^{2g}\omega_i^n\right|\right)^{-1/n} \geq \limsup_{n\to\infty}\left(cq^{n/2}\right)^{-1/n} = q^{-1/2},$$

and hence $R \geq \frac{1}{\sqrt{q}}$.

On the other hand, $D\left(P_K(u)\right) = \sum_{i=1}^{2g}\frac{\omega_i u}{1-\omega_i u}$ implies that the only singularities are $u = \omega_i^{-1}$, $1 \leq i \leq 2g$, so that

$$R = \min_{1\leq i\leq 2g}\left|\omega_i^{-1}\right| \geq q^{-1/2}. \quad \text{Thus} \quad |\omega_i| \leq \sqrt{q} \quad \text{for} \quad 1 \leq i \leq 2g.$$

Finally, by Proposition 7.1.7, $q^g = \prod_{i=1}^{2g}|\omega_i| \leq \prod_{i=1}^{2g}\sqrt{q} = q^g$, which implies that $|\omega_i| = \sqrt{q}$, $1 \leq i \leq 2g$. □

7.2 Proof of the Riemann hypothesis

The purpose of this section is to prove that the conditions of Theorem 7.1.8 hold for any congruence function field K. Let $k = \mathbb{F}_q$ be the field of constants of K.

We first note that in order to prove the Riemann hypothesis, by Proposition 7.1.10 we may assume, extending the field of constants if necessary, that:

(i) $q = a^2$ is a square,
(ii) $q > (g+1)^4$, where g is the genus of K,
(iii) K contains a prime divisor of degree 1.

Indeed, $K_2 = K\mathbb{F}_{q^2}$ has as field of constants $\mathbb{F}_{q^2}$ and q^2 is a square. Since $q^2 > 1$, there exists n such that $q^{2n} = (q^n)^2 > (g+1)^4$, so that $K_{2n} = \mathbb{F}_{q^{2n}} K$ has as field of constants $\mathbb{F}_{q^{2n}}$, the genus of K_{2n} is equal to g (Theorem 6.1.3), and $q^{2n} > (g+1)^4$. Finally, if $\wp$ is a prime divisor of degree m in K, then if $\mathcal{P}$ is above $\wp$ in $K_{2nm} = \mathbb{F}_{q^{2nm}} K$, we have, by Theorem 6.2.1, $d_L(\mathcal{P}) = \frac{m}{(m,2nm)} = \frac{m}{m} = 1$. Then K_{2nm} satisfies (i), (ii), and (iii).

By the above, we may assume that K satisfies (i), (ii), and (iii). Let N be the number of prime divisors of degree 1 in K. If $\sigma \in \operatorname{Aut}\left(K/\mathbb{F}_q\right)$, then for each place $\wp$, $\wp^\sigma$ is a place of K and the respective valuations satisfy $v_{\wp^\sigma}(x) = v_\wp\left(\sigma^{-1}x\right)$.

Let $\tilde{\mathbb{F}}_q$ be an algebraic closure of $k := \mathbb{F}_q$ and let $\tilde{K}$ be an algebraic closure of K. Consider the Frobenius automorphism

$$\varrho\colon \tilde{K} \to \tilde{K}, \quad \text{defined by} \quad \varrho(x) = x^q, \quad \varrho \in \operatorname{Aut}(\tilde{K}/k).$$

Let $\wp$ be a prime divisor of K. For any $\sigma \in \operatorname{Aut}(K/k)$, consider the corresponding prime divisor $\wp^\sigma$. Explicitly, if $\varphi_\wp$ is the place associated to $\wp$, then $\varphi_{\wp^\sigma}$ is the place $\sigma\varphi_\wp$ given by

$$\varphi_{\wp^\sigma}(\alpha) = \sigma\varphi_\wp(\alpha) = \varphi_\wp(\sigma^{-1}\alpha).$$

Define $\wp^q$ as the prime divisor given by the Frobenius automorphism, that is,

$$\varphi_{\wp^q}(\alpha) := \varrho\varphi_\wp(\alpha) = \varphi_\wp(\varrho^{-1}\alpha) = \varphi_\wp(\alpha^{1/q}) = \varphi_\wp(\alpha)^{1/q}.$$

Notice that $\wp^q$ is not the qth power of $\wp$. Now the respective valuation rings of $\wp$ and $\wp^q$ are given by

$$\vartheta_\wp = \{\alpha \in K \mid \varphi_\wp(\alpha) \neq \infty\} \text{ and } \vartheta_{\wp^q} = \{\alpha \in K \mid \varphi_{\wp^q}(x) = \varphi_\wp(x)^{1/q} \neq \infty\}$$

Thus $\vartheta_\wp = \vartheta_{\wp^q}$. Therefore $\varphi_\wp$ and $\varphi_{\wp^q}$ are equivalent (Proposition 2.2.13). We will use the notation $\wp = \wp^q$ to mean that $\varphi_\wp = \varphi_{\wp^q}$ instead of the usual meaning.

Proposition 7.2.1. *We have $\wp = \wp^q$ if and only if $d_K(\wp) = 1$.*

Proof. Clearly, $d_K(\wp) = [\vartheta_\wp/\wp : k]$. Consider $\varphi_\wp\colon K \to (\vartheta_\wp/\wp) \cup \{\infty\}$ and $\varphi_{\wp^q}\colon K \to (\vartheta_{\wp^q}/\wp^q) \cup \{\infty\}$. The following equivalences hold:

$$\begin{aligned} &\varrho\varphi_\wp(y) = \varphi_\wp(y)^{1/q} = \varphi_\wp(y) \text{ for all } y \in K \\ \iff &\varphi_\wp(y) = \varphi_\wp(y)^q \text{ for all } y \in K \iff \varphi_\wp(\alpha) = \infty \text{ or } \varphi_\wp(y) \in \mathbb{F}_q \\ \iff &\vartheta_\wp/\wp = \vartheta_{\wp^q}/\wp^q = \mathbb{F}_q \iff d_K(\wp) = d_K(\wp^q) = 1. \end{aligned}$$

□

Proposition 7.2.1 is one of the main results we will be using in the proof of the Riemann hypothesis. Actually, the $N_1 = N$ prime divisors of degree 1 in K/k are precisely those such that $\wp = \wp^q$. The Riemann hypothesis is equivalent to $|N - (q + 1)| \leq 2g\sqrt{q}$ (Propositions 7.1.9, 7.1.10, and 7.1.11). Therefore it suffices to show that for r large enough and for $K_r := K\mathbb{F}_{q^r}$, if N_r denotes the number of places $\mathfrak{P}$ such that $\mathfrak{P}^{q^r} = \mathfrak{P}$, then N_r satisfies $|N_r - (q^r + 1)| \leq 2gq^{r/2}$.

The proof of the Riemann hypothesis presented here is essentially due to Bombieri [7] (see also [38, 148]). The idea is to construct a function u on K such that every prime divisor of degree 1 but one is a zero of u, and on the other hand, the degree of u is not very large.

We have $q = a^2$. Set $m = a - 1, n = a + 2g$, and $r = m + an$. Then the inequality

$$N - (q + 1) < (2g + 1)\sqrt{q}$$

becomes

$$\begin{aligned} N - 1 &< q + (2g)\sqrt{q} + \sqrt{q} = a^2 + (2g)a + a \\ &= a(a + 2g) + a = an + m + 1 = r + 1. \end{aligned}$$

Thus $N - 1 \leq r$.

Let $\mathfrak{S}$ be a divisor of degree 1 in K/k. We have

$$L\left(\mathfrak{S}^{-1}\right) \subseteq L\left(\mathfrak{S}^{-2}\right) \subseteq \cdots \subseteq L\left(\mathfrak{S}^{-n}\right) \subseteq \cdots.$$

Furthermore, since $\mathfrak{S}^{-n} \mid \mathfrak{S}^{-(n-1)}$, then by Theorem 3.1.11,

$$\ell\left(\mathfrak{S}^{-n}\right) + d\left(\mathfrak{S}^{-n}\right) \leq \ell\left(\mathfrak{S}^{-(n-1)}\right) + d\left(\mathfrak{S}^{-(n-1)}\right).$$

Therefore

$$0 \leq \ell\left(\mathfrak{S}^{-n}\right) - \ell\left(\mathfrak{S}^{-(n-1)}\right) \leq d\left(\mathfrak{S}^{n}\right) + d\left(\mathfrak{S}^{-n+1}\right) = n - (n-1) = 1.$$

Let $t \in \mathbb{N}$ and let I_t be the set of numbers i $(1 \leq i \leq t)$ such that $\ell\left(\mathfrak{S}^{-i}\right) - \ell\left(\mathfrak{S}^{-(i-1)}\right) = 1$. For each $i \in I_t$, let $u_i \in L\left(\mathfrak{S}^{-i}\right) \setminus L\left(\mathfrak{S}^{-(i-1)}\right)$. The pole divisor of u_i is $\mathfrak{N}_{u_i} = \mathfrak{S}^i$.

Proposition 7.2.2. *The system $\{u_i \mid i \in I_t\}$ is a k-base of $L\left(\mathfrak{S}^{-t}\right)$.*

Proof. If $\sum_{i\in I_t} a_i u_i = 0$ with $a_i \in k$ and $a_i \neq 0$ for some i, then $v_{\mathfrak{S}}(a_i u_i) = -i$, so the valuations of the nonzero terms in the sum are all distinct. Therefore $a_i = 0$ for all $i \in I_t$, and the system $\{u_i \mid i \in I_t\}$ is linearly independent.

On the other hand,

$$\ell\left(\mathfrak{S}^{-t}\right) = \sum_{i=1}^{t} \dim_k \frac{L\left(\mathfrak{S}^{-i}\right)}{L\left(\mathfrak{S}^{-(i-1)}\right)} = \sum_{i=1}^{t} \delta_i \quad \text{with} \quad \delta_i = \begin{cases} 0 \text{ if } i \notin I_t, \\ 1 \text{ if } i \in I_t, \end{cases}$$

so $\ell\left(\mathfrak{S}^{-t}\right) = |I_t| = |\{u_i \mid i \in I_t\}|$. Therefore $\{u_i \mid i \in I_t\}$ is a basis of $L\left(\mathfrak{S}^{-t}\right)$. □

As a particular case of Proposition 7.2.2, we take $t = m = a-1 = \sqrt{q}-1$, where a is a power of the characteristic and $n = a + 2g$. The set

$$L\left(\mathfrak{S}^{-n}\right)^a = \left\{y^a \mid y \in L\left(\mathfrak{S}^{-n}\right)\right\} \subseteq K^a$$

is a k-vector space of the same dimension as that of $L\left(\mathfrak{S}^{-n}\right)$.

The space $M = \left\{\sum_{i\in I_m} u_i y_i^a \mid y_i \in L\left(\mathfrak{S}^{-n}\right)\right\}$ is a k-vector space generated by $U = \left\{u_i u_j^a \mid i \in I_m,\ j \in I_n\right\}$. Note that since $a = \sqrt{q}$ is a power of the characteristic, K^a is a field.

Proposition 7.2.3. *The set U is linearly independent over k.*

Proof. Since $u_j^a \in K^a$ and $k \subseteq K^a$, it suffices to prove that $\{u_i \mid i \in I_m\}$ is linearly independent over K^a.

Let $\sum_{i\in I_m} u_i y_i^a = 0$ with some $y_i \neq 0$. This implies that two elements have the same valuation (Proposition 2.2.3 (vi)). Thus there exist $y_i \neq 0$, $y_j \neq 0$, with $i \neq j$ and $v_{\mathfrak{S}}\left(u_i y_i^a\right) = v_{\mathfrak{S}}\left(u_j y_j^a\right)$. Hence,

$$-i + a v_{\mathfrak{S}}(y_i) = -j + a v_{\mathfrak{S}}\left(y_j\right) \quad \text{or} \quad i \equiv j \bmod a.$$

Since $i, j \in I_m$ for $1 \leq i, j \leq m = a - 1 < a$, the latter congruence is impossible. □

As a consequence of Proposition 7.2.3 we obtain $\dim_k M = |U| = |I_m|\,|I_n| = \ell\left(\mathfrak{S}^{-m}\right)\ell\left(\mathfrak{S}^{-n}\right)$. By the Riemann–Roch theorem we have the inequality

$$\begin{aligned} \dim_k M &= \ell\left(\mathfrak{S}^{-m}\right)\ell\left(\mathfrak{S}^{-n}\right) \geq (m-g+1)(n-g+1) \\ &= (a-g)(a+g+1) = a^2 + a - g(g+1) = q + \sqrt{q} - g(g+1). \end{aligned}$$

Now consider the k-vector space

$$M' = \left\{\sum_{i\in I_m} u_i^a y_i \mid y_i \in L\left(\mathfrak{S}^{-n}\right)\right\}.$$

For $i \in I_m$, we have $u_i^a y_i \in L(\mathfrak{S}^{-am}\mathfrak{S}^{-n})$

Again, from the Riemann–Roch theorem and the equality

$$d_K\left(\mathfrak{S}^{am}\mathfrak{S}^n\right) = ma + n = a^2 - a + a + 2g = q + 2g > 2g - 2,$$

we obtain

$$\dim_k M' \leq \ell\left(\mathfrak{S}^{-am}\mathfrak{S}^{-n}\right) = (q + 2g) - g + 1 = q + g + 1.$$

Now, because of our choice of $q > (g+1)^4$, we have

$$\sqrt{q} - g(g+1) > (g+1)^2 - g(g+1) = g + 1.$$

Thus

$$\dim_k M \geq q + \sqrt{q} - g(g+1) > q + g + 1 \geq \dim_k M'.$$

Let

$$\theta\colon M \longrightarrow M' \quad \text{be defined by} \quad \theta\left(\sum_{i\in I_m} u_i y_i^a\right) = \sum_{i\in I_m} u_i^a y_i.$$

Since $k^q = k$, θ is k-linear. Moreover, $\dim_k M > \dim_k M'$ implies that $\ker\theta \neq \{0\}$. Hence, there exist $y_i \in L\left(\mathfrak{S}^{-n}\right)$ $(i \in I_m)$, such that $\sum_{i\in I_m} u_i^a y_i = 0$ and not all y_i are zero. Thus

$$u = \sum_{i\in I_m} u_i y_i^a \in L\left(\mathfrak{S}^{-r}\right) \setminus \{0\} \quad \text{and} \quad u \in \ker\theta.$$

If $\wp$ is any place of K/k distinct from $\mathfrak{S}$, then $\varphi_\wp(y_i) \neq \infty$ and $\varphi_\wp(u_i) \neq \infty$ for all $i \in I_m$.

Furthermore, if $\wp$ satisfies $\wp = \wp^q$, then for all $\alpha \in K$, we have $\varphi_\wp(\alpha) = \varphi_{\wp^q}(\alpha) = \varphi(\alpha)^{1/q}$ or $\varphi_\wp(\alpha) = \varphi(\alpha)^q$, so $\varphi_\wp(\alpha) \in \mathbb{F}_q$. This implies that for $a = \sqrt{q} = p^u$, $\varphi_\wp(\alpha)^a = \varphi_\wp(\alpha)$. From $\sum_{i\in I_m} u_i^a y_i = 0$ we obtain

$$\varphi_\wp(u) = \sum_{i\in I_m} \varphi_\wp(u_i)\,\varphi_\wp(y_i)^a = \sum_{i\in I_m} \varphi_\wp(u_i)^a\,\varphi_\wp(y_i) = 0.$$

Thus $\wp$ belongs to the support of the divisor of zeros of u, $\mathfrak{Z}_u$. Therefore $\prod_{\substack{\wp\neq\mathfrak{S}\\ \wp=\wp^q}} \wp = \prod_{\substack{\wp\neq\mathfrak{S}\\ \deg_K\wp=1}} \wp \,\Big|\, \mathfrak{Z}_u$, and

$$d_K\Big(\prod_{\substack{\wp\neq\mathfrak{S}\\ \wp^\theta=\wp^q}} \wp\Big) = N - 1 \leq d_K(\mathfrak{Z}_u) = d_K(\mathfrak{N}_u) \leq d_K\left(\mathfrak{S}^r\right) = r.$$

This is what we wanted to prove.

Theorem 7.2.4. *We have* $N - (q+1) < (2g+1)\sqrt{q}$. □

To finish the proof of the Riemann hypothesis we must now find a lower bound for $N-(q+1)$. The upper bound we have obtained is not good enough to obtain the Riemann hypothesis. For example, if K is of genus one and ω_1 and ω_2 are the inverses of the roots of $P_K(u)$, then $\omega_1 = q$ and $w_2 = 1$ satisfy $N = q+1-\sum_{i=1}^{2g}\omega_i = q+1-q-1=0$, $\omega_1\omega_2 = q$, but $|\omega_i| \neq \sqrt{q}$.

In order to obtain a lower bound, we consider an automorphism $\theta \in \operatorname{Aut}(K/k)$ and an algebraic closure $\tilde{k}$ of k. Let $\tilde{K} = K\tilde{k}$. We extend θ to $\tilde{\theta} \in \operatorname{Aut}(\tilde{K}/k)$ by defining $\tilde{\theta}(\alpha) = \alpha^q$ for every $\alpha \in \tilde{k}$. Let $\wp$ be any prime divisor of K/k of degree d and $K_d = K\mathbb{F}_{q^d}$. Then by Theorem 6.2.1, $\wp$ decomposes into d prime divisors $\mathfrak{P}_1, \ldots, \mathfrak{P}_d$ of degree one in K_d. Let $\varphi_\wp, \varphi_{\wp^q}, \varphi_{\wp^\theta}, \varphi_{\mathfrak{P}_i}, \varphi_{\mathfrak{P}_i^q}, \varphi_{\mathfrak{P}_i^{\tilde{\theta}}}$ be the places associated to $\wp, \wp^q, \wp^\theta, \mathfrak{P}_i, \mathfrak{P}_i^q$, and $\mathfrak{P}_i^{\tilde{\theta}}$ $(1 \le i \le d)$ respectively.

We have $\varphi_{\wp^\theta}(x) = \varphi(\theta^{-1}x)$ and $\varphi_{\wp^q}(x) = \varphi_\wp(x^{1/q}) = \varphi_\wp(x)^{1/q}$. For $x \in K$ we have

$$\varphi_{\mathfrak{P}_i^{\tilde{\theta}}}(x) = \varphi_{\mathfrak{P}_i}(\tilde{\theta}^{-1}x) = \varphi_\wp(\theta^{-1}x) = \varphi_{\wp^\theta}(x)$$

and

$$\varphi_{\mathfrak{P}_i^q}(x) = \varphi_{\mathfrak{P}_i}(x)^{1/q} = \varphi_\wp(x)^{1/q}.$$

For $\alpha \in \mathbb{F}_{q^d}$,

$$\varphi_{\mathfrak{P}_i^{\tilde{\theta}}}(\alpha) = \varphi_{\mathfrak{P}_i}(\tilde{\theta}^{-1}\alpha) = \varphi_{\mathfrak{P}_i}(\alpha^{1/q}) = \varphi_{\mathfrak{P}_i^q}(\alpha).$$

Therefore $\wp^q = \wp^\theta$ if and only if $\mathfrak{P}_i^q = \mathfrak{P}_i^{\tilde{\theta}}$ for all $1 \le i \le d$.

We define $N^{(\theta)} := \sum_{\wp^\theta = \wp^q} d_K(\wp)$. By the above, $N^{(\theta)}$ is the number of prime divisors $\mathfrak{P}$ of $K\tilde{k}/\tilde{k}$ for which $\mathfrak{P}^{\tilde{\theta}} = \mathfrak{P}^q$. Furthermore, Theorem 7.2.4 can be extended to $N^{(\theta)}$ (see Exercise 7.7.3).

Proposition 7.2.5. *Let K be a function field over k. Let L be a geometric Galois extension of K with Galois group G. If $\theta \in \operatorname{Aut}(L/k)$ is such that $\theta(K) = K$, then*

$$N^{(\theta)}(K) = [L:K]^{-1} \sum_{g \in G} N^{(\theta g)}(L).$$

Proof. Let $\mathcal{P}$ be a prime divisor of L/k and let $\wp = \mathcal{P}|_K$. The places of L over $\wp^\theta$ are the places $\left(\mathcal{P}^\theta\right)^g$ for $g \in G$, and the one over $\wp^q$ is $\mathcal{P}^q$. Thus

$$\wp^\theta = \wp^q \iff \text{there exists } g \in G \text{ such that } \left(\mathcal{P}^\theta\right)^g = \left(\mathcal{P}^{\theta g}\right) = \mathcal{P}^q. \tag{7.3}$$

Now assume that $\mathcal{P}, \mathcal{P}_1, \mathcal{P}_2$ are prime divisors in L over a prime divisor $\wp$ of K. Since G acts transitively in $\{\mathcal{P} \in \mathbb{P}_L \mid \mathcal{P} \mid \wp\}$, then if $\varphi_{\mathcal{P}_1}$ and $\varphi_{\mathcal{P}_2}$ denote the places corresponding to $\mathcal{P}_1, \mathcal{P}_2$ respectively, we have

$$\left|\left\{\sigma \in G \mid \varphi_{\sigma\mathcal{P}_1} = \varphi_{\mathcal{P}_2}\right\}\right| = \left|\left\{\sigma \in G \mid \varphi_{\sigma\mathcal{P}} = \varphi_{\mathcal{P}}\right\}\right|.$$

The following equivalences hold:

$$\begin{aligned} \varphi_{\sigma\mathcal{P}} = \varphi_{\mathcal{P}} &\iff \varphi_{\sigma\mathcal{P}}(x) = \varphi_{\mathcal{P}}\left(\sigma^{-1}x\right) = \varphi_{\mathcal{P}}(x) \quad \text{for all} \quad x \in \vartheta_{\mathcal{P}} \\ &\iff \sigma^{-1}x - x \in \ker \varphi_{\mathcal{P}} = \mathcal{P} \iff \sigma^{-1}x \equiv x \bmod \mathcal{P} \quad \text{for all} \quad x \in \vartheta_{\mathcal{P}} \\ &\iff \sigma^{-1} \in I_{L/K}(\mathcal{P}|\wp) \iff \sigma \in I_{L/K}(\mathcal{P}|\wp). \end{aligned} \tag{7.4}$$

Therefore, $\left|\left\{\sigma \in G \mid \varphi_{\sigma\mathcal{P}_1} = \varphi_{\mathcal{P}_2}\right\}\right| = e_{L/K}(\mathcal{P}|\wp)$. Let $I = I_{L/K}\left(\mathcal{P}^{\theta}|\wp^{\theta}\right)$. We have

$$\begin{aligned} \sum_{g\in G} N^{(\theta g)}(L) &= \sum_{g\in G}\sum_{\mathcal{P}^{\theta g}=\mathcal{P}^q} d_L(\mathcal{P}) = \sum_{\bar\sigma\in G/I}\sum_{g\in I}\sum_{\mathcal{P}^{\theta\sigma g}=\mathcal{P}^q} d_L(\mathcal{P}) \\ &= \sum_{\bar\sigma\in G/I}\sum_{\mathcal{P}^{\theta\sigma}=\mathcal{P}^q} e_{L/K}\left(\mathcal{P}^{\theta}|\wp^{\theta}\right) d_L(\mathcal{P}) && \text{(by (7.2))} \\ &= \sum_{\wp^{\theta}=\wp^q}\sum_{\mathcal{P}|\wp} e_{L/K}(\mathcal{P}|\wp)\, d_{L/K}(\mathcal{P}|\wp)\, d_K(\wp) \\ & && \text{(Proposition 5.1.11 and (7.3))} \\ &= [L:K]\sum_{\wp^{\theta}=\wp^q} d_K(\wp) && \text{(Theorem 5.1.14)} \\ &= [L:K]N^{(\theta)}(K). \end{aligned}$$

□

Let $\theta \in \operatorname{Aut}(K/k)$ be an automorphism of finite order and let $E = K^{(\theta)}$ be the fixed field. Then K/E is a cyclic extension with Galois group $\langle\theta\rangle$.

Proposition 7.2.6. *There exists an element $x \in E \setminus k$ such that $E/k(x)$ is separable.*

Proof. Since there exists a divisor of degree 1 (Theorem 6.3.8), there exists a prime divisor $\wp$ of E of degree t with $(t, p) = 1$ and $p = \operatorname{char} k$. Let $m \in \mathbb{N}$ be such that $m > 2g - 1$ and $(m, p) = 1$. Then by the Riemann–Roch theorem (Corollary 3.5.8), there exists an element x in E such that $\mathfrak{N}_x = \mathcal{P}^m$. Therefore $[E : k(x)] = mt$ and $(mt, p) = 1$ with $p = \operatorname{char} E$, which implies that $E/k(x)$ is separable. □

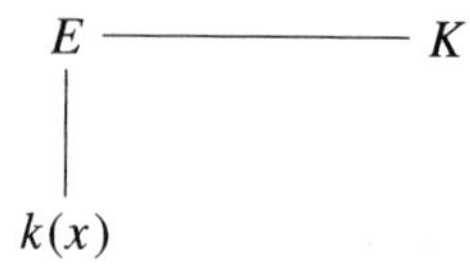

Let $x \in E \setminus k$ enjoy the property of Proposition 7.2.6. Let $\hat{K}$ be the Galois closure of $K/k(x)$ and $\hat{k}$ be the field of constants of $\hat{K}$. Then both $\hat{K}$ and $K\hat{k}$ admit $\hat{k}$ as field of constants. Also, θ is extendable to an element of $\operatorname{Aut}\left(\hat{K}/\hat{k}(x)\right)$.

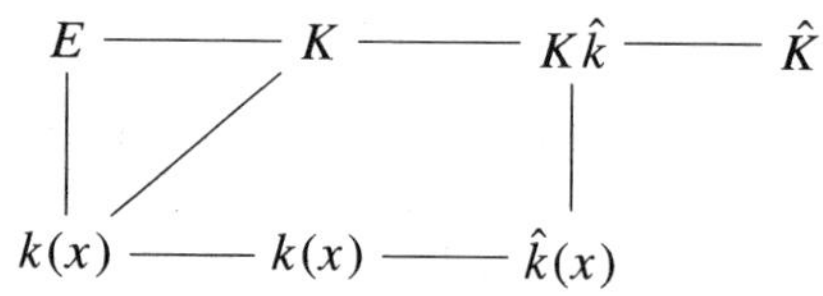

Extending constants of $\hat{K}$ if necessary, we may assume that if $|\hat{k}| = \hat{q}$, then $\hat{q} = a^2$ is a square, $\hat{q} > \left(g_{\hat{K}} + 1\right)^4$, $\hat{q} > \left(g_{K\hat{k}} + 1\right)^4 = (g_K + 1)^4$, and $\hat{K}$ has a prime divisor of degree 1.

Whence, we may assume that K/k satisfies the following conditions:

(1) K/k contains an element $x \in K \setminus k$ such that $K/k(x)$ is separable, and the Galois closure $\hat{K}$ of $K/k(x)$ has as field of constants k,
(2) $|k| = q = a^2$ is a square and $q > (\hat{g} + 1)^4$, $\hat{g} = g_{\hat{K}}$,
(3) $\hat{K}/k$ contains a prime divisor of degree 1.

Proposition 7.2.7. *Let* $m = \left[\hat{K} : K\right]$, $n = \left[\hat{K} : k(x)\right]$, *and* $\theta \in \operatorname{Aut}(K/k)$. *Then* $N^{(\theta)} - (q+1) \geq -\frac{n-m}{m}\left(2\hat{g} + 1\right)\sqrt{q}$.

Proof. Let $H = \operatorname{Gal}\left(\hat{K}/K\right)$ and $G = \operatorname{Gal}\left(\hat{K}/k(x)\right)$. We have $\theta \in G$ and $m = |H|$, $n = |G|$. By Proposition 7.2.5,

$$N^{(\theta)}(K) - \frac{1}{m}\sum_{h\in H} N^{(\theta h)}\left(\hat{K}\right) \quad \text{and} \quad q + 1 = N(k(x)) = \frac{1}{n}\sum_{g\in G} N^{(g)}\left(\hat{K}\right).$$

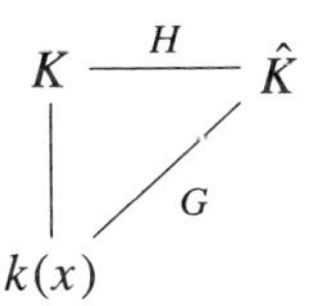

It follows by Theorem 7.2.4 and Exercise 7.7.3 that

$$\begin{aligned}\sum_{g\in G} N^{(g)}\left(\hat{K}\right) &= \sum_{h\in H} N^{(\theta h)}\left(\hat{K}\right) + \sum_{g\in G\setminus\theta H} N^{(g)}\left(\hat{K}\right)\\ &\leq \sum_{h\in H} N^{(\theta h)}\left(\hat{K}\right) + \sum_{g\in G\setminus\theta H}\left((q+1) + \left(2\hat{g}+1\right)\sqrt{q}\right)\\ &= \sum_{h\in H} N^{(\theta h)}\left(\hat{K}\right) + (n-m)\left(q + 1 + \left(2\hat{g}+1\right)\sqrt{q}\right).\end{aligned}$$

Since $\sum_{g\in G} N^{(g)}(\hat{K}) = nN^{(\mathrm{Id})}(k(x)) = n(q+1)$ (Proposition 7.2.5), we have

$$\begin{aligned}\sum_{h\in H} N^{(\theta h)}\left(\hat{K}\right) &\geq n(q+1) - (n-m)\left(q + 1 + \left(2\hat{g}+1\right)\sqrt{q}\right)\\ &= m(q+1) - (n-m)\left(2\hat{g}+1\right)\sqrt{q}.\end{aligned}$$

Finally, by Proposition 7.2.5, we have $\sum_{h\in H} N^{(\theta h)}(\hat{K}) = mN^{(\theta)}(K)$, so

$$N^{(\theta)}(K) \geq (q+1) - \frac{(n-m)}{m}\left(2\hat{g}+1\right)\sqrt{q}.$$

□

Corollary 7.2.8. *Let K/k be a congruence function field and consider an element $\theta \in \operatorname{Aut}(K/k)$ of finite order. Then there exists a finite extension k' of k with q' elements and a constant $c > 0$ such that for all $r \geq 1$ the extension k'_r of degree r of k' satisfies $\left|N^{(\theta)}\left(K'_r\right) - \left((q')^r + 1\right)\right| \leq c(q')^{r/2}$, where $K'_r = Kk'_r$.*

Proof. Let k' be the extension of k satisfying Proposition 7.2.7 and Theorem 7.2.4. The numbers n, m, $\hat{g}$ given in Proposition 7.2.7 are the same for extensions of constants (Theorem 6.1.3). Therefore, for all $r \geq 1$, we have $|k'_r| = (q')^r$ and

$$-\frac{(n-m)}{m}\left(2\hat{g}+1\right)(q')^{r/2} \leq N^{(\theta)}\left(K'_r\right) - \left((q')^r + 1\right) \leq \left(2\hat{g}+1\right)(q')^{r/2}.$$

With $c = \max\left\{\frac{(n-m)}{m}\left(2\hat{g}+1\right), 2\hat{g}+1\right\}$ we obtain the result. □

Finally we have the following theorem:

Theorem 7.2.9 (Riemann hypothesis). *Let K/k be a congruence function field, where $|k| = q$. Then:*

(i) *The zeros of the zeta function $\zeta_K(s)$ belong to the line of equation $\mathbb{R}\mathrm{e}\, s = \frac{1}{2}$.*
(ii) *The zeros of the function $Z_K(u)$ belong to the circle of equation $|u| = q^{-1/2}$.*
(iii) *If $\omega_1, \ldots, \omega_{2g}$ are the inverses of the roots of $P_K(u)$, then $|\omega_i| = \sqrt{q}$, for $i = 1, \ldots, 2g$.*
(iv) *If N_1 denotes the number of prime divisors of degree 1 in K, then $|N_1 - (q+1)| \leq 2g\sqrt{q}$.*

Proof. The statements follow from Theorem 7.1.8, Propositions 7.1.9, 7.1.10, 7.1.11, and Corollary 7.2.8. □

7.3 Consequences of the Riemann Hypothesis

An immediate consequence of the Riemann hypothesis is the following:

Theorem 7.3.1. *Let K/k be a congruence function field of genus 0. Then K is a field of rational functions.*

Proof. If N is the number of prime divisors of degree 1 in K, then by applying Proposition 7.1.9 we get $|N - (q+1)| \leq 2g\sqrt{q} = 0$. Thus $N = q+1$, so K contains prime divisors of degree 1. The result follows by Theorem 4.1.7. □

Our goal is to estimate the number of prime divisors of degree n in K/k.

Theorem 7.3.2. *If $K = \mathbb{F}_q(x)$ is a rational function field over $\mathbb{F}_q$ and if n_i is the number of prime divisors of degree i in K, then $n_1 = q+1$ and $n_i = \frac{1}{i}\sum_{d|i}\mu\left(\frac{i}{d}\right)q^d$ for $i > 1$.*

Proof. The prime divisors different from $\wp_\infty$ are in bijective correspondence with the monic irreducible polynomials (Theorem 2.4.1). Since $\wp_\infty$ is of degree 1, the result follows by Proposition 7.1.5. $\square$

We will generalize the preceding method in order to estimate the number of prime divisors of degree m in any function field K over $k = \mathbb{F}_q$.

Let K/k be a function field and let $x \in K \setminus k$ be such that $[K : k(x)] < \infty$. Let $\zeta_0(s)$ be the zeta function of $k(x)$ and let $\zeta(s)$ be the zeta function of K. Denote by N_m the number of prime divisors of degree m in K. We have, by Theorem 6.3.7,

$$\zeta(s) = \prod_{\mathcal{P}\in\mathbb{P}_K} \left(1 - \frac{1}{(N\mathcal{P})^s}\right)^{-1} = \prod_{m=1}^{\infty} \left(1 - \frac{1}{q^{ms}}\right)^{-N_m} \quad \text{whenever} \quad \mathbb{R}\text{e}\, s > 1.$$

Then

$$\frac{\zeta'(s)}{\zeta(s)} = [\ln \zeta(s)]' = \left[\sum_{m=1}^{\infty} -N_m \left(\ln\left(1 - \frac{1}{q^{ms}}\right)\right)\right]' = -\ln q \left(\sum_{t=1}^{\infty} \frac{c_t}{q^{ts}}\right),$$

where $c_t = \sum_m mN_m$ and where m runs though the natural numbers such that there exists $r \in \mathbb{N}$ with $rm = t$. That is, $c_t = \sum_{m|t} mN_m$.

Therefore we have

$$\frac{\zeta'(s)}{\zeta(s)} = -\ln q \sum_{t=1}^{\infty} \left(\sum_{m|t} mN_m\right) \frac{1}{q^{ts}} \quad \text{whenever} \quad \mathbb{R}\text{e}\, s > 1.$$

In particular, for $K = k(x)$ we have

$$\frac{\zeta_0'(s)}{\zeta_0(s)} = -\ln q \sum_{t=1}^{\infty} \left(\sum_{m|t} mn_m\right) \frac{1}{q^{ts}} \quad \text{whenever} \quad \mathbb{R}\text{e}\, s > 1.$$

On the other hand, $\zeta_0(s) = \frac{1}{(1-q^{1-s})(1-q^{-s})}$, so

$$\frac{\zeta_0'(s)}{\zeta_0(s)} = (\ln \zeta_0(s))' = -\ln q \left(\sum_{n=1}^{\infty} \frac{q^n + 1}{q^{ns}}\right).$$

In particular, equating coefficients we obtain $\sum_{m|t} mn_m = q^t + 1$, and this formula is equivalent to that of Theorem 7.3.2.

Notation 7.3.3. For two real functions $f(x)$, $g(x)$ with $g(x) \geq 0$, we write $f = O(g)$ if there exists a constant $c > 0$ such that $|f(x)| < c|g(x)|$ for x large enough.

Theorem 7.3.4.

$$\frac{\zeta'(s)}{\zeta(s)} = -\ln q \sum_{t=1}^{\infty} \left(\sum_{m|t} mN_m\right) \frac{1}{q^{ts}}, \quad \mathbb{R}\text{e}\, s > 1 \text{ and } n_m = \frac{q^m}{m} + O\left(\frac{q^{m/2}}{m}\right).$$

Proof. The first part was already proved in the course of the previous argument. Since

$$n_m = \frac{1}{m}\sum_{i|m} q^i \mu\left(\frac{m}{i}\right) = \frac{q^m}{m} + \frac{1}{m}\sum_{\substack{i|m\\ i\neq m}} q^i \mu\left(\frac{m}{i}\right),$$

it follows that

$$\left|n_m - \frac{q^m}{m}\right| \leq \frac{1}{m}\sum_{i\leq \frac{m}{2}} q^i = \frac{q^{m/2}}{m}\left(\sum_{i\leq \frac{m}{2}} q^{i-m/2}\right) \leq \frac{q^{m/2}}{m}\sum_{r=0}^{\infty}\frac{1}{q^r} = \frac{q^{m/2}}{m}\frac{1}{1-1/q}.$$

□

On the one hand, we have

$$\frac{\zeta'(s)}{\zeta(s)} = -\ln q \sum_{t=1}^{\infty}\left(\sum_{m|t} m N_m\right)\frac{1}{q^{ts}}$$

and on the other hand

$$\frac{\zeta(s)}{\zeta_0(s)} = P_K(s) = \prod_{i=1}^{2g}\left(1 - \frac{\omega_i}{q^s}\right),$$

where $\omega_1, \ldots, \omega_{2g}$ are the inverses of the roots of $P_K(u)$, $u = q^{-s}$, where $|\omega_i| = \sqrt{q}$ from the Riemann hypothesis.

Now

$$\begin{aligned}\left(\ln\frac{\zeta(s)}{\zeta_0(s)}\right)' &= \frac{\zeta'(s)}{\zeta(s)} - \frac{\zeta_0'(s)}{\zeta_0(s)} = \frac{P_K'(s)}{P_K(s)} = \left(\sum_{i=1}^{2g}\ln q\frac{\omega_i q^{-s}}{1-\omega_i q^{-s}}\right)\\ &= \ln q\sum_{i=1}^{2g}\frac{\omega_i q^{-s}}{1-\omega_i q^{-s}} = \ln q\sum_{i=1}^{2g}\sum_{n=1}^{\infty}\omega_i^n q^{-ns} = \ln q\sum_{n=1}^{\infty}\frac{s_n}{q^{ns}},\end{aligned}$$

where $s_n = \sum_{i=1}^{2g}\omega_i^n$.

We also have

$$\frac{\zeta'(s)}{\zeta(s)} - \frac{\zeta_0'(s)}{\zeta_0(s)} = -\ln q\sum_{t=0}^{\infty}\left(\sum_{m|t} m\,(N_m - n_m)\right)\frac{1}{q^{ts}}.$$

Therefore we obtain $\sum_{m|t} m\,(N_m - n_m) = -s_t$.

From the Möbius inversion formula, we obtain

$$t\,(N_t - n_t) = -\sum_{m|t}\mu\left(\frac{t}{m}\right)s_m,$$

and hence

$$N_t = n_t - \frac{1}{t}\sum_{m|t}\mu\left(\frac{t}{m}\right)s_m$$

with $s_m = \sum_{i=1}^{2g}\omega_i^m$.

Since $|\omega_i| = q^{1/2}$, we deduce

$$t\,|N_t - n_t| \le \sum_{m=1}^{t}|s_m| \le \sum_{m=1}^{t}\sum_{i=1}^{2g}|\omega_i|^m = \sum_{m=1}^{t}2gq^{m/2} = 2gq^{1/2}\frac{q^{t/2}-1}{q^{1/2}-1}.$$

Therefore, $N_t = n_t + O\left(\frac{q^{t/2}}{t}\right)$.

In short we have the following theorem:

Theorem 7.3.5. *Let K/k be a congruence function field with $k = \mathbb{F}_q$. If n_m and N_m denote the prime divisors of degree m in $k(x)$ and K respectively, then*

$$n_m = \frac{1}{m}\sum_{d|m}\mu\left(\frac{m}{d}\right)q^d \text{ for } m > 1 \text{ and } n_1 = q+1,$$

$$N_m = n_m + O\left(\frac{q^{m/2}}{m}\right),$$

$$n_m = \frac{q^m}{m} + O\left(\frac{q^{m/2}}{m}\right).$$

Furthermore,

$$\sum_{d|m} d\,(N_d - n_d) = -s_m$$

and

$$m\,(N_m - n_m) = -\sum_{d|m}\mu\left(\frac{m}{d}\right)s_d,$$

where $s_d = \sum_{i=1}^{2g}\omega_i^d$ and μ is the Möbius function. □

We end this section by relating the number of integral divisors to the number of prime divisors and comparing the number of prime divisors in extensions of constants.

Proposition 7.3.6. *Let K/k be a congruence function field with $k = \mathbb{F}_q$ and for each $n \in \mathbb{N}$, let K_n be the extension of constants of K of degree n. That is, $K_n = K\mathbb{F}_{q^n}$. Let N_j be the number of prime divisors of degree j in K and let $N_1^{(n)}$ be the number of divisors of degree 1 in K_n. Then*

$$N_1^{(n)} = \sum_{d|n} dN_d \quad \text{and} \quad N_n = \frac{1}{n}\sum_{d|n}\mu\left(\frac{n}{d}\right)N_1^{(d)}.$$

Proof. By Theorem 6.2.1, if d divides n and $\wp$ is a prime divisor of degree d in K, then $\wp$ decomposes into $(d, n) = d$ prime divisors of degree $\frac{d}{(d,n)} = 1$ in K_n. Therefore for each prime divisor of degree d in K we obtain d prime divisors of degree 1. Conversely, if $\mathcal{P}$ is a prime divisor of degree 1 in K_n and $\wp = \mathcal{P}|_K$, then by Proposition 5.1.11 we have $1 \cdot n = d_K(\wp)\, d_{K_n/K}(\mathcal{P}|\wp)$. Thus $d_K(\wp)$ divides n and $N_1^{(n)} = \sum_{d|n} dN_d$. By the Möbius inversion formula we obtain $nN_n = \sum_{d|n} \mu\left(\frac{n}{d}\right) N_1^{(d)}$. □

Now as in Chapter 6 we denote by A_n the number of integral divisors of degree n. Recall that $A_n = \sum_{d(C)=n} \frac{q^{N(C)}-1}{q-1}$ and $A_n = h\left(\frac{q^{n-g+1}-1}{q-1}\right)$ for $n > 2g_K - 2$, where h is the class number of K.

Theorem 7.3.7. *We have*

$$A_n = \sum_{\substack{k_1+2k_2+\cdots+nk_n=n \\ k_i \geq 0}} \prod_{i=1}^{n} \binom{k_i + N_i - 1}{k_i},$$

where the sum runs through all partitions of n, i.e., *the n-arrays $(k_1, \ldots, k_n)$ with $k_i \geq 0$ and $\sum_{i=0}^{n} ik_i = n$.*

Proof. We provide two proofs, the first one analytic and the second of combinatorial nature. First recall that $f(x) = \frac{1}{1-x} = \sum_{n=0}^{\infty} x^n$ for $|x| < 1$. Therefore by taking the derivative of both sides $p - 1$ times we obtain

$$\frac{1}{(1-x)^p} = \sum_{n=0}^{\infty} \binom{n+p-1}{p-1} x^n \quad \text{for } |x| < 1.$$

Now, the zeta function is $Z_K(u) = \sum_{i=0}^{\infty} A_n u^n$ for $u = q^{-s}$.
Thus

$$\begin{aligned}
\zeta_K(s) &= \prod_{\mathcal{P} \in \mathbb{P}_K} \left(1 - \frac{1}{(N\mathcal{P})^s}\right)^{-1} = \prod_{n=1}^{\infty} \left(1 - \frac{1}{q^{ns}}\right)^{-N_n} \\
&= \prod_{n=1}^{\infty} \left(\frac{1}{1-u^n}\right)^{N_n} = \prod_{n=1}^{\infty} \left(\sum_{k_n=0}^{\infty} \binom{k_n + N_n - 1}{N_n - 1} u^{nk_n}\right) \\
&= Z_K(u) = 1 + \sum_{t=0}^{\infty} \left(\sum_{\substack{k_1+2k_2+\cdots+tk_t=t \\ k_i \geq 0}} \prod_{i=1}^{t} \binom{k_i + N_i - 1}{N_i - 1} \right) u^t.
\end{aligned}$$

Since $\binom{k_i+N_i-1}{N_i-1} = \binom{k_i+N_i-1}{k_i}$ the equality follows by equating coefficients.

Now we give the combinatorial proof. Let $g(k_1, \ldots, k_n)$ be the number of distinct products of k_1 prime divisors of degree 1, k_2 prime divisors of degree 2, ..., k_n prime divisors of degree n.

We have $g(k_1, \ldots, k_n) = \prod_{i=1}^{n} f_i(k_i)$, where $f_i(k_i)$ is the number of products of k_i prime divisors of degree i.

In general, if $\mathcal{P}_1, \ldots, \mathcal{P}_{N_i}$ are all prime divisors of degree i, a product of k_i of them has the general form $\mathcal{P}_1^{a_1} \cdots \mathcal{P}_{N_i}^{a_{N_i}}$ with $a_1 + \cdots + a_{N_i} = k_i$. These products correspond bijectively to the choices of $N_i - 1$ elements from a set of $k_i + N_i - 1$, namely the elements $a_1 + 1, a_1 + a_2 + 2, \ldots, a_1 + \cdots + a_{N_i} + (N_i - 1)$, as indicated in the following diagram:

$$\begin{array}{ccccccccc}
\mathcal{P}_1 \cdots \mathcal{P}_1 & \sqcup & \mathcal{P}_2 \cdots \mathcal{P}_2 & \sqcup & \cdots & \sqcup & \mathcal{P}_{N_i} \cdots \mathcal{P}_{N_i} \\
\leftrightarrow & \uparrow & \leftrightarrow & \uparrow & & \uparrow & \leftrightarrow \\
a_1 & a_1+1 & a_2 & a_1+a_2+2 & & a_1+\cdots+a_{N_i-1}+(N_i-1) & a_{N_i}
\end{array}$$

Therefore $f_i(k_i) = \binom{k_i+N_i-1}{N_i-1}$.

It follows that

$$\begin{aligned} A_n &= \sum_{k_1+2k_2+\ldots+nk_n=n} g(k_1, \ldots, k_n) = \sum_{k_1+2k_2+\ldots+nk_n=n} \prod_{i=1}^{n} f_i(k_i) \\ &= \sum_{k_1+2k_2+\ldots+nk_n=n} \prod_{i=1}^{n} \binom{k_i + N_i - 1}{N_i - 1}. \end{aligned}$$

□

7.4 Function Fields with Small Class Number

We saw in Chapter 6 that if $P_K(u)$ is the numerator of the zeta function of a congruence function field, then $P_K(u) = a_0 + a_1 u + \cdots + a_{2g} u^{2g}$, $u = q^{-s}$, $a_{2g-i} = a_i q^{g-i}$, and $a_0 = 1$, $a_{2g} = q^g$. Furthermore, $a_i = A_i - (q+1)A_{i-1} + qA_{i-2}$.

On the other hand, $P_K(u) = \prod_{i=1}^{2g} (1 - \omega_i u)$, $|\omega_i| = q^{1/2}$ for $1 \le i \le 2g$.

Finally, $h = P_K(1) = \sum_{i=0}^{2g} a_i = \sum_{i=0}^{g-1} a_i \left(1 + q^{g-i}\right) + a_g = \prod_{i=1}^{2g} (1 - \omega_i)$.

Proposition 7.4.1. *Let $g = g_K$ be the genus of a function field K over $k = \mathbb{F}_q$, and let $h_K = h$ be the class number. Let*

$$S(q, g, r) = (q-1)\left[q^{2g-1} + 1 - 2gq^{(2g-1)/2}\right] - r(2g-1)\left(q^g - 1\right).$$

Then if $S(q, g, r) > 0$, we have $h > r$.

Proof. Let K_{2g-1} be the constant extension of degree $2g-1$ of K. By the Riemann hypothesis applied to K_{2g-1} with field of constants $\mathbb{F}_{q^{2g-1}}$ (K_{2g-1} is also of genus g), if N_1' is the number of prime divisors of degree 1 in K_{2g-1}, then

$$\left|N_1' - \left(q^{2g-1} + 1\right)\right| \le 2gq^{(2g-1)/2}, \quad \text{so} \quad N_1' \ge q^{2g-1} + 1 - 2gq^{(2g-1)/2}.$$

Now if d divides $2g-1$, a prime divisor of degree d in K splits into $(d, 2g-1) = d$ prime divisors of degree $\frac{d}{(d, 2g-1)} = 1$ in K_{2g-1} (Theorem 6.2.1). On the other hand, if a prime divisor of degree 1 in K_{2g-1} restricts to a prime divisor of degree d, then by Proposition 5.1.11, d divides $2g - 1$.

Also, at most $2g-1$ places of degree 1 in K_{2g-1} can restrict to the same place in K. If $\mathcal{P}_1, \ldots, \mathcal{P}_s$ are prime divisors of degree 1 that restrict to the same prime $\wp$ in K with $s \leq 2g-1$, then $\wp^{(2g-1)/d_K(\wp)}$ is an integral divisor of degree $2g-1$ in K. Hence, with at most $2g-1$ divisors of degree one in K_{2g-1}, we obtain an integral divisor of degree $2g-1$ in K. Since there are N_1' places of degree 1 in K_{2g-1}, there exist at least

$$\frac{N_1'}{2g-1} \geq \frac{q^{2g-1}+1-2gq^{(2g-1)/2}}{2g-1}$$

integral divisors of degree $2g-1$ in K.

We have

$$A_{2g-1} = h\frac{q^g-1}{q-1} \geq \frac{q^{2g-1}+1-2gq^{(2g-1)/2}}{2g-1}.$$

Therefore

$$h \geq \frac{\left(q^{2g-1}+1-2gq^{(2g-1)/2}\right)(q-1)}{(2g-1)(q^g-1)} = R.$$

If $S(q,g,r) > 0$, then $R > r$, which implies that $h > r$. □

As an exercise of basic calculus, it can be verified that $S(q,g,1)$ is increasing as a function of g for $q=4$, $g \geq 2$ or $q=3$, $g \geq 3$ or $q=2$, $g \geq 5$.

On the other hand,

$$S(4,2,1) = 3(50-32) = 54 > 0,$$
$$S(3,3,1) = 2\left(179 - 54\sqrt{3}\right) > 0,$$
$$S(2,5,1) = 2\left(117 - 80\sqrt{2}\right) > 0.$$

Hence, we obtain the following result:

Theorem 7.4.2. *We have $h_K > 1$ whenever $q=4$ and $g \geq 2$, $q=3$ and $g \geq 3$, or $q=2$ and $g \geq 5$.* □

On the other hand, we have the following:

Theorem 7.4.3. *If $g \geq 1$, then $h_K > 1$ whenever $q \geq 5$.*

Proof. Let $P_K(u) = \prod_{i=1}^{2g}(1-\omega_i u)$ be the numerator of the zeta function of K. Then by the Riemann hypothesis we have

$$h = P_K(1) = \prod_{i=1}^{2g}(1-\omega_i) = \left|\prod_{i=1}^{2g}(1-\omega_i)\right| = \prod_{i=1}^{2g}|1-\omega_i|$$
$$\geq \prod_{i=1}^{2g}(|\omega_i|-1) = \prod_{i=1}^{2g}\left(\sqrt{q}-1\right) = \left(\sqrt{q}-1\right)^{2g} \geq \left(\sqrt{q}-1\right)^2 \geq \left(\sqrt{5}-1\right)^2 > 1.$$

Thus $h > 1$. □

Thus we see that the number of possibilities for a field K to have class number 1 is very limited. If $g = 0$ then $h = 1$, but if $g \geq 1$, $h = 1$ can hold only in the cases $q = 4, g = 1; q = 3, g = 1, 2; q = 2, g = 1, 2, 3, 4$.

We can study the function $S(q, g, r)$ for several values of r and give criteria in order to have $h > r$. Here we present only the results for $2 \leq r \leq 10$ enumerating the possibilities for g and q. This procedure by no means implies that given (q, g, r) such that $S(q, g, r) < 0$, there necessarily exists a field of genus g with field of constants $\mathbb{F}_q$ and class number $h = r$.

Theorem 7.4.4. *Let K be a congruence function field with field of constants $k = \mathbb{F}_q$, genus $g \geq 1$, and class number h satisfying $2 \leq h \leq 10$. Then we necessarily have*

(i) If $h = 2$, then $q = 2, 3, 4$ and
if $q = 4$, then $g = 1$,
if $q = 3$, then $g \in \{1, 2\}$,
if $q = 2$, then $g \leq 5$.
(ii) If $h = 3$, then $q \leq 7$ and $g \leq 6$.
(iii) If $h = 4$, then $q \leq 8$ and $g \leq 6$.
(iv) If $h = 5$, then $q \leq 9$ and $g \leq 7$.
(v) If $h = 6$, then $q \leq 11$ and $g \leq 7$.
(vi) If $h = 7$, then $q \leq 13$ and $g \leq 7$.
(vii) If $h = 8$, then $q \leq 13$ and $g \leq 8$.
(viii) If $h = 9$, then $q \leq 16$ and $g \leq 8$.
(ix) If $h = 10$, then $q \leq 17$ and $g \leq 8$. □

Remark 7.4.5. Theorem 7.4.4 can be improved by fixing first h, then g, and finally the possible q. For instance, if $h = 10$, and $g = 6$, then q is 2 necessarily, whereas the theorem states only that $q \leq 17$.

Now we state the result that describes all possible fields K with class number 1 (of genus at least 1). The proof is based on a detailed analysis of the function $P_K(u)$.

Theorem 7.4.6 (Leitzel, Madan, Queen [94, 95]). *There exist, up to isomorphism, exactly 7 congruence function fields $K/\mathbb{F}_q$ with class number 1 and genus $g \neq 0$. If $K = \mathbb{F}_q(X, Y)$ is such a field, then the 7 fields are given as follows:*

(i) $q = 2, g = 1, Y^2 + Y = X^3 + X + 1$
(ii) $q = 2, g = 2, Y^2 + Y = X^5 + X^3 + 1$,
(iii) $q = 2, g = 2, Y^2 + Y = \left(X^3 + X^2 + 1\right)\left(X^3 + X + 1\right)^{-1}$,
(iv) $q = 2, g = 3, Y^4 + XY^3 + \left(X^2 + X\right)Y^2 + \left(X^3 + 1\right)Y + \left(X^4 + X + 1\right) = 0$,
(v) $q = 2, g = 3, Y^4 + \left(X^3 + X + 1\right)Y + \left(X^4 + X + 1\right) = 0$,
(vi) $q = 3, g = 1, Y^2 = X^3 + 2X + 2$,
(vii) $q = 4, g = 1, Y^2 + Y = X^3 + \alpha, \alpha \in \mathbb{F}_4 \setminus \{0, 1\}$. □

Now we detail one of the techniques used to prove this kind of result.

Let $1 = h = P_K(1) = \sum_{i=0}^{2g} a_i = \sum_{i=0}^{g-1} \left(q^{g-i}+1\right) a_i + a_g$. Let $S_n = \sum_{i=1}^{2g} \omega_i^n$, where $P_K(u) = \prod_{i=1}^{2g} (1-\omega_i u)$. Then by Theorem 7.3.5, we have $-S_n = \sum_{d|n} d\,(N_d - n_d)$.

Now,

$$u^{-2g} P_K(u) = \prod_{i=1}^{2g} \left(u^{-1} - \omega_i\right) = a_0 u^{-2g} + a_1 u^{-2g+1} + \cdots + a_{2g},$$

that is, $\omega_1, \ldots, \omega_{2g}$ are the roots of $u^{-2g} P_K(u) = P'_K(v)$, with $v = u^{-1}$. Thus

$$P'_K(v) = b_0 + b_1 v + \cdots + b_{2g} v^{2g} = \prod_{i=1}^{2g} (v - \omega_i)$$

with $b_i = a_{2g-i} = q^{g-i} a_i$ and $b_{2g} = a_0 = 1$.

We have

$$b_{2g-i} = a_i = (-1)^{2g-i} \sigma_i = (-1)^i \sigma_i,$$

where σ_i is the ith elementary symmetric function in $\{\omega_1, \ldots, \omega_{2g}\}$, so that by Newton's identities (Theorem 7.1.4)

$$S_m + S_{m-1} a_1 + \cdots + S_1 a_{m-1} + m a_m = 0 \quad \text{for} \quad 0 \le m \le 2g-1.$$

Hence

$$\begin{aligned} S_1 + a_1 = 0, &\quad a_1 = -S_1, \\ S_2 + S_1 a_1 + 2a_2 = 0, &\quad a_2 = \frac{S_1^2 - S_2}{2}, \\ &\quad a_3 = -\frac{S_1^3 - 3S_1 S_2 + 2S_3}{6}, \\ &\quad a_4 = \frac{S_1^4 - 6S_1^2 S_2 + 8S_1 S_2 + 3S_2^2 - 6S_4}{24}, \qquad \text{etc.} \end{aligned}$$

On the other hand, since

$$n_d = \begin{cases} q+1, & d = 1, \\ \frac{1}{d} \sum_{f|d} \mu\left(\frac{d}{f}\right) q^f, & d > 1. \end{cases}$$

and $S_n = -\sum_{d|n} d\,(N_d - n_d)$, we obtain, after making all necessary substitutions,

$$\begin{aligned} a_1 &= N_1 - (q+1), \\ 2a_2 &= N_1^2 - (2q+1)N_1 + 2N_2 + 2q, \\ 6a_3 &= N_1^3 - 3qN_1^2 + (3q-1)N_1 - 6(q+1)N_2 + 6N_1N_2 + 6N_3, \\ 24a_4 &= (4q-2)N_1 - N_1^2 + (2-4q)N_1^3 + (12+24q)N_2 \\ &\quad + 12N_2^2 + N_1^4 - (12+24q)N_1N_2 + 12N_1^2N_2 \\ &\quad - 24(q+1)N_3 + 24N_1N_3 + 24N_4. \end{aligned}$$

For $g \geq 1$, we have $N_1 \leq 1$. Indeed, if there exist two prime divisors of degree 1, say $\mathcal{P}_1, \mathcal{P}_2$, then since $h = 1$, $\frac{\mathcal{P}_1}{\mathcal{P}_2} = (x)$ is a principal divisor. Thus $[K : k(x)] = \deg(\mathfrak{N}_x) = \deg(\mathcal{P}_2) = 1$ (Theorem 3.2.7), so $g = 0$, which is absurd. Therefore $N_1 \leq 1$.

Now if $q = 3$, $g = 2$, we obtain

$$\begin{aligned} P_K(1) = h &= \left(q^2 + 1\right) a_0 + (q+1)a_1 + a_2 \\ &= 10 + 4a_1 + a_2 = \frac{-6 + N_1 + N_1^2 + 2N_2}{2}. \end{aligned}$$

It follows that $h = 1$ if and only if $N_1^2 + N_1 + 2N_2 = 8$.

On the other hand, by the Riemann hypothesis, the inverses of the roots of $P_K(u)$ are $\sqrt{3}e^{\pm i\theta_1}$, $\sqrt{3}e^{\pm i\theta_2}$, so

$$\begin{aligned} P_K(u) &= \left(1 - \sqrt{3}e^{i\theta_1}u\right)\left(1 - \sqrt{3}e^{-i\theta_1}u\right)\left(1 - \sqrt{3}e^{i\theta_2}u\right)\left(1 - \sqrt{3}e^{-i\theta_2}u\right) \\ &= \left(1 - 2\sqrt{3}\cos\theta_1 u + 3u^2\right)\left(1 - 2\sqrt{3}\cos\theta_2 u + 3u^2\right). \end{aligned}$$

Comparing coefficients we obtain

$$\cos\theta_1 + \cos\theta_2 = \frac{(4 - N_1)\sqrt{3}}{6}$$

and

$$\cos\theta_1 cos\theta_2 = \frac{N_1^2 - 7N_1 + 2N_2 - 6}{24}.$$

Since $N_1^2 + N_1 + 2N_2 = 8$, we get $\cos\theta_1\cos\theta_2 = \frac{-7N_1 + 8 - N_1 - 6}{24} = \frac{1 - 4N_1}{12}$.

Let $f(x) = (x - \cos\theta_1)(x - \cos\theta_2) = x^2 + \frac{(N_1 - 4)\sqrt{3}}{6}x + \frac{1 - 4N_1}{12}$. Then $\cos\theta_1$ and $\cos\theta_2$ are roots of $f(x)$. Notice that

$$0 \leq (1 - \cos\theta_1)(1 - \cos\theta_2) = f(1) = \frac{\left(12 + 1 - 8\sqrt{3}\right) + N_1\left(2\sqrt{3} - 4\right)}{12} < 0,$$

which is absurd. Therefore, if $q = 3$ and $g = 2$, then we must have $h > 1$.

7.5 The Class Numbers of Congruence Function Fields

Let $K/\mathbb{F}_q$ be a congruence function field. Its zeta function is given by

$$Z_K(u) = \frac{P_K(u)}{(1-u)(1-qu)},$$

where

$$P_K(u) = \sum_{i=0}^{2g} a_i u^i, \quad a_{2g-i} = a_i q^{g-i} \quad \text{for} \quad 0 \le i \le 2g,$$

and $g = g_K$ is the genus of K (Theorem 6.4.1). Then $P_K(1) = h_K$ is the class number of K (Corollary 6.3.9).

Let $K_n := K\mathbb{F}_{q^{\ell^n}}$ be the constant extension of degree ℓ^n, where ℓ is a rational prime ($q = p^u$, $\ell = p$ or $\ell \neq p$). Then

$$Z_{K_n}\left(u^{\ell^n}\right) = \prod_{j=1}^{\ell^n} Z_K\left(\zeta_{\ell^n}^j u\right),$$

where ζ_{ℓ^n} is any ℓ^nth primitive root of 1 in $\mathbb{C}^*$ (Theorem 7.1.6).

We have

$$P_K(u) = \prod_{i=1}^{2g} \left(1 - \alpha_i^{-1} u\right),$$

where $\alpha_1, \ldots, \alpha_{2g}$ are the roots of $P_K(u)$. Thus

$$P_{K_n}\left(u^{\ell^n}\right) = \prod_{i=1}^{2g} \left(1 - \alpha_i^{-\ell^n} u^{\ell^n}\right).$$

Therefore, if h_n is the class number of K_n, we have

$$\begin{aligned} \frac{h_n}{h} &= \frac{P_{K_n}(1)}{P_K(1)} = \frac{\prod_{i=1}^{2g}\left(1 - \alpha_i^{-\ell^n}\right)}{\prod_{i=1}^{2g}\left(1 - \alpha_i^{-1}\right)} \\ &= \frac{\prod_{i=1}^{2g}\prod_{j=1}^{\ell^n}\left(1 - \zeta_{\ell^n}^j \alpha_i^{-1}\right)}{\prod_{i=1}^{2g}\left(1 - \alpha_i^{-1}\right)} = \prod_{i=1}^{2g}\prod_{j=1}^{\ell^n-1}\left(1 - \zeta_{\ell^n}^j \alpha_i^{-1}\right). \end{aligned}$$

Theorem 7.5.1. *With the above notation, let ℓ^{e_n} be the exact power of ℓ dividing h_n. Then*

$$e_n = \lambda n + \gamma$$

for n sufficiently large, with $0 \le \lambda \le 2g$ and $\gamma \in \mathbb{Z}$.

Proof. We have

$$\frac{h_n}{h} = \prod_{j=1}^{\ell^n-1}\prod_{i=1}^{2g}\left(1 - \zeta_{\ell^n}^j \alpha_i^{-1}\right) = \prod_{j=1}^{\ell^n-1} P_K\left(\zeta_{\ell^n}^j\right).$$

Now, $P_K(T) \in \mathbb{Z}[T]$, so $P_K(T)$ has the form $P_K(T) = 1 + a_1 T + \cdots + q^g T^{2g}$. Let

$$\begin{aligned} R_K(T) = P_K(T+1) &= 1 + a_1(T+1) + \cdots + q^g(T+1)^{2g} \\ &= b_0 + b_1 T + \cdots + b_{2g} T^{2g}. \end{aligned}$$

We have

$$P_K(\zeta_{\ell^n}^j) = R_K(\zeta_{\ell^n}^j - 1) = b_0 + b_1(\zeta_{\ell^n}^j - 1) + \cdots + b_{2g}(\zeta_{\ell^n}^j - 1)^{2g}. \tag{7.5}$$

Note that $R_K(-1) = P_K(0) = a_0 = 1$. Therefore there exists $0 \leq \lambda \leq 2g$ such that $\ell \nmid b_\lambda$. Choose λ to be minimal with this property.

In the cyclotomic number field $\mathbb{Q}(\zeta_{\ell^n})/\mathbb{Q}$, ℓ is fully ramified and $(\ell) = (1 - \zeta_{\ell^n})^{\phi(\ell^n)} = (1 - \zeta_{\ell^n}^j)^{\phi(\ell^n)}$ for all $(j, n) = 1$ and ϕ is the Euler ϕ-function ([156, Proposition 2.1, p. 9]). Let $\mathfrak{L} = (1 - \zeta_{\ell^n})$ be the prime ideal of $\mathbb{Q}(\zeta_{\ell^n})$ above ℓ, i.e., $v_{\mathfrak{L}}(1 - \zeta_{\ell^n}) = 1$. Clearly, if $j = \ell^m j_1$, $m < n$, $(j_1, \ell) = 1$, then

$$1 - \zeta_{\ell^n}^j = 1 - \zeta_{\ell^{n-m}}^{j_1} = (1 - \zeta_{\ell^n})^{\ell^m} u$$

with u a unit in $\mathbb{Q}(\zeta_{\ell^n})$. Hence $v_{\mathfrak{L}}(1 - \zeta_{\ell^n}^j) = \ell^m = v_\ell(j)$. Therefore, in (7.5) we obtain

$$v_{\mathfrak{L}}(b_i(\zeta_{\ell^n}^j - 1)^i) = v_{\mathfrak{L}}(b_i) + i v_{\mathfrak{L}}(\zeta_{\ell^n}^j - 1) = v_\ell(b_i)\phi(\ell^n) + i v_\ell(j).$$

Let ζ be a primitive ℓ^nth root of unity. Then, for $0 \leq i \leq \lambda - 1$,

$$v_{\mathfrak{L}}(b_i(\zeta - 1)^i) \geq \phi(\ell^n) + i > \lambda = v_{\mathfrak{L}}(b_\lambda(\zeta - 1)^\lambda)$$

for n such that $\phi(\ell^n) > \lambda - i$.

For $\lambda < i \leq 2g$,

$$v_{\mathfrak{L}}(b_i(\zeta - 1)^i) \geq i > \lambda = v_{\mathfrak{L}}(b_\lambda(\zeta - 1)^\lambda).$$

Therefore, for a primitive ℓ^nth root ζ of 1 with $\phi(\ell^n) > \lambda$,

$$v_{\mathfrak{L}}(P_K(\zeta)) = v_{\mathfrak{L}}(R_T(\zeta - 1)) = \lambda. \tag{7.6}$$

Let $n_0 \in \mathbb{N}$ be such that $\phi(\ell^{n_0}) > \lambda$. For $n - 1 > \lambda$ we have

$$\frac{h_n}{h_{n-1}} = \frac{h_n}{h} \frac{1}{\left(\frac{h_{n-1}}{h}\right)} = \frac{\prod_{j=1}^{\ell^n - 1} P_K(\zeta_{\ell^n}^j)}{\prod_{j=1}^{\ell^{n-1}-1} P_K(\zeta_{\ell^{n-1}}^j)} = \prod_{\zeta} P_K(\zeta),$$

where the latter product runs through all the primitive ℓ^n th roots of unity. Using (7.6) and the fact that there are $\phi(\ell^n)$ primitive roots of unity, we obtain

$$\begin{aligned} v_\ell(h_n) &= v_\ell(h_{n-1}) + v_\ell\Big(\prod_\zeta P_K(\zeta)\Big) = v_\ell(h_{n-1}) + \frac{1}{\phi(\ell^n)} v_{\mathfrak{L}}\Big(\prod_\zeta P_K(\zeta)\Big) \\ &= v_\ell(h_{n-1}) + \frac{1}{\phi(\ell^n)}\phi(\ell^n)\lambda = v_\ell(h_{n-1}) + \lambda. \end{aligned}$$

Therefore

$$v_\ell(h_n) = \lambda(n - n_0) + v_\ell(h_{n_0}) = \lambda n + (v_\ell(h_{n_0}) - n_0\lambda) = \lambda n + \gamma. \qquad \square$$

Remark 7.5.2. Theorem 7.5.1 states that the Iwasawa μ invariant for congruence function fields is 0 (see [156, Chapter 7]).

7.6 The Analogue of the Brauer–Siegel Theorem

The Brauer–Siegel theorem is a theorem in number fields, that is, finite extensions of $\mathbb{Q}$. For a number field F, let d be its discriminant, R its regulator, and h its class number.

Theorem 7.6.1 (Brauer–Siegel). *We have* $\lim_{|d|\to\infty} \frac{\ln(hR)}{\ln\sqrt{|d|}} = 1$. □

The goal of this section is to present an analogue of the theorem of Brauer and Siegel. Let K/k be a congruence function field with $k = \mathbb{F}_q$. All extensions of K considered in this section have k as their exact field of constants.

If n_m and N_m denote the number of divisors of degree m in the rational function field $k(x)$ and in K respectively, then (Theorem 7.3.5)

$$\begin{aligned}\left|n_m - \frac{q^m}{m}\right| &= \left|\sum_{\substack{d|m\\ d<m}} \mu\left(\frac{m}{d}\right) q^d\right| \leq \sum_{d=1}^{[m/2]} q^d = q\frac{q^{[m/2]}-1}{q-1}\\ &\leq 2\left(q^{[m/2]}-1\right) < 2q^{m/2},\\ |N_m - n_m| &= \frac{1}{m}\left|\sum_{d|m} \mu\left(\frac{m}{d}\right) s_d\right| \leq \frac{1}{m}\left(\sum_{d=1}^{m}\left|\sum_{i=1}^{2g}\omega_i^d\right|\right)\\ &\leq \frac{2g}{m}\sum_{d=1}^{m} q^{d/2} = \frac{2g}{m} q^{1/2}\frac{q^{m/2}-1}{q^{1/2}-1} \leq 4gq^{m/2}.\end{aligned}$$

Now, the number of integral divisors of degree $2g$ is $A_{2g} = h\frac{q^{g+1}-1}{q-1}$, and we have $N_{2g} \geq n_{2g} - 4gq^g > \frac{q^{2g}}{2g} - 2q^g - 4gq^g = \frac{q^{2g}}{2g} - (4g+2)q^g$.
Thus

$$h\frac{q^{g+1}-1}{q-1} = A_{2g} \geq N_{2g} > \frac{q^{2g}}{2g} - (4g+2)q^g.$$

Therefore

$$h > \frac{(q-1)}{\left(q^{g+1}-1\right)}\left(\frac{q^{2g}}{2g} - (4g+2)q^g\right).$$

Theorem 7.6.2. *If k is fixed, then* $\liminf_{g\to\infty} \frac{\ln h}{g \ln q} \geq 1$.

Proof. We have $h \geq q^{g-1}\frac{C}{2g}$, where C is a constant and g is large enough. Therefore

$$\ln h \geq (g-1)\ln q + \ln C - \ln 2g, \qquad \frac{\ln h}{g\ln q} \geq 1 - \frac{1}{g} + \frac{\ln C}{g\ln q} - \frac{\ln 2g}{g\ln q},$$

and the right-hand side goes to 1 when g goes to ∞, which implies the result. $\square$

In order to obtain an analogue to the Brauer–Siegel theorem, we must prove that $\limsup_{g\to\infty} \frac{\ln h}{g \ln q} \leq 1$. This remains an open problem. We will prove that the result holds with a restriction, namely that for K, there exist $x \in K \setminus k$ and m such that $[K : k(x)] \leq m$ with $\frac{m}{g} \to 0$.

Theorem 7.6.3. *We have* $\lim_{\frac{m}{g}\to 0} \frac{\ln h}{g \ln q} = 1$, *where* g *is the genus of* K, h *is the class number of* K, *and* m *is the minimum integer such that there exists* $x \in K \setminus k$ *with* $[K : k(x)] = m$.

Proof. For an integral divisor $\mathfrak{A}$, it follows from the Riemann–Roch theorem that $\ell\left(\mathfrak{A}^{-1}\right) \geq d\left(\mathfrak{A}\right) - g + 1$, so that $A_n \geq h\frac{q^{n-g+1}-1}{q-1}$. Therefore if $\zeta_K(s)$ is the zeta function for $s \in \mathbb{R}$ such that $s > 1$, then

$$\zeta_K(s) = \sum_{n=0}^{\infty} A_n q^{-ns} \geq \sum_{n=g}^{\infty} A_n q^{-ns} \geq \sum_{n=g}^{\infty} h\frac{q^{n-g+1}-1}{q-1}\frac{1}{q^{ns}}$$
$$= \frac{h}{q^{gs}} \sum_{n=g}^{\infty} \frac{q^{n-g+1}-1}{q-1}\frac{1}{q^{(n-g)s}} = \frac{h}{q^{gs}} \sum_{n=0}^{\infty} \frac{q^{n+1}-1}{q-1}\frac{1}{q^{ns}} - \frac{h}{q^{gs}}\zeta_0(s),$$

where $\zeta_0(s)$ is the zeta function of $k(x)$.

Hence $\zeta_K(s) \geq \frac{h}{q^{gs}}\zeta_0(s)$ for $s \in \mathbb{R}$, $s > 1$.

On the other hand, $\zeta_K(s) = \prod_{\mathcal{P}\in\mathbb{P}_K}\left(1 - \frac{1}{N(\mathcal{P})^s}\right)^{-1}$.

Let $\mathcal{P}$ be a divisor of K of relative degree t and $\wp = \mathcal{P}|_{k(x)}$. Then

$$\deg(\wp)\, t = d(\mathcal{P}), \quad N\mathcal{P} = q^{d(\mathcal{P})} = q^{t \deg \wp}$$

and

$$1 - \frac{1}{N(\mathcal{P})^s} = 1 - \frac{1}{q^{d(\mathcal{P})s}} = 1 - \frac{1}{q^{(\deg \wp)ts}} \geq \left(1 - \frac{1}{q^{d(\wp)s}}\right)^t.$$

Therefore if $\mathcal{P}_1, \ldots, \mathcal{P}_r$ are the prime divisors of K over $\wp$ in $k(x)$, $r \leq m = [K : k(x)]$, and each relative degree is t_i, then

$$\prod_{i=1}^{r}\left(1 - \frac{1}{N(\mathcal{P}_i)^s}\right) \geq \prod_{i=1}^{r}\left(1 - \frac{1}{N(\wp)^s}\right)^{t_i} \geq \left(1 - \frac{1}{N(\wp)^s}\right)^m.$$

Thus

$$\zeta_K(s) = \prod_{\mathcal{P}\in\mathbb{P}_K}\left(1 - \frac{1}{N(\mathcal{P})^s}\right)^{-1} \leq \prod_{\wp\in\mathbb{P}_{k(x)}}\left(1 - \frac{1}{N(\wp)^s}\right)^{-m} = \zeta_0(s)^m.$$

It follows that

$$\zeta_0(s)^m \geq \zeta_K(s) \geq \frac{h}{q^{gs}}\zeta_0(s), \quad \text{that is,} \quad \zeta_0(s)^{m-1} \geq \frac{h}{q^{gs}}.$$

Taking logarithms, we obtain

$$(m-1)\ln \zeta_0(s) \geq \ln h - gs \ln q.$$

Therefore

$$s \geq \frac{\ln h}{g \ln q} - \frac{(m-1)\ln \zeta_0(s)}{g \ln q}.$$

Let $\varepsilon > 0$ be fixed and let $s = 1 + \varepsilon$. If $\frac{m}{g} \to 0$, then taking g large enough, we have $1 + \varepsilon \geq \frac{\ln h}{g \ln q} - \varepsilon$, so $\limsup_{\frac{m}{g}\to\infty} \frac{\ln h}{g \ln q} \leq 1$.

The result follows by the above and Theorem 7.6.2. □

An interesting problem that remains open is to determine whether a complete analogue of the Brauer–Siegel theorem holds, that is, $\lim_{g\to\infty} \frac{\ln h}{g \ln q} = 1$ without any restriction. To finish this chapter we present some approximations to this result.

Theorem 7.6.4. *We have* $\left(\sqrt{q}-1\right)^{2g} \leq h \leq \left(\sqrt{q}+1\right)^{2g}$.

Proof. We have $h = P_K(1) = |P_K(1)| = \prod_{i=1}^{2g} |1-\omega_i|$, where $|\omega_i| = \sqrt{q}$. Therefore $\sqrt{q} - 1 \leq |1-\omega_i| \leq \sqrt{q}+1$, from which the result follows. □

Corollary 7.6.5. *We have*

$$\frac{2\,\ln\left(\sqrt{q}-1\right)}{\ln q} \leq \frac{\ln h}{g\,\ln q} \leq \frac{2\,\ln\left(\sqrt{q}+1\right)}{\ln q}.$$ □

Now for $n > 2g-2$, then $A_n = h\left(\frac{q^{n-g+1}-1}{q-1}\right)$ by Theorem 6.2.6.

On the other hand, $A_n = \sum_{p(n)} \prod_{i=1}^{n} \binom{k_i+N_i-1}{k_i}$, where $p(n)$ is the set of partitions of n (Theorem 7.3.7).

Taking $n = 2g-1$, we obtain the equality

$$h\left(\frac{q^g-1}{q-1}\right) = \sum_{p(2g-1)} \prod_{i=1}^{2g-1} \binom{k_i+N_i-1}{k_i}.$$

Let $M = \max_{p(2g-1)} \prod_{i=1}^{2g-1} \binom{k_i+N_i-1}{k_i}$.

Then $M \leq h\left(\frac{q^g-1}{q-1}\right) \leq |p(2g-1)|\,M$.

Furthermore, it is well known that $|p(2g-1)| < e^{T\sqrt{2g-1}}$, where $T = \pi\left(\frac{2}{3}\right)^{1/2}$. Therefore $M \leq h\left(\frac{q^g-1}{q-1}\right) \leq e^{T\sqrt{2g-1}}M$, whence

$$\frac{\ln M}{g\ \ln q} \le \frac{\ln h}{g\ \ln q} + \frac{\ln (q^g - 1) - \ln(q-1)}{g\ \ln q} \le \frac{T\sqrt{2g-1}}{g\ \ln q} + \frac{\ln M}{g\ \ln q}.$$

Now,

$$\lim_{g\to\infty} \frac{\ln (q^g - 1) - \ln(q-1)}{g \ln q} = 1 \quad \text{and} \quad \lim_{g\to\infty} \frac{T\sqrt{2g-1}}{g \ln q} = 0,$$

from which we obtain that

$$\lim_{g\to\infty} \frac{\ln h}{g \ln q} \quad \text{exists} \quad \text{if and only if} \quad \lim_{g\to\infty} \frac{M}{g \ln q} \quad \text{exists.}$$

Furthermore,

$$\lim_{g\to\infty} \frac{\ln h}{g \ln q} = \limsup_{g\to\infty} \frac{M}{g \ln q} - 1.$$

Therefore, proving the analogue of the Brauer–Siegel theorem is equivalent to proving that $\limsup_{g\to\infty} \frac{M}{g \ln q} \le 2$.

7.7 Exercises

Exercise 7.7.1. Prove Lemma 7.1.2.

Exercise 7.7.2. Prove Proposition 7.1.5

Exercise 7.7.3. Prove Theorem 7.2.4 for any $\theta \in \mathrm{Aut}(K/k)$, i.e.,

$$N^{(\theta)} - (q+1) < (2g+1)\sqrt{q}.$$

8

Constant and Separable Extensions

We have seen (Remark 5.2.30 and Example 5.2.31) that the field of constants of a constant extension $K\ell$ can contain ℓ properly . On the other hand, if ℓ is a finite field, the constant field of $K\ell$ is ℓ (Theorem 6.1.2).

Our goal in this chapter is to give a full account on the constant extension $K\ell$. Our main reference is Deuring's monograph [28].

In particular, we shall study the change of genus in extensions of constants; as we shall see, in this case the genus does not increase (Theorem 8.5.3), in contrast to the geometric separable case, in which the genus does not decrease.

At the end of the chapter we present a few results on inseparable extensions.

8.1 Linearly Disjoint Extensions

Definition 8.1.1. Let F and M be two extensions of a field E that are contained in an algebraic closed field Ω. Then F is said to be *linearly disjoint from M over E if every finite set of elements of F that is linearly independent over E is also linearly independent over M.*

$$\begin{array}{ccc} F & \text{---} & FM \\ | & & | \\ E & \text{---} & M \end{array}$$

We can see right away that the relation defined above is symmetric.

Proposition 8.1.2. *Let F be linearly disjoint from M over E. Then M is linearly disjoint from F over E.*

Proof. Let $\alpha_1, \ldots, \alpha_n$ be elements of M that are linearly independent over E. Assume that there exists a nontrivial linear combination

$$a_1\alpha_1 + \cdots + a_m\alpha_m = 0 \tag{8.1}$$

where the elements $a_1, \ldots, a_m$ of F are not all zero.

Suppose that the elements $a_1, \ldots, a_s$ $(s \geq 1)$ are linearly independent over E and $a_{s+1}, \ldots, a_m$ are linear combinations

$$a_i = \sum_{j=1}^{s} \beta_{ij} a_j, \quad \beta_{ij} \in E, \quad i = s+1, \ldots, m.$$

Then (8.1) can be written as

$$\sum_{\ell=1}^{s} a_\ell \alpha_\ell + \sum_{i=s+1}^{m} \Big(\sum_{j=1}^{s} \beta_{ij} a_j \Big) \alpha_i = 0. \tag{8.2}$$

The coefficient of $a_\ell (1 \leq \ell \leq s)$ in (8.2) is $\big(\alpha_\ell + \sum_{i=s+1}^{m} \beta_{i\ell}\alpha_i\big)$.

Therefore

$$\sum_{\ell=1}^{s} \Big(\alpha_\ell + \sum_{i=s+1}^{m} \beta_{i\ell}\alpha_i \Big) a_\ell = 0.$$

Since $\{\alpha_1, \ldots, \alpha_m\}$ is linearly independent over E, it follows that

$$\alpha_\ell + \sum_{i=s+1}^{m} \beta_{i\ell}\alpha_i \neq 0 \quad \text{for} \quad 1 \leq \ell \leq s. \tag{8.3}$$

But (8.3) contradicts the linear independence of $\{a_1, \ldots, a_s\}$ over E. □

Example 8.1.3. We have that $\mathbb{Q}(\sqrt{2})$ and $\mathbb{Q}(\sqrt{3})$ are linearly disjoint over $\mathbb{Q}$.

Example 8.1.4. The fields $\mathbb{Q}(\zeta_3 \sqrt[3]{2})$ and $\mathbb{Q}(\sqrt[3]{2})$ are not linearly disjoint over $\mathbb{Q}$.

Our next result shows that the relation of being linearly disjoint is transitive. More precisely:

Proposition 8.1.5. *Let $E \subseteq F$ and $E \subseteq M$ be two field extensions and let N be an intermediate field, i.e., $E \subseteq N \subseteq M$. Then F and M are linearly disjoint over E if and only if*

(i) *F and N are linearly disjoint over E and*
(ii) *FN and M are linearly disjoint over N.*

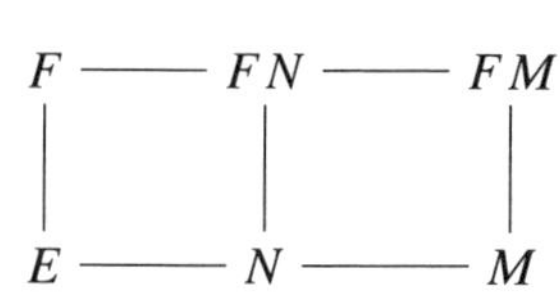

Proof. Assume that F and M are linearly disjoint over E. If $\mathcal{A} \subseteq F$ is any finite set that is linearly independent over E, then it is linearly independent over M. In particular, $\mathcal{A}$ is linearly independent over N. Therefore F and N are linearly disjoint over E.

Now let $\mathcal{A} = \{\alpha_1, \ldots, \alpha_n\} \subseteq M$ be linearly independent over N. Let $\beta_1, \ldots, \beta_n \in FN$ be such that

$$\sum_{i=1}^{n} \beta_i \alpha_i = 0. \tag{8.4}$$

Each β_i is a quotient of elements of the form $\sum_j a_j b_j$ with $a_j \in F$ and $b_j \in N$. Clearing denominators we may assume that $\beta_i = \sum_{j=1}^{m_i} a_{ij} b_{ij}$, with $a_{ij} \in F$ and $b_{ij} \in N$. Furthermore, since we are dealing with a finite number of elements $\{b_{ij}\}_{1\le i\le n}^{1\le j\le m_i}$ in N, we may choose a finite set $\{d_1, \ldots, d_n\} \subseteq N$ that is linearly independent over E and such that $\beta_i = \sum_{j=1}^{m} c_{ij} d_j$ for all $1 \le i \le n$, $c_{ij} \in F$.

Therefore (8.4) becomes

$$\sum_{j=1}^{m} \sum_{i=1}^{n} c_{ij} d_j \alpha_i = 0.$$

Since $\{d_j \alpha_i\}_{1\le j\le m}^{1\le i\le n} \subseteq M$ is linearly independent over E and M and F are linearly disjoint over E, it follows that $c_{ij} = 0$ for all $1 \le i \le n$, $1 \le j \le m$. Therefore $\beta_i = 0$ for $1 \le i \le n$.

Hence M and FN are linearly disjoint over N.

Conversely, assume that N and F are linearly disjoint over E, and M and FN are linearly disjoint over N.

Let $\{\alpha_i\}_{i\in I}$ and $\{\beta_j\}_{j\in J}$ be bases of N over E and of M over N respectively. Then $\{\alpha_i \beta_j\}_{(i,j)\in I\times J}$ is a basis of M/E.

Let $\{\delta_k\}_{k\in K}$ be a basis of F over E. Suppose that we have a relation

$$\sum_{i\in I, j\in J} \left(\sum_{k\in K} a_{kij} \delta_k \right) \left(\alpha_i \beta_j \right) = 0, \tag{8.5}$$

where only finitely many a_{kij}'s in E may be nonzero.

Then

$$\sum_{j\in J} \left(\sum_{\substack{k\in K\\ i\in I}} a_{kij} \delta_k \alpha_i \right) \beta_j = 0. \tag{8.6}$$

Since $\{\beta_j\}_{j\in J}$ is a basis of M over N, M and FN are linearly disjoint over N, and $\sum_{i\in I, k\in K} a_{kij} \delta_k \alpha_i \in FN$, it follows that $\sum_{i\in I, k\in K} a_{kij} \delta_k \alpha_i = 0$ for all j.

Thus $\sum_{i\in I} \left(\sum_{k\in K} a_{kij} \delta_k \right) \alpha_i = 0$ for all $j \in J$. Since $\{\alpha_i\}_{i\in I}$ is a basis of N over E, and N and F are linearly disjoint over E, it follows that $\sum_{k\in K} a_{kij} \delta_k = 0$ for all $i \in I$ and $j \in J$.

Finally, since $\{\delta_k\}_{k\in K}$ is a basis of F over E, we have $a_{k,i,j} = 0$ for all $i \in I$, $j \in J$, and $k \in K$. Hence M and F are linearly disjoint over E. □

For the basic properties we use for tensor products we refer to [89], [69], and [4].

Proposition 8.1.6. *Let F/E and M/E be two field extensions and Ω be an algebraically closed field such that $F, M \subseteq \Omega$. Let $F \otimes_E M$ denote the tensor product of F and M over E. The natural map $\varphi : F \otimes_E M \to FM$ satisfies* $\operatorname{im}\varphi = F[M] = \left\{ \sum_{i=1}^{n} \alpha_i \beta_i \mid n \in \mathbb{N}, \alpha_i \in F, \beta_i \in M \right\}$. *Then F and M are linearly disjoint over E if and only if φ is a monomorphism.*

Proof. Let $\{\beta_i\}_{i\in I}$ be a basis of M over E. Every element of $F \otimes_E M$ can be written as $\sum_{i\in I} \alpha_i \otimes_E \beta_i$ with $\alpha_i = 0$ for almost all i. Since tensor product commutes with direct sum and $A \otimes_R R \cong A$ for any R-module A and R a commutative ring, we have that if $\{\beta_i\}_{i\in I}$ is a basis of M over E, then $\{1 \otimes_E \beta_i\}_{i\in I}$ is a basis of $F \otimes_E M$ over F (with the extension of scalars: $\lambda(a \otimes_E b) = \lambda a \otimes_E b$, $\lambda \in F$). From this we obtain that $\sum_{i\in I} \alpha_i \otimes \beta_i = 0$ if and only if $\alpha_i = 0$ for all i.

Since $\varphi\left(\sum_{i\in I} \alpha_i \otimes \beta_i\right) = \sum_{i\in I} \alpha_i\beta_i$, the result follows. □

We now introduce the concept of a *free* or *algebraically disjoint* set.

Definition 8.1.7. Let F and M be two extensions of a field E. We say that F *is free or* algebraically disjoint from M over E if every finite subset of F that is algebraically independent over E remains algebraically independent over M.

Like linear disjointness, freeness is defined in an asymmetric way. However, as we did for linear disjointness, we shall prove that the relation is in fact symmetric.

Proposition 8.1.8. *If F is free from M over E, then M is free from F over E.*

Proof. Let $y_1, \ldots, y_n$ be elements of M that are algebraically independent over E. If $y_1, \ldots, y_n$ are dependent over F, then they are so in a subfield K of F that is finitely generated over E. Let $\operatorname{tr} K/E = r$. Since F is free from M over E, then $\operatorname{tr}(K(y_1, \ldots, y_n)/E(y_1, \ldots, y_n)) = r$.

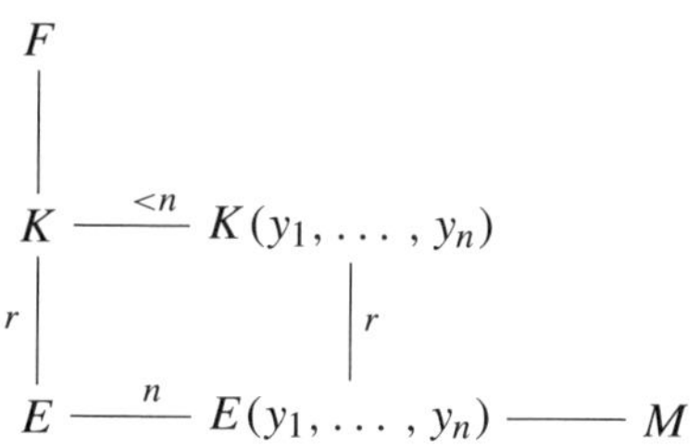

We have, on the one hand,

$$\operatorname{tr}\left(K(y_1, \ldots, y_n)/E\right)$$
$$= \operatorname{tr}\left(K(y_1, \ldots, y_n)/E(y_1, \ldots, y_n)\right) + \operatorname{tr}\left(E(y_1, \ldots, y_n)/E\right) = r + n;$$

on the other hand,

$$\operatorname{tr}\left(K(y_1, \ldots, y_n)/E\right) = \operatorname{tr}\left(K(y_1, \ldots, y_n)/K\right) + \operatorname{tr}(K/E) < n + r.$$

This contradiction shows that M is free from F over E. □

The next proposition proves that linear disjointness implies algebraic disjointness.

Proposition 8.1.9. *If F and M are linearly disjoint over E, then they are algebraically disjoint over E.*

Proof. Let $y_1, \dots, y_n$ be elements of F that are algebraically independent over E. If $y_1, \dots, y_n$ are algebraically dependent over M, then there exists a relation of the type

$$p(y_1, \dots, y_n) = 0 = \sum_{(i_1,\dots,i_n)\in I} a_{i_1\dots i_n} y^{i_1} \cdots y_n^{i_n}, \quad a_{i_1\dots i_n} \in M,$$

where $p(T_1, \dots, T_n) \in M[T_1, \dots, T_n]$ is a nonzero polynomial.

Therefore $\{y_1^{i_1} \cdots y_n^{i_n}\}_{(i_1,\dots,i_n)\in I}$ is linearly dependent over M. On the other hand, since $\{y_1^{i_1} \cdots y_n^{i_n}\}_{(i_1,\dots,i_n)\in I}$ is linearly independent over E this contradicts the linear disjointness of F and M over E. □

An important result that we will need later, when we study the general constant extensions of function fields, is the following:

Proposition 8.1.10. *Let F be a field extension of E and let $\mathcal{A}$ be a set of elements that are algebraically independent over F. Then $E(\mathcal{A})$ is linearly disjoint from F over E.*

Proof. Let $f_1, \dots, f_r \in E(\mathcal{A})$ be linearly independent over E. Then there exists a finite set $\{y_1, \dots, y_n\} \subseteq \mathcal{A}$ such that $f_i = \frac{a_i}{b_i}$, with $a_i, b_i \in E[y_1, \dots, y_n]$. Let $b = \prod_{i=1}^r b_i$. If $\alpha_1, \dots, \alpha_r \in F$ are such that $\sum_{i=1}^r \alpha_i f_i = 0$ then $\sum_{i=1}^r \alpha_i (bf_i) = \sum_{i=1}^r \alpha_i g_i = 0$ with $g_i = bf_i \in E[y_1, \dots, y_n]$, and $\{g_1, \dots, g_r\}$ is linearly independent over E.

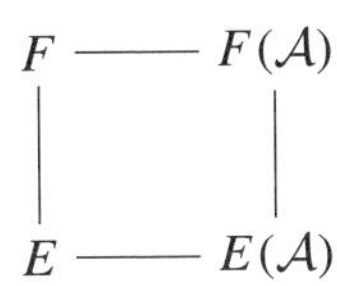

Now if some α_i is nonzero there is a nontrivial algebraic relation of $\{y_1, \dots, y_n\}$ over F. This is impossible since $\{y_1, \dots, y_n\}$ is algebraically independent over F. Therefore $\{f_1, \dots, f_r\}$ is linearly independent over F, and F and $E(\mathcal{A})$ are linearly disjoint over E. □

An observation we shall be using frequently is the following:

Remark 8.1.11. When we need to test whether two fields are either linearly or algebraically disjoint, it suffices to assume that these fields are finitely generated over the base field since in either case the definitions involve only a finite number of elements at a time.

Corollary 8.1.12. *Let F be any purely transcendental extension of E, and let M be any extension of E. If F is algebraically disjoint from M over E, then F is linearly disjoint from M over E.*

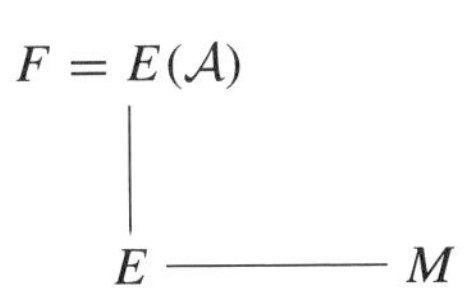

Proof. Let $F = E(\mathcal{A})$, where $\mathcal{A}$ is a transcendence base. Then $\mathcal{A}$ is algebraically independent over M. The result follows immediately by Proposition 8.1.10. $\square$

Corollary 8.1.13. *If F is an algebraic extension of E, and M is a purely transcendental extension of E, then F and M are linearly disjoint over E.*

Proof. Exercise 8.7.7. $\square$

8.2 Separable and Separably Generated Extensions

Definition 8.2.1. A field extension F/E is called *separably generated* if there exists a transcendence basis $\{\alpha_i\}_{i\in I}$ of F over E such that $F/E(\{\alpha\}_{i\in I})$ is algebraic and separable. Such a basis $\{\alpha_i\}_{i\in I}$ is called a *separating transcendence basis for F over E*.

Definition 8.2.2. A field extension F/E is called *separable* if for any subfield $E \subseteq M \subseteq F$ with M/E finitely generated, M/E is separably generated.

Proposition 8.2.3. *If E is a field of characteristic* 0, *any field extension F/E is both separable and separably generated.*

Proof: Let F/E be any field and let $\mathfrak{A} = \{\alpha_i\}_{i\in J}$ be any transcendence basis of F/E. Then $F/E(\mathcal{A})$ is algebraic and therefore separable. Thus F/E is separably generated. Also, if $E \subseteq M \subseteq F$ is any intermediate field with M/E finitely generated, then as before, M/E is separably generated. Hence F/E is separable. $\square$

Remark 8.2.4. We will prove in Theorem 8.2.8 that a separably generated extension is separable. The converse is not true in general (Example 8.2.10). The general definition of separability is compatible with the definition for algebraic extensions. Since every field extension of characteristic 0 is separable and separably generated, in the rest of this section we shall consider fields of characteristic $p > 0$.

Let E be a field of characteristic $p > 0$ and let F/E be an extension. Let $\bar{F}$ be an algebraic closure of F, $n \in \mathbb{N}$, and

$$E^{1/p^n} := \left\{\alpha \in \bar{F} \mid \alpha^{p^n} \in E\right\}. \tag{8.7}$$

Then E^{1/p^n} is a field and $E \subseteq E^{1/p^n} \subseteq \bar{E} \subseteq \bar{F}$. Set

$$E^{1/p^\infty} := \bigcup_{n\geq 0}^{\infty} E^{1/p^n}. \tag{8.8}$$

Then E^{1/p^∞} is also a field.

For algebraic extensions, we have the following proposition:

Proposition 8.2.5. *Let F/E be an algebraic extension of fields of characteristic $p > 0$. Then F/E is separable if and only if F and $E^{1/p}$ are linearly disjoint.*

Proof.
($\Rightarrow$) Let $M = F^pE \subseteq F$. Since F/E is separable, F/M is separable too. If $\alpha \in F$, then $\alpha^p \in F^p \subseteq F^pE$. Therefore F/EF^p is purely inseparable, and $F = EF^p$.

Now let $a_1, \dots, a_n \in F$ be elements that are linearly independent over E. Let $K = E(a_1, \dots, a_n)$. We have $n \leq m = [K : E] < \infty$. We complete $\{a_1, \dots, a_n\}$ to a basis $\{a_1, \dots, a_n, a_{n+1}, \dots, a_m\}$ of K/E.

Clearly, $K = E(a_1, \dots, a_n) = \sum_{i=1}^m Ea_i = \bigoplus_{i=1}^m Ea_i$.

Since K/E is separable, we have $K = EK^p = E(a_1^p, \dots, a_m^p) = \sum_{i=1}^m Ea_i^p$.

It follows from $[K : E] = m$ that $\{a_1^p, \dots, a_m^p\}$ is a basis of K/E. In particular, $\{a_1^p, \dots, a_n^p\}$ is linearly independent over E.

Let $b_1, \dots, b_n \in E^{1/p}$ be such that $\sum_{i=1}^n b_ia_i = 0$. Hence $\sum_{i=1}^n b_i^pa_i^p = 0$ with $b_i^p \in E$. We have $b_i^p = 0$ $(1 \leq i \leq n)$, so $b_i = 0$ $(1 \leq i \leq n)$. It follows that F and $E^{1/p}$ are linearly disjoint over E.

($\Leftarrow$) Let F and $E^{1/p}$ be linearly disjoint over E. Let $\alpha \in F$ and $h(x) = \operatorname{Irr}(\alpha, x, E)$ with $\deg h(x) = n$. We will show that $h(x)$ is separable. It suffices to see that $h(x) \notin E[x^p]$.

The elements $1, \alpha, \dots, \alpha^{n-1}$ are linearly independent over E. Therefore $1, \alpha, \dots, \alpha^{n-1}$ are linearly independent over $E^{1/p}$. This is equivalent to saying that $1, \alpha^p, \alpha^{2p}, \dots, \alpha^{(n-1)p}$ are linearly independent over E. If $h(x) = g(x^p)$, then $\operatorname{Irr}(\alpha^p, x, E) \mid g(x)$ and $[E(\alpha^p) : E] \leq \deg g = \frac{\deg h}{p}$. This contradicts the independence of $\{1, \alpha^p, \dots, \alpha^{(n-1)p}\}$. $\square$

Now we are ready to prove the following result:

Theorem 8.2.6 (MacLane). *Let F/E be a field extension of characteristic $p > 0$. Then the following conditions are equivalent:*

(1) *F/E is separable.*
(2) *F and E^{1/p^n} are linearly disjoint over E for some $n \in \mathbb{N}$.*
(3) *F and E^{1/p^∞} are linearly disjoint over E.*

Proof.
(1)$\Rightarrow$(3): By Remark 8.1.11 we may assume that F/E is finitely generated. Let $\{y_1, \dots, y_m\}$ be a transcendence base of F over E such that $F/E(y_1, \dots, y_m)$ is algebraically separable.

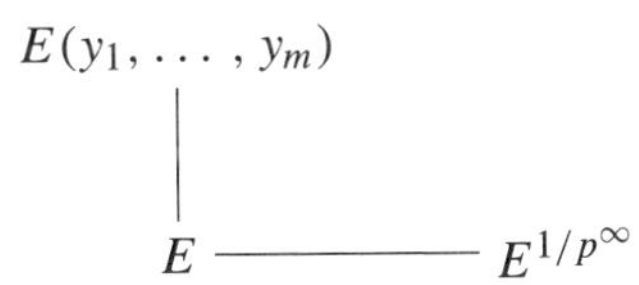

Clearly the set $\{y_1, \dots, y_m\}$ is algebraically independent over E^{1/p^∞}. By Proposition 8.1.10, $E(y_1, \dots, y_m)$ and E^{1/p^∞} are linearly disjoint over E. The composite field $E^{1/p^\infty}E(y_1, \ldots, y_m) = E^{1/p^\infty}(y_1, \ldots, y_m) = K$ is purely inseparable over $E(y_1, \ldots, y_m)$.

We have the following diagram:

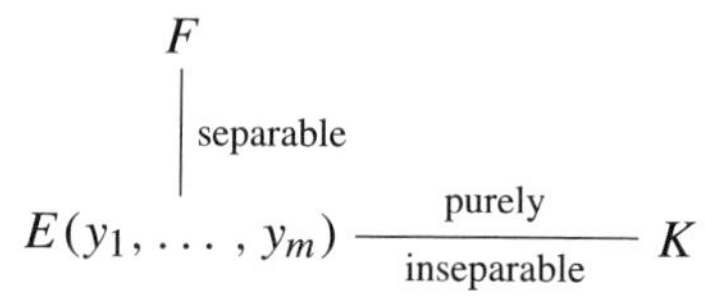

Let $L = E(y_1, \ldots, y_m)$. If $\alpha \in F$, then α is algebraically separable over L. Hence $K(\alpha)/K$ is separable. Let $h(x) = \operatorname{Irr}(\alpha, x, K) \in K'[x]$, where K'/L is a finite purely inseparable extension with $K' \subseteq K$.

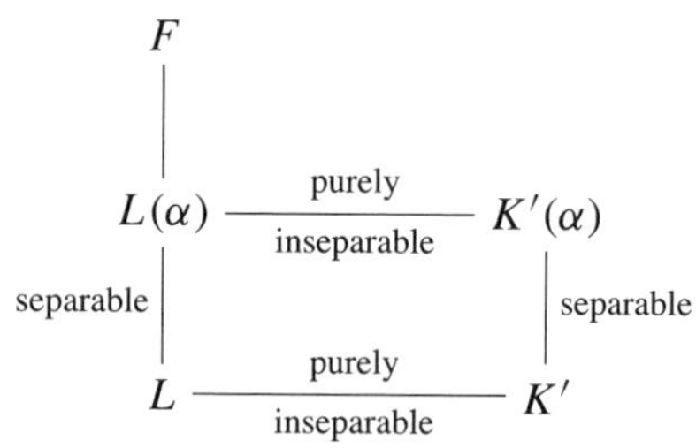

It is easy to see that $\left[K'(\alpha) : L\right]_s = [L(\alpha) : L] = \left[K'(\alpha) : K'\right]$. It follows that F and $K = E^{1/p^\infty}(y_1, \ldots, y_m)$ are linearly disjoint over $L = E(y_1, \ldots, y_m)$. The result follows from Proposition 8.1.5.

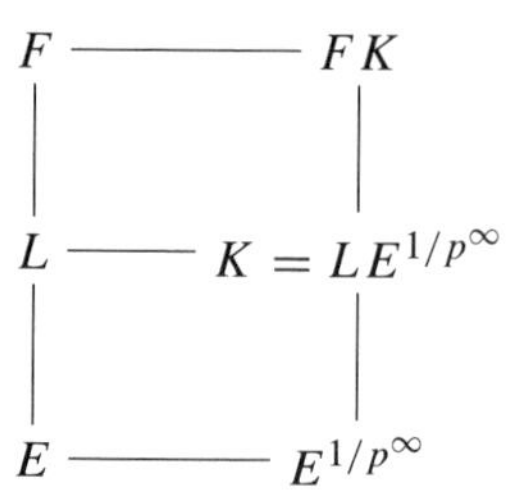

(3) $\Rightarrow$ (2) This implication follows from the fact that $E^{1/p^n} \subseteq E^{1/p^\infty}$.

(2) $\Rightarrow$ (1) By Remark 8.1.11, we may assume that F is finitely generated over E.

Let $F = E(y_1, \ldots, y_m)$ and let r be the transcendence degree of F over E. If $r = m$, the result follows. Otherwise, let $\{y_1, \ldots, y_r\}$ be a transcendence base. Then y_{r+1} is algebraic over $E(y_1, \ldots, y_r)$.

Let $p(T_1, \ldots, T_r, T_{r+1}) \in E[T_1, \ldots, T_r, T_{r+1}]$ be a polynomial of minimum degree such that $p(y_1, \ldots, y_r, y_{r+1}) = 0$.

Clearly, $p(T_1, \ldots, T_r, T_{r+1})$ is irreducible. We shall prove that not all T_i, $1 \leq i \leq r+1$, appear to the pth power throughout. Indeed, assume for the sake of contradiction that

$$p(T_1, \ldots, T_{r+1}) = \sum a_{(i_1, \ldots, i_{r+1})} S_{(i_1, \ldots, i_{r+1})}(T_1, \ldots, T_{r+1})^p, \tag{8.9}$$

where the $S_{(i_1,\dots,i_{r+1})}$'s are monomials and $a_{(i_1,\dots,i_{r+1})} \in E$.

Taking the pth roots in (8.9), we see that the $S_{(i_1,\dots,i_{r+1})}(y_1, \dots, y_{r+1})$ are linearly dependent over $E^{1/p}$. Since $p(T_1, \dots, T_r, T_{r+1})$ is of minimum degree possible, it follows that $\{S_{(i_1,\dots,i_{r+1})}(y_1, \dots, y_{r+1})\}$ is linearly independent over E. This contradicts the linear disjointness of $E^{1/p}$ and $E(y_1, \dots, y_m)$.

Say that T_1 does not appear as a pth root throughout but appears in $p(T_1, \dots, T_{r+1})$. Since $p(T_1, \dots, T_{r+1})$ is irreducible in $E[T_1, \dots, T_{r+1}]$ it follows that the equation $p(T_1, \dots, T_{r+1}) = 0$ is separable for y_1 over $E(y_2, \dots, y_{r+1})$. Hence y_1 is separable and algebraic over $E(y_2, \dots, y_{r+1})$ and over $E(y_2, \dots, y_m)$.

If $\{y_2, \dots, y_m\}$ is a transcendence base, the proof follows immediately. Otherwise, proceeding as before we can show that one y_i, say y_2, is separable and algebraic over $E(y_3, \dots, y_n)$. Therefore F is separable over $E(y_3, \dots, y_m)$.

It is easy to see that we can go on with this process until we find a transcendence base. This proves that (2) $\Rightarrow$ (1). □

Remark 8.2.7. The proof of (2) $\Rightarrow$ (1) in Theorem 8.2.6 shows that a separating transcendence base for $E(y_1, \dots, y_m)$ over E can be selected from a given set of generators $\{y_1, \dots, y_m\}$.

Theorem 8.2.8. *Let F/E be an extension of fields of characteristic p.*

(1) *If F/E is separably generated, then F/E is separable.*
(2) *If F/E is separable and finitely generated, then F/E is separably generated.*

Proof.

(1) Let $\mathcal{A}$ be a transcendence base of F/E such that $F/E(\mathcal{A})$ is an algebraic separable extension.
It is clear that $\mathcal{A}$ is algebraically independent over $E^{1/p}$. Hence, by Proposition 8.1.10, $E^{1/p}$ and $E(\mathcal{A})$ are linearly disjoint.

$$\begin{array}{ccccc} E^{1/p} & \text{———} & E^{1/p}(\mathcal{A}) & & \\ | & & | & & \\ E & \text{———} & E(\mathcal{A}) & \text{———} & F \end{array}$$

Now, $F/E(\mathcal{A})$ is algebraic and separable and $E^{1/p}(\mathcal{A})/E(\mathcal{A})$ is algebraic and purely inseparable. It follows that F and $E^{1/p}(\mathcal{A})$ are linearly disjoint over $E(\mathcal{A})$ (see the proof of (1) $\Rightarrow$ (3) in Theorem 8.2.6). Thus, by Proposition 8.1.5, $E^{1/p}$ and F are linearly disjoint over E. Using MacLane's criterion (Theorem 8.2.6) we obtain that F/E is separable.

(2) Let F/E be a finitely generated separable extension, say $F = E(y_1, \dots, y_m)$ By Remark 8.2.7 we may choose a subset of the set $\{y_1, \dots, y_m\}$ that is a separating transcendence base for F over E. In particular, F/E is separably generated. □

Remark 8.2.9. The hypothesis that F/E is a finitely generated extension cannot be dropped. Indeed, there exists an extension F/E that is separable but not separably generated.

Example 8.2.10. Let E be a perfect field of characteristic $p > 0$. Then $E^{1/p} = E$. In particular, $E^{1/p}$ and F are linearly disjoint over E and F/E is separable for any extension F.

Let x be a transcendental element over E. Let $F = E\left(\{x^{1/p^m}\}_{m=0}^{\infty}\right)$. Then F/E is separable and $\operatorname{tr} F/E = 1$ (actually $\left(x^{1/p^m}\right)^{p^m} = x \in E(x)$; thus $F/E(x)$ is algebraic). Let $\{y\}$ be any transcendence base of F/E. There exist $n \in \mathbb{N}$ and a rational function

$$f(T_1, \ldots, T_n) \in E(T_1, \ldots, T_n)$$

such that $y = f(x, x^{1/p}, \ldots, x^{1/p^{n-1}})$. Then $E(y) \neq F$ since $x^{1/p^n} \notin E(y)$ and $F/E(y)$ is purely inseparable. Therefore $F/E(y)$ is not separable and F/E is not separably generated.

Corollary 8.2.11. *If E is a perfect field, any extension F of E is separable over E.*

Proof. Exercise 8.7.8. □

As a consequence of MacLane's criterion we obtain the following corollaries.

Corollary 8.2.12. *If F is separable over E and $E \subseteq M \subseteq F$, then M in separable over E.*

Proof. Exercise 8.7.9. □

Corollary 8.2.13. *If M/E and F/M are separable field extensions, then F/E is separable.*

Proof. Exercise 8.7.10. □

Proposition 8.2.14. *Let F be a separable extension of E and assume that F is algebraically disjoint from L over E with $E \subseteq L$. Then FL is a separable extension of L.*

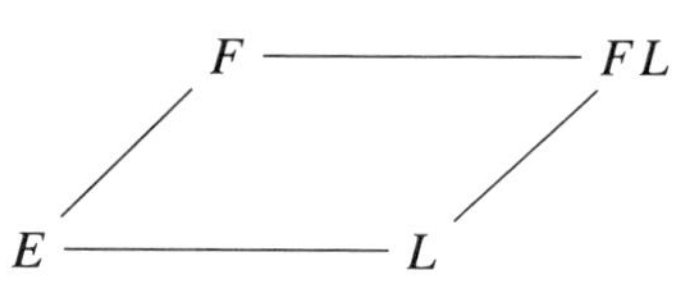

Proof. The elements of FL are of the form $\frac{\sum_{i=1}^{n} a_i b_i}{\sum_{j=1}^{m} c_j d_j}$ with $a_i, c_j \in F$ and $b_i, d_j \in L$. In particular, any finitely generated subfield of FL is contained in a composite ML, where M is a subfield of F that is finitely generated over E. If for any such M we can prove that ML is a separable extension of L, the separability of FL over L will follow by Corollary 8.2.12 and Theorem 8.2.8 (2).

Therefore we may assume that F is finitely generated over E. Let $\{y_1, \ldots, y_m\}$ be a transcendence base of F over E. Since F and L are algebraically disjoint over

E, it follows that $\{y_1, \ldots, y_m\}$ is a transcendence base of FL over L. Every element of F is separable and algebraic over $E(y_1, \ldots, y_m)$, so it is also separable over $L(y_1, \ldots, y_m)$. Thus FL is separably generated over L. The result follows by Theorem 8.2.8. □

Corollary 8.2.15. *Let F and L be two separable extensions of E. If F and L are algebraically disjoint over E, then FL is separable over E.*

Proof. Exercise 8.7.11. □

Proposition 8.2.16. *If F and L are two extensions that are linearly disjoint over E, then F is separable over E if and only if FL is separable over L.*

Proof.
($\Rightarrow$) Proposition 8.2.14 and Proposition 8.1.9.
($\Leftarrow$) If F is not separable over E, then by MacLane's criterion, F is not linearly disjoint from $E^{1/p}$ over E. Hence F is not linearly disjoint from $LE^{1/p}$ over E (Proposition 8.1.5).

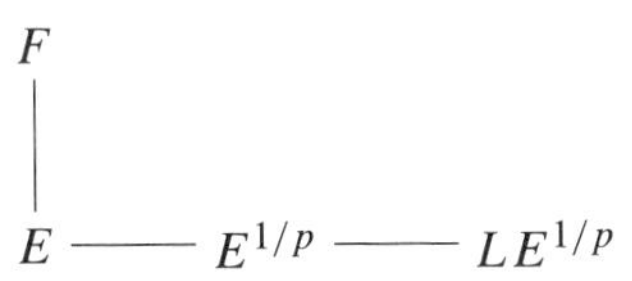

Using MacLane's criterion we obtain that FL is not linearly disjoint from $LE^{1/p}$ over L. Therefore FL and $L^{1/p}$ are not linearly disjoint over L. By Theorem 8.2.6, FL is not separable over L. □

We are now ready to characterize separably algebraic finitely generated extensions.

Proposition 8.2.17. *Let F be a finitely generated extension of E. If $F^{p^m}E = F$ for some $m \in \mathbb{N}$, then F is separably algebraic over E and $F^{p^n}E = F$ for all $n \in \mathbb{N}$. Conversely, if F is separably algebraic over E, then $F^{p^m}E = F$ for all $m \in \mathbb{N}$.*

Proof. If $F^{p^m}E = F$ for some m, then F is an algebraic extension of E (see Exercise 8.7.16). Now $F = F^{p^m}E \subseteq F^pE \subseteq F$. Therefore $F = F^pE$. Furthermore, for all $n \geq 1$, $F^{p^n}E = (F^p)^{p^{n-1}}E = (F^pE)^{p^{n-1}}E = F^{p^{n-1}}E$. Thus $F^{p^n}E = F$ for all $n \in \mathbb{N}$.

Let T be the separable closure of E in F. Then F is a purely inseparable extension of T. Since F is algebraic and finitely generated over E, F is a finite extension of E. In particular, there exists $n \in \mathbb{N}$ such that $F^{p^n} \subseteq T$. It follows that $F = F^{p^n}E \subseteq T \subseteq F$.

Conversely, let F be a separably algebraic extension of E. We have $E \subseteq F^pE \subseteq F$ and F is a purely inseparable extension of F^pE. Hence $F = F^pE$. As before, it follows that $F = F^{p^m}E$ for all $m \in \mathbb{N}$. □

8.3 Regular Extensions

We now study the class of extensions that we will be dealing with when we consider extensions of function fields.

Proposition 8.3.1. *Let k be algebraically closed in an extension K. Let x be an element of the algebraic closure $\bar{k}$ of k. Then $k(x)$ and K are linearly disjoint over k and $[k(x) : k] = [K(x) : K]$.*

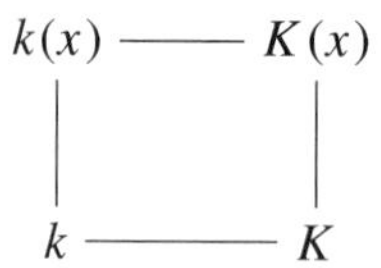

Proof. Let $p(T) = \mathrm{Irr}(x, T, k) \in k[T]$. If $q(T) \in K[T]$ is a nonconstant factor of $p(T)$, then the coefficients of $q(T)$ are algebraic over k. Since k is algebraically closed in K, we have $q(T) \in k[T]$. Hence $p(T)$ is irreducible in $K[T]$. It follows that $[k(x) : k] = [K(x) : K]$ and that $k(x)$ and K are linearly disjoint over k. □

Theorem 8.3.2. *Let K/k be a field extension, and let $\bar{k}$ be an algebraic closure of k. Then the following conditions are equivalent:*

(1) *k is algebraically closed in K and K is separable over k.*
(2) *K and $\bar{k}$ are linearly disjoint over k.*

Proof.
(1) $\Rightarrow$ (2) By Remark 8.1.11 we may assume that K is finitely generated over k, and it suffices to show that K and L are linearly disjoint over k, where L is any finite algebraic extension of k. In this situation, if L is separable over k, then L is of the form $L = k(\alpha)$, with α algebraic over k. The result follows by Proposition 8.3.1.

In general, if L_s is the maximum separable extension of k in L, then L_s and K are linearly disjoint over k. By Proposition 8.1.5, it suffices to show that L and KL_s are linearly disjoint over L_s. Let $\{y_1, \ldots, y_m\}$ be a separating transcendence base for K over k. Then K is separably algebraic over $k(y_1, \ldots, y_m)$.

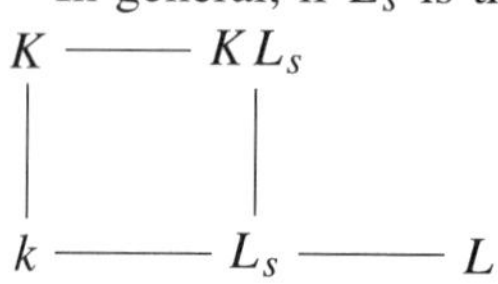

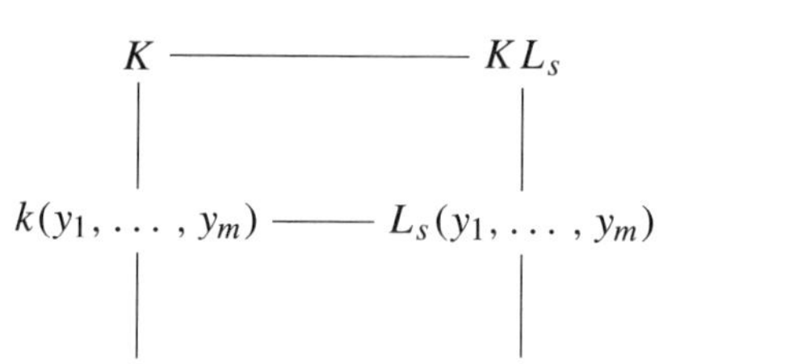

Since $k(y_1, \ldots, y_m)$ and L_s are linearly disjoint over k (Proposition 8.1.10), $\{y_1, \ldots, y_m\}$ is also a separating transcendence basis of KL_s over L_s, and KL_s is separably algebraic over $L_s(y_1, \ldots, y_m)$. Thus KL_s is separable over L_s. Since L/L_s is a purely inseparable extension, it follows that KL_s and L are linearly disjoint over L_s.

(2) $\Rightarrow$ (1) We have $k^{1/p} \subseteq \overline{k}$, so $k^{1/p}$ and K are linearly disjoint over k. By Theorem 8.2.6, K/k is separable. If $\alpha \in \overline{k} \cap K$, then since K and $k(\alpha)$ are linearly disjoint, it follows that $[k(\alpha) : k] = [K(\alpha) : K] = 1$. Hence $\alpha \in k$ and k is algebraically closed in K.

□

Definition 8.3.3. An extension K of k is called *regular* if k is algebraically closed in K and K/k is separable, or equivalently, if K is linearly disjoint from $\overline{k}$ over k.

Remark 8.3.4. In the case of function fields K/k, we are assuming that k is algebraically closed in K. Therefore K/k is regular iff there exists $x \in K$ such that $K/k(x)$ is a finite separable extension.

Proposition 8.3.5. *Let K be a regular extension of k. If $k \subseteq K' \subseteq K$, then K' is a regular extension of k.*

Proof. Since $K' \subseteq K$, K' is linearly disjoint from $\overline{k}$ over k. □

Proposition 8.3.6. *Regularity is transitive, that is, if K is a regular extension of k and L is a regular extension of K, then L is a regular extension of k.*

Proof. k is algebraically closed in K and K is algebraically closed in L. Therefore k is algebraically closed in L. The fact that L is separable over k follows from Corollary 8.2.13. □

Proposition 8.3.7. *If k is algebraically closed, then every extension of k is regular.*

Proof. We have $\overline{k} = k$. If K is any extension of k, then K is linearly disjoint from $\overline{k} = k$ over k. The fact that K is separable over k follows from Corollary 8.2.11 since k is a perfect field. □

The converse of Proposition 8.1.9 holds for regular extensions:

Theorem 8.3.8. *Let F and L be two extensions of a field E such that F and L are contained in some field Ω. If F is a regular extension of E, and F and L are algebraically independent over E, then F and L are linearly disjoint over E.*

Proof. By Remark 8.1.11 we may assume that F is finitely generated over E. Let $\{\alpha_1, \ldots, \alpha_m\}$ be elements of F that are linearly independent over E. If $\{\alpha_1, \ldots, \alpha_m\}$ are not linearly independent over L, let $\beta_1, \ldots, \beta_m \in L$ be such that

$$\beta_1\alpha_1 + \cdots + \beta_m\alpha_m = 0 \tag{8.10}$$

and at least one of the β_i's is nonzero.

Removing the elements that are equal to 0, we may assume that $\beta_i \neq 0$ for all $1 \leq i \leq m$.

Let $\varphi \colon L \to \bar{E} \cup \{\infty\}$ be a place of L such that $\varphi|_E = \mathrm{Id}_E$. Let $\{y_1, \ldots, y_n\}$ be a transcendence base of F over E. Then $\{y_1, \ldots, y_n\}$ is algebraically independent over L. We can extend φ to a place $\widetilde{\varphi}\colon LF \to \bar{F} \cup \{\infty\}$ such that $\widetilde{\varphi}|_{E(y_1,\dots,y_n)} = \mathrm{Id}_{E(y_1,\dots,y_n)}$. Set $\Psi = \widetilde{\varphi}|_F$. If $\xi \in F^*$, then ξ is algebraic over $E(y_1, \dots, y_m) = M$. Hence there exists a relation

$$\xi^t + a_{t-1}\xi^{t-1} + \cdots + a_1\xi + a_0 = 0$$

with $a_0, a_1, \dots, a_{t-1} \in M$, and $a_0 \neq 0$.

Since $a_0 \neq 0$ it follows that $\varphi(\xi) \neq 0$. Similarly, $\varphi\left(\frac{1}{\xi}\right) \neq 0$. Now $1 = \Psi(1) = \Psi\left(\xi\frac{1}{\xi}\right) = \Psi(\xi)\Psi\left(\frac{1}{\xi}\right)$, so $\varphi(\xi) \neq \infty$. Hence Ψ is a field homomorphism and $\varphi(F) = \Psi(F) \cong F$.

By Exercise 8.7.12, there exists an index j_0 (say $j_0 = m$) such that $\varphi\big(\beta_i/\beta_m\big) \neq \infty$ for all i.

Dividing (8.10) by β_m, we obtain

$$\frac{\beta_1}{\beta_m}\alpha_1 + \frac{\beta_2}{\beta_m}\alpha_2 + \cdots + \alpha_m = 0 \tag{8.11}$$

and hence

$$\sum_{i=1}^{m} \varphi\left(\frac{\beta_i}{\beta_m}\right)\varphi(\alpha_i) = 0,$$

where $\varphi\left(\frac{\beta_i}{\beta_m}\right) \in \overline{E}$. Consequently $\{\varphi(\alpha_1), \dots, \varphi(\alpha_m)\}$ are linearly dependent over $\overline{E}$. Since φ is an isomorphism of F onto $\varphi(F)$, it follows that $\{\alpha_1, \dots, \alpha_m\}$ is linearly dependent over $\overline{E}$. This contradicts the regularity of F. Therefore F and L are linearly disjoint over E. □

Theorem 8.3.9. *Let K be a regular extension of k such that K and L are algebraically disjoint over k. Then KL is a regular extension of L.*

Proof. Let $y_1, \dots, y_m$ be elements of K that are algebraically independent over k. Then $\{y_1, \dots, y_m\}$ is algebraically independent over L.

Therefore $m = \operatorname{tr} \bar{L}(y_1, \dots, y_m)/L = \operatorname{tr} \bar{L}(y_1, \dots, y_m)/\bar{L} + \operatorname{tr} \bar{L}/L$ (Proposition 1.1.12). Since $\operatorname{tr} \bar{L}/L = 0$, it follows that $\{y_1, \dots, y_m\}$ is algebraically independent over $\bar{L}$. In particular, K is algebraically disjoint from $\bar{L}$ over k.

By Theorem 8.3.8, K is linearly disjoint from $\overline{L}$ over k. Using Proposition 8.1.5 we deduce that KL is linearly disjoint from $\overline{L}$ over L. Hence KL is regular over L. □

Corollary 8.3.10. *Let K and L be regular extensions of k. If K and L are algebraically independent over k, then KL is a regular extension of k.*

Proof. Exercise 8.7.13. □

8.4 Constant Extensions

Let K/k be an algebraic function field. Given any extension ℓ' of k we wish to obtain the constant extension $K\ell'$. In order to be able to construct $K\ell'$, we need two conditions, first that K and ℓ' be contained in a larger field (see Section 5.4), and second that $K \cap \ell' = k$ (see Definition 5.1.1). Given K and ℓ', both conditions are not always satisfied. However, we can construct a function field L over a constant field ℓ such that ℓ contains a subfield that is k-isomorphic to ℓ'.

Proposition 8.4.1. *If a field k is algebraically closed in K and $\{X_i\}_{i \in \mathcal{A}}$ is an algebraically independent set over K, then $k(\{X_i\}_{i\in\mathcal{A}})$ is algebraically closed in $K(\{X_i\}_{i\in\mathcal{A}})$.*

Proof. Let $\alpha \in K(\{X_i\}_{i\in\mathcal{A}})$ be algebraic over $k(\{X_i\}_{i\in\mathcal{A}})$. There exists a relation

$$\alpha^r + f_{r-1}\alpha^{r-1} + \cdots + f_1\alpha + f_0 = 0 \tag{8.12}$$

with $f_0, \ldots, f_{r-1} \in k(\{X_i\}_{i\in\mathcal{A}})$.

Since only finitely many X_i's appear in (8.12), we may assume that $\mathcal{A}$ is a finite set, say $\mathcal{A} = \{X_1, \ldots, X_n\}$.

We will prove the result by induction on n.

$$\begin{array}{ccc} K & \text{———} & K(x) \\ | & & | \\ k & \text{———} & k(x) \end{array}$$

Assume that $n = 1$ and $X_1 = x$. Let $\alpha \in K(x)$ be a nonzero algebraic element over $k(x)$. We may write $\alpha = A\frac{h(x)}{g(x)}$, where $h(x), g(x) \in K[x]$, $(h(x), g(x)) = 1$, A is a nonzero element of K, and $h(x), g(x)$ are monic.

There exist $f_0, \ldots, f_{r-1}, f_r \in k[x]$ such that $(f_0, \ldots, f_r) = 1$ and

$$f_r(x)\alpha^r + \cdots + f_1(x)\alpha + f_0(x)\alpha = 0. \tag{8.13}$$

Clearing denominators in (8.13) we obtain

$$\begin{aligned} f_r(x)A^r h(x)^r + f_{r-1}(x)A^{r-1}h(x)^{r-1}g(x) + \cdots \\ + f_1(x)Ah(x)g^{r-1}(x) + f_0(x)g^r(x) = 0. \end{aligned} \tag{8.14}$$

Let a be a root of $h(x) \in K[x]$. By means of the substitution $x = a$ in (8.14) we obtain

$$f_0(a)g^r(a) = 0.$$

Then $g(a) \neq 0$, since h and g are relatively prime. It follows that $f_0(a) = 0$.

Thus every root of h is algebraic over k. Because $h(x)$ is monic, the coefficients of h are algebraic over k. Since k is algebraically closed, it follows that $h(x) \in k[x]$. Similarly, $g(x) \in k[x]$.

Now let a be a root of $h(x) - g(x)$. The equality $0 = h(a) - g(a)$ and the fact that h and g are relatively prime imply $h(a) = g(a) \neq 0$.

Substituting x by a in (8.14) we obtain

$$f_r(a)h(a)^r A^r + \cdots + f_1(a)h(a)g^{r-1}(a)A + f_0(a)g^r(a) = 0.$$

Now, $f_r, \ldots, f_1, f_0$ are relatively prime, so there exists i such that $f_i(a) \neq 0$. It follows that A is algebraic over k. Since k is algebraically closed in K, we have $A \in k$.

Therefore $\alpha = A\frac{h(x)}{g(x)} \in k(x)$ and $k(x)$ is algebraically closed in $K(x)$.

Now assume that the result holds for $n-1$. For n, let $\alpha \in K(X_1, \ldots, X_{n-1}, X_n)$. Let $E = k(X_1, \ldots, X_{n-1})$ and $F = K(X_1, \ldots, X_{n-1})$. By the induction hypothesis E is algebraically closed in F. Since X_n is transcendent over F, it follows from the case $n = 1$ that $E(X_n)$ is algebraically closed in $F(X_n)$. Thus if $\alpha \in K(X_1, \ldots, X_n)$ is algebraic over $k(X_1, \ldots, X_n)$, then $\alpha \in E(X_n) = k(X_1, \ldots, X_n)$. □

Theorem 8.4.2. *Let K/k be an algebraic function field and let k' be any extension of k. Then there exists a function field L/ℓ that is an extension of K/k such that:*

(1) There exist a subfield ℓ' such that $k \subseteq \ell' \subseteq \ell$ and a k-isomorphism $\lambda : \ell' \to k'$.
(2) $L = K\ell'$.

Moreover, if M/m is another extension such that there exist a subfield m' of m and a k-isomorphism $\mu : m' \to k'$ satisfying (1) and (2), then there exists a K-isomorphism $\varrho : M \to L$ such that $\varrho|_{m'} = \lambda^{-1} \circ \mu : m' \twoheadrightarrow \ell'$.

Finally, ℓ is a purely inseparable finite extension of ℓ'.

Proof. First we construct a composite field $L = K\ell'$. Let $\{y_\alpha\}_{\alpha \in \mathcal{A}}$ be a transcendence base of k' over k.

Let $\{X_\alpha\}_{\alpha \in \mathcal{A}}$ be an algebraically independent set over K. Then the cardinality of $\{X_\alpha\}_{\alpha \in \mathcal{A}}$ is the same as the cardinality of the transcendence degree of k' over k. Let Ω be an algebraic closure of $K(\{X_\alpha\}_{\alpha \in \mathcal{A}})$.

There exists a k-isomorphism λ_1 from $k(\{y_\alpha\}_{\alpha \in \mathcal{A}})$ to $k(\{X_\alpha\}_{\alpha \in \mathcal{A}})$ such that $\lambda|_k = \mathrm{Id}_k$ and $\lambda(y_\alpha) = X_\alpha$. Since k' is algebraic over $k(\{y_\alpha\}_{\alpha \in \mathcal{A}})$, λ_1 can be extended to an isomorphism of k' onto a subfield $\lambda_2(k') =: \ell'$ of Ω.

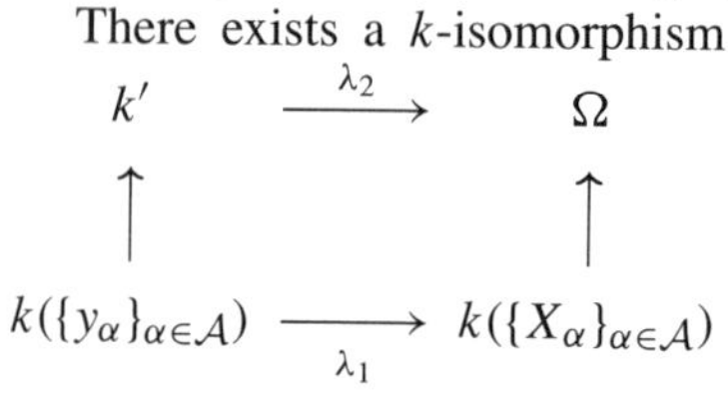

Then Ω contains both K and $\ell' \cong k'$. Therefore we may take the composite field $K\ell'$ in Ω (see Remark 5.4.2).

Let $T \in K \setminus k$ be transcendental over k. If T is not transcendental over ℓ', there exists a finite subset $\{X_1, \ldots, X_m\}$ of the transcendence base $\{X_\alpha\}_{\alpha \in \mathcal{A}}$ such that T is algebraic over $k(X_1, \ldots, X_m)$. Therefore there is a relation $\sum_{i=0}^{r} f_i T^i = 0$ where $f_i \in k[X_1, \ldots, X_m]$ and at least one of the f_i's is a nonconstant polynomial. This implies that $\{X_1, \ldots, X_m\}$ is not algebraically independent over K. Hence T is transcendental over ℓ'. In particular, we have $\ell' \cap K = k$.

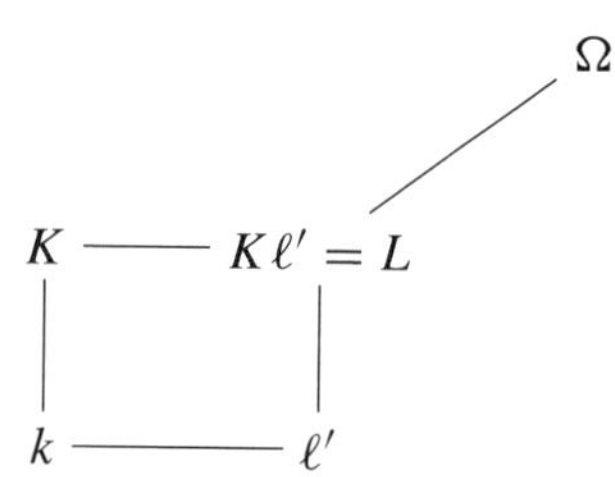

Now let $L = K\ell'$. Then $\ell' \supseteq k$ and $[L : \ell'(T)] \leq [K : k(T)] < \infty$.

$$\begin{array}{ccc}
K & \text{———} & K\ell' = L \\
| & & | \\
k(T) & \text{———} & \ell'(T) \\
| & & | \\
k & \text{———} & \ell'
\end{array}$$

Therefore L/ℓ' is a function field. Let ℓ be the field of constants of L. The field L/ℓ satisfies conditions (1) and (2) of the theorem and $[\ell : \ell'] = [\ell(T) : \ell'(T)] \leq [L : \ell'(T)] < \infty$.

Next, consider another extension M/m of K/k satisfying (1) and (2). We need to find an isomorphism $\varrho : M \to L$ such that $\varrho|_K = \mathrm{Id}_K$ and $\varrho|_{m'} = \lambda^{-1} \circ \mu := \theta$, where $\mu \colon m' \to k'$ is the k-isomorphism of $m' \subseteq m$ onto k'.

Now each element of $M = Km'$ can be written in the form $\frac{\sum_{i=1}^n a_i b_i}{\sum_{j=1}^n c_j d_j}$ with $a_i, c_j \in K$ and $b_i, d_j \in m'$. Therefore ϱ must satisfy

$$\varrho\left(\frac{\sum_{i=1}^n a_i b_i}{\sum_{j=1}^m c_j d_j}\right) = \frac{\sum_{i=1}^n a_i \theta(b_i)}{\sum_{j=1}^m c_j \theta(d_j)}. \tag{8.15}$$

Let $\varrho : M \to L$ be given by (8.15). To prove that ϱ is well defined, we have to verify that if

$$0 = \sum_{i=1}^n a_i b_i, \qquad \text{then} \qquad 0 = \sum_{i=1}^n a_i \theta(b_i).$$

We need to prove that ϱ is an isomorphism and also that if the denominator $\sum_{j=1}^m c_j d_j$ is nonzero in (8.15) then $\varrho\left(\sum_{j=1}^n c_j d_j\right)$ is nonzero. Thus we have to show that

$$\sum_{j=1}^m c_j \theta(d_j) = 0 \qquad \text{implies} \qquad \sum_{j=1}^m c_j d_j = 0.$$

It will suffice to establish that for $\alpha_i \in k$, $\beta_i \in m'$,

$$\sum_{i=1}^n \alpha_i \beta_i = 0 \quad \text{if and only if} \quad \sum_{i=1}^n \alpha_i \theta(\beta_i) = 0. \tag{8.16}$$

Since the expressions in (8.16) involve a finite number of elements, we may assume that ℓ' is finitely generated over k.

Assume ℓ' is a purely transcendental field extension of k, say $\ell' = k(y_1, \ldots, y_n)$. Thus $m' = k(z_1, \ldots, z_n)$ with $z_i = \theta^{-1}(y_i)$. If $X \in K$ is transcendental over k, then X is transcendental over ℓ'. Hence

$$\begin{aligned}
\operatorname{tr} K(y_1, \ldots, y_n)/k &= \operatorname{tr} K(y_1, \ldots, y_n)/k(y_1, \ldots, y_n) + \operatorname{tr} k(y_1, \ldots, y_n)/k \\
&= 1 + n = \operatorname{tr} K/k + \operatorname{tr} K(y_1, \ldots, y_n)/K = 1 + \operatorname{tr}(K(y_1, \ldots, y_n)/K).
\end{aligned}$$

Therefore $\operatorname{tr} K(y_1, \ldots, y_n)/K = n$. It follows that $y_1, \ldots, y_n$ are algebraically independent over K, and so are $z_1, \ldots, z_n$. Thus $M = Km' = K(z_1, \ldots, z_n)$, $L = K\ell' = K(y_1, \ldots, y_n)$ and hence the map $\varrho : M \to L$, such that $\varrho(z_i) = y_i$ for $1 \le i \le n$, is the required isomorphism.

Further, in this case, $L = K\ell' = K(y_1, \ldots, y_n)$ satisfies that the field $\ell' = k(y_1, \ldots, y_n)$ is algebraically closed in L (Proposition 8.4.1) so that the field of constants of L is $\ell = \ell'$.

Also, to prove the pure inseparability of ℓ/ℓ' and m/m', it suffices to assume that ℓ'/k and m'/k are finitely generated. Therefore, to prove the general case we may assume that ℓ'/k and m'/k are finitely generated.

$$\begin{array}{ccccc}
K & \longrightarrow & K\ell'' = K(y_1, \ldots, y_m) & \longrightarrow & K\ell' = L \\
\uparrow & & \uparrow & & \uparrow \\
k & \longrightarrow & \ell'' = k(y_1, \ldots, y_m) & \longrightarrow & \ell'
\end{array}$$

Suppose ℓ' is finitely generated and let $\{y_1, \ldots, y_n\}$ be a transcendence base of ℓ' over k. Consider $\ell'' = k(y_1, \ldots, y_n)$. Then ℓ'/ℓ'' is a finite extension and similarly for M/m'.

Let ϱ be the isomorphism

$$\begin{aligned}
\varrho : Km'' &\longrightarrow K\ell'' \\
z_i &\longmapsto y_i \quad (1 \le i \le n).
\end{aligned}$$

In order to find an isomorphism $\varrho_1 \colon M \to L$ such that $\varrho_1|_{Km''} = \varrho$ will follow from the fact that M/Km' is a finite extension. Thus we may assume that ℓ'/k is a finite extension.

$$\begin{array}{ccccc}
K & \longrightarrow & Km'' = K(z_1, \ldots, z_m) & \longrightarrow & Km' = M \\
\uparrow & & \uparrow & & \uparrow \\
k & \longrightarrow & m'' = k(z_1, \ldots, z_m) & \longrightarrow & m'
\end{array}$$

Let $\ell = k(\alpha_1, \ldots, \alpha_n)$, where α_i algebraic over k for $1 \le i \le n$. Assume that the result holds for $n-1$ and let $\ell_1 = k(\alpha_1, \ldots, \alpha_{n-1})$. We have $L_1 = K\ell_1 = K(\alpha_1, \ldots, \alpha_{n-1})$. Let $\beta_i = \theta^{-1}(\alpha_i)$ $(1 \le i \le n-1)$, $m_1 = k(\beta_1, \ldots, \beta_{n-1})$, and $M_1 = Km_1 = K(\beta_1, \ldots, \beta_{n-1})$. Set $m' = m_1(\beta_n)$ and let ℓ_2 and m_2 be the algebraic closures of ℓ_1 and m_1 in L_1 and M_1 respectively.

$$\begin{array}{ccc}
K & \text{---} & L_1 \\
| & & | \\
| & & \ell_2 \\
| & & | \\
k & \text{---} & \ell_1
\end{array}
\qquad\qquad
\begin{array}{ccc}
K & \text{---} & M_1 \\
| & & | \\
| & & m_2 \\
| & & | \\
k & \text{---} & m_1
\end{array}$$

By the induction hypothesis, there exists an isomorphism $\varrho_1 : M_1 \to L_1$ such that $\varrho_1|_{m_1} = \theta : m_1 \xrightarrow{\cong} \ell_1$ and $\varrho_1|_K = \mathrm{Id}_K$. Also, ℓ_2/ℓ_1 and m_2/m_1 are purely inseparable.

Let $\alpha_n = \alpha$ and $\beta = \theta^{-1}(\alpha)$. Then $L = L_1(\alpha)$ and $M = M_1(\beta)$.

It suffices to extend ϱ_1 to an isomorphism $\varrho : M \to L$ such that $\varrho(\beta) = \alpha$ and that the constant field ℓ of L is purely inseparable over ℓ_2 (and hence over ℓ').

In other words, ϱ_1 can be extended to ϱ if

$$\varrho_1(\mathrm{Irr}(\beta, X, M_1)) = \mathrm{Irr}(\alpha, X, L_1).$$

Let $p(X) = \mathrm{Irr}(\beta, X, M_1) \in M_1[X]$. Since β in algebraic over k, the coefficients of $p(X)$ are algebraic over k. Hence $p(X) \in m_2[X]$. Now, since m_2/m_1 is purely inseparable, it follows that $\mathrm{Irr}(\beta, X, m_1) = \mathrm{Irr}(\beta, X, M_1)^{p^t}$ for some $t \geq 0$ (where p is the characteristic).

Since θ is an isomorphism of $m' = m_1(\beta)$ onto $\ell' = \ell_1(\alpha)$ with $\theta(\beta) = \alpha$, we obtain that $\theta(\mathrm{Irr}(\beta, X, M_1)^{p^t}) = \varrho_1(\mathrm{Irr}(\beta, X, M_1))^{p^t} = \mathrm{Irr}(\alpha, X, \ell_1)$. Since $\varrho_1(m_1) = \ell_1$, we have $\varrho_1(m_2) = \ell_2$. Hence $\varrho_1(\mathrm{Irr}(\beta, X, M_1)) = \mathrm{Irr}(\alpha, X, L_1)$ because $\varrho_1(\mathrm{Irr}(\beta, X, M_1))$ is the only irreducible factor of $\varrho_1(\mathrm{Irr}(\beta, X, M_1))^{p^t}$ over ℓ_2.

This shows that ϱ_1 can be extended to an isomorphism with the required properties.

It remains to prove that the field of constants ℓ of L is purely inseparable over ℓ'. Now since ℓ_2 is purely inseparable over ℓ_1, $\ell_2(\alpha)$ is purely inseparable over $\ell_1(\alpha) = \ell'$. Hence it suffices to prove that ℓ is purely inseparable over $\ell_2(\alpha)$.

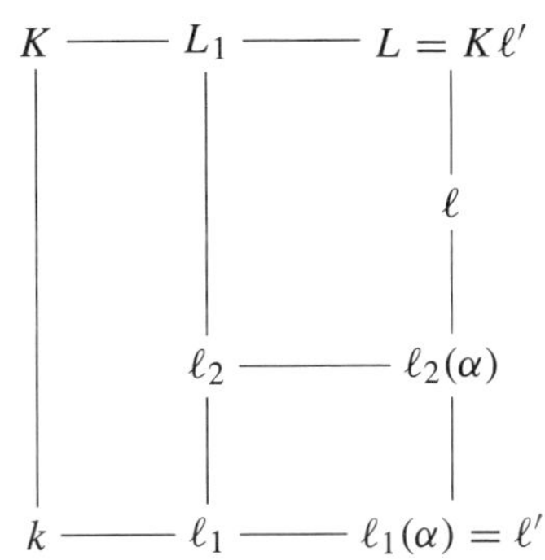

Since ℓ_2 is algebraically closed in L_1, we have $\mathrm{Irr}(\alpha, X, L_1) = \mathrm{Irr}(\alpha, X, \ell_2)$, so

$$[L_1(\alpha) : L_1] = [\ell_2(\alpha) : \ell_2]. \tag{8.17}$$

If $x \in L$ is transcendental over ℓ', we have

$$[\ell_2(\alpha, x) : \ell_2(x)] = [\ell_2(\alpha) : \ell_2] \tag{8.18}$$

(Proposition 2.1.6).

Now

$$[L_1(\alpha) : \ell_2(x)] = [L_1(\alpha) : L_1][L_1 : \ell_2(x)] = [L_1(\alpha) : \ell_2(\alpha, x)][\ell_2(\alpha, x) : \ell_2(x)]. \tag{8.19}$$

From (8.17), (8.18) and (8.19) we obtain

$$\begin{aligned}[L_1 : \ell_2(x)] &= \frac{[L_1(\alpha) : \ell_2(\alpha, x)][\ell_2(\alpha, x) : \ell_2(x)]}{[L_1(\alpha) : L_1]} \\ &= \frac{[L_1(\alpha) : \ell_2(\alpha, x)][\ell_2(\alpha) : \ell_2]}{[\ell_2(\alpha) : \ell_2]} = [L_1(\alpha) : \ell_2(\alpha, x)].\end{aligned} \tag{8.20}$$

Let δ be a constant of $L_1(\alpha)$, that is, $\delta \in \ell$. There exists $t \in \mathbb{N}$ such that δ^{p^t} is separably algebraic over $\ell_2(\alpha)$. Being δ^{p^t} separable over $\ell_2(\alpha)$, $\ell_2(\alpha, \delta^{p^t})$ is a simple extension $\ell_2(\gamma)$ of ℓ_2. We have, by (8.20),

$$[L_1(\alpha) : \ell_2(\alpha, \delta^{p^t}, x)] = [L_1(\gamma) : \ell_2(\gamma, x)] = [L_1 : \ell_2(x)]. \tag{8.21}$$

Using (8.21) with $L_1(\alpha)$ and $\ell_1(\alpha)$, we obtain

$$[L_1(\alpha) : \ell_2(\alpha, \delta^{p^t}, x)] = [L_1(\alpha) : \ell_2(\alpha, x)].$$

Hence $\ell_2(\alpha, \delta^{p^t}, x) = \ell_2(\alpha, x)$ and $\delta^{p^t} \in \ell_2(\alpha, x)$.

Since δ^{p^t} is algebraic over $\ell_2(\alpha)$ and $\ell_2(\alpha)$ is algebraically closed in $\ell_2(\alpha, x)$, it follows that $\delta^{p^t} \in \ell_2(\alpha)$ and δ is purely inseparable over $\ell_2(\alpha)$.

This completes the proof of the theorem. □

Remark 8.4.3. Example 5.2.31 shows that the field of constants of $K\ell'$ can contain ℓ' properly.

Our next result characterizes when the field of constants $K\ell'$ is ℓ'.

Theorem 8.4.4. *Let $L = K\ell'$ be a constant field extension of K such that the field of constants ℓ contains ℓ'. Then the following conditions are equivalent:*

(i) *K and ℓ are linearly disjoint over k.*
(ii) *For every finitely generated field ℓ_0 over k such that $\ell_0 \subseteq \ell'$, the constant field of $L_0 := K\ell_0$ is ℓ_0.*

If these conditions are fulfilled, then for any extension ℓ_0 over k such that $\ell_0 \subseteq \ell'$ (not necessarily finitely generated), the constant field of $L_0 := K\ell_0$ is ℓ_0. In particular, the field of constants of $L = K\ell'$ is ℓ'.

Proof.
(i) ⇒ (ii) Let $k \subseteq \ell_0 \subseteq \ell'$. Let ℓ_0' be the field of constants of $L_0 = L\ell_0$, and we have $\ell_0 \subseteq \ell_0' \subseteq \ell$. It follows from (i) and Proposition 8.1.5 that ℓ_0' and K are linearly disjoint.

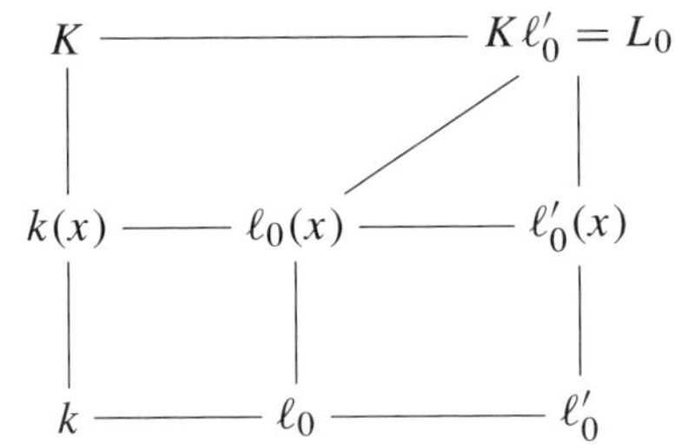

Let $x \in K \setminus k$. By Proposition 8.1.5, $k(x)$ and ℓ_0' are linearly disjoint over k, and K and $\ell_0' k(x) = \ell_0'(x)$ are linearly disjoint over $k(x)$.

Since $\ell_0(x) \subseteq \ell_0'(x)$ and $L_0 = K\ell_0(x) = K\ell_0'(x)$, we have

$$[L_0 : \ell_0'(x)] \leq [L_0 : \ell_0(x)] \leq [K : k(x)]. \tag{8.22}$$

On the other hand, since K and $\ell_0'(x)$ are linearly disjoint over $k(x)$, we obtain

$$[L_0 : \ell_0'(x)] = [K : k(x)]. \tag{8.23}$$

From (8.22) and (8.23), it follows that $\ell_0(x) = \ell_0'(x)$.

Since x is a transcendental element over ℓ_0', using Proposition 2.1.6 we deduce $\ell_0 = \ell_0'$.

(ii) $\Rightarrow$ (i) To prove that K and ℓ are linearly disjoint over k, it is enough to prove that any finitely generated subfield of ℓ over k is linearly disjoint from K over k (Remark 8.1.11).

Let ℓ_0 be a finitely generated subfield of ℓ. We have $\ell_0 \subseteq L = K\ell' = \bigcup_{\substack{\ell_0' \text{ finitely generated over } k \\ k \subseteq \ell_0' \subseteq \ell'}} K\ell_0$. Therefore $\ell_0 \subseteq K\ell_0'$ for some finitely generated extension ℓ_0' of k contained in ℓ'. Since the field of constants of $K\ell_0'$ is ℓ_0', we have $\ell_0 \subseteq \ell_0' \subseteq \ell'$.

Therefore it is enough to prove that any finitely generated subfield ℓ_0 of ℓ' over k is linearly disjoint from K over k.

Let $\ell_0 = k(\alpha_1, \ldots, \alpha_m)$ with $k_i = k(\alpha_1, \ldots, \alpha_i)$ and $K_i = Kk_i$ for $i = 1, \ldots, m$.

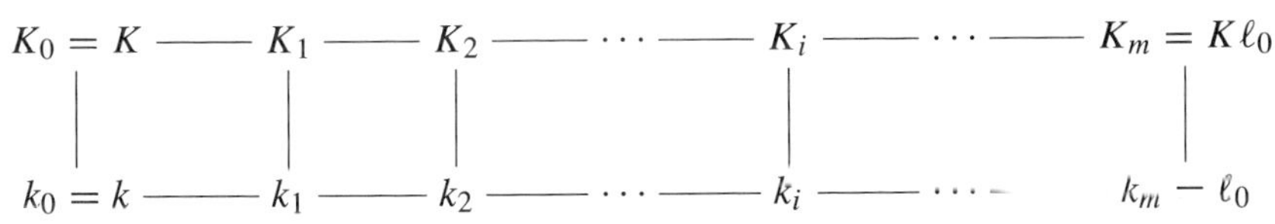

By Proposition 8.1.5, it suffices to show that k_i and K_{i-1} are linearly disjoint over k_{i-1} for $1 \leq i \leq m$.

By hypothesis the field of constants of each K_i is k_i, so that k_i is algebraically closed in K_i for $0 \leq i \leq m$. Since $k_i = k_{i-1}(\alpha_i)$, by Proposition 8.3.1 k_i and K_{i-1} are linearly disjoint over k_{i-1}. This proves (i).

Notice that the proof of (i) $\Rightarrow$ (ii) actually shows a stronger statement, namely that if K and ℓ' are linearly disjoint, then the field of constants of $K\ell_0$ is ℓ_0 for any $k \subseteq \ell_0 \subseteq \ell'$. This finishes the proof of the theorem. $\square$

Remark 8.4.5. The conclusion of Theorem 8.4.4 would not hold under the mere hypothesis that K and ℓ' are linearly disjoint over k.

Example 8.4.6. Let k_0, ℓ_0, u, v, k, and K be as in Example 5.2.31. Since k is algebraically closed in K and $\ell_0 = k(v^{1/p})$ with $v^{1/p}$ algebraic over k, it follows by Proposition 8.3.1 that K and ℓ_0 are linearly disjoint over k. However, the field of constants of K is $\ell = k(u^{1/p}, v^{1/p}) \supsetneqq \ell_0$.

Corollary 8.4.7. *If either K or ℓ' is separable over k, then the field of constant of $L = K\ell'$ is $\ell = \ell'$.*

Proof. By Theorem 8.4.4, we may assume that ℓ' is finitely generated over k. If ℓ' is purely transcendental, the field of constants of $K\ell'$ is $\ell' = \ell$ (Proposition 8.4.1). Therefore we may assume that ℓ'/k is a finite extension.

If ℓ'/k is separable, then $\ell' = k(\alpha)$, where α algebraic and separable over k. Since $\operatorname{Irr}(\alpha, T, K)$ divides $\operatorname{Irr}(\alpha, T, k)$, it follows that $L = K(\alpha)$ is a separable extension of K.

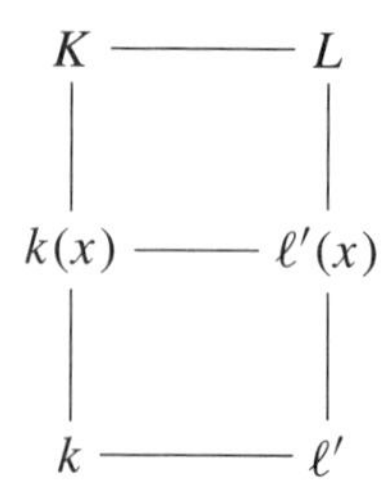

Now if $\beta \in \ell$, we have $\operatorname{Irr}(\beta, T, K) \in k[T]$. Hence β is separable and ℓ/ℓ' is separable. Since by Theorem 8.4.2 ℓ/ℓ' is a purely inseparable extension, it follows that $\ell = \ell'$. Next, assume that K/k is separable. Let $x \in K \setminus k$ be such that $K/k(x)$ is a finite separable extension. Then L is a finite separable extension of $\ell'(x)$, and hence $\ell(x)/\ell'(x)$ is a finite separable extension. Therefore ℓ/ℓ' is separable (Proposition 5.2.20). Again we obtain $\ell = \ell'$. $\square$

Corollary 8.4.8. *If either K or ℓ' is separably generated over k, then the field of constants of $K\ell'$ is $\ell = \ell'$.*

Proof. Since a separably generated extension is separable (Theorem 8.2.8), the result follows by Corollary 8.4.7. $\square$

Remark 8.4.9. If k is a perfect field (for example k algebraically closed, of characteristic 0, finite), then any function field K/k is separable (Corollary 8.2.11). Thus for any extension ℓ of k, the field of constants of the constant extension $L = K\ell$ is ℓ. Hence, Theorem 6.1.2 is a particular case of Corollary 8.4.7

Now we study the constant $\lambda_{L/K}$ introduced in Theorem 5.3.4, that is, if L/K is any function field extension, there exists $\lambda_{L/K} \in \mathbb{Q}^+$ such that $d_K(\mathfrak{A}) = \lambda_{L/K} d_L(\mathfrak{A})$ for any divisor $\mathfrak{A} \in D_K$.

Theorem 8.4.10. *Let $L = K\ell'$ be a constant field extension. If the characteristic of k is 0, then $\lambda_{L/K} = 1$, and if $\operatorname{char} k = p > 0$, then $\lambda_{L/K} = p^t$ for some $t \in \mathbb{N} \cup \{0\}$. Furthermore, if ℓ is the field of constants of L, $\lambda_{L/K} = 1$ if and only if K and ℓ are linearly disjoint over k.*

Proof. Let $x \in K \setminus k$ and $\mathfrak{A} = \mathfrak{Z}_x$. By Theorem 3.2.7 we have

$$d_K(\mathfrak{A}) = [K : k(x)] \quad \text{and} \quad d_L(\mathfrak{A}) = [L : \ell(x)]. \tag{8.24}$$

Hence $\lambda_{L/K} = 1 \iff d_K(\mathfrak{A}) = d_L(\mathfrak{A}) \iff [K : k(x)] = [L : \ell(x)]$.

Now $[K : k(x)] = [L : \ell(x)]$ if and only if K and $\ell(x)$ are linearly disjoint over $k(x)$. Since the field of constants of $\ell(x) = k(x)\ell$ is ℓ, it follows that $k(x)$ and ℓ are linearly disjoint over k (see the proof of Theorem 8.4.4). Therefore $\lambda_{L/K} = 1$ if and only if K and ℓ are linearly disjoint over k.

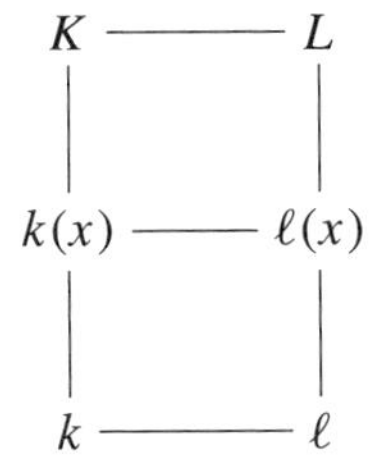

If $\operatorname{char} k = 0$, then K/k is separable, and by Corollary 8.4.7, K and ℓ are linearly disjoint and $\lambda_{L/K} = 1$. Let $\operatorname{char} k = p > 0$ and let K_0 be the separable closure of $k(x)$ in K. Set $L_0 = K_0\ell'$. Since K_0/k is separable, it follows that K_0 and ℓ' are linearly disjoint over k. Thus $[K_0 : k(x)] = [L_0 : \ell'(x)]$.

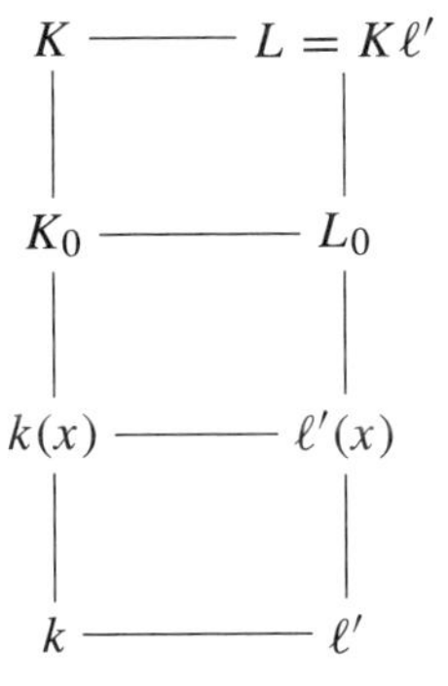

Also, K/K_0 is a purely inseparable extension, say of degree p^s with $s \geq 0$. Hence L/L_0 is a purely inseparable extension, say of degree p^{s_0} with $s_0 \leq s$. We have

$$\begin{aligned}\lambda_{L/K} &= \frac{[K : k(x)]}{[L : \ell(x)]} = \frac{[K : k(x)][\ell(x) : \ell'(x)]}{[L : \ell'(x)]} \\ &= \frac{[K : K_0][K_0 : k(x)]}{[L : L_0][L_0 : \ell'(x)]}[\ell : \ell'] = p^{s-s_0}[\ell : \ell']\end{aligned}$$

Since ℓ/ℓ' is a finite purely inseparable extension, then $\lambda_{L/K} = p^t$ for some $t \geq 0$. □

Assume that k is a finite field, K/k a function field, and $L = K\ell$ a constant extension. If $\mathfrak{P}$ is a place of L and $\mathfrak{p}$ its restriction to K, then the residue fields satisfy $k(\mathfrak{p})\ell = \ell(\mathfrak{P})$ (Theorem 6.1.4). We study this property for arbitrary constant extensions.

Theorem 8.4.11. *Let K/k be a function field and let $L = K\ell$ be an extension of constants. Let $\mathfrak{P}$ be a prime divisor of L lying over the prime divisor $\mathfrak{p}$ of K. If ℓ is a separably generated extension of k, then the residue fields satisfy*

$$\ell(\mathfrak{P}) = k(\mathfrak{p})\ell.$$

Proof. By Proposition 8.2.16 and Corollary 8.4.7, L is a separably generated extension of K. First we assume that ℓ is purely transcendental over k. Since $k(\mathfrak{p})$ is an algebraic extension of k, then $k(\mathfrak{p})$ and ℓ are linearly disjoint over k (Corollary 8.1.13). For any $y \in \vartheta_{\mathfrak{P}}$, put $\overline{y} = y \bmod \mathfrak{P} \in \ell(\mathfrak{P}) = \vartheta_{\mathfrak{P}}/\mathfrak{P}$.

Let $y \in \vartheta_{\mathfrak{P}} \subseteq L$, $y \neq 0$. Then y can be written in the form

$$y = \frac{\sum\limits_{i=1}^{n} a_i b_i}{\sum\limits_{j=1}^{m} a'_j b'_j} \quad \text{for some} \quad a_i, a'_j \in K \quad \text{and} \quad b_i, b'_j \in \ell, \tag{8.25}$$

where $\{b_i\}_{i=1}^n$ and $\{b'_j\}_{j=1}^m$ are chosen to be linearly independent over k.

Let $\alpha, \beta \in K$ be such that

$$v_{\mathfrak{p}}(\alpha) = -\min_{1\leq i\leq n} v_{\mathfrak{p}}(a_i) \quad \text{and} \quad v_{\mathfrak{p}}(\beta) = -\min_{1\leq j\leq m} v_{\mathfrak{p}}(a'_j). \tag{8.26}$$

We have $v_{\mathfrak{p}}(\alpha a_i) = v_{\mathfrak{p}}(\alpha) + v_{\mathfrak{p}}(a_i) \geq v_{\mathfrak{p}}(\alpha) + \min_{1\leq i\leq n} v_{\mathfrak{p}}(a_i) = 0$, so $\alpha a_i \in \vartheta_{\mathfrak{p}}$.

Similarly, $v_{\mathfrak{p}}(\beta a'_j) \geq 0$ for $1 \leq j \leq m$. Also, there exist indices i_0, j_0 such that $1 \leq i_0 \leq n$ and $1 \leq j_0 \leq m$, $v_{\mathfrak{p}}(\alpha a_{i_0}) = 0$, and $v_{\mathfrak{p}}(\beta a'_{j_0}) = 0$. Thus $\overline{\alpha a_{i_0}} \neq 0$ and $\overline{\beta a'_{j_0}} \neq 0$ in $k(\mathfrak{p})$.

It follows that $\sum_{i=1}^n \alpha a_i b_i \in \vartheta_{\mathfrak{P}}$ and $\sum_{j=1}^m \beta a'_j b'_j \in \vartheta_{\mathfrak{P}}$.

We also have $\overline{\sum_{i=1}^n \alpha a_i b_i} = \sum_{i=1}^n \overline{(\alpha a_i)} b_i \neq 0$ and $\overline{\sum_{j=1}^m \beta a'_j b'_j} = \sum_{j=1}^m \overline{(\beta a'_j)} b'_j \neq 0$ since $\{b_i\}_{i=1}^n$ and $\{b'_j\}_{j=1}^m$ are linearly independent over k, $\overline{\beta a'_{j_0}} \neq 0$, and ℓ and $k(\mathfrak{p})$ are linearly disjoint. In particular,

$$v_{\mathfrak{P}}\left(\sum_{i=1}^{n} \alpha a_i \beta_i\right) = 0 \quad \text{and} \quad v_{\mathfrak{P}}\left(\sum_{j=1}^{m} \beta a'_j b'_j\right) = 0.$$

Now

$$\begin{aligned}
v_{\mathfrak{P}}\left(\frac{\alpha}{\beta}y\right) &= v_{\mathfrak{P}}(\alpha) + v_{\mathfrak{P}}\left(\sum_{i=1}^{n} a_i b_i\right) - v_{\mathfrak{P}}\left(\sum_{j=1}^{m} \beta a'_j b'_j\right) \\
&= v_{\mathfrak{P}}(\alpha) + v_{\mathfrak{P}}\left(\sum_{i=1}^{n} a_i b_i\right) \geq v_{\mathfrak{P}}(\alpha) + \min_{1\leq i\leq n}\left\{v_{\mathfrak{P}}(a_i) + v_{\mathfrak{P}}(b_i)\right\} \\
&= v_{\mathfrak{P}}(\alpha) + \min_{1\leq i\leq n}\left\{v_{\mathfrak{P}}(a_i) + 0\right\} = 0.
\end{aligned}$$

Hence $\frac{\alpha}{\beta}y \in \vartheta_{\mathfrak{P}}$,

$$\overline{\left(\frac{\alpha}{\beta}y\right)} = \frac{\sum_{i=1}^{n} \overline{(\alpha a_i)} b_i}{\sum_{j=1}^{m} \overline{(\beta a'_j)} b'_j},$$

and $\sum_{i=1}^{n} \overline{(\alpha a_i)} b_i \neq 0$, $\sum_{j=1}^{m} \overline{(\beta a'_j)} b'_j \neq 0$. Therefore $\overline{\frac{\alpha}{\beta}y} \neq 0$ in $k(\mathfrak{p})\ell$.

Since $y \in \vartheta_{\mathfrak{P}}$, we have $\frac{\beta}{\alpha} \in \vartheta_{\mathfrak{p}}$ and $\overline{y} = \overline{\left(\frac{\beta}{\alpha}\right)\left(\frac{\alpha y}{\beta}\right)} \in k(\mathfrak{p})\ell$.

This shows that $\ell(\mathfrak{P}) \subseteq k(\mathfrak{p})\ell \subseteq \ell(\mathfrak{P})$ or $\ell(\mathfrak{P}) = k(\mathfrak{p})\ell$ when ℓ is a purely transcendental extension of k.

Now assume that ℓ is separably algebraic over k. Any element $\alpha \in \ell(\mathfrak{P}) = \vartheta_{\mathfrak{P}}/\mathfrak{P}$ is the image $\alpha = \overline{y}$ of an element $y \in K\ell'$, where ℓ' a finite extension of k. Therefore, if we prove the theorem for finite separable extensions, it will follow that $\alpha \in k(\mathfrak{p})\ell' \subseteq k(\mathfrak{p})\ell$ and thus $\ell(\mathfrak{P}) \subseteq k(\mathfrak{p})\ell$, so the theorem will be established for any algebraic separable extension of k.

Suppose that ℓ is a finite separable extension of k. Then ℓ is a simple extension of $k : \ell = k(\alpha)$ satisfying $[\ell : k] = n$. Let $\mathfrak{P} = \mathfrak{P}_{1,\ldots,}\mathfrak{P}_h$ be all prime divisors of L lying over $\mathfrak{p}$. Let L' be the Galois closure of L/K and let $\mathfrak{B}$ be a prime divisor of L' lying over $\mathfrak{P}$. For any $\sigma \in \operatorname{Gal}(L'/K)$, we have $\sigma\mathfrak{B}|_L = \mathfrak{P}_j$ for some j.

Pick $\bar{y} \in \ell(\mathfrak{P})$. By the approximation theorem (Theorem 2.5.3) there exists an element $\xi \in L$ such that

$$v_{\mathfrak{P}}(\xi - y) > 0 \quad \text{and} \quad v_{\mathfrak{P}_j}(\xi) \geq 0 \quad \text{for} \quad 2 \leq j \leq h.$$

In particular, $\overline{\xi} = \overline{y}$. By Theorems 5.3.4, 8.4.4, and 8.4.10 and Corollary 8.4.8, we have $\xi \in L = K\ell = Kk(\alpha) = K(\alpha)$ and $\lambda_{L/K} = \frac{[\ell:k]}{[L:K]} = 1$.

Thus

$$[K(\alpha) : K] = [L : K] = [\ell : k] = [k(\alpha) : k].$$

It follows that ξ can be written uniquely in the form

$$\xi = a_0 + a_1\alpha + \cdots + a_{n-1}\alpha^{n-1} \quad \text{with} \quad a_i \in K,\ i = 0, \ldots, n-1. \tag{8.27}$$

Taking a conjugate for ξ in (8.27), we have

$$\xi^{(i)} = a_0 + a_1\alpha^{(i)} + \cdots + a_{n-1}(\alpha^{(i)})^{n-1} \quad \text{for} \quad 1 \leq i \leq n. \tag{8.28}$$

Since α is separable of degree n over K, the Vandermonde determinant

$$\det \begin{bmatrix} 1 & \alpha^{(1)} & \cdots & (\alpha^{(1)})^{n-1} \\ \vdots & \vdots & & \vdots \\ 1 & \alpha^{(n)} & \cdots & (\alpha^{(n)})^{n-1} \end{bmatrix} = \prod_{i>j} \left(\alpha^{(i)} - \alpha^{(j)}\right)$$

is nonzero, so (8.28) has a unique solution $(a_0, \ldots, a_{n-1})$ in K^n, where for $t = 0, \ldots, n-1$,

$$a_t = \frac{\begin{vmatrix} 1 & \alpha^{(1)} & \cdots & \xi^{(1)} & \cdots & (\alpha^{(1)})^{n-1} \\ \vdots & \vdots & & \vdots & & \vdots \\ 1 & \alpha^{(n)} & \cdots & \xi^{(n)} & \cdots & (\alpha^{(n)})^{n-1} \end{vmatrix}}{\begin{vmatrix} 1 & \alpha^{(1)} & \cdots & (\alpha^{(1)})^{t-1} & \cdots & (\alpha^{(1)})^{n-1} \\ \vdots & \vdots & & \vdots & & \vdots \\ 1 & \alpha^{(n)} & \cdots & (\alpha^{(n)})^{t-1} & \cdots & (\alpha^{(n)})^{n-1} \end{vmatrix}} = \frac{c_t}{d} \quad \text{with} \quad d \in \ell \setminus \{0\}.$$

Now

$$v_{\mathfrak{B}}(\xi^{(i)}) = v_{\sigma^{-1}\mathfrak{B}}(\xi) = e_{L'/L}(\sigma^{-1}\mathfrak{B}|\mathfrak{P}_j)v_{\mathfrak{P}_j}(\xi) \geq 0, \tag{8.29}$$

where $\sigma \in \operatorname{Gal}(L'/L)$ is such that $\sigma\xi = \xi^{(i)}$ and $\mathfrak{P}_j = \sigma^{-1}\mathfrak{B}|_L$.

From (8.29) we obtain that

$$v_{\mathfrak{B}}(a_t) \geq 0, \quad v_{\mathfrak{P}}(a_t) \geq 0, \quad \text{and} \quad v_{\mathfrak{p}}(a_t) \geq 0.$$

Thus $a_k \in \vartheta_{\mathfrak{p}}$ and

$$\overline{y} = \overline{\xi} = \overline{a}_0 + \overline{a}_1\alpha + \cdots + \overline{a_{n-1}}\alpha^{n-1} \in k(\mathfrak{p})\ell.$$

It follows that $\ell(\mathfrak{P}) = k(\mathfrak{p})\ell$ when ℓ is separably algebraic over k.

The general case follows immediately since ℓ/k is separably generated. □

In the process of proving Theorem 8.4.11, we have obtained the following:

Proposition 8.4.12. *Let K/k be a function field and ℓ a purely transcendental extension of k. Let $L = K\ell$, $\mathfrak{P}$ a prime divisor of L, and $\mathfrak{p} = \mathfrak{P}|_K$. Let $\{b_1, \ldots, b_n\} \subseteq \ell$ be a system that is linearly independent over k. Then for $a_1, \ldots, a_n \in K$, we have*

$$v_{\mathfrak{P}}\left(\sum_{i=1}^{n} a_i b_i\right) = \min_{1 \leq i \leq n} v_{\mathfrak{p}}(a_i). \tag{8.30}$$

Proof. Let $a_1, \dots, a_n \in K$ with $a_i \neq 0$ for some index i and set $\alpha = -\min_{1\leq i\leq n} v_{\mathfrak{p}}(a_i)$. Then as in the proof of Theorem 8.4.11, we have $v_{\mathfrak{p}}(\alpha a_i) \geq 0$ and there exists an index i_0 such that $1 \leq i_0 \leq n$ and $v_{\mathfrak{p}}(\alpha a_{i_0}) = 0$ and $\overline{\alpha a_{i_0}} \neq 0$ in $k(\mathfrak{p})$.

It follows that $\overline{\sum_{i=1}^n \alpha a_i b_i} = \sum_{i=1}^n \overline{(\alpha a_i)} b_i$, and hence $v_{\mathfrak{P}}\left(\sum_{i=1}^n \alpha a_i b_i\right) \geq 0$. Now ℓ and $k(\mathfrak{p})$ are linearly disjoint over k, $\{b_1, \dots, b_n\} \subseteq \ell$ is linearly independent over k, and hence $\{b_1, \dots, b_n\}$ is linearly independent over $k(\mathfrak{p})$ and $\overline{\alpha a_{i_0}} \neq 0$, so $\sum_{i=1}^n \overline{(\alpha a_i)} b_i \neq 0$. Therefore

$$\begin{aligned} v_{\mathfrak{P}}(\alpha) + v_{\mathfrak{P}}\left(\sum_{i=1}^n a_i b_i\right) &= v_{\mathfrak{P}}\left(\sum_{i=1}^n \alpha a_i b_i\right) = 0 = \min_{1\leq i\leq n} v_{\mathfrak{P}}(\alpha a_i) \\ &= v_{\mathfrak{P}}(\alpha) + \min_{1\leq i\leq n} v_{\mathfrak{P}}(a_i). \qquad \square \end{aligned}$$

We also have the following result:

Proposition 8.4.13. *With the hypotheses of Proposition* 8.4.12, *for each prime divisor* $\mathfrak{p}$ *there exists a unique prime divisor* $\mathfrak{P}$ *in* L *lying over* $\mathfrak{p}$.

Proof. Since $\ell(\mathfrak{P})/\ell$ and $k(\mathfrak{p})/k$ are finite extensions, it follows that $\ell(\mathfrak{P})$ is a purely transcendental extension of $k(\mathfrak{p})$ and any transcendence base of ℓ over k is also a transcendence base of $\ell(\mathfrak{P})$ over $k(\mathfrak{p})$. Also, ℓ and $k(\mathfrak{p})$ are linearly disjoint and the structure of $\ell(\mathfrak{P})$ is uniquely determined; namely, for any transcendence basis $\{\alpha_i\}_{i\in I}$ of ℓ over k and basis $\{\beta_j\}_{j=1}^m$ of $k(\mathfrak{p})$ over k, we have $\ell(\mathfrak{P}) = k\left(\{\alpha_i\}_{i\in I}\right)\left(\{\beta_j\}_{j=1}^m\right)$.

$$\begin{array}{ccc} k(\mathfrak{p}) & \text{———} & k(\mathfrak{p})\ell = \ell(\mathfrak{P}) \\ | & & | \\ k & \text{———} & \ell \end{array}$$

Given any two prime divisors $\mathfrak{P}, \mathfrak{P}'$ of L lying over $\mathfrak{p}$ and using the notation of the proof of Theorem 8.4.11, we have $\overline{Y} = \overline{\left(\frac{\beta}{\alpha}\right)\left(\frac{\alpha Y}{\beta}\right)}$ for any $Y \in \vartheta_{\mathfrak{P}}$, where $\alpha, \beta \in K$, $\frac{\beta}{\alpha} \in \vartheta_{\mathfrak{p}} = K \cap \vartheta_{\mathfrak{P}} = K \cap \vartheta_{\mathfrak{P}'}$, and the definition of $\overline{Y}$ depends only on K, $\mathfrak{p}$, and ℓ and not on $\mathfrak{P}$ and $\mathfrak{P}'$. It follows that $\vartheta_{\mathfrak{P}} = \vartheta_{\mathfrak{P}'}$ and hence $\mathfrak{P} = \mathfrak{P}'$. $\square$

8.5 Genus Change in Constant Extensions

The genus of a geometric extension L/K has been studied in previous chapters, for example in Section 4.3. In Chapter 9 we will examine the general case of the genus of a separable extension L of K (Theorem 9.4.2).

In this section we consider the case of a constant extension $L = K\ell'$ of K, where $\ell \supseteq \ell'$ is the field of constants of L.

Proposition 8.5.1. *If* $\lambda_{L/K} = 1$, *that is,* K *and* ℓ *are linearly disjoint over* k, *then* $g_L \leq g_K$. *For any divisor* $\mathfrak{q} \in D_K$, *any basis of* $L_K(\mathfrak{q})$ *is a subset of a basis of* $L_L(\mathfrak{q})$. *In particular,* $\ell_K(\mathfrak{q}) \leq \ell_L(\mathfrak{q})$.

Proof. We have $L_K(\mathfrak{q}) \subseteq L_L(\mathfrak{q})$. If $\alpha_1, \dots, \alpha_n \in L_K(\mathfrak{q})$ are linearly independent (over k), then $\alpha_1, \dots, \alpha_n$ are linearly independent over ℓ since K and ℓ are linearly disjoint. Hence $\ell_K(\mathfrak{q}) \leq \ell_L(\mathfrak{q})$.

Now choose $\mathfrak{q} \in D_K$ such that $d_K(\mathfrak{q}) > 2g_K - 2$ and $d_L(\mathfrak{q}) > 2g_L - 2$. By Corollary 3.5.6 we have

$$\ell_K(\mathfrak{q}^{-1}) = d_K(\mathfrak{q}) - g_K + 1$$

and

$$\ell_L(\mathfrak{q}^{-1}) = d_L(\mathfrak{q}) - g_L + 1 \tag{8.31}$$

Since $\lambda_{L/K} = 1$, it follows that $d_K(\mathfrak{q}) = d_L(\mathfrak{q})$. Also, $\ell_K(\mathfrak{q}^{-1}) \leq \ell_L(\mathfrak{q}^{-1})$. From (8.31) we obtain

$$-g_K + 1 \leq -g_L + 1. \qquad \square$$

Theorem 8.5.2. *If ℓ' is separably generated over k, then $g_L = g_K$ and any basis of $L_K(\mathfrak{q})$ is also a basis of $L_L(\mathfrak{q})$ for any $\mathfrak{q} \in D_K$. Hence $\ell_K(\mathfrak{q}) = \ell_L(\mathfrak{q})$.*

Proof. Suppose the result has been proved for $\ell' = k(y)$ with y transcendental and for $\ell' = k(\alpha)$, where α is a separable algebraic element. For ℓ' separably generated over k, let $z \in L_L(\mathfrak{q})$. Then z belongs to a field $L_0 = K\ell_0$, where ℓ_0 is a finitely separably generated extension of k, so $z \in L_{L_0}(\mathfrak{q})$. By induction on the transcendence degree of ℓ_0 over k, and using the finite separable case, we obtain $L_{L_0}(\mathfrak{q}) = L_K(\mathfrak{q})\ell_0$. It follows that $L_L(\mathfrak{q}) = L_K(\mathfrak{q})\ell$ and $\ell_L(\mathfrak{q}) = \ell_K(\mathfrak{q})$. The proof of the equality $g_K = g_L$ proceeds along the same lines as that of Proposition 8.5.1.

Therefore we assume first that $\ell' = k(y)$ with y transcendental over k. Let $\xi \in L_L(\mathfrak{q})$. Then ξ can be written uniquely as

$$\xi = \frac{f(y)}{g(y)} = \frac{\sum_{i=0}^{n} a_i y^i}{\sum_{j=0}^{m} b_j y^j} \tag{8.32}$$

with $f(y), g(y) \in k[y]$, $(f, g) = 1$, and $b_m = 1$.

Let $\mathfrak{P}$ be a prime divisor of L lying over an arbitrary prime divisor of K. Using Proposition 8.4.12 we obtain

$$v_{\mathfrak{P}}(g(y)) = v_{\mathfrak{P}}\left(\sum_{j=0}^{m} b_j y^j\right) = \min_{0 \leq j \leq m}\left\{v_{\mathfrak{P}}(b_j)\right\} = \min_{0 \leq j \leq m-1}\left\{0, v_{\mathfrak{P}}(b_j)\right\} \leq 0. \tag{8.33}$$

Statement (8.33) implies that $v_{\mathfrak{P}}(\mathfrak{Z}_{(g(y))}) = 0$ for any place of L that is not variable over K; thus the only possible prime divisors that occur in the zero divisor of $g(y)$ are those that are variable over K.

Now $\xi \in L_L(\mathfrak{q})$, so

$$(\xi)_L \mathfrak{q}^{-1} = \frac{\mathfrak{Z}_{(f(y))}\mathfrak{N}_{(g(y))}}{\mathfrak{N}_{(f(y))}\mathfrak{Z}_{(g(y))}\mathfrak{q}}$$

is an integral divisor in L. Since $\left(\mathfrak{Z}_{(g(y))}, \mathfrak{N}_{(g(y))}\right) = 1$, any prime divisor dividing $\mathfrak{Z}_{(g(y))}$ must divide $\mathfrak{Z}_{(f(y))}$. Moreover, f and g are relatively prime, so

$$\alpha(y)f(y) + \beta(y)g(y) = 1 \tag{8.34}$$

for some $\alpha(y), \beta(y) \in k[y]$.

If $\mathfrak{Q}$ is any prime divisor of L that is variable over K, then if $\alpha(y) = \sum_{\ell=0}^{s} c_\ell y^\ell$, then $v_{\mathfrak{Q}}(c_\ell) = 0$ for $c_\ell \neq 0$ and $v_{\mathfrak{Q}}(y) = 0$. Therefore

$$v_{\mathfrak{Q}}(\alpha(y)) \geq \min_{0 \leq \ell \leq s} v_{\mathfrak{Q}}(c_\ell y^\ell) = \min_{0 \leq \ell \leq s} \left(v_{\mathfrak{Q}}(c_\ell) + \ell v_{\mathfrak{Q}}(y) \right) = 0. \tag{8.35}$$

Similarly,

$$v_{\mathfrak{Q}}(\beta(y)) \geq 0. \tag{8.36}$$

From (8.34), (8.35), and (8.36) we obtain

$$\begin{aligned} 0 = v_{\mathfrak{Q}}(1) &\geq \min\{v_{\mathfrak{Q}}(\alpha(y)) + v_{\mathfrak{Q}}(f(y)), v_{\mathfrak{Q}}(\beta(y)) + v_{\mathfrak{Q}}(g(y))\} \\ &\geq \min\{v_{\mathfrak{Q}}(f(y)), v_{\mathfrak{Q}}(g(y))\}. \end{aligned}$$

It follows that $\mathfrak{Z}_{(f(y))}$ and $\mathfrak{Z}_{(g(y))}$ cannot have a common prime divisor. Thus $\mathfrak{Z}_{(g(y))} = \mathfrak{N}$ and $g(y) = 1$.

Using (8.32) and Proposition 8.4.12 we obtain that for any prime divisor $\mathfrak{p}$ of K,

$$v_{\mathfrak{p}}(\xi) = v_{\mathfrak{p}}(f(y) = \min_{0 \leq i \leq n} \{v_{\mathfrak{p}}(a_i)\} \geq v_{\mathfrak{p}}(\mathfrak{q}). \tag{8.37}$$

Thus $a_i \in L_K(\mathfrak{q})$. It follows that $L_L(\mathfrak{q})$ is the vector space generated over $\ell = \ell'$ by $L_L(\mathfrak{q})$ or, equivalently, $L_K(\mathfrak{q})\ell = L_L(\mathfrak{q})$.

Since K and ℓ are linearly disjoint, we get $\ell_K(\mathfrak{q}) = \ell_L(\mathfrak{q})$. This proves the theorem in the case $\ell' = k(y)$, where y is a transcendental element over k.

Now we consider the case $\ell' = k(\alpha)$ where α is a finite separable element over k. Let $\xi \in L_L(\mathfrak{q})$. Then ξ can be written uniquely in the form

$$\xi = c_0 + c_1\alpha + \cdots + c_{n-1}\alpha^{n-1} \quad \text{where} \quad c_i \in K \ (i = 0, \ldots, n-1) \tag{8.38}$$

and $n = \deg \operatorname{Irr}(\alpha, x, k) = \deg \operatorname{Irr}(\alpha, x, K)$.

Let L_1 be the Galois closure of L/K. By changing each side in (8.38) into its conjugate, we obtain

$$\xi^{(i)} = c_0 + c_1\alpha^{(i)} + \cdots + c_{n-1}(\alpha^{(i)})^{n-1} \quad \text{for} \quad 1 \leq i \leq n. \tag{8.39}$$

Since α is a separable element, we have

$$\Delta = \det \begin{bmatrix} 1 & \alpha^{(1)} & \cdots & (\alpha^{(1)})^{n-1} \\ \vdots & \vdots & & \\ 1 & \alpha^{(n)} & \cdots & (\alpha^{(n)})^{n-1} \end{bmatrix} = \prod_{i>j} (\alpha^{(i)} - \alpha^{(j)}) \neq 0,$$

where $\Delta \in \ell$.

Therefore there exists a unique solution to the system of linear equations (8.39), namely

$$a_t = \frac{\begin{vmatrix} 1 & \alpha^{(1)} & \cdots & (\alpha^{(1)})^{t-1} & \xi^{(1)} & (\alpha^{(1)})^{t+1} & \cdots & (\alpha^{(1)})^{n-1} \\ \vdots & \vdots & & \vdots & \vdots & \vdots & & \vdots \\ 1 & \alpha^{(n)} & \cdots & (\alpha^{(n)})^{t-1} & \xi^{(n)} & (\alpha^{(n)})^{t+1} & \cdots & (\alpha^{(n)})^{n-1} \end{vmatrix}}{\Delta}$$
$$= \frac{b_t}{\Delta} \quad \text{for} \quad 0 \le t \le n-1. \tag{8.40}$$

Each b_t is a linear combination of $\xi^{(i)}$ with coefficients in $\ell' = \ell$. Since $\mathfrak{q} \in D_K$ and $\xi \in L_{L_1}(\mathfrak{q})$, it follows that $\xi^{(i)} = \xi^{\sigma} \in L_{L_1}(\mathfrak{q}^{\sigma}) = L_{L_1}(\mathfrak{q})$ for some $\sigma : L \to L_1$ whose restriction to K is the identity. Thus $a_t \in L_{L_1}(\mathfrak{q}) \cap K$ and $a_t \in L_K(\mathfrak{q})$.

Therefore $L_L(\mathfrak{q}) = L_K(\mathfrak{q})\ell$ and the equality $\ell_L(\mathfrak{q}) = \ell_K(\mathfrak{q})$ follows from the linear disjointness of K and ℓ over k. □

In Proposition 8.5.1 we obtained $g_L \le g_K$ when $\lambda_{L/K} = 1$. This inequality is true for any constant extension. Actually, the following general result holds:

Theorem 8.5.3. *For any constant field extension of function fields L of K, we have*

$$\lambda_{L/K} g_L \le g_K.$$

In particular, $g_L \le g_K$.

Proof. Let $L = K\ell'$ and let $\mathcal{A}$ be a transcendence base of ℓ'/k. Set $\ell_0 = k(\mathcal{A})$ and $L_0 = K\ell_0$. By Corollary 8.4.8 and Theorem 8.4.10, we have proved $\lambda_{L_0/K} = 1$. Hence $g_{L_0} \le g_K$ (Proposition 8.5.1). Since $\lambda_{L/K} = \lambda_{L/L_0}\lambda_{L_0/K}$, if we prove $\lambda_{L/L_0} g_L \le g_{L_0}$, it will follow that

$$\lambda_{L/K} g_L = \lambda_{L/L_0} g_L \le g_{L_0} = \lambda_{L_0/K} g_{L_0} \le g_K.$$

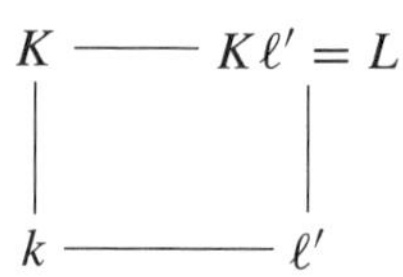

Therefore we may assume that ℓ'/k is an algebraic extension. First consider the case that ℓ' is a finite extension of k. Then $[L : K] = m \le n = [\ell' : k]$. We can take a subset $\{\alpha_1, \ldots, \alpha_m\}$ of a basis $\{\alpha_1, \ldots, \alpha_n\}$ of ℓ' over k that is a basis of L over K. Let $\mathfrak{X}'_K$ be the vector subspace over k of the repartitions of K such that if $\xi \in \mathfrak{X}'_K$ and $\mathfrak{p} \in \mathbb{P}_K$, then $\xi(\mathfrak{p}) \in K \subseteq K_{\mathfrak{p}}$. Similarly, define $\mathfrak{X}'_L$ over ℓ.

Let $\varphi \colon \prod_{i=1}^m \mathfrak{X}'_K \to \mathfrak{X}'_L$ be defined by

$$\varphi(\xi_1, \ldots, \xi_m) = \theta$$

where for any place $\mathfrak{P}$ of L,

$$\theta(\mathfrak{P}) = \sum_{i=1}^{m} \xi_i(\mathfrak{p})\alpha_i \in L \subseteq L_{\mathfrak{P}}$$

and $\mathfrak{P}|_K = \mathfrak{p}$ is the prime divisor in K.

Then θ belongs to $\mathfrak{X}_L$ because $\xi_i(\mathfrak{p}) \in \vartheta_{\mathfrak{p}}$ for almost all $\mathfrak{p} \in \mathbb{P}_K$. Furthermore, since $\{\alpha_1, \ldots, \alpha_m\}$ is a basis of L/K it follows that φ is a k-monomorphism.

Let $\mathfrak{X}_L^0 = \varphi\left(\prod_{i=1}^m \mathfrak{X}_K'\right) \subseteq \mathfrak{X}_L'$. Then $\mathfrak{X}_L^0$ is a vector subspace of $\mathfrak{X}_L'$ (over k).

If $X_1, \ldots, X_m \in K$ and the $\xi_{X_i} = (X_i)_{\mathfrak{p}\in\mathbb{P}_K}$ are the principal repartitions, then $\varphi(\xi_{X_1}, \ldots, \xi_{X_m}) = \xi_y$ whenever $y = \sum_{i=1}^m \alpha_i X_i \in L$.

It follows that $L \subseteq \mathfrak{X}_L^0$. Let $\mathfrak{q}$ be any divisor of K. Then if $\mathfrak{X}_L'(\mathfrak{q}) := \mathfrak{X}_L(\mathfrak{q}) \cap \mathfrak{X}_L'$, we have

$$\mathfrak{X}_L^0 + \mathfrak{X}_L'(\mathfrak{q}) \subseteq \mathfrak{X}_L'. \tag{8.41}$$

Let $\theta \in \mathfrak{X}_L'$ and $\mathfrak{p} \in \mathbb{P}_K$. Let $\mathfrak{P}_1, \ldots, \mathfrak{P}_h$ be the prime divisors of L lying over $\mathfrak{p}$. By the approximation theorem (Theorem 2.5.3), there exists $y_{\mathfrak{p}} \in L$ such that

$$v_{\mathfrak{P}_i}\left(y_{\mathfrak{p}} - \theta(\mathfrak{P}_i)\right) \geq v_{\mathfrak{P}_i}(\mathfrak{q}) \quad \text{for} \quad 1 \leq i \leq h. \tag{8.42}$$

Let $\delta \in \mathfrak{X}_L'$ be defined by

$$\delta(\mathfrak{P}) = \begin{cases} y_{\mathfrak{p}} & \text{if } \mathfrak{P} \mid \mathfrak{p} \text{ and } v_{\mathfrak{P}}(\mathfrak{q}) \neq 0, \\ y_{\mathfrak{p}} & \text{if } \mathfrak{P} \mid \mathfrak{p} \text{ and } v_{\mathfrak{P}'}(\theta(\mathfrak{P}')) < 0 \text{ for some } \mathfrak{P}' \mid \mathfrak{p}, \\ 0 & \text{otherwise.} \end{cases} \tag{8.43}$$

Let $\mathfrak{P} \in \mathbb{P}_L$ and $\mathfrak{p} = \mathfrak{P}|_K$. If $v_{\mathfrak{P}}(\mathfrak{q}) \neq 0$ or $v_{\mathfrak{P}'}(\vartheta(\mathfrak{P}')) < 0$ for some $\mathfrak{P}'$ dividing $\mathfrak{p}$, then $\delta(\mathfrak{P}) = y_{\mathfrak{p}}$, so

$$v_{\mathfrak{P}}(\theta - \delta)(\mathfrak{P}) = v_{\mathfrak{P}}(\theta(\mathfrak{P}) - y_{\mathfrak{p}}) \geq v_{\mathfrak{P}}(\mathfrak{q}).$$

Now if $v_{\mathfrak{P}}(\mathfrak{q}) = 0$ and $v_{\mathfrak{P}'}(\theta(\mathfrak{P}')) \geq 0$ for every $\mathfrak{P}' \mid \mathfrak{p}$, then

$$v_{\mathfrak{P}}((\theta - \delta)(\mathfrak{P})) = v_{\mathfrak{P}}(\theta(\mathfrak{P})) \geq 0 = v_{\mathfrak{P}}(\mathfrak{q}).$$

It follows that $\theta - \delta \in \mathfrak{X}_L'(\mathfrak{q})$.

For any $\mathfrak{p} \in \mathbb{P}_K$, let $y_{\mathfrak{p}} \in L$ be defined as in (8.42).

Let $y_{\mathfrak{p}} = \sum_{i=1}^m \alpha_i X_{i\mathfrak{p}}$, $X_{i\mathfrak{p}} \in K$ and $\delta_i' \in \mathfrak{X}_K'$ be given by

$$\delta_i'(\mathfrak{p}) = \begin{cases} X_{i\mathfrak{p}} & \text{if } v_{\mathfrak{p}}(\mathfrak{q}) \neq 0 \text{ or } v_{\mathfrak{P}}(\mathfrak{q}) < 0 \text{ for some } \mathfrak{P} \mid \mathfrak{p}, \\ 0 & \text{otherwise.} \end{cases}$$

Then $\varphi(\delta_1', \ldots, \delta_m')(\mathfrak{P}) = \sum_{i=1}^m \alpha_i \delta_i'(\mathfrak{p}) = \delta(\mathfrak{P})$. Thus $\delta \in \mathfrak{X}_L^0$, and $\theta = (\theta-\delta)+\delta \in \mathfrak{X}_L'(\mathfrak{q}) + \mathfrak{X}_L^0$. It follows that

$$\mathfrak{X}'_L \subseteq \mathfrak{X}^0_L + \mathfrak{X}'_L(\mathfrak{q}). \tag{8.44}$$

Using (8.41) and (8.44) we obtain

$$\mathfrak{X}'_L = \mathfrak{X}^0_L + \mathfrak{X}'_L(\mathfrak{q}). \tag{8.45}$$

By Exercise 3.6.16 and Corollary 3.4.6 we have

$$\dim_k \frac{\mathfrak{X}'_K}{\mathfrak{X}'_K(\mathfrak{N}) + K} = g_K \tag{8.46}$$

and

$$\dim_\ell \frac{\mathfrak{X}'_L}{\mathfrak{X}'_L(\mathfrak{N}) + K} = g_L. \tag{8.47}$$

Using (8.46) we obtain

$$mg_K = \dim_k \prod_{i=1}^{m} \left(\frac{\mathfrak{X}'_K}{\mathfrak{X}'_K(\mathfrak{N}) + K} \right) = \dim_k \frac{\prod_{i=1}^m \mathfrak{X}'_K}{\prod_{i=1}^m (\mathfrak{X}'_K(\mathfrak{N}) + K)}.$$

Applying the k-monomorphism θ, we get

$$mg_K = \dim_k \frac{\mathfrak{X}^0_L}{\mathfrak{X}^0_L(\mathfrak{N}) + L},$$

where $\mathfrak{X}^0_L(\mathfrak{q}) := \varphi\left(\prod_{i=1}^m \mathfrak{X}'_K(\mathfrak{q})\right) \subseteq \mathfrak{X}'_L$ for any $\mathfrak{q} \in D_K$.

On the other hand, by (8.45),

$$\begin{aligned} ng_L &= n \dim_\ell \frac{\mathfrak{X}'_L}{\mathfrak{X}'_L(\mathfrak{N}) + L} = \dim_k \frac{\mathfrak{X}'_L}{\mathfrak{X}'_L(\mathfrak{N}) + L} \\ &= \dim_k \frac{\mathfrak{X}^0_L + \mathfrak{X}'_L(\mathfrak{N}) + L}{\mathfrak{X}'_L(\mathfrak{N}) + L} = \dim_k \frac{\mathfrak{X}^0_L}{\mathfrak{X}^0_L \cap (\mathfrak{X}'_L(\mathfrak{N}) + L)}. \end{aligned}$$

Now $\mathfrak{X}^0_L(\mathfrak{N}) + L \subseteq \mathfrak{X}^0_L \cap (\mathfrak{X}'_K(\mathfrak{N}) + L)$, so

$$\begin{aligned} ng_L &= \dim_k \frac{\mathfrak{X}^0_L}{\mathfrak{X}^0_L \cap (\mathfrak{X}'_L(\mathfrak{N}) + L)} \\ &= \dim_k \frac{\mathfrak{X}^0_L}{\mathfrak{X}^0_L(\mathfrak{N}) + L} - \dim_k \frac{\mathfrak{X}^0_L \cap (\mathfrak{X}'_L(\mathfrak{N}) + L)}{\mathfrak{X}^0_L(\mathfrak{N}) + L} \\ &= mg_K - \dim_k \frac{\mathfrak{X}^0_L \cap (\mathfrak{X}'_L(\mathfrak{N}) + L)}{\mathfrak{X}^0_L(\mathfrak{N}) + L}. \end{aligned}$$

Therefore

$$ng_L \le mg_K. \tag{8.48}$$

By Theorem 5.3.4, we obtain

$$\lambda_{L/K} = \frac{[\ell : k]}{[L : K]} = \frac{n}{m}. \tag{8.49}$$

Therefore it follows from (8.48) and (8.49) that

$$\lambda_{L/K} g_L = \frac{n}{m} g_L \le g_K.$$

Next, consider ℓ' to be an arbitrary algebraic extension of k. Let $x \in K \setminus k$ and set

$$r := d_K(\mathfrak{N}_x) = [K : k(x)] \qquad \text{and} \qquad s := d_L(\mathfrak{N}_x) = [L : \ell(x)].$$

Any basis $\{\alpha_1, \dots, \alpha_r\}$ of K over $k(x)$ spans L over $\ell(x)$. Thus we obtain $r - s$ relations

$$\sum_{i=1}^{r} \alpha_i c_{ij} = 0 \quad (j = 1, 2, \dots, r - s),$$

with coefficients $c_{ij} \in \ell(x)$ and such that the $r - s$ vectors $(c_{1j}, \dots, c_{nj})$ are linearly independent over $\ell(x)$.

Notice that $c_{ij} \in \ell(x)$, so the coefficients of c_{ij} belong to a finitely generated (and thus finite) extension ℓ_0' of k, with $\ell_0' \subseteq \ell$. Clearly, $L_0 = L\ell_0'$ is spanned by $\{\alpha_1, \dots, \alpha_r\}$ over $\ell_0'(x)$ and $c_{ij} \in \ell_0'(x)$. Therefore if ℓ_0 is the field of constants of L_0, we obtain

$$d_{L_0}(\mathfrak{N}_x) = [L_0 : \ell_0(x)] \le [L_0 : \ell_0'(x)] \le s = d_L(\mathfrak{N}_x).$$

It follows that

$$1 \le \lambda_{L/L_0} = \frac{d_{L_0}(\mathfrak{N}_x)}{d_L(\mathfrak{N}_x)} \le 1,$$

and hence $\lambda_{L/L_0} = 1$. Using the case of a finite extension and Proposition 8.5.1, we deduce that

$$\lambda_{L/K} g_L = \lambda_{L_0/K} \lambda_{L/L_0} g_L \le \lambda_{L_0/K} g_{L_0} \le g_K.$$

□

Corollary 8.5.4. *With the hypotheses of Theorem* 8.5.3, *if* $\lambda_{L/K} > 2$, *then*

$$\lambda_{L/K} g_L < g_K.$$

Proof: Suppose $\lambda_{L/K} g_L = g_K$. Let ω be a nonzero differential of K.

We have

$$d_L((\omega)) = \frac{d_K((\omega))}{\lambda_{L/K}} = \frac{2g_K - 2}{\lambda_{L/K}}.$$

Thus $\lambda_{L/K} \mid 2g_K - 2$. Now $\lambda_{L/K} g_L = g_K$ implies that $\lambda_{L/K}$ divides g_K and therefore $\lambda_{L/K}$ divides 2. □

Remark 8.5.5. If $\lambda_{L/K} = 2$, it is possible to have

$$\lambda_{L/K} g_L = 2g_L = g_K.$$

Example 8.5.6. Let k be a field of characteristic 2 and let α_0, α_1 be elements of k satisfying $[k(\alpha_0^{1/2}, \alpha_1^{1/2}) : k] = 4$ (see Example 5.2.31). Let x be a transcendental element over k and let y be such that

$$y^2 = \alpha_0 + \alpha_1 x^2. \tag{8.50}$$

By Example 5.2.31 (with $p = 2$), if $K = k(x, y)$, then

$$[K : k(x)] = 2 = d(\mathfrak{N}_x) \tag{8.51}$$

and the field of constants of K is k.

If $\mathfrak{P}$ is any place of K such that $v_{\mathfrak{P}}(y) < 0$, we have

$$\begin{aligned} 2v_{\mathfrak{P}}(y) &= v_{\mathfrak{P}}(y^2) = v_{\mathfrak{P}}(\alpha_0 + \alpha_1 x^2) \\ &= \min\left\{ v_{\mathfrak{P}}(\alpha_0), v_{\mathfrak{P}}(\alpha_1) + 2v_{\mathfrak{P}}(x) \right\} = 2v_{\mathfrak{P}}(x). \end{aligned}$$

Similarly, if $v_{\mathfrak{P}}(x) < 0$ then $v_{\mathfrak{P}}(x) = v_{\mathfrak{P}}(y)$. It follows that $\mathfrak{N}_y = \mathfrak{N}_x$. Thus $1, x, x^2, \ldots, x^n, y, yx, \ldots, yx^{n-1} \in L(\mathfrak{N}_x^{-n})$ and these elements are linearly independent. In particular,

$$\ell(\mathfrak{N}_x^{-n}) \geq 2n + 1. \tag{8.52}$$

Using the Riemann–Roch theorem (Corollary 3.5.6), (8.51), and (8.52), we obtain for n large enough

$$2n + 1 \leq \ell(\mathfrak{N}_x^{-n}) = d(\mathfrak{N}_x^n) - g_K + 1 = 2n - g_K + 1.$$

Hence $g_K = 0$.

Now set $\ell' = k(\alpha_0^{1/2})$. We have $[\ell_0'(\alpha_1^{1/2}) : \ell_0'] = 2$. Put $L = K\ell'$. Since $\lambda_{L/K} g_L \leq g_K = 0$, it follows that $g_L = 0$.

By Exercise 5.10.17, the constant field of L is $\ell = k(\alpha_0^{1/2}, \alpha_1^{1/2})$ and $L = \ell'(x, y) = k(\alpha_0^{1/2}, \alpha_1^{1/2})(x, y)$.

Now $y^2 = \alpha_0 + \alpha_1 x^2$, so $y = \alpha_0^{1/2} + \alpha_1^{1/2} x \in k(\alpha_0^{1/2}, \alpha_1^{1/2})(x)$. Consequently $L = k(\alpha_0^{1/2}, \alpha_1^{1/2})(x) = \ell(x)$ and $d_L(\mathfrak{N}_x) = 1$. Therefore

$$\lambda_{L/K} = \frac{d_K(\mathfrak{N}_x)}{d_L(\mathfrak{N}_x)} = \frac{2}{1} = 2.$$

An interesting remark is that this example covers the general case:

Proposition 8.5.7. *Let $L = K\ell'$ be a constant extension such that $g_L = g_K$ and $\lambda_{L/K} > 1$. Then $g_L = g_K = 0$, $\lambda_{L/K} = 2$, $K = k(x, y)$ with $y^2 = \alpha + \beta x^2$, $\alpha, \beta \in k$ such that $[k(\alpha^{1/2}, \beta^{1/2}) : k] = 4$, and $[\ell'(\alpha^{1/2}, \beta^{1/2}) : \ell'] < 4$.*

Proof. If $g_L \neq 0$, then $g_L < \lambda_{L/K} g_L \leq g_K = g_L$. Therefore $g_K = g_L = 0$. By Corollary 8.5.4, we obtain $\lambda_{L/K} = 2$.

Let W be the canonical class of K. By Corollaries 3.5.5 and 3.5.6, we have

$$d_K(W^{-1}) = 2 \qquad \text{and} \qquad N_K(W^{-1}) = 3.$$

Let $\mathfrak{q}$ be a integral divisor in W^{-1} with $d_K(\mathfrak{q}) = 2$ and $\ell_K(\mathfrak{q}^{-1}) = 3$. Let $\{1, x, y\}$ be a basis of $L_K(\mathfrak{q}^{-1})$. Now $x \notin k$ and $x \in L_K(\mathfrak{q}^{-1})$, so $\mathfrak{q}^{-1}$ divides (x), $\mathfrak{N}_x$ divides $\mathfrak{q}$ and $d_L(\mathfrak{q}) = 2$. It follows that $d_K(\mathfrak{N}_x)$ is 1 or 2. If $d_K(\mathfrak{N}_x) = 1$, then $K = k(x)$ (Theorem 3.2.7). Thus $L = K\ell' = \ell'(x)$ and $d_L(\mathfrak{N}_x) = 1$. This is impossible because $\lambda_{L/K} = \frac{d_K(\mathfrak{N}_x)}{d_L(\mathfrak{N}_x)} > 1$.

Therefore we have $d_K(\mathfrak{N}_x) = 2$, $\mathfrak{N}_x = \mathfrak{q}$, and

$$[K : k(x)] = d_K(\mathfrak{N}_x) = 2. \tag{8.53}$$

Now consider y. If $y \in k(x)$, we have $y = \frac{f(x)}{g(x)}$ with $f(x), g(x) \in k[x]$ and $(f, g) = 1$. It follows that

$$(y)_K = \frac{\mathfrak{Z}_{(f)}}{\mathfrak{Z}_{(g)}} \mathfrak{N}_x^{\deg g - \deg f}.$$

Since $y \in L_K(\mathfrak{q}^{-1}) = L_K(\mathfrak{N}_x^{-1})$, $(y)_K \mathfrak{N}_x$ is an integral divisor and $(y)_K = \frac{\mathfrak{B}}{\mathfrak{N}_x}$, where $\mathfrak{B}$ is an integral divisor. Therefore $\mathfrak{Z}_{(g)} = \mathfrak{N}$, $g(x)$ is constant, and $\deg f(x) = 1$.

This is a contradiction to the fact that $1, x$, and y are linearly independent over k. Therefore $y \notin k(x)$ and by (8.53) it follows that $K = k(x, y)$.

Now since $\lambda_{L/K} \neq 1$, using Theorem 8.4.10 and Corollary 8.4.7 we deduce that y is purely inseparable over $k(x)$. Thus

$$y^2 = \frac{h(x)}{m(x)}$$

with $h(x), m(x) \in k[x]$, and $(h(x), m(x)) = 1$. Therefore

$$(y^2)_K = (y)_K^2 = \frac{\mathfrak{Z}_{(h)}}{\mathfrak{Z}_{(m)}} \mathfrak{N}_x^{\deg m - \deg h}.$$

Since $(y)_K \mathfrak{N}_x$ is integral and $(y)_K^2 = \frac{\mathfrak{B}^2}{\mathfrak{N}_x^2}$, it follows that $\mathfrak{Z}_{(m)} = \mathfrak{N}$, $m(x)$ is constant, and $\deg h(x) = 2$ and $\mathfrak{Z}_{(h)} = \mathfrak{B}^2$. Thus

$$h(x) = \alpha + \beta x^2 = y^2.$$

Now $[k(\alpha^{1/2}, \beta^{1/2}) : k]$ divides 4, so $[k(\alpha^{1/2}, \beta^{1/2}) : k]$ is 1, 2, or 4. Assume $[k(\alpha^{1/2}, \beta^{1/2}) : k] \neq 4$. Then $1, \alpha^{1/2}, \beta^{1/2}$ cannot be linearly independent over k and there exist $a, b, c \in k$, not all zero, such that

$$a\alpha^{1/2} + b\beta^{1/2} = c.$$

Say $a \neq 0$. Then $\alpha^{1/2} = \frac{c - b\beta^{1/2}}{a}$. Since $y = \alpha^{1/2} + \beta^{1/2}x \in K$, it follows that

$$y = \frac{c}{a} - \beta^{1/2}\left(\frac{b}{a} + x\right), \quad \beta^{1/2} = \frac{\frac{c}{a} - y}{\frac{b}{a} + x} \in K, \quad \text{and} \quad \alpha^{1/2} = \frac{c}{a} - \frac{b}{a}\beta^{1/2} \in K.$$

Thus $\alpha^{1/2}, \beta^{1/2} \in k$ and $y \in k(x)$, which is absurd, whence $[k(\alpha^{1/2}, \beta^{1/2}) : k] = 4$.

Let ℓ be the field of constants of $L = K\ell'$. Since $\lambda_{L/K} = 2$, it follows that $d_L(\mathfrak{N}_x) = 1$ and $L = \ell(x) = \ell'(x, y)$. Hence $\alpha^{1/2}, \beta^{1/2} \in \ell$.

$$\begin{array}{ccc} K & \text{———} & \ell(x) = L \\ {\scriptstyle 2}\,| & & | \\ k(x) & \text{———} & \ell'(x) \\ | & & | \\ k & \text{———} & \ell' \end{array}$$

Therefore

$$[\ell'(\alpha^{1/2}, \beta^{1/2}) : \ell'] \leq [\ell : \ell'] = [\ell(x) : \ell'(x)] = [L : \ell'(x)] \leq [K : k(x)] = 2. \quad \square$$

Corollary 8.5.8. *If $g_L = g_K > 0$, then $\lambda_{L/K} = 1$.*

Proof: We have $0 \neq g_L \leq \lambda_{L/K} g_L \leq g_K = g_L$. $\square$

We establish the following generalization of Theorem 8.5.2.

Theorem 8.5.9. *Let $L = K\ell$ be a constant extension of K. Then $L_L(\mathfrak{q}) = L_K(\mathfrak{q})\ell$ for any $\mathfrak{q} \in D_K$ if and only if $g_L = g_K$ and $\lambda_{L/K} = 1$.*

If these conditions hold, we have in particular $\ell_K(\mathfrak{q}) = \ell_L(\mathfrak{q})$.

Proof:
($\Rightarrow$) We have $\ell_L(\mathfrak{q}) = \ell_K(\mathfrak{q})$ for all $\mathfrak{q} \in D_K$. Let $\mathfrak{q} \in D_K$ be such that $-d_K(\mathfrak{q}) > 2g_K - 2$ and $-d_L(\mathfrak{q}) > 2g_L - 2$.

By the Riemann–Roch theorem (Corollary 3.5.6) we have

$$\ell_K(\mathfrak{q}) + d_K(\mathfrak{q}) = 1 - g_K \qquad \text{and} \qquad \ell_L(\mathfrak{q}) + d_L(\mathfrak{q}) = 1 - g_L.$$

Thus

$$\lambda_{L/K} = \frac{d_K(\mathfrak{q})}{d_L(\mathfrak{q})} = \frac{1 - g_K - \ell_K(\mathfrak{q})}{1 - g_L - \ell_L(\mathfrak{q})} \xrightarrow[d_K(\mathfrak{q}) \to -\infty]{} 1.$$

Therefore $\lambda_{L/K} = 1$, $d_K(\mathfrak{q}) = d_L(\mathfrak{q})$ for any $\mathfrak{q} \in D_K$ and $g_L = g_K$.
($\Leftarrow$) We have $\ell L_K(\mathfrak{q}) \subseteq L_L(\mathfrak{q})$. Since $\lambda_{L/K} = 1$, it follows by Theorem 8.4.10 that ℓ and K are linearly disjoint over k. Thus

$$\ell_K(\mathfrak{q}) = \dim_\ell \ell L_K(\mathfrak{q}) \leq \dim_\ell L_L(\mathfrak{q}) = \ell_L(\mathfrak{q}). \tag{8.54}$$

Let $\mathfrak{q} \in D_K$ be such that $-d_K(\mathfrak{q}) = -d_L(\mathfrak{q}) > 2g_K - 2$.

Using Corollary 3.5.6 we obtain

$$\ell_K(\mathfrak{q}) + d_K(\mathfrak{q}) = 1 - g_K \quad \text{and} \quad \ell_L(\mathfrak{q}) + d_L(\mathfrak{q}) = 1 - g_L.$$

Since $g_K = g_L$ and $d_K(\mathfrak{q}) = d_L(\mathfrak{q})$, it follows that

$$\ell_K(\mathfrak{q}) = \ell_L(\mathfrak{q}), \quad \ell L_K(\mathfrak{q}) = L_L(\mathfrak{q}).$$

Therefore the result holds for any divisor $\mathfrak{q} \in D_K$ satisfying $-d_K(\mathfrak{q}) > 2g_K - 2$ or $d_K(\mathfrak{q}) < 2 - 2g_K$.

Let $\mathfrak{q} \in D_K$ be an arbitrary divisor and let $\mathfrak{p}_1, \mathfrak{p}_2$ be two prime divisors of K such that $v_{\mathfrak{p}_1}(\mathfrak{q}) = v_{\mathfrak{p}_2}(\mathfrak{q}) = 0$. Let $n, m \in \mathbb{N}$ be large enough so that if $\mathfrak{B} = \mathfrak{p}_1^{-n}\mathfrak{q}$, $\mathfrak{L} = \mathfrak{p}_2^{-m}\mathfrak{q}$, then $d_K(\mathfrak{B}) < 2-2g_K$ and $d_K(\mathfrak{L}) < 2-2g_L$. The least common multiple of $\mathfrak{B}$ and $\mathfrak{L}$ is $\mathfrak{q}$ and therefore

$$L_K(\mathfrak{B}) \cap L_K(\mathfrak{L}) = L_K(\mathfrak{q}) \quad \text{and} \quad L_L(\mathfrak{B}) \cap L_L(\mathfrak{L}) = L_L(\mathfrak{q}).$$

Let $\{\alpha_1, \ldots, \alpha_r\}$ be a basis of $L_K(\mathfrak{q})$. We complete this basis to a basis $\{\beta_1, \ldots, \beta_s, \alpha_1, \ldots, \alpha_r\}$ of $L_K(\mathfrak{B})$ and to a basis $\{\gamma_1, \ldots, \gamma_t, \alpha_1, \ldots, \alpha_r\}$ of $L_K(\mathfrak{L})$.

Now we will prove that $\{\alpha_1, \ldots, \alpha_r, \beta_1, \ldots, \beta_s, \gamma_1, \ldots, \gamma_t\}$ is linearly independent over k. Assume

$$\sum_{i=1}^r a_i\alpha_i + \sum_{j=1}^s b_j\beta_j + \sum_{u=1}^t c_u\gamma_u = 0, \quad \text{with} \quad a_i, b_i, c_u \in k.$$

Notice that $\sum_{i=1}^r a_i\alpha_i + \sum_{j=1}^s b_j\beta_j \in L_K(\mathfrak{B})$ and $-\sum_{u=1}^t c_u\gamma_u \in L_K(\mathfrak{L})$, so $\sum_{u=1}^t c_u\gamma_u \in L_K(\mathfrak{B}) \cap L_K(\mathfrak{L}) = L_K(\mathfrak{q})$. Therefore $c_1 = \cdots = c_t = 0$. Similarly, $b_1 = \cdots = b_s = 0$. It follows that $a_1 = \cdots = a_r = 0$.

Since ℓ and K are linearly disjoint over k, the set

$$\{\alpha_1, \ldots, \alpha_r, \beta_1, \ldots, \beta_s, \gamma_1, \ldots, \gamma_t\}$$

is linearly independent over ℓ.

Let $y \in L_L(\mathfrak{q}) = L_L(\mathfrak{B}) \cap L_L(\mathfrak{L})$. Since $y \in L_L(\mathfrak{B})$, we have

$$y = \sum_{i=1}^r a_i\alpha_i + \sum_{j=1}^s b_j\beta_j, \quad \text{with} \quad a_i, b_j \in \ell. \tag{8.55}$$

Moreover, $y \in L_L(\mathfrak{L})$ implies that

$$y = \sum_{i=1}^r a_i'\alpha_i + \sum_{u=1}^t c_u\gamma_u \quad \text{with all} \quad a_i', c_u \text{ in } \ell. \tag{8.56}$$

It follows from relations (8.55) and (8.56) and the linear independence of the set $\{\alpha_1, \dots, \alpha_r, \beta_1, \dots, \beta_s, \gamma_1, \dots, \gamma_t\}$ that

$$a_i = a_i' \ \ (1 \le i \le r), \ \text{ and } \ b_j = c_u = 0 \ \ (1 \le j \le s, \ \ 1 \le u \le t).$$

Therefore $y \in \ell L_K(\mathfrak{q})$ and $L_L(\mathfrak{q}) = \ell L_K(\mathfrak{q})$. By the linear disjointness of ℓ and K over k, we obtain $\ell_K(\mathfrak{q}) = \ell_L(\mathfrak{q})$. □

A very important corollary is the following:

Corollary 8.5.10. *If $g_K = g_L$ and $\lambda_{L/K} = 1$, the natural homomorphism φ of the class group C_K of K into the class group C_L of L is a monomorphism. We also have $\varphi(W_K) = W_L$.*

Proof. Let $\bar{\mathfrak{q}} \in \ker\varphi$, that is, $\mathfrak{q} \in D_K$ and $\mathfrak{q}$ is principal when considered in L. Then $d_L(\mathfrak{q}) = 0$ and $\ell_L(\mathfrak{q}) = 1$. Using Theorem 8.5.9, we obtain $d_K(\mathfrak{q}) = 0$ and $\ell_K(\mathfrak{q}) = 1$. Therefore $\mathfrak{q} \in P_K$ and φ is an injective homomorphism.

Now $d_L(W_K) = d_K(W_K) = 2g_K - 2 = 2g_L - 2$ and $\ell_L(W_K^{-1}) = \ell_K(W_K^{-1}) = g_K = g_L$. Therefore $\varphi(W_K) = W_L$ (Exercise 3.6.23). □

8.6 Inseparable Function Fields

In this section we recall some of the properties of inseparable function field extensions. In Theorem 5.2.24 we proved that if L/ℓ is a finite purely inseparable extension of K/k, then for each place $\mathfrak{p}$ of K there exists a unique place $\mathfrak{P}$ of L such that $\mathfrak{P} \cap K = \mathfrak{p}$. Furthermore, if k is a perfect field every place of K is fully ramified in L (Corollary 5.2.26).

Now let K/k be a function field of characteristic $p > 0$.

Proposition 8.6.1 (Stichtenoth). *The following conditions are equivalent.*

(i) *K/k is inseparable.*
(ii) *$[K : K^p k] \ge p^2$.*
(iii) *For any place $\mathfrak{P}$ of K, $k(\mathfrak{P})/k$ is inseparable.*

Proof:
(i) $\Rightarrow$ (ii): Let L/k be a subfield of K/k such that $[K : L] = p$ and K/L is inseparable. Then for any $\alpha \in K$, α^p belongs to L. Therefore $K^p k \subseteq L$.

Let $x \in K \setminus k$. Then $K/k(x^p)$ is not a separable extension. Thus there exists an extension $k(x^p) \subseteq L \subseteq K$ such that K/L is of degree p and purely inseparable. We have $K^p k \subseteq L$ and

$$[K : K^p k] \ge [K : L] = p.$$

If $[K : K^p k] = p$, let $y \in K \setminus K^p k$. Then $K/k(y)$ is inseparable since K/k is not separably generated. There exists L_1 such that $k(y) \subseteq L_1$, $[K : L_1] = p$, and K/L_1 is

a purely inseparable extension. Since $K^pk \subseteq L_1$, It follows that $K^pk = L_1$. Therefore $y \in K^pk$. This contradiction shows that $[K : K^pk] \geq p^2$.
(ii) $\Rightarrow$ (iii) Let $\mathfrak{P}$ be a place of K/k and set $\mathfrak{p} = \mathfrak{P} \cap K^pk$. Since K/K^pk is purely inseparable, it follows by Theorem 5.2.24 that $\mathfrak{P}$ is the only place above $\mathfrak{p}$. Let $e = e(\mathfrak{P}|\mathfrak{p})$ and $f = f(\mathfrak{P}|\mathfrak{p})$. Let $z \in K$ be a prime element of $\mathfrak{P}$. Then $z^p \in K^pk$ and $p = v_{\mathfrak{P}}(z^p) = ev_{\mathfrak{p}}(z^p)$. Therefore $e \leq p$. Since $ef = [K : K^pk] \geq p^2$, it follows that $f \geq p$ and $k(\mathfrak{P})/k(\mathfrak{p})$ is inseparable. Thus $k(\mathfrak{P})/k$ is inseparable.
(iii) $\Rightarrow$ (i): Assume that (iii) holds and suppose for the sake of contradiction that K/k is separable. There exists $x \in K$ such that $K/k(x)$ is separable. By Theorem 5.2.33 it follows that all but finitely many places of $k(x)$ are separable. Thus K/k is inseparable. □

Corollary 8.6.2. *K/k is separable if and only if $[K : K^pk] = p$.* □

Corollary 8.6.3. *If K/k is separable then every element x of $K \setminus K^pk$ is a separating element and every subfield L/k of K/k is separable.*

Proof: Assume that there exists $x \in K \setminus K^pk$, such that $K/k(x)$ is not separable. Then $k(x) \subseteq K^pk$. If L/k is a subfield of K/k, then by Corollary 8.2.12, L/k is separable. □

Theorem 8.6.4. *Let K/k be an inseparable extension. Then $[K : K^pk] = p^s$ where s is the minimum number of generators of K/k.*

Proof: Let $[K : K^pk] = p^s$ and let $\{x_1, \ldots, x_t\}$ be a set of generators of K/k, that is, $K = k(x_1, \ldots, x_t)$. Then $K^pk = k(x_1^p, \ldots, x_t^p)$. Thus $[K : K^pk] \leq p^t$ and it follows that $s \leq t$. Since K/K^pk is of degree p^s, there exist $y_1, \ldots, y_s \in K \setminus K^pk$ such that $K^pk(y_1, \ldots, y_s) = K$. Now $s \geq 2$, so y_2 belongs to $K \setminus k$ and $[K : k(y_2)] < \infty$, which implies that $K/k(y_2, \ldots, y_s)$ is a finite extension. Let $L = k(y_1, y_2, \ldots, y_s)$. If K/L is not separable, there exists N such that $L \subseteq N \subseteq K$, K/N is inseparable, and $[K : N] = p$. Thus $K^pk \subseteq N$ and $y_1, \ldots, y_s \in N$, so $N = K$. This contradiction shows that K/L is a separable extension. Let $T = k(y_2, \ldots, y_s)$. Then $T(y_1) = L$ and K/L is separable. Let T_s be the separable closure of T in K and let $z \in T_s \subseteq K$ be such that $T_s = T(z)$. Notice that $T_s(y_1) \supseteq L$, and hence $K/T_s(y_1)$ is separable. Therefore $K = T_s(y_1) = T(z, y_1)$ and $z \in K$ is separable over T. Let $z_1 = z, \ldots, z_m$ be all the roots of $\operatorname{Irr}(z, x, T) = f(x)$, where $z_1, \ldots, z_m$ are all distinct and $\deg \operatorname{Irr}(z, x, T) = m$. Let $y_1 = y^{(1)}, \ldots, y_1^{(n)}$ be all the roots of $\operatorname{Irr}(y_1, x, T) = g(x)$. For all $i = 2, \ldots, n$ and $j = 2, \ldots, m$, choose $\alpha \in T$ such that $\alpha \neq \frac{y_1 - y_1^{(i)}}{z_j - z}$. Set $\omega = y_1 + \alpha z$. Then $\omega \neq y_1^{(i)} + \alpha z_j$ for all $i = 2, \ldots, n$, $j = 2, \ldots, m$. Let $h(x) = g(\omega - \alpha x) \in T(\omega)[x]$. We have $h(z) = g(\omega - \alpha z) = f(y_1) = 0$. Since the roots of $g(x)$ are $y_1, y_1^{(2)}, \ldots, y_1^{(n)}$ (not necessarily distinct), we have $h(z_j) = g(\omega - \alpha z_j) \neq 0$ because $\omega - \alpha z_j \neq y_1^{(i)}$ for $j \geq 2, i \geq 2$.

Now $h(x)$ and $f(x)$ have a common factor $\operatorname{Irr}(z, x, T(\omega))$ in $T(\omega)[x]$, which is linear since z is the only common root of $h(x)$ and $f(x)$. Thus $x - z \in T(\omega)[x]$ and

$z \in T(\omega)$. We also have $y_1 = \omega - \alpha z \in T(\omega)$. Therefore $K = T(y_1, z) \subseteq T(\omega) \subseteq K$, and $K = T(\omega) = k(\omega, y_2, \ldots, y_s)$. In particular, K can be generated by s elements over k.
□

For $n \in \mathbb{N}$, set $K_n = K^{p^n}k$. Then $K_{n+j} = K_n^{p^j}k$. In particular, $K_{m+1} = K_m^p k$. Therefore $[K_m : K_{m+1}] \geq p$ and $[K : K_{m+1}] = [K : K_m][K_m : K_{m+1}] \geq p[K : K_m]$. We obtain

$$1 \leq p^{-1}[K : K_1] \leq p^{-2}[K : K_2] \leq \ldots \leq p^{-m}[K : K_m] \leq \ldots .$$

Note that $p^{-n}[K : K_n] \in \mathbb{N}$.

Proposition 8.6.5. *There exists $n \in \mathbb{N}$ such that K_n/k is separable and for all $m \geq n$,*

$$p^{-m}[K : K_m] = p^{-n}[K : K_n].$$

Proof: Let $M \subseteq K$ be a maximal subfield of K such that M/k is separable. For example, we may choose M to be the separable closure of $k(x)$ in K, where $x \in K \setminus k$. Then K/M is purely inseparable. Since K/M is finitely generated, it follows that $K^{p^n} \subseteq M$ for some $n \in \mathbb{N}$ and $K_n = K^{p^n}k \subseteq M$. In particular, K_n/k is a separable extension. Now M/K_m is a separable extension for all $m \geq n$ and K_n/K_m is separable.

By Corollary 8.6.2 we have $\left[K_{n+i} : K_{n+i+1}\right] = \left[K_{n+i} : K_{n+i}^p k\right] = p$ for all $i \geq 1$. Thus

$$[K_n : K_m] = \prod_{i=0}^{m-n-1} \left[K_{n+i} : K_{n+i+1}\right] = p^{m-n}.$$

It follows that

$$\begin{aligned} p^{-m}\,[K : K_m] &= p^{-m}\,[K : K_n]\,[K_n : K_m] \\ &= p^{-m}\,[K : K_n]\,p^{m-n} = p^{-n}\,[K : K_n]. \end{aligned}$$
□

Proposition 8.6.5 gives an important invariant for any function field.

Definition 8.6.6. Let K/k be any function field of characteristic $p > 0$ and let $n \in \mathbb{N}$ be such that $K_n = K^{p^n}k/k$ is separable. We define the invariant

$$\mu_K := p^{-n}\,[K : K_n].$$

Remark 8.6.7. μ_K is a power of p and provides a measure of the inseparability of K/k. We have $\mu_K = 1$ if and only if K/k is separable. If s is the minimum number of generators of K/k, then

$$\mu_K \geq p^{-1}\,[K : K_1] = p^{-1}p^s = p^{s-1}.$$

Theorem 8.6.8. *Let $\mathfrak{P}$ be a place of K/k. Then μ_K divides $d_K(\mathfrak{P})$.*

Proof: Let $n \in \mathbb{N}$ be such that $\mu_K = p^{-n}[K : K_n]$, $\mathfrak{p} = \mathfrak{P} \cap K_n$, $e = e(\mathfrak{P}|\mathfrak{p})$, $f = f(\mathfrak{P}|\mathfrak{p})$ and let π be a prime element for $\mathfrak{P}$. Since $\pi^{p^n} \in K_n$, we have

$$p^n = v_{\mathfrak{P}}(\pi^{p^n}) = e v_{\mathfrak{p}}(\pi^{p^n}),$$

so $e \leq p^n$. Now, K/K_n is purely inseparable, so it follows by Theorem 5.2.24 that $\mathfrak{P}$ is the only place in K dividing $\mathfrak{p}$ and $ef = [K : K_n]$. Hence $f = e^{-1}[K : K_n] \geq p^{-n}[K : K_n] = \mu_K$. Since f is a power of p, we have $\mu_K \mid f$. Finally, since f divides $d_K(\mathfrak{P})$ it follows that μ_K divides $d_K(\mathfrak{P})$. □

Corollary 8.6.9. *The genus of K/k satisfies*

$$g_K \equiv 1 \bmod \mu_K \quad \text{if} \quad p \neq 2,$$
$$g_K \equiv 1 \bmod \frac{1}{2}\mu_K \quad \text{if} \quad p = 2.$$

Proof. μ_K divides $d_K((\omega)_K) = 2g_K - 2$, where ω is a nonzero differential of K, and hence $2g_K \equiv 2 \bmod \mu_K$.

Since μ_K is a power of p with $p \neq 2$, it follows that 2 is invertible mod p and the statement holds. □

Theorem 8.6.10. *Let K/k be any function field. There exists a finite purely inseparable extension ℓ/k such that ℓ is the field of constants of $L = K\ell$ and L/ℓ is separable.*

Proof. If K/k is separable there is nothing to prove. Let $p > 0$ be the characteristic of K and let $K = k(x_1, \ldots, x_s)$. Let $n \in \mathbb{N}$ be such that $K_n = K^{p^n}k = k(x_1^{p^n}, \ldots, x_s^{p^n})$ is separable over k. Since $\left[K_n : K_{n+1}\right] = \left[K_n : K_n^p k\right] = p$, there exists $1 \leq i \leq s$ such that $x_i^{p^n} \notin K_n^p k$ and thus $x_i^{p^n}$ is a separating element of K_n/k (Corollary 8.6.3). We may assume $i = s$. Therefore, $K_n/k(x_s^{p^n})$ is a separable extension. For $i = 1, \ldots, s-1$, let $f_i(x_i^{p^n}, x_s^{p^n}) = 0$ be a separable equation of $x_i^{p^n}$ over $k(x_s^{p^n})$. Let ℓ be the field obtained by adjoining the p^n roots of the coefficients of each f_i to k. Then ℓ/k is a finite purely inseparable extension.

Considering the equations in ℓ, we have $f_i(x_i^{p^n}, x_s^{p^n}) = g_i(x_i, x_s)^{p^n} = 0$ where $g_i(x_i, x_s) = 0$ is a separable equation of x_i over $\ell(x_s)$. Let $L = K\ell = \ell(x_1, \ldots, x_s)$. Then $L/\ell(x_s)$ is a separable extension. Therefore L/ℓ is separable. Let ℓ_1 be the field of constants of L. Then ℓ_1/ℓ is a purely inseparable extension (Theorem 8.4.2), and hence $\ell_1(x_s)/\ell(x_s)$ is purely inseparable. On the other hand, $\ell_1(x_s)$ is a subset of L and $L/\ell(x_s)$ is separable. It follows that $\ell_1 = \ell$. □

Theorem 8.6.11. *Let K/k be a function field and let $L = K\ell$ be a constant extension of K/k such that the field of constants of L is ℓ and L/ℓ is separable. Then there exists a field m satisfying $k \subseteq m \subseteq \ell$ and*

(i) *m is the field of constants of $M = Km$.*
(ii) *M/m is separable.*
(iii) *If m' is another field such that $k \subseteq m' \subseteq \ell$ and satisfying* (i) *and* (ii), *then $m \subseteq m'$.*

Proof. It suffices to prove that if m_1 and m_2 are two fields such that $k \subseteq m_i \subseteq \ell$ for $i = 1, 2$ and satisfying (i) and (ii), then $m_3 = m_1 \cap m_2$ also satisfies (i) and (ii).

Set $M_i = Km_i$ for $i = 1, 2, 3$. Let m' be the field of constants of M_3. Then $m' \subseteq M_i$ for $i = 1, 2$. Therefore $m' \subseteq m_i$ for $i = 1, 2$, and $m' = m_3$.

Now $L^p\ell = (K\ell)^p\ell = K^p\ell = (K^pk)\ell$. Since K/K^pk is a geometric extension, $L^p\ell/K^pk$ is a constant extension, and $[K : K^pk] \geq p > 1$, it follows that K cannot be contained in $K^p\ell$. Let $x \in K \setminus L^p\ell$. Then $x \in M_i \setminus M_i^p m_i$ for $i = 1, 2$. Therefore $M_i/m_i(x)$ is a separable extension for $i = 1, 2$ (Corollary 8.6.3). We will prove that $M_3/m_3(x)$ is also separable.

Let $y \in M_3$ and consider $F(Y) = \sum_{i=0}^n f_i(x)Y^i \in m_1(x)[Y]$ to be the irreducible polynomial for y over $m_1(x)$.

Since the field of constants of $L = K\ell = M_1\ell$ is ℓ, it follows by Theorem 8.4.4 that M_1 and ℓ are linearly disjoint over m_1. Hence M_1 and $\ell(x)$ are linearly disjoint over $m_1(x)$ (Proposition 8.1.5).

Since $\{1, y, \ldots, y^{m-1}\}$ is linearly independent over $m_1(x)$ where $m = \deg_y F$, it follows that $\{1, y, \ldots, y^{m-1}\}$ is linearly independent over $\ell(x)$ and F is irreducible over $\ell(x)$. Then same thing happens for the irreducible polynomial $G(Y) = \sum_{i=0}^n g_i(x)Y^i$ for y over $m_2(x)$. Thus we obtain that $n = m$ and $g_i(x) = f_i(x) \in m_1(x) \cap m_2(x) = m_3(x)$.

Therefore y is separable over $m_3(x)$, and the result follows. □

Corollary 8.6.12. *Given any function field K/k, there exists a minimal extension ℓ/k such that if $L = K\ell$, the field of constants of L is ℓ and L/ℓ is separable. This extension ℓ/k is a finite purely inseparable extension.*

Proof. Exercise 8.7.17. □

Now we study the relationship between μ_K (Definition 8.6.6) and the invariant $\lambda_{L/K}$ defined in Chapter 5 (Theorem 5.3.4). If $L = K\ell$, with ℓ as in Corollary 8.6.12, then by Theorem 5.3.4 we have $\lambda_{L/K} = \frac{[\ell:k]}{[L:K]}$.

Theorem 8.6.13. *Let L be a finite constant extension of K/k and let ℓ be the field of constants of L. Assume that L/ℓ is separable. Then*

$$\mu_K = \lambda_{L/K}.$$

Proof. Proposition 8.6.5 provides a positive integer n such that $K_n = K^{p^n}k/k$ is separable. Consider the following diagram. Since $L = K\ell$, we have $L^{p^n}\ell = K^{p^n}\ell = K_n\ell$. Since K_n/k is separable, it follows that K_n and ℓ are linearly disjoint over k (Corollary 8.4.7 and Theorem 8.4.4). Hence $\lambda_{K_n\ell/K_n} = 1$ by Theorem 8.4.10. We have $[\ell : k] = [K_n\ell : K_n]$, and hence

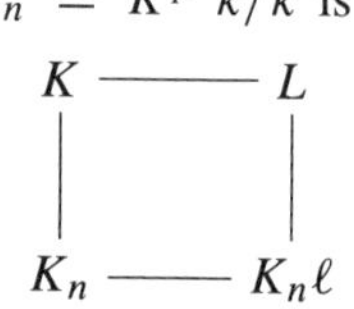

$$
\begin{aligned}
p^n \mu_K[L:K] &= [K:K_n][L:K] = [L:K_n] = [L:K_n\ell][K_n\ell:K_n] \\
&= [L:L^{p^n}\ell][\ell:k] = p^n[\ell:k].
\end{aligned}
$$

Therefore $\mu_K = \frac{[\ell:k]}{[L:K]} = \lambda_{L/K}$. □

Corollary 8.6.14. *If L/ℓ is any constant extension of K/k such that L/ℓ is separable, then $\mu_K = \lambda_{L/K}$.*

Proof. There exists a finite purely inseparable extension ℓ' of k such that $\ell' \subseteq \ell$, $L' = K\ell'$ admits ℓ' as field of constants, and L'/ℓ' is separable (Corollary 8.6.12). Hence $\mu_K = \lambda_{L'/K}$. Since $\lambda_{L/L'} = 1$ (Theorem 8.4.4, Corollary 8.4.7, and Theorem 8.4.10) and $\lambda_{L/K} = \lambda_{L/L'}\lambda_{L'/K}$, the result follows. □

Corollary 8.6.15. *If L/ℓ is a finite constant extension of K/k, we have*

$$\mu_K = \mu_L \lambda_{L/K}.$$

Proof. Using Theorem 8.6.10 we obtain a finite constant extension L'/ℓ' of L/ℓ such that L'/ℓ' is separable. By Theorem 8.6.13 and Corollary 8.6.14, we have

$$\mu_K = \lambda_{L'/K} = \frac{[\ell':k]}{[L':K]} = \frac{[\ell':\ell]}{[L':L]}\frac{[\ell:k]}{[L:K]} = \lambda_{L'/L}\lambda_{L/K} = \mu_L\lambda_{L/K}. \qquad \square$$

Corollary 8.6.16. *If L/ℓ is any constant extension of K/k we have $\mu_K = \mu_L\lambda_{L/K}$.*

Proof. Exercise 8.7.18. □

8.7 Exercises

Exercise 8.7.1. Give an example of a function field K with constant field k such that K/k is not separably generated or show that any function field K is separably generated over its constant field k.

Exercise 8.7.2. Let K/k be a separably generated function field and $K_n = kK^{p^n}$. Prove that K/K_n is a purely inseparable extension of degree p^n.

If $k \subseteq F$ and K/F is a purely inseparable extension of degree p^n, prove that $F = K_n$.

Exercise 8.7.3. Let K/k be a separably generated function field. If $k \subseteq F \subset K$ and K/F is not a separable extension, prove that $F \subseteq K^p k$.

Exercise 8.7.4. Let F/E and M/E be two field extensions with $[F:E] < \infty$. Prove that F and M are linearly disjoint if and only if $[FM:M] = [F:E]$.

Exercise 8.7.5. Give an example of two fields F and M that are not linearly disjoint over $\mathbb{Q}$ such that $F \cap M = \mathbb{Q}$.

Exercise 8.7.6. Assume $[F : \mathbb{Q}] = n$ and $[E : \mathbb{Q}] = m$. Prove that F and E are linearly disjoint over $\mathbb{Q}$ if and only if $[EF : \mathbb{Q}] = nm$.

Exercise 8.7.7. Prove Corollary 8.1.13.

Exercise 8.7.8. Prove Corollary 8.2.11.

Exercise 8.7.9. Prove Corollary 8.2.12.

Exercise 8.7.10. Prove Corollary 8.2.13.

Exercise 8.7.11. Prove Corollary 8.2.15.

Exercise 8.7.12. Let $\varphi\colon K \to E \cup \{\infty\}$ be a place on K. Given a finite number of nonzero elements $\alpha_1, \ldots, \alpha_n \in K$, we define $\alpha_i \leq \alpha_j$ if $\alpha_i \alpha_j^{-1} \in \vartheta_\varphi = \{\xi \in K \mid \varphi(\xi) \neq \infty\}$, where ϑ_ϑ is the valuation ring corresponding to φ. Prove that $\leq$ is transitive. Conclude that there exists an index j_0 such that $\alpha_i \alpha_{j_0}^{-1} \in \vartheta_\varphi$ for all i.

Exercise 8.7.13. Prove Corollary 8.3.10.

Exercise 8.7.14. Let E, K, L be subfields of Ω with $E \subseteq K$, $E \subseteq L$, and $[K : E] = n < \infty$. Show that the composite KL is a finite extension of L and $[KL : L] \leq n$. Furthermore, prove that $[KL : L] = n$ iff K and L are linearly disjoint over E.

Exercise 8.7.15. Let μ_K be given as in Definition 8.6.6. Prove that $\mu_K = 1$ if and only if K/k is separable.

Exercise 8.7.16. Prove that if L/E is a finitely generated extension of fields of characteristic p and $F^{p^m}E = E$, then F/E is an algebraic extension.

Exercise 8.7.17. Prove Corollary 8.6.12.

Exercise 8.7.18. Prove Corollary 8.6.16.

Exercise 8.7.19. Let k be a perfect field of characteristic p. Let K/k be a separably generated function field with $x \in K \setminus k$. Prove that if x is not a separating element, then $x^{1/p} \in K$.

Exercise 8.7.20. Let L/K be a constant extension, $L = K\ell'$ with k the constant field of K. Suppose that ℓ'/k is separably generated. Then

$$\overline{\operatorname{con}}_{K/L}\colon C_{K,0} \to C_{L,0} \quad \text{and} \quad \overline{\operatorname{con}}_{K/L}\colon C_K \to C_L$$

are injective (see Exercise 5.10.21).

9

The Riemann–Hurwitz Formula

Given a function field K/k, the divisor of any nonzero differential ω has degree $2g_K - 2$ (Corollary 3.5.5). Consider an extension L/ℓ of K/k; if we could find a differential Ω of L coming from ω, then we would be able to compare the degrees of Ω and ω, thus obtaining a relation between the respective genera of L and K. In the separable geometric case, we can obtain such a relation between ω and Ω by means of the cotrace of ω, and in this way we get the Riemann–Hurwitz formula.

In the inseparable case, the cotrace does not exist, due to the fact that the trace is trivial. J. Tate [152] discovered a function that is similar to the trace and can substitute it; this led him to prove his genus formula. The two mentioned formulas constitute the body of this chapter.

In the course of this discussion we shall present the Hasse differentials, whose advantage consists in being more natural than the Weil differentials. However, their disadvantage is to be definable only in the case that the field of constants is perfect. In fact, it will be shown that when the constant field is perfect, the Weil and the Hasse differentials are one and the same.

Finally, once the genus formulas have been established, we revisit and characterize fields of genus 0 and 1, now without restriction on their characteristic. On the other hand, we study in detail hyperelliptic function fields, which will be applied in Chapter 10 to cryptography, and in Chapter 14 to Weierstrass points, both in characteristic 0 and in positive characteristic.

9.1 The Differential *dx* in *k*(*x*)

In Section 4.1, we defined the differential dx in $k(x)$ as the differential that vanishes at $\mathfrak{X}\left(\mathfrak{p}_\infty^2\right) + K$ and such thas if ξ is the repartition satisfying $\xi_{\mathfrak{p}_\infty} = \frac{1}{x}$ and $\xi_{\mathfrak{p}} = 0$ for every place $\mathfrak{p} \neq \mathfrak{p}_\infty$, then $dx\,(\xi) = -1$ and $(dx)_{k(x)} = \frac{1}{\mathfrak{p}_\infty^2}$. Here k denotes an arbitrary field.

Throughout this chapter K/k will denote a function field, where k is an arbitrary field of constants.

Let ξ be a repartition and ω a differential.

Definition 9.1.1. For any place $\mathfrak{P}$ of K, we define the *$\mathfrak{P}$th component of* ω as $\omega^{\mathfrak{P}}(\xi) = \omega\left(\xi^{\mathfrak{P}}\right)$, where $\xi^{\mathfrak{P}}$ denotes the repartition whose $\mathfrak{P}$th component is the same as that of ξ (namely $\xi_{\mathfrak{P}}$), and every other component of $\xi^{\mathfrak{P}}$ is zero.

Symbolically we will write $\omega^{\mathfrak{P}}(\xi) = \omega(\xi_{\mathfrak{P}})$. Clearly, $\omega^{\mathfrak{P}}$ is k-linear.

Proposition 9.1.2. *Let ω be any differential and let $\xi \in \mathfrak{X}_K = \mathfrak{X}$. Then $\omega^{\mathfrak{P}}(\xi)$ is zero for all but a finite number of places $\mathfrak{P}$ and $\omega(\xi) = \sum_{\mathfrak{P}\in\mathbb{P}_K} \omega^{\mathfrak{P}}(\xi)$.*

Proof. Let $(\omega)_K = \mathfrak{A} = \prod_{\mathfrak{P}\in\mathbb{P}_K} \mathfrak{P}^{\alpha(\mathfrak{P})}$. All but a finite number of places $\mathfrak{P}$ satisfy the following conditions: $\alpha(\mathfrak{P}) = 0$ and $v_{\mathfrak{P}}(\xi) \geq 0$. Let $\mathfrak{S}_1, \ldots, \mathfrak{S}_s$ be the places that do not satisfy at least one of these two conditions.

If $\mathfrak{P}$ is a place that does not belong to $\{\mathfrak{S}_1, \ldots, \mathfrak{S}_s\}$, then $\xi^{\mathfrak{P}}$ is a repartition that is a multiple of $\mathfrak{A}^{-1}$, and $\xi^{\mathfrak{P}}$ satisfies $v_{\mathfrak{S}}\left(\xi^{\mathfrak{P}}\right) \geq v_{\mathfrak{S}}\left(\mathfrak{A}^{-1}\right)$ for every place $\mathfrak{S}$. Indeed, $v_{\mathfrak{S}}\left(\xi^{\mathfrak{P}}\right) = \infty$ for $\mathfrak{S} \neq \mathfrak{P}$ and $v_{\mathfrak{P}}\left(\xi^{\mathfrak{P}}\right) = v_{\mathfrak{P}}\left(\xi_{\mathfrak{P}}\right) \geq 0 = -\alpha(\mathfrak{P}) = v_{\mathfrak{P}}\left(\mathfrak{A}^{-1}\right)$.

Therefore $\omega\left(\xi^{\mathfrak{P}}\right) = \omega^{\mathfrak{P}}(\xi) = 0$.

Let ξ_i be the repartition such that $(\xi_i)_{\mathfrak{S}_i} = \xi_{\mathfrak{S}_i}$ and $(\xi_i)_{\mathfrak{S}} = 0$ for $\mathfrak{S} \neq \mathfrak{S}_i$. Thus $\xi_i = \xi^{\mathfrak{S}_i}$. Now set $\xi' = \xi_1 + \cdots + \xi_s$. Then $\xi - \xi'$ is a multiple of $\mathfrak{A}^{-1}$, which implies $\omega\left(\xi - \xi'\right) = 0$. It follows that

$$\omega(\xi) = \omega\left(\xi'\right) = \sum_{i=1}^{s} \omega(\xi_i) = \sum_{i=1}^{s} \omega^{\mathfrak{S}_i}(\xi) = \sum_{\mathfrak{P}\in\mathbb{P}_K} \omega^{\mathfrak{P}}(\xi). \qquad \square$$

Remark 9.1.3. In general, $\omega^{\mathfrak{P}}$ is not necessarily a differential.

Example 9.1.4. Let $K = k(x)$, $\omega = dx$, and let ξ be the repartition given by $\xi_{\mathfrak{P}} = \frac{1}{x}$ for all $\mathfrak{P} \in \mathbb{P}_K$. Then

$$\omega^{\mathfrak{P}_\infty}(\xi) = \omega\left(\xi^{\mathfrak{P}_\infty}\right) = \omega\left(\frac{1}{x}\right) = -1 \neq 0.$$

In other words, $\omega^{\mathfrak{P}_\infty}$, does not vanish on K.

Theorem 9.1.5. *Let K/k be a function field and ω a nonzero differential of K. Let $(\omega)_K = \prod_{\mathfrak{P}\in\mathbb{P}_K} \mathfrak{P}^{\beta_{\mathfrak{P}}}$. Then $\beta_{\mathfrak{P}}$ is the largest integer m such that $\omega^{\mathfrak{P}}(\alpha) = 0$ for every $\alpha \in K$ $(K_{\mathfrak{P}})$ satisfying $v_{\mathfrak{P}}(\alpha) \geq -m$. That is, $\beta_{\mathfrak{P}}$ satisfies $\omega^{\mathfrak{P}}(\alpha) = 0$ for all $\alpha \in K$ $(K_{\mathfrak{P}})$ such that $v_{\mathfrak{P}}(x) \geq -\beta_{\mathfrak{P}}$, and there exists $\alpha \in K$ $(K_{\mathfrak{P}})$ such that $v_{\mathfrak{P}}(\alpha) = -\beta_{\mathfrak{P}} - 1$ and $\omega^{\mathfrak{P}}(\alpha) \neq 0$.*

Equivalently, we have

$$\beta_{\mathfrak{P}} = \sup\{m \in \mathbb{Z} \mid \alpha \in K(K_{\mathfrak{P}}),\ v_{\mathfrak{P}}(\alpha) \geq -m \Rightarrow \omega^{\mathfrak{P}}(\alpha) = 0\}.$$

Proof. Let $\alpha \in K$ $(K_{\mathfrak{P}})$ be such that $v_{\mathfrak{P}}(\alpha) \geq -\beta_{\mathfrak{P}}$. Let $\alpha^{\mathfrak{P}}$ be the repartition satisfying $\left(\alpha^{\mathfrak{P}}\right)_{\mathfrak{P}} = \alpha$, and $\left(\alpha^{\mathfrak{P}}\right)_{\mathfrak{q}} = 0$ for all $\mathfrak{q} \neq \mathfrak{P}$. We have

$$\alpha^{\mathfrak{P}} \in \mathfrak{X}_K\left((\omega)_K^{-1}\right),$$

so $\omega^{\mathfrak{P}}(\alpha) = \omega\left(\alpha^{\mathfrak{P}}\right) = 0$.

On the other hand, let $\xi \in \mathfrak{X}_K\left(\mathfrak{P}^{-1}(\omega)_K^{-1}\right)$ be such that $\omega(\xi) \neq 0$. If $\mathfrak{q} \neq \mathfrak{P}$, we have $v_{\mathfrak{q}}\left(\xi_{\mathfrak{q}}\right) \geq -\beta_{\mathfrak{q}}$, and hence

$$\omega^{\mathfrak{q}}(\xi) = \omega^{\mathfrak{q}}\left(\xi_{\mathfrak{q}}\right) = 0.$$

By Proposition 9.1.2 we have $0 \neq \omega(\xi) = \sum_{\mathfrak{q} \in \mathbb{P}_K} \omega^{\mathfrak{q}}(\xi) = \omega^{\mathfrak{P}}(\xi)$, so $\omega^{\mathfrak{P}}(\xi) \neq 0$ and $v_{\mathfrak{P}}\left(\xi_{\mathfrak{P}}\right) \geq -\beta_{\mathfrak{P}} - 1$. □

Corollary 9.1.6. *If $\omega \neq 0$, then $\omega^{\mathfrak{P}} \neq 0$ for all $\mathfrak{P} \in \mathbb{P}_K$.* □

In order to describe completely dx in $k(x)$ we must determine all $\mathfrak{p}$th components $(dx)^{\mathfrak{p}}$. Since $k(x)$ is dense in $k(x)_{\mathfrak{p}}$, it suffices to determine $(dx)^{\mathfrak{p}}(u)$, with $u \in k(x)$. Indeed, if $u' \in k(x)_{\mathfrak{p}}$, let $u \in k(x)$ be such that $v_{\mathfrak{p}}(u' - u) \geq -m$, where m is the exponent of $\mathfrak{p}$ in $(dx)_{k(x)}$. Then $\omega^{\mathfrak{p}}(u' - u) = 0$ and $\omega^{\mathfrak{p}}(u') = \omega^{\mathfrak{p}}(u)$.

Let $\mathfrak{p} \neq \mathfrak{p}_\infty$. Let $f(x) \in k[x]$ be a monic irreducible polynomial such that $(f(x))_{k(x)} = \frac{\mathfrak{p}}{\mathfrak{p}_\infty^{\deg f}}$. For $u \in k(x)$, if $v_{\mathfrak{p}}(u) \geq 0$ and ξ_u denotes the repartition defined by $(\xi_u)_{\mathfrak{p}} = u$, and $(\xi_u)_{\mathfrak{p}'} = 0$ for $\mathfrak{p}' \neq \mathfrak{p}$, then $\xi_u \in \mathfrak{X}\left(\mathfrak{p}_\infty^{-2}\right)$ since $\mathfrak{p} \neq \mathfrak{p}_\infty$ and $(dx)^{\mathfrak{p}}(u) = dx\left(\xi_u\right) = 0$.

Now let $u(x) = \frac{a(x)}{f(x)^r b(x)}$, where $r \geq 1$, $a(x), b(x) \in k[x]$ are relatively prime and each of them is relatively prime to $f(x)$. Since $b(x)$ and $f(x)^r$ are relatively prime, there exist $\alpha(x), \beta(x) \in k[x]$ such that

$$a(x) = \alpha(x) f(x)^r + \beta(x) b(x).$$

Thus

$$u(x) = \frac{\alpha(x) f(x)^r + \beta(x) b(x)}{f(x)^r b(x)} = \frac{\alpha(x)}{b(x)} + \frac{\beta(x)}{f(x)^r}.$$

We may write

$$\beta(x) = g_0(x) + g_1(x) f(x) + \cdots + g_{r-1}(x) f(x)^{r-1} + t(x) f(x)^r,$$

with $g_i(x) \in k[x]$ and $\deg g_i(x) < \deg f(x)$. Therefore

$$u(x) = v(x) + \frac{g_0(x)}{f(x)^r} + \frac{g_1(x)}{f(x)^{r-1}} + \cdots + \frac{g_{r-1}(x)}{f(x)},$$

where $v(x) \in k(x)$, the denominator of $v(x)$ is not divisible by $f(x)$, and $\deg g_i(x) < \deg f(x)$ for $0 \le i \le r-1$.

Since $(dx)_{k(x)} = \mathfrak{p}_\infty^{-2}$, it follows that $\mathfrak{p}$ does not divide $(dx)_{k(x)}$. Therefore $(dx)^{\mathfrak{p}}(v(x)) = 0$ (Theorem 9.1.5). Now if $\mathfrak{S}$ is any place different from $\mathfrak{p}$ and $\mathfrak{p}_\infty$, then $v_{\mathfrak{S}}\left(g_{r-i}(x)f(x)^{-i}\right) \ge 0$, so $(dx)^{\mathfrak{S}}\left(g_{r-i}(x)f(x)^{-i}\right) = 0$.

Since the differentials vanish at the constant repartitions, we obtain

$$\begin{aligned} 0 = (dx)\left(g_{r-i}(x)f(x)^{-i}\right) &= \sum_{\mathfrak{S}\in\mathbb{P}_K} (dx)^{\mathfrak{S}}\left(g_{r-i}(x)f(x)^{-i}\right) \\ &= (dx)^{\mathfrak{p}}\left(g_{r-i}(x)f(x)^{-i}\right) + (dx)^{\mathfrak{p}_\infty}\left(g_{r-i}(x)f(x)^{-i}\right), \end{aligned}$$

so $(dx)^{\mathfrak{p}}\left(g_{r-i}(x)f(x)^{-i}\right) = -(dx)^{\mathfrak{p}_\infty}\left(g_{r-i}(x)f(x)^{-i}\right)$.

Using the fact that $\deg g_{r-i}(x) < d = \deg f(x)$, we deduce that if $i > 1$, then

$$\deg\left(g_{r-i}(x)f(x)^{-i}\right) < d - id.$$

Therefore $v_{\mathfrak{p}_\infty}\left(g_{r-i}(x)f(x)^{-i}\right) > (i-1)d \ge d \ge 1$. It follows that

$$(dx)^{\mathfrak{p}_\infty}\left(g_{r-i}(x)f(x)^{-i}\right) = 0$$

for $i = 2, \ldots, r$ (Theorem 9.1.5).

Let $g_{r-1}(x) = a_0 + a_1x + \cdots + a_{d-1}x^{d-1}$. Then

$$\frac{g_{r-1}(x)}{f(x)} - \frac{a_{d-1}}{x} = \frac{a_0x + \cdots + a_{d-1}x^d - \left(a_{d-1}x^d + \cdots + b_0a_{d-1}\right)}{f(x)x},$$

where $f(x) = x^d + \cdots + b_1x + b_0$.

Hence $\deg\left(g_{r-1}(x)f(x)^{-1} - a_{d-1}x^{-1}\right) \le -2$, and

$$(dx)^{\mathfrak{p}_\infty}\left(\frac{g_{r-1}(x)}{f(x)} - \frac{a_{d-1}}{x}\right) = 0.$$

Thus $(dx)^{\mathfrak{p}_\infty}\left(\frac{g_{r-1}(x)}{f(x)}\right) = (dx)^{\mathfrak{p}_\infty}\left(\frac{a_{d-1}}{x}\right) = -a_{d-1}$ and $(dx)^{\mathfrak{p}}(u) = a_{d-1}$.

We have proved the following result:

Theorem 9.1.7. *Let $f(x) \in k[x]$ be a monic irreducible polynomial of degree d, $\mathfrak{z}_{(f(x))} = \mathfrak{p}$, and let $u \in k(x)$ be represented by*

$$u(x) = v(x) + \frac{g_0(x)}{f(x)^r} + \frac{g_1(x)}{f(x)^{r-1}} + \cdots + \frac{g_{r-1}(x)}{f(x)},$$

where $g_0(x), \ldots, g_{r-1}(x) \in k[x]$ are polynomials of degree at most $d-1$ and $v(x) \in k(x)$ has a denominator that is not divisible by $f(x)$. Then $(dx)^{\mathfrak{p}}(u)$ is the coefficient of x^{d-1} in $g_{r-1}(x)$. □

The simplest case is $d = 1$, i.e., $f(x) = x - a$ with $a \in k$. In this case, the $\mathfrak{p}$-adic completion is

$$k(x)_{\mathfrak{p}} = k((x-a)) = \left\{ \sum_{i=m}^{\infty} a_i (x-a)^i \mid a_i \in k, m \in \mathbb{Z} \right\}.$$

Thus, the completion is the Laurent series in $x - a$ (Theorem 2.5.20). Then Theorem 9.1.7 can be stated as follows:

Theorem 9.1.8. *Let $a \in k$ and $\mathfrak{p}_a = \mathfrak{z}_{(x-a)}$ in $k(x)$. Set $y = \sum_{i=m}^{\infty} c_i (x-a)^i$ with $c_i \in k$, and assume that y belongs to the $\mathfrak{p}_a$-adic completion of $k(x)$. Then*

$$(dx)^{\mathfrak{p}_a}(y) = c_{-1}. \qquad \square$$

Next we find another expression for $(dx)^{\mathfrak{p}}(u)$. Let $f(x)$ be a monic irreducible polynomial that is not necessarily of degree 1, and $\mathfrak{p} = \mathfrak{z}_{(f(x))}$. We will assume that the residue field $k(\mathfrak{p})$ is separable over k. Let $r \in \mathfrak{X} = \mathfrak{X}_{k(x)}$ be a repartition such that $v_{\mathfrak{p}}(r) \geq -1$, and let ξ be the residue class of x in $k(\mathfrak{p})$. Then ξ is a root of $f(x)$. We have

$$k(\mathfrak{p}) = \vartheta_{\mathfrak{p}}/\mathfrak{p} = k[x]_f / f k[x]_f \cong k[x]/(f(x)) \cong k(\xi).$$

Since $f(x)$ is separable, it follows that $f'(\xi) \neq 0$. Now $v_{\mathfrak{p}}(r) \geq -1$ implies $v_{\mathfrak{p}}\left(r_{\mathfrak{p}} f(x)\right) \geq 0$. Let ζ be the residue class of $r_{\mathfrak{p}} f(x)$ in $k(\mathfrak{p}) = k(\xi)$. We write

$$r_{\mathfrak{p}} = \frac{g(x)}{f(x)} + v$$

with $v \in k(x)_{\mathfrak{p}}$, $v_{\mathfrak{p}}(v) \geq 0$, and $g(x) \in k[x]$ has degree less than $d = \deg f(x)$. Then $r_{\mathfrak{p}} f(x) = g(x) + v f(x)$, and therefore $\zeta = g(\xi)$. On the other hand, by Theorem 9.1.7, $(dx)^{\mathfrak{p}}(r)$ is the coefficient of x^{d-1} in $g(x)$.

Proposition 9.1.9. *Let $\ell = k(\xi)$, where ξ is an algebraic separable element over k. Let $f(x)$ be the minimal polynomial of ξ of degree d. Then $\operatorname{Tr}_{\ell/k} \frac{\xi^i}{f'(\xi)} = 0$ for $0 \leq i < d-1$, and $\operatorname{Tr}_{\ell/k} \frac{\xi^{d-1}}{f'(\xi)} = 1$.*

Proof. (See Theorem 5.7.17). Let $\xi = \xi_1, \ldots, \xi_d$ be the d roots of $f(x)$ in an algebraic closure of k. Let $g_i(x) = \frac{f(x)}{(x-\xi_i) f'(\xi_i)}$ for $1 \leq i \leq d$.

Each $g_i(x)$ is a polynomial of degree $d - 1$ and we have

$$g_i(\xi_i) = 1 \quad \text{and} \quad g_i(\xi_j) = 0 \quad \text{for} \quad i \neq j.$$

Let

$$h_j(x) = \sum_{i=1}^{d} \xi_i^j g_i(x) \quad \text{for} \quad 0 \leq j \leq d-1.$$

Clearly, $h_j(x)$ is a polynomial of degree at most $d-1$ and we have $h_j(\xi_i) = \xi_i^j$ for $1 \leq i \leq d$. Therefore $h_j(x) = x^j$. Indeed both polynomials take the same value at d distinct points, and both have degree less than or equal to $d-1$.

Then for $x = 0$, we have

$$\begin{aligned} h_j(0) &= \sum_{i=1}^{d} \xi_i^j g_i(0) = \sum_{i=1}^{d} \xi_i^j \frac{f(0)}{(-\xi_i) f'(\xi_i)} \\ &= -f(0) \sum_{i=1}^{d} \frac{\xi_i^{j-1}}{f'(\xi_i)} = -f(0) \operatorname{Tr}_{\ell/k} \frac{\xi^{j-1}}{f'(\xi)}. \end{aligned}$$

Since

$$h_j(0) = \begin{cases} 1 \text{ if } j = 0, \\ 0 \text{ if } 1 \leq j \leq d-1, \end{cases}$$

we obtain

$$f(0) \operatorname{Tr}_{\ell/k} \frac{\xi^{j-1}}{f'(\xi)} = \begin{cases} -1 \text{ if } j = 0, \\ \ \ 0 \ \text{ if } 1 \leq j \leq d-1. \end{cases}$$

Let

$$f(x) = x^d + a_1 x^{d-1} + \cdots + a_{d-1} x + a_d = x^d + \sum_{t=1}^{d} a_t x^{d-t}.$$

We have

$$f(0) = a_d \quad \text{and} \quad 0 = f(\xi) = \xi^d + \sum_{t=1}^{d} a_t \xi^{d-t}, \quad \text{so} \quad \xi^{d-1} = -\sum_{t=1}^{d} a_t \xi^{d-t-1}.$$

Therefore

$$\operatorname{Tr}_{\ell/k} \frac{\xi^{d-1}}{f'(\xi)} = -\sum_{t=1}^{d} a_t \operatorname{Tr}_{\ell/k} \frac{\xi^{d-t-1}}{f'(\xi)} = -a_d \left(\frac{-1}{f(0)} \right) = 1.$$

Finally, we obtain

$$\operatorname{Tr}_{\ell/k} \frac{\xi^i}{f'(\xi)} = \begin{cases} 0 \text{ if } 0 \leq i < d-1, \\ 1 \text{ if } i = d-1. \end{cases} \qquad \square$$

As an immediate consequence we have the following result:

Theorem 9.1.10. *Let $r \in \mathfrak{X}$ be such that $v_{\mathfrak{P}}(r) \geq -1$, $\mathfrak{P} = \mathfrak{Z}_f$, and $f(x) \in k[x]$ is a monic irreducible polynomial. Let $k(\mathfrak{P})/k$ be separable and ξ be the class of x in $k(\mathfrak{P})$. Then if ζ is the class $r_{\mathfrak{P}} f(x)$ in $k(\mathfrak{P})$, we have*

$$(dx)^{\mathfrak{P}}(r) = \operatorname{Tr}_{k(\mathfrak{P})/k} \frac{\zeta}{f'(\xi)}.$$

Proof. If $r_{\mathfrak{P}} = \frac{g(x)}{f(x)} + v$ with $v_{\mathfrak{P}}(v) \geq 0$ and $\deg g(x) \leq d-1$, then $(dx)^{\mathfrak{P}}(r) = a_{d-1}$, which is the coefficient of x^{d-1} in $g(x)$.

Since $\zeta = g(\xi)$, we have $\frac{\zeta}{f'(\xi)} = \frac{g(\xi)}{f'(\xi)}$, so if $g(x) = a_0 + a_1 x + \cdots + a_{d-1} x^{d-1}$, we have by Proposition 9.1.9,

$$\operatorname{Tr}_{k(\mathfrak{P})/k} \frac{\zeta}{f'(\xi)} = \operatorname{Tr}_{k(\mathfrak{P})/k} \frac{\sum_{i=0}^{d-1} a_i \xi^i}{f'(\xi)} = \sum_{i=0}^{d-1} a_i \operatorname{Tr}_{k(\mathfrak{P})/k} \frac{\xi^i}{f'(\xi)} = a_{d-1}.$$

Therefore $(dx)^{\mathfrak{P}}(r) = a_{d-1} = \operatorname{Tr}_{k(\mathfrak{P})/k} \frac{\zeta}{f'(\xi)}$. □

To conclude our analysis of dx, we state the following result.

Proposition 9.1.11. *Let $u \in k(x)$ be represented by $u = p(x) + a_{-1}x^{-1} + v$, with $p(x) \in k[x]$, $v \in k(x)_{\mathfrak{p}_\infty}$, and $v_{\mathfrak{p}_\infty}(v) \geq 2$. Then $(dx)^{\mathfrak{p}_\infty}(u) = -a_{-1}$.*

Proof. For $i \geq 0$, we have $(dx)^{\mathfrak{p}_\infty}\left(x^i\right) = -\sum_{\mathfrak{p} \neq \mathfrak{p}_\infty} (dx)^{\mathfrak{p}}\left(x^i\right) = 0$ (Theorem 9.1.7). Clearly, $(dx)^{\mathfrak{p}_\infty}\left(x^{-1}\right) = -1$ (Definition 4.1.4), and since $(dx)_{k(x)} = \mathfrak{p}_\infty^{-2}$ we conclude immediately that $(dx)^{\mathfrak{p}_\infty}(v) = 0$. Therefore

$$(dx)^{\mathfrak{p}_\infty}(u) = 0 - a_{-1} + 0 = -a_{-1}.$$ □

9.2 Trace and Cotrace of Differentials

In this section, L/ℓ denotes a finite extension of K/k.

Definition 9.2.1. Let $\xi \in \mathfrak{X}_K$ be a repartition. The *cotrace* of ξ, which we will denote by $\operatorname{cotr}_{K/L} \xi$, is the repartition $\zeta \in \mathfrak{X}_L$ defined as follows: if $\mathfrak{P}$ is a place of L, $\mathfrak{P}|_K = \mathfrak{p}$, and $\xi_{\mathfrak{p}}$ is the $\mathfrak{p}$th component of ξ with $\xi_{\mathfrak{p}} \in K_{\mathfrak{p}} \subseteq L_{\mathfrak{P}}$, then $\zeta_{\mathfrak{P}} := \xi_{\mathfrak{p}}$.

To see that ζ is in fact a repartition, just notice that there exist only finitely many places such that $v_{\mathfrak{p}}\left(\xi_{\mathfrak{p}}\right) < 0$, and above each one of these, there exist finitely many places in L.

The following proposition follows immediately from the definition.

Proposition 9.2.2. *If ξ_x is the principal repartition associated to $x \in K$, i.e., $(\xi_x)_{\mathfrak{p}} = x$ for every place $\mathfrak{p}$ of K, then $\operatorname{cotr}_{K/L} \xi_x = \zeta_x$. Furthermore, if $\lambda, \lambda' \in k$ and $\xi, \xi' \in \mathfrak{X}_K$, we have $\operatorname{cotr}_{K/L}\left(\lambda\xi + \lambda'\xi'\right) = \lambda \operatorname{cotr}_{K/L} \xi + \lambda' \operatorname{cotr}_{K/L} \xi'$, that is, $\operatorname{cotr}_{K/L}$ is k-linear.* □

Definition 9.2.3. We define the *trace* of a repartition $\zeta \in \mathfrak{X}_L$ as $\operatorname{Tr}_{L/K} \zeta = \xi$, where $\xi_{\mathfrak{p}} = \sum_{i=1}^{h} \operatorname{Tr}_{L_{\mathfrak{P}_i}/K_{\mathfrak{p}}} \zeta_{\mathfrak{P}_i}$, and $\mathfrak{P}_1, \ldots, \mathfrak{P}_h$ are the places of L over $\mathfrak{p}$.

It is easy to see that $\operatorname{Tr}_{L/K} \zeta \in \mathfrak{X}_K$. It follows from Corollary 5.5.17 that if $y \in L$, then $\operatorname{Tr}_{L/K} y = \sum_{i=1}^{h} \operatorname{Tr}_{L_{\mathfrak{P}_i}/K_{\mathfrak{p}}} y$. Thus we obtain the following proposition:

Proposition 9.2.4. *If ζ_y is the principal repartition of $\mathfrak{X}_L$ associated to y, then*

$$\operatorname{Tr}_{L/K} \zeta_y = \xi_{\operatorname{Tr}_{L/K} y}$$

is the principal repartition of $\mathfrak{X}_K$ associated to $\operatorname{Tr}_{L/K} y \in K$. Furthermore, if $\lambda, \lambda' \in k$ and $\zeta, \zeta' \in \mathfrak{X}_L$, we have

$$\operatorname{Tr}_{L/K}\left(\lambda\zeta + \lambda'\zeta'\right) = \lambda \operatorname{Tr}_{L/K} \zeta + \lambda' \operatorname{Tr}_{L/K} \zeta'. \qquad \square$$

Theorem 9.2.5. *Let $\xi \in \mathfrak{X}_K$ and $z \in L$. Then $\operatorname{Tr}_{L/K}(z \operatorname{cotr}_{K/L} \xi) = (\operatorname{Tr}_{L/K} z)\xi$.*

Proof. Let $\mathfrak{p}$ be a place of K and let $\mathfrak{P}_1, \ldots, \mathfrak{P}_h$ be the places of L over $\mathfrak{p}$. We have

$$A = \left(\operatorname{Tr}_{L/K}\left(z \operatorname{cotr}_{K/L} \xi\right)\right)(\mathfrak{p}) = \sum_{i=1}^{h} \operatorname{Tr}_{L_{\mathfrak{P}_i}/K_{\mathfrak{p}}}\left(z \operatorname{cotr}_{K/L} \xi\right)(\mathfrak{P}_i).$$

Since $\left(\operatorname{cotr}_{K/L} \xi\right)(\mathfrak{P}_i) = \xi_{\mathfrak{p}} \in K_{\mathfrak{p}}$, it follows by Corollary 5.5.17 that

$$A = \left(\sum_{i=1}^{h} \operatorname{Tr}_{L_{\mathfrak{P}_i}/K_{\mathfrak{p}}}(z)\right) \xi_{\mathfrak{p}} = \left(\left(\operatorname{Tr}_{L/K} z\right) \xi\right)(\mathfrak{p}). \qquad \square$$

Definition 9.2.6. Let Ω be a differential of L and $\xi \in \mathfrak{X}_K$. The function ω defined by

$$\omega(\xi) = \Omega\left(\operatorname{cotr}_{K/L} \xi\right)$$

is called the *trace* of Ω and it is denoted by $\omega = \operatorname{Tr}_{L/K} \Omega$.

Theorem 9.2.7. *$\omega = \operatorname{Tr}_{L/K} \Omega$ is a differential of K.*

Proof. By Proposition 9.2.2, ω is k-linear. Now if $\xi_x \in \mathfrak{X}_K$, we have $\operatorname{cotr}_{K/L} \xi_x = \zeta_x \in \mathfrak{X}_L$, from which we obtain that $\omega(\xi_x) = \operatorname{Tr}_{L/K} \Omega(\zeta_x) = 0$. Thus $K \subseteq \ker \omega$.

If $\Omega = 0$, it follows at once that $\omega = 0$. If $\Omega \neq 0$, let $(\Omega)_L = \prod_{\mathfrak{P} \in \mathbb{P}_L} \mathfrak{P}^{a(\mathfrak{P})}$ be its divisor. Let $\mathfrak{p}$ be a place of K and let $\mathfrak{P}_1, \ldots, \mathfrak{P}_h$ be the places of L above $\mathfrak{p}$ with respective ramification indices e_i $(1 \leq i \leq h)$. Let $a'(\mathfrak{p})$ be the greatest integer such that $e_i a'(\mathfrak{p}) \leq a(\mathfrak{P}_i)$ for $1 \leq i \leq h$. Then $a'(\mathfrak{p}) = 0$ for all but a finite number of places.

Let $\mathfrak{A} = \prod_{\mathfrak{p} \in \mathbb{P}_K} \mathfrak{p}^{a'(\mathfrak{p})}$ be a divisor of K. Let $\xi \in \mathfrak{X}_K$ be such that $\xi \equiv 0 \bmod \mathfrak{A}^{-1}$. Thus, $\xi \in \mathfrak{X}_K\left(\mathfrak{A}^{-1}\right)$ and $v_{\mathfrak{p}}(\xi) \geq -v_{\mathfrak{p}}(\mathfrak{A}) = -a'(\mathfrak{p})$ for every place $\mathfrak{p}$ of K. If $\mathfrak{P}$ is a place of L above $\mathfrak{p}$, we have

$$v_{\mathfrak{P}}\left(\left(\operatorname{cotr}_{K/L} \xi\right)(\mathfrak{P})\right) = v_{\mathfrak{P}}\left(\xi_{\mathfrak{p}}\right) = e_{L/K}(\mathfrak{P}|\mathfrak{p})\, v_{\mathfrak{p}}\left(\xi_{\mathfrak{p}}\right) \geq -ea'(\mathfrak{p}) \geq -a(\mathfrak{P}),$$

where $e = e_{L/K}(\mathfrak{P}|\mathfrak{p})$. Therefore we have $\operatorname{cotr}_{K/L} \xi \in \mathfrak{X}_L((\Omega_L)^{-1})$ and $\Omega\left(\operatorname{cotr}_{K/L} \xi\right) = 0$, which implies $\omega(\xi) = 0$. Thus $\mathfrak{X}_K\left(\mathfrak{A}^{-1}\right) \subseteq \ker \omega$, which proves that ω is a differential of K. $\square$

Proposition 9.2.8. *If Ω, Ω' are two differentials of L and x an element of K, then*

$$\operatorname{Tr}_{L/K}(\Omega + \Omega') = \operatorname{Tr}_{L/K}(\Omega) + \operatorname{Tr}_{L/K}(\Omega') \quad \text{and} \quad \operatorname{Tr}_{L/K}(x\Omega) = x \operatorname{Tr}_{L/K}(\Omega).$$

Proof. The first formula is obvious. For the second one, consider a repartition $\xi \in \mathfrak{X}_K$. We have

$$\begin{aligned}\left(\operatorname{Tr}_{L/K}(x\Omega)\right)(\xi) &= \operatorname{Tr}_{L/K}\left(x\Omega\left(\operatorname{cotr}_{K/L}\xi\right)\right) = \operatorname{Tr}_{L/K}\left(\Omega\left(\operatorname{cotr}_{K/L}x\xi\right)\right) \\ &= \left(\operatorname{Tr}_{L/K}\Omega\right)(x\xi) = x(\operatorname{Tr}_{L/K}\Omega)\xi.\end{aligned}$$ □

According to the Proposition 9.2.8, an operation of trace of differentials corresponds to the cotrace operation on repartitions. Conversely, we wish to associate an operation of cotrace on differentials corresponding to the operation of trace on repartitions. However, at this point a difficulty arises with respect to linearity, for we have only k-linearity. This forces us to consider only geometric extensions, i.e., the case $\ell = k$. The general case can be solved using Theorem 9.5.17.

Thus, we consider a finite geometric extension L/K of function fields.

Definition 9.2.9. Let ω be a differential in K. For $\zeta \in \mathfrak{X}_L$ we define

$$\Omega(\zeta) = \omega\left(\operatorname{Tr}_{L/K}\zeta\right).$$

We say that Ω is the *cotrace* of ω and we denote it by $\Omega = \operatorname{cotr}_{K/L}\omega$.

Theorem 9.2.10. *In the geometric case $\ell = k$, the cotrace Ω is a differential of L.*

Proof. By Proposition 9.2.4, Ω is k-linear. On the other hand, if ζ_y is the principal repartition in L corresponding to y, it follows by Proposition 9.2.4 that $\operatorname{Tr}_{L/K}\zeta_y = \xi_{\operatorname{Tr}_{L/K}y}$ is the principal repartition in K associated to $\operatorname{Tr}_{L/K}y$, so $\Omega\left(\zeta_y\right) = 0$.

Now, if L/K is inseparable, we have $\operatorname{Tr}_{L/K} \equiv 0$. Thus $\Omega = 0$ and Ω is a differential. Assume that L/K is a separable extension. Let $\omega \neq 0$ and let $(\omega)_K = \prod_{\mathfrak{p}\in\mathbb{P}_K}\mathfrak{p}^{a(\mathfrak{p})}$ be its divisor.

For each divisor $\mathfrak{P}$ of L, let $\mathfrak{p} = \mathfrak{P}|_K$, $e(\mathfrak{P}) = e_{L/K}(\mathfrak{P}|\mathfrak{p})$ be the ramification index of $\mathfrak{P}$ over $\mathfrak{p}$, and let $m(\mathfrak{P})$ be the exponent of $\mathfrak{P}$ in the different $\mathfrak{D}_{L/K}$.

Let $u \in L_{\mathfrak{P}}$ and let $\mathfrak{M}$ be the repartition that takes the value u at $\mathfrak{P}$, and 0 at every other place. Then $\operatorname{Tr}_{L/K}\mathfrak{M}$ is the repartition that takes the value 0 at any place other than $\mathfrak{p}$. At $\mathfrak{p}$ we have

$$\left(\operatorname{Tr}_{L/K}\mathfrak{M}\right)_{\mathfrak{p}} = \sum_{i=1}^{h}\operatorname{Tr}_{L_{\mathfrak{P}_i}/K_{\mathfrak{p}}}\mathfrak{M}_{\mathfrak{P}_i} = \operatorname{Tr}_{L_{\mathfrak{P}}/K_{\mathfrak{p}}}u,$$

so

$$\Omega^{\mathfrak{P}}(u) = \Omega(\mathfrak{M}) = \omega\left(\operatorname{Tr}_{L/K}\mathfrak{M}\right) = \omega^{\mathfrak{p}}\left(\operatorname{Tr}_{L_{\mathfrak{P}}/K_{\mathfrak{p}}}u\right).$$

Let $\pi \in K$ be such that $v_{\mathfrak{p}}(\pi) = 1$. If $v_{\mathfrak{P}}(u) \geq -e(\mathfrak{P})a(\mathfrak{p}) - m(\mathfrak{P})$, then

$$v_{\mathfrak{P}}\left(\pi^{a(\mathfrak{p})}u\right) = v_{\mathfrak{P}}\left(\pi^{a(\mathfrak{p})}\right) + v_{\mathfrak{P}}(u) = e(\mathfrak{P})\,a(\mathfrak{p}) + v_{\mathfrak{P}}(u) \geq -m(\mathfrak{P}).$$

Thus, by Theorem 5.6.1 and Definition 5.6.2, we have

$$v_{\mathfrak{p}}\left(\mathrm{Tr}_{L_{\mathfrak{P}}/K_{\mathfrak{p}}}\,\pi^{a(\mathfrak{p})}u\right) = v_{\mathfrak{p}}\left(\pi^{a(\mathfrak{p})}\,\mathrm{Tr}_{L_{\mathfrak{P}}/K_{\mathfrak{p}}}\,u\right) \geq 0.$$

Therefore $v_{\mathfrak{p}}\left(\mathrm{Tr}_{L_{\mathfrak{P}}/K_{\mathfrak{p}}}\,u\right) \geq -v_{\mathfrak{p}}\left(\pi^{a(\mathfrak{p})}\right) = -a(\mathfrak{p})$. Hence,

$$\Omega^{\mathfrak{P}}(u) = \omega^{\mathfrak{p}}\left(\mathrm{Tr}_{L_{\mathfrak{P}}/K_{\mathfrak{p}}}\,u\right) = 0.$$

On the other hand, there exists an element $z \in L_{\mathfrak{P}}$ such that

$$v_{\mathfrak{P}}(z) = -m(\mathfrak{P}) - 1 \quad \text{with} \quad v_{\mathfrak{p}}\left(\mathrm{Tr}_{L_{\mathfrak{P}}/K_{\mathfrak{p}}}\,z\right) < 0.$$

Since $v_{\mathfrak{P}}(\pi z) \geq -m(\mathfrak{P})$, we have $v_{\mathfrak{p}}\left(\pi\,\mathrm{Tr}_{L_{\mathfrak{P}}/K_{\mathfrak{p}}}\,z\right) \geq 0$. It follows that $v_{\mathfrak{p}}\left(\mathrm{Tr}_{L_{\mathfrak{P}}/K_{\mathfrak{p}}}\,z\right) = -1$. Now, $a(\mathfrak{p})$ is the exponent of the divisor of ω, so by Theorem 9.1.5 there exists an element $y \in K$ such that

$$v_{\mathfrak{p}}(y) = -a(\mathfrak{p}) - 1 \quad \text{and} \quad \omega^{\mathfrak{p}}(y) \neq 0.$$

Then

$$\begin{aligned} v_{\mathfrak{P}}\left(yz\left(\mathrm{Tr}_{L_{\mathfrak{P}}/K_{\mathfrak{p}}}\,z\right)^{-1}\right) &= v_{\mathfrak{P}}(y) + v_{\mathfrak{P}}(z) - v_{\mathfrak{P}}\left(\mathrm{Tr}_{L_{\mathfrak{P}}/K_{\mathfrak{p}}}\,z\right) \\ &= e(\mathfrak{P})(-a(\mathfrak{p}) - 1) - m(\mathfrak{P}) - 1 - e(\mathfrak{P})(-1) \\ &= -e(\mathfrak{P})\,a(\mathfrak{p}) - m(\mathfrak{P}) - 1. \end{aligned}$$

Furthermore,

$$\Omega^{\mathfrak{P}}\left(yz\left(\mathrm{Tr}_{L_{\mathfrak{P}}/K_{\mathfrak{p}}}\,z\right)^{-1}\right) = \omega^{\mathfrak{p}}\left(\mathrm{Tr}_{L_{\mathfrak{P}}/K_{\mathfrak{p}}}\left(yz\left(\mathrm{Tr}_{L_{\mathfrak{P}}/K_{\mathfrak{p}}}\,z\right)^{-1}\right)\right) = \omega^{\mathfrak{p}}(y) \neq 0.$$

Thus Ω is a k-linear function from $\mathfrak{X}_L$ to k vanishing in L as well as in $\mathfrak{X}_L\left(\mathfrak{D}_{L/K}^{-1}\left(\mathrm{con}_{K/L}(\omega)_K\right)^{-1}\right)$. Therefore Ω is a differential of L when $\ell = k$. □

9.3 Hasse Differentials and Residues

In Section 3.4 we gave the definition of differential based on the "usual" differentials in the complex plane. The differentials defined in Section 3.4 are due to A. Weil. Helmut Hasse ([53, 54]) established a theory of differentials for function fields whose field of constants is a perfect field, which constitutes a natural extension of the classical notion. We will see that this new concept of differentials (which we will call H-differentials) is essentially the same as that of the Weil differentials. Actually, the differentials presented in Section 3.4 for the sake of motivation are the Hasse differentials.

Let K/k be a function field, where k is a perfect field. Let $\mathfrak{P}$ be a place of K and let $K_{\mathfrak{P}}$ be the completion of K at $\mathfrak{P}$. Let π be a prime element of $\mathfrak{P}$. Then by Proposition 2.3.13 and Theorem 2.5.20 an arbitrary element $\alpha \in K_{\mathfrak{P}}$ can be uniquely expanded as

$$\alpha = \sum_{i=v_{\mathfrak{P}}(\alpha)}^{\infty} s_i \pi^i, \quad \text{where} \quad s_i \in k(\mathfrak{P}) \subseteq K_{\mathfrak{P}}.$$

Definition 9.3.1. The derivative $\frac{d\alpha}{d\pi}$, or differentiation with respect to π, is defined by

$$\frac{d\alpha}{d\pi} = \sum_{i=v_{\mathfrak{P}}(\alpha)}^{\infty} i s_i \pi^{i-1}.$$

Proposition 9.3.2. *The derivative* $\frac{d}{d\pi} : K_{\mathfrak{P}} \to K_{\mathfrak{P}}$ *is continuous and satisfies*

(1) $\frac{d}{d\pi}(a\alpha + b\beta) = a\frac{d\alpha}{d\pi} + b\frac{d\beta}{d\pi}$ for all $a, b \in k(\mathfrak{P})$ and $\alpha, \beta \in K_{\mathfrak{P}}$.
(2) $\frac{d}{d\pi}(\alpha\beta) = \alpha\frac{d\beta}{d\pi} + \beta\frac{d\alpha}{d\pi}$ for all $a, b \in k(\mathfrak{P})$ and $\alpha, \beta \in K_{\mathfrak{P}}$.
(3) $\frac{d}{d\pi}(\alpha^n) = n\alpha^{n-1}\frac{d\alpha}{d\pi}$ for all $n \in \mathbb{Z}$.

Proof: Exercise 9.7.1. □

Now let π_1 be another prime element for $\mathfrak{P}$. Since $\frac{d}{d\pi}$ and $\frac{d}{d\pi_1}$ are continuous, the derivative of a convergent power series can be carried out term by term.

Let $\alpha \in K_{\mathfrak{P}}$, $\alpha = \sum_{i=v_{\mathfrak{P}}(\alpha)}^{\infty} s_i' \pi_1^i$, $s_i' \in k(\mathfrak{P})$. Then

$$\frac{d\alpha}{d\pi_1} = \sum_{i=v_{\mathfrak{P}}(\alpha)}^{\infty} i s_i' \pi_1^{i-1}.$$

On the other hand,

$$\frac{d\alpha}{d\pi} = \sum_{i=v_{\mathfrak{P}}(\alpha)}^{\infty} \frac{d}{d\pi}(s_i'\pi_1^i) = \sum_{i=v_{\mathfrak{P}}(\alpha)}^{\infty} s_i' \frac{d}{d\pi}(\pi_1^i) = \sum_{i=v_{\mathfrak{P}}(\alpha)}^{\infty} s_i' i \pi_1^{i-1}\frac{d\pi_1}{d\pi} = \frac{d\alpha}{d\pi_1}\frac{d\pi_1}{d\pi}.$$

Proposition 9.3.3. *The differentiation with respect to prime elements* π, π_1 *of* $\mathfrak{P}$ *satisfies*

$$\frac{d\alpha}{d\pi} = \frac{d\alpha}{d\pi_1}\frac{d\pi_1}{d\pi}. \tag{9.1}$$

□

Let $A_{\mathfrak{P}} = \{(a, b) \mid a, b \in K_{\mathfrak{P}}\}$.

Definition 9.3.4. Put $(\alpha, \beta) \sim_H (\alpha', \beta')$ if for a prime element π of $\mathfrak{P}$ on $K_{\mathfrak{P}}$ the equality

$$\alpha \frac{d\beta}{d\pi} = \alpha' \frac{d\beta'}{d\pi} \tag{9.2}$$

holds. Clearly, $\sim_H$ is an equivalence relation on $A_{\mathfrak{P}}$.

Proposition 9.3.5. *The class does not depend on the prime element.*

Proof. If $(\alpha, \beta) \sim_H (\alpha', \beta')$ with respect to the prime element π, then

$$\alpha \frac{d\beta}{d\pi} = \alpha' \frac{d\beta'}{d\pi}.$$

It follows that $\alpha \frac{d\beta}{d\pi_1} = \alpha \frac{d\beta}{d\pi} \frac{d\pi}{d\pi_1} = \alpha' \frac{d\beta'}{d\pi} \frac{d\pi}{d\pi_1} = \alpha' \frac{d\beta'}{d\pi_1}$.

Thus the equivalence classes do not depend on the prime element. □

Definition 9.3.6. The classes in $A_{\mathfrak{P}}/\sim_H$ are called the local Hasse differentials of $K_{\mathfrak{P}}$. The class of (α, β) is denoted by $\alpha\, d\beta$ and we will use the notation $\sim$ instead of $\sim_H$.

If $(\alpha, \beta) \sim (\alpha', \beta')$, then for any $\gamma \in K_{\mathfrak{P}}$ we have $(\gamma\alpha, \beta) \sim (\gamma\alpha', \beta')$. It follows that we can define the product $\gamma\alpha\, d\beta$ as the class of $(\gamma\alpha, \beta)$, i.e.,

$$\gamma\alpha\, d\beta = (\gamma\alpha)\, d\beta. \tag{9.3}$$

In particular, $\alpha\, d\beta$ is the product of α and $d\beta = 1\, d\beta$.

Proposition 9.3.7. *For any two prime elements π and π_1 for $\mathfrak{P}$ we have*

$$v_{\mathfrak{P}}\left(\alpha \frac{d\beta}{d\pi}\right) = v_{\mathfrak{P}}\left(\alpha \frac{d\beta}{d\pi_1}\right).$$

Proof. Since π and π_1 are prime elements we have, $v_{\mathfrak{P}}\left(\frac{d\pi_1}{d\pi}\right) = 0$, and

$$v_{\mathfrak{P}}\left(\alpha \frac{d\beta}{d\pi}\right) = v_{\mathfrak{P}}\left(\alpha \frac{d\beta}{d\pi_1} \frac{d\pi_1}{d\pi}\right) = v_{\mathfrak{P}}\left(\alpha \frac{d\beta}{d\pi_1}\right) + v_{\mathfrak{P}}\left(\frac{d\pi_1}{d\pi}\right) = v_{\mathfrak{P}}\left(\alpha \frac{d\beta}{d\pi_1}\right). \quad \square$$

Definition 9.3.8. We define the *order* of $\alpha\, d\beta$ at $\mathfrak{P}$ by

$$v_{\mathfrak{P}}(\alpha\, d\beta) := v_{\mathfrak{P}}\left(\alpha \frac{d\beta}{d\pi}\right),$$

where π is any prime element for $\mathfrak{P}$.

If $v_{\mathfrak{P}}\left(\alpha \frac{d\beta}{d\pi}\right) = m > 0$, $\mathfrak{P}$ is called a *zero* of order m of $\alpha\, d\beta$. If $m < 0$, $\mathfrak{P}$ is called a *pole* of order $-m$.

The following result establishes that the "residue" of a differential does not depend on the prime element considered.

Theorem 9.3.9. *Let π and π_1 be prime elements in $K_{\mathfrak{P}}$ for $\mathfrak{P}$. Let $\alpha, \beta \in K_{\mathfrak{P}}$ and*

$$\alpha \frac{d\beta}{d\pi} = \sum_i s_i \pi^i, \qquad \alpha \frac{d\beta}{d\pi_1} = \sum_i s_i' \pi_1^i.$$

Then $s_{-1} = s_{-1}'$.

Proof. Let

$$\pi_1 = \sum_{i=1}^{\infty} a_i \pi^i, \quad \text{where} \quad a_1 \neq 0 \quad \text{and} \quad a_i \in k(\mathfrak{P}), \quad i = 1, 2, \ldots, \infty.$$

If $m = v_{\mathfrak{P}}\left(\alpha \frac{d\beta}{d\pi}\right)$, then

$$\begin{aligned}\sum_{i=m}^{\infty} s_i \pi^i &= \alpha \frac{d\beta}{d\pi} = \alpha \frac{d\beta}{d\pi_1} \frac{d\pi_1}{d\pi} = \left(\sum_{i=m}^{\infty} s_i' \pi_1^i\right) \frac{d\pi_1}{d\pi} \\ &= \sum_{i=m}^{\infty} s_1' \left(\sum_{j=1}^{\infty} a_j \pi^j\right)^i \left(\sum_{j=1}^{\infty} j a_j \pi^{j-1}\right). \end{aligned} \tag{9.4}$$

For $i = -1$ we obtain

$$\begin{aligned} & s_{-1}' \left(\sum_{j=1}^{\infty} a_j \pi^j\right)^{-1} \left(\sum_{j=1}^{\infty} j a_j \pi^{j-1}\right) \\ &= s_{-1}' \left(a_1^{-1} \pi^{-1}\right) \left(1 + \sum_{\ell=1}^{\infty} a_1^{-1} a_{\ell+1} \pi^{\ell}\right)^{-1} \left(\sum_{j=1}^{\infty} j a_j \pi^{j-1}\right) \\ &= s_1' a_1^{-1} \pi^{-1} \left(1 - a_1^{-1} a_{\ell+1} \pi + \cdots\right) (a_1 + 2a_2 \pi + \cdots) \\ &= \frac{s_{-1}'}{\pi} + \sum_{\ell=0}^{\infty} s_\ell'' \pi^\ell. \end{aligned}$$

To prove the theorem it suffices to show that for $i \neq -1$, the expansion of

$$\left(\sum_{j=1}^{\infty} a_j \pi^j\right)^i \left(\sum_{j=1}^{\infty} j a_j \pi^{j-1}\right) \tag{9.5}$$

does not contain the term π^{-1}.

First we consider the case $\operatorname{char} k = 0$. Let $\pi_1^{i+1} = \sum_{\ell=i+1}^{\infty} \varepsilon_\ell \pi^\ell$. Then for any $i \neq -1$, we have

$$\left(\sum_{j=1}^{\infty} a_j \pi^j\right)^i \left(\sum_{j=1}^{\infty} j a_j \pi^{j-1}\right) = \pi_1^i \frac{d\pi_1}{d\pi} = \frac{1}{i+1}\frac{d\pi_1^{i+1}}{d\pi} = \frac{1}{i+1}\sum_{\ell=i+1}^{\infty} \ell \varepsilon_\ell \pi^{\ell-1}. \tag{9.6}$$

The coefficient of π^{-1} in (9.6) is $\frac{0\varepsilon_0}{1+i} = 0$.

Now consider the case $\operatorname{char} k = p > 0$. Let $\{y_n\}_{n=1}^{\infty}$ be an algebraically independent set that replaces the above set of coefficients $\{a_n\}_{n=1}^{\infty}$. Let $M = \mathbb{Q}\left(\{y_n\}_{n=1}^{\infty}\right)$. Then (9.5) takes the form

$$\sum_{\ell} w_\ell \pi^\ell = \left(\sum_{j=1}^{\infty} y_j \pi^j\right)^i \left(\sum_{j=1}^{\infty} j y_j \pi^{j-1}\right), \quad \text{with} \quad w_n \in M \quad \text{and} \quad i \neq -1. \tag{9.7}$$

By the characteristic 0 case, the coefficient ω_{-1} of π^{-1} in (9.7) is 0.

Notice that w_ℓ is a rational function on a finite subset of $\{y_n\}_{n=1}^{\infty}$ whose denominator is at most a power of y_1 and whose numerator is a polynomial with coefficients in $\mathbb{Z}$. When we take the numerator modulo p, we obtain a rational function $\overline{w}_i \in \mathbb{F}_p(\{y_n\}_{n=1}^{\infty}) = \overline{M}$. Thus, by viewing (9.7) as a power series in π with coefficients in $\overline{M}$, we obtain

$$\sum_{\ell} \overline{w}_\ell t^\ell = \left(\sum_{j=1}^{\infty} y_j \pi^j\right)^i \left(\sum_{j=1}^{\infty} j y_j \pi^{j-1}\right) \bmod p. \tag{9.8}$$

We have $a_1 \neq 0$. Let $\xi_\ell = \overline{w}_\ell(a_1, a_2, \ldots) \in k(\mathfrak{P})$. From (9.8) we obtain

$$\sum_{\ell} \xi_\ell t^\ell = \left(\sum_{j=1}^{\infty} a_j \pi^j\right)^i \left(\sum_{j=1}^{\infty} j a_j \pi^{j-1}\right).$$

Since $w_{-1} = 0$, it follows that $\overline{w}_{-1} = 0$ and $\xi_{-1} = 0$. □

Definition 9.3.10. Let $\alpha\, d\beta$ be a local Hasse differential, π a prime element, and

$$\alpha\, d\beta = \sum_{i=m}^{\infty} s_i \pi^i \in K_{\mathfrak{P}}.$$

Then the *residue of $\alpha\, d\beta$ is defined by*

$$\operatorname*{Res}_{\mathfrak{P}} \alpha\, d\beta := \operatorname{Tr}_{k(\mathfrak{P})/k} s_{-1} \in k.$$

Theorem 9.3.9 proves that the residue is independent from the prime element. Recall that we are considering a perfect field k, so $\operatorname{Tr}_{k(\mathfrak{P})/k} \not\equiv 0$.

To define the global Hasse differential, we consider an arbitrary function field K/k, where k is a perfect field. Set $A = K \times K$. For $(\alpha, \beta) \in A$, $\mathfrak{P} \in \mathbb{P}_K$, and $\alpha, \beta \in K_{\mathfrak{P}}$, let $(\alpha\, d\beta)_{\mathfrak{P}}$ be the local Hasse differential at $\mathfrak{P}$. We define

$$(\alpha, \beta) \sim_H (\alpha_1', \beta')$$

if $(\alpha\, d\beta)_{\mathfrak{P}} = (\alpha'\, d\beta')_{\mathfrak{P}}$ for all $\mathfrak{P} \in \mathbb{P}_K$.

It is easy to see that $\sim_H$ defines an equivalence relation in A.

Definition 9.3.11. The equivalence class corresponding to $(\alpha, \beta) \in A$ is called a *Hasse differential* or *H-differential*, and the class of (α, β) is denoted by $\alpha\, d\beta$.

Since k is a perfect field, it follows by Corollary 8.2.11 that K/k is separable. The separating elements of K are characterized by the following theorem.

Theorem 9.3.12. *An element x of K is a separating element if and only if $dx \neq 0$. Furthermore, when x is a separating element we have $(dx)_{\mathfrak{P}} \neq 0$ for all $\mathfrak{P} \in \mathbb{P}_K$.*

Proof. If K is of characteristic 0, every x in $K \setminus k$ is a separating element. Since if for some prime divisor $\mathfrak{P}$ and some prime element π at $\mathfrak{P}$, $\frac{dx}{d\pi} = 0$ implies $x \in k$, the result follows.

Consider k to be of characteristic $p > 0$. Let $\mathfrak{P} \in \mathbb{P}_K$ and let π be a prime element at $\mathfrak{P}$. Let $x \in K$. If x is not a separating element, then $y = x^{1/p} \in K$ (Exercise 8.7.19). Hence

$$\frac{dx}{d\pi} = \frac{dy^p}{d\pi} = py^{p-1}\frac{dy}{d\pi} = 0. \tag{9.9}$$

Since (9.9) holds for any $\mathfrak{P} \in \mathbb{P}_K$ it follows that $dx = 0$.

Conversely, let $x \in K$ be a separating element. Let $K = k(x, y)$ with $f(x, y) = 0$, where $f(T_1, T_2) \in k[T_1, T_2]$ is an irreducible polynomial. Using the chain rule, we obtain

$$f_x(x, y)\frac{dx}{d\pi} + f_y(x, y)\frac{dy}{d\pi} = 0, \tag{9.10}$$

where f_x and f_y denote the usual partial derivatives.

Since $f(T_1, T_2)$ is irreducible and y is separable over $k(x)$ it follows that

$$f_y(x, y) \neq 0. \tag{9.11}$$

Suppose that $\frac{dx}{d\pi} = 0$. From (9.10) and (9.11) we obtain

$$\frac{dy}{d\pi} = 0.$$

Let $x = \sum_i s_i \pi^i$ and $y = \sum_i t_i \pi^i$, with $s_i, t_i \in k(\mathfrak{P})$. Since $\frac{dx}{d\pi} = 0 = \frac{dy}{d\pi}$ it follows that $s_i = t_i = 0$ for $i \not\equiv 0 \mod p$. Therefore $x = \sum_j s_{pj}\pi^{pj}$ and $y = \sum_j t_{pj}\pi^{pj}$, that is, x and y are power series in π^p. Since $K = k(x, y)$, every element of K is a power series of π^p. We may assume without loss of generality that $\pi \in K$. This contradiction proves that $\frac{dx}{d\pi} \neq 0$, i.e., $(dx)_{\mathfrak{P}} \neq 0$. Furthermore, this holds for any $\mathfrak{P} \in \mathbb{P}_K$. □

Now we prove the analogue of Theorem 3.4.9 for H-differentials.

Theorem 9.3.13. *Let $\beta \in K$ be such that $d\beta \neq 0$. Then any H-differential in K can be written uniquely as $\alpha\, d\beta$ for some $\alpha \in K$.*

Proof. Let $x \in K$ be arbitrary. To prove the theorem it suffices to prove that there exists a unique $\alpha \in K$ such that $dx = \alpha\, d\beta$.

It is clear that $K/k(\beta)$ is a finite separable extension. Thus there exists an irreducible polynomial g such that

$$g(x, \beta) = 0.$$

Let $\mathfrak{P}$ be an arbitrary place and let π be a prime element at $\mathfrak{P}$. Using the chain rule, we obtain

$$g_x(x, \beta)\frac{dx}{d\pi} + g_\beta(x, \beta)\frac{d\beta}{d\pi} = 0.$$

Now β is a separating element, x is separable over $k(\beta)$, and g is irreducible, so we have $\frac{d\beta}{d\pi} \neq 0$ and $g_x(x, \beta) \neq 0$.

Let $\alpha = -\frac{g_\beta(x,\beta)}{g_x(x,\beta)} \in K$. Then

$$\frac{dx}{d\pi} = \alpha\frac{d\beta}{d\pi}$$

for any $\mathfrak{P}$. It follows that $dx = \alpha\, d\beta$.

The uniqueness is a consequence of the fact that the H-differentials form a K-vector space. □

Theorem 9.3.14 (Residue Theorem). *Let $\alpha\, d\beta$ be any H-differential. Then* $\mathrm{Res}_{\mathfrak{P}}\, \alpha d\beta = 0$ *for almost all places $\mathfrak{P}$. Furthermore,*

$$\sum_{\mathfrak{P}\in\mathbb{P}_K} \operatorname*{Res}_{\mathfrak{P}}(\alpha\, d\beta) = 0. \tag{9.12}$$

Proof. For $\mathfrak{P}$ such that $v_{\mathfrak{P}}(\alpha) \geq 0$ and $v_{\mathfrak{P}}(\beta) \geq 0$, we have

$$\alpha\frac{d\beta}{d\pi} = \left(\sum_{i=0}^{\infty} a_i\pi^i\right)\left(\sum_{j=1}^{\infty} jb_j\pi^{j-1}\right) = \sum_{i=0}^{\infty} c_i\pi^i,$$

so $\mathrm{Res}_{\mathfrak{P}}(\alpha\, d\beta) = 0$.

Since $v_{\mathfrak{P}}(\alpha) \geq 0$ and $v_{\mathfrak{P}}(\beta) \geq 0$ hold for almost all $\mathfrak{P}$, we obtain the first part of the theorem.

For any perfect field k, if x is a separating element of K and L/K is a finite separable extension, then if $\mathfrak{p}$ is a place of K we have

$$\operatorname*{Res}_{\mathfrak{p}}(\mathrm{Tr}_{L_{\mathfrak{P}}/K_{\mathfrak{p}}}(y)dx) = \operatorname*{Res}_{\mathfrak{P}}(y\, dx), \tag{9.13}$$

where $\mathfrak{P}$ is place of L dividing $\mathfrak{p}$. It follows that

$$\operatorname{Res}_{\mathfrak{p}}(\operatorname{Tr}_{L/K}(y)dx) = \sum_{\mathfrak{P}|\mathfrak{p}} \operatorname{Res}_{\mathfrak{P}}(y\,dx). \tag{9.14}$$

In particular,

$$\operatorname{Res}_{\mathfrak{p}_0}(\operatorname{Tr}_{K/k(x)}(y)dx) = \sum_{\mathfrak{p}|\mathfrak{p}_0} \operatorname{Res}_{\mathfrak{p}}(y\,dx), \tag{9.15}$$

where $\mathfrak{p}_0 = \mathfrak{p} \cap k(x)$. For a proof of (9.13), (9.14), and (9.15) see Exercises 9.7.18 and 9.7.19 as well as the above proof of the case in which k is algebraically closed.

By the above argument we may assume that k is algebraically closed.

If $d\beta = 0$, (9.12) follows. Now assume that $d\beta \neq 0$, i.e., β is a separating element of K.

For $K = k(\beta)$ we leave the verification of (9.12) to the reader (Exercise 9.7.20).

For the case $K \neq k(\beta)$, it suffices to show that if $\mathfrak{P}$ is an arbitrary place on $k(\beta)$ and $\wp_1, \ldots, \wp_h$ are the places on K above $\mathfrak{P}$, we have

$$\sum_{i=1}^{h} \operatorname{Res}_{\wp_i}(\alpha\, d\beta) = \operatorname{Res}_{\mathfrak{P}}(\operatorname{Tr}_{K/k(\beta)}(\alpha)d\beta). \tag{9.16}$$

Indeed, from (9.16) and the case $K = k(\beta)$ we obtain

$$\sum_{\wp\in\mathbb{P}_K} \operatorname{Res}_{\wp}(\alpha\, d\beta) = \sum_{\mathfrak{P}\in\mathbb{P}_{k(\beta)}} \operatorname{Res}_{\mathfrak{P}}\left(\operatorname{Tr}_{K/k(\beta)}\alpha)d\beta\right) = 0.$$

Since β is a separating element of K, let y be such that

$$K = k(\beta, y) \quad \text{and} \quad f(\beta, y) = 0, \tag{9.17}$$

where y is separable over $k(\beta)$.

Set $F(T) := f(\beta, T) = \prod_{i=1}^{h} p_i(T)$ in $k(\beta)_{\mathfrak{P}}[T]$.

By Corollary 5.4.9,

$$K \otimes_{k(\beta)} k(\beta)_{\mathfrak{P}} \cong \bigoplus_{i=1}^{h} K_{\wp_i}.$$

Indeed, we have

$$\begin{aligned} K \otimes_{k(\beta)} k(\beta)_{\mathfrak{P}} &= k(\beta)[T]/(F(T)) \otimes_{k(\beta)} k(\beta)_{\mathfrak{P}} \\ &\cong k(\beta)_{\mathfrak{P}}[T]/(F(T)) \cong \prod_{i=1}^{h} k(\beta)_{\mathfrak{P}}/(p_i(T)) \cong \bigoplus_{i=1}^{h} K_{\wp_i}. \end{aligned}$$

By Corollary 5.5.17, we have

$$\operatorname{Tr}_{K/k(\beta)} y = \sum_{i=1}^{h} \operatorname{Tr}_{K_{\wp_i}/k(\beta)_{\mathfrak{P}}} y$$

and

$$\operatorname*{Res}_{\mathfrak{P}}(\operatorname{Tr}_{K/k(\beta)} y\, d\beta) = \sum_{i=1}^{h} \operatorname*{Res}_{\mathfrak{P}}((\operatorname{Tr}_{K_{\wp_i}/k(\beta)_{\mathfrak{P}}} y) d\beta).$$

Thus, to prove (9.16), it suffices to show that

$$\operatorname*{Res}_{\wp_i}(y\, d\beta) = \operatorname*{Res}_{\mathfrak{P}} \left((\operatorname{Tr}_{K_{\wp_i}/k(\beta)_{\mathfrak{P}}} y) d\beta \right).$$

In other words, we need to prove that if $L/k(\beta)_{\mathfrak{P}}$ is a finite extension and $\wp$ is the extension of $\mathfrak{P}$ to L, then for any $\alpha \in L$,

$$\operatorname*{Res}_{\wp}(\alpha\, d\beta) = \operatorname*{Res}_{\mathfrak{P}}((\operatorname{Tr}_{L/k(\beta)_{\mathfrak{P}}} \alpha) d\beta). \tag{9.18}$$

Let π be a prime element for $\mathfrak{P}$. Then if $\operatorname{Tr} = \operatorname{Tr}_{L/k(\beta)_{\mathfrak{P}}}$, we have

$$\operatorname{Tr}(\alpha) d\beta = \operatorname{Tr}(\alpha) \frac{d\beta}{d\pi} d\pi = \operatorname{Tr}\left(y \frac{d\beta}{d\pi} \right) d\pi$$

because $\frac{d\beta}{d\pi} \in k(\beta)_{\mathfrak{P}}$. Thus it suffices to prove

$$\operatorname*{Res}_{\wp}(\alpha d\pi) = \operatorname*{Res}_{\mathfrak{P}}((\operatorname{Tr} \alpha) d\pi). \tag{9.19}$$

We know that Tr is a linear and continuous map. Furthermore, any α has a unique expansion

$$\alpha = \sum_{i=m}^{\infty} s_i t^i \quad \text{with} \quad s_i \in k,$$

where t is any prime element of $\wp$. Thus it suffices to prove that

$$\operatorname*{Res}_{\wp}(t^n d\pi) = \operatorname*{Res}_{\mathfrak{P}}(\operatorname{Tr}(t^n) d\pi) \quad \text{for} \quad n \in \mathbb{Z}. \tag{9.20}$$

Since k is algebraically closed, it follows that $\left[L : k(\beta)_{\mathfrak{P}}\right] = e$ is the ramification index of $\mathfrak{P}$. If k is of characteristic 0, we use Proposition 5.5.12. That is, we may assume that $t^e = \pi$.

Using Newton's identities (Theorem 7.1.4) it is easy to see that

$$\operatorname{Tr}(t^n) = \begin{cases} 0 & \text{for } e \nmid n, \\ e\pi^m & \text{for } n = me. \end{cases} \tag{9.21}$$

It follows that

$$\operatorname*{Res}_{\mathfrak{P}}(\operatorname{Tr}(t^n) d\pi) = \begin{cases} 0 \text{ for } n \neq -e, \\ e \text{ for } n = -e. \end{cases}$$

Now, since $t^n \frac{d\pi}{dt} = t^n(et^{e-1}) = et^{n+e-1}$, we have

$$\operatorname*{Res}_{\wp}(t^n d\pi) = \begin{cases} 0 \text{ for } n \neq -e, \\ e \text{ for } n = -e. \end{cases}$$

Note that this proves (9.20) in the case $t^e = \pi$. But for $t^e = \pi$, (9.20) implies (9.19). Therefore, we obtain (9.20) for arbitrary prime elements t and π.

Thus (9.20) holds when k has characteristic 0.

Now we consider k to be algebraically closed of characteristic $p > 0$. We have $\left[L : k(\beta)_{\mathfrak{P}}\right] = e$ and $L = k(\beta)_{\mathfrak{P}}(t)$. Let

$$\frac{t^e}{\pi} = a_0 + a_1 t + \cdots + a_{e-1}t^{e-1} \quad \text{with} \quad a_i \in k(\beta)_{\mathfrak{P}} \quad \text{and} \quad a_0 \neq 0.$$

We have $v_{\wp}(a_i t^i) = ev_{\mathfrak{P}}(a_i) + i \neq ev_{\mathfrak{P}}(a_j) + j$ whenever $0 \leq i, j \leq e-1, i \neq j$, $a_i \neq 0, a_j \neq 0$.

It follows that $0 = v_{\wp}\left(\frac{t^e}{\pi}\right) = \min_{0 \leq j \leq e-1}\left\{ev_{\mathfrak{P}}(a_j) + j\right\}$. Thus $v_{\mathfrak{P}}(a_j) \geq 0$ and $v_{\mathfrak{P}}(a_0) = 0$. Since $a_0\pi$ is a prime element for $\mathfrak{P}$ in $(k(\beta))_{\mathfrak{P}}$, we rewrite $a_0\pi$ as π again. We have $k(\beta)_{\mathfrak{P}} = k((\pi))$. Hence

$$t^e = \pi(1 + A_1(\pi)t + \cdots + A_{e-1}(\pi)t^{e-1}),$$

where

$$A_i(\pi) = \sum_{j=0}^{\infty} a_{ij}\pi^j, \quad a_{ij} \in k,$$

and $A_i(\pi) \in k((\pi))$ is considered as a power series.

Let $M = \mathbb{Q}(z_{ij})$ for $1 \leq i \leq e-1$ and $j \in \mathbb{N}$, where $\{z_{ij}\}$ is a set of variables corresponding to a_{ij}. Let $\overline{M}$ be an algebraic closure of M and

$$A_i^*(\pi) = \sum_{j=0}^{\infty} z_{ij}\pi^j \in \overline{M}((\pi))$$

corresponding to $A_i(\pi)$. Set $\overline{L} = \overline{M}((\pi))(t)$, where

$$t^e = \pi(1 + A_1^*(\pi)t + \cdots + A_{e-1}^*(\pi)t^{e-1}). \tag{9.22}$$

Let $\wp_0$ be the zero divisor of π (considered as a variable).

Since $\overline{M}((\pi))$ is a complete field, there exists a unique prime divisor $\mathfrak{P}_0$ above $\wp_0$ (Theorem 5.4.7).

By (9.22) we have

$$v_{\mathfrak{P}_0}(t^e) = ev_{\mathfrak{P}_0}(t) = v_{\mathfrak{P}_0}(\pi) + 0 = e(\mathfrak{P}_0|\wp_0)v_{\mathfrak{P}_0}(\pi).$$

It follows that $e(\mathfrak{P}_0|\wp_0) = e$, $\left[\overline{L} : \overline{M}((\pi))\right] = e$, and the equation (9.22) is irreducible in t.

Since the characteristic of $\overline{M}$ is zero, we have by (9.20),

$$\operatorname*{Res}_{\mathfrak{P}_0}(t^n d\pi) = \operatorname*{Res}_{\wp_0}(\operatorname{Tr}(t^n)d\pi). \tag{9.23}$$

Now we obtain from (9.22) that

$$\pi = t^e - \pi(A_1^*(\pi)t + \cdots + A_{e-1}^*(\pi)t^{e-1}). \tag{9.24}$$

If we substitute the expression of π again in the right-hand side of (9.24), we conclude that the terms containing π contain π^2. Repeating this process and using $\lim_{m\to\infty}\pi^m = 0$, we obtain

$$\pi = \sum_{\ell=e}^{\infty} \beta_\ell t^\ell \quad \text{with} \quad \beta_\ell \in \overline{M} \quad \text{and} \quad \beta_e = 1. \tag{9.25}$$

We already knew the existence of an expression such as (9.25), but with this method of computation we obtain the additional information that the β_ℓ are all polynomials in z_{ij} with coefficients in $\mathbb{Z}$. Thus

$$t^n \frac{d\pi}{dt} = \sum_{\ell=e}^{\infty} \ell\beta_\ell t^{\ell-1+n}$$

is also a polynomial in z_{ij} with coefficients in $\mathbb{Z}$.

On the other hand, by (9.24) we have

$$\frac{1}{t} = -A^*(\pi) - A_2^*(\pi)t - \cdots - A_{a-1}^*(\pi)t^{e-2} + \frac{t^{e-1}}{\pi}.$$

Let

$$\operatorname{Tr}(t^n) = \sum_m c_{nm}(z_{ij})\pi^m, \quad \text{with} \quad c_{nm}(z_{ij}) \in \overline{M}.$$

Then each c_{nm} belongs to $\mathbb{Z}[z_{ij}]$. In particular,

$$\operatorname*{Res}_{\wp_0}(\operatorname{Tr}(t^n)d\pi) = c_{n,-1}(z_{ij}) \in \mathbb{Z}[z_{ij}].$$

It follows that (9.23) is a polynomial identity. Let $\bar{c}_{nm} = c_{nm} \bmod p \in \mathbb{F}_p[z_{ij}]$ and substitute z_{ij} by a_{ij}. Then the equation (9.23) holds mod p, which implies that (9.18) holds for the extension $L/k(\beta)_\mathfrak{p}$. This completes the proof. □

With Theorem 9.3.14 at hand, we can now see that Weil differentials and Hasse differentials are the same when the ground field k is perfect.

Theorem 9.3.15. *Let K/k be an algebraic function field where k is a perfect field. Let $\alpha\, d\beta$ be an arbitrary H-differential in K. Define*

$$w\colon \mathfrak{X}_K \mapsto k$$

by

$$w(\xi) = \sum_{\mathfrak{P}\in\mathbb{P}_K} \operatorname*{Res}_{\mathfrak{P}}(\xi_{\mathfrak{P}}\alpha\, d\beta). \tag{9.26}$$

Then w is a differential in K. Furthermore, the correspondence $y\,dx \leftrightarrow w$ is a K-module isomorphism.

Proof. We denote by Dif_H and Dif_W the Hasse and the Weil differentials respectively.

Let $\varphi\colon \mathrm{Dif}_H \mapsto \mathrm{Dif}_W$ be the function given in (9.26), that is,

$$\varphi(\alpha\, d\beta)(\xi) = \sum_{\mathfrak{P}\in\mathbb{P}_K} \operatorname*{Res}_{\mathfrak{P}}(\xi_{\mathfrak{P}}\alpha\, d\beta).$$

For any $\xi \in \mathfrak{X}_K$ there are only finitely many elements $\mathfrak{P}$ of $\mathbb{P}_K$ such that $v_{\mathfrak{P}}(\xi_{\mathfrak{P}}) < 0$. It follows that $\mathrm{Res}_{\mathfrak{P}}(\xi_{\mathfrak{P}}\alpha\, d\beta)$ is equal to zero except for finitely many $\mathfrak{P} \in \mathbb{P}_K$. Thus the sum in (9.26) is well defined.

Now we will see that w is a differential. Since $\mathrm{Res}_{\mathfrak{P}}$ and $\mathrm{Tr}_{k(\mathfrak{P})/k}$ are linear, it follows that w is k-linear. Let $\mathfrak{A} = \prod_{\mathfrak{P}} \mathfrak{P}^{a(\mathfrak{P})}$, where

$$a(\mathfrak{P}) = \begin{cases} v_{\mathfrak{P}}(\alpha\, d\beta) & \text{if } v_{\mathfrak{P}}(\alpha\, d\beta) < 0,\\ 0 & \text{otherwise.}\end{cases}$$

If $\xi \in \mathfrak{X}_K$ is such that $\mathfrak{A}^{-1}$ divides ξ and $v_{\mathfrak{P}}(\xi_{\mathfrak{P}}) \geq -v_{\mathfrak{P}}(\mathfrak{A})$, then

$$v_{\mathfrak{P}}(\xi_{\mathfrak{P}}\alpha\, d\beta) = v_{\mathfrak{P}}(\xi_{\mathfrak{P}}) + v_{\mathfrak{P}}(\alpha\, d\beta) \geq -v_{\mathfrak{P}}(\mathfrak{A}) + v_{\mathfrak{P}}(\alpha\, d\beta) \geq 0$$

for all $\mathfrak{P} \in \mathbb{P}_K$. Thus $\mathrm{Res}_{\mathfrak{P}}(\xi_{\mathfrak{P}}\alpha\, d\beta) = 0$ for all $\mathfrak{P} \in \mathbb{P}_K$. If $x \in K$, by the residue theorem (Theorem 9.3.14) we have $w(x) = \sum_{\mathfrak{P}\in\mathbb{P}_K} \mathrm{Res}_{\mathfrak{P}}(x\alpha\, d\beta) = 0$. Hence $\mathfrak{X}(\mathfrak{A}^{-1}) + K \subseteq \ker w$ and w is a differential.

Next, if $\alpha\, d\beta = 0$, then $w = 0$. If $\alpha\, d\beta \neq 0$, let $\mathfrak{P} \in \mathbb{P}_K$ be such that $(\alpha\, d\beta)_{\mathfrak{P}} \neq 0$. Let $a \in K_{\mathfrak{P}}$ be such that

$$\operatorname*{Res}_{\mathfrak{P}}(a\alpha\, d\beta) \neq 0 \quad \text{and} \quad \mathrm{Tr}_{k(\mathfrak{P})/k} \operatorname*{Res}_{\mathfrak{P}}(a\alpha\, d\beta) \neq 0.$$

Such an a exists since $k(\mathfrak{P})/k$ is separable.

Let $\xi \in \mathfrak{X}_K$ be defined by

$$\xi_{\mathfrak{q}} = \begin{cases} a \text{ if } & \mathfrak{q} = \mathfrak{P},\\ 0 \text{ otherwise.}\end{cases}$$

Then $w(\xi) = \mathrm{Res}_{\mathfrak{P}}(a\alpha\, d\beta) \neq 0$, so φ is one-to-one.

Finally, if $\varphi(\alpha_1\, d\beta) = w_1$, $\varphi(\alpha_2\, d\beta) = w_2$ and $x \in K$, then

$$\varphi(\alpha_1\, d\beta + \alpha_2\, d\beta) = w_1 + w_2$$

and

$$\varphi(z\alpha_1\, d\beta) = zw_1. \tag{9.27}$$

Thus φ is a one-to-one K-linear homomorphism and since both Dif_H and Dif_W are one-dimensional K-modules, it follows that φ is a K-isomorphism. □

Corollary 9.3.16. *With the hypotheses of Theorem* 9.3.15, *we have*

$$w^{\mathfrak{P}}(\xi) = \operatorname*{Res}_{\mathfrak{P}}(\xi_{\mathfrak{P}}\alpha\, d\beta). \qquad \square$$

Finally, we have the following theorem:

Theorem 9.3.17. *Let $\alpha\, d\beta$ be a nonzero H-differential in K and let w be the corresponding W-differential given in Theorem* 9.3.15. *Then the divisor of w is given by*

$$v_{\mathfrak{P}}((w)_K) = v_{\mathfrak{P}}(\alpha\, d\beta).$$

Proof. Let $\mathfrak{A}$ be a divisor such that $v_{\mathfrak{P}}(\alpha\, d\beta) \geq v_{\mathfrak{P}}(\mathfrak{A})$ for all $\mathfrak{P} \in \mathbb{P}_K$. Let $\xi \in \mathfrak{X}(\mathfrak{A}^{-1})$. Then

$$v_{\mathfrak{P}}(\xi_{\mathfrak{P}}) \geq -v_{\mathfrak{P}}(\mathfrak{A}) \geq -v_{\mathfrak{P}}(\alpha\, d\beta).$$

Hence $v_{\mathfrak{P}}(\xi_{\mathfrak{P}}\alpha\, d\beta) \geq 0$ and $\mathrm{Res}_{\mathfrak{P}}(\xi_{\mathfrak{P}}\alpha\, d\beta) = 0$. It follows that $w(\xi) = 0$ and $\mathfrak{A}$ divides w.

Now let $\mathfrak{B}$ be a divisor such that for a $\mathfrak{P} \in \mathbb{P}_K$, $v_{\mathfrak{P}}(\mathfrak{B}) > v_{\mathfrak{P}}(\alpha\, d\beta)$. Let $a \in K_{\mathfrak{P}}$ be such that $\mathrm{Res}_{\mathfrak{P}}(a\alpha\, d\beta) \neq 0$. Such an a exists since $(\alpha\, d\beta)_{\mathfrak{P}} \neq 0$ and $k(\mathfrak{P})/k$ is separable. Furthermore, we may choose a such that $v_{\mathfrak{P}}(a\alpha d\beta) = v_{\mathfrak{P}}(a) + v_{\mathfrak{P}}(\alpha\, d\beta) = -1$. Thus

$$v_{\mathfrak{P}}(a) = -1 - v_{\mathfrak{P}}(\alpha\, d\beta) > -1 - v_{\mathfrak{P}}(\mathfrak{B}).$$

Hence $v_{\mathfrak{P}}(a) \geq -v_{\mathfrak{P}}(\mathfrak{B})$. Let $\xi \in \mathfrak{X}_K$ be given by

$$\xi_{\mathfrak{q}} = \begin{cases} a \text{ if } & \mathfrak{q} = \mathfrak{P}, \\ 0 \text{ otherwise.} \end{cases}$$

Then $w(\xi) = \mathrm{Res}_{\mathfrak{P}}\, a\alpha\, d\beta \neq 0$, and the result follows. □

Let w be any (Weil) differential over an algebraic function field K/k, where k is a perfect field. Let $\alpha\, d\beta$ be the corresponding Hasse differential. Then

$$\operatorname*{Res}_{\mathfrak{P}}(\alpha\, d\beta) = w(\xi),$$

where

$$\xi_{\mathfrak{q}} = \begin{cases} 0 \text{ if } & \mathfrak{q} \neq \mathfrak{P}, \\ 1 \text{ if } & \mathfrak{A} = \mathfrak{P}. \end{cases}$$

Therefore $\mathrm{Res}_{\mathfrak{P}}(\alpha\, d\beta) = w^{\mathfrak{P}}(1)$.

We have not defined H-differentials in the case of an imperfect field, but we may define the residue of a differential. We use the idea of the H-differentials. First we recall a basic result from basic algebra.

Proposition 9.3.18. *Let E be any field and let V be a finite-dimensional E-vector space. Let $V^* = \operatorname{Hom}_k(V, E)$. Then V^* and V are isomorphic as E-vector spaces. Furthermore, if $\phi : V \times V \to E$ is a nondegenerate bilinear form (that is, for any nonzero $v \in V$, there exists $w \in V$ such that $\phi(v, w) \neq 0$), then for any $T \in V^*$ there exists a unique $v \in V$ such that $T(w) = \phi(v, w)$ for all $w \in V$.*

Proof: Let $\{e_1, \ldots, e_n\}$ be a basis of V over E. Let $v \in V$ be written as $v = \sum_{i=1}^{h} x_i e_i$, $x_i \in E$.

Define $f_i : V \to E$ by $f_i(v) = x_i$. Then $f_i \in V^*$, $\{f_1, \ldots, f_n\}$ is a basis of V^* and $\dim_E V^* = \dim_E V$. Next, let $\phi : V \times V \to E$ be a nondegenerate bilinear form. Let $T_i \in V^*$ be defined by $T_i(w) := \phi(e_i, w)$. Then $\{T_1, \ldots, T_n\}$ is linearly independent over k and since $\dim_E V^* = n$, given $T \in V^*$, there exist $a_1, \ldots, a_n \in E$ such that

$$T = \sum_{i=1}^{n} a_i T_i.$$

Thus

$$T(w) = \sum_{i=1}^{n} a_i T_i(w) = \sum_{i=1}^{a} a_i \phi(e_i, w) = \phi(\sum_{i=1}^{n} a_i e_i, w).$$

It follows that $T(w) = \phi(v, w)$ for all $w \in V$ with $v = \sum_{i=1}^{n} a_i e_i$. Clearly v is unique. □

Now let K/k be an arbitrary function field. Let w be any (Weil) differential. Then if $\mathfrak{P} \in \mathbb{P}_K$, the local component $w^{\mathfrak{P}}$ of w is a function

$$w^{\mathfrak{P}} : K_{\mathfrak{P}} \to k.$$

Since k is not necessarily perfect, we consider the separable closure $k(\mathfrak{P})_s$ of k in the residue field $k(\mathfrak{P})$.

Then the function

$$\varphi : k(\mathfrak{P})_s \times k(\mathfrak{P})_s \to k$$

defined by $\varphi(a, b) = \operatorname{Tr}_{k(\mathfrak{P})_s/k}(ab)$ is a nondegenerate bilinear pairing. It follows by Proposition 9.3.18 that for the k-linear map

$$w^{\mathfrak{P}}|_{k(\mathfrak{P})_s} : k(\mathfrak{P})_s \to k$$

there exists a unique $\varrho \in k(\mathfrak{P})_s$ such that

$$w^{\mathfrak{P}}|_{k(\mathfrak{P})_s} = \varphi(-, \varrho).$$

Thus

$$w^{\mathfrak{P}}(\alpha) = \operatorname{Tr}_{k(\mathfrak{P})_s/k}(\alpha\varrho) \quad \text{for all} \quad \alpha \in k(\mathfrak{P})_s.$$

Definition 9.3.19. Let K/k be an arbitrary function field. Let w be a (Weil) differential on K. For $\mathfrak{P} \in \mathbb{P}_K$, we define the *residue of* w *at* $\mathfrak{P}$ as the element $\operatorname{Tr}_{k(\mathfrak{P})_s/k} \varrho \in k$ satisfying $w^{\mathfrak{P}}(\alpha) = \operatorname{Tr}_{k(\mathfrak{P})_s/k}(\alpha\varrho)$ for all $\alpha \in k(\mathfrak{P})_s$. We use the notation

$$\operatorname{Tr}_{k(\mathfrak{P})_s/k} \varrho = \operatorname*{Res}_{\mathfrak{P}} w.$$

We have

$$w^{\mathfrak{P}}(1) = \operatorname*{Res}_{\mathfrak{P}} w. \tag{9.28}$$

Proposition 9.3.20. *Let* w *be a differential in a function field* K/k. *If* $\mathfrak{P} \in \mathbb{P}_K$ *is not a pole of* w, *then*

$$\operatorname*{Res}_{\mathfrak{P}} w = 0.$$

In particular, $\operatorname{Res}_{\mathfrak{P}} w \neq 0$ *for only finitely many* $\mathfrak{P} \in \mathbb{P}_K$.

Proof: If $\mathfrak{P}$ is not a pole of w, i.e., $v_{\mathfrak{P}}((w)_K) \geq 0$, then $w^{\mathfrak{P}}(\alpha) = 0$ for any $\alpha \in K_{\mathfrak{P}}$ such that $v_{\mathfrak{P}}(\alpha) \geq 0$ (Theorem 9.1.5). In particular, $w^{\mathfrak{P}}(\alpha) = 0$ for all $\alpha \in k(\mathfrak{P})_s$. The result follows. □

Definition 9.3.21. A differential w is said to be of the *second kind* if $\operatorname{Res}_{\mathfrak{P}} w = 0$ for all $\mathfrak{P} \in \mathbb{P}_K$.

It is easy to see that if w is of the first kind (that is, holomorphic), then w is of the second kind.

Theorem 9.3.22 (Residue Theorem). *For any differential* w *of a function field* K/k, *we have*

$$\sum_{\mathfrak{P} \in \mathbb{P}_k} \operatorname*{Res}_{\mathfrak{P}} w = 0. \tag{9.29}$$

Proof: By (9.28), we have

$$0 = w(1) = \sum_{\mathfrak{P} \in \mathbb{P}_K} w^{\mathfrak{P}}(1) = \sum_{\mathfrak{P} \in \mathbb{P}_K} \operatorname*{Res}_{\mathfrak{P}} w. \qquad \square$$

Let K/k be any function field, and let $\mathfrak{A}$ be any divisor. If $D_K(\mathfrak{A}) = \{w \mid \mathfrak{A} \mid w\}$, then

$$D_K(\mathfrak{A}) \cong \left(\frac{\mathfrak{X}_K}{\mathfrak{X}_K(\mathfrak{A}^{-1}) + K}\right)^*,$$

where $*$ denotes the dual k-vector space (Proposition 3.4.5).

This isomorphism can be obtained from the k-bilinear pairing

$$\begin{aligned} \varphi\colon \operatorname{Dif}_K \times \mathfrak{X}_K &\to k \\ \varphi(w, \xi) &= w(\xi). \end{aligned} \tag{9.30}$$

Thus (9.30) can be written as

$$w(\xi) = \sum_{\mathfrak{P}\in\mathbb{P}_K} w^{\mathfrak{P}}(\xi) = \sum_{\mathfrak{P}} \operatorname{Res}_{\mathfrak{P}}(\xi_{\mathfrak{P}} w)$$

(Corollary 9.3.16).

We have obtained the following result:

Proposition 9.3.23. *For any function field K/k, $\frac{\mathfrak{X}_K}{\mathfrak{X}(\mathfrak{A}^{-1})+K}$ and $D_K(\mathfrak{A})$ are dual k-vector spaces obtained from the bilinear pairing defined by*

$$\begin{aligned} \varphi \colon \operatorname{Dif}_K \times \mathfrak{X}_K &\to k \\ \varphi(w, \xi) &= \sum_{\mathfrak{P}\in\mathbb{P}_K} \operatorname{Res}_{\mathfrak{P}}(\xi_{\mathfrak{P}} w). \end{aligned} \tag{9.31}$$

□

9.4 The Genus Formula

We begin this section by observing that in the last part of the proof of Theorem 9.2.10 we have shown more than is stated. Indeed, assume that L/K is a finite separable geometric extension. If $\mathfrak{A}$ denotes the divisor $\operatorname{con}_{K/L}(\omega)_K$, where ω is a nonzero differential of K and $\Omega = \operatorname{cotr}_{K/L}\omega$, then $\Omega^{\mathfrak{P}}$ vanishes at every $u \in L_{\mathfrak{P}}$ with

$$v_{\mathfrak{P}}(u) \geq -e(\mathfrak{P})\,a(\mathfrak{p}) - m(\mathfrak{P})$$

and there exists an element $u \in L_{\mathfrak{P}}$ such that

$$v_{\mathfrak{P}}(u) = -e(\mathfrak{P})\,a(\mathfrak{p}) - m(\mathfrak{P}) - 1 \quad \text{and} \quad \Omega^{\mathfrak{P}}(u) \neq 0.$$

As an immediate consequence of what was proved in Theorems 9.2.10 and 9.1.5, we have the following theorem:

Theorem 9.4.1. *Let L/K be a finite separable geometric extension of function fields, ω a nonzero differential of K, and $\Omega = \operatorname{cotr}_{K/L}\omega$. Then $\Omega \neq 0$ and $(\Omega)_L = \mathfrak{D}_{L/K}\operatorname{con}_{K/L}(\omega)_K$.*

Proof According to what was seen in Theorem 9.2.10, the exponent of $\mathfrak{P}$ appearing in $(\Omega)_L$ is

$$e(\mathfrak{P})\,a(\mathfrak{p}) + m(\mathfrak{P}),$$

where $\mathfrak{p} = \mathfrak{P}|_K$, $m(\mathfrak{P})$ is the exponent of $\mathfrak{P}$ in $\mathfrak{D}_{L/K}$, $a(\mathfrak{p})$ is the exponent of $\mathfrak{p}$ in $(\omega)_K$, and $e(\mathfrak{P})$ is the ramification index of $\mathfrak{P}$ over K. On the other hand, $e(\mathfrak{P})\,a(\mathfrak{p}) + m(\mathfrak{P})$ is the exponent of $\mathfrak{P}$ appearing in the divisor $\mathfrak{D}_{L/K}\operatorname{con}_{K/L}(\omega)_K$. This proves the result. □

As a corollary we obtain the Riemann–Hurwitz genus formula:

Theorem 9.4.2 (Riemann–Hurwitz Genus Formula). *Let L/K be a finite geometric separable extension of function fields and g_L, g_K the genera of L and K respectively. If $d_L\left(\mathfrak{D}_{L/K}\right)$ denotes the degree of the different of the extension, we have*

$$g_L = 1 + [L : K]\,(g_K - 1) + \frac{1}{2} d_L\left(\mathfrak{D}_{L/K}\right).$$

Proof. By Corollary 3.5.5 the degree of the divisor of any nonzero differential in a field E is $2g_E - 2$. On the other hand, by Theorem 5.3.4 we have $d_L\left(\operatorname{con}_{K/L}(\omega)_K\right) = [L : K] d_K\left((\omega)_K\right)$. Finally, using Theorem 9.4.1 we get

$$\begin{aligned} 2g_L - 2 = d_L\left((\Omega)_L\right) &= d_L\left(\mathfrak{D}_{L/K} \operatorname{con}_{K/L}(\omega)_K\right) \\ &= d_L\left(\mathfrak{D}_{L/K}\right) + d_L\left(\operatorname{con}_{K/L}(\omega)_K\right) \\ &= d_L\left(\mathfrak{D}_{L/K}\right) + [L : K] d_K\left((\omega)_K\right) \\ &= d_L\left(\mathfrak{D}_{L/K}\right) + [L : K]\,(2g_K - 2)\,, \end{aligned}$$

from which the result follows. □

Now we consider L/K to be an arbitrary finite separable extension of function fields. Let ℓ and k be the fields of constants of L and K respectively. Then by Proposition 5.2.20 and Corollary 8.4.7, ℓ is the field of constants of $K\ell$ and

$$[K\ell : K] = [\ell : k]. \tag{9.32}$$

Now, by Proposition 5.2.32, $K\ell/K$ is unramified and every place is separable. Hence $\mathfrak{D}_{K\ell/K} = \mathfrak{N}$ (Proposition 5.6.7). Using Theorem 5.7.15 we get

$$\mathfrak{D}_{L/K} = \mathfrak{D}_{L/K\ell} \tag{9.33}$$

and by Theorem 8.5.2 we have

$$g_{K\ell} = g_K. \tag{9.34}$$

Since $L/K\ell$ is a geometric extension we obtain from Theorem 9.4.2, (9.32), (9.33), and (9.34) that

$$\begin{aligned} g_L &= 1 + [L : K\ell](g_{K\ell} - 1) + \frac{1}{2} d_L(\mathfrak{D}_{L/K\ell}) \\ &= 1 + \frac{[L : K]}{[K\ell : K]}(g_K - 1) + \frac{1}{2} d_L(\mathfrak{D}_{L/K}) \\ &= 1 + \frac{[L : K]}{[\ell : k]}(g_K - 1) + \frac{1}{2} d_L(\mathfrak{D}_{L/K}). \end{aligned}$$

Thus we have proved the following generalization of the Riemann–Hurwitz genus formula.

Corollary 9.4.3 (Riemann–Hurwitz Genus Formula). *Let L/K be a finite separable extension of function fields. If ℓ and k denote the fields of constants of L and K respectively, then*

$$g_L = 1 + \frac{[L:K]}{[\ell:k]}(g_K - 1) + \frac{1}{2}d_L\big(\mathfrak{D}_{L/K}\big). \qquad \square$$

Example 9.4.4. Here we will apply the genus formula to obtain g_K, where $K = k(x, y)$, $y^2 = f(x)$, $f(x) \in k[x]$ is square-free, and char $K \neq 2$ (that is, what we have already done in Section 4.3). Let $f(x) = p_1(x) \ldots p_r(x)$, $m = \deg f = \sum_{i=1}^{r} \deg p_i$. Set $\mathfrak{Z}_{p_i(x)} = \mathfrak{p}_i$. By Example 5.8.9, the ramified prime divisors are $\mathfrak{p}_1 \ldots, \mathfrak{p}_r$ and possibly $\mathfrak{p}_\infty$. Moreover, $\mathfrak{p}_\infty$ is ramified if and only if m is odd.

Since char $K \neq 2$, it follows that $\mathfrak{D}_{K/k(x)} = \mathfrak{p}_1 \cdots \mathfrak{p}_r \mathfrak{p}_\infty^{\varepsilon}$ with

$$\varepsilon = \begin{cases} 0 \text{ if m is even,} \\ 1 \text{ if m is odd} \end{cases}$$

(Theorem 5.6.3). Therefore $d\left(\mathfrak{D}_{L/K}\right) = m+\varepsilon$. Now $g_{k(x)} = 0$, so using the Riemann–Hurwitz formula we obtain

$$\begin{aligned} g_K &= 1 + 2(0-1) + \frac{1}{2}(m+\varepsilon) = \frac{m+\varepsilon-2}{2} \\ &= \begin{cases} \frac{m}{2} - 1 & \text{if m is even,} \\ \frac{m+1}{2} - 1 = \frac{m-1}{2} & \text{if m is odd,} \end{cases} \end{aligned}$$

which coincides with Corollary 4.3.7.

Example 9.4.5. Let $y^n = f(x) \in k[x]$, where k is a perfect field and

$$f(x) = p_1(x)^{\lambda_1} \cdots p_r(x)^{\lambda_r}, \quad \text{with} \quad 0 < \lambda_i < n \quad \text{for} \quad 1 \leq i \leq r,$$

and $p_1(x), \ldots, p_r(x)$ are distinct irreducible polynomials.

Let $K = k(x, y)$, and assume that the nth roots of 1 are contained in k, and that char $K \nmid n$ or char $K = 0$. Set $(p_i(x))_{k(x)} = \frac{\mathfrak{p}_i}{\mathfrak{p}_\infty^{\deg p_i}}$ and $m_i = \deg p_i(x)$. Let

$$\operatorname{con}_{k(x)/K}(\mathfrak{p}_i) = \left(\mathfrak{P}_i^{(1)} \cdots \mathfrak{P}_i^{(g_i)}\right)^{n/d_i}$$

with $d_i = (\lambda_i, n)$. For convenience we will assume that $\mathfrak{p}_\infty$ is not ramified, that is, n divides $\deg f(x)$ (Example 5.8.9).

Finally, let f_i be the relative degree of $\mathfrak{P}_i^{(j)}$ over $\mathfrak{p}_i$, so that $f_i m_i$ is equal to the degree of $\mathfrak{P}_i^{(j)}$.

Then

$$\mathfrak{D}_{K/k(x)} = \prod_{i=1}^{r} \left(\mathfrak{P}_i^{(1)} \cdots \mathfrak{P}_i^{(g_i)}\right)^{(n/d_i)-1}$$

and

$$d\left(\mathfrak{D}_{K/k(x)}\right)=\sum_{i=1}^{r}\left(\frac{n}{d_i}-1\right)(g_i)\,d_K\left(\mathfrak{P}_i^{(j)}\right)=\sum_{i=1}^{r}\left(\frac{n}{d_i}-1\right)g_i f_i m_i$$
$$=\sum_{i=1}^{r}\frac{n}{d_i}g_i f_i m_i-\sum_{i=1}^{r}g_i f_i m_i.$$

We have $\frac{n}{d_i}g_i f_i m_i=[K:k(x)]m_i=nm_i$, so

$$d\left(\mathfrak{D}_{K/k(x)}\right)=n\sum_{i=1}^{r}m_i-\sum_{i=1}^{r}d_i m_i=n\deg f(x)-\sum_{i=1}^{r}d_i m_i.$$

Therefore

$$g_K=1+n(0-1)+\frac{1}{2}\left(n\deg f(x)-\sum_{i=1}^{r}d_i m_i\right)$$
$$=\frac{1}{2}\left(n\deg f(x)+2-2n-\sum_{i=1}^{r}(\lambda_i,n)\deg p_i(x)\right).$$

Example 9.4.6. Let $K=k(x,y)$, where

$$y^p-y=\frac{f(x)}{p_1(x)^{\lambda_1}\cdots p_r(x)^{\lambda_r}},$$

$f(x),\ p_i(x)\in k[x]$, $p_1(x),\ldots,p_r(x)$ are distinct irreducible polynomials, $\lambda_i>0$, $p\nmid\lambda_i$, $\operatorname{char}k=p$, and k is a perfect field. For convenience we will assume that $\mathfrak{p}_\infty$ is not ramified. By Example 5.8.8, if

$$(p_i(x))_{k(x)}=\frac{\mathfrak{p}_i}{\mathfrak{p}_\infty^{\deg p_i}}\quad\text{and}\quad \operatorname{con}_{k(x)/K}(\mathfrak{p}_i)=\mathfrak{P}_i^p,$$

then $\mathfrak{D}_{K/k(x)}=\prod_{i=1}^{r}\mathfrak{P}_i^{(\lambda_i+1)(p-1)}$. It follows that

$$d\left(\mathfrak{D}_{K/k(x)}\right)=\sum_{i=1}^{r}(\lambda_i+1)(p-1)d_K(\mathfrak{P}_i)$$
$$=\sum_{i=1}^{r}(\lambda_i+1)(p-1)d_{k(x)}(\mathfrak{p}_i)$$
$$=\sum_{i=1}^{r}(\lambda_i+1)(p-1)m_i,$$

where $m_i=\deg p_i(x)$. Therefore

$$\begin{aligned} g_K &= 1 + [K : k(x)]\left(g_{k(x)} - 1\right) + \frac{1}{2}\left(d\left(\mathfrak{D}_{K/k(x)}\right)\right) \\ &= 1 + p(0-1) + \frac{1}{2}\sum_{i=1}^{r}(\lambda_i + 1)(p-1)m_i \\ &= \frac{1}{2}(p-1)\left\{\sum_{i=1}^{r}(\lambda_i + 1)\, m_i - 2\right\}. \end{aligned}$$

9.5 Genus Change in Inseparable Extensions

We have studied the genus change in constant extensions and in finite separable extensions. In the latter case, the trace was used to find differentials in a subfield. Since we were considering separable extensions, the trace was nontrivial. When we consider inseparable extensions, the trace is the trivial map and we cannot use the trace map any longer to find nontrivial differentials.

In this section we present a substitute for the trace map due to John Tate [152].

Let E be a field of characteristic $p > 0$ and let F be an inseparable extension of E of degree p. Let α be any generator of F over E, that is, $F = E(\alpha)$. Let $\xi \in F$. Then ξ can be expressed uniquely in terms of α as

$$\xi = a_0 + a_1\alpha + \cdots + a_{p-1}\alpha^{p-1}, \quad \text{with} \quad a_i \in E. \tag{9.35}$$

Definition 9.5.1. We define the nontrivial E-map

$$\begin{aligned} &S_\alpha : F \to E \quad \text{by putting} \\ &S_\alpha(\xi) = a_{p-1} \quad \text{for all} \quad \xi \in F. \end{aligned} \tag{9.36}$$

Proposition 9.5.2. *We have*

$$\xi = \sum_{i=0}^{p-1} S_\alpha(\xi\alpha^{p-1-j})\alpha^j.$$

Proof: Let $X^p - b = \mathrm{Irr}(\alpha, X, E)$. Then

$$\xi\alpha^{p-1-j} = a_0\alpha^{p-1-j} + \cdots + a_j\alpha^{p-1} + a_{j+1}b + \cdots + a_{p-1}b\alpha^{p-1-j-1}.$$

It follows that $S_\alpha(\xi\alpha^{p-1-j}) = a_j$, and the result follows by (9.35) □

Since the map S_α depends on the generator α, the question that arises is how S_α changes when α is replaced by another generator β. First we note that S_α is E-linear.

Let $\phi : F \times F \to E$ be given by

$$\phi(x, y) = S_\alpha(xy).$$

Then ϕ is E-bilinear and if $S_\alpha(z) \neq 0$, then for any $x \neq 0$, $\phi(x, x^{-1}z) = S_\alpha(z) \neq 0$. Thus ϕ is a nondegenerate bilinear form on F. In particular, for any E linear map

$S : F \to E$, there exists an element γ in F uniquely determined by S and such that $S(\xi) = S_\alpha(\xi\gamma)$ for all $\xi \in F$ (Proposition 9.3.18).

In particular, there exists a unique $\gamma \in F$ such that

$$S_\beta(\xi) = S_\alpha(\xi\gamma) \tag{9.37}$$

for all $\xi \in F$.

Definition 9.5.3. Let R be a commutative ring. A *derivation* D of R is a mapping $D : R \to R$ such that

$$D(x+y) = Dx + Dy \quad \text{and} \quad D(xy) = x\,Dy + y\,Dx$$

for all $x, y \in R$.

Example 9.5.4. Let $R = k[x]$, where k is a field. Then for $f(x) = \sum_{i=0}^{n} a_i x^i$, the mapping D defined by $Df(x) = f'(x) = \sum_{i=1}^{n} i a_i x^{i-1}$ is a derivation.

Example 9.5.5. Let $R = k[x_1, \ldots, x_n]$, where k is a field. Then the usual partial derivative $\frac{\partial}{\partial x_i}$ is a derivation of $k[x_1, \ldots, x_n]$.

Given any derivation D of R, $x \in R$, and $n \in \mathbb{N}$, we have $D(x^n) = nx^{n-1}Dx$.

In our case, F is an inseparable extension of E of degree p. Let

$$\begin{aligned} D : E[x] &\to E[x] \\ f(x) &\mapsto f'(x). \end{aligned}$$

We have $((x^p - b)f(x))' = (x^p - b)'f(x) + (x^p - b)f'(x) = (x^p - b)f'(x)$. Thus D maps the principal ideal $x^p - b$ into itself. Since F is inseparable over E, it follows that F is isomorphic to $E[x]/((x^p - b))$ for some $b \in E$. Let α be the root of $x^p - b$ and set $F = E(\alpha)$. Then the kernel of the epimorphism

$$\begin{aligned} \phi : E[x] &\to E(\alpha) = F \\ f(x) &\mapsto f(\alpha) \end{aligned}$$

is the ideal $(x^p - b)$.

It is easy to see that D induces a well-defined derivation in F,

$$\begin{array}{ccc} E[x] & \xrightarrow{\;D\;} & E[x] \\ \phi\big\downarrow & & \big\downarrow\phi \\ E(\alpha) & \xrightarrow[D_\alpha]{} & E(\alpha) \end{array}$$

which will be denoted by D_α. Notice that D_α is given by the formula $D_\alpha(f(\alpha)) = f'(\alpha)$.

If $\xi = a_0 + a_1\alpha + \cdots + a_{p-1}\alpha^{p-1}$, then

$$D_\alpha(\xi) = a_1 + 2a_2\alpha + \cdots + (p-1)a_{p-1}\alpha^{p-2}.$$

It follows that $D_\alpha(\xi) = 0$ if and only if $a_1 = a_2 = \cdots = a_{p-1} = 0$ if and only if $\xi = a_0 \in E$. Also, D_α is E-linear.

Proposition 9.5.6. *We have*

$$S_\alpha(D_\alpha(\xi)) = 0$$

for all $\xi \in F$.

Proof: Let $\xi = a_0 + a_1\alpha + \cdots + a_{p-1}\alpha^{p-1}$. Then

$$S_\alpha(D_\alpha(\xi)) = S_\alpha(a_1 + 2a_2\alpha + \cdots + (p-1)a_{p-1}\alpha^{p-2}) = 0$$

by (9.36). □

Proposition 9.5.7. *The map* S_α *satisfies* $S_\alpha(\xi^{p-1}D_\alpha\xi) = (D_\alpha\xi)^p$ *for all* $\xi \in F$. *Equivalently,*

$$S_\alpha\left(\frac{D_\alpha\xi}{\xi}\right) = \left(\frac{D_\alpha\xi}{\xi}\right)^p$$

for all $\xi \in F \setminus \{0\}$.

Proof: For any $\xi \in F$, ξ^p belongs to E, so if $\xi \neq 0$ we have

$$S_\alpha(\xi^{p-1}D_\alpha\xi) = S_\alpha\left(\frac{\xi^p D_\alpha(\xi)}{\xi}\right) = \xi^p S_\alpha\left(\frac{D_\alpha\xi}{\xi}\right)$$

and $(D_\alpha\xi)^p = \left(\frac{D_\alpha\xi}{\xi}\right)^p \xi^p$. The stated equivalence follows.

Let $R = \left\{\xi \in F \mid S_\alpha(\xi^{p-1}D_\alpha\xi) = (D_\alpha\xi)^p\right\}$. Let $T : F \setminus \{0\} \to E$ be defined by $T\xi = S_\alpha\left(\frac{D_\alpha\xi}{\xi}\right) - \left(\frac{D_\alpha\xi}{\xi}\right)^p$. We have

$$\begin{aligned}
T(\xi\xi_1) &= S_\alpha\left(\frac{D_\alpha(\xi\xi_1)}{\xi\xi_1}\right) - \left(\frac{D_\alpha(\xi\xi_1)}{\xi\xi_1}\right)^p \\
&= S_\alpha\left(\frac{\xi D_\alpha\xi_1 + \xi_1 D_\alpha\xi}{\xi\xi_1}\right) - \left(\frac{\xi D_\alpha\xi_1 + \xi_1 D_\alpha\xi}{\xi\xi_1}\right)^p \\
&= S_\alpha\left(\frac{D_\alpha\xi_1}{\xi_1} + \frac{D_\alpha\xi}{\xi}\right) - \left(\frac{D_\alpha\xi_1}{\xi_1} + \frac{D_\alpha\xi}{\xi}\right)^p = T(\xi) + T(\xi_1).
\end{aligned}$$

Thus T is a group homomorphism of $F \setminus \{0\}$ into E. The kernel of T is $R \setminus \{0\}$, so $R \setminus \{0\}$ is a multiplicative subgroup of $F \setminus \{0\}$.

Now if $\xi \in R$ we have $D_\alpha(\xi + 1) = D_\alpha\xi$ and $((\xi + 1)^{p-1} - \xi^{p-1})D_\alpha\xi = \sum_{i=0}^{p-2} a_i\xi^i D_\alpha\xi$ for some $a_i \in E$.

Also, $\xi^i D_\alpha\xi = D_\alpha\left(\frac{\xi^{i+1}}{i+1}\right)$. Hence, using Proposition 9.5.6 we obtain

$$S_\alpha(((\xi + 1)^{p-1} - \xi^{p-1})D_\alpha\xi) = \sum_{i=0}^{p-1} a_i S_\alpha\left(D_\alpha\left(\frac{\xi^{i+1}}{i+1}\right)\right) = 0.$$

In particular, since $\xi \in R$ we have

$$\begin{aligned} S_\alpha((\xi+1)^{p-1}D_\alpha(\xi+1)) &= S_\alpha((\xi+1)^{p-1}D_\alpha\xi) \\ &= S_\alpha(\xi^{p-1}D_\alpha\xi) = (D_\alpha\xi)^p = (D_\alpha(\xi+1))^p. \end{aligned}$$

It follows that $\xi+1 \in R$. Finally, if $\xi \in R$ and η is a nonzero element of R, we have $\xi+\eta = \eta(\eta^{-1}\xi+1) \in R$. Thus $R\backslash\{0\}$ is a multiplicative group and R is closed under addition. Hence $E \subseteq R$ and $\alpha \in R$, so R is a subfield of F containing E and α. Therefore $E(\alpha) \subseteq R \subseteq F = E(\alpha)$. □

Now we can find the relationship between two generators α and β.

Theorem 9.5.8. *If α and β are two generators of F over E, then*

$$S_\beta(\xi) = S_\alpha(\xi(D_\alpha\beta)^{1-p}) \quad \text{for all} \quad \xi \in F. \tag{9.38}$$

Proof: Since both sides of (9.38) are E-linear, it suffices to prove (9.38) for $\xi = \beta^i$ $(0 \le i \le p-1)$.

Multiplying both sides by $(D_\alpha\beta)^p \in E$, the equality becomes

$$(D_\alpha\beta)^p S_\beta(\beta^i) = S_\alpha(\beta^i D_\alpha\beta) \quad (0 \le i \le p-1). \tag{9.39}$$

For $i < p-1$, we have $\beta^i D_\alpha\beta = D_\alpha\left(\frac{\beta^{i+1}}{i+1}\right)$, so by Proposition 9.5.6 and (9.35), both sides of (9.39) are equal to zero.

For $i = p-1$, we have $(D_\alpha\beta)^p S_\beta(\beta^{p-1}) = (D_\alpha\beta)^p$. Therefore by (9.35) and Proposition 9.5.7 we have

$$S_\alpha(\beta^{p-1}D_\alpha\beta) = (D_\alpha\beta)^p.$$

Thus (9.38) holds also for $i = p-1$. □

Now we establish some basic facts about an inseparable extension L/K of function fields of degree p^n.

Proposition 9.5.9. *Let K be a function field, L a purely inseparable extension of K, and $\mathfrak{P}$ a place of L lying over the place $\wp$ of K. Then the local degree satisfies*

$$[L_\mathfrak{P} : K_\wp] = [L : K].$$

Proof: By Theorem 5.2.24, $\mathfrak{P}$ is the only place above $\wp$. On the other hand, using Theorem 5.1.14, the proof of Corollary 5.4.6, and Theorem 5.4.10 we obtain

$$[L : K] = \dim_K L = \sum_{\mathfrak{P}|\wp}[L_\mathfrak{P} : K_\wp] = [L_\mathfrak{P} : K_\wp]. \qquad \square$$

Corollary 9.5.10. *Any repartition $\xi \in \mathfrak{X}_L$ of L can be written uniquely in the form*

$$\xi = \xi_0 + \xi_1\alpha + \cdots + \xi_{p^n-1}\alpha^{p^n-1},$$

where $\xi_0, \ldots, \xi_{p^n-1} \in \mathfrak{X}_K$ are repartitions of K, $L = K(\alpha)$, and $[L : K] = p^n$.

Proof: The statement follows immediately from Corollary 5.5.8, Proposition 9.5.9, and the fact that $\{1, \alpha, \ldots, \alpha^{p^n-1}\}$ is a basis of $L_{\mathfrak{P}}$ over $K_{\wp}$. □

Proposition 9.5.11. *If L is a purely inseparable extension of K of degree p, then for any place $\mathfrak{P}$ of L lying over the place $\wp$ of K, there exists $\beta \in \vartheta_{\mathfrak{P}}$ such that*

$$\vartheta_{\mathfrak{P}} = \vartheta_{\wp}[\beta]. \tag{9.40}$$

Thus $\vartheta_{\mathfrak{P}}$ has a power basis over $\vartheta_{\wp}$.

Proof: We have $[L_{\mathfrak{P}} : K_{\wp}] = p$. If $L_{\mathfrak{P}}$ is unramified over $K_{\wp}$ let $\beta \in \vartheta_{\mathfrak{P}}$ be such that $\overline{\beta}$ generates $\ell(\mathfrak{P})$ over $k(\wp)$ (i.e., $\beta \in \vartheta_{\mathfrak{P}} \backslash (\vartheta_{\wp} + \mathfrak{P})$). Since

$$\vartheta_{\mathfrak{P}} \cong \ell(\mathfrak{P})[[\pi]] \quad \text{and} \quad \vartheta_{\wp} \cong k(\wp)[[\pi]],$$

where $\pi \in \vartheta_{\wp}$ satisfies $v_{\mathfrak{P}}(\pi) = v_{\wp}(\pi) = 1$ (Theorem 2.5.20), it follows that $\vartheta_{\mathfrak{P}} = \vartheta_{\wp}[\beta]$.

If $L_{\mathfrak{P}}$ is ramified over $K_{\wp}$ and π_L is a prime element for $\mathfrak{P}$, then $\pi_L^p = \pi_K \in K_{\wp}$ is a prime element for $\wp$. We have $\ell(\mathfrak{P}) = k(\wp) = m$, $\vartheta_{\mathfrak{P}}$ is isomorphic to $m[[\pi_L]]$, and $\vartheta_{\wp}$ to $m[[\pi_K]]$. It follows that $\vartheta_{\mathfrak{P}} = \vartheta_{\wp}[\pi_L]$. □

The values $v_{\mathfrak{P}}(D_\alpha \beta)$ are fundamental for the genus formula we will establish below.

Proposition 9.5.12. *Let L be a purely inseparable extension of K of degree p, and $L = K(\alpha)$. Let $\wp$ be the place of K that lies below the place $\mathfrak{P}$ of L. Set*

$$r_{\wp} = r = \max_{x \in K_{\wp}} \{v_{\wp}(a - x^p)\},$$

where $\alpha^p = a \in K$. Then

$$p^n v_{\mathfrak{P}}(D_\beta \alpha) \deg_L \mathfrak{P} = \begin{cases} r \deg_K \wp & \text{if} \quad p \mid r, \\ (r-1) \deg_K \wp & \text{if} \quad p \nmid r, \end{cases}$$

where $\beta \in \vartheta_{\mathfrak{P}}$ satisfies $\vartheta_{\mathfrak{P}} = \vartheta_{\wp}[\beta]$ and $p^n = [\ell : k]$.

Proof: Since $L_{\mathfrak{P}} = K_{\wp}(\alpha) = K_{\wp}(a^{1/p})$ and $[L_{\mathfrak{P}} : K_{\wp}] = p$, it follows that a is not a pth power in $K_{\wp}$. Therefore $a - x^p \neq 0$ for all $x \in K_{\wp}$ and r is finite.

Let $b \in K_{\mathfrak{P}}$ be such that $r = v_{\wp}(a - b^p)$.

If p divides r, put $r = sp$. Let π be a prime element in $K_{\wp}$ such that $v_{\wp}(\pi) = 1$ and set $\tau = (\alpha - b)\pi^{-s} \in L_{\mathfrak{P}}$. Then $\tau^p = (\alpha^p - b^p)\pi^{-sp} = (a - b^p)\pi^{-sp}$ satisfies $v_{\mathfrak{P}}(\tau^p) = r - sp = 0$. Therefore τ^p is a unit in $\vartheta_{\wp}$.

If the residue class of τ^p were a pth power of a residue class in $K_{\wp}$, there would exist $c \in K_{\wp}$ such that

$$c^p \equiv (a - b^p)\pi^{-sp} \bmod \wp.$$

Then if $x = b + \pi^s c$, x would satisfy

$$\begin{aligned} v_{\mathfrak{P}}(a - x^p) &= v_{\wp}(a - b^p - \pi^{sp} c^p) = v_{\wp}(\pi^{sp}((a - b^p)\pi^{-sp} - c^p)) \\ &= sp + v_{\wp}((a - b^p)\pi^{-sp} - c^p) \geq sp + 1 = r + 1. \end{aligned}$$

This contradicts the maximality of r. It follows that

$$[k(\wp)(\overline{\tau}) : k(\wp)] = p \quad \text{and} \quad \ell(\mathfrak{P}) = k(\wp)(\overline{\tau}).$$

In this case Proposition 9.5.11 yields $\vartheta_{\mathfrak{P}} = \vartheta_{\wp}[\beta]$ with $\beta = \tau$.

We have $D_\beta \alpha = (D_\alpha \beta)^{-1} = \pi^s$ (see Exercise 9.7.7). Thus $v_{\mathfrak{P}}(D_\beta \alpha) = s$ and by Theorem 5.3.4,

$$\begin{aligned} v_{\mathfrak{P}}(D_\beta \alpha) p^n \deg_L \mathfrak{P} = sp^n \deg_L \mathfrak{P} &= sp^n \frac{\deg_K \wp}{\lambda_{L/K}} \\ &= sp^n \frac{\deg_K \wp}{[\ell : k]}[L : K] = sp^n \frac{\deg_K \wp}{p^n} p \\ &= ps \deg_K \wp = r \deg_K \wp. \end{aligned}$$

Now assume that p does not divide r. Let $u, v \in \mathbb{Z}$ be such that $ru - pv = 1$. Let π be a prime element in $K_\wp$ and $\tau = (\alpha - b)^u \pi^{-v} \in L_{\mathfrak{P}}$. Then $\tau^p = (a - b^p)^u \pi^{-vp}$ satisfies

$$v_\wp(\tau^p) = ru - vp = 1.$$

It follows that τ is a prime element in $L_{\mathfrak{P}}$ and by Proposition 9.5.11, $\vartheta_{\mathfrak{P}} = \vartheta_\wp[\beta]$ with $\beta = \tau$, and $L_{\mathfrak{P}}/K_\wp$ is a ramified extension.

We have

$$D_\alpha \beta = D_\alpha \tau = u(\alpha - b)^{u-1}\pi^{-v} = u(\alpha - b)^u \pi^{-v}(\alpha - b)^{-1} = u(\alpha - b)^{-1}\tau.$$

Hence $D_\beta \alpha = D_\tau \alpha = (D_\alpha \tau)^{-1} = u^{-1}(\alpha - b)\tau^{-1}$ and since $(u, p) = 1$, it follows that

$$v_{\mathfrak{P}}(D_\beta \alpha) = v_{\mathfrak{P}}((a - b^p)^{1/p}\tau^{-1}) = r - 1.$$

We have $\wp = \mathfrak{P}^p$. Moreover, $\deg_L \wp = p \deg_L \mathfrak{P} = \frac{\deg_K \wp}{\lambda_{L/K}} = p \deg_K \wp$ because $\lambda_{L/K} = \frac{[\ell:k]}{[L:K]} = \frac{1}{p}$ $(n = 0)$.

Thus $v_{\mathfrak{P}}(D_\beta \alpha) p^n \deg_L \mathfrak{P} = (r - 1) \deg_K \wp$. □

The map S_α given in Definition 9.5.1 can be extended to a K-linear map of $\mathfrak{X}_L$ into $\mathfrak{X}_K$ as follows.

Definition 9.5.13. Let K be a function field, L a purely inseparable extension of K of degree p, and α a generator of L over K. For $\xi \in \mathfrak{X}_L$, ξ can be written as

$$\xi = \xi_0 + \xi_1 \alpha + \cdots + \xi_{p-1}\alpha^{p-1} \quad \text{with} \quad \xi_0, \ldots, \xi_{p-1} \in \mathfrak{X}_K. \tag{9.41}$$

We define the K-linear map

$$S_\alpha : \mathfrak{X}_L \to \mathfrak{X}_K$$

by

$$S_\alpha(\xi) = \xi_{p-1}.$$

Proposition 9.5.14. *Let S_α be the K-linear map that we have just defined. Given a divisor $\mathfrak{A} \in D_K$ there exists $\mathfrak{U} \in D_L$ such that $\mathfrak{A} \mid S_\alpha(\xi) = \xi_{p-1}$ whenever $\mathfrak{U}$ divides ξ.*

Proof: Let $\mathfrak{A} \in D_K$ be an arbitrary divisor. For any place $\mathfrak{P}$ of L lying over the place $\wp$ of K, let $\beta_{\mathfrak{P}} \in \vartheta_{\mathfrak{P}}$ be such that $\vartheta_{\mathfrak{P}} = \vartheta_\wp[\beta_{\mathfrak{P}}]$ (Proposition 9.5.11).

Let $\pi_\wp$ be a prime element for $\wp$. Set $\mathfrak{U} = \prod_{\mathfrak{P}} \mathfrak{P}^{c_{\mathfrak{P}}}$, where

$$c_{\mathfrak{P}} = e(\mathfrak{P}|\wp)v_\wp(\mathfrak{A}) + v_{\mathfrak{P}}(D_{\beta_{\mathfrak{P}}}\alpha)^{p-1}. \tag{9.42}$$

According to Corollary 5.5.9 we may choose $\beta_{\mathfrak{P}} = \alpha$ for almost all $\mathfrak{P} \in D_L$. In particular, $v_{\mathfrak{P}}((D_{\beta_{\mathfrak{P}}}\alpha)^{p-1}) = 0$ for almost all $\mathfrak{P}$. Thus $\mathfrak{U}$ is a divisor in L. Further, the $\mathfrak{P}$th component in (9.41) is

$$\xi_{\mathfrak{P}} = \xi_{0,\mathfrak{P}} + \xi_{1,\mathfrak{P}}\alpha + \cdots + \xi_{p-1,\mathfrak{P}}\alpha^{p-1} \in L_{\mathfrak{P}}, \text{ with } \xi_{i,\mathfrak{P}} \in K_\wp \text{ for } 0 \le i \le p-1.$$

If $\mathfrak{U}$ divides ξ, then $v_{\mathfrak{P}}(\xi_{\mathfrak{P}}) \ge v_{\mathfrak{P}}(\mathfrak{U}) = c_{\mathfrak{P}}$. Therefore

$$v_{\mathfrak{P}}(\xi_{\mathfrak{P}}(D_{\beta_{\mathfrak{P}}}\alpha)^{1-p}\pi_\wp^{-v_\wp(\mathfrak{A})}) \ge c_{\mathfrak{P}} + v_{\mathfrak{P}}((D_{\beta_{\mathfrak{P}}}\alpha)^{1-p}) - e(\mathfrak{P}|\wp)v_\wp(\mathfrak{A}) \ge 0.$$

It follows that $\xi_{\mathfrak{P}}(D_{\beta_{\mathfrak{P}}}\alpha)^{1-p}\pi_\wp^{v_\wp(\mathfrak{A})} \in \vartheta_{\mathfrak{P}}$. By Proposition 9.5.2 we have

$$\xi_{\mathfrak{P}}(D_{\beta_{\mathfrak{P}}}\alpha)^{1-p}\pi_\wp^{-v_\wp(\mathfrak{A})} = \sum_{i=0}^{p-1} S_{\beta_{\mathfrak{P}}}(\xi_{\mathfrak{P}}(D_{\beta_{\mathfrak{P}}}\alpha)^{1-p}\pi_\wp^{-v_\wp(\mathfrak{A})}\beta_{\mathfrak{P}}^{p-1-i})\beta_{\mathfrak{P}}^i.$$

Now, $\{1, \beta_{\mathfrak{P}}, \ldots, \beta_{\mathfrak{P}}^{p-1}\}$ is an integral basis of $\vartheta_{\mathfrak{P}}$ over $\vartheta_\wp$, so

$$S_{\beta_{\mathfrak{P}}}(\xi_{\mathfrak{P}}(D_{\beta_{\mathfrak{P}}}\alpha)^{1-p}\pi_\wp^{-v_\wp(\mathfrak{A})}\beta_{\mathfrak{P}}^{p-1-i}) \in \vartheta_\wp.$$

In particular, for $i = p-1$ we have

$$S_{\beta_{\mathfrak{P}}}(\xi_{\mathfrak{P}}(D_{\beta_{\mathfrak{P}}}\alpha)^{1-p}\pi_\wp^{-v_\wp(\mathfrak{A})}) \in \vartheta_\wp. \tag{9.43}$$

We obtain from Theorem 9.5.8 that

$$S_\alpha(\xi_{\mathfrak{P}}\pi_\wp^{-v_\wp(\mathfrak{A})}) = S_{\beta_{\mathfrak{P}}}(\xi_{\mathfrak{P}}(D_{\beta_{\mathfrak{P}}}\alpha)^{1-p}\pi_\wp^{-v_\wp(\mathfrak{A})}). \tag{9.44}$$

Using (9.43) and (9.44), it follows that

$$v_\wp(S_\alpha(\xi_{\mathfrak{P}}\pi_\wp^{-v_\wp(\mathfrak{A})})) = v_\wp(\pi_\wp^{-v_\wp(\mathfrak{A})}S_\alpha(\xi_{\mathfrak{P}})) = -v_\wp(\mathfrak{A}) + v_\wp(S_\alpha(\xi_{\mathfrak{P}})) \ge 0.$$

Therefore $\mathfrak{A}$ divides $S_\alpha(\xi_{\mathfrak{P}})$. $\square$

Definition 9.5.15. Let w be a nontrivial differential of K. We define

$$\Omega\colon \mathfrak{X}_L \to k \quad \text{by} \quad \Omega(\xi) = w(S_\alpha(\xi)).$$

Assume that $y \in L$ and ξ_y is the principal repartition (i.e., $(\xi_y)_{\mathfrak{P}} = y$ for $\mathfrak{P} \in \mathbb{P}_L$). Then if

$$y = a_0 + \cdots + a_{p-1}\alpha^{p-1}$$

and $\xi_{p-1,a_{p-1}}$ is the principal repartition of K (i.e., $\left(\xi_{p-1,a_{p-1}}\right)_{\wp} = a_{p-1}$), we have $\Omega(\xi_y) = w(S_\alpha(\xi_y)) = w(\xi_{p-1,a_{p-1}}) = 0$.

By Proposition 9.5.14, there exists a divisor $\mathfrak{U}$ in L such that if $\mathfrak{U}$ divides ξ then (w) divides $S_\alpha(\xi)$; so if $\mathfrak{U}$ divides ξ we have

$$\Omega(\xi) = w(S_\alpha(\xi)) = 0.$$

In particular we have the following proposition:

Proposition 9.5.16. *If $\ell = k$, i.e., L is a geometric extension, then the map Ω given in Definition 9.5.15 is a nontrivial differential in L.* □

We are interested in the genus change from K to L. We might proceed as at the end of Section 9.4, namely assuming first that L/K is a geometric extension and finding a formula relating g_L to g_K. Then we would apply the constant field extension and use the results of Chapter 8.

Instead of this approach we prove in general that the map Ω given in Definition 9.5.15 can be replaced by a true differential of L. For this purpose we prove the following theorem:

Theorem 9.5.17. *Let k be any field and let ℓ be a finite extension of k. Let $T : \ell \to k$ be a nontrivial k-linear map of ℓ into k. Then if V is any vector space over ℓ and Ω is any k-linear map of V into k, there exists a uniquely determined ℓ-map Λ from V into ℓ such that*

$$\Omega = T\Lambda, \quad \text{i.e.,} \quad \Omega(\xi) = T(\Lambda(\xi))$$

for all $\xi \in V$.

Proof:

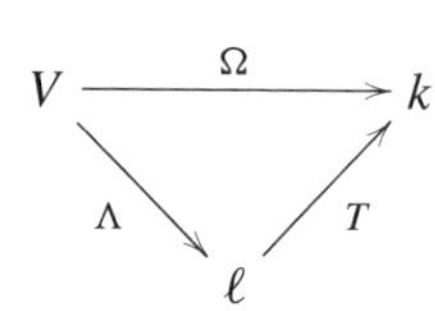

If such a map Λ actually exists, it must satisfy

$$T(\alpha\Lambda(\xi)) = T(\Lambda(\alpha\xi)) = \Omega(\alpha\xi)$$

for all $\alpha \in \ell$. If we fix $\xi \in V$, let $\varphi_\xi : \ell \to k$ be defined by $\varphi_\xi(\alpha) = \Omega(\alpha\xi)$. Then φ_ξ is a linear map from ℓ into k.

Since T is nontrivial, there exists a unique element α_ξ in ℓ such that

$$\varphi_\xi(\alpha) = T(\alpha\alpha_\xi) \quad \text{for all} \quad \alpha \in \ell$$

(apply Proposition 9.3.18 to the nondegenerate form

$$\phi : \ell \times \ell \to k \quad \text{such that} \quad \phi(a, b) = T(ab)).$$

Let $\Lambda : V \to \ell$ be defined by $\Lambda(\xi) = \alpha_\xi$. Then

$$T(\Lambda(\xi)) = T(\alpha_\xi) = \varphi_\xi(1) = \Omega(1 \times \xi) = \Omega(\xi)$$

and $T(\alpha\Lambda(\xi)) = \Omega(\alpha\xi)$.

Given $\alpha \in \ell$, $a, b \in \ell$, and $\xi, \xi \in V$, we have

$$\begin{aligned} T(\alpha\Lambda(a\xi + b\xi_1)) &= \Omega(\alpha(a\xi + b\xi_1)) = \Omega(\alpha a\xi) + \Omega(\alpha b\xi_1) \\ &= T(\alpha a\Lambda(\xi)) + T(\alpha b\Lambda(\xi_1)) = T(\alpha(a\Lambda(\xi) + b\Lambda(\xi_1))). \end{aligned}$$

Therefore $\alpha(\Lambda(a\xi + b\xi_1) - a\Lambda(\xi) - b\Lambda(\xi)) \in \ker T$ for all $\alpha \in \ell$.

Since T is nontrivial, there exists $w \in \ell$ such that $T(w) \neq 0$. Given any nonzero $v \in \ell$, let $\alpha = wv^{-1} \in \ell$ be such that $T(\alpha v) = T(w) \neq 0$. We have

$$\Lambda(a\xi + b\xi_1) = a\Lambda(\xi) + b\Lambda(\xi) \quad \text{for all} \quad a, b \in \ell \quad \text{and} \quad \xi, \xi, \in V_1. \qquad \square$$

Returning to our case, let $T : \ell \to k$ be an arbitrary but fixed nontrivial map from ℓ into k (where ℓ and k are the constant fields of L and K respectively).

Given a nontrivial differential w of K, let Ω be given as in Definition 9.5.15, that is, $\Omega(\xi) = w(S_\alpha(\xi))$ for all $\xi \in \mathfrak{X}_L$.

Consider the ℓ-linear map $\Lambda : \mathfrak{X}_L \to \ell$ defined in Theorem 9.5.17 and satisfying

$$T(\Lambda(\xi)) = \Omega(\xi) = w(S_\alpha(\xi)). \tag{9.45}$$

Corollary 9.5.18. *The map Λ satisfying* (9.45) *is a nontrivial differential of L.* $\square$

Recall that Λ depends on the choice of α. In order to compute the divisor of Λ, we define

$$\mathfrak{D}_\alpha = \prod_{\mathfrak{P}} \mathfrak{P}^{\gamma_\mathfrak{P}}, \tag{9.46}$$

where $\gamma_\mathfrak{P} = v_\mathfrak{P}((D_{\beta_\mathfrak{P}}\alpha)^{1-p})$, $\vartheta_\mathfrak{P} = \vartheta_\wp[\beta_\mathfrak{P}]$, and $\wp = \mathfrak{P}|_K$.

Theorem 9.5.19. *Let L/K be a purely inseparable extension of degree p of function fields with $L = K(\alpha)$. If w is a nontrivial differential of K and Λ is given as in* (9.45), *then the divisors of Λ and w are related by the formula*

$$(\Lambda)_L = (\operatorname{con}_{K/L}(w)_K)\mathfrak{D}_\alpha,$$

where $\mathfrak{D}_\alpha$ is defined as in (9.46).

Proof: Let $\mathfrak{U} = \operatorname{con}_{K/L}(w)_K \mathfrak{D}_\alpha = \prod_{\mathfrak{P}} \mathfrak{P}^{c_{\mathfrak{P}}}$, where

$$c_{\mathfrak{P}} = e(\mathfrak{P}|\wp) v_\wp((w)_K) + v_{\mathfrak{P}}((D_{\beta_{\mathfrak{P}}}\alpha)^{1-p}).$$

Let $\xi \in \mathfrak{X}_L$ and

$$\xi = \xi_0 + \xi_1\alpha + \cdots + \xi_{p-1}\alpha^{p-1}, \quad \text{with} \quad \xi_0, \ldots, \xi_{p-1} \in \mathfrak{X}_K.$$

Each component $\xi_{\mathfrak{P}} \in L_{\mathfrak{P}}$ ($\mathfrak{P} \in \mathbb{P}_L$) satisfies

$$\xi_{\mathfrak{P}} = \xi_{0,\mathfrak{P}} + \xi_{1,\mathfrak{P}}\alpha + \cdots + \xi_{p-1,\mathfrak{P}}\alpha^{p-1} \in L_{\mathfrak{P}}, \quad \text{with} \quad \xi_{0,\mathfrak{P}}, \ldots, \xi_{p-1,\mathfrak{P}} \in K_\wp.$$

If $\mathfrak{U}^{-1}$ divides ξ, then for any $a \in \ell$, $\mathfrak{U}^{-1}$ divides $a\xi$ and by Proposition 9.5.14 (see (9.42)), $(w)_K^{-1}$ divides $S_\alpha(a\xi)$. It follows that

$$T(a\Lambda(\xi)) = \Omega(a\xi) = w(S_\alpha(a\xi)) = 0$$

for all $a \in \ell$. Therefore $\Lambda(\xi) = 0$, and $\mathfrak{U}$ divides $(\Lambda)_L$. We have

$$v_{\mathfrak{P}}((\Lambda)_L) \geq c_{\mathfrak{P}}. \tag{9.47}$$

Now let $\vartheta_{\mathfrak{P}} = \vartheta_\wp[\beta_{\mathfrak{P}}]$ and $\beta_{\mathfrak{P}} \in \vartheta_{\mathfrak{P}}$.

Let $\xi_{\mathfrak{P}} \in L \subseteqq L_{\mathfrak{P}}$ be such that $v_{\mathfrak{P}}(\xi_{\mathfrak{P}}) = -v_{\mathfrak{P}}((D_{\beta_{\mathfrak{P}}}\alpha)^{1-p}) - 1$.

Then $v_{\mathfrak{P}}(\xi_{\mathfrak{P}}(D_{\beta_{\mathfrak{P}}}\alpha)^{1-p}) = -1$ and by Proposition 9.5.2 and Theorem 9.5.8, we have

$$\begin{aligned}
\xi_{\mathfrak{P}}(D_{\beta_{\mathfrak{P}}}\alpha)^{1-p} &= \sum_{i=0}^{p-1} S_{\beta_{\mathfrak{P}}}(\xi_{\mathfrak{P}}(D_{\beta_{\mathfrak{P}}}\alpha)^{1-p}\beta_{\mathfrak{P}}^{p-1-i})\beta_{\mathfrak{P}}^i \\
&= \sum_{i=0}^{p-1} S_\alpha(\xi_{\mathfrak{P}}\beta_{\mathfrak{P}}^{p-1-i})\beta_{\mathfrak{P}}^i \notin \vartheta_{\mathfrak{P}}.
\end{aligned}$$

Therefore there exists i such that $0 \leq i \leq p-1$ and

$$S_\alpha(\xi_{\mathfrak{P}}\beta_{\mathfrak{P}}^{p-1-i}) \notin \vartheta_\wp.$$

Also, there exists $y \in K$ such that $v_\wp(y) = -v_\wp((w)_K) - 1$ and $w^\wp(y) \neq 0$. The definition of Λ establishes that in local components

$$T(\Lambda^{\mathfrak{P}}(\xi)) = w^\wp(S_\alpha(\xi)) \quad \text{for all} \quad \xi \in L_{\mathfrak{P}}.$$

Let $z = (S_\alpha(\xi_{\mathfrak{P}}\beta_{\mathfrak{P}}^{p-1-i}))^{-1} \in \vartheta_\wp$ be such that $v_\wp(z) \geq 1$. Then

$$\begin{aligned}
T(\Lambda^{\mathfrak{P}}(yz\xi_{\mathfrak{P}}\beta_{\mathfrak{P}}^{p-1-i})) &= w^\wp(S_\alpha(yz\xi_{\mathfrak{P}}\beta_{\mathfrak{P}}^{p-1-i})) \\
&= w^\wp(yzS_\alpha(\xi_{\mathfrak{P}}\beta_{\mathfrak{P}}^{p-1-i})) = w^\wp(yzz^{-1}) = w^\wp(y) \neq 0.
\end{aligned}$$

Thus $\Lambda^{\mathfrak{P}}(yz\xi_{\mathfrak{P}}\beta_{\mathfrak{P}}^{p-1-i}) \neq 0$ and

$$\begin{aligned} v_{\mathfrak{P}}(yz\xi_{\mathfrak{P}}\beta_{\mathfrak{P}}^{p-1-i}) &= e(\mathfrak{P}|\wp)(v_{\wp}(y) + v_{\mathfrak{P}}(z)) + v_{\mathfrak{P}}(\xi_{\mathfrak{P}}\beta_{\mathfrak{P}}^{p-1-i}) \\ &= e(\mathfrak{P}|\wp)(-v_{\wp}((w)_K) - 1) + e(\mathfrak{P}|\wp)v_{\wp}(z) + v_{\mathfrak{P}}(\xi_{\mathfrak{P}}\beta_{\mathfrak{P}}^{p-1-i}) \\ &\geq -e(\mathfrak{P}|\wp)v_{\wp}((w)_K) - e(\mathfrak{P}|\wp) + e(\mathfrak{P}|\wp) - v_{\mathfrak{P}}((D_{\beta_{\mathfrak{P}}}\alpha)^{1-p}) - 1 \\ &= -e(\mathfrak{P}|\wp)v_{\wp}((w)_K) - v_{\mathfrak{P}}((D_{\beta_{\mathfrak{P}}}\alpha)^{1-p}) - 1. \end{aligned}$$

It follows that

$$v_{\mathfrak{P}}((\Lambda)_L) \leq c_{\mathfrak{P}}. \tag{9.48}$$

We deduce the result from (9.47) and (9.48) . □

Corollary 9.5.20 (Tate Genus Formula). *The genera of L and K are related by the formula*

$$2g_L - 2 = p^{1-n}(2g_K - 2) + (1-p)\sum_{\mathfrak{P}\in\mathbb{P}_L} v_{\mathfrak{P}}(D_{\beta_{\mathfrak{P}}}\alpha)\deg_L \mathfrak{P},$$

where for each $\mathfrak{P} \in \mathbb{P}_L$, we have $\wp = \mathfrak{P}|_K$, $\vartheta_{\mathfrak{P}} = \vartheta_{\wp}[\beta_{\mathfrak{P}}]$, and $[\ell : k] = p^n$ for some $n \geq 0$.

Proof: By Corollary 3.5.5 we have $d_L((\Lambda)_L) = 2g_L - 2$ and $d_K((w)_K) = 2g_K - 2$. On the other hand, using Theorem 5.3.4 we obtain

$$d_L(\operatorname{con}_{K/L}(w)_K) = \frac{d_K((w)_K)}{\lambda_{L/K}} = \frac{[L:K]}{[\ell:k]}d_K((w)_K) = p^{1-n}(2g_K - 2).$$

The results follows immediately by (9.46) and Theorem 9.5.19. □

Corollary 9.5.21. *Let K be a function field of characteristic $p > 2$. Let L be a purely inseparable finite extension of K. Then $g_L - g_K$ is divisible by $\frac{p-1}{2}$.*

Proof: Since the extension is obtained from a finite number of successive extensions of degree p, it suffices to consider the case $[L : K] = p$. Multiplying the formula of Corollary 9.5.20 by p^n and using the fact that $p^n \equiv 1 \bmod (p-1)$, we obtain

$$2g_L - 2 \equiv 2g_K - 2 \bmod (p-1).$$

The result follows. □

Example 9.5.22. Let k be a field of characteristic $p > 0$ such that $p \neq 2$. Let $K = k(x, y)$ be the hyperelliptic function field generated by

$$y^2 = x^p - a, \quad \text{with} \quad a \notin k^p$$

By Corollary 4.3.7 we have

$$g_K = \left[\frac{p+1}{2}\right] - 1 = \frac{p+1}{2} - 1 = \frac{p-1}{2}.$$

Let $L = K(\alpha)$ with $\alpha = a^{1/p}$. Then $L = k(x,y)(a^{1/p}) = k(a^{1/p})(x,y) = k(a^{1/p})(z)$, where $z = \frac{y}{(x-\alpha)^{(p-1)/2}}$.

Then $g_L = 0$ and $g_L - g_K = -\frac{p-1}{2}$.

Remark 9.5.23. Example 9.5.22 shows again that even though $K_{\wp}$ is isomorphic to $k'((\pi))$, k is not contained in k' (see Example 2.5.23).

Proposition 9.5.24. *Let K be a function field of characteristic $p > 0$ such that $g_K < \frac{p-1}{2}$. Then for any constant extension $L = K\ell'$, we have $g_L = g_K$.*

Proof: If $\mathcal{A}$ is a transcendence basis of ℓ' over k and if $L_1 = Kk(\mathcal{A})$, we have $g_{L_1} = g_K$ (Theorem 8.5.2).

Therefore we may assume that ℓ'/k is algebraic. If ℓ'_s is the separable closure of k in ℓ', then if $L_2 = K\ell'_s$ we have $g_{L_2} = g_K$ (Theorem 8.5.2).

Thus we may assume that ℓ'/k is purely inseparable. We have $g_L \leq \lambda_{L/K} g_K$. If $g_L < g_K$, the change of genus can be obtained in a finite extension ℓ'/k (see the proof of Theorem 8.5.3).

Hence, we may assume that ℓ'/k is a finite purely inseparable extension.

By Corollary 9.5.21,

$$\frac{p-1}{2} \mid g_K - g_L \quad \text{and} \quad g_K - g_L \geq 0.$$

It follows that $0 \leq g_K - g_L \leq g_K < \frac{p-1}{2}$. Therefore $g_K = g_L$. □

Definition 9.5.25. A function field K is called *conservative* if any constant extension $L = K\ell$ satisfies $g_L = g_K$.

Example 9.5.26. K is conservative in the following two cases:

(i) char $K = 0$ (Theorem 8.5.2)
(ii) $g_K < \frac{p-1}{2}$ and char $K = p$ (Proposition 9.5.24).

For constant extensions we have the following result:

Theorem 9.5.27. *Let L be a finite purely inseparable constant extension of K/k. Then*

$$2g_K - 2 = \lambda_{L/K}(2g_L - 2) + \mu_K(p-1)A,$$

where A is a nonnegative integer and μ_K is the invariant given in Definition 8.6.6. *If $\lambda_{L/K} > 1$, we have $A > 0$.*

Proof: We proceed by induction on $[L : K]$. If $L = K$, there is nothing to prove. Assume $[L : K] \geq p$ (where p is the characteristic). Since L/K is purely inseparable, there exists L' such that $K \subseteq L' \subseteq L$ and $[L : L'] = p$. By the induction hypothesis we have

$$2g_K - 2 = \lambda_{L'/K}(2g_{L'} - 2) + \mu_K(p-1)A'. \tag{9.49}$$

Applying Tate's genus formula (Corollary 9.5.20) to the purely inseparable extension L/L', we get

$$2g_L - 2 = p^{1-n}(2g_{L'} - 2) + (1-p) \sum_{\mathfrak{P} \in \mathbb{P}_L} v_{\mathfrak{P}}(D_{\beta_{\mathfrak{P}}}\alpha) \deg_L \mathfrak{P},$$

where $p^n = [\ell : \ell']$. Therefore $p^{1-n} = \frac{p}{p^n} = \frac{[L:L']}{[\ell:\ell']} = \lambda_{L/L'}^{-1}$.

Let $a_{\mathfrak{P}} = v_{\mathfrak{P}}(D_{\beta_{\mathfrak{P}}}\alpha)$. We obtain

$$2g_{L'} - 2 = \lambda_{L/L'}(2g_L - 2) + \lambda_{L/L'}(p-1) \sum_{\mathfrak{P} \in \mathbb{P}_L} a_{\mathfrak{P}} \deg_L \mathfrak{P}. \tag{9.50}$$

Notice that $a_{\mathfrak{P}}$ belongs to $\mathbb{Z}$ and $a_{\mathfrak{P}} = 0$ for almost all $\mathfrak{P}$. Since L/L' is a constant extension, it follows that $a_{\mathfrak{P}} \geq 0$ (Proposition 9.5.12).

By Theorem 8.6.8, μ_L divides $\deg_L \mathfrak{P}$, and by Corollary 8.6.15, $\mu_L \lambda_{L/L'} = \mu_{L'}$. Using (9.50) we obtain

$$2g_{L'} - 2 = \lambda_{L/L'}(2g_L - 2) + \mu_{L'}(p-1)A'' \tag{9.51}$$

with $A'' \in \mathbb{Z}$ and $A'' \geq 0$.

It follows from (9.49) and (9.51) that

$$\begin{aligned} 2g_K - 2 &= \lambda_{L'/K}(2g_{L'} - 2) + \mu_K(p-1)A' \\ &= \lambda_{L'/K}\lambda_{L/L'}(2g_L - 2) + \lambda_{L'/K}\mu_{L'}(p-1)A'' + \mu_K(p-1)A' \\ &= \lambda_{L/K}(2g_L - 2) + \mu_K(p-1)A, \end{aligned}$$

where $A = A' + A'' \geq 0$ (here we have used the facts that $\lambda_{L/K} = \lambda_{L'/K}\lambda_{L/L'}$ and $\lambda_{L'/K}\mu_{L'} = \mu_K$).

Finally, assume $\lambda_{L/K} > 1$. Then if $\lambda_{L'/K} > 1$ it follows by the induction hypothesis that $A' > 0$ and $A \geq A' > 0$.

If $\lambda_{L'/K} = 1$, then $\lambda_{L/L'} > 1$ and $\lambda_{L/L'} g_L \leq g_{L'}$ (Theorem 8.5.3). We have

$$2g_{L'} - 2 \geq 2\lambda_{L/L'} g_L - 2\lambda_{L/L'} = \lambda_{L/L'}(2g_L - 2).$$

Using (9.51), we conclude that $A'' > 0$ and $A \geq A'' > 0$. □

Theorem 9.5.28. *Let K/k be an inseparable function field and L/ℓ the minimum constant extension of K/k such that L/ℓ is separable. Then*

$$g_K = \mu_K\left(g_L - 1 + \frac{1}{2}(p-1)A\right) + 1$$

with $A \in \mathbb{N}$.

Proof: By Theorem 8.6.13 we have $\mu_K = \lambda_{L/K}$. The result is a consequence of Theorem 9.5.27 since $\mu_K = \lambda_{L/K} > 1$ (see Remark 8.6.7). □

Corollary 9.5.29. *If K/k is any inseparable function field we have*

$$g_K \geq \frac{(p-1)(p-2)}{2}.$$

Proof: Since μ_K is a power of p and $\mu_K > 1$, we have $\mu_K \geq p$.
Therefore

$$\begin{aligned} g_K &\geq p\left(0 - 1 + \frac{1}{2}(p-1)\times 1\right) + 1 = \frac{1}{2}p(p-1) + (1-p) \\ &= \frac{(p-1)}{2}(p-2). \end{aligned}$$

□

Remark 9.5.30. There exist examples where the equality $g_K = \frac{1}{2}(p-1)(p-2)$ holds.

Example 9.5.31. Let $K = k(x, y)$ be the function field given in Example 5.2.31. Recall that $k = k_0(u, v)$, where k_0 is a field of characteristic p, $\left[k(u^{1/p}, v^{1/p}) : k\right] = p^2$, and $y^p = ux^p + v$. By Corollary 9.5.29 we have

$$g_K \geq \frac{(p-1)(p-2)}{2}$$

since K/k is not separable. Indeed, the field of constants of $L = Kk(u^{1/p})$ is $k(u^{1/p}, v^{1/p}) \neq k(u^{1/p})$. Let $\mathfrak{N}_x$ be the pole divisor of x in K. For t large enough, we have

$$\ell(\mathfrak{N}_x^{-t}) = t\deg_K(\mathfrak{N}_x) - g_K + 1 = pt - g_K + 1$$

(because $[K : k(x)] = p = d_K(\mathfrak{N}_x)$). We have $\mathfrak{N}_x = \mathfrak{N}_y = \mathfrak{A}$ and $\mathfrak{N}_{x^i y^j} = \mathfrak{A}^{i+j}$. Therefore $x^i y^j \in L(\mathfrak{A}^{-t})$ for $i \geq 0, 0 \leq j \leq p-1$, and $i + j \leq t$, and these elements are k-linearly independent (since $j \leq p-1$). Let $t \geq p-1$. Then

$$\begin{aligned} &|\{(i,j) \mid i \geq 0, 0 \leq j \leq p-1, i+j \leq t\}| \\ &\qquad = \sum_{j=0}^{p-1}(t-j+1) = pt - \frac{p(p-1)}{2} + p = pt - \frac{p}{2}(p-3). \end{aligned}$$

Thus $pt - g_K + 1 = \ell(\mathfrak{A}^{-t}) \geq pt - \frac{p}{2}(p-3)$.
Therefore $g_K \leq 1 + \frac{p}{2}(p-3) = \frac{(p-1)(p-2)}{2}$, and it follows that

$$g_K = \frac{(p-1)(p-2)}{2}.$$

9.6 Examples

9.6.1 Function Fields of Genus 0

Let K/k be a function field of genus 0 that is not rational, and let $\mathfrak{A}$ be any divisor. We will prove that $d_K(\mathfrak{A})$ is even. First suppose that $d_K(\mathfrak{A}) = 1$. Then

$$d_K(\mathfrak{A}) = 1 > 2g_K - 2 = -2.$$

Using Corollary 3.5.6 we obtain that

$$\ell_K(\mathfrak{A}^{-1}) = d_K(\mathfrak{A}) - g_K + 1 = 2.$$

According to Exercise 3.6.22, there exists an integral divisor $\mathfrak{P}$ of degree 1. Thus $\mathfrak{P}$ must be a prime divisor. By Theorem 4.1.7, it follows that K is a rational function field. This contradiction shows that $d_K(\mathfrak{A})$ must be different from 1.

Now assume that there exists a divisor $\mathfrak{A}$ of odd degree, say $d_K(\mathfrak{A}) = 2n + 1$ with $n \in \mathbb{N}$. By Proposition 4.1.6 there exists a prime divisor $\mathfrak{P}$ of degree 2. Thus $\mathfrak{A}\mathfrak{P}^{-n}$ has degree 1.

In particular, it follows that $d_K(D_K) = 2\mathbb{Z}$.

Proposition 9.6.1. *A function field K/k of genus 0 is a rational function field if and only if K contains a divisor of degree 1.* □

Let $x \in K \setminus k$ be such that $[K : k(x)] = 2$. Then

$$(x)_K = \frac{\mathfrak{P}_1}{\mathfrak{P}},$$

where $\mathfrak{P}_1, \mathfrak{P}$ are prime divisors of degree 2.

Since $d_K(\mathfrak{P}) = 2 > 2g_K - 2 = -2$, it follows by Corollary 3.5.6 that $\ell_K(\mathfrak{P}^{-1}) = d_K(\mathfrak{P}) - g_K + 1 = 3$. Let $\{1, x, y\}$ be a basis of $L_K(\mathfrak{P}^{-1})$. If $y = f(x) \in k[x]$, we have $(y)_K = \frac{\mathfrak{P}_2}{\mathfrak{P}} = (f(x))_K$, with $\mathfrak{P}_2 \neq \mathfrak{P}_1$, so $[K : k(f(x))] = 2$.

Consequently we have $k(f(x)) = k(x)$, where $f(x) \in k[x]$ has degree 1. This contradicts the independence of $\{1, x, y\}$. Thus $K = k(x, y)$. We also have $\ell_K(\mathfrak{P}^{-2}) = 5$. Since $\{1, x, y, x^2, y^2, xy\} \subseteq L_K(\mathfrak{P}^{-2})$ it follows that there is an equation

$$ax^2 + by^2 + cxy + dx + ey + f = 0 \tag{9.52}$$

with $a, b, c, d, e, f \in k$ not all zero.

If $a = 0$, (9.52) reduces to $by^2 + (cy + d)x + ey + f = 0$.

We have $cy + d \neq 0$ since otherwise $by^2 + ey + f = 0$. Hence $cy + d \neq 0$.

It follows that $x \in k(y)$. In particular, $K = k(x, y) = k(y)$ and K is rational. This proves that $a \neq 0$. Similarly, we obtain $b \neq 0$. Therefore $F(X, Y) = aX^2 + bY^2 + cXY + dX + eY + f$ is an irreducible polynomial in $k[X, Y]$.

We may assume $b = 1$. In this case, (9.52) can be written as

$$y^2 + (cx + e)y + (ax^2 + dx + f) = 0. \tag{9.53}$$

If $\operatorname{char} k \neq 2$, then (9.53) can be reduced to

$$y^2 = h(x),$$

where $h(x) \in k[x]$ and $h(x)$ has degree 1 or 2. The degree 1 case is not possible because otherwise K would be a rational function field.

Furthermore, $h(x)$ must be irreducible, since otherwise, if $h(x) = (Ax + \alpha)(x + \beta)$, $\alpha, \beta \in k$, then if $\alpha \neq 0$ or $\beta \neq 0$, we have

$$\frac{y^2}{(x+\beta)^2} = \left(\frac{y}{x+\beta}\right)^2 = \left(\frac{Ax+\alpha}{x+\beta}\right).$$

Let $z = \sqrt{\frac{Ax+\alpha}{x+\beta}}$. Then $y = \pm(x+\beta)z \in k(x, z)$ and

$$x = \frac{\alpha - \beta z^2}{z^2 - A} \in k(z).$$

If $\alpha = \beta = 0$, then $y^2 = Ax^2$ and $y = \sqrt{A}x \in k(\sqrt{A})(x)$. Thus $\sqrt{A} = \frac{y}{x} \in K$ and K is rational. Therefore $f(x)$ is irreducible.

Let us now consider the case $\operatorname{char} k = 2$. If $cx + e = 0$, the extension $K/k(x)$ is inseparable of degree 2. Assume that $K/k(y)$ is also an inseparable extension, that is,

$$x^2 = g(y) \in k(y).$$

As before, $g(y) \in k[y]$ is a polynomial of degree 2. Thus

$$g(y) = \alpha y^2 + \beta y + \gamma, \quad \text{with} \quad \alpha, \beta, \gamma \in k.$$

We have $x^4 = \alpha^2 y^4 + \beta^2 y^2 + \gamma^2 = \alpha^2(ax^2 + bx + c)^2 + \beta^2(ax^2 + bx + c) + \gamma^2 = \alpha^2 a^2 x^4 + (\alpha^2 b^2 + \beta^2 a)x^2 + \beta^2 bx + (\alpha^2 c^2 + \beta^2 c + \gamma^2)$.

It follows that

$$\alpha^2 a^2 = 1, \quad \alpha^2 b^2 + \beta^2 a = 0, \quad \beta^2 b = 0, \quad \text{and} \quad \alpha^2 c^2 + \beta^2 c + \gamma^2 = 0.$$

Thus $b = 0$, $\beta = 0$, $\alpha a = 1$, and $\gamma = \alpha c = \frac{c}{a}$.

The latter imply that $x^2 = \frac{1}{a}y^2 + \frac{c}{a}$, or

$$y^2 = ax^2 + c.$$

Note that $a^{1/2}$ and $c^{1/2} \in k$ cannot occur since in this case $y = a^{1/2}x + c^{1/2}$ and $K = k(x) = k(y)$.

Now in the case that $K/k(y)$ is separable we may assume, by exchanging the roles of and x and y, that $K/k(x)$ is separable and $cx + e \neq 0$ in (9.53).

Let $z = \frac{y}{cx+e}$. Then $K = k(x, z)$ and

$$z^2 - z = \frac{ax^2 + dx + f}{(cx + e)^2} = h(x). \tag{9.54}$$

Note that $ax^2 + dx + f$ and $cx + e$ are relatively prime, since otherwise $z^2 - z = \frac{Ax+B}{cx+e}$ and $z \in \overline{k}$, or $x \in k(z)$ and $K = k(z)$.

If $c \neq 0$, setting $x_1 = \frac{1}{cx+e}$, (9.54) reduces to

$$z^2 - z = h_1(x_1),$$

where $h_1(x_1)$ is a polynomial of degree 2.

Therefore, when $K/k(x)$ is separable, x and y can be chosen such that

$$y^2 - y = f(x) \in k[x] \quad \text{with} \quad \deg f(x) = 2.$$

We have proved the following theorem:

Proposition 9.6.2. *Let K/k be a function field of genus 0 that is not a rational function field. Then there exist $x, y \in K$ such that $K = k(x, y)$, $[K : k(x)] = 2 = [K : k(y)]$, and x, y satisfy*

$$ax^2 + y^2 + cxy + dx + ey + f = 0 \quad \text{for some} \quad a \neq 0, \tag{9.55}$$

where $F(X, Y) = aX^2 + Y^2 + cXY + dX + eY + f$ is an irreducible polynomial in $k[X, Y]$.

Furthermore:

(a) *If* char $k \neq 2$, *then* (9.55) *can be reduced to*

$$y^2 = f(x) \in k[x], \quad \text{where} \quad \deg f(x) = 2 \quad \text{and} \quad f(x) \text{ is irreducible.} \tag{9.56}$$

(b) *If* char $k = 2$ *and either $K/k(x)$ or $K/k(y)$ is separable,* (9.55) *can be reduced to*

$$y^2 - y = f(x) \in k[x], \quad \text{with} \quad \deg f(x) = 2. \tag{9.57}$$

(c) *If* char $k = 2$ *and both $K/k(x)$ and $K/k(y)$ are purely inseparable, then*

$$y^2 = ax^2 + c \in k[x] \tag{9.58}$$

with $a^{1/2} \notin k$ or $c^{1/2} \notin k$. □

To see which conditions (9.55), (9.56), (9.57), and (9.58) must satisfy in order for K to be or not be a rational function field, first consider the case $K = k(x, y)$ with x, y satisfying (9.55).

If this equation is of the first degree ($a = b = c = 0$), then K is rational. If the equation is reducible, then again K is rational. If there is an algebraic element α in $\overline{k} \backslash k$ such that (α, x) is a solution of (9.55), then $k(x) \subseteq k'(x) \subseteq K$ and $\left[k'(x) : k(x)\right] =$

$\left[k' : k\right] \geq 2 = [K : k(x)]$. Thus $K = k'(x)$ is a rational function field. Therefore we may assume that K is not rational. We will show that K has genus 0. As before, we have $a \neq 0$ and we may assume $b = 1$.

Let $\wp_\infty$ be the pole divisor of x in $k(x)$. If $\mathfrak{P}_\infty$ is a prime divisor in K satisfying $\mathfrak{P}_\infty|_{k(x)} = \wp_\infty$, then $d_K(\mathfrak{P}_\infty) = 2$ since K is not rational. Thus $d_K(\mathfrak{P}_\infty^{-s}) = -2s$ for all $s \in \mathbb{N}$.

Let $\mathcal{A} = \{a(x) + yb(x) \mid a(x), b(x) \in k[x], \deg a(x) \leq s, \deg b(x) \leq s - 1\}$.

Then $\mathcal{A} \subseteqq L_K(\mathfrak{P}_\infty^{-s})$ by Proposition 4.3.5 (see Exercise 9.7.9). Thus $\ell_K(\mathfrak{P}_\infty^{-s}) \geq 2s+1$. Let $s \in \mathbb{N}$ be such that $2s > 2g_K - 2$. We have $\ell_K(\mathfrak{P}_\infty^{-s}) = d_K(\mathfrak{P}_\infty^{s}) - g_K + 1 = 2s + 1 - g_K \geq 2s + 1$. It follows that $g_K = 0$.

Now we consider (9.56) ($\operatorname{char} k \neq 2$). In this case, $K/k(x)$ is a separable extension. If K is not rational, then $f(x)$ is not a square and for any place

$$\varphi : K \to k(\mathfrak{P}) \cup \{\infty\},$$

we have $k(\mathfrak{P}) \neq k$. This means that the prime divisor $\mathfrak{P}$ is of degree larger than 1. Assume that $\mathfrak{P}$ is such that $\varphi(x) \neq \infty$. This is equivalent to $v_{\mathfrak{P}}(x) \geq 0$, i.e., $\mathfrak{P} \neq \mathfrak{P}_\infty$, which implies $\varphi(x) \notin k$ or $\varphi(y) \notin k$. If $\varphi(y) \in k$ then $\varphi(x) \notin k$, and

$$\varphi(y)^2 = f(\varphi(x)).$$

It follows that $f(x) - \alpha^2$ is irreducible for any $\alpha \in k$.

Conversely, if $f(x) - \alpha^2 \in k[x]$ is irreducible for all $\alpha \in k$, then for any place φ of K such that $\varphi(x) \neq \infty$, we have $k(\mathfrak{P}) \neq k$.

Now if $\operatorname{char} k = 2$ and $K = k(x, y)$ is given by (9.57), then $k(\mathfrak{P}) \neq k$ if and only if $f(x) - (\alpha^2 - \alpha) \in k[x]$ is irreducible for all $\alpha \in k$.

Next, assume $\operatorname{char} k = 2$ and let $K = k(x, y)$ be as in (9.58).

If $k(a^{1/2}) = k(c^{1/2}) = k'$, then $y \in k'(x)$ and $k'K = k'(x, y) = k'(x)$, with $[k'(x) : k(x)] = [k' : k] = 2$. Hence $K = k'(x)$ and K is rational. Therefore if K is not rational, we have $[k(a^{1/2}, c^{1/2}) : k] = 4$.

Conversely, if $[k(a^{1/2}, c^{1/2}) : k] = 4$ we will prove that K is not rational. Set $k' = k(a^{1/2})$. Then $Kk' = k(a^{1/2}, c^{1/2})(y)$ and

$$\begin{aligned} 4 &= [k(a^{1/2}, c^{1/2}) : k] = [k(a^{1/2}, c^{1/2})(x) : k(x)] = [Kk' : k(x)] \\ &= [Kk' : k'(x)][k'(x) : k(x)] = 2[Kk' : k'(x)]. \end{aligned}$$

Thus $\left[Kk' : k'(x)\right] = 2$.

If $\varphi : K \to k(\mathfrak{P}) \cup \{\infty\}$ is any place of K the restriction of φ to k is the identity (see the discussion in Section 2.2). Thus

$$\varphi(y)^2 = a\varphi(x)^2 + c \quad \text{or} \quad \varphi(y) = a^{1/2}\varphi(x) + c^{1/2}.$$

It follows that $\varphi(x) \notin k$ or $\varphi(y) \notin k$, and hence there is no place of degree 1 in K.

We have proved the following theorem:

Theorem 9.6.3. *Let K/k be a function field. Then K is of genus 0 iff $K = k(x, y)$ with*

$$ax^2 + by^2 + cxy + dx + ey + f = 0. \tag{9.59}$$

Furthermore, (9.59) *can be reduced to:*

(a) $y^2 = f(x)$ *if* $\operatorname{char} k \neq 2$*, where* $f(x) \in k[x]$ *is a polynomial of degree* 2. *In this case* K *is a rational function field if and only if there exists* $\alpha \in k$ *such that* $f(x) - \alpha^2 \in k[x]$ *is reducible.*
(b) $y^2 - y = f(x)$*, where* $f(x)$ *has degree* 2 *if* $\operatorname{char} k = 2$*, and* $K/k(x)$ *is separable. In this case,* K *is a rational function field if and only if there exists* $\alpha \in k$ *such that* $f(x) - (\alpha^2 - \alpha) \in k[x]$ *is reducible.*
(c) $y^2 = ax^2 + c$ *for some* $a \neq 0$ *if* $\operatorname{char} k = 2$*, and* $K/k(x)$ *and* $K/k(y)$ *are purely inseparable. In this case, we have that* K *is a rational function field if and only if* $\left[k(a^{1/2}, c^{1/2}) : k\right] < 4$. □

We end the discussion with a result on the different $\mathfrak{D}_{K/k(x)}$.

Theorem 9.6.4. *Let* K/k *be a function field of genus* 0.

(a) Assume that $K = k(x, y)$, $\operatorname{char} k \neq 2$, $y^2 = f(x)$, $f(x) = p_1(x)^{e_1} \cdots p_r(x)^{e_r}$ *with* $r = 1$ *or* $r = 2$, $e_i = 1$*, and* $\sum_{i=1}^{r} e_i \deg p_i = 2$. *If* $(p_i(x))_{k(x)} = \wp_i \wp_\infty^{-\deg p_i}$*, then* $\mathfrak{D}_{K/k(x)} = \prod_{i=1}^{r} \mathfrak{P}_i \mathfrak{P}_\infty^{\varepsilon}$*, where the* $\mathfrak{P}_i$*'s are the prime divisors in* K *lying above* $\wp_i$*,* $\mathfrak{B}_\infty$ *is a prime divisor above* $\wp_\infty$*, and* ε *is* 0 *or* 1.
(b) Assume that $y^2 - y = f(x) \in k(x)$, $\deg f = 2$, $\mathfrak{D}_{K/k(x)} = \mathfrak{P}_\infty^{\delta}$*, and* δ *is* 0, 1, *or* 2*, where* $\mathfrak{P}_\infty$ *is the prime divisor in* K *above the pole divisor of* x *in* $k(x)$*, and* $d_K(\mathfrak{P}_\infty)$ *is* 1 *or* 2.

Proof: (a) This is just Example 5.8.9.

(b) Since k is not a perfect field we cannot apply directly Example 5.8.8. Clearly $K/k(x)$ is a separable extension. If $K/k(x)$ is a constant extension, then $K = k'(x)$, $[k' : k] = 2$, k'/k is a separable extension and for any place $\mathfrak{P}$ we have $k'(\mathfrak{P}) = k'k(\wp)$ (Theorem 8.4.11). Thus there are no inseparable or ramified places (Theorem 5.2.32), and $\mathfrak{D}_{K/k(x)} = \mathfrak{N}$. If $K/k(x)$ is a geometric extension, then since $K/k(x)$ is separable, we may apply the genus formula and we obtain

$$0 = g_K = 1 + (g_{k(x)} - 1)[K : k(x)] + \frac{1}{2} d_K(\mathfrak{D}_{K/k(x)}) = \frac{1}{2} d_K(\mathfrak{D}_{K/k(x)}) - 1.$$

It follows that $d_K(\mathfrak{D}_{K/k(x)}) = 2$. On the other hand, by Example 5.8.8 the only ramified prime of $k(x)$ in K can be the pole divisor $\wp_\infty$ of x in $k(x)$. Therefore $\wp_\infty$ ramifies or is inert in K and we have

$$\mathfrak{P}_\infty = \wp_\infty^{\delta}$$

with $\delta = 1$ if and only if $\wp_\infty$ is inseparable or $\delta = 2$, if and only if $\wp_\infty$ is ramified. It follows that $\mathfrak{D}_{K/k(x)} = \mathfrak{P}_\infty^{s}$ with $s d_K(\mathfrak{P}_\infty) = 2$. □

Note that $\mathfrak{P}_\infty | \wp_\infty$ may be inseparable (see Exercise 5.10.18).

9.6.2 Function Fields of Genus 1

Let K/k be a field of genus 1. Let W_K denote the canonical class of K. Then $d_K(W_K) = 2g_K - 2 = 0$, and we have

$$N(W_K) = d_K(W_K) - g_K + 1 + N(W_K^{-1}W_K) = 0 - 1 + 1 + 1 = 1.$$

It follows that $W_K = P_K$ is the principal class.

For a function field of genus 0, there exist divisors of degree 2. This is not the case for function fields of genus 1.

Proposition 9.6.5. *Let $n \in \mathbb{N}$. Then there exists a function field K/k with $g_K = 1$ such that $d_K(D_K) = n\mathbb{Z}$.*

Proof: Corollary to Theorem 7 of [91], [133, Theorem 2]. □

In Section 4.2 we studied elliptic function fields K/k such that $\operatorname{char} k \neq 2$. In this section we will consider the case $\operatorname{char} k = 2$.

By (4.1) we have $K = k(x, y)$ with

$$y^2 + \gamma xy + \delta y = \alpha_3 x^3 + \alpha_2 x^2 + \alpha_1 x + \alpha_0. \tag{9.60}$$

As in Section 4.2, we also have $\mathfrak{N}_x = \mathfrak{P}^2$ and $\mathfrak{N}_y = \mathfrak{P}^3$, where $\mathfrak{P}$ denotes a prime divisor of degree 1. Thus $[K : k(y)] = 3$. It follows that $\alpha_3 \neq 0$ (since otherwise x satisfies an equation of degree 2 over $k(y)$ and then $[K : k(y)] = [k(x, y) : k(y)] \leq 2$). Multiplying by α_3^2 and putting $y_1 = \alpha_3 y, x_1 = \alpha_3 x$, we may assume $\alpha_3 = 1$. Hence $K = k(x, y)$ with

$$y^2 + (\gamma x + \delta)y = x^3 + \alpha_2 x^2 + \alpha_1 x + \alpha_0. \tag{9.61}$$

First we handle the case $\gamma x + \delta = 0$ in (9.61). In this case $K/k(x)$ is a purely inseparable extension of degree 2. By means of the substitution $x_1 = x + \alpha_2$, (9.60) reduces to

$$y^2 = x^3 + ax + b, \quad \text{with} \quad a, b \in k. \tag{9.62}$$

Consider any function field $K = k(x, y)$ satisfying (9.62). Let

$$(x)_{k(x)} = \frac{\wp_0}{\wp_\infty}, \quad \text{where} \quad v_{\wp_\infty}(x) = -1.$$

Let $\mathfrak{P}$ be any prime divisor in K that lies above $\wp_\infty$. Since $v_{\mathfrak{P}}(x) < 0$, we have

$$\begin{aligned} v_{\mathfrak{P}}(x^3 + ax + b) &= \min\{v_{\mathfrak{P}}(x^3), v_{\mathfrak{P}}(ax), v_{\mathfrak{P}}(b)\} = v_{\mathfrak{P}}(x^3) \\ &= 3v_{\mathfrak{P}}(x) = 3e(\mathfrak{P}|\wp_\infty)v_{\wp_\infty}(x) = -3e(\mathfrak{P}|\wp_\infty). \end{aligned}$$

Thus

$$v_{\mathfrak{P}}(y^2) = 2v_{\mathfrak{P}}(y) = -3e(\mathfrak{P}|\wp_\infty).$$

Therefore 3 divides $v_{\mathfrak{P}}(y)$ and $\mathfrak{P}^3$ divides $\mathfrak{N}_y$.

Since $3 \geq [K : k(y)] = d_K(\mathfrak{N}_y) \geq 3$, it follows that

$$\mathfrak{N}_y = \mathfrak{P}^3 \quad \text{and} \quad e(\mathfrak{P}|\wp_\infty) = 2.$$

In particular, $d_K(\mathfrak{P}) = 1$ and k is the field of constants of K.

Let n be such that $n > 2g_K - 2$. For $2 \leq m \leq n$, we write $m = 3t + r, r \in \{0, 1, 2\}$. If $r = 0$, consider the element y^t. If $r = 1$, then $m = 3t + 1 = 3(t-1) + 4 \geq 2$ and hence $t \geq 1$. In this case we can work with $y^{t-1}x^2$. If $r = 2$, consider the element $y^t x$. In any case, for any $2 \leq m \leq n$, there exist i and j such that $0 \leq i, 0 \leq j$, and $v_{\mathfrak{P}}(y^i x^j) = -(3i + 2j) = -m$. It follows that $\ell_K(\mathfrak{P}^{-n}) \geq n$. Therefore

$$n \leq \ell_K(\mathfrak{P}^{-n}) = d_K(\mathfrak{P}^{-n}) - g_K + 1 = n - g_K + 1.$$

Thus $g_K \leq 1$.

Proposition 9.6.6. *Assume that $K = k(x, y)$ has characteristic 2 and is given by (9.62). Then K contains a prime divisor of degree 1 and $g_K \leq 1$.* □

In order to study the situation in which $g_K = 0$ and $g_K = 1$, consider the equation $y^2 = x^3 + ax + b$, and let $k' = k(\sqrt{a}, \sqrt{b})$. In $K' = Kk'$, we have

$$y^2 = x^3 + a_1^2 x + b_1^2 = x(x^2 + a_1^2) + b_1^2,$$

where $a_1^2 = a$, $b_1^2 = b$, and $a_1, b_1 \in k'$.

It follows that

$$x = \left(\frac{y + b_1}{x + a_1}\right)^2.$$

Therefore, if $z = \sqrt{x} = \frac{y+b_1}{x+a_1}$, the field $K' = k'(z)$ is a rational function field. Assume that $\mathfrak{P}'$ is a prime divisor of K' above $\mathfrak{P}$; then $v_{\mathfrak{P}'}(z) = -1$. In Kk' we have $\mathfrak{N}_x = (\mathfrak{P}')^2$ and $\mathfrak{N}_y = (\mathfrak{P}')^3$.

It is easy to see that $g_K = 0$ if and only if $\ell_K(\mathfrak{P}^{-1}) = 2$ (and $g_K = 1$ if and only if $\ell_K(\mathfrak{P}^{-1}) = 1$).

Now, in K', we have

$$y^2 = x^3 + ax + b = z^6 + a_1^2 z^2 + b_1^2 = (z^3 + a_1 z + b_1)^2,$$

that is,

$$y = z^3 + a_1 z + b_1. \tag{9.63}$$

Assume that $g_K = 0$. Since K contains a prime divisor of degree 1, it follows that K is a rational function field. In this case we have $\ell_K(\mathfrak{P}^{-1}) = 2$, or equivalently, there exists $w \in K \setminus k$ such that $\mathfrak{N}_w = \mathfrak{P}$.

Since $\{1, z\}$ is a basis of $L_{K'}((\mathfrak{P}')^{-1})$ and $w \in L_K(\mathfrak{P}^{-1}) \subseteq L_{K'}((\mathfrak{P}')^{-1})$, there exist $\alpha, \beta \in k'$ such that

$$w = \alpha + \beta z \in K. \tag{9.64}$$

Also, $\{1, w, w^2, w^3\}$ is a basis of $L_K(\mathfrak{P}^{-3})$. Therefore there exist A, B, C, D in k such that

$$y = A + Bw + Cw^2 + Dw^3. \tag{9.65}$$

Taking squares in (9.65) and substituting w by its value given by (9.64), we obtain

$$\begin{aligned} x^3 + ax + b = y^2 &= A^2 + B^2(\alpha^2 + \beta^2 x) + C^2(\alpha^4 + \beta^4 x^2) + D^2(\alpha^2 + \beta^2 x)^3 \\ &= A^2 + B^2\alpha^2 + B^2\beta^2 x + C^2\alpha^4 + C^2\beta^4 x^2 + D^2\alpha^6 \\ &\quad + D^2\alpha^4\beta^2 x + D^2\alpha^2\beta^4 x^2 + D^2\beta^6 x^3 \\ &= (A^2 + B^2\alpha^2 + C^2\alpha^4 + D^2\alpha^6) + (B^2\beta^2 + D^2\alpha^4\beta^2)x \\ &\quad + (C^2\beta^4 + D^2\alpha^2\beta^4)x^2 + D^2\beta^6 x^3. \end{aligned}$$

It follows that

$$\begin{aligned} A^2 + B^2\alpha^2 + C^2\alpha^4 + D^2\alpha^6 &= b, \\ B^2\beta^2 + D^2\alpha^4\beta^2 &= a, \\ C^2\beta^4 + D^2\alpha^2\beta^4 &= 0, \\ D^2\beta^6 &= 1. \end{aligned}$$

Therefore $\beta^3 = 1/D \in k$. In particular, $k(\beta)/k$ is separable. Since k'/k is purely inseparable, it follows that $\beta \in k$.

We also have $C^2 + D^2\alpha^2 = 0$, and hence $\alpha = \frac{C}{D} \in k$.

Thus $g_K = 0$ implies $z \in K$ and by (9.63), we have $a_1, b_1 \in k$. The converse is clear. We have proved the following proposition:

Proposition 9.6.7. *If $K = k(x, y)$ is given by (9.62), then $g_K = 0$ (and K is a rational function field) if and only if $\sqrt{a}, \sqrt{b} \in k$. In this case, we have $K = k(\sqrt{x})$.* □

As an application of Tate's genus formula (Corollary 9.5.20) we present another proof that if K/k is a function field such that $K = k(x, y)$ and

$$y^2 = x^3 + ax + b = f(x)$$

with $k(\sqrt{a}, \sqrt{b}) \neq k$, then $g_K = 1$.

We have already proved that if $\mathfrak{P}$ divides $\wp_\infty$ where $(x)_{k(x)} = \frac{\wp_0}{\wp_\infty}$, then $\wp_\infty = \mathfrak{P}^2$, $d_K(\mathfrak{P}) = 1$, and k is the field of constants of K.

We now compute the numbers $r_\wp$ given in Proposition 9.5.12. Let $\mathfrak{B}$ be a prime divisor in K, $\wp = \mathfrak{B}|_{k(x)}$, and $\wp \neq \wp_\infty$. Let $h(x) \in k[x]$ be a prime element for $\wp$. Then $h(x)$ satisfies

$$(h(x))_{k(x)} = \frac{\wp}{\wp_\infty^{\deg h}}.$$

Notice that since $v_\wp(f(x))$ is nonnegative, $r_\wp$ is nonnegative too. We wish to show that $r_\wp = 0$ or $r_\wp = 1$. Assume for the time being that $r_\wp \geq 2$. Let $\xi \in k(x)_\wp$ be such that

$$v_\wp(f(x) - \xi^2) = r_\wp.$$

Since $k(x)$ is dense in $k(x)_\wp$, there exists $t(x) \in k(x)$ such that

$$v_\wp(\xi - t(x)) > r_\wp.$$

We have $v_\wp(\xi^2 - t(x)^2) = 2v_\wp(\xi - t(x)) > 2r_\wp$. Thus

$$v_\wp(f(x) - t(x)^2) = v_\wp(f(x) - \xi^2) = r_\wp.$$

Let $t(x) = \frac{p(x)}{q(x)}$, where $p(x), q(x) \in k[x]$ and $(p(x), q(x)) = 1$. Since $v_\wp(t(x)) \geq 0$, it follows that $(h(x), q(x)) = 1$. Let $c(x), d(x) \in k[x]$ be such that $h^n(x)c(x) + q(x)d(x) = p(x)$ for a given $n \in \mathbb{N}$. Then

$$t(x) - d(x) = \frac{p(x)}{q(x)} - d(x) = \frac{p(x) - d(x)q(x)}{q(x)} = \frac{h^n(x)c(x)}{q(x)}.$$

Thus $v_\wp(t(x) - d(x)) \geq n$ and $c(x) \in k[x]$. If we take $n > r_\wp$, it follows that $v_\wp(f(x) - c(x)^2) = r_\wp$, and we may assume that $t(x) \in k[x]$.

Since $r_\wp \geq 2$, we have

$$f(x) - t(x)^2 = h(x)^2 s(x) \tag{9.66}$$

with $s(x) \in k[x]$. Taking the usual derivative in (9.66), it follows that

$$x^2 + a = f'(x) = h(x)^2 s'(x).$$

Hence $s'(x) \in k[x]$, $\deg s'(x) = 0$ and $h(x) = x + \sqrt{a}$. Thus $\sqrt{a} \in k$ and $s(x) = \ell(x^2) + x + \beta$ with $\ell(0) = 0$.

Substituting in (9.66) we obtain

$$x^3 + ax + b = f(x) = (x^2 + a)(\ell(x^2) + x + \beta) + t(x)^2.$$

Therefore

$$b = x^2\ell(x^2) + \beta x^2 + a\ell(x^2) + a\beta + t(x)^2. \tag{9.67}$$

Let $\ell(x) = d_m x^m + \cdots + d_1 x$, $t(x) = c_n x^n + \cdots + c_1 x + c_0$.

It follows from (9.67) that $n = m + 1$ and

$$\left.\begin{array}{rl} d_m + c_{m+1}^2 & = 0 \\ \vdots & \vdots\ \vdots \\ d_i + ad_{i+1} + c_{i+1}^2 & = 0 \\ \vdots & \vdots\ \vdots \\ d_1 + ad_2 + c_2^2 & = 0 \\ \beta + ad_1 + c_1^2 & = 0 \\ a\beta + c_0^2 & = b \end{array}\right\}. \tag{9.68}$$

From (9.68) we deduce that b belongs to k^2, which is a contradiction. Thus for all $\wp \neq \wp_\infty$, we have $r_\wp = 0$ or $r_\wp = 1$. Since k is the field of constants of K, it follows by Proposition 9.5.12 that

$$v_{\mathfrak{B}}(D_{\tau_{\mathfrak{B}}}\alpha) \deg_K \mathfrak{B} = 0.$$

Now assume $\wp = \wp_\infty$. Then

$$r_{\wp_\infty} \geq v_{\wp_\infty}(f(x)) = -3.$$

For any $\xi \in k(x)_{\wp_\infty}$,

$$v_\wp(f(x) - \xi^2) = \min\{v_\wp(f(x)), 2v_\wp(\xi)\} = \min\{-3, 2v_\wp(\xi)\} \leq -3.$$

Hence $r_{\wp_\infty} = -3$ and since $2 = p \nmid -3$, we have $v_\wp(D_{\tau_{\mathfrak{P}}}\alpha) \deg_K \mathfrak{P} = -4$. Therefore

$$v_{\mathfrak{B}}(D_{\tau_{\mathfrak{B}}}\alpha) \deg_K \mathfrak{B} = \begin{cases} 0 & \text{if } \mathfrak{B}|_{k(x)} = \wp \neq \wp_\infty, \\ -3 - 1 = -4 & \text{if } \mathfrak{B} = \mathfrak{P}. \end{cases}$$

Using Tate's genus formula, we obtain

$$\begin{aligned} 2g_K - 2 &= 2^{1-0}(2g_{k(x)} - 2) + (1-2) \sum_{\mathfrak{B}} v_{\mathfrak{B}}(D_{\tau_{\mathfrak{B}}}\alpha) \deg_K \mathfrak{B} \\ &= 2(-2) - (-4) = 0. \end{aligned}$$

It follows that $g_K = 1$, which was to be shown.

Now we consider the case $\gamma x + \delta \neq 0$ in (9.61).

Let $y_1 = \frac{y}{\gamma x + \delta}$. Then $K = k(y_1, x)$ and

$$\begin{aligned} y_1^2 - y_1 = y_1^2 + y = \frac{y^2}{(\gamma x + \delta)^2} + \frac{y}{\gamma x + \delta} &= \frac{y^2 + (\gamma x + \delta) y}{(\gamma x + \delta)^2} \\ &= \frac{x^3 + \alpha_2 x^2 + \alpha_1 x + \alpha_0}{(\gamma x + \delta)^2}. \end{aligned}$$

Clearly, $K/k(y)$ is a separable extension of degree 2. We distinguish two subcases. If $\gamma = 0$, then denoting y_1 by y, we have

$$y^2 - y = f(x) \in k[x], \tag{9.69}$$

where $f(x)$ is a polynomial of degree 3.

If $\gamma \neq 0$, let $x_1 = x + \delta/\gamma$. Then

$$x^3 + \alpha_2 x^2 + \alpha_1 x + \alpha_0 = x_1^3 + \varepsilon_2 x_1^2 + \varepsilon_1 x_1 + \varepsilon_0$$

and

$$y^2 - y = \frac{x_1}{\gamma^2} + \frac{\varepsilon_2}{\gamma^2} + \frac{\varepsilon_1}{\gamma^2 x_1} + \frac{\varepsilon_0}{\gamma^2 x_1^2}. \tag{9.70}$$

Note that if $g_K = 1$, $\varepsilon_0 = \varepsilon_1 = 0$ does not hold. Indeed, if this were the case, x_1 would belong to $k(y)$ and K would be a rational function field.

Assuming that ε_0 is a square in k (which happens when k is a perfect field), then if $\beta_0 \in k$ is such that $\beta_0^2 = \varepsilon_0$, let

$$y_1 = y + \frac{\beta_0}{\gamma x_1}.$$

We have

$$y_1^2 - y_1 = y_1^2 + y_1 = y^2 - y + \frac{\beta_0^2}{\gamma^2 x_1^2} - \frac{\beta_0}{\gamma x_1} = \frac{x_1}{\gamma^2} + \frac{\varepsilon_2}{\gamma_2} + \frac{\varepsilon_3}{\gamma^2 x_1}.$$

Let $x_2 = \frac{x_1}{\gamma^2} + \frac{\varepsilon_2}{\gamma^2}$. Then

$$y_1^2 - y_1 = x_2 + \frac{1}{\alpha x_2 + \beta} \quad \text{for some} \quad \alpha \in k^*. \tag{9.71}$$

If ε_0 is not a square in k, the substitution $x_2 = \frac{x_1}{\gamma^2} + \frac{\varepsilon_2}{\gamma^2}$ reduces (9.70) to

$$y_1^2 - y_1 = x_2 + \frac{\alpha x_2 + \beta}{\delta x_2^2 + \varepsilon} \quad \text{for some} \quad \delta \neq 0. \tag{9.72}$$

If $(\alpha x_2 + \beta, \delta x_2^2 + \varepsilon) \neq 1$, (9.72) reduces to (9.71). We now assume that $(\alpha x_2 + \beta, \delta x_2^2 + \varepsilon) = 1$, with $\delta \neq 0$. Then (9.72) can be written as

$$y^2 - y = x + \frac{a'x + b'}{x^2 + c'}, \tag{9.73}$$

where $(x^2 + c', a'x + b') = 1$. Then

$$\left(x + \frac{a'x + b'}{x^2 + c'}\right)_{k(x)} = \frac{\mathfrak{A}}{\wp_\infty \wp} \quad \text{or} \quad \frac{\mathfrak{A}}{\wp_\infty \wp^2}$$

according to whether $\sqrt{c} \notin k$ and $d_{k(x)}(\wp) = 2$ or $\sqrt{c} \in k$ and $d_{k(x)}(\wp) = 1$ respectively. Even though k is not a perfect field we can use Example 5.8.8 to see that $\wp_\infty$ is ramified in K. Furthermore, $d_K(\mathfrak{P}) = 1$, $\mathfrak{P}$ divides $\wp_\infty$, and k is the field of constants of K. If $\sqrt{c} \notin k$, then by Example 5.8.8 we have

$$\mathfrak{P}^2 \mathfrak{P}_1^2 \mid \mathfrak{D}_{K/k(x)},$$

where $d_K(\mathfrak{P}_1) \geq 2$. In this case $g_K = 2$. Thus $\sqrt{c} \in k$.

We are now ready to prove the following theorem:

Theorem 9.6.8. *Suppose that k has characteristic 2. Then an elliptic function field K/k is given by $K = k(x, y)$, where:*

(1) *If $K/k(x)$ is purely inseparable, then*

$$y^2 = x^3 + ax + b$$

with $k(\sqrt{a}, \sqrt{b}) \neq k$.

(2) *If $K/k(x)$ is separable, then K is given by one of the following equations:*

(a) $y^2 - y = f(x) \in k[x]$, *where* $\deg f(x) = 3$ *and* $f(x)$ *is irreducible.*

(b) $y^2 - y = x + \frac{1}{ax+b}$, *with* $a \in k^*$.

(c) $y^2 - y = x + \frac{\alpha x+\beta}{(x+\varepsilon)^2}$, *where* $(\alpha x + \beta, x + \varepsilon) = 1$.

When k is a perfect field, K/k is given by either (a) *or* (b).

Conversely, any of the above equations defines an elliptic function field.

Proof: We have already proved (1) (Proposition 9.6.7). On the other hand, any elliptic function field K such that $K/k(x)$ is separable is given by (a), (b), or (c).

Now if K/k is defined by (a) or (b), let $(x)_{k(x)} = \frac{\wp_0}{\wp_\infty}$. Then either

$$(f(x))_{k(x)} = \frac{\mathfrak{A}}{\wp_\infty^3}, \quad \text{for some integral divisor } \mathfrak{A}$$

or

$$\left(x + \frac{1}{ax+b}\right)_{k(x)} = \frac{\mathfrak{B}}{\wp_1 \wp_\infty},$$

where $\mathfrak{B}$ is an integral divisor and $\wp_1$ is a prime divisor of degree 1.

It follows by Example 5.8.8 that

$$\mathfrak{D}_{K/k(x)} = \mathfrak{P}^4 \quad \text{or} \quad \mathfrak{D}_{K/k(x)} = \mathfrak{P}_1^2 \mathfrak{P}^2,$$

where $\mathfrak{P}_1, \mathfrak{P}$ are prime divisors in K that lie above $\wp_1$ and $\wp_\infty$ respectively. In particular, $\wp_\infty$ is ramified, k is the field of constants of K, and $d_K(\mathfrak{P}) = 1$. Using the genus formula we obtain in both cases that

$$g_K = 1 + (g_{k(x)} - 1)[K : k(x)] + \frac{1}{2} d_K(\mathfrak{D}_{K/k(x)}) = 1 - 2 + \frac{1}{2}(4) = 1.$$

Finally, consider (c). We have

$$\left(x + \frac{\alpha x + \beta}{(x+\varepsilon)^2}\right)_{k(x)} = \frac{\mathfrak{A}}{\wp_\infty \wp_\varepsilon^2},$$

where $\mathfrak{A}$ is an integral divisor relatively prime to $\wp_\infty \wp_\varepsilon$.

It follows by Example 5.8.8 that $\wp_\infty$ is ramified. Moreover, if $\mathfrak{P}$ divides $\wp_\infty$, then $d_K(\mathfrak{P}) = 1$ and k is the field of constants of K. Also, $\mathfrak{N}_x = \mathfrak{P}^2$ and $\mathfrak{N}_y = \mathfrak{P}^3$. Let

$n > 2g_K - 2$. For any m satisfying $2 \leq m \leq n$, there exists an element $y^i x^j$ such that $\eta_{y^j x^i} = \mathfrak{P}^m$. Hence

$$n \leq \ell_K(\mathfrak{P}^{-n}) = d_K(\mathfrak{P}^n) - g_K + 1 = n + 1 - g_K.$$

Thus $g_K \leq 1$. Let $k' = k(\sqrt{\beta})$. Then, as before, $K' = Kk'$ can be given by

$$y^2 - y = x + \frac{1}{ax+b}, \quad \text{where} \quad a \in k' \setminus \{0\}$$

and $g_{K'} = 1$. Since K' is a constant extension of K, by Theorem 8.5.3 we have

$$1 = g_{K'} \leq g_K \leq 1.$$

Thus $g_K = 1$ and K is an elliptic function field (and as a corollary we obtain that $\mathfrak{D}_{K/k(x)} = \mathfrak{P}^2 \mathfrak{P}_\varepsilon^2$). □

9.6.3 The Automorphism Group of an Elliptic Function Field

Now we study the automorphism group of an elliptic function field.

Let K/k be an arbitrary elliptic function field, C_K the divisor class group of K, and $C_{K,0}$ its subgroup of divisor classes of degree 0 (see Section 3.2).

Set $M_K = \{\mathfrak{P} \in \mathbb{P}_K \mid d_K(\mathfrak{P}) = 1\}$. Let $\mathfrak{P}_0 \in M_K$ be fixed and $K = k(x, y)$ with $\mathfrak{N}_x = \mathfrak{P}_0^2$, $\mathfrak{N}_y = \mathfrak{P}_0^3$. Let

$$\varphi : M_K \to C_{K,0}$$

be defined by

$$\varphi(\mathfrak{P}) = \left[\frac{\mathfrak{P}}{\mathfrak{P}_0}\right]. \tag{9.74}$$

Proposition 9.6.9. *The function φ given in* (9.74) *is bijective.*

Proof: Let $\mathfrak{B}$ be a divisor of degree 0. Then $d_K(\mathfrak{B}\mathfrak{P}_0) = 1 > 0 = 2g_K - 2$. By Corollary 3.5.6 we have

$$\ell_K(\mathfrak{B}^{-1}\mathfrak{P}_0^{-1}) = d_K(\mathfrak{B}\mathfrak{P}_0) - g_K + 1 = 1.$$

If $\alpha \in L_K(\mathfrak{B}^{-1}\mathfrak{P}_0^{-1})$, α is nonzero and satisfies $(\alpha)_K = \frac{\mathfrak{P}}{\mathfrak{B}\mathfrak{P}_0}$, where $\mathfrak{P}$ is an integral divisor of degree 1. Thus $\mathfrak{P}$ is a prime divisor and $\overline{\mathfrak{B}} = \left(\frac{\mathfrak{P}}{\mathfrak{P}_0}\right) = \varphi(\mathfrak{P})$.

Now if $\varphi(\mathfrak{P}) = \varphi(\mathfrak{P}_1)$, then $\frac{\mathfrak{P}}{\mathfrak{P}_0}$ and $\frac{\mathfrak{P}_1}{\mathfrak{P}_0}$ define the same class. Therefore $\frac{\mathfrak{P}}{\mathfrak{P}_0}\left(\frac{\mathfrak{P}_1}{\mathfrak{P}_0}\right)^{-1} = \frac{\mathfrak{P}}{\mathfrak{P}_1}$, which is principal. Let $(x)_K = \frac{\mathfrak{P}}{\mathfrak{P}_1}$. If $\mathfrak{P} \neq \mathfrak{P}_1$, we have $[K : k(x)] = d_K(\mathfrak{Z}_x) = d_K(\mathfrak{P}) = 1$, which contradicts the fact that K is of genus 1. It follows that $\mathfrak{P} = \mathfrak{P}_1$, and φ is bijective. □

Remark 9.6.10. The bijection φ provides M_K with an additive group structure whose operation $\oplus$ is defined by

$$\mathfrak{P} \oplus \mathfrak{P}_1 := \varphi^{-1}\left(\varphi(\mathfrak{P})\varphi(\mathfrak{P}_1)\right) = \varphi^{-1}\left(\left[\frac{\mathfrak{P}\mathfrak{P}_1}{\mathfrak{P}_0^2}\right]\right).$$

In the other words, $\mathfrak{P}\oplus\mathfrak{P}_1 = \mathfrak{P}_2$, where $\left(\overline{\frac{\mathfrak{P}\mathfrak{P}_1}{\mathfrak{P}_0^2}}\right) = \left(\overline{\frac{\mathfrak{P}_2}{\mathfrak{P}_0}}\right)$. We have $\mathfrak{P}\oplus\mathfrak{P}_0 = \mathfrak{P}$ and $\mathfrak{P} \oplus \mathfrak{P}_1 = \mathfrak{P}_2$ if and only if $\frac{\mathfrak{P}\mathfrak{P}_1}{\mathfrak{P}_2\mathfrak{P}_0}$ is principal. With this structure, M_K is isomorphic to $C_{K,0}$.

Now consider

$$\operatorname{Aut}_k(K) = \{\sigma : K \to K \mid \sigma \text{ is an automorphism of } K \text{ and } \sigma|_k = \operatorname{Id}_k\}.$$

Proposition 9.6.11. *Let K/k be any function field and let $\sigma, \theta \in \operatorname{Aut}_k(K)$ be such that $\wp^\sigma = \wp^\theta$ for all $\wp \in \mathbb{P}_K$. Then $\sigma = \theta$.*

Proof: Put $\varphi = \sigma\theta^{-1}$. It follows from the choice of σ, θ that $\wp^\varphi = \wp$ for all $\wp \in \mathbb{P}_K$. Let $z \in K$ be such that $(z)_K = \wp_1^{a_1} \cdots \wp_r^{a_r}$. Then

$$(z^\varphi)_K = (z)_K^\varphi = (\wp_1^\varphi)^{a_1} \cdots (\wp_r^\varphi)^{a_r} = \wp_1^{a_1} \cdots \wp_r^{a_r} = (z)_K.$$

Thus there exists $C_z \in k$ such that

$$z^\varphi = C_z z.$$

If z_1 and z_2 are linearly independent over k, we have

$$(z_1 + z_2)^\varphi = C_{z_1+z_2}(z_1 + z_2) = z_1^\varphi + z_2^\varphi = C_{z_1}z_1 + C_{z_2}z_2.$$

Therefore $C_{z_1+z_2} = C_{z_1} = C_{z_2}$. Since $C_1 = 1$, it follows that $C_z = 1$ for all $z \in K$ and $\varphi = \operatorname{Id}_K$. Hence $\sigma = \theta$. □

Now we return to the case of an elliptic function field K/k. Let $\mathfrak{P}$ and $\mathfrak{P}_1$ be two prime divisors of degree 1, not necessarily distinct. We choose $\mathfrak{P}_0 = \mathfrak{P}$ as in Remark 9.6.10. We have $\ell_K((\mathfrak{P}\mathfrak{P}_1)^{-1}) = d_K(\mathfrak{P}\mathfrak{P}_1) - g_K + 1 = 2$.

Let $z \in K \setminus k$ be such that $z \in L_K((\mathfrak{P}\mathfrak{P}_1)^{-1})$. Then $(z)_K = \frac{\mathfrak{A}}{\mathfrak{P}\mathfrak{P}_1}$ for some integral divisor $\mathfrak{A}$ such that $\mathfrak{A} \neq \mathfrak{P}\mathfrak{P}_1$. We have $[K : k(z)] = 2 = d(\mathfrak{N}_z)$.

Next, assume that $K/k(z)$ is a separable extension.

Let $\operatorname{Gal}(K/k(z)) = \{1, \sigma\}$, where $\sigma \neq \operatorname{Id}$ and $\sigma(z) = z$. Notice that σ fixes $\mathfrak{A}(\mathfrak{P}\mathfrak{P}_1)^{-1}$ (such an automorphism is called a *reflection automorphism* of $\mathfrak{P}$ and $\mathfrak{P}_1$ in K). If $\mathfrak{P} \neq \mathfrak{P}_1$, then $\mathfrak{P}$ and $\mathfrak{P}_1$ are the prime divisors above the pole divisor $\wp_\infty$ of z in $k(z)$, and thus $\sigma\mathfrak{P} = \mathfrak{P}_1$, $\sigma\mathfrak{P}_1 = \mathfrak{P}$ (because $\operatorname{Gal}(K/k(z)) = \{1, \sigma\}$).

Let $\mathfrak{q}$ be a divisor of degree 1 in K. Then

$$\left(\frac{\mathfrak{q}}{\mathfrak{P}}\right)\left(\frac{\mathfrak{q}}{\mathfrak{P}}\right)^\sigma = \frac{\mathfrak{q}\mathfrak{q}^\sigma}{\mathfrak{P}\mathfrak{P}^\sigma},$$

and the latter is a divisor of degree 0 in $k(z)$, and hence a principal divisor. This means that

$$\mathfrak{q} \oplus \mathfrak{q}^{\sigma} = \mathfrak{P}^{\sigma} = \mathfrak{P}_1 \quad \text{or} \quad \mathfrak{q}^{\sigma} = \mathfrak{P}^{\sigma} \ominus \mathfrak{q}.$$

In particular, taking $\mathfrak{P}_1 = \mathfrak{P}$, we get $\mathfrak{q}^{\sigma} = \ominus\mathfrak{q}$.

Let us denote

$$\sigma_{\mathfrak{P},\mathfrak{P}_1} \tag{9.75}$$

by σ for $\mathfrak{P}$ and $\mathfrak{P}_1$. Then

$$\mathfrak{q}^{\sigma_{\mathfrak{P},\mathfrak{P}_1}} = \mathfrak{P}_1 \ominus \mathfrak{q}, \quad \text{and so} \quad \mathfrak{q}^{\sigma_{\mathfrak{P},\mathfrak{P}}} = \ominus\mathfrak{q}.$$

Set

$$\tau = \tau_{\mathfrak{P},\mathfrak{P}_1} = \sigma_{\mathfrak{P},\mathfrak{P}} \circ \sigma_{\mathfrak{P},\mathfrak{P}_1}. \tag{9.76}$$

For any prime divisor $\mathfrak{q}$ of degree 1, the divisor $\frac{\mathfrak{q}^{\sigma_{\mathfrak{P},\mathfrak{P}}}}{\mathfrak{P}} \frac{\mathfrak{q}^{\tau}}{\mathfrak{P}_1} = (z)$ is principal. Thus $\mathfrak{q}^{\tau} \oplus \mathfrak{q}^{\sigma_{\mathfrak{P},\mathfrak{P}}} = \mathfrak{q}^{\tau} \ominus \mathfrak{q} = \mathfrak{P}_1$, and we have

$$\mathfrak{q}^{\tau_{\mathfrak{P},\mathfrak{P}_1}} = \mathfrak{q} \oplus \mathfrak{P}_1. \tag{9.77}$$

Because of (9.77) $\tau_{\mathfrak{P},\mathfrak{P}_1}$ is called the *translation automorphism* from $\mathfrak{P}$ to $\mathfrak{P}_1$.

Assume that $\tau' = \tau_{\mathfrak{P},\mathfrak{P}_2}$ is another translation automorphism. Then if $\mathfrak{q} \oplus \mathfrak{P}_1 = \mathfrak{q}_1 = \mathfrak{q}^{\tau}$, it follows that $\mathfrak{q}_1$ is a prime divisor of degree 1 and

$$\mathfrak{q}_1^{\tau'} = \mathfrak{q}_1 \oplus \mathfrak{P}_2.$$

Thus

$$\mathfrak{q}^{\tau\tau'} = (\mathfrak{q}^{\tau})^{\tau'} = \mathfrak{q}_1^{\tau'} = \mathfrak{q}_1 \oplus \mathfrak{P}_2 = \mathfrak{q} \oplus \mathfrak{P}_1 \oplus \mathfrak{P}_2.$$

Therefore $\tau_{\mathfrak{P},\mathfrak{P}_1} \circ \tau_{\mathfrak{P},\mathfrak{P}_2}$ has the same effect as $\tau_{\mathfrak{P},\mathfrak{P}_1 \oplus \mathfrak{P}_2}$ on prime divisors of degree one.

Proposition 9.6.12. *With the above notation we have*

$$\tau_{\mathfrak{P},\mathfrak{P}_1} \circ \tau_{\mathfrak{P},\mathfrak{P}_2} = \tau_{\mathfrak{P},\mathfrak{P}_1 \oplus \mathfrak{P}_2}.$$

Proof: Let $G = \operatorname{Aut}_k(K)$ and $\overline{G} = \operatorname{Aut}_{\overline{k}}(\overline{K})$, where $\overline{k}$ is an algebraic closure of k and $\overline{K} = K\overline{k}$ is the constant field extension.

The natural map from G to $\overline{G}$ is a monomorphism of groups. (If $\sigma \in G$, the extension of σ to $\overline{K}$ is defined as follows: if $\alpha = \sum_{i=1}^{n} \alpha_i x_i$ with $\alpha_i \in \overline{k}$ and $x_i \in K$, then $\sigma(\alpha) = \sum_{i=1}^{n} \alpha_i \sigma(x_i)$. See the proof of Corollary 14.3.9.)

All the prime divisors of $\overline{K}$ are of degree 1 since $\overline{k}$ is algebraically closed, and $\tau_{\mathfrak{P},\mathfrak{P}_1} \circ \tau_{\mathfrak{P},\mathfrak{P}_2}$ and $\tau_{\mathfrak{P},\mathfrak{P}_1 \oplus \mathfrak{P}_2}$ have the same effect on all prime divisors of $\overline{K}$. The statement follows by Proposition 9.6.11. $\square$

Theorem 9.6.13. *Let*

$$G = \left\{ \tau_{\mathfrak{P},\mathfrak{P}_1} \mid \mathfrak{P}_1 \in \mathbb{P}_K \text{ of degree } 1 \right\}$$

be the set of all translation automorphisms. Then G is a group that is isomorphic to M_K and to $C_{K,0}$.

Proof: Let $\varphi : G \to M_K$ be given by $\varphi(\tau_{\mathfrak{P},\mathfrak{P}_1}) = \mathfrak{P}_1$ (respectively, let $\widetilde{\varphi} : G \to C_{K,0}$ be given by $\widetilde{\varphi}(\tau_{\mathfrak{P},\mathfrak{P}_1}) = \left[\frac{\mathfrak{P}_1}{\mathfrak{P}}\right]$). Then $\varphi(\tau_{\mathfrak{P},\mathfrak{P}_1} \circ \tau_{\mathfrak{P},\mathfrak{P}_2}) = \varphi(\tau_{\mathfrak{P},\mathfrak{P}_1 \oplus \mathfrak{P}_2}) = \mathfrak{P}_1 \oplus \mathfrak{P}_2 = \varphi(\tau_{\mathfrak{P},\mathfrak{P}_1})\varphi(\tau_{\mathfrak{P},\mathfrak{P}_2})$.

Hence φ is a group homomorphism. Clearly φ is bijective. □

Theorem 9.6.14. *Let $\mathfrak{G} = \operatorname{Aut}_k(K)$ and set $G = \{\tau_{\mathfrak{P},\mathfrak{P}_1} \mid d_K(\mathfrak{P}_1) = 1, \mathfrak{P}_1 \in \mathbb{P}_K\}$.*

Then G is a normal subgroup of $\mathfrak{G}$ that satisfies $|\mathfrak{G}/G| < \infty$.

Proof: Let $\sigma \in \mathfrak{G}$ and $\tau = \tau_{\mathfrak{P},\mathfrak{P}_1} \in G$. Let $\mathfrak{q}$ be a prime divisor of degree 1. Set $\varphi = \sigma\tau\sigma^{-1}$. Using (9.77) we see that $\mathfrak{q}^\tau = \mathfrak{q} \oplus \mathfrak{P}_1$ is equivalent to

$$\frac{\mathfrak{q}\mathfrak{P}_1}{\mathfrak{q}^\tau\mathfrak{P}} = (z)_K \quad \text{being principal or} \quad \mathfrak{q}^\tau = (z^{-1})_K \frac{\mathfrak{q}\mathfrak{P}_1}{\mathfrak{P}} = \frac{\mathfrak{q}\mathfrak{P}_1}{(z)_K\mathfrak{P}}.$$

It follows that

$$\begin{aligned}
\left(\frac{\mathfrak{q}^\varphi}{\mathfrak{P}^\varphi}\right) &= \left(\frac{\mathfrak{q}}{\mathfrak{P}}\right)^\varphi = \left(\frac{\mathfrak{q}}{\mathfrak{P}}\right)^{\sigma\tau\sigma^{-1}} = \left(\frac{\mathfrak{q}^\sigma}{\mathfrak{P}^\sigma}\right)^{\tau\sigma^{-1}} = \left(\frac{(\mathfrak{q}^\sigma)^\tau}{(\mathfrak{P}^\sigma)^\tau}\right)^{\sigma^{-1}} \\
&= \left(\frac{\mathfrak{q}^\sigma\mathfrak{P}_1}{\mathfrak{P}(z_1)_K}\frac{\mathfrak{P}(z_2)_K}{\mathfrak{P}^\sigma\mathfrak{P}_1}\right)^{\sigma^{-1}} = \left(\frac{\mathfrak{q}^\sigma}{\mathfrak{P}^\sigma}\right)^{\sigma^{-1}}\left(\frac{z_2^{\sigma^{-1}}}{z_1^{\sigma^{-1}}}\right)_K \\
&= \left(\frac{\mathfrak{q}}{\mathfrak{P}}\right)\left(\frac{z_2^{\sigma^{-1}}}{z_1^{\sigma^{-1}}}\right)_K.
\end{aligned}$$

Thus $\frac{\mathfrak{P}^\varphi\mathfrak{q}}{\mathfrak{P}\mathfrak{q}^\varphi}$ is principal, and $\mathfrak{q}^\varphi = \mathfrak{q} \oplus \mathfrak{P}^\varphi$.

Therefore $\varphi = \tau_{\mathfrak{P},\mathfrak{P}^\varphi} = \sigma\tau_{\mathfrak{P},\mathfrak{P}_1}\sigma^{-1}$ and G is a normal subgroup of $\mathfrak{G}$. Furthermore, we have $\mathfrak{P}^{\sigma\tau} = \mathfrak{P}^\sigma \oplus \mathfrak{P}_1$ and $\frac{\mathfrak{P}^\sigma\mathfrak{P}_1}{\mathfrak{P}\mathfrak{P}^{\sigma\tau}}$ is principal, and so is $\left(\frac{\mathfrak{P}^\sigma\mathfrak{P}_1}{\mathfrak{P}\mathfrak{P}^{\sigma\tau}}\right)^{\sigma^{-1}} = \frac{\mathfrak{P}\mathfrak{P}_1^{\sigma^{-1}}}{\mathfrak{P}^{\sigma^{-1}}\mathfrak{P}^\varphi}$. It follows that

$$\mathfrak{P}^\varphi \oplus \mathfrak{P}^{\sigma^{-1}} = \mathfrak{P}_1^{\sigma^{-1}}$$

or

$$\mathfrak{P}^\varphi = \mathfrak{P}_1^{\sigma^{-1}} \ominus \mathfrak{P}^{\sigma^{-1}}.$$

Now let $\sigma \in \mathfrak{G}$ and set $\mathfrak{P}_1 = \mathfrak{P}^\sigma$. Then

$$\mathfrak{P}^{\tau_{\mathfrak{P},\mathfrak{P}_1}} = \mathfrak{P} \oplus \mathfrak{P}_1 = \mathfrak{P}_1 = \mathfrak{P}^{\sigma}.$$

Therefore $\tau_{\mathfrak{P},\mathfrak{P}_1}\sigma^{-1} = \sigma^{-1}\tau_{\mathfrak{P},\mathfrak{P}^{\varphi}}$ fixes $\mathfrak{P}$. Thus if we show that $\mathrm{Stat}_{\mathfrak{G}}(\mathfrak{P}) = \{\theta \in \mathfrak{G} \mid \mathfrak{P}^{\theta} = \mathfrak{P}\}$ is finite, it will follow that

$$|\mathfrak{G}/G| \leq |\mathrm{Stat}_{\mathfrak{G}}(\mathfrak{P})| < \infty.$$

We have $\ell_K(\mathfrak{P}^{-n}) = d_K(\mathfrak{P}^n) - g_K + 1 = n$ for every $n \geq 1$.

Let $\{1, x\}$ be a basis of $L_K(\mathfrak{P}^{-2})$ and $\{1, x, y\}$ a basis of $L_K(\mathfrak{P}^{-3})$ (with $\mathfrak{N}_x = \mathfrak{P}^2, \mathfrak{N}_y = \mathfrak{P}^3$). Let $\sigma \in \mathrm{Stat}_{\mathfrak{G}}(\mathfrak{P})$. We have

$$L_K(\mathfrak{P}^{-2})^{\sigma} = L_K(\mathfrak{P}^{-2}) \quad \text{and} \quad L_K(\mathfrak{P}^{-3})^{\sigma} = L_K(\mathfrak{P}^{-3}).$$

It follows that

$$\sigma x = ax + b \quad \text{for some} \quad a \neq 0, \quad \text{and} \quad \sigma y = c + dx + ey \quad \text{for some} \quad e \neq 0. \tag{9.78}$$

If char $k \neq 2, 3$, then by (4.6) we have $K = k(x, y)$ with

$$y^2 = 4x^3 - g_2x - g_3. \tag{9.79}$$

Substituting (9.78) in (9.79) we obtain

$$(c + dx + y)^2 = 4(ax + b)^3 - g_2(ax + b) - g_3.$$

Hence

$$c^2 + d^2x^2 + e^2y^2 + 2cdx + 2cey + 2dexy$$
$$= 4a^3x^3 + 12a^2bx^2 + 12ab^2x + 4b^3 - g_2ax - g_2b - g_3.$$

It follows that

$$\frac{2d}{e} = 0, \quad \frac{2c}{e} = 0, \quad \text{and} \quad \frac{d^2}{e^2} = \frac{12a^2b}{e^2}, \quad \frac{4a^3}{e^2} = 4. \tag{9.80}$$

Therefore $d = c = 0$, $b = 0$ and $e^2 = a^3$.

If $\lambda = \frac{e}{a}$, we have $a = \frac{e^2}{a^2} = \lambda^2$, so $e = a\lambda = \lambda^3$.

Therefore

$$\sigma x = \lambda^2 x \quad \text{and} \quad \sigma y = \lambda^3 y. \tag{9.81}$$

If we substitute (9.81) in (9.79) we obtain

$$\lambda^6 y^2 = 4\lambda^6 x^3 - g_2\lambda^2 x - g_3.$$

Hence

$$y^2 = 4x^3 - \frac{g_2}{\lambda^4}x - \frac{g_3}{\lambda^6} = 4x^3 - g_2x - g_3.$$

If g_2 and g_3 are nonzero we have $\lambda^4 = \lambda^6 = 1$, so $\lambda^2 = 1$, i.e., $\lambda = \pm 1$. If $g_2 = 0$, then clearly $\lambda^6 = 1$. If $g_3 = 0$, then $\lambda^4 = 1$. Therefore $\mathrm{Stab}_{\mathfrak{G}}(\mathfrak{P})$ is isomorphic to C_2, C_4, or C_6. In any case, it is finite.

If $\operatorname{char} k = 3$, by (4.3) we have

$$y^2 = x^3 + \alpha_2 x^2 + \alpha_3 x + \alpha_4. \tag{9.82}$$

Substituting (9.78) in (9.82), we obtain

$$(c + dx + ey)^2 = (ax + b)^3 + \alpha_2(ax + b)^2 + \alpha_3(ax + b) + \alpha_4,$$

so

$$\begin{aligned} &c^2 + d^2x^2 + e^2y^2 + 2cdx + 2cey + 2dexy \\ &\qquad = a^3x^3 + b^3 + \alpha_2 a^2 x^2 + 2\alpha_2 abx + \alpha_2 b^2 + \alpha_3 ax + \alpha_3 b + \alpha_4. \end{aligned}$$

Hence

$$\frac{a^3}{e^2} = 1, \quad \frac{2d}{e} = 0, \quad \text{and} \quad \frac{2c}{e} = 0.$$

Thus $c = d = 0$, $e = \lambda^3$, $a = \lambda^2$, and $\lambda \in k^*$.

It follows that $\sigma x = \lambda^2 x + b$ and $\sigma y = \lambda^3 y$. We have

$$\begin{aligned} \lambda^6 y^2 &= \lambda^6 x^3 + b^3 + \alpha_2 \lambda^4 x^2 + 2\alpha_2 \lambda^2 bx + \alpha_2 b^2 + \alpha_3 \lambda^2 x + \alpha_3 b + \alpha_4 \\ &= \lambda^6 x^3 + \alpha_2 \lambda^4 x^2 + x(2\alpha_2 \lambda^2 b + \alpha_3 \lambda^2) + (b^3 + \alpha_2 b^2 + \alpha_3 b + \alpha_4). \end{aligned}$$

Therefore

$$b^3 + \alpha_2 b^2 + \alpha_3 b + \alpha_4(1 - \lambda^6) = 0, \quad \frac{\alpha_2}{\lambda^2} = \alpha_2, \quad \text{and} \quad \frac{2\alpha_2 b + \alpha_3}{\lambda^4} = \alpha_3.$$

If $\alpha_2 \neq 0$ then $\lambda^2 = 1$ and $\lambda^6 = 1$, so $b^3 + \alpha_2 b^2 + \alpha_3 b = 0$.

Hence λ can take at most two values (± 1) and b can take at most three values; it follows that $\mathrm{Stab}_{\mathfrak{G}}(\mathfrak{P})$ is a finite group. If $\alpha_2 = 0$, then $\alpha_3 \neq 0$ since $x^3 + \alpha_2 x^2 + \alpha_3 x + \alpha_4$ is a separable polynomial. Therefore $\lambda^4 = 1$, and thus the possible number of λ's and b's is finite.

Finally we consider $\operatorname{char} k = 2$. Since we are assuming that $K/k(x)$ is separable, by Theorem 9.6.8 we have

$$y^2 - y = f(x) \in k[x], \quad \text{with} \quad \deg f(x) = 3, \tag{9.83}$$

$$y^2 - y = x + \frac{1}{Ax + B}, \quad \text{with} \quad A \in k^*, \tag{9.84}$$

or

$$y^2 - y = x + \frac{\alpha x + \beta}{(x+\varepsilon)^2}, \quad \text{where} \quad (\alpha x + \beta, x + \varepsilon) = 1. \tag{9.85}$$

Note that in the proof of Theorem 9.6.8, we showed that in a quadratic constant extension, (9.85) reduces to (9.84). Thus if k' is this quadratic extension of k and $K' = Kk'$, we have $g_{K'} = 1$ and the stabilizer $\mathrm{Stab}_{\mathfrak{G}}(\mathfrak{P})$ in k is contained in the stabilizer in K'. Consequently we may assume that K is given by (9.83) or (9.84) and also that $f(x)$ is monic.

If $K = k(x, y)$ is given by (9.83) and $f(x) = x^3 + \alpha_2 x^2 + \alpha_1 x + \alpha_0$, we have

$$\begin{aligned}(c + dx + ey)^2 - (c + dx + ey) &= (ax + b)^3 + \alpha_2(ax + b)^2 \\ &\quad + \alpha_1(ax + b) + \alpha_0, \\ c^2 + d^2x^2 + e^2y^2 - c - dx - ey &= a^3x^3 + a^2bx^2 + ab^2x + b^3 + \alpha_2 a^2 x^2 \\ &\quad + \alpha_2 b^2 + \alpha_1 ax + \alpha_1 b + \alpha_0.\end{aligned}$$

Hence

$$\begin{aligned}\frac{1}{e} &= 1, \\ a^3 &= 1, \\ a^2b + \alpha_2 a^2 - d^2 &= \alpha_2, \\ ab^2 + \alpha_1 a + d &= \alpha_1, \\ b^3 + \alpha_2 b^2 + \alpha_1 b + \alpha_0 - (c^2 - c) &= 0, \\ d^2 &= a^2b + \alpha_2(a^2 + 1), \\ d &= ab^2 + \alpha_1(a + 1), \\ d^2 &= a^2b^4 + \alpha_1^2(a^2 + 1), \\ a^2b + \alpha_2(a^2 + 1) &= a^2b^4 + \alpha_1^2(a^2 + 1), \\ a^2b(1 - b^3) &= (\alpha_1^2 + \alpha_2)(a^2 + 1).\end{aligned}$$

Therefore there is a finite number of choices for b, and thus for d and c also. Thus $\mathrm{Stat}_{\mathfrak{G}}(\mathfrak{P})$ is finite.

Finally, if K is given by (9.84), we have

$$(c^2 + d^2x^2 + e^2y^2) - (c + dx + ey) = ax + b + \frac{1}{A(ax + b) + B}.$$

Then $\frac{1}{e} = 1$, so $e = 1$ and

$$y^2 - y - (a + d)x + (b + c^2 - c) + d^2x^2 + \frac{1}{Aax + (Ab + B)} = x + \frac{1}{Ax + B}.$$

Hence $d = 0$ and

$$(a + d - 1)x + (b + c^2 - c) = \frac{1}{Ax + B} - \frac{1}{Aax + (Ab + B)}.$$

It follows that $a = 1$, $b + c^2 - c = 0$, $Ab + B = B$, $b = 0$, $c^2 = c$, and therefore $c = 0$ or 1.

This proves that in any case $\mathrm{Stab}_{\mathfrak{G}}(\mathfrak{P})$ is finite. □

9.6.4 Hyperelliptic Function Fields

Definition 9.6.15. A function field K/k is called *hyperelliptic* if $g_K \geq 2$ and K is a quadratic extension of a field of genus 0.

First we consider the special case of K/k a quadratic extension of a rational function field. Assume $[K : k(x)] = 2$. If $\operatorname{char} k \neq 2$, then $K = k(x, y)$ with

$$y^2 = f(x) \in k[x].$$

Recall that $f(x)$ is a square-free polynomial of degree m, and by Corollary 4.3.7, $g_K = \left[\frac{m+1}{2}\right] - 1$. Thus K is hyperelliptic if and only if $m \geq 5$.

Proposition 9.6.16. *A function field K/k is a hyperelliptic function field that is a quadratic extension of a rational function field if and only if $g_K \geq 2$ and there exists $\mathfrak{A} \in D_K$ such that $d(\mathfrak{A}) = 2$ and $\ell(\mathfrak{A}^{-1}) \geq 2$.*

Proof: Assume $[K : k(x)] = 2$ and $g_K \geq 2$. Let $\mathfrak{N}_x$ be the pole divisor of x. By Theorem 3.2.7, $d(\mathfrak{N}_x) = [K : k(x)] = 2$. Since 1 and x belong to $L_K(\mathfrak{N}_x^{-1})$ and are linearly independent, it follows that $\ell(\mathfrak{N}_x^{-1}) \geq 2$.

Conversely, if $g_K \geq 2$ and $\mathfrak{A}$ is a divisor of degree 2 such that $\ell_K(\mathfrak{A}^{-1}) \geq 2$, let $y \in L_K(\mathfrak{A}^{-1}) \setminus k$. Then $(y)_K = \mathfrak{A}^{-1}\mathfrak{B}$ for some integral divisor $\mathfrak{B}$. Since $y \notin k$, it follows that $d(\mathfrak{B}) = d(\mathfrak{A}) = 2$ and $\ell_K(\mathfrak{B}^{-1}) = \ell_K(\mathfrak{A}^{-1}) \geq 2$ (see the proof of Theorem 3.3.2).

Let $x \in L_K(\mathfrak{B}^{-1}) \setminus k$. Then $\mathfrak{N}_x$ divides $\mathfrak{B}$ and $d(\mathfrak{N}_x) = [K : k(x)] \leq 2$. Since $K \neq k(x)$, we have $[K : k(x)] = 2$ and K is hyperelliptic. □

Corollary 9.6.17. *If K is any function field of genus 2, then K is hyperelliptic.*

Proof: Exercise. □

Example 9.6.18. Let $n \in \mathbb{N}$ be any positive integer, and let k be any field. We consider extensions $K/k(x)$ such that the field of constants of K is k and $[K : k(x)] = 2$.

(1) If $K/k(x)$ is separable, then we distinguish two subcases:

(a) $\operatorname{char} K \neq 2$. Then $K = k(x, y)$ and $y^2 = f(x) \in k[x]$, where $f(x)$ is a separable polynomial of degree m. By Corollary 4.3.7, $g_K = \left[\frac{m+1}{2}\right] - 1$. Let $m = 2n + 1$. Thus $g_K = n$.

(b) char $K = 2$. Then $K = k(x, y)$ and $y^2 - y = f(x) \in k(x)$. Let $f(x) = \frac{1}{x^\lambda}$, where $(2, \lambda) = 1$ and $\lambda \in \mathbb{N}$. By Example 5.8.8 we have

$$\mathfrak{D}_{K/k(x)} = \mathfrak{P}^{(\lambda+1)(2-1)} = \mathfrak{P}^{\lambda+1},$$

where $\mathfrak{P}$ is the prime divisor of K that lies above the pole divisor of x in $k(x)$. Using the genus formula we obtain

$$g_K = 1 + 2(g_{k(x)} - 1) + \frac{1}{2}(\lambda + 1) = \frac{\lambda - 1}{2}.$$

Let $\lambda = 2n + 1$. Then $g_K = n$.

(2) Now we consider $K/k(x)$ to be purely inseparable. In this case we have $K = k(x, y)$ and

$$y^2 = f(x) \in k[x],$$

where $f(x)$ is a separable polynomial of degree m and char $k = 2$.
First we will see that if $g_K \neq 0$, then k is an imperfect field (see also the proof of Proposition 9.6.7). Assume for the sake of contradiction that k is perfect and let $f(x) = a_m x^m + a_{m-1}x^{m-1} + \cdots + a_1 x + a_0$.
Let $b_i \in k$ be such that $b_i^2 = a_i$ for $0 \leq i \leq m$. Then

$$y^2 = f(x) = b_m^2 x^m + b_{m-1}^2 x^{m-1} + \cdots + b_1^2 x + b_0 = xg(x)^2 + h(x)^2$$

for some $g(x), h(x) \in k[x]$, where $g(x) \neq 0$ since $f(x)$ is assumed to be a separable polynomial. Hence

$$x = \left[\frac{y - h(x)}{g(x)}\right]^2 = z^2$$

with $z = \frac{y-h(x)}{g(x)} \in K$. Therefore $K = k(z)$, where $z = \sqrt{x}$, K is a rational function field, and $g_K = 0$.
Now assume that k is an imperfect field and let $\alpha \in k \setminus k^2$. Let m be an odd positive integer and

$$f(x) = x^m - \alpha \in k[x].$$

We will calculate g_K using Tate's genus formula. Consider $(x)_{k(x)} = \frac{\wp_0}{\wp_\infty}$ and let $\mathfrak{P}$ be a prime divisor in K such that $\mathfrak{P}$ divides $\wp_\infty$. Then

$$v_\mathfrak{P}(y^2) = 2v_\mathfrak{P}(y) = v_\mathfrak{P}(f(x)) = e(\mathfrak{P}|\wp_\infty)v_{\wp_\infty}(f(x)) = -e(\mathfrak{P}|\wp_\infty)m.$$

Since m is odd, we have $e(\mathfrak{P}|\wp_\infty) = 2$, $d_K(\mathfrak{P}) = 1$, and the field of constants of K is k.
Let $\wp$ be any divisor of $k(x)$ distinct from $\wp_\infty$. Set

$$r_\wp = \max_{\xi \in k(x)_\wp} \left\{ v_\wp(f(x) - \xi^2) \right\}.$$

Let $\xi \in k(x)_\wp$ be such that $v_\wp(f(x) - \xi^2) = r_\wp$.
We have $r_\wp \geq v_\wp(f(x)) \geq 0$. Since $k(x)$ is dense in $k(x)_\wp$, there exists $h(x) \in k[x]$ such that

$$v_\wp(h(x) - \xi) > r_\wp.$$

Thus $v_\wp(h(x)^2 - \xi^2) = 2v_\wp(h(x) - \xi) > 2r_\wp \geq r_\wp$.
It follows that

$$v_\wp(f(x) - h(x)^2) = r_\wp. \tag{9.86}$$

Let $h(x) = \frac{p(x)}{q(x)}$ with $p(x), q(x) \in k[x]$ and $(p(x), q(x)) = 1$.
We have $v_\wp(q(x)) = 0$ since otherwise, $v_\wp(q(x)) > 0$ and

$$\begin{aligned} 0 \leq r_\wp &= v_\wp(f(x) - h(x)^2) = \min\{v_\wp(f(x)), 2v_\wp(h(x))\} \\ &= 2v_\wp(h(x)) = -2v_\wp(q(x)) < 0. \end{aligned}$$

Assume that $r_\wp \geq 2$. From (9.86) we obtain

$$f(x) - a(x)^2 = \ell(x)^2 s(x), \; s(x) \in k[x] \tag{9.87}$$

with $\ell(x) \in k[x]$ a prime element for $\wp$. Taking the derivative with respect to x in (9.87) we get

$$x^{m-1} = mx^{m-1} = \ell(x)^2 s'(x).$$

Since $\ell(x)$ is a prime element for $\wp$, it follows that $\ell(x) = x$ and $s'(x) = x^{m-3}$. Thus

$$s(x) = x^{m-2} + r(x^2) \quad \text{for some} \quad r(x) \in k[x].$$

Using (9.87) we deduce that

$$x^m - \alpha = a(x)^2 + x^2(x^{m-3} + r(x^2)).$$

Furthermore, we obtain

$$\alpha = a(0)^2 \in k^2,$$

which is a contradiction. Thus

$$r_\wp = 0 \quad \text{or} \quad r_\wp = 1 \quad \text{for all} \quad \wp \neq \wp_\infty.$$

For $\wp = \wp_\infty$, we have $r_{\wp_\infty} \geq v_{\wp_\infty}(f(x)) = -m$, which is an odd number. For any $\xi \in k(x)_{\wp_\infty}$, we have $v_{\wp_\infty}(f(x)-\xi^2) = \min\{v_{\wp_\infty}(f(x)), 2v_{\wp_\infty}(\xi)\} \leq v_{\wp_\infty}(f(x))$. Thus $r_{\wp_\infty} = -m$.

Since the field of constants of K is k, Tate's genus formula yields

$$\begin{aligned} 2g_K - 2 &= p^1(2g_{k(x)} - 2) + (1-p)\sum_{\mathfrak{B}} v_{\mathfrak{B}}(D_{\tau_{\mathfrak{B}}}\alpha)\deg_K \mathfrak{B} \\ &= 2(0-2) - 1((r_{\wp_\infty} - 1)\deg \wp_\infty) \\ &= -4 - 1(-m-1) = -4 + m + 1 = m - 3. \end{aligned}$$

Therefore $g_K = \frac{m-1}{2}$. If we set $m = 2n+1$, we get $g_K = n$.

In any case we have obtained a hyperelliptic function field of genus n for any $n \in \mathbb{N}$.

By Example 9.6.18, if K/k is a hyperelliptic function field with $[K : k(x)] = 2$ and $K/k(x)$ separable (for instance if $\operatorname{char} k \neq 2$ or $\operatorname{char} k = 2$ and k a perfect field), then K is given by $K = k(x, y)$ with

$$y^2 = f(x) \in k[x], \tag{9.88}$$

if $\operatorname{char} k \neq 2$ and $\deg f = m$. In this case $g_K = \left[\frac{m+1}{2}\right] - 1 \geq 2$. That is,

$$m = \begin{cases} 2g_K + 1 & \text{if } m \text{ is odd,} \\ 2g_K + 2 & \text{if } m \text{ is even.} \end{cases}$$

Now if $\operatorname{char} k = 2$, then $K/k(x)$ is an Artin–Schreier extension and $K = k(x, y)$ can be given by

$$y^2 - y = \frac{a(x)}{b(x)}, \quad \text{where} \quad a(x), b(x) \in k[x] \quad \text{and} \quad (a(x), b(x)) = 1. \tag{9.89}$$

When k is a perfect field, we know from Example 5.8.8 that we may modify y in such a way that

$$y^2 - y = r(x) \in k(x)$$

with $(r(x))_{k(x)} = \frac{\mathfrak{A}}{\wp_1^{\lambda_1} \cdots \wp_m^{\lambda_m}}$, where $\wp_1, \ldots, \wp_m$ are prime divisors, $\mathfrak{A}$ is an integral divisor relatively prime to $\wp_1, \ldots, \wp_m$, $\lambda_i > 0$, and $(\lambda_i, 2) = 1$.

The genus of K is given by the equation

$$g_K = \frac{1}{2}\sum_{i=1}^{m}(\lambda_i + 1)\deg \wp_i - 1$$

(see Example 5.8.8 and Theorem 9.4.2).

Now we study an important characterization of hyperelliptic function fields.

Let K/k be any function field of genus $g > 0$. If w_0 is a nonzero differential, then by Theorem 3.4.9, for any differential w there exists a unique $z \in K$ such that $w = zw_0$.

Definition 9.6.19. The element z defined above is called the *ratio* of the differentials w and w_0 and it is denoted by $z = \frac{w}{w_0}$.

Now we consider a function field K/k of genus $g \geq 2$. Let $\{w_1, \dots, w_g\}$ be a basis of the holomorphic differentials. Assume that $z_i = \frac{w_i}{w_1} \in K$ for $i = 2, \dots, g$, and $z_1 = 1$. Let $L = k(1, z_2, \dots, z_g)$.

Proposition 9.6.20. *If $L \neq K$, then K is a hyperelliptic function field.*

Proof: Let $[K : L] = m \geq 2$ and let $\mathfrak{A}$ be a canonical divisor in W_K. We may choose $\mathfrak{A}$ to be integral since $g = g_K \geq 2$. By Corollary 3.5.5,

$$\ell(\mathfrak{A}^{-1}) = N(W_K) = g.$$

Let $\{y_1, \dots, y_g\}$ be a basis of $L_K(\mathfrak{A}^{-1})$. For any nonzero x in K, the set $\{xy_1, \dots, xy_g\}$ is a basis of $L_K((x)_K\mathfrak{A}^{-1})$ and $(x)_K\mathfrak{A}^{-1} \in W_K$.

We may assume that $z_1 = 1, z_2, \dots, z_g$ form a basis of $L_K(\mathfrak{A}^{-1})$.

Let $\mathfrak{P}$ be an arbitrary prime divisor of K. Since $z_i \in L_K(\mathfrak{A}^{-1})$, we have $v_{\mathfrak{P}}(z_i) \geq v_{\mathfrak{P}}(\mathfrak{A}^{-1})$ for $2 \leq i \leq g$. Thus

$$v_{\mathfrak{P}}(\mathfrak{A}^{-1}) \leq \min_{2 \leq i \leq g} v_{\mathfrak{P}}(z_i).$$

On the other hand, by Exercise 3.6.19, there exists an index i_0 such that $2 \leq i_0 \leq g$ and $z_{i_0} \notin L(\mathfrak{A}^{-1}\mathfrak{P})$.

Thus $v_{\mathfrak{P}}(\mathfrak{A}^{-1}) = v_{\mathfrak{P}}(z_{i_0})$. We have

$$v_{\mathfrak{P}}(\mathfrak{A}^{-1}) = \min_{2 \leq i \leq g} v_{\mathfrak{P}}(z_i), \quad \text{with} \quad \mathfrak{P} \in \mathbb{P}_K. \tag{9.90}$$

It follows that $\mathfrak{A}^{-1} \in D_L$ is a divisor of L.

By Theorem 5.3.4, we have

$$d_L(\mathfrak{A}^{-1}) = \lambda_{K/L} d_K(\mathfrak{A}^{-1}) = \frac{2 - 2g}{[K : L]} = \frac{2 - 2g}{m}.$$

Since $z_1, \dots, z_g \in L$ and $L_L(\mathfrak{A}^{-1}) \subseteq L_K(\mathfrak{A}^{-1})$, it follows that

$$\ell_L(\mathfrak{A}^{-1}) = \ell_K(\mathfrak{A}^{-1}) = g.$$

Using the Riemann–Roch theorem we obtain

$$\begin{aligned} \ell_L(\mathfrak{A}^{-1}) &= d_L(\mathfrak{A}) - g_L + 1 + \ell_L(W_L^{-1}\mathfrak{A}), \\ g_K &= -\frac{2 - 2g_K}{m} - g_L + 1 + \ell_L(W_L^{-1}\mathfrak{A}). \end{aligned}$$

Therefore

$$m(\ell_L(W_L^{-1}\mathfrak{A}) - g_L) = mg_K + 2 - 2g_K - m = (m - 2)(g_K - 1) \geq 0.$$

It follows that $\ell_L(W_L^{-1}\mathfrak{A}) \geq g_L = \ell_L(W_L^{-1})$. But W_L^{-1} divides $W_L^{-1}\mathfrak{A}$, so

$$\ell_L(W_L^{-1}\mathfrak{A}) \leq \ell_L(W_L^{-1}).$$

Thus $(m-2)(g_K-1) = (m-2)(g-1) = 0$, and $m = 2$.

The case $g_L \neq 0$ is not possible (Exercise 3.6.19), so $g_L = 0$ and K is a hyperelliptic function field. □

We will prove that the converse of Proposition 9.6.20 holds.

Proposition 9.6.21. *If K/k is any function field of genus $g = g_K \geq 2$ and $L \subseteq K$ is such that $g_L = 0$, then*

$$\{\alpha_1 z_1 + \cdots + \alpha_g z_g \mid \alpha_i \in L, 1 \leq i \leq g\} \neq K,$$

where $z_1, \ldots, z_g$ are as in Proposition 9.6.20.

Proof: By Proposition 3.4.5 and Corollary 3.4.6, we have

$$\dim_k \frac{\mathfrak{X}_L}{\mathfrak{X}_L(\mathfrak{N}) + L} = g_L = 0.$$

It follows that $\mathfrak{X}_L = \mathfrak{X}_L(\mathfrak{N}) + L$. Let $\mathfrak{A}$ be a canonical divisor such that $\{z_1, \ldots, z_g\}$ is a basis of $L_K(\mathfrak{A}^{-1})$. If $\xi \in \mathfrak{X}_L(\mathfrak{N})$, then $\xi z_i \in \mathfrak{X}_L(\mathfrak{N})\mathfrak{X}_K(\mathfrak{A}^{-1}) = \mathfrak{X}_K(\mathfrak{A}^{-1})$, which implies

$$\sum_{i=1}^{g} \mathfrak{X}_L(\mathfrak{N}) z_i \subseteq \mathfrak{X}_K(\mathfrak{A}^{-1}). \tag{9.91}$$

It follows that

$$\sum_{i=1}^{g} (\mathfrak{X}_L(\mathfrak{N}) + K) z_i \subseteq \mathfrak{X}_K(\mathfrak{A}^{-1}) + K.$$

On the other hand, we have

$$\dim_k \frac{\mathfrak{X}_K}{\mathfrak{X}_K(\mathfrak{A}^{-1}) + K} = \delta_K(\mathfrak{A}) = \ell_K(W_K^{-1}\mathfrak{A}) = \ell_K(\mathfrak{N}) = 1.$$

In particular,

$$\mathfrak{X}_K(\mathfrak{A}^{-1}) + K \neq \mathfrak{X}_K.$$

Therefore

$$\begin{aligned}\sum_{i=1}^{g} \mathfrak{X}_L z_i = \sum_{i=1}^{g} (\mathfrak{X}_L(\mathfrak{N}) + L) z_i &\subseteq \sum_{i=1}^{g} (\mathfrak{X}_L(\mathfrak{N}) + K) z_i \\ &\subseteq \mathfrak{X}_K(\mathfrak{A}^{-1}) + K \subsetneqq \mathfrak{X}_K.\end{aligned}$$

Now if $Lz_1 + \cdots + Lz_g = K$ held, then by Corollary 5.5.8 it would follow that

$$\mathfrak{X}_{Lz_1} + \cdots + \mathfrak{X}_{Lz_g} = \mathfrak{X}_K.$$

This contradiction shows that $Lz_1 + \cdots + Lz_g \neq K$. □

The converse of Proposition 9.6.20 is also true:

Theorem 9.6.22. *Let $z_1, \ldots, z_g$ be as before. If K/k is a hyperelliptic function field, then*

$$L = k\left(\frac{z_2}{z_1}, \ldots, \frac{z_g}{z_1}\right)$$

is the only quadratic subfield of K of genus 0.

Proof: If E is a quadratic subfield of K of genus 0, we have $[K : E] = 2$, and by Proposition 9.6.21,

$$E \subseteq Ez_1 + \cdots + Ez_g \neq K.$$

Thus $E = Ez_1 + \cdots + Ez_g$, $z_i \in E$, and

$$F = k\left(1, \frac{z_2}{z_1}, \ldots, \frac{z_g}{z_1}\right) \subseteq E \neq K.$$

By Proposition 9.6.20, we have

$$[K : F] = 2.$$

Therefore $F = E = k\left(\frac{z_2}{z_1}, \ldots, \frac{z_g}{z_1}\right)$ and $k\left(\frac{z_2}{z_1}, \ldots, \frac{z_g}{z_1}\right)$ is the only quadratic subfield of genus 0 of K. □

Remark 9.6.23. The above results are no longer true for elliptic function fields. Clearly the explicit construction of $k\left(\frac{z_2}{z_1}, \ldots, \frac{z_g}{z_1}\right) = E$ implies $g \geq 2$. When $g = 1$ we have $E = k$. The uniqueness of the quadratic subfield does not hold when $g_K = 1$. For instance, if k is an algebraically closed field, and K/k is an elliptic function field, then $K = k(x, y)$ with $\mathfrak{N}_x = \mathfrak{P}^2$, and $\mathfrak{N}_y = \mathfrak{P}^3$ for some prime divisor. If we choose a prime divisor $\mathfrak{q}$ such that $\mathfrak{q} \neq \mathfrak{P}$ and $\mathfrak{q}$ is not ramified in $K/k(x)$, we have $\ell_K(\mathfrak{q}^{-2}) = 2$. If $z \in L_K(\mathfrak{q}^{-2}) \setminus k$, then $\mathfrak{N}_z = \mathfrak{q}^2$ and $[K : k(z)] = 2$. Thus $z \notin k(x)$, since otherwise $k(x) = k(z)$ and

$$z = \frac{ax + b}{cx + d}.$$

This is impossible since $\mathfrak{N}_{\frac{ax+b}{cx+d}} \neq \mathfrak{q}^2$.

9.7 Exercises

Exercise 9.7.1. Prove Proposition 9.3.2.

Exercise 9.7.2. With the notation of Section 9.3, prove that if $(\alpha, \beta) \sim (\alpha', \beta')$, then for any $\gamma \in K_{\mathfrak{p}}$ we have $(\gamma\alpha, \beta) \sim (\gamma\alpha', \beta')$.

Exercise 9.7.3. Prove Equation (9.21).

Exercise 9.7.4. Let $K = k(x)$. Prove that the Hasse differential dx corresponds to the differential dx given in Definition 4.1.4.

Exercise 9.7.5. Prove Equation (9.27).

Exercise 9.7.6. Prove that the differentials of the second kind form a k-vector space of infinite dimension.

Exercise 9.7.7. If L/K is a purely inseparable extension of function fields of degree p, prove that for all $\alpha, \beta \in L \setminus K$, $(D_\alpha\beta)(D_\beta\alpha) = 1$.

Exercise 9.7.8. Using Theorem 9.5.17 give a new proof of Corollary 9.4.3.

Exercise 9.7.9. With the notation of Section 9.6.1, prove that $\Lambda \subseteq L_K(\mathfrak{P}_\infty^{-s})$.

Exercise 9.7.10. Let $L/k(x)$ be a geometric separable proper extension and let k be a perfect field. Prove that there exists at least one prime divisor in $k(x)$ that is ramified in K.

Exercise 9.7.11. Show that Exercise 9.7.10 is no longer true if we do not assume k to be perfect.

Exercise 9.7.12. Let K/k be a function field with $g_K > 1$. Prove that $d_K(D_K) = n\mathbb{Z}$ with $n \leq 2g_K - 2$. Compare with Proposition 9.6.5.

Exercise 9.7.13. Let K be a field of genus 2. Prove that K is a hyperelliptic function field.

Exercise 9.7.14. Let K/k be any function field such that $\operatorname{char} k = 2$, given by $K = k(x, y)$, with $y^2 = f(x) \in k[x]$. Prove that there exists a constant extension k' of k such that K' is a rational function field where $K' = Kk'$.

Exercise 9.7.15. Let $K = k(x, y)$, where

$$y^2 = (x - \alpha_1)(x - \alpha_2) \cdots (x - \alpha_{2n+1}), \qquad n \geq 2,$$

and $\alpha_1, \ldots, \alpha_{2n+1}$ are distinct elements of k. If $\operatorname{char} k \neq 2$ then the places $\mathfrak{p}_{\alpha_i}$ $(i = 1, \ldots, n+1)$ and $\mathfrak{p}_\infty$ of $k(x)$ are ramified in $K/k(x)$ with ramification index 2. Let $\mathfrak{P}_1, \ldots, \mathfrak{P}_{2n+1}$ and $\mathfrak{P}_\infty$ be the prime divisors in K above $\mathfrak{p}_{\alpha_1}, \ldots, \mathfrak{p}_{\alpha_{2n+1}}$, and $\mathfrak{p}_\infty$ respectively. Prove that

$$(dx)_K = \frac{\mathfrak{P}_1 \cdots \mathfrak{P}_{2n+1}}{\mathfrak{P}_\infty^3} \quad \text{and} \quad (y)_K = \frac{\mathfrak{P}_1 \cdots \mathfrak{P}_{2n+1}}{\mathfrak{P}_\infty^{2n+1}}.$$

From the above, deduce that $g_K = n$ and that the holomorphic differentials can be written as

$$\frac{\beta_0 + \beta_1 x + \cdots + \beta_{n-1} x^{n-1}}{y} dx, \quad \text{with} \quad \beta_i \in k.$$

Exercise 9.7.16. Assume $\operatorname{char} k = 2$ and let $\alpha_1, \ldots, \alpha_{n+1}$ be distinct elements of k. Let $K = k(x, y)$ be such that

$$y^3 = (x - \alpha_1)(x - \alpha_2) \cdots (x - \alpha_{n+1}).$$

Prove that the places $\mathfrak{p}_{\alpha_i}$ and $\mathfrak{p}_\infty$ can be extended to $\mathfrak{P}_i$ and $\mathfrak{P}_\infty$ in K in such a way that the ramification indices are 3. Prove that

$$(dx)_K = \frac{\left(\mathfrak{P}_1 \mathfrak{P}_2 \cdots \mathfrak{P}_{n+1}\right)^2}{\mathfrak{P}_\infty^4}$$

and that $g_K = n$.

Exercise 9.7.17. Let L/K be a geometric extension of function fields. Let ω be a nonzero differential in K. Prove that L/K is separable if and only if $\operatorname{cotr}_{K/L} \omega \neq 0$.

Exercise 9.7.18. Let F be a perfect field and consider a finite separable extension L/K of formal series $L = F((T))$, with $K = F((t))$. For $\alpha \in L$, set $\alpha dt = \alpha \frac{dt}{dT} dT$ and $\alpha \frac{dt}{dT} = \sum_{n=m}^{\infty} c_n T^n$, with $\operatorname{Res}_T \alpha \, dt = c_{-1}$.

(i) Show that if $\operatorname{char} F = 0$ and $\alpha = T^m$ for some $m \in \mathbb{Z}$, then $\operatorname{Res}_T \alpha dt = \operatorname{Res}_t (\operatorname{Tr}_{L/K} \alpha) \, dt$.
(ii) Prove that the same holds for $\operatorname{char} F = p$ using formally what was obtained in (i).
(iii) Prove that for any $\alpha \in L$, $\operatorname{Res}_T \alpha dt = \operatorname{Res}_t (\operatorname{Tr}_{L/K} \alpha) dt$.

Exercise 9.7.19. Let k be a perfect field. Let K/k be a function field over k and let L/K be a finite separable extension. Let $\mathfrak{p}$ be a place of K and $\mathfrak{P}|\mathfrak{p}$ a place of L. Prove that $\operatorname{Res}_{\mathfrak{p}} \left(\operatorname{Tr}_{L_{\mathfrak{P}}/K_{\mathfrak{p}}}(y) dx \right) = \operatorname{Res}_{\mathfrak{P}}(y \, dx)$.

Deduce that $\operatorname{Res}_{\mathfrak{p}} \left(\operatorname{Tr}_{L/K}(y) dx \right) = \sum_{\mathfrak{P}|\mathfrak{p}} \operatorname{Res}_{\mathfrak{P}}(y \, dx)$.

Exercise 9.7.20. Prove the residue theorem for $k(x)$ when k is an algebraically closed field.

Exercise 9.7.21. Prove Lüroth's theorem: Let $K = k(x)$ be a rational function field. Let $k \subsetneq T \subseteq K$ be any intermediate field other than k. Then $T = k(t)$ for some $t \in K \setminus k$.

10

Cryptography and Function Fields

10.1 Introduction

The term *cryptography* comes from the two Greek words: *kryptós* (hidden, secret) and *gráphein* (to write). In this way, cryptography may be understood as a method of writing in a secret way. More precisely, it is the art of transforming written information from its original or standard form to one that cannot be understood unless one knows a secret key.

Cryptography consists of two processes. The first one, called *encryption*, is a way of codifying the information, which means concealing it in a such a way that it becomes unintelligible to persons that are not authorized to read it; various methods are known for keeping messages or data secret. The second and inverse process is the *decryption* of the codified message; in order to decode or decipher the codified message, one needs special knowledge.

Let us assume that a person, who from now on will be called Arnold, wishes to share a given piece of information with another person, say Charlotte, in such a way that no one other than Charlotte can understand it. We will say that Arnold wants to send a message, which we shall call *plaintext*, to Charlotte. In order to keep the message inaccessible to eavesdroppers and understandable by Charlotte only, Arnold codifies it, obtaining in this way a new message, which will be called *ciphertext*. Once Charlotte receives the message, she decodes it, obtains the plaintext, and reads it.

How does such a process work? First of all, Arnold needs to use an *encryption key* in order to obtain the ciphertext from the original message; second, Charlotte must use a *decryption key* to be able to decipher the message and obtain the plaintext. The decryption key must be kept secret from everyone else so that the method can work properly.

There are two basic types of codification: *symmetric* and *asymmetric*. Let us assume that the encryption and decryption keys are called a and b respectively. We say that the codification system is symmetric if $a = b$ or b can be computed easily from a. Observe that if Arnold and Charlotte are using a symmetric system, they need to exchange the secret key before they begin sending each other information.

In symmetric cryptography, exchanging keys is a process of capital importance, since if a is not kept secret anyone could deduce b from a and then decipher the message.

In the case of an asymmetric cryptosystem, a and b are distinct and the computation of b from a is not achievable. The advantage of such a system is that a may be made public without danger. Asymmetric systems work as follows: If Charlotte wishes to receive an encrypted message, she publishes the encryption key a while keeping b secret. When Arnold sends a message to Charlotte, he uses a to obtain the ciphertext. Only Charlotte can decipher the message, since she is the only one who knows b; not even Arnold would be able to obtain the original message from the encrypted one.

For the mentioned reason, asymmetric cryptosystems are called *public-key cryptosystems*. Some of the most popular public-key systems will be described in Section 10.2.

Symmetric cryptosystems used to work efficiently when communication systems were still restricted, for instance, between spies and intelligence and counterintelligence agencies (if one may call that intelligence). In these cases a small number of select persons know the keys from the beginning, and they are the only ones who use them.

Nowadays the situation has changed drastically; all kinds of persons, and not only at governmental levels, use cipher systems to exchange information. This is done in big businesses such as banks and credit card companies, etc. At the personal level, cryptosystems are used for various purposes, for example exchanging scientific papers between various collaborators who prefer to keep their work unpublicized in order to avoid plagiarism. It is in such cases that public-key cryptosystems are useful; indeed, sometimes it is not possible for several persons who live at a distance from one another to get together and agree on a secret key.

10.2 Symmetric and Asymmetric Cryptosystems

One of the simplest cryptosystems is the so-called *Caesar cipher*. In this case, the plaintext is written using the twenty-six usual letters of the alphabet $\Sigma = \{A, B, C, \dots, Z\}$. The encryption and decryption keys are one and the same, namely $a = b \in \Omega = \{0, 1, \dots, 25\}$, where each letter of the alphabet is identified with a member of Ω. The codification scheme is

$$\begin{aligned} \varphi\colon \Sigma &\to \Sigma, \\ x &\mapsto x + a \bmod 26. \end{aligned}$$

The decoding function is

$$\begin{aligned} \psi\colon \Sigma &\to \Sigma, \\ y &\mapsto y - b \bmod 26. \end{aligned}$$

Arnold and Charlotte just need to know a in order to exchange information. Since we have only 26 choices for a, it is easy to guess its value and thus to obtain the plaintext from the ciphertext. This shows that the Caesar cipher is quite unsecured.

A major problem with symmetric cryptosystems is key distribution and key management. If Arnold and Charlotte use such a system, they must exchange the secret key before exchanging messages.

In public-key systems, key exchange is no longer a problem. Charlotte makes public the encryption key a so that anyone who wants to send a message to her uses a. When Charlotte receives a ciphertext, she uses the decryption key b that she has kept secret.

The most popular public-key cryptosystem is the *RSA cryptosystem*, named after Ron Rivest, Adi Shamir, and Len Adleman, created in 1978 [123]. In fact, this was one of the first public-key cryptosystems to be invented, and nowadays it remains the most important one. The security of this cryptosystem is due to the difficulty of finding the factorization of a composite positive integer that is the product of two large primes. Let us see how it works.

Charlotte finds two large prime numbers p and q and computes $n = pq$. Then she chooses any integer a such that $1 < a < \varphi(n) = (p-1)(q-1)$ and $\gcd(a, \varphi(n)) = 1$. Because of this choice, there exists $b \in \{0, \ldots, \varphi(n)-1\}$ such that $ab \equiv 1 \bmod \varphi(n)$. The number b can be computed using the extended Euclidean algorithm ([12]).

Charlotte publishes the pair (n, a) and her private key is b. Note that if an attacker or an eavesdropper is able to find the prime factorization of n, then (s)he can easily find b and the system breaks down. Therefore the security of the system depends on making the factorization of n infeasible. If p and q are sufficiently large, it seems that nobody yet knows how to factor n.

Let the plaintext be an integer m such that $0 \leq m < n$. The ciphertext is $c := m^a \bmod n$. If Arnold wants to send the message m to Charlotte, then since he knows a and n, he can encrypt m and send c.

Example 10.2.1. Let $p = 17$ and $q = 29$. Then $n = 17 \times 29 = 493$ and $\varphi(n) = (p-1)(q-1) = 16 \times 28 = 448$. Let $a = 5$. Then $b = 269$. If $m = 75$ is the plaintext, then $c = 75^5 \bmod 493 = 249$ is the ciphertext.

Note that $c^{269} \bmod 493 = 75 = m$.

Now, the way Arnold sends a message is as follows. Assume that the alphabet contains N letters and he assigns to each letter a unique number between 0 and $N-1$. Set $t := [\log_N n]$ and assume that Arnold has a text $m_1 m_2 \ldots m_k$, where each m_i is the number corresponding to a letter. Then he defines

$$m := \sum_{i=1}^{t} m_i N^{t-i}.$$

We have $0 \leq m \leq (N-1)\sum_{i=1}^{t} N^{t-i} = N^t - 1 < n$. Let $c := m^a \bmod n$ be the ciphertext, and write c in base N.

Since $0 \leq c < n < N^{t+1}$, the N-adic expansion of c has length at most $t+1$, that is,

$$c = \sum_{i=0}^{t} c_i N^{t-i} \quad \text{with} \quad c_i \in \{0, 1, \ldots, N-1\} \quad \text{for} \quad 0 \leq i \leq t.$$

Therefore the encrypted message consists of the integer $c = c_0 c_1 \dots c_k$.

Example 10.2.2. Suppose that our alphabet consists of the set of vowels $\{a, e, i, o, u\}$, that is, $N = 5$. In the setting of Example 10.2.1 we have $k = [\log_5 493] = 3$ since $5^3 = 125 < 493 < 625 = 5^4$. The numerical assignment of our alphabet is

$$\begin{aligned} a &\mapsto 0, \\ e &\mapsto 1, \\ i &\mapsto 2, \\ o &\mapsto 3, \\ u &\mapsto 4. \end{aligned}$$

If Arnold encrypts eio, which corresponds to 123, he obtains

$$m = 1 \times 5^2 + 2 \times 5 + 3 \times 5^0 = 25 + 10 + 3 = 38.$$

The encrypted integer is

$$c = 38^5 \bmod 493 = 208.$$

Writing 208 in its 5-adic expansion, we obtain

$$208 = 1 \times 5^3 + 3 \times 5^2 + 1 \times 5 + 3 \times 5^0.$$

Therefore the ciphertext is *eoeo*, which corresponds to 1313. Again note that $208^{269} \bmod 493 = 38$.

The reason why the RSA system works is the following elementary result.

Theorem 10.2.3. *If p and q are distinct prime numbers, $n = pq$, $\varphi(n) = (p-1)(q-1)$, and a is such that $(a, \varphi(n)) = 1$, whenever $0 \leq m < n$ we have*

$$(m^a)^b \bmod n = m,$$

where b is such that $ab \equiv 1 \bmod \varphi(n)$.

Proof. Exercise 10.9.1. □

10.3 Finite Field Cryptosystems

As we already mentioned in Section 10.2, the RSA cryptosystem is the most important public-key cryptosystem. The concept of a public key was defined by Diffie and Hellman in 1976 ([29]); the difference with respect to symmetric cryptosystems lies in the idea of using a one-way function for encryption.

There are several public-key cryptosystems. We are interested in elliptic and hyperelliptic cryptosystems, which are applications of elliptic and hyperelliptic function fields. We will study these cryptosystems later on.

First we introduce some concepts that are necessary in studying the feasibility, security, and efficiency of a cryptosystem.

10.3.1 The Discrete Logarithm Problem

Let $\mathbb{F}_p^* = \{1, 2, \ldots, p-1\}$ be the multiplicative group of the finite field of p elements. We choose an element g of $\mathbb{F}_p^*$, which will be called the "*base*." The *discrete logarithm problem* in $\mathbb{F}_p^*$ with respect to the base g is that of, given $y \in \mathbb{F}_p^*$, determining an integer x such that $y = g^x$ (that is, x" $=$ "$\log_g y$). Of course, the existence of x is equivalent to y belonging to the subgroup of $\mathbb{F}_p^*$ generated by g.

The discrete logarithm problem can be defined for any finite group. More precisely:

Definition 10.3.1. The *discrete logarithm problem for the finite group* G is the following: given a *base* $g \in G$ and $y \in G$, find $x \in \mathbb{Z}$ such that $g^x = y$ if such an x exists, that is, if $y \in \langle g \rangle$. In other words, the discrete logarithm problem consists in finding $x = \log_g y$.

Another useful concept for making a cryptosystem realizable is that of a *hash function*. The idea behind these functions is that in order to make a cryptosystem secure, we need keys that require a lot of space, often much more than what is realistically possible. For instance, we frequently need several numbers, each of which has several thousand digits. To be able to reduce the quantity of space, we use a function, say $H\colon \mathbb{Z}/s\mathbb{Z} \to \mathbb{Z}/t\mathbb{Z}$, where s is much larger than t. Usually s is of the order of several millions of bits and t is smaller than 200 bits. Since $t < s$, the function H is not injective. We say that H is a hash function if its values can be computed in an easy and efficient way, and if on the other hand it is not computationally feasible to find two distinct elements x_1, x_2 such that $H(x_1) = H(x_2)$.

Definition 10.3.2. A *cryptographic hash function* is a function $H\colon \mathbb{Z}/s\mathbb{Z} \to \mathbb{Z}/t\mathbb{Z}$ such that $s > t$ and:

(i) Given m, $H(m)$ can be easily computed.
(ii) Given n, it is not computationally feasible to find m such that $H(m) = n$. We say that H is *preimage resistant*.
(iii) It is not computationally feasible to find $x_1, x_2 \in \mathbb{Z}/s\mathbb{Z}$ such that $x_1 \neq x_2$ and $H(x_1) = H(x_2)$. We say that H is *collision resistant*.

There exist several good hash functions. For a complete discussion see [110].

Another issue to be considered in cryptography is that concerning the signature of the message. When Charlotte receives a message that supposedly comes from Arnold, she must make sure with a reasonable degree of certainty that Arnold is really the one signing the message. Whenever one sends a message, it must be sent together with a digital and nonfalsifiable signature; that is what we mean by a *digital signature*.

10.3.2 The Diffie–Hellman Key Exchange Method and the Digital Signature Algorithm (DSA)

Assume that Arnold and Charlotte want to agree upon an integer to be used as a key for their private-key cryptosystem. They must use some public communication channel like the Internet, telephone, e-mail, or regular mail in order to achieve this agreement. First of all, both of them agree on a large prime number p and a base $g \in \mathbb{F}_p^*$. This is agreed publicly, so any eavesdropper knows p and g. Second, Arnold secretly chooses a large number $a < p$, computes $g^a \bmod p$, and communicates his result to Charlotte. Meanwhile, Charlotte does the same: she secretly chooses a large integer $b < p$ and communicates $g^b \bmod p$ to Arnold. Finally, they agree upon a key, which will be the integer $g^{ab} \in \mathbb{F}_p^*$.

The eavesdropper knows g, g^a, and $g^b \in \mathbb{F}_p^*$, and faces the problem of finding g^{ab}. This is the *Diffie–Hellman problem.* It is known that anyone who can solve the discrete logarithm problem in $\mathbb{F}_p^*$ can solve the Diffie–Hellman problem as well. The converse is still an open question ([84]).

Now we present a digital signature public-key cryptosystem that was proposed in 1991. It is the analogue to the older Data Encryption Standard, which is a private-key cryptosystem. This cryptosystem is called the *Digital Signature Algorithm* (DSA). Let us see how it works.

Arnold chooses a large prime number p, say that p is of order about 10^{50}. This can be achieved using a random number generator and a primality test (see [12]). Secondly, he chooses a second prime number $\ell \equiv 1 \bmod p$ of more than 512 bits and whose number of bits is a multiple of 64. Hence ℓ is larger than 10^{154}.

Thirdly, Arnold chooses a generator of the unique cyclic subgroup of $\mathbb{F}_\ell^*$ of order p by computing $y = g_0^{(q-1)/p} \bmod \ell$ for a random integer g_0; note that if $y \neq 1$, then g_0 is a generator.

Finally Arnold takes a random integer x such that $0 < x < p$ as his secret key, and sets as his public key $z = g^x \bmod p$.

If Arnold sends a message, he first applies a hash function to the plaintext, obtaining an integer H such that $0 < H < p$. Next, he chooses an integer k, computes $g^k \bmod \ell = A$, and sets $r = A \bmod p$. Finally, let $tk \equiv H + xr (\mathrm{mod} p)$. Arnold's signature is then the pair $(r, t) \bmod p$.

Charlotte verifies the signature as follows. Let $\alpha = t^{-1} H \bmod p$ and $\beta = t^{-1} r \bmod p$, and consider $g^\alpha z^\beta \bmod \ell$. If $g^\alpha z^\beta \equiv r \bmod p$, then Charlotte is reasonably satisfied.

The DSA signature scheme uses relatively short signatures, since they consist of numbers of order about 10^{50}. The security of the system depends on the nontreatability of the discrete logarithm problem in the large-order field $\mathbb{F}_\ell$. The DSA seems to have attained a fairly high level of security without sacrificing small signature storage and implementation time.

We are interested in a variant of DSA using elliptic function fields, which is even harder to break than the DSA described in this subsection.

10.4 Elliptic Function Fields Cryptosystems

Elliptic curves and elliptic function fields can be used to implement public-key cryptosystems. The Diffie–Hellman key exchange described in Section 10.3.2 can be implemented in this case if instead of using finite fields we use elliptic function fields over finite fields. We will also present a variant of the DSA given in Section 10.3.2 using elliptic function fields.

Elliptic cryptosystems were first proposed in 1985 by Neal Koblitz [81] and Victor Miller [111]. There are two good reasons for using these cryptosystems. The first one is that there exists only one finite field of q elements, whereas there are many elliptic function fields over $\mathbb{F}_q$. The second and more important one is the absence of subexponential-time algorithms to break the system if the elliptic function field is chosen to be nonsupersingular. In fact, Menezes, Okamoto, and Varistone [109] found a way to tackle the discrete logarithm problem using the Weil pairing in elliptic curves to embed them in $\mathbb{F}_{q^k}^*$, thus reducing the discrete logarithm problem to the discrete logarithm problem in $\mathbb{F}_{q^k}^*$. This is useful only if k is small; in fact, the only elliptic curves for which k is small are essentially the supersingular ones. The supersingular elliptic function fields are those such that $C_{0,K\overline{\mathbb{F}}_q}(p) = \{1\}$, where p is the characteristic.

Now, K is a supersingular elliptic function field over $\mathbb{F}_q$ if and only if $N_1(\mathbb{F}_q) \equiv 1 \bmod p$, where $N_1(\mathbb{F}_q)$ denotes the number of prime divisors of degree 1 (see [157, Proposition 4.29]). Moreover, as a consequence of the Riemann hypothesis (Theorem 7.2.9 (iv)), if $p \geq 5$, then K is supersingular if and only if $N_1(\mathbb{F}_p) = p + 1$.

Therefore we must choose a nonsupersingular elliptic function field. Even though nobody seems to know how to find a subexponential-time algorithm for the discrete logarithm problem on nonsupersingular elliptic function fields, the progress made in computing discrete logarithms for finite fields and in factoring integers implies that the key sizes necessary for the public-key systems to be secure grow every single day.

10.4.1 Key Exchange Elliptic Cryptosystems

In this subsection we present the Diffie–Hellman key exchange adapted for elliptic function fields. Let $\mathbb{F}_q$ be a finite field and let K be an elliptic function field with exact field of constants $\mathbb{F}_q$. Let M_K be the set of prime divisors of K of degree 1. Choose $\mathfrak{P}_0 \in M_K$ such that $K = \mathbb{F}_q(x, y)$ with $\mathfrak{N}_x = \mathfrak{P}_0^2$, $\mathfrak{N}_y = \mathfrak{P}_0^3$ and let

$$\varphi\colon M_K \to C_{K,0}$$
$$\mathfrak{P} \mapsto \left[\frac{\mathfrak{P}}{\mathfrak{P}_0}\right]$$

be the bijection given in Proposition 9.6.9 and Equation (9.74). Therefore the set of prime divisors of degree 1 forms an abelian group.

To any prime divisor $\mathfrak{P} \neq \mathfrak{P}_\infty$, where $\mathfrak{P}_\infty = \mathfrak{P}_0$ is the infinite prime, corresponds a unique rational point $(a, b) \in \mathbb{F}_q^2$ satisfying the defining equation

$$y^2 - h(x)y = f(x)$$

of the elliptic function fields, where $f(x)$ is a polynomial of degree 3 (see Exercise 10.9.1). Here $h(x) = 0$, $f(x)$ is square-free if $\operatorname{char} k \neq 2$ and $h(x) \neq 0$, and $\deg h(x) \leq 1$ if $\operatorname{char} k = 2$ (see Exercise 10.9.2). The infinite prime $\mathfrak{P}_\infty$ corresponds to the point at infinity (∞, ∞).

First we choose a random prime divisor of degree one in an elliptic function field K as the key. Of course, Arnold and Charlotte have agreed in advance on a method to convert an arbitrary point on an elliptic curve or a prime divisor of degree one on an elliptic function field into an integer. One way to do this is to use the fact that to any prime divisor of degree one corresponds a rational point $(a, b) \in \mathbb{F}_q^2$ of the corresponding elliptic curve over $\mathbb{F}_q$ and then to convert $a \in \mathbb{F}_q$ into an integer after choosing a suitable map from $\mathbb{F}_q$ to $\mathbb{Z}$.

Next, Arnold and Charlotte choose an elliptic function field K over $\mathbb{F}_q$ where the discrete logarithm problem is hard, and a prime divisor $\mathfrak{p} \in \mathbb{P}_K$ of degree one. Now Arnold chooses an integer α, computes $\mathfrak{p}_\alpha := \mathfrak{p}^\alpha$, and sends $\mathfrak{p}^\alpha$ to Charlotte. In the same way, Charlotte chooses a secret integer β, computes $\mathfrak{p}_\beta := \mathfrak{p}^\beta$, and sends it to Arnold. Now Arnold and Charlotte compute

$$\mathfrak{p}_{\alpha\beta} = \mathfrak{p}_{\beta\alpha} = \mathfrak{p}_\alpha^\beta = \mathfrak{p}_\beta^\alpha = \mathfrak{p}^{\alpha\beta} = \mathfrak{p}^{\beta\alpha}.$$

Suppose that the eavesdropper John is spying on Arnold and Charlotte. Then John has to find $\mathfrak{P} = \mathfrak{p}^{\alpha\beta}$ knowing $\mathfrak{p}$, $\mathfrak{p}^\alpha$, and $\mathfrak{p}^\beta$, but neither α nor β. John's task is what is called the *Diffie–Hellman problem for elliptic curves or elliptic function fields*. That is, he has to solve the

Diffie–Hellman problem for elliptic function fields:

Given $\mathfrak{p}$, $\mathfrak{p}^\alpha$, and $\mathfrak{p}^\beta$ in D_K, compute $\mathfrak{p}^{\alpha\beta}$.

Note that if John solves the discrete logarithm problem in elliptic function fields, he can obtain α using $\mathfrak{p}$ and $\mathfrak{p}^\alpha$. Thus he can find $\mathfrak{p}^{\alpha\beta} = (\mathfrak{p}^\beta)^\alpha$. That is, the elliptic function field discrete logarithm problem with respect to the base $\mathfrak{A} \in D_K$ is, given $\mathfrak{B} \in D_K$, to find $a \in \mathbb{Z}$ such that $\mathfrak{B} = \mathfrak{A}^a$ if such an a exists. Therefore, if John can solve the discrete logarithm problem, then he can solve the Diffie–Hellman problem.

10.5 The ElGamal Cryptosystem

The ElGamal cryptosystem [33] is quite close to the Diffie–Hellman key exchange, and its security is based on the difficulty of solving the Diffie–Hellman problem. Let us first consider its implementation in the finite field $\mathbb{F}_p^*$.

Let p be a prime number and let g be an element of $\mathbb{F}_p^*$, preferably but not necessarily a generator. Arnold chooses a random exponent $\alpha \in \{0, 1, \ldots, p-2\}$ and computes $a = g^\alpha \bmod p$. Arnold's public key is (p, g, a) and his secret key is α. Note that in the setting of the Diffie–Hellman protocol, a is Arnold's key, which is fixed in the ElGamal cryptosystem.

When Charlotte wants to encrypt a plaintext m, which we will assume, as usual, is an integer in $\{1, \ldots, p-1\}$, she obtains (p, g, a) from Arnold. Then she chooses a random exponent $\beta \in \{1, \ldots, p-2\}$ and computes $b = g^\beta \bmod p$.

Again b is Charlotte's key in the Diffie–Hellman cryptosystem. Charlotte finds

$$c = a^\beta m \bmod p.$$

That is, Charlotte encrypts the message m by multiplying it mod p by the Diffie–Hellman key. The ElGamal ciphertext is (b, c).

Once Arnold gets (b, c), he computes

$$\frac{c}{b^{\alpha}} \equiv cb^{p-1-\alpha} \bmod p.$$

We have

$$\begin{aligned} cb^{p-1-\alpha} &\equiv a^{\beta} m g^{\beta(p-1-\alpha)} \equiv a^{\beta} m (g^{p-1})^{\beta} g^{-\alpha\beta} \\ &\equiv a^{\beta} m (1) a^{-\beta} \equiv m \bmod p. \end{aligned}$$

The implementation of the ElGamal cryptosystem for elliptic function fields runs as follows.

Charlotte chooses an elliptic function field K over the finite field $\mathbb{F}_q$ such that the discrete logarithm problem is infeasible for C_{0K}. Then she picks a prime divisor $\mathfrak{p}$ of degree one such that the order of the class of $\bar{\mathfrak{p}}$ is a large prime number. Next, she selects a secret integer α and computes $\mathfrak{A} = \mathfrak{p}^{\alpha}$. The elliptic function field K, $\mathbb{F}_q$, $\mathfrak{p}$, and $\mathfrak{A}$ constitute Charlotte's public key. Her private key is α.

Now when Arnold wants to send a message to Charlotte, say that it corresponds to a prime divisor of degree one $\mathfrak{q}$, he selects a secret random integer β and computes $\mathfrak{B} = \mathfrak{p}^{\beta}$ and $\mathfrak{C} = \mathfrak{q}\mathfrak{A}^{\beta}$. Finally, Arnold sends $(\mathfrak{B}, \mathfrak{C})$ to Charlotte. Charlotte simply computes $\mathfrak{C}\mathfrak{B}^{-\alpha}$. This method works since

$$\mathfrak{C}\mathfrak{B}^{-\alpha} = \mathfrak{q}\mathfrak{A}^{\beta}\mathfrak{p}^{-\alpha\beta} = \mathfrak{q}\mathfrak{p}^{\alpha\beta}\mathfrak{p}^{-\alpha\beta} = \mathfrak{q}.$$

The eavesdropper John knows Charlotte's public key, namely K, $\mathbb{F}_q$, $\mathfrak{p}$, and $\mathfrak{A} = \mathfrak{p}^{\alpha}$ and also $\mathfrak{B}$ and $\mathfrak{C}$. If he could solve the discrete logarithm problem, he could get α from $\mathfrak{p}$ and $\mathfrak{A}$, where $\mathfrak{A} = \mathfrak{p}^{\alpha}$ and $\alpha = \log_{\mathfrak{p}} \mathfrak{A}$, and use α to find $\mathfrak{q} = \mathfrak{C}\mathfrak{B}^{-\alpha}$. The same result is obtained if John obtains β from $\mathfrak{p}$ and $\mathfrak{B}$, $\beta = \log_{\mathfrak{p}} \mathfrak{B}$, and computes $\mathfrak{q} = \mathfrak{C}\mathfrak{A}^{-\beta}$ (where $\mathfrak{C}\mathfrak{A}^{-\beta} = \mathfrak{q}\mathfrak{A}^{\beta}\mathfrak{A}^{-\beta} = \mathfrak{q}$).

Thus the security of this method relies on the infeasibility of solving the discrete logarithm problem.

Note that if Arnold chooses β all the time, then when he sends two different messages $\mathfrak{q}$ and $\mathfrak{q}_1$, we have $\mathfrak{B} = \mathfrak{B}_1 = \mathfrak{p}^{\beta}$, and hence

$$\mathfrak{C}_1\mathfrak{C}^{-1} = \mathfrak{q}_1\mathfrak{p}^{\beta}\mathfrak{q}^{-1}\mathfrak{p}^{-\beta} = \mathfrak{q}_1\mathfrak{q}^{-1}.$$

Now, depending on the kind of message, sooner or later $\mathfrak{q}$ is made public (say that the message informing about the status of the stock market has to be published some days later) and John then knows $\mathfrak{q}$, $\mathfrak{C}_1$, and $\mathfrak{C}$, so he knows $\mathfrak{q}_1 = \mathfrak{C}_1\mathfrak{C}^{-1}\mathfrak{q}$.

10.5.1 Digital Signatures

As we established in Section 10.3.2, digital signatures are used to legitimate a message or a document. The traditional way, which we use in everyday life, is the written signature; but when we send a message, secret or not, and the addressee of our message needs to be reasonably sure that the message comes from us, it is necessary to use another kind of signature, namely a *digital signature*. Here we present the *ElGamal digital signature* method using elliptic function fields.

Again, our old friends Arnold and Charlotte wish to share some information without the knowledge of John. As before, for several good reasons, Arnold and Charlotte have to use public key exchange. The digital signature must satisfy the following conditions:

(i) The signature must depend on the document or message in such a way that nobody can use it in another message.
(ii) It should be possible for Charlotte to find out that Arnold has sent the message.

First, Arnold must select a public key. He uses an elliptic function field K over $\mathbb{F}_q$ such that the discrete logarithm problem cannot be solved (at least for now) for K. Let $\mathfrak{p} \in \mathbb{P}_K$ be of order ℓ, usually a very large prime number although this is not necessary. Then Arnold chooses a secret integer α and computes $\mathfrak{A} = \mathfrak{p}^{\alpha}$. As explained in Section 10.4.1, he chooses a function from $\mathbb{P}_K$ to $\mathbb{Z}$ (say $f : \mathbb{P}_K \to \mathbb{Z}$, $f(\mathfrak{q}) = \varphi_{\mathfrak{q}}(x)$, where $\varphi_{\mathfrak{q}}$ is the place corresponding to $\mathfrak{q}$, that is, $f(\mathfrak{q}) = \varphi_{\mathfrak{q}}(x) = x \bmod q$ where $K = k(x, y)$, $y^2 = u(x)$ or $y^2 + y = u(x)$).

The public information given by Arnold is K, f, $\mathfrak{p}$, and $\mathfrak{A}$. Now when Arnold sends a message, he first represents it as an integer m (see Section 10.2) and selects an integer β that is relatively prime to ℓ. Next he computes $\mathfrak{B} = \mathfrak{p}^{\beta}$ and takes $\gamma \equiv \beta^{-1}(m - \alpha f(\mathfrak{B})) \bmod \ell$. Recall that $\mathfrak{B}$ is represented by a pair (a, b) satisfying the equation that defines K (see Exercise 10.9.3).

The signed ciphertext is $(m, \mathfrak{B}, \gamma)$. In this way m is not kept secret. If Arnold wants to make m secret, he may use any cryptosystem to perform this task. The main point is that Charlotte receives $(m, \mathfrak{B}, \gamma)$ or $(m', \mathfrak{B}, \gamma)$; in the former case, m is not secret, and in the latter m' is the encryption of m and Charlotte wants to verify that Arnold is sending the message.

Charlotte computes $\mathfrak{C} = \mathfrak{A}^{f(\mathfrak{B})}\mathfrak{B}^{\gamma}$ and $\mathfrak{D} = \mathfrak{p}^m$. If the signature is valid then

$$\mathfrak{C} = \mathfrak{A}^{f(\mathfrak{B})}\mathfrak{B}^{\gamma} = \mathfrak{p}^{\alpha f(\mathfrak{B})}\mathfrak{p}^{\beta\gamma} = \mathfrak{p}^{\alpha f(\mathfrak{B}) + (m - \alpha f(\mathfrak{B}))} = \mathfrak{p}^m = \mathfrak{D}.$$

Therefore, if $\mathfrak{C} = \mathfrak{D}$ Charlotte can be reasonably sure that the signature is valid.

Again we see that if John is able to compute discrete logarithms, then he can use $\mathfrak{p}$ and $\mathfrak{A}$ to find $\alpha = \log_{\mathfrak{p}} \mathfrak{A}$, and this enables him to sign any message as if he were Arnold.

Now, Arnold's secret keys are α and β and he must use a different β for every document. Indeed, assume he keeps the same β every time, say that he sends two messages m and m' with $\beta = \beta'$. Then John gets $(m, \mathfrak{B}, \gamma)$ and $(m', \mathfrak{B}', \gamma')$ but $\mathfrak{B} = \mathfrak{p}^{\beta} = \mathfrak{p}^{\beta'} = \mathfrak{B}'$, so he recognizes that the same key has been used. Thus, John obtains

$$\beta\gamma \equiv (m - \alpha f(\mathfrak{B})) \bmod \ell$$

and

$$\beta\gamma' \equiv (m' - \alpha f(\mathfrak{B})) \bmod \ell.$$

He deduces that $\beta(\gamma - \gamma') \equiv (m - m') \bmod \ell$, which implies that if r is the greatest common divisor of ℓ and $\gamma - \gamma'$ ($r = 1$ if ℓ was chosen to be prime), then r divides $m - m'$ and

$$\beta \equiv \left(\frac{\gamma - \gamma'}{r}\right)^{-1} \left(\frac{m - m'}{r}\right) \bmod \frac{\ell}{r} \equiv A \bmod \frac{\ell}{r}, \quad 0 < A \leq \frac{\ell}{r} - 1.$$

Thus $\beta \in \{iA \mid 1 \leq i \leq r\}$. Then John tries these r values and obtains β (that is, until he gets $\mathfrak{B} = \mathfrak{p}^{\beta}$). Once he knows β he can obtain α as follows. He knows γ, $f(\mathfrak{B})$, and m. From

$$\alpha f(\mathfrak{B}) \equiv (m - \beta\gamma) \bmod \ell$$

he obtains, as before, $r = \gcd(f(\mathfrak{B}), \ell)$ possible values for α. Each candidate can be tested until $\mathfrak{A} = \mathfrak{p}^{\alpha}$ is reached.

10.6 Hyperelliptic Cryptosystems

In 1989, Koblitz [83] generalized the use of elliptic curve cryptosystems to the use of hyperelliptic curves. In this section we show how hyperelliptic function fields may be used in cryptography. We shall see that among all function fields, the hyperelliptic ones are differentiated by some special properties. Of course, one can consider elliptic fields as forming part of the class of hyperelliptic fields although they are formally defined otherwise. Everything presented in the rest of the chapter is valid for fields of elliptic functions.

The main reason why hyperelliptic fields may be used in cryptography is that their group of divisor classes of degree 0 has some special representatives that can be operated within a computationally feasible algorithmic form. This does not happen with other function fields.

Let $K = \mathbb{F}_q(x, y)$ be a hyperelliptic function field where $K/\mathbb{F}_q(x)$ is a quadratic separable extension. Thus the defining equation of K is

$$y^2 = g(x) \in \mathbb{F}_q[x] \quad \text{if} \quad \operatorname{char} K \neq 2 \tag{10.1}$$

and

$$y^2 - y = g_1(x) \in \mathbb{F}_q(x) \quad \text{if} \quad \operatorname{char} K = 2, \tag{10.2}$$

where $g(x)$ is square-free, $g_1(x) = \frac{\alpha(x)}{\beta(x)}$, $\alpha(x)$, $\beta(x)$ are relatively prime elements of $\mathbb{F}_q[x]$, and if $p(x)$ is an irreducible polynomial dividing $\beta(x)$, then the power of $p(x)$ dividing $\beta(x)$ is odd.

Assume that the infinite prime of $\mathbb{F}_q(x)$ or, more precisely, the pole divisor of x in $\mathbb{F}_q(x)$, ramifies in K. Let g be the genus of K. Then the defining equation of $K = \mathbb{F}_q(x, y)$ can be written as

$$y^2 - h(x)y = f(x), \tag{10.3}$$

where $h(x)$ is a polynomial of degree at most g, $h(x) = 0$ if char $K \neq 2$, $h(x)$ is nonzero and relatively prime to $f(x)$ if char $K = 2$, and $f(x)$ is a polynomial of degree $2g+1$. Furthermore, we may choose $h(x)$ and $f(x)$ as follows. If char $K = 2$, the ramified primes in $K/k(x)$ are precisely the infinite prime and the prime divisors of $h(x)$; if char $K \neq 2$, then $f(x)$ is square-free and the ramified primes in $K/k(x)$ are the infinite prime and the prime divisors of $f(x)$ (see Exercise 10.9.2). We will denote the infinite prime in K by $\mathfrak{P}_\infty$ and the infinite prime in $k(x)$ by $\mathfrak{p}_\infty$.

The following definition is standard in algebraic geometry.

Definition 10.6.1. Given any function field K/k, the group $C_{K,0}$ of divisor classes of degree 0 is called the *Jacobian* of K. It will be also denoted by $\mathbb{J}_K$, or simply $\mathbb{J}$ if the underlying field K is implicitly known.

In the case of a hyperelliptic function field over an algebraically closed field, there is a way to represent every member of $\mathbb{J}$: every class C contains a unique reduced divisor. That is, there is a correspondence between reduced divisors and the Jacobian of K. Furthermore, there are algorithms that are computationally feasible that multiply two reduced divisors and provide the reduced divisor in the class of the product.

In the rest of this section, $K = k(x, y)$ will be a hyperelliptic function field over an algebraically closed field of constants k.

Definition 10.6.2. Let $K = k(x, y)$ be a hyperelliptic function field over an algebraically closed field k (usually $k = \overline{\mathbb{F}}_q$) given by Equation 10.3. A divisor $\mathfrak{A} \in D_{K,0}$ of degree 0 is called *reduced* if:

(1) $\mathfrak{A} = \frac{\mathfrak{B}}{\mathfrak{P}_\infty^n}$, where $\mathfrak{B}$ is an integral divisor of degree n that is relatively prime to $\mathfrak{P}_\infty$.
(2) If $\mathfrak{p} \in \mathbb{P}_{k(x)}$ is not ramified and $\operatorname{con}_{k(x)/K} \mathfrak{p} = \mathfrak{P}\mathfrak{P}'$, then $v_{\mathfrak{P}}(\mathfrak{B}) > 0$ implies that $v_{\mathfrak{P}'}(\mathfrak{B}) = 0$.
(3) If $\mathfrak{p} \in \mathbb{P}_{k(x)}$ is ramified, $\mathfrak{p} \neq \mathfrak{p}_\infty$, and $\operatorname{con}_{k(x)/K} \mathfrak{p} = \mathfrak{P}^2$, then $v_{\mathfrak{P}}(\mathfrak{B}) \in \{0, 1\}$.
(4) $\deg_K \mathfrak{B} = n \leq g = g_K$.

If $\mathfrak{A}$ satisfies (1)–(3), then $\mathfrak{A}$ is said to be *semireduced*.

The reasons to consider hyperelliptic function fields and not a general function field for cryptosystem issues are the following:

(i) Every class divisor of degree 0 can be represented in a unique way by a reduced divisor.
(ii) Every reduced divisor can be represented by two explicit functions.
(iii) The sum of two reduced divisors can be effectively computed.

Before proving these facts, we give the following notation and definition.

Definition 10.6.3. Given any two divisors $\mathfrak{A}, \mathfrak{A}_1 \in D_{K,0}$, we define the 0-*greatest common divisor* of $\mathfrak{A}$ and $\mathfrak{A}_1$ as

$$\begin{aligned} [\mathfrak{A}, \mathfrak{A}_1]_0 &:= \mathfrak{A}_2, \quad \text{where} \\ v_{\mathfrak{P}}(\mathfrak{A}_2) &= \min\{v_{\mathfrak{P}}(\mathfrak{A}), v_{\mathfrak{P}}(\mathfrak{A}_1)\} \quad \text{for} \quad \mathfrak{P} \neq \mathfrak{P}_\infty \quad \text{and} \\ v_{\mathfrak{P}_\infty}(\mathfrak{A}_2) &= -\sum_{\mathfrak{P}\neq\mathfrak{P}_\infty} v_{\mathfrak{P}}(\mathfrak{A}_2). \end{aligned}$$

Notice that $\mathfrak{A}_2 \in D_{K,0}$.

The following result is due to Mumford [114].

Theorem 10.6.4. *Let $\mathfrak{A} = \prod_{\mathfrak{P}} \mathfrak{P}^{v_{\mathfrak{P}}(\mathfrak{A})} = \prod_{i=1}^r \mathfrak{P}_i^{\alpha_i} \cdot \mathfrak{P}_\infty^\beta$ be a semireduced divisor and assume that for $1 \leq i \leq r$, we have, $\mathfrak{P}_i \cap k[x] = \mathfrak{p}_i$, $(x - a_i)_{k(x)} = \frac{\mathfrak{p}_i}{\mathfrak{p}_\infty}$, $\mathfrak{P}_i \cap k[y] = \mathfrak{q}_i$, and $(y - b_i)_{k(x)} = \frac{\mathfrak{q}_i}{\mathfrak{q}_\infty}$. In other words, if $\varphi_{\mathfrak{P}_i}$ is the place corresponding to $\mathfrak{P}_i$, then $\varphi_{\mathfrak{P}_i}(x) = a_i$ and $\varphi_{\mathfrak{P}_i}(y) = b_i$. If $p(x) = \prod_{i=1}^r (x - a_i)^{\alpha_i}$, then there exists a unique polynomial $q(x)$ such that:*

(1) $\deg q(x) < \deg p(x)$,
(2) $q(a_i) = b_i$ *for* $1 \leq i \leq r$,
(3) $p(x) \mid (q(x)^2 - h(x)q(x) - f(x))$ *where $h(x)$ and $f(x)$ are as in Equation* (10.3).

Furthermore, we have $\mathfrak{A} = [(p(x))_K, (q(x) - y)_K]_0$.

Proof. Assume that $1 \leq i \leq r$ and $\mathfrak{p}_i$ is unramified. Consider $y \in K_{\mathfrak{P}_i} = k(x)_{\mathfrak{p}_i}$. Since $x - a_i$ is a prime element for $\mathfrak{P}_i$, then $y = \sum_{j=0}^\infty c_j (x - a_i)^j$ with $c_0 = b_i$. Define $q_i(x) := \sum_{j=0}^{\alpha_i - 1} c_j (x - a_i)^j$. We have:

(1) $\deg q_i(x) \leq \alpha_i - 1 < \alpha_i = \deg(x - a_i)^{\alpha_i}$.
(2) $q_i(a_i) = c_0 = b_i$.
(3) Reducing the equation $y^2 - h(x)y = f(x)$ modulo $(x - a_i)^{\alpha_i}$ and using the fact that $y \bmod (x - a_i)^{\alpha_i} = q_i(x)$, we obtain

$$q_i^2(x) - h(x)q_i(x) \equiv f(x) \bmod (x - a_i)^{\alpha_i}.$$

Hence $(x - a_i)^{\alpha_i}$ divides $q_i(x)^2 - h(x)q_i(x) - f(x)$.

Now if $t(x)$ is another polynomial satisfying (1)–(3), then

$$(x - a_i)^{\alpha_i} \quad \text{divides} \quad (q_i(x) - t(x))(q_i(x) + t(x) - h(x)).$$

In case $h(x) = 0$, that is, char $K \neq 2$, we have $y^2 = f(x)$, $b_i^2 = f(a_i)$. If $(x - a_i)$ divides $q(x) + t(x)$, then $q(a_i) + t(a_i) = 2b_i = 0$, so $b_i = 0$, $(x - a_i) \mid f(x)$ and $(x - a_i) \mid q_i(x)$. Since $f(x)$ is square-free, $(x - a_i)^2$ does not divide $f(x)$ and $(x - a)^2$ divides $q_i(x)^2$. It follows that $\alpha_i = 1$ and $q(x) = g(x) = c_0 = b_i = 0$ (in fact this case is impossible since $\mathfrak{P}_i$ would be ramified)

Now if char $K = 2$, we have $h(x) \neq 0$. Because of (2), $(x - a_i)$ divides $(q_i(x) - t(x))$; then since the ramified primes are precisely those dividing $h(x)$, it follows that

$h(a_i) \neq 0$, so $(x-a_i) \nmid (q_i(x)-t(x)-h(x))$. Hence $(x-a_i)^{\alpha_i}$ divides $(q_i(x)-t(x))$. Since $\deg(q_i(x)-t(x)) \leq \alpha_i - 1$, we conclude that $q_i(x) = t(x)$.

We have shown that in any case, $q_i(x)$ is the unique polynomial satisfying (1)–(3).

Now we study the case of $\mathfrak{p}_i$ ramified. Then $\alpha_i = 1$, and hence $q_i(x) = b_i$ is the unique polynomial satisfying (1)–(3). It follows by the Chinese remainder theorem that there exists a unique polynomial $q(x)$ such that $q(x) \equiv q_i(x) \bmod (x-a_i)^{\alpha_i}$ for $1 \leq i \leq r$ and $\deg q(x) < \sum_{i=1}^r \alpha_i$. It is easy to verify that $q(x)$ is the unique polynomial satisfying statements (1)–(3) of the theorem.

Now let $\mathfrak{P}_1, \ldots, \mathfrak{P}_s$ be the unramified prime divisors and let $\mathfrak{P}_{s+1}, \ldots, \mathfrak{P}_r$ be the ramified ones. Set $\mathfrak{p}_i = \mathfrak{P}_i \cap k[x]$ and $\operatorname{con}_{k(x)/K} \mathfrak{p}_i = \mathfrak{P}_i\mathfrak{P}_i'$ for $1 \leq i \leq s$. Then

$$(p(x))_K = \frac{\prod_{i=1}^s (\mathfrak{P}_i\mathfrak{P}_i')^{\alpha_i} \prod_{i=s+1}^r \mathfrak{P}_i^{2\alpha_i}}{\mathfrak{P}_\infty^\alpha} \quad \text{for some} \quad \alpha \geq 0.$$

Now for $q(x) - y$, if $\mathfrak{P}$ is distinct from $\mathfrak{P}_1, \ldots, \mathfrak{P}_r, \mathfrak{P}_\infty$, then $v_{\mathfrak{P}}(q(x)-y) \geq 0$.

For $1 \leq i \leq s$ we have $y \equiv q(x) \bmod (x-a_i)^{\alpha_i}$, so $v_{\mathfrak{P}_i}(y-q(x)) \geq \alpha_i$.

Finally, for $s+1 \leq i \leq r$, the conjugate of $y - q(x)$ is $-y - q(x)$ if $\operatorname{char} k \neq 2$, and $y + h(x) - q(x)$ if $\operatorname{char} k = 2$. Now the product of $y - q(x)$ and its conjugate is $-y^2 + q(x)^2$ or $y^2 + h(x)y - h(x)q(x) + q(x)^2$, that is, $f(x) - h(x)q(x) + q(x)^2$. It is easy to verify that $(x-a_i)^2 \nmid q(x)^2 - h(x)q(x) - f(x)$. Therefore $v_{\mathfrak{P}_i}(y-q(x)) = 1$.

We have proved that $[(p(x))_K, (y-q(x))_K]_0 = \mathfrak{A}$. □

Definition 10.6.5. The divisor $[(p(x))_K, (y-q(x))_K]_0$ will be denoted by $\operatorname{div}(p, q)$.

Another key fact concerning the use of hyperelliptic function fields in cryptography is the following.

Theorem 10.6.6. *Let $C \in \mathbb{J}_K$ be any element of the Jacobian of K. Then there exists a unique reduced divisor $\mathfrak{B}$ in C.*

In other words, every divisor of degree 0 is equivalent to a unique reduced divisor.

Proof. Let C be any class of degree zero and let $\mathfrak{C} \in C$ be any arbitrary divisor in C. Then $\deg_K(\mathfrak{C}\mathfrak{P}_\infty^g) = g$. By the Riemann–Roch theorem it follows that

$$\ell(\mathfrak{C}^{-1}\mathfrak{P}_\infty^{-g}) \geq g - g + 1 = 1.$$

Therefore there exists an integral divisor $\mathfrak{A}$ of degree g such that $\frac{\mathfrak{A}}{\mathfrak{P}_\infty^g} \in C$. Let $\mathfrak{A}_1$ be of degree $n \leq g$, such that $(\mathfrak{A}_1, \mathfrak{P}_\infty) = 1$ and $\frac{\mathfrak{A}_1}{\mathfrak{P}_\infty^n} \in C$. Note that such an element exists for any function field (such that $\deg_K \mathfrak{P}_\infty = 1$).

Next we consider M to be the set of prime divisors other than $\mathfrak{P}_\infty$ that are not ramified in $K/k(x)$. Consider the partition $M_1 \cup M_2$ of M. If $\mathfrak{p} \in \mathbb{P}_{k(x)}$ splits as $\operatorname{con}_{k(x)/K} \mathfrak{p} = \mathfrak{P}\mathfrak{P}'$ and if $v_{\mathfrak{P}}(\mathfrak{A}_1) \geq v_{\mathfrak{P}'}(\mathfrak{A}_1)$, then $\mathfrak{P} \in M_1$ and $\mathfrak{P}' \in M_2$. Define

$$\mathfrak{B} := \frac{\mathfrak{A}_1}{\mathfrak{P}_\infty^n} \prod_{\mathfrak{P} \in M_2} (\alpha_{\mathfrak{P}})_K^{-v_{\mathfrak{P}}(\mathfrak{A}_1)} \prod_{\substack{\mathfrak{P} \text{ ramified} \\ \mathfrak{P} \neq \mathfrak{P}_\infty}} (\alpha_{\mathfrak{P}})_K^{-[v_{\mathfrak{P}}(\mathfrak{A}_1)/2]},$$

where $(\alpha_{\mathfrak{P}})_{k(x)} = \frac{\mathfrak{P}_{|k(x)}}{\mathfrak{p}_\infty}$. Note that

$$(\alpha_{\mathfrak{P}})_K = \begin{cases} \frac{\mathfrak{P}^2}{\mathfrak{P}_\infty^2} & \text{if } \mathfrak{P} \text{ is ramified,} \\ \frac{\mathfrak{P}\mathfrak{P}'}{\mathfrak{P}_\infty^2} & \text{if } \mathfrak{P} \text{ is not ramified.} \end{cases}$$

Thus $\mathfrak{B} = \prod_{\mathfrak{P}\in M_1} \mathfrak{P}^{s_{\mathfrak{P}}} \prod_{\substack{\mathfrak{P} \text{ ramified} \\ \mathfrak{P}\neq\mathfrak{P}_\infty}} \mathfrak{P}^{t_{\mathfrak{P}}} \cdot \mathfrak{P}_\infty^u$, where $s_{\mathfrak{P}} = v_{\mathfrak{P}}(\mathfrak{A}_1) - v_{\mathfrak{P}'}(\mathfrak{A}_1) \geq 0$, $t_{\mathfrak{P}} = v_{\mathfrak{P}}(\mathfrak{A}_1) - 2\left[\frac{v_{\mathfrak{P}}(\mathfrak{A})}{2}\right] \in \{0, 1\}$, and $u \leq 0$. Clearly $\mathfrak{B}$ is a reduced divisor and $\mathfrak{B} \in C$.

It remains to prove that $\mathfrak{B}$ is unique. Now, since K is hyperelliptic and $\mathfrak{P}_\infty$ is ramified, it can be shown that $\ell_K(\mathfrak{P}_\infty^{-2t}) = \ell_K(\mathfrak{P}_\infty^{-(2t+1)}) = t + 1$ for $0 \leq t \leq g - 1$ and that there is no $\alpha \in K^*$ such that $\mathfrak{N}_\alpha = \mathfrak{P}_\infty^{2t+1}$ for $0 \leq t \leq g - 1$ (see Corollary 14.2.72).

Now for $t \geq 2g - 1$, using the Riemann–Roch theorem we obtain $\ell_K(\mathfrak{P}_\infty^{-t}) = t - g + 1$.

Next, let $\mathfrak{B}$ be a principal semireduced divisor. Say $(\alpha)_K = \mathfrak{B} = \frac{\prod_{i=1}^t \mathfrak{P}_i}{\mathfrak{P}_\infty^t}$. Then t is even. Let $\ell_K(\mathfrak{P}_\infty^{-t}) = t/2 + 1$ and notice that $t/2 \geq 0$. We have $\ell_K(\mathfrak{P}_\infty^{-t}) = \ell_{k(x)}(\mathfrak{p}_\infty^{-t/2})$. For each $\mathfrak{P}_i$ such that $1 \leq i \leq t$, we set $(x - a_i)_K = \frac{\mathfrak{P}_i\mathfrak{P}_i'}{\mathfrak{P}_\infty^2}$, where $\mathfrak{P}_i = \mathfrak{P}_i'$ if $\mathfrak{P}_i$ is ramified and $\mathfrak{P}_i \neq \mathfrak{P}_i'$ otherwise. Thus a basis of $L_K(\mathfrak{P}_\infty^{-t})$ is $\{1 = \alpha_0, \alpha_1, \ldots, \alpha_{t/2}\}$, where $\alpha_i = \prod_{j=1}^i (x - a_j)$.

Therefore $\alpha = \sum_{t=0}^{t/2} \lambda_i\alpha_i \in k[x]$. Assume $t \geq 2$. Then $\alpha(a_1) = 0$ and $\lambda_0 = 0$, so $\mathfrak{P}_1$ and $\mathfrak{P}_1'$ divide $\mathfrak{N}_\alpha$. But this contradicts the fact that $\mathfrak{B}$ is semireduced; it follows that $t = 0$ and $\mathfrak{B} = \mathfrak{N}$.

Now let $\mathfrak{B}_1$ and $\mathfrak{B}_2$ be two reduced divisors in the same class, i.e., $\mathfrak{B}_1 P_K = \mathfrak{B}_2 P_K$. Say $\mathfrak{B}_1 = \mathfrak{A}_1\mathfrak{P}_\infty^{-a_1}$ and $\mathfrak{B}_2 = \mathfrak{A}_2\mathfrak{P}_\infty^{-a_2}$, $\deg(\mathfrak{A}_1\mathfrak{A}_2) \leq 2g$, and as in the first part of the proof, we construct a semireduced divisor $\mathfrak{B}_3$ such that $\mathfrak{B}_3 P_K = \mathfrak{B}_1\mathfrak{B}_2^{-1} P_K$.

If we assume that $\mathfrak{B}_1 \neq \mathfrak{B}_2$, there exists a prime divisor $\mathfrak{T}_1 \neq \mathfrak{P}_\infty$ such that $v_{\mathfrak{T}_1}(\mathfrak{B}_1) \neq v_{\mathfrak{T}_1}(\mathfrak{B}_2)$. We may assume that $v_{\mathfrak{T}_1}(\mathfrak{B}_1) > v_{\mathfrak{T}_1}(\mathfrak{B}_2)$ and $v_{\mathfrak{T}_1}(\mathfrak{B}_1) \geq v_{\mathfrak{T}_1'}(\mathfrak{B}_2)$ if $\mathfrak{T}_1 \neq \mathfrak{T}_1'$. It is easy to see that $v_{\mathfrak{T}_1}(\mathfrak{B}_3) > 0$, so $\mathfrak{B}_3 \neq \mathfrak{N}$. This contradicts the equalities $\mathfrak{B}_1\mathfrak{B}_2^{-1} P_K = P_k = \mathfrak{B}_3 P_K$. Hence $\mathfrak{B}_1 = \mathfrak{B}_2$. □

10.7 Reduced Divisors over Finite Fields

We apply the results of Section 10.6 to finite fields. Let $K = k(x, y)$ be the hyperelliptic function field given by Equation (10.3), where k is a finite field. Then $\mathbb{J}_K = C_{K,0}$ is a finite group (Theorem 6.2.2). Now if $\bar{k}$ is an algebraic closure of k and $\bar{K} = K\bar{k}$, then by Exercise 8.7.20 we have $C_{K,0} \subseteq C_{\bar{K},0}$ and each element of $C_{K,0}$ admits a unique representation as a reduced division $\operatorname{div}(p, q)$, where $p, q \in k[x]$, $\deg p(x) \leq g$, and $\deg q(x) < \deg p(x)$.

Given two reduced divisors $\mathfrak{A}_1 = \operatorname{div}(p_1, q_1)$ and $\mathfrak{A}_2 = \operatorname{div}(p_2, q_2)$, Koblitz [83] presented an algorithm to find the reduced divisor $\mathfrak{A}_3 = \operatorname{div}(p_3, q_3)$ such that $\overline{\mathfrak{A}_1}\,\overline{\mathfrak{A}_2} = \overline{\mathfrak{A}_3}$. In this way it is possible to compute the Jacobian of a hyperelliptic function field. For general function fields it is difficult to compute the Jacobian.

The first part of the algorithm is as follows:

Let $\mathfrak{A}_1 = \operatorname{div}(p_1, q_1)$ and $\mathfrak{A}_2 = \operatorname{div}(p_2, q_2)$. Set $d_1 = (p_1, p_2)$ and let $\alpha_1, \alpha_2 \in k[x]$ be such that $d_1 = \alpha_1 p_1 + \alpha_2 p_2$. Set $d_2 = (d_1, q_1 + q_2 - h)$, $d_2 = \beta_1 d_1 + \beta_2(q_1 + q_2 - h)$, $\gamma_1 = \alpha_1\beta_1$, and $\gamma_2 = \alpha_2\beta_1$, $\gamma_3 = \beta_2$. We have

$$d_2 = \gamma_1 p_1 + \gamma_2 p_2 + \gamma_3(q_1 + q_2 - h).$$

Next, put $p := \frac{p_1 p_2}{d_2^2}$ and $q := \frac{\gamma_1 p_1 q_2 + \gamma_2 p_2 q_1 + \gamma_3(q_1 q_2 + f)}{d_2} \bmod p$. We obtain the following theorem:

Theorem 10.7.1. $\mathfrak{A} = \operatorname{div}(p, q)$ *is a semireduced divisor that satisfies* $\overline{\mathfrak{A}} = \overline{\mathfrak{A}_1}\,\overline{\mathfrak{A}_2}$.

Proof. [84, Page 173, Theorem 7.1]. □

The second part of the algorithm starts with a given semireduced divisor $\mathfrak{A} = \operatorname{div}(p, q)$. The task is to find the reduced divisor $\mathfrak{A}_3 = \operatorname{div}(p_3, q_3)$ such that $\overline{\mathfrak{A}} = \overline{\mathfrak{A}_3}$ in $C_{K,0}$.

Let $p_3' := \frac{f + hq - q^2}{p}$ and $q_3' = (h - q) \bmod p_3'$. If $\deg p_3' > g$ we repeat the process. Once we get $\deg p_3' \le g$, we finally set $p_3 := a^{-1} p_3'$ and $q_3 = q_3'$, where a is the leading coefficient of p_3'. Then $\mathfrak{A}_3 = \operatorname{div}(p_3, q_3)$ is reduced and $\overline{\mathfrak{A}_3} = \overline{\mathfrak{A}}$ ([84, p. 176, Theorem 7.2]).

Remark 10.7.2. Note that the computations take place in the field k.

Example 10.7.3. Consider the hyperelliptic curve of equation $K = \mathbb{F}_{2^4}(x, y)$ over $\mathbb{F}_{2^4} = \mathbb{F}_{16}$, where $y^2 + x(x + \beta)y = x^5 + 1$ and $\beta \in \mathbb{F}_{2^2} \setminus \mathbb{F}_2$ so $\beta^2 = \beta + 1$. Consider $p(x) = x^4 + x^3 + x^2 + x^1 + 1 = \frac{x^5+1}{x+1} \in \mathbb{F}_2[x]$. Notice that $p(x)$ is irreducible over $\mathbb{F}_2[x]$ and let $\xi_1, \xi_2, \xi_3, \xi_4 \in \mathbb{F}_{16}$ be its roots. Fix one of them, say $\xi = \xi_1$.

The ramified prime divisors of $k(x)$ in K, where $k := \mathbb{F}_{2^4}$, are $\mathfrak{p}_0$ and $\mathfrak{p}_\beta$, which correspond to x and $(x + \beta)$ respectively, and the infinite prime $\mathfrak{p}_\infty$. Let $\mathfrak{P}_0$, $\mathfrak{P}_\beta$, and $\mathfrak{P}_\infty$ be the prime divisors in K that lie above $\mathfrak{p}_0$, $\mathfrak{p}_\beta$, and $\mathfrak{p}_\infty$ respectively. Define $h(x) := x(x + \beta)$ and $f(x) := x^5 + 1 = (x + 1)p(x)$. If $\mathfrak{p}_1$ and $\mathfrak{p}_\xi$ are the prime divisors in $k(x)$ corresponding to $x + 1$ and $x + \xi$ respectively, then $\mathfrak{p}_1$ and $\mathfrak{p}_\xi$ split in $K/k(x)$. If $\mathfrak{p}_1 = \mathfrak{P}_1\mathfrak{P}_1'$ and $\mathfrak{p}_\xi = \mathfrak{P}_\xi\mathfrak{P}_\xi'$, then from the defining equation of K we obtain that

$$\varphi_{\mathfrak{P}_1}(y) = 0; \quad \varphi_{\mathfrak{P}_1'}(y) = 1 + \beta; \quad \varphi_{\mathfrak{P}_\xi}(y) = 0; \quad \varphi_{\mathfrak{P}_\xi'}(y) = \xi.$$

In this way we deduce that the divisor of y in K is

$$(y)_K = \frac{\mathfrak{P}_1 \prod_{i=1}^4 \mathfrak{P}_{\xi_i}}{\mathfrak{P}_\infty^5}.$$

Now we apply the algorithm to the divisors $\mathfrak{A}_1 = \frac{\mathfrak{P}_1\mathfrak{P}_\beta\mathfrak{P}_\xi}{\mathfrak{P}_\infty^3}$ and $\mathfrak{A}_2 = \frac{\mathfrak{P}_1'\mathfrak{P}_0\mathfrak{P}_\xi'}{\mathfrak{P}_\infty^3}$.

Applying the first part of the algorithm we obtain p_1, q_1, p_2, q_2 such that $\mathfrak{A}_j = \operatorname{div}(p_j, q_j)$ for $j = 1, 2$. We have

$$p_1(x) = (x-1)(x-\beta)(x-\xi); \quad q_1(x) = \frac{1}{\beta+\xi}(x-1)(x-\xi)$$

and

$$p_2(x) = (x-1)x(x-\xi); q_2(x) = (1+\xi^{-1})x^2 + (1+\xi^{-1}+\beta)x + 1$$

Therefore $d_1 = (p_1, p_2) = (x-1)(x-\xi) = \alpha_1 p_1 + \alpha_2 p_2$, from which we obtain $\alpha_1 = \alpha_2 = \beta + 1$. Next, we have $d_2 = (d_1, q_1 + q_2 + h) = (x-1)(x-\xi)$. Thus $d_2 = \beta_1 d_1 + \beta_2(q_1 + q_2 + h) = 1 \times d_1 + 0 \times (q_1 + q_2 + h)$, that is, $\beta_1 = 1$ and $\beta_2 = 0$. Hence we have $\gamma_1 = \alpha_1\beta_1 = \alpha_1 = \beta + 1$, $\gamma_2 = \alpha_2\beta_1 = \alpha_2 = \beta + 1$, and $\gamma_3 = \beta_2 = 0$.

In this way, we get

$$p(x) = \frac{p_1 p_2}{d_2^2} = x(x-\beta)$$

and

$$q(x) = \frac{\gamma_1 p_1 q_2 + \gamma_2 p_2 q_1 + \gamma_3(q_1 + q_2 + f)}{d_2} \bmod p = x + 1.$$

Note that if $\varphi_{\mathfrak{P}_0}$ and $\varphi_{\mathfrak{P}_\beta}$ are the places associated to $\mathfrak{P}_0$ and $\mathfrak{P}_\beta$ respectively, we have

$$\varphi_{\mathfrak{P}_0}(y-q) = \varphi_{\mathfrak{P}_0}(y) - q(0) = 1 - 1 = 0$$

and

$$\varphi_{\mathfrak{P}_\beta}(y-q) = \varphi_{\mathfrak{P}_\beta}(y) - q(\beta) = \beta + 1 - (\beta+1) = 0.$$

Using this valuation it is easy to check that $v_{\mathfrak{P}_0}(y-q) = v_{\mathfrak{P}_\beta}(y-q) = 1$.

Therefore the semireduced divisor in the class of $\mathfrak{A}_1\mathfrak{A}_2$ is

$$\mathfrak{B} = \operatorname{div}(p, q) = [(p(x))_K, (y - q(x))_K]_0 = \left[\frac{\mathfrak{P}_0^2\mathfrak{P}_\beta^2}{\mathfrak{P}_\infty^4}, \frac{\mathfrak{P}_0\mathfrak{P}_\beta\mathfrak{Q}}{\mathfrak{P}_\infty^5}\right]_0 = \frac{\mathfrak{P}_0\mathfrak{P}_\beta}{\mathfrak{P}_\infty^2},$$

where $\mathfrak{Q}$ is an integral divisor relatively prime to $\mathfrak{P}_0\mathfrak{P}_\infty$.

Observe that since $\mathfrak{B}$ is already a reduced divisor, the second part of the algorithm is not necessary.

Example 10.7.4. Consider again the hyperelliptic curve of equation $y^2+x(x+\beta)y = x^5+1$ over $\mathbb{F}_{2^4}$ as in Example 10.7.3, and the divisors of degree zero $\mathfrak{A}_1 = \frac{\mathfrak{P}_1\mathfrak{P}_\beta\mathfrak{P}_\xi}{\mathfrak{P}_\infty^3}$ and $\mathfrak{A}_2 = \frac{\mathfrak{P}_1'\mathfrak{P}_0}{\mathfrak{P}_\infty^2}$. Then $\mathfrak{A}_1\mathfrak{A}_2 = \frac{\mathfrak{P}_1\mathfrak{P}_1'\mathfrak{P}_0\mathfrak{P}_\beta\mathfrak{P}_\xi}{\mathfrak{P}_\infty^5}$. Using the first part of the algorithm as in Example 10.7.3, we obtain that the semireduced divisor that belongs to the same class as $\mathfrak{A}_1\mathfrak{A}_2$ is (see Exercise 10.9.5)

$$\mathfrak{B} = \frac{\mathfrak{P}_0\mathfrak{P}_\beta\mathfrak{P}_\xi}{\mathfrak{P}_\infty^5}.$$

Now we use the second part of the algorithm to find the reduced divisor $\mathfrak{A}_3$ that belongs to the same class as $\mathfrak{B}$.

Let $\mathfrak{B} = \operatorname{div}(p, q)$, where

$$p(x) = x(x+\beta)(x+\xi), \quad q(x) \le 2;$$
$$q(0) = 1, \quad q(\beta) = \beta + 1, \quad \text{and} \quad q(\xi) = 0.$$

It is easy to see that $q(x) = \frac{\xi^4+1}{\xi+\beta}\left(x + \frac{\beta+\xi}{\xi+1}\right)(x+\xi)$.

To simplify the notation, we set $\mu = \xi + \beta \in \mathbb{F}_{2^4} \setminus \mathbb{F}_{2^2}$. Then we have $p(x) = x^3 + \mu x^2 + \mu^{13}x$ and $q(x) = x^2 + \mu^5 x + 1$.

Using the algorithm we obtain

$$p_3'(x) = \frac{f + hq - q^2}{p} = (x+1)(x+\mu^{12}) = x^2 + \mu x + \mu^{12} \quad (\xi^4 = \mu^{12})$$

and

$$q_3'(x) = h - q \bmod p_3' = (x^2 + \mu^{10}x) - (x^2 + \mu^5 x + 1) \bmod p_3' = x + 1.$$

It follows that $(p_3'(x))_K = \frac{\mathfrak{P}_1\mathfrak{P}_1'\mathfrak{P}_{\xi^4}\mathfrak{P}_{\xi^4}'}{\mathfrak{P}_\infty^4}$. We have $y - q_3' = y - x - 1$, $v_{\mathfrak{P}_\infty}(y - x - 1) = -5$, and $v_{\mathfrak{P}_1}(y-x-1) \ge 0$, $v_{\mathfrak{P}_1'}(y-x-1) = v_{\mathfrak{P}_{\xi^4}}(y-x-1) = v_{\mathfrak{P}_{\xi^4}'}(y-x-1) = 0$.

Therefore $\mathfrak{A}_3 = \operatorname{div}(p_3', q_3') = [(p_3'(x))_K, (y - q_3'(x))_K]_0 = \frac{\mathfrak{P}_1}{\mathfrak{P}_\infty}$.

10.8 Implementation of Hyperelliptic Cryptosystems

The advantage of using hyperelliptic function fields cryptosystems as compared to elliptic ones is that we can construct such a cryptosystem at the same security level as the elliptic one using a smaller defining field. More precisely, the order of the Jacobian of a hyperelliptic function field of genus g over a field of q elements is approximately q^g. This means that if we have an elliptic function field, i.e., of genus one, with a field size of q of order 2^{200}, then a hyperelliptic curve of genus two, three, or four can have field size of order 2^{100}, 2^{67}, or 2^{50} respectively.

The Diffie–Hellman key exchange and the ElGamal message transmission can be implemented in the Jacobian of a hyperelliptic function field. We need to choose $k = \mathbb{F}_q$ and a suitable K for the implementation.

Now K must satisfy several conditions to be suitable for implementation. We summarize the main security requirements for our function field. First, given the current state of computing power, the class number h over $\mathbb{F}_q$ must be divisible by a large prime p of order larger than $2^{160} \approx 1.47 \times 10^{50}$ in order to avoid Pollard-rho ([118, 119]), Shanks's Baby-Step/Giant-Step, and Pohlig–Hellman ([116]) attacks. These attacks are discrete logarithm problem algorithms. Second, after Gaudry [47] it is recommended that the genus should be less than four so that one can construct a secure hyperelliptic cryptosystem. Next, the order of the field base should be a prime power of two in order to protect the cryptosystem against Weil descent on the Jacobian of K (for instance see [40]). Finally, Frey and Rück [37] reduced the discrete logarithm problem in $C_{K,0}$ to the discrete logarithm problem in $\mathbb{F}^*_{q^m}$. Therefore to avoid the Frey–Rück attack, p must not divide $q^m - 1$ for "small" m, say of order about $m \approx 2000/\log_2 q$, that is, p must not divide $q^t - 1$ for $1 \leq t \leq 2000/\log_2 q$.

In short, assume that $K/\mathbb{F}_q$ is a hyperelliptic function field of genus g suitable for implementation in cryptography. If p is a prime dividing the order of the class group of K, then K, g, q, and p must satisfy:

- $p > 2^{160}$,
- $g = 2$ or $g = 3$,
- $q = 2^r$ with r a prime number,
- The smallest $s \geq 1$ such that $q^s \equiv 1 \bmod p$ should be greater than $2000/\log_2 q$

In order to determine the class group, we use the Riemann zeta function. Let K be a congruence function field over $\mathbb{F}_q$ and let $K_r := K\mathbb{F}_{q^r}$ with $r \geq 1$. If $P_r(u) = P_{K_r}(u)$ is the numerator of the zeta function of K_r and if h_r denotes the class number of K_r, we have

$$h := h_1 = P_1(u) = \prod_{i=1}^{g} |1 - \alpha_i|^2,$$

where $\{\alpha_i, \overline{\alpha}_i\}$ are the roots of $P(u)$. We also have for any $r \geq 1$ (see Theorem 7.1.6),

$$h_r = \prod_{i=1}^{g} |1 - \alpha_i^r|^2.$$

We write $P(T) = P_1(T) = 1 + a_1 T + \cdots + a_g T^g + q a_{g-1} T^{g+1} + \cdots + q^g T^{2g}$. Denote by N_r the number of divisors of degree 1 in $K_r = K\mathbb{F}_{q^r}$. Then $a_1 = N_1 - 1 - q$ and $a_2 = (N_2 - 1 - q^2 - a_1^2)/2$.

To compute h_r in the case of genus $g = 2$, we may use Exercise 10.9.4.

Example 10.8.1. Consider $y^2 - y = x^5 + x$. Here the only ramified prime is $\mathfrak{p}_\infty$, and the genus is 2.

Using Exercises 10.9.3 and 10.9.4 we find that $N_1 = 5$ and $N_2 = 9$. Then $a_1 = 0$ and $a_2 = 2$. The solutions of the equations $T^2 + (-2) = 0$ are $\sqrt{2}$ and $-\sqrt{2}$. Finally we obtain $\alpha_1 = -\sqrt{2}\bar{\zeta}_3$ and $\alpha_2 = \sqrt{2}\bar{\zeta}_3$, where $\zeta_3 = \frac{-1+\sqrt{3}i}{2}$ is a primitive third root of unity.

It follows that

$$h_r = \begin{cases} (2^{r/2} - 1)^4 & \text{if } r \equiv 0 \bmod 6, \\ 1 + 2^r + 2^{2r} & \text{if } r \equiv 1, 5 \bmod 6, \\ (1 + 2^{r/2} + 2^r)^2 & \text{if } r \equiv 2, 4 \bmod 6, \\ (2^r - 1)^2 & \text{if } r \equiv 3 \bmod 6. \end{cases} \tag{10.4}$$

For $r \leq 666$ all these hyperelliptic function fields satisfy that if a prime number p divides h_r then it divides $2^i - 1$ for some $i \leq 2000$. This follows from Equation (10.4), since $2^{3r} - 1 = (2^r - 1)(2^{2r} + 2^r + 1)$. Therefore all such function fields are vulnerable to the Frey–Rück attack for $r \leq 666$. That is, the discrete logarithm problem can be solved in $\mathbb{F}_{2^i}^*$ for some $i \leq 2000$ and therefore all these hyperelliptic function fields offer no security and are not suitable for cryptography.

Note that this example is quite similar to that of Koblitz [84, Example 6.1, p. 149].

Example 10.8.2. Consider the equation $y^2 + x(x+\beta)y = x^5 + 1$ over $\mathbb{F}_4 = \mathbb{F}_{2^2}$, where $\beta^2 = \beta + 1$. Let $\xi^5 = 1$ be such that $\xi \in \mathbb{F}_{16} \setminus \mathbb{F}_4$. We use the element $\mu = \xi + \beta$ for the explicit computations described below. Note that $\mu^3 = \xi$, $\mu^5 = \beta^2$, etc. We have $\mathbb{F}_4 = \{0, 1, \mu^{10}, \mu^5\}$ and $\mathbb{F}_{16}^* = \{\mu^i \mid 0 \leq i \leq 14\}$, $\mu^{15} = \mu^0 = 1$. Using Exercises 10.9.3 and 10.9.4 we find by direct computation $N_1 = 5$ and $N_2 = 23$. Therefore $a_1 = 0$ and $a_2 = 3$.

The solutions of the equation $x^2 + a_1 x + (a_2 - 2q) = 0$ are $\gamma_1 = \sqrt{5}$ and $\gamma_2 = -\sqrt{5}$. Finally, one of the roots of $x^2 - \sqrt{5}x + 4$ is $\alpha_1 = \frac{\sqrt{5} - \sqrt{11}i}{2}$ and a root of $x^2 + \sqrt{5}x + 4$ is $\alpha_2 = \frac{-\sqrt{5} + \sqrt{11}i}{2}$. We have $\alpha_2 = -\alpha_1$. It follows that

$$\begin{aligned} h_r &= |1 - \alpha_1^r|^2 |1 - \alpha_2^r|^2 = |1 - \alpha_1^r|^2 |1 - (-1)^r \alpha_1^r|^2 \\ &= \begin{cases} |\alpha_1^r - 1|^4 & \text{if } r \text{ is even}, \\ \\ |1 - \alpha_1^{2r}|^2 & \text{if } r \text{ is odd}. \end{cases} \end{aligned}$$

For instance, for $r = 61$ we obtain $h_r = (271)^2 p^2$ where p is the fifty-three-digit prime number

$$44947399259371741314172478713222775636987866517942801 \approx 4.5 \times 10^{52}.$$

Furthermore, p does not divide $2^i - 1$ for $1 \leq i \leq 1000 = 2000/\log_2 q$. However, K might not be completely suitable for cryptography purposes because the base field is of order 4^{61}, which is not a prime power of 2 and thus is vulnerable to the Weil descent on the Jacobian.

In the next examples we present some hyperelliptic function fields. For the algorithms used to compute the order of the Jacobian we refer to the original papers.

Let $p(t) \in \mathbb{F}_2[t]$ be a monic irreducible polynomial of degree m and let λ be a root of $p(t)$. Then $\mathbb{F}_{2^m} = \mathbb{F}_2(\lambda)$.

For any element $\alpha = \sum_{i=0}^{m-1} \alpha_i \lambda^i \in \mathbb{F}_{2^m}$, $\alpha_i \in \mathbb{F}_2$, we represent α by the integer $\sum_{i=0}^{m-1} \alpha_i 2^i$ written in hexadecimal notation. For instance, the hexadecimal number $C1$ represents the element $\alpha = \lambda^7 + \lambda^6 + 1$.

We will use the above notation in the following examples.

Example 10.8.3 ([64]). Let $p = 10001300006400014200121$ and consider the genus-2 hyperelliptic function field defined by

$$y^2 + y = \alpha x^5, \quad \text{or equivalently} \quad (y')^2 = \alpha x^5 + 4^{-1},$$

over $\mathbb{F}_p$, where $\alpha \in \mathbb{F}_p$ and α is not a 5th power. Then the class group h is of order $h = 5 \times \ell$, where

$$\ell = 20005200592038621583241900701806833029 81.$$

This cryptosystem is not secure since K is defined over $\mathbb{F}_p$ where p is a large prime.

Example 10.8.4 ([24]). Let $\mathbb{F}_{2^{83}} = \mathbb{F}_2(\lambda)$, where λ is a root of $p(t) = t^{83} + t^7 + t^4 + t^2 + 1$ and let $K/\mathbb{F}_{2^{83}}$ be the genus-2 hyperelliptic function field given by the equation

$$y^2 + (\alpha_0 + \alpha_1 x + \alpha_2 x^2) = x^5 + \beta_4 x^4 + \beta_3 x^3 + \beta_2 x^2 + \beta_1 x + \beta_0,$$

where

$$\begin{aligned}
\alpha_0 &= 4D168CAB78F1F7EB78D54, & \alpha_1 &= 3B167A2F520486B2A8A60,\\
\alpha_2 &= 507FC6D8D98A1411D1F24, & &\\
\beta_0 &= 6ABF379716E615F0997AF, & \beta_1 &= 1D13C5C10A58A238681F3,\\
\beta_2 &= 3ACC287DAA28D01EDDB58, & \beta_3 &= 74BF8FFD1A04B1E8B845B,\\
\beta_4 &= 10046A0ED36CF3B146071. & &
\end{aligned}$$

The order of the class group of $K/\mathbb{F}_{2^{83}}$ is $2p$, where

$$p = 46768052394612054553468807679365619497317916118893 \approx 4.68 \times 10^{49}.$$

Now p does not divide $2^i - 1$ for $i = 1, 2, \ldots, 25$. Moreover, we have $2000/\log_2 q = 2000/83 \approx 24.0964 < 25$. Hence the system is reasonably secure and therefore suitable for being used in cryptography.

Example 10.8.5 ([24]). Let $\mathbb{F}_{2^{59}} = \mathbb{F}_2(\lambda)$, where λ is a root of $p(t) = t^{59} + t^7 + t^4 + t^2 + 1$, and consider the genus-3 hyperelliptic curve given by

$$\begin{aligned}
&y^2 + (\alpha_0 + \alpha_1 x + \alpha_2 x^2 + \alpha_3 x^3)\\
&\quad = x^7 + \beta_6 x^6 + \beta_5 x^5 + \beta_4 x^4 + \beta_3 x^3 + \beta_2 x^2 + \beta_1 x + \beta_0,
\end{aligned}$$

where

$$\begin{aligned} \alpha_0 &= 44EC0A3F607D5FE, & \alpha_1 &= 183AFFC60B6C97A, \\ \alpha_2 &= 5E8C286F052173E, & \alpha_3 &= 39BFF4C327D0FCC, \\ \beta_0 &= 2CE03A6BD01418F, & \beta_1 &= 15160EE501EA31D, \\ \beta_2 &= 2DDF3B805A56673, & \beta_3 &= 72EAAC2B03D6F33, \\ \beta_4 &= 30BF8CAF4CF398A, & \beta_5 &= 288F45CEB700047, \\ \beta_6 &= 692BDF3913214F7. \end{aligned}$$

The order of the class group of $K/\mathbb{F}_{2^{59}}$ is $2p$, where

$$\begin{aligned} p &= 95780971232851005943503002779523943538413536699032693 \\ &\approx 9.58 \times 10^{52}. \end{aligned}$$

Now, $2000/\log_2 q = 2000/59 \approx 33.9 < 34$ and p does not divide $2^i - 1$ for $1 \leq i \leq 34$. Thus K is suitable for cryptography purposes.

Example 10.8.6 ([80, 67]). Assume $\mathbb{F}_{2^{59}} = \mathbb{F}_2(\lambda)$, where λ is a root of $p(t) = t^{59} + t^6 + t^5 + t^4 + t^3 + t + 1$. Let $K/\mathbb{F}_{2^{59}}$ be the hyperelliptic genus-3 function field given by

$$y^2 + (x^3 + x^2 + ax + b)y = x^7 + x^6 + cx^5 + dx^4 + ex^3 + f,$$

where

$$\begin{aligned} a &= 6723B8D13BC30C7, & b &= 72D7EE15A5C9CF5, \\ c &= 6723B8D13BC30C7, & d &= 72D7EE15A5C9CF4, \\ e &= 24198E10C3B7566, & f &= 1EB9AF07BD3B303. \end{aligned}$$

The order of the Jacobian of $K/\mathbb{F}_{2^{59}}$ is $2p$, where

$$\begin{aligned} p &= 95780971304118053647396689122057683977359360476125197 \\ &\approx 9.58 \times 10^{52}. \end{aligned}$$

Finally, $2000/\log_2 q = 2000/59 \approx 33.9 < 34$ and p does not divide $2^i - 1$ for $1 \leq i \leq 34$. It follows that the hyperelliptic function field K is suitable for cryptography purposes.

10.9 Exercises

Exercise 10.9.1. Prove Theorem 10.2.3.

Exercise 10.9.2. Let $K = k(x, y)$ be a hyperelliptic function field of genus g over an arbitrary constant field k, and assume $[K : k(x)] = 2$. Show that if the pole divisor of x in $k(x)$ is ramified, then the defining equation of K can be given as follows:

(i) If char $K \neq 2$ then $y^2 = f(x) \in k(x)$, where $f(x)$ is a square-free polynomial of degree $2g + 1$ and the ramified primes in $K/k(x)$ are precisely the prime divisors of $f(x)$ and the pole divisor of x.

(ii) If char $K = 2$ then $y^2 - h(x)y = f(x)$, where $f(x)$ is a polynomial of degree $2g+1$, $h(x)$ is a nonzero polynomial of degree at most g that is relatively prime to $f(x)$, and the ramified primes in $K/k(x)$ are precisely the prime divisors of $h(x)$ and the pole divisor of x.

Exercise 10.9.3. Let $K = k(x, y)$ be a hyperelliptic or an elliptic function field of genus $g \geq 1$ given by

$$y^2 - h(x)y = f(x),$$

where

$$\deg h(x) \leq g, \quad \deg f(x) = 2g + 1,$$

$h(x) = 0$, $f(x)$ is square-free if char $k \neq 2$ and $h(x) \neq 0$ if char $k = 2$.

Let $\mathfrak{p}$ be a prime divisor of degree 1 and let $\varphi_{\mathfrak{P}}$ be the associated place. If $\mathfrak{P} \neq \mathfrak{P}_\infty$ then $\varphi_{\mathfrak{P}}(x), \varphi_{\mathfrak{P}}(y) \in k \cong \vartheta_{\mathfrak{P}}/\mathfrak{P}$ and $\varphi_{\mathfrak{P}_\infty}(x) = \varphi_{\mathfrak{P}_\infty}(y) = \infty$. Prove that

$$\begin{aligned} \varphi_{\mathfrak{P}} &\mapsto (\varphi_{\mathfrak{P}}(x), \varphi_{\mathfrak{P}}(y) \quad \text{if} \quad \mathfrak{P} \neq \mathfrak{P}_\infty \\ \varphi_{\mathfrak{P}_\infty} &\mapsto (\infty, \infty) \end{aligned}$$

defines a 1-to-1 correspondence between the set of places of degree 1 in K and the set of "rational points": $A = \{(a, b) \in k^2 \mid b^2 - h(a)b = f(a)\} \cup \{(\infty, \infty)\}$.

Exercise 10.9.4. Let $K/\mathbb{F}_q$ be a hyperelliptic function field of genus 2 and let h_r, $r \geq 1$, be the class number of $K_r = K\mathbb{F}_{q^r}$. Show that the following procedure works for finding h_r:

(i) Let N_r be the number of prime divisors of degree 1 in K_r. Find by direct computation N_1 and N_2 (you may use Exercise 10.9.3).
(ii) The coefficients of the numerator $P(u)$ of the zeta function of K are given by $a_1 = N_1 - 1 - q$ and $a_2 = (N_2 - 1 - q^2 + a_1^2)/2$.
(iii) Solve the equation $T^2 + a_1 T + (a_2 - 2q) = 0$. Let b_1 and b_2 be its roots.
(iv) Solve $T^2 - b_i T + q = 0$ for $i = 1, 2$ to obtain $\alpha_1, \overline{\alpha}_1, \alpha_2$, and $\overline{\alpha}_2$.
(v) Finally, obtain $h_r = |1 - \alpha_1^r|^2 |1 - \alpha_2^r|^2$.

Exercise 10.9.5. Let $K = \mathbb{F}_{2^4}(x, y)$ be the hyperelliptic function field given by

$$y^2 + x(x + \beta)y = x^5 + 1,$$

where $\beta^2 + \beta = 1$. Set $\mathfrak{A}_1 = \frac{\mathfrak{P}_1 \mathfrak{P}_\beta \mathfrak{P}_\xi}{\mathfrak{P}_\infty^3}$ and $\mathfrak{A}_2 = \frac{\mathfrak{P}'_1 \mathfrak{P}_0}{\mathfrak{P}_\infty^2}$. Using Koblitz's algorithm, show that the semireduced divisor in the class of $\mathfrak{A}_1 \mathfrak{A}_2$ is $\mathfrak{B} = \frac{\mathfrak{P}_0 \mathfrak{P}_\beta \mathfrak{P}_\xi}{\mathfrak{P}_\infty^3}$, where $\xi \neq 1$ is a root of $x^5 + 1$.

11

Introduction to Class Field Theory

11.1 Introduction

The notion of class fields is usually attributed to Hilbert, but the concept was already in the mind of Kronecker and the term was used by Weber before the appearance of the fundamental papers of Hilbert.

During the years 1880 to 1927, class field theory developed into three topics: prime decomposition, abelian extensions, and class groups of ideals.

In 1936 Chevalley introduced the concept of idele in order to formulate a class field theory for abelian extensions.

There is another way to study class fields, given by Hasse at the beginning of the of the 1930s. This approach uses the theory of simple algebras, which belongs to the area of noncommutative algebra.

Generally speaking, class field theory is the study of extensions where the prime divisors of degree 1 decompose totally. Particular features of the theory are the study of abelian extensions of $k(x)$ and of $\mathbb{Q}$, where k denotes a finite field, as well as the "reciprocity law."

There are several approaches to the theory of class fields:

(1) Relations between groups of congruence classes and abelian extensions (Weber).
(2) Theory of adeles (repartitions) and ideles (Chevalley and Weil).
(3) Theory of simple algebras (Hasse, Noether, Witt).
(4) Nonabelian L series (Artin).
(5) Providing natural generators for class fields as values of transcendental functions (Kronecker).

Unfortunately, a systematic treatment of class field theory would be too long and technical for our goals, so we have to confine ourselves to an explicit description of the abelian extensions of $k(x)$, where k is a finite field. This work is due to Carlitz and Hayes and is the objective of Chapter 12. The study of abelian extensions of a congruence function field K can be done by means of the so-called elliptic modules or Drinfeld modules. We will discuss Drinfeld modules in Chapter 13. In this chapter we present the Čebotarev density theorem, profinite groups and infinite Galois theory.

We will end this chapter with the principal results, without proof, of the theory of class fields for local as well as for global fields.

11.2 Čebotarev's Density Theorem

The proof we present here of Čebotarev's density theorem is based on [38]. In the rest of this chapter, the fields under consideration are congruence function fields. Let L/ℓ be a Galois extension of K/k with Galois group G. Let $\mathcal{P}$ be a place of L, and $\wp = \mathcal{P}|_K$. If D and I are the decomposition and inertia groups of $\mathcal{P}$ over $\wp$ respectively, then by Corollary 5.2.12, $\operatorname{Gal}(\ell(\mathcal{P})/k(\wp))$ is isomorphic to D/I. Since $\ell(\mathcal{P})$ and $k(\wp)$ are finite fields, it follows that D/I is a cyclic group generated by the *Frobenius automorphism*

$$\sigma : \ell(\mathcal{P}) \longrightarrow \ell(\mathcal{P}), \quad \text{defined by} \quad \sigma(x) = x^{q^f},$$

where $|k| = q$ and $f = [k(\wp) : k]$, i.e., $|k(\wp)| = q^f = N\wp$.

If $\wp$ is not ramified, then $I = \{1\}$. Therefore D is generated by the Frobenius automorphism.

Definition 11.2.1. Let $\mathcal{P}$ be a place in L and $\wp = \mathcal{P}|_K$, where $\wp$ is not ramified. Then $\left[\frac{L/K}{\mathcal{P}}\right]$ denotes the Frobenius automorphism of $\ell(\mathcal{P})/k(\wp)$.

Whenever we use the symbol $\left[\frac{L/K}{\mathcal{P}}\right]$ we will understand that $\mathcal{P}$ is not ramified.

Proposition 11.2.2. *The Frobenius automorphism is characterized by the property*

$$\left[\frac{L/K}{\mathcal{P}}\right](x) \equiv x^{N(\wp)} \bmod \mathcal{P} \text{ for all } x \in \vartheta_{\mathcal{P}},$$

where $\wp = \mathcal{P}|_K$.

Proof. Let $\sigma = \left[\frac{L/K}{\mathcal{P}}\right]$. If $\bar{\sigma}$ is the image of σ in $\operatorname{Gal}(\ell(\mathcal{P})/k(\wp))$, then $\bar{\sigma}x = x^{N(\wp)}$ for $x \in \ell(\mathcal{P}) = \vartheta_{\mathcal{P}}/\mathcal{P}$. The result follows. □

Proposition 11.2.3. *We have* $\left[\frac{L/K}{\sigma(\mathcal{P})}\right] = \sigma\left[\frac{L/K}{\mathcal{P}}\right]\sigma^{-1}$ *for all* $\sigma \in G$.

Proof. Let $\sigma \in G$ and put $\theta = \sigma\left[\frac{L/K}{\mathcal{P}}\right]\sigma^{-1}$. Pick $x \in \vartheta_{\sigma\mathcal{P}} = \sigma(\vartheta_{\mathcal{P}})$. Then $\sigma^{-1}x \in \vartheta_{\mathcal{P}}$, which implies that

$$\left[\frac{L/K}{\mathcal{P}}\right]\sigma^{-1}x \equiv \left(\sigma^{-1}x\right)^{N(\wp)} \bmod \mathcal{P}.$$

From the latter we obtain

$$\theta x = \sigma \left[\frac{L/K}{\mathcal{P}}\right] \sigma^{-1} x \equiv \sigma \left(\left(\sigma^{-1} x \right)^{N(\wp)} \right) \bmod \sigma \mathcal{P} = x^{N(\wp)} \bmod \sigma \mathcal{P}.$$

Therefore

$$\sigma \left[\frac{L/K}{\mathcal{P}}\right] \sigma^{-1} = \left[\frac{L/K}{\sigma(\mathcal{P})}\right]. \qquad \square$$

Proposition 11.2.4. *Assume $K \subseteq E \subseteq L$, where E/K is also a Galois extension. Then*

$$\operatorname{res}|_E \left[\frac{L/K}{\mathcal{P}}\right] = \left[\frac{E/K}{\mathcal{P} \cap E}\right].$$

Proof. Let $\theta = \left[\frac{L/K}{\mathcal{P}}\right]$, $\wp = \mathcal{P}|_K$, and $x \in \vartheta_{\mathcal{P} \cap E} = \vartheta_{\mathcal{P}} \cap E$. Then $\theta x - x^{N(\wp)} \in \mathcal{P} \cap E$. $\square$

When $\mathcal{P}$ run through the prime divisors above $\wp$, the Frobenius automorphisms runs through a conjugation class of G (Proposition 11.2.3).

Definition 11.2.5. The *Artin's symbol* $\left(\frac{L/K}{\wp}\right)$ of a place $\wp$ of K is the conjugation class

$$\left(\frac{L/K}{\wp}\right) = \left\{ \sigma \left[\frac{L/K}{\mathcal{P}}\right] \sigma^{-1} \mid \sigma \in G \right\}, \text{ with } \mathcal{P}|_K = \wp.$$

Definition 11.2.6. Let A be a set of places of K. Then the limit ($s \in \mathbb{R}$, $s > 1$)

$$\delta(A) = \lim_{s \to 1^+} \frac{\sum_{\mathcal{P} \in A} (N\mathcal{P})^{-s}}{\sum_{\mathcal{P} \in \mathbb{P}_K} (N\mathcal{P})^{-s}},$$

is called *Dirichlet's density of A*, in case this limit exists.

Proposition 11.2.7. *If A is finite, then $\delta(A) = 0$.*

Proof. Let $\zeta_K(s) = \prod_{\mathcal{P} \in \mathbb{P}_K} \left(1 - (N\mathcal{P})^{-s}\right)^{-1}$. Then $\zeta_K(s)$ has a pole at $s = 1$, so

$$\lim_{s \to 1^+} \prod_{\mathcal{P} \in \mathbb{P}_K} \left(1 - \frac{1}{(N\mathcal{P})^s}\right)^{-1} = \lim_{s \to 1^+} \prod_{\mathcal{P} \in \mathbb{P}_K} \frac{1}{1 - \left(\frac{1}{N\mathcal{P}}\right)^s} = \infty.$$

Therefore

$$\lim_{s \to 1^+} \prod_{\mathcal{P} \in \mathbb{P}_K} \left(1 - \frac{1}{(N\mathcal{P})^s}\right) = 0,$$

which implies that

$$\lim_{s\to 1^+} \sum_{\mathcal{P}\in\mathbb{P}_K} (N\mathcal{P})^{-s} = \infty.$$

Now if A is finite, then $\sum_{\mathcal{P}\in A} (N\mathcal{P})^{-s}$ is uniformly bounded, and we have

$$\delta(A) = \lim_{s\to 1^+} \frac{\sum_{\mathcal{P}\in A} (N\mathcal{P})^{-s}}{\sum_{\mathcal{P}\in\mathbb{P}_K} (N\mathcal{P})^{-s}} = 0. \qquad \square$$

Proposition 11.2.8. *Assume that A, B are disjoint sets of prime divisors such that $\delta(A)$ and $\delta(B)$ exist. Then $\delta(A \cup B) = \delta(A) + \delta(B)$.*

Proof. The statement is an immediate consequence of the definition. □

In what remains of this section, we will use the following notation. Let L/ℓ be a finite Galois extension of K/k with Galois group G and $|k| = q$. Let $x \in K \setminus k$, where $K/k(x)$ is a finite separable extension. Set

$$n = [\ell : k] = [K\ell : K], \quad d = [K : k(x)], \quad m = [L : K\ell],$$
$$P(K) = \left\{\wp \in \mathbb{P}_K \mid \wp|_{k(x)} \neq \wp_\infty\right\}, \quad \operatorname{ord}\tau = o(\tau), \quad \text{for} \quad \tau \in G.$$

Define

$$P_{nr}(K) = \{\wp \in P(K) \mid \wp \text{ is not ramified over } k(x)\}.$$

For $i \in \mathbb{N}$, let

$$P_i(K) = \{\wp \in P_{nr}(K) \mid d_K(\wp) = i\},$$
$$C_i(L/K, \mathfrak{C}) = \left\{\wp \in P_i(K) \mid \left(\frac{L/K}{\wp}\right) = \mathfrak{C}\right\},$$

where $\mathfrak{C}$ is a given conjugation class of G. For $\tau \in G$, let

$$D_i(L/K, \tau) = \left\{\mathcal{P} \in P(L) \mid \left[\frac{L/K}{\mathcal{P}}\right] = \tau, \mathcal{P} \cap K \in P_i(K)\right\}.$$

The Frobenius automorphism of the algebraic closure $\bar{k}$ of k will be denoted by φ. Thus $\varphi : \bar{k} \longrightarrow \bar{k}$ is defined by $\varphi(x) = x^q$.

Let $C = \bigcup_{i=1}^{\infty} C_i(L/K, \mathfrak{C}) = \left\{\wp \in P_{nr}(K) \mid \left(\frac{L/K}{\wp}\right) = \mathfrak{C}\right\}$.

The Čebotarev density theorem states that $\delta(C) = |\mathfrak{C}|/|G|$.

Proposition 11.2.9. *Let $i \in \mathbb{N}$, $\wp \in C_i(L/K, \mathfrak{C})$, and $\tau, \tau' \in \mathfrak{C}$.*

(1) *There are exactly $[L : K]/\operatorname{ord}(\tau)$ prime divisors of $P_{nr}(K)$ that lie above $\wp$.*
(2) *If $C_i' \subseteq C_i(L/K, \mathfrak{C})$ and $D_i'(\tau)$ is the set of prime divisors in $D_i(L/K, \tau)$ lying above C_i', then $|C_i'| = |\mathfrak{C}| \operatorname{ord}(\tau) |D_i'(\tau)| [L : K]^{-1}$.*

Proof.

(1) Let h be the number of prime divisors over $\wp$. We have $d_{L/K}(\mathcal{P}|\wp) = \operatorname{ord}(\tau)$ since $e_{L/K}(\mathcal{P}|\wp) = 1$. Furthermore, by Theorem 5.1.14 $[L:K] = efh = fh = \operatorname{ord}(\tau)\,h$. Hence $h = \frac{[L:K]}{\operatorname{ord}(\tau)}$.

(2) For $\sigma \in G = \operatorname{Gal}(L/K)$, we have $D'_i\left(\sigma\tau\sigma^{-1}\right) = \sigma D'_i(\tau)$. If $\tau' \in \mathfrak{C}$ is distinct from τ, then $D'_i(\tau)$ and $D'_i(\tau')$ are disjoint. Therefore $\bigcup_{\tau'\in\mathfrak{C}} D'_i(\tau')$ is the set of prime divisors of $P_{nr}(L)$ over C'_i. By (1),

$$\left|C'_i\right| \frac{[L:K]}{\operatorname{ord}(\tau)} = \sum_{\tau'\in\mathfrak{C}} \left|D'_i(\tau')\right| = |\mathfrak{C}|\left|D'_i(\tau)\right|. \qquad \square$$

Proposition 11.2.10. *Let T be an intermediate field, i.e., $K \subseteq T \subseteq L$, and let t be the field of constants of T. Let $\tau \in \operatorname{Gal}(L/T)$. Set $|t| = q^r$. If r divides i then*

$$D_i(L/K,\tau) = D_{i/r}(L/T,\tau) \cap \{\mathcal{P} \in P(L) \mid d_K(\mathcal{P}\cap K) = i\}.$$

Proof. Let $\mathcal{P} \in P_{nr}(L)$ be such that $\wp = \mathcal{P}\cap K$ is of degree i. Thus $N\wp = q^i$. Let $\mathfrak{S} = \mathcal{P}\cap T$ be of degree s, that is, $N\mathfrak{S} = (q^r)^s = q^{rs}$. By definition,

$$\left[\frac{L/K}{\mathcal{P}}\right] = \tau \iff \tau x \equiv x^{q^i} \bmod \mathcal{P} \text{ for all } x \in \vartheta_{\mathcal{P}} \tag{11.1}$$

and

$$\left[\frac{L/T}{\mathcal{P}}\right] = \tau \iff \tau x \equiv x^{q^{rs}} \bmod \mathcal{P} \text{ for all } x \in \vartheta_{\mathcal{P}}. \tag{11.2}$$

Thus, it suffices to prove that $\left[\frac{L/K}{\mathcal{P}}\right] = \tau$ implies $rs = i$. Since $\tau \in \operatorname{Gal}(L/T)$, (11.1) implies that $x \equiv x^{q^i} \bmod \mathcal{P}$ for all $x \in \vartheta_{\mathfrak{S}}$. Hence $t(\mathfrak{S}) \subseteq \mathbb{F}_{q^i}$. On the other hand, $t(\mathfrak{S}) \supseteq k(\wp) = \mathbb{F}_{q^i}$, so $t(\mathfrak{S}) = \mathbb{F}_{q^i}$. Finally, we have $t(\mathfrak{S}) = \mathbb{F}_{(q^r)^s}$, i.e., $i = rs$. $\square$

Corollary 11.2.11. *With the hypotheses of Proposition* 11.2.10, *let $\mathfrak{C}$, $\mathfrak{C}'$ be the conjugation classes of $\tau \in G$ and of $\tau \in \operatorname{Gal}(L/T)$ respectively. Assume that r divides i and let*

$$C'_{i/r} = C_{i/r}\left(L/T,\mathfrak{C}'\right) \setminus \{\mathfrak{S} \in P(T) \mid d_K(\mathfrak{S}\cap K) \le i/2\}.$$

Then $|C_i(L/K,\mathfrak{C})| = \frac{|\mathfrak{C}|\left|C'_{i/r}\right|}{|\mathfrak{C}'|[T:K]}$.

Proof. Put $s = \frac{i}{r}$. The set

$$D'_i(\tau) = D_s(L/T,\tau) \cap \{\mathcal{P} \in P(L) \mid d_K(\mathcal{P}\cap K) = i\}$$

is the set of prime divisors in $D_s(L/T,\tau)$ that lie above

$$C_i'' = C_s\left(L/T, \mathfrak{C}'\right) \cap \{\mathfrak{S} \in P(T) \mid d_K(\mathfrak{S} \cap K) = i\}.$$

We have

$$\begin{aligned}
\frac{[L:K]}{|\mathfrak{C}| \operatorname{ord}(\tau)} |C_i(L/K, \mathfrak{C})| &= |D_i(L/K, \tau)| && \text{(Proposition 11.2.9)}\\
&= \left|D_i'(\tau)\right| && \text{(Proposition 11.2.10)}\\
&= \frac{[L:T]}{|\mathfrak{C}'| \operatorname{ord}(\tau)} \left|C_i''\right| && \text{(Proposition 11.2.9).}
\end{aligned}$$

Therefore, $\left|C_i''\right| = \frac{[T:K]|\mathfrak{C}'|}{|\mathfrak{C}|} |C_i(L/K, \mathfrak{C})|$.

By the above, it suffices to prove that $C_i'' = C_{i/r}'$. If $\mathfrak{S} \in P_{nr}(T)$ is of degree s and $\wp = \mathfrak{S} \cap K$, then $k \subseteq k(\wp) \subseteq t(\mathfrak{S}) = \mathbb{F}_{(q^r)^s} = \mathbb{F}_{q^{rs}} = \mathbb{F}_{q^i}$. Therefore $d_K(\wp)$ divides i. It follows that $d_K(\wp) = i$ or $d_K(\wp) \leq \frac{i}{2}$. □

Proposition 11.2.12. *Let $i \in \mathbb{N}$ be such that $\tau|_\ell = \varphi^i|_\ell$ (where φ is the Frobenius automorphism) for all $\tau \in \mathfrak{C}$. Let ℓ' be a finite extension of ℓ and let $L' = L\ell'$. Then L'/K is a Galois extension, the field of constants of L' is ℓ', $g_L = g_{L'}$ and for each $\tau \in \mathfrak{C}$ there exists a unique $\tau' \in \operatorname{Gal}(L'/K)$ such that $\tau'|_L = \tau$ and $\tau'|_{\ell'} = \varphi^i|_{\ell'}$. Furthermore:*

(i) $\operatorname{ord}\left(\tau'\right)$ *is the least common multiple of* $\operatorname{ord}(\tau)$ *and* $\left[\ell' : \ell' \cap \mathbb{F}_{q^i}\right]$,
(ii) $\mathfrak{C}' = \left\{\tau' \mid \tau \in \mathfrak{C}\right\}$ *is a conjugation class of* $\operatorname{Gal}(L'/K)$,
(iii) $C_i\left(L'/K, \mathfrak{C}'\right) = C_i(L/K, \mathfrak{C})$.

Proof. Since L/K and ℓ'/k are Galois extensions and $L' = L\ell'$, it follows that L'/K is a Galois extension. By Theorem 6.1.2, ℓ' is the field of constants of L' and by Theorem 6.1.3, $g_L = g_{L'}$. Now assume that $\tau \in \mathfrak{C}$ and $G' = \operatorname{Gal}(L'/K)$; since $K\ell' \cap L = K\ell$, we have $G = G'/H$, where $H = \operatorname{Gal}(L'/L) \cong \operatorname{Gal}\left(\ell'/\ell\right)$.

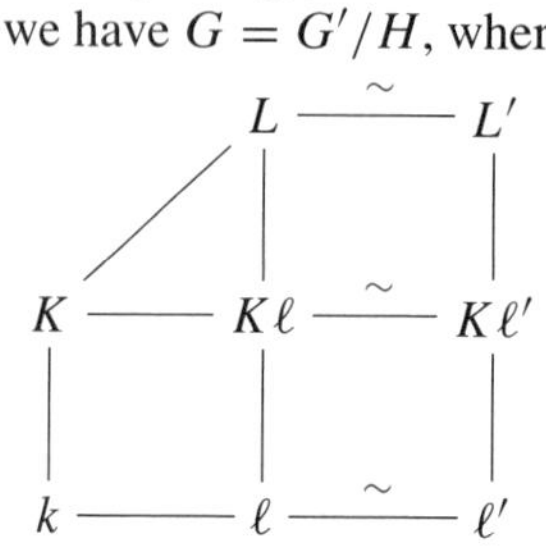

If $\tau_1 \in G'$ is such that $\tau_1|_L = \tau$, then $\tau_1|_\ell = \tau|_\ell = \varphi^i|_\ell$. It follows that $\theta = (\tau_1)^{-1}\varphi^i|_{\ell'} \in \operatorname{Gal}\left(\ell'/\ell\right) \cong H$. Therefore $\tau' = \tau_1\theta$ and $\tau'|_L = \tau_1|_L\theta|_L = \tau_1|_L = \tau$, $\tau'|_{\ell'} = \tau_1|_{\ell'}\theta|_{\ell'} = \varphi^i|_{\ell'}$. This proves the existence of τ'. The uniqueness follows immediately from $L' = L\ell'$. Now, $\operatorname{ord}\left(\tau'|_{\ell'}\right) = \operatorname{ord}\left(\varphi^i|_{\ell'}\right) = \left[\ell' : \ell' \cap \mathbb{F}_{q^i}\right]$ and

$$\operatorname{ord}\left(\tau'\right) = \left[\operatorname{ord}\left(\tau'|_L\right), \operatorname{ord}\left(\tau'|_{\ell'}\right)\right] = \left[\operatorname{ord}(\tau), \left[\ell' : \ell' \cap \mathbb{F}_{q^i}\right]\right],$$

which proves (i).

To establish (ii), let $\mathfrak{C}' = \left\{\tau' \mid \tau \in \mathfrak{C}\right\}$ and $\theta' \in \operatorname{Gal}(L'/K)$. Then

$$\theta'\tau'\left(\theta'\right)^{-1}|_L = \theta'|_L\tau'|_L\left(\theta'\right)^{-1}|_L = \theta\tau\theta^{-1},$$

where $\theta = \theta'|_L$ and $\theta'\tau'\left(\theta'\right)^{-1}|_{\ell'} = \theta'|_{\ell'}\tau'|_{\ell'}\left(\theta'\right)^{-1}|_{\ell'} = \tau'|_{\ell'} = \varphi^i|_{\ell'}$, since $\operatorname{Gal}(\ell'/k)$ is a cyclic group. Now $\theta\tau\theta^{-1} \in \mathfrak{C}$ implies $\theta'\tau'\left(\theta'\right)^{-1} \in \mathfrak{C}'$.

Finally, in order to verify (iii) it suffices to demonstrate the following: Assume that $\mathfrak{S} \in P_{nr}(L')$, $\wp = \mathfrak{S} \cap K$ is of degree i, and $\mathcal{P} = \mathfrak{S}|_L$. Then $\left[\frac{L/K}{\mathcal{P}}\right] = \tau$ if and only if $\left[\frac{L'/K}{\mathfrak{S}}\right] = \tau'$.

To prove this, suppose that $\left[\frac{L/K}{\mathcal{P}}\right] = \tau$. Then $\tau x \equiv x^{q^i} \bmod \mathcal{P}$ for $x \in \vartheta_{\mathcal{P}}$. If $x \in \ell'$, we have $\varphi^i(x) = x^{q^i}$. Furthermore, $\vartheta_{\mathfrak{S}} = \ell' \vartheta_{\mathcal{P}}$ since $L' = L\ell'$ (Exercise 11.7.1). It follows that $\tau' x \equiv x^{q^i} \bmod \mathfrak{S}$ for all $x \in \vartheta_{\mathfrak{S}}$. Therefore $\left[\frac{L'/K}{\mathfrak{S}}\right] = \tau'$.

Conversely, if $\left[\frac{L'/K}{\mathfrak{S}}\right] = \tau'$, then by Proposition 11.2.4,

$$\left[\frac{L'/K}{\mathfrak{S}}\right]\bigg|_L = \tau'|_L = \left[\frac{L/K}{\mathfrak{S} \cap L}\right] = \left[\frac{L/K}{\mathcal{P}}\right]. \qquad \square$$

Corollary 11.2.13. *If $L = K\ell$ is the extension of constants and $\tau \in \mathrm{Gal}(L/K)$ satisfies $\tau|_\ell = \varphi^i|_\ell$, then $C_i\,(K/K, \mathrm{Id}) = C_i\,(L/K, \{\tau\})$.*

Proof. Notice that in the context of Proposition 11.2.12, K plays the role of L and L plays the role of L'. We have $\mathfrak{C} = \{\mathrm{Id}\}$ and $\mathfrak{C}' = \{\tau\}$, so the result follows. $\square$

Proposition 11.2.14. *Suppose that $K\ell = L$ and that $\tau|_\ell = \varphi|_\ell$, $\mathfrak{C} = \{\tau\}$, $\tau \in G$. Then*

$$||C_1\,(L/K, \mathfrak{C})| - q| < 2\left(g_L + g_L d + d^2\right) q^{1/2}$$

Proof. Taking $i = 1$ in the previous corollary, we have $C_1\,(L/K, \{\tau\}) = C_1\,(K/K, \{\mathrm{Id}\}) = P_1(K)$. Here $\mathfrak{C} = \{\tau\}$ since $G = \mathrm{Gal}\,(\ell/k)$ is a cyclic group. We will denote by $\bar{P}_1(K)$ the set of all prime divisors of degree 1. By Theorem 6.1.3 we have $g_L = g_K$, so using the Riemann hypothesis (Theorem 7.2.9 (iv)) we obtain that

$$\left||\bar{P}_1(K)| - (q+1)\right| \leq 2 g_K q^{1/2}.$$

Now by Theorem 9.4.2, we have

$$g_K = 1 + \left(g_{k(x)} - 1\right)[K : k(x)] + \frac{1}{2} d\left(\mathfrak{D}_{K/k(x)}\right) = 1 - d + \frac{1}{2} d\left(\mathfrak{D}_{K/k(x)}\right),$$

and hence $d\left(\mathfrak{D}_{K/k(x)}\right) = 2g_K - 2 + 2d$. This implies that there are at most $2g_L - 2 + 2d$ prime divisors of $k(x)$ that are ramified in K. On the other hand, there exist at most d elements in $\bar{P}_1(K)$ above each element of $\bar{P}_1(k(x))$. Thus there are at most d prime divisors of K above $\mathfrak{N}_x = \wp_\infty$ in $k(x)$. Clearly, none of these divisors belongs to $P_1(K)$, but they could belong to $\bar{P}_1(K)$. Then

$$\bar{P}_1(K) \setminus P_1(K) = \left\{\wp \mid d_K\,(\wp) = 1, \wp \text{ is ramified in } K/k(x) \text{ or } \wp|_{k(x)} = \wp_\infty\right\},$$

so

$$\left|\bar{P}_1(K) \setminus P_1(K)\right| \leq d\,(2g_L - 2 + 2d) + d = d\,(2g_L - 1 + 2d)\,.$$

Therefore

$$\begin{aligned}|P_1(K) - q| &\leq \big||P_1(K)| - |\bar{P}_1(K)|\big| + \big||\bar{P}_1(K)| - q\big| \\ &\leq d\,(2g_L - 1 + 2d) + 2g_L q^{1/2} + 1 < 2\sqrt{q}\left(g_L + g_L d + d^2\right). \quad \square\end{aligned}$$

Proposition 11.2.15. *For each finite extension M of K and for each natural number i, we have*

$$|\{\mathfrak{S} \in P_{nr}(M) \mid d_K\,(\mathfrak{S} \cap K) \leq i/2\}| \leq 4[M : K]\,(g_K + 1)\, q^{i/2}.$$

Proof. For each prime divisor in $P(K)$ there exist at most $[M : K]$ places in $P(M)$. Therefore

$$|\{\mathfrak{S} \in P_{nr}(M) \mid d_K\,(\mathfrak{S} \cap K) \leq i/2\}| \leq [M : K] \sum_{j \leq i/2} \left|P_j(K)\right|.$$

By Theorem 6.2.1, for each $\wp \in P_j(K)$ there exist precisely j divisors of $\mathbb{F}_{q^j} K$ of degree 1. Hence, using the Riemann hypothesis (Theorem 7.2.9), we obtain that

$$\left|P_j(K)\right| \leq \frac{1}{j}\left|\bar{P}_1\left(\mathbb{F}_{q^j} K\right)\right| \leq \frac{1}{j}\left(2g_K q^{j/2} + q^j + 1\right).$$

For $i \geq 4$ we have

$$\sum_{j=1}^{[i/2]} \frac{1}{j} q^j \leq \frac{2}{i} q^{i/2} + \sum_{j=0}^{[i/2]-1} q^j = \frac{2}{i} q^{i/2} + \frac{q^{[i/2]} - 1}{q - 1} \leq 2q^{i/2}.$$

For $i = 1, 2, 3$ we also obtain the inequality.
Similarly, $\sum_{j=1}^{[i/2]} \frac{1}{j}\left(q^j + 1\right) \leq 4q^{i/2}$. Combining all these inequalities, we obtain

$$\begin{aligned}&\left|\{\mathfrak{S} \in P_{nr}(M) \mid d_K\,(\mathfrak{S} \cap K) \leq i/2\}\right| \\ &\quad\leq [M : K] \sum_{j \leq i/2} \frac{1}{j}\left(2g_K q^{j/2} + q^j + 1\right) \\ &\quad\leq [M : K]\left\{2g_K\left(2q^{i/2}\right) + 4q^{i/2}\right\} = 4[M : K] q^{i/2}\,(g_K + 1)\,. \quad \square\end{aligned}$$

Now we will prove the following result, from which the Čebotarev density theorem will be an immediate consequence.

Proposition 11.2.16. *Let $a \in \mathbb{N}$ be such that $\tau|_\ell = \varphi^a|_\ell$ for all $\tau \in \mathfrak{C}$. If $i \not\equiv a \bmod n$, then $C_i\,(L/K, \mathfrak{C}) = \emptyset$. If $i \equiv a \bmod n$, then*

$$\left|C_i\,(L/K, \mathfrak{C}) - \frac{|\mathfrak{C}|}{im} q^i\right| < 4\,|\mathfrak{C}|\left(d^2 + \frac{1}{2} g_L d + \frac{1}{2} g_L + g_K + 1\right) q^{i/2}.$$

Proof. Since $\tau|_\ell = \varphi^a|_\ell$, if $\mathcal{P} \in P(L)$ is above $\wp \in C_i\,(L/K, \mathfrak{C})$, we have

$$\varphi^a|_\ell = \left[\frac{L/K}{\mathcal{P}}\right]\Bigg|_\ell = \varphi^i|_\ell.$$

This shows that if $C_i\,(L/K, \mathfrak{C}) \neq \emptyset$, then we necessarily have $i \equiv a \bmod n$.

Now assume that $i \equiv a \bmod n$. We substitute ℓ by a finite extension ℓ' such that $i\ \operatorname{ord}(\tau)$ divides $[\ell' : k]$. Set $L' = L\ell'$. Since L'/L is an extension of constants, we have $K\ell' \cap L = K\ell$. Therefore $[L' : K\ell'] = [L : K\ell] = m$ and $g_{L'} = g_L$. Furthermore, by Proposition 11.2.12 there exists a unique $\tau' \in \operatorname{Gal}(L'/K)$ such that $\tau'|_L = \tau$, $\tau'|_{\ell'} = \varphi^i|_{\ell'}$,

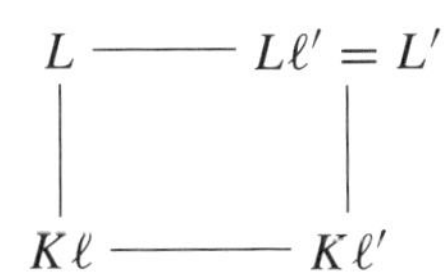

$$\operatorname{ord}(\tau') \text{ is the least common multiple of } \{\operatorname{ord}(\tau), [\ell' : \mathbb{F}_{q^i}]\} = [\ell' : \mathbb{F}_{q^i}],$$

and

$$C_i\,(L'/K, \mathfrak{C}'') = C_i\,(L/K, \mathfrak{C}), \tag{11.3}$$

where $\mathfrak{C}''$ is the conjugacy class of τ' in $\operatorname{Gal}(L'/K)$ and $|\mathfrak{C}''| = |\mathfrak{C}|$.

We substitute L by L' and take T to be the fixed field of L' under $\langle \tau' \rangle$, as in Proposition 11.2.10. Then $K \subseteq T \subseteq L'$ and

$$D_j\,(L'/K, \tau') = D_{j/r}\,(L'/T, \tau) \cap \{\mathcal{P} \in P(L') \mid d_K\,(\mathcal{P} \cap K) = j\}.$$

Here t is the field of constants of T, $|t| = q^r$, and r divides i. Observe that $t = \ell' \cap T$ is the fixed field of ℓ' under φ^i and therefore equal to $\mathbb{F}_{q^i}$.

Then $[\ell' : \mathbb{F}_{q^i}] = [L' : T] = \operatorname{ord}(\tau')$. In particular, $T\ell' = L'$, so $[T : \mathbb{F}_{q^i}K] = [L' : \ell' K] = m$, and $[T : K] = [T : \mathbb{F}_{q^i}K][\mathbb{F}_{q^i}K : K] = mi$.

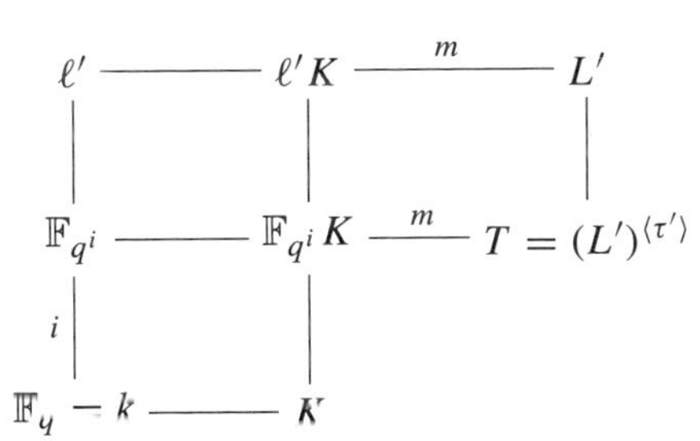

Now $T = L'^{\langle \tau' \rangle}$, so if we substitute L by L' in Corollary 11.2.11, we obtain $|\mathfrak{C}'| = 1$ for $r = i$. By Proposition 11.2.15 and Corollary 11.2.11,

$$\begin{aligned}
&\left|\frac{|\mathfrak{C}|}{[T:K]}\left|C_1\left(L'/T,\{\tau\}\right)\right|-\left|C_i\left(L'/K,\mathfrak{C}''\right)\right|\right| \\
&\quad=\frac{|\mathfrak{C}|}{[T:K]}\left(\left|C_1\left(L'/T,\{\tau\}\right)\right|-\left|C_1'\right|\right) \\
&\quad\le\frac{|\mathfrak{C}|}{[T:K]}\left|\{\mathfrak{S}\in P_{nr}(T)\mid d_K\left(\mathfrak{S}\cap K\right)\le i/2\}\right|\le 4\,|\mathfrak{C}|\,(g_K+1)\,q^{i/2}.
\end{aligned} \tag{11.4}$$

By Proposition 11.2.14,

$$\left|\left|C_1\left(L'/T,\{\tau\}\right)\right|-q^i\right|<2\left(g_L+g_Ld+d^2\right)q^{i/2}.$$

Multiplying the last inequality by $|\mathfrak{C}|/(im)$, where $im=[T:K]$, we obtain

$$\left|\frac{|\mathfrak{C}|}{[T:K]}\left|C_1\left(L'/T,\{\tau\}\right)\right|-\frac{|\mathfrak{C}|}{im}q^i\right|\le\frac{2\,|\mathfrak{C}|}{im}\left(g_L+g_Ld+d^2\right)q^{i/2}. \tag{11.5}$$

Hence by (11.3), (11.4), and (11.5) we get

$$\begin{aligned}
&\left|\,|C_i\left(L/K,\mathfrak{C}\right)|-\frac{|\mathfrak{C}|}{im}q^i\right| \\
&\qquad\le\left|\left|C_i\left(L'/K,\mathfrak{C}''\right)\right|-\frac{|\mathfrak{C}|}{[T:K]}\left|C_1\left(L'/T,\{\tau\}\right)\right|\right| \\
&\qquad\quad+\left|\frac{|\mathfrak{C}|}{[T:K]}\left|C_1\left(L'/T,\{\tau\}\right)\right|-\frac{|\mathfrak{C}|}{im}q^i\right| \\
&\qquad\le 4\,|\mathfrak{C}|\,(g_K+1)\,q^{i/2}+\frac{2\,|\mathfrak{C}|}{im}\left(g_L+g_Ld+d^2\right)q^{i/2} \\
&\qquad=4\,|\mathfrak{C}|\left(g_K+1+\frac{g_L}{2im}+\frac{g_Ld}{2im}+\frac{d^2}{2im}\right)q^{i/2} \\
&\qquad<4\,|\mathfrak{C}|\left(d^2+\frac{1}{2}g_Ld+\frac{1}{2}g_L+g_K+1\right)q^{i/2}.
\end{aligned}$$

□

Notation 11.2.17. For two functions $f(x)$ and $g(x)$ of a real variable, we will write $f(x)=O(g(x))$ as $x\to c$ to express the fact that $|f(x)|\le M|g(x)|$ when x is in a neighborhood of c. In particular, if $g(x)=1$, $f(x)=O(1)$ means that $f(x)$ is bounded in a neighborhood of c (see Notation 7.3.3).

Proposition 11.2.18.

$$\sum_{j=1}^{\infty}\frac{x^{a+jn}}{a+jn}=-\frac{1}{n}\ln(1-x)+O(1)\quad\text{when}\quad x\to 1^-.$$

Proof. If ξ is an nth root of 1, then ξ is distinct from 1 and satisfies $1+\xi+\xi^2+\cdots+\xi^{n-1}=0$. Therefore

$$
\begin{aligned}
-\frac{1}{n}\sum_{i=0}^{n-1}\ln\left(1-\xi^i x\right)\xi^{-ia} &= \frac{1}{n}\sum_{i=0}^{n-1}\sum_{j=1}^{\infty}\frac{\left(\xi^i x\right)^j}{j}\xi^{-ia}\\
&= \frac{1}{n}\sum_{j=1}^{\infty}\frac{x^j}{j}\left(\sum_{i=0}^{n-1}\xi^{i(j-a)}\right)\\
&= \frac{1}{n}\sum_{j\equiv a \bmod n}\frac{x^j}{j}n = \sum_{j\equiv a \bmod n}\frac{x^j}{j} = \sum_{t=1}^{\infty}\frac{x^{a+tn}}{a+tn}.
\end{aligned}
$$

Since for $1 < i \leq n-1$, $\ln\left(1-\xi^i x\right)$ is bounded in a neighborhood of 1, the result follows. □

Proposition 11.2.19. *If $a \in \mathbb{N}$ is such that $0 < a \leq n$ and $\tau|_\ell = \varphi^a|_\ell$ for all $\tau \in \mathfrak{C}$, then*

$$
\sum_{\mathcal{P}\in C}(N_\mathcal{P})^{-s} = -\frac{|\mathfrak{C}|}{[L:K]}\ln\left(1-q^{1-s}\right)+O(1), \quad s\to 1^+.
$$

Proof. Recall that $C = \bigcup_{i=1}^{\infty} C_i\,(L/K,\mathfrak{C})$. We have

$$
\begin{aligned}
\sum_{\mathcal{P}\in C}\frac{1}{(N\mathcal{P})^s} &= \sum_{j=0}^{\infty}\ \sum_{\mathcal{P}\in C_{a+jn}(L/K,\mathfrak{C})}(N\mathcal{P})^{-s}\\
&= \sum_{j=0}^{\infty}\left(\frac{|\mathfrak{C}|}{m(a+jn)}q^{a+jn}+O\left(q^{\frac{1}{2}(a+jn)}\right)\right)q^{-(a+jn)s}\\
&\qquad\qquad \text{(Proposition 11.2.16)}\\
&= \frac{|\mathfrak{C}|}{m}\sum_{j=0}^{\infty}\frac{q^{(1-s)(a+jn)}}{a+jn}+O\left(q^{(\frac{1}{2}-s)a}\sum_{j=0}^{\infty}q^{(\frac{1}{2}-s)jn}\right)\\
&= -\frac{|\mathfrak{C}|}{mn}\ln\left(1-q^{1-s}\right)+O(1)+O\left(\frac{q^{(\frac{1}{2}-s)a}}{1-q^{(\frac{1}{2}-s)n}}\right)\\
&\qquad\qquad \text{(Proposition 11.2.18 with } x = q^{1-s})\\
&= -\frac{|\mathfrak{C}|}{[L:K]}\ln\left(1-q^{1-s}\right)+O(1), \quad s\to 1^+.
\end{aligned}
$$

□

Theorem 11.2.20 (Čebotarev's Density Theorem). *Let L/K be a finite Galois extension of congruence function fields and let $\mathfrak{C}$ be a conjugacy class of* $\operatorname{Gal}(L/K)$. *Then the Dirichlet density of the set*

$$
\left\{\wp\in\mathbb{P}_K \mid \left(\frac{L/K}{\wp}\right)=\mathfrak{C}\right\}
$$

exists and is equal to $\frac{|\mathfrak{C}|}{[L:K]}$.

Proof. In Proposition 11.2.19 we take $L = K$ and obtain

$$\sum_{\wp \in \mathbb{P}_K} (N\wp)^{-s} = -\ln\left(1 - q^{1-s}\right) + O(1), \quad s \to 1^+.$$

Since the number of prime divisors of $k(x)$ above $\wp_\infty$ is finite and so is the number of ramified prime divisors, then the Dirichlet density of the set $\left\{\wp \in \mathbb{P}_K \mid \left(\frac{L/K}{\wp}\right) = \mathfrak{C}\right\}$ is equal to the density of $C = \bigcup_{i=1}^{\infty} C_i\,(L/K, \mathfrak{C})$. Hence by Propositions 11.2.7 and 11.2.8 we have

$$\begin{aligned}\delta\left(\left\{\wp \in \mathbb{P}_K \mid \left(\frac{L/K}{\wp}\right) = \mathfrak{C}\right\}\right) &= \delta\left(C\right) = \lim_{s \to 1^+} \frac{\sum_{\wp \in C} (N\wp)^{-s}}{\sum_{\wp \in \mathbb{P}_K} (N\wp)^{-s}} \\ &= \lim_{s \to 1^+} \frac{-\frac{|\mathfrak{C}|}{[L:K]}\ln\left(1 - q^{1-s}\right) + O(1)}{-\ln\left(1 - q^{1-s}\right) + O(1)} \\ &= \frac{|\mathfrak{C}|}{[L:K]}.\end{aligned}$$

□

11.3 Inverse Limits and Profinite Groups

Definition 11.3.1. By a *directed partially ordered set* or a *directed poset* we understand a nonempty partially ordered set I such that if $i, j \in I$, there exists $k \in I$ satisfying $i \le k$ and $j \le k$.

Now suppose that I is an ordered set such that to any $i \in I$ is associated a set A_i (which might be just a set, a group, a ring, a field, a topological space, etc.) in such a way that whenever $i \le j$, there exists a map

$$\phi_{ji} : A_j \longrightarrow A_i$$

which, depending on A_i, is a map, a group homomorphism, a ring homomorphism, a continuous map, etc., such that

(i) $\phi_{ii} = \mathrm{Id}_{A_i}$,
(ii) $\phi_{ji} \circ \phi_{kj} = \phi_{ki}$ for $i \le j \le k$.

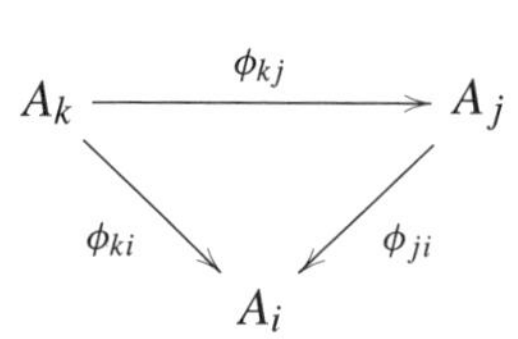

Definition 11.3.2. The system $\{A_i, \phi_{ji}, I\}_{\substack{i,j \in I\\ i \le j}}$ above is called an *inverse system* or a *projective system*.

Definition 11.3.3. Given an inverse system $\{A_i, \phi_{ji}, I\}$ we say that $(X, \varphi_i)_{i \in I}$ is an *inverse limit* of the system if there exist maps (group homomorphisms, continuous maps, etc.)

$$\varphi_i : X \longrightarrow A_i$$

for all $i \in I$ such that $\phi_{ji} \circ \varphi_j = \varphi_i$ whenever $i \leq j$

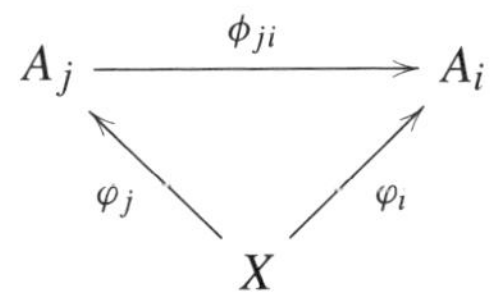

and such that if $(Y_i, \xi_i)_{i \in I}$ is any other object with maps

$$\xi_i : Y \longrightarrow A_i$$

for all $i \in I$ such that $\phi_{ji} \circ \xi_j = \xi_i$ whenever $i \leq j$, then there exists a unique map (group homomorphism, continuous map, etc.)

$$\xi : Y \longrightarrow X$$

such that $\varphi_i \circ \xi = \xi_i$ for all $i \in I$.

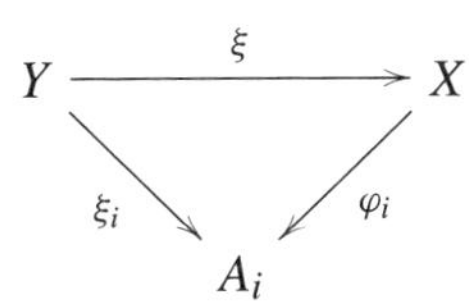

We write $X = \varprojlim_{i \in I} A_i = \varprojlim_{i} A_i = \varprojlim A_i$.

Theorem 11.3.4. *Given an inverse system $\{A_i, \phi_{ji}, I\}$, there exists an inverse limit $(X, \varphi_i)_{i \in I}$, $X = \varprojlim_{i} A_i$. Furthermore, $(X, \varphi_i)_{i \in I}$ is unique in the following sense: if $(Z, \theta_i)_{i \in I}$ is another inverse limit, there exists a unique map $\alpha : X \to Z$ (α group homomorphism, continuous map, etc.) such that α is an isomorphism satisfying $\theta_i \circ \alpha = \varphi_i$ for all $i \in I$.*

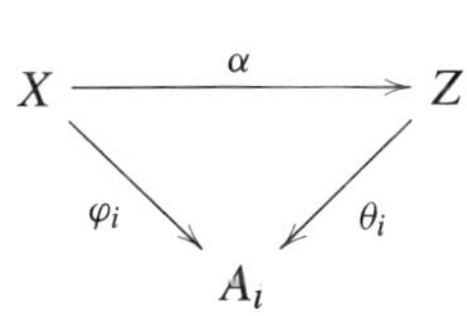

Proof: First we prove uniqueness. Since X and Z are both inverse limits, there exist unique maps $\alpha : X \to Z$ and $\beta : Z \to X$ such that the following diagrams commute:

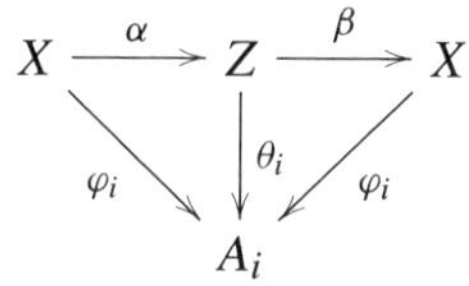

Thus $\beta \circ \alpha$ and Id_X satisfy $\varphi_i \circ (\beta \circ \alpha) = \varphi_i = \varphi_i \circ (\mathrm{Id}_X)$. By the uniqueness, we have $\beta \circ \alpha = \mathrm{Id}_X$. Similarly, $\alpha \circ \beta = \mathrm{Id}_Z$. It follows that α and β are inverse isomorphisms (of groups, rings, topological spaces, etc.).

To see the existence, let $B = \prod_{i \in I} A_i$ be the direct product, considered with the product topology (and with the algebraic operations defined componentwise).

Let $X = \{(a_i)_{i \in I} \in B \mid a_i = \varphi_{ji}(a_j) \text{ for all } i \leq j\}$. Let $\varphi_i : X \to A_i$ be the map induced by the projection ($\varphi_i = \pi_i|_X$):

$$\begin{aligned} \pi_i : \prod_{j \in I} A_j &\longrightarrow A_i \\ (a_j)_{j \in I} &\longmapsto a_i . \end{aligned}$$

Then $(\phi_{ji} \circ \varphi_j)((a_k)_{k \in I}) = \phi_{ji}(a_j) = a_i = \varphi_i((a_k)_{k \in I})$ for all $(a_k)_{k \in I} \in X$. Assume that $(Y, \xi_i)_{i \in I}$ is another object such that the maps $\xi_i : Y \to A_i$ satisfy $\phi_{ji} \circ \xi_j = \xi_i$ for all $i \leq j$. Let

$$\xi : Y \to X$$

be defined by

$$\xi(y) = (\xi_i(y))_{i \in I}.$$

Notice that ξ is well defined since $(\phi_{ji})(\xi_j(y)) = \xi_i(y)$ and we have

$$(\varphi_i \circ \xi)(y) = \varphi_i((\xi_k(y)_{k \in I})) = \xi_i(y),$$

so $(\xi(y))_{i \in I} \in X$. Thus X is an inverse limit of $\{A_i, \phi_{ji}, I\}$. □

Remark 11.3.5. Given an inverse system $\{A_i, \phi_{ij}, I\}$, we denote by $A := \prod_{i \in I} A_i$ the direct product. Then

$$\varprojlim_{i \in I} A_i = \left\{ (\ldots, a_i, \ldots) \in A \mid \phi_{kj}(a_k) = a_j \text{ for all } j \leq k \right\}$$

is the inverse limit or the projective limit.

Given an inverse system $\{A_i, \phi_{ji}, I\}$, let

$$\begin{aligned} \pi_i : A &\longrightarrow A_i \\ (a_j)_{j\in I} &\longmapsto a_i \end{aligned}$$

be the natural projection. For each $i \in I$, let

$$\phi_i := \pi_i|_{\varprojlim_i A_i} : \varprojlim_i A_i \longrightarrow A_i$$

be the map induced by the projection. We have $\phi_{jk} \circ \phi_j = \phi_k$ for $k \le j$.

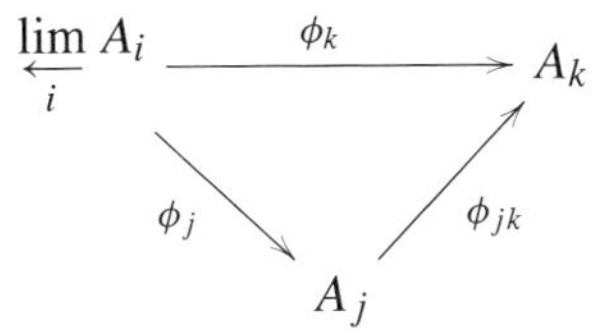

Now if for each $i \in I$, A_i is a topological Hausdorff space, we provide A with the product topology and $\varprojlim_i A_i$ is a topological space with the induced topology. We always assume that the maps ϕ_{ji} are continuous.

Notice that the maps ϕ_i are always continuous; indeed, if U is an open set of A_i, we have

$$\phi_i^{-1}(U) = \pi_i^{-1}(U) \cap \varprojlim_i A_i,$$

where $\pi_i^{-1}(U)$ is an open set by definition of the product topology. In fact, the topology of $\varprojlim_i A_i$ is generated by unions and finite intersections of the sets $\phi_i^{-1}(U_i)$ such that U_i is open in A_i. Furthermore, if T is open in $\varprojlim_i A_i$, we shall see that T contains some $\phi_k^{-1}(U_k)$ for some k and some U_k that is open in A_k. Since T is generated by unions and finite intersections of sets of the form

$$\pi_j^{-1}(U_j) \cap \varprojlim A_i,$$

it suffices to see that

$$\phi_i^{-1}(U_i) \cap \phi_j^{-1}(U_j) = \phi_k^{-1}(U_k) \quad \text{for some} \quad k.$$

Choose $k \ge i, j$ and let

$$U_k := \phi_{kj}^{-1}(U_j) \cap \phi_{ki}^{-1}(U_i).$$

Then

$$\phi_k^{-1}(U_k) = \phi_k^{-1}(\phi_{kj}^{-1}U_j)) \cap \phi_k^{-1}(\phi_{ki}^{-1}(U_i)) = \phi_j^{-1}(U_j) \cap \phi_i^{-1}(U_i).$$

Definition 11.3.6. Let I be a directed poset. Let I' be a subset such that I' is also a directed poset with the order induced by the one in I. We say that I' is *cofinal* in I if for every $i \in I$, there exists $i' \in I'$ such that $i \leq i'$.

If $\{A_i, \phi_{ji}, I\}$ is an inverse system, then $\{A_i, \phi_{ji}, I'\}$ becomes an inverse system and we say that $\{A_i, \phi_{ji}, I'\}$ is a cofinal subsystem of $\{A_i, \phi_{ji}, I\}$.

Theorem 11.3.7. *If $\{A_i, \phi_{ji}, I\}$ is an inverse system of groups, compact topological spaces, or compact topological groups, and $I' \subseteq I$ is cofinal in I, then*

$$\varprojlim_{i \in I} A_i \cong \varprojlim_{i \in I'} A_i .$$

Proof: Let $X := \left(\varprojlim_{i \in I} A_i, \varphi_i\right)$ and $Y := \left(\varprojlim_{i' \in I'} A_{i'}, \varphi'_{i'}\right)$. For $j \in I$, let $j' \in I'$ be such that $j \leq j'$. We define

$$\widetilde{\varphi}_j : Y \to A_j$$

by $\widetilde{\varphi}_j := \phi_{j'j} \circ \varphi'_{j'}$.

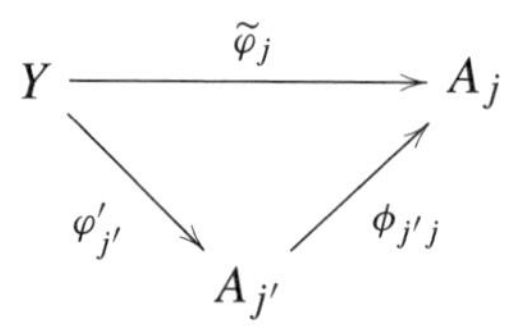

If $k \in I'$ satisfies $j \leq k$, let $\ell \in I$ be such that $j', k \leq \ell$. Then

$$\phi_{j'j}\varphi'_{j'} = \phi_{j'j}\phi_{\ell j'}\varphi'_{\ell} = \phi_{\ell j}\varphi'_{\ell} = \phi_{kj}\phi_{\ell k}\varphi'_{\ell} = \phi_{kj}\varphi'_{k}.$$

Thus $\widetilde{\varphi}_j$ is independent of the choice of $j' \in I'$. Furthermore, if $i, j \in I$ and $i \leq j$, then if $k \in I'$ satisfies $j \leq k$, we have

$$\phi_{ji}\widetilde{\varphi}_j = \phi_{ji}\phi_{kj}\varphi'_k = \phi_{ki}\varphi'_k = \widetilde{\varphi}_i .$$

Therefore, there exists a unique map

$$\overline{\varphi} : Y \longrightarrow X$$

such that $\varphi_j\overline{\varphi} = \widetilde{\varphi}_j$ for all $j \in I$. If $(a'_i)_{i' \in I'} \in Y$ and $\overline{\varphi}((a'_{i'})_{i' \in I'}) = (b_i)_{i \in I}$, then $b_{i'} = a_{i'}$ for $i' \in I'$. It follows that $\overline{\varphi}$ is an injection.

Now if $(b_i)_{i \in I} \in X$, we define $(a_{i'})_{i' \in I'} \in Y$ by $a_{i'} = b_{i'}$ for all $i' \in I'$. Then $\overline{\varphi}((a_{i'})_{i' \in I'}) = (b_i)_{i \in I}$ since I' is cofinal in I and $\overline{\varphi}$ is a surjection. In the case of an algebraic structure, $\overline{\varphi}$ is an isomorphism. In the case of compact topological spaces, $\overline{\varphi}$ is a continuous bijection and since X and Y are compact spaces, it follows that $\overline{\varphi}$ is a closed map and that X and Y are homeomorphic. □

Theorem 11.3.8. *Let $\{A_i, \phi_{ji}, I\}$ be an inverse system of nonempty compact Hausdorff topological spaces A_i over a directed poset I. Then the set $\varprojlim A_i$ is nonempty. In particular, the inverse limit of an inverse system of nonempty finite sets is nonempty.*

Proof: For each $j \in I$, let $Y_j = \{(a_i) \in \prod A_i \mid \phi_{jk}(a_j) = a_k \text{ for all } k \leq j\}$.

By the axiom of choice, Y_j is nonempty. Note that $Y_j \supseteq Y_{j'}$ for $j \leq j'$. In particular, the intersection of finitely many Y_j's is a nonempty set. Since $\prod_{i \in I} A_i$ is a compact topological space, $\bigcap_{j \in I} Y_j$ is nonempty. Now

$$\bigcap_{j \in I} Y_j = \varprojlim_i A_i,$$

so the result follows. □

Proposition 11.3.9. *The set $\varprojlim_i A_i$ is closed in $A = \prod_{i \in I} A_i$.*

Proof: Let $(a_i)_{i \in I} \in A \setminus \varprojlim_i A_i$. There exist $i \leq j$ such that $\phi_{ji}(a_j) \neq a_i$. Since A_i is Hausdorff, we can find open neighborhoods U of $\phi_{ji}(a_j)$ and V of a_i such that $U \cap V = \emptyset$. Set $W := \phi_{ji}^{-1}(U)$. Then W is an open set of A_j. Let $\widetilde{U} = V \times W \times \prod_{k \neq i, j} A_k \subseteq A$. Clearly, $\widetilde{U}$ is an open set of A, and $(a_i)_{i \in I} \in \widetilde{U}$. Moreover, since $\phi_{ji}(W) \subseteq U$ and $U \cap V = \emptyset$, we have $\widetilde{U} \cap \varprojlim_i A_i = \emptyset$. It follows that $\varprojlim_i A_i$ is closed in A. □

Definition 11.3.10. A group G is called a *topological group* if G is a topological space such that the group operations

$$\begin{aligned} i \colon G &\longrightarrow G \\ x &\longmapsto x^{-1} \end{aligned} \qquad \text{and} \qquad \begin{aligned} \cdot \colon G \times G &\longrightarrow G \\ (x, y) &\longmapsto x \cdot y \end{aligned}$$

are continuous.

Proposition 11.3.11. *Let G be a topological group. Then G is Hausdorff if and only if $\{e\}$ is closed in G, where e denotes the identity of G.*

Proof:
($\Rightarrow$) Since G is T_2, it is T_1.
($\Leftarrow$) Let

$$\begin{aligned} \varphi \colon G \times G &\longrightarrow G \\ (x, y) &\longmapsto xy^{-1}. \end{aligned}$$

Since $\varphi = \cdot(\mathrm{Id}, i)$, it follows that φ is continuous. Furthermore,

$$\varphi^{-1}(\{e\}) = \{(x, y) | xy^{-1} = e\} = \{(x, x) | x \in G\} = \Delta,$$

and therefore the diagonal Δ is closed in $G \times G$. Thus G is a Hausdorff space. □

Now for each $x \in G$, the map

$$\begin{aligned} \xi_x : G &\longrightarrow G \\ y &\longmapsto xy \end{aligned}$$

is continuous and satisfies $\xi_x^{-1} = \xi_{x^{-1}}$ (because $(\xi_{x^{-1}} \circ \xi_x)(y) = x^{-1}(xy) = y$). Thus ξ_x is a homeomorphism and V is a open neighborhood of e if and only if $\xi_x(V) = xV$ is an open neighborhood of $\{x\}$. This means that the topology of G is determined by the neighborhoods of $\{e\}$.

Definition 11.3.12. A *profinite group* is a topological group G that is Hausdorff, compact, and contains a basis of open neighborhoods of $\{e\}$ that consists of normal subgroups of G.

Theorem 11.3.13. *Let G be a compact Hausdorff topological group. Then G contains a basis of open neighborhoods of $\{e\}$ consisting of normal subgroups if and only if G is totally disconnected (that is, every element of G is its own connected component).*

Proof:
($\Rightarrow$) Let $x \neq e$. Since G is a Hausdorff space, there exist open sets U and V such that $e \in U$, $x \in V$, and $U \cap V = \emptyset$. Let N be a normal subgroup of G. Then N is open and contained in U. We have

$$G = \left(\bigcup_{g \notin N} gN \right) \cup N.$$

Thus $x \in \bigcup_{g \in N} gN = W$, which is an open set. Moreover, $W \cap N = \emptyset$ and $W \cup N = G$. Thus the connected component of $\{e\}$ is $\{e\}$.

Now for any $y \in G$, the map

$$\begin{aligned} \xi_y : G &\longrightarrow G \\ Z &\longmapsto yZ \end{aligned}$$

is a homeomorphism. Therefore the connected component of y is the image under ξ_y of the connected component of $\{e\}$, namely $\xi_y(\{e\}) = \{y\}$. It follows that G is totally disconnected.
($\Leftarrow$) Assume that G is a totally disconnected topological group. Let V be an open set of G containing e. Then $V^c := G \setminus V$ is a closed set and $e \notin V^c$. Since G is a compact space, it follows that V^c is also compact. On the other hand, G is a Hausdorff space, so for each $x \in V^c$ there exist open sets W_x, U_x, such that $e \in W_x$, $x \in U_x$, and $W_x \cap U_x = \emptyset$. Thus $V^c \subseteq \bigcup_{x \in V^c} U_x$. Since V^c is a compact set, there exist $x_1, \ldots, x_n \in V^c$ such that $V^c \subseteq U := \bigcup_{i=1}^n U_{x_i}$.

Let $W := \bigcap_{i=1}^n W_{x_i}$. Then $e \in W$ and $W \cap U = \emptyset$, so $W \subseteq U^c$ and U^c is a closed set. It follows that $\overline{W} \subseteq U^c$.

Therefore

$$\emptyset = \overline{W} \cap U \supseteq \overline{W} \cap V^c,$$

and $\overline{W} \subseteq V$. Thus, there exists an open neighborhood W of e such that $\overline{W} \subseteq V$ and $\overline{W}$ is a compact set.

Next we show that $\{e\} = \bigcap_{U \in \mathcal{A}} U$, where

$$\mathcal{A} = \{U \mid e \in U \text{ and } U \text{ is open and closed in } G\}.$$

Let $A = \bigcap_{U \in \mathcal{A}} U \supseteq \{e\}$. It suffices to show that A is connected. Assume that $A = C \cup D$, $C \cap D = \emptyset$, and C and D are closed in A (and therefore closed in G). Since G is Hausdorff (therefore a normal space) and C and D are disjoint compact subsets, there exists open subsets C' and D' in G such that $C' \supset C$, $D' \supseteq D$, and $C' \cap D' = \emptyset$. Now $A \subseteq C' \cup D'$, so $(C' \cup D')^c \subseteq A^c = \bigcup_{U \in \mathcal{A}} U^c$. Now since $(C' \cup D')^c$ is closed and compact and U^c is open, $U \in \mathcal{A}$, it follows that there exist finitely many $U_1, \ldots, U_n \in \mathcal{A}$ such that

$$(C' \cup D')^c \subseteq \bigcup_{i=1}^{n} U_i^c \quad \text{or} \quad \bigcap_{i=1}^{n} U_i = P \subseteq C' \cup D',$$

P is open and closed in G. Now $x \in P = (P \cap C') \cup (P \cap D')$, say $x \in P \cap C'$, which is open. Also $P \cap D'$ is open. Since $C' \cap D' = \emptyset$, we have $P \cap C' = P \setminus (P \cap D') = P \cap (P \cap D')^c$. Hence $P \cap C'$ is also a closed subset of G. It follows that $P \cap C' \in \mathcal{A}$ and $A \subseteq P \cap C'$. Therefore $A \cap D \subseteq A \cap D' = \emptyset$. Then A is connected and $A = \bigcap_{U \in \mathcal{A}} U = \{e\}$.

Next we show that if W is an open neighborhood of x, there exists a closed domain P (that is, P is an open and closed set) such that $\{e\} \subseteq P \subseteq W$. Now W is closed and $W^c \subseteq \{e\}^c = \bigcup_{U \in \mathcal{A}} U^c$, with U^c an open set. Since W^c is compact, there exist finitely many $U_1, \ldots, U_n$ of $\mathcal{A}$ such that $W^c \subseteq \bigcup_{U \in \mathcal{A}} U_i^c$. Thus $P' := \bigcap_{i=1}^n U_i \subseteq W$ is a closed domain and $x \in P' \subseteq W$.

Let $Q = \{q \in G \mid P'q \subseteq P'\}$ and $H = Q \cap Q^{-1}$, take $q \in Q$ and $x \in P'$. Then $xq \in P'$ and since P' is open, it follows by the continuity of the product that there exist open sets U_x and V_x containing x and q respectively, $U_x, V_x \subseteq P'$ such that $U_x V_x \subseteq P'$. Since P' is closed and thus compact, and $P' = \bigcup_{x \in P'} U_x$, there exist $x_1, \ldots, x_m \in P'$ such that $P' = \bigcup_{i=1}^m U_{x_i}$. Let $V' = \bigcap_{i=1}^m V_{x_i}$. Then $q \in V'$ and $P'V' \subseteq P'$, so $V' \subseteq Q$. It follows that Q is open.

Now let $r \in G \setminus Q$. There exists $p \in P'$ such that $pr \notin P'$. Since $G \setminus P'$ is an open set and the product is a continuous map, there exists an open neighborhood W of r such that $pW' \subseteq G \setminus P'$. Therefore $W' \subseteq G \setminus Q$, $G \setminus Q$ is open, and hence Q is closed. Since Q^{-1} is homeomorphic to Q, it follows that $H = Q \cap Q^{-1}$ is an open and closed set of G.

For $y \in Q$, we have $y = ey \in P'$, so $Q \subseteq P'$. Also, $P'e = P' \subseteq P'$, and hence $e \in Q$.

Let $h_1, h_2 \in H$. Then $h_1 \in Q$, $h_2^{-1} \in Q$, and

$$P'(h_1 h_2^{-1}) = (P'h_1)h_2^{-1} \subseteq P'h_2^{-1} \subseteq P'.$$

Therefore $h_1h_2^{-1} \in Q$. Similarly, $(h_1h_2^{-1})^{-1} = h_2h_1^{-1} \in Q$, so $h_1h_2^{-1} \in Q^{-1}$. It follows that $h_1h_2^{-1} \in H$ and H is a subgroup of G.

We have shown that H is an open and closed subgroup of G. Finally, since G is compact and $G = \bigcup_{x \in G} Hx$, where Hx is open for each $x \in G$, it follows that H is of finite index in G, $[G : H] = t < \infty$, and $G = \bigcup_{i=1}^{t} Hx_i$. Let $N = \bigcap_{x \in G} xHx^{-1} = \bigcap_{i=1}^{t} x_i H x_i^{-1}$. Then N is a normal subgroup of G, and we have $e \in N \subseteq H \subseteq W \subseteq \overline{W} \subseteq V$. Furthermore, N is an open and closed normal subgroup of G of finite index. This proves the theorem. □

Remark 11.3.14. If G is a finite group, then G is a topological group with the discrete topology. Clearly G is a profinite group.

The term *profinite group* comes from the following theorem.

Theorem 11.3.15. *Let G be a profinite group. Then if N runs through all open normal subgroups of G, we have*

$$G \cong \varprojlim_{N} G/N$$

algebraically and topologically (note that G/N is finite), that is, G is the inverse limit of finite groups.

Conversely, if $\{G_i, \phi_{ji}\}$ is a projective system of finite groups G_i with the discrete topology, then the group $G := \varprojlim G_i$ is a profinite group.

Proof: First, let G be a profinite group. Let N be an open and normal subgroup of G. Then $G = \bigcup_{\overline{x} \in G/N} xN$, where xN is homeomorphic to N for all $x \in G$. Since G is a compact space, we have $[G : N] < \infty$. Thus G/N is a finite group and since $N = G \setminus \bigcup_{\substack{x \in G/N \\ x \notin N}} xN$ and $\bigcup_{\substack{x \in G/N \\ x \notin N}} xN$ is open, it follows that N is a closed subgroup of G.

Let $\mathcal{A} = \{N_i \mid i \in I\}$ be the set of all open normal subgroups of G and let $G_i := G/N_i$ for each $i \in I$. We define a partial order on I by setting $i \leq j \iff N_i \supseteq N_j$. Now for $i \leq j$, let

$$\begin{aligned} f_{ji} : G_j = G/N_j &\longrightarrow G/N_i = G_i \\ x \bmod N_j &\longmapsto x \bmod N_i \end{aligned}$$

be the natural projection.

Given $i, j \in I$, let $N_k := N_i \cap N_j$. Then $N_k \in \mathcal{A}$ and $i \leq k$, $j \leq k$. Therefore $\{G_i, f_{ji}\}$ is a projective system. Let

$$\begin{aligned} f : G &\longrightarrow \varprojlim_{i} G_i \\ \sigma &\longmapsto \prod_{i \in I} \sigma_i, \quad \text{where} \quad \sigma_i := \sigma \bmod N_i. \end{aligned}$$

Then f is a group homomorphism whose kernel is $\bigcap_{i\in I} N_i = \{e\}$. Indeed, $\{N_i \mid i \in I\}$ is a fundamental system of open neighborhoods of $\{e\}$ and G is a Hausdorff space. Therefore f is a monomorphism of groups.

Now $\{U_S := \prod_{i\notin S} G_i \times \prod_{i\in S}\{e_{G_i}\} \mid S \subseteq I,\, S \text{ finite}\}$ is a subbasis of neighborhoods of $e \in \prod_{i\in I} G_i$. We have

$$f^{-1}\Big(U_S \cap \varprojlim_i G_i\Big) = \bigcap_{i\in S} N_i.$$

Since the latter is open, it follows that f is a continuous map.

Now, G is compact, so $f(G)$ is a compact space too. Thus $f(G)$ is a closed subset of $\varprojlim_i G_i$. Let $\varphi = (\varphi_i)_{i\in I} \in \varprojlim_i G_i$. Then $\varphi(U_S \cap \varprojlim_i G_i)$ is a basic open neighborhood of φ. Let $N_k := \bigcap_{i\in S} N_i$ and let $\sigma \in G$ be such that $\sigma \bmod N_k = \varphi_k \in G/N_k$.

Then the diagram

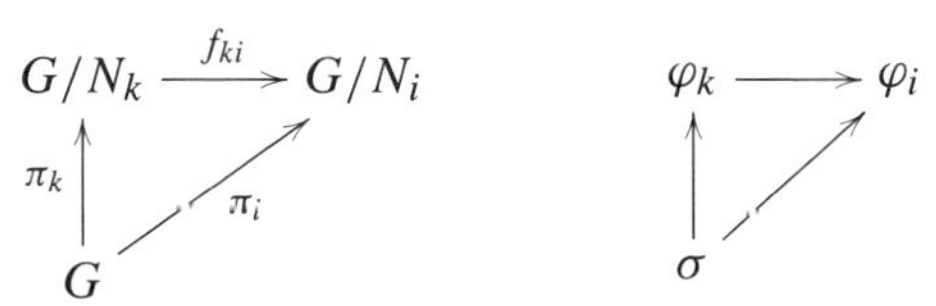

commutes for $i \in S$. Therefore $\sigma \bmod N_i = \varphi_i$. It follows that $f(\sigma) \in \varphi(U_S \cap \varprojlim_i G_i)$. Hence $f(G)$ is dense in $\varprojlim_i G_i$ and since $f(G)$ is closed, we conclude that f is onto. In particular, f is an algebraic isomorphism.

Finally, if $T \subseteq G$ is closed, then T and $f(T)$ are compact. Therefore $f(T)$ is a closed set in $\varprojlim_i G_i$. It follows that f is a closed map and f is a homeomorphism.

Conversely, let $\{G_i, f_{ji}\}$ be a projective system where for all $i \in I$, G_i is a finite group considered with the discrete topology.

Let $G := \varprojlim_i G_i$. Then G is closed in $\prod_{i\in I} G_i$. Since each G_i is compact, it follows by Tychonov's theorem that $\prod_{i\in I} G_i$ is a compact space. Therefore G is a compact group. Also, since each G_i is a Hausdorff space, so is $\prod_{i\in I} G_i$, and G is Hausdorff too.

Let V be an open neighborhood of $e \in G$. Then $V = V' \cap \varprojlim_i G_i$, where V' an open neighborhood of $e \in \prod_{i\in I} G_i$. Therefore there exists a finite subset $S \subseteq I$ such that $U_S = \prod_{i\notin S} G_i \times \prod_{i\in S} H_i \subseteq V'$ with $H_i \lhd G_i$ for each $i \in S$. Thus $e \in U_S \cap \varprojlim_i G_i \subseteq V$. It follows that $\big\{U_S \cap \varprojlim_i G_i \mid S \subseteq I \text{ finite}\big\}$ is a basis of neighborhoods of $e \in G$ and since U_S is a normal subgroup of $\prod_{i\in I} G_i$, we have $U_S \cap \varprojlim_i G_i \lhd \varprojlim_i G_i$. By Theorem 11.3.13, it follows that G is a profinite group.

$\square$

We have proved the following theorem:

Theorem 11.3.16. *Let G be a topological group. The following conditions are equivalent*

(i) *G is a profinite group.*
(ii) *G is the inverse limit of finite groups.*
(iii) *G is a topological group that is Hausdorff, compact, and totally disconnected.*
(iv) *G is a topological group that is a Hausdorff compact space that contains a basis of neighborhoods of e consisting of open normal subgroups of G.* □

Example 11.3.17. If G is a finite group, then G is a profinite group.

Example 11.3.18. Let $I = \mathbb{N} = \{1, 2, \dots, \}$. We define $n \leq m \iff n \mid m$.

Let $f_{m,n} : \mathbb{Z}/m\mathbb{Z} \longrightarrow \mathbb{Z}/n\mathbb{Z}$ be the natural projection

$$a \bmod m \longmapsto a \bmod n.$$

Set $\hat{\mathbb{Z}} := \varprojlim_{n\in\mathbb{N}} \mathbb{Z}/n\mathbb{Z}$. Then $\hat{\mathbb{Z}}$ is called the *Prüfer ring*. We have $\hat{\mathbb{Z}} < \prod_{n=1}^{\infty} \mathbb{Z}/n\mathbb{Z}$. Let

$$\begin{aligned} \varphi\colon \mathbb{Z} &\longrightarrow \hat{\mathbb{Z}} \\ a &\longmapsto (a \bmod n)_{n\in\mathbb{N}}. \end{aligned} \tag{11.6}$$

Let $\alpha = (\alpha_n)_{n\in\mathbb{Z}} \in \hat{\mathbb{Z}}$ and let V be an open neighborhood of α. Then there exists a finite set $S \subseteq \mathbb{N}$ such that $W = \alpha\Big(\Big(\prod_{n\in S}\{1\} \times \prod_{n\notin S} \mathbb{Z}/n\mathbb{Z}\Big) \cap \hat{\mathbb{Z}}\Big) \subseteq V$. Let $m = \prod_{s\in S} s$. Then $s \leq m$ for all $s \in S$. Let $a \in \mathbb{Z}$ be such that $a \equiv \alpha_m \bmod m$.

Then $a \bmod s \equiv \alpha_s \bmod s$ for all $s \in S$. Hence $\varphi(a) \in W$, and $\varphi(\mathbb{Z})$ is dense in $\hat{\mathbb{Z}}$.

For $n \in \mathbb{N}$, the map

$$\begin{aligned} \theta_n : \hat{\mathbb{Z}} &\longrightarrow n\hat{\mathbb{Z}} \\ x &\longmapsto nx := x + \cdots + x \end{aligned}$$

is an algebraic and topological isomorphism. Thus $n\hat{\mathbb{Z}} \cong \hat{\mathbb{Z}}$ and therefore $n\hat{\mathbb{Z}}$ is open and closed in $\hat{\mathbb{Z}}$.

Conversely, let $H < \hat{\mathbb{Z}}$ be an open subgroup. Since $\hat{\mathbb{Z}}$ is compact, H is a closed subgroup and $[\hat{\mathbb{Z}} : H] = n < \infty$.

In particular, $n\hat{\mathbb{Z}} \subseteq H$. Now the map φ given in (11.6) satisfies $\varphi(n\mathbb{Z}) \subseteq n\hat{\mathbb{Z}}$ and induces

$$\mathbb{Z} \xrightarrow{\varphi} \hat{\mathbb{Z}} \xrightarrow{\pi} \hat{\mathbb{Z}}/n\hat{\mathbb{Z}}$$
$$\widetilde{\varphi} : \mathbb{Z} \longrightarrow \hat{\mathbb{Z}}/n\hat{\mathbb{Z}};$$

$\widetilde{\varphi}$ is dense and $\hat{\mathbb{Z}}/n\hat{\mathbb{Z}}$ is finite. Hence $\widetilde{\varphi}$ is onto and $\ker \varphi = n\mathbb{Z}$, so $\mathbb{Z}/n\mathbb{Z} \cong \hat{\mathbb{Z}}/n\hat{\mathbb{Z}}$.

Therefore $[\hat{\mathbb{Z}} : n\hat{\mathbb{Z}}] = [\mathbb{Z} : n\mathbb{Z}] = n = [\hat{\mathbb{Z}} : H]$,

$$\begin{aligned} \hat{\mathbb{Z}}/n\hat{\mathbb{Z}} &\longrightarrow \hat{\mathbb{Z}}/H \\ x \bmod n\hat{\mathbb{Z}} &\longmapsto x \bmod H \end{aligned}$$

is an epimorphism, and $n\hat{\mathbb{Z}} = H$.

Therefore the open subgroups of $\hat{\mathbb{Z}}$ are the subgroups $n\hat{\mathbb{Z}}$ with $n \in \mathbb{N}$.

Example 11.3.19. Let $p \subset \mathbb{N}$ be a rational prime. For $n \in \mathbb{N} \cup \{0\} := \mathbb{N}_0$ and $m \leq n$, the natural projection

$$\begin{aligned} f_{n,m} : \mathbb{Z}/p^n\mathbb{Z} &\longrightarrow \mathbb{Z}/p^m\mathbb{Z} \\ x \bmod p^n &\longmapsto x \bmod p^m \end{aligned}$$

defines an inverse system. Set

$$\mathcal{Y} := \varprojlim_n \mathbb{Z}/p^n\mathbb{Z}.$$

Let $\mathbb{Z}_p = \left\{ \sum_{n=0}^{\infty} a_n p^n | a_n \in \{0, 1, \ldots, p-1\} \right\}$ (see Example 2.3.7) and let

$$\begin{aligned} \varphi : \mathbb{Z}_p &\longrightarrow \mathcal{Y} \\ \sum_{n=0}^{\infty} a_n p^n &\longmapsto \left(\sum_{n=0}^{i} a_n p^n \right)_{i \in \mathbb{N}_0}. \end{aligned}$$

Clearly, φ is a monomorphism of groups. If $(\alpha_i)_{i \in \mathbb{N}_0} \in \mathcal{Y}$, then the class of α_i in $\mathbb{Z}/p^i\mathbb{Z}$ contains an element x_i such that $0 \leq x_i \leq p^i - 1$. Put

$$x_i = \sum_{n=0}^{i-1} a_{in} p^n$$

with $a_{in} \in \{0, 1, \ldots, p-1\}$. Since for $i \geq j$, $f_{ji}(x_j) = x_i$, it follows that $a_{in} = a_{jn}$ for $0 \leq n \leq j$. Set

$$a_n := a_{in} \quad \text{for} \quad n \leq i.$$

Then $(\alpha_i)_{i \in \mathbb{N}_0} = \left(\sum_{n=0}^{i-1} a_n p^n \right)_{i \in \mathbb{N}_0} = \varphi \left(\sum_{n=0}^{\infty} a_n p^n \right)$ and φ is a group isomorphism.

Let $V = \prod_{s \in S} U_S \times \prod_{n \notin S} \mathbb{Z}/p^n\mathbb{Z}$ be a basic open neighborhood, and $S \subseteq \mathbb{N}_0$ a finite set. Let $t = \sup S$. Then if $\alpha = (\alpha_i)_{i \in \mathbb{N}_0} \in V \cap \mathcal{Y}$, we have

$$\begin{aligned} \alpha_t \bmod p^t &\equiv a_0 + a_1 p + \cdots + a_{t-1} p^{t-1}, \\ \varphi^{-1}\Big(\prod_{s \in S} \{\alpha_s\} \times \prod_{n \notin S} \mathbb{Z}/p^n\mathbb{Z}\Big) &= (a_0 + a_1 p + \cdots + a_{t-1} p^{t-1}) + p^t \mathbb{Z}_p \end{aligned}$$

is open in $\mathbb{Z}_p$, and $\varphi^{-1}(V \cap \mathcal{Y})$ is a finite union of such subsets. Thus φ is continuous.

Finally, φ is closed. Indeed, if $T \subseteq \mathbb{Z}_p$ is closed, then T is compact and so is $\varphi(T)$. Thus $\varphi(T)$ is closed in $\mathcal{Y}$. We have

$$\mathbb{Z}_p \cong \varprojlim_{n \in \mathbb{N}_0} \mathbb{Z}/p^n\mathbb{Z}$$

algebraically and topologically.

As in Example 11.3.18, the open subgroups of $\mathbb{Z}_p$ are precisely those of the form $p^n\mathbb{Z}_p$ with $n \in \mathbb{N}_0$.

Now let H be a closed subgroup of $\mathbb{Z}_p$.

If $H \neq (0)$, let $x \in H$ be such that $v_p(x)$ is minimal and put $v_p(x) = n$.

We have $\mathbb{Z}x = \{mx \mid m \in \mathbb{Z}\} \subseteq H$. Since H is closed, it follows that $\overline{\mathbb{Z}x} = \mathbb{Z}_p x \subseteq H$. We have $x = a_0 p^n$ with $v_p(a_0) = 0$. Hence $a_0^{-1} \in \mathbb{Z}_p$ and $p^n = a_0^{-1}x \in \mathbb{Z}_p x$. Thus $p^n\mathbb{Z}_p \subseteq H$. On the other hand, if $y \in H \setminus \{0\}$, we have $v_p(y) = m \geq n$, so $y = p^m b_0 = p^n(p^{m-n}b_0) \in p^n\mathbb{Z}_p$. Consequently $H = p^n\mathbb{Z}_p$. In particular, the closed subgroups of $\mathbb{Z}_p$ are $\{0\}$ and $p^n\mathbb{Z}_p$ for $n \in \mathbb{N} \cup \{0\}$.

Example 11.3.20. Let A be an abelian torsion group. Then for any $a \in A$ there exists $n \in \mathbb{N}$ such that $na = 0$. Let $\mathbb{Q}/\mathbb{Z} = \{\overline{x} = x + \mathbb{Z} \mid x \in \mathbb{Q}\}$ (we have $\mathbb{Q}/\mathbb{Z} \cong \{\xi \in \mathbb{C} \mid \xi^m = 1 \text{ for some } m \in \mathbb{N}\}$). We define the *Pontryagin dual* of A as

$$\chi(A) = \operatorname{Hom}(A, \mathbb{Q}/\mathbb{Z}).$$

Then $A = \bigcup_{i \in I} A_i$, where the union runs through all finite subgroups A_i of A.

We define $i \leq j \iff A_j \supseteq A_i$. For $i \leq j$, let

$$f_{ij} \colon A_i \longrightarrow A_j$$

be the natural injection

Let

$$\phi_{ji} \colon \chi(A_j) \to \chi(A_i)$$

be given by $\phi_{ji}(\sigma) = \sigma \circ f_{ij}$. Then $\left\{\chi(A_i), \phi_{ji}, I\right\}$ is an inverse system. Note that $\chi(A_i) \cong \hat{A}_i \cong A_i$, where $\hat{A}_i$ denotes the group of characters of A_i.

Then $\chi(A)$ is isomorphic to $\varprojlim_i \chi(A_i)$ (see Exercise 11.7.14).

11.4 Infinite Galois Theory

Definition 11.4.1. Let k be any field and $\bar{k}$ the separable algebraic closure of k. The Galois group $\operatorname{Gal}(\bar{k}/k) =: G_k$ is called the *absolute Galois group* of k.

In general, G_k is an infinite group and the usual main theorem of Galois theory does not hold anymore in the usual sense. The next example explains this difference.

Example 11.4.2. Let $\mathbb{F}_p$ be the finite field of p elements, and $G = G_{\mathbb{F}_p} = \operatorname{Gal}(\overline{\mathbb{F}}_p/\mathbb{F}_p)$. Let

$$\begin{aligned} \varphi : \overline{\mathbb{F}}_p &\longrightarrow \overline{\mathbb{F}}_p \\ x &\longmapsto x^p \end{aligned}$$

be the Frobenius automorphism. Let $H = (\varphi) = \{\varphi^n \mid n \in \mathbb{Z}\}$. Note that if $x \in \overline{\mathbb{F}}_p^H$, then $\varphi(x) = x^p = x$, so $x \in \mathbb{F}_p$. Therefore

$$\mathbb{F}_p = \overline{\mathbb{F}}_p^H = \overline{\mathbb{F}}_p^G.$$

We will now see that $H \neq G$.

Let $n \in \mathbb{N}$ and write $n = b_n p^{v_p(n)}$, where $(b_n, p) = 1$. Let $x_n, y_n \in \mathbb{Z}$ be such that

$$1 = b_n x_n + p^{v_p(n)} y_n.$$

Define $a_n = b_n x_n \in \mathbb{Z}$. If m divides n, then

$$m = b_m p^{v_p(m)} \mid b_n p^{v_p(n)} = n,$$

so

$$b_m \mid b_n \quad \text{and} \quad v_p(m) \leq v_p(n).$$

Now, $a_n - a_m = b_n x_n - b_m x_m$. Hence b_m divides $a_n - a_m$ and

$$a_n - a_m - \left(1 - p^{v_p(n)} y_n\right) - \left(1 - p^{v_p(m)} y_m\right) = p^{v_p(m)} y_m - p^{v_p(n)} y_n.$$

It follows that $p^{v_p(m)}$ divides $a_n - a_m$, and

$$a_n \equiv a_m \bmod m \quad \text{whenever} \quad m \text{ divides } n.$$

Now assume that there exists an integer a such that $a_n \equiv a \bmod n$ for all n. If q is any prime other than p and $\alpha \in \mathbb{N}$ is arbitrary, consider $n = q^\alpha$. Then

$$a_n = q^\alpha x_n \equiv a \bmod q^\alpha.$$

Thus q^α divides a for all α, so $a = 0$. But

$$a_p = p - 1 \not\equiv 0 \bmod p.$$

This contradiction shows that there does not exist $a \in \mathbb{Z}$ such that $a_n \equiv a \bmod n$ for all n.

Let $\psi_n = \varphi^{a_n}|_{\mathbb{F}_{p^n}} \in \mathrm{Gal}(\mathbb{F}_{p^n}/\mathbb{F}_p)$. If $\mathbb{F}_{p^m} \subseteq \mathbb{F}_{p^n}$, then m divides n, so $a_n \equiv a_m \bmod m$. Since $o(\varphi|_{\mathbb{F}_{p^m}}) = m$ we have

$$\psi_n|_{\mathbb{F}_{p^m}} = \varphi^{a_n}|_{\mathbb{F}_{p^m}} = \varphi^{a_m}|_{\mathbb{F}_{p^m}} = \psi_m.$$

Let $\psi \in G$ be defined as follows. If $x \in \overline{\mathbb{F}}_p$, then $x \in \mathbb{F}_{p^n}$ for some n, and we put $\psi(x) = \psi_n(x)$. Clearly, ψ is a well-defined element of G. If $\psi \in H = (\varphi)$, then $\psi = \varphi^a$ for some $a \in \mathbb{Z}$. Then $\psi|_{\mathbb{F}_{p^n}} = \varphi^{a_n}|_{\mathbb{F}_{p^n}} = \varphi^a|_{\mathbb{F}_{p^n}}$. Hence $a_n \equiv a \bmod n$ for all n. This contradiction shows that $H \neq G$ but

$$\overline{\mathbb{F}}_p^H = \overline{\mathbb{F}}_p^G.$$

In order to establish the "right" main theorem of Galois theory we must take into account the topological nature of the Galois group of an arbitrary Galois extension.

Let K/F be an algebraic, normal, and separable extension of fields, that is, a Galois extension. Let

$$\mathcal{K} = \{K_i \mid i \in I\}$$

be the collection of all intermediate subfields K_i ($F \subseteq K_i \subseteq K$) such that K_i/F is a finite Galois extension. Then

$$K = \bigcup_{i \in I} K_i.$$

Let $G := \mathrm{Gal}(K/F)$ and $N_i = \mathrm{Gal}(K/K_i)$. We have $K_i = K^{N_i} = \{\alpha \in K \mid \sigma\alpha = \alpha \;\forall\, \sigma \in N_i\}$. Then:

(1) For $i \in I$, $N_i \lhd G$ and $G/N_i \cong \mathrm{Gal}(K_i/F)$ is a finite group.
(2) For every $i, j \in I$, $N_k := N_i \cap N_j$ satisfies that $N_k \lhd G$ and G/N_k is a finite group (in fact, if $K_i = K^{N_i}$ and $K_j = K^{N_j}$, then $K_k = K^{N_i} K^{N_j} = K^{N_i \cap N_j}$).
(3) $\bigcap_{i \in I} N_i = \{1\}$.

We define a topology on G by taking the cosets

$$\sigma N_i, \quad i \in I,$$

as a basis of neighborhoods of σ for each $\sigma \in G$.

Proposition 11.4.3. *For the topology defined above, the multiplication and the inversion maps*

$$\begin{array}{ccc} G \times G \xrightarrow{\varphi} G & \qquad & G \xrightarrow{i} G \\ (\sigma, \varphi) \mapsto \sigma\psi & \qquad & \sigma \mapsto \sigma^{-1} \end{array}$$

are continuous.

Proof: The statement follows from the facts that $\varphi^{-1}(\sigma\psi N_j) \supseteq \sigma N_j \times \psi N_j$ and

$$i^{-1}(\sigma^{-1} N_j) = \sigma N_j$$

for all $j \in I$. □

Definition 11.4.4. The topology defined above on G is called the *Krull topology* and with this topology G becomes a topological group.

Theorem 11.4.5. *The Galois group $G = \mathrm{Gal}(K/F)$ endowed with the Krull topology is a profinite group. Moreover, we have*

$$G \cong \varprojlim_{i \in I} G/N_i \cong \varprojlim_{i \in I} \mathrm{Gal}(K_i/F)$$

algebraically and topologically, where $N_i = \mathrm{Gal}(K/K_i)$ and K_i runs through the set $\{K_i \mid F \subseteq K_i \subseteq K$, and K_i/F is a finite Galois extension$\}$.

Proof: For each $i \in I$, denote by G_i the group $\mathrm{Gal}(K_i/F)$, which is isomorphic to G/N_i.

We define a partial order $\leq$ in I by

$$i \leq j \iff K_i \subseteq K_j \quad \text{or equivalently,} \quad i \leq j \iff N_i \supseteq N_j.$$

Then I is a directed poset since if $i, j \in I$, the composite $K_k := K_i K_j$ is a finite Galois extension of F and $K_i, K_j \subseteq K_k$.

Now, if $i \leq j$, let

$$\begin{aligned} \phi_{ji} : G_j &\to G_i \\ \sigma &\mapsto \sigma|_{K_i}. \end{aligned}$$

We have obtained an inverse system $\{G_i, \phi_{ji}, I\}$ of finite Galois groups. Let

$$\begin{aligned} \Phi : G &\longrightarrow \varprojlim_{i \in I} G_i \subseteq \prod_{i \in I} G_i \\ \sigma &\longmapsto (\sigma|_{K_i})_{i \in I}. \end{aligned}$$

Clearly, Φ is a group homomorphism whose kernel is $\bigcap_{i \in I} G_i = \{1\}$.

Now consider the following composition:

$$G \xrightarrow{\Phi} \varprojlim_{i \in I} G_i \xrightarrow{\phi_i} G_i.$$

For each $i \in I$, $\phi_i \circ \Phi$ is continuous. Indeed, G_i is a finite group with the discrete topology, so if $A \subseteq G_i$, we have

$$\begin{aligned} (\phi_i \circ \Phi)^{-1}(A) &= \bigcup_{a \in A} (\phi_i \circ \Phi)^{-1}(a) = \bigcup_{a \in A} \Phi^{-1}(\phi_i^{-1}(a)) \\ &= \bigcup_{a \in A} \{\sigma \in G \mid \sigma|_{K_i} = a\} = \bigcup_{a \in A} a \,\mathrm{Gal}(K/K_i) = \bigcup_{a \in A} a N_i, \end{aligned}$$

which is open. It follows that if $S \subseteq I$ is a finite set, then

$$\Phi^{-1}\Big(\Big(\prod_{i \in S} A_i \times \prod_{i \notin S} G_i\Big) \cap \varprojlim_{j \in I} G/N_j\Big) = \bigcap_{i \in S} (\phi_i \circ \Phi)^{-1}(A_i)$$

is open. Therefore Φ is continuous.

Now we have

$$\Phi(N_i) = \Big(\varprojlim_{j \in I} G_i\Big) \bigcap \Big[\Big(\prod_{K_j \not\subseteq K_i} G_i\Big) \times \Big(\prod_{K_j \subseteq K_i} \{1\}_j\Big)\Big]$$

and $\{j \in I \mid K_j \subseteq K_i\}$ is finite, so $\Phi(N_i)$ is an open set. Therefore Φ is an open map.

Finally, if $(\sigma_i)_{i \in I} \in \varprojlim G_i$, let $\sigma : K \to K$ be such that $\sigma(\alpha) = \sigma_i(\alpha)$ for $\alpha \in K_i$. Then σ is a well-defined element of G and $\Phi(\sigma) = (\sigma_i)_{i \in I}$. Thus Φ is a group epimorphism. The result follows. □

Example 11.4.6. Assume $q = p^u$ for some prime number p and some $u \in \mathbb{N}$. Let $\mathbb{F}_q$ be the finite field of q elements. For each n, there exists a unique extension $\mathbb{F}_{q^n}$ of $\mathbb{F}_q$, and $\mathbb{F}_{q^n}/\mathbb{F}_q$ is a cyclic extension. It follows that $\overline{\mathbb{F}}_q = \bigcup_{n=1}^{\infty} \mathbb{F}_{q^n}$ and $G_n := \operatorname{Gal}(\mathbb{F}_{q^n}/\mathbb{F}_q) \cong \mathbb{Z}/n\mathbb{Z}$. Therefore

$$\operatorname{Gal}(\overline{\mathbb{F}}_q/\mathbb{F}_q) \cong \varprojlim_n \mathbb{Z}/n\mathbb{Z} \cong \hat{\mathbb{Z}}.$$

Example 11.4.7. Let $q = p^u$ as in Example 11.4.6 and let ℓ be any prime number ($\ell = p$ or $\ell \neq p$). Let $T_n := \mathbb{F}_{q^{\ell^n}}$.

Then $H_n := \operatorname{Gal}(\mathbb{F}_{q^{\ell^n}}/\mathbb{F}_q) \cong \mathbb{Z}/\ell^n\mathbb{Z}$. If $T_\ell = \bigcup_{n=0}^{\infty} T_n$, then $T_\ell/\mathbb{F}_q$ is a Galois extension and

$$\operatorname{Gal}(T_\ell/\mathbb{F}_q) \cong \varprojlim \operatorname{Gal}(T_n/\mathbb{F}_q) \cong \varprojlim \mathbb{Z}/\ell^n\mathbb{Z} \cong \mathbb{Z}_\ell.$$

Since $T_\ell \subseteq \overline{\mathbb{F}}_q$, if $N_\ell = \operatorname{Gal}(\overline{\mathbb{F}}_q/T_\ell)$, then $\mathbb{Z}_\ell \cong \hat{\mathbb{Z}}/N_\ell$. By Exercise 11.7.16,

$$T_\ell \cap \left(\prod_{\ell' \neq \ell} T_{\ell'} \right) = \mathbb{F}_q \quad \text{and} \quad \prod_{\ell \text{ prime}} T_\ell = \overline{\mathbb{F}}_q.$$

Therefore

$$\hat{\mathbb{Z}} \cong \operatorname{Gal}(\overline{\mathbb{F}}_q/\mathbb{F}_q) \cong \prod_{\ell \text{ prime}} \operatorname{Gal}(T_\ell/\mathbb{F}_q) \cong \prod_{\ell \text{ prime}} \mathbb{Z}_\ell.$$

Example 11.4.8. For each $n \in \mathbb{N}$, let ζ_n denote a primitive nth root of 1 in $\mathbb{C}$ (for example $\zeta_n = e^{2\pi i/n}$). Let $\mathbb{Q}(\zeta_n)$ be the nth cyclotomic number field. Then

$$\operatorname{Gal}(\mathbb{Q}(\zeta_n)/\mathbb{Q}) \cong \mathcal{U}_n = (\mathbb{Z}/n\mathbb{Z})^*.$$

If $n = p_1^{\alpha_1} \cdots p_r^{\alpha_r}$, then $\mathcal{U}_n \cong \prod_{i=1}^{r} \mathcal{U}_{p_i^{\alpha_i}}$. We have

$$\mathcal{U}_2 = \{1\}, \quad \mathcal{U}_{2^2} \cong \mathbb{Z}/2\mathbb{Z}, \quad \mathcal{U}_{2^\alpha} \cong \mathbb{Z}/2\mathbb{Z} \times \mathbb{Z}/2^{\alpha-2}\mathbb{Z}$$

for $\alpha \geq 3$ and

$$\mathcal{U}_{p^n} \cong \mathbb{Z}/(p-1)\mathbb{Z} \times \mathbb{Z}/p^{n-1}\mathbb{Z}$$

for each odd prime p.

Let $\mathbb{Q}(\zeta_\infty) := \bigcup_{n=1}^{\infty} \mathbb{Q}(\zeta_n)$. Then

$$G_\infty := \operatorname{Gal}(\mathbb{Q}(\zeta_\infty)/\mathbb{Q}) \cong \varprojlim_n \operatorname{Gal}(\mathbb{Q}(\zeta_n)/\mathbb{Q}).$$

If $\mathbb{Q}(\zeta_{p^\infty}) := \bigcup_{n=1}^{\infty} \mathbb{Q}(\zeta_{p^n})$, where p is any prime, then

$$\mathbb{Q}(\zeta_\infty) = \prod_{p \text{ prime}} \mathbb{Q}(\zeta_{p^\infty}) \qquad \text{and} \qquad \mathbb{Q}(\zeta_{p^\infty}) \cap \prod_{q \neq p} \mathbb{Q}(\zeta_{q^\infty}) = \mathbb{Q}.$$

Therefore G_∞ is isomorphic to $\prod_{p \text{ prime}} \mathrm{Gal}(\mathbb{Q}(\zeta_{p^\infty})/\mathbb{Q})$.

Now

$$\begin{aligned}\mathrm{Gal}(\mathbb{Q}(\zeta_{p^\infty})/\mathbb{Q}) &\cong \varprojlim_n \mathrm{Gal}(\mathbb{Q}(\zeta_{p^n})/\mathbb{Q}) \cong \varprojlim_n \mathcal{U}_{p^n} \\ &\cong \begin{cases} \mathbb{Z}/2\mathbb{Z} \times \mathbb{Z}_2 & \text{if } \quad p = 2, \\ \mathbb{Z}/(p-1)\mathbb{Z} \times \mathbb{Z}_p & \text{if } \quad p > 2. \end{cases}\end{aligned}$$

From Example 11.4.7, we obtain that

$$G_\infty \cong (\mathbb{Z}/2\mathbb{Z} \times \prod_{p>2} \mathbb{Z}/(p-1)\mathbb{Z}) \times \hat{\mathbb{Z}}.$$

Now we are ready to state the main theorem in Galois theory

Theorem 11.4.9 (Fundamental Theorem in Galois Theory). *Let K/F be a Galois extension of fields with Galois group $G = \mathrm{Gal}(K/F)$. Set*

$$\mathcal{F}(K/F) = \{L \mid L \text{ is a field and } F \subseteq L \subseteq K\}$$

and

$$S(G) = \{H \mid H \text{ is a closed subgroup of } G\}.$$

Let

$$\Phi : \mathcal{F}(K/F) \to S(G) \quad \text{and} \quad \Psi : S(G) \to \mathcal{F}(K/F)$$

be defined by

$$\Phi(L) = \{\sigma \in G \mid \sigma|_L = \mathrm{Id}_L\} = \mathrm{Gal}(K/L)$$

and

$$\Psi(H) = \{\alpha \in K | \sigma\alpha = \alpha \, \forall \, \sigma \in H\} = K^H.$$

Then Φ and Ψ are mutually inverse bijections. Furthermore, we have $L_1 \subseteq L_2$ if and only if $\Phi(L_1) \geq \Phi(L_2)$, and $H_1 \leq H_2$ if and only if $\Psi(H_1) \supseteq \Psi(H_2)$.

Finally, if $\sigma \in G$ and $L \in \mathcal{F}(K|F)$, then

$$\mathrm{Gal}(K/\sigma L) = \Phi(\sigma L) = \sigma\Phi(L)\sigma^{-1} = \sigma \, \mathrm{Gal}(K/L)\sigma^{-1}.$$

In particular, $L \in \mathcal{F}(K/F)$ is a normal extension of F if and only if $\mathrm{Gal}(K/L)$ is normal in G, and in this case, $\mathrm{Gal}(L/F) \cong \frac{\mathrm{Gal}(K/F)}{\mathrm{Gal}(K/L)}$.

The open subgroups of G correspond to the finite subextensions of K/F.

Proof: It is easy to see that Φ and Ψ reverse inclusions. By Theorem 11.4.5, $\Phi(L) = \operatorname{Gal}(K/L)$ is a profinite group, so $\Phi(L)$ is closed in G. Hence $\Phi(L) \in S(G)$.

Let $L \in \mathcal{F}(K/F)$. Then $\Psi\Phi(L) = \Psi(\operatorname{Gal}(K/L)) = K^{\operatorname{Gal}(K/L)} \supseteq L$. Suppose that $y \in K^{\operatorname{Gal}(K/L)}$. Then if $f(x) = \operatorname{Irr}(y, x, L)$, every root of $f(x)$ is of the form σy for some $\sigma \in \operatorname{Gal}(K/L)$. Thus

$$f(x) = (x - y)^n \in L[x].$$

Since K/F is a separable extension, we have $n = 1$ and $y \in L$. This shows that $\Psi\Phi(L) = L$.

Conversely, pick $H \in S(G)$. Let $L = \Psi(H) = K^H$. Then $\Phi\Psi(H) = \operatorname{Gal}(K/K^H) \supseteq H$. To see that $\Phi\Psi(H) = H$, it suffices to show that H is dense in $\operatorname{Gal}(K/L)$ since H is closed.

Let $L \subseteq N \subseteq K$ be such that N/L is a finite Galois extension, and let $\tau \in \operatorname{Gal}(K/L)$. We wish to show that

$$\tau \operatorname{Gal}(K/N) \cap H \neq \emptyset.$$

If $\sigma \in H$, since $\sigma|_L = \operatorname{Id}_L$ and N/L is normal, we have $\sigma(N) = N$.

Let $H_1 = \{\sigma|_N \mid \sigma \in H\} \leq \operatorname{Gal}(N/L)$. Then $N^{H_1} \supseteq N^{\operatorname{Gal}(N/L)} = L$. If $\alpha \in N^{H_1}$, then $\sigma\alpha = \alpha$ for all $\sigma \in H$. Hence $\alpha \in K^H = L$, and we have $N^{H_1} = L$. Using finite Galois theory, we obtain

$$H_1 = \operatorname{Gal}(N/L).$$

In particular, there exists $\sigma \in H$ such that $\sigma|_N = \tau|_N$, i.e., $\sigma \in \tau \operatorname{Gal}(K/N) \cap H \neq \emptyset$. Therefore $\Phi\Psi(H) = H$. This shows that Φ and Ψ are inverse bijections.

Now consider $\sigma \in G$ and $L \in \mathcal{F}(K/F)$. Let $\Phi(L) = H = \operatorname{Gal}(K/L)$ and $\Phi(\sigma L) = H_1 = \operatorname{Gal}(K/\sigma L)$. We have $\theta \in H_1 \Leftrightarrow \theta(\sigma\alpha) = \sigma\alpha \ \forall\, \alpha \in L \Leftrightarrow (\sigma^{-1}\theta\sigma)(\alpha) = \alpha \ \forall\, \alpha \in L \Leftrightarrow \sigma^{-1}\theta\sigma \in H \Leftrightarrow \theta \in \sigma H \sigma^{-1}$. Thus $H_1 = \sigma H \sigma^{-1}$.

When L/F is normal the group homomorphism

$$\begin{aligned} G = \operatorname{Gal}(K/F) &\overset{\Theta}{\to} \operatorname{Gal}(L/F) \\ \sigma &\mapsto \sigma|_L \end{aligned}$$

is onto because every F-automorphism of L can be extended to any algebraic extension. Since

$$\ker \Theta = \{\sigma \in G \mid \sigma|_L = \operatorname{Id}_L\} = \operatorname{Gal}(K/L)$$

we obtain that

$$\operatorname{Gal}(L/F) \cong \frac{\operatorname{Gal}(K/F)}{\operatorname{Gal}(K/L)}.$$

Finally if H is an open subgroup, H is also closed and of finite index. □

We have shown that the Galois group of any G extension is a profinite group. (Theorem 11.4.5). We also know that any finite group G is the Galois group of a certain Galois extension. Next we show that this is also true for an arbitrary profinite group, or in other words, that the converse of Theorem 11.4.5 also holds.

Theorem 11.4.10 (Leptin). *Let G be any profinite group. Then there exists a Galois extension of fields K/F such that*

$$G \cong \mathrm{Gal}(K/F).$$

Proof: Consider any field E. Let T be the disjoint union of all the sets G/N, where N runs through the collection of all open normal subgroups of G. We have

$$T := \dot{\bigcup_{\substack{N \lhd G \\ [G:N]<\infty}}} G/N = \dot{\bigcup_{\substack{N \lhd G \\ [G:N]<\infty}}} \left(\bigcup_{\theta \in G/N} \theta N \right).$$

For each $t \in T$, define x_t such that $\{x_t\}_{t\in T} = \mathcal{T}$ is an algebraically independent set over E. Let $K = E(\mathcal{T})$ be the field of rational functions with indeterminates in $\mathcal{T}$ and coefficients in E. Notice that G acts on $\mathcal{T}$ in a natural way: if $\sigma \in G$ and $\theta N \in G/N$, then $\sigma(\theta N) = (\sigma\theta)N$ or $\sigma(x_t) = x_{\sigma t}$, where $t = \theta N$ and $\sigma t = (\sigma\theta)N$.

This action induces an action on K in a natural manner: if $f \in K$, then in the expression of f appear only finitely many variables $x_t \in \mathcal{T}$. Then if $\sigma \in G$ and $f = f(x_{t_1}, \ldots, x_{t_n})$, put

$$\sigma f = f(x_{\sigma t_1}, \ldots, x_{\sigma t_n}).$$

Let $F := K^G = \{\alpha \in K \mid \sigma\alpha = \alpha \text{ for all } \sigma \in G\}$. Let $\alpha \in K$ and

$$G_\alpha = \{\sigma \in G \mid \sigma\alpha = \alpha\}.$$

Then G_α is a subgroup of G and if the indeterminates that appear in the expression of α are $\{x_{t_i} | t_i \in G/N_i, 1 \le i \le m\}$, we have

$$G_\alpha \supseteq \bigcap_{i=1}^{n} N_i = N.$$

Since each N_i is open, N is open too and thus $[G : N] < \infty$. It follows that $G_\alpha = \bigcup_{g\in G_\alpha} gN$ is open and $[G : G_\alpha] < \infty$.

The orbit of α is the finite set $C(\alpha) = \{\sigma\alpha \mid \sigma \in G\}$ containing $[G : G_\alpha]$ elements (it is well known that

$$\begin{aligned} G/G_\alpha &\to C(\alpha) \\ gG_\alpha &\mapsto g\alpha \end{aligned}$$

is a well defined bijection). Let $f_\alpha(x) = \prod_{\bar{\sigma}\in G/G_\alpha} (x - \bar{\sigma}\alpha)$.

Clearly, $\tau f_\alpha = f_\alpha$ for all $\tau \in G$ and thus $f_\alpha(x) \in F[x]$. It follows that α is algebraic over F and since the roots of f_α are all distinct, K/F is an algebraic separable extension. Now $\mathrm{Irr}(\alpha, x, F)$ divides $f_\alpha(x)$ and all the roots of $f_\alpha(x)$ belong to K. Thus K/F is a normal extension. (Furthermore, $\sigma\alpha$ is a conjugate of α for all $\sigma \in G$, so $f_\alpha(x) = \mathrm{Irr}(\alpha, x, F)$ although we do not need this fact.)

Let $H = \mathrm{Gal}(K/F)$ and notice that $G \subseteq H$. Consider the natural injection

$$i : G \hookrightarrow H.$$

Let N be an open normal subgroup of H and let $K^N = \{\alpha \in K | \sigma\alpha = \alpha$ for all $\sigma \in N\}$. By the fundamental theorem of Galois theory (Theorem 11.4.9), K^N/F is a finite Galois extension, say, $K^N = F(\alpha_1, \dots, \alpha_m)$. Then

$$i^{-1}(N) = G \cap N \supseteq \bigcap_{j=1}^{m} G_{\alpha_j}$$

is an open set in G. It follows that i is continuous. Since G is compact, $i(G) = G$ is compact. Hence G is closed in H. Finally, $K^H = K^G$, so by Theorem 11.4.9, $H = G$. □

Remark 11.4.11. Artin's theorem establishes that if G is a finite group of automorphisms of a field L, then L/L^G is a Galois extension with Galois group G. This theorem is no longer true for a profinite group.

Example 11.4.12. Let G be any infinite profinite group and let F be any field. For each $g \in G$, consider an indeterminate x_g such that $\{x_g\}_{g\in G}$ is algebraically independent over F. Let $E = F(x_g \mid g \in G)$ be the rational function field on the variables $\{x_g\}_{g\in G}$ over F.

Then G acts on E naturally: if $f(x_{g_1}, \dots, x_{g_n}) \in E$ and $h \in G$, then

$$h \circ f(x_{g_1}, \dots, x_{g_n}) = f(x_{hg_1}, \dots, x_{hg_n}).$$

If $\alpha \in E \setminus F$, we have $\alpha = f(x_{g_1}, \dots, x_{g_n})$. Let $h \in G \setminus \{g_i g_1^{-1} \mid 1 \leq i \leq n\}$. Then $hg_1 \notin \{g_1, \dots, g_n\}$ and $h \circ \alpha \neq \alpha$. Thus $E^G = F$. Clearly E/F is not a Galois extension.

In any case we establish a light version of Artin's theorem for profinite groups.

Theorem 11.4.13 (Artin). *Let L be any field and G any profinite group of automorphisms of L, i.e., G is a subgroup of* $\{\sigma : L \to L \mid \sigma$ is a field automorphism$\}$.

Assume that for any $\alpha \in L$, the stabilizer

$$G_\alpha = \{\sigma \in G \mid \sigma\alpha = \alpha\}$$

is of finite index in G. Then L/L^G is a Galois extension with Galois group G.

Proof: The orbit of α is $C(\alpha) = \{\tau\alpha \mid \tau \in G\}$, which is a finite set with $[G : G_\alpha] = n$ elements. Let $C(\alpha) = \{\alpha_1, \dots, \alpha_n\} = \{\sigma_1\alpha, \dots, \sigma_n\alpha\}$ and let $f(x) = \prod_{i=1}^n (x - \sigma_i\alpha)$.

Since $\tau f(x) = f(x)$ for all $\tau \in G$ we have $f(x) \in K[x]$, where $K = L^G$. Then $f(x)$ is a separable polynomial and all the conjugates of α are in L. It follows that L/K is a Galois extension.

Let $H = \mathrm{Gal}(L/L^G)$. Then $G \subseteq H$. Let $i : G \hookrightarrow H$ be the natural embedding. If N is a normal subgroup of H, then $[H : N] < \infty$ and $\left[L^N : K\right]$ is a finite extension, say $L^N = K(\alpha)$. Thus $i^{-1}(N) = N \cap G \supseteq \bigcap_{j=1}^{n} G_{\alpha_j}$, where $\alpha_1, \ldots, \alpha_n$ are the conjugates of α. By Exercise 11.7.6, G_{α_j} is open for all j and so is $i^{-1}(N)$. Therefore i is a continuous map. Since G is compact, it follows that G is closed in H and $K = L^G = L^H$. By Theorem 11.4.9, we have $G = H$. □

11.5 Results on Global Class Field Theory

In this section and the next, we will not present the proofs of the stated results. We present only the main results, since a systematic treatment is beyond the scope of this book. The principal references are [17, 76, 90, 115].

In what follows K/k is a function field with $k = \mathbb{F}_q$. Let L/K be a finite Galois extension and $S(L/K) = \{\wp \mid \wp \in \mathbb{P}_K, \wp \text{ is totally decomposed in } L\}$. Then $\wp \in S(L/K)$ if and only if $\left(\frac{L/K}{\wp}\right) = \{1\}$ (see Exercise 11.7.2).

Theorem 11.5.1 (Bauer). *For two finite Galois extensions L_1 and L_2 of K, we have $S(L_1/K) \subseteq S(L_2/K)$ if and only if $L_2 \subseteq L_1$.*

Proof.
($\Longleftarrow$) This is immediate.
($\Longrightarrow$) Let $L = L_1L_2$. Then $S(L/K) = S(L_1/K)$ and by the Čebotarev density theorem (Theorem 11.2.20),

$$\delta(S(L/K)) = \frac{1}{[L:K]} = \frac{1}{[L_1:K]} = \delta(S(L_1/K)).$$

This implies that $[L : K] = [L_1 : K]$ and since $L_1 \subseteq L = L_1L_2$, it follows that $L_1 = L_1L_2$, or, equivalently, $L_2 \subseteq L_1$. □

Definition 11.5.2. The *idele group* J_K of K is defined as

$$J_K := \left\{ (\ldots, x_\wp, \ldots) \in \prod_{\wp \in \mathbb{P}_K} K_\wp^* \mid x_\wp \in \vartheta_\wp^* \text{ for almost all } \wp \right\}.$$

The group J_K is provided with the following topology: a basis of open sets consists of the subsets of the form $\prod_{\wp \in \mathbb{P}_K} A_\wp$, where $A_\wp \subseteq K_\wp^*$ is open for all $\wp$ and $A_\wp = \vartheta_\wp^*$ for almost all $\wp \in \mathbb{P}_K$ ([17, p. 62]). In other words, the topology of J_K is generated by the open sets

$$U_S = \prod_{\wp \in S} A_\wp \times \prod_{\wp \notin S} \vartheta_\wp^*,$$

where S is a finite set and $S \subseteq \mathbb{P}_K$, $A_\wp \subseteq K_\wp^*$ is open.

We have $K^* \subseteq J_K$ under the diagonal embedding and K^* is a discrete subgroup of J_K.

Definition 11.5.3. We define the *idele class group* of K as $\mathfrak{C}_K = J_K/K^*$.

Let S be a finite set of prime divisors of K such that for some extension L/K, S contains all the ramified prime divisors. Let I^S be the free abelian group generated by the prime divisors $\wp \notin S$. In other words, $I^S = D_K/\langle S \rangle$.

If L/K is an abelian extension with Galois group G and $\wp \notin S$, $\left(\frac{L/K}{\wp}\right)$ consists of a unique element. This defines a function

$$\psi_{L/K}(\wp) = \left(\frac{L/K}{\wp}\right) \quad \text{from} \quad \mathbb{P}_K \setminus S \quad \text{into} \quad G.$$

Then $\psi_{L/K}$ can be extended to

$$\psi_{L/K} : I^S \longrightarrow G, \quad \psi_{L/K}\left(\wp_1^{a_1} \cdots \wp_r^{a_r}\right) = \psi_{L/K}(\wp_1)^{a_1} \cdots \psi_{L/K}(\wp_r)^{a_r}.$$

For $x \in J_K$, we write $(x)^S = \prod_{\wp \notin S} \wp^{v_\wp(x_\wp)} \in I^S$.

Definition 11.5.4. We say that the *reciprocity law* holds for an abelian extension L of K if there exists a homomorphism $\psi : J_K \longrightarrow \operatorname{Gal}(L/K)$ such that:

(i) ψ is continuous,
(ii) $\psi(K^*) = 1$,
(iii) $\psi(x) = \psi_{L/K}\left((x)^S\right)$ for $x \in J_K^S = \left\{(x_\wp)_{\wp \in \mathbb{P}_K} \mid x_\wp = 1, \wp \in S\right\}$, where S consists of the ramified prime divisors in L/K.

In this case $K^* \subseteq \ker \psi$. Therefore ψ can be viewed as $\psi : \mathfrak{C}_K = J_K/K^* \longrightarrow \operatorname{Gal}(L/K)$.

Theorem 11.5.5. *When there exists a map ψ satisfying the three conditions of Definition* 11.5.4, *it is unique.*

Proof. See [17, Chapter 7, Section 4, Proposition 4.1, p. 169]. □

The next theorem describes the global class field theory.

Theorem 11.5.6 (Takagi–Artin).

(i) *Every finite abelian extension L/K satisfies the reciprocity law.*
(ii) *The Artin map $\psi_{L/K}$ is surjective and its kernel is $K^* N_{L/K}(J_L)$, where $N_{L/K}$ is the norm map. Therefore $\psi_{L/K}$ induces an isomorphism from $\mathfrak{C}_K/N_{L/K}\mathfrak{C}_L$ onto* $\operatorname{Gal}(L/K)$.
(iii) *(Existence Theorem) For each open subgroup N of finite index in $\mathfrak{C}_K$, there exists a unique finite abelian extension L/K such that $N_{L/K}\mathfrak{C}_L = N$.*

Proof. [17, Chapter 7, Section 5, Theorem 5.1, p. 172]. □

Remark 11.5.7. Since the reciprocity law holds for any finite extension L/K we have the map

$$\phi_{L/K}\colon J \to \operatorname{Gal}(L/K).$$

By the universal property of inverse limits, we have the reciprocity law homomorphism ϕ:

$$\phi\colon J \to \operatorname{Gal}\left(K^{ab}/K\right),$$

where K^{ab} is the maximal abelian extension of K. Thus

$$K^{ab} = \bigcup_{\substack{L/K \text{ finite} \\ \text{abelian}}} L, \qquad \operatorname{Gal}\left(K^{ab}/K\right) \cong \varprojlim_{L} \operatorname{Gal}(L/K),$$

where ϕ is the unique homomorphism given by $\phi_{L/K}$. We have $\ker\phi = K^*$.

11.6 Results on Local Class Field Theory

Here we consider the completion $K_\wp$ of a congruence function field K at a prime divisor $\wp$. Recall that $K_\wp$ is of the form $k((\pi))$ for some finite field k. In this section K will denote a field of the form $k((\pi))$, where k is a finite field.

Theorem 11.6.1. *If L/K is a finite abelian extension, there exists a function $\psi_{L/K}$: $K^* \longrightarrow \operatorname{Gal}(L/K)$, $\psi_{L/K}(a) = (a, L/K)$, that induces an isomorphism between $K^*/N_{L/K}L^*$ and* $\operatorname{Gal}(L/K)$. □

Definition 11.6.2. The map $\psi_{L/K}$ of Theorem 11.6.1 is called *Artin's local map*.

Theorem 11.6.3 (Existence Theorem). *If $H \subseteq K^*$ is an open subgroup of finite index, then there exists a unique abelian extension L/K such that $H = N_{L/K}L^*$. Furthermore, if L_1 and L_2 are finite extensions of K^* we have $N_{L/K}L_1^* \supseteq N_{L/K}L_2^*$ if and only if $L_1 \subseteq L_2$.* □

11.7 Exercises

Exercise 11.7.1. Let K/k be a congruence function field, ℓ/k a finite extension, $L = K\ell$. If $\mathfrak{P} \in \mathbb{P}_L$ and $\mathfrak{p} = \mathfrak{P} \cap K$, prove that $\vartheta_{\mathfrak{P}} = \vartheta_{\mathfrak{p}}\ell$.

Exercise 11.7.2. Let L/K be a finite Galois extension of congruence function fields. Let $\mathfrak{p} \in \mathbb{P}_K$ be an unramified prime divisor. Show that $\mathfrak{p}$ splits completely in L/K if and only if $\left(\frac{L/K}{\mathfrak{p}}\right) = 1$.

Exercise 11.7.3. For a finite Galois extension of congruence function fields L/K set $S(L/K) = \{\mathfrak{p} \in \mathbb{P}_K \mid \mathfrak{p} \text{ splits completely in } L/K\}$. Prove that if L and L' are two finite Galois extensions of a congruence function field K such that $S(L/K)$ and $S(L'/K)$ differ by only finitely many elements, then $L = L'$.

Exercise 11.7.4. With the notation of Exercise 11.7.3, prove that the Dirichlet density of $S(L/K)$ is equal to $\frac{1}{[L:K]}$.

Exercise 11.7.5. If U is an open subgroup of a profinite group G, show that U is closed.

Exercise 11.7.6. Let G be a profinite group and let $[G : H] < \infty$. Prove that H is open and closed in G.

Exercise 11.7.7. Give an example of nonempty topological spaces A_i such that $\varprojlim_i A_i = \emptyset$.

Exercise 11.7.8. Prove that if A_i is a group for all i, and $\varphi_{ji} \colon A_j \to A_i$ is a group homomorphism, then $\varprojlim_i A_i \neq \emptyset$.

Exercise 11.7.9. Let $\left(A_i, \varphi_{ji}, I\right)$ be such that each A_i is a nonempty compact Hausdorff topological space and φ_{ji} is a surjective morphism for each $i, j \in I$. Prove that

$$\varphi_j \colon \varprojlim_i A_i \to A_j$$

is a surjection for all $j \in I$.

Exercise 11.7.10. Let G be any group. Let $\mathcal{A} := \left\{N \mid N \lhd G, \left|G/N\right| < \infty\right\}$. If $N, M \in \mathcal{A}$ we define

$$N \leq M \iff M \subseteq N.$$

Put $G_N := G/N$. Define

$$\begin{array}{rccc} \varphi_{MN} \colon & G_M & \to & G_N \\ & g \bmod M & \mapsto & g \bmod N. \end{array}$$

Then $\left\{G_N, \varphi_{MN}, \mathcal{A}\right\}$ is an inverse system. Let

$$\hat{G} := \varprojlim_N G_N = \varprojlim_N G/N.$$

$\hat{G}$ is called the *completion* of G. Show that there exists a canonical group homomorphism $\phi \colon G \to \hat{G}$ and that $\hat{G}$ is a complete topological space. Show that $\phi(G)$ is dense in $\hat{G}$. Is ϕ necessarily a monomorphism?

Exercise 11.7.11. Prove that if G is a finite group, then G is also a profinite group that is isomorphic to its own completion.

Exercise 11.7.12. Let G be any group and p a prime number. Set

$$\mathcal{A}_p := \left\{ N \mid N \lhd G, \left|G/N\right| = p^n < \infty, n \in \mathbb{N} \cup \{0\} \right\}.$$

Let $\hat{G}_p := \varprojlim_{N \in \mathcal{A}_p} G/N$. Is it true that $\hat{G} \cong \prod_{p \text{ prime}} \hat{G}_p$?

Exercise 11.7.13. If $G = \mathbb{Z}_p$, what is $\hat{G}_\ell$ for ℓ a prime number? Consider the cases $\ell = p$ and $\ell \neq p$ (see Exercise 11.7.12).

Exercise 11.7.14. In Example 11.3.20 show that $\chi(A) \cong \varprojlim_i \chi(A_i)$.

Exercise 11.7.15. Prove that $\chi(\mathbb{Q}/\mathbb{Z}) \cong \hat{\mathbb{Z}}$ and that $\chi(\mathbb{Q}_p/\mathbb{Z}_p) \cong \mathbb{Z}_p$. Also show that $\chi(\mathbb{Z}_p) \cong \mathbb{Q}_p/\mathbb{Z}_p$.

Exercise 11.7.16. Let p be a prime number and $q = p^u$ for some $u \in \mathbb{N}$. Let ℓ be another prime number, not necessarily distinct from p, and set $T_\ell := \bigcup_{n=0}^{\infty} \mathbb{F}_{q^{\ell^n}}$. Prove that

$$T_\ell \cap \Big(\prod_{\ell' \neq \ell} T_{\ell'} \Big) = \mathbb{F}_q \qquad \text{and} \qquad \bar{\mathbb{F}}_q = \prod_{\ell \text{ prime}} T_\ell.$$

Exercise 11.7.17. Let $G_n := \dfrac{\left(R_x/\left(x^{n+1}\right)\right)^*}{\mathbb{F}_q^*}$ where $R_x = \mathbb{F}_q[x]$ is the ring of polynomials in one variable. For $n \leq m$, consider the natural epimorphism

$$\varphi_{m,n}\colon G_m \to G_n.$$

Then $\left\{G_n, \varphi_{m,n}, \mathbb{N}\right\}$ is an inverse system. Prove that

$$G_\infty := \varprojlim_n G_n \cong \left\{ f(x) \in \mathbb{F}_q[[x]] \mid f(0) = 1 \right\},$$

where $\mathbb{F}_q[[x]]$ is the formal power series in one variable over $\mathbb{F}_q$.

Exercise 11.7.18. Let K be a local field that is complete with respect to a discrete valuation v whose residue class field is finite. Let ϑ be the ring of integers and $\mathfrak{p}$ the maximal ideal. Prove that

$$\begin{aligned} \vartheta &\overset{\varphi}{\cong} \varprojlim_n \vartheta/\mathfrak{p}^n \\ a &\mapsto \Big(\prod_n a \bmod \mathfrak{p}^n\Big) \end{aligned}$$

where

$$\begin{aligned} \varphi_{m,n}\colon \quad \vartheta/\mathfrak{p}^m &\to \vartheta/\mathfrak{p}^n \\ a \bmod \mathfrak{p}^m &\mapsto a \bmod \mathfrak{p}^n \end{aligned}$$

is the natural map for $m \geq n$. In particular, ϑ is a profinite ring.

Exercise 11.7.19. With the notation of Exercise 11.7.18, the group of units U of ϑ is closed in ϑ, hence Hausdorff and compact. Furthermore, the subgroups $U^{(n)} := 1+\mathfrak{p}^n$ form a basis of neighborhoods of $1 \in U$. Prove that

$$U \cong \varprojlim U/U^{(n)}$$

and conclude that U is a profinite group.

Exercise 11.7.20. Let K/F be any Galois extension of fields with Galois group $G = \mathrm{Gal}(K/F)$. Let H be a subgroup of G. Prove that $K^H = K^{\bar{H}}$, where $K^A := \{\alpha \in K \mid \sigma\alpha = \alpha \ \forall\, \sigma \in A\}$ and $\bar{H}$ denotes the closure of H.

Exercise 11.7.21. In this exercise, the M_i's could be other structures such as groups or fields. Let I be a direct poset and $(M_i)_{i\in I}$ a family of A-modules, where A is a commutative ring with unit. For $i \leq j$, let $\mu_{ij}\colon M_i \to M_j$ be an A-homomorphism and assume that the set of μ_{ij}'s satisfies:

(i) $\mu_{ii} = \mathrm{Id}_{M_i}$ for all $i \in I$.
(ii) $\mu_{ik} = \mu_{jk} \circ \mu_{ij}$ whenever $i \leq j \leq k$.

Then (M_i, μ_{ij}, I) is a *direct system*. Set $C = \bigoplus_{i\in I} M_i$ and let D be the submodule of C generated by the elements of the form $x_i - \mu_{ij}(x_i)$ with $i \leq j$. Let $M = C/D$. Let $\mu\colon C \to M$ be the projection and let $\mu_i = \mu|_{M_i}$. Then (M_i, μ_i, I), $\mu_i\colon M_i \to M$, is called the *direct limit* of the system (M_i, μ_{ij}, I) and we write $M := \varinjlim_i M_i$. We have $\mu_i = \mu_j \circ \mu_{ij}$ if $i \leq j$.

Prove that every element M can be written as $\mu_i(x_i)$ for some $i \in I$ and some $x_i \in M_i$.

Exercise 11.7.22. Prove that if $\mu_i(x_i) = 0$, there exists $j \geq i$ such that $\mu_{ij}(x_i) = 0$ in M_i.

Exercise 11.7.23. Show that the direct limit satisfies the following universal property. Let P be an A-module such that for each $i \in I$, there exists an A-module homomorphism $\alpha_i\colon M_i \to P$ such that $\alpha_i = \alpha_j \circ \mu_{ij}$ whenever $i \leq j$. Then there exists a unique homomorphism $\alpha\colon M \to P$ satisfying $\alpha_i = \alpha \circ \mu_i$ for all $i \in I$.

Conclude that the direct limit is unique up to isomorphism.

Exercise 11.7.24. Let $(M_i)_{i\in I}$ be a family of A-submodules of an A-module such that for every $i, j \in I$, there exists $k \in I$ such that $M_i + M_j \subseteq M_k$. Define $i \leq j$ to mean $M_i \subseteq M_j$ and let $\mu_{ij}\colon M_i \to M_j$ be the natural embedding. Show that

$$\varinjlim_{i\in I} M_i = \sum_{i\in I} M_i = \bigcup_{i\in I} M_i.$$

Exercise 11.7.25. Assume that L/K is a Galois extension, $L = \bigcup_{i\in I} K_i$, where $[K_i : K] < \infty$, K_i/K is a Galois extension, and $L = \varinjlim_{i\in I} K_i$. Prove that

$$\mathrm{Gal}(L/K) = \mathrm{Gal}(\varinjlim_{i\in I} K_i/K) \cong \varprojlim_{i\in I} \mathrm{Gal}(K_i/K).$$

12

Cyclotomic Function Fields

12.1 Introduction

As we have seen, there is a close analogy between algebraic number fields and algebraic functions, and this analogy is even more pronounced if we consider the case of congruence function fields, that is, when the field of constants is finite.

Since the nineteenth century, it is well known that every abelian extension of $\mathbb{Q}$ is contained in a cyclotomic extension. This result is known as the Kronecker–Weber theorem. In other words, the maximal abelian extension of $\mathbb{Q}$ is $\bigcup_{n=1}^{\infty} \mathbb{Q}(\zeta_n)$, where $\zeta_n = e^{e\pi i/n}$. Note that ζ_n is a torsion element of $\mathbb{Z}$ acting on $\overline{\mathbb{Q}}^*$, where $\overline{\mathbb{Q}}$ denotes an algebraic closure of $\mathbb{Q}$. More precisely, $\overline{\mathbb{Q}}^*$ is a multiplicative abelian group, that is, a $\mathbb{Z}$-module. The torsion of $\overline{\mathbb{Q}}$ is $M = \operatorname{tor} \overline{\mathbb{Q}}^* = \{\zeta \in \overline{\mathbb{Q}}^* \mid \zeta^n = 1, \text{ some } n \in \mathbb{N}\} =$ roots of 1. Therefore $\mathbb{Q}(M)$ is the maximal abelian extension of $\mathbb{Q}$.

If we want to describe something similar for function fields, the role of $\mathbb{Q}$ must be played by $k(T)$, where k is a finite field, $|k| = q$, and T is a variable. The role of $\mathbb{Z}$ will then be played by $k[T]$. This choice is not canonical since $k(T) = k\big(\frac{aT+b}{cT+d}\big)$, $ad - bc \neq 0$, $a, b, c, d \in k$, and the corresponding ring of polynomials is $k\big[\frac{aT+b}{cT+d}\big]$. Here the infinite prime is different in each case. In the case of $\mathbb{Z}$, the infinite prime is canonical and it corresponds to the unique archimedean valuation of $\mathbb{Q}$. Furthermore, for $n \in \mathbb{N}$, $k(T^{1/n})$ and $k(T^n)$ are rational function fields over k and $\big[k(T^{1/n}) : k(T)\big] = n = \big[k(T) : k(T^n)\big]$. Notice that the case of a rational congruence function field $k(T)$ differs from that of $\mathbb{Q}$ in the following sense: If $A \subseteq \mathbb{Q}$ is a field, then $A = \mathbb{Q}$ and if B is an overfield of $\mathbb{Q}$ strictly containing $\mathbb{Q}$, then $\mathbb{Q}$ is not isomorphic to B. This is not the case for $k(T)$.

Using the ideas of Carlitz [14], Hayes [61] gave a description for the class field theory of a rational function field over the finite field k similar to that of $\mathbb{Q}$. In the rest of this chapter we describe the work of Carlitz and Hayes.

12.2 Basic Facts

As usual, let $k = \mathbb{F}_q$ be the finite field of cardinality q. Let K be a rational function field over $\mathbb{F}_q$, $K = \mathbb{F}_q(T)$, and let $R_T = \mathbb{F}_q[T]$. Here K will play the role of $\mathbb{Q}$ and R_T the role of $\mathbb{Z}$. Let $\bar{K}$ be an algebraic closure of K and set

$$\begin{aligned} A = \operatorname{End}_{\mathbb{F}_q}(\bar{K}) =\{\varphi : \bar{K} \to \bar{K} \mid \varphi(a+b) = \varphi(a) + \varphi(b), \\ \varphi(\alpha a) = \alpha\varphi(a) \ \forall\, \alpha \in \mathbb{F}_q \text{ and } \forall\, a, b \in \bar{K}\}. \end{aligned}$$

Thus, A is the $\mathbb{F}_q$-algebra (meaning an $\mathbb{F}_q$-module that has a ring structure) consisting of the $\mathbb{F}_q$-endomorphisms of the abelian additive group of $\bar{K}$.

We consider two special elements of A.

Definition 12.2.1.

(i) Let $\varphi \in A$ be the Frobenius automorphism of $\bar{K}/\mathbb{F}_q$, that is, $\varphi\colon \bar{K} \to \bar{K}$ is given by $u \mapsto u^q$.

(ii) Denote by μ_T the element of A that acts as multiplication by T, that is, $\mu_T : \bar{K} \to \bar{K}$ is given by $u \mapsto Tu$.

Given any $f(T) \in R_T$, the substitution $T \to \varphi + \mu_T$ in f gives an element of A. In other words, if $f(T) = a_n T^n + \cdots + a_1 T + a_0$ then

$$f(\varphi + \mu_T)(u) = a_n(\varphi + \mu_T)^n(u) + \cdots + a_1(\varphi + \mu_T)(u) + a_0(u)$$

for all $u \in \overline{K}$. Thus we obtain a map $\xi : R_T \to A$ given by $\xi(T) = \varphi + \mu_T$, and $\xi(f(T)) = f(\varphi + \mu_T)$. It is easy to see that ξ is a ring homomorphism. Therefore ξ provides $\bar{K}$ with the structure of a R_T-module.

Remark 12.2.2. We have

$$\begin{aligned} (\varphi \circ \mu_T)(u) &= \varphi(Tu) = T^q u^q, \\ (\mu_T^q \circ \varphi)(u) &= \mu_T^q(u^q) = T^q u^q. \end{aligned}$$

Therefore $\varphi \circ \mu_T = \mu_T^q \circ \varphi$. In particular, $\varphi \circ \mu_T \neq \mu_T \circ \varphi$.

Notation 12.2.3. If $u \in \bar{K}$ and $M \in R_T$ we write $u^M = M(\varphi + \mu_T)(u)$. That is, $M \circ u = \xi(M)(u) = M(\varphi + \mu_T)(u)$.

Remark 12.2.4. For $\alpha \in \mathbb{F}_q$, we have $u^\alpha = \alpha u$, so the R_T-action preserves the $\mathbb{F}_q$-algebra structure of the algebraic closure of K.

For $u \in \overline{K}$ and $M, N \in R_T$, we have

$$u^{M+N} = u^M + u^N \quad \text{and} \quad u^{MN} = (u^M)^N.$$

Theorem 12.2.5. *If $M = a_d T^d + a_{d-1}T^{d-1} + \cdots + a_1 T + a_0$ with $a_d \neq 0$, then*

$$u^M = \sum_{i=0}^{d} \begin{bmatrix} M \\ i \end{bmatrix} u^{q^i},$$

where $\begin{bmatrix} M \\ i \end{bmatrix}$ is a polynomial of R_T of degree $(d-i)q^i$. Furthermore, we have

$$\begin{bmatrix} M \\ 0 \end{bmatrix} = M, \quad \begin{bmatrix} M \\ d \end{bmatrix} = a_d \quad \text{and} \quad \begin{bmatrix} M \\ i \end{bmatrix} = a_i + \sum_{n=i+1}^{d} a_n h_n(i,T)$$

where each $h_n(i,T) = \sum_{0 \le j_1 \le j_2 \le \cdots \le j_{n-i} \le i} T^{q^{j_1}+q^{j_2}+\cdots+q^{j_{n-i}}}$ is a polynomial of degree $(n-i)q^i$ (here we put $j_0 = 0$).

Proof. First we consider the case u^{T^n}. We will prove by induction on n that

$$u^{T^n} = \sum_{i=0}^{n-i} \left(\sum_{0 \le j_1 \le j_2 \le \ldots \le j_{n-i} \le i} T^{q^{j_1}+q^{j_2}+\cdots+q^{j_{n-i}}} \right) u^{q^i} + u^{q^n},$$

i.e.,

$$u^{T^n} = \sum_{i=0}^{n-1} h_n(i,T)\, u^{q^i} + u^{q^n}. \tag{12.1}$$

For $n = 1$ we have $u^T = (\varphi + \mu_T)(u) = u^q + Tu = Tu + u^q$ and

$$\sum_{i=0}^{n-1} h_n(i,T) u^{q^i} + u^{q^n} = h_1(0,T) u^{q^0} + u^q,$$

$$h_1(0,T) = \sum_{0 \le j_1 \le \cdots \le j_1 = j_{1-0} \le 0} T^{(q^{j_1}+\cdots+q^{j_{1-0}})} = T^{q^0} = T^1 = T.$$

Thus (12.1) holds for $n = 1$. Assume that (12.1) holds for a given $n \ge 1$. For $n+1$ we have

$$\begin{aligned} u^{T^{n+1}} &= (u^{T^n})^T = (\mu_T + \varphi)(u^{T^n}) = Tu^{T^n} + (u^{T^n})^q \\ &= \sum_{i=0}^{n} \left(\sum_{0 \le j_1 \le \cdots \le j_{n-i+1} \le i} T^{q^{j_1}+\cdots+q^{j_{n-i+1}}} \right) u^{q^i} + u^{q^{n+1}}. \end{aligned}$$

Thus $u^{T^{n+1}} = \sum_{i=0}^{n} h_{n+1}(i,T) u^{q^i} + u^{q^{n+1}}$ and (12.1) holds for $u^{T^{n+1}}$. Define $h_n(i,T) = 1$ if $i = n$ and $h_n(i,T) = 0$ if $i > n$.

Now $M = a_0 + a_1 T + \cdots + a_d T^d = \sum_{n=0}^{d} a_n T^n$, where $a_d \neq 0$. Hence

$$u^M = u^{\sum_{n=0}^{d} a_n T^n} = \sum_{n=0}^{d} a_n u^{T^n} = \sum_{i=0}^{d} \left(\sum_{n=0}^{d} a_n h_n(i, T) \right) u^{q^i}.$$

Therefore for $0 \leq i \leq d-1$, we have

$$\begin{aligned} \begin{bmatrix} M \\ i \end{bmatrix} &= \sum_{n=0}^{d} a_n h_n(i, T) = \sum_{n=i}^{d} a_n h_n(i, T) \\ &= a_i + \sum_{n=i+1}^{d} a_n h_n(i, T). \end{aligned}$$

Finally,

$$\begin{bmatrix} M \\ 0 \end{bmatrix} = \sum_{n=0}^{d} a_n h_n(0, T) = \sum_{n=0}^{d} a_n T^n = M,$$

$$\begin{bmatrix} M \\ d \end{bmatrix} = \sum_{n=0}^{d} a_n h_n(d, T) = a_d h_d(d, T) = a_d. \qquad \square$$

Remark 12.2.6. It is easy to see that if $\begin{bmatrix} M \\ i \end{bmatrix} = 0$ for $i < 0$ and $i > \deg M$, then

$$\begin{bmatrix} \alpha M + \beta N \\ i \end{bmatrix} = \alpha \begin{bmatrix} M \\ i \end{bmatrix} + \beta \begin{bmatrix} N \\ i \end{bmatrix} \quad \text{for} \quad \alpha, \beta \in \mathbb{F}_q,$$

$$\begin{bmatrix} T^{d+1} \\ i \end{bmatrix} = T \begin{bmatrix} T^d \\ i \end{bmatrix} + \begin{bmatrix} T^d \\ i-1 \end{bmatrix}^q.$$

Remark 12.2.7. It turns out that in spite of the fact that the action u^M is technically complicated, it is the counterpart in $\overline{\mathbb{Q}}^*$ to exponentiation.

More precisely, $\mathbb{Z}$ acts on $\overline{\mathbb{Q}}^* = \overline{\mathbb{Q}} \setminus \{0\}$ as follows: For $n \in \mathbb{Z}$ and $u \in \overline{\mathbb{Q}}^*$, put $nu = u^n$. The cyclotomic number fields correspond to $\{u \in \bar{\mathbb{Q}}^* \mid u^n = 1\} = \{\zeta_n^a\}_{a=0}^{n-1}$, where $\zeta_n = e^{2\pi i/n}$.

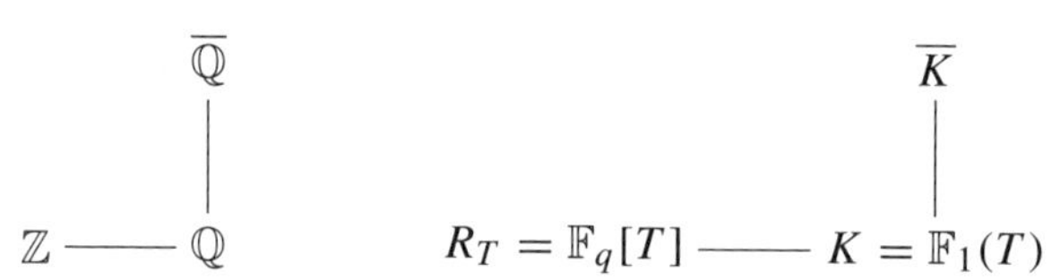

In our case R_T acts on $\overline{K}$ by exponentiation: For $M \in R_T$ and $u \in \overline{K}$, we have $M \circ u = u^M$. The cyclotomic function fields will correspond to $\{u \in \overline{K} \mid u^M = 0\}$.

Definition 12.2.8. Let Λ_M be the set of elements in $\bar{K}$ corresponding to the M-torsion of $\overline{K}$. Thus

$$\Lambda_M = \{u \in \bar{K} \mid u^M = 0\} \text{ is the set of zeros of the polynomial } u^M \text{ in } u.$$

Λ_M is called the *Carlitz–Hayes module of M*.

Now R_T is a commutative ring, so if $u \in \Lambda_M$ and $N \in R_T$, we have

$$N \circ u = u^N \in \Lambda_M$$

since $M \circ u^N = (u^N)^M = u^{NM} = (u^M)^N = 0^N = 0$. Therefore we obtain the following result:

Proposition 12.2.9. *Λ_M is an R_T-submodule of $\overline{K}$.* □

Remark 12.2.10. If $\alpha \in \mathbb{F}_q \setminus \{0\}$, we have $\Lambda_M = \Lambda_{\alpha M}$ since

$$\lambda^{\alpha M} = (\lambda^M)^\alpha = \alpha\lambda^M = 0 \Longleftrightarrow \lambda^M = 0.$$

Proposition 12.2.11. *Considered as a polynomial in u over K, u^M is a separable polynomial of degree q^d, where $d = \deg M$. Therefore Λ_M is a finite set with q^d elements. Furthermore, Λ_M is a vector space of dimension d over $\mathbb{F}_q$.*

Proof. We have $u^M = \sum_{i=0}^{d} \begin{bmatrix} M \\ i \end{bmatrix} u^{q^i}$. Thus $\frac{d}{du}(u^M) = \begin{bmatrix} M \\ 0 \end{bmatrix} = M \neq 0$, where $\frac{d}{du}(u^M)$ is constant with respect to u. It follows that u^M is a separable polynomial of degree q^d, and $|\Lambda_M| = \deg_u u^M = q^d$. Finally, since Λ_M is an $\mathbb{F}_q$-module, we have $\dim_{\mathbb{F}_q} \Lambda_M = d$. □

Remark 12.2.12. Over $\mathbb{Q}$ we have $\Lambda_n = \{\xi \in \overline{\mathbb{Q}}^* \mid \xi^n = 1\} = W_n \cong \prod_{i=1}^r W_{p_i^{\alpha_i}}$, where $n = p_1^{\alpha_1} \cdots p_r^{\alpha_r}$, $p_1, \dots, p_r$ are rational primes, and W_s denotes the group of sth roots of 1. Thus Λ_n is $\mathbb{Z}$-cyclic.

One would think intuitively that the same happens over $\bar{K}$, i.e.,

$$\Lambda_M = \{u \in \overline{K} \mid u^M = 0\} \cong \prod_{i=1}^{r} \Lambda_{P_i^{\alpha_i}},$$

where $M = \prod_{i=1}^r P_i^{\alpha_i}$, $P_1, \dots, P_r$ are irreducible polynomials in R_T, and Λ_M is R_T-cyclic. It turns out that this is true.

Proposition 12.2.13. *If $M = \prod_{i=1}^i P_i^{\alpha_i}$, then $\Lambda_M \cong \bigoplus_{i=1}^r \Lambda_{P_i^{\alpha_i}}$ as R_T-modules.*

Proof. We know that Λ_M is an R_T-module and R_T is a principal ideal domain. Every torsion R_T-module A decomposes as $A = \bigoplus_P A(P)$, where the sum runs over all prime elements of R_T and $A(P) = \{a \in A \mid P^n \circ a = 0 \text{ for some } n \in \mathbb{N}\}$.

For $A = \Lambda_M$, we have (see Exercise 12.10.4)

$$A(P) = \begin{cases} 0 & \text{if } P \notin \{P_1, \dots, P_r\}, \\ \Lambda_{P_i^{\alpha_i}} & \text{if } P = P_i. \end{cases}$$

Thus $\Lambda_M \cong \bigoplus_{i=1}^r \Lambda_{P_i^{\alpha_i}}$. □

Proposition 12.2.14. *Assume that $M = P^n$ for some irreducible polynomial $P \in R_T$ and some positive integer n. Then Λ_M is a cyclic R_T-module.*

Proof. We proceed by induction on n. For $n = 1$, let ξ be a nonzero element of Λ_P. Define $\phi : R_T \to \Lambda_P$ given by $N \mapsto \xi^N$. Notice that $\phi \neq 0$ since $\phi(1) = \xi^1 = \xi \neq 0$. On the other hand, $\phi(P) = \xi^P = 0$, so $P \in \ker\phi$ and $(P) \subseteq \ker\phi$. Now since R_T is a principal ideal domain and P is a nonzero irreducible polynomial, it follows that (P) is a maximal ideal. Hence

$$(P) \subseteq \ker\phi \subsetneqq R_T \quad \text{and} \quad (P) = \ker\phi.$$

(We might also proceed as follows: If $N \notin (P)$, we have $(P, N) = 1$. Let $A, B \in R_T$ be such that $1 = AP + BN$. If $\phi(N) = 0 = \xi^N$ we have $\xi = \xi^1 = \xi^{PA+NB} = (\xi^P)^A + (\xi^N)^B = 0 + 0 = 0$.) Returning to our proof, we obtain

$$R_T/(P) \cong R_T/\ker\phi \cong \phi(R_T).$$

On the other hand, we have $|\phi(R_T)| = |R_T/(P)| = q^d = |\Lambda_P|$. Hence $\phi(R_T) = \Lambda_P$ and Λ_P is isomorphic to $R_T/(P)$. Finally, for any $S \in R_T$, $R_T/(S)$ is a cyclic R_T-module (because 1 is a generator). Thus Λ_P is R_T-cyclic (or simply $\Lambda_P = \phi(R_T) = \{\xi^N | N \in R_T\} = \langle\xi\rangle$).

Now for any $n \in \mathbb{N}$ we consider

$$\begin{aligned} \theta : \Lambda_{P^{n+1}} &\to \Lambda_{P^n} \\ u &\mapsto u^P. \end{aligned}$$

Then θ is an R_T-homomorphism and $\ker\theta = \Lambda_P$.

It follows that $\Lambda_{P^{n+1}}/\Lambda_P$ is isomorphic to $\theta(\Lambda_{P^{n+1}})$ and

$$|\theta(\Lambda_{P^{n+1}})| = \left|\Lambda_{P^{n+1}}/\Lambda_P\right| = \frac{q^{d(n+1)}}{q^d} = q^{nd} = |\Lambda_{P^n}|.$$

Therefore θ is onto and $\Lambda_{P^{n+1}}/\Lambda_P \cong \Lambda_{P^n}$.

Let $\lambda \in \Lambda_{P^{n+1}}$ be such that $\lambda^P = \theta(\lambda)$ generates Λ_{P^n}. We will prove that λ generates $\Lambda_{P^{n+1}}$.

Let $u \in \Lambda_{P^{n+1}}$. Then $\theta(u) = u^P = \theta(\lambda)^A = \lambda^{PA}$ for some $A \in R_T$. It follows that $u - \lambda^A \in \Lambda_P = \ker\theta$. Since $\theta(\lambda^{P^n}) = \lambda^{P^{n+1}} = 0$, λ^{P^n} belongs to Λ_P. Now λ^P generates Λ_{P^n}, so $(\lambda^P)^{P^{n-1}} = \lambda^{P^n} \neq 0$. It follows from the case $n = 1$ (or the fact that Λ_P is a 1-dimensional $R_T/(P)$-vector space) that λ^{P^n} is a generator of Λ_P. Therefore there exists $B \in R_T$ such that $u - \lambda^A = \lambda^{P^n B}$, so $u = \lambda^{A+P^n B} \in \langle\lambda\rangle$. Thus λ generates $\Lambda_{P^{n+1}}$ as an R_T-module and $\Lambda_{P^{n+1}}$ is a cyclic R_T-module. □

Corollary 12.2.15. *Let P be an irreducible polynomial in R_T. Then:*

(i) *Λ_P is a one-dimensional $R_T/(P)$-vector space whose scalar product is given by $u^{N+(P)} = u^N$ for each $u \in \Lambda_P$ and $N \in R_T$.*
(ii) *For $n \in \mathbb{N}$, we have $\Lambda_{P^n} \subseteq \Lambda_{P^{n+1}}$ and $\Lambda_{P^{n+1}}/\Lambda_P \cong \Lambda_{P^n}$.*

(iii) *Given $n \in \mathbb{N}$, if $\lambda \in \Lambda_{P^{n+1}}$ is such that λ^P generates Λ_{P^n}, then λ generates $\Lambda_{P^{n+1}}$. Conversely, if λ generates $\Lambda_{P^{n+1}}$, then λ^P generates Λ_{P^n}.* □

Corollary 12.2.16. *Let M be a nonzero element of R_T and let $M = \alpha P_1^{n_1} \cdots P_r^{n_r}$ be its factorization in R_T in terms of irreducible monic polynomials. For each $i = 1, \ldots, r$, let λ_i be a generator of $\Lambda_{P_i^n}$. Then Λ_M is a cyclic R_T-module and $\lambda_1 + \cdots + \lambda_r$ is a generator of Λ_M.*

Proof. We have $\Lambda_M = \bigoplus_{i=1}^r \Lambda_{P_i^{\alpha_i}}$. Each $\Lambda_{P_i^{\alpha_i}}$ is a cyclic R_T–module and $\Lambda_{P_i^{\alpha_i}}$ is the P_ith primary component of Λ_M. The result follows. □

A more precise version of Corollary 12.2.16 is the following.

Theorem 12.2.17. *For each $M \in R_T \setminus \{0\}$, the R_T-module Λ_M is canonically isomorphic to $R_T/(M)$. In particular, Λ_M is a cyclic R_T-module.*

Proof. If λ is a generator of Λ_M, define $\theta\colon R_T \to \Lambda_M$ given by $A \mapsto \lambda^A$. Then θ is an epimorphism of R_T-modules and $\Lambda_M \cong R_T/\ker\theta$, where $\ker\theta = \{A \in R_T \mid \lambda^A = 0\} = \operatorname{ann}(\lambda) = \operatorname{ann}(\Lambda_M)$.

Clearly, $M \in \ker\theta$ since $\lambda^M = 0 \quad (\lambda \in \Lambda_M)$. Thus $(M) \subseteq \ker\theta$. On the other hand, $|\Lambda_M| = |R_T/M| = q^d$, where $d = \deg M$. Therefore $\ker\theta = (M)$ and Λ_M is isomorphic to $R_T/(M)$. □

Definition 12.2.18. For $M \in R_T \setminus \{0\}$ we define $\Phi(M)$ as the order of the group of units of $R_T/(M)$, that is, $\Phi(M) = \left|(R_T/(M))^*\right|$. Equivalently $\Phi(M) = |\{N \in R_T | (N, M) = 1, \deg N < \deg M\}|$.

Remark 12.2.19. Φ is the analogue of the Euler function ϕ on $\mathbb{N}$, defined for $n \in \mathbb{N}$ by $\phi(n) = |\{m \in \mathbb{N} | (m, n) = 1, m < n\}|$.

Proposition 12.2.20. *For $M, N \in R_T$, we have:*

(i) *If $(M, N) = 1$, then $\Phi(MN) = \Phi(M)\Phi(N)$.*
(ii) *If $P \in R_T$ is irreducible, then $\Phi(P) = q^d - 1$, where $d = \deg P$.*
(iii) *If $P \in R_T$ is irreducible, then*

$$\Phi(P^n) = |R_T/(P^{n-1})|\Phi(P) = q^{nd} - q^{(n-1)d},$$

where $d = \deg P$.

Proof. Exercise 12.10.5. □

Proposition 12.2.21. *The R_T-cyclic module Λ_M contains precisely $\Phi(M)$ generators. In fact, if λ is any generator of Λ_M, then for $A \in R_T$, λ^A is a generator if and only if $(A, M) = 1$.*

Proof. Let λ be a generator of Λ_M. If $(A, M) = 1$, let $\xi \in \Lambda_M$ and let $B \in R_T$ be such that $\xi = \lambda^B$. Let $S, U \in R_T$ be such that $SA + UM = 1$. Then $B = SAB + UMB$. It follows that

$$\xi = \lambda^B = \lambda^{SAB+UMB} = \lambda^{SAB} + (\lambda^M)^{UB} = (\lambda^A)^{SB} + 0 = (\lambda^A)^{SB}.$$

Thus λ^A is a generator of Λ_M.

Conversely, if λ^A is a generator of Λ_M, then there exists $B \in R_T$ such that $\lambda^{AB} = \lambda$. Hence $\lambda^{AB-1} = 0$. Since λ is a generator it follows that if $\lambda^C = 0$ for some $C \in R_T$, then M divides C. Therefore M divides $AB - 1$. Thus $AB \equiv 1 \bmod M$ and $(A, M) = 1$. □

12.3 Cyclotomic Function Fields

Let $R_T = \mathbb{F}_q[T]$ and $K = \mathbb{F}_q(T)$ as before.

Definition 12.3.1. The pole divisor $\mathfrak{p}_\infty$ of T in K, defined by $(T)_K = \frac{\mathfrak{p}_0}{\mathfrak{p}_\infty}$, is called the *infinite prime* in K.

Definition 12.3.2. Let $M \in R_T \setminus \{0\}$. The field $K(\Lambda_M)$ generated over K by adjoining $\Lambda_M = \{u \in \bar{K} \mid u^M = 0\}$ is called the *cyclotomic function field* determined by M over K.

Proposition 12.3.3. *$K(\Lambda_M)/K$ is a Galois extension.*

Proof. Since $\Lambda_M \cong R_T/(M)$, which is a cyclic R_T-module generated by λ, we have $\lambda^{R_T} = \Lambda_M = \{\lambda^A | A \in R_T\}$, so $K(\Lambda_M) = K(\lambda)$. Indeed, any element $\xi \in \Lambda_M$ is of the form λ^A for some $A \in R_T$ and

$$\xi = A(\mu_T + \varphi)(\lambda) \in K(\lambda^q, \{T^S\lambda\}) = K(\lambda).$$

Finally, since $K(\Lambda_M)$ is the decomposition field of the separable polynomial $F(u) = u^M \in K[u]$, it follows that $K(\Lambda_M)/K$ is a Galois extension. □

Remark 12.3.4. Let $M(T) = a_d T^d + \cdots + a_1 T + a_0$. Then

$$\begin{aligned} u^M &= a_d u^{q^d} + \begin{bmatrix} M \\ d-1 \end{bmatrix} u^{q^{d-1}} + \cdots + \begin{bmatrix} M \\ 1 \end{bmatrix} u^q + Mu \in R_T[u], \\ u^M &= a_d \left[u^{q^d} + \cdots + a_d^{-1} Mu \right] \end{aligned}$$

with $u^{q^d} + \cdots + a_d^{-1} Mu \in R_T[u]$ and the leading coefficient is 1. It follows that the elements of Λ_M are integral over R_T.

Definition 12.3.5. We will denote the Galois group of $K(\Lambda_M)/K$ by G_M, i.e., $G_M = \mathrm{Gal}\,(K(\Lambda_M)/K)$.

Proposition 12.3.6. *The action of G_M over $K(\Lambda_M)$ commutes with the action of R_T. In other words, if $u \in K(\Lambda_M)$, $\sigma \in G_M$, and $N \in R_T$, then $\sigma(u^N) = \sigma(u)^N$.*

Proof. Let $u \in K(\Lambda_M)$. First note that $u^N \in K(\Lambda_M)$ since if $u = \sum_{i=1}^r a_i u_i$ with $a_i \in K$ and $u_i \in \Lambda_M$, we have $u^N = \sum_{i=1}^r a_i^N u_i^N$, where $a_i^N \in K$ and $u_i^N \in \Lambda_M$. Therefore $u^N \in K(\Lambda_M)$. Now

$$\sigma(u^N) = \sigma\left(\sum_{i=0}^{\deg N} \begin{bmatrix} N \\ i \end{bmatrix} u^{q_i}\right) = \sum_{i=0}^{\deg N} \begin{bmatrix} N \\ i \end{bmatrix} \sigma(u)^{q^i} = \sigma(u)^N. \qquad \square$$

When the fields under consideration are number fields, if $\mathbb{Q}(\zeta_n)/\mathbb{Q}$ is the cyclotomic extension, we have $\mathrm{Gal}\,(\mathbb{Q}(\zeta_n)/\mathbb{Q}) \cong U_n = (\mathbb{Z}/n\mathbb{Z})^*$. The analogue for function fields would be

$$\mathrm{Gal}(K(\Lambda_M)/K) \cong (R_T/(M))^* = G_M.$$

We will see that this is indeed the case.

Proposition 12.3.7. *The group G_M is a subgroup of $(R_T/(M))^*$. In particular, $K(\Lambda_M)/K$ is an abelian extension and*

$$[K(\Lambda_M) : K] \leq \Phi(M) = \left|(R_T/(M))^*\right|.$$

Proof. Since $K(\Lambda_M) = K(\lambda)$, an element σ of G_M is determined by its action on λ. Now, $\sigma\lambda$ is a conjugate of λ, so $\sigma(\lambda) \in \Lambda_M$ and $\sigma(\lambda) = \lambda^A$ for some $A \in R_T$. We will show that $\sigma\lambda$ must be a generator of Λ_M.

If $\xi \in \Lambda_M$, then $\sigma^{-1}(\xi) \in \Lambda_M$, so $\sigma^{-1}(\xi) = \lambda^B$ for some $B \in R_T$. Hence $\xi = (\sigma\lambda)^B$. Therefore $\sigma\lambda$ is a generator of Λ_M and it follows that $(A, M) = 1$. Thus $A \bmod M \in (R_T/(M))^*$. To see that A does not depend on λ, let λ_1 be another generator of Λ_M, say $\lambda_1 = \lambda^B$ for some $B \in R_T$. Then

$$\sigma\lambda_1 = \sigma(\lambda^B) = \sigma(\lambda)^B = \lambda^{AB} = (\lambda^B)^A = \lambda_1^A.$$

Now, if $\sigma(\lambda) = \lambda^A = \lambda^{A_1}$, we have $\lambda^{A-A_1} = 0$. Thus $A - A_1 \in (M)$ and $A \equiv A_1 \bmod M$.

Define $\theta : G_M \to (R_T/(M))^*$ given by $\sigma \mapsto A \bmod M$ where $\sigma\lambda = \lambda^A$.

If $\Psi \in G_M$, we have $\Psi(\lambda) = \lambda^B$ and $(\Psi \circ \sigma)(\lambda) = \Psi(\lambda^A) = \lambda^{AB}$. Hence $\theta(\Psi\sigma) = AB \bmod M = \theta(\Psi)\theta(\sigma)$. Therefore θ is a group homomorphism.

Finally, if $\theta(\sigma) - 1 \bmod M$, we have $\sigma \in \ker\theta$ and $\sigma\lambda = \lambda^1 = \lambda$, so $\sigma = \mathrm{Id}$ and θ is a monomorphism. It follows that

$$G_M \subseteq (R_T/(M))^* \quad \text{and} \quad |G_M| = [K(\Lambda_M) : K] \leq \left|(R_T/(M))^*\right| = \Phi(M).$$

Since $(R_T/(M))^*$ is abelian, G_M is abelian too and the proof is complete. $\square$

Definition 12.3.8. Let $S \in R_T$ be a monic polynomial. We define the *S-cyclotomic polynomial* or *the cyclotomic polynomial with respect to S* by

$$\Psi_S(u) = \prod_{\substack{(B,S)=1 \\ \deg B < \deg S}} (u - \lambda_S^B),$$

where λ_S is a generator of Λ_S. We have $\Psi_S(u) \in K(\Lambda_S)[u]$.

Proposition 12.3.9. *For any monic polynomial $S \in R_T$ we have $\Psi_S(u) \in K[u]$.*

Proof. Let $\sigma \in G_S = \mathrm{Gal}(K(\Lambda_S)/K)$. Then $\sigma(\lambda_S) = \lambda_S^A$, with $(A, S) = 1$. Therefore $\sigma(\Psi_S(u)) = \prod_{\substack{(B,S)=1 \\ \deg B < \deg S}} (u - \lambda_S^{AB})$. Now $(A, S) = 1$ and $(B, S) = 1$ imply $(AB, S) = 1$. If $AB = QS + B_1$ with $\deg B_1 < \deg S$, then $\lambda_S^{AB} = \lambda_S^{B_1}$. Similarly, if $AB = Q_1 S + B_1$ and $AC = Q_2 S + C_1$ with $\deg B_1 < \deg S$ and $\deg C_1 < \deg S$, then $AB \equiv AC \bmod S$ implies $B_1 \equiv C_1 \bmod S$.

Therefore $\prod_{\substack{(B,S)=1 \\ \deg B < \deg S}} (u - \lambda_S^{AB}) = \prod_{\substack{(B_1,S)=1 \\ \deg B_1 < \deg S}} (u - \lambda_S^{B_1}) = \Psi_S(u)$. It follows that $\sigma(\Psi_S(u)) = \Psi_S(u)$ for all $\sigma \in G_S$, and hence $\Psi_S(u) \in K[u]$. □

Remark 12.3.10. We have $\deg \Psi_S(u) = \Phi(S)$. For $R, S \in R_T$ we choose generators $\lambda_R, \lambda_S, \lambda_{RS}$ of Λ_R, Λ_S, and Λ_{RS} such that $\lambda_{RS}^R = \lambda_S$ and $\lambda_{RS}^S = \lambda_R$.

We wish to prove that we may choose such generators for all $M \in R_T$. More precisely:

Proposition 12.3.11. *There exists a system $\{\lambda_M\}_{\substack{M \in R_T \\ M \text{ monic}}}$ such that λ_M generates Λ_M as an R_T-module and for all $N, M \in R_T$ such that N divides M, we have $\lambda_M^N = \lambda_{M/N}$.*

Proof. We call a subset I of R_T *admissible* if for all $A \in I$, A is a monic polynomial and there exists $\{\lambda_A\}_{A \in I} \subseteq \overline{K}$ such that for all $A \in I$, λ_A generates Λ_A and if B is an element of I that divides A, we have $\lambda_A^B = \lambda_{A/B}$.

Let $\mathcal{A} = \{I | I \text{ is admissible}\}$. Then $\mathcal{A}$ is nonempty since $I = \{P, 1\} \in \mathcal{A}$, where P is a monic irreducible polynomial (here we choose λ_P to be any generator of Λ_P and $\lambda_1 = 0$).

We define a relation $\leq$ in $\mathcal{A}$ as follows: $I \leq J$ if $I \subseteq J$ and for all $A \in I$, $\lambda_{A,I} = \lambda_{A,J}$. Clearly $\leq$ is a partial order in $\mathcal{A}$ and if $\{I\}_{I \in \mathcal{A}}$ is a chain in $\mathcal{A}$, $\overline{I} = \bigcup_{I \in \mathcal{A}} I$ is an upper bound of $\{I\}_{I \in \mathcal{A}}$.

Let I_0 be a maximal element of $\mathcal{A}$. If I_0 does not contain all monic polynomials of R_T, there exists a monic polynomial M in $R_T \setminus I_0$.

Let $M = P_1^{\alpha_1} \cdots P_r^{\alpha_r}$. Note that if N is monic and M divides N, then $N \notin I_0$ since otherwise, $N \in I_0$ and $\lambda_M := \lambda_N^{N/M}$ would satisfy all the conditions.

Let $M \in R_T$ be a monic polynomial of minimal degree such that $M \notin I_0$. Then $P_1^{\beta_1} \cdots P_r^{\beta_r} \in I_0$ for all $\sum_{i=1}^r \beta_i \deg P_i < \sum_{i=1}^r \alpha_i \deg P_i$. Let $H_i = \frac{M}{P_i} \in I_0$, and let

λ_{H_i} be the generator of Λ_{H_i}. Since $\{P_i\}_{i=1}^r$ are relatively prime, there exist elements $\gamma_i \in R_T$ satisfying $1 = \sum_{i=1}^r \gamma_i P_i$. Let $\lambda_M := \lambda_{H_1}^{\gamma_1} + \cdots + \lambda_{H_r}^{\gamma_r}$. Then

$$\lambda_M^{P_i} = \lambda_{H_1}^{\gamma_1 P_i} + \cdots + \lambda_{H_i}^{\gamma_i P_i} + \cdots + \lambda_{H_r}^{\gamma_r P_i}, \quad \text{and} \quad \gamma_i P_i = 1 - \sum_{j \neq i} \gamma_j P_j.$$

Hence

$$\lambda_M^{P_i} = \left(\lambda_{H_1}^{P_i} - \lambda_{H_i}^{P_1}\right)^{\gamma_1} + \cdots + \left(\gamma_{H_{i-1}}^{P_i} - \lambda_{H_i}^{P_{i-1}}\right)^{\gamma_{i-1}} + \lambda_{H_i}$$
$$+ \left(\lambda_{H_{i+1}}^{P_i} - \lambda_{H_i}^{P_{i+1}}\right)^{\gamma_{i+1}} + \cdots + \left(\lambda_{H_r}^{P_i} - \lambda_{H_i}^{P_r}\right)^{\gamma_r}.$$

Now for all $j \neq i$, we have $\lambda_{H_j}^{P_i} = \lambda_{H_j/P_i} = \lambda_{M/P_iP_j} = \lambda_{H_i/P_j} = \lambda_{H_i}^{P_j}$.

Therefore $\lambda_M^{P_i} = \lambda_{H_i}$ and λ_M satisfies $\lambda_M^S = \lambda_{M/S}$ for all $S \in R_T$ such that $S \mid M$. In particular, $I_1 = I_0 \cup \{M\}$ is an element of $\mathcal{A}$ that is strictly larger than I_0. This contradicts the maximality of I_0 and proves the proposition. □

Remark 12.3.12. Since Λ_M is isomorphic to $R_T/(M)$ we may take $\lambda_M = 1 \bmod M$ for all M. However, Proposition 12.3.11 provides a system that does not depend on the identification $\Lambda_M \cong R_T/(M)$.

Proposition 12.3.13. *We have*

(1) *If N and M are two distinct monic polynomials in R_T, then $(\Psi_N(u), \Psi_M(u)) = 1$.*
(2) $u^M = \prod_{\substack{N \mid M \\ N \text{ monic}}} \Psi_N(u)$*, where M is a monic polynomial in R_T.*
(3) $\Psi_M(u) = \prod_{\substack{N \mid M \\ N \text{ monic}}} (u^N)^{\mu(M/N)}$*, where*

$$\mu(D) = \begin{cases} 1 & \text{if } D = 1, \\ (-1)^s & \text{if } D = P_1 \cdots P_s, \text{ where the } P_1, P_2, \ldots, P_s \text{ are} \\ & \quad \text{distinct irreducible monic polynomials of } R_T, \\ 0 & \text{otherwise,} \end{cases}$$

and M is a monic polynomial.

Proof. Exercises 12.10.6, 12.10.8, and 12.10.12. □

Proposition 12.3.14. *Let $P \in R_T$ be a monic irreducible polynomial of degree d and let $M = P^n$ with $n \in \mathbb{N}$. Then:*

(1) *No divisor in K other than $\mathfrak{p}_\infty$ and $\mathfrak{p}$ is ramified in $K(\Lambda_M)/K$. Here $(P)_K = \frac{\mathfrak{p}}{\mathfrak{p}_\infty^{\deg \mathfrak{p}}}$.*
(2) *The ramification index of $\mathfrak{p}$ in $K(\Lambda_M)/K$ is*

$$e(\mathfrak{p}) = \Phi(M) = q^{dn} - q^{d(n-1)} = [K(\Lambda_M) : K].$$

Proof. Let ϑ_M be the integral closure of R_T in $K(\Lambda_M)$. Since R_T is a Dedekind domain, then ϑ_M is a Dedekind domain (Theorem 5.7.7). The ramified primes in $K(\Lambda_M)/K$ other than the infinite prime $\mathfrak{p}_\infty$ are those appearing in the discriminant $\partial_{\vartheta_M/R_T}$.

$$\begin{array}{ccc} \vartheta_M & \text{———} & K(\Lambda_M) \\ | & & | \\ R_T & \text{———} & K \end{array}$$

Let λ be a generator of Λ_M. Then $R_T[\lambda] \subseteq \vartheta_M$. Set $g(u) := \operatorname{Irr}(\lambda, u, K) \in K[u]$. Let $f(u) = u^M$. Since $f(\lambda) = 0$, there exists $h(u) \in K[u]$ such that $f(u) = h(u)g(u)$. Therefore

$$M = f'(u) = h'(u)g(u) + h(u)g'(u). \tag{12.2}$$

Substituting u by λ in (12.2) we obtain

$$M = f'(\lambda) = h'(\lambda)g(\lambda) + h(\lambda)g'(\lambda) = h(\lambda)g'(\lambda).$$

It follows that $(g'(\lambda))_{\vartheta_M} \mid (M)_{\vartheta_M} = P^n\vartheta_M$. By Theorem 5.7.21, the different $\mathfrak{D}_{\vartheta_M/R_T}$ satisfies

$$\mathfrak{D}_{\vartheta_M/R_T} = \gcd\{(F'(\alpha)) \mid \alpha \text{ is integral}, K(\Lambda_M) = K(\alpha),\ F(u) = \operatorname{Irr}(\alpha, u, K)\}.$$

Therefore $\mathfrak{D}_{\vartheta_M/R_T} \mid (g'(\lambda))_{K(\Lambda_M)} = P^n = (\mathfrak{p}_1 \cdots \mathfrak{p}_h)^{en}$, where

$$P\vartheta_M = (\mathfrak{p}_1 \cdots \mathfrak{p}_h)^e. \tag{12.3}$$

It follows that the only possible ramified prime divisors in $K(\Lambda_M)/K$ are $\mathfrak{p}$ and $\mathfrak{p}_\infty$. This proves (1).

Next, we calculate $e = e_{K(\Lambda_M)/K}(\mathfrak{p}_i|P)$. Let $d = \deg P$. We have

$$\begin{aligned} u^{P^n} &= (u^{P^{n-1}})^P = \sum_{i=0}^{d} \begin{bmatrix} P \\ i \end{bmatrix} (u^{P^{n-1}})^{q^i} \\ &= u^{P^{n-1}} \left(\sum_{i=0}^{d} \begin{bmatrix} P \\ u \end{bmatrix} (u^{P^{n-1}})^{q^i - 1} \right) = u^{p^{n-1}} t(u) \end{aligned}$$

with $t(u) \in R_T[u]$ and

$$t(u) = \frac{u^{P^n}}{u^{P^{n-1}}} = \sum_{i=0}^{d} \begin{bmatrix} P \\ i \end{bmatrix} (u^{P^{n-1}})^{q^i - 1}.$$

Therefore $t(\alpha) = 0 \iff \alpha \in \Lambda_{P^n} \setminus \Lambda_{P^{n-1}}$, or in other words, $t(\alpha) = 0 \iff \alpha$ is generator of Λ_{P^n}. Recall that $\Lambda_{P^n}/\Lambda_{P^{n-1}} \cong \Lambda_P$ (see Exercise 12.10.3).

Therefore

$$t(u) = \prod_{(A,M)=1} (u - \lambda^A) = \begin{bmatrix} P \\ 0 \end{bmatrix} + \sum_{i=1}^{d} \begin{bmatrix} P \\ i \end{bmatrix} (u^{P^{n-1}})^{q^i-1}$$
$$= P + \sum_{i=1}^{d} \begin{bmatrix} P \\ i \end{bmatrix} (u^{P^{n-1}})^{q^i-1}.$$

For $u = 0$, we have

$$t(0) = \pm \prod_{(A,M)-1} \lambda^A = P. \tag{12.4}$$

Now by Theorem 12.2.5, $u^A = u(F(u))$ for some $F(u) \in R_T[u]$.

Thus $\lambda^A = \lambda F(\lambda)$ and λ divides λ^A in ϑ_M. If $(A, M) = 1$, then λ^A is a generator and by symmetry we obtain $\lambda^A \mid \lambda$, so

$$\lambda = \beta_A \lambda^A \tag{12.5}$$

with $\beta_A \in \vartheta_M^*$.

Using Equation (12.4) we obtain $\pm P = \beta_0 \lambda^{\Phi(M)}$ for some $\beta_0 \in \vartheta_M^*$. Hence (12.3) yields $(\mathfrak{p}_1 \cdots \mathfrak{p}_n)^e = (P)_{\vartheta_M} = (\lambda)^{\Phi(M)}$. Now $v_{\mathfrak{p}_i}(\lambda) \geq 1$, so $e = v_{\mathfrak{p}_i}((\mathfrak{p}_1 \ldots \mathfrak{p}_n)^e) = v_{\mathfrak{p}_i}\left(\lambda^{\Phi(M)}\right) \geq \Phi(M)$. Therefore $e \geq \Phi(M) = \left|(R_T/(M))^*\right| \geq [K(\Lambda_M) : K] \geq e$. It follows that

$$e = \Phi(M) = [K(\Lambda_M) : K] = q^{dn} - q^{d(n-1)}.$$

This proves (2) and the proposition. □

Remark 12.3.15. We have

$$t(u) = \frac{u^{P^n}}{u^{P^{n-1}}} = \frac{\prod_{N \mid P^n} \Psi_N(u)}{\prod_{N \mid P^{n-1}} \Psi_N(u)} = \Psi_{P^n}(u) = \prod_{(A,M)=1} (u - \lambda^A).$$

Thus the polynomial $t(u)$ found in the proof of Proposition 12.3.14 is nothing other than the P^n-cyclotomic polynomial.

Theorem 12.3.16. *Let $M \in R_T \setminus \{0\}$ be a monic polynomial. Then*

(1) $t(u) = \Psi_M(u) = \operatorname{Irr}(\lambda, u, K)$. *In particular, $\Psi_M(u)$ is an irreducible polynomial.*
(2) $G_M = \operatorname{Gal}(K(\Lambda_M)/K) \cong (R_T/(M))^*$.
(3) $[K(\Lambda_M) : K] = \Phi(M)$.
(4) *If $M = P^n$ for some irreducible polynomial P, then $\mathfrak{p}$ is totally ramified in $K(\Lambda_M)/K$, where $(P)_K = \frac{\mathfrak{p}}{\mathfrak{p}_\infty^{\deg P}}$.*

Proof. If $M = P^n$, where P is an irreducible polynomial, we have

$$[K(\Lambda_M) : K] = \Phi(M) = \left|(R_T/(M))^*\right| = |G_M|$$

By Proposition 12.3.7, G_M is a subset of $(R_T/(M))^*$. Since both sets have the same order, they must be isomorphic. Further, P is totally ramified since $e = \Phi(M) = [K(\Lambda_M) : K]$. From the latter we obtain (4).

Now let $M = P^{\alpha_1} \cdots P_r^{\alpha_r}$, where $P_i, \ldots, P_r$ are distinct irreducible polynomials in R_T. Then $\Lambda_M \cong \bigoplus_{i=1}^r \Lambda_{P_i^{\alpha_i}}$.

If we prove that $[K(\Lambda_M) : K] = \Phi(M)$ we will be able to deduce that $G_M \cong (R_T/(M))^*$ since $G_M \subseteq (R_T/(M))^*$ and both sets have the same order $\Phi(M)$. Then (2) and (3) will follow, and then (1) will follow too from the facts that $t(\lambda) = 0$, $\deg(t(u)) = \Phi(u) = \deg \operatorname{Irr}(\lambda, u, K)$, and $\operatorname{Irr}(\lambda, u, K)$ divides $t(u)$, so $\Psi_M(u) = t(u) = \operatorname{Irr}(\lambda, u, K)$. To prove that $[K(\Lambda_M) : K] = \Phi(M)$, notice that $K(\Lambda_{P_1^{\alpha_1}}), \ldots, K(\Lambda_{P_r^{\alpha_r}})$ are pairwise linearly disjoint because each $\mathfrak{p}_i$ is totally ramified in $K(\Lambda_{P_i^{\alpha_i}})/K$ and unramified in $\prod_{j \neq i} K(\Lambda_{P^{\alpha_i}})/K$.

It follows that

$$[K(\Lambda_M) : K] = \prod_{i=1}^r \left[K(\Lambda_{P_i^{\alpha_i}}) : K\right] = \prod_{i=1}^r \Phi(P_i^{\alpha_i}) = \Phi(M). \qquad \square$$

Corollary 12.3.17. *For any $M \in R_T \setminus \{0\}$, the extension $K(\Lambda_M)/K$ is geometric, that is, the field of constants of $K(\Lambda_M)$ is the same as that of K.*

Proof. Let $M = P_1^{\alpha_1} \cdots P_r^{\alpha_r}$, where $P_1, \ldots, P_r$ are distinct irreducible polynomials of R_T. Then

$$K(\Lambda_M) = \prod_{i=1}^r K\left(\Lambda_{P_i^{\alpha_i}}\right).$$

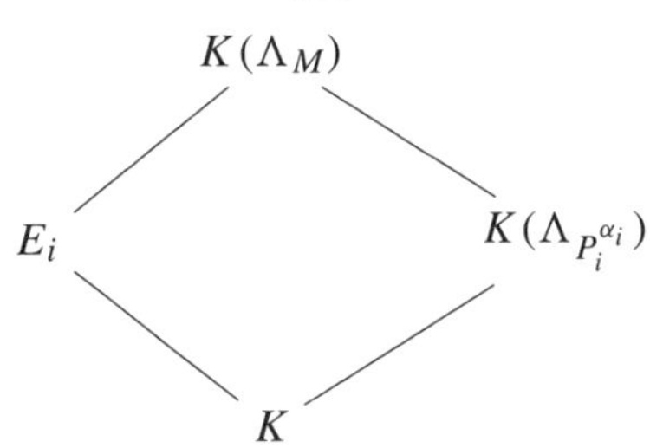

For each $i = 1, \ldots, r$, let $E_i = K(\Lambda_{M/P_i^{\alpha_i}})$. Then $\operatorname{Gal}(K(\Lambda_M)/E_i)$ is isomorphic to $\operatorname{Gal}(K(\Lambda_{P_i^{\alpha_i}}/K))$. Let L be the maximal unramified extension of K contained in $K(\Lambda_M)$, $K \subseteq L \subseteq K(\Lambda_M)$. Since $K(\Lambda_M)/E_i$ is totally ramified at the prime divisors above $\mathfrak{p}_i$ and E_iL/E_i is unramified, it follows that $E_iL = E_i$. Thus $L \subseteq E_i$ for $1 \leq i \leq r$.

Therefore $K \subseteq L \subseteq \bigcap_{i=1}^r E_i = K$, and $L = K$. In particular, it follows that every extension S/K such that $K \subsetneqq S \subseteq K(\Lambda_M)$ is ramified. If $\mathbb{F}_{q^s}$ is the field of constants of $K(\Lambda_M)$ and $\mathbb{F}_q$ is the field of constants of K, then $\mathbb{F}_q(T) = K \subseteq \mathbb{F}_{q^s}(T) \subseteq K(\Lambda_M)$ and $\mathbb{F}_{q^s}(T)/\mathbb{F}_q(T)$ is unramified (Theorem 5.2.32). Thus $\mathbb{F}_{q^s}(T) = \mathbb{F}_q(T)$ and by Proposition 2.1.6,

$$1 = \left[\mathbb{F}_{q^s}(T) : \mathbb{F}_q(T)\right] = \left[\mathbb{F}_{q^s} : \mathbb{F}_q\right] = s. \qquad \square$$

Proposition 12.3.18. *Let P be a monic irreducible polynomial in R_T and $M = P^n$ for some $n \geq 1$. Then*

$$\Psi_{P^n}(u) = \frac{u^{P^n}}{u^{P^{n-1}}}$$

is an Eisenstein polynomial over R_T at P. In other words, if

$$\Psi_{P^n}(u) = u^d + a_{d-1}u^{d-1} + \cdots + a_0 \in R_T[u],$$

then P divides a_i for $0 \leq i \leq d-1$, and P^2 does not divide a_0.

Proof. We have $\Psi_{P^n}(u) = \prod_{(A,P^n)=1}(u - \lambda_{P^n}^A)$, and P is totally ramified.

Let $\mathfrak{p}^{\Phi(M)} = P\vartheta_M$. We have $\Psi_{P^n}(0) = P = \pm\prod_{(A,P^n)=1}\lambda_{P^n}^A$. It follows that

$$\begin{aligned} v_{\mathfrak{p}}(P) = \Phi(M) &= \sum_A v_{\mathfrak{p}}(\lambda^A) = \sum_A v_{\mathfrak{p}^{A^{-1}}}(\lambda) \\ &= \sum_A v_{\mathfrak{p}}(\lambda) = \Phi(M)v_{\mathfrak{p}}(\lambda). \end{aligned}$$

Thus $v_{\mathfrak{p}}(\lambda^A) = v_{\mathfrak{p}}(\lambda) = 1$, so

$$\begin{aligned} \Psi_{P^n}(u) = u^{\Phi(P^n)} &- f_{\Phi(P^n)-1}(\{\lambda^A\}_A)u^{\Phi(P^n)-1} + \cdots \\ &+ f_1\left(\{\lambda^A\}_A\right)u + (-1)^{\Phi(P^n)} f_0\left(\{\lambda^A\}_A\right), \end{aligned}$$

where the $f_i(\{\lambda^A\}_A)$ are the elementary symmetric polynomials in $\{\lambda^A\}_A$ and $f_0(\{\lambda^A\}_A) = \Psi_{P^n}(0) = P$. Hence

$$\Psi_{P^n}(u) = u^{\Phi(M)} + \beta_{\Phi(M)-1}u^{\Phi(M)-1} + \cdots + \beta_1 u + \beta_0 \in R_T[u],$$

P divides β_i for $1 \leq i \leq \Phi(M)-1$, $\beta_0 = \pm P$, and $\beta_{\Phi(M)} = 1$. □

As a corollary we recover the irreducibility of $\Psi_{P^n}(u)$.

Corollary 12.3.19. *The polynomial $\Psi_{P^n}(u) \in R_T$ is irreducible.*

Proof. The statement is an application of Eisenstein's criterion. □

12.4 Arithmetic of Cyclotomic Function Fields

In the case of number fields, assume that $\mathbb{Q}(\zeta_n)/\mathbb{Q}$ is a cyclotomic extension, where $n \in \mathbb{N}$ is such that $n \not\equiv 2 \bmod 4$. Then a rational prime p is ramified in $\mathbb{Q}(\zeta_n)/\mathbb{Q}$ if and only if p divides n and the infinite prime is ramified. Furthermore, if p is a finite prime not dividing n, then

$$p\vartheta_{\mathbb{Q}(\zeta_n)} = \mathfrak{P}_1 \cdots \mathfrak{P}_g,$$

where $[\vartheta_{\mathbb{Q}(\zeta_n)}/\mathfrak{P}_i : \mathbb{Z}/p] = f$, $fg = \phi(n)$, and $f = o(p \bmod n)$, that is

$$f = \min\{m \in \mathbb{N} \mid p^m \equiv 1 \bmod n\}.$$

We will see that the same statements hold in the function field case. The key result is that $\mathfrak{p}_\infty$ is tamely ramified in $K(\Lambda_M)/K$. We need two general facts: Newton's method (Section 12.4.1) and Abhyankar's lemma (Section 12.4.2).

12.4.1 Newton Polygons

Let F be a complete field with respect to a discrete valuation v with place $\mathfrak{p}$. Let Ω be an algebraic closure of F and

$$f(x) = a_0 + a_1 x + \cdots + a_n x^n \in F[x], \quad \text{where} \quad a_0 a_n \neq 0.$$

We associate to each term of $f(x)$ a point in $\mathbb{R} \times (\mathbb{R} \cup \{\infty\})$ as follows:

If $a_i x^i \neq 0$, i.e., if $a_i \neq 0$, we take the point $(i, v(a_i))$.

If $a_i x^i = 0$, i.e., if $a_i = 0$, we take the formal point $(i, \infty) = (i, v(a_i))$ (which is the same as not taking any point of $\mathbb{R} \times \mathbb{R}$).

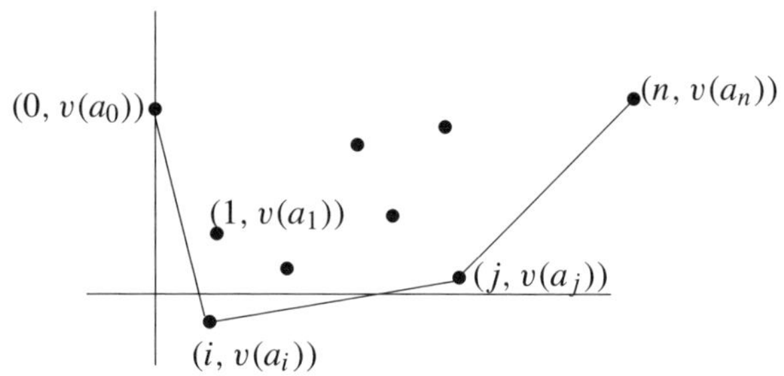

Consider the bottom convex cover of the set

$$\{(i, v(a_i)) | i = 0, 1, \ldots, n, |a_i \neq 0\}.$$

Definition 12.4.1. This cover is called a *Newton polygon*.

More precisely, the set of vertices of this bottom cover is

$$\left\{(0 = i_0, v(a_0)), (i_1, v(a_{i_1})), \ldots, (i_m = n, v(a_n))\right\},$$

where $a_0, a_{i_1}, \ldots, a_n$ satisfy the following. First, consider $S = \{i > 0 \mid a_i \neq 0\}$ and let i_1 be maximum such that

$$\frac{v(a_{i_1}) - v(a_0)}{i_1 - 0} = \min\left\{\frac{v(a_j) - v(a_0)}{j - 0} \mid j \in S\right\}.$$

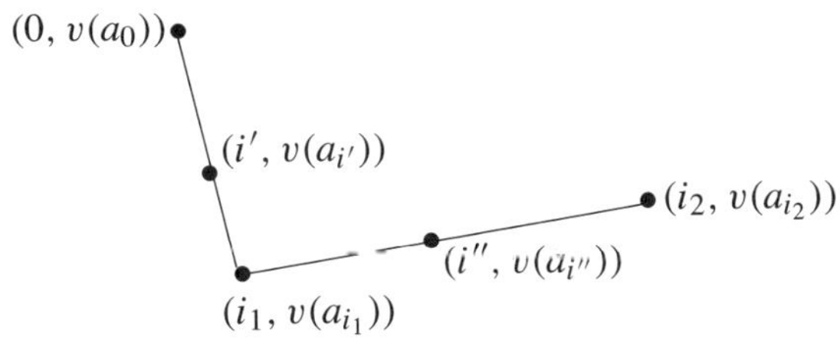

Now let i_2 be maximum such that

$$\frac{v(a_{i_2}) - v(a_{i_1})}{i_2 - i_1} = \min\left\{\frac{v(a_j) - v(a_{i_1})}{j - a_{i_1}} \mid j \in S, j > i_1\right\}$$

and so on.

Theorem 12.4.2. *Let* $[(r, v(a_r)), (s, v(a_s))]$ *be any segment of the Newton polygon corresponding to* $f(x)$. *Let* $\frac{v(a_s)-v(a_r)}{s-r} = -m$ *be its slope. Then* $f(x)$ *has exactly* $s-r$ *roots* $\alpha_1, \ldots, \alpha_{s-r}$ *satisfying* $v(\alpha_1) = \cdots = v(\alpha_{s-r}) = m$.

Furthermore, define $f_m(x) = \prod_{i=1}^{s-r}(x - \alpha_i)$. *Then* $f_m(x) \in F[x]$ *and* $f_m(x)$ *divides* $f(x)$.

Proof. Let $f(x) = a_n^{-1} f(x) = a_n^{-1}a_n x^n + a_n^{-1}a_{n-1}x^{n-1} + \cdots + a_n^{-1}a_1 x + a_n^{-1}a_0$.

Then

$$\frac{v(a_i a_n^{-1}) - v(a_j a_n^{-1})}{i - j} = \frac{v(a_i) - v(a_j)}{i - j}$$

and the Newton polygon corresponding to $g(x)$ is obtained from the one corresponding to $f(x)$ by a translation of $-v(a_n)$ in the y-direction, as follows:

$$(i, v(a_i a_n^{-1})) = (i, v(a_i) - v(a_n)) = (i, v(a_i)) - (0, v(a_n)).$$

Moreover, the roots of $g(x)$ and $f(x)$ are the same. Thus we may assume that $a_n = 1$. Let $\alpha_1, \ldots, \alpha_n \in \Omega$ be the roots of $f(x)$. We partition the set of α_i's according to the value $v(\alpha_i)$, obtaining

$$\begin{aligned} v(\alpha_1) &= \cdots = v(\alpha_{s_1}) = m_1, \\ v(\alpha_{s_1+1}) &= \cdots = v(\alpha_{s_2}) = m_2, \\ &\vdots \\ v(\alpha_{s_t+1}) &= \cdots = v(\alpha_{s_{t+1}}) = m_{t+1}, \end{aligned}$$

with $m_1 < m_2 < \cdots < m_t < m_{t+1}$.

We have

$$f(x) = \prod_{i=1}^{n}(x-\alpha_i) = x^n - h_1(\alpha_1,\dots,\alpha_n)x^{n-1} + h_2(\alpha_1,\dots,\alpha_n)x^{n-2} - \cdots$$
$$+ (-1)^{n-1}h_{n-1}(\alpha_1,\dots,\alpha_n)x + (-1)^n h_n(\alpha_1,\dots,\alpha_n),$$

where $h_j(\alpha_1,\dots,\alpha_n) = \sum_{i_1<\dots<i_j} \alpha_{i_1},\dots\alpha_{i_j} = (-1)^j a_{n-j}$, $1 \le j \le n$.

Also $v(a_n) = v(1) = 0$.

For $0 \le u < s_{j+1} - s_j$, we have $n - s_j \ge n - s_j - u > n - s_{j+1}$, so

$$\begin{aligned} v(a_{n-s_j-u}) &= v(h_{s_j+u}(\alpha_1,\dots,\alpha_n)) = v\left(\sum_{i_1<\dots<i_{s_j+u}} \alpha_{i_1}\dots\alpha_{i_{s_j+u}}\right) \\ &\ge \min_{i_1,\dots,i_{s_j+u}} \{v(\alpha_{i_1}\dots\alpha_{i_{s_j+u}})\} \\ &= v(\alpha_1\cdots\alpha_{s_1}\alpha_{s_1+1}\cdots\alpha_{s_2}\alpha_{s_2+1}\cdots\alpha_{s_j+1}\alpha_{s_j+2}\cdots\alpha_{s_j+u}) \\ &= s_1m_1 + (s_2-s_1)m_2 + \cdots + (s_j - s_{j-1})m_j + um_j. \end{aligned} \tag{12.6}$$

For $a_{n-s_{j+1}}$ there is a single term with minimum valuation such that

$$\begin{aligned} v\left(a_{n-s_{j+1}}\right) &= v\Big(\sum_{i_1<\dots<i_{s_{j+1}}} \alpha_{i_1}\cdots\alpha_{i_{s_{j+1}}}\Big) = v(\alpha_1\cdots\alpha_{s_1}\cdots\alpha_{s_j+1}\cdots\alpha_{s_{j+1}}) \\ &= s_1m_1 + (s_2-s_1)m_2 + \cdots + (s_{j+1}-s_j)m_{j+1}. \end{aligned} \tag{12.7}$$

We will deduce from (12.6) and (12.7) that the vertices of the Newton polygon of $f(x)$ are

$$\begin{aligned} &(0, v(a_0)) = (0, v(a_{n-s_{t+1}})) \\ &\quad = (n - s_{t+1}, s_1m_1 + (s_2-s_1)m_2 + \cdots + (s_{t+1}-s_t)m_{t+1}), \\ &(n-s_t, v(a_{n-s_t})) = (n-s_t, s_1m_1 + (s_2-s_1)m_2 + \cdots + (s_t - s_{t-1})m_t), \\ &\qquad \vdots \\ &(n-s_2, s_1m_1 + (s_2-s_1)m_2), \\ &(n-s_1, s_1m_1), \\ &(n, 0). \end{aligned}$$

Now the slope between $(n - s_{j+1}, v(a_{n-s_{j+1}}))$ and $(n - s_j, v(a_n - s_j))$ is given by

$$\begin{aligned} &\frac{v(a_{n-s_{j+1}}) - v(a_{n-s_j})}{(n-s_{j+1}) - (n-s_j)} \\ &\qquad = \frac{\left[m_1s_1 + (s_2-s_1)m_2 + \cdots + (s_{j+1}-s_j)m_{j+1}\right] - \cdots}{-(s_{j+1}-s_j)} \cdots \\ &\qquad \cdots \frac{\cdots - \left[m_1s_1 + (s_2-s_1)m_2 + \cdots + (s_j - s_{j-1})m_j\right]}{-(s_{j+1}-s_j)} \\ &\qquad\qquad = -\frac{s_{j+1}-s_j}{s_{j+1}-s_j}m_{j+1} = -m_{j+1}. \end{aligned}$$

Thus the slope is $s_{j+1} - s_j$, which is the number of roots of f with valuation m_{j+1}. This proves the first part.

For the second part, we proceed by induction on n to show that $f_m(x) = \prod_{i=1}^{s-r}(x - \alpha_i) \in F[x]$. Clearly, $f_m(x)$ divides $f(x)$.

For $n = 1$, $f_0(x) = x + a_0$ and there is nothing to prove.

For $n = 2$, we consider two cases. If $f(x)$ is irreducible, assume that E is the decomposition field of $f(x)$; then the other root of $f(x)$ is either α (if E/F is inseparable) or $\sigma\alpha$, where $\mathrm{Gal}(E/F) = \{1, \sigma\}$. In any case we obtain

$$v(\sigma\alpha) = v_{\mathfrak{p}}(\sigma\alpha) = v_{\sigma^{-1}\mathfrak{p}}(\alpha) = v_{\mathfrak{p}}(\alpha) = v(\alpha)$$

because F is a complete field. Therefore all the roots have the same valuation and the Newton polygon is a segment.

Suppose that $f(x)$ is reducible. If both roots have the same valuation, there is nothing to prove and if the two roots have different valuation, we have $f(x) = (x - a)(x - b)$, with $a, b \in F$, so we are done.

Now assume that $f_m(x) \in F[x]$ and that $f(x)$ is any polynomial of degree less than n. For n, let

$$f_{s_j}(x) = \prod_{i=s_j+1}^{s_{j+1}} (x - \alpha_i), \quad j = 0, 1, \ldots, t \text{ (with } s_0 = 0), \qquad f(x) = \prod_{j=0}^{t} f_{s_j}(x).$$

Let $g(x) = \frac{f(x)}{\mathrm{Irr}(\alpha_1, x, F)}$. Then $g(x) \in F[x]$. Since every conjugate of α_1 has the same valuation, it follows that $\mathrm{Irr}(\alpha_1, x, F) \mid f_{s_0}(x)$. Let $g_0(x) = \frac{f_{s_0}(x)}{\mathrm{Irr}(\alpha_1, x, F)}$. Then $g(x) = g_0(x) \prod_{j=1}^{t} f_{s_j}(x)$. Since $\deg g(x) < \deg f(x) = n$, we use induction on $\deg g(x)$ to conclude that $f_{s_j}(x) \in F[x]$, for $j = 1, \ldots, t$, and $g_0(x) = g_{s_0}(x) \in F[x]$.

Therefore $f_{s_0}(x) = g_0(x)\,\mathrm{Irr}(\alpha_1, x, F) \in F[x]$. ◻

12.4.2 Abhyankar's Lemma

The other ingredient needed to determine the ramification type of $\mathfrak{p}_\infty$ in $K(\Lambda_M)/K$ is Abhyankar's lemma. First we establish a result on finite groups.

Proposition 12.4.3. *Let G be a finite group and let U be a normal subgroup of G of order p^n, where p a rational prime or $p = 1$. Let G/U be a cyclic group of order relatively prime to p.*

Assume that H_1 is a subgroup of G whose order is a multiple of p^n. Then for any subgroup H_2 of G we have $|H_1 \cap H_2| = (|H_1|, |H_2|)$.

Proof. Since $|H_1 \cap H_2|$ divides $|H_i|$ for $i = 1, 2$, it follows that $|H_1 \cap H_2| \mid (|H_1|, |H_2|)$. Put $|H_1| = a_1 p^n$ and $|H_2| = a_2 p^m$ with $(a_1, p) = (a_2, p) = 1$, and let $d = (a_1, a_2)$. Then $(|H_1|, |H_2|) = dp^m$ with $(d, p) = 1$. In particular, $|H_1 \cap H_2| \leq dp^m$.

By hypothesis, the normal subgroup U is the p-Sylow subgroup of G (or $U = \{e\}$). Thus U contains any subgroup of G of order p^m. Therefore, if W is a p-Sylow subgroup of H_2, of order p^m, then $W \subseteq H_2$ and $W \subseteq U \subseteq H_1$. It follows that

$$W \subseteq H_1 \cap H_2 \quad \text{and} \quad p^m \mid |H_1 \cap H_2| \,. \tag{12.8}$$

Let $\pi : G \longrightarrow G/U$ be the canonical epimorphism.

We have $\pi(H_i) = \frac{H_i U}{U} \cong \frac{H_i}{U \cap H_i}$. Hence $|\pi(H_i)| = \frac{|H_i|}{|U \cap H_i|} = a_i$ for $i = 1, 2$.

Since G/U is a cyclic group, $\pi(H_1) \cap \pi(H_2)$ is a cyclic group of order $d = (a_1, a_2)$. In particular, there exists $x \in H_1 \cap H_2$ such that d divides $o(x)$. Since $(d, p) = 1$ it follows by (12.8) that

$$dp^m \mid |H_1 \cap H_2| \quad \text{and} \quad |H_1 \cap H_2| = dp^m = (|H_1|, |H_2|) \,. \qquad \square$$

Theorem 12.4.4 (Abhyankar's Lemma). *Let L/K be a finite separable extension of function fields. Suppose that $L = K_1 K_2$ with $K \subseteq K_i \subseteq L$. Let $\mathfrak{p}$ be a prime divisor of K and $\mathfrak{P}$ a prime divisor in L above $\mathfrak{p}$. Let $\mathfrak{P}_i = \mathfrak{P} \cap K_i$ for $i = 1, 2$. If at least one of the extensions K_i/K, $i = 1, 2$, is tamely ramified at $\mathfrak{p}$, then*

$$e_{L/K}(\mathfrak{P}|\mathfrak{p}) = \left[e_{K_1/K}(\mathfrak{P}_1|\mathfrak{p}), e_{K_2/K}(\mathfrak{P}_2|\mathfrak{p}) \right].$$

Proof. Let $\widetilde{L}$ be the Galois closure of L/K and let $\mathfrak{B}$ be a prime divisor in $\widetilde{L}$ such that $\mathfrak{B}|_L = \mathfrak{P}$.

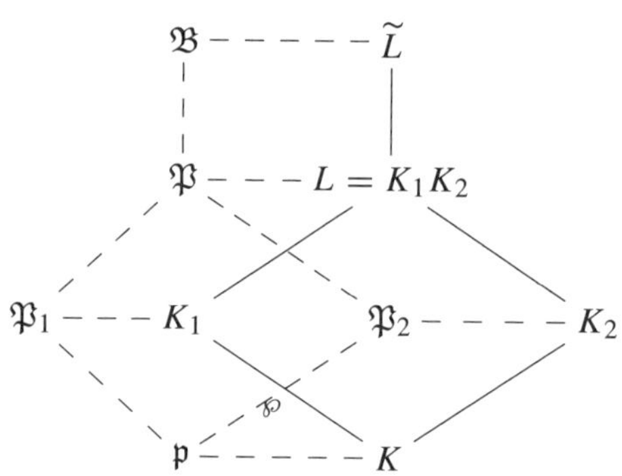

Let $G = I(\mathfrak{B}|\mathfrak{p})$ and let $H_i = I(\mathfrak{B}|\mathfrak{P}_i)$, $i = 1, 2$, be the inertia groups. Define

$$p = \begin{cases} \text{char}\, K & \text{if char}\, K \neq 0, \\ 1 & \text{if char}\, K = 0. \end{cases}$$

We may assume without loss of generality that K_1/K is tamely ramified at $\mathfrak{P}_1$. Then $(e(\mathfrak{P}_1|\mathfrak{p}), p) = 1$.

Let U be a p-Sylow subgroup of G. Then U corresponds to the wild ramification of $\mathfrak{p}$ in $\widetilde{L}/K$. Thus U is the first ramification group (Corollary 5.9.10) and $U \lhd G$ (Theorem 5.9.4). Set $|U| = p^n$. Since the ramification in $\mathfrak{P}_1|\mathfrak{p}$ is tame, it follows by Corollary 5.9.17 that $U \subseteq H_1$ and G/U is a cyclic group.

Therefore H_1 and H_2 satisfy the conditions of Proposition 12.4.3, and we have $|H_1 \cap H_2| = (|H_1|, |H_2|)$. Now since $L = K_1 K_2$, it follows that $\text{Gal}(\widetilde{L}/L) = \text{Gal}(\widetilde{L}/K_1) \cap \text{Gal}(\widetilde{L}/K_2)$ and $I(\mathfrak{B}|\mathfrak{P}) = I(\mathfrak{B}|\mathfrak{P}_1) \cap I(\mathfrak{B}|\mathfrak{P}_2) = H_1 \cap H_2$. Therefore

$$\begin{aligned} e(\mathfrak{B}|\mathfrak{P}) &= |I(\mathfrak{B}|\mathfrak{P})| = |H_1 \cap H_2| = (|H_1|, |H_2|) \\ &= (e(\mathfrak{B}|\mathfrak{P}_1), e(\mathfrak{B}|\mathfrak{P}_2)) \\ &= (e(\mathfrak{B}|\mathfrak{P})e(\mathfrak{P}|\mathfrak{P}_1), e(\mathfrak{B}|\mathfrak{P})e(\mathfrak{P}|\mathfrak{P}_2)) \\ &= e(\mathfrak{B}|\mathfrak{P})(e(\mathfrak{P}|\mathfrak{P}_1), e(\mathfrak{P}|\mathfrak{P}_2)). \end{aligned}$$

Hence $(e(\mathfrak{P}|\mathfrak{P}_1), e(\mathfrak{P}|\mathfrak{P}_2)) = 1$. We have

$$e(\mathfrak{P}|\mathfrak{p}) = e(\mathfrak{P}|\mathfrak{P}_1)e(\mathfrak{P}_1|\mathfrak{p}) = e(\mathfrak{P}|\mathfrak{P}_2)e(\mathfrak{P}_2|\mathfrak{p}),$$

on the other hand. If $a, b, x, y \in \mathbb{Z} \setminus \{0\}$ satisfy $ax = by$ and $(x, y) = 1$, then $[a, b] = ax = by$ (see Exercise 12.10.16).

Therefore $e(\mathfrak{P}|\mathfrak{p}) = [(e(\mathfrak{P}_1|\mathfrak{p}), e(\mathfrak{P}_2|\mathfrak{p})]$. □

12.4.3 Ramification at $\mathfrak{p}_\infty$

The main objective of this subsection is to prove that for any $M \in R_T \setminus \{0\}$, with $R_T = \mathbb{F}_q[T]$, the infinite prime of $K = \mathbb{F}_q(T)$, where $(T)_K = \frac{\mathfrak{p}_0}{\mathfrak{p}_\infty}$, is tamely ramified in $K(\Lambda_M)/K$.

Proposition 12.4.5. *Assume $M = P^n \in R_T$, where P is a monic irreducible polynomial of degree d. Then $\mathfrak{p}_\infty$ decomposes into $\Phi(M)/(q-1)$ prime divisors in $K(\Lambda_M)$. The ramification index of $\mathfrak{p}_\infty$ in $K(\Lambda_M)$ is $e_\infty = q - 1$ and each prime divisor in $K(\Lambda_M)$ is of degree 1, so the relative inertia degree f_∞ is 1.*

$\mathfrak{B} - K(\Lambda_M)$
$\mathfrak{P} - K(\Lambda_P)$
$\mathfrak{p}_\infty - K$

Proof: Let $\mathfrak{B}$ be a prime divisor of $K(\Lambda_M)$ that lies above $\mathfrak{p}_\infty$. Since $K(\Lambda_M)/K$ is a Galois extension of degree $\Phi(M)$, it suffices to prove that $e_\infty = e(\mathfrak{B}|\mathfrak{p}_\infty) = q - 1$ and $f_\infty = f(\mathfrak{B}|\mathfrak{p}_\infty) = 1$. Let $\mathfrak{P} := \mathfrak{B} \cap K(\Lambda_P)$. First we will prove that $e_{\mathfrak{P}} = e(\mathfrak{P}|\mathfrak{p}_\infty) = q - 1$, $f_{\mathfrak{P}} = f(\mathfrak{P}|\mathfrak{p}_\infty) = 1$, and that $\mathfrak{P}$ decomposes fully in $K(\Lambda_M)/K(\Lambda_P)$. Let $g(u) = u^P/u = \Psi_P(u)$. Then $K(\Lambda_P)$ is obtained by adjoining the roots of $g(u)$ to K.

We have $g(u) = \sum_{i=0}^{d} \begin{bmatrix} P \\ i \end{bmatrix} u^{q^i - 1} = h(u^{q-1})$ where $h(u) = \sum_{i=0}^{d} \begin{bmatrix} P \\ i \end{bmatrix} u^{\frac{q^i-1}{q-1}}$ and $\deg_T \begin{bmatrix} P \\ i \end{bmatrix} = (d-1)q^i$.

Let K_∞ be the completion of K at $\mathfrak{p}_\infty$ and denote by v_∞ the corresponding valuation. Clearly, $v_\infty\left(\begin{bmatrix} P \\ i \end{bmatrix}\right) = -(d-i)q^i = -\deg_T\left(\begin{bmatrix} P \\ i \end{bmatrix}\right)$. We write $h(u) = \sum_{j=0}^{\frac{q^d-1}{q-1}} f_j(T)u^j$ where $f_j(T) \neq 0 \iff j = \frac{q^i-1}{q-1}$ for some $0 \leq i \leq d$.

We draw the Newton polygon corresponding to $h(u)$ in K_∞. The vertices of the coefficients are given by

$$\big(j, v_\infty(f_j(T))\big) = \left(\frac{q^i - 1}{q - 1}, -(d-i)q^i\right) = \beta_i$$

for $j = \frac{q^i-1}{q-1}$.

The slope from β_i to β_{i+1} is

$$s_i = \frac{-(d-(i+1))q^{i+1} + (d-i)q^i}{\frac{q^{i+1}-1}{q-1} - \frac{q^i-1}{q-1}} = -d(q-1) + q + i(q-1) < s_{i+1}.$$

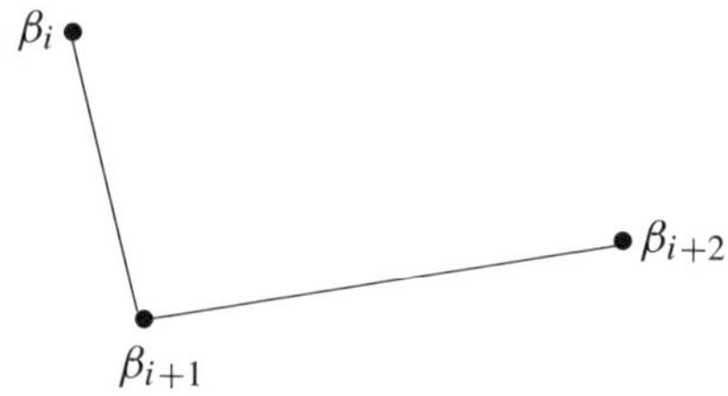

Thus the slopes increase with i and therefore $\beta_0, \beta_1, \ldots, \beta_d$ are the vertices of the Newton polygon of $h(u)$.

The slope from β_0 to β_1 is $s_0 = -d(q-1) + q$. Hence $h(u)$ contains $\frac{q^1-1}{q-1} - 0 = 1 - 0 = 1$ root θ in K_∞ such that $v_\infty(\theta) = d(q-1) - q$. Now since $g(u) = h(u^{q-1})$, it follows that $K(\Lambda_P)_{\mathfrak{P}} = K_\infty(\lambda)$, where λ is a root of $u^{q-1} - \theta$. Thus $\lambda^{q-1} = \theta$.

Let $v_{\mathfrak{P}}$ be the valuation above v_∞. We have

$$v_{\mathfrak{P}}(\lambda^{q-1}) = (q-1)v_{\mathfrak{P}}(\lambda) = v_{\mathfrak{P}}(\theta) = e_\infty v_\infty(\theta) = e_\infty(d(q-1) - q).$$

Since $(d(q-1) - q, q-1) = 1$, it follows that $(q-1)$ divides e_∞ and

$$e_\infty \leq e_\infty f_\infty = \left[K(\Lambda_P)_{\mathfrak{P}} : K_\infty\right] = [K_\infty(\lambda) : K_\infty] \leq q - 1 \leq e_\infty.$$

Therefore $e_\infty = q-1$ and $f_\infty = 1$, so $K(\Lambda_P)_{\mathfrak{P}}/K_\infty$ is totally ramified.

Now we will prove that $\mathfrak{P}$ decomposes fully in $K(\Lambda_{P^n})/K(\Lambda_P)$. Let λ be a root of $g(u)$, and $v_{\mathfrak{P}}(\lambda) = d(q-1) - q$. We have $u^P = ug(u)$. Then $u^M = u^{P^n} = (u^P)^{P^{n-1}} = u^{P^{n-1}} g(u^{P^n-1})$ (in other words, $\Psi_{P^n}(u) = \Psi_P(u^{P^{n-1}}) = u^{P^n}/u^{P^{n-1}}$).

The field $K(\Lambda_M)$ is obtained by adjoining any root of $g(u^{P^{n-1}})$ to $K(\Lambda_P)$. If λ_{P^n} is a generator of $\Lambda_{P^n} = \Lambda_M$, then $\lambda_{P^n}^{P^{n-1}} = \lambda_{P^n/P^{n-1}} = \lambda_P = \lambda$ is a generator of Λ_P. Therefore $K(\Lambda_M)$ is obtained from $K(\Lambda_P)$ by adjoining a root of $u^{P^{n-1}} - \lambda$.

Next, we determine the Newton polygon of $u^{P^{n-1}} - \lambda$. We have

$$u^{P^{n-1}} - \lambda = \sum_{i=0}^{d(n-1)} \begin{bmatrix} P^{n-1} \\ i \end{bmatrix} u^{q^i} - \lambda.$$

Define

$$\gamma_{-1} = (0, v_{\mathfrak{P}}(-\lambda)) = (0, d(q-1) - q),$$

and

$$\begin{aligned}\gamma_i &= \left(q^i, v_{\mathfrak{P}}\left(\begin{bmatrix} P^{n-1} \\ i \end{bmatrix}\right)\right) = \left(q^i, e(\mathfrak{P}|\mathfrak{p}_\infty)v_\infty\left(\begin{bmatrix} P^{n-1} \\ i \end{bmatrix}\right)\right) \\ &= (q^i, -(q-1)(d(n-1)-i)q^i) \quad \text{for} \quad 0 \le i \le d(n-1).\end{aligned}$$

The slope from γ_{-1} to γ_0 is

$$\begin{aligned}\frac{-(q-1)(d(n-1)) - (d(q-1)-q)}{1-0} &= -(q-1)(d(n-1)+d)+q \\ &= -dn(q-1)+q = t_{-1}.\end{aligned}$$

Next, for $0 \le i \le d(n-1)$ the slope from γ_i to γ_{i+1} is given by

$$\begin{aligned}t_i &= \frac{-(q-1)(d(n-1)-(i+1))q^{i+1} + (q-1)(d(n-1)-i)q^i}{q^{i+1}-q^i} \\ &= -q(d(n-1)-(i+1)) + (d(n-1)-i) = -(q-1)(d(n-1)-i)+q \\ &= -(q-1)d(n-1) + i(q-1) + q.\end{aligned}$$

Therefore $t_i < t_{i+1}$. Similarly $t_{-1} = -dn(q-1)+q < -(q-1)d(n-1)+q = t_0$, so t_j is an increasing function of j.

It follows that $\gamma_{-1}, \gamma_0, \ldots, \gamma_{d(n-1)}$ are precisely the vertices of the Newton polygon of $u^{p^{n-1}} - \lambda$. Now the segment from γ_{-1} to γ_0 shows that $u^{p^{n-1}} - \lambda$ has a root in $K(\Lambda_P)_{\mathfrak{P}}$. Since the extension $K(\Lambda_M)_{\mathfrak{B}}/K(\Lambda_P)_{\mathfrak{P}}$ is Galois, it follows that $K(\Lambda_M)_{\mathfrak{B}} = K(\Lambda_P)_{\mathfrak{P}}$. Thus $u^{p^{n-1}} - \lambda$ decomposes in $K(\Lambda_P)_{\mathfrak{P}}[u]$ and $f(\mathfrak{B}|\mathfrak{P}) = e(\mathfrak{B}|\mathfrak{P}) = 1$. This proves the proposition. □

Theorem 12.4.6. *Let M be a nonzero polynomial of R_T. Then $\mathfrak{p}_\infty$ is tamely ramified in $K(\Lambda_M)/K$. Furthermore, we have $e_\infty = q-1$ and $f_\infty = 1$, and there are exactly $h_\infty = \Phi(M)/(q-1)$ prime divisors of $K(\Lambda_M)$ above $\mathfrak{p}_\infty$.*

Proof. Let $M = P_1^{\alpha_1} \cdots P_r^{\alpha_r}$ and $K(\Lambda_M) = \prod_{i=1}^r K(\Lambda_{P_i^{\alpha_i}})$. By Proposition 12.4.5, $e_{K(\Lambda_{P_i}^{\alpha_i})} = q-1$. Moreover, $\mathfrak{p}_\infty$ is tamely ramified in $K(\Lambda_{P_i^{\alpha_i}})/K$ for every i. Indeed, set $p = \operatorname{char} K$, where $q = p^n$ for some $n \ge 1$; then p does not divide $q-1$.

We obtain from Abhyankar's lemma that

$$e_\infty = \left[e_{K(\Lambda_{P_1^{\alpha_1}})}, \ldots, e_{K(\Lambda_{P_r^{\alpha_r}})}\right] = [q-1, \ldots, q-1] = q-1.$$

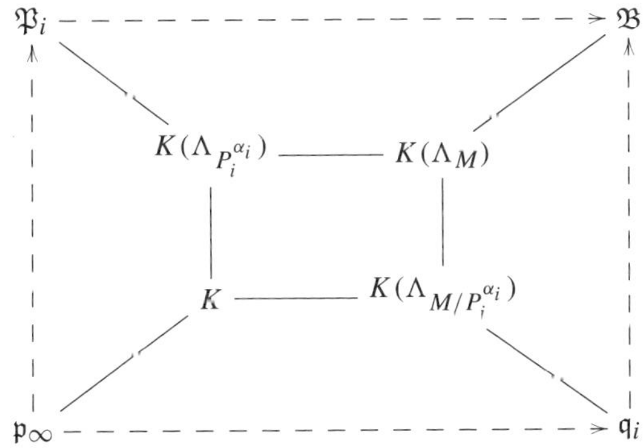

We wish to prove by induction on r that $f_\infty = 1$. The case $r = 1$ is a consequence of Proposition 12.4.5. For the general case, let $\mathfrak{B}$ be a prime divisor in $K(\Lambda_M)$ that lies above $\mathfrak{p}_\infty$. Let $\mathfrak{P}_i = \mathfrak{B} \cap K(\Lambda_{p_i^{\alpha_i}})$ and $\mathfrak{q}_i = \mathfrak{B} \cap K(\Lambda_{M/p_i^{\alpha_i}})$. Then $1 = f(\mathfrak{P}_i|\mathfrak{p}_\infty) \geq f(\mathfrak{B}|\mathfrak{q}_i)$ and $f(\mathfrak{B}|\mathfrak{p}_\infty) = f(\mathfrak{B}|\mathfrak{q}_i) f(\mathfrak{q}_i|\mathfrak{p}_\infty) = f(\mathfrak{B}|\mathfrak{q}_i)$. By the induction hypothesis $f(\mathfrak{q}_i|\mathfrak{p}_\infty) = 1$. It follows that $f_\infty = f(\mathfrak{B}|\mathfrak{p}_\infty) = 1$. Finally, the equality $h_\infty = \Phi(M)/(q-1)$ follows from Corollary 5.2.17 and the facts that $e_\infty = q-1$, $f_\infty = 1$, and $[K(\Lambda_M) : K] = \Phi(M)$. □

12.5 The Artin Symbol in Cyclotomic Function Fields

First we determine the Artin symbol in an extension $K(\Lambda_M)/K$ (see Definition 11.2.5).

Theorem 12.5.1. *Let $M \in R_T \setminus \{0\}$ and let P be an irreducible polynomial that does not divide M. Then the map*

$$\begin{aligned} \varphi_P \colon \Lambda_M &\longrightarrow \Lambda_M \\ \lambda &\longmapsto \lambda^P \end{aligned}$$

corresponds to the Artin symbol $\left[\frac{K(\Lambda_M)/K}{P}\right]$.

Proof. Let $(R_T)_P$ denote the localization of P, i.e.,

$$(R_T)_P = \left\{ \frac{f}{g} \mid f, g \in R_T,\ P \nmid g \right\}.$$

If $(P)_K = \frac{\mathfrak{p}}{\mathfrak{p}_\infty^{\deg P}}$, then $k(\mathfrak{p}) = (R_T)_P / P(R_T)_P \cong R_T/(P) \cong \mathbb{F}_{q^d}$, where $d = \deg P$ (see Section 2.4).

Let $\mathfrak{P}$ be a prime divisor in $K(\Lambda_M)$ that divides $\mathfrak{p}$.

Clearly, $N(\mathfrak{p}) = |\mathbb{F}_{q^d}| = q^d$ and $\Lambda_M \subseteq \vartheta_{\mathfrak{P}}$. It follows by Proposition 11.2.2 that

$$\left[\frac{K(\Lambda_M)/K}{P}\right](\lambda) \equiv \lambda^{q^d} \bmod \mathfrak{P}.$$

We have $u^P = u\Psi_P(u) = u\left(u^{q^d-1} + \beta_{q^d-2}u^{q^d-2} + \cdots + \beta_1 u + \beta_0\right)$. Moreover, by Proposition 12.3.18, P divides β_i for all $0 \leq i \leq q^d - 2$. Hence $\lambda^P \equiv \lambda^{q^d} \bmod \mathfrak{P}$. Now

$$u^M = \prod_{A \bmod M} (u - \lambda^A), \tag{12.9}$$

so taking the derivative with respect to u in (12.9) we obtain, using Proposition 12.2.11,

$$M = \sum_{A \bmod M} \Bigg(\prod_{\substack{B \neq A \\ B \bmod M}} (u - \lambda^{\beta}) \Bigg), \tag{12.10}$$

which is constant with respect to u.

Taking $u = \lambda^C$ in (12.10), we obtain $M = \prod_{C \neq B} (\lambda^C - \lambda^B)$. Since P does not divide M, it follows that $\lambda^C \not\equiv \lambda^B \bmod \mathfrak{P}$ whenever $C \not\equiv B \bmod M$.

Hence $\lambda^P \equiv \lambda^Q \bmod \mathfrak{P}$ implies $\lambda^P = \lambda^Q$.

Finally, from $\lambda^P \equiv \left[\frac{K(\Lambda_M)/K}{P}\right](\lambda) \equiv \lambda^{q^d}$, we conclude that $\varphi_P = \left[\frac{K(\Lambda_K)/K}{P}\right]$. □

Proposition 12.5.2. *Let $M \in R_T \setminus \{0\}$ and let P be an irreducible polynomial that does not divide M. In $K(\Lambda_M)/K$ we have*

$$e_P = 1, \quad f_P = o(P \bmod M), \quad \text{and} \quad h_P = \Phi(M)/f_P.$$

Proof. Let $\lambda = \lambda_M$ be a generator of Λ_M. Then $K(\Lambda_M) = K(\lambda)$.

Let $\mathfrak{P}$ be a prime divisor in $K(\Lambda_M)$ dividing $\mathfrak{p}$, where $(P)_K = \frac{\mathfrak{p}}{\mathfrak{p}_\infty^{\deg P}}$. Then

$$\vartheta_{\mathfrak{P}} = \left\{\xi \in K(\Lambda_M) | v_{\mathfrak{P}}(\xi) \geq 0\right\}$$

and

$$\begin{aligned} f_P &= \left[\vartheta_{\mathfrak{P}}/\mathfrak{P} : (R_T)_P/P(R_T)_P\right] = \left[(\vartheta_M)_{\mathfrak{P}}/\mathfrak{P}(\vartheta_M)_{\mathfrak{P}} : R_T/(P)\right] \\ &= \left[\vartheta_M/\mathfrak{P}\vartheta_M : R_T/(P)\right], \end{aligned}$$

where ϑ_M denotes the integral closure of R_T in $K(\Lambda_M)$.

Set $d = \deg P$. By Proposition 12.3.14, $\mathfrak{p}$ is not ramified in $K(\Lambda_M)/K$. Furthermore, the Artin symbol $\varphi_P = \left[\frac{K(\Lambda_M)/K}{P}\right]$ at P is given by $\varphi_P(\lambda) = \lambda^P$. Then $e_P = 1$ and $h_P = [K(\Lambda_M) : K]/f_P = \Phi(M)/f_P$.

Now $f_P = o(\varphi_P)$, so f_P is the minimum natural number such that

$$\varphi_P^{f_P} = \mathrm{Id} \in G_M = \mathrm{Gal}\,(K(\Lambda_M)/K).$$

We have

$$\begin{aligned} \varphi_P^f = \mathrm{Id} &\iff \varphi_P^f(\lambda) = \lambda^{P^f} = \lambda \\ &\iff \lambda^{P^f - 1} = 0 \iff M \mid P^f - 1 \\ &\iff P^f \equiv 1 \bmod M. \end{aligned}$$

Thus $f_P = o(P \bmod M)$. □

We are ready to state the general theorem about the behavior of prime divisors in cyclotomic extensions.

Theorem 12.5.3. *Let $M = P_1^{\alpha_1} \cdots P_r^{\alpha_r} \in R_T$, where $P_1, \ldots, P_r$ are irreducible polynomials, and let $K(\Lambda_M)/K$ be a cyclotomic extension. If $P \in R_T$ is distinct from $P_1, \ldots, P_r, P_\infty$, then*

$$e_P = 1, \quad f_P = o(P \bmod M), \quad \text{and} \quad h_P = \Phi(M)/f_P.$$

If $P = P_i$, then

$$e_P = \Phi(P^{\alpha_i}), \quad f_P = o\left(P_i \bmod \frac{M}{P_i^{\alpha_i}}\right),$$

and

$$h_P = \frac{\Phi(M)}{\Phi(P_i^{\alpha_i}) f_P} = \frac{\Phi(M/P_i^{\alpha_i})}{o(P_i \bmod M/P_i^{\alpha_i})}.$$

If $P = P_\infty$, then

$$e_\infty = q - 1, \quad f_\infty = 1, \quad \text{and} \quad h_\infty = \Phi(M)/(q-1).$$

Proof. The statement follows from Proposition 12.3.14, Theorem 12.4.6, and Proposition 12.5.2. □

Next we determine the inertia group of the infinite prime.

Proposition 12.5.4. *We have $\mathbb{F}_q^* = G_0$, where G_0 denotes the inertia group of any prime divisor of $K(\Lambda_M)$ above $\mathfrak{p}_\infty$.*

Proof. Let $\mathfrak{P}$ be a prime divisor of $K(\Lambda_M)$ above $\mathfrak{p}_\infty$. If M is a nonzero element of R_T, then for $A = \alpha \in \mathbb{F}_q^* \subseteq (R_T/(M))^*$ we have $\sigma_A(\lambda) = \sigma_\alpha(\lambda) = \lambda^\alpha = \alpha\lambda$, where $\lambda = \lambda_M$ is a generator of Λ_M.

Since $f_\infty = f(\mathfrak{P}|\mathfrak{p}_\infty) = 1$, it follows that G_0 is equal to the decomposition group of $\mathfrak{P}$. Assume that $M = P^n$ for some irreducible polynomial P. Then

$$G_M = \text{Gal }(K(\Lambda_M)/K) \cong \left(R_T/(P^n)\right)^*$$

and

$$|G_M| = \Phi(P^n) = q^{dn} - q^{d(n-1)} = q^{d(n-1)}(q-1),$$

where $d = \deg P$.

It follows by the decomposition law for abelian groups that G_M contains a unique subgroup or order $(q-1)$, and this can only be $\mathbb{F}_q^*$.

On the other hand, we have $|G_0| = e_\infty f_\infty = (q-1)1 = q-1$. Thus $G_0 \cong \mathbb{F}_q^*$. Now let $M \in R_T \setminus \{0\}$ be arbitrary. Assume that P divides M. First we will see that there exists $\lambda \in \Lambda_P \subseteq \Lambda_M$ such that $v_{\mathfrak{P}'}(\lambda) = -1$, where $\mathfrak{P}' = \mathfrak{P} \cap K(\Lambda_P)$.

For $\lambda \in \Lambda_P \setminus \{0\}$,

$$\frac{\lambda^P}{\lambda} = \lambda^{q^d-1} + \begin{bmatrix} P \\ d-1 \end{bmatrix} \lambda^{q^{d-1}-1} + \cdots + \begin{bmatrix} P \\ 1 \end{bmatrix} \lambda^{q-1} + P = 0, \tag{12.11}$$

where $d = \deg P$ and $\begin{bmatrix} P \\ i \end{bmatrix} \in R_T$ is of degree $(d-i)q^i$.

Dividing by T^{q^d-1} in (12.11) we obtain

$$\left(\frac{\lambda}{T}\right)^{q^d-1} + g_{d-1}\left(\frac{1}{T}\right)\left(\frac{\lambda}{T}\right)^{q^{d-1}-1} + \cdots + g_1\left(\frac{1}{T}\right)\left(\frac{\lambda}{T}\right)^{q-1} + g_0\left(\frac{1}{T}\right) = 0,$$

where $g_i\left(\frac{1}{T}\right) \in \mathbb{F}_q\left[\frac{1}{T}\right]$,

$$g_{d-i}\left(\frac{1}{T}\right) = \frac{1}{T^{(q^d-1)-(q^{d-i}-1)}} \begin{bmatrix} P \\ d-i \end{bmatrix} = \frac{1}{T^{q^d-q^{d-i}}} \begin{bmatrix} P \\ d-i \end{bmatrix},$$

and

$$\begin{aligned} v_\infty\left(g_{d-i}\left(\frac{1}{T}\right)\right) &= v_\infty\left(\begin{bmatrix} P \\ d-i \end{bmatrix}\right) - v_\infty\left(T^{q^d-q^{d-i}}\right) \\ &= -iq^{d-i} + q^d - q^{d-i} = q^d - (i+1)q^{d-i}. \end{aligned}$$

Therefore

$$\begin{aligned} v_\infty\left(g_{d-1}\left(\frac{1}{T}\right)\right) &= q^d - 2q^{d-1} < q^d - (i+1)q^{d-i} \\ &= v_\infty\left(g_{d-i}\left(\frac{1}{T}\right)\right) \quad \text{for all} \quad i > 1. \end{aligned}$$

Now $\frac{\lambda}{T}$ is an integral element with respect to $\mathfrak{P}' \mid \mathfrak{p}_\infty$, since it satisfies a monic polynomial with coefficients in $\mathbb{F}_q\left[\frac{1}{T}\right]$. Since $\frac{1}{T}$ is a prime element at $\mathfrak{p}_\infty$ it follows that $\frac{\lambda}{T}$ is integral with respect to $\frac{1}{T}$. Thus $v_{\mathfrak{P}'}\left(\frac{\lambda}{T}\right) \geq 0$. Assume for the sake of contradiction that $v_{\mathfrak{P}'}\left(\left(\frac{\lambda}{T}\right)^{q^d-1}\right) < v_{\mathfrak{P}'}\left(g_{d-1}\left(\frac{1}{T}\right)\right)$. Then

$$v_{\mathfrak{P}'}\left(\left(\frac{\lambda}{T}\right)^{q^d-1}\right) < v_{\mathfrak{P}'}\left(g_{d-i}\left(\frac{1}{T}\right)\right) < v_{\mathfrak{P}'}\left(g_{d-i}\left(\frac{1}{T}\right)\left(\frac{\lambda}{T}\right)^{q^{d-i}-1}\right)$$

for all $i > 0$.

Thus,

$$\begin{aligned} \infty = v_{\mathfrak{P}'}(0) = v_{\mathfrak{P}'}\Bigg(\left(\frac{\lambda}{T}\right)^{q^d-1} &+ g_{d-1}\left(\frac{1}{T}\right)\left(\frac{\lambda}{T}\right)^{q^{d-1}-1} + \cdots \\ &+ g_1\left(\frac{1}{T}\right)\left(\frac{\lambda}{T}\right)^{q-1} + g_0\left(\frac{1}{T}\right)\Bigg) = v_{\mathfrak{P}'}\left(\left(\frac{\lambda}{T}\right)^{q^d-1}\right) \neq \infty. \end{aligned}$$

This shows that

$$(q^d - 1)(v_{\mathfrak{P}'}(\lambda) - v_{\mathfrak{P}'}(T)) = v_{\mathfrak{P}'}\left(\left(\frac{\lambda}{T}\right)^{q^d-1}\right)$$
$$\geq v_{\mathfrak{P}'}\left(g_{d-1}\left(\frac{1}{T}\right)\right) = e(\mathfrak{P}'|\mathfrak{p}_\infty)v_\infty\left(g_{d-1}\left(\frac{1}{T}\right)\right) = (q-1)(q^d - 2q^{d-1}).$$

Therefore

$$v_{\mathfrak{P}'}(\lambda) \geq \frac{(q-1)(q^d - 2q^{d-1})}{q^d - 1} + v_{\mathfrak{P}'}(T) \geq -1.$$

In particular, $v_{\mathfrak{P}'}(\lambda) < 0 \Rightarrow v_{\mathfrak{P}'}(\lambda) = -1$. By Exercise 12.10.19, $[K(\Lambda_P) : \mathbb{F}_q(\lambda)] = q^{d-1}$.

Therefore $\deg \mathfrak{Z}_\lambda = \deg \mathfrak{N}_\lambda = q^{d-1}$ and $e_\infty = q - 1$, where $(q-1, q^{d-1}) = 1$.

It follows that there are q^{d-1} prime divisors $\mathfrak{q}$ of $K(\Lambda_P)$ such that $v_\mathfrak{q}(\lambda) = -1$ in the pole divisor of λ.

Note that since $\lambda \in \Lambda_P$, λ belongs to ϑ_P. Thus the pole divisor of λ consists of prime divisors dividing $\mathfrak{p}_\infty$. Therefore if $\mathfrak{q}$ is any prime divisor in $K(\Lambda_P)$ that divides $\mathfrak{Z}_\lambda$, then $v_\mathfrak{q}(\lambda) = -1$. Let $A \in R_T$ be such that $\sigma_{A^{-1}}(\mathfrak{q}) = \mathfrak{P}'$, $\sigma_A(\lambda) = \lambda^A$, and $\sigma_A \in G_P$.

Thus $v_{\mathfrak{P}'}(\lambda^A) = v_{\mathfrak{P}'^{\sigma_{A^{-1}}}}(\lambda) = v_\mathfrak{q}(\lambda) = -1$. We may assume $\lambda = \lambda^A$.

Our claim is thereby proved. Now since $\mathfrak{P} \mid \mathfrak{P}'$ is unramified (Theorem 12.4.6), then $v_\mathfrak{P}(\lambda) = e(\mathfrak{P}|\mathfrak{P}')v_{\mathfrak{P}'}(\lambda) = 1 \times v_{\mathfrak{P}'}(\lambda) = -1$. In short, there exists an element $\lambda \in \Lambda_{P^n} \subseteq \Lambda_M$ such that $v_\mathfrak{P}(\lambda) = -1$. Then $\frac{1}{\lambda}$ is a prime element for $\mathfrak{P} \mid \mathfrak{p}_\infty$, that is, $v_\mathfrak{P}\left(\frac{1}{\lambda}\right) = 1$. Finally, if $\alpha \in \mathbb{F}_q^*$, then $\sigma_\alpha\left(\frac{1}{\lambda}\right) = \alpha\frac{1}{\lambda}$. Therefore $\sigma_\alpha(K(\Lambda_M)_\mathfrak{P}) = \sigma_\alpha(\mathbb{F}_{q^d}((\lambda))) = \mathbb{F}_{q^d}((\lambda)) = K(\lambda_M)_\mathfrak{P}$. Thus $\sigma_\alpha \in \mathrm{Gal}(K(\lambda_M)_\mathfrak{P}/K_\mathfrak{p})$, so $\mathfrak{P}^{\sigma_\alpha} = \mathfrak{P}$ and $\mathbb{F}_q^* \subseteq D(\mathfrak{P} \mid \mathfrak{p}_\infty) = G_0$. Since $\mathbb{F}_q^*$ and G_0 are of order $q - 1$, the result follows. □

Definition 12.5.5. Let M be a nonzero element of R_T, and

$$K(\Lambda_M)^+ := K(\Lambda_M)^{G_0}.$$

$K(\Lambda_M)^+$ is called the *maximal real subfield* of $K(\Lambda_M)$.

Remark 12.5.6. We have $[K(\Lambda_M) : K(\Lambda_M)^+] = |G_0| = q - 1$ and $\mathfrak{p}_\infty$ decomposes totally into $\Phi(M)/(q-1)$ prime divisors in $K(\Lambda_M)^+/K$.

Remark 12.5.7. The inertia group of the infinite prime divisors in the cyclotomic number field $\mathbb{Q}(\zeta_n)/\mathbb{Q}$ is $G_0 = \{1, J\}$, where J denotes complex conjugation, and $\mathbb{Q}(\zeta_n)^+ = \mathbb{Q}(\zeta_n) \cap \mathbb{R} = \mathbb{Q}(\zeta_n)^{\{1,J\}}$. The above equality motivates Definition 12.5.5.

For any $M \in R_T$, denote by ϑ_M the integral closure of R_T in $K_M = K(\Lambda_M)$.

Proposition 12.5.8. *Assume that $M = P^n$ for some irreducible polynomial P. Then $\vartheta_M = R_T[\lambda_M]$, where λ_M is a generator of Λ_M.*

Proof. Set $\lambda = \lambda_M$. Since λ is integral, we have $R_T[\lambda] \subseteq \vartheta_M$. Let $\alpha \in \vartheta_M$. Since $\{1, \lambda, \ldots, \lambda^{\Phi(M)-1}\}$ is a basis of K_M/K, there exist $a_1, a_2, \ldots, a_r \in K$ such that $\alpha = a_0 + a_1\lambda + \cdots + a_r\lambda^r$, where $r = \Phi(M) - 1$. We wish to show that $a_i \in R_T$ for $i = 1, 2, \ldots, r$. By the proof of Proposition 12.3.14 we have $v_{\mathfrak{P}}(\lambda) = 1$, where $\mathfrak{P}$ is the (unique) prime divisor of K_M above $\mathfrak{p}$ and $(P)_K = \frac{\mathfrak{p}}{\mathfrak{p}_\infty^{\deg P}}$.

Clearly, $v_{\mathfrak{P}}(a_i\lambda^i) = i + \Phi(M)v_{\mathfrak{p}}(a_i) \equiv i \bmod \Phi(M)$. Thus, whenever $i \neq j$, $a_i \neq 0$, and $a_j \neq 0$, we have $v_{\mathfrak{P}}(a_i\lambda^i) \neq v_{\mathfrak{P}}(a_j\lambda^j)$. It follows that

$$0 \leq v_{\mathfrak{P}}(\alpha) = \min_{a_i \neq 0}\left\{v_{\mathfrak{P}}(a_i\lambda^i)\right\} = \min_{a_i \neq 0}\left\{i + \Phi(M)v_{\mathfrak{P}}(a_i)\right\}.$$

Hence $v_{\mathfrak{P}}(a_i) \geq 0$ for all i. Now for any $\sigma_A \in G_M = \operatorname{Gal}(K_M/K)$ such that $\sigma_A(\lambda) = \lambda^A$, we have

$$\alpha_A = \sigma_A(\alpha) = a_0 + a_1\lambda^A + \cdots + a_r(\lambda^A)^r, \tag{12.12}$$

where $A \bmod M \in (R_T/(M))^*$. If $\{\overline{A}_1, \ldots, \overline{A}_{\Phi(M)}\}$ is a set of representatives of $(R_T/(M))^*$ we obtain from (12.12), writing $\alpha_i = \alpha^{A_i}$, $\lambda_i = \lambda^{A_i}$, that

$$\begin{pmatrix} \alpha_1 \\ \vdots \\ \alpha_{\Phi(M)} \end{pmatrix} = \begin{pmatrix} 1 & \lambda_1 & \lambda_1^2 & \cdots & \lambda_1^r \\ \vdots & \vdots & \vdots & \vdots & \vdots \\ 1 & \lambda_{r+1} & \lambda_{r+1}^2 & \cdots & \lambda_{r+1}^r \end{pmatrix} \begin{pmatrix} a_0 \\ \vdots \\ a_r \end{pmatrix}.$$

The determinant of the matrix $\left[\lambda_i^j\right]_{\substack{0\leq j\leq r \\ 1\leq i\leq r+1}}$ is a Vandermonde determinant, so that $\det\left[\lambda_i^j\right] = \prod_{1\leq t\leq \ell\leq r+1}(\lambda_\ell - \lambda_t) = d$ (see Exercise 12.10.22). Therefore

$$a_i = \frac{\det\begin{bmatrix} 1 & \lambda_1 & \cdots & \lambda_1^{i-1} & \alpha_1 & \lambda_1^{i+1} & \cdots & \lambda_1^r \\ \vdots & \vdots & & \vdots & \vdots & \vdots & & \vdots \\ 1 & \lambda_{r+1} & \cdots & \lambda_{r+1}^{i-1} & \alpha_{r+1} & \lambda_{r+1}^{i+1} & \cdots & \lambda_{r+1}^r \end{bmatrix}}{\det\begin{bmatrix} 1 & \lambda_1 & \cdots\cdots & \lambda_1^r \\ \vdots & \vdots & \cdots\cdots & \vdots \\ 1 & \lambda_{r+1} & \cdots\cdots & \lambda_{r+1}^r \end{bmatrix}} = \frac{b_i}{d},$$

where $b_i \in \vartheta_M$.

By the proof of Proposition 12.3.14 ((12.5)), for all $A \bmod (R_T/(M))^*$, we have $\lambda = \beta_A\lambda^A$ and $P = \beta_0\lambda^{\Phi(M)}$ for some $\beta_A, \beta_0 \in \vartheta_M^*$.

Then for any prime divisor $\mathfrak{q}$ in $K(\Lambda_M)$ dividing neither $\mathfrak{p}$ nor $\mathfrak{p}_\infty$, we have $v_{\mathfrak{q}}(\lambda) = v_{\mathfrak{q}}(\lambda^A) = 0$. It follows that the support of the pole divisor of a_i can consist only of $\mathfrak{p}$ and $\mathfrak{p}_\infty$. Since $v_{\mathfrak{p}}(a_i) \geq 0$, we have $a_i \in R_T$. Thus $\vartheta_M = R_T[\lambda]$. □

Proposition 12.5.8 holds for any $M \in R_T \setminus \{0\}$. To see this fact, first we prove the following proposition:

Proposition 12.5.9. *Let $M, N \in R_T \setminus \{0\}$ be two relatively prime polynomials. Then $\vartheta_{MN} = \vartheta_M \vartheta_N$.*

Proof. By Theorem 5.7.15,

$$\mathfrak{D}_{\vartheta_{MN}/R_T} = \mathfrak{D}_{\vartheta_{MN}/\vartheta_M} \operatorname{con}_{M/MN} \mathfrak{D}_{\vartheta_M/R_T} = \mathfrak{D}_{\vartheta_{MN}/\vartheta_N} \operatorname{con}_{N/MN} \mathfrak{D}_{\vartheta_N/R_T}.$$

Since $\mathfrak{p}_\infty$ is not being considered in the Dedekind domain ϑ_E $(E \in \{N, M, NM\})$ and M and N are relatively prime, it follows by Theorem 12.5.3 and Proposition 5.6.7 that $\operatorname{con}_{M/MN} \mathfrak{D}_{\vartheta_M/R_T}$ and $\operatorname{con}_{N/MN} \mathfrak{D}_{\vartheta_N/R_T}$ have no common factor, and neither do $\mathfrak{D}_{\vartheta_{MN}/\vartheta_M}$ and $\mathfrak{D}_{\vartheta_{MN}/\vartheta_N}$.

$$\begin{array}{ccc} K_M & \text{———} & K_{MN} \\ | & & | \\ K & \text{———} & K_N \end{array}$$

Thus

$$\operatorname{con}_{M/MN} \mathfrak{D}_{\vartheta_M/R_T} = \mathfrak{D}_{\vartheta_{MN}/\vartheta_N} \quad \text{and} \quad \operatorname{con}_{N/MN} \mathfrak{D}_{\vartheta_N/R_T} = \mathfrak{D}_{\vartheta_{MN}/\vartheta_M}. \qquad (12.13)$$

Now since R_T is a principal ideal domain and ϑ_E is a torsion-free R_T-module, it follows using the theory of finitely generated modules over principal ideal domains that ϑ_E is R_T-free. Let V be a basis for ϑ_M/R_T and V^* the dual basis of V with respect to the trace map. Then V^* generates $\mathfrak{D}^{-1}_{\vartheta_M/R_T}$ as an R_T-module. By (12.13) it follows that V^* generates $\mathfrak{D}^{-1}_{\vartheta_{MN}/\vartheta_N}$. Hence $V^{**} = V$ generates ϑ_{MN} over ϑ_N, and therefore $\vartheta_{MN} = \vartheta_M \vartheta_N$. □

As a corollary, we obtain the following theorem:

Theorem 12.5.10. *For any $M \in R_T \setminus \{0\}$, let $\lambda = \lambda_M$ be a generator of the Carlitz–Hayes module Λ_M. Then $\vartheta_M = R_T[\lambda]$.*

Proof. Let $M = \alpha P_1^{\alpha_1} \cdots P_r^{\alpha_r}$, where $P_1, \dots, P_r$ are distinct monic irreducible polynomials in R_T. Using Propositions 12.5.8 and 12.5.9 we obtain

$$\vartheta_M = \prod_{i=1}^r \vartheta_{P_i^{\alpha_i}} = \prod_{i=1}^r R_T[\lambda_{P_i^{\alpha_i}}] = R_T[\lambda]. \qquad \square$$

Next we present a particular case of an analogue of Dirichlet theorem on distribution of primes in arithmetic progressions without using the Čebotarev density theorem.

Proposition 12.5.11. *Let $M \in R_T \setminus \{0\}$. If M is not a prime power, then $\Psi_M(u) \equiv 1 + Cu^{q-1} \pmod{u^{2(q-1)}}$, where $C \in R_T$ and $\deg C = (\deg M - 1)(q-1) - 1$.*

Proof. By Exercise 12.10.12, $\Psi_M(u) = \prod_{A|M}(u^A)^{\mu(M/A)}$. Since M is not a prime power, it follows that $\sum_{A|M}\mu(M/A) = 0$ (Exercise 12.10.9). Hence $\Psi_M(u) = \prod_{A|M}(u^A/u)^{\mu(M/A)}$. Let $y = u^{q-1}$. By Theorem 12.2.5, $u^A/u \equiv A + \begin{bmatrix} A \\ 1 \end{bmatrix} y \bmod y^2$. Thus

$$\begin{aligned}\Psi_M(u) &\equiv \prod_{A|M}\left(A + \begin{bmatrix} A \\ 1 \end{bmatrix} y\right)^{\mu(M/A)} \bmod y^2 \\ &\equiv \left(\prod_{A|M} A^{\mu(M/A)}\right)\prod_{A|M}\left(1 + \left(\begin{bmatrix} A \\ 1 \end{bmatrix}/A\right) y\right)^{\mu(M/A)} \bmod y^2.\end{aligned}$$

By Exercise 12.10.23,

$$\begin{aligned}\Psi_M(u) &\equiv \prod_{A|M}\left(1 + \left(\begin{bmatrix} A \\ 1 \end{bmatrix}/A\right) y\right)^{\mu(M/A)} \bmod y^2 \\ &\equiv 1 + C(T)y \bmod y^2,\end{aligned}$$

where $C(T) = \sum_{A|M}\mu(M/A)\left(\begin{bmatrix} A \\ 1 \end{bmatrix}/A\right)$. Therefore

$$M(T)C(T) = \sum_{A|M}\mu(M/A)\left(\frac{M}{A}\right)\begin{bmatrix} A \\ 1 \end{bmatrix}.$$

Set $d_1 = \deg A$. Then

$$\begin{aligned}\deg\left(\frac{M}{A}\right)\begin{bmatrix} A \\ 1 \end{bmatrix} &= \deg M - d_1 + (d_1 - 1)q \\ &= (\deg M - 1)q + (d_1 - \deg M)(q-1) \le (\deg M - 1)q,\end{aligned}$$

and we have equality if and only if $d_1 = \deg M$, i.e., $A = M$. Hence

$$\deg C = (\deg M - 1)q - \deg M = (\deg M - 1)(q-1) - 1. \qquad \square$$

Corollary 12.5.12. *If $M \in R_T \setminus \{0\}$ is not a prime power and $\lambda \in \Lambda_M$ is a generator, then λ is a unit in ϑ_M.*

Proof. We have $0 = \Psi_M(\lambda) = 1 + C(T)\lambda^{q-1} \bmod \lambda^{2(q-1)}$. Therefore $1 = \lambda(-C_1(T)\lambda^{q-2} + \lambda^{2q-3}\alpha)$ for some $\alpha \in \vartheta_M$. It follows that λ is invertible in ϑ_M. $\square$

Definition 12.5.13. Let $P \in R_T$ be a monic irreducible polynomial and let $A \in R_T$. We say that

$$\hat{o}(A \bmod P) = M \in R_T$$

if M is monic and of minimal degree satisfying $A^M \equiv 0 \bmod P$.

Remark 12.5.14. The notation given in Definition 12.5.13 is $\hat{o}(A \bmod P)$ instead of $o(A \bmod P)$, that is, the one that denotes the order of an element in a quotient group.

Remark 12.5.15. Assume that $N \in R_T$ satisfies $A^N \equiv 0 \bmod P$, and let $N = QM+R$ with Q, $R \in R_T$ and $R = 0$ or $\deg R < \deg M$. Then $A^N = (A^M)^Q + A^R$. It follows that $A^R \equiv 0 \bmod P$. Therefore $R = 0$ and M divides N. In particular, the polynomial M given in Definition 12.5.13 is unique.

Remark 12.5.16. Since $R_T/(P)$ is finite, $\{A^M \bmod P \mid M \in R_T\}$ is finite too, and there exist two distinct elements M_1, M_2 in R_T such that $A^{M_1} \equiv A^{M_2} \bmod P$. Hence $A^{M_1-M_2} \equiv 0 \bmod P$.

Proposition 12.5.17. *Let $P \in R_T$ be an irreducible polynomial and $M \in R_T$ monic polynomial not divisible by P. If $A \in R_T$, then*

$$P \mid \Psi_M(A) \iff \hat{o}(A \bmod P) = M.$$

Proof. First assume that P divides $\Psi_M(A)$. Since $u^M = \prod_{D|M} \Psi_D(u)$ it follows that $A^M = \prod_{D|M} \Psi_D(A) \equiv 0 \bmod P$.

Let $\hat{o}(A \bmod P) = N$. Then N divides M. Hence $A^N = \prod_{D|N} \Psi_D(A) \equiv 0 \bmod P$. Therefore, there exists D_0 dividing N such that $P \mid \Psi_{D_0}(A)$.

Suppose that $D_0 \neq M$. Then $A^M = \Psi_M(A)\Psi_{D_0}(A) \prod_{\substack{D|M \\ D\neq D_0, D\neq M}} \Psi_D(A) \equiv 0 \bmod P^2$. Now $\Psi_M(A+P) \equiv \Psi_M(A) \bmod P \equiv 0 \bmod P$ and $\Psi_{D_0}(A+P) \equiv \Psi_{D_0}(A) \bmod P \equiv 0 \bmod P$. Hence $0 \equiv (A+P)^M = A^M + P^M \equiv P^M \bmod P^2$. We have

$$P^M = \sum_{i=0}^{\deg M} \begin{bmatrix} M \\ i \end{bmatrix} P^{q^i} = MP + P^2 C \equiv MP \bmod P^2.$$

But this is impossible since $P \nmid M$. It follows that $\hat{o}(A \bmod P) = M$.

Conversely, let $\hat{o}(A \bmod P) = M$, where $A^M = \prod_{D|M} \Psi_D(A) \equiv 0 \bmod P$. Thus P divides $\Psi_D(A)$ for some D dividing M. If $D \neq M$, then $A^D = \prod_{D'|D} \Psi_{D'}(A) \equiv 0 \bmod P$, which contradicts the fact that $\hat{o}(A \bmod P) = M$. Hence $D = M$ and P divides $\Psi_M(A)$. □

Proposition 12.5.18. *Let $P \in R_T$ be an irreducible polynomial, and $M \in R_T$ a monic polynomial such that $P \nmid M$. Then P divides $\Psi_M(A)$ for some $A \in R_T$ if and only if $P \equiv 1 \bmod M$.*

Proof: If P divides $\Psi_M(A)$ for some $A \in R_T$, then by Proposition 12.5.17, $\hat{o}(A \bmod P) = M$. By Proposition 12.3.18, the polynomial $\Psi_P(u) = u^P/u$ is Eisenstein. Thus $u^P = u\Psi_P(u) \equiv u^{q^d} \bmod P$, where $d = \deg P$. In particular, we have $A^P \equiv A^{q^d} \bmod P$.

Since $\Phi(P) = q^d - 1 = \left|(R_T/(P))^*\right|$, it follows that if $P \nmid A$, we have $A^{q^d-1} \equiv 1 \bmod P$, so that $A^{q^d} \equiv A \bmod P$. If P divides A, we have $A^{q^d} \equiv 0 \equiv A \bmod P$.

In any case we obtain $A^{q^d} \equiv A \bmod P$. Therefore $A^P \equiv A \bmod P$, or $A^P - A = A^{P-1} \equiv 0 \bmod P$. Since $\hat{o}(A \bmod P) = M$, it follows by Remark 12.5.15 that M divides $(P-1)$. Thus $P \equiv 1 \bmod M$.

Conversely, assume that $P \equiv 1 \bmod M$. Then $d = \deg(P-1) = \deg P$ and $u^{P-1} = \sum_{i=0}^{d} \begin{bmatrix} P-1 \\ i \end{bmatrix} u^{q^i}$. Hence $(u^{P-1})' \bmod P \equiv (P-1) \bmod P \equiv -1 \bmod P \not\equiv 0$.

Therefore the polynomial $u^{P-1} \bmod P \in (R_T/(P))\,[u]$ is separable.

Since $\deg_u u^{P-1} = q^d = |R_T/(P)|$ and $A^{P-1} \equiv 0 \bmod P$ for all $A \in R_T$, it follows that

$$u^{P-1} \bmod P = \prod_{D|P-1} \Psi_D(u) \bmod P = \prod_{\substack{A \bmod P \\ A \in R_T}} (u-A) \bmod P.$$

Therefore there exists $A \in R_T$ such that $\psi_M(A) \equiv 0 \bmod P$. Thus P divides $\Psi_M(A)$ and $\hat{o}(A \bmod P) = M$. $\square$

Corollary 12.5.19. *For any nonconstant polynomial $M \in R_T$, there exist infinitely many irreducible polynomials P in R_T such that $P \equiv 1 \bmod M$.*

Proof. Let $\{P_1, \ldots, P_r\}$ be any finite set of irreducible polynomials satisfying $P_i \equiv 1 \bmod M$. Set $N = MP_1 \cdots P_r$ and let $Q \in R_T$ be arbitrary. Then $\Psi_M(NQ) \equiv \Psi_M(0) \bmod N$. Since we may take $r \geq 1$ and P_1 not dividing M, it follows that M is not a prime power. By Exercise 12.10.26, we have $\Psi_M(0) = 1$. Thus $\Psi_M(NQ) \equiv 1 \bmod N$. In particular,

$$\Psi_M(NQ) \equiv 1 \bmod M \quad \text{and} \quad \Psi_M(NQ) \equiv 1 \bmod P_i \quad \text{for} \quad 1 \leq i \leq r.$$

It follows from the above that if P is any irreducible polynomial dividing $\Psi_M(NQ)$, we have $P \equiv 1 \bmod M$ by Proposition 12.5.18, and $P \neq P_i$ for $1 \leq i \leq r$. $\square$

Remark 12.5.20. Corollary 12.5.19 is a particular case of Dirichlet's theorem (Theorem 12.5.21 above), which is an easy consequence of Čebotarev's density theorem (Theorem 11.2.20). However, the proof we provided for Corollary 12.5.19 does not use Čebotarev's density theorem.

Theorem 12.5.21 (Dirichlet). *Let $M, N \in R_T$ be two nonconstant monic polynomials such that $(M, N) = 1$. Then there exist infinitely many irreducible polynomials $P \in R_T$ such that $P \equiv N \bmod M$.*

Proof. Consider the extension $K(\Lambda_M)/K$ with $\operatorname{Gal}(K(\Lambda_M)/K) \cong (R_T/(M))^*$. Let $\sigma \in \operatorname{Gal}(K(\Lambda_M)/K)$ be the element of the Galois group corresponding to the element $N \bmod M \in (R_T/(M))^*$. Then $\sigma(\lambda_M) = \lambda_M^N$, where λ_M is a generator of Λ_M.

By Theorem 12.5.1, the Artin symbol $\left[\frac{K(\Lambda_M)/K}{P}\right]$ corresponds to the map

$$\begin{aligned} \varphi_P : K(\Lambda_M) &\to K(\Lambda_M) \\ \lambda_M &\mapsto \lambda_M^P. \end{aligned}$$

By Čebotarev's density theorem, there exist infinitely many irreducible polynomials $P \in R_T$ such that $\left[\frac{K(\Lambda_M)/K}{P}\right] = \sigma$. Therefore $\sigma = \varphi_P$ for infinitely many irreducible polynomials $P \in R_T$. Now

$$\sigma = \varphi_P \iff \lambda_M^N = \lambda_M^P \iff N \equiv P \bmod M. \qquad \square$$

12.6 Dirichlet Characters

Definition 12.6.1. Let $M \in R_T \setminus \{0\}$ be a monic polynomial. A *Dirichlet character* mod M is a homomorphism

$$\mathcal{X} : (R_T/(M))^* \to \mathbb{C}^*.$$

Remark 12.6.2. Assume that M divides an element N of R_T and consider the canonical homomorphism

$$\begin{aligned} \varphi_{N,M}\colon\ (R_T/(N))^* &\to (R_T/(M))^* \\ A \bmod N &\mapsto A \bmod M. \end{aligned}$$

Then for any Dirichlet character mod M, $\mathcal{X} : (R_T/(M))^* \to \mathbb{C}^*$, $\varphi_{N,M}$ induces a Dirichlet character mod N, namely $\mathcal{X} \circ \varphi_{N,M} : (R_T/(N))^* \to \mathbb{C}^*$.

$$\begin{array}{ccc} (R_T/(M))^* & \xrightarrow{\quad \mathcal{X} \quad} & \mathbb{C}^* \\ & {\scriptstyle \varphi_{M,F}} \searrow \quad \nearrow {\scriptstyle \xi} & \\ & (R_T/(F))^* & \end{array}$$

Conversely, if $\mathcal{X}$ is a Dirichlet character mod M, we say that we may define $\mathcal{X}$ mod F for $F \mid M$ if there exists $\xi : (R_T/(F))^* \to \mathbb{C}^*$ such that $\xi \circ \varphi_{M,F} = \mathcal{X}$.

Next we show the existence of the *conductor*. Let $\mathcal{X}\colon (R_T/M)^* \to \mathbb{C}^*$ be a Dirichlet character and A and B such that $A \mid M$, $B \mid M$, and $\mathcal{X} = \mathcal{X}_A \circ \varphi_{M,A}$ and $\mathcal{X} = \mathcal{X}_B \circ \varphi_{M,B}$. Consider $C = (A, B)$ and set D as the product of all the monic irreducible polynomials dividing M but not dividing B. It follows that $C = (DA, B)$. Consider any $U, V \in R_T$ such that $(UV, M) = 1$ and $U \equiv V \bmod C$. By the Chinese remainder theorem, there exists $S \in R_T$ such that $S \equiv U \bmod DA$ and $S \equiv V \bmod B$.

If P is any irreducible polynomial such that $P \mid S$ and $P \mid M$, then writing $S = V + QB$, we deduce that $P \nmid B$, since otherwise $P \mid V$ and then $P \mid (V, M) = 1$. Now since $P \mid M$ and $P \nmid B$, it follows that $P \mid D$. Therefore $P \mid DA$ and $P \mid S$. Hence $P \mid U$ and $P \mid (U, M) = 1$. This contradiction shows that $(S, M) = 1$. It follows that

$$\mathcal{X}(S) = \mathcal{X}_A \circ \varphi_{M,A}(S) = \mathcal{X}_{DA} \circ \varphi_{M,DA}(S) = \mathcal{X}_{DA} \circ \varphi_{M,DA}(U) = \mathcal{X}(U)$$

and

$$\mathcal{X}(S) = \mathcal{X}_B \circ \varphi_{M,B}(S) = \mathcal{X}_B \circ \varphi_{M,B}(V) = \mathcal{X}(V).$$

Thus $\mathcal{X}(S) = \mathcal{X}(U) = \mathcal{X}(V)$. Therefore $\mathcal{X}$ can be defined mod C.

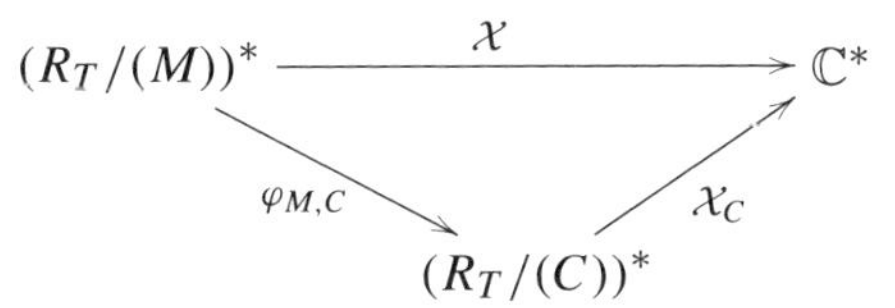

In particular, if $\mathcal{X}$ can be defined mod F_1 and mod F_2 with F_1 and F_2 monic of minimal degree, then since it can be defined mod C, $C = (F_1, F_2)$ and $C \mid F_1$ and $C \mid F_2$, it follows that $C = F_1 = F_2$.

Theorem 12.6.3. *Given a Dirichlet character $\mathcal{X}$, there exists a unique monic polynomial F in R_T of minimal degree dividing M such that $\mathcal{X}$ can be defined* mod F. □

Definition 12.6.4. Given a Dirichlet character $\mathcal{X}$ mod M the *conductor of $\mathcal{X}$* is F if $F \in R_T$ is a monic polynomial of minimal degree dividing M such that $\mathcal{X}$ can be defined mod F. We denote the conductor of $\mathcal{X}$ by $F_{\mathcal{X}}$.

Example 12.6.5. Let $\mathcal{X} : \left(R_T/(T^3)\right)^* \to \mathbb{C}^*$ (with $q = 2$) be given by $\mathcal{X}(1) = 1$, $\mathcal{X}(T+1) = -1$, $\mathcal{X}(T^2+T+1) = -1$, and $\mathcal{X}(T^2+1) = 1$.

Let $\xi : \left(R_T/(T^2)\right)^* \to \mathbb{C}^*$ be defined by $\xi(1) = 1$ and $\xi(T+1) = -1$.

Then $\varphi_{T^3,T^2} : \left(R_T/(T^3)\right)^* \to \left(R_T/(T^2)\right)^*$ is given by

$$\varphi_{T^3,T^2}(1) = \varphi_{T^3,T^2}(T^2+1) = 1$$

and

$$\varphi_{T^3,T^2}(T+1) = \varphi_{T^3,T^2}(T^2+T+1) = T+1.$$

Hence $\xi \circ \varphi_{T^3,T^2} = \mathcal{X}$. Clearly T^2 is minimal since $(R_T/(T))^* = \{1\}$. Therefore $F_{\mathcal{X}} = T^2$.

Example 12.6.6. Let $\mathcal{X} : \left(R_T/(T^2(T+1))\right)^* \to \mathbb{C}^*$ with $q = 2$ given by

$$\mathcal{X}(1) = 1 \quad \text{and} \quad \mathcal{X}(T^2+T+1) = -1.$$

Then $\xi \circ \varphi_{T^2(T+1),T^2} = \mathcal{X}$ where $\xi(1) = 1$ and $\xi(T+1) = -1$. Hence $F_{\mathcal{X}} = T^2$.

Remark 12.6.7. Given a Dirichlet character $\mathcal{X}$ we may regard $\mathcal{X}$ as a map $\mathcal{X}\colon R_T \to \mathbb{C}$ by defining $\mathcal{X}(Q) = 0$ if $(Q, F_{\mathcal{X}}) \neq 1$. Unless otherwise specified, we will always view $\mathcal{X}$ as being defined modulo its conductor.

Definition 12.6.8. A Dirichlet character $\mathcal{X}$ defined modulo its conductor is called *primitive*. In this case $\mathcal{X}(Q) = 0$ as infrequently as possible. Also notice that when $\mathcal{X}$ is defined modulo its conductor, we have $\mathcal{X}(A + F_{\mathcal{X}}) = \mathcal{X}(A)$. Thus $\mathcal{X}$ is periodic of period $F_{\mathcal{X}}$.

Notation 12.6.9. Whenever we mention the characters of $(R_T/(M))^*$ for $M \in R_T$ or characters mod M, we will be including all characters whose conductor divides M and the trivial character of conductor 1. The trivial character ε satisfies $\varepsilon(Q) = 1$ for all $Q \in R_T$.

Definition 12.6.10. Let $\mathcal{X}$ and ϕ be two Dirichlet characters of conductors $F_{\mathcal{X}}$ and F_ϕ respectively. We define the product of $\mathcal{X}$ and ϕ as follows. First let

$$Q = \left[F_{\mathcal{X}}, F_\phi\right] \quad \text{and define} \quad \gamma\colon\ (R_T/(Q))^* \to \mathbb{C}^*$$

by $\gamma(\overline{A}) = \mathcal{X}(\overline{A})\phi(\overline{A})$. Then the product $\mathcal{X}\phi$ is defined as the primitive character associated to γ.

Remark 12.6.11. It is not true in general that $(\mathcal{X}\phi)(\overline{A}) = \mathcal{X}(\overline{A})\phi(\overline{A})$.

Example 12.6.12. Let $q = 2$, and $\mathcal{X}$ mod $T^2(T^2+1)$ be given by

$$\mathcal{X}(1) = 1, \quad \mathcal{X}(T^2+T+1) = 1, \quad \mathcal{X}(T^3+T^2+1) = -1,$$

and

$$\mathcal{X}(T^3+T+1) = -1.$$

If $F_{\mathcal{X}}$ is the conductor of $\mathcal{X}$, then

$$F_{\mathcal{X}} \in \left\{1, T, T+1, T(T+1), T^2, T^2(T+1), T^2+1, T(T^2+1), T^2(T^2+1)\right\}.$$

Note that $\left|(R_T/(T))^*\right| = \left|(R_T/(T+1))^*\right| = \left|(R_T/(T(T+1)))^*\right| = 1$. Thus $F_{\mathcal{X}} \neq 1, T, T+1, T(T+1)$.

Now $T^3+T^2+1 \bmod T^2 = 1$, $\mathcal{X}(T^3+T^2+1) = -1 \neq 1$, T^3+T+1 mod $(T^2+1) = 1$ and $\mathcal{X}(T^3+T+1) = -1 \neq 1$. Thus $F_{\mathcal{X}} \neq T^2, T^2+1$. Finally we have

$$\begin{aligned} T^3+T^2+1 \bmod T^2(T+1) &= 1, \quad \mathcal{X}(T^3+T^2+1) = -1 \neq 1,\\ T^3+T+1 \bmod T(T^2+1) &= 1, \quad \mathcal{X}(T^3+T+1) = -1 \neq 1. \end{aligned}$$

Hence $F_{\mathcal{X}} \neq T^2(T+1), T(T^2+1)$. It follows that $F_{\mathcal{X}} = T^2(T^2+1)$. Now let φ mod (T^2) be given by $\varphi(1) = 1$ and $\varphi(T+1) = -1$. Then $F_\varphi = T^2$.

Consider the product $\mathcal{X}\varphi$. We have $\left[F_{\mathcal{X}}, F_\varphi\right] = \left[T^2(T^2+1), T^2\right] = T^2(T^2+1)$. Define $\gamma : \left(R_T/T^2(T^2+1)\right)^* \to \mathbb{C}^*$ by $\gamma(A) = \mathcal{X}(A)\varphi(A)$. Then

$$\begin{aligned} \gamma(1) &= \mathcal{X}(1)\varphi(1) = 1 \times 1 = 1,\\ \gamma(T^2+T+1) &= \mathcal{X}(T^2+T+1)\varphi(T^2+T+1) = (1)(-1) = -1,\\ \gamma(T^3+T^2+1) &= \mathcal{X}(T^3+T^2+1)\varphi(T^3+T^2+1) = (-1)(1) = -1,\\ \gamma(T^3+T+1) &= \mathcal{X}(T^3+T+1)\varphi(T^3+T+1) = (-1)(-1) = 1. \end{aligned}$$

Let $\xi\colon \left(R_T/(T^2+1)\right)^* \to \mathbb{C}^*$ be such that $\xi(1) = 1$ and $\xi(T) = -1$.

Then $\xi \circ \varphi_{T^2(T^2+1), T^2+1} = \gamma$. Thus $F_\gamma = T^2+1$ and $\xi = \mathcal{X}\varphi$. Notice that $\xi(T) = -1 \neq 0 = \varphi(T) = \mathcal{X}(T)\varphi(T)$.

Definition 12.6.13. If $\mathcal{X}$ is any Dirichlet character, we define the conjugate $\overline{\mathcal{X}}$ of $\mathcal{X}$ by $\overline{\mathcal{X}}(A) = \overline{\mathcal{X}(A)}$. Notice that $\overline{\mathcal{X}}(A) = \mathcal{X}(A)^{-1}$ for any A such that $(A, F_{\mathcal{X}}) = 1$. Hence $\mathcal{X}\overline{\mathcal{X}}$ is the trivial character defined by

$$\mathcal{X}\overline{\mathcal{X}}(A) \equiv 1 \quad \text{for all} \quad A \in R_T.$$

Remark 12.6.14. We have $G_M = \operatorname{Gal}(K(\Lambda_M)/K) \cong (R_T/(M))^*$, where $R_T = \mathbb{F}_q[T]$, $K = \mathbb{F}_q(T)$, and $\Lambda_M = \{\lambda \in \overline{K} \mid \lambda^M = 0\}$. Then a Dirichlet character is a character of G_M for some $M \in R_T$. In this case the Dirichlet character may be considered as a *Galois character*.

Example 12.6.15. Let $\mathcal{X}$ be as in Example 12.6.5. Then

$$\mathcal{X}\colon \left(R_T/(T^3)\right)^* \cong G_{T^3} = \operatorname{Gal}\left(K(\Lambda_{T^3})/K\right) \to \mathbb{C}^*$$

and $\ker \mathcal{X} = \{1 \bmod T^3, (T^2+1) \bmod T^3\}$. Therefore $\mathcal{X}$ is a character of $\left(R_T/(T^3)\right)^* / \ker \mathcal{X} \cong \left(R_T/(T^2)\right)^* \cong \operatorname{Gal}(K(\Lambda_{T^2})/K)$ and it may be considered as a character of $\operatorname{Gal}(K(\Lambda_{T^2})/K)$.

Example 12.6.16. Let $\mathcal{X}$ be as in Example 12.6.6. Then $\left(R_T/T^2(T+1)\right)^* \cong \left(R_T/T^2\right)^*$, and since any character mod $T^2(T+1)$ or mod T^2 is the same character, it follows that $K(\Lambda_{T^2(T+1)}) = K(\Lambda_{T^2})$.

Our main interest in the topic of Dirichlet characters is the study of some arithmetic properties of cyclotomic function fields. For this purpose, we need some general facts on group characters, which we now review.

Definition 12.6.17. Let G be any finite group. The *character group of* G is

$$\hat{G} = \operatorname{Hom}(G, \mathbb{C}^*).$$

Assume that $\mathcal{X} \in \operatorname{Hom}(G, \mathbb{C}^*)$. Since $\mathbb{C}^*$ is an abelian group, we have $\mathcal{X}([a,b]) = 1$ for any $a, b \in G$, where $[a,b] = aba^{-1}b^{-1}$ is the commutator of a and b. Therefore we can factor $\mathcal{X}$ through $[G,G] = \langle [a,b] | a, b \in G\rangle$ by defining $\widetilde{\mathcal{X}} : G/[G,G] = G^{ab} \longrightarrow \mathbb{C}^*$. In particular, $\hat{G} = \operatorname{Hom}(G, \mathbb{C}^*) \cong \operatorname{Hom}\ (G^{ab}, \mathbb{C}^*) = \widehat{G^{ab}}$. For instance, if G is a simple nonabelian group, we have $[G,G] = G$ and $\hat{G} = \{\mathrm{Id}\}$.

From now on, all groups considered will be abelian (and finite).

Proposition 12.6.18. *Any abelian group G is isomorphic to its character group $\hat{G}$.*

Proof. If G is a cyclic group of order m and if a is a generator of G, let $\mathcal{X} \in \hat{G}$ be given by $\mathcal{X}(a) = \zeta_m$, where ζ_m is a generator of the mth roots of 1 in $\mathbb{C}^*$. We have $\mathcal{X}^n(a) = \mathcal{X}(a)^n = \zeta_m^n$. Hence $o(\mathcal{X}) = m$.

Now let $\varphi \in \hat{G}$ be arbitrary. Then $\varphi(a) \in \mathbb{C}$ and since $1 = \varphi(1) = \varphi(a^m) = \varphi(a)^m$, it follows that $\varphi(a) = \zeta_m^i$ for some $0 \leq i < m-1$. Thus $\varphi = \mathcal{X}^i$ and $\hat{G} = \langle \mathcal{X} \rangle \cong \mathbb{Z}/m\mathbb{Z} \cong G$. In general, let $G \cong \prod_{i=1}^r \mathbb{Z}/m_i\mathbb{Z}$. If $\mathcal{X} \in \hat{G}$, let $\mathcal{X}_i : \mathbb{Z}/m_i\mathbb{Z} \to$

$\mathbb{C}^*$ be given by $\mathcal{X}_i(a) = \mathcal{X}(0, \ldots, 0, a, 0, \ldots, 0)$. It is clear that $\mathcal{X} = \prod_{i=1}^r \mathcal{X}_i$ and this factorization is unique. Moreover,

$$\hat{G} \cong \prod_{i=1}^{r} \left(\widehat{\mathbb{Z}a/m_i\mathbb{Z}}\right) \cong \prod_{i=1}^{r} (\mathbb{Z}/m_i\mathbb{Z}) \cong G. \qquad \square$$

Now we consider the pairing $\Psi : G \times \hat{G} \to \mathbb{C}^*$, $(g, \mathcal{X}) \mapsto \mathcal{X}(g)$.

Proposition 12.6.19. *Ψ is a perfect pairing, which means that Ψ is not degenerate. In other words, if $g \in G$ is such that $\mathcal{X}(g) = 1$ for all $\mathcal{X} \in \hat{G}$, then $g = 1$. Conversely, if $\mathcal{X} \in \hat{G}$ is such that $\mathcal{X}(g) = 1$ for all $g \in G$, then $\mathcal{X} = 1$ (by definition).*

Proof. If $g \neq 1$, it follows by Proposition 12.6.18 that there exists $\mathcal{X} \in \hat{G}$ such that $\mathcal{X}(g) \neq 1$. $\square$

Proposition 12.6.20. *There is a canonical isomorphism between G and $\hat{\hat{G}}$.*

Proof. We have $\hat{\hat{G}} = (\hat{\hat{G}}) \cong \hat{G} \cong G$. Furthermore, if $g \in G$, let $\hat{g} \in \hat{\hat{G}}$ be defined by $\hat{g}(\mathcal{X}) = \mathcal{X}(g) = \Psi(g, \mathcal{X})$ for all $\mathcal{X} \in \hat{G}$. Then $\theta : G \longrightarrow \hat{\hat{G}}$ is a natural group homomorphism. It follows by Proposition 12.6.19 that θ is an isomorphism. $\square$

Definition 12.6.21. Let G be an abelian group, and H a subgroup of G. We define

$$H^\perp = \left\{\mathcal{X} \in \hat{G} \mid \mathcal{X}(h) = 1 \text{ for all } h \in H\right\} = \left\{\mathcal{X} \in \hat{G} \mid H \subseteq \ker \mathcal{X}\right\}.$$

If M is a subgroup of $\hat{G}$, let

$$\begin{aligned} M^\perp &= \{g \in G \mid \mathcal{X}(g) = 1 \text{ for all } \mathcal{X} \in M\} \\ &= \left\{\hat{g} \in \hat{\hat{G}} \mid \hat{g}(\mathcal{X}) = 1 \text{ for all } \mathcal{X} \in M\right\}. \end{aligned}$$

Proposition 12.6.22. *For any $H < G$ and any $M < \hat{G}$ we have*

$$H^\perp \cong \left(\widehat{G/H}\right) \qquad \text{and} \qquad M^\perp = \left(\widehat{\hat{G}/M}\right).$$

Proof. If suffices to exhibit an isomorphism between $H^\perp$ and $\widehat{G/H}$. If $\mathcal{X} \in H^\perp$ then $\mathcal{X}(h) = 1$ for all $h \in H$ and $\mathcal{X}$ can be factored as $\tilde{\mathcal{X}} \circ \pi$.

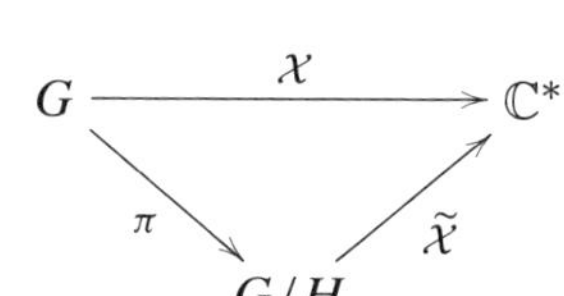

Thus $H^\perp \to \widehat{G/H}$, $\mathcal{X} \mapsto \tilde{\mathcal{X}}$ is a group isomorphism. $\square$

Proposition 12.6.23. *For any subgroup H of G, $\hat{H}$ is isomorphic to $\hat{G}/H^\perp$.*

Proof. Let $\mathcal{X} \in \hat{G}$. Then the restriction map $\hat{G} \to \hat{H}$, $\mathcal{X} \mapsto \mathcal{X}|_{\hat{H}}$ is a group homomorphism with kernel $H^{\perp}$. Thus $\hat{G}/H^{\perp} \subseteq \hat{H}$. On the other hand,

$$|\hat{G}/H^{\perp}| = \frac{|\hat{G}|}{|H^{\perp}|} = \frac{|G|}{|\widehat{(G/H)}|} = \frac{|G|}{|G/H|} = |H| = |\hat{H}|.$$

It follows that $\hat{G}/H^{\perp} \cong \hat{H}$. □

Proposition 12.6.24. *With the identification $G = \hat{\hat{G}}$, we have $(H^{\perp})^{\perp} = H$.*

Proof. If $h \in H$, then for any $\mathcal{X} \in H^{\perp}$ we have $\mathcal{X}(h) = 1$. Hence $\hat{h}(\mathcal{X}) = \mathcal{X}(h) = 1$, so $\hat{h} \in (H^{\perp})^{\perp}$. Thus $H \subseteq (H^{\perp})^{\perp}$. Finally, by Proposition 12.6.23,

$$|(H^{\perp})^{\perp}| = \left|\left(\widehat{\hat{G}/(H^{\perp})}\right)\right| = \frac{|G|}{|H^{\perp}|} = \frac{|G|}{\left|\widehat{G/H}\right|} = \frac{|G|}{|G|/|H|} = |H|.$$

Therefore $H = (H^{\perp})^{\perp}$. □

Definition 12.6.25. Let $M \in R_T \setminus \{0\}$ and let

$$\mathcal{X} \in \widehat{(R_T/(M))^*} \cong \hat{G}_M = \widehat{\mathrm{Gal}(K(\Lambda_M)/K)}$$

be a Dirichlet character mod M. We have $\mathbb{F}_q^* \subseteq G_M$. We say that $\mathcal{X}$ is *even* if $\mathcal{X}(\alpha) = 1$ for $\alpha \in \mathbb{F}_q^*$, and *odd* otherwise.

Definition 12.6.26. Let $\mathcal{X}$ be any Dirichlet character mod M with conductor M, that is, $\mathcal{X} \in \widehat{G}_M$. Let $\ker \mathcal{X} \subseteq G_M$ and

$$K_{\mathcal{X}} = K(\Lambda_M)^{\ker \mathcal{X}}.$$

Then $K_{\mathcal{X}}$ is called the *field belonging to $\mathcal{X}$* or *the field associated to $\mathcal{X}$*.

Remark 12.6.27. We have that $\mathcal{X}$ is even iff $\mathfrak{p}_{\infty}$ decomposes totally in $K_{\mathcal{X}}/K$.

Remark 12.6.28. Let $\mathcal{X}$ be a Dirichlet character mod M and let $N \in R_T \setminus \{0\}$ be a multiple of M. Consider the Dirichlet character $\widetilde{\mathcal{X}}$ defined mod N, that is,

$$\begin{array}{ccc} (R_T/(N))^* & \xrightarrow{\quad\widetilde{\mathcal{X}}\quad} & \mathbb{C}^* \\ & {\scriptstyle\varphi_{N,M}}\searrow \quad \nearrow{\scriptstyle\mathcal{X}} & \\ & (R_T/(M))^* & \end{array} \qquad \tilde{\mathcal{X}} = \mathcal{X} \circ \varphi_{N,M}.$$

Let $K_1 = K(\Lambda_M)^{\ker \mathcal{X}}$ and $K_2 = K(\Lambda_N)^{\ker \widetilde{\mathcal{X}}}$. Then

$$\ker \varphi_{N,M} = \{A \bmod N \mid A \equiv 1 \bmod M\}$$

and

$$(R_T/(N))^* \,/ \ker \varphi_{N,M} \cong (R_T/(M))^* \,.$$

Since $G_N \cong (R_T/(N))^*$ and $G_M \cong (R_T/(M))^*$,

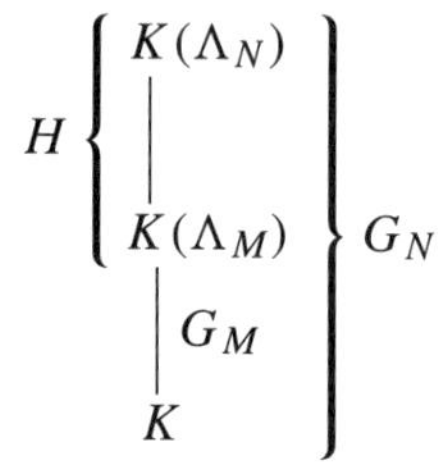

it follows that $K(\Lambda_M) = K(\Lambda_N)^H$, where $H = \operatorname{Gal}(K(\Lambda_N)/K(\Lambda_M))$. Hence $\ker \varphi_{N,M} \cong \operatorname{Gal}(K(\Lambda_N)/K(\Lambda_M))$.

Now, $\ker \widetilde{\mathcal{X}} = \varphi_{N,M}^{-1}(\ker \mathcal{X})$, and since $\varphi_{N,M}^{-1}(\ker \mathcal{X}) \supseteqq \varphi_{N,M}^{-1}(\{1\}) = \ker \varphi_{N,M}$, we have $K_2 = K(\Lambda_N)^{\ker \widetilde{\mathcal{X}}} \subseteq K(\Lambda_N)^{\ker \varphi_{N,M}} = K(\Lambda_M)$. Thus $K_2 \subseteq K(\Lambda_M)^{\ker \mathcal{X}} = K_1$. On the other hand,

$$\begin{aligned} |\ker \widetilde{\mathcal{X}}| &= \left|\varphi_{N,M}^{-1}(\ker \mathcal{X})\right| = \left|\ker \varphi_{N,M}\right| |\ker \mathcal{X}| \\ &= [K(\Lambda_N) : K(\Lambda_M)]\,[K(\Lambda_M) : K_1] \\ &= [K(\Lambda_N) : K_1] \end{aligned}$$

and $|\ker \widetilde{\mathcal{X}}| = [K(\Lambda_N) : K_2]$. Therefore $K_1 = K_2$.

Thus, given any Dirichlet character $\mathcal{X}$ defined mod M (the conductor does not matter) the field $K_{\mathcal{X},M} = K(\Lambda_M)^{\ker \mathcal{X}}$ depends only on $\mathcal{X}$ and not on M.

Definition 12.6.29. Let X be any finite group of Dirichlet characters. Let M be the least common multiple of $\{F_{\mathcal{X}} \mid \mathcal{X} \in X\}$. Then $X \subseteq \widehat{G_M}$. Set $H = \bigcap_{\mathcal{X} \in X} \ker \mathcal{X}$ and $K_X = K(\Lambda_M)^H$; K_X is called the *field belonging to X* or *the field associated to X*.

When $X = \langle \mathcal{X} \rangle$, we have $K_X = K_{\mathcal{X}}$.

Remark 12.6.30. H is a subgroup of G_M and $G_M/H \cong \operatorname{Gal}(K_X/K)$. Therefore, by Proposition 12.6.22, $H^{\perp} \cong \widehat{\operatorname{Gal}(K_X/K)}$. Since G_M is abelian, it follows that $H^{\perp} \cong \widehat{\operatorname{Gal}(K_X/K)} \cong \operatorname{Gal}(K_X/K)$.

Also, if $\mathcal{X} \in X < \hat{G}_M$, then since $\ker \mathcal{X} \supseteq H$, we can consider the induced map $\widetilde{\mathcal{X}} : G_M/H \to \mathbb{C}^*$. Therefore $X \subseteq \widehat{G_M/H} \cong H^{\perp}$. Now $X^{\perp} < G_M$ and if $\alpha \in X^{\perp}$, then $\mathcal{X}(\alpha) = 1$ for all $\mathcal{X} \in X$. Hence $\alpha \in H$ and $X^{\perp} \subseteq H$, so $H^{\perp} \subseteq X^{\perp\perp} = X$.

It follows that

$$X = H^{\perp} \cong \operatorname{Gal} \widehat{K_X/K}) \cong \operatorname{Gal}(K_X/K). \tag{12.14}$$

Let X be any finite group of Dirichlet characters. Since $X \cong \widehat{\operatorname{Gal}(K_X|K)}$, we can consider the natural pairing

$$\begin{aligned}\Psi : \mathrm{Gal}(K_X/K) \times X &\longrightarrow \mathbb{C}^* \\ (g, \mathcal{X}) &\longmapsto \mathcal{X}(g).\end{aligned}$$

Under Ψ we have that if L is a subfield of K_X, let

$$Y_L = \mathrm{Gal}(K_X/L)^{\perp} \cong \frac{\widehat{\mathrm{Gal}\,(K_X/K)}}{\mathrm{Gal}(K_X/L)} \cong \widehat{\mathrm{Gal}(L/K)}.$$

Conversely, if $Y \subseteq X$ is a subgroup of X, let $L_Y = K_X^{Y^{\perp}}$. Then L_Y is the fixed subfield of $\{g \in \mathrm{Gal}(K_X/K) \mid \mathcal{X}(g) = 1 \text{ for all } \mathcal{X} \in Y\}$.

We have $Y^{\perp} = \mathrm{Gal}(K_X/L_Y)$, so $Y = Y^{\perp\perp} = \mathrm{Gal}(K_X/L_Y)^{\perp} = Y_{L_Y}$. Conversely, $L_{Y_L} = K_X^{Y_L^{\perp}} = K_X^{(\mathrm{Gal}(K_X/L)^{\perp})^{\perp}} = K_X^{\mathrm{Gal}(K_X/L)} = L$. In other words, we have the following theorem:

Theorem 12.6.31. *There is a bijective correspondence between $\mathcal{A} = \{Y \mid Y < X\}$ and $\mathcal{B} = \{L \mid L \subseteq K_X\}$ given by*

$$\begin{aligned}\mathcal{A} &\longleftrightarrow \mathcal{B} \\ Y &\longrightarrow L_Y = K_X^{Y^{\perp}} \\ \widehat{\mathrm{Gal}(L/K)} \cong \mathrm{Gal}(K_X/L)^{\perp} = Y_L &\longleftarrow L.\end{aligned}$$

In particular, we obtain a one-to-one correspondence between all groups of Dirichlet characters and subfields of cyclotomic function fields. □

Remark 12.6.32. Since $\mathrm{Gal}(L/K)$ is a finite group, we have $\mathrm{Gal}(L/K) \cong \widehat{\mathrm{Gal}(L/K)} \cong Y_L$. This may be expressed by means of the natural nondegenerate pairing

$$\begin{aligned}\mathrm{Gal}(L/K) \times Y_L &\longrightarrow \mathbb{C}^* \\ (g, \mathcal{X}) &\longmapsto \mathcal{X}(g).\end{aligned}$$

Proposition 12.6.33. *Let X_1, X_2 be two groups of Dirichlet characters and let $K_i = K_{X_i}$ $(i = 1, 2)$ be the field belonging to X_i. Then*

(1) $X_1 \subseteq X_2 \iff K_1 \subseteq K_2$,
(2) $K_{\langle X_1, X_2\rangle} = K_1 K_2$.

Proof. See Exercise 12.10.28. □

Now we shall see the way Dirichlet characters may be applied to study some arithmetic properties of cyclotomic function fields.

Let $M \in R_T \setminus \{0\}$ and let $M = \prod_{i=1}^{r} P_i^{\alpha_i}$ be its decomposition as a product of irreducible polynomials. Then

$$(R_T/(M))^* \cong \prod_{i=1}^{r} \left(R_T/(P_i^{\alpha_i})\right)^*. \tag{12.15}$$

If $\mathcal{X}$ is a Dirichlet character mod M, then corresponding to (12.15) let $\mathcal{X} = \prod_{i=1}^{r} \mathcal{X}_{P_i}$ where $\mathcal{X}_{P_i}$ is a character mod $P_i^{\alpha_i}$. In other words,

$$\mathcal{X}(A \bmod M) = \prod_{i=1}^{r} \mathcal{X}_{P_i}\left(A \bmod P_i^{\alpha_i}\right).$$

Example 12.6.34. Let $\mathcal{X}$ and φ be as in Example 12.6.12. Then $\mathcal{X}$ is defined mod $T^2(T^2+1)$ and φ is defined mod T^2. Let $\phi := \mathcal{X}\varphi$, where ϕ is defined mod T^2+1. We have $\mathcal{X} = (\mathcal{X}\varphi)\varphi^{-1} = \phi\varphi^{-1}$ and ϕ is defined mod T^2+1, so $\mathcal{X}_{T^2} = \varphi^{-1} = \varphi$, and $\mathcal{X}_{T^2+1} = \phi$.

Definition 12.6.35. Let X be any finite group of Dirichlet characters. Then for a monic irreducible polynomial P in R_T we set

$$X_P = \{\mathcal{X}_P \mid \mathcal{X} \in X\}.$$

Theorem 12.6.36. *Let X be a finite group of Dirichlet characters and K_X its associated field. Let $P \in R_T \setminus \{0\}$ be an irreducible polynomial and set $(P)_K = \frac{\mathfrak{p}}{\mathfrak{p}_\infty^{\deg P}}$. Let $\mathfrak{P}$ be a prime divisor of K_X that lies above $\mathfrak{p}$ and set $e = e(\mathfrak{P}|\mathfrak{p})$. Then $e = |X_P|$.*

Proof. Let M be the least common multiple of $\{F_{\mathcal{X}} | \mathcal{X} \in X\}$. Then $K_X \subseteq K(\Lambda_M)$. Let $M = P^a A$, where $A \in R_T$ and P does not divide A. Let $L = K_X(\Lambda_A) = K_X K(\Lambda_A)$.

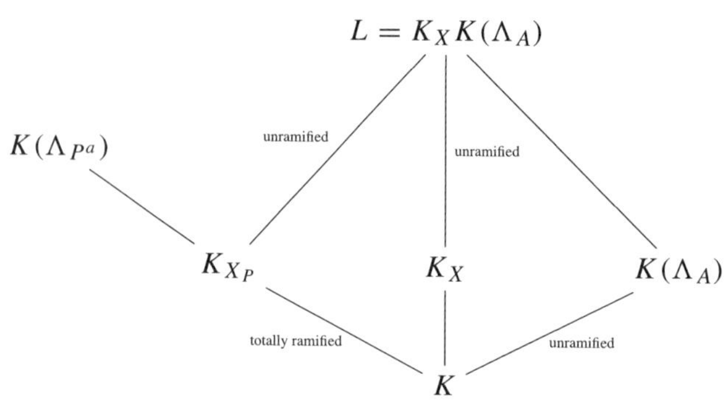

By Proposition 12.6.33,

$$L = K_X K(\Lambda_A) = K_X K_{\widehat{G_A}} = K_{\langle X, \widehat{G_A}\rangle}.$$

Thus L is the field belonging to the group generated by X and $\hat{G}_A$. Equivalently, the group of characters of L is generated by X and each Dirichlet character of G_M whose conductor is prime with P. Thus

$$\langle X, \hat{G}_A\rangle \cong X_P \times \hat{G}_A.$$

We have $K_{X_P} \subseteq K(\Lambda_{P^a})$ and $L = K_{X_P} K(\Lambda_A)$.

Notice that $\mathfrak{p}$ is unramified in $K(\Lambda_A)/K$. It follows that the ramification index of $\mathfrak{p}$ in K_X/K is the same as that of L/K. Since L/K_{X_P} is not ramified in the prime divisors above $\mathfrak{p}$ and $\mathfrak{p}$ is fully ramified in K_{X_P}/K (Proposition 12.3.14), we conclude that $e = \left[K_{X_P} : K\right] = |X_P|$. □

Example 12.6.37. Set $q = 2$ and consider the character $\mathcal{X}$ given in Example 12.6.12. The conductor of $\mathcal{X}$ is $T^2(T^2+1)$. By Example 12.6.34 we have $\mathcal{X}_{T^2} = \varphi$ and $\mathcal{X}_{T^2+1} = \phi$. Note that

$$\Phi(T^2) = \Phi(T^2+1) = q^{dn} - q^{d(n-1)} = 2^{1\times 2} - 2^{1\times(2-1)} = 2^2 - 2 = 4 - 2 = 2.$$

Hence $\left[K(\Lambda_{T^2}) : K\right] = \left[K(\Lambda_{T^2+1}) : K\right] = 2$. We have

$$u^{T^2} = \sum_{i=0}^{2} \begin{bmatrix} T^2 \\ i \end{bmatrix} u^{q^i} = T^2 u + \begin{bmatrix} T^2 \\ 1 \end{bmatrix} u^q + u^{q^2}.$$

Now $\begin{bmatrix} T^2 \\ 1 \end{bmatrix} = T\begin{bmatrix} T \\ 1 \end{bmatrix} + \begin{bmatrix} T \\ 0 \end{bmatrix}^q = T + T^q = T + T^2$, where $\begin{bmatrix} T \\ 1 \end{bmatrix} = a_1 = 1$. Thus $u^{T^2} = T^2u + (T+T^2)u^2 + u^4$. We also have

$$\Psi_{T^2}(u) = u^{T^2}/u^T = \frac{T^2u + (T+T^2)u^2 + u^4}{Tu + u^2} = u^2 + Tu + T.$$

Hence each root α of $\Psi_{T^2}(u)$ is of the form $\left(\frac{\alpha}{T}\right)^2 + \left(\frac{\alpha}{T}\right) = -\frac{1}{T} = \frac{1}{T}$. Hence $K(\Lambda_{T^2}) = K(\beta)$, where β is a root of the Artin–Schreier extension satisfying $\beta^2 - \beta = \frac{1}{T}$. Similarly, $K(\Lambda_{T^2+1}) = K(\gamma)$ where $\gamma^2 - \gamma = \frac{1}{T+1}$. It follows that $K_{\mathcal{X}} = K(\varepsilon)$ with $\varepsilon^2 - \varepsilon = \frac{1}{T(T+1)}$ and we have the following diagram:

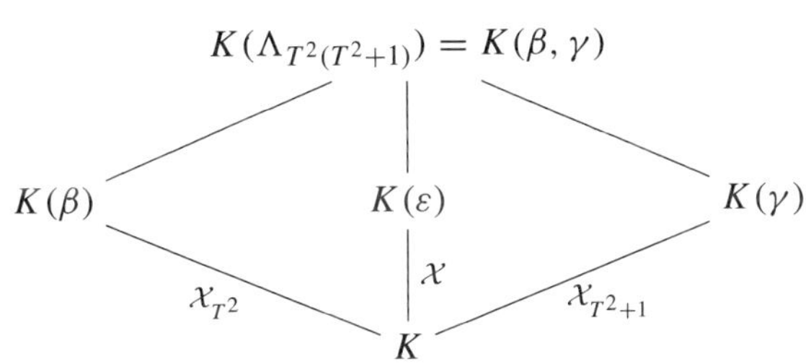

In $K(\varepsilon)/K$, T and $T+1$ are the ramified primes; in $K(\beta)/K$, T is the only ramified prime and in $K(\gamma)/K$, $T+1$ is the only ramified prime.

Corollary 12.6.38. *Let $\mathcal{X}$ be a Dirichlet character. Then P ramifies in $K_{\mathcal{X}}/K$ if and only if $\mathcal{X}(P) = 0$ (or equivalently P divides $F_{\mathcal{X}}$). If X is any finite group of Dirichlet characters, then P is unramified in K_X/K if and only if $\mathcal{X}(P) \neq 0$ for all $\mathcal{X} \in X$.*

Proof. We have the following equivalences: P is ramified in $K_X/K \Leftrightarrow X_P \neq 1 \Leftrightarrow \exists\, \mathcal{X} \in X$ such that $\mathcal{X}_P \neq 1 \Leftrightarrow \exists\, \mathcal{X} \in X$ with $P \mid F_{\mathcal{X}} \Leftrightarrow \exists\, \mathcal{X} \in X$ with $\mathcal{X}(P) = 0$. $\square$

The inertia group and the decomposition group are related to Dirichlet characters in the following manner:

Theorem 12.6.39. *Let X be a finite group of Dirichlet characters and let K_X be its associated field. Let $P \in R_T$, and $Y = \{\mathcal{X} \in X \mid \mathcal{X}(P) \neq 0\}$, $Z = \{\mathcal{X} \in X \mid \mathcal{X}(P) = 1\}$. Set $(P)_K = \frac{\mathfrak{p}}{\mathfrak{p}_\infty^{\deg P}}$ and consider $\mathfrak{P}$ to be a prime divisor in K_X lying above $\mathfrak{p}$. Then*

$$X/Y \cong \widehat{I(\mathfrak{P}|\mathfrak{p})} \cong I(\mathfrak{P}|\mathfrak{p}) \quad \text{and} \quad X/Z \cong D(\mathfrak{P}|\mathfrak{p}).$$

In particular, $e = e(\mathfrak{P}|\mathfrak{p}) = [X : Y]$, $f = f(\mathfrak{P}|\mathfrak{p}) = [Y : Z]$, and $h = [Z : 1] = |Z|$, where h is the number of prime divisors in K_X above $\mathfrak{p}$. Finally, the group Y/Z is cyclic of order f.

Proof. Let K_Y be the field corresponding to Y. Since $Y \subseteq X$ we have $K_Y \subseteq K_X$. By Corollary 12.6.38, K_Y/K is the maximal extension in which $\mathfrak{p}$ is unramified. Since K_X/K is an abelian extension, it follows that any place in K_Y above $\mathfrak{p}$ is fully ramified in K_X/K_Y and $K_Y = K_X^{I(\mathfrak{P}|\mathfrak{p})}$. By Theorem 12.6.31, $K_Y = K_X^{Y^\perp}$. Thus $Y^\perp = I(\mathfrak{P}|\mathfrak{p}) = \operatorname{Gal}(K_X/K_Y)$. Therefore, using Proposition 12.6.23 we obtain

$$X/Y = \widehat{\operatorname{Gal}(K_X/K)}/\operatorname{Gal}(K_X/K_Y)^\perp \cong \widehat{\operatorname{Gal}(K_X/K_Y)} = \widehat{I(\mathfrak{P}|\mathfrak{p})} \cong I(\mathfrak{P}|\mathfrak{p}).$$

Now, $\mathfrak{p}$ is unramified in K_Y/K. Let M be the least common multiple of $\{F_\mathcal{X} \mid \mathcal{X} \in Y\}$.

By Corollary 12.6.38, P does not divide M and $Y \subseteq \hat{G}_M$, so by Proposition 12.6.33, $K_Y \subseteq K(\Lambda_M)$. Clearly, the Frobenius map φ_P for $K(\Lambda_M)$ corresponds to the map $\lambda_M \longrightarrow \lambda_M^P$, where λ_M is a generator of Λ_M. Thus $\varphi_P \in G_M \cong (R_T/(M))^*$ corresponds to $P \bmod M$. Since

$$\operatorname{Gal}(K_Y/K) \cong \frac{\operatorname{Gal}(K(\Lambda_M)/K)}{\operatorname{Gal}(K(\Lambda_M)/K_Y)},$$

the Frobenius map $\widetilde{\varphi}_P$ for K_Y/K corresponds to the coset of P in this quotient.

If $\mathcal{X} \in Y$, $\ker \mathcal{X} \supseteq \operatorname{Gal}(K(\Lambda_M)/K_Y)$. Then $\mathcal{X}(\widetilde{\varphi}_P) = \mathcal{X}(P)$. In particular, $\mathcal{X}(\widetilde{\varphi}_P) = 1$ if and only if $\mathcal{X}(P) = 1$. Thus $Z = \langle\widetilde{\varphi}_P\rangle^\perp$ under the pairing $\operatorname{Gal}(K_Y/K) \times Y \to \mathbb{C}^*$.

Using Propositions 12.6.18 and 12.6.23 we obtain

$$Y/\langle\widetilde{\varphi}_P\rangle^\perp = Y/Z \cong \widehat{\langle\widetilde{\varphi}_P\rangle} \cong \langle\widetilde{\varphi}_P\rangle.$$

The latter is a cyclic group of order $f = [Y : Z]$.

Set $L = K_Y^{\langle\widetilde{\varphi}_P\rangle}$. Then L is the decomposition field of $\mathfrak{p}$, so $\mathfrak{p}$ is fully decomposed in L/K and any prime divisor in L above $\mathfrak{p}$ is inert in K_Y/L (because $[\ell(\mathfrak{q}) : k(\mathfrak{p})] = 1$ and $[k_Y(\mathfrak{P}) : k(\mathfrak{p})] = [k_Y(\mathfrak{P}) : \ell(\mathfrak{q})] = [K_Y : L] = o(\widetilde{\varphi}_P)$, where $\mathfrak{q}$ is a prime divisor of L above $\mathfrak{p}$ and ℓ, k, and k_Y are the fields of constants of L, K, and K_Y respectively). Therefore if h is the number of prime divisors in K_X above $\mathfrak{p}$, we have

$$h = [L : K] = \frac{[K_Y : K]}{[K_Y : L]} = \frac{|Y|}{|\langle\widetilde{\varphi}_P\rangle|} = |Z|.$$

In K_X/K, the splitting field of $\mathfrak{p}$ is the fixed field of the decomposition group, and this is L. Thus L is the field corresponding to Z, and $Z = \widehat{\operatorname{Gal}(L|K)}$.

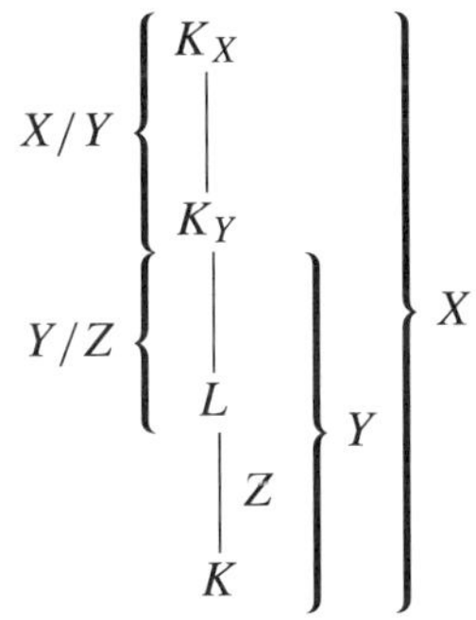

By Exercise 12.10.29,

$$X/Z = \frac{\widehat{\mathrm{Gal}(K_X/K)}}{\widehat{\mathrm{Gal}(L/K)}} \cong \frac{\widehat{\mathrm{Gal}(K_X/K)}}{\widehat{\left(\frac{\mathrm{Gal}(K_X/K)}{\mathrm{Gal}(K_X/L)}\right)}} \cong \widehat{\mathrm{Gal}(K_X/L)} \cong \widehat{D(\mathfrak{P}|\mathfrak{p})}. \qquad \square$$

Lemma 12.6.40. *Let $P \in R_T$ be a monic irreducible polynomial of degree d and let $n = p^t$. Then $(R_T/(P^n))^*$ contains a cyclic subgroup of order $p^t a$ for any a dividing $q^d - 1$.*

Proof. We have $\left|(R_T/(P^n))^*\right| = \Phi(P^n) = q^{dn} - q^{d(n-1)} = q^{d(n-1)}(q^d - 1)$. Therefore the groups $(R_T/(P^n))^*$ is isomorphic to a direct sum $H \oplus A$ where $|H| = q^{d(n-1)}$ and $|A| = q^d - 1$. Note that A is the only subgroup of $(R_T/(P^n))^*$ of order $q^d - 1$. Define

$$\begin{aligned} \theta : \left(R_T/(P^n)\right)^* &\longrightarrow (R_T/P)^* \\ B \bmod P^n &\longmapsto B \bmod P. \end{aligned}$$

Then θ is an epimorphism and $(R_T/(P^n))^* / \ker\theta \cong (R_T/(P))^*$.

Since $\left|(R_T/(P))^*\right| = \Phi(P) = q^d - 1$, it follows that

$$A \cong (R_T/(P))^* \quad \text{and} \quad H \cong \ker\theta \cong \left\{B \bmod P^n | B \equiv 1 \bmod P\right\}.$$

But $R_T/(P)$ and $\mathbb{F}_{q^d}$ are isomorphic, so A must be the multiplicative group of nonzero elements of a field, and therefore A is a cyclic group.

Now let $B = 1 + P$. We wish to determine the order of $B \bmod P^n$ in the quotient R_T/P^n Since B belongs to $\ker\theta$, we have $B \in H$ and $o(B) = p^s$ for some $s \geq 0$. Then

$$B^{p^s} = 1 + P^{p^s} \equiv 1 \bmod P^n \Leftrightarrow p^s \geq n = p^t \Leftrightarrow s \geq t.$$

Thus $o(B) = p^t$, and the result follows. $\square$

Theorem 12.6.41. *Let G be any finite abelian group. There exist fields E and F such that:*

(i) $\mathrm{Gal}(F/E) \cong G$.

(ii) *F/E is unramified at all prime divisors.*

(iii) *F/K is an abelian extension and E/K is a cyclic extension, where, as usual, $K = \mathbb{F}_q(T)$.*

(iv) *The field of constants of E and F is $\mathbb{F}_q$.*

Proof. Assume $G \cong \mathbb{Z}/m_1\mathbb{Z} \times \cdots \times \mathbb{Z}/m_r\mathbb{Z}$. Let $m_i = p^{t_i} a_i$ with $(a_i, p) = 1$ for $t_i \geq 0$ and $1 \leq i \leq r$. Let $d_i' = o(p \bmod a_i)$ for $1 \leq i \leq r$, that is, $p^{d_i'} \equiv 1 \bmod a_i$. Suppose $d_1 < d_2 < \cdots < d_r$, where each d_i' divides d_i (for instance, take $d_1 = d_1'$, $d_i = 2d_{i-1}d_i'$, $i = 2, \ldots, r$). Let $P_i \in R_T$ be a monic irreducible polynomial of degree d_i. Such a P_i exists since $\mathbb{F}_{q^{d_i}} = \mathbb{F}_q(\alpha_i)$ for some α_i and if $P_i = \operatorname{Irr}(\alpha_i, T, \mathbb{F}_q)$, then $\mathbb{F}_q(\alpha_i) \cong R_T/(P_i)$. By Lemma 12.6.40, $\left(R_T/(P_i^{p^{t_i}})\right)^*$ contains an element of order $p^{t_i} a_i = m_i$. Since the character group of $\left(R_T/(P_i^{p^{t_i}})\right)^*$ is isomorphic to the group, there exists a character $\mathcal{X}_i \bmod P_i^{p^{t_i}}$ and order m_i. Thus, $\mathcal{X}_i$ satisfies $o(\mathcal{X}_i) = m_i$ and $F_{\mathcal{X}_i} = P_i^{s_i}$ with $s_i \leq p^{t_i}$.

Let P_{r+1} be another monic irreducible polynomial of degree $d_{r+1} > d_r$ such that $a_1 \cdots a_r \mid q^{d_{r+1}} - 1$. Such d_{r+1} exists since $(a_1 \cdots a_r, q) = 1$. Let $\mathcal{X}_{r+1}$ be a Dirichlet character defined mod $P_{r+1}^{p^t}$ for $t = t_1 + \cdots + t_r$ and order $m_{r+1} = p^t(q^{d_{r+1}} - 1)$ (Lemma 12.6.40). Then $m_1 \cdots m_r = a_1 \cdots a_r p^{t_1 + \cdots + t_r} \mid m_{r+1}$. Let $\mathcal{X} = \mathcal{X}_1 \cdots \mathcal{X}_r \mathcal{X}_{r+1}$ and $E = K_X$ be the field corresponding to $X = \langle \mathcal{X} \rangle$. Let $Y = \langle \mathcal{X}_1, \ldots, \mathcal{X}_r, \mathcal{X}_{r+1} \rangle$ and $F = K_Y$ be the field corresponding to Y. We have $K \subseteq E = K_X \subseteq K_Y = F \subseteq K(\Lambda_M)$, where

$$M = P_1^{\alpha_1} \cdots P_r^{\alpha_r} P_{r+1}^{\alpha_{r+1}} \quad \text{with} \quad \alpha_i = p^{t_i}, \quad 1 \leq i \leq r, \quad \text{and} \quad \alpha_{r+1} = p^t.$$

Thus the field of constants of E and F is $\mathbb{F}_q$. This proves (d). Also, F/K is an abelian extension.

Now by (12.14), the group $\operatorname{Gal}(E/K) \cong X = \langle \mathcal{X} \rangle$ is cyclic. This proves (c).

We have $Y = \langle \mathcal{X}_1, \ldots, \mathcal{X}_r, \mathcal{X}_{r+1} \rangle = \langle \mathcal{X}_1, \ldots, \mathcal{X}_r, \mathcal{X} \rangle$. Moreover, $o(\mathcal{X}) = o(\mathcal{X}_{r+1}) = m_{r+1}$ and since $m_1 \cdots m_r$ divides m_{r+1}, $\mathcal{X}$ is of maximal order. It follows that on the one hand, $Y/X = Y/\langle \mathcal{X} \rangle \cong \langle \mathcal{X}_1, \ldots, \mathcal{X}_r \rangle$, and on the other hand, by Exercise 12.10.29,

$$\begin{aligned} Y/X = \frac{\widehat{\operatorname{Gal}(K_Y/K)}}{\widehat{\operatorname{Gal}(K_X/K)}} &\cong \frac{\widehat{\operatorname{Gal}(K_Y/K)}}{\widehat{\left(\frac{\operatorname{Gal}(K_Y/K)}{\operatorname{Gal}(K_Y/K_X)}\right)}} \cong \widehat{\operatorname{Gal}(K_Y/K_X)} \\ &\cong \widehat{\operatorname{Gal}(F/E)} \cong \operatorname{Gal}(F/E). \end{aligned}$$

Thus $\mathrm{Gal}(F/E) \cong \langle \mathcal{X}_1, \dots, \mathcal{X}_r \rangle \cong \mathbb{Z}/m_1\mathbb{Z} \times \cdots \times \mathbb{Z}/m_r\mathbb{Z} \cong G$. This proves (a).

Finally, by Theorem 12.5.3 the ramified primes in F/K are $\mathfrak{p}_1, \dots, \mathfrak{p}_r, \mathfrak{p}_{r+1}$, and $\mathfrak{p}_\infty$, where $(P_i)_K = \frac{\mathfrak{p}_i}{\mathfrak{p}_\infty^{\deg P_i}}$. Notice that the ramification index of $\mathfrak{p}_\infty$ in E/K is $q-1$ as well as in F/K since E is the field belonging to $\mathcal{X}$, $q-1 \mid o(\mathcal{X})$, $\left(R_T/(P_{r+1}^{p^t})\right)^*$ contains a unique subgroup of order $q-1$ (Lemma 12.6.40), and this subgroup is precisely the inertia group of $\mathfrak{p}_\infty$ (Proposition 12.5.4). Therefore $\mathfrak{p}_\infty$ is unramified in F/E. Finally, we have $Y_{P_i} = \langle \mathcal{X}_i \rangle = X_{P_i}$. By Theorem 12.6.36, the ramification index in F/E of any prime divisor in F above $\mathfrak{p}_i$ is $\frac{|Y_{P_i}|}{|X_{P_i}|} = 1$. Thus F/E is unramified for every prime divisor. This proves (b) and the theorem. □

12.7 Different and Genus

Let $M \in R_T \setminus \{0\}$ be a monic nonconstant polynomial. We denote by $\mathfrak{D}_M$ the different of the extension $K(\Lambda_M)/K$ and by g_M the genus of $K(\Lambda_M)$. Since the extension $K(\Lambda_M)/K$ is geometric and separable we may apply the Riemann–Hurwitz genus formula. For an irreducible polynomial $P \in R_T$, we write $(P)_K = \frac{\mathfrak{p}}{\mathfrak{p}_\infty^{\deg P}}$.

Proposition 12.7.1. *Let P be a monic irreducible polynomial of degree d and let $n \in \mathbb{N}$. If $M = P^n$, then*

$$\mathfrak{D}_M = \mathfrak{P}^s \prod_{\mathfrak{B}|\mathfrak{p}_\infty} \mathfrak{B}^{q-2},$$

where $\mathfrak{P}$ is the only prime divisor in $K(\Lambda_M)$ above $\mathfrak{p}$,

$$s = n\Phi(M) - q^{d(n-1)} = nq^{dn} - nq^{d(n-1)} - q^{d(n-1)} = nq^{dn} - (n+1)q^{d(n-1)},$$

and

$$2g_M - 2 = (dqn - dn - q)\frac{\Phi(P^n)}{q-1} - dq^{d(n-1)}.$$

Proof. By Theorem 12.5.3 any prime divisor other than $\mathfrak{p}$ and $\mathfrak{p}_\infty$ is unramified in $K(\Lambda_M)/K$. Also, $\mathfrak{p}$ is fully ramified, $e_\infty = q-1$, and $\mathfrak{p}_\infty$ is tamely ramified. Thus $\mathfrak{D}_M = \mathfrak{P}^s \prod_{\mathfrak{B}|\mathfrak{p}_\infty} \mathfrak{B}^{q-2}$.

Now we shall find s. To this end we calculate $(\mathfrak{D}_M)_{\mathfrak{p}} = \mathfrak{D}_{(K(\Lambda_M)_{\mathfrak{P}}/K_{\mathfrak{p}})} = \mathfrak{P}^s$. Clearly, $K(\Lambda_M)_{\mathfrak{P}}$ is generated over $K_{\mathfrak{p}}$ for a root λ of $\Psi_{P^n}(u) = \frac{u^{P^n}}{u^{P^{n-1}}}$.

By Proposition 12.5.8, $\{\lambda^i\}_{i=0}^{\Phi(M)-1}$ is an integral basis of the extension $K(\Lambda_M)_{\mathfrak{P}}/K_{\mathfrak{p}}$. By Theorem 5.7.17, we have $(\mathfrak{D}_M)_{\mathfrak{P}} = \left(\Psi'_{P^n}(\lambda)\right)_{\mathfrak{P}}$. Now $u^{P^n} = u^{P^{n-1}}\Psi_{P^n}(u)$, so

$$\begin{aligned} P^n &= (u^{P^n})' = (u^{P^{n-1}})'\Psi_{P^n}(u) + u^{P^{n-1}}\Psi'_{P^n}(u) \\ &= P^{n-1}\Psi_{P^n}(u) + u^{P^{n-1}}\Psi'_{P^n}(u). \end{aligned}$$

Therefore $P^n = \lambda^{P^{n-1}} \Psi'_{P^n}(\lambda)$ and $\left(\Psi'_{P^n}(\lambda)\right) = \left(\frac{P^n}{\lambda^{P^{n-1}}}\right)$.

Since $\lambda^{P^{n-1}} \in \Lambda_P$ and $\Psi_P(u) = \prod_{(A,P)=1}(u - \lambda_P^A)$, it follows that

$$\Psi_P(0) = P = \pm \prod_{(S,P)=1} \lambda_P^S = (\text{unity}) \times \lambda_P^{\Phi(P)}.$$

We obtain that $\left(\left(\lambda^{P^{n-1}}\right)^{\Phi(P)}\right) = (P)$ and if $\mathfrak{q}$ is the prime divisor of $K(\Lambda_P)$ above $\mathfrak{p}$, we have $v_{\mathfrak{q}}\left(\lambda^{P^{n-1}}\right) = \frac{v_{\mathfrak{q}}(P)}{\Phi(P)} = \frac{e(\mathfrak{q}|\mathfrak{p})v_{\mathfrak{p}}(P)}{\Phi(P)} = 1$ because $\mathfrak{q}|\mathfrak{p}$ is totally ramified in $K(\Lambda_P)/K$. Then $v_{\mathfrak{P}}\left(\lambda^{P^{n-1}}\right) = e\left(\mathfrak{P}|\mathfrak{q}\right) v_{\mathfrak{q}}\left(\lambda^{P^{n-1}}\right) = \frac{\Phi(P^n)}{\Phi(P)}$. Consequently,

$$s = v_{\mathfrak{P}}\left(\Psi'_{P^n}(\lambda)\right) = v_{\mathfrak{P}}\left(\frac{P^n}{\lambda^{P^{n-1}}}\right) = n\Phi(P^n) - q^{d(n-1)}.$$

Finally, by Theorem 9.4.2 we have

$$2g_M - 2 = (dnq - dn - q)\left(\Phi(P^n)/(q-1)\right) - dq^{d(n-1)}. \qquad \square$$

Now we state the general result.

Theorem 12.7.2 (Genus and Different formulas). *Let $M \in R_T \setminus \mathbb{F}_q$ be a monic polynomial of the form $M = P_1^{\alpha_1} \cdots P_r^{\alpha_r}$, where $P_1, \ldots, P_r$ are distinct irreducible polynomials. Set $d_i = \deg P_i$. Then*

$$\mathfrak{D}_M = \prod_{i=1}^{r} \left(\prod_{\mathfrak{P}|\mathfrak{p}_i} \mathfrak{P}\right)^{s_i} \prod_{\mathfrak{B}|\mathfrak{p}_\infty} \mathfrak{B}^{q-2},$$

where $(P_i)_K = \frac{\mathfrak{p}_i}{\mathfrak{p}_\infty^{\deg P_i}}$, $s_i = \alpha_i \Phi\left(P_i^{\alpha_i}\right) - q^{d_i(\alpha_i - 1)}$, and

$$2g_M - 2 = -2\Phi(M) + \sum_{i=1}^{r} d_i s_i \frac{\Phi(M)}{\Phi(P_i^{\alpha_i})} + (q-2)\frac{\Phi(M)}{q-1}.$$

Proof. For each $i \in \{1, \ldots, r\}$, $\mathfrak{p}_i$ is fully ramified in $K(\Lambda_{P_i^{\alpha_i}})/K$ and unramified in $K(\Lambda_M)/K(\Lambda_{P_i^{\alpha_i}})$.

Now, for each $\mathfrak{q}$ prime divisor in $K(\Lambda_{P_i^{\alpha_i}})$ that lies above $\mathfrak{p}_i$, there are $\frac{\Phi(M)/\Phi(P_i^{\alpha_i})}{f_i}$ prime divisors in $K(\Lambda_M)$ above $\mathfrak{q}$, each of them of relative degree f_i. Therefore the contribution to $\mathfrak{D}_M$ of $\mathfrak{p}_i$ is $\left(\prod_{\mathfrak{P}|\mathfrak{p}_i} \mathfrak{P}\right)^{s_i}$, where s_i is as in Proposition 12.7.1. We have $\deg_{K(\Lambda_M)}\left(\prod_{\mathfrak{P}|\mathfrak{p}_i} \mathfrak{P}\right) = d_i \frac{\Phi(M)/\Phi(P_i^{\alpha_i})}{f_i} f_i = d_i \frac{\Phi(M)}{\Phi(P_i^{\alpha_i})}$. Thus $\mathfrak{D}_M = \prod_{i=1}^{r} \left(\prod_{\mathfrak{P}|\mathfrak{p}_i} \mathfrak{P}\right)^{s_i} \prod_{\mathfrak{B}|\mathfrak{p}_\infty} \mathfrak{B}^{q-2}$ and

$$\begin{aligned} 2g_M - 2 &= (2g_K - 2)\,[K(\Lambda_M) : K] + \deg_{K(\Lambda_M)} \mathfrak{D}_M \\ &= -2\Phi(M) + \sum_{i=1}^{r} s_i d_i \frac{\Phi(M)}{\Phi(P_i^{\alpha_i})} + (q-2)\frac{\Phi(M)}{q-1} \end{aligned}$$

with $s_i = \alpha_i \Phi(P_i^{\alpha_i}) - q^{d_i(\alpha_i - 1)}$. □

12.8 The Maximal Abelian Extension of K

We will denote by A the maximal abelian extension of K. We will construct A explicitly, namely A is generated by certain extensions of finite degree over K, each one of which is generated by roots of a polynomial that can be given explicitly. We can also describe Artin's reciprocity law over these roots (Theorem 11.5.6).

It turns out that A is the composition of three pairwise linearly disjoint extensions E/K, K_T/K, and L_∞/K.

12.8.1 E/K

Let E be the union of constant extensions of K. More precisely, $E = \bigcup_{n=1}^{\infty} \mathbb{F}_{q^n}(T)$. By Theorem 6.1.3, $\left[\mathbb{F}_{q^n}(T) : \mathbb{F}_q(T)\right] = \left[\mathbb{F}_{q^n} : \mathbb{F}_q\right] = n$ and

$$\operatorname{Gal}\left(\mathbb{F}_{q^n}(T)/\mathbb{F}_q(T)\right) \cong \operatorname{Gal}\left(\mathbb{F}_{q^n}/\mathbb{F}_q\right) \cong \mathbb{Z}/n\mathbb{Z}.$$

Recall that $\mathbb{F}_{q^n}(T)$ is obtained by adding the roots of $u^{q^n} - u = f(u)$ to K.

We have

$$\begin{aligned} \operatorname{Gal}(E/K) &= \operatorname{Gal}\left(\bigcup_{n=1}^{\infty} \mathbb{F}_{q^n}(T)/\mathbb{F}_q(T)\right) \\ &= \operatorname{Gal}\left(\varinjlim_{n} \mathbb{F}_{q^n}(T)/\mathbb{F}_q(T)\right) \\ &\cong \varprojlim_{n} \operatorname{Gal}(\mathbb{F}_{q^n}/\mathbb{F}_q) = \varprojlim_{n} \mathbb{Z}/n\mathbb{Z}. \end{aligned} \tag{12.16}$$

The inverse limit in (12.16) is given by maps $\pi_{m,n} : C_m := \mathbb{Z}/m\mathbb{Z} \to \mathbb{Z}/n\mathbb{Z} = C_n$, $x \bmod m \mapsto x \bmod n$ for n dividing m.

Thus $\operatorname{Gal}(E/K) \cong \varprojlim_{n} \mathbb{Z}/n\mathbb{Z} \cong \hat{\mathbb{Z}}$, where $\hat{\mathbb{Z}}$ denotes the Prüfer ring, which is the completion of $\mathbb{Z}$.

More precisely, assume that n divides m. Then $n = P_1^{\alpha_1} \cdots P_r^{\alpha_r}$ and $m = P_1^{\beta_1} \cdots P_r^{\beta_r}$ for some $\alpha_i \le \beta_i$ and $1 \le i \le r$. We have $\pi_{m,n} = \pi_{P_1^{\beta_1}, P_1^{\alpha_1}} \times \cdots \times \pi_{P_r^{\beta_r}, P_r^{\alpha_r}}$ where $\pi_{P_i^{\beta_i}, P_i^{\alpha_i}} : \mathbb{Z}/P_i^{\beta_i}\mathbb{Z} \to \mathbb{Z}/P_i^{\alpha_i}\mathbb{Z}$, $x \bmod P_i^{\beta_i} \mapsto x \bmod P_i^{\alpha_i}$.

Therefore $\varprojlim_{n} \mathbb{Z}/n\mathbb{Z} \simeq \prod_{p \text{ prime}} \varprojlim_{\alpha} \mathbb{Z}/p^\alpha\mathbb{Z}$. Now $\varprojlim_{\alpha} \mathbb{Z}/p^\alpha\mathbb{Z} \simeq \mathbb{Z}_p$, where $\mathbb{Z}_p$ is the ring of p-adic integers.

Theorem 12.8.1. $G_E := \mathrm{Gal}(E/K) \cong \hat{\mathbb{Z}} \cong \prod_{p \text{ prime}} \mathbb{Z}_p$. □

Remark 12.8.2. Topologically, the group $G_E = \mathrm{Gal}(E/K)$ is generated by the Frobenius automorphism $\sigma : E \to E, u \mapsto u^q$.

12.8.2 K_T/K

Put $K_T = \bigcup_{M \in R_T} K(\Lambda_M)$. Now $\mathrm{Gal}\,(K(\Lambda_M)/K) \cong (R_T/(M))^*$, and we have

$$\begin{aligned} G_T := \mathrm{Gal}(K_T/K) &= \mathrm{Gal}(\varinjlim_M K(\Lambda_M)/K) \\ &\cong \varprojlim_M \mathrm{Gal}(K(\Lambda_M)/K) \cong \varprojlim_M (R_T/(M))^* . \end{aligned}$$

We define an action of G_T on K_T as follows. If $u \in K_T$, then $u \in K(\Lambda_M)$ for some $M \in R_T$. For $\sigma \in G_T$, we have $\sigma|_{K(\Lambda_M)} = \phi$ with $\phi(\lambda_M) = \lambda_M^A$ and $(A, M) = 1$. In fact, if $\mathcal{G} := \varprojlim_M (R_T/(M))^*$, there exists a natural projection $\mathcal{G} \xrightarrow{\pi_M} (R_T/(M))^*$, which corresponds to the restriction $\sigma(\lambda_M) = \pi_M(\sigma)(\lambda_M)$.

We shall now describe explicitly $G_T \cong \mathcal{G}$.

Proposition 12.8.3. *Let $M = P^n \in R_T$, where $n \geq 1$ and P is a monic irreducible polynomial. Then $(R_T/(M))^* \cong H_M \oplus C_{q^d-1}$, where H_M is a p-group of order $q^{d(n-1)}$ and C_{q^d-1} is a cyclic group of order $q^d - 1$ with $d = \deg P$.*

Proof. The group $(R_T/(M))^*$ is abelian of order $\Phi(M) = q^{dn} - q^{d(n-1)} = q^{d(n-1)}(q^d - 1)$. It follows that $(R_T/(M))^* \cong H_M \oplus B$ with $|H_M| = q^{d(n-1)}$ and $|B| = q^d - 1$. Finally, since

$$\begin{aligned} \theta : (R_T/(M))^* &\longrightarrow (R_T/(P))^* \\ C \bmod M &\longmapsto C \bmod P \end{aligned}$$

is an epimorphism, it follows that B is isomorphic to $(R_T/(P))^*$, which is a cyclic group because it is the multiplicative group of the nonzero elements of a finite field. □

Remark 12.8.4. In fact we have

$$H_M \cong \ker\theta = \left\{D \bmod P^n | D \equiv 1 \bmod P\right\}.$$

That is, if $D \in H_M$, then $D \equiv 1 + CP^s \bmod P^n$ with $C \in R_T$, $(C, P) = 1$, and $1 \leq s \leq n$. Now the elements of H_M of the form $D \equiv 1 + CP^s \bmod P^n$, where $1 \leq s \leq n - 1$ and $(C, P) = 1$, are in correspondence with $\left(R_T/(P^{n-s})\right)^*$. Therefore H_M contains $\Phi(P^{n-s}) = q^{d(n-s)} - q^{d(n-s-1)}$ elements of the form $1 + CP^s \bmod P^n$ with $1 \leq s \leq n - 1$, and $(C, P) = 1$.

Proposition 12.8.5. *Set $M = P^n$ and let t be the positive integer satisfying $p^{t-1} < n \le p^t$. Let $n_0 = \left[\frac{n}{p^{t-1}}\right]$ be the integral part of n/p^{t-1}. Then the elements of maximum order in H_M are those of order p^t. Furthermore*

(i) *If $n_0 = n/p^{t-1}$, then the number of elements of order p^t in H_M is*

$$q^{d(n-1)} - q^{d(n-n_0)}.$$

(ii) *If $n_0 < n/p^{t-1}$, then the number of elements of order p^t in H_M is*

$$q^{d(n-1)} - q^{d(n-n_0-1)}.$$

Proof. We have

$$(1 + CP^s)^{p^m} \equiv 1 + C^{P^m} P^{sp^m} \bmod P^{p^{m+1}}. \tag{12.17}$$

Thus $o(1 + CP^s) = p^m \Leftrightarrow sp^m \ge n$ and $sp^{m-1} < n$.

Since any $s \ge 1$ satisfies $sp^t \ge n$, we have $H_M^{p^t} = \{1\}$ and $(1 + P)^{p^{t-1}} \equiv 1 + P^{p^{t-1}} \not\equiv 1 \bmod P^n$. Therefore H_M is exactly of exponent p^t.

It follows from (12.17) that the elements of order p^t are those such that $sp^t \ge n$ and $sp^{t-1} < n$. Since $sp^t \ge n$ for any $s \ge 1$, we get $o(1 + CP^s) = p^t$ if and only if $1 \le s < \frac{n}{p^{t-1}}$.

If $n_0 = \frac{n}{p^{t-1}}$, then $1 \le s \le n_0 - 1$ and there exist

$$\sum_{s=1}^{n_0-1} \left(q^{d(n-s)} - q^{d(n-s-1)}\right) = q^{d(n-1)} - q^{d(n-n_0)}$$

elements of order p^t in H_M.

If $n_0 < n/p^{t-1}$, there are

$$\sum_{s=1}^{n_0} \left(q^{d(n-s)} - q^{d(n-s-1)}\right) = q^{d(n-1)} - q^{d(n-n_0-1)}$$

elements of order p^t in H_M. □

Corollary 12.8.6. *With the notation of Proposition* 12.8.5 *assume*

$$H_M \cong \left(\mathbb{Z}/p^t\mathbb{Z}\right)^\alpha \times \mathbb{Z}/p^{n_1}\mathbb{Z} \times \cdots \times \mathbb{Z}/p^{n_s}\mathbb{Z} = G$$

with $t > n_1 \ge \cdots \ge n_s \ge 0$. Then

(i) $\alpha = ud(n_0 - 1)$ *if* $n_0 = n/p^{t-1}$,
(ii) $\alpha = udn_0$ *if* $n_0 < n/p^{t-1}$,

where $q = p^u$. In particular, if $n = p^t$, then $\alpha = ud(p-1)$.

Proof. The element $(a_1, \dots, a_\alpha, b_1, \dots, b_s)$ of G is of order p^t if and only if $(a_i, p) = 1$ for some $i \in \{1, \dots, \alpha\}$. Therefore G contains

$$\left(p^{\alpha t} - p^{\alpha(t-1)}\right) p^{n_1+\cdots+n_s} = (p^\alpha - 1)p^{\alpha(t-1)+m} = p^{\alpha t+m} - p^{\alpha(t-1)+m}$$

elements of order p^t, where $m = n_1 + \cdots + n_s$.

Thus if $n_0 = n/p^{t-1}$, we obtain using Proposition 12.8.5 that

$$\begin{aligned} q^{d(n-1)} - q^{d(n-n_0)} &= p^{ud(n-1)} - p^{ud(n-n_0)} = p^{ud(n-n_0)}\left(p^{ud(n_0-1)} - 1\right) \\ &= p^{\alpha(t-1)+m}\left(p^\alpha - 1\right). \end{aligned}$$

Hence $\alpha = ud(n_0 - 1)$. Now if $n_0 < \frac{n}{p^t}$, we have

$$p^{ud(n-n_0-1)}\left(p^{udn_0} - 1\right) = p^{\alpha(t-1)+m}(p^\alpha - 1).$$

So $\alpha = udn_0$ in this case. □

Now for each $t \in \mathbb{N}$, let $pH_{P^{p^t}}$ denote the subgroup of $H_{P^{p^t}}$ consisting of all elements of the form v^p with $v \in H_{P^{p^t}}$.

Proposition 12.8.7. *For every integer $t \ge 2$, the map $\Psi\colon H_{P^{p^{t-1}}} \longrightarrow pH_{P^{p^t}}$ defined by*

$$\Psi\left((1 + CP^s) \bmod P^{p^{t-1}}\right) = \left((1 + CP^s) \bmod P^{p^t}\right)^p,$$

with $(C, P) = 1$ and $1 \le s \le p^{t-1}$, is a group isomorphism.

Proof. Clearly, Ψ is a well-defined epimorphism. Now

$$\begin{aligned} &(1 + CP^s) \bmod P^{p^{t-1}} \in \ker \Psi \\ \iff &\left((1 + CP^s) \bmod P^{p^t}\right)^p = \left(1 + C^p P^{sp}\right)\left(\bmod P^{p^t}\right) \equiv 1 \bmod P^{p^t} \\ \iff &0 \equiv C^p P^{sp} \bmod P^{p^t} \iff s = p^{t-1} \iff 1 + CP^s \equiv 1 \bmod P^{p^{t-1}}. \end{aligned}$$

Thus Ψ is a monomorphism and consequently a group isomorphism. □

Theorem 12.8.8. *We have*

(i) $H_{P^p} \cong (\mathbb{Z}/p\mathbb{Z})^{\alpha_1}$ *with* $\alpha_1 = ud(p-1)$
(ii) *For* $t \ge 2$, $H_{P^{p^t}} \cong \prod_{i=1}^t \left(\mathbb{Z}/p^i\mathbb{Z}\right)^{\alpha_i}$,

where $\alpha_i = udp^{t-i-1}(p-1)^2$ if $1 \le i \le t-1$ and $\alpha_t = ud(p-1)$.

Proof. Each element of $H_{P^p}\setminus\{0\}$ is of order p, and since $|H_{P^p}| = q^{d(p-1)} = p^{ud(p-1)}$, statement (i) follows. We shall prove (ii) by induction on t for $t \ge 2$.

We have

$$H_{P_{p^2}} \cong \left(\mathbb{Z}/p^2\mathbb{Z}\right)^{\alpha_2} \times (\mathbb{Z}/p\mathbb{Z})^{\alpha_1}$$

and $|H_{P_{p^2}}| = p^{ud(p^2-1)}$, where $\alpha_2 = ud(p-1)$ and $\alpha_1 \geq 0$.

Thus $ud(p^2-1) = 2ud(p-1) + \alpha_1$, and $\alpha_1 = ud(p-1)^2$. It follows that (ii) holds for $t = 2$. Now by Proposition 12.8.7, $pH_{P_{p^{t+1}}}$ and $H_{P_{p^t}}$ are isomorphic. Hence $pH_{P_{p^{t+1}}} \cong \prod_{i=1}^{t} \left(\mathbb{Z}/p^i\mathbb{Z}\right)^{\alpha_{i+1}}$, where $\alpha_{i+1} = udp^{t-i-1}(p-1)^2$ if $1 \leq i \leq t-1$ and $\alpha_{t+1} = ud(p-1)$.

Therefore $H_{P_{p^{t+1}}} \cong \prod_{i=1}^{t+1} \left(\mathbb{Z}/p^i\mathbb{Z}\right)^{\alpha_i}$ for some $\alpha_1 \geq 0$. Since $|H_{P_{p^{t+1}}}| = p^{ud(p^{t+1}-1)}$, we have

$$ud(p^{t+1}-1) = \sum_{i=1}^{t+1} i\alpha_i = (t+1)ud(p-1) + \sum_{i=2}^{t} i\alpha_i + \alpha_1.$$

Thus $\alpha_1 = udp^{(t+1)-2}(p-1)^2$. This proves (ii). □

Theorem 12.8.9. *If P is an irreducible polynomial of degree d in R_T and $q = p^u$, then:*

(i) $\mathrm{Gal}(K(\Lambda_{P^p})/K) \cong (\mathbb{Z}/p\mathbb{Z})^{\alpha_1} \times (\mathbb{Z}/(q^d-1)\mathbb{Z})$ *with* $\alpha_1 = ud(p-1)$.
(ii) *For each positive integer* $t \geq 2$,

$$\mathrm{Gal}(K(\Lambda_{P_{p^t}})/K) \cong \prod_{i=1}^{t} \left(\mathbb{Z}/p^i\mathbb{Z}\right)^{\alpha_i} \times \mathbb{Z}/(q^d-1)\mathbb{Z}$$

where $\alpha_i = udp^{t-i-1}(p-1)^2$ *if* $1 \leq i \leq t-1$ *and* $\alpha_t = ud(p-1)$.

Proof. The statements follow immediately by Proposition 12.8.3 and Theorem 12.8.8. □

We have $\Lambda_P \subseteq \Lambda_{P^2} \subseteq \cdots \subseteq \Lambda_{P^n} \subseteq \cdots$, so $K \subseteq K(\Lambda_P) \subseteq \cdots \subseteq K(\Lambda_{P^n}) \subseteq \cdots$ is a tower of field extensions. In particular, for each $n \geq 1$ there exists $t \geq 1$ such that $K(\Lambda_{P^n}) \subseteq K(\Lambda_{P^t})$. Let $K(\Lambda_{P^\infty}) := \bigcup_{n=1}^{\infty} K(\Lambda_{P^n}) = \bigcup_{t=1}^{\infty} K(\Lambda_{P_{p^t}})$.

Theorem 12.8.10. *With the previous notation, we have*

$$\mathrm{Gal}\left(K(\Lambda_{P^\infty})/K(\Lambda_P)\right) \simeq \varprojlim_t H_{P_{p^t}} \cong \varprojlim_t \left(\prod_{i=1}^{t} \left(\mathbb{Z}/p^i\mathbb{Z}\right)^{\alpha_i}\right),$$

where $\alpha_i = udp^{t-i-1}(p-1)^2$ *if* $1 \leq i \leq t-1$, $\alpha_t = ud(p-1)$, *and* $q = p^u$.

Proof. For each $t \geq 2$, denote by Ψ_t the composition of the homomorphisms

$$\begin{array}{ccccc} H_{P_{p^t}} & \longrightarrow & pH_{P_{p^t}} & \longrightarrow & H_{P_{p^{t-1}}} \\ (1+CP^s) \bmod P^{p^t} & \longmapsto & \left((1+CP^s) \bmod P^{p^t}\right)^p & \longmapsto & (1+CP^s) \bmod P^{p^{t-1}}. \end{array}$$

Let λ be a generator of $\Lambda_{P^{p^t}}$. Then $\lambda^{P^{(p^t-p^{t-1})}}$ is a generator of $\Lambda_{P^{p^{t-1}}}$. Let $\sigma \in H_{P^{p^t}}$, and let $C \in R_T$ be such that $(C, P) = 1$ and $\sigma(\lambda) = \lambda^C$. We have $\sigma\left(\lambda^{P^{(p^t-p^{t-1})}}\right) = \left(\lambda^{P^{(p^t-p^{t-1})}}\right)^C$. Therefore, Ψ_t is the homomorphism $H_{P^{p^t}} \to H_{P^{p^{t-1}}}$, $\sigma \mapsto \sigma|_{K(\Lambda_{P^{p^{t-1}}})}$. Hence the homomorphisms Ψ_t $(t \geq 2)$ induce the projective system of the groups $\mathrm{Gal}(K(\Lambda_{P^{p^t}})/K(\Lambda_P))$, and consequently

$$\mathrm{Gal}\,(K(\Lambda_{P^\infty})/K(\Lambda_P)) = \varprojlim_t \mathrm{Gal}\left(K(\Lambda_{P^{p^t}})/K(\Lambda_P)\right) \cong \varprojlim_t H_{P^{p^t}}.$$

The second isomorphism follows using Theorem 12.8.9. □

The next result is a corollary of Theorem 12.8.10.

Theorem 12.8.11. *We have*

$$\begin{aligned}\mathrm{Gal}\,(K(\Lambda_{P^\infty})/K) &\cong \varprojlim_t \mathrm{Gal}\left((\Lambda_{P^{p^t}})/K(\Lambda_P)\right) \times \left(\mathbb{Z}/(q^d-1)\mathbb{Z}\right)\\ &\cong \varprojlim_t \left(\prod_{i=1}^{t}\left(\mathbb{Z}/p^i\mathbb{Z}\right)^{\alpha_i}\right) \times \left(\mathbb{Z}/(q^d-1)\mathbb{Z}\right)\\ &\cong \mathbb{Z}_p^\infty \times \left(\mathbb{Z}/(q^d-1)\mathbb{Z}\right),\end{aligned}$$

where $\alpha_i = udp^{t-i-1}(p-1)^2$ *if* $1 \leq i \leq t-1$, $\alpha_t = ud(p-1)$, *and* $\mathbb{Z}_p^\infty$ *denotes the direct product of a countable number of copies of the ring of* p*-adic integers.*

Proof. The result is a consequence of the fact that the inverse limit commutes with direct product and the isomorphism between $\varprojlim_i \left(\mathbb{Z}/p^i\mathbb{Z}\right)$ and $\mathbb{Z}_p$. □

Theorem 12.8.12. *Let* $\mathcal{M}$ *be the set of all monic irreducible polynomials in* R_T. *Then*

$$\mathrm{Gal}\,(K_T/K) \cong \mathbb{Z}_p^\infty \times \prod_{p\in\mathcal{M}} \left(\mathbb{Z}/(q^{d_P}-1)\mathbb{Z}\right),$$

where $d_P = \deg P$ *for each* $P \in \mathcal{M}$ *and* $K_T = \bigcup_{M\in R_T} K(\Lambda_M)$.

Proof. Let $M \in R_T$ be a nonconstant polynomial, and let $M = \alpha P_1^{n_1} \cdots P_r^{n_r}$ be its factorization into powers of monic irreducible polynomials.

We have $K(\Lambda_M) = K\left(\Lambda_{P_1^{n_1}}, \ldots, \Lambda_{P_r^{n_r}}\right) = \prod_{i=1}^r K(\Lambda_{P_i^{n_i}})$. Therefore $K_T = \prod_{P\in\mathcal{M}} K(\Lambda_{P^\infty})$.

For each $P \in \mathcal{M}$, if $(P)_K = \frac{\mathfrak{p}}{\mathfrak{p}_\infty^{\deg P}}$, then $\mathfrak{p}$ is fully ramified in $K(\Lambda_{P^\infty})/K$ and unramified in $\prod_{Q\in\mathcal{M}\setminus\{P\}} K(\Lambda_{Q^\infty})/K$. In particular, if P, Q are distinct elements of $\mathcal{M}$, then $K(\Lambda_{P^\infty})$ and $K(\Lambda_{Q^\infty})$ are linearly disjoint over K.

Thus $\mathrm{Gal}(K_T/K) \cong \prod_{P\in\mathcal{M}} \mathrm{Gal}\,(K(\Lambda_{P^\infty})/K)$ and the result follows by Theorem 12.8.11. □

12.8.3 L_∞/K

Note that EK_T cannot be the maximal abelian extension of K because $\mathfrak{p}_\infty$ is tamely ramified in EK_T/K. We need certain extensions for which $\mathfrak{p}_\infty$ is wildly ramified. For instance consider the Artin–Schreier extension $K(y)$, where

$$y^p - y = T.$$

Since $(T)_K = \frac{\mathfrak{p}}{\mathfrak{p}_\infty}$, it follows by Example 5.8.8 that $\mathfrak{p}_\infty$ is the only ramified prime in $K(y)/K$ and the index of ramification of $\mathfrak{p}_\infty$ is $e = p = [K(y) : K]$. Thus $\mathfrak{p}_\infty$ is wildly ramified in $K(y)/K$.

Let $X = \frac{1}{T}$. Then $K = \mathbb{F}_q(X)$ and $R_X = \mathbb{F}_q[X] = \mathbb{F}_q\left[\frac{1}{T}\right]$. For $n \geq 1$, put $F_n = K(\Lambda_{X^{n+1}})$ and let $\lambda_{X^{n+1}}$ be a generator of the cyclic R_X-module $\Lambda_{X^{n+1}} = \Lambda_{T^{-n-1}}$.

Any polynomial $N \in R_X$ acts on F_n. Furthermore, we have

$$\begin{aligned}\operatorname{Gal}(F_n/K) &\cong \left(R_X/(X^{n+1}\right)^* \\ &\cong \left\{f(X) \bmod X^{n+1} | f(X) \in R_X \text{ and } f(0) \neq 0\right\}.\end{aligned}$$

If $N = \beta \in \mathbb{F}_q^*$, then $\beta \in \left(R_X/X^{n+1}\right)^*$, so $\mathbb{F}_q^* \subseteq \operatorname{Gal}(F_n/K)$.

We have $\lambda^\beta = \beta\lambda$, where $\lambda = \lambda_{X^{n+1}}$. Let L_n be the subfield of F_n fixed by $\mathbb{F}_q^*$: $L_n := F_h^{\mathbb{F}_q^*}$. Then

$$[L_n : K] = \frac{[F_n : K]}{[F_n : L_n]} = \frac{q^n(q-1)}{q-1} = q^n.$$

It is easy to see that $\mathfrak{p}_\infty$ is totally ramified in F_n/L. The only other ramified prime in F_n/K is $\mathfrak{p}_0$, where $(T)_K = \frac{\mathfrak{p}_0}{\mathfrak{p}_\infty}$. Here $\mathfrak{p}_0$ is the infinite prime in R_X. Now, $\mathfrak{p}_0$ is tamely ramified in F_n/K with ramification index $q-1$. The decomposition group corresponds to $\mathbb{F}_q^*$ (Proposition 12.5.4).

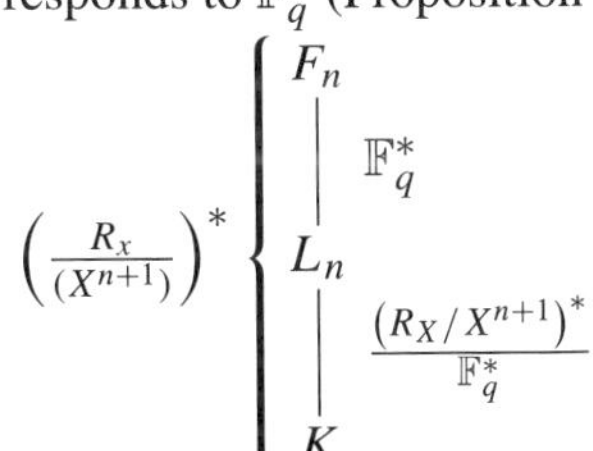

In the extension L_n/K the only ramified prime is $\mathfrak{p}_\infty$ and it is totally and wildly ramified.

Theorem 12.8.13. *Let G_n be the group of polynomials in R_X mod X^{n+1} with constant term equal to* 1, *namely*

$$G_n := \left\{\overline{f(X)} \in \left(R_X/(X^{n+1}\right)^* | f(0) = 1\right\}.$$

Then for each $n \geq 1$, $\operatorname{Gal}(L_n/K)$ and G_n are isomorphic.

Proof. Define $\phi : \left(\frac{R_X}{X^{n+1}}\right)^* \to G_n, \overline{f(X)} \mapsto f^{-1}(0)\left(\overline{f(X)}\right)$. Notice that if $\overline{f(X)} = b_0 + b_1X + \cdots + b_nX^n$ with $b_0 \neq 0$, then

$$\phi(\overline{f(X)}) = 1 + (b_0^{-1}b_1)X + \cdots + (b_0^{-1}b_n)X^n.$$

Clearly ϕ is a group epimorphism and

$$\ker\phi = \left\{\overline{f(x)} \mid f(0)^{-1}\overline{f(X)} = 1\right\} = \left\{\overline{f(X)} \mid \overline{f(X)} = f(0)\right\} = \mathbb{F}_q^*.$$

Thus $G_n \cong \frac{(R_X/(X^{n+1}))^*}{\mathbb{F}_q^*} \cong \operatorname{Gal}(L_n/K)$. □

Now we have $L_n \subseteq L_{n+1}$ for all $n \geq 1$. Let $L_\infty := \bigcup_{n=1}^\infty L_n$.

Theorem 12.8.14. *We have that L_∞/K is an abelian extension, where $\mathfrak{p}_\infty$ is the only ramified prime and it is totally and wildly ramified. Furthermore,*

$$G_\infty := \operatorname{Gal}(L_\infty/K) = \varprojlim_n G_n \cong \left\{f\left(\frac{1}{T}\right) \in \mathbb{F}_q\left[\left[\frac{1}{T}\right]\right] \mid f(0) = 1\right\}.$$

Proof: For each positive integer n, the extension L_n/K is abelian, where $\mathfrak{p}_\infty$ is totally and wildly ramified and there is no other ramified prime. Thus the same holds for L_∞/K. We also have $\operatorname{Gal}(L_\infty/K) = \varprojlim_n \operatorname{Gal}(L_n/K) \cong \varprojlim_n G_n$. Finally, if $H = \{f(X) \in \mathbb{F}_q[[X]] | f(0) = 1\}$, define $\Psi\colon H \to G_n$, $f(X) \mapsto f(X) \bmod X^{n+1}$. Then Ψ is a group epimorphism that satisfies the universal property of the inverse limit (see Exercise 11.7.17). Hence H is isomorphic to $\varprojlim_n G_n$. □

12.8.4 $A = EK_TL_\infty$

Let A be the composite of E, K_T, and L_∞. Since E/K, K_T/K, and L_∞/K are abelian extensions, it follows that A/K is an abelian extension too.

Theorem 12.8.15. *The extensions E/K, K_T/K, and L_∞/K are pairwise linearly disjoint over K. Therefore the Galois group of A/K is naturally isomorphic to*

$$G_E \times G_T \times G_\infty.$$

Proof. First note that for any finite subextension of EK_T/K, $\mathfrak{p}_\infty$ is tamely ramified; indeed, the extension is contained in the composition of a finite subextension of E with one of K_T (and therefore in $K(\Lambda_M)$ for some $M \in R_T$). In both subextensions $\mathfrak{p}_\infty$ is tamely ramified. On the other hand, in any subextension of L_∞, $\mathfrak{p}_\infty$ is totally and wildly ramified since any finite subextension of L_∞/K is contained in some L_n. It follows that EK_T and L_∞ are linearly disjoint over K. In particular, E and L_∞ as well as K_T and L_∞ are linearly disjoint over K.

To prove that E and K_T are linearly disjoint over K, it suffices to show that for any $M \in R_T$, $K(\Lambda_M) \cap E = K$.

If M is constant, then $K(\Lambda_M) = K$ and there is nothing to prove. Now if $R = E \cap K(\Lambda_M)$ and $R \neq K$, then R/K is ramified (see the proof of Corollary 12.3.17 or Remark 12.6.30 together with Theorem 12.6.36).

On the other hand, R/K is unramified by Theorem 5.2.32. Thus $R = K$. This proves the theorem. □

Now we will prove that A is the maximal abelian extension of K. For this purpose, we consider first (see Definition 11.5.2)

$$J = J_K = \big\{\alpha = (\alpha_{\mathfrak{p}})_{\mathfrak{p}\in\mathbb{P}_K} | \alpha_{\mathfrak{p}} \in K_{\mathfrak{p}}^* \text{ for all } \mathfrak{p} \text{ and } v_{\mathfrak{p}}(\alpha_{\mathfrak{p}}) = 0 \text{ for almost all } \mathfrak{p}\big\}.$$

Thus, J consists of all sequences $(\alpha_{\mathfrak{p}})_{\mathfrak{p}\in\mathbb{P}_K}$ such that $\alpha_{\mathfrak{p}} \neq 0$ for all $\mathfrak{p}$, $\alpha_{\mathfrak{p}}$ belongs to the completion $K_{\mathfrak{p}}$ of K at $\mathfrak{p}$ and such that all but finitely many $\alpha_{\mathfrak{p}}$ are units, $\alpha_{\mathfrak{p}} \in \vartheta_{\mathfrak{p}}^* := U_{\mathfrak{p}}$. The topology of J is given in Definition 11.5.2.

Our next task is to construct a group homomorphism

$$\Psi : J \longrightarrow \operatorname{Gal}(A/K).$$

This will be done by writing J as a direct product of four subgroups of J and then defining Ψ on each factor separately. The map will be trivial on one factor and the other three factors map into the Galois groups of E/K, K_T/K, and L_∞/K respectively.

We choose a canonical prime element $\pi_{\mathfrak{p}}$ for $\mathfrak{p}$ defined by:

(a) $\pi_{\mathfrak{p}} = P$ if $\mathfrak{p}$ is not the infinite prime $\mathfrak{p}_\infty$ and P is the monic irreducible polynomial in R_T such that $(P)_K = \frac{\mathfrak{p}}{\mathfrak{p}_\infty^{\deg P}}$.

(b) $\pi_{\mathfrak{p}} = \frac{1}{T}$ if $\mathfrak{p} = \mathfrak{p}_\infty$ is the infinite prime.

Every element $\zeta \in K_{\mathfrak{p}}^*$ can be written uniquely as

$$\zeta = u\pi_{\mathfrak{p}}^n \tag{12.18}$$

for some $u \in U_{\mathfrak{p}} = \vartheta_{\mathfrak{p}}^*$ and $n \in \mathbb{Z}$.

Definition 12.8.16. For $\zeta \in K_{\mathfrak{p}}^*$ given by (12.18) we define

$$\operatorname{sgn}_{\mathfrak{p}}\zeta := \overline{u},$$

where $\overline{u}$ is the residue class of u in the class field $k(\mathfrak{p})$.

Remark 12.8.17. The map $\operatorname{sgn}: K_{\mathfrak{p}}^* \longrightarrow k(\mathfrak{p})^*$ is a multiplicative epimorphism.

For $\alpha \in \mathbb{F}_q^*$, we identify α with $\operatorname{sgn}_{\mathfrak{p}}(\alpha)$.

Definition 12.8.18. We define the groups

$$V_{\mathfrak{p}} := \ker(\operatorname{sgn}_{\mathfrak{p}}) \quad \text{and} \quad K_{\mathfrak{p}}^{(1)} := V_{\mathfrak{p}} \cap U_{\mathfrak{p}} = V_{\mathfrak{p}} \cap \vartheta_{\mathfrak{p}}^*.$$

Proposition 12.8.19. *As a topological group, $V_{\mathfrak{p}}$ is isomorphic to $K_{\mathfrak{p}}^{(1)} \times \mathbb{Z}$.*

Proof: Using (12.18) we obtain a map

$$\begin{aligned} K_{\mathfrak{p}}^{(1)} \times \mathbb{Z} &\xrightarrow{\phi} V_{\mathfrak{p}} \\ (\alpha, n) &\longmapsto \alpha\pi_{\mathfrak{p}}^{n}. \end{aligned}$$

Since $\operatorname{sgn}_{\mathfrak{p}}(\alpha\pi_{\mathfrak{p}}^{n}) = \operatorname{sgn}_{\mathfrak{p}}(\alpha)$, ϕ is a well-defined epimorphism of groups. Now if $\zeta \notin \vartheta_{\mathfrak{p}}^{*}$, then $v_{\mathfrak{p}}(\zeta) = n \in \mathbb{Z} \setminus \{0\}$. Thus $\|\zeta\|_{\mathfrak{p}} \neq 1$, where $\|\ \|_{\mathfrak{p}}$ denotes the absolute value associated to $\mathfrak{p}$. Therefore if $\varepsilon = |1 - \|\zeta\|_{\mathfrak{p}}| > 0$ and $\zeta' \in B(\zeta, \varepsilon) = \{x \in K_{\mathfrak{p}} \mid \|x - \zeta\|_{\mathfrak{p}} < \varepsilon\}$, we have

$$\left|\|\zeta'\|_{\mathfrak{p}} - \|\zeta\|_{\mathfrak{p}}\right| \leq \|\zeta - \zeta'\|_{\mathfrak{p}} < \varepsilon = \left|1 - \|\zeta\|_{\mathfrak{p}}\right|.$$

Hence $\|\zeta'\|_{\mathfrak{p}} \neq 1$, and $\vartheta_{\mathfrak{p}}^{*}$ is open in $K_{\mathfrak{p}}$. It follows that $K_{\mathfrak{p}}^{(1)} = V_{\mathfrak{p}} \cap \vartheta_{\mathfrak{p}}^{*}$ is open in $V_{\mathfrak{p}}$.

Let C be an open set in $K_{\mathfrak{p}}^{(1)}$ and $B \subseteq \mathbb{Z}$. Then $\phi(C \times B) = \bigcup_{n \in B} C\pi_{\mathfrak{p}}^{n}$, which is open in $V_{\mathfrak{p}}$ since C is open and so is $C\pi_{\mathfrak{p}}^{n}$. Thus ϕ is an open map.

Now if $\mathcal{U}$ is any open set of $V_{\mathfrak{p}}$, then since for all $n \in \mathbb{Z}$, $\pi_{\mathfrak{p}}^{n}\mathcal{U}$ is homeomorphic to $\mathcal{U}$, the set $\pi_{\mathfrak{p}}^{n}\mathcal{U} \cap K_{\mathfrak{p}}^{(1)}$ is open in $K_{\mathfrak{p}}^{(1)}$. Therefore

$$\begin{aligned} \phi^{-1}(\mathcal{U}) &= \phi^{-1}\left(\mathcal{U} \bigcap \left(\bigcup_{n \in \mathbb{Z}} \pi_{\mathfrak{p}}^{n} K_{\mathfrak{p}}^{(1)}\right)\right) = \bigcup_{n \in \mathbb{Z}} \phi^{-1}\left(\mathcal{U} \cap \pi_{\mathfrak{p}}^{n} K_{\mathfrak{p}}^{(1)}\right) \\ &= \bigcup_{n \in \mathbb{Z}} \left(\left(\pi_{\mathfrak{p}}^{-n}\mathcal{U} \cap K_{\mathfrak{p}}^{(1)}\right) \times \{n\}\right) \end{aligned}$$

is open in $K_{\mathfrak{p}}^{(1)} \times \mathbb{Z}$. The result follows. □

Definition 12.8.20. Let $\xi \in J$. We define

$$\partial\xi = \prod_{\mathfrak{p} \in \mathbb{P}_K} \mathfrak{p}^{v_{\mathfrak{p}}(\xi_{\mathfrak{p}})} \quad \text{and} \quad d_T\xi = \operatorname{sgn}_{\mathfrak{p}_\infty}(\xi_{\mathfrak{p}_\infty}) \prod_{\substack{\mathfrak{p} \in \mathbb{P}_K \\ \mathfrak{p} \neq \mathfrak{p}_\infty}} \pi_{\mathfrak{p}}^{v_{\mathfrak{p}}(\xi_{\mathfrak{p}})}.$$

Since $\xi \in J$, $\pi_{\mathfrak{p}} \in K$ we have $\partial\xi \in D_K$ and $d_T\xi \in K^{*}$.

Lemma 12.8.21. *The maps $\partial\colon J \to D_K$ and $d_T\colon J \to K^{*}$ are group epimorphisms.*

Proof. This is clear. □

Definition 12.8.22. Consider $K^{*} \subseteq J$ with the discrete topology ($K^{*} \subseteq J$ along the diagonal $x \mapsto (x)_{\mathfrak{p} \in \mathbb{P}_K}$) (see Exercise 12.10.35). We can view $V_\infty = K_{\mathfrak{p}_\infty}^{(1)} \times \mathbb{Z}$ as a subgroup of $K_{\mathfrak{p}_\infty}^{*} \subseteq J$, by identifying V_∞ with the group of ideles with all components equal to 1 except the component corresponding to $\mathfrak{p}_\infty$.

Finally, consider the subgroup $\mathcal{U}_T$ of J consisting of all ideles whose $\mathfrak{p}_\infty$-component is 1 and whose other components are elements of $\vartheta_{\mathfrak{p}}^{*}$.

Remark 12.8.23. The topology of $K^*_{\mathfrak{p}_\infty}$ considered as a subgroup of J is the same as its usual topology (if $C \subseteq K^*_{\mathfrak{p}_\infty}$ is open in the usual topology, then $C \times \prod_{\mathfrak{p}\neq\mathfrak{p}_\infty} U_\mathfrak{p} = \mathcal{U}$ is open in J and $\mathcal{U} \cap K^*_{\mathfrak{p}_\infty} = C$). Also the topological groups $\mathcal{U}_T$ and $\prod_{\mathfrak{p}\neq\mathfrak{p}_\infty} U_\mathfrak{p}$ are equal.

Theorem 12.8.24. *We have*

$$J \cong K^* \times \mathcal{U}_T \times K^{(1)}_{\mathfrak{p}_\infty} \times \mathbb{Z}$$

both algebraically and topologically.

Proof. Let $\xi \in J$ and consider

$$d_T(\xi) = \operatorname{sgn}_{\mathfrak{p}_\infty}(i_{\mathfrak{p}_\infty}) \prod_{\mathfrak{p}\neq\mathfrak{p}_\infty} \pi_\mathfrak{p}^{v_\mathfrak{p}(\xi_\mathfrak{p})} \in K^* \subseteq J \quad \text{and let} \quad \xi^* = d_T(\xi)^{-1}\xi.$$

For $\mathfrak{p} \neq \mathfrak{p}_\infty$, we have

$$\xi^*_\mathfrak{p} = \left(\operatorname{sgn}_{\mathfrak{p}_\infty}\left(\xi_{\mathfrak{p}_\infty}\right) \prod_{\mathfrak{q}\neq\mathfrak{p},\mathfrak{p}_\infty} \pi_\mathfrak{q}^{v_\mathfrak{q}(\xi_\mathfrak{q})}\right)^{-1} \pi_\mathfrak{p}^{-v_\mathfrak{p}(\xi_\mathfrak{p})} u \pi_\mathfrak{p}^{v_\mathfrak{p}(\xi_\mathfrak{p})},$$

where $u \in U_\mathfrak{p}$. Thus $\xi^*_\mathfrak{p} \in U_\mathfrak{p}$ for all $\mathfrak{p} \neq \mathfrak{p}_\infty$.

For $\mathfrak{p} = \mathfrak{p}_\infty$ we have

$$\xi^*_{\mathfrak{p}_\infty} = \operatorname{sgn}_{\mathfrak{p}_\infty}(\xi_{\mathfrak{p}_\infty})^{-1}\xi_{\mathfrak{p}_\infty} u \in V_{\mathfrak{p}_\infty} = K^{(1)}_{\mathfrak{p}_\infty} \times \mathbb{Z} \quad \text{for some} \quad u \in U_{\mathfrak{p}_\infty}.$$

Thus $\xi^* \in \mathcal{U}_T \times V_{\mathfrak{p}_\infty}$ and

$$\xi = d_T(\xi)\xi^*. \tag{12.19}$$

The decomposition of $\xi \in J$ as a product of an element of K^* and an element of $\mathcal{U}_T \times V_{\mathfrak{p}_\infty}$ is unique since if $\xi = \alpha\theta$ is another such decomposition with $\alpha \in K^*$ and $\theta \in V_{\mathfrak{p}_\infty}$, then for all $\mathfrak{p} \neq \mathfrak{p}_\infty$,

$$v_\mathfrak{p}(\xi) = v_\mathfrak{p}(\xi_\mathfrak{p}) = v_\mathfrak{p}(\alpha) + v_\mathfrak{p}(\theta_\mathfrak{p}) = v_\mathfrak{p}(\alpha) = v_\mathfrak{p}(d_T\xi).$$

Now $\deg_K((\alpha)_K) = \deg_K((d_T\xi)_K) = 0$, so $(\alpha)_K = (d_T\xi)_K$. Thus $\alpha = Cd_T\xi$ with $C \in k^*$.

Since $C = \alpha(d_T\xi)^{-1} = \theta^{-1}_{\mathfrak{p}_\infty}\xi^*_{\mathfrak{p}_\infty} \in K^* \cap V_\infty = \{1\}$, it follows that $\alpha = d_T\xi$ and $\theta = \xi^*$.

In particular, J and $K^* \times \mathcal{U}_T \times V_{\mathfrak{p}_\infty}$ are isomorphic as groups. Now since $V_{\mathfrak{p}_\infty}$ is an open subgroup of $K^*_{\mathfrak{p}_\infty}$ (because $V_{\mathfrak{p}_\infty} = \operatorname{sgn}^{-1}_{\mathfrak{p}_\infty}(\{1\})$ and $\{1\}$ is open in $k(\mathfrak{p}_\infty) \subseteq K_{\mathfrak{p}_\infty}$), it follows that $\mathcal{U}_T \times V_{\mathfrak{p}_\infty}$ is open in J. Using the fact that K^* is a discrete subspace of J, we obtain

$$J \cong K^* \times \mathcal{U}_T \times V_{\mathfrak{p}_\infty}.$$

Finally, since $V_{\mathfrak{p}_\infty} \cong K^{(1)}_\mathfrak{p} \times \mathbb{Z}$, we obtain the result using Proposition 12.8.19. $\square$

Theorem 12.8.25. *The group $\mathcal{U}_T$ is isomorphic to $G_T = \mathrm{Gal}(K_T/K)$ in a natural way. The isomorphism will be denoted by ψ_T.*

Proof. Let $\xi \in \mathcal{U}_T$ and let $M \in R_T$ be a monic polynomial. Suppose $M = \prod P^n$ is the factorization of M. By the Chinese remainder theorem, there exists $C \in R_T$ such that $C \equiv \xi_{\mathfrak{p}} \bmod P^n$ for every P dividing M, where $(P)_K = \frac{\mathfrak{p}}{\mathfrak{p}_\infty^{\deg P}}$. Then C is unique mod M and C mod M determines a unique automorphism σ_C of $K(\Lambda_M)/K$ such that $\sigma_C(\lambda) = \lambda^C$ or all $\lambda \in \Lambda_M$.

Define $\Psi_T^M : \mathcal{U}_T \to \mathrm{Gal}(K(\Lambda_M)/K)$, $\xi \mapsto \sigma_C$. Then $\left(\Psi_T^M\right)^{-1}\left(\{\sigma_C\}\right) = \Big\{\xi \in \mathcal{U}_T \mid \xi_{\mathfrak{p}} \equiv C \bmod P^n \ \forall\, P \mid M\Big\}$.

For each P dividing M, let

$$T_{\mathfrak{p}} = \Big\{X \in \mathcal{U}_{\mathfrak{p}} \mid X \equiv C \bmod P^n\Big\} = \Big\{X \in K_{\mathfrak{p}}^* \mid ||X - C||_{\mathfrak{p}} < ||P||_{\mathfrak{p}}^{n-1}\Big\}$$

and notice that $T_{\mathfrak{p}}$ is open in $K_{\mathfrak{p}}^*$.

Set $S = \Big\{\mathfrak{p} \in \mathbb{P}_K \setminus \{\mathfrak{p}_\infty\} \mid v_{\mathfrak{p}}(M) \neq 0\Big\}$. Then

$$\left(\Psi_T^M\right)^{-1}(\{\sigma_C\}) = \prod_{\mathfrak{p}\in S} T_{\mathfrak{p}} \times \prod_{\mathfrak{p}' \in \mathbb{P}_K \setminus (S \cup \{\mathfrak{p}_\infty\})} U_{\mathfrak{p}'},$$

which is open in $\mathcal{U}_T$. It follows that Ψ_T^M is a continuous epimorphism.

On the other hand, if M divides N the restriction of Ψ_T^N to $K(\Lambda_M)$ is just Ψ_T^M. Using the universal property of inverse limits ($G_T = \varprojlim_M \mathrm{Gal}(K(\Lambda_M)/K)$) we obtain a continuous homomorphism

$$\Psi_T : \mathcal{U}_T \longrightarrow G_T.$$

If $\xi \in \ker \Psi_T$, then for every $\mathfrak{p} \neq \mathfrak{p}_\infty$, $\xi_{\mathfrak{p}} \equiv 1 \bmod P^n$ for all $n \in \mathbb{N}$ and $v_{\mathfrak{p}}(\xi_{\mathfrak{p}} - 1) \geq n$ for all n. Thus $\xi_{\mathfrak{p}} = 1$ and ξ is the unit idele.

Now let $\tau \in G_T$ and let N be an open normal subgroup of G_T. Since $K_T = \bigcup_{M \in R_T} K(\Lambda_M)$, if $L = K_T^N$ we have $L \subseteq K(\Lambda_M)$ for some M. Let $\xi \in \mathcal{U}_T$ be such that $\Psi_T(\xi)|_{K(\Lambda_M)} = \tau|_{K(\Lambda_M)}$. Then $\tau^{-1}\Psi_T(\xi)|_{K(\Lambda_M)} = \mathrm{Id}_{K(\Lambda_M)}$ and $\tau^{-1}\Psi_T(\xi)|_L = \mathrm{Id}_L$. Hence $\tau^{-1}\Psi_T(\xi) \in N$ and $\Psi_T(\xi) \in \tau N$. Therefore $\Psi_T(\mathcal{U}_T)$ is dense.

Since $\mathcal{U}_T$ is compact, it follows that Ψ_T is onto and hence an isomorphism of topological groups. □

Theorem 12.8.26. *As a topological group, $K_{\mathfrak{p}_\infty}^{(1)}$ is naturally isomorphic to $G_\infty = \mathrm{Gal}(L_\infty/K)$. The corresponding isomorphism from $K_{\mathfrak{p}_\infty}^{(1)}$ to G_∞ will be denoted by Ψ_∞.*

Proof. By Theorem 12.8.14, $G_\infty \cong \Big\{f(X) \in \mathbb{F}_q[[X]] \mid f(0) = 1\Big\}$. On the other hand, by Theorem 2.5.20, $\vartheta_{\mathfrak{p}_\infty} \cong \mathbb{F}_q[[X]]$ since $\deg_K \mathfrak{p}_\infty = 1$. Now $K_{\mathfrak{p}_\infty}^{(1)} = V_{\mathfrak{p}_\infty} \cap U_{\mathfrak{p}_\infty} =$

$\{f(x) \in \mathbb{F}_q[[x]] \mid f(0) = 1\}$. The action of $K_{\mathfrak{p}_\infty}^{(1)}$ on L_∞ is described in Section 12.8.3. □

Finally we have the monomorphism

$$\Psi_{\mathbb{Z}} : \mathbb{Z} \longrightarrow G_E = \mathrm{Gal}(E/K) \cong \widehat{\mathbb{Z}}$$

defined in such a way that $\Psi_{\mathbb{Z}}(1)$ is the Frobenius automorphism.

Since $\mathbb{Z}$ has the discrete topology, $\Psi_{\mathbb{Z}}$ is a dense continuous monomorphism.

By Theorem 12.8.24, any element ξ in J can be written uniquely as

$$\xi = d_T(\xi)\xi_T\xi_\infty\xi_{\mathbb{Z}} \tag{12.20}$$

with $d_T(\xi) \in K^*$, $\xi_T \in \mathcal{U}_T$, $\xi_\infty \in K_{\mathfrak{p}_\infty}^{(1)}$, and $\xi_{\mathbb{Z}} \in \mathbb{Z}$. Note that $\xi_\infty \neq \xi_{\mathfrak{p}_\infty}$.

Definition 12.8.27. We define a homomorphism of topological groups

$$\Psi\colon J \longrightarrow \mathrm{Gal}(A/K) \cong G_T \times G_\infty \times G_E$$

as follows: If $\xi \in J$, then ξ can be written as in (12.20), and we put

$$\Psi(\xi) = \Psi_T(\xi_T^{-1})\Psi_\infty(\xi_\infty^{-1})\Psi_{\mathbb{Z}}(\xi_{\mathbb{Z}}). \tag{12.21}$$

Since Ψ_T and Ψ_∞ are isomorphisms and $\Psi_{\mathbb{Z}}$ is a monomorphism, it follows that $\ker\Psi = K^*$ $(= \{d_T(\xi) \mid \xi \in J\})$ and that Ψ is continuous.

Therefore we have proved the following result:

Theorem 12.8.28. *The map Ψ defined by* (12.21) *is a continuous dense homomorphism from J into* $\mathrm{Gal}(A/K)$ *whose kernel is K^*.* □

Remark 12.8.29. Our reason for defining

$$\Psi(\xi) = \Psi_T(\xi_T^{-1})\Psi_\infty(\xi_\infty^{-1})\Psi_{\mathbb{Z}}(\xi_{\mathbb{Z}})$$

instead of

$$\Psi(\xi) = \Psi_T(\xi_T)\Psi_\infty(\xi_\infty)\Psi_{\mathbb{Z}}(\xi_{\mathbb{Z}})$$

is that the former yields Artin's reciprocity law homomorphism for K, as will now be seen.

Let A^* be the maximal abelian extension of K. Since A/K is abelian, $A \subseteq A^*$. Let $\Psi^*\colon J \longrightarrow A^*$ be the reciprocity law homomorphism (see Remark 11.5.7).

Let res: $\mathrm{Gal}(A^*/K) \longrightarrow \mathrm{Gal}(A/K)$ be the restriction map.

We will prove that $\mathrm{res} \circ \Psi^* = \Psi$ and since $\ker\Psi = \ker\Psi^* = K^*$ it will follow that

$$K^* = \ker\Psi = (\Psi^*)^{-1}(\ker \mathrm{res}) = (\Psi^*)^{-1}(\{1\}).$$

Thus ker res $= 1$, res is an isomorphism, and res $=$ Id. Therefore $A = A^*$.

Now in order to show that res $\circ \Psi^* = \Psi$, it suffices to prove that for any idele $\xi \in J$, $\Psi^*(\xi)|_F = \Psi(\xi)|_F$ for all $K \subseteq F \subseteq A$ such that $[F : K] < \infty$. Any such extension L is contained in the composite $\mathbb{F}_{q^m}K(\Lambda_M)L_n$ for some $m, n \in \mathbb{N}$, $M \in R_T$, and $K(\Lambda_M) = \prod_{P|M} K(\Lambda_{P^t})$. Thus it will be sufficient to show that $\Psi^*(\xi)\,|_F = \Psi(\xi)|_F$ for any F that has one of the following forms:

$$\begin{aligned} &\text{(i)}\quad F = \mathbb{F}_{q^m} \text{ for some } m \geq 1,\\ &\text{(ii)}\quad F = K(\Lambda_{P^t}) \text{ for some monic irreducible polynomial } P \in R_T \text{ and } t \geq 1,\\ &\text{(iii)}\quad F = L_n \text{ for some } n \geq 1. \end{aligned} \tag{12.22}$$

Let F/K be a finite extension of type (i), (ii), or (iii). The restriction of $\Psi^*(\xi)$ from $\mathrm{Gal}(A^*/K)$ to $\mathrm{Gal}(F/K)$ induces

$$\Psi_F^* : J \longrightarrow \mathrm{Gal}(F/K).$$

The Takagi–Artin theorem (Theorem 11.5.6) yields the following characterization of Ψ_F^*: For any finite set S of prime divisors containing all those prime divisors that ramify in F/K, then Ψ_F is the unique homomorphism $J \to \mathrm{Gal}(F/K)$ such that

(a) Ψ_F^* is continuous,
(b) $\Psi_F^*(K^*) = 1$,
(c) $\Psi_F^*(\xi) = \prod_{\mathfrak{p}\notin S} \left(\frac{F/K}{\mathfrak{p}}\right)^{v_{\mathfrak{p}}(\xi_{\mathfrak{p}})} = \left(\frac{F/K}{\partial\xi}\right)$, where $\left(\frac{F/K}{}\right)$ is the Artin symbol (Definition 11.2.5).

In short, we need to verify that $\Psi_F : J \to \mathrm{Gal}(F/K)$ satisfies (a), (b), and (c) on all extensions of type (i), (ii), or (iii) of (12.22).

By (12.21), Ψ_F satisfies (a) and (b), so we only need to prove (c). For $\mathfrak{p} \in \mathbb{P}_K$ we call an idele a $\mathfrak{p}$-idele if $\xi \in J$ is such that for some $\mathfrak{p} \in \mathbb{P}_K$,

$$\xi_{\mathfrak{p}'} = \begin{cases} \mu_{\mathfrak{p}'} & \text{if } \mathfrak{p}' \neq \mathfrak{p},\\ P = \pi_{\mathfrak{p}} & \text{if } \mathfrak{p}' = \mathfrak{p}, \end{cases}$$

where $(P)_K = \frac{\mathfrak{p}}{\mathfrak{p}_\infty^{\deg P}}$ (or if $\mathfrak{p} = \mathfrak{p}_\infty$, $\xi_{\mathfrak{p}_\infty} = \frac{1}{T}$) and $\mu_{\mathfrak{p}'} \in U_{\mathfrak{p}'}$.

Now every $\xi \in J^S$ can be written as the finite product of $\mathfrak{p}$-ideles and inverses of $\mathfrak{p}$-ideles for various $\mathfrak{p} \notin S$, so it suffices to prove (c) for a $\mathfrak{p}$-idele ξ.

Proposition 12.8.30. *If F is of any of the three types of* (12.22) *and ξ is a $\mathfrak{p}$-idele, then $\Psi_F = \Psi_F^*$.*

Proof.
Case 1: F/K is a finite constant field extension. The extension F/K is unramified. Let $S = \{\mathfrak{p}_\infty\}$. Let ξ be a $\mathfrak{p}$-idele with $\mathfrak{p} \neq \mathfrak{p}_\infty$. Then $\partial\xi = \mathfrak{p}$ and by Proposition 11.2.2, $\left(\frac{F/K}{\mathfrak{p}}\right) = \sigma^{\deg \mathfrak{p}}$, where σ is the Frobenius automorphism.

On the other hand, if $\theta = \xi d_T(\xi)^{-1}$ then

$$\theta_{\mathfrak{p}_\infty} = \xi_{\mathfrak{p}_\infty}\left(\operatorname{sgn}_{\mathfrak{p}_\infty}\xi_{\mathfrak{p}_\infty}\right)^{-1}\prod_{\substack{\mathfrak{p}'\in\mathbb{P}_K\\ \mathfrak{p}'\neq\mathfrak{p}_\infty}}\pi_{\mathfrak{p}'}^{-v_{\mathfrak{p}'}(\xi_{\mathfrak{p}'})} = 1\times 1\times\pi_{\mathfrak{p}}^{-1} = P^{-1}\in V_{\mathfrak{p}_\infty}.$$

Thus $v_{\mathfrak{p}_\infty}(\theta_{\mathfrak{p}_\infty}) = \deg_{R_T} P = \deg\mathfrak{p}$. Therefore $\xi_{\mathbb{Z}} = \deg\mathfrak{p}$.

Since F/K is a constant extension, it follows that

$$\Psi_F(\xi) = \Psi_{\mathbb{Z}}(\xi_{\mathbb{Z}}) = \sigma^{\deg\mathfrak{p}}.$$

Hence in this case we obtain $\Psi_F = \Psi_F^*$.

Case 2: $F = K(\Lambda_{P^t})$ for a monic irreducible polynomial $P\in R_T$ and some $t\geq 1$. In this case the only ramified primes are $\mathfrak{p}$ and $\mathfrak{p}_\infty$, where $(P)_K = \frac{\mathfrak{p}}{\mathfrak{p}_\infty^{\deg P}}$ (Theorem 12.5.3).

Set $S = \{\mathfrak{p},\mathfrak{p}_\infty\}$. Let $\mathfrak{q}$ be a prime divisor such that $\mathfrak{q}\notin S$ and let $\xi\in J^S$ be a $\mathfrak{q}$-idele. Then $d_T(\xi) = Q = \pi_{\mathfrak{q}}$. We write $\xi = d_T(\xi)\xi^*\in J^S$. Therefore $\xi_{\mathfrak{p}} = 1 = Q\xi_{\mathfrak{p}}^*$, and $\xi_{\mathfrak{p}}^* = Q^{-1}$.

Now ξ acts on $K(\Lambda_{P^t})/K$ via the $\mathfrak{p}$th component of ξ^*, so on $K(\Lambda_{P^t})$ $\Psi(\xi) = \Psi_T(\xi_T^{-1})$ is the automorphism $\Lambda_{P^t}\to\Lambda_{P^t}$, $\lambda\mapsto\lambda^Q$. By Theorem 12.5.1 this corresponds to the Artin symbol at $Q = \partial\xi$. This proves that $\Psi_F = \Psi_F^*$ in this case.

Case 3: $F = L_n$ for some $n\geq 1$. Set $S = \{\mathfrak{p}_0,\mathfrak{p}_\infty\}$ where $(T)_K = \frac{\mathfrak{p}_0}{\mathfrak{p}_\infty}$. The only ramified prime is $\mathfrak{p}_\infty$ Let $\xi\in J^S$ be a $\mathfrak{p}$-idele where $\mathfrak{p}$ corresponds to some $P\in R_T$ distinct from T. Note that

$$d_T(\xi)^{-1} = \pi_{\mathfrak{p}}^{-1} = P^{-1} = (P^{-1}T^d)(1/T)^d,$$

where $d = \deg P$ and $P^{-1}T^d$ is a unit at $\mathfrak{p}_\infty$. Hence the $\mathfrak{p}_\infty$-coordinate of $\xi^* = \xi d_T(\xi)^{-1}$ is $\xi_{\mathfrak{p}_\infty}^* = P^{-1}$. Therefore $\xi_\infty = P^{-1}T^d$. Now if $P = T^d + a_{d-1}T^{d-1} + \cdots + a_1T + a_0$ with $a_0\neq 0$, then

$$\xi_\infty^{-1} = \frac{P(T)}{T^d} = 1 + \frac{a_{d-1}}{T} + \cdots + \frac{a_1}{T^{d-1}} + \frac{a_0}{T^d} = a_0P_1(1/T),$$

where P_1 is a monic polynomial. We have $v_{\mathfrak{p}}(P_1) = v_{\mathfrak{p}}\left(\frac{P}{T^d}\right) = 1$. Thus P_1 is the canonical uniformizer when we consider $K = \mathbb{F}_q(1/T)$, i.e., $1/T$ is a generator of K. By definition ξ acts on L_n via the component of ξ_∞, that is, $\Psi(\xi) = \Psi_\infty(\xi_\infty^{-1})$. Considered on L_n/K this is the restriction of the automorphism of $F_n = K(\Lambda_{T^{-n-1}})$ such that $\Lambda_{T^{-n-1}}\to\Lambda_{T^{-n-1}}$, $\lambda\mapsto\lambda^{a_0P_1}$. The restriction of this automorphism to L_n is the same as the restriction of the automorphism

$$\begin{aligned}\Lambda_{T^{-n-1}} &\longrightarrow \Lambda_{T^{-n-1}}\\ \lambda &\longmapsto \lambda^{P_1}\end{aligned}\tag{12.23}$$

because the automorphism of F_n associated to $a_0\in\mathbb{F}_q^*$ fixes L_n. By Theorem 12.5.1 the automorphism defined by (12.23) corresponds to the Artin symbol in F_n at $\mathfrak{p}$ and therefore its restriction to L_n is the Artin symbol in L_n at $\mathfrak{p}$.

This shows that $\Psi_F = \Psi_F^*$ in this case, and the proof is complete. □

We have obtained the analogue of the Kronecker–Weber theorem for function fields:

Theorem 12.8.31. *The extension A/K constructed in Section* 12.8.4 *is the maximal abelian extension of K, and the homomorphism*

$$\Psi\colon J \longrightarrow \operatorname{Gal}(A/K)$$

given in (12.21) *is the Artin reciprocity law homomorphism.* □

In particular, A and Ψ do not depend upon the original choice of the generator T.

As a corollary of Theorem 12.8.31 we have the following:

Theorem 12.8.32. *The maximal abelian extension of K is $K_T K_{1/T}$.*

Proof. According to the construction of Ψ, the group of ideles fixing K_T is $K^* V_{\mathfrak{p}_\infty} = K^* K_{\mathfrak{p}_\infty}$. Similarly, the group of ideles fixing $K_{1/T}$ is $K^* K_{\mathfrak{p}_0}$ (where $(T)_K = \frac{\mathfrak{p}_0}{\mathfrak{p}_\infty}$). The intersection of these two groups is K^*, so the kernel of the map

$$J \longrightarrow \operatorname{Gal}(K_T K_{1/T}/K),$$

induced by restriction, is $K^* = \ker \Psi$. It follows that $A = K_T K_{1/T}$. □

12.9 The Analogue of the Brauer–Siegel Theorem

As we saw in Section 7.6, the analogue of the Brauer–Siegel theorem for function fields is the limit

$$\lim_{g\to\infty} \frac{\ln h}{g \ln q} = 1, \tag{12.24}$$

where g is the genus, h is the class number, and q is the cardinality of the constant field. In this section we prove that the analogue of the Brauer–Siegel theorem holds for the class of cyclotomic function fields. We shall prove that in the class of cyclotomic function fields over the finite field of constants $\mathbb{F}_q$, we have

$$\lim_{g\to\infty} \frac{\Phi(M)}{g} = 0,$$

where $g = g_M$ is the genus of $K(\Lambda_M)$ and $\Phi(M) = [K(\Lambda_M) : K] = \left|(R_T/(M))^*\right|$.

Therefore, in this class of function fields, the conditions of Theorem 7.6.3 are satisfied and we have

$$\lim_{g\to\infty} \frac{\ln h}{g \ln q} = 1.$$

Let $M = \prod_{i=1}^{t} P_i^{n_i}$ be the factorization of $M \in R_T \setminus \mathbb{F}_q$ into powers of irreducible polynomials with $n_i \geq 1$ and $d_i = \deg(P_i) \geq 1$ for $i = 1, \ldots, t$. Let g_M be the genus of $K(\Lambda_M)$. Then by Theorem 12.7.2,

$$g_M = \frac{\Phi(M)}{2}\left(\sum_{i=1}^{t}\left(n_i d_i - \frac{d_i}{q^{d_i}-1}\right) - \frac{q}{q-1}\right) + 1. \tag{12.25}$$

Now if $d = \deg M$, using (12.25) we obtain

$$\begin{aligned} g_M &\leq \Phi(M)d + 1 = d\prod_{i=1}^{t} q^{d_i(n_i-1)}(q^{d_i}-1) + 1 \\ &\leq d\prod_{i=1}^{d} q^{d_i n_i} + 1 = dq^d + 1. \end{aligned} \tag{12.26}$$

Suppose that d is sufficiently large so as to satisfy $d \geq \frac{4q}{q-1}$. If $n_i = d_i = 1$ for some $i \in \{1, \ldots, t\}$, we have $\Phi(M) = (q-1)\prod_{\substack{j=1 \\ j\neq i}}^{t} \Phi\left(P_j^{n_j}\right)$.

Since we want to estimate the quotient $\Phi(M)/g_M$ when g_M is sufficiently large and the number of irreducible polynomials of degree one in R_T is finite, we may assume that $n_i \geq 2$ or $d_i \geq 2$ for $i = 1, \ldots, t$. Hence for $i = 1, \ldots, t$, we have $n_i(q^{d_i} - 1) \geq 2$. Therefore

$$n_i d_i - \frac{d_i}{q^{d_i}-1} \geq \frac{n_i d_i}{2} \tag{12.27}$$

for $i = 1, \ldots, t$.

Using (12.25) and (12.27) we obtain

$$\begin{aligned} g_M &\geq \frac{g_M}{\Phi(M)} \geq \frac{1}{2}\left(\sum_{i=1}^{t}(n_i d_i - \frac{d_i}{q^{d_i}-1}) - \frac{q}{q-1}\right) \\ &\geq \frac{1}{2}\left(\sum_{i=1}^{t}\frac{n_i d_i}{2} - \frac{q}{q-1}\right) = \frac{1}{2}\left(\frac{d}{2} - \frac{q}{q-1}\right) \geq \frac{d}{8}. \end{aligned}$$

Therefore we obtain the following

Proposition 12.9.1. *In the class of cyclotomic function fields $K(\Lambda_M)$ over the finite field of constants $\mathbb{F}_q$, we have*

$$g_M \longrightarrow \infty \iff d \longrightarrow \infty,$$

where $M \in R_T \setminus \mathbb{F}_q$, $d = \deg M$ and g_M is the genus of $K(\Lambda_M)$. Furthermore,

$$\lim_{g_M \to \infty} \frac{\Phi(M)}{g_M} = 0. \qquad \square$$

As a corollary we get the following theorem:

Theorem 12.9.2. *In the class of cyclotomic function fields $K(\Lambda_M)$ over the finite field of constants $\mathbb{F}_q$, we have*

$$\lim_{g_M\to\infty} \frac{\ln h_M}{g_M \ln q} = 1,$$

where h_M is the class number of $K(\Lambda_M)/\mathbb{F}_q$.

Proof. The statement is an immediate consequence of Theorem 7.6.3 and Proposition 12.9.1. □

12.10 Exercises

Exercise 12.10.1. Let $M, N \in R_T = \mathbb{F}_q[T]$, $u \in \bar{K}$, and $K = \mathbb{F}_q(T)$. Prove that

$$u^{M+N} = u^M + u^N \quad \text{and} \quad u^{MN} = (u^M)^N.$$

Exercise 12.10.2. Let k be any field and let T be an indeterminate over k. Prove that for all $n \in \mathbb{N}$,

$$\left[k(T^{1/n}) : k(T)\right] = \left[k(T) : k(T^n)\right] = n.$$

Exercise 12.10.3. Prove that if $P \in R_T$ is an irreducible polynomial and $n \in \mathbb{N}$, then $\Lambda_{P^n}/\Lambda_{P^{n-1}} \cong \Lambda_P$.

Exercise 12.10.4. Let $M = \prod_{i=1}^r P_i^{\alpha_i} \in R_T$, where $P_1, P_2, \ldots, P_r$ are irreducible polynomials. Let $A = \Lambda_M$. Prove that

$$A(P) = \begin{cases} 0 & \text{if } P \notin \{P_1, \ldots, P_r\}, \\ \Lambda_{P_i^{\alpha_i}} & \text{if } P = P_i, \end{cases}$$

where $A(P)$ denotes the P-torsion of A.

Exercise 12.10.5. Let $\Phi(M) = \left|\left(R_T/M\right)^*\right|$ for $M \in R_T$. Prove that:

(i) If $M = P$ is irreducible with $d = \deg P$, then $\Phi(P) = q^d - 1$.
(ii) If $M, N \in R_T$ are relatively prime, then

$$\left(R_T/(MN)\right)^* \cong \left(R_T/M\right)^* \times \left(R_T/N\right)^*.$$

(iii) If $M, N \in R_T$ are relatively prime, then

$$\Phi(MN) = \Phi(M)\Phi(N).$$

(iv) If $M = P^n$ with P irreducible of degree d, then

$$\Phi(P^n) = \left|\left(R_T/(P^{n-1})\right)^*\right|\Phi(P) = q^{dn} - q^{d(n-1)}.$$

Exercise 12.10.6. If M and N are distinct elements of $\in R_T \setminus \{0\}$, prove that $\big(\Psi_M(u), \Psi_N(u)\big) = 1$.

Exercise 12.10.7. If $M \in R_T \setminus \{0\}$ is of degree d, prove that

$$\sum_{\substack{D|M \\ D \text{ monic}}} \Phi(D) = q^d.$$

Exercise 12.10.8. Let $M \in R_T \setminus \{0\}$ be a monic polynomial. Prove that

$$u^M = \prod_{\substack{D|M \\ D \text{ monic}}} \Psi_D(u).$$

Exercise 12.10.9. Let $\mu\colon R_T \setminus \{0\} \to \mathbb{Q}$ be given by

$$\mu(D) = \begin{cases} 1 & \text{if } D = 1, \\ (-1)^t & \text{if } D = P_1 \cdots P_t, \quad P_i \in R_T \quad \begin{array}{l}\text{distinct irreducible,} \\ \text{polynomials,}\end{array} \\ 0 & \text{otherwise.} \end{cases}$$

Prove that

$$\sum_{\substack{D|M \\ D \text{ monic}}} \mu(D) = \varepsilon(M) = \begin{cases} 1 & \text{if } M \text{ is a nonzero constant,} \\ 0 & \text{otherwise.} \end{cases}$$

Exercise 12.10.10. Let E be a field and let $\mathcal{A} = \{\xi\colon R_T \setminus \{0\} \to E\}$. We define the *convolution product* $*$ in $\mathcal{A}$ by

$$(\xi * \phi)(M) = \sum_{\substack{D|M \\ M \text{ monic}}} \xi(D)\phi(M/D).$$

Prove that $(\mathcal{A}, *, +)$ is a commutative ring with unit $1_{\mathcal{A}} = \varepsilon$, where $\varepsilon\colon R_T \setminus \{0\} \to E$ is defined by

$$\varepsilon(M) = \begin{cases} 1 & \text{if } M \text{ is a nonzero constant,} \\ 0 & \text{otherwise.} \end{cases}$$

Exercise 12.10.11. Prove that if $f, g \in \mathcal{A}$ are such that

$$g(M) = \sum_{\substack{D|M \\ D \text{ monic}}} f(D)$$

for all $M \in R_T \setminus \{0\}$, then

$$f(M) = \sum_{\substack{D|M \\ D \text{ monic}}} g(D)\mu\left(\frac{M}{D}\right),$$

where μ is as in Exercise 12.10.9.

Exercise 12.10.12. Prove that

$$\Psi_M(u) = \prod_{\substack{D|M \\ D \text{ monic}}} \left(u^D\right)^{\mu(M/D)}.$$

Exercise 12.10.13. Let $M, N \in R_T \setminus \{0\}$. Prove that

$$\left(u^M, u^N\right) = u^{(M,N)},$$

where $(\_, \_)$ denotes the greatest common divisor.

Exercise 12.10.14. Show that if $P \in R_T$ is a monic polynomial and $n \in \mathbb{N}$, then

$$\Psi_{P^n}(u) = \Psi_P\left(u^{P^{n-1}}\right).$$

Exercise 12.10.15. Show that the vertices of the Newton polygon of $f(x)$ given in the proof of Theorem 12.4.2 are

$$(n - s_{t+1}, v(a_{n-s_{t+1}})), (n - s_t, v(a_{n-s_t})), \ldots ,$$
$$\ldots , (n - s_2, s_1m_1 + (s_2 - s_1)m_2), (n - s_1, s_1m_1), (n, 0).$$

Exercise 12.10.16. Let $a, b, x, y \in \mathbb{Z} \setminus \{0\}$ be such that $(x, y) = 1$ and $ax = by$. Prove that $[a, b] = ax = by$ where $[\_, \_]$ denotes the greatest common divisor.

Exercise 12.10.17. Let F be a complete field with respect to a discrete valuation v. Prove that if $f(x) \in F[x]$ has all its roots with distinct valuations, then $f(x)$ is a product of linear factors in $F[x]$.

Exercise 12.10.18. Let F and v be as in Exercise 12.10.17. Let $f(x) \in F[x]$. Prove that if $f(x)$ is irreducible, then the Newton polygon of $f(x)$ is a segment.

Exercise 12.10.19. Let $M = P^n$, where $P \in R_T = \mathbb{F}_q[T]$ is a monic irreducible polynomial of degree d. Let λ be a generator of the Carlitz–Hayes module Λ_M. Let N_1 and N_2 be the zero and pole divisors of λ respectively.

(i) Let $g_n(T, u) = \Psi_{P^n}(u)$ considered as a polynomial in two variables T and u. Prove that

$$\deg_T g_n(T, u) = \begin{cases} q^{d-1} & \text{if } n = 1, \\ q^{d(n-1)-1}(q^d - 1) & \text{if } n > 1. \end{cases}$$

(ii) Let $h_n(z) = g_n(z, \lambda)$. Deduce from (i) that

$$\deg h_n(z) = \begin{cases} q^{d-1} & \text{if } n = 1, \\ q^{d(n-1)-1}(q^d - 1) & \text{if } n > 1, \end{cases}$$

(iii) Using the irreducibility of $\Psi_{P^n}(u)$, prove that $h_n(z)$ is an irreducible polynomial in z with coefficients in $\mathbb{F}_q[\lambda]$.

(iv) Show that if T is a root of $h_n(z)$ and $L = \mathbb{F}_q(\lambda)$, then $K(\Lambda_M) = L(T)$, where $K = \mathbb{F}_q(T)$. Therefore $[K(\Lambda_M) : \mathbb{F}_q(\lambda)] = \deg_z h_n$.

(v) Deduce that

$$\deg N_1 = \deg N_2 = \deg h_n = \begin{cases} q^{d-1} & \text{if } n = 1, \\ q^{d(n-1)-1}(q^d - 1) & \text{if } n > 1. \end{cases}$$

Exercise 12.10.20. Let k be a field of characteristic $p > 0$ such that $\mathbb{F}_{p^2} \subseteq k$, and let $K = k(x)$ be a rational function field over k. Let $L = K(y) = k(x, y)$ where

$$y^{p^2} - y = x. \tag{12.28}$$

Let $\wp_\infty$ be the infinite prime divisor in K and $\mathfrak{p}$ a prime divisor in L above $\wp_\infty$.

(i) Show that $v_{\mathfrak{p}}(y) < 0$ using equation (12.28).

(ii) Deduce from (i) that $p^2 \mid e(\mathfrak{p}|\wp_\infty)$.

(iii) Deduce from (ii) that $[L : K] = p^2$ and that $\wp_\infty$ is totally ramified in L/K.

(iv) Prove that

$$T^{p^2} - T - x = \prod_{\alpha \in \mathbb{F}_{p^2}} (T - (y + \alpha)).$$

(v) Prove that L/K is a Galois extension with Galois group

$$\operatorname{Gal}(L/K) \cong (\mathbb{F}_{p^2}, +) \cong C_p \times C_p.$$

(vi) Using two subextensions F_1, F_2 such that $K \subseteq F_i \subseteq L$ and $[F_i : K] = p$, $i = 1, 2$, deduce that Abhyankar's lemma does not hold for two wildly ramified extensions.

Exercise 12.10.21. Let $\lambda \in \Lambda_P \setminus \{0\}$. Prove that $[K(\Lambda_P) : \mathbb{F}_q(\lambda)] = q - 1$.

Exercise 12.10.22. Let E be a field, $\beta_1, \ldots, \beta_n \in E$, and

$$A = \begin{bmatrix} 1 & \beta_1 & \cdots & \beta_1^{n-1} \\ \vdots & \vdots & \vdots & \vdots \\ 1 & \beta_n & \cdots & \beta_n^{n-1} \end{bmatrix}.$$

Prove that $\det A = \prod_{1 \le i < j \le n} (\beta_j - \beta_i)$.

Exercise 12.10.23. Prove that if $M \in R_T \setminus \{0\}$ and μ is the function given in Exercise 12.10.9, then

$$\prod_{A|M} A^{\mu(M/A)} = \begin{cases} P & \text{if } M = P^n,\ P \text{ an irreducible polynomial,} \\ 1 & \text{otherwise.} \end{cases}$$

Exercise 12.10.24. Let $P \in R_T$ be an irreducible polynomial. Prove that

$$\Psi_P(u) = 1 + \prod_{\substack{D|(P-1) \\ D\neq 1}} \Psi_D(u).$$

Exercise 12.10.25. Let $P \in R_T$ be an irreducible polynomial. Let $\mathcal{M} := R_T/(P)$. We define an action of R_T on $\mathcal{M}$ as follows: if $A \bmod P \in \mathcal{M}$ and $Q \in R_T$, then $Q \circ (A \bmod P) := A^Q \bmod P$.

Prove that this action is well defined, that is, if $A \equiv B \bmod P$ then $A^Q \equiv B^Q \bmod P$.

Show that $A^P \equiv A \bmod P$ for all $A \in R_T$ and deduce that $\mathcal{M}$ is an $R_T/(P-1)$-module. Does it hold that $\mathcal{M} \cong R_T/(P-1)$ as modules?

Exercise 12.10.26. Prove that $\Psi_M(0) = \begin{cases} P & \text{if } M = P^n \text{ for some } n \in \mathbb{N}, \\ 1 & \text{otherwise.} \end{cases}$

Exercise 12.10.27. If χ, ϕ are two Dirichlet characters such that $(F_\chi, F_\phi) = 1$, prove that $F_{\chi\phi} = F_\chi F_\phi$.

Exercise 12.10.28. Prove Proposition 12.6.33.

Exercise 12.10.29. Let G be a finite abelian group and let $H < G$. Prove that there exists an exact sequence of groups

$$1 \longrightarrow \widehat{(G/H)} \longrightarrow \hat{G} \xrightarrow{\phi} \hat{H} \longrightarrow 1$$

where $\phi(\sigma) = \sigma|_H$. In particular, $\hat{G}/\widehat{(G/H)} \cong \hat{H}$.

Exercise 12.10.30. Let G be a finite abelian group and let $H < G$. Prove that G contains a subgroup isomorphic to G/H.

Exercise 12.10.31. Let X be a finite group of Dirichlet characters. Describe in terms of X the maximal abelian extensions L of K_X such that L is abelian over K, the field of constants of L is $\mathbb{F}_q$, and L/K_X is unramified at every prime divisor.

Exercise 12.10.32. Let χ be a Dirichlet character and let $\chi = \prod_P \chi_P$ be its decomposition.

(i) Prove that $(\chi\psi)_P = \chi_P \psi_P$.
(ii) Prove that if $(F_\chi, F_\phi) = 1$ then $\chi(A)\psi(A) = (\chi\psi)(A)$ for all $A \in R_T$.

(iii) Prove that if χ and ψ are two arbitrary Dirichlet characters, then $\chi(A)\psi(A) = (\chi\psi)(A)$ unless $\chi(A) = \psi(A) = 0$.

Exercise 12.10.33. Prove that if χ is any nontrivial character of conductor $F_\chi = F$, then $\sum_{A \bmod F} \chi(A) = 0$.

Exercise 12.10.34. Let $M \in R_T$ and $A \in R_T$ be such that $A \not\equiv 1 \bmod M$ and $(A, M) = 1$. Prove that there exists a character χ defined modulo M and of conductor $F = F_\chi \mid M$ such that $\chi(A) \neq 1$.

Conclude that $\sum_{\chi \bmod M} \chi(A) = 0$.

Show that if $q = 2$ and $M = T^2(T^2+1)$, then $\sum_{\chi \bmod M} \chi\left(T^2\right) \neq 0$.

Exercise 12.10.35. Show that the subgroup K^* of J_K is discrete.

Exercise 12.10.36. Let $P \in R_T$ be an irreducible monic polynomial and let $\lambda \in \Lambda_P$ be a generator. Prove that $P = \prod_{\deg M < \deg P} \lambda^M$, where the product is over all nonzero polynomials of degree less than $\deg P$.

Exercise 12.10.37. Let $\mathcal{M} := \{M \mid M \in R_T, \deg M < \deg P \text{ monic}\}$ and let $\varrho = \prod_{M \in \mathcal{M}} \lambda^M$. Using Exercise 12.10.36, obtain that $P = (-1)^{\deg P} \varrho^{q-1}$.

Exercise 12.10.38. Assume that d divides $q-1$. If $M \in R_T$ is such that $P \nmid M$, prove that $N^d \equiv M \bmod P$ is solvable if and only if $M^{\frac{q^d-1}{d}} \equiv 1 \bmod P$, where $d = \deg P$. The order of the element $M^{\frac{q^d-1}{d}}$ is a divisor of d in $(R_T/P)^*$. In particular, $M^{\frac{q^d-1}{d}} \equiv \alpha \bmod P$ for a unique $\alpha \in \mathbb{F}_q^*$. We write $M^{\frac{q^d-1}{d}} \equiv \left(\frac{M}{P}\right)_d \bmod P$ and call $\left(\frac{M}{P}\right)_d$ the *dth power residue symbol*. Set $\left(\frac{M}{P}\right)_d = 0$ if $P \mid M$. Also define $\left(\frac{M}{P}\right) := \left(\frac{M}{P}\right)_{q-1}$. Thus $\left(\frac{M}{P}\right)_d = \left(\frac{M}{P}\right)^{\frac{q-1}{d}}$.

Exercise 12.10.39. If Q and P are two distinct monic irreducible polynomials, prove that $\varphi_Q(\varrho) = \left(\frac{P^*}{Q}\right)\varrho$, where ϱ is given in Exercise 12.10.37, φ_Q is the Artin automorphism (Theorem 12.5.1), and $P^* := (-1)^{\deg P} P = \varrho^{q-1}$.

Exercise 12.10.40. Show that every nonzero residue class module P has a unique representative of the form μM, where $\mu \in \mathbb{F}_q^*$ and $M \in \mathcal{M}$. Let $S \in R_T$ be such that $P \nmid S$. For $M \in \mathcal{M}$ write $SM = \mu_M M' \bmod P$ with $\mu_M \in \mathbb{F}_q^*$ and $M' \in \mathcal{M}$. Show that $\left(\frac{S}{P}\right)_{q-1} = \left(\frac{S}{P}\right) = \prod_{M \in \mathcal{M}} \mu_M$.

Exercise 12.10.41. Using Theorems 12.5.1 and 12.5.3, show that $\varphi_Q(\varrho) = \prod_{M \in \mathcal{M}} \lambda^{QM}$. Use Exercise 12.10.40 to show that $\varphi_Q(\varrho) = \left(\frac{Q}{P}\right)\varrho$.

Exercise 12.10.42. Combine Exercises 12.10.39 and 12.10.41 to show that if P and Q are two distinct monic irreducible polynomials, then $\left(\frac{P}{Q}\right)\left(\frac{Q}{P}\right)^{-1} = (-1)^{\deg P \deg Q}$.

13

Drinfeld Modules

13.1 Introduction

In this chapter we present a brief introduction to *Drinfeld modules* or, as they were called by Drinfeld himself, *elliptic modules*. The main goal of V. G. Drinfeld [30] was to generalize three classical results: a) the Kronecker–Weber Theorem; b) the Eichler–Shimura Theorem on ζ functions of modular curves and c) the fundamental theorem on complex multiplication.

In Chapter 12 we studied cyclotomic function fields, that is, the Carlitz–Hayes theory. These fields are the analogue to the classical cyclotomic fields $\mathbb{Q}(\zeta_n)$, and, as we saw, this analogy provides an explicit class field theory for congruence rational function fields (Theorem 12.8.31). The Drinfeld paper cited above provides an explicit class field theory for arbitrary congruence function fields.

Independently, D. Hayes [62] applied one rank Drinfeld modules in order to develop explicit class fields for global fields in characteristic p. His method does not require the use of the scheme-theoretic machinery used by Drinfeld; instead, Hayes used methods which are similar to Deuring's complex multiplication theory of elliptic curves.

With these results of Drinfeld and Hayes, Hilbert's problem 12 becomes completely solved for the function field case. Note that there is no similar explicit class field theory for number fields except for $\mathbb{Q}$ and the imaginary quadratic extensions of $\mathbb{Q}$, which are in some way similar to one rank Drinfeld modules.

The case considered in Chapter 12 is a particular case of Drinfeld modules, namely, the *Carlitz module*. The study of this module, provides explicit class fields, namely, the cyclotomic function fields. Drinfeld modules are one-dimensional objects and their rank can be any positive integer. When such a module is of rank one, as we mentioned before, there are some analogies with number fields, whereas in rank two there are analogies with the theory of elliptic curves. Nothing analogous to the classical case is known for Drinfeld modules of rank larger than or equal to three.

In Section 13.5 we apply the theory of rank one Drinfeld modules over the analogue in characteristic p of the field of complex numbers and, as in Chapter 12, we find explicit class fields over an arbitrary congruence function field K and we give explicitly the maximal abelian extension of K.

We follow very closely the seminal paper of D. Hayes [63]. Other important sources are the class field paper of Hayes [62] and the books of Goss [51] and of Thakur [151].

13.2 Additive Polynomials and the Carlitz Module

Our first goal in this section will be to define an exponential function in characteristic $p > 0$. Assume $R_T = \mathbb{F}_q[T]$ and $K = \mathbb{F}_q(T)$. The usual power series for $e^z = \sum_{n=0}^{\infty} \frac{z^n}{n!}$ does not make sense in positive characteristic, in which case we do not even know what e means. Recall that the classical exponential function e^z is multiplicative, that is, $e^{z+w} = e^z e^w$ for all $z, w \in \mathbb{C}$. Consider a multiplicative function f in characteristic p, that is, $f(x+y) = f(x)f(y)$. Assume that f is defined on some integral domain of characteristic p. Then $f(x)^p = f(px) = f(0) = f(0+0) = f(0)^2$. Therefore $f(0)$ is 0 or 1, and $f(x)$ is identically 0 or 1.

On the other hand, there exist several additive functions in characteristic p; indeed, any polynomial of the form $f(x) = \sum_{i=0}^{n} a_i x^{p^i}$ is additive: $f(x+y) = f(x) + f(y)$. Moreover, in the zero characteristic case, any additive function f satisfies $f(x) = cx$ for some constant c.

Now let $\mathbb{C}_\infty$ be the completion of an algebraic closure of $K_{\mathfrak{p}_\infty}$, where $\mathfrak{p}_\infty$ is the pole divisor of T. We want to define an *additive* exponential $\operatorname{ex}\colon \mathbb{C}_\infty \to \mathbb{C}_\infty$.

In the classical case we have $e^z = 1$ if and only if z is of the form $2n\pi i$ with $n \in \mathbb{Z}$, and such elements are zeros of multiplicity one of the equation $e^z - 1 = 0$. Therefore the analogous situation in positive characteristic would be a function $\operatorname{ex}(u)$ satisfying $\operatorname{ex}(u) = 0$ if and only if $u = \tilde{\pi} M$ with $M \in R_T$ ($u \in \mathbb{C}_\infty$) for some $\tilde{\pi} \in \mathbb{C}_\infty$, which would be similar to the classical $2\pi i$.

Considering $\operatorname{ex}(u)$ as an infinite product, we obtain

$$\operatorname{ex}(u) = cu \prod_{\lambda \in \tilde{\pi} R_T \setminus \{0\}} \left(1 - \frac{u}{\lambda}\right). \tag{13.1}$$

We normalize (13.1) by taking $c = 1$. Observe that since the zeros of $\operatorname{ex}(Tu)$ and $\prod_{\lambda \in (\tilde{\pi}/T) R_T / \tilde{\pi} R_T} (\operatorname{ex}(u) - \operatorname{ex}(\lambda))$ are the same, it follows that

$$\operatorname{ex}(Tu) = \alpha \prod_{\lambda \in (\tilde{\pi}/T) R_T / \tilde{\pi} R_T} (\operatorname{ex}(u) - \operatorname{ex}(\lambda)) \tag{13.2}$$

for some $\alpha \neq 0$. We normalize (13.2) by taking $\alpha = 1$. It follows that $\operatorname{ex}(Tu)$ is an $\mathbb{F}_q$-linear polynomial in $\operatorname{ex}(u)$ of degree $q = |R_T/TR_T| = |(\tilde{\pi}/T)R_T/\tilde{\pi} R_T|$.

From Corollary 13.2.5 below, we obtain

$$\operatorname{ex}(Tu) = \operatorname{ex}(u)^q + T\operatorname{ex}(u). \tag{13.3}$$

Note that (13.3) corresponds to the action given in Definition 12.2.1:

$$u^T = (\varphi + \mu_T)(u) = \varphi(u) + \mu_T(u) = u^q + Tu.$$

For this reason (13.3) is called the *Carlitz exponential* and $u^T = u^q + Tu$ defines the *Carlitz module*. It is clear that if $M \in R_T$, then $\mathrm{ex}(Mu)$ is a polynomial $C_M(\mathrm{ex}(u))$ in $\mathrm{ex}(u)$. In fact, it can be shown that $\mathrm{ex}(u) = C_M(\mathrm{ex}(u/M)) = \lim_{\deg M \to \infty} C_M(u/M)$, which is the analogue of $e^z = \lim_{n\to\infty}\left(1 + \frac{z}{n}\right)^n$ in the classical case.

Definition 13.2.1. Let E be a field of characteristic $p > 0$ and $p(x) \in E[x]$. We say that $p(x)$ is *additive* if $p(x+y) = p(x) + p(y)$ whenever $x + y \in E[x, y]$ in the polynomial ring of two variables.

Example 13.2.2. Let $\tau_p(x) = x^p$. Then τ_p is additive.

Example 13.2.3. If $p(x)$ and $h(x)$ are additive, then $p(x) + h(x)$, $\alpha p(x)$ and $p(h(x))$ are additive for any $\alpha \in E$.

Proposition 13.2.4. *If $p(x) \in E[x]$ is an additive polynomial, then $p(x) = \sum_{i=0}^n a_i x^{p^i}$ for some $a_0, \dots, a_n$.*

Proof. Consider the equality $p(x+y) = p(x) + p(y)$ and take the formal derivative of both sides with respect to x. We obtain $p'(x+y) = p'(x)$ and $p(0) = 0$. It follows that $p'(y) = p'(0) = c_0 \in E$. Thus

$$p(x) = c_0 x + \sum_{j=1}^m c_j x^{pj} = c_0 x + p_1(x)^p,$$

where $p_1(x) \in E[c_1^{1/p}, \dots, c_m^{1/p}][x]$ and $p_1(x)$ is additive. By induction on $\deg p(x)$, we obtain $p_1(x) = \sum_{t=0}^{m_1} b_t x^{p^t}$. Thus $p(x) = c_0 x + \sum_{t=0}^{m_1} b_t^p x^{p^{t+1}} \in E[x]$. □

Corollary 13.2.5. *If E contains $\mathbb{F}_q$ ($q = p^u$) and $p(x) \in E[x]$ is $\mathbb{F}_q$-linear, that is, $p(x)$ is additive and satisfies $p(\alpha x) = \alpha p(x)$ with $\alpha \in \mathbb{F}_q$, then $p(x)$ is of the form $p(x) = \sum_{i=0}^n a_i x^{q^i}$.*

Proof. Since $p(x) = \sum_{j=0}^m b_j x^{p^j}$ and $p(\alpha x) = \alpha p(x)$ for all $\alpha \in \mathbb{F}_q$, we have $\sum_{j=0}^m b_j \alpha^{p^j} x^{p^j} = \sum_{j=0}^m b_j \alpha x^{p^j}$. Thus $\alpha^{p^j} = \alpha$ for every j such that $b_j \neq 0$, and the result follows. □

It follows from Example 13.2.3 and Corollary 13.2.5 that the set of $\mathbb{F}_q$-linear maps in $E[x]$, where $\mathbb{F}_q$ is contained in E, forms a ring under composition.

Definition 13.2.6. Let E be a field containing $\mathbb{F}_q$, and let R be the ring of $\mathbb{F}_q$-linear polynomials in $E[x]$. Set $\tau(x) = x^q$. Then $R \cong E\langle\tau\rangle$, where $E\langle\tau\rangle$ is the *twisted polynomial ring* consisting of the $\mathbb{F}_q$-algebra generated by E and the element τ such that

$$\tau u = u^q \tau \tag{13.4}$$

for all $u \in E$.

In other words, $E\langle\tau\rangle$ is similar to a polynomial ring except that the multiplication of τ by elements of E is given by (13.4).

Definition 13.2.7. Let $R_T = \mathbb{F}_q[T]$ and $K = \mathbb{F}_q(T)$ as usual. The *Carlitz module* for R_T defined over K is the $\mathbb{F}_q$-algebra homomorphism

$$\begin{aligned} C\colon R_T &\to K\langle\tau\rangle \\ M &\mapsto C_M \end{aligned}$$

such that $C_T = T + \tau$.

Note that Definition 13.2.7 is the same as Definition 12.2.1. Also, Definition 13.2.7 provides an $\mathbb{F}_q$-algebra homomorphism such that the constant term of C_M is M and there exists $M \in R_T$, for instance $M = T$, such that $C_M \notin K$.

13.3 Characteristic, Rank, and Height of Drinfeld Modules

In this section we generalize the definition of a Carlitz module, which is the simplest example of a *Drinfeld module*.

Let K be a congruence function field whose exact field of constants is the finite field $\mathbb{F}_q$ of q elements. We fix a prime divisor $\mathfrak{P}_\infty$ of K, which will be called the *infinite prime*. Let $A \subset K$ be the ring of elements in K whose only poles are at $\mathfrak{P}_\infty$. That is, $A = \bigcup_{t=1}^{\infty} L(\mathfrak{P}_\infty^{-t})$.

Now, A is the integral closure of some $\mathbb{F}_q[T]$ with $T \in K$ (choose T such that $\mathfrak{N}_T = \mathfrak{P}_\infty^t$ for some positive t). By Theorems 5.7.7 and 5.7.9, A is a Dedekind domain whose prime ideals other than zero are in one-to-one correspondence with the prime divisors of K other than $\mathfrak{P}_\infty$: if $\mathfrak{P}$ is a prime divisor, distinct from $\mathfrak{P}_\infty$, then $A \subset \vartheta_{\mathfrak{P}}$ and $\mathfrak{P} \cap A$ is the corresponding nonzero prime ideal of A.

We set $d_\infty = \deg_K \mathfrak{P}_\infty \geq 1$. Let k be any field containing $\mathbb{F}_q$ and consider the twisted polynomial ring $k\langle\tau\rangle$, where $\tau(u) = u^q$, $u \in \Omega$, and Ω is any k-algebra. The action of $k\langle\tau\rangle$ on Ω is given by

$$\Big(\sum_{i=0}^{n} a_i\tau^i\Big)(u) = \sum_{i=0}^{n} a_i u^{q^i} \in k[u],$$

where $\sum_{i=0}^{n} a_i u^{q^i}$ is an additive polynomial. Let $D\colon k\langle\tau\rangle \to k$ be the *augmentation homomorphism*, that is, $D\Big(\sum_{i=0}^{n} a_i\tau^i\Big) = a_0$.

Definition 13.3.1. Let $\iota\colon k \to k\langle\tau\rangle$ be the inclusion map defined by $\iota(\alpha) = \alpha (= \alpha\tau^0)$. A *Drinfeld module over* k is a homomorphism $\rho\colon A \to k\langle\tau\rangle$ of $\mathbb{F}_q$-algebras such that $\rho \neq \iota \circ D \circ \rho$. We denote $\rho(a)$ by ρ_a.

Remark 13.3.2. $\delta := D\rho\colon A \to k$ is a homomorphism of $\mathbb{F}_q$-algebras. We say that k *is an A-field*. Also, notice that $D(\rho_a) = \delta(a)$ for $a \in A$. The condition $\rho \neq \iota \circ D \circ \rho = \iota \circ \delta$ means that ρ does not factor through k via ι.

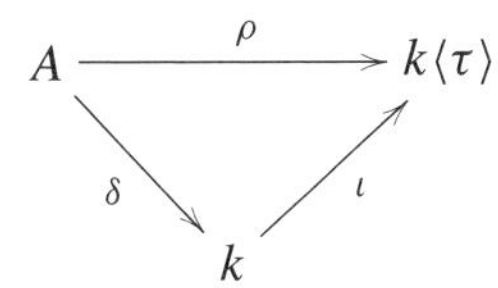

Alternatively, if we fix a homomorphism $\delta\colon A \to k$, a Drinfeld A-module over k is a homomorphism $\rho\colon A \to k\langle\tau\rangle$ of $\mathbb{F}_q$-algebras such that $D \circ \rho = \delta$ and $\rho_a \neq \delta(a)\tau^0$ for some $a \in A$.

Example 13.3.3. Assume that $A = R_T$, k is any field containing A, and $\delta\colon A \to k$ is any $\mathbb{F}_q$-algebra homomorphism. Let n be an integer greater than or equal than one and a_n a nonzero element of k. Let $\rho_T = \delta(T) + \sum_{i=1}^{n} a_i\tau^i$ be arbitrary with $a_n \neq 0$, $n \geq 1$. Then ρ can be extended in a unique way to a homomorphism $\rho\colon A \to k\langle\tau\rangle$, and ρ is a Drinfeld A-module.

Definition 13.3.4. The kernel $\mathfrak{P}$ of the map $\delta\colon A \to k$ is called the *characteristic of* ρ. If $\mathfrak{P} = (0)$, we say that ρ has *generic characteristic* or *infinite characteristic* in order to avoid confusion with the usual 0 characteristic. If $\mathfrak{P} \neq (0)$ we say that ρ has *finite characteristic*. We denote the characteristic of a Drinfeld A-module by $\operatorname{char}(\rho)$.

Proposition 13.3.5. *The map ρ given above is injective.*

Proof. Exercise 13.7.1. □

Note that Drinfeld A-modules are essentially nontrivial embeddings of A into $k\langle\tau\rangle$.

Example 13.3.6. Assume $A = R_T$, $k = K = \mathbb{F}_q(T)$, $\delta = \mathrm{id}\colon R_T \xrightarrow{\mathrm{id}} K$, $\rho = C\colon R_T \to K\langle\tau\rangle$, $\rho(M) = \rho_M = C_M$, and $C_T = T + \tau$ is the Carlitz module. Then C is a Drinfeld module.

Notation 13.3.7. We use the notation $\operatorname{Drin}_A(k)$ for the set of all Drinfeld A-modules over k once the map $\delta = D\rho\colon A \to k$ has been fixed.

In practice δ is either an inclusion or a reduction map module over some nonzero prime ideal of A.

Now, given any k-algebra V, A acts con V via δ if we define

$$a \circ v = \delta(a)v \quad \text{for all} \quad v \in V \quad \text{and} \quad a \in A. \tag{13.5}$$

In this way V is an A-module. However, if we consider ρ, V is also an A-module under the operation defined by

$$a * v = \rho_a(v) \quad \text{for all} \quad v \in V \quad \text{and} \quad a \in A. \tag{13.6}$$

The linear term of $a * v$ is $\delta(a)v = a \circ v$ but by the definition of a Drinfeld module, there exists $a \in A$ such that $a * v \neq a \circ v$. Thus, the idea of a Drinfeld module may be understood as the deformation of a standard A-module. The A-module structure of V given by (13.6) will be denoted by V_ρ.

Definition 13.3.8. Given two Drinfeld A-modules ρ, ρ' over k, an *isogeny* from ρ to ρ' is a twisted polynomial $f \in k\langle\tau\rangle$ such that $f\rho_a = \rho'_a f$ for all $a \in A$.

The product of two isogenies is easily seen to be again an isogeny. In this way, using the language of categories, we may say that $\mathrm{Drin}_A(k)$ is a category whose morphisms are the isogenies. The isogenies from ρ to ρ' will be denoted by $\mathrm{Isog}(\rho, \rho')$.

In particular, the isomorphisms in $\mathrm{Drin}_A(k)$ are the invertible twisted polynomials in $k\langle\tau\rangle$ and these polynomials are precisely the nonzero constant polynomials k^*. Therefore ρ and ρ' are isomorphic if and only if there exists an element $\alpha \in k^*$ such that $\alpha\rho_a = \rho'_a\alpha$ for all $a \in A$.

Example 13.3.9. Assume as usual $A = R_T$ and $K = \mathbb{F}_q(T)$, and consider the following two Drinfeld A-modules, where $\delta\colon A \to K$ is the inclusion map:

$$\begin{aligned} \rho := C\colon A &\to K\langle\tau\rangle \\ T &\mapsto C_T = T + \tau \end{aligned} \qquad \text{and} \qquad \begin{aligned} \rho' := C'\colon A &\to K\langle\tau\rangle \\ T &\mapsto T - \tau. \end{aligned}$$

Then C is the Carlitz module. Now ρ and ρ' are isomorphic if and only if there exists $\alpha \in k^*$ such that $\alpha C_M = C'_M\alpha$ for all $M \in R_T$. It is easily seen that α must be a $(q-1)$th root of -1 (see Exercise 13.7.2).

For $p \neq 2$, K does not contain any $(q-1)$th root of -1 and therefore C and C' are not isomorphic over K. However, they are isomorphic over any overfield of K that contains the $(q-1)$th roots of -1.

More generally, if $\alpha_1 \in A$ and k is a field containing $\mathbb{F}_q\big(T, \alpha_1^{1/(q-1)}\big)$ then the module $\rho_T = T + \alpha_1\tau$ is isomorphic to the Carlitz module over K.

Now we will define the rank of a Drinfeld A-module $\rho \in \mathrm{Drin}_A(k)$. Let $\phi\colon A \to \mathbb{Z}$ be defined by $\phi(a) := -\deg \rho_a$ (in τ). Then ϕ is a nontrivial valuation on A (Exercise 13.7.3).

Now the unique extension of ϕ to $K = \mathrm{quot}\, A$ defines the prime divisor $\mathfrak{P}_\infty$. Therefore there exists a unique rational number r_ρ such that

$$\deg \rho_a = -d_\infty \times r_\rho \times v_{\mathfrak{P}_\infty}(a) \tag{13.7}$$

for all $a \in A$.

Definition 13.3.10. The number r_ρ is called the *rank* of ρ.

We will see (Theorem 13.3.19) that r_ρ is a positive integer.

Example 13.3.11. Assume that C is the Carlitz module. Then $d_\infty = 1$ and $\deg C_T = 1 = -d_\infty \times r_C \times v_{\mathfrak{P}_\infty}(T) = -1 \times r_C \times (-1) = r_C$. Therefore the Carlitz module is of rank one.

Now we define another number attached to a Drinfeld A-module ρ. If $\mathrm{char}(\rho) = 0$, define the *height* of ρ by $h_\rho = 0$.

Assume that $\mathfrak{P} = \mathrm{char}(\rho) \neq 0$ and let $v_{\mathfrak{P}}$ be the valuation associated to the place $\mathfrak{P}$. Let a be a nonzero element of A, and $\rho_a = \sum_{i=0}^n \alpha_i\tau^i$. Pick i_0 such that $\alpha_{i_0} \neq 0$ and $\alpha_j = 0$ whenever $0 \leq j \leq i_0 - 1$. We define

$$j_\rho(a) = \operatorname{ord}(\rho_a) = i_0.$$

Note that $\operatorname{ord}(\rho_a) > 0$ if and only if $a \in \mathfrak{P}$. Furthermore, j_ρ defines a nontrivial valuation on A (Exercise 13.7.3) that is equivalent to $v_{\mathfrak{P}}$.

Hence, there exists a positive rational number h_ρ such that

$$j_\rho(a) = \operatorname{ord}(\rho_a) = h_\rho \times v_{\mathfrak{P}}(a) \times \deg_K \mathfrak{P} \tag{13.8}$$

for all $a \in A$.

Definition 13.3.12. The number h_ρ defined above is called the *height* of the Drinfeld A-module ρ.

We will prove (Theorem 13.3.19) that h_ρ is a nonnegative integer.

Example 13.3.13. If C is the Carlitz module, the structural map δ is injective, so the height of C is $h_C = 0$.

Example 13.3.14. Let $A = R_T$, k be any field containing $\mathbb{F}_q$, and $\rho\colon A \to k\langle\tau\rangle$ a Drinfeld module of rank r and height h. Then $\rho_T = \sum_{i=0}^{r} \alpha_i \tau^i$ with $\alpha_0, \ldots, \alpha_r \in k$ and $\alpha_r \neq 0$, since

$$\deg \rho_T = -d_{\mathfrak{P}_\infty} \times r_\rho \times v_{\mathfrak{P}_\infty}(T) = r_\rho = r.$$

Now $\delta(T) = \alpha_0$, so $\delta(f(T)) = f(\delta(T)) = f(\alpha_0)$. Therefore

$$\operatorname{char}(\rho) = \begin{cases} (0) & \text{if } \alpha_0 \text{ is transcendental over } \mathbb{F}_q \\ (\operatorname{Irr}(\alpha_0, T, \mathbb{F}_q)) & \text{if } \alpha_0 \text{ is algebraic over } \mathbb{F}_q. \end{cases}$$

If $\mathfrak{P}$ is any nonzero prime ideal of A, $k = A/\mathfrak{P}$, and δ is the canonical projection, then $\operatorname{char}(\rho) = \mathfrak{P}$.

In general, assume that $\operatorname{char}(\rho) = \mathfrak{P} \neq (0)$. Then if $\alpha_0 = 0$, we have $\mathfrak{P} = (T)$ and $\operatorname{ord}(\rho_T) = i_0 = h_\rho \times v_{\mathfrak{P}}(T) \times \deg_K(T) = h_\rho \times 1 \times 1 = h_\rho = h$.

Therefore if $k = \mathbb{F}_q = A/(T)$, $\delta\colon A \to k$ is the canonical projection and $\rho\colon A \to k\langle\tau\rangle$, then $\rho_T = \tau^h + \tau^r$ is a Drinfeld A-module of height h and rank r. Note that if $\alpha_0 \neq 0$, α_0 algebraic over $\mathbb{F}_q$, and $\mathfrak{P} = (\operatorname{Irr}(\alpha_0, T, \mathbb{F}_q)) = (f(T))$, then

$$\operatorname{ord}(\rho_{f(T)}) = h_\rho \times v_{\mathfrak{P}}(f(T)) \times \deg_K(f(T)) = h_\rho \times \deg_T f(T) = h \times \deg_T f(T),$$

that is, $h = \frac{\operatorname{ord}(\rho_{f(T)})}{\deg_T f(T)}$.

In order to show that the rank r_ρ and the height h_ρ of a Drinfeld A-module are integers, we need the basic general results on finitely generated modules over Dedekind domains. The structure of these modules is similar to that of finitely generated modules over principal ideal domains.

Let ρ be a Drinfeld A-module over k. If $\mathfrak{A}$ is an integral ideal of A, then $\mathfrak{A}$ can be generated by at most two elements (Exercise 13.7.4). Let k be any field containing $\mathbb{F}_q$. Given any two twisted polynomials $f(\tau), g(\tau) \in k\langle\tau\rangle$ with $g(\tau) \neq 0$ there exists a unique pair of twisted polynomials $q(\tau)$ (the *right quotient*) and $r(\tau)$ (the *right residue*) such that $\deg r(\tau) < \deg g(\tau)$ and

$$f(\tau) = q(\tau)g(\tau) + r(\tau). \tag{13.9}$$

The proof of (13.9) is similar to that in the case of a polynomial ring $k[x]$. As in that case, we deduce that every left ideal of $k\langle\tau\rangle$ is principal. Now if we assume that k is a perfect field, we obtain the left analogue of (13.9), namely, if $f(\tau), g(\tau) \in k\langle\tau\rangle$ and $g(\tau) \neq 0$, then there exists a unique pair $q_1(\tau)$ and $r_1(\tau)$, consisting of the *left quotient* and the *left residue*, such that $\deg r_1(\tau) < \deg g(\tau)$ and

$$f(\tau) = g(\tau)q_1(\tau) + r_1(\tau). \tag{13.10}$$

Again, the proof is similar to the polynomial ring case but here we need the fact that $k^q = k$. As a consequence we obtain that when k is perfect, every right ideal of $k\langle\tau\rangle$ is principal.

Example 13.3.15. If $k = \mathbb{F}_q(T)$, $f(\tau) = T + \tau - \tau^2$, and $g(\tau) = \tau + T^2$, then using the same Euclidean algorithm and the relation (13.4) ($\tau a = a^q\tau$ for $a \in k$), we obtain

$$-\tau^2 + \tau + T = (-\tau + (1 + T^2)^q)(\tau + T^2) + (T - (1 + T^2)^q T^2).$$

Therefore $q(\tau) = -\tau + (1 + T^2)^q$ and $r(\tau) = T - (1 + T^2)^q T^2 = T - T^2 - T^{2q+2}$.

In the algebraic closure $\bar{k}$ of k we have

$$-\tau^2 + \tau + T = (\tau + T^2)(-\tau + (1 + T^2)^{1/q}) + T + T^2(1 + T^2)^{1/q}.$$

Therefore $q_1(\tau) = -\tau + (1+T^2)^{1/q}$ and $r(\tau) = T + T^2(1+T^2)^{1/q} = T + T^2 + T^{2+2/q}$.

Definition 13.3.16. Given $f(\tau), g(\tau) \in k\langle\tau\rangle$, the *right greatest common divisor* of $f(\tau)$ and $g(\tau)$ is the monic generator of the left ideal of $k\langle\tau\rangle$ generated by $f(\tau)$ and $g(\tau)$. We will denote it by $\operatorname{rgcd}(f(\tau), g(\tau))$.

If $h(\tau) = \operatorname{rgcd}(f(\tau), g(\tau))$, the left ideal of $k\langle\tau\rangle$ generated by $f(\tau)$ and $g(\tau)$ is $k\langle\tau\rangle h(\tau)$.

Example 13.3.17. We have $\operatorname{rgcd}(T + \tau - \tau^2, \tau + T^2) = 1$. In fact,

$$\operatorname{rgcd}(f(\tau), g(\tau)) = \gcd(f(x), g(x))_{x=\tau}$$

(see Exercise 13.7.5).

Now let $\rho \in \operatorname{Drin}_A(k)$, and let $\mathfrak{A} = (a, b)$ be an integral ideal of A. Let $\rho_a, \rho_b \in k\langle\tau\rangle$, and consider $\operatorname{rgcd}(\rho_a, \rho_b) := \rho_{\mathfrak{A}}$. That is, $\rho_{\mathfrak{A}}$ denotes the monic generator of the left ideal of $k\langle\tau\rangle$ generated by ρ_a and ρ_b.

Definition 13.3.18. Let $\bar{k}$ be an algebraic closure of k and $\rho \in \operatorname{Drin}_A(k)$. We define $\rho[\mathfrak{A}] \subseteq \bar{k}$ as the set of roots of $\rho_{\mathfrak{A}}$ in $\bar{k}$. Note that

$$\rho[\mathfrak{A}] = \left\{u \in \bar{k} \mid \rho_a(u) = 0 \;\forall\, a \in \mathfrak{A}\right\}$$

(this corresponds to the Λ_M given in Definition 12.2.8).

Note that $\rho[\mathfrak{A}]$ is a finite additive subgroup of $\bar{k}$ since ρ_a is an additive polynomial. Furthermore, if $a \in A$ and $u \in \rho[\mathfrak{A}]$, let $\rho_a(u) \in \bar{k}^*$. Then for $f \in \mathfrak{A}$, we have $\rho_f \circ \rho_a(u) = \rho_{fa}(u) = \rho_{af}(u) = \rho_a \circ \rho_f(u) = \rho_a(0) = 0$.

In other words, $\rho[\mathfrak{A}]$ is also a finite A-module under the action given in (13.6). This is the natural generalization of Proposition 12.3.6.

Theorem 13.3.19. *Let $\rho \in \mathrm{Drin}_A(k)$ be any Drinfeld A-module of rank r and height h. Then r is a positive integer and h is a nonnegative integer.*

Proof. Let $\rho_{\mathfrak{A}} = a_{i_0}\tau^{i_0} + \cdots + a_n\tau^n$, where $i_0 := \operatorname{ord}\rho_{\mathfrak{A}}$, $a_{i_0}, a_n \neq 0$, and $n = \deg \rho_{\mathfrak{A}}$. Then $|\rho[\mathfrak{A}]| = q^{n-i_0} = q^{\deg\rho_{\mathfrak{A}} - \operatorname{ord}\rho_{\mathfrak{A}}}$. Let $\mathfrak{P}$ be any nonzero prime ideal of A. The sequence

$$0 \to \rho[\mathfrak{P}] \to \rho[\mathfrak{P}^m] \xrightarrow{\varphi} \rho[\mathfrak{P}^{m-1}] \to 0, \tag{13.11}$$

where $\varphi\colon \rho[\mathfrak{P}^m] \to \rho[\mathfrak{P}^{m-1}]$ is defined by $\varphi(u) = \pi \times u$ with $\pi \in \mathfrak{P} \setminus \mathfrak{P}^2$, is exact (this result corresponds to the proof of Proposition 12.2.14). Now $\rho[\mathfrak{P}]$ is an A-module that is annihilated by $\mathfrak{P}$. Thus $\rho[\mathfrak{P}]$ is a finite $A/\mathfrak{P}$ vector space, say of dimension $d_{\mathfrak{P}}$. From (13.11) we obtain that

$$|\rho[\mathfrak{P}^m]| = |\rho[\mathfrak{P}^{m-1}]||\rho[\mathfrak{P}]| = |A/\mathfrak{P}|^{md_{\mathfrak{P}}} = q^{md_{\mathfrak{P}}\deg_K \mathfrak{P}}.$$

By Exercise 13.7.6, there exists $m \in \mathbb{N}$ such that $\mathfrak{P}^m = (a)$ is principal. Therefore $\rho[a] := \rho[(a)] = \rho[\mathfrak{P}^m]$. Since $|\rho[a]| = q^{\deg\rho_a - \operatorname{ord}\rho_a}$, it follows that

$$md_{\mathfrak{P}} \deg_K \mathfrak{P} = \deg \rho_a - \operatorname{ord}\rho_a, \tag{13.12}$$

where $\mathfrak{P}^m = (a)$ and $d_{\mathfrak{P}} = \dim_{A/\mathfrak{P}} \rho[\mathfrak{P}]$.

If $\mathfrak{P} \neq \operatorname{char}(\rho)$, then ρ_a is separable since $\delta(a) \neq 0$. Hence $|\rho[a]| = q^{\deg\rho_a}$ and we obtain

$$m \times d_{\mathfrak{P}} \times \deg_K \mathfrak{P} = \deg\rho_a = -d_\infty \times r \times v_{\mathfrak{P}_\infty}(a).$$

Now in K, we have $(a)_K = \frac{\mathfrak{P}^m}{\mathfrak{P}_\infty^{-v_{\mathfrak{P}_\infty}(a)}}$. Thus $m \deg_K \mathfrak{P} = -v_{\mathfrak{P}_\infty}(a) d_\infty$, and therefore $m \times d_{\mathfrak{P}} \times \deg_K \mathfrak{P} = r \times m \times \deg_K \mathfrak{P}$. It follows that $r = d_{\mathfrak{P}} = \dim_{A/\mathfrak{P}} \rho[\mathfrak{P}] \in \mathbb{N}$.

In the case that ρ is of generic characteristic, we have $h_\rho = 0$. Otherwise, assume $\mathfrak{P} = \operatorname{char}(\rho)$ and that ρ is of finite characteristic. Then $\operatorname{ord}\rho_a = j_\rho(a) = hv_{\mathfrak{P}}(a)\deg_K \mathfrak{P} = hm \deg_K \mathfrak{P}$.

From (13.12) we obtain

$$md_{\mathfrak{P}} \deg_K \mathfrak{P} = rm \deg_K \mathfrak{P} - hm \deg_K \mathfrak{P}.$$

Therefore $d_{\mathfrak{P}} = r - h$, and h is an integer. ◻

Remark 13.3.20. We have obtained that

$$\rho[\mathfrak{P}] \cong \begin{cases} (A/\mathfrak{P})^r & \text{if } \mathfrak{P} \neq \operatorname{char}(\rho), \\ (A/\mathfrak{P})^{r-h} & \text{if } \mathfrak{P} = \operatorname{char}(\rho). \end{cases}$$

Using the theory of Dedekind domains, it can be proved that for any $m \geq 1$,

$$\rho[\mathfrak{P}^m] \cong \begin{cases} (A/\mathfrak{P}^m)^r & \text{if } \mathfrak{P} \neq \operatorname{char}(\rho), \\ (A/\mathfrak{P}^m)^{r-h} & \text{if } \mathfrak{P} = \operatorname{char}(\rho). \end{cases}$$

For the reader who is familiar with the torsion of the Jacobian (which is isomorphic to $C_{K,0}$) of a function field over an algebraically closed field, we observe that if $p = \operatorname{char} k$ and ℓ is any prime number, then

$$C_{K,0}(\ell) \cong \begin{cases} R^{2g_K} & \text{if } \ell \neq p, \\ R^{\lambda_K} & \text{if } \ell = p, \end{cases}$$

where $R = \mathbb{Q}_\ell/\mathbb{Z}_\ell$, g_K is the genus of K, and λ_K is the *Hasse–Witt invariant* of K. Thus the rank r of a Drinfeld module is the analogue of $2g_K$ and $r-h$ is the analogue of λ_K.

13.4 Existence of Drinfeld Modules. Lattices

As we saw in Example 13.3.3, if $A = \mathbb{F}_q(T)$ and $\delta\colon A \to k$ is any $\mathbb{F}_q$-algebra homomorphism, then any assignment $T \mapsto f(\tau) \in k\langle\tau\rangle \setminus k$, where $f(0) = \delta(T)$, can be extended to a Drinfeld A-module.

Now we want to construct Drinfeld A-modules for a general A. The method for achieving this goal is due to Drinfeld, and we will follow D. Hayes's papers [62, 63] and M. Rosen's book [128] to present the construction.

The idea is to make an analogous construction to the one used in the classical case of elliptic curves over $\mathbb{C}$. More precisely, consider a lattice $\Gamma = \mathbb{Z}\omega_1 + \mathbb{Z}\omega_2$ in $\mathbb{C}$, with $\operatorname{im}\frac{\omega_1}{\omega_2} > 0$. Then $\mathbb{C}/\Gamma$ corresponds to an elliptic curve over $\mathbb{C}$, and Γ and Γ' give $\mathbb{C}$-isomorphic elliptic curves if and only if there exists a nonzero complex number α such that $\alpha\Gamma = \Gamma'$.

Thus the procedure to obtain a Drinfeld A-module ρ over $\mathbb{C}_\infty$ is to define an *A-lattice* Γ in $\mathbb{C}_\infty$ and to find a Drinfeld A-module ρ^Γ attached to Γ. Finally, we will see that every Drinfeld A-module ρ over $\mathbb{C}_\infty$ is of the form ρ^Γ for some lattice Γ.

Let $K_\infty = K_{\mathfrak{P}_\infty} \cong \mathbb{F}_{q^{d_\infty}}[[\pi]]$ be the completion of K at $\mathfrak{P}_\infty$ and let π be a uniformizer at $\mathfrak{P}_\infty$. Let $\bar{K}_\infty$ be an algebraic closure of K_∞. Then $\bar{K}_\infty$ is not a complete field, but its completion $\mathbb{C}_\infty$ is algebraically closed. We consider $\mathbb{C}_\infty$ as the function field analogue of $\mathbb{C}$. The analytic theory of power series and infinite products can be developed similarly to the way it is done in $\mathbb{C}$ (see [51, Chapter 2]).

Let $\delta\colon A \to k$ be any $\mathbb{F}_q$-algebra monomorphism. By abuse of language we also use the notation δ for the extension $\delta\colon K \to k$. Let $k\langle\langle\tau\rangle\rangle$ by the ring of *left twisted power series* generated over k by τ. Thus the relation (13.4), $\tau\alpha = \alpha^q\tau$, holds for all $\alpha \in k$. Finally, let D be the *derivative at* 0 or the *augmentation homomorphism* $D\colon k\langle\langle\tau\rangle\rangle \to k$, defined by $D(f(\tau)) := f(0)$.

Definition 13.4.1. Any ring homomorphism $\rho\colon K \to k\langle\langle\tau\rangle\rangle$ such that $D \circ \rho = \delta$ is called a *formal K-module over k*. We assume that ρ is nontrivial, or in other words, that $\rho(K)$ is not contained in k.

Note that if $\rho \in \mathrm{Drin}_A(k)$ satisfies $\mathrm{char}(\rho) = 0$, then ρ_a is invertible in $k\langle\langle\tau\rangle\rangle$ for each $a \in A\setminus\{0\}$. The proof is the same as in the case of formal power series. Therefore ρ extends to a nontrivial K-module. This extension is also called ρ.

For the rest of this section, $\delta\colon A \to \mathbb{C}_\infty$ will denote the inclusion map. As we mentioned before, the exponential map $e^z\colon \mathbb{C} \to \mathbb{C}$ is a fundamental entire function on $\mathbb{C}$. The exponential functions associated to lattices in $\mathbb{C}_\infty$ have turned out to be an important source for the construction of Drinfeld A-modules of arbitrary rank. Our main goal in this section is to sketch a proof of one of the fundamental results in the analytic theory of Drinfeld modules. The result is the *analytic uniformization theorem* for Drinfeld A-modules over $\mathbb{C}_\infty$.

Theorem 13.4.2 (Analytic Uniformization Theorem). *Let ρ be a Drinfeld A-module over $\mathbb{C}_\infty$. There exists a unique lattice Γ in $\mathbb{C}_\infty$ such that $\rho = \rho^\Gamma$.*

We will see the meaning of ρ^Γ soon.

Definition 13.4.3. A *lattice* Γ is a discrete finitely generated A-submodule of $\mathbb{C}_\infty$.

In other words, Γ is discrete in the topology of $\mathbb{C}_\infty$ and the action of A on Γ is multiplication in $\mathbb{C}_\infty$.

Definition 13.4.4. If Γ is a lattice, then the dimension over K_∞ of the K_∞ vector space $K_\infty\Gamma$ is called the rank of Γ and will be denoted by $r_\Gamma := \dim_{K_\infty} K_\infty\Gamma$.

Example 13.4.5. Let $\{\alpha_1, \ldots, \alpha_r\} \subseteq \mathbb{C}_\infty$ be linearly independent over K_∞ (we have $[\mathbb{C}_\infty : K_\infty] = \infty$, so r can be chosen arbitrarily). Pick any nonzero elements $a_1, \ldots, a_r$ in A. Then

$$\Gamma := A\frac{\alpha_1}{a_1} + \cdots + A\frac{\alpha_r}{a_r}$$

is a lattice of rank $r_\Gamma = r$.

The following result is the nonarchimedean analogue of the Weierstrass factorization theorem.

Theorem 13.4.6. *Let $f\colon \mathbb{C}_\infty \to \mathbb{C}_\infty$ be an entire function, that is, a function that can be represented as a power series $f(u) = \sum_{n=0}^{\infty} a_n u^n \in \mathbb{C}_\infty[[u]]$ that is convergent everywhere. Let $\{\lambda\}_{\lambda\in I}$ be the nonzero roots of f in $\mathbb{C}_\infty$, where each λ is of multiplicity m_λ. Then I is at most countable, $\{\lambda\}_{\lambda\in I} = \{\lambda_1, \ldots, \lambda_t, \ldots\}$, $\lim_{t\to\infty} v_{\mathfrak{P}_\infty}(\lambda_t) = -\infty$, and if n is the multiplicity of the zero of f at $z = 0$, we have*

$$f(u) = cu^n \prod_{t=1}^{\infty} \left(1 - \frac{u}{\lambda_t}\right)^{m_t} \tag{13.13}$$

for some constant $c \in \mathbb{C}_\infty$ and where $m_t = m_{\lambda_t}$. Conversely (13.13) defines an entire function on $\mathbb{C}_\infty$.

Proof. See Goss [51]. □

An entire function $f(u) = \sum_{n=0}^{\infty} a_n u^n$ that is $\mathbb{F}_q$-linear satisfies that if $q \nmid n$ then $a_n = 0$. Thus f must be of the form $f(u) = \sum_{n=0}^{\infty} b_{nq} \alpha^{nq}$. We define the derivative at 0 by $Df = b_0 (= f(0))$.

As a corollary of Theorem 13.4.6, we obtain the following:

Corollary 13.4.7. *Assume that $f_1(u)$, $f_2(u)$ are two $\mathbb{F}_q$-linear entire functions with $Df_1 = Df_2 \neq 0$, and $f_1(u)$ and $f_2(u)$ have the same set of roots with the same multiplicities. Then $f_1(u) = f_2(u)$.* □

In order to define the exponential function of a lattice, we need the following result, whose proof is left to the reader.

Proposition 13.4.8. *If Γ is a lattice, then $\sum_{\gamma \in \Gamma \setminus \{0\}} \frac{1}{\gamma}$ is absolutely convergent in $\mathbb{C}_\infty$.* □

Definition 13.4.9. Let Γ be a lattice. We define the *exponential function associated to* Γ as the entire function defined by

$$e_\Gamma(u) = u \prod_{\gamma \in \Gamma \setminus \{0\}} \left(1 - \frac{u}{\gamma}\right). \tag{13.14}$$

The usual exponential function is multiplicative; indeed, we have $e^{z_1 + z_2} = e^{z_1} e^{z_2}$ for all $z_1, z_2 \in \mathbb{C}$. As expected, we have the following result:

Proposition 13.4.10. *The exponential function e_Γ associated to any lattice Γ is $\mathbb{F}_q$-linear, that is,*

$$e_\Gamma(\alpha u + \beta w) = \alpha e_\Gamma(u) + \beta e_\Gamma(w)$$

for all $\alpha, \beta \in \mathbb{F}_q$ and $u, w \in \mathbb{C}_\infty$.

Proof. Let N be a positive integer and let $\Gamma_N := \{\lambda \in \Gamma \mid |\lambda|_{\mathfrak{P}_\infty} \leq N\}$. Since $\lim_{\gamma \in \Gamma} |\gamma|_\infty = \infty$, Γ_N is finite and clearly it is an $\mathbb{F}_q$-linear space.

Let $p_N(u) = u \prod_{\gamma \in \Gamma_N} \left(1 - \frac{u}{\gamma}\right) \in \mathbb{C}_\infty[u]$. Since $\lim_{N \to \infty} p_N(u) = e_\Gamma(u)$, it suffices to show that $p_N(u)$ is $\mathbb{F}_q$-linear. More generally, if V is a finite $\mathbb{F}_q$-linear space and we define $f_V(u) = A \prod_{v \in V} (u - v)$ for a constant $A \in \mathbb{C}_\infty$, then $f_V(u)$ is an $\mathbb{F}_q$-linear polynomial.

We will prove the latter statement by induction on $\dim_{\mathbb{F}_q} V = n$. For $n = 0$, we have $f_V(u) = Au$ and the result follows. Assume $n \geq 1$ and let W be an $(n-1)$-dimensional subspace of V. Then for $v_0 \in V \setminus W$, we have $V = W + \mathbb{F}_q v_0$.

Therefore $f_V(u) = A \prod_{\substack{w\in W\\ \mu\in\mathbb{F}_q}} (u - (w + \mu v_0))$. Let $f_W(u) = \prod_{w\in W} (u - w)$. Then $f_W(u)$ is $\mathbb{F}_q$-linear and

$$\begin{aligned} f_V(u) &= A \prod_{w\in W} (u - w) \times \prod_{\substack{w\in W\\ \mu\in\mathbb{F}_q^*}} ((u - \mu v_0) - w) \\ &= A f_W(u) \times \prod_{\mu\in\mathbb{F}_q^*} f_W(u - \mu v_0) \\ &= A f_W(u) \times \prod_{\mu\in\mathbb{F}_q^*} ((f_W(u)) - \mu f_W(v_0)) \\ &= A \times \prod_{\mu\in\mathbb{F}_q^*} (f_w(u) - \mu f_W(v_0)) \\ &= A[f_W(u)][f_W(u)^{q-1} - f_W(v_0)^{q-1}]. \end{aligned}$$

Thus $f_V(u)$ is $\mathbb{F}_q$-linear. $\square$

Note that since the polynomial $p_N(u)$ defined in the proof of Proposition 13.4.10 is $\mathbb{F}_q$-linear, it follows by Corollary 13.2.5 that $p_N(u) = u + \sum_{i=1}^{n} a_i u^{q^i}$. Therefore, the power series extension of $e_\Gamma(u)$ is of the form $e_\Gamma(u) = u + \sum_{i=1}^{\infty} c_i u^{q^i}$. Since e_Γ is a nonconstant entire function, it follows that $e_\Gamma : \mathbb{C}_\infty \to \mathbb{C}_\infty$ is an epimorphism. In fact, any nonconstant entire function has a zero (see [51, Proposition 2.13]). Therefore if f is any nonconstant entire function and $c \in \mathbb{C}_\infty$, then $g = -c + f$ has a zero. Thus f is onto.

The importance of the lattices and exponential functions is that for any lattice Γ of rank r we will obtain a Drinfeld A-module ρ^Γ over $\mathbb{C}_\infty$ of rank r.

Now we consider two lattices Γ, Γ' such that $\Gamma \subseteq \Gamma'$ and Γ is of finite index in Γ'. Since $e_\Gamma(u)$ is periodic with group of periods Γ, it follows that $e_\Gamma(\Gamma')$ and Γ'/Γ are isomorphic as $\mathbb{F}_q$-vector spaces. In particular, $e_\Gamma(\Gamma')$ is a finite set.

Definition 13.4.11. Let Γ, Γ' be two lattices such that $\Gamma \subseteq \Gamma'$ and Γ has finite index in Γ'. We define,

$$P(\Gamma'/\Gamma; u) = u \prod_{\lambda\in e_\Gamma(\Gamma')\setminus\{0\}} \left(1 - \frac{u}{\lambda}\right) \tag{13.15}$$

which is an $\mathbb{F}_q$-linear polynomial of degree $|\Gamma'/\Gamma|$ associated to Γ'/Γ.

Proposition 13.4.12. *Let $\Gamma, \Gamma', \Gamma''$ be three lattices such that $\Gamma'' \supseteq \Gamma' \supseteq \Gamma$ and Γ has finite index in Γ''. Then*

$$e_{\Gamma'}(u) = P(\Gamma'/\Gamma; e_\Gamma(u)) \quad \text{with} \quad u \in \mathbb{C}_\infty \tag{13.16}$$

and

$$P(\Gamma''/\Gamma; u) = P(\Gamma''/\Gamma'; P(\Gamma'/\Gamma; u)). \tag{13.17}$$

Proof. The roots of the left side of (13.16) are elements λ of Γ', and the roots of the right side are precisely the elements u such that $e_\Gamma(u) \in e_\Gamma(\Gamma')$. Therefore both sides of (13.16) are entire functions with the same roots and $D(e_{\Gamma'}(u)) = D(P(\Gamma'/\Gamma; e_\Gamma(u)) = 1$. Thus (13.16) is a consequence of Corollary 13.4.7.

Finally, (13.17) follows from (13.16) and (13.15). □

Next, we will see that the lattice Γ provides a Drinfeld A-module $\rho_\Gamma \in \mathrm{Drin}_A(\mathbb{C}_\infty)$. First note that if $a \in A \setminus \{0\}$, then $a^{-1}\Gamma \supseteq \Gamma$.

Theorem 13.4.13. *Let Γ be a lattice of rank r. For $a \in A \setminus \{0\}$, let $\rho_a^\Gamma : \mathbb{C}_\infty \to \mathbb{C}_\infty$ be given by*

$$\rho_a^\Gamma(u) := aP(a^{-1}\Gamma/\Gamma; u). \tag{13.18}$$

Then $\rho_a^\Gamma \in \mathbb{C}_\infty\langle\tau\rangle$. Let $\rho^\Gamma : A \to \mathbb{C}_\infty\langle\tau\rangle$ be defined by $\rho^\Gamma(a) = \rho_a^\Gamma$ if $a \neq 0$ and $\rho^\Gamma(0) = 0$. Then ρ^Γ is a Drinfeld A-module of rank r over $\mathbb{C}_\infty$.

Proof. We have $\rho_a^\Gamma \in \mathbb{C}_\infty\langle\tau\rangle$ by Definition 13.4.11. Now

$$\rho_a^\Gamma(u) = aP(a^{-1}\Gamma/\Gamma; u) = au \prod_{\lambda \in e_\Gamma(a^{-1}\Gamma)\setminus\{0\}} \left(1 - \frac{u}{\lambda}\right).$$

Therefore $D(\rho_a^\Gamma) = a = \delta(a)$, where $\delta\colon A \to \mathbb{C}_\infty$ is the inclusion map. Thus $D \circ \rho^\Gamma = \delta$. Now we have

$$\begin{aligned} e_{a^{-1}\Gamma}(u) &= u \prod_{\lambda \in a^{-1}\Gamma\setminus\{0\}} \left(1 - \frac{u}{\lambda}\right) = u \prod_{\mu \in \Gamma\setminus\{0\}} \left(1 - \frac{u}{a^{-1}\mu}\right) \\ &= a^{-1}(au) \prod_{\mu \in \Gamma\setminus\{0\}} \left(1 - \frac{au}{\mu}\right) = a^{-1}e_\Gamma(au). \end{aligned} \tag{13.19}$$

Using (13.16) and (13.19), we obtain

$$e_\Gamma(au) = ae_{a^{-1}\Gamma}(u) = aP(a^{-1}\Gamma/\Gamma; e_\Gamma(u)) = \rho_a^\Gamma(e_\Gamma(u)). \tag{13.20}$$

If $a, b \in A$, we have

$$\begin{aligned} \rho_{a+b}^\Gamma(e_\Gamma(u)) &= e_\Gamma((a+b)u) = e_\Gamma(au) + e_\Gamma(bu) = \rho_a^\Gamma(e_\Gamma(u)) + \rho_b^\Gamma(e_\Gamma(u)), \\ \rho_{ab}^\Gamma(e_\Gamma(u)) &= e_\Gamma(abu) = \rho_a^\Gamma(e_\Gamma(bu)) = \rho_a^\Gamma(\rho_b^\Gamma(e_\Gamma(u)). \end{aligned}$$

Since the exponential map is onto, it follows that

$$\rho_{a+b}^\Gamma = \rho_a^\Gamma + \rho_b^\Gamma \quad \text{and} \quad \rho_{ab}^\Gamma = \rho_a^\Gamma \rho_b^\Gamma \quad \text{for all} \quad a, b \in A.$$

Therefore $\rho^\Gamma \in \mathrm{Drin}_A(\mathbb{C}_\infty)$. It remains to show that the rank of ρ^Γ is r. Now $\rho_a^\Gamma(u) = aP(a^{-1}\Gamma/\Gamma; u)$, so we have $\deg_u \rho_a^\Gamma(u) = |a^{-1}\Gamma/\Gamma|$. Since Γ is of rank r, Γ is isomorphic to a sum of r fractional ideals of A. Moreover, if $\mathfrak{A}$ is any fractional ideal, we have $a^{-1}\mathfrak{A}/\mathfrak{A} \cong a^{-1}A/A \cong A/aA$, and hence $|a^{-1}\Gamma/\Gamma| = |A/aA|^r = q^{r \deg_K a}$. It follows that

$$r \deg_K a = \log_q |a^{-1}\Gamma/\Gamma| = \deg_\tau \rho_a^\Gamma = -d_\infty r_{\rho^\Gamma} v_{\mathfrak{P}_\infty}(a) = r_{\rho^\Gamma} \deg_K a.$$

Hence $r_{\rho^\Gamma} = r$. □

Now we are ready to present a sketch of the proof of Theorem 13.4.2.

Proof of Theorem 13.4.2 (sketch). First pick an element ϕ in the left twisted power series $\mathbb{C}_\infty\langle\langle\tau\rangle\rangle$ such that $D(\phi) = \alpha$ is a transcendental element over $\mathbb{F}_q$. By equating coefficients, we obtain a unique power series $\lambda_\phi = \sum_{i=0}^\infty c_i\tau^i \in \mathbb{C}_\infty\langle\langle\tau\rangle\rangle$ such that

$$\lambda_\phi \alpha = \phi\lambda_\phi \tag{13.21}$$

and $c_0 = 1$.

Next, we show that for each $\beta \in \mathbb{C}_\infty$,

$$\tau_\beta = \lambda_\phi \beta \lambda_\phi^{-1}$$

is the unique power series $\mathbb{C}_\infty\langle\langle\tau\rangle\rangle$ with constant term β that commutes with ϕ. Finally, we obtain that if $\Lambda\colon K \to \mathbb{C}_\infty\langle\langle\tau\rangle\rangle$ is a formal K-module over $\mathbb{C}_\infty$, then there exists a unique power series $\lambda_\Lambda = \sum_{i=0}^\infty c_i\tau^i$ such that $c_0 = 1$ and

$$\Lambda_a = \lambda_\Lambda \delta(a) \lambda_\Lambda^{-1} \tag{13.22}$$

for all $a \in K$.

For the proofs of the above statements see [62, 63] and [51, Chapter 4].

Now for the given Drinfeld A-module ρ, let $\lambda_\rho = \sum_{i=0}^\infty c_i\tau^i$ be the twisted power series defined by (13.22). Then $\lambda_\rho(u) = \sum_{i=0}^\infty c_i u^{q^i}$ converges for all $u \in \mathbb{C}_\infty$. This can be proved using (13.21). Now $\lambda_\rho\delta(a) = \rho_a\lambda_\rho$ is equivalent to the relation given in (13.20) with $\lambda_\rho(u)$ replacing $e_\Gamma(u)$. Finally, it can be shown that the roots of λ_ρ form a lattice Γ. Hence $\rho = \rho^\Gamma$. □

Example 13.4.14. Consider the Carlitz module $C \in \mathrm{Drin}_A(\mathbb{C}_\infty)$, where $A = R_T = \mathbb{F}_q[T]$. Since C has rank 1, if Γ is the lattice such that $C = \rho^\Gamma$, then Γ is of rank 1 over A. It follows that there exists $\pi \in \mathbb{C}_\infty$ such that $\Gamma = A\tilde{\pi}$. Γ is the set of roots of the Carlitz exponential. Note that $\tilde{\pi}$ plays the role of $2\pi i \in \mathbb{C}$ since the lattice of zeros of the complex function $e^z - 1$ is $\{2n\pi i \mid n \in \mathbb{Z}\} = (2\pi i)\mathbb{Z}$.

To compute $\tilde{\pi}$, notice that the Carlitz exponential function ex is given by (13.3),

$$\mathrm{ex}(Tu) = T\,\mathrm{ex}(u) + \mathrm{ex}(u)^q.$$

Consider the power series expansion of ex:

$$\operatorname{ex}(u) = \sum_{i=0}^{\infty} \frac{u^{q^i}}{D_i}. \tag{13.23}$$

It follows by (13.3) that the coefficients in (13.23) are given by

$$D_0 = 1 \quad \text{and} \quad D_i = (T^{q^i} - T)D_{I-1}^q \quad \text{for} \quad i > 0. \tag{13.24}$$

We write $[i] := T^{q^i} - T$. Then

$$D_i = [i][i-1]^q \cdots [1]^{q^{i-1}}.$$

Thus we have $\deg_T D_i = iq^i$, $v_{\mathfrak{P}_\infty}(D_i) = -iq^i$, and $v_{\mathfrak{P}_\infty}(u^{q^i}/D_i) = -q^i v_{\mathfrak{P}_\infty} + iq^i > 0$ for i large enough. Therefore $\lim\limits_{\substack{i\to\infty \\ \|\,\|_\infty}} u^{q^i}/D_i = 0$ for all $u \in \mathbb{C}_\infty$ and (13.23) is convergent for all $u \in \mathbb{C}_\infty$.

Since ex is periodic, it is clearly not injective. However, we may define an inverse function, called *logarithm*, in a neighborhood of 0. Let

$$L(u) = \sum_{i=0}^{\infty} \frac{(-1)^i u^{q^i}}{L_i} \tag{13.25}$$

be the logarithm. It follows from (13.3) that

$$TL(u) = L(Tu + u^q).$$

On the other hand, (13.25) yields

$$L_i = (T^{q^i} - T)(T^{q^{i-1}} - T) \cdots (T^q - T) = [i][i-1] \cdots [1].$$

Now $\deg_T L_i = q(q^i - 1)/(q-1)$ and $v_{\mathfrak{P}_\infty}(u^{q^i}/L_i) = -q^i v_{\mathfrak{P}_\infty}(u) + q(q^i-1)/(q-1)$. Therefore $v_{\mathfrak{P}_\infty}(u^{q^i}/L_i) > 0$ for i large enough if and only if $d_\infty(u) < q/(q-1)$. Thus (13.25) is convergent for u of degree less than $q/(q-1)$.

Our goal is to find an expression for $\tilde{\pi}$ and find its degree (which must be $q/(q-1)$ since the set of zeros of $\operatorname{ex}(u)$ is $A\tilde{\pi}$, and the inverse around 0 is defined for elements of degree less than $q/(q-1)$). For $x \in \mathbb{C}_\infty$ we write

$$\operatorname{ex}(xL(u)) = \sum_{j=0}^{\infty} \begin{bmatrix} x \\ j \end{bmatrix} u^{q^j}.$$

Equating coefficients, we obtain

$$\begin{bmatrix} x \\ j \end{bmatrix} = \sum_{i=0}^{j} (-1)^{j-i} \frac{x^{q^i}}{D_i L_{j-i}^{q^i}}.$$

If we write the Carlitz module as

$$C_M = \sum_{i=0}^{d} C_{M,i}\tau^i$$

where $M \in R_T \setminus \{0\}$ is of degree d, then $C_{M,i} = \begin{bmatrix} M \\ i \end{bmatrix}$. Note that $C_{M,i}$ is the same as in Theorem 12.2.5.

If $d = \deg M < t \in \mathbb{N}$, then $C_{M,t} = 0$. Therefore every polynomial of degree less than t is a zero of $\begin{bmatrix} x \\ t \end{bmatrix}$, and there are q^t polynomials of degree less than t. Define

$$\mathrm{ex}_t(x) := \prod_{\substack{M \in R_T \\ \deg M < t}} (x - M) = A_t x \prod_{\substack{M \in R_T \setminus \{0\} \\ \deg M < t}} \left(1 - \frac{x}{M}\right),$$

where $A_t = \pm \prod_{\substack{M \in R_T \setminus \{0\} \\ \deg M < t}} M$. Then $A_t = (-1)^t D_t / L_t$, and $\begin{bmatrix} x \\ t \end{bmatrix} = \frac{\mathrm{ex}_t(x)}{\alpha_t}$ for some constant $\alpha_t \in K = \mathbb{F}_q(T)$.

By Theorem 12.2.5, we have $\begin{bmatrix} T^t \\ t \end{bmatrix} = 1$. Thus $\alpha_t = \mathrm{ex}_t(T^t) = \prod_{\substack{M \in R_T \\ \deg M < t}} (T^t - M)$. The reader should try to conclude that in fact, $\alpha_t = D_t$ and hence D_t is the product of all monic polynomials of degree t. In short, we have

$$\begin{bmatrix} x \\ t \end{bmatrix} = \frac{(-1)^t}{L_t} x \prod_{\substack{M \in R_T \setminus \{0\} \\ \deg M < t}} \left(1 - \frac{x}{M}\right).$$

Thus

$$x \prod_{\substack{M \in R_T \setminus \{0\} \\ \deg M < t}} \left(1 - \frac{x}{M}\right) = (-1)^t L_t \begin{bmatrix} x \\ t \end{bmatrix} = (-1)^t L_t \sum_{i=0}^{t} (-1)^{t-i} \frac{x^{q^i}}{D_i L_{t-i}^{q^i}}$$

$$= \sum_{i=0}^{t} (-1)^i \frac{L_t}{D_i L_{t-i}^{q^i}} x^{q^i}.$$

It follows that

$$\sum_{i=0}^{\infty} \frac{\tilde{\pi}^{q^i} u^{q^i}}{D_i} = \mathrm{ex}(\tilde{\pi} u) = \tilde{\pi} u \prod_{M \in R_T \setminus \{0\}} \left(1 - \frac{u}{M}\right)$$

$$= \tilde{\pi} \lim_{d \to \infty} \frac{\mathrm{ex}_d(u)}{A_d} = \tilde{\pi} \lim_{d \to \infty} \left(\sum_{i=0}^{d} (-1)^i \frac{L_d}{D_i L_{d-i}^{q^i}} u^{q^i} \right)$$

$$= \tilde{\pi} \sum_{i=0}^{\infty} \left\{ \lim_{d \to \infty} (\ 1)^i \frac{L_d}{L_{d-i}^{q^i}} \right\} \frac{u^{q^i}}{D_i}.$$

Therefore

$$\tilde{\pi}^{q^i-1} = \lim_{d\to\infty} (-1)^i \frac{L_d}{L_{d-i}^{q^i}}. \tag{13.26}$$

It follows from (13.26) that $\deg \tilde{\pi} = q/(q-1)$ and that $\tilde{\pi}$ is a $(q-1)$th root of $\lim_{d\to\infty} \frac{(-L_d)}{L_{d-1}^q}$.

L. Carlitz [14] found an explicit expression for $\tilde{\pi}$ in the form of an infinite product. First note that $[i+1]-[i]=[1]^{q^i}$. Let

$$\alpha_i := \prod_{j=2}^{i}\left(1-\frac{[j-1]}{[j]}\right) = \frac{[1]^{(q^i-1)/(q-1)}}{L_j}.$$

Since $\sum_{j=2}^{\infty}\frac{[j-1]}{[j]}$ is convergent, each α_i is convergent. Furthermore, $|\alpha_i|_\infty = 1$ since $\deg \alpha_i = 0$. Let $\alpha = \lim_{i\to\infty}\alpha_i \in \mathbb{C}_\infty$, where $|\alpha| = 1$. Notice that $\alpha_{i+1}-\alpha_i = -\frac{[i]}{[i+1]}\alpha_i$ and $\deg(\alpha_{i+1}-\alpha_i) = -q^i(q-1)$.

Let $\delta_i = \alpha_i - \alpha$, where $\deg \delta_i = -q^i$. From this expression Carlitz deduced that

$$\lim_{d\to\infty}\sum_{i=0}^{d}(-1)^i\frac{L_d}{D_i L_{d-i}^{q^i}}u^{q^i} = \sum_{i=0}^{\infty}\frac{(-1)^i}{D_i}u^{q^i}\alpha^{q^i-1}x_i,$$

where $x_i = [1]^{(q^i-1)/(q-1)}$. In particular,

$$\lim_{d\to\infty}\left(\frac{-L_d}{D_1 L_{d-1}^q}\right) = \frac{(-1)}{D_1}\alpha^{q-1}x_1 = (-1)\alpha^{q-1}.$$

Therefore $\lim_{d\to\infty}\left(\frac{-L_d}{L_{d-1}^q}\right) = (-[1])\alpha^{q-1} = \tilde{\pi}^{q-1}$.

Choose a fixed $(q-1)$th root ξ of $-[1] = T - T^q$. We have

$$\tilde{\pi} = \xi\alpha = \xi\prod_{i=1}^{\infty}\left(1-\frac{[i-1]}{[i]}\right). \tag{13.27}$$

The arbitrary character of the choice of a $(q-1)$th root of $-[1]$ is the analogue of the fact that in the classical case we may choose $2\pi i$ or $-2\pi i$.

13.5 Explicit Class Field Theory

As we saw in Chapter 12, the maximal abelian extension of $K = \mathbb{F}_q(T)$ is obtained by considering the torsion of the Carlitz module, first for $A = \mathbb{F}_q[T]$ and then for $A' = \mathbb{F}_q[1/T]$ (Theorem 12.8.32). D. Hayes [62] developed an explicit class field theory for a general A using Drinfeld modules.

In fact, this explicit class field theory uses the theory of rank-one Drinfeld modules and provides explicit abelian extensions. Finally, using the general theory of class fields, it is shown that the abelian extensions found are the ones prescribed by the reciprocity map. In this section we present an introduction to explicit class field theory of general congruence function fields. For a more complete history and proofs see [62, 63, 128, 151].

We use the same notations as in Sections 13.3 and 13.4. We set $\mathbb{F}_\infty := \mathbb{F}_{q^{d_\infty}}$ as the residue field of K_∞ at $\mathfrak{P}_\infty$. The *class group* of A is denoted by Pic A (see Exercise 13.7.6) and h_A denotes the cardinality of Pic A. The group Pic A also receives the name of *Picard group* of A. For a nonzero ideal $\mathfrak{A}$ we use the notation $N\mathfrak{A}$ or $\Phi(\mathfrak{A})$ for the cardinality of the group of units of A mod $\mathfrak{A}$ $(A/\mathfrak{A})^*$. For any nonzero element $x \in K$, we define $\deg x := -d_\infty v_{\mathfrak{P}_\infty}(x)$, and we put $N(x) = q^{\deg x}$.

Note that the Drinfeld A-module ρ is of rank one if $\deg \rho_a = -d_\infty v_{\mathfrak{P}_\infty}(a) = \deg a$ for all $a \in A$.

For an ideal $\mathfrak{A}$ of A and a Drinfeld A-module $\rho \in \operatorname{Drin}_A(k)$, recall that $\rho[\mathfrak{A}]$ denotes the $\mathfrak{A}$-torsion of $\bar{k}$, that is, $\rho[\mathfrak{A}] = \{u \in \bar{k} \mid \rho_a(u) = 0 \ \forall\, a \in \mathfrak{A}\} = \{u \in \bar{k} \mid \rho_{\mathfrak{A}}(u) = 0\}$. The next result is completely similar to Proposition 12.3.7.

Proposition 13.5.1. *If ρ is a Drinfeld A-module of rank one over K and if $a \in A\setminus\{0\}$, then $K(\rho[a])/K$ is an abelian extension and* $\operatorname{Gal}(K(\rho[a])/K)$ *is isomorphic in a natural way to a subgroup of $(A/(a))^*$.*

Proof. Let $\mathfrak{P} = \operatorname{char}(\rho)$. If (a) is relatively prime to $\mathfrak{P}$ then $\rho[a]$ and $A/(a)$ are isomorphic as A-modules (see Exercise 13.7.10). Choose a generator λ of $\rho[a]$. If $\sigma \in G := \operatorname{Gal}(K(\rho[a])/K)$, then $\sigma\lambda \in \rho[a]$, so that $\sigma\lambda = \rho_{a_\sigma}(\lambda)$ for a unique $a_\sigma \in (A/(a))$. Now, since $\sigma\lambda$ is also a generator of $\rho[a]$, we have $a_\sigma \in (A/(a))^*$ and the correspondence $\sigma \mapsto a_\sigma$ is a group monomorphism of G into $(A/(a))^*$.

Suppose that $\mathfrak{P} \neq 0$. Then $h_\rho = 1$. If $(a) = \mathfrak{P}$, then $\rho[a] = 0$, so that $K(\rho[a]) = K$. This case is analogous to the situation in which in characteristic p we adjoin pth roots of unity, which can only be 1.

More generally, if $a \in \mathfrak{P}$, then $(a) = \mathfrak{C}\mathfrak{P}^n$ for some $(\mathfrak{C}, \mathfrak{P}) = 1$. Thus $\rho[a] = \rho[\mathfrak{C}\mathfrak{P}^n] = \rho[\mathfrak{C}]$, and we are in the first case. Therefore $G \subseteq (A/(\mathfrak{C}))^* \subseteq (A/(a))^*$. □

13.5.1 Class Number One Case

Now we generalize the results of Sections 12.3, 12.4, and 12.5 when A has class number one. We will use the following notations. We take $k = K$ and the Drinfeld modules $\rho \in \operatorname{Drin}_A(K)$ under consideration will be of rank one unless otherwise stated. In case $h_A = 1$, any nonzero ideal $\mathfrak{A}$ of A is principal and $\alpha_{\mathfrak{A}}$ will denote a generator of $\mathfrak{A}$ such that the leading coefficient of $\rho_{\alpha_{\mathfrak{A}}}$ is one. Note that $\rho_{\alpha_{\mathfrak{A}}} = \rho_{\mathfrak{A}}$. Let $\lambda_{\mathfrak{A}}$ denote a generator of the A-module $\rho[\mathfrak{A}]$. The Galois group of the extension $K(\rho[\mathfrak{A}])/K$ will be denoted by $G_{\mathfrak{A}}$ (we will prove shortly that the extension $K(\rho[\mathfrak{A}])/K$ is an abelian extension).

Definition 13.5.2. If A has class number one and $\mathfrak{A}$ is a nonzero ideal of A we define the *cyclotomic polynomial* with respect to $\mathfrak{A}$ by

$$\Psi_{\mathfrak{A}}(u) = \prod_{\alpha_{\mathfrak{B}} \in (A/\mathfrak{A})^*} (u - \rho_{\mathfrak{B}}(\lambda_{\mathfrak{A}}))$$

where the product runs through a set of representatives $\alpha_{\mathfrak{B}} \in A$ of $(A/\mathfrak{A})^*$. We have $\Psi_{\mathfrak{A}}(u) \in K(\rho[\mathfrak{A}])[u]$.

Note that if $\alpha_{\mathfrak{B}} \neq \alpha_{\mathfrak{B}'}$ then $\mathfrak{B} \neq \mathfrak{B}'$ since otherwise $\alpha_{\mathfrak{B}} = \xi \alpha_{\mathfrak{B}'}$ with ξ a unit of A and $\rho_{\alpha_{\mathfrak{B}}} = \rho_\xi \rho_{\alpha_{\mathfrak{B}'}}$ has leading coefficient $\xi = 1$.

Proposition 13.5.3. *If A has class number one, then for any nonzero ideal $\mathfrak{A}$ of A we have, $\Psi_{\mathfrak{A}}(u) \in K[u]$.*

Proof. Let $\sigma \in G_{\mathfrak{A}}$. Then $\sigma(\lambda_{\mathfrak{A}})$ is a generator of $\rho[\mathfrak{A}] = \rho[\alpha_{\mathfrak{A}}] = \ker \rho_{\alpha_{\mathfrak{A}}} \cong A/\mathfrak{A}$ (see Remark 13.3.20). The argument follows as in Proposition 12.3.9. □

Proposition 13.5.4. *Let $\mathfrak{A}$ be a prime power ideal of A, $\mathfrak{A} = \mathfrak{P}^m$ with $\mathfrak{P}$ a nonzero prime ideal of A and $m \in \mathbb{N}$. Then*

(1) *$\mathfrak{P}$ is fully ramified in $K(\rho[\mathfrak{P}^m])/K$.*
(2) *The ramification index of $\mathfrak{P}$ in $K(\rho[\mathfrak{P}^m])/K$ is $[K(\rho[\mathfrak{P}^m]) : K]$.*
(3) *If $\mathfrak{T}$ is any prime divisor in K other than $\mathfrak{P}_\infty$ and $\mathfrak{P}$, then $\mathfrak{T}$ is not ramified in $K(\rho[\mathfrak{P}^m])/K$.*

In particular we have $G_{\mathfrak{P}^m} \cong (A/\mathfrak{P}^m)^*$.

Proof. It is analogous to the proof of Proposition 12.3.14 and the details are left to the reader (see Exercise 13.7.12). □

Note that $\rho_{\mathfrak{P}^m}(u) = \rho_{\mathfrak{P}}(\rho_{\mathfrak{P}^{m-1}}(u))$. Now we have $\rho_{\mathfrak{P}} = \alpha_{\mathfrak{P}} u + a_1 u^q + \cdots + a_d u^{q^d}$. Hence $\rho_{\mathfrak{P}^m}(u) = \alpha_{\mathfrak{P}} \rho_{\mathfrak{P}^{m-1}}(u) + a_1 \rho_{\mathfrak{P}^{m-1}}(u)^q + \cdots + a_d \rho_{\mathfrak{P}^{m-1}}(u)^{q^d}$. That is,

$$\rho_{\mathfrak{P}^m}(u) = \rho_{\mathfrak{P}^{m-1}}(u) H(u)$$

with $H(u)$ a polynomial in $K[u]$ and $H(0) = \alpha_{\mathfrak{P}}$. Furthermore, the roots of $H(u)$ are precisely the elements of $\rho[\mathfrak{P}^m] \setminus \rho[\mathfrak{P}^{m-1}]$. Hence $H(u) = \Psi_{\mathfrak{P}^m}(u)$, $\alpha_{\mathfrak{P}} = H(0) = \pm \prod_{\alpha_{\mathfrak{C}} \in (A/\mathfrak{P}^m)^*} \rho_{\mathfrak{C}}(\lambda_{\mathfrak{P}^m})$ and $\Psi_{\mathfrak{P}^m}(u) = \operatorname{Irr}(\lambda_{\mathfrak{P}^m}, u, K)$.

Assume that A is of class number one and let a be a nonzero element of A. We write $(a) = aA = \mathfrak{A} = \prod_{i=1}^t \mathfrak{P}_i^{s_i}$. Then $(A/(a))^* \cong \prod_{i=1}^t (A/\mathfrak{P}_i^{s_i})^*$. Let $\mathfrak{P}_i^{s_i} = (a_i)$ for $i = 1, \ldots, t$. It follows that the fields $K(\lambda_{\mathfrak{P}_i^{s_i}})$ are pairwise linearly disjoint over K since $\mathfrak{P}_i$ is fully ramified in $K(\lambda_{\mathfrak{P}_i^{s_i}})/K$ and unramified in $\prod_{\substack{j=1\\ j\neq i}}^t K(\lambda_{\mathfrak{P}_j^{s_j}})/K$. We have obtained

Theorem 13.5.5. *Assume A has class number one, that is, $h_A = 1$. For a nonzero element a of A, let $\mathfrak{A} = (a)$. Then*

(1) *$\Psi_{\mathfrak{A}}(u) = \operatorname{Irr}(\lambda_{\mathfrak{A}}, u, K)$. In particular $\Psi_{\mathfrak{A}}(u)$ is an irreducible polynomial.*
(2) *$G_{\mathfrak{A}} = \operatorname{Gal}(K(\lambda_{\mathfrak{A}})/K) = \operatorname{Gal}(K(\rho[\mathfrak{A}])/K) = \operatorname{Gal}(K(\rho[a])/K) \cong (A/(a))^* = (A/\mathfrak{A})^*$.*

(3) $[K(\rho[a]) : K] = \Phi((a))$. □

Next, we determine the Artin symbol of the extension $K(\rho[a])/K$ with a any nonzero element of A.

Theorem 13.5.6. *If A has class number one and $\rho \in \mathrm{Drin}_A(K)$ is a Drinfeld A-module of rank one over K, then if $\mathfrak{P}$ is a prime ideal of A not dividing (a), we have for $\lambda \in \rho[a]$*

$$\lambda^{\varphi_{\mathfrak{P}}} = \rho_{\mathfrak{P}}(\lambda)$$

where $\varphi_{\mathfrak{P}}$ denotes the Frobenius automorphism $\left[\frac{K(\rho[a])/K}{\mathfrak{P}}\right]$.

Proof. Let $\lambda = \lambda_{(a)}$ and $\mathfrak{Q}$ be a prime divisor of $K(\rho[a])$ above $\mathfrak{P}$. Then $\rho_{\mathfrak{P}}(u)/u$ is Eisenstein at $\mathfrak{P}$. The proof goes as in Proposition 12.3.18. It follows that $\rho_{\mathfrak{P}}(\lambda) \equiv \lambda^{q^d} \bmod \mathfrak{Q}$. It follows that $\rho_{\mathfrak{P}}(\lambda) \equiv \left[\frac{K(\rho[a])/K}{\mathfrak{P}}\right](\lambda) \bmod \mathfrak{Q}$. To prove the equality and not just the congruence, consider the derivative of $\rho_a(x) = \prod_{b\in A/(a)}(x - \rho_b(\lambda))$, that is, $\rho'_a(\lambda) = a = \prod_{\substack{b\in A/(a)\\ b\neq c}}(\rho_c(\lambda) - \rho_b(\lambda))$. Since $v_{\mathfrak{Q}}(a) = 0$, we have $\rho_c(\lambda) \not\equiv \rho_b(\lambda) \bmod \mathfrak{Q}$ for all $c \not\equiv b \bmod (a)$. It follows that $\rho_{\mathfrak{P}}(\lambda) = \left[\frac{K(\rho[a])/K}{\mathfrak{P}}\right](\lambda)$. □

Since the Frobenius automorphism at $\mathfrak{P}$ acts as $\mathfrak{P}$, it follows that the decomposition of prime divisors in $K(\rho[a])/K$ is analogous to the cyclotomic cases, both the classic and function field one (see Theorem 12.5.3).

13.5.2 General Class Number Case

Now we try to generalize the results of Section 13.5.1. Here A will be arbitrary and $\rho \in \mathrm{Drin}_A(\mathbb{C}_\infty)$ a Drinfeld A-module of rank one. By a *class field* of A we mean a finite abelian extension field of K on which $\mathfrak{P}_\infty$ splits completely. By a *narrow class field* of A we mean a finite abelian extension of K.

Definition 13.5.7. Let ρ be a Drinfeld A-module over $\mathbb{C}_\infty$ such that $\delta(a) = a$ for all $a \in A$. Let E be a subfield of $\mathbb{C}_\infty$ containing K. We say that ρ is *defined over* E or that E is a *field of definition for* ρ if ρ is isomorphic over $\mathbb{C}_\infty$ to a Drinfeld A-module ρ' such that $\rho'_a \in E\langle\tau\rangle$ for all $a \in A$.

Example 13.5.8. If ρ is a Drinfeld module of rank one, then K_∞ is a field of definition for ρ (see Exercise 13.7.11).

Next result proves that there always exists a smallest field of definition for ρ.

Theorem 13.5.9. *Let ρ be a Drinfeld A-module over $\mathbb{C}_\infty$ of any rank. Then there exists a field of definition K_ρ, finitely generated over K, which is contained in every field of definition for ρ.*

Proof. For $a \in A$, let $\rho_a = a + \sum_{i=1}^{r_\rho \deg a} c_i \tau^i$. For any $\xi \in \mathbb{C}_\infty$ we have

$$(\xi \rho \xi^{-1})_a = a + \sum_{i=1}^{r_\rho \deg a} \xi^{1-q^i} c_i \tau^i.$$

Fix a nonconstant $a \in A$ and consider the set S of indices such that $c_i \neq 0$. Let g be the greatest common divisor of the set $\{q^i - 1 \mid i \in S\}$ and let $g = \sum_{j \in S} \alpha_j (q^j - 1)$ with $\alpha_j \in \mathbb{Z}$. For each $i \in S$ consider the element

$$I_i := c_i \Big(\prod_{j \in S} c_j^{\alpha_j} \Big)^{(1-q^i)/g} \in \mathbb{C}_\infty, \tag{13.28}$$

which is invariant under the map $c_j \mapsto \xi^{1-q^j} c_j$. Therefore the elements $\{I_j \mid j \in S\}$ belong to any field of definition of ρ. Let K_ρ be the field generated over K by the elements $I_i, i \in S$.

Let $\xi \in \mathbb{C}_\infty$ be chosen such that

$$\xi^g = \prod_{i \in S} c_i^{\alpha_i}. \tag{13.29}$$

Then $I_i = \xi^{1-q^i} c_i$. It follows that $\xi \rho_a \xi^{-1}$ has coefficients in K_ρ. By (13.22) we obtain that $\xi \rho_x \xi^{-1}$ has coefficients in K_ρ for all $x \in A$. Therefore ρ is defined over K_ρ. □

Definition 13.5.10. The field K_ρ is called the *smallest field of definition* for ρ or the *field of invariants* of ρ.

We will show in Section 13.5.4 that for ρ of rank one, K_ρ is the maximal unramified abelian extension of K in which $\mathfrak{P}_\infty$ splits completely. Thus for rank one Drinfeld A-modules over $\mathbb{C}_\infty$, K_ρ is independent of the choice of ρ.

To see the role that K_ρ plays in class field theory, we consider an action of $\mathrm{Pic}(A)$ (see Exercise 13.7.6) on $\mathrm{Drin}_A(k)$. Let $\mathfrak{A}$ be an integral ideal of A and $\rho \in \mathrm{Drin}_A(k)$. Consider the left ideal $I_{\mathfrak{A}}$ of $k\langle\tau\rangle$ generated by $\{\rho_a\}_{a \in \mathfrak{A}}$ and let $\rho_{\mathfrak{A}} \in k\langle\tau\rangle$ be a generator: $I_{\mathfrak{A}} = k\langle\tau\rangle \rho_{\mathfrak{A}}$. We write

$$\rho_{\mathfrak{A}} = f_1(\tau)\rho_{a_1} + \cdots + f_m(\tau)\rho_{a_m}$$

for some $f_i(\tau) \in k\langle\tau\rangle$ where $a_i \in A$ for $1 \leq i \leq m$.

Since $\mathfrak{A}$ is an ideal, we have $I_{\mathfrak{A}} \rho_a \subseteq I_{\mathfrak{A}}$ for $a \in A$. Therefore, for any $a \in A$ there exists a unique $\rho'_a \in k\langle\tau\rangle$ such that

$$\rho_{\mathfrak{A}} \rho_a = \rho'_a \rho_{\mathfrak{A}}. \tag{13.30}$$

Then $\rho' \colon A \to k\langle\tau\rangle, a \mapsto \rho'_a$, is a Drinfeld A-module over k. We denote this module by

$$\rho' := \mathfrak{A} * \rho. \tag{13.31}$$

Clearly, if E is a field of definition for ρ, then E is also a field of definition for $\mathfrak{A} * \rho$ for any nonzero ideal $\mathfrak{A}$ in A.

Now if $\mathfrak{A} = (a)$ is principal with $a \neq 0$, let α be the leading coefficient of ρ_a. Then $\rho_{\mathfrak{A}} = \alpha^{-1}\rho_a$ and $(\mathfrak{A} * \rho)_b = \alpha^{-1}\rho_b\alpha$ for all $b \in A$. It follows that $\mathfrak{A} * \rho$ and ρ are isomorphic. Thus $\operatorname{Pic}(A)$ acts on the isomorphism classes of Drinfeld A-modules ρ over K such that $r_\rho = r$ and $D \circ \rho = \delta$.

We will now see that this action defines the *Hilbert class field* H_A of A.

Definition 13.5.11. The *Hilbert class field* H_A of A is the maximal abelian extension of K in which the infinite prime $\mathfrak{P}_\infty$ splits completely.

For a rank one Drinfeld A-module ρ, we will show that $K_\rho = H_A$. In particular, for rank one A-modules K_ρ is independent of ρ.

To show $K_\rho = H_A$ we use a sign function sgn (see Definition 12.8.16) and use it to define $\operatorname{Pic}^+ A$ which is an extension of $\operatorname{Pic} A$ and corresponds to the extension H_A^+ of K, where every finite place of A is unramified. With the addition of the sign function we control the top coefficient of ρ_a and this turns out to be more efficient than controlling K_ρ. In this way we do not have to deal with isomorphism classes of rank one A-modules.

For the rest of this section we will consider only rank one Drinfeld A-modules over $\mathbb{C}_\infty$: $\rho \in \operatorname{Drin}_A(\mathbb{C}_\infty)$. We recall the definition of a sign function.

Definition 13.5.12. A *sign function* $\operatorname{sgn}\colon K_\infty^* \to \mathbb{F}_\infty^*$ is a homomorphism which is the identity on $\mathbb{F}_\infty^*$ and trivial on $U^{(1)} := U_{K_\infty}^{(1)} = 1 + \mathfrak{P}_\infty$ the group of units congruent to one mod $\mathfrak{P}_\infty$ (see Definition 5.9.13). We also use the convention $\operatorname{sgn}(0) = 0$.

For $\sigma \in \operatorname{Gal}(\mathbb{F}_\infty/\mathbb{F}_q)$, the composite map $\sigma \circ \operatorname{sgn}$ is called *twisting of the sign function* sgn.

Note that there are $|\mathbb{F}_\infty^*| = q^{d_\infty} - 1$ sign functions, depending on the choice of the prime element at $\mathfrak{P}_\infty$. In fact, if sgn and sgn$'$ are two sign functions, the map $x \mapsto \operatorname{sgn}(x)/\operatorname{sgn}'(x)$ with x a nonzero element of K_∞, is trivial on the units $U_{\mathfrak{P}_\infty}$ of K_∞ so it factors through $v_{\mathfrak{P}_\infty}\colon K_\infty^* \to \mathbb{Z}$. Thus

$$\operatorname{sgn}(x) = \operatorname{sgn}'(x)\xi^{\deg x/d_\infty}$$

for some $\xi \in \mathbb{F}_\infty^*$.

Definition 13.5.13. A Drinfeld A-module over $\mathbb{C}_\infty$, $\rho \in \operatorname{Drin}_A(\mathbb{C})_\infty$ is called *normalized* if the leading coefficient $\mu_\rho(x)$ of ρ_x belongs to $\mathbb{F}_\infty$ for all $x \in A$. If for some sign function sgn, the map $x \mapsto \mu_\rho(x)$ is a twisting of sgn, ρ is called sgn–*normalized.*

We have defined the map $x \mapsto \mu_\rho(x)$ as the leading coefficient ρ_x, $x \in A$. Now we will show that the map μ_ρ can be extended to $K = \operatorname{quot} A$.

For $x, y \in A$ we have (see Exercise 13.7.13)

$$\mu_\rho(xy) = \mu_\rho(x)\mu_\rho(y)^{r\deg x} = \mu_\rho(y)\mu_\rho(x)^{r\deg y}, \tag{13.32}$$

where A is of rank r

Fix a prime element $\pi \in K$ at $\mathfrak{P}_\infty$. Let $x \in K_\infty^*$. Then x can be written uniquely as

$$x = c\xi\pi^m = \operatorname{sgn}(x)\xi\pi^m$$

with $c \in \mathbb{F}_\infty^*$, $\xi \in U^{(1)}$ and $m \in \mathbb{Z}$.

Let $b \in A$ be chosen such that $a = bx \in A$. In fact we can make this choice since if $\mathfrak{N}_{\pi^m} = \mathfrak{P}_1^{\alpha_1} \cdots \mathfrak{P}_t^{\alpha_t}$, by the Riemann–Roch Theorem, there exists $b \in K$ such that $\mathfrak{P}_1^{\alpha_1} \cdots \mathfrak{P}_t^{\alpha_t} \mid \mathfrak{Z}_b$ and $\mathfrak{N}_b = \mathfrak{P}_\infty^u$ for some u large enough. That is, $b \in L_K(\mathfrak{P}_1^{\alpha_1} \cdots \mathfrak{P}_t^{\alpha_t}\mathfrak{P}_\infty^{-u})$: just take $u > 2g_K - 1 + \sum_{i=1}^t \alpha_i \deg_K \mathfrak{P}_i$. Therefore $a = bx$ and we may define $\mu_\rho(x)$ by the rule given in (13.32), that is, $\mu_\rho(a) = \mu_\rho(x)\mu_\rho(b)^{r \deg x}$ or

$$\mu_\rho(x) := \mu_\rho(a)\mu_\rho(b)^{-r \deg x}. \tag{13.33}$$

We leave to the reader to verify that this definition is independent of a and b and satisfies (13.32) (see Exercise 13.7.14).

We fix a sign function sgn and the object we study will be denoted by $(K, \mathfrak{P}_\infty, \operatorname{sgn})$. In this way, this object is analogous to $\mathbb{Q}$ with its archimedean place and the usual sign function on $\mathbb{R}$. Therefore an element a of A is called *positive* if $\operatorname{sgn}(x) = 1$.

One key result is the following theorem.

Theorem 13.5.14. *Every Drinfeld A-module* $\rho \in \operatorname{Drin}_A(\mathbb{C}_\infty)$ *is isomorphic over* $\mathbb{C}_\infty$ *to a* sgn*–normalized A-module* ρ'.

Proof. Let π be a prime element at $\mathfrak{P}_\infty$ which is positive for sgn. Let $\xi \in \mathbb{C}_\infty$ be such that $\xi^{q^{d_\infty}-1} = \mu_\rho(\pi^{-1})^{-1}$. Let $\rho' := \xi\rho\xi^{-1}$. Then $\mu_{\rho'}(\pi^{-1}) = 1$ (see Exercise 13.7.13).

For $x \in K_\infty^*$ we write $x = c\xi\pi^n$ with $c \in \mathbb{F}_\infty^*$, $\xi \in U^{(1)}$ and $n \in \mathbb{Z}$. Then $\operatorname{sgn}(x) = c \in \mathbb{F}_\infty^*$. In particular for $x = a \in A$, by Exercise 13.7.13, and since $\mu_{\rho'}(\xi\pi^n) = 1$, we obtain

$$\mu_{\rho'}(a) = \mu_{\rho'}(c\xi\pi^n) = \mu_{\rho'}(c) = \mu_{\rho'}(\operatorname{sgn}(a)).$$

Now the restriction of $\mu_{\rho'}$ to the residue field $\mathbb{F}_\infty$ of K_∞, is an automorphism $\iota_{\rho'}\colon \mathbb{F}_\infty \to \mathbb{F}_\infty$ fixing pointwise $\mathbb{F}_q$. Therefore $\mu_{\rho'}(a) = \iota_{\rho'}(\operatorname{sgn}(a))$. Therefore ρ s isomorphic to ρ' which is a sgn–normalized A-module. □

The next step is to find how many sgn–normalized A-modules are there in each isomorphism class. We restrict ourselves to rank one modules. First we give the definition of *Hayes modules*.

Definition 13.5.15. A *Hayes A-module* is a sgn normalized rank one Drinfeld A-module over $\mathbb{C}_\infty$.

The set of Hayes modules will be denoted by $\mathfrak{H}$.

The Carlitz module is a Hayes module.

Proposition 13.5.16. *If ρ and $\rho' = \xi\rho\xi^{-1}$ are* sgn*–normalized rank one Drinfeld A-modules over $\mathbb{C}_\infty$, then $\xi \in \mathbb{F}_\infty^*$ and $\mu_\rho = \mu_{\rho'}$.*

Proof. Since $\mu_\rho(\pi^{-1}) = \mu_{\rho'}(\pi^{-1}) = 1$ by Exercise 13.7.13 (4), we obtain that $\xi^{1-q^{d_\infty}} = 1$. Hence $\xi \in \mathbb{F}_\infty$. Finally we obtain for any $a \in A$

$$\mu_{\rho'}(a) = \xi^{1-q^{\deg a}}\mu_\rho(a) = \mu_\rho(a). \qquad \square$$

Corollary 13.5.17. *Each isomorphism class of Drinfeld A-modules or rank one over $\mathbb{C}_\infty$ contains exactly $(q^{d_\infty}-1)/(q-1)$* sgn*–normalized A-modules.*

Proof. Given any $\rho \in \operatorname{Drin}_A(\mathbb{C}_\infty)$ of rank one, ρ is isomorphic to a sgn–normalized one ρ'. Now, every A-module ρ'' isomorphic to ρ' is given as $\rho'' = \xi\rho'\xi^{-1}$. By Proposition 13.5.16 if ρ'' is also a sgn–normalized A-module, then $\xi \in \mathbb{F}_\infty^*$. Finally, since $\operatorname{Aut}(\rho) \cong \mathbb{F}_q^*$ we obtain exactly $|\mathbb{F}_\infty^*|/|\mathbb{F}_q^*| = (q^{d_\infty}-1)/(q-1)$ sgn–normalized modules isomorphic to ρ'. $\square$

Now, we consider the set of Hayes modules $\mathfrak{H}$. If $\rho \in \mathfrak{H}$ and $\mathfrak{A}$ is a nonzero ideal of A, let $\rho' = \mathfrak{A} * \rho$ (see (13.31)) Then ρ' satisfies $\rho_{\mathfrak{A}}\rho_a = \rho'_a\rho_{\mathfrak{A}}$ for all $a \in A$.

Since ρ is sgn–normalized, for each $\alpha \in \mathfrak{A}$ we have $\mu_\rho(\alpha) \in \mathbb{F}_\infty$, so $\xi := \mu(\rho_{\mathfrak{A}}) \in \mathbb{F}_\infty$. It follows that $\rho' = \mathfrak{A} * \rho$ is also sgn–normalized.

The Galois group $\operatorname{Gal}(\mathbb{C}_\infty/K_\infty)$ acts naturally on $\mathbb{C}_\infty\langle\tau\rangle$ and hence if $\rho \in \operatorname{Drin}_A(\mathbb{C}_\infty)$ and $\sigma \in \operatorname{Gal}(\mathbb{C}_\infty/K_\infty)$, $\sigma\rho$ is also a Drinfeld A-module. See Exercise 13.7.15. Clearly, if $\rho \in \mathfrak{H}$ then $\sigma\rho \in \mathfrak{H}$. Furthermore, if $\mathfrak{A}$ is a nonzero ideal and $\sigma \in \operatorname{Gal}(\mathbb{C}_\infty/K_\infty)$, then, from the definition we obtain that

$$\mathfrak{A} * \sigma\rho = \sigma(\mathfrak{A} * \rho).$$

Since every Drinfeld A-module ρ is obtained from a unique lattice Γ in $\mathbb{C}_\infty$: $\rho = \rho^\Gamma$ (Theorem 13.4.2), it can be proven that the fractional ideals of A act transitively on the isomorphism classes of Drinfeld A-modules over $\mathbb{C}_\infty$ such that the nonzero principal ideals operate trivially on these classes since if $\mathfrak{A} = aA$ is principal and $\rho \in \operatorname{Drin}_A(\mathbb{C}_\infty)$, then

$$\mathfrak{A} * \rho = \mu_\rho(a)^{-1}\rho\mu_\rho(a) \tag{13.34}$$

(see [63, §§8-9]).

It follows that Pic A acts on the isomorphism classes of Drinfeld A-modules. Furthermore, the set of isomorphism classes of rank one A-modules over $\mathbb{C}_\infty$, $\mathfrak{D}_1$, is a principal homogeneous space for Pic A, that is, the action of Pic A on $\mathfrak{D}_1$ is faithful. We recall that faithful means that if $\bar{\mathfrak{A}} \in \operatorname{Pic} A$ is such that $\mathfrak{A} * \rho = \rho$ for all $\rho \in \mathfrak{D}_1$, then $\bar{\mathfrak{A}} = 0$. In particular $|\mathfrak{D}_1| = |\operatorname{Pic} A|$ (see [63, Proposition 9.2]). That is

Theorem 13.5.18. *There are exactly h_A isomorphism classes of rank one Drinfeld A-modules over $\mathbb{C}_\infty$, and $\mathfrak{D}_1$ is a principal homogeneous space for* Pic A *under the $*$ action* $\square$

Now let $\rho \in \mathfrak{H}$ be a Hayes module and let $\mathfrak{A}$ be a nonzero ideal of A such that $\mathfrak{A} * \rho = \rho$. In particular the class $\bar{\mathfrak{A}}$ of $\mathfrak{A}$ in Pic A and $\bar{\rho}$ the isomorphism class of ρ, satisfy $\bar{\mathfrak{A}} \circ \bar{\rho} = \bar{\rho}$. Since the action of Pic A is transitive on $\mathfrak{D}_1$, the stabilizer of $\bar{\rho}$ is the set of nonzero principal ideals of A. Therefore $\mathfrak{A} = xA$ is principal. Now, by (13.34) we have

$$\mathfrak{A} * \rho = \mu_\rho(x)^{-1} \rho \mu_\rho(x) = \rho. \tag{13.35}$$

That is $\mu_\rho(x) \in \operatorname{Aut}(\rho) = \mathbb{F}_q^*$.

Therefore the stabilizer of ρ is

$$\{xA \mid x \in A, \mu_\rho(x) \in \mathbb{F}_q^*\} = \{xA \mid x \in A, \operatorname{sgn}(x) = 1\}.$$

Let

$$P_A^+ = \{xA \mid x \in K, \operatorname{sgn}(x) = 1\} \quad \text{and} \quad \operatorname{Pic}^+ A = \frac{M_A}{P_A^+}, \tag{13.36}$$

where M_A denotes the set of fractional ideals of A.

Definition 13.5.19. The group $\operatorname{Pic}^+ A$ is called the *narrow class group of A relative to* sgn.

The induced sgn function provides an isomorphism between P_A/P_A^+ and $\mathbb{F}_\infty^*/\mathbb{F}_q^*$. Therefore

$$h_A^+ := |\operatorname{Pic}^+ A| = \frac{q^{d_\infty} - 1}{q - 1} |\operatorname{Pic} A| = \frac{q^{d_\infty} - 1}{q - 1} h_A = |\mathfrak{H}|. \tag{13.37}$$

It follows

Theorem 13.5.20. *The set $\mathfrak{H}$ of Hayes modules is a principal homogeneous space for* $\operatorname{Pic}^+ A$ *under the $*$ action and* $|\mathfrak{H}| = \frac{q^{d_\infty}-1}{q-1} h_A = h_A^+$. □

13.5.3 The Narrow Class Field H_A^+

In this section we study a normalized field, H_A^+ over K that is a narrow class field where all finite prime divisors are unramified. Let $\rho \in \mathfrak{H}$ be a Hayes module and let α be a nonconstant element of A. Let H_A^+ be the field generated over K by the coefficients of ρ_α. Note that ρ_β is uniquely determined by ρ_α since if $\rho_\alpha = \sum_{i=0}^n a_i \tau^i$ and $\rho_\beta = \sum_{j=0}^m b_j \tau^j$, with $a_0 = \alpha$, then the equality

$$\rho_\alpha \rho_\beta = \rho_\beta \rho_\alpha$$

is equivalent to the recurrences

$$(\alpha^{q^i} - \alpha) b_i = \sum_{j=1}^{i} (a_j b_{i-j}^{q^j} - a_j^{q^{i-j}} b_{i-j}). \tag{13.38}$$

Since α is nonconstant, α is transcendental over $\mathbb{F}_q$ and so $\alpha^{q^i} - \alpha \neq 0$ for all $i \geq 1$. It follows that every b_i is uniquely determined.

It follows from (13.38) that H_A^+ is independent of α. By (13.30) and (13.31) we also have that all Hayes A-modules $\mathfrak{A} * \rho$ for nonzero ideals $\mathfrak{A}$ in A are defined over H_A^+. Thus, H_A^+/K is independent of ρ. However, it does depend upon choice of the sign function sgn.

Definition 13.5.21. The field H_A^+ is called the *normalizing field* for rank one Drinfeld A-modules over $(K, \mathfrak{P}_\infty, \mathrm{sgn})$.

Proposition 13.5.22. *The extension H_A^+/K is a finite abelian extension with Galois group isomorphic to a subgroup of* $\operatorname{Pic}^+ A$.

Proof. Fix $\rho \in \mathfrak{H}$, a Hayes module. For any $\sigma \in \operatorname{Aut}(\mathbb{C}_\infty/K)$, $\sigma\rho$ is a sgn–normalized Drinfeld A-modules, so $\sigma\rho$ is defined over H_A^+. In particular H_A^+ contains all the conjugates of its generators. It follows that H_A^+ is a finite normal extension of K.

Now, H_A^+ contains the smallest field of definition K_ρ of ρ. By Exercise 13.7.11, K_ρ/K is a finite subextension of K_∞/K. It follows that K_ρ/K is a separable extension. Let $\xi \in \mathbb{C}_\infty$ and $\rho' = \xi\rho\xi^{-1}$ be such that ρ' is defined over K_ρ. Let a be any nonconstant positive element of A. Then by Exercise 13.7.13 (3)

$$\xi^{1-q^{r \deg a}} \mu_\rho(a) = \mu_{\rho'}(a) \in K_\rho,$$

which implies that $K_\rho(\xi)/K_\rho$ and $K(\xi)/K$ are separable extensions. It follows that H_A^+/K is separable and therefore Galois since H_A^+ is a subextension of $K_\rho(\xi)$.

By Exercise 13.7.15 we have $\mathfrak{A} * \sigma\rho = \sigma(\mathfrak{A} * \rho)$ for any $\sigma \in \operatorname{Gal}(\mathbb{C}_\infty/K)$, $\rho \in \operatorname{Drin}_A(\mathbb{C}_\infty)$ and $\mathfrak{A}$ any nonzero ideal of A. Therefore the action of $\operatorname{Gal}(H_A^+/K)$ commutes with the action of $\operatorname{Pic}^+ A$.

Define $\theta\colon \operatorname{Gal}(H_A^+/K) \to \operatorname{Pic}^+ A$ as follows. If $\sigma \in \operatorname{Gal}(H_A^+/K)$, then $\theta(\sigma) = \bar{\mathfrak{A}}_\sigma$ where $\mathfrak{A}_\sigma$ satisfies $\sigma\rho = \mathfrak{A}_\sigma * \rho$. The homomorphism θ is injective by Theorem 13.5.20 since $\sigma\rho \neq \rho$ for $\sigma \neq \mathrm{Id}$. □

In order to study ramification in H_A^+/K we need to consider the inertia group which is related to the reduction map mod $\mathfrak{P}$ for a place $\mathfrak{P}$. Therefore we study the reduction map. Let B^+ be the integral closure of A in H_A^+.

At this point we make a detour to discuss how a Drinfeld module can be reduced to some residue fields.

In general, let $\rho \in \operatorname{Drin}_A(k)$ of rank r. Suppose that the field k has a discrete valuation v and with all the coefficients of ρ_a integral at v. Let ϑ_v be the valuation ring at v with maximal ideal $\mathfrak{p}$ and residue field $k(\mathfrak{p})$. We take the coefficients of ρ_a mod $\mathfrak{p}$ and denote this reduction by $\rho^{(\mathfrak{p})}$. In general $\rho^{(\mathfrak{p})}$ is not a Drinfeld A-module if all the nonconstant terms are congruent to 0 mod $\mathfrak{p}$.

Definition 13.5.23. We say that ρ has *stable reduction* at $\mathfrak{p}$, if there exists a Drinfeld A-module $\rho' \in \operatorname{Drin}_A(k)$ isomorphic to ρ such that the coefficients of ρ'_a are integral at v for all $a \in A$ and $\rho'^{(\mathfrak{p})}$ is a Drinfeld A-module $\rho'^{(\mathfrak{p})} \in \operatorname{Drin}_A(k(\mathfrak{p}))$.

We say that ρ has *good reduction* at $\mathfrak{p}$ if in addition $\rho'^{(\mathfrak{p})}$ has rank r.

Remark 13.5.24. If ρ has rank one, then every stable reduction is good.

The key fact is that even if ρ has no stable reduction at $\mathfrak{p}$, there exists an extension k' of k such that ρ has stable reduction over k'.

Definition 13.5.25. We say that ρ has *potential stable reduction* (resp. *potential good reduction*) if there exists an extension k' of k such that ρ has stable reduction (resp. good reduction) over k'.

Example 13.5.26. If $A = \mathbb{F}_q[T]$, then for any $r > 1$ the Drinfeld A-module $\rho_T = T + \tau + a_2\tau^2 + \cdots + a_{r-1}\tau^{r-1} + T\tau^r$ has stable reduction but not good reduction. Also, the Drinfeld A-module $\phi_T = T + T\tau + T\tau^2 + \cdots + T\tau^r$ does not have stable reduction.

Theorem 13.5.27. *Every Drinfeld module ρ over a field k with a discrete valuation v has potential stable reduction. In particular if ρ is of rank one, ρ has potential good reduction.*

Proof. For $a \in A$ let $\rho_a = \sum a_i\tau^i$, and set

$$\gamma_a = \min_{i>0} \frac{v(a_i)}{q^i - 1}.$$

Let $x_1, \ldots, x_s$ be a set of generators of A as $\mathbb{F}_q$–algebra and set $\gamma := \min_{1\leq j\leq s} \gamma_{x_j}$. There exists a finite extension k' of k with a valuation v' extending v and an element $x \in k'$ such that $v'(x) = \gamma$. Then it is easy to verify that conjugation by x gives a Drinfeld module isomorphic to ρ which has stable reduction. □

We return to our discussion. In our case a sgn-normalized A-module ρ of rank one is defined over B^+ by Theorem 13.5.27 and ρ may be reduced at every nonzero prime ideal $\mathfrak{T}$ of B^+. Let $\pi_{\mathfrak{T}}\colon B^+ \to B^+/\mathfrak{T}$ be the reduction map and let $\mathfrak{P} := \mathfrak{T} \cap A$.

Proposition 13.5.28. *The reduction map $\rho \mapsto \pi_{\mathfrak{T}} \circ \rho$ is injective on $\mathfrak{H}$.*

Proof. Assume that ρ and ρ' belonging to $\mathfrak{H}$ reduce modulo $\mathfrak{T}$ to the same $\phi \in \operatorname{Drin}_A((B^+/\mathfrak{T}))$, that is, $\phi = \rho^{(\mathfrak{T})} = \rho'^{(\mathfrak{T})}$. By Theorem 13.5.20 there exists an ideal $\mathfrak{A}$ of A such that $\rho' = \mathfrak{A} * \rho$. Using an argument similar to that of Exercise 5.10.36, we may assume that $\mathfrak{A}$ is relatively prime to $\mathfrak{T}$.

Reducing the defining equation $\rho_{\mathfrak{A}}\rho_x = \rho'_x\rho_{\mathfrak{A}}$ modulo $\mathfrak{T}$, we obtain $\rho_{\mathfrak{A}}\rho_x \equiv \rho_x\rho_{\mathfrak{A}} \bmod \mathfrak{T}$ for all $x \in A$. It follows that $\pi_{\mathfrak{T}}(\rho_{\mathfrak{A}}) \in \operatorname{End}(\pi_{\mathfrak{T}} \circ \rho)$. Now we know that $\operatorname{End}(\pi_{\mathfrak{T}} \circ \rho) = A$ (see [63, Corollary 5.14] or [151, Theorem 2.7.2]). Therefore there exists $a \in A$ such that

$$\rho_{\mathfrak{A}} \equiv \rho_a \bmod \mathfrak{T}. \tag{13.39}$$

Since the leading coefficient of $\rho_{\mathfrak{A}}$ is one, $\mu_\rho(a) = 1$. Thus a is a positive element of A. The proof will follow if we prove that $\mathfrak{A} = aA$ (see (13.34)). Define $\mathfrak{B} := \mathfrak{A} + aA$. By (13.39) the torsion modules $\rho^{(\mathfrak{T})}[\mathfrak{B}]$, $\rho^{(\mathfrak{T})}[\mathfrak{A}]$ and $\rho^{(\mathfrak{T})}[aA]$ in an algebraic closure $\overline{B^+/\mathfrak{T}}$ of $B^+/\mathfrak{T}$ are the same. From the proof of Theorem 13.3.19 we obtain that $|A/\mathfrak{A}| = |A/\mathfrak{B}| = |A/aA|$. It follows that $\mathfrak{B} = \mathfrak{A} = aA$. □

Now we can prove our claim about the ramification of H_A^+/K.

Proposition 13.5.29. *The extension H_A^+/K is unramified at every finite place $\mathfrak{P}$ of A.*

Proof. Let σ be an element of the inertia group of $\mathfrak{P}$. Therefore $\sigma\rho \equiv \rho \bmod \mathfrak{Q}$, where $\mathfrak{Q}$ is the prime ideal of B^+ above $\mathfrak{P}$.

From Proposition 13.5.28, it follows that $\sigma\rho = \rho$. Since H_A^+ is generated by the coefficients of ρ, it follows that $\sigma = \mathrm{Id}$. The result is now a consequence of Corollary 5.2.19. □

For a nonzero ideal $\mathfrak{A}$ in A, let $\sigma_{\mathfrak{A}} = \left(\frac{H_A^+/K}{\mathfrak{A}}\right)$ be the Artin automorphism associated to $\mathfrak{A}$. That is, $\sigma_{\mathfrak{A}} = \prod_{i=1}^s \sigma_{\mathfrak{P}_i}^{\alpha_i}$ where $\mathfrak{A} = \prod_{i=1}^s \mathfrak{P}_i^{\alpha_i}$ is the prime decomposition of $\mathfrak{A}$ and $\sigma_{\mathfrak{P}_i} = \left(\frac{H_A^+/K}{\mathfrak{P}_i}\right)$ is the Artin symbol for the prime $\mathfrak{P}_i$.

One of the main results in the theory of Drinfeld A-modules of rank one over $\mathbb{C}_\infty$ is the following theorem.

Theorem 13.5.30. *For every Hayes A-module ρ, we have*

$$\sigma_{\mathfrak{A}}\rho = \mathfrak{A} * \rho. \tag{13.40}$$

In particular $\mathrm{Gal}(H_A^+/K)$ *is isomorphic to* $\mathrm{Pic}^+ A$ *and* $[H_A^+ : K] = \frac{q^{d_\infty}-1}{q-1} h_A = \frac{q^{d_\infty}-1}{q-1} d_\infty h_K$.

Proof. Since for any nonzero ideals $\mathfrak{A}$, $\mathfrak{B}$ of A we have

$$\mathfrak{A} * (\mathfrak{B} * \rho) = (\mathfrak{A}\mathfrak{B}) * \rho,$$

it suffices to show (13.40) for $\mathfrak{A} = \mathfrak{P}$ a nonzero prime ideal of A.

Let $\mathfrak{Q}$ be a prime divisor of B^+ above $\mathfrak{P}$ and consider the Frobenius automorphism $\sigma_{\mathfrak{Q}}$ at $\mathfrak{Q}$ where $\sigma_{\mathfrak{Q}} = \sigma_{\mathfrak{P}}$ is the Artin symbol at $\mathfrak{P}$. Then

$$\sigma_{\mathfrak{P}}(x) \equiv x^{N\mathfrak{P}} \bmod \mathfrak{Q} \quad \text{for all} \quad x \in B^+.$$

Let $\rho' = \mathfrak{P} * \rho$. then for any $y \in A$ we have

$$\rho_{\mathfrak{P}}\rho_y = \rho'_y \rho_{\mathfrak{P}}. \tag{13.41}$$

Now the reduction $\rho \bmod \mathfrak{Q} := \phi$ satisfies that $r_\phi = 1$. Since $\operatorname{char}\phi = \mathfrak{Q}$ and $1 \le h_\phi < r_\phi = 1$, we have $h_\phi = 1$. By the proof of Theorem 13.3.19 we obtain that $\phi_{\mathfrak{P}} = \tau^{\deg \mathfrak{P}}$. Reducing (13.41) mod $\mathfrak{Q}$ we obtain

$$\tau^{\deg \mathfrak{P}} \rho_y = \rho'_y \tau^{\deg \mathfrak{P}} \bmod \mathfrak{Q}. \tag{13.42}$$

Let $\rho_y = \sum_{i=0}^{\deg y} a_i \tau^i$, $\rho'_y = \sum_{j=0}^{\deg y} b_j \tau^j$. Then (13.42) implies

$$a_i^{N(\mathfrak{P})} \equiv b_i \bmod \mathfrak{Q}.$$

Therefore

$$\sigma_{\mathfrak{P}}\rho_y = \sum_{i=0}^{\deg y}(\sigma_{\mathfrak{P}}a_i)\tau^i \equiv \sum_{i=0}^{\deg y} a_i^{N(\mathfrak{P})}\tau^i \equiv \sum_{i=0}^{\deg y} b_i\tau^i = (\mathfrak{P} * \rho)_y \bmod \mathfrak{Q}.$$

Since reduction mod $\mathfrak{Q}$ is injective, it follows that $\sigma_{\mathfrak{P}}\rho = \mathfrak{P} * \rho$. □

Corollary 13.5.31. *For $x \in K^*$, define σ_x as the Artin symbol σ_{xA} corresponding to the principal ideal xA. Then $\sigma_x\rho = \mu_\rho(x)^{-1}\rho\mu_\rho(x)$.*

Proof. Exercise 13.7.17. □

Finally we prove the *Principal Ideal Theorem* for H_A^+.

Theorem 13.5.32. *Let $\mathfrak{A}$ be any nonzero ideal of A. Then $\operatorname{con}_{A/B^+}\mathfrak{A} = \mathfrak{A}B^+ = D(\rho_{\mathfrak{A}})B^+$ where $D(\rho_{\mathfrak{A}})$ is the constant term of $\rho_{\mathfrak{A}}$.*

Proof. It is easy to see that for any two nonzero ideals $\mathfrak{A}$ and $\mathfrak{B}$ of A we have $D(\rho_{\mathfrak{A}\mathfrak{B}}) = D((\mathfrak{B} * \rho)_{\mathfrak{A}})D(\rho_{\mathfrak{B}})$. Thus it suffices to consider $\mathfrak{A} = \mathfrak{P}$ any nonzero prime ideal.

We have that all the coefficients of $\rho_{\mathfrak{P}}$ other than the leading coefficient belong to any ideal $\mathfrak{Q}$ above $\mathfrak{P}$ (see (13.42)). Choose $x \in A$ such that $v_{\mathfrak{Q}}(x) = 1$. We write $xA = \mathfrak{P}\mathfrak{C}$. Then by Exercise 13.7.16, we obtain $\rho_x = \mu_\rho(x)(\mathfrak{P} * \rho)_{\mathfrak{C}}\rho_{\mathfrak{P}}$. It follows that

$$1 = v_{\mathfrak{Q}}(D(\rho_x)) = v_{\mathfrak{Q}}(D(\mathfrak{P} * \rho)_{\mathfrak{C}}) + v_{\mathfrak{Q}}(D(\rho_{\mathfrak{P}})) \geq v_{\mathfrak{Q}}(D(\rho_{\mathfrak{P}})).$$

Therefore $v_{\mathfrak{Q}}(D(\rho_{\mathfrak{P}})) = 1$ for any ideal $\mathfrak{Q}$ of B^+ dividing $\mathfrak{P}$. The result will follow by showing that no other nonzero prime ideal of B^+ divides $D(\rho_{\mathfrak{P}})$. Let $\mathfrak{T}$ be another prime ideal of B^+, $\mathfrak{T} \nmid \mathfrak{P}$. Let $e \geq 1$ be such that $\mathfrak{P}^e = yA$ is principal. Set $\mathfrak{D} = \mathfrak{P}^{e-1}$. Then

$$v_{\mathfrak{T}}(D((\mathfrak{P} * \rho)_{\mathfrak{D}})) + v_{\mathfrak{T}}(D(\rho_{\mathfrak{P}})) = v_{\mathfrak{T}}(\mu_\rho(y)^{-1}y) = 0.$$

Since both ρ and $\mathfrak{P} * \rho$ are defined over B^+, it follows that all the above valuations are nonnegative. Hence $v_{\mathfrak{T}}(D(\rho_{\mathfrak{P}})) = 0$. □

13.5.4 The Hilbert Class Field H_A

We return to K_ρ, the field of definition of a Drinfeld A-module $\rho \in \operatorname{Drin}_A(\mathbb{C}_\infty)$ of rank one. We may assume that ρ is sgn–normalized. We have $K\mathbb{F}_\infty \subseteq K_\rho \subseteq H_A^+$.

Let $\xi \in \mathbb{C}_\infty$ be so that $\rho' = \xi\rho\xi^{-1}$ is defined over K_ρ. Since the group of automorphisms of ρ is $\mathbb{F}_q^*$, the greatest common divisor g given in the proof of Theorem 13.5.9 is $g = q - 1$. From (13.29) we obtain that $\xi_0 := \xi^{q-1} \in H_A^+$, because H_A^+ is the field generated by the coefficients of ρ. Since ρ is sgn–normalized, $\mu_\rho(\pi^{-1}) = 1$ and by Exercise 13.7.13 (4) we obtain that

$$\xi_0^{(q^{d_\infty}-1)/(q-1)} = \xi^{q^{d_\infty}-1} = \mu_{\rho'}(\pi^{-1})^{-1} \in K_\rho.$$

Furthermore, $H_A^+ = K_\rho(\xi_0)$ since the coefficients of $\rho_x = \xi\rho'_x\xi^{-1}$ generate H_A^+ for any nonconstant $x \in A$. Therefore $[H_A^+ : K_\rho] \leq (q^{d_\infty} - 1)/(q-1)$.

Now we consider the exact sequence

$$1 \longrightarrow \frac{P_A}{P_A^+} \longrightarrow \text{Pic}^+ A \xrightarrow{\theta} \text{Pic } A \to 0 \tag{13.43}$$

where θ is the natural map.

We identify $\text{Pic}^+ A$ with $\text{Gal}(H_A^+/K)$ ((13.40)). Under this identification, we have

Proposition 13.5.33. *The subfield K_ρ of H_A^+ is the fixed field of H_A^+ of the subgroup of* $\text{Pic}^+ A$ *generated by σ_x, $x \in K^*$. Furthermore, the extension $H_A^+/K_\rho = K_\rho(\xi_0)/K_\rho$ is a cyclic Kummer extension of degree $(q^{d_\infty} - 1)/(q-1)$ and for any $x \in K^*$ we have*

$$\xi_0^{\sigma_x} = \mu_\rho(x)^{q-1}\xi_0. \tag{13.44}$$

In particular, K_ρ is independent of the choice of ρ.

Proof. From Corollary 13.5.31, σ_x fixes all the invariants I_i in (13.28) which generate K_ρ. Therefore each σ_x fixes K_ρ. Denote again by σ_x some extension of σ_x to a monomorphism of $H_A^+(\xi)$ into $\mathbb{C}_\infty$. Then, by Corollary 13.5.31, we obtain

$$\begin{aligned}\rho' = \sigma_x\rho' &= \xi^{\sigma_x}\sigma_x\rho\xi^{-\sigma_x} = \xi^{\sigma_x}\mu_\rho(x)^{-1}\rho\mu_\rho(x)\xi^{-\sigma_x}\\ &= (\xi^{\sigma_x-1}\mu_\rho(x)^{-1})\rho'(\xi^{\sigma_x-1}\mu_\rho(x)^{-1})^{-1}\end{aligned}$$

where $\rho_x = \xi\rho'_x\xi^{-1}$ for $x \in A$.

Therefore $\xi^{\sigma_x-1}\mu_\rho(x)^{-1}$ is an automorphism of ρ' and so it is an element of $\mathbb{F}_q^*$. The definition of ξ_0 implies (13.44). It follows that $[H_A^+ : K_\rho] \geq (q^{d_\infty} - 1)/(q-1)$ and therefore $[H_A^+ : K_\rho] = (q^{d_\infty} - 1)/(q-1)$. Since P_A/P_A^+ is isomorphic to $\mathbb{F}_\infty^*/\mathbb{F}_q^*$ and thus of order $(q^{d_\infty} - 1)/(q-1)$, we have that K_ρ is indeed the fixed field of the subgroup $\{\sigma_x \mid x \in K^*\}$ of $\text{Pic}^+ A$. □

Definition 13.5.34. The common field of definition of the rank one Drinfeld A-modules over $\mathbb{C}_\infty$ is called the *Hilbert class field* of A and it is denoted by H_A.

One of the main results in class field theory is next theorem.

Theorem 13.5.35. *The prime $\mathfrak{P}_\infty$ splits completely in the extension H_A/K and every prime divisor $\mathfrak{P}$ of K is unramified in H_A/K. The extension H_A/K is of degree h_A with Galois group isomorphic to* $\text{Pic } A$ *under the Artin map. If ρ is a Drinfeld A-module defined over H_A, then*

$$\sigma_{\mathfrak{A}}\rho = \mathfrak{A} * \rho \tag{13.45}$$

for any nonzero ideal $\mathfrak{A}$ in A, where $\sigma_{\mathfrak{A}}$ is the Artin map.

Proof. Since H_A/K is a Galois subextension of K_∞/K, it follows that $\mathfrak{P}_\infty$ splits completely in H_A/K. Since $H_A \subseteq H_A^+$ every finite place of K is unramified in H_A.

By (13.43) and Proposition 13.5.33 we obtain that $\mathrm{Gal}(H_A/K) \cong \operatorname{Pic} A$. Finally (13.45) is an immediate consequence of (13.40). □

Now we have that the maximal unramified abelian extension of K such that $\mathfrak{P}_\infty$ splits completely has Galois group isomorphic to $\operatorname{Pic} A$ (see [127]). Thus H_A is precisely this field. That is, H_A is the maximal unramified extension of K in which $\mathfrak{P}_\infty$ splits completely.

The proof of the next result can be found in [63, Theorem 15.8] or [151, Theorem 3.5.1].

Theorem 13.5.36. *Let B be the integral closure of A in H_A. Then every rank one Drinfeld A-module ρ is isomorphic to an A-module ρ' which is defined over B and where $\mu_{\rho'}(a)$ is a unit in B.* □

As a consequence we obtain the following theorem:

Theorem 13.5.37 (Principal Ideal Theorem). *Let ρ be a rank one A-module which is defined over B. If $\mathfrak{A}$ is any nonzero ideal in A, then $\mathfrak{A}B = D(\rho_{\mathfrak{A}})B$ is principal generated by $D(\rho_{\mathfrak{A}})$.*

Proof. Similar to that of Theorem 13.5.32. □

13.5.5 Explicit Class Fields and Ray Class Fields

Now we construct the maximal abelian extension of a congruence function field K. This construction is analogous to that for cyclotomic function fields. We fix a sgn function. Let $\mathfrak{m}$ be any nonzero proper ideal of A. Let $K_\mathfrak{m} := K(\rho[\mathfrak{m}])$. Exactly as in Section 13.5.3 we will see that $K_\mathfrak{m}$ is a Galois extension of K unramified away from the prime ideals dividing $\mathfrak{m}$ and $\mathfrak{P}_\infty$. We have that $K_\mathfrak{m}$ is a narrow ray class field of conductor $\mathfrak{m}$ (see [78]). We define $K_\mathfrak{m}^+$ as the maximal extension of K contained in $K_\mathfrak{m}$ in which $\mathfrak{P}_\infty$ splits completely. It turns out that $K_\mathfrak{m}^+$ is the ray class field modulo $\mathfrak{m}$. In this way, we will obtain an explicit description of the maximal abelian extension of K in which $\mathfrak{P}_\infty$ splits completely. One obtains all class fields by varying $\mathfrak{P}_\infty$. The techniques to study $K_\mathfrak{m}$ are similar to those of H_A^+ (see Section 13.5.3).

To begin with, let $M_{\mathfrak{m},A}$ be the fractional ideals of A generated by the prime ideals $\mathfrak{P}$ not dividing $\mathfrak{m}$ and let

$$P_{\mathfrak{m},A}^+ = \{xA \mid x \in K^*, x \text{ positive}, x \equiv 1 \bmod \mathfrak{m}\}.$$

Definition 13.5.38. The quotient group $\operatorname{Pic}_\mathfrak{m}^+ A = M_{\mathfrak{m},A}/P_{\mathfrak{m},A}^+$ is called the *narrow ray class group* modulo $\mathfrak{m}$ relative to sgn.

We consider the set of Hayes modules $\mathfrak{H}$. We have $\rho[\mathfrak{m}] \cong A/\mathfrak{m}$ as A-modules (Remark 13.3.20). Recall that $\Phi(\mathfrak{m}) = \left|(A/\mathfrak{m})^*\right|$. Then $\rho[\mathfrak{m}]$ has $\Phi(\mathfrak{m})$ generators as an A-module. Consider the set $X_{\mathfrak{m}}$ consisting of the pairs (ρ, λ), where $\rho \in X$ and λ is a generator of $\rho[\mathfrak{m}]$. We define the action

$$\mathfrak{A} * (\rho, \lambda) := (\mathfrak{A} * \rho, \rho_{\mathfrak{A}}(\lambda)) \tag{13.46}$$

of $M_{\mathfrak{m},A}$ on $X_{\mathfrak{m}}$. As in Section 13.5.3, we have that the stabilizer of any point is $P^+_{\mathfrak{m},A}$. Now, on the one hand we have $|X_{\mathfrak{m}}| = |\mathfrak{H}|\Phi(\mathfrak{m}) = |\operatorname{Pic}^+ A|\Phi(\mathfrak{m})$ and on the other hand we have the exact sequence

$$0 \longrightarrow \frac{M_{\mathfrak{m},A} \cap P_A^+}{P^+_{\mathfrak{m},A}} \longrightarrow \operatorname{Pic}^+_{\mathfrak{m}} A \stackrel{\theta}{\longrightarrow} \operatorname{Pic}^+ A \longrightarrow 0.$$

We have $\left|\frac{M_{\mathfrak{m},A} \cap P_A^+}{P^+_{\mathfrak{m},A}}\right| = \Phi(\mathfrak{m})$ (see [78, Chapter IV]). Hence $X_{\mathfrak{m}}$ and $\operatorname{Pic}^+_{\mathfrak{m}} A$ have the same cardinality and therefore we obtain the analogue of Theorem 13.5.20.

Theorem 13.5.39. *The set $X_{\mathfrak{m}}$ is a principal homogeneous space for* $\operatorname{Pic}^+_{\mathfrak{m}} A$ *under the action $*$ given in (13.46).* □

Let $K(\mathfrak{m}) := H_A^+(\rho[\mathfrak{m}])$. As in the case of H_A^+ it will be proved that $K(\mathfrak{m})/K$ is a Galois extension and unramified away from $\mathfrak{P}_\infty$ and the prime ideals dividing $\mathfrak{m}$. First we prove the analogue of Proposition 12.3.18

Proposition 13.5.40. *Let L/K be a finite extension and let $\rho \in$* $\operatorname{Drin}_A(\mathbb{C}_\infty)$ *of rank one which is defined over a finite valuation ring $\vartheta_{\mathfrak{T}}$ in L where $\mathfrak{T}$ is unramified in L/K. Let $\mathfrak{P} := \mathfrak{T} \cap A$. Set $\mathfrak{A} = \mathfrak{P}^e$ and $\mathfrak{B} = \mathfrak{P}^{e-1}$. Then $\rho_{\mathfrak{B}}(t)$ divides $\rho_{\mathfrak{A}}(t)$ in $\vartheta_{\mathfrak{T}}[t]$ and the quotient is Eisenstein at $\mathfrak{P}$.*

Proof. The proof for the case $e = 1$ is similar to that given in the proof of Theorem 13.5.32.

For $e > 1$, let $f(t) := (\mathfrak{B} * \rho)_{\mathfrak{P}}(t)/t$. Then

$$\rho_{\mathfrak{A}}(t) = f(\rho_{\mathfrak{B}}(t))\rho_{\mathfrak{B}}(t).$$

By case $e = 1$, we know that $f(t)$ is Eisenstein at $\mathfrak{P}$ and $\rho_{\mathfrak{B}}(t) \equiv t^{N(\mathfrak{B})} \bmod \mathfrak{T}$ (see Theorem 13.3.19). □

By Proposition 13.5.40 it follows that in the extension $K(\mathfrak{m})/K$ we have the same type of ramification as in the cyclotomic case. More precisely, we have

Proposition 13.5.41. *Let $\mathfrak{m} = \mathfrak{P}^e$ where $\mathfrak{P}$ is a prime ideal of A. Then the extension $K(\mathfrak{P}^e) = H_A^+(\rho[\mathfrak{P}^e])/H_A^+$ is fully ramified at $\mathfrak{T}$, where $\mathfrak{T}$ is a prime divisor of H_A^+ above $\mathfrak{P}$ and the ramification index is $\Phi(\mathfrak{P}^e)$. Furthermore, the extension $K(\mathfrak{P}^e)/H_A^+$ is unramified at every finite prime ideal $\mathfrak{P}_1 \neq \mathfrak{P}$ and at $\mathfrak{P}_\infty$. Finally, we have $[K(\mathfrak{P}^e) : H_A^+] = \Phi(\mathfrak{P}^e)$.*

Proof. The polynomial $f(u) := \rho_{\mathfrak{P}^e}(u)/\rho_{\mathfrak{P}^{e-1}}(u)$ is Eisenstein and $f(u) = \prod(u-\lambda)$ where the product runs over the set of generators of the A-cyclic module $\rho[\mathfrak{P}^e]$. We also have that $\deg_u f(u) = \Phi(\mathfrak{P}^e)$. The proof follows as in the one of Proposition 12.3.14. □

Corollary 13.5.42. *For any nonzero ideal* $\mathfrak{m}$ *of* A, $K(\mathfrak{m})/H_A^+$ *is a Galois extension with Galois group isomorphic to* $(A/\mathfrak{m})^*$. *The ramified primes are the prime ideals* $\mathfrak{P}$ *dividing* $\mathfrak{m}$ *with ramification index* $\Phi(\mathfrak{P}^e)$ *where* $\mathfrak{P}^e$ *is the exact power of* $\mathfrak{P}$ *dividing* $\mathfrak{m}$.

Proof. Similar to that of Theorem 12.5.3. □

Now $K_\mathfrak{m} := K(\rho[\mathfrak{m}])$ is a normal extension and therefore $\sigma(K_\mathfrak{m}) = K(\sigma\rho[\mathfrak{m}]) = K_\mathfrak{m}$ for any $\sigma \in \mathrm{Gal}(\mathbb{C}_\infty/K)$.

Since H_A^+ is generated by the coefficients of ρ_a, with $\rho \in \mathfrak{H}$ and a a nonconstant element of A, it follows that $K_\mathfrak{m} = K(\mathfrak{m})$.

The next result is a complement of Theorem 13.5.30.

Theorem 13.5.43. *Let* $\mathfrak{A}$ *be any nonzero ideal of* A *which is prime to* $\mathfrak{m}$ *and let* $\lambda \in \rho[\mathfrak{m}]$. *Then if* $\sigma_\mathfrak{A}$ *is the Artin automorphism, we have*

$$\lambda^{\sigma_\mathfrak{A}} := \sigma_\mathfrak{A}\lambda = \rho_\mathfrak{A}(\lambda). \tag{13.47}$$

Proof. If $\mathfrak{A}$ and $\mathfrak{B}$ are two nonzero ideals of A, prime to $\mathfrak{m}$ and $\sigma_\mathfrak{A}$ and $\sigma_\mathfrak{B}$ satisfy (13.47), then by Theorem 13.5.30, and Exercise 13.7.16 we have

$$\begin{aligned}\sigma_{\mathfrak{A}\mathfrak{B}}(\lambda) &= \sigma_\mathfrak{A}\sigma_\mathfrak{B}\lambda = \sigma_\mathfrak{B}\sigma_\mathfrak{A}\lambda = \sigma_\mathfrak{B}(\rho_\mathfrak{A}(\lambda)) = (\sigma_\mathfrak{B}\rho_\mathfrak{A})(\sigma_\mathfrak{B}\lambda)\\ &= (\sigma_\mathfrak{B}\rho_\mathfrak{A})(\rho_\mathfrak{B}(\lambda)) = (\mathfrak{B}*\rho)_\mathfrak{A}(\rho_\mathfrak{B}(\lambda)) = \rho_{\mathfrak{A}\mathfrak{B}}(\lambda).\end{aligned}$$

Thus, we may assume $\mathfrak{A} = \mathfrak{P}$ to be prime. If $\mathfrak{T}$ in $K_\mathfrak{m}$ is a prime ideal above $\mathfrak{P}$, $\sigma_\mathfrak{P}$ satisfies $\sigma_\mathfrak{P}\lambda \equiv \lambda^{N\mathfrak{P}} \bmod \mathfrak{T}$.

Since $\rho \bmod \mathfrak{T} = \phi$ satisfies $\phi_\mathfrak{P} = \tau^{\deg \mathfrak{P}}$, it follows as in the proof of Theorem 13.5.30 that $\rho_\mathfrak{P}(\lambda) \equiv \lambda^{N(\mathfrak{P})} \bmod \mathfrak{T}$, and therefore $\sigma_\mathfrak{P}(\lambda) = \rho_\mathfrak{P}(\lambda)$. □

As a consequence of Theorem 13.5.43 we see that $K_\mathfrak{m}$ is independent of ρ (see (13.38)). We also have that $\mathrm{Pic}_\mathfrak{m}^+ A$ acts on $K_\mathfrak{m}$ as automorphisms via (13.47).

Since $\mathrm{Pic}_\mathfrak{m}^+ A$ and $[K_\mathfrak{m} : K]$ have the same cardinality, it follows that $\mathrm{Pic}_\mathfrak{m}^+ A \cong \mathrm{Gal}(K_\mathfrak{m}/K)$ as in Theorem 13.5.30.

Now, the positive elements of A generate $A/\mathfrak{m}$, so the map $a \mapsto \sigma_a := \sigma_{aA}$, where $a \in A$ is a nonzero element of A prime to $\mathfrak{m}$, induces an isomorphism between $(A/\mathfrak{m})^*$ and $\mathrm{Gal}(K_\mathfrak{m}/H_A^+)$. For an element $\lambda \in \rho[\mathfrak{m}]$ and $x \in A$ congruent to 1 mod $\mathfrak{m}$ in K^*, by (13.47) we obtain

$$\sigma_x(\lambda) = \rho_{xA}(\lambda) = \mu_\rho(x)^{-1}\lambda$$

and $\mu_\rho(x)^{-1} \in \mathbb{F}_\infty^*$. Therefore $\mathrm{Gal}(K_\mathfrak{m}/K)$ contains a subgroup $I_{\mathfrak{P}_\infty}$ isomorphic to $\mathbb{F}_\infty^*$ and in fact Hayes has shown that $I_{\mathfrak{P}_\infty}$ is both the decomposition and the inertia groups at $\mathfrak{P}_\infty$.

Definition 13.5.44. The fixed field of $K_{\mathfrak{m}}$ under $I_{\mathfrak{P}_\infty}$, $K_{\mathfrak{m}}^+ := K_{\mathfrak{m}}^{I_{\mathfrak{P}_\infty}}$ is called the *ray class field of conductor* $\mathfrak{m}$.

We have that $K_{\mathfrak{m}}^+$ is the ray class field of K of conductor $\mathfrak{m}$ that is completely split over $\mathfrak{P}_\infty$. This situation is analogous to the familiar situation of cyclotomic fields, where $K_{\mathfrak{m}}$ plays the role of the usual cyclotomic number field $\mathbb{Q}(\zeta_m)$ and $K_{\mathfrak{m}}^+$ that of $\mathbb{Q}(\zeta_m)^+$, the maximal real subfield of $\mathbb{Q}(\zeta_m)$. Note that the ramification index of $\mathfrak{P}_\infty$ in $K_{\mathfrak{m}}/K_{\mathfrak{m}}^+$ and of $K_{\mathfrak{m}}/K$ is $(q^{d_\infty} - 1)$.

Now if K_∞^+ is the union of all $K_{\mathfrak{m}}^+$ where $\mathfrak{m}$ runs through all nonzero proper ideals $\mathfrak{m}$ of A. Then K_∞^+ is the maximal abelian extension of K in which $\mathfrak{P}_\infty$ splits completely. If $K_\infty = \bigcup_{\mathfrak{m}} K_{\mathfrak{m}}$, then K_∞ is the fixed field of K_∞^+ under $I_{\mathfrak{P}_\infty}$.

We have (Theorem 12.8.25) $\operatorname{Gal}(K_\infty/K) \cong \mathcal{U}_{\mathfrak{P}_\infty}$ where $\mathcal{U}_{\mathfrak{P}_\infty}$ corresponds to the $\mathcal{U}_T$ defined in Chapter 12. That is, $\mathcal{U}_{\mathfrak{P}_\infty}$ is the subgroup of the idele group whose $\mathfrak{P}_\infty$ component is 1 and whose other components are elements of $\vartheta_{\mathfrak{P}}^*$. More precisely, $\operatorname{Gal}(K_\infty/K) \cong J_K/(K^* \times K_\infty^{(1)} \times \mathbb{Z})$ and K_∞ corresponds to $K^* \times K_\infty^{(1)} \times \mathbb{Z}$ (see Theorem 12.8.25).

If we take $\mathfrak{P}'_\infty \neq \mathfrak{P}_\infty$, $\mathfrak{P}'_\infty$ a prime divisor of K and we consider K'_∞, the intersection of the corresponding idele subgroups is $\{1\}$ so $K_\infty K'_\infty$ is the maximal abelian extension of K.

Theorem 13.5.45. *Let $\mathfrak{P}_\infty$, $\mathfrak{P}'_\infty$ be two different prime divisors of K. If K_∞ and K'_∞ are as above, then $K_\infty K'_\infty$ is the maximal abelian extension of K.* □

13.6 Drinfeld Modules and Cryptography

The similarity between elliptic curves and Drinfeld modules of rank two allows us to define a cryptosystem based on Drinfeld modules. In 2001, T. Scanlon [132] used this idea. Unfortunately, he showed that his approach was insecure. In 2003, R. Gillard et al. [49] proposed a new public-key cryptosystem based on Drinfeld modules. However, S. R. Blackburn et al. [6] showed that this cryptosystem is also insecure. In this section we present these Drinfeld-module-based cryptosystems.

Definition 13.6.1. The *discrete logarithm problem for a Drinfeld module* $\rho\colon A \to k\langle\tau\rangle$ is as follows: given $a \in A$ such that $\rho_a\colon k \to k$ is bijective, find $b \in A$ such that $\rho_b\colon k \to k$ is the inverse of ρ_a.

Note that public-key cryptosystems based on the intractability of the discrete algorithm problem for certain groups, for instance, Diffie–Hellman, ElGamal, have natural Drinfeld module analogues. Likewise, cryptosystems such as RSA admit Drinfeld module versions. Here we present briefly the Drinfeld module version of the Diffie–Hellman cryptosystem.

13.6.1 Drinfeld Module Version of the Diffie–Hellman Cryptosystem

Let $A = \mathbb{F}_q[T]$ and $k = \mathbb{F}_q$. Let $\rho\colon A \to k\langle\tau\rangle$ be any Drinfeld module, where $\delta\colon A \to \mathbb{F}_q$ is the natural projection. Fix an arbitrary element ξ in k. Then A, k, ρ, δ, and ξ constitute the public key. Arnold and Charlotte choose a and b in A respectively. Arnold transmits $\rho_a(\xi)$ to Charlotte, while Charlotte transmits $\rho_b(\xi)$ to Arnold. The common private key is

$$\rho_b(\rho_a(\xi)) = \rho_{ba}(\xi) = \rho_{ab}(\xi) = \rho_a(\rho_b(\xi)). \tag{13.48}$$

Now a possible attack to this cryptosystem would come from the fact that the ring of functions induced by a Drinfeld module on a finite field is isomorphic to a ring of linear functions (here we are talking of $\mathbb{F}_p$-linearity and not k-linearity).

Proposition 13.6.2. *For any Drinfeld module $\rho\colon A \to k\langle\tau\rangle$ and any $a \in A$ such that $\rho_a\colon k \to k$ is bijective, there exist real numbers C and r and an algorithm to find an inverse to ρ_a using at most $C(\log_p |k|)^r$ field operations in $\mathbb{F}_p$.*

Proof. [132, Proposition 2]. □

Proposition 13.6.2 proves that cryptosystems based on probable intractability of inverting the action of a Drinfeld module, for example the Drinfeld version or RSA, are insecure.

Each polynomial $f(\tau)$ in τ corresponds to an additive map of k, i.e., $f(\tau) = \sum_{i=0}^{n} a_i\tau^i$, $f(\tau)(\alpha) = \sum_{i=0}^{n} a_i\alpha^{q^i}$. Let $\varphi\colon k\langle\tau\rangle \to \mathrm{Hom}(k, +)$ be this assignment.

Notice that knowing the inverse of $\varphi \circ \rho_a$ is not enough to find b. However, with some additional effort we can find $b \in A$ such that $\varphi \circ \rho_b = (\varphi \circ \rho_a)^{-1}$ (see [132, Proposition 3]).

The techniques of Proposition 3 of [132] extend to the discrete logarithm problem for Drinfeld modules. In fact, we have the following:

Proposition 13.6.3. *For any Drinfeld module $\rho \in \mathrm{Drin}_A(k)$ and any elements $\alpha, \beta \in k$, there exist real numbers C', r' and an algorithm that computes $a \in A$ with $\rho_a(\alpha) = \beta$, if such an a exists, using at most $C'(\log_p |k|)^{r'}$ field operations in $\mathbb{F}_p$.*

Proof. [132, Proposition 4]. □

Proposition 13.6.3 proves that no public key cryptosystem based on the supposed infeasibility of solving the discrete logarithm problem for Drinfeld modules, such as the Drinfeld module versions of Diffie–Hellman and ElGamal cryptosystems, is secure.

13.6.2 The Gillard et al. Drinfeld Cryptosystem

In this section, we shall define again the cryptosystem proposed by Gillard et al. [49]. Take $q = p$, $A = R_T = \mathbb{F}_q[T]$, $k = K = \mathbb{F}_q(T)$, and let ρ be a Drinfeld A-module over k such that $\rho_a \in A\langle\tau\rangle$ for all $a \in A$. Let $p(T)$ be a monic irreducible polynomial of degree d larger than one. Let $\mathfrak{B} = A/(p(T)) \cong \mathbb{F}_{q^d}$. For $a \in A$ we write $\bar{a}$ for the class $a + (p(T))$ in $\mathfrak{B}$.

Now the ideal $(p(T))$ is an $A\langle\tau\rangle$-submodule of A. Hence the relation

$$f(\tau)(\bar{a}) := \overline{f(\tau)(a)} \quad \text{for} \quad a \in A$$

defines an $A\langle\tau\rangle$-module structure on $\mathfrak{B}$ in a natural way. That is, if $f(\tau) = \sum_{i=0}^{m} \alpha_i \tau^i \in A\langle\tau\rangle$, we have

$$f(\tau)\bar{a} = \overline{\sum_{i=0}^{m} \alpha_i a^{p^i}} = \sum_{i=0}^{m} \bar{\alpha}_i \bar{a}^{p^i}.$$

In particular, the map ψ from $\mathfrak{B}$ to $\mathfrak{B}$: $\bar{a} \stackrel{\psi}{\longmapsto} f(\tau)(\bar{a})$ is $\mathbb{F}_p$-linear. Furthermore, we have $\bar{a}^{p^d} = \bar{a}$. In particular, if we define $b_i = \sum_{j \equiv i \bmod d} \bar{\alpha}_j$ for $i \in \{0, 1, \ldots, d-1\}$, the map ψ is of the form

$$\psi = b_0 + b_1\tau + \cdots + b_{d-1}\tau^{d-1} \quad \text{with} \quad b_i \in \mathfrak{B}. \tag{13.49}$$

For $x \in A$, we write $\bar{\rho}_x$ for the map $\bar{a} \mapsto \rho_x(\bar{a})$ discussed above. Now choose secretly c_1 and $c_2 \in A$ such that the maps $\lambda_1 := \bar{\rho}_{c_1}$ and $\lambda_2 := \bar{\rho}_{c_2}$ are bijective. The private key is then given by the function $\varphi\colon \mathbb{F}_{p^d} \to \mathbb{F}_{p^d}$ defined by

$$\varphi(z) = \lambda_1((\lambda_2(z))^e + \delta), \tag{13.50}$$

where $\delta \in \mathbb{F}_{p^d}$ and $e \in \mathbb{N}$ are secret.

The public key of the system consists of the prime p, the integer d, and some information about how to compute φ (see [49]).

Note that for any $y \in \mathbb{F}_{p^d}$ and any $j \in \{0, 1, 2, \ldots, d-1\}$, the private key $(\lambda_1\tau^{-j}y^{-e}, b\tau^j\lambda_2, e, b^e\tau^j\delta)$ gives the same function φ in (13.50) as the original private key $(\lambda_1, \lambda_2, e, \delta)$. Thus any of these solutions can be used as a private key for φ.

S. R. Backburn et al. [6] showed how to recover a private key from a public key, proving in this way that the Gillard et al. cryptosystem is insecure. We refer the reader to the original paper for details.

13.7 Exercises

Exercise 13.7.1. Let $\delta\colon A \to k\langle\tau\rangle$ be a Drinfeld A-module. Show that ρ is an injective map.

Exercise 13.7.2. Let $A = R_T$. Let k be a field containing $\mathbb{F}_q(T)$ and $\delta\colon A \to k$ the inclusion. Let $C\colon R_T \to k\langle\tau\rangle$ and $C'\colon R_T \to k\langle\tau\rangle$ be the Drinfeld R_T-modules given by $C_T = T + \tau$ and $C'_T = T - \tau$. Prove that C_T and C'_T are isomorphic if and only if there exists a $(q-1)$th root of -1 in k.

More generally, assume that ρ, ρ' are Drinfeld R_T-modules given by $\rho_T = T + \tau$, $\rho'_T = T + f_1\tau$, where $f_1 \in k$ and $k \supseteq \mathbb{F}_q(T, \sqrt[q-1]{f_1})$. Show that ρ and ρ' are isomorphic over k.

Exercise 13.7.3. Let $\rho \in \mathrm{Drin}_A(k)$ and let $\phi\colon A \to \mathbb{Z}$ be given by $\phi(a) := -\deg_\tau \rho_a$. Show that ϕ is a nontrivial valuation on A equivalent to $v_{\mathfrak{P}_\infty}$.

If $\mathrm{char}(\rho) = \mathfrak{P} \neq 0$, consider the map $j_\rho = \mathrm{ord}\colon A \to \mathbb{Z}$ defined for $a \neq 0$ by $j_\rho(a) = i_0$, where $\rho_a = \sum_{i=0}^{n} \alpha_i \tau^i$, $\alpha_{i_0} \neq 0$, and $\alpha_j = 0$ for all $0 \leq j \leq i_0 - 1$. For $a = 0$ we put $j_\rho(0) = \infty$. Prove that j_ρ is a nontrivial valuation on A equivalent to $v_{\mathfrak{P}}$.

Exercise 13.7.4. Let R be any Dedekind domain, and let I be an integral ideal of R. Prove that I can be generated by at most two elements.

Exercise 13.7.5. Let k be a field of characteristic p. Let $k\langle\tau\rangle$ the ring of twisted polynomials and $k[x]$ be the ring of polynomials. Prove that if $f(\tau), g(\tau) \in k\langle\tau\rangle$ then $\mathrm{rgcd}(f(\tau), g(\tau)) = \gcd(f(x), g(x))_{x=\tau}$. That is, if $h(x)$ denotes the greatest common divisor of $f(x)$ and $g(x)$, then $h(\tau) = \mathrm{rgcd}(f(\tau), g(\tau))$.

Exercise 13.7.6. Let K be a congruence function field and let A be the Dedekind domain consisting of the elements in K whose only poles are at a fixed place $\mathfrak{P}_\infty$ of K. Let M_A be the abelian group consisting of the fractional ideals of A, and P_A be the subgroup consisting of the principal fractional ideals. The abelian group M_A/P_A is called the *Picard group* of A and is denoted by Pic A. Show that Pic A is a finite group. We denote the cardinality of Pic A by h_A. This is similar to the case of rings of integers in number fields.

Hint: Consider the class group of K. For a divisor $\mathfrak{A} \in D_K$, write $\mathfrak{A} = \mathfrak{A}_0 \mathfrak{P}_\infty^t$ with $(\mathfrak{A}, \mathfrak{P}_\infty) = 1$. Then $\begin{array}{c}\varphi\colon C_K \to \mathrm{Pic}\, A \\ \mathfrak{A} \mapsto \mathfrak{A}_0 \cap A\end{array}$ is an epimorphism. Consider the restriction ψ of φ to $C_{K,0}$. Then show that ψ has finite cokernel. Since $C_{K,0}$ is a finite group, it follows that Pic A is also finite.

Note: For an arbitrary Dedekind domain D, Pic D is not necessarily finite.

Exercise 13.7.7. Let A be as in Exercise 13.7.6. Prove that $h_A = d_\infty h_K$, where h_K is the class number of K and $h_A = |\,\mathrm{Pic}\, A|$.

Exercise 13.7.8. Let Γ be a lattice. Prove that the series $\sum_{\gamma \in \Gamma\setminus\{0\}} \frac{1}{\gamma}$ is absolutely convergent in $\mathbb{C}_\infty$. It follows that the infinite product $\prod_{\gamma \in \Gamma\setminus\{0\}} \left(1 - \frac{1}{\gamma}\right)$ is convergent.

In fact, prove that a series $\sum_{n=0}^{\infty} a_n$ in $\mathbb{C}_\infty$ converges if and only if $\lim_{n\to\infty} a_n = 0$. In this case both products $\prod_{n=0}^{\infty}(1 \pm a_n)$ converge.

Exercise 13.7.9. Let Γ be a lattice. Prove that $e_\Gamma(u)$ is a periodic function with group of periods Γ. In particular, $\mathbb{C}_\infty/\Gamma$ and $\mathbb{C}_\infty$ are isomorphic as $\mathbb{F}_q$-vector spaces.

Exercise 13.7.10. Let ρ be a Drinfeld A-module of rank one and let $a \in A \setminus \{0\}$. If (a) is relatively prime to $\mathrm{char}(\rho)$, then $\rho[a]$ and $A/(a)$ are isomorphic as A-modules.

Exercise 13.7.11. Let ρ be any Drinfeld A-module of rank one. Assume that $\delta(a) = a$ for all $a \in A$. Prove that K_∞ is a field of definition for ρ.

Exercise 13.7.12. Prove Proposition 13.5.4.

Exercise 13.7.13. Let ρ be a Drinfeld A-module over k of rank r. For each $x \in A$, let $\mu_\rho(x)$ be the leading coefficient of ρ_x. Prove that

(i) $\mu_\rho(xy) = \mu_\rho(x)\mu_\rho(y)^{r \deg x} = \mu_\rho(y)\mu_\rho(x)^{r \deg y}$ for $x, y \in A$.
(ii) If $\deg x = \deg y$, then $\mu_\rho(x + y) = \mu_\rho(x) + \mu_\rho(y)$.
(iii) If $\rho' = \xi\rho\xi^{-1}$ for some $\xi \in K_\infty^*$, then

$$\mu_{\rho'}(a) = \xi^{1-q^{r \deg a}} \mu_\rho(a).$$

(iv) If π is a prime element for $\mathfrak{P}_\infty$, then

$$\mu_{\rho'}(\pi^{-1}) = \xi^{(1-q^{d_\infty r})} \mu_\rho(\pi^{-1}).$$

Exercise 13.7.14. Verify that (13.33) is independent of a and b.

Exercise 13.7.15. Consider $\sigma \in \mathrm{Gal}(\mathbb{C}_\infty/K)$, $\rho \in \mathrm{Drin}_A(\mathbb{C}_\infty)$. Define $\sigma\rho$ as the map $x \mapsto \rho_x$ followed by the action of σ. Prove that $\sigma\rho \in \mathrm{Drin}_A(\mathbb{C}_\infty)$ and that for any nonzero ideal $\mathfrak{A}$ of A

$$\mathfrak{A} * \sigma\rho = \sigma(\mathfrak{A} * \rho).$$

Exercise 13.7.16. Prove that for any nonzero ideals $\mathfrak{A}$, $\mathfrak{B}$ of A and any Drinfeld A-module $\rho \in \mathrm{Drin}_A(k)$ we have

$$\rho_{\mathfrak{A}\mathfrak{B}} = (\mathfrak{B} * \rho)_{\mathfrak{A}}\rho_{\mathfrak{B}} \quad \text{and} \quad \mathfrak{A} * (\mathfrak{B} * \rho) = (\mathfrak{A}\mathfrak{B}) * \rho.$$

Exercise 13.7.17. Prove Corollary 13.5.31.

14

Automorphisms and Galois Theory

In this chapter we continue our study of the arithmetic of extensions in function fields. We study the group

$$G = \operatorname{Aut}_k K = \{\sigma : K \to K \mid \sigma \text{ is an automorphism and } \sigma|_k = \operatorname{Id}_k\},$$

where K/k is an arbitrary function field. When g_K is 0 or 1, the group G is infinite, except in the case that k is a finite field. For $g_K \geq 2$, G is almost always a finite group. In order to investigate G, we need to consider some special points in K called the Weierstrass points. We also need to know the genus g_K of K. It is often difficult to determine precisely the genus of a function field, so we will derive some bounds for the genus in special cases. This result is the *Castelnuovo–Severi inequality*.

14.1 The Castelnuovo–Severi Inequality

In this section we consider a separably generated function field K/k, that is, K/k is a separably generated extension (Definition 8.2.1).

The proof of the Castelnuovo–Severi inequality that we present here is due to Stichtenoth [148, Chapter III.10.3] and [147].

Proposition 14.1.1. *Let K'/k be a subfield of K/k and $[K : K'] = n$. Assume that $\{y_1, \ldots, y_n\}$ is a basis of K/K' such that $y_i \in L_K(\mathfrak{C}^{-1})$ for some $\mathfrak{C} \in D_K$. Then*

$$g_K \leq 1 + n(g_{K'} - 1) + d_K(\mathfrak{C}). \tag{14.1}$$

Proof. Let $\mathfrak{A}_1 \in D_{K'}$ be of sufficiently large degree such that

$$\ell_{K'}(\mathfrak{A}_1^{-1}) =: t = d_{K'}(\mathfrak{A}_1) + 1 - g_{K'}$$

(Corollary 3.5.6).

Let $\{x_1, \ldots, x_t\}$ be a basis of $L_{K'}(\mathfrak{A}_1^{-1})$ and $\mathfrak{A} = \operatorname{con}_{K'/K} \mathfrak{A}_1 \in D_K$. Then

$$\mathcal{A} = \{x_i y_j \mid 1 \leq i \leq t, 1 \leq j \leq n\} \subseteq L_K(\mathfrak{A}^{-1}\mathfrak{C}^{-1}).$$

Clearly $\mathcal{A}$ is linearly independent over k. Thus

$$\ell_K(\mathfrak{A}^{-1}\mathfrak{C}^{-1}) \geq nt = n(d_{K'}(\mathfrak{A}_1) + 1 - g_{K'}). \tag{14.2}$$

Since we may assume that $d_K(\mathfrak{A}\mathfrak{C})$ is of sufficiently large degree, we obtain using Corollary 3.5.6 and Theorem 5.3.4 that

$$\begin{aligned} \ell_K(\mathfrak{A}^{-1}\mathfrak{C}^{-1}) &= d_K(\mathfrak{A}\mathfrak{C}) + 1 - g_K \\ &= d_K(\mathfrak{A}) + d_K(\mathfrak{C}) + 1 - g_K \\ &= nd_{K'}(\mathfrak{A}_1) + d_K(\mathfrak{C}) + 1 - g_K. \end{aligned} \tag{14.3}$$

Substituting (14.3) in (14.2) we get

$$nd_{K'}(\mathfrak{A}_1) + n - ng_{K'} \leq nd_{K'}(\mathfrak{A}_1) + d_K(\mathfrak{C}) + 1 - g_K.$$

This is (14.1) □

One of the key points in the proof of the Castelnuovo–Severi inequality is the following:

Lemma 14.1.2. *Let k be a separably closed field, K/k a separably generated function field, and K_1/k, K_2/k two subfields of K/k such that $K = K_1K_2$ and each K/K_i is a finite extension. Then:*

(i) At least one of the extensions K/K_i, $i = 1, 2$, is separable.
(ii) K/k (and thus K_1/k and K_2/k) contains infinitely many places of degree 1.
(iii) If K/K_1 is separable and $n = [K : K_1]$, then for almost all places $\wp \in \mathbb{P}_{K_1}$ of degree 1, we have:
  (a) $\wp$ decomposes fully in K/K_1, that is, $\wp$ has n distinct extensions $\mathfrak{P}_1, \ldots, \mathfrak{P}_n$ in K/K_1.
  (b) The restrictions $\mathfrak{P}_i|_{K_2} = \mathfrak{P}_i \cap K_2$, $1 \leq i \leq n$, are distinct places of K_2.

Proof.

(i) If both K/K_1 and K/K_2 are inseparable, then by Exercise 8.7.3 we have $K_i \subseteq K^pk$ for $i = 1, 2$. Thus $K_1K_2 \subseteq K^pk$. On the other hand by Exercise 8.7.2 we have $[K : K^pk] = p$, so $K \neq K^pk$. Therefore K/K_1 or K/K_2 is separable.
(ii) This is just Corollary 5.2.35.
(iii) Since $K = K_1K_2$ and K/K_1 is separable, there exist $y_1, \ldots, y_s \in K_2$ and $\alpha_1, \ldots, \alpha_s \in k$ such that $K = K_1(y_1, \ldots, y_s)$ and $y := \alpha_1y_1 + \cdots + \alpha_sy_s \in K_2$, where $K = K_1(y)$.

Let $\varphi(T) = \operatorname{Irr}(y, T, K_1) = T^n + a_{n-1}T^{n-1} + \cdots + a_1T + a_0 \in K_1[T]$ be the minimal polynomial of y over K_1. Since φ is separable, its discriminant $d = \operatorname{disc}(\varphi) \in K_1$ is nonzero. Let $\wp \in \mathbb{P}_{K_1}$ be any place satisfying

$$d_{K_1}(\wp) = 1, \quad a_0, \ldots, a_{n-1} \in \vartheta_\wp, \quad \text{and} \quad v_\wp(d) = 0. \tag{14.4}$$

It is easy to see that almost all $\wp \in \mathbb{P}_{K_1}$ of degree 1 satisfy (14.4). For $a \in \vartheta_\wp$, we denote by $\overline{a}$ its residue module $\wp$, $\overline{a} \in \vartheta_\wp/\wp \cong k$. The polynomial

$$\overline{\varphi}(T) = T^n + \overline{a}_{n-1}T^{n-1} + \cdots + \overline{a}_1 T + \overline{a}_0 \in k[T]$$

is separable because $\overline{d} = \operatorname{disc}(\overline{\varphi}) \neq 0$ in $\vartheta_\wp/\wp$. Since k is separably closed we have $\overline{\varphi}(T) = \prod_{i=1}^n (T - \gamma_i)$, where the elements $\gamma_1, \ldots, \gamma_n$ of k are distinct. For $j = 1, \ldots, n$ we define the homomorphism

$$\begin{aligned} \tau_j \colon \vartheta_\wp[y] &\to k \\ \sum c_i y^i &\mapsto \sum \overline{c}_i \gamma_j^i. \end{aligned}$$

By Theorem 2.4.4, τ_j can be extended to a place $\mathfrak{P}_j$ of K/k (the extension is not a homomorphism of fields since $\tau_j(y - \gamma_j) = 0$ and $y - \gamma_j \neq 0$). Note that each $\mathfrak{P}_j$, $1 \leq j \leq n$, is above $\wp$ (because $\vartheta_\wp \subseteq \vartheta_\wp[y]$). Now the places $\mathfrak{P}_j$ are distinct and since $[K : K_1] = n$, it follows that $\wp$ is fully decomposed in K/K_1. Finally, if the restriction of $\mathfrak{P}_j$ to $k(y)$ is $\mathfrak{P}_j \cap k(y) = \mathfrak{q}_j$, this satisfies $\tau_j(y - \gamma_j) = 0$ and $\tau_i(y - \gamma_j) = \gamma_i - \gamma_j \neq 0$, so $\mathfrak{q}_1, \ldots, \mathfrak{q}_n$ are distinct in $k(y)$. Therefore the restrictions of the $\mathfrak{B}_j$'s to K_2 are distinct. □

Theorem 14.1.3 (Castelnuovo–Severi Inequality). *Let K/k be a function field such that K/k is separably generated. Let K_1/k and K_2/k be two subfields of K/k satisfying $K = K_1K_2$. Put $[K : K_i] = n_i$ and $g_i = g_{K_i}$ for $i = 1, 2$. If $g = g_K$, then*

$$g \leq n_1 g_1 + n_2 g_2 + (n_1 - 1)(n_2 - 1). \tag{14.5}$$

Proof. First assume that k is separably closed. By Lemma 14.1.2, K/K_1 or K/K_2 is separable. Say that K/K_1 is separable and let $K = K_1(y)$ with $y \in K_2$. By Corollary 5.2.35 there is an integral divisor $\mathfrak{A} \in D_{K_2}$ such that

$$d_{K_2}(\mathfrak{A}) = g_2 \quad \text{and} \quad \ell_{K_2}(\mathfrak{A}^{-1}) = 1.$$

Let $\wp_0 \in \mathbb{P}_{K_2}$ be of degree 1 and relatively prime to $\mathfrak{A}$ and let $\mathfrak{B} = \frac{\mathfrak{A}}{\wp_0}$. Since $\mathfrak{A}$ is integral and $\ell_{K_2}(\mathfrak{A}^{-1}) = 1$, it follows that $L_{K_2}(\mathfrak{A}^{-1}) = k$. Thus

$$d_{K_2}(\mathfrak{B}) = d_{K_2}(\mathfrak{A}) - d_{K_2}(\wp_0) = g_2 - 1$$

and if $\xi \in L_{K_2}(\mathfrak{B}^{-1}) \setminus \{0\}$, then $\xi \in L_{K_2}(\mathfrak{A}^{-1}) = k$. We have

$$(\xi)_{K_2} = \frac{\mathfrak{C}}{\mathfrak{B}} = \frac{\wp_0 \mathfrak{C}}{\mathfrak{A}} = \mathfrak{N} \text{ for some integral divisor } \mathfrak{C}.$$

It follows that $\mathfrak{C} = \mathfrak{B}$. Since $\mathfrak{B}$ is not integral this is impossible and thus $L_{K_2}(\mathfrak{B}^{-1}) = \{0\}$. In particular, $\ell_{K_2}(\mathfrak{B}^{-1}) = 0$.

According to Lemma 14.1.2 we may choose $\mathfrak{P} \in \mathbb{P}_{K_1}$ of degree 1 satisfying the following: $\mathfrak{P}$ has n_1 extensions $\mathfrak{q}_1, \ldots, \mathfrak{q}_{n_1}$ in K/k such that the restrictions

$$\mathfrak{Q}_i = \mathfrak{q}_i \cap K_2 \in \mathbb{P}_{K_2}$$

are distinct and $\mathfrak{Q}_i$ is relatively prime to $\mathfrak{B}$ for $i = 1, \ldots, n_1$.

Using the Riemann–Roch theorem we obtain

$$\ell_{K_2}(\mathfrak{B}^{-1}\mathfrak{Q}_i^{-1}) \geq d_{K_2}(\mathfrak{B}\mathfrak{Q}_i) + 1 - g_2 = g_2 + 1 - g_2 = 1.$$

Let $\xi \in L_{K_2}(\mathfrak{B}^{-1}\mathfrak{Q}_i^{-1}) \setminus \{0\}$. Then $(\xi)_{K_2} = \frac{\mathfrak{C}}{\mathfrak{B}\mathfrak{Q}_i}$. If $\mathfrak{Q}_i$ divides $\mathfrak{C}$, we have $\xi \in L_{K_2}(\mathfrak{B}^{-1}) \setminus \{0\}$, which is impossible. Therefore $\mathfrak{Q}_i \nmid \mathfrak{C}$ and $v_{\mathfrak{Q}_i}(\xi) = -1$.

It follows that for $1 \leq i \leq n_1$, there exists $u_i \in L_{K_2}(\mathfrak{B}^{-1}\mathfrak{Q}_i^{-1})$ such that $v_{\mathfrak{Q}_i}(u_i) = -1$ and $v_{\mathfrak{Q}_j}(u_i) \geq 0$ whenever $i \neq j$. We will see that $\{u_1, \ldots, u_{n_1}\}$ is a linearly independent system over K_1. Assume that there exist $x_1, \ldots, x_n \in K_1$ not all zero such that $\sum_{i=1}^n x_i u_i = 0$. We may assume that $x_i \neq 0$ for all $1 \leq i \leq n$. Let $j \in \{1, \ldots, n_1\}$ be such that $v_{\mathfrak{P}}(x_j) \leq v_{\mathfrak{P}}(x_i)$ for $1 \leq i \leq n$.

Since $\mathfrak{q}_j \mid \mathfrak{P}$ is unramified in K/K_1, we have $v_{\mathfrak{q}_j}(x_j) = v_{\mathfrak{P}}(x_j)$ and $v_{\mathfrak{q}_j}(u_j) \leq v_{\mathfrak{Q}_j}(u_j) = -1$. Therefore $v_{\mathfrak{q}_j}(x_j u_j) \leq v_{\mathfrak{P}}(x_j) - 1$.

For $i \neq j$, we have

$$v_{\mathfrak{q}_j}(x_i u_i) = v_{\mathfrak{q}_j}(x_i) + v_{\mathfrak{q}_j}(u_i) \geq v_{\mathfrak{P}}(x_i) \geq v_{\mathfrak{P}}(x_j).$$

In particular, $v_{\mathfrak{q}_j}(x_j u_j) = v_{\mathfrak{q}_j}(x_j) + v_{\mathfrak{q}_j}(u_j) \leq v_{\mathfrak{q}_j}(x_j) - 1 < v_{\mathfrak{q}_j}(x_i) + v_{\mathfrak{q}_j}(u_i) = v_{\mathfrak{q}_j}(x_i u_i)$ for $i \neq j$. Therefore, by Proposition 2.2.3 (v) we have

$$\infty = v_{\mathfrak{q}_j}(0) = v_{\mathfrak{q}_j}\left(\sum_{i=1}^n x_i u_i\right) = v_{\mathfrak{q}_j}(x_j u_j) < \infty.$$

This contradiction shows that $\{u_1, \ldots, u_{n_1}\}$ is linearly independent over K_1 and therefore a basis of K/K_1. Let $\mathfrak{D} = \operatorname{con}_{K_2/K}\left(\mathfrak{B} \prod_{i=1}^{n_1} \mathfrak{Q}_i\right) \in D_K$.

We have $d_K(\mathfrak{D}) = n_2 d_{K_2}\left(\mathfrak{B} \prod_{i=1}^{n_1} \mathfrak{Q}_i\right) = n_2(g_2 - 1 + n_1)$.

Since $u_i \in L_{K_2}\left(\mathfrak{B}^{-1}\mathfrak{Q}_i^{-1}\right) \subseteq L_{K_2}\left(\mathfrak{B}^{-1} \prod_{j=1}^{n_1} \mathfrak{Q}_j^{-1}\right)$, it follows that $u_i \in L_K\left(\mathfrak{D}^{-1}\right)$. Using Proposition 14.1.1, we obtain

$$\begin{aligned} g = g_K &\leq 1 + n_1(g_1 - 1) + d_K(\mathfrak{D}) \\ &= 1 + n_1(g_1 - 1) + n_2(g_2 - 1 + n_1) \\ &= n_1 g_1 + n_2 g_2 + (n_1 - 1)(n_2 - 1). \end{aligned}$$

This is (14.5) for k separably closed.

Now if k is arbitrary, denote by $\bar{k}$ its separable closure. Set $\bar{K} = K\bar{k}$, $\bar{K}_i = K_i k$, $g_{\bar{K}} = \bar{g}$, $g_{\bar{K}_i} = \bar{g}_i$, and $\left[\bar{K} : \bar{K}_i\right] = \bar{n}_i$, $i = 1, 2$.

Since $\bar{k}/k$ is separable it follows by Theorem 8.5.2 that $\bar{g} = g$ and $\bar{g}_i = g_i$ for $i = 1, 2$. By Theorem 8.4.10 and Corollary 8.5.8, $\bar{k}$ and K are linearly disjoint over k. By Proposition 8.1.5 it follows that K and $\bar{K}_i$ are linearly disjoint over K_i for $i = 1, 2$.

Therefore $n_i = [K : K_i] = \left[\overline{K} : \overline{K}_i\right] = \overline{n_i}$ for $i = 1, 2$. Thus

$$g = \overline{g} \leq \overline{n}_1 \overline{g}_1 + \overline{n}_2 \overline{g}_2 + (\overline{n}_1 - 1)(\overline{n}_2 - 1)$$
$$= n_1 g_1 + n_2 g_2 + (n_1 - 1)(n_2 - 1).$$

This proves the theorem. □

Corollary 14.1.4 (Riemann's Inequality). *Let $K = k(x, y)$ be any function field such that K/k is separably generated. Then*

$$g_K \leq ([K : k(x)] - 1)([K : k(y)] - 1).$$

Proof. Clearly $K = K_1 K_2$ with $K_1 = k(x)$ and $K_2 = k(y)$. Thus $g_{K_1} = g_{K_2} = 0$, and the result follows immediately by (14.5). □

Example 14.1.5. Let p be an odd prime and k any field of characteristic p.

Let $K = k(x, y)$ with $y^p - y = \frac{x^2}{x+1}$.

Then $[K : k(x)] = p$ and $[K : k(y)] = 2$. By Example 5.8.8 we have $\mathfrak{D}_{K/k(x)} = (\mathfrak{P}_\infty \mathfrak{P}_1)^{2(p-1)}$ where $(x+1)_{k(x)} = \frac{\wp_1}{\wp_\infty}$ and $\mathfrak{P}_1, \mathfrak{P}_\infty$ are the prime divisors above $\wp_1$ and $\wp_\infty$ respectively.

Using Theorem 9.4.2, we obtain

$$g_K = 1 + [K : k(x)](g_{k(x)} - 1) + \frac{1}{2} d_K(\mathfrak{D}_{K/k(x)})$$
$$= 1 - p + \frac{1}{2}(2(p-1) + 2(p-1)) = (p-1)$$
$$= (2-1)(p-1) = ([K : k(y)] - 1)([K : k(x)] - 1).$$

Example 14.1.5 shows that Castelnuovo's inequality cannot be improved in general.

Proposition 14.1.6. *Let K/k be a separably generated function field with $K = k(x, y)$, where* $\operatorname{Irr}(y, T, k(x)) = \sum_{j=0}^{n-1} f_j(x) T^j + T^n$, $f_j(x) \in k[x]$, *and* $\deg f_j(x) \leq n - j$ *for* $0 \leq j \leq n-1$. *Then*

$$g_K \leq \frac{1}{2}(n-1)(n-2).$$

Proof. Let $\mathfrak{A} := \operatorname{con}_{k(x)/K} \wp_\infty = \mathfrak{N}_x$ where $(x)_{k(x)} = \frac{\wp_0}{\wp_\infty}$. We have $\deg_K \mathfrak{A} = [K : k(x)] = n = \operatorname{Irr}(y, T, k(x))$. Furthermore, $\mathfrak{A}$ is an integral divisor.

Let $\mathfrak{P} \in \mathbb{P}_K$. If $v_{\mathfrak{P}}(x) \geq 0$, then $v_{\mathfrak{P}}(f_j(x)) \geq 0$ and since y is integral over $k(x)$ it follows that $v_{\mathfrak{P}}(y) \geq 0$ (see the proof of Theorem 3.2.7). Thus $v_{\mathfrak{P}}(y) \geq 0 = -v_{\mathfrak{P}}(\mathfrak{A})$.

If $v_{\mathfrak{P}}(x) < 0$, then $\mathfrak{P}$ divides $\wp_\infty$ and $v_{\mathfrak{P}}(x) = -v_{\mathfrak{P}}(\mathfrak{A})$.

Now $v_{\mathfrak{P}}(f_j(x)) = \deg f_j(x) v_{\mathfrak{P}}(x) \geq (n-j) v_{\mathfrak{P}}(x)$ and

$$v_{\mathfrak{P}}(y^j f_j(x)) = j v_{\mathfrak{P}}(y) + v_{\mathfrak{P}}(f_j(x)) \geq j v_{\mathfrak{P}}(y) + (n-j) v_{\mathfrak{P}}(x)$$

In particular, we have $v_{\mathfrak{P}}(y) \geq -v_{\mathfrak{P}}(\mathfrak{A}) = v_{\mathfrak{P}}(x)$. Indeed, assume otherwise, i.e., $v_{\mathfrak{P}}(y) < v_{\mathfrak{P}}(x)$. Then for $j = 0, 1, \ldots, n-1$,

$$v_{\mathfrak{P}}(y^j f_j(x)) > j v_{\mathfrak{P}}(y) + (n-j) v_{\mathfrak{P}}(y) = n v_{\mathfrak{P}}(y) = v_{\mathfrak{P}}(y^n)$$

and hence $\infty = v_{\mathfrak{P}}(0) = v_{\mathfrak{P}}\left(\sum_{j=0}^{n-1} f_j(x) y^j + y^n\right) = v_{\mathfrak{P}}(y^n)$.

Therefore, we have

$$v_{\mathfrak{P}}(y) \geq -v_{\mathfrak{P}}(\mathfrak{A}) \quad \text{for all} \quad \mathfrak{P} \in \mathbb{P}_K. \tag{14.6}$$

It follows that $\mathfrak{A}^{-1} \mid (x)_K$ and $\mathfrak{A}^{-1} \mid (y)_K$. In particular, for any $m \geq n$ and $0 \leq j \leq n-1, 0 \leq i \leq m-j$, we have $x^i y^j \in L_K(\mathfrak{A}^{-m})$. Since $\deg \operatorname{Irr}(y, T, k(x)) = n$, $\{x^i y^j\}_{\substack{0 \leq j \leq n-1 \\ 0 \leq i \leq m-j}}$ are linearly independent over k. Thus

$$\ell_K(\mathfrak{A}^{-m}) \geq \sum_{j=0}^{n-1} (m - j + 1) = n(m+1) - \frac{1}{2} n(n-1). \tag{14.7}$$

If m is large enough, we obtain using the Riemann–Roch theorem

$$\ell_K(\mathfrak{A}^{-m}) = d_K(\mathfrak{A}^m) - g_K + 1 = m d_K(\mathfrak{A}) - g_K + 1 = mn - g_K + 1.$$

By (14.7) we have $mn - g_K + 1 \geq n(m+1) - \frac{1}{2} n(n-1)$, i.e., $g_K \leq \frac{(n-1)(n-2)}{2}$. ☐

14.2 Weierstrass Points

In the case of compact Riemann surfaces, there exists a finite number of special points. Here the term special means being the unique pole of a certain order for some element of the field. Since these points are special, and consequently invariants of the field, they become permuted under the action of a field automorphism. Therefore, they provide information about such automorphisms and the arithmetic of the field.

In characteristic $p > 0$ and algebraically closed field of constants, special points exist and may be used for the study of the given field. Such points are the *Weierstrass points*, which will be considered below.

Definition 14.2.1. Let K/k be any function field and let $\mathfrak{P}$ be a prime divisor of K. A natural number n is called a *pole number* of $\mathfrak{P}$ if there exists $x \in K$ such that $\mathfrak{N}_x = \mathfrak{P}^n$. Notice that the pole divisor of x is precisely $\mathfrak{P}^n$. If n is not a pole number of $\mathfrak{P}$, n is called a *gap number of* $\mathfrak{P}$.

Remark 14.2.2. A natural number n is a pole number of $\mathfrak{P}$ iff there exists $x \in L_K(\mathfrak{P}^{-n}) \backslash L_K(\mathfrak{P}^{-(n-1)})$. In other words, n is a pole number if and only if $\ell_K(\mathfrak{P}^{-n}) > \ell_K(\mathfrak{P}^{-n+1})$. Furthermore, if n and m are pole numbers of $\mathfrak{P}$ then $n+m$ is a pole number of $\mathfrak{P}$ (since if $\mathfrak{N}_x = \mathfrak{P}^n$ and $\mathfrak{N}_y = \mathfrak{P}^m$, then $\mathfrak{N}_{xy} = \mathfrak{P}^{n+m}$).

By the Riemann–Roch theorem, we have

$$\ell_K(\mathfrak{P}^{-n}) = d_K(\mathfrak{P}^n) - g_K + 1 + \delta_K(\mathfrak{P}^n)$$

and

$$\ell_K(\mathfrak{P}^{-n+1}) = d_K(\mathfrak{P}^{n-1}) - g_K + 1 + \delta_K(\mathfrak{P}^{n-1}).$$

Therefore $\ell_K(\mathfrak{P}^{-n}) - \ell_K(\mathfrak{P}^{-n+1}) = d_K(\mathfrak{P}) + \delta_K(\mathfrak{P}^n) - \delta_K(\mathfrak{P}^{n-1})$. Thus, using Remark 14.2.2 we obtain the following:

Proposition 14.2.3. *A number $n \in \mathbb{N}$ is a gap number of $\mathfrak{P}$ iff $\ell_K(\mathfrak{P}^{-n}) = \ell_K(\mathfrak{P}^{-n+1})$ iff $\delta_K(\mathfrak{P}^{n-1}) - \delta_K(\mathfrak{P}^n) = d_K(\mathfrak{P})$.* □

Let $\mathfrak{P}$ be any prime divisor. By Corollary 3.5.8, if $n > 2g_K - 1$ then n is a pole number of $\mathfrak{P}$.

Now we consider a prime divisor $\mathfrak{P}$ of degree 1, and $g_K = g > 0$. By Proposition 3.1.13, $L(\mathfrak{P}^0) = L(\mathfrak{N}) = k$, and by Corollary 3.5.6,

$$\ell_K(\mathfrak{P}^{-(2g-1)}) = d_K(\mathfrak{P}^{2g-1}) - g + 1 = 2g - 1 - g + 1 = g.$$

We have $k = L_K(\mathfrak{P}^0) \subseteq L_K(\mathfrak{P}^{-1}) \subseteq \ldots \subseteq L_K(\mathfrak{P}^{-(2g-1)})$ and

$$g = \dim_k L_K(\mathfrak{P}^{-(2g-1)}) = \sum_{i=1}^{2g-1} \dim_k \frac{L_K(\mathfrak{P}^{-i})}{L_K(\mathfrak{P}^{-i+1})} + \dim_k L_K(\mathfrak{P}^0). \tag{14.8}$$

For any $n \in \mathbb{N}$, we have

$$\ell_K(\mathfrak{P}^{-n}) = d_K(\mathfrak{P}^n) - g + 1 + \delta_K(\mathfrak{P}^n),$$
$$\ell_K(\mathfrak{P}^{-n+1}) = d_K(\mathfrak{P}^{n-1}) - g + 1 + \delta_K(\mathfrak{P}^{n-1}).$$

Hence $\ell_K(\mathfrak{P}^{-n}) - \ell_K(\mathfrak{P}^{-n+1}) = 1 + \delta_K(\mathfrak{P}^n) - \delta_K(\mathfrak{P}^{n-1})$. By Theorem 3.4.11 $D_K(\mathfrak{P}^n) \subseteq D_K(\mathfrak{P}^{n-1})$ and $L_K(\mathfrak{P}^{-n+1}) \subseteq L_K(\mathfrak{P}^{-n})$. It follows that

$$0 \le \ell_K(\mathfrak{P}^{-n}) - \ell_K(\mathfrak{P}^{-n+1}) \le 1. \tag{14.9}$$

Let $t_i = \dim_k \frac{L_K(\mathfrak{P}^{-i})}{L_K(\mathfrak{P}^{-i+1})} = \ell_K(\mathfrak{P}^{-i}) - \ell_K(\mathfrak{P}^{-i+1}) \in \{0, 1\}$.

Using (14.8) we obtain $g - 1 = \sum_{i=1}^{2g-1} t_i$.

In particular there are exactly $g - 1$ indices i such that $1 \le i \le 2g - 1$ and $t_i = 1$. The remaining g indices between 1 and $2g - 1$ such that $t_i = 0$ are gap numbers of $\mathfrak{P}$.

Theorem 14.2.4 (Weierstrass Gap Theorem). *Let K/k be a function field of genus $g_K = g > 0$. Let $\mathfrak{P}$ be a prime divisor of K of degree 1. Then there exist exactly g gap numbers $j_1, \ldots, j_g$ of $\mathfrak{P}$ such that $1 = j_1 < j_2 < \cdots < j_g \le 2g - 1$. The set $\{j_1, \ldots, j_g\}$ is called the* gap sequence of $\mathfrak{P}$.

Proof. We have $(1)_K = \frac{\mathfrak{N}}{\mathfrak{N}} = \frac{\mathfrak{N}}{\mathfrak{P}^0}$, so 0 is not a gap of $\mathfrak{P}$. If $n \geq 2g-1$, then n is a pole number of $\mathfrak{P}$. Finally, if 1 is a pole number, there exists $x \in K$ such that $\mathfrak{N}_x = \mathfrak{P}$. Then $[K : k(x)] = \deg_K \mathfrak{N}_x = 1$ implies that $K = k(x)$ and $g = 0$, a contradiction. Thus 1 is not a pole number. □

Corollary 14.2.5. *The number $n \in \mathbb{N}$ is a gap number of the prime divisor $\mathfrak{P}$ of degree 1 if and only if there exists a holomorphic differential w such that $\mathfrak{P}^{n-1} \mid w$ and $\mathfrak{P}^n \nmid w$. Equivalently, n is a gap number of $\mathfrak{P}$ if and only if there exists a holomorphic differential w such that $v_{\mathfrak{P}}((w)_K) = n-1$.*

Proof. We have

$$\ell_K(\mathfrak{P}^{-n}) - \ell_K(\mathfrak{P}^{-n+1}) = 1 + \delta_K(\mathfrak{P}^n) - \delta_K(\mathfrak{P}^{n-1}).$$

Hence n is a gap number if and only if $\delta_K(\mathfrak{P}^{n-1}) - \delta_K(\mathfrak{P}^n) = 1$, that is, there exists $w \in D_K(\mathfrak{P}^{n-1})$ such that $w \notin D_K(\mathfrak{P}^n)$. □

Example 14.2.6. If K is a function field of genus $g_K = 0$ and $\mathfrak{P}$ is a prime divisor of degree 1, then K is a rational function field and every $n \in \mathbb{N}$ is a pole number of $\mathfrak{P}$.

Example 14.2.7. If K is a function field of genus $g_K = 1$ and $\mathfrak{P}$ is a prime divisor of degree 1, then K is a field of elliptic functions and $n = 1$ is the only gap number of $\mathfrak{P}$.

Remark 14.2.8. For any function field of genus $g_K > 0$ and any prime divisor of degree 1, $n = 1$ is a gap number of $\mathfrak{P}$.

In the rest of this section we consider a function field K/k where k is an algebraically closed field.

14.2.1 Hasse–Schmidt Differentials

For a rational function field $k(x)$ we consider the usual derivative, that is, the one induced by

$$f(x) = a_0 + a_1 x + \cdots + a_n x^n \quad \text{and} \quad f'(x) = a_1 + 2a_2 x + \cdots + n a_n x^{n-1}$$

with $f(x) \in k[x]$. We repeat the process with $f'(x)$ and we obtain $f''(x)$ and so on. Unfortunately, in characteristic $p > 0$, the nonconstant function $f(x) = x^p$ satisfies $f^{(n)}(x) = 0$ for all $n \geq 1$, where $f^{(n)}$ denotes the nth derivative. If we want to obtain for function fields of characteristic $p > 0$ information similar to that obtained in characteristic 0, we must modify the usual definition of derivative. This was done by H. Hasse and F. K. Schmidt [58].

In this section we present the work of Hasse and Schmidt. We will use this new definition of differentiation to study the Wronskian determinant and the arithmetic theory of Weierstrass points.

Definition 14.2.9. A sequence $\{D^{(n)}\}_{n\in\mathbb{N}\cup\{0\}}$ of maps $D^{(n)}\colon K \to K$ is called a *differentiation* of K/k if

(i) $D^{(0)} = \mathrm{Id}_K$.
(ii) $D^{(n)}|_k = 0$ for all $n \geq 1$.
(iii) For $x, y \in K$,

$$D^{(n)}(x+y) = D^{(n)}(x) + D^{(n)}(y) \qquad \text{(sum rule)}$$

and

$$D^{(n)}(xy) = \sum_{m=0}^{n} D^{(m)}(x)D^{(n-m)}(y) \qquad \text{(product rule).}$$

The differentiation $\{D^{(n)}\}_{n=0}^{\infty}$ is called *iterative* if
(iv) For all $n, m \in \mathbb{N} \cup \{0\}$,

$$D^{(n)} \circ D^{(m)} = \binom{n+m}{n} D^{(n+m)}.$$

Remark 14.2.10. Consider the local field $K_\wp = k((\pi))$ of characteristic $p > 0$, and let $\alpha \in K$. We can express α in $K_\wp$ as $\alpha = \sum_{i=m}^{\infty} a_i \pi^i$.

Then the usual derivative with respect to π yields

$$\frac{d^n\alpha}{d\pi^n} = \sum_{i=m}^{\infty} i(i-1)\cdots(i-n+1)a_i\pi^{i-n}.$$

Therefore if $n \geq p$, we have $i(i-1)\cdots(i-n+1) \equiv 0 \bmod p$ for all i. Thus $\frac{d^n}{d\pi^n} \equiv 0$ for $n \geq p$.

If instead of $\frac{d^n}{d\pi^n}$ we define

$$D_\pi^{(n)}(\alpha) := \sum_{i=m}^{\infty} \binom{i}{n} a_i \pi^{i-n},$$

then $D_\pi^{(n)}$ is nonzero. In fact, it is easy to see that $\{D_\pi^{(n)}\}_{n=0}^{\infty}$ satisfies the iterative rule (iv) of Definition 14.2.9:

$$D_\pi^{(n)} \circ D_\pi^{(m)} = \binom{n+m}{m} D^{(n+m)}.$$

This is the motivation for constructing differentiations that satisfy the iterative rule.

Note that the product rule defined in Definition 14.2.9 is different from the classical case. We also have

$$D^{(n)}(y+c) = D^{(n)}y \ (n \geq 1) \quad \text{and} \quad D^{(n)}(cy) = cD^{(n)}y \ (n \geq 0) \tag{14.10}$$

for $y \in k$ and $c \in k$.

Finally, in characteristic 0 the iterative rule translates into

$$D^{(n)}y = \frac{D^{(1)}(\cdots D^{(1)}(y)\cdots)}{n!}.$$

Let K/k be a function field and $D = \{D^{(n)}\}_{n=0}^{\infty}$ a differentiation on K. Let $M = K[[u]]$ be the power series in u with coefficients in K. Define

$$\begin{aligned} \phi\colon K &\overset{\phi}{\to} M \\ y &\mapsto \phi(y) = \sum_{n=0}^{\infty}(D^{(n)}y)u^n. \end{aligned} \tag{14.11}$$

Proposition 14.2.11. *ϕ is a ring monomorphism.*

Proof. Clear. □

If $\phi : K \to K[[u]]$ is a ring homomorphism such that for all $y \in K, \phi(y) = \sum_{n=0}^{\infty} a_n u^n$, $a_0 = y$, then $D^{(n)}y := a_n$ is a differentiation on K. This is a consequence of the proof of Proposition 14.2.11 and Definition 14.2.9.

Now assume that there exists an iterative differentiation $\{D^{(n)}\}_{n\in\mathbb{N}\cup\{0\}}$ on K and let $\phi\colon K \to K[[u]]$, $\phi(y) = \sum_{n=0}^{\infty} D^{(n)}(y)u^n$. Then

$$\phi(D^{(m)}y) = \sum_{n=0}^{\infty}(D^{(n)} \circ D^{(m)})(y)u^n = \sum_{n=0}^{\infty}\binom{n+m}{m}D^{(n+m)}(y)u^n.$$

We write $D_u^{(m)}u^n = \binom{n}{m}u^{n-m}$ ($D_u^{(m)}u^n = 0$ for $m > n$).
With this notation we have

$$\begin{aligned} D_u^{(m)}(\phi(y)) &= D_u^{(m)}\left(\sum_{n=0}^{\infty} D^{(n)}(y)u^n\right) = \sum_{n=0}^{\infty} D^{(n)}(y)D_u^{(m)}(u^n) \\ &= \sum_{n=0}^{\infty} D^{(n)}(y)\binom{n}{m}u^{n-m} = \sum_{n=m}^{\infty}\binom{n}{m}D^{(n)}(y)u^{n-m} \\ &= \sum_{t=0}^{\infty}\binom{t+m}{m}D^{(t+m)}(y)u^t = \phi(D^{(m)}y). \end{aligned}$$

Thus we obtain the following result:

Proposition 14.2.12. *A derivation D is iterative if and only if*

$$D_u^{(m)}(\phi(y)) = \phi(D^{(m)}(y)).$$ □

Theorem 14.2.13. *Given a differential D on K, then D can be extended in a unique way to a finite separable extension $L = K(w)$. If D is iterative, then the extension of D to L is also iterative.*

Proof. Consider the ring of power series $L[[u]] \supseteq K[[u]]$. Let

$$g(t) = \operatorname{Irr}(w, t, K) = t^n + a_{n-1}t^{n-1} + \cdots + a_1 t + a_0 \in K[t]$$

be the irreducible polynomial of w. Let $A_i = \phi(a_i) \in K[[u]] \subseteq L[[u]]$ be the power series corresponding to each coefficient, i.e., $A_i = \sum_{n=0}^{\infty} \left(D^{(n)} a_i\right) u^n$.

Set $G(t) = t^n + A_{n-1}t^{n-1} + \cdots + A_1 t + A_0 \in K[[u]][t]$. If we find $B \in L[[u]]$ such that $G(B) = 0$, and if $B = \sum_{n=0}^{\infty} b_n u^n$, $b_0 = w$, then $\phi : K \to K[[u]]$ can be extended to $\widetilde{\phi} : L \to L[[u]]$ ($\widetilde{\phi}|_K = \phi$) by defining $\widetilde{\phi}(w) = B$. Thus $b_n = D^{(n)} w$ is the required extension.

Now $L[[u]]$ is a complete field whose absolute value is given by the valuation

$$v\left(\sum_{n=0}^{\infty} c_n u^n\right) = n_0, \quad \text{where} \quad c_i = 0, \quad 0 \le i \le n_0 - 1, \quad \text{and} \quad c_{n_0} \neq 0.$$

The residue field is L. Since $g(t) = (t - w)h(t) \in L[t]$ and $G(t) \equiv g(t) \bmod u$, it follows by Hensel's lemma (Theorem 2.3.14) that G has a root B in $L[[u]]$ and $b_0 = w$.

The uniqueness of the extension of D to L is also a consequence of Hensel's lemma since $g(t)$ is separable and then G has a unique root with constant term w.

Finally, assume that D is iterative. If $z \in L$, write $Z = \phi(z)$ and $Z^{(m)} = \phi(D^{(m)} z)$, so that $Z^{(m)} = \sum_{n=0}^{\infty} D^{(n)} \left(D^{(m)} z\right)$. We also have

$$D_u^{(m)} Z = \sum_{n=0}^{\infty} D^{(n)}(z) D_u^{(m)} u^n.$$

There are two differentiations in L given by the corresponding Taylor series, namely $\phi_1(D^{(m)} z) = Z^{(m)}$ and $\phi_2(D^{(m)} z) = D_u^{(m)} Z$. Since both homomorphisms when restricted to K are the same, they yield extensions of D to L. But the extension is unique, so it follows that $\phi(D^{(m)} z) = D_u^{(m)} \phi(z)$. Thus the extension of D is iterative. □

Proposition 14.2.14. *For each separating element $x \in K \setminus k$, there exists one and only one differentiation $D_x := \left\{D_x^{(n)}\right\}_{n=0}^{\infty}$ of K/k such that*

$$D_x^{(1)}(x) = 1 \quad \text{and} \quad D_x^{(n)}(x) = 0 \quad \text{for} \quad n \ge 2.$$

Notice that this differentiation is iterative; it is called differentiation with respect to x *and will be denoted by $D_x^{(1)} = \frac{d}{dx}$.*

Proof. Let $F = k(x)$, where $K/k(x)$ is separable. If D is any differential satisfying $D^{(1)} x = 1$ and $D^{(n)} x = 0$ for $n \ge 2$, then it is easy to verify, using induction and the product rule, that

$$D^{(n)} x^m = \binom{m}{n} x^{m-n} \tag{14.12}$$

for $n, m \geq 0$. In particular, $D^{(n)}x^m = 0$ for $n > m$, and $D^{(n)}x^n = 1$. For any $f(x) = a_m x^m + \cdots + a_1 x + a_0 \in k[x]$, we have

$$D^{(n)} f(x) = a_m \binom{m}{n} x^{m-n} + \cdots + a_1 \binom{1}{n} x^{1-n} + D^{(n)}(a_0).$$

Also, if $g(x) = \frac{1}{f(x)}$ with $f(x) \in k[x]$, then for $n \geq 1$,

$$0 = D^{(n)}(1) = D^{(n)}(fg) = \sum_{i=0}^{n} D^{(n-i)}(f) D^{(i)}(g),$$

so $D^{(n)}(g)$ is uniquely defined. Now formula (14.12) defines a differential D_x on F satisfying $D_x^{(1)} x = 1$, and $D_x^{(n)} x = 0$ for $n \geq 2$.

We have

$$\begin{aligned}(D_x^{(n)} \circ D_x^{(m)})(x^t) &= D_x^{(n)} \binom{t}{m} x^{t-m} = \binom{t}{m} D_x^{(n)}(x^{t-m}) \\ &= \binom{t}{m}\binom{t-m}{n} x^{t-m-n} \\ &= \binom{m+n}{m}\binom{t}{m+n} x^{t-(n+m)} = \binom{m+n}{m} D_x^{(n+m)}(x^t).\end{aligned}$$

Therefore D_x is iterative. By Theorem 14.2.13 there exists a unique extension of D_x to K, and this extension is iterative. □

Proposition 14.2.15. *Let $\wp$ be a place of K/k and π a prime element of $\wp$. For $\alpha \in K$, consider its power series expansion $\alpha = \sum_{n=n_0}^{\infty} a_n \pi^n$ in $K_\wp$. Then*

$$D_\pi^{(m)} \alpha = \sum_{m=n_0}^{\infty} a_n \binom{n}{m} \pi^{n-m}. \tag{14.13}$$

Proof. We have that $K_\wp$ is isomorphic to $k((\pi))$ and contains K. For $f(\pi) = a_n \pi^n + \cdots + a_1 \pi + a_0 \in k[\pi]$, we obtain $D_\pi^{(m)} f(\pi) = \sum_{i=0}^{n} a_i \binom{i}{m} \pi^{i-m}$.

On the other hand, (14.13) defines an iterative differential D on $k((\pi))$. Since D and D_π agree on $k(\pi)$, the result follows. □

Lemma 14.2.16. *Let F be any field and let $M = F[[u]]$ be the field of power series in u with coefficients in F. Let $v = h(u) = \sum_{n=1}^{\infty} a_n u^n \in M$ with $a_1 \neq 0$. Then there exists g in $M_1 = F[[v]]$, which is the field of power series in v with coefficients in F, such that $(g \circ h)(u) = u$ and $(h \circ g)(v) = v$.*

Proof. If such g exists, let $g(v) = \sum_{n=1}^{\infty} b_n v^n$. Then

$$v = h(g(v)) = \sum_{n=1}^{\infty} a_n \left(\sum_{m=1}^{\infty} b_m v^m\right)^n = \sum_{n=1}^{\infty} a_n v^n \left(\sum_{m=1}^{\infty} b_m v^{m-1}\right)^n.$$

Hence,

$$\begin{aligned} a_1b_1 = 1, &\quad \text{so} \quad b_1 = a_1^{-1}; \\ a_1b_2 + a_2b_1 = 0, &\quad \text{so} \quad b_2 = -a_1^{-1}a_2b_1 = -a_2b_1^2; \\ a_1b_3 + 2a_2b_1b_2 + a_3b_1 = 0, &\quad \text{so} \quad b_3 = -a_1^{-1}(2a_2b_1b_2 + a_3b_1) \\ &\qquad\qquad = -b_1(2a_2b_1b_2 + a_3b_1). \end{aligned}$$

In general, we have

$$a_1b_n + a_2p_1^{(n)}(b_1, b_2, \dots, b_{n-1}) + \cdots + a_np_{n-1}^{(n)}(b_1, \dots, b_{n-1}) = 0, \tag{14.14}$$

where $p_1^{(n)}, \dots, p_{n-1}^{(n)}$ are polynomials in $\mathbb{Z}[b_1, \dots, b_{n-1}]$.

Then $g(v) = \sum_{n=1}^{\infty} b_nv^n$, where b_n is as in (14.14) and satisfies $v = h(g(v))$. Now since $b_1 \neq 0$, there exists $h_1(u) \in M$ such that $u = g(h_1(u))$. Thus

$$h(u) = h(g(h_1(u))) = (h \circ g)(h_1(u)) = h_1(u). \qquad \square$$

Proposition 14.2.17. *Let D be an iterative differential defined on K and let $x \in K$ be such that $D^{(1)}x \neq 0$. Let $\phi : K \to M = K[[u]]$ be defined by $\phi(y) = \sum_{n=0}^{\infty} D^{(n)}(y)u^n$ and let $v = h(u) = \sum_{n=1}^{\infty} D^{(n)}(x)u^n$.*

Then if $\psi : K \to M_1 = K[[v]]$ is defined by

$$\psi(y)(v) = \phi(y)(g(v)) = \sum_{m=0}^{\infty} b_mv^m,$$

where $h(g(v)) = v$, $g(h(u)) = u$, the formula

$$D_1^{(m)}(y) := b_m$$

defines a differentiation on K. We also have $\phi(y)(u) = \psi(y)(h(u))$.

Proof. We have $b_0 = \psi(y)(0) = \phi(y)(g(0)) = \phi(y)(0) = D^{(0)}(y) = y$ and

$$\begin{aligned} \psi(y_1 + y_2)(v) &= \phi(y_1 + y_2)(g(v)) = \phi(y_1)(g(v)) + \phi(y_2)(g(v)) \\ &= \psi(y_1)(v) + \psi(y_2)(v), \\ \psi(y_1y_2)(v) &= \phi(y_1y_2)(g(v)) = \phi(y_1)(g(v))\phi(y_2))(g(v)) = \psi(y_1)(v)\psi(y_2)(v). \end{aligned}$$

The result follows by Proposition 14.2.11. $\square$

Corollary 14.2.18. *The new differential obtained in Proposition 14.2.17 satisfies $D_1^{(1)}x = x$ and $D_1^{(n)}x = 0$ for $n \geq 2$. That is, $D_1 = D_x$ is the derivative with respect to x.*

Proof. We have

$$\begin{aligned}\psi(x)(v) = \phi(x)(g(v)) &= \sum_{n=0}^{\infty} D^{(n)}(x)(g(v))^n \\ &= D^0 x + \sum_{n=1}^{\infty} D^{(n)} x(g(v))^n = x + h(g(v)) = x + v.\end{aligned}$$

Thus $D_1^{(0)}(x) = x$, $D_1^{(1)}(x) = 1$, and $D_1^{(n)}(x) = 0$ for $n \geq 2$. $\square$

Now we consider a separably generated function field K/k.

Theorem 14.2.19. *Let D be a differential defined over K such that $D^{(1)}c = 0$ for all $c \in k$. Let $x \in K$ be such that $D^{(1)}x \neq 0$. Then x is a separating element, that is, $K/k(x)$ is separable.*

Proof. Let t be a separating element. To show that x is a separating element, it suffices to see that $k(x,t)/k(x)$ is separable, or equivalently that t is separable over $k(x)$. Assume that t is not separable over $k(x)$. Then if

$$p(T) = \operatorname{Irr}(t, T, k(x)) \in k(x)[T],$$

there exists $\ell(T) \in k(x)[T]$ such that $p(T) = \ell(T^p)$. In other words, there is an irreducible equation

$$\sum_{n,m} c_{nm} t^n x^m = 0 \quad \text{with} \quad c_{n,m} \in k$$

such that if $c_{nm} \neq 0$, then $p \mid n$. On the other hand, since $k(t,x)/k(t)$ is separable, there exists $c_{nm} \neq 0$ such that $p \nmid m$. Therefore x is separable over $k(t^p)$. Clearly we have $D^{(1)}\alpha = 0$ for all $\alpha \in k(t^p)$. By Theorem 14.2.13, $D|_{k(t^p)}$ can be extended uniquely to $k(x,t^p)$. Since $D^{(1)}|_{k(x,t^p)} = 0$ is one such extension, it follows that $D^{(1)}x = 0$. This contradiction proves the theorem. $\square$

Theorem 14.2.20. *Let D and F be two iterative differentials on the separably generated function field K/k. Assume that $D^{(1)} \not\equiv 0$ and $F^{(1)} \not\equiv 0$. Then there exists $z \in K$ such that F is obtained from D as in Proposition* 14.2.17. *More precisely, define*

$$\phi : K \to M = K[[u]]$$

and

$$\psi : K \to M = K[[v]]$$

by $\phi(\alpha)(u) = \sum_{n=0}^{\infty} D^{(n)}(\alpha)u^n$ and $\psi(\alpha)(v) = \sum_{n=0}^{\infty} F^{(n)}(\alpha)v^n$. Then there exists $z \in K$ such that $D^{(1)}z \neq 0$ and if $v = h(u) = \sum_{n=1}^{\infty} D^{(n)}(z)u^n$, then $\psi(\alpha)(v) = \phi(\alpha)(g(v))$, where $h(g(v)) = v$, $g(h(u)) = u$.

Proof. Let $x, y \in K$ be such that $D^{(1)}x \neq 0$ and $F^{(1)}y \neq 0$. Let D_x and D_y be the differentiations with respect to x and y respectively. By Theorem 14.2.19, x and y are separating elements of K/k. Thereby $k(x, y)/k(x)$ and $k(x, y)/k(y)$ are separable extensions.

Let $\sum_{n,m} c_{n,m}x^n y^m = 0$ be an irreducible equation. Then there exist $c_{n,m} \neq 0$, $c_{n',m'} \neq 0$ such that $p \nmid n$, $p \nmid m'$, where $\operatorname{char} K = p \geq 0$ (if $p = 0$, the above condition is vacuous). Thus

$$\sum_{n,m} nc_{nm}x^{n-1}y^m + D_x^{(1)}(y) \sum_{n,m} mc_{n,m}x^n y^{m-1} = 0.$$

Since $\sum_{n,m} c_{nm}x^n y^m = 0$ is irreducible, we have

$$\sum_{n,m} nc_{nm}x^{n-1}y^m \neq 0 \quad \text{and} \quad \sum_{n,m} mc_{n,m}x^n y^{m-1} \neq 0.$$

Therefore $D_x^{(1)}(y) = -\frac{\sum nc_{nm}x^{n-1}y^m}{\sum mc_{nm}x^n y^{m-1}} \neq 0$.

Let $\theta : K \to M_2 = K[[w]]$ and $\delta : K \to M_3 = K[[t]]$ be defined by

$$\theta(\alpha)(w) = \sum_{n=0}^{\infty} D_x^{(n)}(\alpha)w^n \quad \text{and} \quad \delta(\alpha)(t) = \sum_{n=0}^{\infty} D_y^{(n)}(\alpha)t^n.$$

Set $p(w) = \sum_{n=1}^{\infty} D_x^{(n)}(y)w^n$. Since $D_x^{(1)}(y) \neq 0$, by Lemma 14.2.16 there exists $\ell(t) \in M_3$ such that $(\ell \circ p)w = w$ and $(p \circ \ell)(t) = t$. By Proposition 14.2.17 we have $\delta(\alpha)(t) = \theta(\alpha)(\ell(t))$ and $\theta(\alpha)(w) = \delta(\alpha)(p(w))$.

Since $D^{(1)}x \neq 0$ and $F^{(1)}y \neq 0$, then by Lemma 14.2.16 and Proposition 14.2.17 we have the following: Assume that $w = h(u) = \sum_{n=1}^{\infty} D^{(n)}(x)u^n$ and $t = h_1(v) = \sum_{n=1}^{\infty} F^{(n)}(y)v^n$; then for g and g_1 such that $g(h(u)) = u$, $h(g(w)) = w$ and $g_1(h_1(v)) = v$, $h_1(g_1(t)) = t$, we obtain

$$\theta(\alpha)(w) = \phi(\alpha)(g(w)) \quad \text{and} \quad \delta(\alpha)(t) = \psi(\alpha)(g_1(t)).$$

Therefore

$$\psi(\alpha)(v) = \delta(\alpha)(h_1(v)) = \theta(\alpha)(\ell(h_1(v))) = \phi(\alpha)(g(\ell(h_1(v)))) = \phi(\alpha)(g_2(v)),$$

where $g_2 = g \circ \ell \circ h_1$ and $h_2 = g_1 \circ h \circ p$, then $(g_2 \circ h_2)(u) = u$ and $(h_2 \circ g_2)(v) = v$. $\square$

Corollary 14.2.21. *If K/k is a separably generated function field and D is any iterative differential such that $D^{(1)} \not\equiv 0$, there exists $x \in K \setminus k$ such that $D = D_x$.*

Proof. Let $z \in K$ be such that $D^{(1)}z \neq 0$. Assume that $v = h(u) = \sum_{n=1}^{\infty} D^{(n)}(z)u^n$ and $g \in K[[v]]$ satisfies $g(h(u)) = u$ and $h(g(v)) = v$. Then if $\phi : K \to K[[u]]$ is defined by $\phi(\alpha)(u) = \sum_{n=0}^{\infty} D^{(n)}(\alpha)u^n$, the differential $\psi : K \to K[[v]]$ given by

$$\psi(\alpha)(v) = \phi(\alpha)(g(v))$$

is of the form D_y for some $y \in K$ (Corollary 14.2.18). Thus $\phi(\alpha)(u) = \psi(\alpha)(h(u))$ is also of the form D_x for some $x \in K$ $\square$

Remark 14.2.22. If $D_x y \neq 0$, then $D_y x \neq 0$. Furthermore, $D_x^{(1)}(y) D_y^{(1)}(x) = 1$. This follows from Lemma 14.2.16 since if

$$\phi\colon K \to K[[u]] \quad \text{is defined by} \quad \phi(\alpha)(u) = \sum_{n=0}^{\infty} D_x^{(n)}(\alpha) u^n,$$

then assuming $v = h(u) = \sum_{n=1}^{\infty} D_x^{(n)}(y) u^n$, we obtain

$$g(v) = \sum_{n=1}^{\infty} a_n v^n, \quad \text{where} \quad a_1 = \left(D_x^{(1)} y\right)^{-1} = D_y^{(1)} x.$$

Notation 14.2.23. If $D_x y \neq 0$, we write $\frac{dy}{dx} := D_x^{(1)} y$, so that

$$\frac{dx}{dy} = D_y^{(1)}(x) = \left(D_x^{(1)}(y)\right)^{-1} = \left(\frac{dy}{dx}\right)^{-1}.$$

Remark 14.2.24. If x, y are two separating elements of K/k and if $F(x, y) = 0$ is an irreducible equation, then using the proof of Theorem 14.2.20 we obtain

$$D_x^{(1)} y = -\frac{\frac{\partial F}{\partial x}(x, y)}{\frac{\partial F}{\partial y}(x, y)},$$

where $\frac{\partial F}{\partial x}, \frac{\partial F}{\partial y}$ denote the usual partial derivatives. That is, if $F(x, y) = \sum_{n,m} c_{n,m} x^n y^m$, then

$$\frac{\partial F}{\partial x}(x, y) = \sum_{n,m} n c_{n,m} x^{n-1} y^m \quad \text{and} \quad \frac{\partial F}{\partial y}(x, y) = \sum_{n,m} m c_{n,m} x^n y^{m-1}.$$

14.2.2 The Wronskian

In the classical case, given a basis of the holomorphic differentials, the zeros of the Wronskian determinant are the so-called Weierstrass points. These points depend only on the function field, or equivalently, on the Riemann surface [34, 36]. Since the Weierstrass points are field invariants, they were used by Weierstrass and others to study the group of automorphisms of a function field over $\mathbb{C}$ [70, 117, 162]. It was noticed that the Weierstrass points were often related to the branch (ramified) points. H. L. Schmid [135] assumed that the behavior of the Weierstrass points was the same in characteristic $p > 0$ as in characteristic 0. Then he deduced that any ramified prime divisor in a Galois extension of degree p of a rational function field $k(x)$, where k is algebraically closed of characteristic p, is a Weierstrass point. However, the behavior in characteristic p of the Wronskian determinant differs from the characteristic 0 case.

F. K. Schmidt was the first to study the Wronskian and the Weierstrass points in characteristic greater than 0. Here we present the theory of the Wronskian and Weierstrass points for any function field over an algebraically closed field of constants. We follow very closely the original papers of Schmidt [138, 139].

In this section we consider an iterative differentiation D on K/k such that if $D^{(n)} a = 0$ for all $n \geq 1$, then $a \in k$ and $D^{(1)} \neq 0$.

Proposition 14.2.25. *Let $\{y_0, \dots, y_n\} \subseteq K$ be linearly independent over k. For $0 \leq i \leq n$, put*

$$Y_i := \phi(y_i) = \sum_{n=0}^{\infty} D^{(n)}(y_i)u^n.$$

Then $\{Y_0, \dots, Y_n\} \subseteq K[[u]]$ is linearly independent over K.

Proof. Suppose for the sake of contradiction that $\{Y_0, \dots, Y_n\}$ is linearly dependent over K. Then we may assume that $\{Y_1, \dots, Y_r\}$ is linearly independent over K and $\{Y_0, Y_1, \dots, Y_r\}$ is linearly dependent over K. Let $a_i \in K$ be such that $Y_0 = \sum_{i=1}^r a_i Y_i$. In particular, we have

$$D^{(n)} y_0 = \sum_{i=1}^{r} a_i D^{(n)} y_i \quad \text{for} \quad n \geq 0. \tag{14.15}$$

Applying the operator $D^{(m)}$ to (14.15) we obtain

$$D^{(m)} \circ D^{(n)} y_0 = \sum_{i=1}^{r} D^{(m)}(a_i D^{(n)} y_i) = \sum_{i=1}^{r}\sum_{j=0}^{m} D^{(m-j)}(a_i)D^{(j)}D^{(n)}(y_i).$$

Using the iterative rule, we obtain

$$\begin{aligned}\binom{n+m}{n} D^{(n+m)}(y_0) &= \sum_{i=1}^{r}\sum_{j=0}^{m}\binom{n+j}{n} D^{(m-j)}(a_i)D^{(n+j)}(y_i)\\ &= \sum_{i=1}^{r}\sum_{j=0}^{m-1}\binom{n+j}{n} D^{(n-j)}(a_i)D^{(n+j)}(y_i) + \binom{n+m}{n}\sum_{i=1}^{r} a_i D^{(n+m)}(y_i).\end{aligned}$$

Applying (14.15) to $n+m$, we get

$$0 = \sum_{i=1}^{r}\sum_{j=0}^{m-1}\binom{n+j}{n} D^{(m-j)}(a_i)D^{(n+j)}(y_i). \tag{14.16}$$

For $m = 1, 2, \dots$ in (14.16) we obtain

$$\begin{aligned} m=1: \quad 0 &= \sum_{i=1}^{r} D^{(1)}(a_i)D^{(n)}(y_i) \quad \text{for} \quad n \geq 0,\\ m=2: \quad 0 &= \sum_{l=1}^{r} D^{(2)}(a_i)D^{(n)}(y_i) + \binom{n+1}{n}\sum_{i=1}^{r} D^{(1)}(a_i)D^{(n+1)}(y_i)\\ &= \sum_{i=1}^{r} D^{(2)}(a_i)D^{(n)}(y_i), \quad \text{for} \quad n \geq 0,\\ \dots \quad & \quad \dots \end{aligned}$$

It follows by induction that $0 = \sum_{i=1}^{r} D^{(m)}(a_i)D^{(n)}(y_i)$ for any $m \geq 1$ and any $n \geq 0$. Therefore we have

$$\left(D^{(m)}(a_1)\right) Y_1 + \cdots + \left(D^{(m)}(a_r)\right) Y_r = 0 \quad \text{for each} \quad m \geq 1.$$

Since $\{Y_1, \ldots, Y_r\}$ is linearly independent over K, we have $D^{(m)}(a_i) = 0$ for all $m \geq 1$ and all i. Thus $a_i \in k$, and since $Y_0 = \sum_{i=1}^{r} a_i Y_i$ it follows that $y_0 = \sum_{i=1}^{r} a_i y_i$, $a_i \in k$. Hence $\{y_0, \ldots, y_r\}$ is not linearly independent over k. This contradiction shows that $\{Y_0, \ldots, Y_n\}$ is linearly independent over K. □

Theorem 14.2.26. *If $\{y_0, \ldots, y_n\}$ is linearly independent over k, then there exist n integer numbers $m_1, \ldots, m_n$ such that $0 < m_1 < m_2 < \cdots < m_n$ and*

$$\Delta_{m_1,\ldots,m_n}(y_0, \ldots, y_n) = \det \begin{bmatrix} y_0 & \cdots & y_n \\ D^{(m_1)}(y_0) & \cdots & D^{(m_1)}(y_n) \\ \vdots & \cdots & \vdots \\ D^{(m_n)}(y_0) & \cdots & D^{(m_n)}(y_n) \end{bmatrix} \neq 0. \tag{14.17}$$

Proof. By Proposition 14.2.25 the $(n+1)$ power series

$$\begin{array}{ccccccccc} Y_0 &=& y_0 &+& D^{(1)}(y_0)u &+ \cdots +& D^{(m)}(y_0)u^m &+& \cdots \\ \vdots && \vdots && \vdots && \vdots && \vdots \\ Y_n &=& y_n &+& D^{(1)}(y_n)u &+ \cdots +& D^{(m)}(y_n)u^m &+& \cdots \end{array}$$

form a linearly independent set over K. Thus the rank of the matrix $\left[D^{(m)}(y_i)\right]_{0 \leq m \leq \infty}^{0 \leq i \leq n}$ is $n+1$. The result follows. □

We write $\widetilde{y} = \begin{pmatrix} y_0 \\ \vdots \\ y_n \end{pmatrix}$ and $D^{(m)}\widetilde{y} := \begin{pmatrix} D^{(m)}(y_0) \\ \vdots \\ D^{(m)}(y_n) \end{pmatrix}$. Define integers $\varepsilon_0, \ldots, \varepsilon_n$ as follows: Set $\varepsilon_0 = 0$ and if $\varepsilon_1, \ldots, \varepsilon_i$ have been defined and $i \leq n-1$, let $\varepsilon_{i+1} = \min\{j \in \mathbb{N} \mid D^{(\varepsilon_{i+1})}(\widetilde{y})$ is linearly independent from $D^{(\varepsilon_0)}\widetilde{y}, \ldots, D^{(\varepsilon_i)}\widetilde{y}\}$. Thus $\varepsilon_0 < \varepsilon_1 < \ldots < \varepsilon_n$, and $\{\varepsilon_0, \ldots, \varepsilon_n\}$ is minimal satisfying Theorem 14.2.26.

Definition 14.2.27. Let $\{\varepsilon_0, \ldots, \varepsilon_n\}$ and $\{y_0, \ldots, y_n\}$ be as above. Then

$$W := \Delta_{\varepsilon_1,\ldots,\varepsilon_n}(y_0, \ldots, y_n)$$

is called the *Wronskian determinant*. The set $\{\varepsilon_0, \varepsilon_1, \ldots, \varepsilon_n\}$ is called the *order* of $\widetilde{y} = \begin{pmatrix} y_0 \\ \vdots \\ y_n \end{pmatrix}$ with respect to D and each ε_i is called an *order* of $\widetilde{y}$.

Proposition 14.2.28 (Matzat). *Let $\{\alpha_0, \ldots, \alpha_n\}$ be natural numbers such that $0 \leq \alpha_0 < \alpha_1 < \cdots < \alpha_n$ and $\Delta_{\alpha_0,\ldots,\alpha_n}(y_0, \ldots, y_n) \neq 0$. Then $\varepsilon_i \leq \alpha_i$ for $0 \leq i \leq n$.*

Proof. Assume that there exists $r \in \{1, \dots, n\}$ such that $\varepsilon_i \leq \alpha_i, 0 \leq i \leq r-1$, and $\alpha_r < \varepsilon_r$. By the definition of ε_i the rank of

$$[D^{(0)}(\widetilde{y}), \dots, D^{(\varepsilon_1)}(\widetilde{y})]^T$$

is 2, the rank of

$$\left[D^{(0)}(\widetilde{y}), \dots, D^{(\varepsilon_1)}(\widetilde{y}), D^{(\varepsilon_1+1)}(\widetilde{y}), \dots, D^{(\varepsilon_2)}(\widetilde{y})\right]^T$$

is 3, and so on. In particular, the rank of

$$\left[D^{(0)}(\widetilde{y}), D^{(1)}(\widetilde{y}), \dots, D^{(\varepsilon_r-1)}(\widetilde{y})\right]^T$$

is r. It follows that the rank of $\left[D^{(\alpha_0)}(\widetilde{y}), \dots, D^{(\alpha_r)}(\widetilde{y})\right]$ is at most r, a contradiction. Thus $\varepsilon_i \leq \alpha_i$ for $0 \leq i \leq n$. □

Now assume that $\{y_0, \dots, y_n\}$ is a linearly independent set over k and V is the k-vector space generated by $\{y_0, \dots, y_n\}$. If $\{z_0, \dots, z_n\}$ is another basis of V, consider the matrix A defined by

$$z_i = \sum_{j=0}^{n} a_{ij} y_j \quad \text{for} \quad i = 0, \dots, n \quad \text{and} \quad a_{ij} \in k.$$

Then $D^{(m)} z_i = \sum_{j=0}^{n} a_{ij} D^{(m)} y_j$ for all $m \geq 0$. Hence

Proposition 14.2.29. *Whenever* $0 = \alpha_0 < \alpha_1 < \cdots < \alpha_n$, *we have*

$$\Delta_{\alpha_1,\dots,\alpha_n}(z_0, \dots, z_n) = (\det A)\Delta_{\alpha_1,\dots,\alpha_n}(y_0, \dots, y_n). \qquad \square$$

A consequence of the latter is that the Wronskian determinant is an invariant of the space $V = \sum_{i=1}^{n} k y_i$.

The Wronskian determinant can be determined with the help of the power series (14.11).

Definition 14.2.30. Two integral domains P, P_1 with iterative differentiations D and D_1 respectively are called *differentially isomorphic* if there exists a ring isomorphism $\theta\colon P \to P_1$ such that $\theta(D^{(n)} y) = D_1^{(n)}(\theta(y))$ for all $y \in P, n \in \mathbb{Z}, n \geq 0$.

Now for K/k and $M = K[[u]]$, let $T = \phi(K) \subseteq M$, where ϕ is given by (14.11), that is, $\phi(y) = \sum_{n=0}^{\infty} (D^{(n)} y) u^n$. Define $D_u^{(n)}$ in T by

$$D_u^{(n)}\left(\sum_{m=0}^{\infty} a_m u^m\right) = \sum_{m=n}^{\infty} \binom{m}{n} a_m u^{m-n}.$$

Then by Proposition 14.2.12, K and T are differentially isomorphic (recall that we are assuming D to be iterative). Note that if

$$D_u^{(n)}(z) = 0 \quad \text{for all} \quad n \geq 1, \quad \text{then} \quad z \in K, \tag{14.18}$$

Definition 14.2.31. Let $z_0, \ldots, z_n \in M$. We define the *Wronskian determinant of* $\{z_0, \ldots, z_n\}$ by

$$\Delta_{\varepsilon_1,\ldots,\varepsilon_n}(z_0, \ldots, z_n) := \det \begin{bmatrix} D_u^{(0)}(z_0) & \cdots & D_u^{(0)}(z_n) \\ D_u^{(\varepsilon_1)}(z_0) & \cdots & D_u^{(\varepsilon_1)}(z_n) \\ \vdots & & \vdots \\ D_u^{(\varepsilon_n)}(z_0) & \cdots & D_u^{(\varepsilon_n)}(z_n) \end{bmatrix}.$$

For $n+1$ linearly independent power series $\{z_0, \ldots, z_n\}$ over K, the Wronskian determinant of $z_0, \ldots, z_n$ will be denoted by $\Delta(z_0, \ldots, z_n)$.

Now, T and K are differentially isomorphic, so if $\{y_0, \ldots, y_n\} \subseteq K$ and $\phi(y_i) = Y_i$ for $0 \le i \le n$, then since D is iterative we have

$$\phi(\Delta_{\varepsilon_0,\ldots\varepsilon_n}(y_0, \ldots, y_n)) = \Delta_{\varepsilon_1,\ldots,\varepsilon_n}(Y_0, \ldots, Y_n).$$

Thus

$$\Delta_{\varepsilon_1,\ldots,\varepsilon_n}(Y_0, \ldots, Y_n) \equiv \Delta_{\varepsilon_1,\ldots,\varepsilon_n}(y_0, \ldots, y_n) \mod u \tag{14.19}$$

whenever $0 < \varepsilon_1 < \cdots < \varepsilon_n$.

If $\{y_0, \ldots, y_n\}$ is a linearly independent set over k, then by Proposition 14.2.25, $\{Y_0, \ldots, Y_n\}$ is linearly independent. It follows that $\Delta(y_0, \ldots, y_n)$ and $\Delta(Y_0, \ldots, Y_n)$ have the same orders. Thus $\Delta(Y_0, \ldots, Y_n)$ is the minimal set, ordered in lexicographic order, such that

$$\Delta_{\varepsilon_1,\ldots,\varepsilon_n}(Y_0, \ldots, Y_n) \not\equiv 0 \mod u.$$

Now we define the K-vector space $\mathcal{U}$ generated by $\{Y_0, \ldots, Y_n\}$. For any other basis $\{Z_0, \ldots, Z_n\}$ of $\mathcal{U}$, it follows from Proposition 14.2.29 that the Wronskian determinants $\Delta(Y_0, \ldots, Y_n)$ and $\Delta(Z_0, \ldots, Z_n)$ have the same orders. We may choose a basis $\{Z_0, \ldots, Z_n\}$ of $\mathcal{U}$ such that

$$Z_j = u^{h_j} + \sum_{n=h_j+1}^{\infty} a_n^{(j)} u^n \quad \text{for} \quad 0 \le j \le n \tag{14.20}$$

with $0 \le h_0 < h_1 < \cdots < h_n$. Note that h_0 is the greatest integer such that u^{h_0} divides every element of $\mathcal{U}$, and in general, h_{i+1} is the maximum integer such that $u^{h_{i+1}}$ divides every element of $\mathcal{U}$ that is divisible by u^{h_i+1}. Therefore $h_0, \ldots, h_n$ are invariants of the vector space $\mathcal{U}$.

Definition 14.2.32. The powers u^{h_i}, $0 \le i \le n$, are called the *Hermitian invariants* of $\mathcal{U}$ and the basis given in (14.20) is called a *Hermitian basis* of $\mathcal{U}$ over K.

Since the Y_i's are the power series corresponding to the y_i's, it follows that $h_0 = 0$.

Proposition 14.2.33. *If $\{Z_0, \dots, Z_n\}$ is a Hermitian basis of $\mathcal{U}/K$, with respective coefficients of highest degree satisfying $0 < h_1 < \cdots < h_n$, then*

$$\Delta_{h_1,\dots,h_n}(Z_0, \dots, Z_n) \equiv 1 \mod u.$$

Proof. Using the definition of the Hermitian basis and $D_u^{(n)}$ given in (14.18) we obtain immediately that $D_u^{(h_i)} Z_j$ is congruent to 1 or 0 modulo u, depending on whether $i = j$ or $i < j$ respectively. Therefore

$$\Delta_{h_1,\dots,h_n}(Z_0, \dots, Z_n) \equiv \det \begin{bmatrix} 1 & 0 & \dots & 0 \\ 0 & 1 & \dots & 0 \\ \vdots & \vdots & \ddots & \vdots \\ 0 & 0 & \dots & 1 \end{bmatrix} \mod u \equiv 1 \mod u. \qquad \square$$

Proposition 14.2.34. *If $\{Z_0, \dots, Z_n\}$ is a Hermitian basis of $\mathcal{U}$ over K with leading coefficients $0 = h_0 < h_1 < \cdots < h_n$, then if $0 = v_0 < v_1 < \cdots < v_n$ are such that for some $1 \le r \le n$, $v_j - h_j = 0$ for $0 \le j \le r-1$ and $v_r - h_r < 0$, we have*

$$D_{v_1,\dots,v_n}(Z_0, \dots, Z_n) \equiv 0 \mod u.$$

Proof. It follows from the form of Z_r and the definition of $D_u^{(v_r)}$ that $D_u^{(v_r)}(z_r) \equiv 0 \mod u$. Thus

$$\Delta_{v_1,\dots,v_n}(z_0, \dots, z_n) \equiv \det \begin{bmatrix} 1 & 0 & & 0\,0 & 0 & \cdots & 0 \\ 0 & 1 & & 0\,0 & 0 & \cdots & 0 \\ & \ddots & \ddots & & \ddots & & \\ 0 & 0 & & 1\,0 & 0 & \cdots & 0 \\ 0 & 0 & & 0\,0 & 0 & \cdots & 0 \\ & \cdots & \cdots & & \cdots & & \end{bmatrix} \equiv 0 \mod u. \qquad \square$$

Theorem 14.2.35. *Let $Y_0, \dots, Y_n$ be the power series associated to the functions $y_0, \dots, y_n$, that is, $Y_i := \phi(y_i)$ for $0 \le i \le n$. Then the orders of the Wronskian determinant $\Delta(y_0, \dots, y_n)$ are precisely the Hermitian invariants of $\mathcal{U}$ the K-vector space generated by $\{Y_0, \cdots, Y_n\}$.*

Furthermore, assume that $\{Z_0, \dots, Z_n\}$ is a Hermitian basis of $\mathcal{U}$. Let $A \in M_{n+1}(K)$ be the change of basis matrix, i.e., $Y_i = AZ_i$ for $0 \le i \le n$. Then

$$\Delta(y_0, \dots, y_n) = \det A.$$

Proof. By (14.19), the orders of the Wronskian determinant $\Delta(y_0, \dots, y_n)$ are the same as those of $\Delta(Y_0, \dots, Y_n)$. Moreover, by Proposition 14.2.29, these orders are equal to those of $\Delta(Z_0, \dots, Z_n)$. Finally, by Propositions 14.2.33 and 14.2.34 the orders of the Wronskian determinant are equal to the Hermitian invariants of $\mathcal{U}$.

Since $\begin{pmatrix} Y_0 \\ \vdots \\ Y_n \end{pmatrix} = A \begin{pmatrix} Z_0 \\ \vdots \\ Z_n \end{pmatrix}$, we have

$$\Delta(Y_0, \ldots, Y_n) = \det(A)\Delta(Z_0, \ldots, Z_n).$$

It follows by Proposition 14.2.33 that $\Delta(Y_0, \ldots, Y_n) \equiv \det(A) \bmod u$. On the other hand, using (14.19) we obtain

$$\Delta(Y_0, \ldots, Y_n) \equiv \Delta(y_0, \ldots, y_n) \bmod u.$$

These two congruence relations yield $\Delta(y_0, \ldots, y_n) = \det A$. □

Now we study the Wronskian determinant relative to two (iterative) differentiations on K/k. By Corollary 14.2.21 these differentials are of the forms D_x and D_y. Let $\{z_0, \ldots, z_n\}$ be a linearly independent set over k. Let $W_x(z_0, \ldots, z_n)$ and $W_y(z_0, \ldots, z_n)$ be the Wronskian determinants with respect to D_x and D_y respectively.

Theorem 14.2.36. *$W_x(z_0, \ldots, z_n)$ and $W_y(z_0, \ldots, z_n)$ have the same set of orders $\{\mu_0, \mu_1, \ldots, \mu_n\}$ with $0 = \mu_0 < \mu_1 < \cdots < \mu_n$. Furthermore,*

$$W_x(z_0, \ldots, z_n) = W_y(z_0, \ldots, z_n)(D_x^{(1)}(w))^{\mu_0+\cdots+\mu_n}$$

for some $w \in K$ such that $D_x^{(1)}(w) \neq 0$.

Proof. Let $\phi : K \to M = K[[u]]$ and $\psi : K \to M_1 = K[[v]]$ be given by

$$\phi(\alpha)(u) = \sum_{n=0}^{\infty} D_x^{(n)}(\alpha)u^n \quad \text{and} \quad \psi(\alpha)(v) = \sum_{v=0}^{\infty} D_y^{(n)}(\alpha)v^n.$$

For $0 \le i \le n$, put $Z_{iu} := \phi(z_i)$ and $Z_{iv} := \psi(z_i)$. Consider the K-vector spaces $\mathfrak{M}$ and $\mathfrak{N}$ generated by $\{Z_{iu}\}_{i=0}^n$ and $\{Z_{iv}\}_{i=0}^n$ respectively.

Let $w \in K$ be such that $D_x^{(1)}w \neq 0$, given by Theorem 14.2.20 and $v = h(u) = \sum_{n=1}^{\infty} D_x^{(n)}(w)u^n$ be such that $\psi(\alpha)(v) = \phi(\alpha)(g(v))$ with $g(v) = u$.

Clearly $\mathfrak{N}$ is obtained from $\mathfrak{M}$ by means of the substitution $u = g(v)$.

Let

$$\begin{aligned}
\mathcal{U}_0 &= 1 + \sum_{m=1}^{\infty} a_m^{(0)} u^m, \\
\mathcal{U}_1 &= u^{\mu_1} + \sum_{m=\mu_1+1}^{\infty} a_m^{(1)} u^m, \\
&\cdots \quad \cdots \\
\mathcal{U}_i &= u^{\mu_i} + \sum_{m=\mu_i+1}^{\infty} a_m^{(i)} u^m, \\
&\cdots \quad \cdots \\
\mathcal{U}_n &= u^{\mu_n} + \sum_{m=\mu_n+1}^{\infty} a_m^{(n)} u^m
\end{aligned}$$

be the elements of a Hermitian basis of $\mathfrak{M}$ over K. Using the substitution $u = g(v)$, $v = h(u)$, we obtain the Hermitian basis $\{W_0, \dots, W_n\}$ of $\mathfrak{N}/K$ with $W_i = \left(D_x^{(1)}(w)\right)^{\mu_i} \mathcal{U}_i(g(v))$ for $0 \le i \le n$. Let A be the matrix defined by $Z_{iu} = A\mathcal{U}_i$ for $0 \le i \le n$, and B the matrix defined by $Z_{iv} = BW_i$ for $0 \le i \le n$. Then $BW_i = Z_{iv} = Z_{iu}(g(v)) = A\mathcal{U}_i(g(v)) = A\left(D_x^{(1)}(w)\right)^{-\mu_i} W_i$. Hence

$$B = \begin{bmatrix} (D_x^{(1)}(w))^{-\mu_0} & & 0 \\ & \ddots & \\ 0 & & (D_x^{(1)}(w))^{-\mu_n} \end{bmatrix} A.$$

By Theorem 14.2.35 we have

$$W_x(z_0, \dots, z_n) = \det A \quad \text{and} \quad W_y(z_0, \dots, z_n) = \det B.$$

Thus

$$\begin{aligned} W_y(z_0, \dots, z_n) = \det B &= \left(D_x^{(1)}(w)\right)^{-(\mu_0+\cdots+\mu_n)} \det A \\ &= \left(D_x^{(1)}(w)\right)^{-(\mu_0+\cdots+\mu_n)} W_x(z_0, \dots, z_n). \end{aligned}$$ □

Remark 14.2.37. Note that x and y are separating elements (because $D_x^{(1)}x = D_y^{(1)}y = 1$). In particular, $D_y^{(1)}x$ is nonzero. Indeed, assume otherwise. Then $D_y^{(1)}|_{k(x)} = 0$. It follows that $D_y^{(n)}|_{k(x)} = 0$ for all $n \ge 1$. Thus the extension of $D_y|_{k(x)}$ to $k(x, y)$ satisfies $D_y^{(1)}(y) = 0$. The element w given in the proof of Theorem 14.2.36 is the one that transforms D_x into D_y. This element is $w = D_x^{(1)}y = \frac{dy}{dx}$. In particular, we have

$$W_x(z_0, \dots, z_n) = \left(\frac{dy}{dx}\right)^{\mu_0+\mu_1+\cdots+\mu_n} W_y(z_0, \dots, z_n). \tag{14.21}$$

Now we investigate the arithmetic of the orders of the Wronskian determinant. We fix an iterative differentiation D on K/k such that $D^{(n)}a = 0$ for all $n \ge 1$ and $a \in k$, and such that there exists $x \in K$ satisfying $D^{(1)}x \ne 0$.

We need to consider the cases of characteristic 0 and $p > 0$.

Definition 14.2.38. Let p be a rational prime and $n, m \in \mathbb{N} \cup \{0\}$. We define the p-adic order in $\mathbb{N} \cup \{0\}$ by setting $n \le_p m$ if and only if the p-adic coefficients of n are less than or equal to those of m. More precisely, let

$$n = \sum_{i=0}^{r} a_i p^i \quad \text{and} \quad m = \sum_{i=0}^{r} b_i p^i \quad \text{for} \quad 0 \le a_i, b_i \le p-1 \quad \text{and} \quad 0 \le i \le r.$$

Then $n \le_p m$ if and only if $a_i \le b_i$ for all $i = 0, \dots, r$.

For $p = 0$, we may define $n \le_0 m$ as the usual order in $\mathbb{N} \cup \{0\}$.

Lemma 14.2.39. *Let $n, m \in \mathbb{N} \cup \{0\}$. Then p does not divide $\binom{n}{m}$ if and only if $m \leq_p n$ (for $p = 0$ this may be viewed as the equivalence between $\binom{n}{m} \neq 0$ and $m \leq n$).*

Proof. Let $p(t) = (1+t)^n \in F(t)$, where F is any field of characteristic p.

We have $p(t) = (1+t)^n = \sum_{m=0}^{n} \binom{n}{m} t^m$. Then $p \nmid \binom{n}{m}$ if and only if t^m appears in the expansion of $p(t)$.

Let $n = a_0 + a_1 p + \cdots + a_r p^r$ for $0 \leq a_j \leq p-1$ and $0 \leq j \leq r$. Then

$$p(t) = (1+t)^n = \prod_{j=0}^{r} (1+t)^{a_j p^j} = \prod_{j=0}^{r} (1+t^{p^j})^{a_j}$$

and $(1+t^{p_j})^{a_j} = \sum_{i_j=0}^{a_j} \binom{a_j}{i_j} t^{i_j p^j}$. Therefore the powers t^m with nonzero coefficient in $p(t)$ are those of the form $i_0 + i_1 p + \cdots + i_r p^r$ with $0 \leq i_j \leq a_j$ and $0 \leq j \leq r$. This proves the lemma. □

Theorem 14.2.40. *Let* $\operatorname{char} k = p \geq 0$ *and let ε be an order of $W_x(z_0, \ldots, z_n)$. Then if $\mu \leq_p \varepsilon$, μ is an order of $W_x(z_0, \ldots, z_n)$.*

Proof. Let $\mathfrak{M} = Z_0 K + \cdots + Z_n K$, where as usual, each $Z_i = \phi(z_i)$ is the power series associated to z_i. For $0 \leq i \leq n$, let

$$\mathfrak{W}_i = u^{h_i} + \sum_{m=h_i+1}^{\infty} a_m^{(i)} u^m$$

be a Hermitian basis of $\mathfrak{M}$. The orders of $\Delta(\mathfrak{W}_0, \ldots, \mathfrak{W}_n)$ are precisely $h_0, \ldots, h_n$ with $0 = h_0 < h_1 < \cdots < h_n$ and these are also the orders of $W_x(z_0, \ldots, z_n)$. We write $\mathcal{U}_j^{(m)} = D_u^{(m)} Z_j$ and $U^{(m)} = (\mathcal{U}_0^{(m)}, \ldots, \mathcal{U}_n^{(m)})$.

Suppose that $\varepsilon \in \{h_1, \ldots, h_n\}$, $0 \neq \mu \leq_p \varepsilon$ and $\mu \notin \{h_1, \ldots, h_n\}$. Then $\{U^{(0)}, \ldots, U^{(h_r)}, U^{(\mu)}\}$ is linearly dependent over K, where $h_r < \mu < h_{r+1} = \varepsilon$. Now let

$$\mathcal{U} = u^{\varepsilon} + \sum_{m=\varepsilon+1}^{\infty} b_m u^m$$

be any power series starting at u^{ε}. Then

$$\mathcal{U}^{(\mu)} = \binom{\varepsilon}{\mu} u^{\varepsilon-\mu} + \sum_{m=\varepsilon+1}^{\infty} \binom{m}{\mu} b_m u^{m-\mu}.$$

By Lemma 14.2.39, p does not divide $\binom{\varepsilon}{\mu}$. Hence $\mathcal{U}^{(\mu)} = \binom{\varepsilon}{\mu} u^{\varepsilon-\mu} W$ with $W \equiv 1 \bmod u$. On the other hand, $u^{\varepsilon-\mu+1}$ divides $\mathcal{U}^{(h_i)}$ for $0 \leq i \leq r$. Hence $\mathcal{U}^{(h_i)} = \binom{\varepsilon}{\mu} u^{\varepsilon-\mu} W_i$ with $W_i \equiv 0 \mod u$ for $0 \leq i \leq r$.

Therefore we have

$$\det \begin{bmatrix} \mathcal{U}_0 & \cdots & \mathcal{U}_r & \mathcal{U} \\ \mathcal{U}_0^{(h_1)} & \cdots & \mathcal{U}_r^{(h_1)} & \mathcal{U}^{(h_1)} \\ \vdots & & \vdots & \vdots \\ \mathcal{U}_0^{(h_r)} & \cdots & \mathcal{U}_r^{(h_r)} & \mathcal{U}^{(h_r)} \\ \mathcal{U}_0^{(\mu)} & \cdots & \mathcal{U}_r^{(\mu)} & \mathcal{U}^{(\mu)} \end{bmatrix} = \binom{\varepsilon}{\mu} u^{\varepsilon-\mu} \det \begin{bmatrix} \mathcal{U}_0 & \cdots & \mathcal{U}_r & W_0 \\ \mathcal{U}_0^{(h_1)} & \cdots & \mathcal{U}_r^{(h_1)} & W_1 \\ \vdots & & \vdots & \vdots \\ \mathcal{U}_0^{(h_r)} & \cdots & \mathcal{U}_r^{(h_r)} & W_r \\ \mathcal{U}_0^{(\mu)} & \cdots & \mathcal{U}_r^{(\mu)} & W \end{bmatrix}$$

and

$$\det \begin{bmatrix} \mathcal{U}_0 & \cdots & \mathcal{U}_r & W_0 \\ \mathcal{U}_0^{(h_1)} & \cdots & \mathcal{U}_r^{(h_1)} & W_1 \\ \vdots & & \vdots & \vdots \\ \mathcal{U}_0^{(h_r)} & \cdots & \mathcal{U}_r^{(h_r)} & W_r \\ \mathcal{U}_0^{(\mu)} & \cdots & \mathcal{U}_r^{(\mu)} & W \end{bmatrix} \equiv \begin{bmatrix} 1 & & & \\ & 1 & 0 & \\ & & \ddots & \\ & * & 1 & \\ & & & 1 \end{bmatrix} \bmod u \equiv 1 \bmod u.$$

Hence

$$\det \begin{bmatrix} \mathcal{U}_0 & \cdots & \mathcal{U}_r & \mathcal{U} \\ \mathcal{U}_0^{(h_1)} & \cdots & \mathcal{U}_r^{(h_1)} & \mathcal{U}^{(h_1)} \\ \vdots & & \vdots & \vdots \\ \mathcal{U}_0^{(h_r)} & \cdots & \mathcal{U}_r^{(h_r)} & \mathcal{U}^{(h_r)} \\ \mathcal{U}_0^{(\mu)} & \cdots & \mathcal{U}_r^{(\mu)} & \mathcal{U}^{(\mu)} \end{bmatrix} \neq 0.$$

It follows that $\{U^{(0)}, \ldots, U^{(h_r)}, U^{(\mu)}\}$ is linearly independent. This contradiction proves the theorem. □

Corollary 14.2.41. *If* char $k = 0$, *then for any system* $\{y_0, \ldots, y_n\}$ *in* K *that is linearly independent over* k, *the numbers* $0, 1, \ldots, n$ *are the orders of the Wronskian determinant* $\Delta(y_0, \ldots, y_n)$. □

In the characteristic 0 case, we have $D_x^{(1)} x = 1$. Moreover, for any n the set $\{1, x, \ldots, x^n\}$ is linearly independent and the orders of the Wronskian determinant are $0, 1, \ldots, n$.

Now assume that char $k = p > 0$ and we are given $0 = \mu_0 < \mu_1 < \cdots < \mu_n$ such that whenever $\varepsilon \in \{\mu_0, \ldots, \mu_n\}$ and $\mu \leq_p \varepsilon$, then $\mu \in \{\mu_0, \ldots, \mu_n\}$. In this case $\mu_0, \ldots, \mu_n$ are precisely the orders of the Wronskian determinant $\Delta\left(x^{\mu_0}, \ldots, x^{\mu_n}\right)$.

14.2.3 Arithmetic Theory of Weierstrass Points

In this subsection we consider a function field K/k where k is an algebraically closed field of characteristic $p \geq 0$.

Let $\mathfrak{P}$ be any prime divisor of K. Since k is algebraically closed, $\mathfrak{P}$ is of degree 1. Thus by Theorem 14.2.4 there exist exactly $g = g_K$ gap numbers $j_1, \dots, j_g$ of $\mathfrak{P}$ such that $1 = j_1 < j_2 < \cdots < j_g \leq 2g - 1$. The sequence $\{1 = j_1, j_2, \dots, j_g\}$ depends on $\mathfrak{P}$. In the classical case, that is, when $k = \mathbb{C}$ is the field of complex numbers, the gap sequence of $\mathfrak{P}$ is $\{1, 2, \dots, g\}$ for almost all $\mathfrak{P}$. Every prime divisor with gap sequence $\{1, 2, \dots, g\}$ is called an *ordinary point* and the finite set of prime divisors of K with distinct gap sequences is called the set of *Weierstrass points*.

Now if $g \geq 2$, there exist at least $2g + 2$ Weierstrass points and furthermore, the automorphism group $\operatorname{Aut}_k(K) = \{\sigma : K \to K | \sigma|_k = \operatorname{Id}_k\}$ is finite. Also, $|\operatorname{Aut}_k(K)| \leq 84(g - 1)$. This can be proved using the Weierstrass points of the field.

In the arithmetic case, that is, $\operatorname{char} k = p > 0$, some of the above results are no longer true. We need to change the definition of Weierstrass points since there exist fields such that for every prime divisor $\mathfrak{P}$, its gap sequence $\{j_1, \dots, j_g\}$ is different from $\{1, \dots, g\}$. In this case, almost all prime divisors have the same gap sequence (not necessarily equal to $\{1, 2, \dots, g\}$). This is our main result. The proof of this important fact relies on the theory of the Wronskian determinant. When k is not algebraically closed, it is possible to have two infinite disjoint sets A and B of prime divisors, such that every element of A has the same gap sequence $\{i_1, \dots, i_r\}$ and every element of B has the same gap sequence $\{j_1, \dots, j_s\}$ but $\{i_1, \dots, i_r\} \neq \{j_1, \dots, j_r\}$. It can also happen that every possible gap sequence appears for infinitely many prime divisors (that is, there are no Weierstrass points).

For all the reasons stated above, we will consider k to be algebraically closed in the rest of this subsection. In particular, K/k is separably generated (Corollary 8.2.11).

When k is of characteristic $p > 0$, it is possible that there is only one Weierstrass point for arbitrarily large genus. This is in contrast to characteristic 0, where there exist at least $2g + 2$ distinct Weierstrass points.

For $g = 0$, we have $K = k(x)$ and the gap sequence of any prime divisor is empty. For $g = 1$, the gap sequence of every prime divisor is $\{1\}$. Assume $g \geq 2$. Let $W = W_K$ be the canonical class in K and w a nonzero differential, $(w)_K \in W$. By Corollary 3.5.5, $\ell_K((w)_K^{-1}) = N(W) = g$. Let $\{y_0, \dots, y_{g-1}\}$ be a basis of $\ell_K((w)_K^{-1})$. Then $\{y_0, \dots, y_{g-1}\}$ is a linearly independent set over k. Let x be any separating element and let $W_x(y_0, \dots, y_{g-1})$ be the Wronskian determinant. Denote by $0, \mu_1, \dots, \mu_{g-1}$ the orders of $W_x(y_0, \dots, y_{g-1})$.

Definition 14.2.42. We define the *branch divisor* by

$$\mathfrak{B}_K := (w)_K^g (W_x(y_0, \dots, y_{g-1}))_K (dx)_K^{\sum_{i=0}^{g-1} \mu_i}. \tag{14.22}$$

Remark 14.2.43. Write $\frac{dy}{dx} = D_x^{(1)} y$. We have

$$dy = \frac{dy}{dx} dx.$$

Here dx and dy denote the Weil differentials. To prove that $dy = \frac{dy}{dx} dx$, we use the equivalence between Hasse and Weil differentials (Theorem 9.3.15). In fact, assuming that $\mathfrak{P}$ is any place of K and π is a prime element for $\mathfrak{P}$, then if $F(x, y) = 0$ is irreducible we have

$$\frac{dy}{dx} = D_x^{(1)} y = -\frac{\frac{\partial F}{\partial x}(x, y)}{\frac{\partial F}{\partial y}(x, y)}$$

(see Remark 14.2.24).

Let $x = \sum_n a_n \pi^n$ and $y = \sum_n b_n \pi^n$ be the power series expansions of x and y in $K_{\mathfrak{P}}$. Since $F(x, y) = 0$, we have $0 = F\left(\sum_n a_n \pi^n, \sum_n b_n \pi^n\right)$. Using the chain rule we obtain

$$\begin{aligned} 0 = \frac{d(0)}{d\pi} &= \frac{\partial F}{\partial x}\left(\sum_n a_n \pi^n, \sum_n b_n \pi^n\right) \sum_n n a_n \pi^{n-1} \\ &\quad + \frac{\partial F}{\partial y}\left(\sum_n a_n \pi^n, \sum_n b_n \pi^n\right) \sum_n n b_n \pi^{n-1} \\ &= \frac{\partial F}{\partial x}(x, y)\frac{dx}{d\pi} + \frac{\partial F}{\partial y}(x, y)\frac{dy}{d\pi}. \end{aligned}$$

Hence

$$\frac{dy}{d\pi} = -\frac{\frac{\partial F}{\partial x}(x, y)}{\frac{\partial F}{\partial y}(x, y)} \frac{dx}{d\pi} = D_x^{(1)} y \frac{dx}{d\pi}. \tag{14.23}$$

Since (14.23) holds for any place $\mathfrak{P}$, we have

$$dy = D_x^{(1)}(y) dx = \frac{dy}{dx} dx.$$

Theorem 14.2.44. *The branch divisor $\mathfrak{B}_K$ is independent of the differential w, of the basis $y_0, \ldots, y_{g-1}$ of $L_K((w)_K^{-1})$, and of the separating element x. Thus $\mathfrak{B}_K$ is an invariant of the field.*

Proof. If t is another separating element, then by (14.21) and Remark 14.2.43 we have

$$\left(W_x(y_0, \ldots, y_{g-1})\right)_K (dx)_K^{\sum_{i=0}^{g-1} \mu_i} = \left(W_t(y_0, \ldots, y_{g-1})\right)_K (dt)_K^{\sum_{i=0}^{g-1} \mu_i}.$$

Next, if $\{z_0, \ldots, z_{g-1}\}$ is another basis of $L_k\left((w)_K^{-1}\right)$, then

$$z_i = \sum_{j=0}^{g-1} a_{ij} y_j \quad \text{for} \quad 0 \le i \le g-1,$$

where $a_{ij} \in k$. Thus $\begin{pmatrix} z_0 \\ \vdots \\ z_{g-1} \end{pmatrix} = A \begin{pmatrix} y_0 \\ \vdots \\ y_{g-1} \end{pmatrix}$, where $A = (a_{ij})_{0 \le i, j \le g-1}$ is an invertible $g \times g$ matrix with coefficients in k. Therefore by Proposition 14.2.29 we have

$$W_x(z_0, \dots, z_{g-1}) = \det A W_x(y_0, \dots, y_{g-1})$$

and $\det A \in k^*$, so $(\det A)_K = \mathfrak{N}$.

Finally, if $w' \neq 0$ is another differential, put $w' = aw$ for some $a \in K^*$ (Theorem 3.4.9). Then, if $\{y'_0, \dots, y'_{g-1}\}$ is a basis of $L_K((w')^{-1})$, then $\{ay'_0, \dots, ay'_{g-1}\}$ is a basis of $L_K((w)_K^{-1})$. Set $y_i = ay'_i$, $0 \leq i \leq g-1$. Then

$$W_x(y_0, \dots, y_{g-1}) = W_x(ay'_0, \dots, ay'_{g-1}) = a^g W_x(y'_0, \dots, y'_{g-1}).$$

Since $(a^g) = (a)^g = \frac{(w')_K^g}{(w)_K^g}$, we obtain that

$$(w)_K^g (W_x(y_0, \dots, y_{g-1}))_K = (w')_K^g (W_x(y'_0, \dots, y'_{g-1}))_K.$$

The result follows. □

Remark 14.2.45. The degree of $\mathfrak{B}_K$ is

$$\begin{aligned} d_K(\mathfrak{B})_K &= d_K((w)_K^g) + d_K(W_x(y_0, \dots, y_{g-1})_K) + d_K\left((dx)_K^{\sum_{i=0}^{g-1}\mu_i}\right) \\ &= g(2g-2) + 0 + \left(\sum_{i=0}^{g-1}\mu_i\right)(2g-2) \\ &= (2g-2)\left(g + \sum_{i=0}^{g-1}\mu_i\right). \end{aligned}$$

Therefore for $g = 1$, we have $d_K(\mathfrak{B}_K) = 0$ and $d_K(\mathfrak{B}) > 0$ for $g \geq 2$. We will prove that $\mathfrak{B}$ is an integral divisor.

Let $\mathfrak{P}$ be any prime divisor. We may choose $w \neq 0$ such that $\mathfrak{P}$ and $(w)_K$ are relatively prime. Indeed, assume $(w)_K = \mathfrak{P}^i\mathfrak{A}$ with $i \in \mathbb{Z} \setminus \{0\}$ and $(\mathfrak{A}, \mathfrak{P}) = 1$. Then for any other prime $\mathfrak{q} \neq \mathfrak{P}$ and $n \in \mathbb{N}$ large enough, we have, by Corollary 3.5.6, $\ell(\mathfrak{P}^{-i}\mathfrak{q}^{-n}) - \ell(\mathfrak{P}^{-i+1}\mathfrak{q}^{-n}) = 1$. So there exists $z \in K$ such that $(z)_K = \frac{\mathfrak{A}'}{\mathfrak{P}^i\mathfrak{q}^n}$, where $(\mathfrak{P}, \mathfrak{A}') = 1$ and $\mathfrak{A}'$ is integral. Thus $z \notin k$, $(zw)_K = (z)_K(w)_K = \frac{\mathfrak{A}'\mathfrak{A}}{\mathfrak{q}^n}$ and $v_{\mathfrak{P}}((z)_K(w)_K) = 0$.

Let π be a prime element for $\mathfrak{P}$, that is, $v_{\mathfrak{P}}(\pi) = 1$. Then $\mathfrak{P}$ is unramified in $K/k(\pi)$ and $\mathfrak{P}|_{k(\pi)}$ is relatively prime to $(d\pi)_{k(\pi)}$ in $k(\pi)$. Therefore $\mathfrak{P}$ is relatively prime to $(d\pi)_K = \operatorname{con}_{k(\pi)/K}(d\pi)_{k(\pi)}\mathfrak{D}_{K/k(\pi)}$.

It follows that $v_{\mathfrak{P}}(\mathfrak{B}_K) = v_{\mathfrak{P}}\left((W_\pi(y_0, \dots, y_{g-1}))_K\right)$.

Now choose a Hermitian basis of $L_K\left((w)_K^{-1}\right)$ with respect to $\mathfrak{P}$. Then for any nonzero element z of $L_K\left((w)_K^{-1}\right)$ we have $v_{\mathfrak{P}}(z) \geq 0$. So when we consider $z \in K_{\mathfrak{P}}$, we have

$$z = a_z \pi^{n_z} + \sum_{m=n_z+1}^{\infty} a_m \pi^m, \quad \text{with} \quad a_z, a_m \in k, \quad n_z \geq 0, \quad \text{and} \quad a_z \neq 0.$$

Thus $a_z^{-1} z = \pi^{n_z} + \sum_{m=n_z+1}^{\infty} b_m \pi^m$. Let h_0 be the minimum nonnegative integer such that

$$z = \pi^{h_0} + \sum_{m-h_0+1}^{\infty} a_m \pi^m \in L_K((w)^{-1}).$$

If z' is another such z, we have $(z - z')a = \pi^{h_1} + \sum_{m=h_1+1}^{\infty} b_m \pi^m$ for some $a \notin k^*$ and $h_1 > h_0$. Continuing in this way, we conclude that there exists a basis $\{z_0, z_1, \ldots, z_{g-1}\}$ of $L_K((w)_K^{-1})$ such that

$$z_i = \pi^{h_i} + \sum_{m=h_i+1}^{\infty} a_m^{(i)} \pi^m \quad \text{for} \quad 0 \leq i \leq g-1 \tag{14.24}$$

and $h_0 < h_1 < \cdots < h_{g-1}$. Since $\mathfrak{P}$ is relatively prime to $(w)_K$, we have $h_0 = 0$. By Proposition 14.2.33, $\Delta_{h_1,\ldots,h_{g-1}}(z_0, \ldots, z_{g-1}) \equiv 1 \bmod \pi$ and in particular, $\Delta_{h_1,\ldots,h_{g-1}}(z_0, \ldots, z_{g-1}) \neq 0$. By Proposition 14.2.28 we have $\mu_i \leq h_i$ whenever $0 \leq i \leq g-1$. If $\mu_i = h_i$ for all i, then

$$W_\pi(z_0, \ldots, z_{g-1}) = \Delta_{\mu_1,\ldots,\mu_{g-1}}(z_0, \ldots, z_{g-1}) \equiv 1 \bmod \pi$$

and $v_{\mathfrak{P}}(\mathfrak{B}_K) = 0$.

Conversely, if $\mu_i < h_i$ for some i, then $v_{\mathfrak{P}}(\mathfrak{B}_K) \geq \sum_{i=0}^{g-1}(h_i - \mu_i) > 0$. Therefore we have the following theorem:

Theorem 14.2.46. *The branch divisor $\mathfrak{B}_K$ is an integral divisor. Furthermore, $\mathfrak{P}$ divides $\mathfrak{B}$ if and only if the Hermitian powers $h_0, \ldots, h_{g-1}$ described in* (14.24) *satisfy $h_i > \mu_i$ for some i such that* $0 \leq i \leq g-1$. □

Theorem 14.2.47. *Let $\mathfrak{P}$ be any prime divisor. If $h_0, \ldots, h_{g-1}$ are the Hermitian powers associated to $\mathfrak{P}$, then $\{h_i + 1 \mid 0 \leq i \leq g-1\}$ is precisely the gap sequence of $\mathfrak{P}$.*

Proof. Let $z_i = \pi^{h_i} + \sum_{m=h_i+1}^{\infty} a_m^{(i)} \pi^m \in L_K\left((w)_K^{-1}\right)$. Then $(z_i)_K = \frac{\mathfrak{A}_i}{(w)_K}$ for some integral divisor $\mathfrak{A}_i$.

Therefore $(z_i w)_K = \mathfrak{A}_i$ and $z_i w$ is a holomorphic differential. On the other hand, $v_{\mathfrak{P}}(z_i) = h_i$ and $v_{\mathfrak{P}}((w)_K) = 0$, so $v_{\mathfrak{P}}(\mathfrak{A}_i) = h_i$. By Corollary 14.2.5, $h_i + 1$ is a gap number of $\mathfrak{P}$ and conversely. □

We have obtained the main result of this section:

Theorem 14.2.48. *Let K/k be a function field where k is an algebraically closed field. Let w be any nonzero differential of K. Let $\{y_0, \ldots, y_{g-1}\}$ be a basis of $L_K((w)_K^{-1})$*

and x a separating element of K/k. Denote by $\mu_0, \dots, \mu_{g-1}$ the orders of the Wronskian determinant $W_x(y_0, \dots, y_{g-1})$ and by $\mathfrak{B}_K$ the branch divisor of K. Then for any prime divisor $\mathfrak{P}$, the gap sequence of $\mathfrak{P}$ is $\mu_0 + 1, \dots, \mu_{g-1} + 1$ if and only if $\mathfrak{P} \nmid \mathfrak{B}_K$. In particular, all but finitely many divisors have the same gap sequence

$$\mu_0 + 1, \dots, \mu_{g-1} + 1. \qquad \square$$

Definition 14.2.49. Put $\varphi_i = \mu_{i-1} + 1$ for $1 \le i \le g$. The sequence $\{\varphi_1, \dots, \varphi_g\}$ is called the *gap sequence of the field K/k*.

Remark 14.2.50. If $\operatorname{char} k = 0$, then the gap sequence of the function field K/k is $\{1, 2, \dots, g\}$. This is called the *classical gap sequence*.

Definition 14.2.51. A prime divisor $\mathfrak{P}$ of K is called a *Weierstrass point* if its gap sequence is different from the gap sequence of the field. A prime $\mathfrak{P}$ whose gap sequence is equal to the gap sequence of the field is called an *ordinary point*.

Corollary 14.2.52. *If g_K is 0 or 1, then K has no Weierstrass points. If $g \ge 2$, the number of Weierstrass points is at least 1 and at most $(g-1)g(3g-1)$.*

Proof. By Remark 14.2.45 we have

$$d_K(\mathfrak{B}_K) = (2g-2)\left(g + \sum_{i=0}^{g-1} \mu_i\right).$$

Now $\mu_0 = 0 < \mu_1 < \mu_2 < \cdots < \mu_{g-1} \le 2g-1$. Therefore $i \le \mu_i \le g + i$. Hence

$$\sum_{i=0}^{g-1} \mu_i = \sum_{i=1}^{g-1} \mu_i \le \sum_{i=1}^{g-1} (g+i) = g(g-1) + \frac{g(g-1)}{2} = \frac{3}{2} g(g-1)$$

and

$$\sum_{i=1}^{g-1} \mu_i \ge \sum_{i=1}^{g-1} i = \frac{g(g-1)}{2}.$$

Thus

$$\begin{aligned} 0 < (g-1)g(g+1) &= (2g-2)\left(g + \frac{g(g-1)}{2}\right) \le d_K(\mathfrak{B}) \\ &\le (2g-2)\left(g + \frac{3}{2}g(g-1)\right) = (g-1)g(3g-1). \end{aligned} \qquad \square$$

Corollary 14.2.53. *If* $\operatorname{char} k = 0$*, the gap sequence of the function field K/k is $\{1, 2, \dots, g\}$ and*

$$d_K(\mathfrak{B}_K) = g^3 - g.$$

Proof. By Corollary 14.2.41 and Theorem 14.2.46, the sequence $\{1, 2, \ldots, g\}$ is the gap sequence of the field and we have $\mu_i = i$ for $0 \le i \le g-1$. Thus

$$\begin{aligned} d_K(\mathfrak{B}_K) &= (2g-2)\left(g + \sum_{i=0}^{g-1} i\right) = 2(g-1)\left(g + \frac{g(g-1)}{2}\right) \\ &= (g-1)(2g + g^2 - g) = g^3 - g. \end{aligned}$$

□

Definition 14.2.54. The *weight* of a Weierstrass point $\mathfrak{P}$ is defined by $v_{\mathfrak{P}}(\mathfrak{B}_K)$, where $\mathfrak{B}_K$ is the branch divisor.

Remark 14.2.55. Assume that $\varphi_{i+1} := \mu_i + 1$, $0 \le i \le g-1$, is the gap sequence of the field, and $\{\alpha_1, \ldots, \alpha_g\}$ is the gap sequence of a prime divisor $\mathfrak{P}$. Then if $\{h_{0_1}, \ldots, h_{g-1}\}$ is the set of Hermitian powers associated to $\mathfrak{P}$, we have

$$\alpha_{i+1} = h_i + 1 \quad \text{for} \quad 0 \le i \le g-1$$

and

$$\sum_{i=0}^{g-1}(h_i - \mu_i) = \sum_{j=1}^{g}(\alpha_j - \varphi_j).$$

If $\operatorname{char} k = 0$, then

$$v_{\mathfrak{P}}(\mathfrak{B}_K) = \sum_{i=0}^{g-1}(h_i - \mu_i) = \sum_{j=1}^{g}(\alpha_j - \varphi_j).$$

Now for $\operatorname{char} k = p$ we might have strict inequality

$$v_{\mathfrak{P}}(\mathfrak{B}_K) > \sum_{i=0}^{g-1}(h_i - \mu_i) = \sum_{j=0}^{g}(\alpha_j - \varphi_j).$$

Let $\mathfrak{P}$ be a prime divisor,

$$W(\mathfrak{P}) = \{n \in \mathbb{N} \mid n \text{ is a gap of } \mathfrak{P}\},$$

and

$$\mathbb{N} \setminus W(\mathfrak{P}) = P(\mathfrak{P}) = \{n \in \mathbb{N} \mid n \text{ is a pole of } \mathfrak{P}\}.$$

If $n, m \in P(\mathfrak{P})$ then there exist $f, g \in K$ such that $\mathfrak{N}_f = \mathfrak{P}^n$ and $\mathfrak{N}_g = \mathfrak{P}^m$. Hence $\mathfrak{N}_{fg} = \mathfrak{P}^{n+m}$ and $n + m \in P(\mathfrak{P})$. Therefore $P(\mathfrak{P})$ is a semigroup. Now we have

$$|W(\mathfrak{P})| = g \quad \text{and} \quad W(\mathfrak{P}) \subseteq \{1, 2, \ldots, 2g-1\}.$$

Thus

$$|P(\mathfrak{P}) \cap \{1, 2, \dots, 2g\}| = g.$$

Set $P(\mathfrak{P}) \cap \{1, 2, \dots, 2g\} = \{\beta_1, \dots, \beta_g\}$ with $1 < \beta_1 < \cdots < \beta_g = 2g$.

Our next task is to find a lower bound for the number of Weierstrass points of a function field of characteristic 0.

Lemma 14.2.56. *Whenever* $0 < i < g$ *we have* $\beta_i + \beta_{g-i} \geq 2g$.

Proof. Assume for the sake of contradiction that some i satisfies $\beta_i + \beta_{g-i} < 2g$. For $0 < j \leq i$, we have $\beta_j \leq \beta_i$, so $\beta_j + \beta_{g-i} \leq \beta_i + \beta_{g-i} < 2g$. Therefore

$$\beta_{g-i} < \beta_1 + \beta_{g-i} < \beta_2 + \beta_{g-i} < \cdots < \beta_i + \beta_{g-i} < 2g$$

and the subset

$$\{\beta_{g-i}, \beta_1 + \beta_{g-i}, \dots, \beta_i + \beta_{g-i}\} \subseteq P(\mathfrak{P})$$

has cardinality $1 + i$. Thus

$$\{\beta_1 < \beta_2 < \cdots < \beta_{g-i-1} < \beta_{g-i} < \beta_1 + \beta_{g-i} < \cdots < \beta_i + \beta_{g-i} < 2g = \beta_g\} \subseteq P(\mathfrak{P}).$$

It follows that $g = |P(\mathfrak{P}) \cap \{1, 2, \dots, 2g\}| \geq g - i - 1 + i + 1 + 1 = g + 1$, which constitutes the desired contradiction. □

Lemma 14.2.57. $\beta_1 = 2$ *iff* $W(\mathfrak{P}) = \{1, 3, 5, \dots, 2g - 1\}$.

Proof. We have $n\beta_1 = \beta_1 + \cdots + \beta_1 = 2n \in P(\mathfrak{P})$. □

Theorem 14.2.58. *We have* $\beta_1 = 2$ *for some* $\mathfrak{P}$ *if and only if* K/k *is a hyperelliptic function field.*

Proof. If $\beta_1 = 2$, there exists $x \in K$ such that $\mathfrak{N}_x = \mathfrak{P}^2$ and $g_K \geq 2$. Thus $[K : k(x)] = 2$, and by Definition, 9.6.15, K/k is hyperelliptic.

Conversely, if K/k is hyperelliptic, we have $g_K \geq 2$ and there exists $x \in K$ such that $[K : k(x)] = 2$. Then $\deg_K \mathfrak{N}_x = 2$. Now $K = k(x, y)$ where $y^2 = h(x) \in k(x)$ (when $\operatorname{char} k \neq 2$ or $K/k(x)$ inseparable) or $y^2 - y = h(x)$ (when $\operatorname{char} k = 2$ and $K/k(x)$ separable). In either case there exists a prime divisor $\mathfrak{p}$ of $k(x)$ that is ramified in $K/k(x)$, that is, $\mathfrak{p} = \mathfrak{P}^2$. If $\mathfrak{p}'$ is another prime divisor in $k(x)$, then $\frac{\mathfrak{p}'}{\mathfrak{p}}$ is principal. Therefore 2 is a pole number of $\mathfrak{P}$. □

Theorem 14.2.59. *Assume that* $\operatorname{char} k = 0$ *and* $g = g_K \geq 2$. *Then* K/k *is a hyperelliptic function field iff* K *has exactly* $2g+2$ *Weierstrass points, each of them with gap sequence* $\{1, 3, 5, \ldots, 2g-1\}$ *and weight* $\frac{g(g-1)}{2}$. *Furthermore, if* K/k *is hyperelliptic, then the Weierstrass points are precisely the ramified prime divisors in* $K/k(x)$, *where* $k(x)$ *is the unique quadratic rational subfield of* K.

Proof. If K/k is a hyperelliptic function field, there exists $x \in K$ such that $[K : k(x)] = 2$. Let $K = k(x, y)$, $y^2 = f(x)$ where $f(x)$ is a separable polynomial of degree m. We may assume without loss of generality that the infinite prime is unramified, so $m = 2g+2$ (see Corollary 4.3.7). Therefore every prime divisor dividing $\mathfrak{z}_f$ is a ramified prime with the first pole number 2. It follows that all these $2g+2$ prime divisors are Weierstrass points with gap sequence $\{1, 3, \ldots, 2g-1\}$ and weight $\frac{g(g-1)}{2}$. Thus

$$(g-1)g(g+1) = d_K(\mathfrak{B}_K) \leq (2g+2)\frac{g(g-1)}{2} = (g-1)g(g+1).$$

Hence K/k contains $2g+2$ Weierstrass points, and these are precisely the ramified primes.

The converse is immediate. □

Proposition 14.2.60. *With the above notation, if* $\beta_1 > 2$ *there exists* j *such that* $0 < j < g$ *and* $\beta_j + \beta_{g-j} > 2g$.

Proof. If $g = 2$, then $\beta_1 = 3$, $\beta_2 = 4$, and there is nothing to prove. If $g = 3$, then $\{\beta_1, \beta_2, \beta_3\} = \{3, 4, 6\}$ or $\{3, 5, 6\}$ or $\{4, 5, 6\}$ and $\beta_1 + \beta_2 \geq 3 + 4 = 7 > 2(3) = 6$.

Now assume $g \geq 4$ and suppose that $\beta_j + \beta_{g-j} = 2g$ for all $0 < j < g$. If $[x]$ denotes the greatest integer less than or equal to x, then

$$\beta_1, 2\beta_1, \cdots, \left[\frac{2g}{\beta_1}\right]\beta_1 \in \{\beta_1, \ldots, \beta_g\} = P(\mathfrak{P}) \cap \{1, \ldots, 2g\}.$$

Since $\beta_1 > 2$, we have $\beta_1 \geq 3$ and $\frac{2g}{\beta_1} \leq \frac{2}{3}g < g$.

Therefore $\left[\frac{2g}{\beta_1}\right] \leq \frac{2}{3}g < g$, and there exists a pole number of $\mathfrak{P}$ that is smaller than $2g$ and distinct from $\beta_1, 2\beta_1, \ldots, \left[\frac{2g}{\beta_1}\right]\beta_1$. Let β be the first pole number of this type. Then there exists an integer r such that $1 \leq r \leq \left[\frac{2g}{\beta_1}\right] < g-1$ and $r\beta_1 < \beta < (r+1)\beta_1$. Hence $\beta_1, \beta_2 = 2\beta_1, \ldots, \beta_r = r\beta_1, \beta_{r+1} = \beta$.

Since $\beta_j + \beta_{g-j} = 2g$, we have

$$\beta_{g-1} = 2g - \beta_1, \ldots, \beta_{g-r} = 2g - r\beta_1, \beta_{g-(r+1)} = 2g - \beta,$$

whence $\{\beta_{g-(r+1)}, \beta_{g-r}, \ldots, \beta_{g-1}\} = \{a \in P(\mathfrak{P}) | \beta_{g-(r+1)} \leq a \leq \beta_g = 2g\}$.

Now $\beta_1 + \beta_{g-(r+1)} = \beta_1 + 2g - \beta = 2g - (\beta - \beta_1) > 2g \quad r\beta_1 = \beta_{g-r}$ and $\beta_1 + \beta_{g-(r+1)} = 2g - (\beta - \beta_1) < 2g$. Therefore

$$\beta_1 + \beta_{g-(r+1)} \in \{a \in P(\mathfrak{P}) | \beta_{g-(r+1)} \leq a \leq \beta_g = 2g\},$$

but $\beta_1 + \beta_{g-(r+1)} \notin \{\beta_{g-(r+1)}, \ldots, \beta_{g-1}\}$. This contradiction proves the proposition. □

Corollary 14.2.61. *For any prime divisor $\mathfrak{P}$, we have*

$$\sum_{i=1}^{g} \alpha_i = \sum_{\alpha \in W(\mathfrak{P})} \alpha \leq g^2,$$

with equality if and only if the first nongap β_1 of $\mathfrak{P}$ is 2.

Proof. By Lemma 14.2.56 we have

$$2\sum_{i=1}^{g-1} \beta_i = \sum_{i=1}^{g-1} (\beta_i + \beta_{g-i}) \geq 2g(g-1).$$

Thus $\sum_{i=1}^{g} \beta_i = \beta_g + \sum_{i=1}^{g-1} \beta_i \geq 2g + g(g-1) = g(g+1)$, whence

$$\begin{aligned}\sum_{i=1}^{g} \alpha_i &= \sum_{j=1}^{2g} j - \sum_{i=1}^{g} \beta_i \leq \frac{2g(2g+1)}{2} - g(g+1)\\ &= g(2g+1) - g(g+1) = g^2.\end{aligned}$$

Furthermore, by Lemma 14.2.57 and Proposition 14.2.60 we have

$$\sum_{i=1}^{g} \alpha_i = g^2 \iff \sum_{i=1}^{g-1} \beta_i = g(g-1) \iff \beta_1 = 2. \qquad \square$$

Theorem 14.2.62. *Assume* $\operatorname{char} k = 0$. *Then there exist at least $2g+2$ Weierstrass points in K/k. Furthermore, there are exactly $2g+2$ Weierstrass points if and only if K/k is a hyperelliptic function field.*

Proof. We have $d_K(\mathfrak{B}_K) = (g-1)g(g+1) = g^3 - g$. Moreover, the gap sequence of the field is $\{\varphi_1, \ldots, \varphi_g\} = \{1, 2, \ldots, g\}$. If $\mathfrak{P}$ is a Weierstrass point and $v_{\mathfrak{P}}(\mathfrak{B}_K) = S_{\mathfrak{P}}$, then

$$S_{\mathfrak{P}} = \sum_{i=1}^{g} (\alpha_i - \varphi_i) = \sum_{i=1}^{g} \alpha_i - \sum_{i=1}^{g} \varphi_i \leq g^2 - \frac{g(g+1)}{2} = \frac{g(g-1)}{2}.$$

Thus we have at least

$$\frac{(g-1)g(g+1)}{g(g-1)/2} = 2(g+1) = 2g+2$$

distinct prime divisors dividing the branch divisor $\mathfrak{B}_K$. These are precisely the Weierstrass points.

There are exactly $2g+2$ Weierstrass points if $S_{\mathfrak{P}} = \frac{g(g-1)}{2}$ for all $\mathfrak{P}$ dividing $\mathfrak{B}_K$. By Corollary 14.2.61 this happens if and only if $\beta_1 = 2$. The result follows using Theorem 14.2.59. $\square$

Remark 14.2.63. Assume $\operatorname{char} k = 0$. Since $d_K(\mathfrak{B}) = (g-1)g(g+1)$ there exist at most $g^3 - g$ Weierstrass points. This happens exactly in the case that every Weierstrass point has weight 1, and this is possible only if the gap sequence of every Weierstrass point is $\{1, 2, \ldots, g-1, g+1\}$.

In case $\operatorname{char} k = p > 0$, we have

$$d_K(\mathfrak{B}_K) = (2g-2)\left(g + \sum_{i=0}^{g-1} \mu_i\right) \leq (g-1)g(3g-1),$$

and there may exist a single Weierstrass point for arbitrarily large genus.

Proposition 14.2.64. *If the gap sequence of a function field K/k is nonclassical, then $p+1 < 2g$ where* $\operatorname{char} k = p > 0$.

Proof. Suppose that $p+1 \geq 2g$. Then $g < p$. Thus, if n is the first pole number of an ordinary point, we have $n < g < p$ since the gap sequence of the field is nonclassical. If m is the next gap number, then $n < m$ and $m-1$ is an order of the field. Now $n-1$ is not an order, so by Theorem 14.2.40, $n-1 \not\leq_p m-1$. Therefore $m-1 \geq p$ or $m \geq p+1 \geq 2g$, which is impossible since $m \leq 2g-1$. □

14.2.4 Gap Sequences of Hyperelliptic Function Fields

Now we consider an arbitrary field k, not necessarily algebraically closed. Let K/k be any hyperelliptic function field, not necessarily separably generated. Let $g = g_K$ be the genus of K and $W = W_K$ the canonical class of K. Consider any prime divisor $\mathfrak{P}$ of degree f. Then n is a gap number iff $\ell_K(\mathfrak{P}^{-n}) - \ell_K(\mathfrak{P}^{(n-1)}) = 0$. Equivalently, n is a gap number iff $\delta_K(\mathfrak{P}^{n-1}) - \delta_K(\mathfrak{P}^n) = f = d_K(\mathfrak{P})$. We assume that the unique genus-zero subfield of K is a rational function field $k(x)$.

Lemma 14.2.65. *Let $x \in K$ be such that $[K : k(x)] = 2$. Then $\mathfrak{N}_x^{g-1} \in W$ and $\{1, x, \ldots, x^{g-1}\}$ is a basis of $L_K\left(\mathfrak{N}_x^{-(g-1)}\right)$.*

Proof. Clearly $1, x, \ldots, x^{g-1}$ belong to $L_K\left(\mathfrak{N}_x^{-(g-1)}\right)$ and $\ell_K\left(\mathfrak{N}_x^{-(g-1)}\right) \geq g$. On the other hand, $d_K\left(\mathfrak{N}_x^{g-1}\right) = 2(g-1) = 2g-2$, so $d_K\left(W^{-1}\mathfrak{N}_x^{g-1}\right) = 0$ and $\ell_K\left(W^{-1}\mathfrak{N}_x^{g-1}\right) \leq 1$. By the Riemann–Roch theorem, we have

$$\begin{aligned}\ell_K\left(\mathfrak{N}_x^{-(g-1)}\right) &= d_K\left(\mathfrak{N}_x^{g-1}\right) - g + 1 + \ell_K\left(W^{-1}\mathfrak{N}_x^{g-1}\right)\\ &\leq 2g - 2 - g + 1 + 1 = g.\end{aligned}$$

Thus $\ell_K\left(\mathfrak{N}_x^{-(g-1)}\right) = g$ and $\ell_K\left(W^{-1}\mathfrak{N}_x^{g-1}\right) = 1$. It follows that $\mathfrak{N}_x^{g-1} \in W$. □

Lemma 14.2.66. *For any integral divisor* $\mathfrak{A}$ *in* K, *we have*

$$\ell_K(\mathfrak{A}W^{-1}) = \delta(\mathfrak{A}) = g - \mu,$$

where $\mu = \min\{d_{k(x)}(\mathfrak{A} \cap k(x)), g\}$.

Proof. First assume that $\mathfrak{A}$ and $\mathfrak{N}_x$ are relatively prime. Then

$$\delta(\mathfrak{A}) = \ell_K(\mathfrak{A}W^{-1}) = \ell_K\left(\mathfrak{A}\mathfrak{N}_x^{-(g-1)}\right).$$

Therefore

$$y \in L_K\left(\mathfrak{A}\mathfrak{N}_x^{-(g-1)}\right) \iff (y)_K = \frac{\mathfrak{A}\mathfrak{C}}{\mathfrak{N}_x^{g-1}}$$

for some integral divisor $\mathfrak{C}$. Since $(\mathfrak{A}, \mathfrak{N}_x) = 1$, we have

$$y \in L_K\left(\mathfrak{A}\mathfrak{N}_x^{-(g-1)}\right) \iff y \in L_K\left(\mathfrak{N}_x^{-(g-1)}\right) \quad \text{and} \quad \mathfrak{A} \mid \mathfrak{Z}_y.$$

Since $\{1, x, \ldots, x^{g-1}\}$ is a basis of $L_K\left(\mathfrak{N}_x^{-(g-1)}\right)$ we have

$$L_K\left(\mathfrak{N}_x^{-(g-1)}\right) = \{a_0 + a_1x + \cdots + a_{g-1}x^{g-1} \mid a_i \in k\} \subseteq k[x].$$

In particular, $L_K(\mathfrak{N}_x^{-(g-1)}) = L_{k(x)}(\mathfrak{N}_x^{-(g-1)})$ and $\mathfrak{A} \mid \mathfrak{Z}_y$ if and only if $\mathfrak{A} \cap k(x) \mid \mathfrak{Z}_y$. The divisor $\mathfrak{A} \cap k(x)$ corresponds to a polynomial $f(x) \in k[x]$, that is, $(f(x))_{k(x)} = \frac{\mathfrak{A} \cap k(x)}{\wp_\infty^{\deg f}}$, where $\wp_\infty = \mathfrak{N}_x \cap k(x)$.

Therefore $L_K\left(\mathfrak{A}\mathfrak{N}_x^{-(g-1)}\right) = \{p(x) \in k[x] | \deg p(x) \leq g - 1$ and $f(x)$ divides $p(x)\}$. This proves the statement in this case.

Finally, if $\mathfrak{N}_x$ and $\mathfrak{A}$ are not relatively prime and k is infinite, then we can take $x' = \frac{ax+b}{cx+d} \in k(x)$ such that $k(x') = k(x)$ and $(\mathfrak{N}_{x'}, \mathfrak{A}) = 1$. If k is finite, k is a perfect field and we can consider a constant field extension $K' = Kk'$ such that $(\mathfrak{A}', \mathfrak{N}_x) = 1$. The result follows since $[K' : k'(x)] = 2 = [K : k(x)]$ and $g_{K'} = g_K$ (Theorem 8.5.2). □

Theorem 14.2.67. *Let* K/k *be a hyperelliptic function field, and* $\mathfrak{P}$ *be a prime divisor of* K *of degree* f. *Let* $\wp = \mathfrak{P} \cap k(x)$, *where* $k(x)$ *is the unique quadratic subfield of* K. *Then if*

(i) $\wp$ *decomposes in* K, $\wp = \mathfrak{P}\mathfrak{P}'$, *then the gap sequence of* $\mathfrak{P}$ *is* $\left\{1, 2, \ldots, \left[\frac{g}{f}\right]\right\}$.
(ii) $\wp$ *is inert in* K, $\wp = \mathfrak{P}$, *then* $\mathfrak{P}$ *has no gap numbers.*
(iii) $\wp$ *ramifies in* K, $\wp = \mathfrak{P}^2$, *then the gap sequence of* $\mathfrak{P}$ *is equal to* $\{2n - 1 \mid 1 \leq n \leq \left[\frac{g}{f}\right]\}$.

Proof. By Lemma 14.2.66 we have

$$\delta(\mathfrak{P}^n) = g - \mu_n, \tag{14.25}$$

where $\mu_n = \min\left\{d_{k(x)}\left(\mathfrak{P}^n \cap k(x)\right), g\right\}$.

(i) If $\wp$ decomposes, then $\mathfrak{P}^n \cap k(x) = \wp^n$ for all $n \geq 0$ and $d_{k(x)}(\wp) = d_K(\mathfrak{P}) = f$.

Hence $\mu_n = \min\left\{d_{k(x)}(\mathfrak{P}^n), g\right\} = \min\left\{n d_{k(x)}(\wp), g\right\} = \min\{nf, g\}$.

It follows that $\mu_n = nf \iff nf \leq g \iff n \leq \left[\frac{g}{f}\right]$.

For $1 \leq n \leq \left[\frac{g}{f}\right]$, we have

$$\delta\left(\mathfrak{P}^{n-1}\right) - \delta\left(\mathfrak{P}^n\right) = (g-(n-1)f) - (g-nf) = f$$

and n is a gap number for $\mathfrak{P}$.

For $n = \left[\frac{g}{f}\right]+1$, $\delta\left(\mathfrak{P}^{n-1}\right)-\delta\left(\mathfrak{P}^n\right) = (g-(n-1)f)-(g-g) = g-\left[\frac{g}{f}\right]f < f$.

For $n > \left[\frac{g}{f}\right]+1$, $\delta(\mathfrak{P}^{n-1}) - \delta(\mathfrak{P}^n) = 0 - 0 = 0$.

Therefore the gap sequence of $\mathfrak{P}$ is $\left\{1, 2, \ldots, \left[\frac{g}{f}\right]\right\}$.

(ii) If $\wp$ is inert and $\wp = \mathfrak{P}$, then $d_{k(x)}(\wp) = \frac{f}{2}$ and

$$\mathfrak{P}^n \cap k(x) - \wp^n, \deg_{k(x)}(\mathfrak{P}^n \cap k(x)) = n\frac{f}{2}.$$

We have

$$\mu_n = \min\left\{n\frac{f}{2}, g\right\} = \frac{nf}{2} \iff \frac{nf}{2} \leq g \iff n \leq \left[\frac{2g}{f}\right].$$

For $1 \leq n \leq \left[\frac{2g}{f}\right]$ we have

$$\delta(\mathfrak{P}^{n-1}) - \delta(\mathfrak{P}^n) = \left(g-(n-1)\frac{f}{2}\right) - \left(g - n\frac{f}{2}\right) = \frac{f}{2} < f.$$

For $n > \left[\frac{2g}{f}\right]$, we have $\delta(\mathfrak{P}^{n-1}) - \delta(\mathfrak{P}^n) \leq g - \left(\left[\frac{2g}{f}\right]\right)\frac{f}{2} < \frac{f}{2} < f$. Thus $\mathfrak{P}$ has no gap numbers.

(iii) If $\wp$ is ramified and $\wp = \mathfrak{P}^2$, then $\deg_{k(x)}(\wp) = f$ and

$$\mathfrak{P}^{2m-1} \cap k(x) = \mathfrak{P}^{2m} \cap k(x) = \wp^m, \quad \text{or} \quad \mathfrak{P}^n \cap k(x) = \wp^{\left[\frac{n+1}{2}\right]}.$$

We have

$$\mu_n = \min\left\{\left[\frac{n+1}{2}\right]f, g\right\} = \left[\frac{n+1}{2}\right]f \leq g \iff \left[\frac{n+1}{2}\right] \leq \left[\frac{g}{f}\right].$$

Let $n = 2m - 1$; then $\frac{n+1}{2} = m \leq \left[\frac{g}{f}\right]$, and

$$\delta\left(\mathfrak{P}^{n-1}\right) - \delta(\mathfrak{P}^n) = (g - (m-1)f) - (g - mf) = f.$$

Thus each $n = 2m - 1$ such that $1 \leq m \leq \left[\frac{g}{f}\right]$ is a gap number.

Let $n = 2m$, where $\left[\frac{n+1}{2}\right] = m \leq \left[\frac{g}{f}\right]$. We have

$$\delta\left(\mathfrak{P}^{n-1}\right) - \delta(\mathfrak{P}^n) = (g - mf) - (g - mf) = 0.$$

Finally, assume n is $2m - 1$ or $2m$ with $m \geq \left[\frac{g}{f}\right]$. Then

$$\delta\left(\mathfrak{P}^{n-1}\right) - \delta(\mathfrak{P}^n) \leq \left(g - \left[\frac{g}{f}\right] f\right) < f.$$

Thus the gap sequence of $\mathfrak{P}$ is $\left\{2m - 1 \mid 1 \leq m \leq \left[\frac{g}{f}\right]\right\}$. □

Remark 14.2.68. Part (ii) of Theorem 14.2.67 can also be deduced from the fact that $\wp = \mathfrak{P}$ and $\frac{\wp}{\wp_\infty^f} = (f(x))_{k(x)}$. Thus $\mathfrak{N}_{(f(x)^{-1})} = \mathfrak{P}$ and 1 is a nongap number for $\mathfrak{P}$. It follows that $\mathfrak{P}$ has no gap numbers.

Remark 14.2.69. Theorem 14.2.67 shows that when k is not an algebraically closed field, there might exist several gap sequences of the field K/k. Thus there exist two infinite disjoint sets A and B of prime divisors such that every $\mathfrak{P} \in A$ has the same gap sequence $\{i_1, \ldots, i_r\}$ and every $\mathfrak{P} \in B$ has the same gap sequence $\{j_1, \ldots, j_s\}$, and $\{i_1, \ldots, i_r\} \neq \{j_1, \ldots, j_r\}$. Furthermore we may define Weierstrass points as those prime divisors $\mathfrak{P}$ such that only finitely many prime divisors have the same gap sequence as $\mathfrak{P}$. Corollary 4.3.7 shows that there exist function fields of arbitrarily large genus without Weierstrass points, even in characteristic 0.

Example 14.2.70. Let $k = \mathbb{Q}_p$ be the field of p-adic numbers $(p > 2)$ and let $g \geq 2$. Then by Eisenstein's criterion, the polynomial $x^{2g+2} + p$ is irreducible in $\mathbb{Q}_p[x]$. Let $K = k(x, y)$ with $y^2 = x^{2g+2} + p$. By Corollary 4.3.7, $g_K = g$. If $\left(x^{2g+2} + p\right)_{k(x)} = \frac{\wp}{\wp_\infty^{2g+2}}$, then $\wp$ is the only ramified prime in $K/k(x)$ and $f = d_{k(x)}(\wp) = 2g + 2 > g$. Thus $\left[\frac{g}{f}\right] = 0$.

Next we will see that for each f satisfying $1 \leq f \leq g$, there are infinitely many prime divisors in $\mathbb{Q}_p(x)$ that have degree f and decompose in $K/\mathbb{Q}_p(x)$. Using Theorem 14.2.67 we will deduce that K has no Weierstrass points.

Assume $1 \leq f \leq g$, and set $\mathbb{F}_{p^f} = \mathbb{F}_p(\alpha)$. Suppose that $\ell(x) = \operatorname{Irr}(\alpha, x, \mathbb{F}_p)$, $\deg \ell(x) = f$, and $\ell(x) \neq x$. Let $h(x) \in \mathbb{Q}_p[x]$ be such that $\overline{h(x)} = -x^{g+1} + \ell(x)$ where $\deg h(x) = g + 1$. We have

$$\left(x^{2g+2}+p-h(x)^2\right) \bmod p = \left(x^{g+1}-\overline{h(x)}\right)\left(x^{g+1}+\overline{h(x)}\right)$$
$$= \left(x^{g+1}-\overline{h(x)}\right)\ell(x).$$

If $\ell(x) \mid x^{g+1}-\overline{h(x)}$, then $\ell(x) \mid \ell(x)+x^{g+1}-\overline{h(x)} = 2x^{g+1}$. This contradiction shows that $\ell(x)$ and $x^{g+1}+\overline{h(x)}$ are relatively prime. By Hensel's lemma (Theorem 2.3.14), we have $x^{2g+2}+p-h(x)^2 = m(x)t(x)$, where $m(x)$ has degree f and $\overline{m(x)} = \ell(x)$. Thus $m(x)$ is irreducible since $\ell(x)$ is irreducible. On the other hand, if we fix $m(x)$, we have $m_\mu(x) := m(x)+p\mu$, where $\mu \in \mathbb{Z}_p$ is another irreducible polynomial of degree f in $\mathbb{Q}_p[x]$. In short, there exist infinitely many monic irreducible polynomials $m(x)$ of degree f and $\overline{m(x)} = \ell(x)$.

Let φ be the place associated with a given such $m(x)$. Let ϑ and $\wp$ be the valuation and the maximal ideal associated to $m(x)$. We have

$$\vartheta/\wp \cong \mathbb{Q}_p[x]/(m(x)) = F \quad \text{where} \quad \left[F:\mathbb{Q}_p\right] = f.$$

Consider the function

$$\varphi\colon \mathbb{Q}_p(x) \to F\cup\{\infty\} \quad \text{defined by} \quad \varphi(x) := x \bmod \wp, \quad \varphi|_{\mathbb{Q}_p} = \mathrm{Id}_{\mathbb{Q}_p}.$$

Let $\sigma : K \to F_1 \cup \{\infty\}$ be an extension of φ to K.

Then $[F_1 : F]$ is 1 or 2. In fact, we have $F_1 = F$ if and only if $\sigma(y) \in F$ and $\mathfrak{p}$ decomposes in $K/\mathbb{Q}_p(x)$. Now we have

$$y^2 = x^{2g+2}+p$$
$$\Longrightarrow \sigma(y^2) = \sigma(y)^2 = \sigma(x^{2g+2}+p) = \sigma(x)^{2g+2}+\sigma(p) = \left(x^{2g+2}+p\right) \bmod \wp.$$

That is, $\sigma(y) \in F \iff x^{2g+2}+p-\beta^2 \equiv 0 \mod m(x)$ has a solution β in $\mathbb{Q}_p[x]$. If $h(x)$ is as before, then $m(x)$ divides $x^{2g+2}+p-h(x)^2$. Thus $\sigma(y) \in F$. This shows that K has no Weierstrass points.

Example 14.2.71. If $K/\mathbb{F}_q$ is a hyperelliptic function field, then there are finitely many prime divisors of degree 1 in $\mathbb{F}_q(x)$ that decompose in $K/\mathbb{F}_q(x)$. Therefore these points are Weierstrass points.

The next result is a generalization of Theorem 14.2.59.

Corollary 14.2.72. *If K/k is a hyperelliptic function field for an algebraically closed field k of characteristic $p \geq 0$, then the gap sequence of the field is classical, that is, equal to $\{1, 2, \ldots, g\}$. The Weierstrass points correspond to the ramified prime divisors of $K/k(x)$ and their gap sequence is $\{1, 3, \ldots, 2g-1\}$.*

Proof. This is an immediate consequence of Theorem 14.2.67, since in this case $f = 1$. ◻

Example 14.2.73. Let k be an algebraically closed field of characteristic 2. Let $K = k(x, y)$ where

$$y^2 - y = x^m, \quad m \geq 3, \quad \text{and} \quad 2 \nmid m.$$

Then the only ramified prime in $K/k(x)$ is $\wp_\infty$ and

$$g_K = 1 + \left(g_{k(x)} - 1\right)[K : k(x)] + \frac{1}{2}(m+1)(2-1) = \frac{m+1}{2} - 1 = \frac{m-1}{2}$$

(see Example 5.8.8 and Theorem 9.4.2).

Since K is a hyperelliptic field, there exist fields of arbitrarily large genus g ($g = \frac{m-1}{2}$ or $m = 2g+1$) with a single Weierstrass point.

Corollary 14.2.74. *If K/k is a hyperelliptic function field for some algebraically closed field k of characteristic distinct from 2, then K/k has $2g+2$ Weierstrass points.*

Proof. The result follows from Corollary 14.2.72 since there are $2g+2$ ramified primes in $K/k(x)$. □

14.2.5 Fields with Nonclassical Gap Sequence

In this subsection we consider an algebraically closed field k of characteristic $p > 0$.

Theorem 14.2.75. *Let $K = k(x, y)$ be the function field defined by the equation*

$$y^q - y = x^m,$$

where $q = p^u$, $m > 1$, and m divides $q+1$. Set $q + 1 = mn$. Then

$$g_K = \frac{(m-1)(q-1)}{2}.$$

Furthermore, the gap sequence of the field is

$$\{rq + s + 1 \mid r, s \geq 0, (r+1)n + (s+1) \leq q\}.$$

Proof. From Example 5.9.12 we have $g_K = \frac{(m-1)(q-1)}{2}$.

Now we compute the cardinality of the set

$$A = \{rq + s + 1 \mid r, s \geq 0, (r+1)n + (s+1) \leq q\}.$$

For $0 \leq r$, we have

$$(r+1)n + (s+1) \leq q \iff s \leq q - (r+1)n - 1.$$

Set $a_r = \max\{q - (r+1)n, 0\}$. Then $0 \leq s \leq a_r - 1$.

Also, $(r+1)n \leq q = nm - 1$. Hence $r \leq m - \frac{1}{n} - 1$ and $r \leq m - 2$. It follows that

$$\begin{aligned} |A| &= \sum_{r=0}^{m-2} a_r = \sum_{r=0}^{m-2} \{q - (r+1)n\} = (m-1)q - n\frac{m(m-1)}{2} \\ &= (m-1)q - \frac{(q+1)(m-1)}{2} = (m-1)\left(\frac{2q-q-1}{2}\right) \\ &= \frac{(m-1)(q-1)}{2} = g_K. \end{aligned}$$

By Theorem 9.4.1 we have

$$(dx)_K = \operatorname{con}_{k(x)/K}(dx)_{k(x)}\mathfrak{D}_{K/k(x)} = \mathfrak{B}^{-2q+(m+1)(q-1)} = \mathfrak{B}^{m(q-n-1)}.$$

Denote by $\wp$ the zero of x in $k(x)$. Let $r, s \geq 0$ be such that $(r+1)n + (s+1) \leq q$. Set $w = x^r y^s dx$. Clearly $v_{\mathfrak{P}}((w)_K) \geq 0$ for all $\mathfrak{P} \neq \mathfrak{B}$. Now we have

$$\begin{aligned} v_{\mathfrak{B}}((w)_K) &= r v_{\mathfrak{B}}((x)_K) + s v_{\mathfrak{B}}((y)_K) + v_{\mathfrak{B}}((dx)_K) \\ &= -rq - sm + m(q-n-1) = -r(nm-1) - sm + m(q-n-1) \\ &= m(-rn - s + q - n - 1) + r = m(q - n(r+1) - (s+1)) + r \\ &\geq m(0) + r = r \geq 0. \end{aligned}$$

Now the set $\{x^r y^s dx \mid r \geq 0, s \geq 0, (r+1)n + (s+1) \leq q\}$ is linearly independent over k. Moreover, this set is of cardinality g and consists of holomorphic differentials, so it is a basis of holomorphic differentials.

Therefore the gap sequence of the field is

$$\{\mu_i + 1 \mid 0 \leq i \leq g-1\},$$

where $\mu_0, \ldots, \mu_{g-1}$ are the orders of the Wronskian determinant

$$W_y(x^r y^s \mid r, s \geq 0, (r+1)n + (s+1) \leq q)$$

(Theorem 14.2.48).

We associate the corresponding power series for D_y for each $x^r y^s$: $\phi\colon K \to K[[u]]$, where $Y = \phi(y) = y + u$ and $X = \phi(x) = x + \sum_{n=1}^{\infty} D_y^{(n)}(x)u^n$. Now $y^q - y = x^m = x^{\frac{q+1}{n}}$, so $x = (y^q - y)^{\frac{n}{q+1}}$. Since $\frac{1}{q+1} = q^2 - q + 1 - \frac{q^3}{q+1}$ it follows that

$$x = \left(y^2 - y\right)^{\frac{n}{q+1}} = \left(y^q - y\right)^{n(q^2-q+1)}\left(y^q - y\right)^{-\frac{n}{q+1}q^3} = \left(y^q - y\right)^{n(q^2-q+1)} x^{-q^3}.$$

Therefore

$$X = \left(Y^q - Y\right)^{n(q^2-q+1)} X^{-q^3}. \tag{14.26}$$

Next, $X^{q^3} = x^{q^3} + u^{q^3} R(u)$ for some $R(u) \in K[[u]]$. Thus

$$X^{-q^3} = x^{-q^3} - u^{q^3} R_1(u) \quad \text{for some} \quad R_1(u) \in K[[u]].$$

Using (14.26) we obtain that

$$X \equiv \left(y^q - y + u^q - u\right)^{n(q^2-q+1)} x^{-q^3} \mod (u^{q+1}).$$

We have

$$\begin{aligned}(y^q - y - u + u^q)^{n(q^2-q)} &= \left(y^{q^2} - y^q - u^q + u^{q^2}\right)^{n(q-1)} \\ &\equiv \left(y^{q^2} - y^q\right)^{n(q-1)} - n(q-1)\left(y^{q^2} - y^q\right)^{n(q-1)-1} u^q \mod u^{q+1}\end{aligned}$$

and

$$\begin{aligned}(y^2 - y - u + u^q)^n &\equiv (y^q - y - u)^n + n(y^q - y - u)^{n-1} u^q \\ &\equiv (y^q - y - u)^n + n(y^q - y)^{n-1} u^q \mod u^{q+1}.\end{aligned}$$

Therefore

$$\begin{aligned}X &\equiv (y^q - y - u + u^q)^{n(q^2-q)} (y^q - y - u + u^q)^n x^{-q^3} \\ &\equiv \left[\left(y^{q^2} - y^q\right)^{n(q-1)} - n(q-1)\left(y^{q^2} - y^q\right)^{n(q-1)-1} u^q\right] \\ &\qquad \left[(y^q - y - u)^n + n(y^q - y)^{n-1} u^q\right] x^{-q^3} \\ &\equiv a(y^q - y - u)^n + b u^q \mod u^{q+1}\end{aligned}$$

with

$$\begin{aligned}a &= x^{-q^3}\left[\left(y^{q^2} - y^q\right)^{n(q-1)} - n(q-1)\left(y^{q^2} - y^q\right)^{n(q-1)-1} u^q\right] \\ &= x^{-q^3}\left(y^q - y\right)^{n(q^2-q)-q}\left[(y^q - y)^{q^2} + n u^q\right] \\ &= x^{-q^3} x^{mn(q^2-q)-mq}\left(x^{mq^2} + n u^q\right) \neq 0\end{aligned}$$

and

$$b = x^{-q^3}\left(y^{q^2} - y^q\right)^{n(q-1)} n(y^q - y)^{n-1} = n x^{-q^3} x^{mnq(q-1)} x^{m(n-1)} \neq 0.$$

Thus

$$X \equiv a(y^q - y - u)^n + b u^q \mod u^{q+1}, \tag{14.27}$$

where a and b are two nonzero elements of K.

Consider the k-vector space

$$V = L_K\left((dx)_K^{-1}\right) = \bigoplus_{r,s} kx^r y^s \quad \text{with} \quad (r+1)n + (s+1) \le q,$$

and the respective K-vector space consisting of the power series of the form

$$U = \bigoplus_{r,s} KX^r Y^s, \quad \text{with} \quad (r+1)n + (s+1) \le q.$$

We will find the Hermitian powers associated to U. We have

$$u^s = (Y-y)^s \equiv 0 \mod \left(\bigoplus_{i=0}^{s} KY^i\right), \tag{14.28}$$

and

$$\begin{aligned}(y^q - y - u)^n &= \sum_{m=0}^{n} \binom{n}{m} (y^q - y)^m (-1)^{n-m} u^{n-m} \\ &\in \sum_{m=0}^{n} \bigoplus_{i=0}^{n-m} KY^i = \bigoplus_{j=1}^{n} KY^j.\end{aligned}$$

Therefore, using (14.27) we obtain that $X \equiv bu^q \mod M_1$, where M_1 is the K-vector space $\langle u^{q+1}, Y^i | 0 \le i \le n\rangle$. Since b is a nonzero element of K, it follows that $u^q \in \langle u^{q+1}, X, Y^i | 0 \le i \le n\rangle \subseteq \langle u^{q+1}, X^j Y^i | j + \frac{i}{n} \le 1\rangle$. Then the K-vector space $\bigoplus_{j+\frac{i}{n}\le 1} KX^jY^i = M$ contains a power series of the form $P = u^q + u^{q+1}P_1 \in M$. Thus $P \equiv 0 \mod \left(\bigoplus_{j+\frac{i}{n}\le r} KX^jY^i\right)$ and if we choose $s \ge 0$ such that $(r+1)n + (s+1) \le q$, by (14.28) we have $P^r u^s \equiv 0 \mod \left(\bigoplus_{\substack{j+\frac{i}{n}\le r \\ t\le s}} KX^jY^{i+t}\right)$. Now $nj + i \le nr$, so

$$(j+1)n + (i+t+1) \le nr + n + s + 1 = (r+1)n + (s+1) \le q.$$

Hence $P^r u^s \in U$ for all r, s such that $(r+1)n + (s+1) \le q$.

There are $g = \frac{(m-1)(q-1)}{2}$ power series of the form $P^r u^s$ from U.

Note that

$$P^r u^s = u^{rq+s} + \sum_{n=rq+s+1}^{\infty} a_n^{(r,s)} u^n.$$

That is, $\{P^r u^s \mid r, s \ge 0, (r+1)n + (s+1) \le q\}$ is a Hermitian basis of U whose set of Hermitian invariants is

$$\{rq + s \mid r, s \ge 0, (r+1)n + (s+1) \le q\}.$$

By Theorem 14.2.35 the orders of the Wronskian

$$W_y\left(x^r y^s \mid r, s \geq 0, (r+1)n + (s+1) \leq q\right)$$

are precisely $\{rq+s\}_{r,s}$. It follows by Theorem 14.2.48 that the gap sequence of the field is $\{rq+s+1\}_{r,s}$ □

Remark 14.2.76. Theorem 14.2.75 provides us with examples of function fields K/k where k is algebraically closed of characteristic $p > 0$ such that the gap sequence of K is nonclassical.

Example 14.2.77. In the setting of Theorem 14.2.75, let $p = q = 3$ and $m = 4$. Then $K = k(x, y)$ with

$$y^3 - y = x^4.$$

We have $g = \frac{(m-1)(q-1)}{2} = 3$ and $q + 1 = 4 = mn = 4n$, so $n = 1$. Thus the gap sequence of the field is

$$\begin{aligned}
&\{rq+s+1 \mid r, s \geq 0, (r+1)n + (s+1) \leq q\} \\
&= \{3r+s+1 \mid r, s \geq 0, r+1+s+1 \leq 3\} \\
&= \{3r+s+1 \mid r, s \geq 0, r+s \leq 1\} \\
&= \{1, 2, 4\} \neq \{1, 2, 3\}.
\end{aligned}$$

14.3 Automorphism Groups of Algebraic Function Fields

Let K/k be a function field, and

$$\operatorname{Aut}_k K := \{\sigma : K \to K \mid \sigma \text{ is a field automorphism } \sigma|_k = \operatorname{Id}_k\}.$$

If $K = k(x)$, then

$$\operatorname{Aut}_k(k(x)) \cong GL(2,k)/k^* \cong PGL(2,k) = \{\sigma \mid \sigma x = \frac{ax+b}{cx+d}, ad - bc \neq 0\}.$$

In particular, if k is infinite, then $\operatorname{Aut}_k(k(x))$ is infinite too. If K is an elliptic function field and $|k| = \infty$, then $|\operatorname{Aut}_k(K)| = \infty$ by Theorem 9.6.13. Klein and Poincaré [117] proved using the analytic theory of Riemann surfaces that when $g = g_K \geq 2$ and $k = \mathbb{C}$ is the field of complex numbers, then $\operatorname{Aut}_k(K)$ is finite. On the other hand, Weierstrass and Hurwitz gave algebraic proofs of the same result [70, 162]. Because of its algebraic nature, the latter method is applicable to the case of an arbitrary constant field k of characteristic 0. In the case of characteristic $p \neq 0$, H. L. Schmid [136] proved the theorem using Weierstrass points, in a way similar to Hurwitz's proof. On the other hand, K. Iwasawa and T. Tamagawa [73, 74, 75] gave another proof of Schmid's theorem using the representation of $\operatorname{Aut}_k(K)$ in the k-vector space $D_K(0)$ of holomorphic differentials of K instead of Weierstrass points.

In this section we present the proof of Schmid's theorem following the ideas of Iwasawa and Tamagawa and of Schmid himself. At the end of the section we give a proof or Hurwitz's theorem. Let k be an algebraically closed field of characteristic $p \geq 0$, K/k a function field over k, and $G = \operatorname{Aut}_k(K)$. We assume $g = g_K \geq 1$ and $g \geq 2$ for our main result.

Proposition 14.3.1. *Let $\sigma \in G$ be such that $\sigma(k(x)) = k(x)$ for some $x \in K \setminus k$. Let $n = [K : k(x)]$. Assume that $p \nmid n$ whenever $p > 0$ and n is arbitrary whenever $p = 0$. Then*

$$o(\sigma) \leq \max\{n(2n+2g-2)(2n+2g-3)(2n+2g-4),\, pn(g+1)\} < \infty.$$

Proof. Set $\mathfrak{D}_{K/k(x)} = \mathfrak{P}_1^{a_1} \cdots \mathfrak{P}_s^{a_s}$ and

$$\{\mathfrak{P}_i \cap k(x) \mid 1 \leq i \leq s\} = \{\wp_1, \dots, \wp_r\},$$

where the latter has cardinality $r \leq s$. For each $\wp_i$, let $\wp = \wp_i$ and let

$$\wp = \mathfrak{B}_1^{e_1} \cdots \mathfrak{B}_t^{e_t}, \quad \mathfrak{B}_j \in \{\mathfrak{P}_1, \dots, \mathfrak{P}_s\}$$

be the decomposition of $\wp$ in K. First assume that p does not divide e_j for all $1 \leq i \leq r$ and $1 \leq j \leq t$ (if $p = 0$ this condition is automatically satisfied). In this case the contribution of each $\wp$ to the different is

$$\mathfrak{B}_1^{e_1-1} \cdots \mathfrak{B}_t^{e_t-1},$$

whose degree is equal to

$$\sum_{j=1}^{t}(e_j - 1) = \sum_{j=1}^{t} e_j - t = n - t \leq n - 1.$$

By the Riemann–Hurwitz genus formula, we have

$$2(g-1) = 2(g_{k(x)} - 1)n + d_K(\mathfrak{D}_{K/k(x)}).$$

Thus $d_K(\mathfrak{D}_{K/k(x)}) := d = 2n + 2(g-1) > 2(n-1)$. In particular, $3 \leq r \leq d$. Now since $\sigma(k(x)) = k(x)$, it follows that $\sigma(\mathfrak{D}_{K/k(x)}) = \mathfrak{D}_{K/k(x)}$ and σ permutes the sets $\{\mathfrak{P}_1, \dots, \mathfrak{P}_s\}$ and $\{\wp_1, \dots, \wp_r\}$. With the identification $\sigma \in S_r$, where S_r denotes the symmetric group, if

$$\sigma = (\alpha_1^{(1)}, \dots, \alpha_{i_1}^{(1)})(\alpha_1^{(2)}, \dots, \alpha_{i_2}^{(2)}) \cdots (\alpha_1^{(u)}, \dots, \alpha_{i_u}^{(u)})$$

is the cyclic decomposition of σ, where $i_1 + \cdots + i_u = r$, some power σ^ℓ with $\ell \leq i_1 i_2 i_3 \leq r(r-1)(r-2) \leq d(d-1)(d-2)$ fixes at least 3 distinct prime divisors from the set $\{\wp_1, \dots, \wp_r\}$. By Exercise 5.10.14, $\sigma^\ell|_{k(x)} = \mathrm{Id}_{k(x)}$ We have $|\operatorname{Aut}_{k(x)}(K)| \mid [K : k(x)] = n$. Thus $\sigma^{n\ell} = \mathrm{Id}_K$. Therefore

$$\begin{aligned} o(\sigma) \leq n\ell &\leq nd(d-1)(d-2) \\ &= n(2n+2g-2)(2n+2g-3)(2n+2g-4). \end{aligned}$$

Now assume that p divides some e_j. In particular, we have $p > 0$.

Notice that $\sigma|_{k(x)} \in \operatorname{Aut}_k(k(x))$, so that $\sigma x = \frac{ax+b}{cx+d}$ with $ad - bc \neq 0$.

If $c = 0$, then $\sigma x = \frac{a}{d}x + \frac{b}{d} = \alpha x + \beta$ for some $\alpha \neq 0$. If $\alpha = 1$, we have $\sigma x = x + \beta$. If $\alpha \neq 1$,

$$\begin{aligned} \sigma\left(x + \frac{\beta}{\alpha - 1}\right) &= \sigma x + \frac{\beta}{\alpha - 1} \\ &= \alpha x + \beta + \frac{\beta}{\alpha - 1} = \alpha x + \frac{\alpha\beta}{\alpha - 1} = \alpha\left(x + \frac{\beta}{\alpha - 1}\right). \end{aligned}$$

Let $y = x + \frac{\beta}{\alpha-1}$; then $k(x) = k(y)$ and $\sigma y = \alpha y$.

If $c \neq 0$, we may assume that $c = 1$ and $\sigma x = \frac{ax+b}{x+d} = a + \frac{b-ad}{x+d}$.

Therefore $\sigma(x - a) = \sigma x - a = \frac{b-ad}{x+d} = \frac{b-ad}{x-a+(a+d)}$. Put $y = x - a$. Then $k(x) = k(y)$ and $\sigma y = \frac{\alpha}{y+\beta}$ for some $\alpha \neq 0$. Let $\lambda \in k$ be a solution of the equation $\lambda^2 - \beta\lambda - \alpha = 0$. Note that $\lambda \neq 0$ and $\lambda \neq \beta$ since $\alpha \neq 0$.

Let $\delta = \frac{\beta-\lambda}{\lambda} = \frac{\beta}{\lambda} - 1 \neq 0$ and let $z = \frac{y+\beta}{y+\lambda}$. Then $\det\begin{pmatrix} 1 & \beta \\ 1 & \lambda \end{pmatrix} = \lambda - \beta \neq 0$. Therefore $k(x) = k(y) = k(z)$ and

$$\sigma z = \frac{\sigma y + \beta}{\sigma y + \lambda} = \frac{\frac{\alpha}{y+\beta} + \beta}{\frac{\alpha}{y+\beta} + \lambda} = \frac{\beta y + (\alpha + \beta^2)}{\lambda y + (\alpha + \lambda\beta)} = \delta z + 1.$$

From this point on we can proceed as above.

In short, we may assume without loss of generality that

$$\sigma x = \alpha x \quad \text{or} \quad \sigma x = x + \alpha, \quad \text{with} \quad \alpha \in k.$$

If $\sigma x = x + \alpha$, then $\sigma^p x = x + p\alpha = x$. Therefore $o(\sigma^p) \leq n$ and

$$o(\sigma) \leq pn \leq pn(g+1).$$

Now assume $\sigma x = \alpha x$ with $\alpha \in k^*$. If the divisor of x in K is of the form $(x)_K = \frac{\mathfrak{Q}_1^n}{\mathfrak{Q}_2^n}$, where $\mathfrak{Q}_1, \mathfrak{Q}_2$ are distinct prime divisors, then $\mathfrak{Q}_1$ and $\mathfrak{Q}_2$ are fully ramified prime divisors in $K/k(x)$ and $v_{\mathfrak{Q}_1}(\mathfrak{D}_{K/k(x)}) = v_{\mathfrak{Q}_2}(\mathfrak{D}_{K/k(x)}) = n - 1$. Thus we have

$$\begin{aligned} g &= 1 + (0-1)n + \frac{1}{2}d_K(\mathfrak{D}_{K/k(x)}) \\ &= 1 - n + \frac{1}{2}(2(n-1)) + \frac{1}{2}d_K\left(\frac{\mathfrak{D}_{K/k(x)}}{\mathfrak{Q}_1^{n-1}\mathfrak{Q}_2^{n-1}}\right) \\ &= \frac{1}{2}d_K\left(\frac{\mathfrak{D}_{K/k(x)}}{\mathfrak{Q}_1^{n-1}\mathfrak{Q}_2^{n-1}}\right) > 0. \end{aligned}$$

Therefore $\mathfrak{D}_{K/k(x)}$ is divided by at least three different prime divisors of $k(x)$ and the first case of the proof can be applied to this situation. We obtain

$$o(\sigma) \leq n(2n + 2g - 2)(2n + 2g - 3)(2n + 2g - 4).$$

Next, suppose that either $\mathfrak{N}_x$ or $\mathfrak{Z}_x$, say $\mathfrak{N}_x$, is divisible by at least two distinct prime divisors $\mathfrak{Q}_1, \mathfrak{Q}_2$ of K. Since $\sigma x = \alpha x$, we have $\mathfrak{N}_x^\sigma = \mathfrak{N}_x$ and $\mathfrak{Z}_x^\sigma = \mathfrak{Z}_x$. Hence there exists some $\ell \leq n$ such that $\mathfrak{Q}_1^{\sigma^\ell} = \mathfrak{Q}_1$. Let $\tau = \sigma^\ell$. We have

$$\ell_K({\mathfrak{Q}_1}^r) \geq d_K(\mathfrak{Q}_1^r) - g + 1 = r - (g - 1).$$

It follows that $\ell_K(\mathfrak{Q}_1^{-r}) = 2$ for some $r \leq g + 1$. Let $y \in L_K(\mathfrak{Q}_1^{-r}) \setminus k$ satisfy $\mathfrak{N}_y = \mathfrak{Q}_1^s$ and $1 \leq s \leq r \leq g + 1$. Since $\mathfrak{Q}_1^\tau = \mathfrak{Q}_1$, there exist $\beta \in k \setminus \{0\}$, $\gamma \in k$, such that $\tau(y) = \beta y + \gamma$. If $\beta = 1$, then $\tau^p(y) = y + p\gamma = y$. Since $[K : k(y)] = d_K(\mathfrak{N}_y) = s \leq g + 1$, we have $o(\tau^p) \leq g + 1$. Hence

$$o(\tau) \leq p(g + 1) \quad \text{and} \quad o(\sigma) \leq \ell p(g + 1) \leq np(g + 1).$$

We may assume that $\beta \neq 1$, and using the same argument as above, we may assume that $\tau y = \beta y$. Let $F(X, Y) = \sum_{i,j} a_{ij} X^i Y^j$ be an irreducible polynomial over k such that $F(x, y) = 0$ ($[k(x, y) : k(x)] \leq [K : k(x)] = n$). Since $\tau = \sigma^\ell$, $\tau x = \alpha^\ell x$ and $\tau y = \beta y$, we have

$$0 = \tau F(x, y) = F(\tau x, \tau y) = F(\alpha^\ell x, \beta y).$$

Therefore

$$F(\alpha^\ell X, \beta Y) = \xi F(X, Y) \tag{14.29}$$

for some nonzero ξ in k.

Now suppose $a_{ij} \neq 0$ and $a_{i',j'} \neq 0$ for some $(i, j) \neq (i', j')$, and consider $z := x^{i-i'} y^{j-j'}$. Then $x^{i'} y^{j'} z = x^i y^j$. By (14.29), we have

$$\alpha^{\ell i'} x^{i'} \beta^{j'} y^{j'} = \xi x^{i'} y^{j'} \quad \text{and} \quad \alpha^{\ell i} x^i \beta^j y^j = \xi x^i y^j.$$

Thus

$$\begin{aligned} \tau(x^{i'} y^{j'} z) &= \alpha^{\ell i'} x^{i'} \beta^{j'} y^{j'} \tau z = \xi x^{i'} y^{j'} = \tau(z) = \tau(x^i y^j) = \alpha^{i\ell} x^i \beta^j y^j \\ &= \xi x^i y^j = \xi x^{i'} y^{j'} z. \end{aligned}$$

It follows that $\tau z = x^{i-i'} y^{j-j'} = z$. Since $\mathfrak{Q}_1, \mathfrak{Q}_2 \mid \mathfrak{N}_x, \mathfrak{N}_y = \mathfrak{Q}_1^s$, it is easy to see that $z \notin k$. Now

$$\deg_X F \leq [K : k(y)] = s \leq g + 1 \quad \text{and} \quad \deg_Y F \leq [K : k(x)] = n,$$

so $|i - i'| \leq g + 1$ and $|j - j'| \leq n$. Therefore

$$\begin{aligned}[K : k(z)] = d_K(\mathfrak{N}_z) &\le |i - i'| d_K(\mathfrak{N}_x) + |j - j'| d_K(\mathfrak{N}_y) \\ &\le (g+1)n + n(g+1) = 2n(g+1).\end{aligned}$$

Finally, $\tau(z) = z$ implies that $o(\tau) \le 2n(g+1)$. Hence $o(\sigma^\ell) \le 2n(g+1)$ and since $g \ge 1$,

$$\begin{aligned}o(\sigma) &\le 2\ell n(g+1) \le 2n^2(g+1) \\ &\le n(2n+2g-2)(2n+2g-3)(2n+2g-4).\end{aligned}$$

This completes the proof. □

Let $\mathfrak{P}$ be a fixed prime divisor in K. Our next step is to prove that the group $G_\mathfrak{P} = \{\sigma \in G \mid \mathfrak{P}^\sigma = \mathfrak{P}\}$ is finite. We already know this for $g = 1$ (see the proof of Theorem 9.6.14). Here we present another proof.

First we prove that the elements σ of $G_\mathfrak{P}$ have finite order. In characteristic 0 this is easy to see. Choose r to be the minimum pole number of $\mathfrak{P}$, that is, $\ell_K(\mathfrak{P}^{-r}) = 2$ and $r \le g+1$. If $x \in L_K(\mathfrak{P}^{-r}) \setminus k$, then $\sigma x = ax + b$ with $a \ne 0$. Then $[K : k(x)] \le g+1$ and by Proposition 14.3.1, $o(\sigma) < \infty$. The same argument may be applied in case k has characteristic p where $p \nmid r$.

Now we consider the case $\operatorname{char} k = p \ge 0$. Let $n_1, \dots, n_g, n_{g+1}$ be the first $g+1$ pole numbers of $\mathfrak{P}$ with $1 < n_1 < n_2 < \cdots < n_g = 2g < n_{g+1} = 2g+1$. Choose $x_0 = 1$ and $x_i \in L_K\left(\mathfrak{P}^{-n_i}\right) \setminus L_K\left(\mathfrak{P}^{-(n_i-1)}\right)$ for $1 \le i \le g+1$. Then

$$\ell_K\left(\mathfrak{P}^{-n_i}\right) = i+1 \quad \text{and} \quad \{x_0, x_1, \dots, x_i\} \text{ a } k\text{-basis of } L_K(\mathfrak{P}^{-n_i})$$

for $1 \le i \le g+1$. We have $\mathfrak{N}_{x_i} = \mathfrak{P}^{n_i}$. Every $\sigma \in G_\mathfrak{P}$ induces a k-linear map $\sigma\colon L_K(\mathfrak{P}^{-n_i}) \to L_K(\mathfrak{P}^{-n_i})$ for each i. Therefore $\sigma\colon L_K(\mathfrak{P}^{-(2g+1)}) \to L_K(\mathfrak{P}^{-(2g+1)})$ satisfies $\sigma x_j = \sum_{i=1}^{j} a_{ij} x_i$. Thus the matrix A_σ of σ with respect to the basis $\{x_1, \dots, x_{g+1}\}$ is triangular, that is,

$$A_\sigma = \begin{pmatrix} a_1 & & * \\ & \ddots & \\ 0 & & a_{g+1} \end{pmatrix} \tag{14.30}$$

with $a_i = a_{ii}$ for $1 \le i \le g+1$. If $A_\sigma = \mathrm{Id}_{g+1}$, then $\sigma x_g = x_g$ and $\sigma x_{g+1} = x_{g+1}$. Since $\left[K : k(x_g, x_{g+1})\right]$ divides $\left[K : k(x_g)\right] = d_K(\mathfrak{N}_{x_g}) = d_K(\mathfrak{P}^{2g}) = 2g$ and $\left[K : k(x_{g+1})\right] = d_K(\mathfrak{N}_{x_{g+1}}) = d_K(\mathfrak{P}^{2g+1}) = 2g+1$, it follows that $\left[K : k(x_g, x_{g+1})\right] = 1$ and $K = k(x_g, x_{g+1})$. Hence $\sigma = 1$.

Proposition 14.3.2. *Assume that k has characteristic $p \ge 0$ and let $g \ge 1$. For any $\sigma \in G_\mathfrak{P}$, $o(\sigma)$ is finite and $o(\sigma)$ has an upper bound depending only on g and p.*

Proof. The characteristic values of A_σ are $\{a_1, \dots, a_g, a_{g+1}\}$. If A_σ is diagonalizable, that is, there exists a basis $\{y_1, \dots, y_g, y_{g+1}\}$ of $L_K\left(\mathfrak{P}^{-(2g+1)}\right)$ such that

$$\sigma y_i = a_i y_i \quad \text{for} \quad 1 \le i \le g+1$$

and $y_i \in L_K\left(\mathfrak{P}^{-n_i}\right) \setminus L_K\left(\mathfrak{P}^{-n_i+1}\right)$, then

$$\sigma(k(y_g)) = k(y_g) \quad \text{and} \quad \sigma(k(y_{g+1})) = k(y_{g+1}).$$

Furthermore, one of the degrees $\left[K : k(y_g)\right] = 2g$ and $\left[K : k(y_{g+1})\right] = 2g + 1$ is relatively prime to p. It follows by Proposition 14.3.1 that

$$o(\sigma) \leq \max\{n(2n + 2g - 2)(2n + 2g - 3)(2n + 2g - 4), pn(g + 1)\}$$

with $n = 2g$ or $2g + 1$.

Now assume that A_σ is not diagonalizable. Then the minimum polynomial of A_σ contains a quadratic linear divisor. Using the Jordan canonical form for A_σ, we see that there exist two k-linearly independent elements y_1, y_2 in $L_K\left(\mathfrak{P}^{-(2g+1)}\right)$ such that $\sigma y_1 = a y_1, \sigma y_2 = y_1 + a y_2$ with $0 \neq a = a_i = a_j$ and $i \neq j$. Set $y_1' = y_1, y_2' = a y_2$. Then

$$\sigma y_1' = \sigma y_1 = a y_1 = a y_1'$$

and

$$\sigma y_2' = \sigma(a y_2) = a \sigma y_2 = a(y_1 + a y_2) = a(y_1' + y_2').$$

Put $z = \frac{y_2'}{y_1'}$. We have $\sigma z = \frac{\sigma y_2'}{\sigma y_1'} = \frac{a(y_1' + y_2')}{a y_1'} = z + 1$ and $\sigma(k(z)) = k(z)$. We have $[K : k(z)] = n = d_K(\mathfrak{N}_z) \leq d_K(\mathfrak{N}_{y_2'}) + d_K(\mathfrak{Z}_{y_1'}) \leq 2(2g + 1)$.

If $\operatorname{char} k = 0$, then by Proposition 14.3.1, $o(\sigma)$ is finite and has a bound depending only on g. If $p > 0$, then $\sigma^p(z) = z$. Assuming $E = K^{\langle \sigma^p \rangle}$, we get $k(z) \subseteq E$ and $\operatorname{Gal}(K/E) = \langle \sigma^p \rangle$. Therefore

$$o(\sigma^p) = [K : E] \leq n \leq 2(2g + 1).$$

Thus $o(\sigma) \leq 2p(2g + 1)$. □

Let π be a prime element for $\mathfrak{P}$ such that $v_{\mathfrak{P}}(\pi) = 1$. For $\sigma \in G_{\mathfrak{P}}$, $\sigma\pi$ is also a prime element for $\mathfrak{P}$ and we have

$$\sigma\pi \equiv \gamma_\sigma \pi \mod \mathfrak{P}^2,$$

where γ_σ is a $\mathfrak{P}$-unit, that is, $\gamma \in \left(\vartheta_{\mathfrak{P}}/\mathfrak{P}\right)^* = k^*$. Define

$$\begin{aligned} \phi : G_{\mathfrak{P}} &\to k^* \\ \text{by} \qquad \phi(\sigma) &= \gamma_\sigma. \end{aligned} \tag{14.31}$$

Clearly, ϕ is a group homomorphism. Let $N = \ker \phi$. Thus $G_{\mathfrak{P}}/N \cong \Gamma = \{\gamma_\sigma | \sigma \in G\} < k^*$. Since the orders of the elements in $G_{\mathfrak{P}}$ are bounded, it follows that the orders of the elements of $G_{\mathfrak{P}}/N \cong \Gamma$ are bounded too. In particular, $\Gamma < k^*$ is the cyclic group consisting of the mth roots of unity in k^* for some m satisfying $(p, m) = 1$.

Let $\varrho \in G_{\mathfrak{P}}$, $o(\varrho) = m$, and let E be the fixed field of ϱ. Then $\mathfrak{P}$ is fully ramified in K/E and since $p \nmid m$, $v_{\mathfrak{P}}(\mathfrak{D}_{K/E}) = m - 1$. Let $\wp_1 = \mathfrak{P} \cap E$ and $\wp_2, \ldots, \wp_r$ be the prime divisors in E that are ramified in K and set

$$\wp_1 = \mathfrak{P}^m, \wp_i = (\mathfrak{P}_1^{(i)} \cdots \mathfrak{P}_{g_i}^{(i)})^{e_i} \quad \text{for} \quad 2 \le i \le r.$$

Then

$$\begin{aligned} d := d_K(\mathfrak{D}_{K/E}) &= (m-1) + \sum_{i=2}^{r} (e_i - 1) g_i = (m-1) + \sum_{i=2}^{r} (e_i g_i - g_i) \\ &= (m-1) + \sum_{i=2}^{r} \left(m - \frac{m}{e_i}\right) = m - 1 + m\left(\sum_{i=2}^{r}\left(1 - \frac{1}{e_i}\right)\right). \end{aligned}$$

Using the genus formula we obtain $2(g-1) = 2(g_E - 1)m + d$. If $g_E = 0$, then $2(g-1) = -2m + d$. Now $d = (m-1) + \sum_{i=2}^{r}(m - g_i) \le r(m-1)$ and $-2m + d \ge 0$, so $r \ge 3$. If $r = 3$, then $2(g-1) = -2m + (m-1) + m\left(1 - \frac{1}{e_2}\right) + m\left(1 - \frac{1}{e_3}\right) = m - 1 - m\left(\frac{1}{e_2} + \frac{1}{e_3}\right)$.

Thus $2g - 1 = m\left(1 - \frac{1}{e_2} - \frac{1}{e_3}\right)$. The case $e_2 = e_3 = 2$ is impossible, so we conclude that $e_2 \ge 2$ and $e_3 \ge 3$. Therefore $2g - 1 = m\left(1 - \frac{1}{e_2} - \frac{1}{e_3}\right) \ge m\left(1 - \frac{1}{2} - \frac{1}{3}\right) = \frac{m}{6}$. It follows that $m \le 6(2g-1)$.

Next, if $r \ge 4$ we have

$$\begin{aligned} 2(g-1) &= -2m + (m-1) + m\sum_{i=2}^{r}\left(1 - \frac{1}{e_i}\right) \\ &\ge -m - 1 + m\left(1 - \frac{1}{2}\right)(r-1) \ge -m - 1 + \frac{3m}{2} = \frac{m}{2} - 1. \end{aligned}$$

Thus $m \le 2(2g-1)$ in this case.

Finally, if $g_E \ge 1$, we have $2(g-1) \ge d \ge m - 1$. Hence $m \le 2g - 1$. In any case, we obtain that

$$|\Gamma| = |G_{\mathfrak{P}}/N| = m \le 6(2g-1). \tag{14.32}$$

In order to study N, we consider the basis $\{x_1, \ldots, x_{g+1}\}$ given in (14.30). We have $\sigma\pi = \gamma\pi \bmod \pi^2$ with $\gamma = \gamma_\sigma \in k^*$. Moreover, for $x_i \in L_K\left(\mathfrak{P}^{-n_i}\right) \setminus L_K\left(\mathfrak{P}^{-(n_i-1)}\right)$, we obtain $x_i \equiv c\pi^{-n_i} \bmod \pi^{-n_i+1}$ with $c \in k^*$, so

$$\sigma x_i \equiv c(\sigma\pi)^{-n_i} \bmod \pi^{-n_i+1} \equiv c\gamma^{-n_i}\pi^{-n_i} \bmod \pi^{-n_i+1}.$$

On the other hand, since $\sigma x_i = a_i x_i + \sum_{j<i} a_{ij} x_j$, we have

$$a_i c\pi^{-n_i} \equiv c\gamma^{-n_i}\pi^{-n_i} \bmod \pi^{-n_i+1}.$$

It follows that $a_i = \gamma^{-n_i}$ for $1 \leq i \leq g+1$. In particular,

$$a_g = \gamma^{-n_g} = \gamma^{-2g} \quad \text{and} \quad a_{g+1} = \gamma^{-n_{g+1}} = \gamma^{-(2g+1)}.$$

Now N is the kernel of the map ϕ given in (14.31), so N consists of the elements σ in $G_{\mathfrak{P}}$ for which the matrix A_σ has the form

$$\begin{pmatrix} 1 & & * \\ & \ddots & \\ 0 & & 1 \end{pmatrix}. \tag{14.33}$$

If $p = \operatorname{char} k = 0$, any such matrix is of infinite order unless it is the unit matrix. Therefore, using Proposition 14.3.2, we get $N = \{\mathrm{Id}\}$ in characteristic 0. Assume $p > 0$. Then any element of N corresponds to a matrix of the form $B = \mathrm{Id} + A$, where A is of the form $A = \begin{pmatrix} 0 & & * \\ & \ddots & * \\ 0 & & 0 \end{pmatrix}$. Clearly, A is nilpotent and $B^{p^n} = \mathrm{Id} + A^{p^n}$. Therefore the order of any element of N is a power of p. We will prove that N is a finite p-group. If $C, D \in N$, it is easy to verify that $CDC^{-1}D^{-1}$ is of the form $\begin{pmatrix} 1 & 0 & * & * \\ 0 & \ddots & \ddots & * \\ 0 & \cdots & 1 & 0 \\ 0 & \cdots & 0 & 1 \end{pmatrix}$, so that

$$N' = [N, N] \subseteq \left\{ \begin{pmatrix} 1 & 0 & * & * \\ 0 & \ddots & \ddots & * \\ 0 & \cdots & 1 & 0 \\ 0 & \cdots & 0 & 1 \end{pmatrix} \right\}.$$

In this way, we obtain

$$N^{(i)} = \left[N^{(i-1)}, N^{(i-1)}\right] \subseteq \left\{ \begin{pmatrix} 1 & \overbrace{0 \cdots 0}^{i} & * \\ & \ddots \quad \ddots & \\ 0 & \cdots \quad \cdots & 0 \\ \vdots & \vdots \quad \vdots & \vdots \\ 0 & \cdots \quad \cdots & 0 \\ \cdots & \cdots \quad \cdots & \cdots \\ 0 & \cdots \quad \cdots & 0 \\ 0 & \cdots \quad \cdots & 1 \end{pmatrix} \right\}.$$

(where the rows of zeros below the first row span i rows)

Therefore $N^{(g)} = \{\mathrm{Id}\}$ and N is a nilpotent group.

Proposition 14.3.3. *Assume that k has characteristic $p > 0$, $H < G_{\mathfrak{P}}$, and $g_K > 0$. If H is an abelian group such that the order of any element of H is a power of p and for any nontrivial finite subgroup U of H, K^U is a rational function field, then H is a cyclic group of order 1, p, or p^2.*

Proof. Assume that U is a cyclic subgroup of H generated by an element σ of order p. Then $K^U = k(x)$. Since $p = [K : k(x)] = |U|$ we may assume that $\mathfrak{N}_x = \mathfrak{P}^p$. For any $\tau \in H$, we have $K^{\langle \sigma, \tau \rangle} \subseteq k(x)$ and since $\langle \sigma, \tau \rangle$ is abelian, $k(x)/K^{\langle \sigma, \tau \rangle}$ is normal. Therefore $\tau(k(x)) = k(x)$ and $\mathfrak{P}^\tau = \mathfrak{P}$. Since $o(\tau)$ is a power of p, it follows that $\tau x = x + a$, $a \in k$. Thus $\tau^p(x) = x$, $\tau^p \in U$, and $\tau^{p^2} = e$. That is, the order of any element τ of H is 1, p, or p^2. The result will follow if we prove that the only subgroup of H of order p is U. Assume that there exists $V < H$, $V = \langle \tau \rangle$, such that $o(\tau) = p$ and $U \neq V$. Then $\tau x = x + a$ and since $\tau \notin U$, a is nonzero. Thus

$$\tau\left(\frac{x}{a}\right) = \frac{\tau(x)}{a} = \left(\frac{x}{a}\right) + 1.$$

Setting $y = \frac{x}{a}$, we have $k(x) = k(y)$ and $\tau(y) = y + 1, \sigma y = y, \mathfrak{N}_y = \mathfrak{P}^p$. Applying the same to V instead of U, we obtain the existence of an element z in K such that $\mathfrak{N}_z = \mathfrak{P}^p$, $\tau(z) = z$, and $\sigma z = z + 1$.

Now $z \notin k(y)$ since $\sigma z \neq z$ and $[K : k(y)] = p$. Therefore $K = k(y, z)$. Consider the subgroup UV of H, and notice that $|UV| = p^2$. Set $E = K^{UV}$; then $[K : E] = p^2$ and E a rational function field. Clearly, $v_{\mathfrak{P}}(y^p - y) = -p^2$ and $v_{\mathfrak{P}}(z^p - z) = -p^2$. We have

$$\sigma(y^p - y) = y^p - y, \quad \tau(y^p - y) = (\tau y)^p - \tau(y) = y^p + 1 - (y + 1) = y^p - y.$$

Hence $y^p - y \in E$. Similarly, $z^p - z \in E$ and

$$\left[K : k(y^p - y)\right] = d_K(\mathfrak{N}_y) = p^2 = [K : E].$$

Therefore $E = k(y^p - y) = k(z^p - z) = k(w)$, where $(w)_E = \frac{\wp_0}{\wp_\infty}$ and $\wp_\infty = \mathfrak{P}^{p^2}$ in K. Since $\mathfrak{N}_y = \mathfrak{N}_z = \mathfrak{N}_w$, we have $y^p - y = Aw + B$ and $z^p - z = Cw + D$. In particular, $y^p - y = \beta(z^p - z) + \gamma$ for some $\beta, \gamma \in k$ such that $\beta \neq 0$. Let $u = y - \beta^{1/p} z$. Then

$$\begin{aligned} u^p - u - \gamma &= y^p - \beta z^p - y + \beta^{1/p} z - \gamma \\ &= \beta(z^p - z) - \beta z^p + \beta^{1/p} z = \left(\beta^{1/p} - \beta\right) z. \end{aligned}$$

If $\beta^{1/p} - \beta = 0$, then u is a constant and therefore $k(y) = k(z)$, which contradicts the fact that $U \neq V$. If $\beta^{1/p} - \beta \neq 0$, then $z \in k(u)$ and $y \in k(u)$. Thus $K = k(y, z) \subseteq k(u)$, which contradicts $g_K > 0$. This completes the proof. □

Proposition 14.3.4. *Assume* char $k = p > 0$ *and let $g_K > 0$. Let $H < G_{\mathfrak{P}}$ be such that H is abelian and every element of H is a power of p. Then H is a finite group such that $|H| \leq p^2(2g - 1)$.*

Proof. Let U be any finite subgroup of H, say of order n (a power of p). Putting $E = K^U$, we obtain using the Riemann–Hurwitz genus formula

$$2(g-1) = 2n(g_E - 1) + d,$$

where $d = d_K(\mathfrak{D}_{K/E})$. Since $\mathfrak{P}^\sigma = \mathfrak{P}$ for all $\sigma \in U$, $\mathfrak{P}$ is fully ramified in K/E and $\mathfrak{P}^{n-1} \mid \mathfrak{D}_{K/E}$. Therefore $d \geq n-1$. If $2g \leq n$, then

$$g_E = \frac{2(g-1)-d}{2n} + 1 \leq \frac{n-2-(n-1)}{2n} + 1 = -\frac{1}{2n} + 1 < 1.$$

Thus, in this case, $g_E = 0$. Let U be a maximal finite subgroup of H such that $g_E > 0$. Then $|U| < 2g$. We have $H/U < \operatorname{Aut}_k(E)$, so H/U satisfies the condition of Proposition 14.3.3. Therefore $|H/U| \leq p^2$, and

$$|H| \leq p^2|U| \leq p^2(2g-1). \qquad \square$$

Proposition 14.3.5. *Let G be a group with at least n elements (G may be infinite) and a subgroup H contained in the center of G such that $|H| = p$ and G/H is an elementary abelian p-group (that is, $\sigma^p = e$ for all $\sigma \in G/H$ and G/H is abelian). Then G contains an abelian subgroup with at least $\sqrt{pn}$ elements.*

Proof. If G is infinite, then since every element of G is of order at most p^2 ($\sigma \in G$, $\sigma^p \in H$, $\sigma^{p^2} = e$), we may replace G by a finite subgroup of order at least n and assume that G is finite. Let U be a maximal abelian normal subgroup of G. Then $H \subseteq U$, where U/H is an elementary abelian p-group. Let $\sigma_1, \ldots, \sigma_s \in U$ be such that the elements $\overline{\sigma}_i = \sigma_i \bmod H$ form a basis of U/H. Let $\sigma \in G$ be arbitrary and let

$$\varrho_i(\sigma) = \varrho_i := \sigma\sigma_i\sigma^{-1}\sigma_i^{-1}, \quad 1 \leq i \leq s.$$

Since G/H is an abelian group, we have $\varrho_i(\sigma) \in H$. Define

$$\phi\colon G \to H^s$$

by $$\phi(\sigma) = (\varrho_1(\sigma), \ldots, \varrho_s(\sigma)).$$

It is easy to verify that ϕ is a group homomorphism whose kernel is

$$\ker\phi = \{\sigma \in G \mid \varrho_i(\sigma) = 1, 1 \leq i \leq s\} = \{\sigma \in G \mid \sigma\sigma_i = \sigma_i\sigma \text{ for all } i\}.$$

Therefore $\ker\phi \supseteq U$ and $\ker\phi \lhd G$. If $\sigma \in \ker\phi$, then σ commutes with U and $\langle\sigma, U\rangle$ is an abelian subgroup containing U. Since $G/U \cong \frac{G/H}{U/H}$ is abelian, then $\langle\sigma, U\rangle$ is a normal subgroup of G. It follows that $\langle\sigma, U\rangle = U$ and $\sigma \in U$. Therefore $\ker\phi = U$. Now $|G/U| \leq |H^s| = p^s$ and $|U/H| = p^s$, so $|U| = p^{s+1}$. Hence

$$n \leq |G| = |G/U||U| \leq p^s p^{s+1} \quad \text{and} \quad \sqrt{pn} \leq |U|. \qquad \square$$

Now we return to $G_{\mathfrak{P}}$. The group $G_{\mathfrak{P}}/N = \Gamma$ is finite of order at most $6(2g-1)$. Consider again the basis $\{x_1, \ldots, x_g, x_{g+1}\}$ as in (14.30). Set $x = x_1$, that is, $\mathfrak{N}_x =$

$\mathfrak{P}^{-n_1}$, where $\{1, x\}$ is a basis of $L_K(\mathfrak{P}^{-n_1})$. By (14.33) $\sigma x = x + \alpha_\sigma$ where $\alpha_\sigma \in k$ for $\sigma \in N$. Since for $\psi, \sigma \in N$, we have $(\psi\sigma)(x + a_\sigma) = x + a_\sigma + a_\psi = x + a_{\psi\sigma}$, it follows that $\Lambda \colon N \to k$, $\Lambda(\sigma) = a_\sigma$, is a group homomorphism. Let $N_1 = \ker \Lambda$. Then N/N_1 is a subgroup of k. Therefore N/N_1 is an elementary abelian p-group. If $N_1 = \{\mathrm{Id}\}$, N is abelian and by Proposition 14.3.4 we have $|N| \leq p^2(2g-1) < p^3(2g-1)$. Assume $N_1 \neq \{\mathrm{Id}\}$. If $\sigma \in N_1$ then $\sigma \in \mathrm{Aut}_{k(x)}(K)$. Thus

$$|N_1| \leq [K : k(x)] = d_k(\mathfrak{N}_x) = n_1.$$

Since N is nilpotent, there exists N_2 such that $[N_1 : N_2] = p$, $N_2 \lhd N$ and N_1/N_2 is contained in the center of N/N_2. Let $E = K^{N_2}$. Then

$$p\,[K : E] = [N_1 : N_2]\,|N_2| = |N_1| \leq [K : k(x)] = n_1.$$

We have $N_2 \subseteq G_{\mathfrak{P}}$, so $\mathfrak{P}$ is fully ramified in K/E. Let $\wp = \mathfrak{P} \cap E$. If $g_E = 0$ there exists $z \in E$ such that $(z)_E = \frac{\wp'}{\wp}$ and $(z)_K = \frac{\mathfrak{q}}{\mathfrak{P}^{[K:E]}}$ with $[K : E] < n_1$. This contradicts the fact that n_1 is the first pole number of $\mathfrak{P}$. Thus $g_E > 0$.

Next, since K/E is normal and N_2 is trivial on E, we may consider N/N_2 as a subgroup of $\mathrm{Aut}_k(E)$. By Proposition 14.3.4 any abelian subgroup of N/N_2 is of order at most $p^2(2g_E - 1)$.

Let $H = N_1/N_2 < N/N_2 = \overline{N}$. Then H is contained in the center of $\overline{N}$, $|H| = p$, and $\overline{N}/H$ is elementary abelian. By Proposition 14.3.5, $\overline{N}$ contains an abelian subgroup of order at least $\sqrt{pn'}$ if $|\overline{N}| \geq n'$. It follows that

$$\sqrt{pn'} \leq p^2(2g_E - 1).$$

Thus $pn' \leq p^4(2g_E - 1)^2$, so $|N/N_2| \leq p^3(2g_E - 1)^2$ and

$$|N| \leq |N_2| p^3 (2g_E - 1)^2 = |N_1| p^{-1} p^3 (2g_E - 1)^2 = |N_1| p^2 (2g_E - 1)^2.$$

Finally, using the genus formula and the facts that $\mathfrak{P}$ is fully ramified in K/E and $[K : E] = |N_2|$, we obtain

$$2(g_K - 1) = 2[K : E](g_E - 1) + d(\mathfrak{D}_{K/E}) \geq 2|N_2|(g_E - 1) + (|N_2| - 1).$$

Hence $(2g_K - 1)^2 \geq |N_2|^2(2g_E - 1)^2 \geq |N_2|(2g_E - 1)^2$. It follows that

$$|N| \leq |N_1| p^2 (2g_E - 1)^2 \leq |N_1| p^2 \frac{(2g_K - 1)^2}{|N_2|} = p^3 (2g_K - 1)^2.$$

Since $|G_{\mathfrak{P}}/N| \leq 6(2g-1)$, we have $|G_{\mathfrak{P}}| \leq 6p^3(2g-1)^3$. In characteristic 0, we have $|N| = 1$ and $|G_{\mathfrak{P}}| \leq 6(2g-1)$. We have proved the following theorem:

Theorem 14.3.6. *Let K/k be a function field of genus $g \geq 1$ where k is algebraically closed of characteristic $p \geq 0$. Let $\mathfrak{P}$ be any prime divisor of K and $G_{\mathfrak{P}}$ its decomposition group. Then:*

(i) *If $p = 0$, $G_{\mathfrak{P}}$ is a cyclic group of order at most* $6(2g-1)$.

(ii) *If $p > 0$, the p-Sylow subgroup N of $G_{\mathfrak{P}}$ is normal,*

$$|N| \leq p^3(2g-1)^2,$$

and $G_{\mathfrak{P}}/N$ is a cyclic group of order $\leq 6(2g-1)$. Finally,

$$|G_{\mathfrak{P}}| \leq 6p^3(2g-1)^3. \qquad \square$$

As a corollary of Theorem 14.3.6, we obtain our main result.

Theorem 14.3.7. *Let K/k be a function field where k is algebraically closed and of genus $g \geq 2$. Then $\operatorname{Aut}_k(K)$ is a finite group.*

Proof. Define $W = \{\mathfrak{P} \mid \mathfrak{P}$ is a Weierstrass point of $K\}$. By Corollary 14.2.52, we have $1 \leq |W| \leq (g-1)g(3g-1)$.

Set $G = \operatorname{Aut}_k(K)$. If $\sigma \in G$ then $W^\sigma = W$ since the gap sequences $\mathfrak{P}$ and $\mathfrak{P}^\sigma$ are the same. Thus G acts on W and if $\mathfrak{P} \in W$ we have

$$[G : G_{\mathfrak{P}}] = |\text{orbit of } \mathfrak{P}| = |\{\mathfrak{P}^\sigma \mid \sigma \in G\}| \leq |W|.$$

Hence $|G| \leq |W||G_{\mathfrak{P}}| < \infty$. $\square$

Remark 14.3.8. If $\operatorname{char} k = 0$, we have $|W| \leq g^3 - g = (g-1)g(g+1)$ and $|G_{\mathfrak{P}}| \leq 6(2g-1)$. Thus $|G| \leq 6(g-1)g(g+1)(2g-1)$. This bound is much larger than Hurwitz's bound, since Hurwitz [70] proved that $|G| \leq 84(g-1)$. We present a proof of Hurwitz's theorem above (Theorem 14.3.13). If $\operatorname{char} k = p > 0$, P. Roquette [126] showed that Hurwitz's bound is valid if $p > g+1$ with one exception. Henn [66] proved that $|G| \leq 3(2g)^{5/2}$ when K does not belong to one of four exceptional classes.

Corollary 14.3.9. *Let k be an arbitrary field and K/k any function field. Let $\bar{k}$ be the algebraic closure of k and $\bar{K} = K\bar{k}$. If $g_{\bar{K}} \geq 2$, then $\operatorname{Aut}_k(K)$ is a finite group.*

Proof. Let $\sigma \in \operatorname{Aut}_k(K)$. Then σ can be extended to $\widetilde{\sigma} \in \operatorname{Aut}_{\bar{k}}(\bar{K})$ and $\widetilde{\sigma}|_{\bar{k}} = \operatorname{Id}_{\bar{k}}$.

In fact, if we consider $A = \{(\varphi, E)\}$, where E is a function field such that $K \subseteq E \subseteq \bar{K}$ and whose field of constants k_E satisfies $k \subseteq k_E \subseteq \bar{k}$, and $\varphi \in \operatorname{Aut}_{k_E}(E)$ is such that $\varphi|_K = \sigma$, then $A \neq \emptyset$. Indeed, $(\sigma, K) \in A$ and the relation $(\varphi_1, E_1) \leq (\varphi_2, E_2) \iff E_1 \subseteq E_2, \varphi_2|_{E_1} = \varphi_1$ defines a partial order in A. By Zorn's lemma, A contains a maximal element $(\widetilde{\sigma}, F)$. If $F \neq \bar{K}$, there exists $\alpha \in \bar{K} \setminus F$. Let $f(x) = \operatorname{Irr}(\alpha, x, F)$. Since $\alpha \in \bar{k}$, we have $f(x) \in k_F[x]$. Thus $\widetilde{\sigma}(f(x)) = f(x)$ and $\widetilde{\sigma}$ can be extended to $F(\alpha)$ by defining $\widetilde{\sigma}\alpha = \alpha$. Therefore $F = \bar{K}$.

This proves that the function $\varphi : \operatorname{Aut}_k(K) \to \operatorname{Aut}_{\bar{k}}(\bar{K})$, defined by $\varphi(\sigma) = \widetilde{\sigma}$, $\widetilde{\sigma}|_K = \sigma, \widetilde{\sigma}|_k = \operatorname{Id}$, is a group monomorphism and $|\operatorname{Aut}_k(K)| \leq |\operatorname{Aut}_{\bar{k}}(\bar{K}| < \infty$. $\square$

Corollary 14.3.10. *For any function field K/k of genus g such that $g_{\bar{K}} \geq 2$ we have*

$$|\operatorname{Aut}_k(K)| \leq \begin{cases} 6(2g-1)(g-1)g(g+1) & \text{if } \operatorname{char} k = 0, \\ \\ 6p^3(2g-1)^3(g-1)g(3g-1) & \text{if } \operatorname{char} k = p. \end{cases}$$

Proof. By Theorem 14.3.7 we have $|\operatorname{Aut}_{\bar{K}}(\bar{K})| \leq 6(2g_{\bar{K}}-1)(g_{\bar{K}}^3 - g_{\bar{K}})$ if $\operatorname{char} k = 0$, and $|\operatorname{Aut}_{\bar{k}}(\bar{K})| \leq 6p^3(2g_{\bar{K}}-1)^3(g_{\bar{K}}-1)g_{\bar{K}}(3g_{\bar{K}}-1)$ if $\operatorname{char} k = p > 0$. Since $\bar{K}$ is an extension of constants, Theorem 8.5.3 yields $g_{\bar{K}} \leq g$. The result follows. □

Remark 14.3.11. If k is not an algebraically closed field, then $\operatorname{Aut}_k(K)$ may be infinite even though $g_K \geq 2$.

Example 14.3.12 (Rosenlicht). Let k be an imperfect separably closed field, that is, $k \neq k^p$, and if ℓ/k is an algebraic separable extension, then $\ell = k$. Let $a \in k$ be such that $a \notin k^p$ and set $K = k(x, y)$, where $y^p - y = ax^p$.

For each $\beta \in k$, let α satisfy $\alpha^p - \alpha = a\beta^p$. Since $T^p - T - a\beta^p$ is a separable polynomial, we have $\alpha \in k$. Now let $\sigma \in \operatorname{Aut}_k(k(x))$ be defined by $\sigma x = x + \beta$. Since $K/k(x)$ is a cyclic extension, σ can be extended to K by putting $\sigma y = y + \alpha$ (because $(\sigma y)^p - (\sigma y) = (y+\alpha)^p - (y+\alpha) = y^p - y + (\alpha^p - \alpha) = ax^p + a\beta^p = a(x+\beta)^p = a\sigma x^p$).

Therefore there are infinitely many automorphisms in $\operatorname{Aut}_k(K)$ defined by $\sigma x = x + \beta, \sigma y = y + \alpha$, where $\alpha, \beta \in k, \alpha^p - \alpha = a\beta^p$.

Using Tate's genus formula we can prove that $g_K = \frac{(p-1)(p-2)}{2}$ (see Exercise 14.5.16). In particular, if $p \geq 5$, we get $g_K \geq 6$ and $|\operatorname{Aut}_k(K)| = \infty$.

Theorem 14.3.13 (Hurwitz). *Let K/k be a function field of genus $g \geq 2$ where k is algebraically closed of characteristic either* 0 *or p with $p > 2g+1$. Then $|\operatorname{Aut}_k K| \leq 84(g-1)$.*

Proof. Let $G = \operatorname{Aut}_k K$ and let $F := K^G$ be the fixed field of K under G. Then K/F is a finite Galois extension of degree $m = |\operatorname{Aut}_k K|$. Let g' be the genus of F. By the Riemann–Hurwitz genus formula (Theorem 9.4.2) we have

$$g = 1 + m(g'-1) + 1/2 \deg_K(\mathfrak{D}_{K/F}). \tag{14.34}$$

Let $\mathfrak{p}_1, \ldots, \mathfrak{p}_t$ be the prime divisors of F that are ramified in K. Then $\mathfrak{D}_{K/F} = \prod_{i=1}^t (\operatorname{con}_{F/K} \mathfrak{p}_i)^{e_i'-1}$, where $e_i' = e_i + \varepsilon_i$, each e_i is the ramification index of $\mathfrak{p}_i$ in K/F, and $\varepsilon_i \geq 0$. Therefore $\deg_K(\operatorname{con}_{F/K} \mathfrak{p}_i)^{e_i'-1} = m + h_i(\varepsilon_i - 1)$, where h_i is the number of prime divisors in K above $\mathfrak{p}_i$. Hence (14.34) simplifies to

$$2g - 2 = m(2g' - 2 + \delta), \tag{14.35}$$

where $\delta = \sum_{i=1}^t \delta_i$, $\delta_i = 1 + \frac{h_i(\varepsilon_i - 1)}{m} = 1 + \frac{\varepsilon_i - 1}{e_i} = \frac{e_i - 1 + \varepsilon_i}{e_i}$. Notice that $\delta_i \geq \frac{e_i - 1}{e_i} \geq \frac{1}{2}$ for all $i = 1, 2, \ldots, t$.

The proof of $m \leq 84(g-1)$ consists of a detailed case-by-case analysis of (14.35). First we study the possible genus g'. If $g' \geq 2$, it follows from (14.35) that $m \leq g - 1$ and we are done. If $g' = 1$, then $m\delta = 2g - 2 > 0$. Therefore $t > 0$, $\delta \geq \delta_i \geq 1/2$, and $m \leq 4(g-1)$.

Finally we consider the case $g' = 0$. In this situation we obtain $2g - 2 = m(\delta - 2) > 0$, so $\delta > 2$ and $m = \frac{2g-2}{\delta-2}$. Now we consider all the choices for t. If $t \geq 5$, then since $\delta_i \geq 1/2$, we obtain $\delta \geq 5/2$ and thus $m \leq 4(g-1)$. For the case $t = 4$, one of the δ_i's must satisfy $\delta_i > 1/2$, say δ_1. Hence $\delta_1 \geq 2/3$ and $\delta \geq 2/3 + 3/2$, so $\delta - 2 \geq 1/6$ and $m \leq 12(g-1)$.

The next case is $t \leq 3$. First we consider the case that K/F is tamely ramified, that is, $\varepsilon_i = 0$ for $i = 1, \ldots, t$. In this situation and since $\delta_1 < 1$, $\delta_2 < 1$, and $\delta - 2 > 0$, we have $t = 3$. Assume that $e_1 \leq e_2 \leq e_3$. It is straightforward to verify that

- if $e_1 \geq 3$ then $m \leq 24(g-1)$,
- if $e_1 = 2$ and $e_2 \geq 4$, then $m \leq 40(g-1)$,
- if $e_1 = 2$ and $e_2 = 3$, then $e_3 \geq 7$ and $m \leq 84(g-1)$,
- if $e_1 = e_2 = 2$, then $\delta_3 > 1$, which is not possible.

This finishes the tamely ramified case.

It remains to consider the case that $g' = 0$, $t \leq 3$, and K/F is wildly ramified. In particular, we have $\operatorname{char} k = p > 0$. We will show that under the hypothesis $p > 2g+1$ this case does not occur.

Let $\mathfrak{p} = \mathfrak{p}_1$ be a wildly ramified prime divisor of F and $\mathfrak{P} = \mathfrak{P}_1$ a prime divisor in K above $\mathfrak{p}$. There exists a subgroup H of order p of the inertia group of $\mathfrak{P}/\mathfrak{p}$. Let $E = K^H$ be the fixed field of K under H. Let $\mathfrak{q} := \mathfrak{P} \cap E$ be the prime divisor of E below $\mathfrak{P}$. Then $e_{K/K}(\mathfrak{P}|\mathfrak{p}) = p$. By Example 5.8.8 we know that the power of $\mathfrak{P}$ appearing in $\mathfrak{D}_{K/E}$ is equal to $(\lambda+1)(p-1)$ for some integer $\lambda \geq 1$. In particular, this power is greater than or equal to $2(p-1)$.

Let $d := \deg_K(\mathfrak{D}_{K/E})$. Then by the genus formula, we have

$$2g - 2 = p(2g'' - 2) + d, \tag{14.36}$$

where g'' denotes the genus of E. If $g'' \geq 1$, then we obtain from (14.36) and from $d \geq 2(p-1)$ that $2g - 2 \geq 2p - 2$, contrary to our assumption on p.

Thus $g'' = 0$ and (14.36) becomes $2g - 2 = -2p + d$. Let $r \geq 1$ be the number of ramified prime divisors in K/E. Then, if $r = 1$, by Example 5.8.8 we have $(\mathfrak{D}_{K/E}) = \mathfrak{P}^{(\lambda+1)(p-1)}$ with $\lambda \geq 1$. Therefore $2g - 2 = -2p + (\lambda+1)(p-1)$.

The case $\lambda = 1$ is not possible since in this case we would obtain $g = 0$. Thus $\lambda \geq 2$ and then we obtain $2g - 2 \geq -2p + 3(p-1) = p - 3$. This is contrary to the hypothesis $p > 2g + 1$.

Therefore $r \geq 2$. This case is also impossible since in this situation, by (14.36) we have

$$2g - 2 = -2p + d \geq -2p + 2r(p-1)$$

and this implies $g \geq (p-1)(r-1) \geq (p-1)$. □

14.4 Properties of Automorphisms of Function Fields

Theorem 14.4.1 (Schmid). *Let K/k be an algebraic function field such that k is algebraically closed. Let $\sigma \in \operatorname{Aut}_k(K)$ be such that $\sigma \neq \operatorname{Id}$. Then σ fixes at most $2g+2$ distinct prime divisors of K, where g denotes the genus of K.*

Proof. If $g = 0$, it follows by Exercise 5.10.14 that there are at most two places fixed under σ. Denote by n the order of σ. Since σ is not the identity, we have $n > 1$. If $g = 1$ and σ has at least one fixed point, then $o(\sigma) < \infty$ by Theorem 14.3.6. Assume $g \geq 1$ and let $E = K^{\langle\sigma\rangle}$. Then K/E is a Galois extension with Galois group $\langle\sigma\rangle$. By the genus formula,

$$2(g_K - 1) = 2n(g_E - 1) + d, \tag{14.37}$$

where $d = d_K(\mathfrak{D}_{K/E})$. Let $\mathfrak{P}_1, \ldots, \mathfrak{P}_r$ be r distinct places of K fixed by σ. Since $\mathfrak{P}_1^\sigma = \mathfrak{P}_i$ and k is algebraically closed, $\wp_i := \mathfrak{P}_i \cap E$ is fully ramified in K/E. Therefore $(\mathfrak{P}_1 \cdots \mathfrak{P}_r)^{n-1} \mid \mathfrak{D}_{K/E}$ and $d \geq r(n-1)$.

It follows by (14.37) that

$$2(g_K - 1) \geq -2n + r(n-1),$$

or

$$r \leq \frac{2(g_K + n - 1)}{n - 1} = 2\left(\frac{g_K}{n-1} + 1\right) \leq 2(g_K + 1). \qquad \square$$

Remark 14.4.2. Theorem 14.4.1 is no longer true if we do not assume that k is algebraically closed.

Example 14.4.3. Let $K = \mathbb{F}_p(x)$ and let $\sigma \in \operatorname{Aut}_{\mathbb{F}_p}(K)$ be different from the identity (for instance, $\sigma(x) = x + 1$). Let $E = \mathbb{F}_p(x)^{\langle\sigma\rangle}$. By Čebotarev's density theorem (Theorem 11.2.20), there exist infinitely prime divisors in E such that $\left(\frac{K/E}{\wp}\right) = \langle\sigma\rangle$. All these primes correspond to fully ramified or fully inert prime divisors, that is, $\wp = \mathfrak{P}^e$ in K. Since at most two prime divisors of E are ramified in K (Exercise 14.5.6), there are infinitely many inert prime divisors and $\mathfrak{P}^\sigma = \mathfrak{P}$ for all such prime divisors.

Using Theorem 14.4.1 we can provide a proof of Theorem 14.3.7 when $\operatorname{char} k = 0$ and K is not a hyperelliptic function field (of course we have used Theorem 14.3.6 to prove Theorem 14.4.1).

Proposition 14.4.4. *Assume that K/k is a function field of genus $g \geq 2$ and k an algebraically closed field of characteristic* 0. *If K is not a hyperelliptic function field, then* $\operatorname{Aut}_k(K)$ *is finite.*

Proof. Let $W = \{\mathfrak{P} \in \mathbb{P}_K \mid \mathfrak{P} \text{ is a Weierstrass point}\}$. By Theorem 14.2.62, $|W| > 2g + 2$. For any $\sigma \in G = \operatorname{Aut}_k(K)$, we have $\sigma(W) = W$. Thus there is a group homomorphism ϕ from G to the symmetric group S_W. We have $\ker\phi = \{\sigma \in G \mid \mathfrak{P}^\sigma = \mathfrak{P} \text{ for all } \mathfrak{P} \in W\}$. By Theorem 14.4.1, $\ker\phi = \{\mathrm{Id}\}$ and $|G| \leq |S_W| < \infty$. $\square$

Theorem 14.4.5 (Madden–Valentini). *Let k be an algebraically closed field and let L/K be a finite extension of function fields over k. Suppose that for every intermediate extension M such that $K \subsetneqq M \subseteq L$, we have*

$$g_M > [M:K]^2 + 2[M:K](g_K - 1) + 1.$$

Then for every $\sigma \in \operatorname{Aut}_k(L)$, *we have* $\sigma(K) = K$.

Proof. If $\sigma \in \operatorname{Aut}_k(L)$ is such that $\sigma(K) \neq K$, let $M = K\sigma(K)$. By the Castelnuovo–Severi inequality (Theorem 14.1.3) we have

$$\begin{aligned} g_M &\leq [M:K]g_K + [M:\sigma(K)]g_{\sigma(K)} + ([M:K]-1)([M:\sigma(K)]-1) \\ &= 2[M:K](g_K - 1) + [M:K]^2 - 1, \end{aligned}$$

which contradicts the hypothesis. □

Theorem 14.4.6 (Valentini–Madan [154]). *Let K/k be an algebraic function field of genus g, for an algebraically closed field k. Let $T = \{\mathfrak{P}_1, \ldots, \mathfrak{P}_t\}$ be a set of prime divisors of K with $t > 2g+3$. Then for all but finitely many prime divisors $\mathfrak{Q}$, the set $T' = T \cup \{\mathfrak{Q}\}$ has the property that the identity is the only element of* $\operatorname{Aut}_k(K)$ *that maps T' into itself.*

Proof. By Theorem 14.4.1, if $\theta, \sigma \in \operatorname{Aut}_k(K)$ satisfy $\sigma(\wp) = \theta(\wp)$ for $2g+3$ distinct prime divisors, then $\sigma = \theta$. Put

$$\Gamma = \{\sigma \in \operatorname{Aut}_k(K) \mid |\sigma(T) \cap T| \geq t-1\}.$$

Let $A_1, \ldots, A_t$ be the subsets of T of cardinality $t-1$. Any $\sigma \in \operatorname{Aut}_k(K)$ is determined by its action on each A_i. Now if $\Gamma_i = \{\sigma \in \operatorname{Aut}_k(K) \mid A_i^\sigma \subseteq T\}$, then $|\Gamma_i| \leq t!$ and $\Gamma \subseteq \bigcup_{i=1}^t \Gamma_i$. Therefore $|\Gamma| \leq tt! < \infty$. For each $\gamma \in \Gamma$ such that $\gamma \neq \mathrm{Id}$, let

$$W_\gamma = \{\mathfrak{Q} \in \mathbb{P}_K, \mathfrak{Q} \notin T \mid \mathfrak{Q}^\gamma = \mathfrak{Q} \text{ or } \mathfrak{Q}^\gamma \in T\}.$$

If $|W_\gamma| > 2g+3$, then since $\gamma \neq \mathrm{Id}$, γ can fix at most $2g+2$ prime divisors. Thus there exist $\mathfrak{Q}$ and $\mathfrak{Q}' \in W_\gamma$ such that $\mathfrak{Q}^\gamma$ and $(\mathfrak{Q}')^\gamma \in T$. Since $|\gamma(T) \cap T| \geq t-1$, either $\mathfrak{Q}^\gamma$ or $(\mathfrak{Q}')^\gamma \in \gamma(T) \cap T$. This contradiction shows that $|W_\gamma| \leq 2g+3$.

Let $W = \bigcup_{\substack{\gamma \in \Gamma \\ \gamma \neq \mathrm{Id}}} W_\gamma$. Then $|W| \leq (2g+3)|\Gamma| < \infty$. Let $\mathfrak{Q} \notin W \cup T$ and let $T' = T \cup \{\mathfrak{Q}\}$. Suppose that $\sigma \in \operatorname{Aut}_k(K)$ satisfies $\sigma(T') = T'$. Then $|\sigma(T) \cap T| \geq t-1$. Therefore $\mathfrak{Q}^\sigma = \mathfrak{Q}$ or $\mathfrak{Q}^\sigma \in T$, and $\sigma \in \Gamma$. Since $\mathfrak{Q} \notin W = \bigcup_{\substack{\gamma \in \Gamma \\ \gamma \neq \mathrm{Id}}} W_\gamma$, it follows that if $\sigma \neq \mathrm{Id}$, we have $\mathfrak{Q} \notin W_\sigma$, so $\mathfrak{Q}^\sigma \neq \mathfrak{Q}$ and $\mathfrak{Q}^\sigma \notin T$. This contradiction shows that $\sigma = \mathrm{Id}$ and proves the theorem. □

Definition 14.4.7. Let G be a finite group. Then G is called *realizable* over a function field K/k, with k algebraically closed, if there exists a Galois extension L/K such that

$$\operatorname{Aut}_K(L) = \operatorname{Gal}(L/K) \cong G.$$

The group G is called *exactly realizable* over K if the Galois extension L/K satisfies

$$\operatorname{Aut}_k(L) = \operatorname{Aut}_K(L) = \operatorname{Gal}(L/K) \cong G.$$

Our next goal is to prove that given any finite separable extension $K/k(x)$ (k algebraically closed), and any function field F over k of genus at least two, there exists a separable extension L/F such that

$$\operatorname{Aut}_k(L) = \operatorname{Aut}_F(L) \cong \operatorname{Aut}_{k(x)}(K)$$

and such that $[L : F] = [K : k(x)]$.

Proposition 14.4.8. *Let $K/k(x)$ be a finite separable extension and let $M \in \mathbb{R}$ satisfy $M > 0$. Then there exists a separable extension $K_1/k(x)$ such that*

(i) $[K_1 : k(x)] = [K : k(x)]$,
(ii) $\operatorname{Aut}_{k(x)}(K_1) \cong \operatorname{Aut}_{k(x)}(K)$,
(iii) *For any field E such that $k(x) \subsetneqq E \subseteqq K_1$, we have $g_E \geq M$.*

Proof. Let L be the Galois closure of $K/k(x)$, and $G = \operatorname{Gal}(L/k(x))$. Since there exist finitely many ramified prime divisors in $L/k(x)$, by means of a variable substitution $x \mapsto \frac{ax+b}{cx+d}$ with $ad - bc \neq 0$, we may assume that if $\mathfrak{P} \mid \mathfrak{N}_x$ or $\mathfrak{P} \mid \mathfrak{Z}_x$, then $\mathfrak{P}$ is not ramified in $L/k(x)$. Choose $m \in \mathbb{N}$ such that $m \geq M + 1$ and $\operatorname{char} k \nmid m$. Let $t = x^{1/m}$. Then $k(t)/k(x)$ is a cyclic extension of degree m (because the primitive mth roots of unity are in k) such that the primes $\wp_0, \wp_\infty$ are the only prime divisors of $k(x)$ that are ramified in $k(t)$, and they are fully ramified, where $(x)_{k(x)} = \frac{\wp_0}{\wp_\infty}$ (see Example 5.8.9).

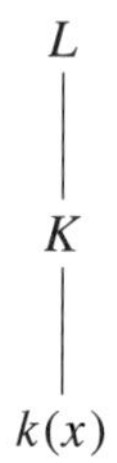

Since $\wp_0$ and $\wp_\infty$ are not ramified in $L/k(x)$ and they are fully ramified in $k(t)/k(x)$, then L and $k(t)$ are linearly disjoint over $k(x)$.

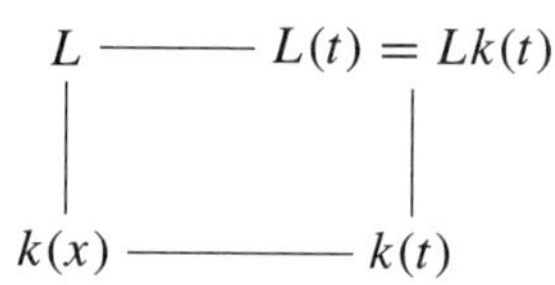

Using Galois theory, it follows that $L(t)/k(t)$ is a Galois extension and $\operatorname{Gal}(L(t)/k(t)) \cong \operatorname{Gal}(L/k(x))$.

Also, $K(t)/k(t)$ is a separable extension satisfying $[K(t) : k(t)] = [K : k(x)]$ and $\operatorname{Aut}_{k(t)}(K(t)) \cong \operatorname{Aut}_{k(x)}(K)$. In fact, we can exhibit an isomorphism

$$\begin{aligned} G_1 = \operatorname{Gal}(L(t)/k(t)) &\xrightarrow{\varphi} \operatorname{Gal}(L/k(x)) = G \\ \sigma &\longmapsto \sigma|_L. \end{aligned}$$

If $K = L^H$, then $K(t) = L(t)^{\varphi^{-1}(H)}$. In particular, $[K(t) : k(t)] = \frac{|G_1|}{|\varphi^{-1}(H)|} = \frac{|G|}{|H|} = [K : k(x)]$.

Finally, $\operatorname{Aut}_{k(x)}(K) = \{\sigma \in G \mid \sigma(K) = K\} = \varphi(\{\theta \in G_1 \mid \theta(K(t)) = K(t)\}) = \varphi(\operatorname{Aut}_{k(t)} K(t)) \cong \operatorname{Aut}_{k(t)}(K(t))$.

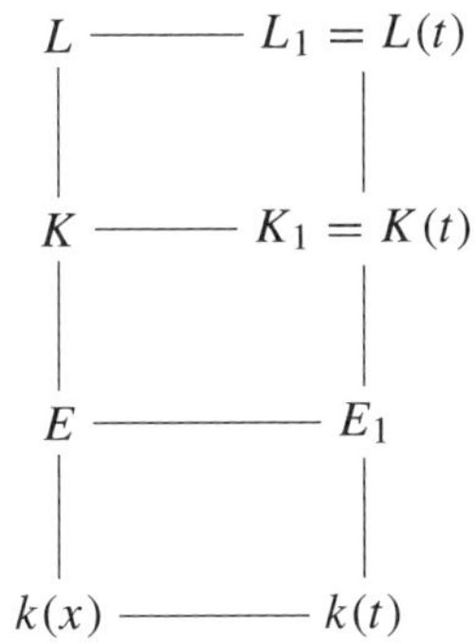

Let E_1 be any intermediate field such that $k(t) \subsetneqq E_1 \subseteqq K_1 = K(t)$. Let $E = E_1 \cap L$. Since k is algebraically closed, E_1/E is a cyclic extension of degree m with $[E : k(x)] = [E_1 : k(t)] \geq 2$. Since $\wp_0$ and $\wp_\infty$ decompose in $E/k(x)$, the prime divisors in E above $\wp_0$ and $\wp_\infty$ are totally ramified in E_1/E. It follows that $d_{E_1}(\mathfrak{D}_{E_1/E}) \geq 4(m-1)$. Using the genus formula we obtain

$$g_{E_1} = 1 + m(g_E - 1) + \frac{1}{2} d_{E_1}(\mathfrak{D}_{E_1/E}) \geq 1 - m + 2(m-1) = m - 1 \geq M.$$

Therefore $K_1/k(t)$ satisfies the conditions of the proposition. □

Remark 14.4.9. The field extension constructed in the proof of Proposition 14.4.8 is an extension $K_1/k(t)$, where t is not necessarily the same element x given by the field extension $K/k(x)$. In order that $K_1/k(x)$ satisfy the same conditions as in the proposition, first notice that the map

$$\begin{aligned} \psi : k(t) &\to k(x) \\ \text{defined by} \qquad \psi(f(t)) &= f(x) \end{aligned}$$

is a field isomorphism. Since $K_1/k(t)$ is an algebraic extension, ψ can be extended to a field monomorphism

$$\widetilde{\psi} : K_1 \to \overline{k(x)},$$

where $\overline{k(x)}$ is an algebraic closure of $k(x)$. Therefore if $K_2 = \widetilde{\psi}(K_1)$, then $K_2/k(x)$ satisfies the same properties as $K_1/k(t)$. Furthermore, since $[K_1 : k(t)] > 1$, there exists at least one prime divisor in $k(t)$ that is ramified in K_1 (Exercise 9.7.10). We may fix this prime in advance using the change of variables

$$t \mapsto \frac{at+b}{ct+d},$$

and therefore we may choose in advance a prime divisor of $k(x)$ that is ramified in $K_2/k(x)$ as well as a prime divisor that is not ramified in $K_2/k(x)$. This observation will be used in our next result.

Theorem 14.4.10 (Stichtenoth). *Let $K/k(x)$ be a finite separable extension for an algebraically closed field k such that $[K : k(x)] > 1$. Let F/k be any function field over k with $g_F \geq 2$. Then there exists a separable extension L/F such that*

$$[L : F] = [K : k(x)] \quad \text{and} \quad \operatorname{Aut}_k(L) = \operatorname{Aut}_F(L) \cong \operatorname{Aut}_{k(x)}(K).$$

Proof. Let $H := \operatorname{Aut}_k(F)$. By Theorem 14.3.7, H is finite. Let $n = |H|$ and let $T = F^H$ be the fixed field. Then F/T is a Galois extension and $\operatorname{Gal}(F/T) = H$. Let q be a rational prime number such that $q \neq \operatorname{char} k$, $q \geq 2g_F$, and $(q, n) = 1$. If $\mathfrak{B}$ is a prime divisor of F, then by Corollary 3.5.8 there exists $z \in F$ such that $\mathfrak{N}_z = \mathfrak{B}^q$. Thus we have:

(a) $F/k(z)$ is a separable extension of degree q (Theorem 3.2.7).
(b) If $\mathfrak{P} = \mathfrak{B} \cap k(z)$, then $\mathfrak{P}$ is the pole divisor of z in $k(z)$ and $\mathfrak{P}$ is fully ramified in $F/k(z)$.
(c) Since $[F : T] = n$, $[F : k(z)] = q$, and $q \nmid n$, then $k(z) \subsetneqq T(z) \subseteqq F$ and since q is prime, $T(z) = F$.

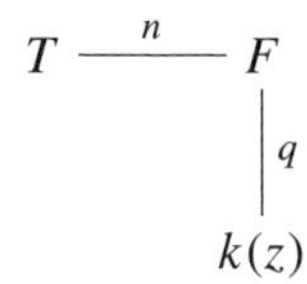

Now by Lemma 14.1.2, there exists a place $\mathfrak{Q}$ in T such that $\mathfrak{Q}$ decomposes fully in F/T, that is, there exist n different places $\mathfrak{B}_1, \ldots, \mathfrak{B}_n$ of F dividing $\mathfrak{Q}$, and the restrictors $\mathfrak{P}_i := \mathfrak{B}_i \cap k(z)$ are distinct places of $k(z)$.

Furthermore, we may choose $\mathfrak{P}_1$ to be unramified in $F/k(z)$.

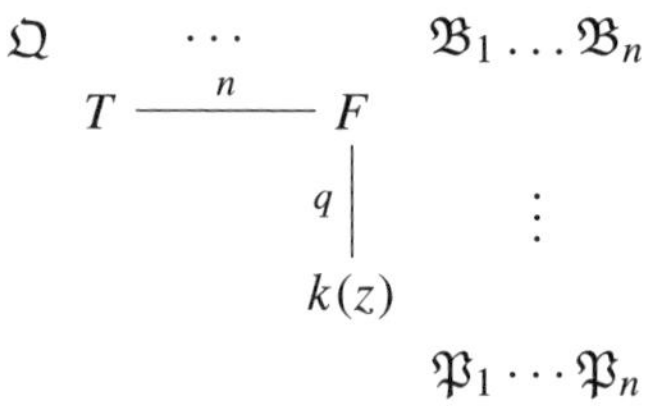

By Proposition 14.4.8 there exists a separable extension $K_1/k(z)$ such that $[K_1 : k(z)] = [K : k(x)]$, $\operatorname{Aut}_{k(z)} K_1 \cong \operatorname{Aut}_{k(x)} K$, and for any intermediate field M satisfying $k(z) \subsetneqq M \subseteqq K_1$, we have $g_M > 2g_F[K : k(x)] + ([K : k(x)] - 1)^2$. Since $K_1 \neq k(z)$ there exists at least one and only finitely many ramified primes in $K_1/k(z)$ (Exercise 9.7.10 and Theorem 5.2.33). By Remark 14.4.9 we may assume that $\mathfrak{P}_1 = \mathfrak{B}_1 \cap k(z)$ is ramified in $K_1/k(z)$ and $\mathfrak{P}_2, \ldots, \mathfrak{P}_n$ are unramified in $K_1/k(z)$. We may also assume that $\mathfrak{P} = \mathfrak{B} \cap k(z)$ is unramified in $K_1/k(z)$. In short, $\mathfrak{Q} = \mathfrak{B}_1 \cdots \mathfrak{B}_n$ is fully decomposed in F/T, $\mathfrak{P}_i = \mathfrak{B}_i \cap k(z)$, $\mathfrak{P} = \mathfrak{B} \cap k(z)$, $\mathfrak{P} = \mathfrak{B}^q$, $\mathfrak{N}_z = \mathfrak{B}^q$, $\mathfrak{P}$ not ramified in $K_1/k(z)$, $\mathfrak{P}_2, \ldots, \mathfrak{P}_n$ not ramified in $K_1/k(z)$, $\mathfrak{P}_1$ ramified in $K_1/k(z)$, $[K_1 : k(z)] = [K : k(x)]$, $\operatorname{Aut}_{k(z)} K_1 \cong \operatorname{Aut}_{k(x)} K$, and if $k(z) \subsetneqq M \subseteqq K_1$, $g_M > 2g_F[K : k(x)] + ([K : K(x)] - 1)^2$. Let L be the Galois closure of $K_1/k(z)$. Thus, by Exercise 5.10.13, $\mathfrak{P}$ is not ramified in $L/k(z)$ because $\mathfrak{P}$ is not ramified in $K_1/k(z)$. Since $\mathfrak{P}$ is fully ramified in $F/k(z)$ then F and L are linearly disjoint over $k(z)$.

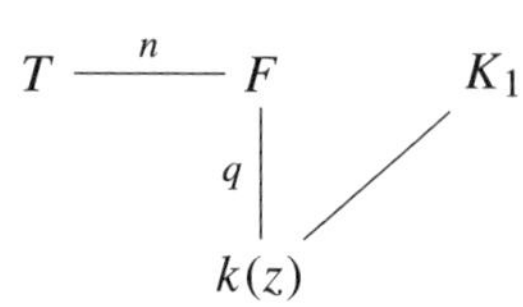

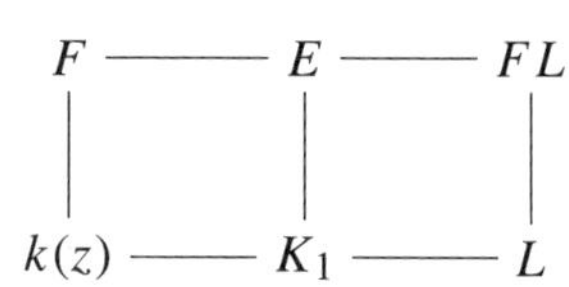

Using basic Galois theory we obtain $\operatorname{Gal}(FL/F) \cong \operatorname{Gal}(L/F \cap L) = \operatorname{Gal}(L/k(z))$. Let $E := FK_1$. Then E/F is a separable extension such that $[E : F] = [K_1 :$

$k(z)] = [K : k(x)]$ and $\mathrm{Aut}_F(E) \cong \mathrm{Aut}_{k(z)}(K_1) \cong \mathrm{Aut}_{k(x)}(K)$. Obviously, $\mathrm{Aut}_F(E) \subseteqq \mathrm{Aut}_k(E)$. Let $\sigma \in \mathrm{Aut}_k(E)$ and let M be an intermediate field such that $F \subsetneqq M \subseteqq E$. Let $M_1 = M \cap K_1$. Then $[M : F] = [M_1 : k(z)]$. It follows that

$$\begin{aligned} g_M \geq g_{M_1} &> 2[K : k(x)]g_F + ([K : k(x)] - 1)^2 \\ &\geq 2[M : F]g_F + ([M : F] - 1)^2. \end{aligned}$$

Using Theorem 14.4.5 we get $\sigma(F) = F$. Thus $\sigma_0 = \sigma|_F \in \mathrm{Aut}_k(F) = H$.

Since $\mathfrak{Q} = \mathfrak{B}_1 \cdots \mathfrak{B}_n$, where $n = |H|$, we have that the decomposition group $D(\mathfrak{B}_1|\mathfrak{Q})$ is the identity. Now $\sigma_0(\mathfrak{B}_1) = \mathfrak{B}_i$ for some i. Since $\mathfrak{P}_1$ is ramified in $K_1/k(z)$ and $\mathfrak{P}_1$ is not ramified in $F/k(z)$, it follows that $\mathfrak{B}_1$ is ramified in E/F. Therefore $\sigma_0(\mathfrak{B}_1) = \mathfrak{B}_i$ is also ramified in E/F. On the other hand, for $j \geq 2$, $\mathfrak{B}_j$ is not ramified in E/F since $\mathfrak{P}_j = \mathfrak{B}_j \cap k(z)$ is not ramified in $K_1/k(z)$ (Exercise 5.10.12). It follows that $\sigma_0(\mathfrak{B}_1) = \mathfrak{B}_1$ and $\sigma_0 = \mathrm{Id}_F$. Therefore $\sigma \in \mathrm{Aut}_F(E)$. This completes the proof of the theorem. □

Theorem 14.4.10 has several interesting consequences.

Theorem 14.4.11. *Let G be any nontrivial finite group, $|G| > 1$. If G is realizable over a rational function field, then G is exactly realizable over any function field K where $g_K \geq 2$ and k is an algebraically closed field.* □

Theorem 14.4.12. *For each function field K/k where k is algebraically closed, $g_K \geq 2$ and for each $n \in \mathbb{N}$, $n \geq 3$, there exists an extension L/K such that $[L : K] = n$ and* $\mathrm{Aut}_k(L) = \{\mathrm{Id}\}$.

Proof. Let $E/k(x)$ be the extension $E = k(x, y)$ where $y^{n-1}(y - 1) = x$. Then $E = k(y)$. The pole divisor of x in $k(x)$ has ramification index n, so $[E : k(x)] = n$. Since y satisfies

$$f(T) = T^n - T^{n-1} - x,$$

we have $f'(T) = nT^{n-1} - (n-1)T^{n-2} = T^{n-2}(nT - (n-1))$. If the characteristic of k divides n, the root of $f'(T)$ is 0, which is not a root of $f(T)$. If the characteristic of k does not divide n, the roots of $f'(T)$ are 0 and $\frac{n-1}{n}$. Therefore $E/k(x)$ is a separable extension.

Let $\sigma \in \mathrm{Aut}_{k(x)} k(y)$. We have $(x)_E = \frac{\mathfrak{P}_0^{n-1}\mathfrak{P}_1}{\mathfrak{P}_\infty^n}$. Since $\sigma x = x$ and $n - 1 \geq 2$, it follows that $\mathfrak{P}_0^\sigma = \mathfrak{P}_0$, $\mathfrak{P}_1^\sigma = \mathfrak{P}_1$ and $\mathfrak{P}_\infty^\sigma = \mathfrak{P}_\infty$. Using Theorem 14.4.1 or Exercise 5.10.14 we conclude that $\sigma = \mathrm{Id}$. Therefore $\mathrm{Aut}_{k(x)} E \cong \{\mathrm{Id}\}$. The result follows by Theorem 14.4.10. □

Lemma 14.4.13. *Let $n \geq 2$ and let G be the transitive subgroup of S_n generated by transpositions. Then $G = S_n$.*

Proof. It suffices to show that at least one transposition belongs to G. We will show that $(1, 2) \in G$. Since G is transitive, there exists $\sigma \in G$ such that $\sigma(1) = 2$. Choose

$\sigma \in G$ such that $\sigma(1) = 2$ and t is minimum, where $\sigma = \varepsilon_1 \cdots \varepsilon_t$, $\varepsilon_i \in G$, and ε_i a transposition. If $t = 1$, then $\sigma = (1, 2)$ and we are done. If $t > 1$, then we have $\varepsilon_1 = (1, a_1)$ and $a_1 \neq 2$; indeed, assume otherwise. Then $\varepsilon_1 = (x, y)$ with $x \neq 1 \neq y$. Then since $\sigma(1) \neq 1$, there exists $\varepsilon_i = (1, x)$ with $i > 1$. Therefore if $\sigma' = \varepsilon_1\sigma$, we have $\sigma'(1) = 2$ (because $\varepsilon_1(1) = 1$) and σ' is a product of $t - 1$ transpositions of G. This contradicts the minimality of t. Hence $\varepsilon_1 = (1, a_1)$. If $\varepsilon_2 = (x, y)$, then $a_1 \in \{x, y\}$ since otherwise, $\sigma' = \varepsilon_1\varepsilon_3 \cdots \varepsilon_t$ satisfies $\sigma'(1) = 2$ (because $\varepsilon_2(a_1) = a_1$) and σ' is a product of $t - 1$ transpositions of G. Therefore $\varepsilon_2 = (a_1, a_2)$. In this way we obtain $\sigma = (1, a_1)(a_1, a_2)\varepsilon_3 \cdots \varepsilon_t$. Finally, $(1, a_1)(a_1, a_2) = (1, a_2, a_1) \in G$, $(1, a_1)(1, a_2, a_1)^2 = (1, a_1)(1, a_1, a_2) = (1, a_2) \in G$ and $\sigma' = (1, a_2)\varepsilon_3 \cdots \varepsilon_t$ satisfies $\sigma'(1) = 2$. This shows that $t = 1$ and $(1, 2) \in G$. □

Now we will prove that if $\operatorname{char} k = p \geq 0$ and $p \nmid n(n-1)$ (n arbitrary for $p = 0$) then the equation $h(T) = T^n + T - x \in k(x)[T]$ has Galois group S_n. For this purpose we first prove two lemmas.

Lemma 14.4.14. *Let K be the splitting field of $h(T)$ over $k(x)$ and let $\wp_\infty$ be the pole divisor of x in $k(x)$. Then $\wp_\infty$ ramifies in $K/k(x)$. Further, if $p \nmid n$, (n arbitrary for $p = 0$) then the ramification index of any prime $\mathfrak{P}$ of K dividing $\wp_\infty$ is n.*

Proof. First note that $h(T)$ is separable since the roots of $h'(T) = nT^{n-1} + 1$ belong to k (if $p \mid n$, $h'(T) = 1$; if $p \nmid n$, the roots of $h'(T)$ are $\sqrt[n-1]{-1/n} \in k$). On the other hand, if y is any root of $h(T)$,

$$v_{\mathfrak{P}}(y^n + y) = v_{\mathfrak{P}}(x) = e(\mathfrak{P}|\wp_\infty)v_{\wp_\infty}(x) = -e(\mathfrak{P}|\wp_\infty) \neq 0.$$

Hence $v_{\mathfrak{P}}(y^n+y) = nv_{\mathfrak{P}}(y) = -e(\mathfrak{P} \mid \wp_\infty)$. Therefore $v_{\mathfrak{P}}(y) = -1, n = e(\mathfrak{P}|\wp_\infty)$, and $\wp_\infty$ is fully ramified in $k(y)/k(x)$, so $y \notin k$ (unless $n = 1$). Thus $K/k(x)$ is a Galois extension. If $p \nmid n$, then $\wp_\infty$ is tamely ramified in $k(y)/k(x)$ (y any root of $h(T)$). Now $K = \prod_{i=1}^n k(y_i)$, where $y_1, \ldots, y_n$ are the roots of $h(T)$. By Abhyankar's lemma (Theorem 12.4.4) we have

$$e(\mathfrak{P}|\wp_\infty) = \left[e(\mathfrak{q}_i|\wp_\infty) \mid 1 \leq i \leq n\right] = n,$$

where $\mathfrak{q}_i = \mathfrak{P} \cap k(y_i)$. □

Lemma 14.4.15. *If $p \nmid n(n-1)$, then for any prime divisor $\wp \neq \wp_\infty$ of $k(x)$ ramified in $K/k(x)$, the decomposition group $D = D(\mathfrak{P}|\wp)$ of any prime $\mathfrak{P}$ of K dividing $\wp$ is cyclic of order* 2. *Moreover, if $\sigma \in \operatorname{Gal}(K/k(x))$ generates D, then the permutation induced by σ on the roots of $h(T)$ is a transposition.*

Proof. Let $x - \beta$ be a prime element for $\wp$ in $k(x)$. Consider $h(T) \bmod \wp \in \left(k[x]_\wp/\wp k[x]_\wp\right)[T] \cong k[T]$.

We have $h(T) \bmod \wp = T^n + T - \beta$. If $\wp$ is ramified in $K/k(x)$, then $\wp$ is ramified in the completions $K_{\mathfrak{P}}/k(x)_\wp$ (Theorem 5.6.3 and Propositions 5.6.7 and 5.6.9). On the other hand, by Theorems 5.7.18 and 5.8.1, $\wp$ is ramified if and only if $h(T) \in k(x)_\wp[T]$ has a multiple root. By Hensel's lemma (Theorem 2.3.14), it follows that $\wp$

is ramified only when $h(T)$ mod $\wp$ has multiple roots. Finally, if $\overline{h(T)} = h(T)$ mod $\wp = T^n + T - \beta$ has a multiple root α, then

$$\overline{h(\alpha)} = \alpha^n + \alpha - \beta = 0 \quad \text{and} \quad \overline{h'(\alpha)} = n\alpha^{n-1} + 1 = 0.$$

Thus $\alpha \neq 0$ and $\alpha^n = -\frac{\alpha}{n} = -\alpha + \beta$, and $\left(1 - \frac{1}{n}\right)\alpha = \beta$, that is, $\alpha = \frac{n\beta}{n-1}$. Since $\overline{h''(\alpha)} = n(n-1)\alpha^{n-2} \neq 0$, α is of multiplicity two and it is the only multiple root of $\overline{h(T)}$. Hence

$$h(T) = (T - z)^2 \prod_{i=1}^{n-2}(T - z_i) \in k(x)_\wp[T]$$

with $z_i \neq z_j, i \neq j$, and $z_i \neq z, i = 1, \ldots, n-2$. Since $D = \operatorname{Gal}(K_{\mathfrak{P}}/k(x)_\wp)$, we have $|D| = \left[K_{\mathfrak{P}} : k(x)_\wp\right] = 2$ and if $\sigma \in D$ and $\sigma \neq \operatorname{Id}$, then σ fixes $z_1, \ldots, z_{n-2}$ and thus σ is a transposition. □

Theorem 14.4.16 (Hayes). *Let $h(T) = T^n + T - x \in k(x)[T]$. If K is the splitting field of $h(T)$ over $k(x)$, and if $p = \operatorname{char} k$ and $p \nmid n(n-1)$ (n arbitrary for $p = 0$), then $K/k(x)$ is a Galois extension and $\operatorname{Gal}(K/k(x))$ is isomorphic to the symmetric group S_n on n elements.*

Proof. By Lemma 14.4.14, $h(T)$ is separable and thus $K/k(x)$ is a Galois extension. Let $G = \operatorname{Gal}(K/k(x))$. Let H be the subgroup of G generated by the decomposition groups of all ramified prime divisors $\mathfrak{P}$ of K that do not divide the pole divisor $\wp_\infty$ of x in $k(x)$. Set $F = K^H$. In $F/k(x)$, the only prime divisor $k(x)$ that can ramify is $\wp_\infty$.

If $\mathfrak{B}_1, \ldots, \mathfrak{B}_r$ are the prime divisors of F dividing $\wp_\infty$, it follows by Lemma 14.4.14 that $\wp_\infty$ is tamely ramified in $F/k(x)$. Thus $\mathfrak{D}_{F/k(x)} = \mathfrak{B}_1^{e_1-1} \cdots \mathfrak{B}_r^{e_r-1}$, where e_i is the ramification index $e(\mathfrak{B}_i|\wp_\infty)$. Therefore

$$d = d_K(\mathfrak{D}_{F/k(x)}) = \sum_{i=1}^{r}(e_i - 1) = [F : k(x)] - r.$$

By the Riemann–Hurwitz formula we have

$$2g_F - 2 = -2[F : k(x)] + d = -[F : k(x)] - r.$$

Hence $r + [F : k(x)] \leq 2$ and $F = k(x)$. Thus $H = G$. By Lemma 14.4.15, G is generated by transpositions. Since G is a transitive subgroup of S_n, using Lemma 14.4.13 we obtain $G = S_n$. □

Theorem 14.4.17 (Madden–Valentini, Stichtenoth). *Let G be any finite group and let k be an algebraically closed field. There exists a separable extension $K/k(x)$ such that $\operatorname{Aut}_{k(x)}(K) \cong G$.*

Proof. Choose $n \in \mathbb{N}$ such that $p \nmid n(n-1)$ (we assume $p \neq 2$), and $n \geq |G|$. Then $G < S_n$. Let $L/k(x)$ be such that $\mathrm{Gal}(L/k(x)) \cong S_n$. Let $E = L^G$. Then $\mathrm{Gal}(L/E) \cong G$. Choose a prime number q such that $2 < q \neq \operatorname{char} k$ and $q \geq 2g_E$ and choose two places $\mathfrak{P}_1, \mathfrak{P}_2$ in E that are unramified in L/E. Let $\mathfrak{A} = \mathfrak{P}_1\mathfrak{P}_2^{q-1}$. Since $d_E(\mathfrak{A}\mathfrak{P}_1^{-1}) = d_E(\mathfrak{A}\mathfrak{P}_2^{-1}) = q-1 \geq 2g_E - 1$ and $d_E(\mathfrak{A}) \geq 2g_E$, it follows by the Riemann–Roch theorem that

$$\ell_E(\mathfrak{A}^{-1}) = q - g_E + 1 = 1 + \ell_E(\mathfrak{A}^{-1}\mathfrak{P}_1) = 1 + \ell_E(\mathfrak{A}^{-1}\mathfrak{P}_2).$$

Therefore $L_E(\mathfrak{A}^{-1}\mathfrak{P}_2^{-1})$ and $L_E(\mathfrak{A}^{-1}\mathfrak{P}_1)$ are proper subspaces of $L_E(\mathfrak{A}^{-1})$ and since k is infinite, there exists

$$x \in L_E(\mathfrak{A}^{-1}) \backslash (L_E(\mathfrak{A}^{-1}\mathfrak{P}_1) \cup L_E(\mathfrak{A}^{-1}\mathfrak{P}_2)).$$

It follows that $\mathfrak{N}_x = \mathfrak{P}_1\mathfrak{P}_2^{q-1}$. Then $[E : k(x)] = d(\mathfrak{N}_x) = q$. Since $q \neq \operatorname{char} k$, $E/k(x)$ is a separable extension of degree q.

We have $G = \mathrm{Gal}(L/E) = \mathrm{Aut}_E(L) < \mathrm{Aut}_{k(x)}(L)$. Let $\sigma \in \mathrm{Aut}_{k(x)}(L)$, and consider $T = L^{\langle\sigma\rangle} \supseteq k(x)$. Then $k(x) \subseteq T \cap E \subseteq E$. Since $[E : k(x)] = q$ is prime and $[T \cap E : k(x)]$ divides q, it follows that $[T \cap E : k(x)]$ is 1 or q and $T \cap E = k(x)$ or $T \cap E = E$. Assume that $T \cap E = k(x)$. Since L/E and L/T are normal extensions, $L/k(x)$ is a normal extension too.

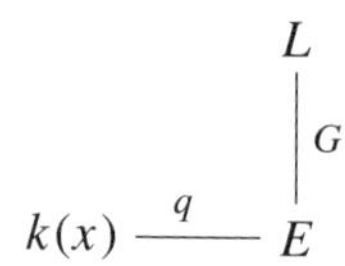

This is impossible since $(x)_L = \mathrm{con}_{E/L}((x)_E) = \mathrm{con}_{E/L}\left(\frac{\mathfrak{A}}{\mathfrak{P}_1\mathfrak{P}_2^{q-1}}\right)$. Now $\mathfrak{P}_1$ and $\mathfrak{P}_2$ are unramified in L/E, so we have

$$\mathrm{con}_{E/L}\, \mathfrak{P}_1\mathfrak{P}_2^{q-1} = (\mathfrak{Q}_1 \cdots \mathfrak{Q}_h)(\mathfrak{Q}_1' \cdots \mathfrak{Q}_h')^{q-1}.$$

This contradicts Proposition 5.2.16. It follows that $T \cap E = E$ and $E \subseteq T$. Therefore $\sigma \in \mathrm{Aut}_E(L)$, and

$$\mathrm{Aut}_{k(x)}(L) = \mathrm{Aut}_E(L) = \mathrm{Gal}(L/E) \cong G.$$

This proves the theorem for $p \neq 2$. For $p = 2$ we consider the equation $h(T) = T^n + T^2 - x \in k(x)[T]$, where $n \geq 3$ and $2 \nmid n$. Then $h(T)$ is separable and since $2 \nmid n$, Lemma 14.4.14 holds for $h(T)$. Also, the conclusion of Lemma 14.4.15 holds since the only possible multiple root of $h(T) \bmod \wp = T^n + T^2 - \beta$ holds when $\beta = 0$ and this is a root of multiplicity 2. Therefore the Galois group of the splitting field of $h(T)$ is S_n. The rest of the proof is the same as in the case $p \neq 2$. □

Remark 14.4.18. One of the key points in the proof of Theorem 14.4.17 is the fact that for infinitely many $n \in \mathbb{N}$, S_n is the Galois group of some extension $K/k(x)$. In fact, for every $n \in \mathbb{N}$ there exists a Galois extension $K/k(x)$ such that $\mathrm{Gal}(K/k(x) \cong S_n$.

Proposition 14.4.19 (Stichtenoth). *Let k be any algebraically closed field and let $n \in \mathbb{N}$. Let $K = k(x, y)$ be given by*

$$y^2 \prod_{i=1}^{n-2} (y - a_i) - x(y - a_{n-1})^{n-1}(y - a_n) = 0,$$

where $a_1, \ldots, a_n$ are n distinct elements of $k \setminus \{0\}$. If $\widetilde{K}$ is the Galois closure of $K/k(x)$, then $\mathrm{Gal}(\widetilde{K}/k(x)) \cong S_n$.

Proof. See Exercise 14.5.8. □

14.5 Exercises

Exercise 14.5.1. Let $\mathfrak{B}$ be an integral divisor, $\mathfrak{A}$ any divisor, and $n \geq 0$ an integer such that $\mathfrak{B}^{-1}$ divides $\mathfrak{A}^n$. Prove that $\mathfrak{B}^{-1}$ divides $\mathfrak{A}^m$ for all m satisfying $0 \leq m \leq n$.

Exercise 14.5.2. If k is an arbitrary field, prove that for a prime divisor $\mathfrak{p}$, n is a gap of $\mathfrak{p} \iff \delta(\mathfrak{p}^{n-1}) - \delta(\mathfrak{p}^n) = f$, where $f = d_K(\mathfrak{p})$.

Exercise 14.5.3. Prove that $\mathrm{Aut}_k(K) = \{\mathrm{Id}\}$ where K/k is the function field given in Example 5.2.31.

Exercise 14.5.4. Give an example where $|\mathrm{Aut}_k(K)| < \infty$, but $\left|\mathrm{Aut}_{\bar{k}}(\bar{K})\right| = \infty$, where $\bar{k}$ is an algebraic closure of k and $\bar{K} = K\bar{k}$.

Exercise 14.5.5. Prove that if F is a field such that the only derivative on F is the 0-derivative, then F is an algebraic extension of $\mathbb{Q}$ or $\mathbb{F}_p$. Here a derivative on a ring R is a function $D \colon R \to R$ such that

$$D(x + y) = D(x) + D(y) \quad \text{and} \quad D(xy) = D(x)y + xD(y)$$

for all $x, y \in R$.

Exercise 14.5.6. Let $K = k(x)$ for some arbitrary field k, and let $\sigma \in \mathrm{Aut}_k(K)$ be such that $\sigma \neq \mathrm{Id}$ and $o(\sigma) < \infty$. Prove that if $E = k(x)^{\langle \sigma \rangle}$, then there are at most two distinct divisors in E that are ramified in K.

Exercise 14.5.7. Let K/k be an algebraic function field for some algebraically closed field k. If the genus of $g_K = g$ of K is nonzero, prove that any $\sigma \in \mathrm{Aut}_k(K) \setminus \{\mathrm{Id}\}$ has $2g + 2$ fixed points if and only if $p = \operatorname{char} k \neq 2$, $o(\sigma) = 2$, and $K^{\langle \sigma \rangle}$ is a rational function field. In particular, K/k is an elliptic or a hyperelliptic function field.

Exercise 14.5.8. Let k be an algebraically closed field. Let $\tilde{K}$ be the normal closure of $K/k(x)$, where $K = k(x, y) = k(y)$ is given by

$$y^2 \prod_{i=1}^{n-2} (y - a_i) - x(y - a_{n-1})^{n-1}(y - a_n)$$

and $a_1, \ldots, a_n \in k$ are distinct elements of $k \setminus \{0\}$. Prove the following statements:

(i) $K/k(x)$ is separable, that is,

$$f(T) = T^2 \prod_{i=1}^{n-2} (T - a_i) - x(T - a_{n-1})^{n-1}(T - a_n),$$

where $f(T) \in k(x)[T]$ is a separable polynomial. Thus $\tilde{K}/k(x)$ is a Galois extension.

(ii) Set $(x)_{k(x)} = \frac{\mathfrak{p}_0}{\mathfrak{p}_\infty}$. If $\mathfrak{P}_0$ is a prime divisor of $\tilde{K}$ above $\mathfrak{p}_0$, then the decomposition group $D(\mathfrak{P}_0|\mathfrak{p}_0)$ is a transposition in $G = \mathrm{Gal}(\tilde{K}/k(x)) < S_n$.
Hint: See the proof of Lemma 14.4.15.

(iii) Let $H < G$ be such that $K = \tilde{K}^H$, that is, H is a stabilizer in G. Then H is a maximal subgroup of G. Equivalently, there is no field F such that $k(x) \subsetneqq F \subsetneqq K$. This is the same as saying that G is a *primitive* subgroup of S_n.

(iv) Any primitive subgroup of S_n that contains a transposition is S_n itself.
Hint: Consider a subgroup G of S_n that contains $(1, 2)$ and let $H = \mathrm{Stab}_G(1) = \{\sigma \in G \mid \sigma(1) = 1\}$. If there exists $r \geq 2$ such that $2^h = h(2) \neq r$ for all $h \in H$, that is, H is not transitive on $\{2, \ldots, n\}$, put $M = \langle \{(1, 2^h) \mid h \in H \rangle$. Prove that MH is a subgroup of G satisfying $H \subsetneqq MH \subsetneqq G$ by showing that there is no $\psi \in MH$ such that $\psi(1) = r$.

(v) Conclude that $\mathrm{Gal}(K/k(x))$ is S_n.

Exercise 14.5.9. Let k be an algebraically closed field of characteristic $p > 0$. Let L/K be a finite separable extension of function fields over k. Let $\mathfrak{P}$ be a prime divisor of L that is either unramified or tamely ramified over K. Set $\mathfrak{p} := \mathfrak{P} \cap K$. If λ is a gap number of $\mathfrak{p}$, then prove that $j\lambda$ is a gap number of $\mathfrak{P}$ for any positive integer j dividing the ramification index e of $\mathfrak{P}$ in L/K.

Exercise 14.5.10. Let k be an algebraically closed field of characteristic $p > 0$. Let L/K be a finite separable extension, $\mathfrak{P}$ a prime divisor in L that is unramified over K, and $\mathfrak{p} := \mathfrak{P} \cap K$. Let $\mathfrak{A}$ be a divisor of K such that $\mathfrak{D}_{L/K} \operatorname{con}_{K/L} \mathfrak{A}$ is an integral divisor of L that is relatively prime to $\mathfrak{P}$. If a positive integer λ satisfies

$$\delta\left(\mathfrak{p}^{\lambda-1}\mathfrak{A}\right) - \delta\left(\mathfrak{p}^{\lambda}\mathfrak{A}\right) = 1,$$

prove that λ is a gap number of $\mathfrak{P}$.

Exercise 14.5.11. Let k be an algebraically closed field of characteristic $p > 0$. Let $K/k(x)$ be a cyclic extension of degree m with $(m, p) = 1$. Show that if at least $m+3$ prime divisors of K are fully ramified, then every fully ramified prime is a Weierstrass point.

Hint: We have $K = k(x, y)$, $y^m = \prod_{i=1}^{s}(x - a_i)^{\lambda_i}$ for $0 < \lambda_i < m$ and $\lambda_1, \ldots, \lambda_{m+3}$ are relatively prime to m (see Example 5.8.9). Prove that m is not a gap number of $\mathfrak{P}_i$ where $(x - a_i)_{k(x)} = \frac{\mathfrak{p}_i}{\mathfrak{p}_\infty}$, $\operatorname{con}_{k(x)/K} \mathfrak{p}_i = \mathfrak{P}_i$, and $\mathfrak{p}_i = \mathfrak{P}_i^m$. Show that $v_{\mathfrak{P}_i}(\omega) = m$ where $\omega = (x - a_i)\prod_{j=1}^{s}(x - a_j)^{b_j} y^{-a} dx$, $\frac{a\lambda_j}{(m,\lambda_j)} = b_j e_j + c_j$ for $0 \leq c_j < e_j$, and $0 < a < m$ is such that $(a, m) = 1$ and $a\lambda_i \equiv m - 1 \bmod m$.

Conclude that the gap sequence of $\mathfrak{P}_i$ does not satisfy the condition of Theorem 14.2.40.

Exercise 14.5.12. Let k be an algebraically closed field of characteristic 2 and let $K/k(x)$ be a cyclic extension of degree 2 with $g_K \geq 2$. Prove that K is classical, that is, the gap sequence of K is $\{1, 2, \ldots, g_K\}$, and prove that the Weierstrass points of K are precisely the ramified prime divisors of K over $k(x)$.

Hint: Use Theorem 14.1.3 or Corollary 14.1.4.

Exercise 14.5.13. Let L/K be an extension of function fields of degree n over an algebraically closed field of constants k. If $g_L > n^2 g_K + (n-1)^2$, prove that the fully ramified prime divisors are Weierstrass points of L.

Exercise 14.5.14. Let L/K be as in Exercise 14.5.13. Let r be the number of fully ramified prime divisors of L/K and assume that n is relatively prime to the characteristic. Prove that if $r > 2n(g_K + 1)$, then the fully ramified prime divisors are Weierstrass points.

Exercise 14.5.15. Let L/K, r, and n be as in Exercise 14.5.14. If $r > 4$ and L is classical, prove that the fully ramified prime divisors are Weierstrass points.

Exercise 14.5.16. Let k be any nonperfect field of characteristic $p > 0$, and let $a \in k$ be such that $a \notin k^p$. Let $K = k(x, y)$ be the function field defined by

$$y^p - y = ax^p.$$

Prove that $g_K = \frac{(p-1)(p-2)}{2}$.

Hint: Set $k' := k\left(a^{1/p}\right)$. Then $K' := Kk' = k'\left(y - a^{1/p}x\right)$ is a rational function field. Since $x^p = \frac{y^p - y}{a}$ belongs to both $k(y)$ and $k'(y)$, it follows that $K/k(y)$ and $K'/k'(y)$ are purely inseparable extensions of degree p. Using the Tate genus formula for $K'/k'(y)$ show that for any place $\mathfrak{p}'$ of $k'(y)$ distinct from the infinite prime divisor $\mathfrak{p}'_\infty$ of $k'(y)$, we have

$$r_{\mathfrak{p}'_\infty} = -1 \quad \text{and} \quad r_{\mathfrak{p}'} = 0 \text{ or } 1.$$

Deduce that $r_\mathfrak{p}$ is 0 or 1 whenever $\mathfrak{p}$ is a place of $k(y)$ that is distinct from the infinite prime $\mathfrak{p}_\infty$ of $k(y)$. Finally, calculate $r_{\mathfrak{p}_\infty}$.

Exercise 14.5.17. Assume that $\operatorname{char} k = p > 0$. Let $a \in k$ be such that $a^{1/p} \notin k$, and $K = k(x, y)$ with $y^2 = x^p - a$. Prove that $g_K = \frac{p-1}{2}$ and $\operatorname{Aut}_k(K) = \operatorname{Aut}_{k(x)}(K) = \{1, \sigma\}$ with $\sigma(y) = -y$. Conclude that $Kk\left(a^{1/p}\right)$ is a rational function field.

A

Cohomology of Groups

In this appendix we present a brief introduction to the cohomology of groups. This topic is independent of the rest of the material contained in the book. The reason why we decided to include it is that in order to continue the study of arithmetic properties of function fields, it is absolutely necessary to master group cohomology as a tool.

In this spirit, Theorem A.3.6 is especially useful. Also, notice that Hilbert's Theorem 90 (Theorem A.2.16) was used in Chapter 5 for the study of Kummer and Artin–Schreier extensions.

A.1 Definitions and Basic Results

For the results and definitions on modules and rings that we will be using in this chapter, we refer to [4] and [9].

Definition A.1.1. For a group G we define the *integral group ring* as

$$\mathbb{Z}[G] = \Big\{ \sum_{\sigma\in G} a_\sigma \sigma \mid a_\sigma \in \mathbb{Z} \text{ y } a_\sigma = 0 \text{ for all but a finite number of } \sigma \Big\},$$

with the operations

$$\left(\sum_{\sigma\in G} a_\sigma\sigma\right) + \left(\sum_{\sigma\in G} b_\sigma\sigma\right) = \sum_{\sigma\in G} (a_\sigma + b_\sigma)\,\sigma,$$

and

$$\left(\sum_{\sigma\in G} a_\sigma\sigma\right)\left(\sum_{\sigma\in G} b_\sigma\sigma\right) = \sum_{\sigma\in G}\left(\sum_{\theta\psi=\sigma} a_\theta b_\psi\right)\sigma.$$

Proposition A.1.2. *For any group G, $\mathbb{Z}[G]$ is a ring with unity, where the 1 corresponds to $\sum_{\sigma\in G} a_\sigma\sigma$ with $a_{\mathrm{Id}} = 1$ and $a_\sigma = 0$ for all $\sigma \neq \mathrm{Id}$. Furthermore, $\mathbb{Z}[G]$ is commutative if and only if G is abelian.*

Proof. We leave the proof to the reader (see Exercise A.5.1). □

Definition A.1.3. Let A be an abelian group written additively and let G be an arbitrary group. We say that A is a *(left) G-module* if there exists a group homomorphism $\varphi : G \longrightarrow \operatorname{Aut} A$, where $\operatorname{Aut} A$ is the automorphism group of A.

This definition is equivalent to the existence of a function

$$\psi : G \times A \longrightarrow A, \quad \text{denoted by} \quad \psi(g, a) = ga,$$

such that

(i) $1a = a$ for all $a \in A$,
(ii) $(gh)a = g(ha)$ for all $g, h \in G$ and $a \in A$,
(iii) $g(a + b) = ga + gb$ for all $g \in G$ and $a, b \in A$.

A similar definition is made for a right G-module.

Observe that if A is a G-module, then A is a $\Lambda = \mathbb{Z}[G]$-module in a natural way, that is,

$$\left(\sum_{\sigma \in G} a_\sigma \sigma\right)(x) = \sum_{\sigma \in G} a_\sigma (\sigma x) \quad \text{for} \quad \sum_{\sigma \in G} a_\sigma \sigma \in \Lambda \text{ and } x \in A.$$

Conversely, if A is a $\mathbb{Z}[G]$-module, then A is an abelian group and we consider the function

$$\varphi : G \longrightarrow \operatorname{Aut} A$$

given by

$$\theta \in G, \quad \varphi(\theta) : A \longrightarrow A, \quad \varphi(\theta)\, a = \theta a,$$

where θ is viewed as the element $\sum_{\sigma \in G} a_\sigma \sigma$ of $\mathbb{Z}[G]$ given by

$$a_\sigma = \begin{cases} 0 \text{ if } \sigma \neq \theta, \\ 1 \text{ if } \sigma = \theta. \end{cases}$$

It is easy to see that $\varphi(\theta) \in \operatorname{Aut} A$ and that φ is a group homomorphism.

Therefore a left (right) G-module is the same as a left (right) $\mathbb{Z}[G]$-module.

Example A.1.4. If A is any abelian group, we can give a G-module structure to A by defining the trivial action; that is, $ga = a$ for all $a \in A$ and all $g \in G$. In this case we say that G acts *trivially* on A or that A is a *trivial G-module*. The fact that A is a trivial G-module is equivalent to the fact that

$$\varphi\colon G \longrightarrow \operatorname{Aut} A \quad \text{satisfies} \quad \varphi(G) = 1.$$

Definition A.1.5. If A and B are G-modules, a *G-homomorphism* is a group homomorphism

$$\varphi\colon A \longrightarrow B \quad \text{such that} \quad \varphi(ga) = g\varphi(a) \quad \text{for all} \quad g \in G \quad \text{and} \quad a \in A.$$

Notation A.1.6. For two G-modules A and B we define

$$\begin{aligned} \operatorname{Hom}(A, B) &= \text{ group of all group homomorphisms from } A \text{ to } B, \\ \operatorname{Hom}_G(A, B) &= \text{ group of } G\text{-homomorphisms from } A \text{ to } B. \end{aligned}$$

$\operatorname{Hom}_G(A, B)$ will be considered only with its group structure.

The proof of the following proposition is easy.

Proposition A.1.7. $\operatorname{Hom}(A, B)$ *can be given a G-module structure as follows: for all $\varphi \in \operatorname{Hom}(A, B)$ and all $g \in G$, let $g \circ \varphi \in \operatorname{Hom}(A, B)$ be defined by*

$$(g \circ \varphi)(a) = g\varphi\left(g^{-1}a\right). \qquad \square$$

Definition A.1.8. If A is a G-module, A^G denotes the maximum G-trivial submodule of A, i.e., $A^G = \{a \in A \mid g \circ a = a \text{ for all } g \in G\}$.

Example A.1.9. If L/K is a finite Galois extension of fields with Galois group, L is a G-module and $L^G = K$.

Proposition A.1.10. *We have* $\operatorname{Hom}_G(A, B) = (\operatorname{Hom}(A, B))^G$. *In particular,* $\operatorname{Hom}_G(\mathbb{Z}, A) = (\operatorname{Hom}(\mathbb{Z}, A))^G \cong A^G$.

Proof. If $\varphi \in \operatorname{Hom}_G(A, B)$, then $\varphi \in \operatorname{Hom}(A, B)$. Now if $g \in G$,

$$(g \circ \varphi)(a) = g \circ \varphi\left(g^{-1}a\right) = gg^{-1}\varphi(a) = \varphi(a).$$

Therefore $g \circ \varphi = \varphi$ for all $g \in G$. Hence $\varphi \in (\operatorname{Hom}(A, B))^G$.

Conversely, if $\varphi \in (\operatorname{Hom}(A, B))^G$, let $a \in A$ and $g \in G$; we have

$$\varphi(ga) = (g \circ \varphi)(ga) = g\varphi\left(g^{-1}(ga)\right) = g\varphi(1a) = g\varphi(a).$$

Thus $\varphi \in \operatorname{Hom}_G(A, B)$ and this proves the first part of the proposition.

The last part of the proposition follows from the fact that $\operatorname{Hom}(\mathbb{Z}, A)$ is isomorphic to A under the G-isomorphism of modules

$$\theta : \operatorname{Hom}(\mathbb{Z}, A) \longrightarrow A \quad \text{defined by} \quad \theta(\varphi) = \varphi(1). \qquad \square$$

Theorem A.1.11. *Let $0 \longrightarrow A \stackrel{f}{\longrightarrow} B \stackrel{g}{\longrightarrow} C \longrightarrow 0$ be an exact sequence of G-modules and let P be a projective G-module. Then*

$$0 \longrightarrow \operatorname{Hom}_G(P, A) \stackrel{f^*}{\longrightarrow} \operatorname{Hom}_G(P, B) \stackrel{g^*}{\longrightarrow} \operatorname{Hom}_G(P, C) \longrightarrow 0$$

is an exact sequence of groups, where $f^*(\varphi) = f \circ \varphi$, $g^*(\theta) = g \circ \theta$.

Proof. If $f^*(\varphi) = f \circ \varphi = 0$, then since f is injective we have $\varphi = 0$, so f^* is injective.

Next, we have $g^* \circ f^* = (g \circ f)^* = 0^* = 0$, and hence $\operatorname{im} f^* \subseteq \ker g^*$.

Now if $\varphi \in \ker g^*$, then $g^*(\varphi) = g \circ \varphi = 0$ (see diagram).

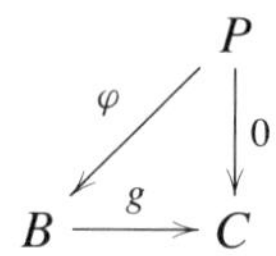

It follows that $\varphi(P) \subseteq \ker g = \operatorname{im} f = A$, so

$$f^{-1} \circ \varphi \in \operatorname{Hom}_G(P, A) \quad \text{and} \quad f^*\left(f^{-1} \circ \varphi\right) = f \circ f^{-1} \circ \varphi = \varphi,$$

that is, $\operatorname{im} f^* = \ker g^*$.

Finally, if $\varphi \in \operatorname{Hom}_G(P, C)$, then since the module P is projective, there exists $\theta \in \operatorname{Hom}_G(P, B)$ such that $g \circ \theta = g^*(\theta) = \varphi$. Therefore g^* is surjective. □

Note A.1.12. If P is an arbitrary G-module and

$$0 \longrightarrow A \longrightarrow B \longrightarrow C$$

is an exact G-sequence, then

$$0 \longrightarrow \operatorname{Hom}_G(P, A) \longrightarrow \operatorname{Hom}_G(P, B) \longrightarrow \operatorname{Hom}_G(P, C)$$

is exact, as follows immediately from the previous proof. In fact, the projectivity of P is equivalent to the exactness of the sequence in Theorem A.1.11 (see [9, Chapter II, Proposition 4, page 231]).

Note A.1.13. For the definition and basic properties of tensor products, we refer to [4, Chapter 2], [9, Chapter 2, §3] and [11, Chapter III, §0].

For any ring R, M a right R-module, and N a left R-module, the tensor product of M and N will be denoted by $M \otimes_R N$. The tensor product is obtained as the quotient of $M \otimes_{\mathbb{Z}} N$ obtained by the relations $mr \otimes_{\mathbb{Z}} n = m \otimes_{\mathbb{Z}} rn$, $m \in M, n \in N, r \in R$. That is, $mr \otimes_{\mathbb{Z}} n = m \otimes_{\mathbb{Z}} rn$ for all $m \in M, n \in N, r \in R$.

In case $R = \mathbb{Z}[G]$, the right module can be made a left module by setting $gm := mg^{-1}$, $g \in G$, and conversely. In this way we define the tensor product of two left $\mathbb{Z}[G]$-modules M and N. Note that for two left $\mathbb{Z}[G]$-modules M and N we have

$$gm \otimes_R gn = mg^{-1} \otimes_R gn = m \otimes_R g^{-1}(gn) = m \otimes_R n$$

for all $g \in G$, $m \in M$, and $n \in N$.

In other words, if we define an action of $\mathbb{Z}[G]$ on $M \otimes_{\mathbb{Z}} N$ by setting the diagonal action

$$g \circ (m \otimes_{\mathbb{Z}} n) := gm \otimes_{\mathbb{Z}} gn,$$

then for any left $\mathbb{Z}[G]$-modules, $M \otimes_{\mathbb{Z}[G]} N \cong (M \otimes_{\mathbb{Z}} N)_G$, the quotient of $M \otimes_{\mathbb{Z}} N$ modulo the elements $m \otimes_{\mathbb{Z}} n$ satisfying $gm \otimes_{\mathbb{Z}} gn = m \otimes_{\mathbb{Z}} n$ for all $g \in G$, the latter with the diagonal action. In particular, $M \otimes_{\mathbb{Z}[G]} N \cong N \otimes_{\mathbb{Z}[G]} M$.

We will denote the tensor product by $M \otimes_{\mathbb{Z}[G]} N = M \otimes_G N = M \otimes N$.

Theorem A.1.14. *Let*

$$0 \longrightarrow A \xrightarrow{f} B \xrightarrow{g} C \longrightarrow 0$$

be an exact sequence of G-modules and let P be a projective G-module. Then

$$0 \longrightarrow P \otimes A \xrightarrow{1 \otimes f} P \otimes B \xrightarrow{1 \otimes g} P \otimes C \longrightarrow 0$$

is exact. Here $P \otimes X$ denotes the tensor product of the G-modules P and X.

Proof. Since P is projective, P is a direct summand of a free G-module, say

$$P \oplus R \cong T = \bigoplus_{i \in I} \mathbb{Z}[G].$$

Recall that the tensor product commutes with the direct sum. Furthermore, for any G-module M, $\mathbb{Z}[G] \otimes M$ is isomorphic to M. Therefore we have

$$(P \oplus R) \otimes M \cong (P \otimes M) \oplus (R \otimes M) \cong \bigoplus_{i \in I} M.$$

Now consider

$$1_T \otimes f : T \otimes A \longrightarrow T \otimes B, \quad \text{defined by} \quad (1_T \otimes f)\,(e_i \otimes a) = e_i \otimes f(a),$$

where e_i is the generator of $\mathbb{Z}[G]$ viewed as the ith component of $T = \bigoplus_{i \in I} \mathbb{Z}[G]$. Then $1_T \otimes f$ is injective since f is. Finally,

$$(1_T \otimes f) \mid_{P \otimes A} = 1_P \otimes f,$$

so that the latter map is injective.

We will see that $1 \otimes g$ is surjective. Given $p \otimes c \in P \otimes C$, there exists $b \in B$ such that $g(b) = c$, so that $(1 \otimes g)\,(p \otimes b) = p \otimes g(b) = p \otimes c$.

We have

$$(1 \otimes g) \circ (1 \otimes f) = 1 \otimes g \circ f = 1 \otimes 0 = 0,$$

hence $\operatorname{im}(1 \otimes f) \subset \ker(1 \otimes g)$.

Now let

$$\varphi\colon\ (P \otimes B)\,/\,(\ker(1 \otimes g)) \longrightarrow P \otimes C$$

be the isomorphism induced by $1 \otimes g$. Since $\operatorname{im}(1 \otimes f) \subseteq \ker(1 \otimes g)$, we can consider the epimorphism

$$\psi\colon\ (P \otimes B)\,/\,(\operatorname{im}(1 \otimes f)) \longrightarrow P \otimes C$$

induced by φ. We have

$$\ker \psi = \ker(1 \otimes g)/\operatorname{im}(1 \otimes f).$$

Let

$$\theta\colon P \times C \longrightarrow (P \otimes B)\,/\,(\operatorname{im}(1 \otimes f))$$

be defined by

$$\theta\,(p, c) = p \otimes b + \operatorname{im}(1 \otimes f) \quad \text{for} \quad c = g(b) \in C.$$

To see that θ is well defined, assume that $g\,(b_1) = g\,(b_2) = c$. Then $g\,(b_1 - b_2) = 0$, so $b_1 - b_2 \in \ker g = \operatorname{im} f$. Thus $b_1 - b_2 = f(a)$ for some $a \in A$. Therefore

$$p \otimes b_1 = p \otimes b_2 + p \otimes f(a) \quad \text{and} \quad p \otimes f(a) \in \operatorname{im}(1 \otimes f)\,,$$

whence

$$p \otimes b_1 \bmod (\operatorname{im}(1 \otimes f)) = p \otimes b_2 \bmod (\operatorname{im}(1 \otimes f))\,.$$

Thus θ is well defined and it is clearly $\mathbb{Z}$-bilinear. Let

$$\tilde{\theta}: P \otimes C \longrightarrow (P \otimes B)\,/\,(\operatorname{im}(1 \otimes f))$$

be the homomorphism induced. It is easy to verify that $\tilde{\theta} \circ \psi = \mathrm{Id}$ and $\psi \circ \tilde{\theta} = \mathrm{Id}$, so ψ is an isomorphism. This proves that $\ker(1 \otimes g) = \operatorname{im}(1 \otimes f)$. □

Remark A.1.15. The projectivity of P was used only once in the proof of Theorem A.1.14, namely to show the injectivity of $1 \otimes f$. A module that satisfies this property is called *flat*, and what we have proved is that any projective module is flat.

Theorem A.1.16 (Snake Lemma). *Let*

$$\begin{array}{ccccccc} & & A & \xrightarrow{f} & B & \xrightarrow{g} & C & \longrightarrow & 0 \\ & & \downarrow{\scriptstyle\alpha} & & \downarrow{\scriptstyle\beta} & & \downarrow{\scriptstyle\gamma} & & \\ 0 & \longrightarrow & A' & \xrightarrow{f'} & B' & \xrightarrow{g'} & C' & & \end{array}$$

be a commutative diagram of G-modules, where the rows are exact. Then there exists a connecting homomorphism $\delta : \ker \gamma \longrightarrow \operatorname{coker} \alpha$ such that

$$\ker\alpha \xrightarrow{\tilde{f}} \ker\beta \xrightarrow{\tilde{g}} \ker\gamma \xrightarrow{\delta} \operatorname{coker}\alpha \xrightarrow{\tilde{f}'} \operatorname{coker}\beta \xrightarrow{\tilde{g}'} \operatorname{coker}\gamma$$

is an exact sequence, where $\tilde{f}'$ and $\tilde{g}'$ are the induced homomorphisms from f' and g' respectively and $\tilde{f}$ and $\tilde{g}$ are the restrictions of f and g respectively.

If in addition f is injective, then $\tilde{f}$ is injective and if g' is surjective, $\tilde{g}'$ is surjective.

Proof. Let f be injective. If $x \in \ker\alpha$, then

$$(\beta \circ f)(x) = \left(f' \circ \alpha\right)(x) = 0.$$

Thus $\tilde{f}(x) \in \ker\beta$ and since f is injective,

$$\tilde{f} = f\mid_{\ker\alpha} \colon \ker\alpha \longrightarrow \ker\beta$$

is injective too. It is easy to see that sequence is exact at $\ker\beta$ and at $\operatorname{coker}\beta$.

Now if g' is surjective, let us see that $\tilde{g}'$ is surjective too. Let $c + \operatorname{im}\gamma \in \operatorname{coker}\gamma$ and let $b \in B$ be such that $g(b) = c$. Then $g'(b + \operatorname{im}\beta) = c + \operatorname{im}\gamma$.

It remains to define $\delta : \ker\gamma \longrightarrow \operatorname{coker}\alpha$ and to demonstrate that $\operatorname{im}\tilde{g} = \ker\delta$ and $\operatorname{im}\delta = \ker\tilde{f}'$. Let $z \in \ker\gamma$ be of the form $g(y)$ with $y \in B$. Then

$$\gamma(z) = (\gamma g)(y) = 0 = \left(g' \circ \beta\right)(y).$$

Therefore we have $\beta(y) \in \ker g' = \operatorname{im} f'$, so $\beta(y) = f'(a)$ for some $a \in A'$. Let $\delta(z) = a + \operatorname{im}\alpha$. We will see that δ is well defined. If $z = g(y) = g(y_1)$, then $y - y_1 \in \ker g = \operatorname{im} f$. Therefore

$$y = y_1 + f(x) \quad \text{for some} \quad x \in A.$$

Since $\beta(y_1) = f'(a_1)$, we have

$$\beta(y) = f'(a) = \beta(y_1) + \beta(f(x)) = f'(a_1) + \beta f(x) = f'(a_1) + f'(\alpha(x)).$$

It follows from the injectivity of f' that $a = a_1 + \alpha(x)$, so that $a + \operatorname{im}\alpha = a_1 + \operatorname{im}\alpha$. Clearly δ is a G-homomorphism.

Now let $z \in \ker\gamma$. Since $z \in \operatorname{im}\tilde{g}$, there exists $y \in \ker\beta$ such that $g(y) = z$. Then

$$\beta(y) = 0 = f'(0), \quad \text{that is} \quad (\delta\tilde{g})(y) = \delta(z) = 0 + \operatorname{im}\alpha.$$

Therefore $\operatorname{im}\tilde{g} \subseteq \ker\delta$. Let $z \in \ker\delta$. Since $\delta(z) = 0$, it follows that if $z = g(y)$ then $\beta(y) = f'(x)$ for some $x \in \operatorname{im}\alpha$. In other words,

$$x = \alpha(a) \quad \text{and} \quad \beta(y) = \left(f' \circ \alpha\right)(a) = \beta(f(a)).$$

Thus

$$y - f(a) \in \ker\beta \quad \text{and} \quad \tilde{g}(y - f(a)) = g(y) - (gf)(a) = g(y) = z.$$

It follows that the sequence is exact at $\ker \gamma$.

Finally,

$$\big(\tilde{f}' \circ \delta\big)(z) = \tilde{f}'\,(a + \operatorname{im}\alpha) = f'(a) + \operatorname{im}\beta, \text{ where } z = g(y) \text{ and } \beta(y) = f'(a).$$

Therefore

$$\left(\tilde{f}' \circ \delta\right)(z) = \beta(y) + \operatorname{im}\beta = 0, \quad \text{i.e.,} \quad \operatorname{im}\delta \subseteq \ker \tilde{f}'.$$

Finally, if $a + \operatorname{im}\alpha \in \ker \tilde{f}'$, then

$$f'(a) \in \operatorname{im}\beta, \quad \text{so} \quad f'(a) = \beta(y) \quad \text{for some} \quad y \in B.$$

If $z = g(y)$, then $\delta(z) = a + \operatorname{im}\alpha$. Therefore the sequence is exact at $\operatorname{coker}\alpha$. □

Definition A.1.17. A *projective resolution* P of $\mathbb{Z}$ is an exact sequence of G-modules of the form

$$P: \qquad \cdots \longrightarrow P_n \xrightarrow{\partial_n} P_{n-1} \longrightarrow \cdots \longrightarrow P_1 \xrightarrow{\partial_1} P_0 \xrightarrow{\partial_0} \mathbb{Z} \longrightarrow 0,$$

where $\mathbb{Z}$ is the trivial G-module and each P_i is projective. In particular, $\partial_n \circ \partial_{n+1} = 0$ for all n.

Lemma A.1.18. *If P, P' are two projective resolutions with respective homomorphisms ∂_n $(n \geq 0)$ and ∂_n' $(n \geq 0)$, then there exist homomorphisms*

$$\varepsilon_n \colon P_n' \longrightarrow P_n (n \geq -1)$$

such that $\partial_n \circ \varepsilon_n = \varepsilon_{n-1} \circ \partial_n'$ for all $n \geq 0$ and $\varepsilon_{-1} = \mathrm{Id}_{\mathbb{Z}}$.

Proof. The proof will be done by induction on n. Let $\varepsilon_{-1} = \mathrm{Id}_{\mathbb{Z}}$. Since P_0' is projective, there exists $\varepsilon_0 : P_0' \longrightarrow P_0$ such that $\partial_0' = \partial_0 \circ \varepsilon_0 = \mathrm{Id}_{\mathbb{Z}} \circ \partial_0' = \varepsilon_{-1} \circ \partial_0'$. Assume that we have constructed $\varepsilon_0, \varepsilon_1, \ldots, \varepsilon_n$ such that $\varepsilon_i : P_i' \longrightarrow P_i$ satisfies $\partial_i \circ \varepsilon_i = \varepsilon_{i-1} \circ \partial_i'$ for $i = 0, 1, \ldots, n$.

$$\begin{array}{ccc} & & P_0' \\ & \overset{\varepsilon_0}{\swarrow} & \downarrow{\scriptstyle \partial_0'} \\ P_0 & \xrightarrow{\partial_0} & \mathbb{Z} \end{array}$$

$$\begin{array}{ccccc} P_{n+1} & \xrightarrow{\partial_{n+1}} & P_n & \xrightarrow{\partial_n} & P_{n-1} \\ & & \uparrow{\scriptstyle \varepsilon_n} & & \uparrow{\scriptstyle \varepsilon_{n-1}} \\ P_{n+1}' & \xrightarrow{\partial_{n+1}'} & P_n' & \xrightarrow{\partial_n'} & P_{n-1}' \end{array} \qquad \begin{array}{ccccc} P_{n+1} & \xrightarrow{\partial_{n+1}} & \operatorname{im}\partial_{n+1} & \longrightarrow & 0 \\ & \overset{\varepsilon_{n+1}}{\nwarrow} & \uparrow{\scriptstyle \varepsilon_n \circ \partial_{n+1}'} & & \\ & & P_{n+1}' & & \end{array}$$

Let $x \in P_{n+1}'$ and notice that $(\varepsilon_n \circ \partial_{n+1}')x \in P_n$. Since

$$\partial_n \left(\varepsilon_n \circ \partial'_{n+1}\right)(x) = \partial_n \varepsilon_n \partial'_{n+1}(x) = \varepsilon_{n-1} \partial'_n \partial'_{n+1}(x) = 0,$$

we have

$$\left(\varepsilon_n \circ \partial'_{n+1}\right)(P_{n+1}) \subseteq \ker \partial_n = \operatorname{im} \partial_{n+1}.$$

Since P'_{n+1} is projective, there exists

$$\varepsilon_{n+1} \colon P'_{n+1} \longrightarrow P_{n+1} \quad \text{such that} \quad \partial_{n+1} \circ \varepsilon_{n+1} = \varepsilon_n \circ \partial'_{n+1}. \qquad \square$$

Now given a projective resolution P and a G-module A, let

$$K_i = \operatorname{Hom}_G(P_i, A) \quad \text{and} \quad R_i = P_i \otimes_G A = P_i \otimes_{\mathbb{Z}[G]} A,$$

where P_i can be made a right G-module by defining the action

$$x \circ g = g^{-1} x \quad \text{for all} \quad g \in G \quad \text{and} \quad x \in P_i.$$

Consider the sequences

$$0 \longrightarrow K_0 \xrightarrow{\partial_1^*} K_1 \xrightarrow{\partial_2^*} \cdots \longrightarrow K_{n-1} \xrightarrow{\partial_n^*} K_n \longrightarrow \cdots$$

and

$$\cdots \longrightarrow R_n \xrightarrow{\partial_n^+} R_{n-1} \longrightarrow \cdots \longrightarrow R_1 \xrightarrow{\partial_1^+} R_0 \longrightarrow 0,$$

where $\partial_n^*(\varphi) = \varphi \circ \partial_n$ and $\partial_n^+(x \otimes a) = \partial_n x \otimes a$.

We have

$$\partial_{n+1}^* \circ \partial_n^* = (\partial_n \circ \partial_{n+1})^* = 0^* = 0 \text{ and } \partial_n^+ \circ \partial_{n+1}^+ = (\partial_n \circ \partial_{n+1})^+ = 0^+ = 0,$$

so $\operatorname{im} \partial_n^* \subseteq \ker \partial_{n+1}^*$ and $\operatorname{im} \partial_{n+1}^+ \subseteq \ker \partial_n^+$.

Definition A.1.19. We define for $n = 0, 1, \ldots,$ the *nth cohomology group of A with respect to P* to be the group

$$H^n(P, A) := \ker \partial_{n+1}^* / \operatorname{im} \partial_n^*,$$

and the *nth homology group* to be the group

$$H_n(P, A) := \ker \partial_n^+ / \operatorname{im} \partial_{n+1}^+.$$

Here we define $\partial_0^* = 0$; $\quad \partial_0^+ = 0$ (see Remark A.2.6).

Theorem A.1.20. *If P and P′ are two projective resolutions, then*

$$H^n(P, A) \cong H^n\left(P', A\right) \quad \text{and} \quad H_n(P, A) \cong H_n\left(P', A\right) \text{ for all } n = 0, 1, \ldots.$$

Proof. Let $\varepsilon_n : P'_n \longrightarrow P_n$ and $\delta_n : P_n \longrightarrow P'_n$ be given by Lemma A.1.18, that is, $\partial_n \circ \varepsilon_n = \varepsilon_{n-1} \circ \partial'_n$ and $\partial'_n \circ \delta_n = \delta_{n-1} \circ \partial_n$. We will construct homomorphisms $h_n : P_n \longrightarrow P_{n+1}$ such that

$$\partial_{n+1} h_n + h_{n-1} \partial_n = \operatorname{Id} - \varepsilon_n \delta_n \tag{A.1}$$

and similarly, $f_n : P'_n \longrightarrow P'_{n+1}$ such that

$$\partial'_{n+1} f_n + f_{n-1} \partial'_n = \operatorname{Id} - \delta_n \varepsilon_n. \tag{A.2}$$

Let $h_{-1} : \mathbb{Z} \longrightarrow P_0$ be such that $h_{-1} = 0$. We wish to find $h_0 : P_0 \longrightarrow P_1$ such that $\partial_1 h_0 + h_{-1} \partial_0 = \partial_1 h_0 = \operatorname{Id} - \varepsilon_0 \delta_0$.

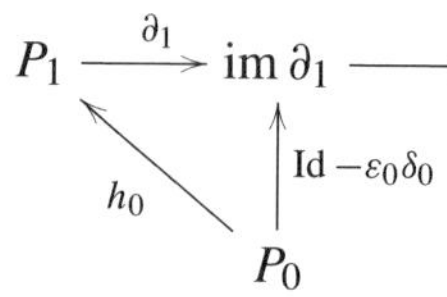

For $x \in P_0$, we obtain

$$\partial_0 (\operatorname{Id} - \varepsilon_0 \delta_0)(x) = \partial_0(x) - \partial_0 \varepsilon_0 \delta_0(x) = \partial_0(x) - \partial'_0 \delta_0(x) = \partial_0(x) - \partial_0(x) = 0.$$

Therefore $x \in \ker \partial_0 = \operatorname{im} \partial_1$. Since P_0 is projective, there exists

$$h_0 : P_0 \longrightarrow P_1 \quad \text{such that} \quad \partial_1 \circ h_0 = \operatorname{Id} - \varepsilon_0 \delta_0.$$

Assume that we have constructed $h_0, h_1, \ldots, h_n$ with property (A.1). If $x \in P_{n+1}$, we have

$$\partial_{n+1} \big(\operatorname{Id} - \varepsilon_{n+1} \delta_{n+1} - h_n \partial_{n+1} \big)(x) = 0,$$

and thus $\operatorname{im} (\operatorname{Id} - \varepsilon_{n+1} \delta_{n+1} - h_n \partial_{n+1}) \subseteq \ker \partial_{n+1} = \operatorname{im} \partial_{n+2}$.

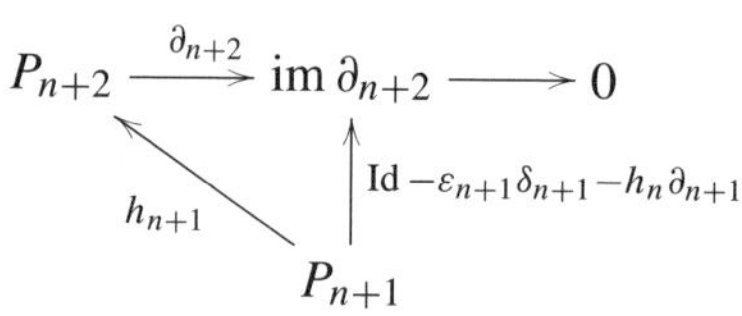

Since P_{n+1} is projective, there exists

$$h_{n+1} : P_{n+1} \longrightarrow P_{n+2} \quad \text{such that} \quad \partial_{n+2} h_{n+1} = \operatorname{Id} - \varepsilon_{n+1} \delta_{n+1} - h_n \partial_{n+1}.$$

Similarly for $f_n : P'_n \longrightarrow P'_{n+1}$.

For $\varepsilon_n : P'_n \longrightarrow P_n$, let

$$\varepsilon_n^* : \operatorname{Hom}_G (P_n, A) \longrightarrow \operatorname{Hom}_G \left(P'_n, A \right)$$

and

$$\varepsilon_n^+ = \varepsilon_n \otimes \mathrm{Id}_A : P'_n \otimes A \longrightarrow P_n \otimes A$$

be defined by

$$\varepsilon_n^* (\varphi) = \varphi \circ \varepsilon_n \quad \text{and} \quad \varepsilon_n^+ (x \otimes a) = \varepsilon_n(x) \otimes a.$$

If $\varphi \in \ker \partial_{n+1}^*$, we have

$$\partial'^*_{n+1} \left(\varepsilon_n^* (\varphi)\right) = \varphi \circ \varepsilon_n \circ \partial'_{n+1} = \varphi \circ \partial_{n+1} \circ \varepsilon_{n+1} = \varepsilon_{n+1}^* \left(\partial_{n+1}^* \circ \varphi\right) = 0,$$

so $\varepsilon_n^* \left(\ker \partial_{n+1}^*\right) \subseteq \ker \partial'^*_{n+1}$. Similarly we have

$$\varepsilon_n^* \left(\operatorname{im} \partial_n^*\right) \subseteq \operatorname{im} \partial'^*_n; \quad \varepsilon_n^+ \left(\ker \partial'^+_n\right) \subseteq \ker \partial_n^+; \quad \varepsilon_n^+ \left(\operatorname{im} \partial'^+_{n+1}\right) \subseteq \operatorname{im} \partial_{n+1}^+.$$

Therefore we have the following induced homomorphisms:

$$\tilde{\varepsilon}_n^* : H^n (P, A) \longrightarrow H^n \left(P', A\right) \quad \text{and} \quad \tilde{\varepsilon}_n^+ : H_n \left(P', A\right) \longrightarrow H_n (P, A).$$

We proceed in a similar way for $\tilde{\delta}_n^*$ and $\tilde{\delta}_n^+$.
Now if $\varphi \in \ker \partial_{n+1}^*$, we have

$$\begin{aligned}(\partial_{n+1} h_n + h_{n-1}\partial_n)^* \varphi &= \varphi \partial_{n+1} h_n + \varphi h_{n-1} \partial_n \\ &= 0 + \varphi h_{n-1} \partial_n = \partial_n^* (\varphi h_{n-1}) \in \operatorname{im} \partial_n^*.\end{aligned}$$

Therefore

$$\overline{(\partial_{n+1} h_n + h_{n-1}\partial_n)^*} = 0 = \overline{(\mathrm{Id} - \varepsilon_n \delta_n)^*} = \overline{\mathrm{Id}^*} - \bar{\delta}_n^* \bar{\varepsilon}_n^*,$$

from which it follows that $\overline{\mathrm{Id}^*} = \mathrm{Id} = \bar{\delta}_n^* \bar{\varepsilon}_n^*$. Similarly we have $\mathrm{Id} = \bar{\varepsilon}_n^* \bar{\delta}_n^*$.

We can show analogously that $\bar{\varepsilon}_n^+ \bar{\delta}_n^+ = \mathrm{Id}$ and $\bar{\delta}_n^+ \bar{\varepsilon}_n^+ = \mathrm{Id}$. □

Definition A.1.21. For an arbitrary G-module A and for $n = 0, 1, \ldots,$ we define the *cohomology groups* $H^n (G, A)$ as $H^n (P, A)$ and the *homology groups* $H_n (G, A)$ as $H_n (P, A)$, where P is any projective resolution.

By Theorem A.1.20 the above definition depends only on G and on A and does not depend on the resolution. On the other hand, to see that Definition A.1.21 is not vacuous, we need to exhibit at least one projective resolution of A.

Let $G^{n+1} = G \times \cdots \times G$ ($n+1$ copies) and let $A_n = \mathbb{Z}\left[G^{n+1}\right]$ be the group ring. Then A_n is an abelian group and G acts on A_n as follows:

$$x \circ (g_0, \ldots, g_n) = (xg_0, \ldots, xg_n) \text{ for all } x \in G \text{ and } (g_0, \ldots, g_n) \in G^{n+1}.$$

Thus A_n is a free $\mathbb{Z}$-module with basis $\{(g_0, \ldots, g_n) \mid g_i \in G\}$.
The proof of the following proposition is straightforward.

Proposition A.1.22. *For each $n \geq 0$, A_n is a free $\mathbb{Z}[G]$-module with basis $\{(1, x_1, ldots, x_n) \mid x_i \in G\}$.* □

Now put $P_i = A_i$ and let $\partial_n : P_n \longrightarrow P_{n-1}$ be defined by

$$\partial_n (g_0, g_1, \dots, g_n) = \sum_{i=0}^{n} (-1)^i \left(g_0, g_1, \dots, \hat{g_i}, \dots, g_n\right),$$

where the symbol $\hat{g_i}$ means that the element g_i does not appear, that is,

$$\left(g_0, g_1, \dots, \hat{g_i}, \dots, g_n\right) = (g_0, g_1, \dots, g_{i-1}, g_{i+1}, \dots, g_n).$$

If $g \in G$, then

$$\begin{aligned} g \circ (\partial_n (g_0, g_1, \dots, g_n)) &= g \circ \sum_{i=0}^{n} (-1)^i \left(g_0, g_1, \dots, \hat{g_i}, \dots, g_n\right) \\ &= \sum_{i=0}^{n} (-1)^i (gg_0, gg_1, \dots, \widehat{gg_i}, \dots, gg_n) \\ &= \partial_n (gg_0, gg_1, \dots, gg_n) = \partial_n (g \circ (g_0, g_1, \dots, g_n)), \end{aligned}$$

so ∂_n is a G-homomorphism.

Now $\partial_0 : P_0 = A_0 = \mathbb{Z}[G] \longrightarrow \mathbb{Z}$ is defined by $\partial_0(g) = 1$ for all $g \in G$.

Proposition A.1.23. *The sequence*

$$\cdots \longrightarrow P_n \xrightarrow{\partial_n} P_{n-1} \longrightarrow \cdots \longrightarrow P_1 \xrightarrow{\partial_1} P_0 \xrightarrow{\partial_0} \mathbb{Z} \longrightarrow 0$$

is G-exact.

Proof. For $n = 0, 1, \dots$ we have

$$\begin{aligned} \partial_{n-1} \circ \partial_n (g_0, g_1, \dots, g_n) &= \partial_{n-1} \left(\sum_{i=0}^{n} (-1)^i \left(g_0, g_1, \dots, \hat{g_i}, \dots, g_n\right) \right) \\ = \sum_{i=0}^{n} (-1)^i & \left(\sum_{j=0}^{i-1} (-1)^j \left(g_0, g_1, \dots, \hat{g_j}, \dots, \hat{g_i}, \dots, g_n\right) \right. \\ & \left. + \sum_{j=i+1}^{n} (-1)^{j-1} \left(g_0, g_1, \dots, \hat{g_i}, \dots, \hat{g_j}, \dots, g_n\right) \right). \end{aligned} \tag{A.3}$$

For any two indices $0 \leq r < s \leq n$, the element $\left(g_0, \dots, \hat{g_r}, \dots, \hat{g_s}, \dots, g_n\right)$ appears exactly twice in (A.3) and its coefficient is $(-1)^{r+s} + (-1)^{r+s-1} = 0$, which proves that $\partial_{n-1} \circ \partial_n = 0$. Therefore $\operatorname{im} \partial_n \subseteq \ker \partial_{n-1}$. Now let

$$h_n : P_{n-1} \longrightarrow P_n, \quad h_n (g_0, \dots, g_{n-1}) = (1, g_0, \dots, g_{n-1}) \quad \text{for} \quad n = 1, 2, \dots.$$

We also define $h_0 \colon P_{-1} = \mathbb{Z} \longrightarrow P_0$ by $h_0(1) = 1 \in \mathbb{Z}[G] = P_0$. We have

$$\begin{aligned}
&(\partial_n h_n + h_{n-1}\partial_{n-1})(g_0, \ldots, g_{n-1}) \\
&\quad = \partial_n (1, g_0, \ldots, g_{n-1}) + h_{n-1}\left(\sum_{i=0}^{n-1}(-1)^i \left(g_0, g_1, \ldots, \hat{g}_i, \ldots, g_{n-1}\right)\right) \\
&\qquad = (g_0, \ldots, g_{n-1}) + \sum_{i=0}^{n-1}(-1)^{i+1}\left(1, g_0, \ldots, \hat{g}_i, \ldots, g_{n-1}\right) \\
&\qquad\qquad + \sum_{i=0}^{n-1}(-1)^i \left(1, g_0, \ldots, \hat{g}_i, \ldots, g_{n-1}\right) = (g_0, \ldots, g_{n-1}).
\end{aligned}$$

Thus

$$\partial_n h_n + h_{n-1}\partial_{n-1} = \mathrm{Id}_{P_{n-1}} \quad \text{for} \quad n = 1, 2, \ldots .$$

Observe that h_n has been defined as a $\mathbb{Z}$-homomorphism but not as a G-homomorphism.

Now if $x \in \ker \partial_{n-1}$, we have

$$x = \mathrm{Id}_{P_{n-1}}(x) = \partial_n h_n(x) + h_{n-1}\partial_{n-1}(x) = \partial_n (h_n(x)) + h_{n-1}(0) = \partial_n (h_n(x)).$$

Thus $x = \partial_n (h_n(x)) \in \operatorname{im} \partial_n$, which proves the exactness of the sequence. □

The resolution defined in Proposition A.1.23 is called the *canonical resolution* or *bar resolution*.

From now on, unless otherwise stated, by resolution we will mean the canonical resolution.

We have proved the existence of the homology and cohomology groups for any G-module A. Now if A and B are two G-modules and $f : A \longrightarrow B$ is a G-homomorphism, we will define in a natural way group homomorphisms

$$H^n(f) : H^n(G, A) \longrightarrow H^n(G, B) \quad \text{and} \quad H_n(f) : H_n(G, A) \longrightarrow H_n(G, B).$$

Let

$$P: \qquad \cdots \longrightarrow P_n \xrightarrow{\partial_n} P_{n-1} \longrightarrow \cdots \longrightarrow P_1 \xrightarrow{\partial_1} P_0 \xrightarrow{\partial_0} \mathbb{Z} \longrightarrow 0$$

be a projective resolution. We have

$$P \otimes_G A: \quad \cdots P_n \otimes A \xrightarrow{\partial_n \otimes 1_A} P_{n-1} \otimes A \longrightarrow \cdots \longrightarrow P_1 \otimes A \xrightarrow{\partial_1 \otimes 1_A} P_0 \otimes A \longrightarrow 0,$$

where $P_i \otimes A$ means $P_i \otimes_{\mathbb{Z}[G]} A$.

Let

$$\begin{gathered}
f_n \colon P_n \otimes A \to P_n \otimes B, \\
f_n(x \otimes a) = x \otimes f(a) = \left(1_{P_n} \otimes f\right)(a).
\end{gathered}$$

We have

$$f_{n-1} \circ (\partial_n \otimes 1_A) = \partial_n \otimes f = (\partial_n \otimes 1_B) \circ \left(1_{P_n} \otimes f\right) = (\partial_n \otimes 1_B) \circ f_n.$$

If $\alpha \in \ker(\partial_n \otimes 1_A)$, then

$$f_{n-1} \circ (\partial_n \otimes 1_A)(\alpha) = 0 = (\partial_n \otimes 1_B) \circ (1 \otimes f)(\alpha) = (\partial_n \otimes 1_B) f_n(\alpha),$$

so $f_n(\alpha) \in \ker(\partial_n \otimes 1_B)$.

If $\alpha \in \operatorname{im}(\partial_{n+1} \otimes 1_A)$, then $\alpha = (\partial_{n+1} \otimes 1_A)(\beta)$. Hence

$$f_n(\alpha) = f_n \circ (\partial_{n+1} \otimes 1_A)(\beta) = ((\partial_{n+1} \otimes 1_B) \circ f_{n+1})(\beta) \in \operatorname{im}(\partial_{n+1} \otimes 1_B).$$

Therefore f_n induces in a natural way the group homomorphisms

$$H_n(f) : H_n(G, A) \longrightarrow H_n(G, B), \qquad n = 0, 1, \ldots.$$

We now consider the sequence

$$\operatorname{Hom}_G(P, A) : \quad \begin{array}{l} 0 \longrightarrow \operatorname{Hom}_G(P_0, A) \xrightarrow{\partial_1^*} \operatorname{Hom}_G(P_1, A) \xrightarrow{\partial_2^*} \cdots \\ \cdots \longrightarrow \operatorname{Hom}_G(P_{n-1}, A) \xrightarrow{\partial_n^*} \operatorname{Hom}_G(P_n, A) \longrightarrow \cdots \end{array}$$

Let $f_n^* : \operatorname{Hom}_G(P_n, A) \longrightarrow \operatorname{Hom}_G(P_n, B)$ be given by $f_n^*(\varphi) = f \circ \varphi$. We have

$$\left.\begin{array}{l} \left(f_n^* \circ \partial_n^*\right)(\varphi) = f \circ \varphi \circ \partial_n \\ \left(\partial_n^* \circ f_{n-1}^*\right)(\varphi) = f \circ \varphi \circ \partial_n \end{array}\right\} \Longrightarrow f_n^* \circ \partial_n^* = \partial_n^* \circ f_{n-1}^*.$$

If $\varphi \in \ker \partial_{n+1}^*$, then

$$\left(\partial_{n+1}^* \circ f_n^*\right)(\varphi) = \left(f_{n+1}^* \circ \partial_{n+1}^*\right)(\varphi) = 0.$$

Therefore $f_n^*(\varphi) \in \ker \partial_{n+1}^*$.

If $\varphi \in \operatorname{im} \partial_n^*$, then $\partial_n^*(\theta) = \theta \circ \partial_n = \varphi$. It follows that

$$f_n^*(\varphi) = \left(f_n^* \circ \partial_n^*\right)(\theta) = \left(\partial_n^* \circ f_{n-1}^*\right)(\theta) \in \operatorname{im} \partial_n^*.$$

Hence f_n^* induces in a natural way a group homomorphism

$$H^n(f) : H^n(G, A) \longrightarrow H^n(G, B), \quad n = 0, 1, \ldots.$$

The following result is a powerful tool for studying the arithmetic of fields by means of the cohomology and the homology groups.

Theorem A.1.24. *Let*

$$0 \longrightarrow A \xrightarrow{f} B \xrightarrow{g} C \longrightarrow 0$$

be an exact sequence of G-modules. Then there exist group homomorphisms

$$\varepsilon_n : H_{n+1}(G, C) \to H_n(G, A)$$

and

$$\delta_n : H^n(G, C) \to H^{n+1}(G, A), \quad n = 0, 1, \ldots,$$

such that the homology sequence

$$\begin{aligned} \cdots \longrightarrow H_{n+1}(G, B) &\xrightarrow{H_{n+1}(g)} H_{n+1}(G, C) \xrightarrow{\varepsilon_n} H_n(G, A) \xrightarrow{H_n(f)} H_n(G, B) \longrightarrow \\ \cdots &\xrightarrow{\varepsilon_0} H_0(G, A) \xrightarrow{H_0(f)} H_0(G, B) \xrightarrow{H_0(g)} H_0(G, C) \longrightarrow 0, \end{aligned}$$

and the cohomology sequence

$$\begin{aligned} 0 \longrightarrow H^0(G, A) &\xrightarrow{H^0(f)} H^0(G, B) \xrightarrow{H^0(g)} H^0(G, C) \xrightarrow{\delta_0} H^1(G, A) \longrightarrow \cdots \\ \xrightarrow{\delta_{n-1}} H^n(G, A) &\xrightarrow{H^n(f)} H^n(G, B) \xrightarrow{H^n(g)} H^n(G, C) \xrightarrow{\delta_n} H^{n+1}(G, A) \longrightarrow \cdots, \end{aligned}$$

are exact sequences of groups.

Proof. Let

$$P: \qquad \cdots \longrightarrow P_n \xrightarrow{\partial_n} P_{n-1} \longrightarrow \cdots \longrightarrow P_1 \xrightarrow{\partial_1} P_0 \xrightarrow{\partial_0} \mathbb{Z} \longrightarrow 0$$

be a projective resolution.

We will use the notation

$$\left.\begin{aligned} X_n &= P_n \otimes X \\ X^n &= \mathrm{Hom}_G(P_n, X) \end{aligned}\right\},$$

$X = A, B,$ or C and $k_n = H_n(k)$, $k^n = H^n(k)$, $k = f$ or g.

Consider the following commutative diagrams of groups:

$$\begin{array}{ccccccccc} 0 & \longrightarrow & A_n & \xrightarrow{f_n} & B_n & \xrightarrow{g_n} & C_n & \longrightarrow & 0 \\ & & \downarrow{\scriptstyle \partial_n} & & \downarrow{\scriptstyle \partial_n} & & \downarrow{\scriptstyle \partial_n} & & \\ 0 & \longrightarrow & A_{n-1} & \xrightarrow{f_{n-1}} & B_{n-1} & \xrightarrow{g_{n-1}} & C_{n-1} & \longrightarrow & 0 \end{array} \tag{A.4}$$

and

$$\begin{array}{ccccccccc} 0 & \longrightarrow & A^{n-1} & \xrightarrow{f^{n-1}} & B^{n-1} & \xrightarrow{g^{n-1}} & C^{n-1} & \longrightarrow & 0 \\ & & \downarrow{\scriptstyle \partial^n} & & \downarrow{\scriptstyle \partial^n} & & \downarrow{\scriptstyle \partial^n} & & \\ 0 & \longrightarrow & A^n & \xrightarrow{f^n} & B^n & \xrightarrow{g^n} & C^n & \longrightarrow & 0 \end{array} \tag{A.5}$$

Since $\operatorname{im}\partial_{n+1} \subseteq \ker\partial_n$, the map $\partial_n : X_n \longrightarrow X_{n-1}$ induces the natural map $\tilde{\partial_n} : \operatorname{coker}\partial_{n+1} \longrightarrow \ker\partial_{n-1}$,

$$\operatorname{coker}\partial_{n+1} = X_n/\operatorname{im}\partial_{n+1} \overset{\tilde{\partial_n}}{\twoheadrightarrow} X_n/\ker\partial_n \cong \operatorname{im}\partial_n \subseteq \ker\partial_{n-1}.$$

We have

$$\ker\tilde{\partial_n} = \ker\partial_n/\operatorname{im}\partial_{n+1} = H_n(G, X)$$

and

$$\operatorname{coker}\tilde{\partial_n} = \ker\partial_{n-1}/\operatorname{im}\partial_n = H_{n-1}(G, X).$$

Similarly, consider

$$\partial^n : X^{n-1} \longrightarrow X^n.$$

We have $\operatorname{im}\partial^{n-1} \subseteq \ker\partial^n$, so we obtain the natural map

$$X^{n-1}/\operatorname{im}\partial^{n-1} \twoheadrightarrow X^{n-1}/\ker\partial^n \cong \operatorname{im}\partial^n \subseteq \ker\partial^{n+1};$$

that is,

$$\tilde{\partial^n} : \operatorname{coker}\partial^{n-1} \longrightarrow \ker\partial^{n+1},$$
$$\ker\tilde{\partial^n} = \ker\partial^n/\operatorname{im}\partial^{n-1} = H^{n-1}(G, X)$$

and

$$\operatorname{coker}\tilde{\partial^n} = \ker\partial^{n+1}/\operatorname{im}\partial^n = H^n(G, X).$$

Consider the commutative diagram

$$\begin{array}{ccccccccc}
 & & 0 & & 0 & & 0 & & \\
 & & \downarrow & & \downarrow & & \downarrow & & \\
0 & \longrightarrow & \ker\partial_n & \longrightarrow & \ker\partial_n & \longrightarrow & \ker\partial_n & & \\
 & & \downarrow & & \downarrow & & \downarrow & & \\
0 & \longrightarrow & A_n & \xrightarrow{f_n} & B_n & \xrightarrow{g_n} & C_n & \longrightarrow & 0 \\
 & & \downarrow{\scriptstyle\partial_n} & & \downarrow{\scriptstyle\partial_n} & & \downarrow{\scriptstyle\partial_n} & & \\
0 & \longrightarrow & A_{n-1} & \xrightarrow{f_{n-1}} & B_{n-1} & \xrightarrow{g_{n-1}} & C_{n-1} & \longrightarrow & 0 \\
 & & \downarrow & & \downarrow & & \downarrow & & \\
 & & \operatorname{coker}\partial_n & \longrightarrow & \operatorname{coker}\partial_n & \longrightarrow & \operatorname{coker}\partial_n & \longrightarrow & 0 \\
 & & \downarrow & & \downarrow & & \downarrow & & \\
 & & 0 & & 0 & & 0 & &
\end{array} \tag{A.6}$$

By the snake lemma (Theorem A.1.16), the rows are exact. Now

$$\partial_n : X_n \longrightarrow X_{n-1}$$

induces

$$\begin{aligned} 0 \to H_n\,(G, X) = \ker \tilde{\partial}_n \to \operatorname{coker} \partial_{n+1} \overset{\tilde{\partial}_n}{\to} \ker \partial_{n-1} \\ \to \operatorname{coker} \tilde{\partial}_n = H_{n-1}\,(G, X)\,. \end{aligned}$$

We obtain the diagram

$$\begin{array}{ccccccccc}
 & & 0 & & 0 & & 0 & & \\
 & & \downarrow & & \downarrow & & \downarrow & & \\
 & & H_n(G, A) & \xrightarrow{f_n} & H_n(G, B) & \xrightarrow{g_n} & H_n(G, C) & & \\
 & & \downarrow & & \downarrow & & \downarrow & & \\
 & & \operatorname{coker} \partial_{n+1} & \longrightarrow & \operatorname{coker} \partial_{n+1} & \longrightarrow & \operatorname{coker} \partial_{n+1} & \longrightarrow & 0 \\
 & & \downarrow{\scriptstyle \tilde{\partial}_n} & & \downarrow{\scriptstyle \partial_n} & & \downarrow{\scriptstyle \tilde{\partial}_n} & & \\
0 & \longrightarrow & \ker \partial_{n-1} & \longrightarrow & \ker \partial_{n-1} & \longrightarrow & \ker \partial_{n-1} & & \\
 & & \downarrow & & \downarrow & & \downarrow & & \\
 & & H_{n-1}(G, A) & \xrightarrow{f_{n-1}} & H_{n-1}(G, B) & \xrightarrow{g_{n-1}} & H_{n-1}(G, C) & & \\
 & & \downarrow & & \downarrow & & \downarrow & & \\
 & & 0 & & 0 & & 0 & &
\end{array} \tag{A.7}$$

Again by the snake lemma, there exists a group homomorphism $\varepsilon_{n-1} : H_n\,(G, C) \longrightarrow H_{n-1}\,(G, A)$ such that

$$H_n\,(G, A) \to H_n\,(G, B) \to H_n\,(G, C) \overset{\varepsilon_{n-1}}{\to} H_{n-1}\,(G, A) \to H_{n-1}\,(G, B) \to \cdots$$

is exact.

Similarly, for the cohomology groups we have diagrams

$$
\begin{array}{ccccccccc}
 & & 0 & & 0 & & 0 & & \\
 & & \downarrow & & \downarrow & & \downarrow & & \\
0 & \longrightarrow & \ker \partial^n & \longrightarrow & \ker \partial^n & \longrightarrow & \ker \partial^n & & \\
 & & \downarrow & & \downarrow & & \downarrow & & \\
0 & \longrightarrow & A^n & \xrightarrow{f^n} & B^n & \xrightarrow{g^n} & C^n & \longrightarrow & 0 \\
 & & \downarrow{\scriptstyle \partial^n} & & \downarrow{\scriptstyle \partial^n} & & \downarrow{\scriptstyle \partial^n} & & \\
0 & \longrightarrow & A^{n+1} & \xrightarrow{f^{n+1}} & B^{n+1} & \xrightarrow{g^{n+1}} & C^{n+1} & \longrightarrow & 0 \\
 & & \downarrow & & \downarrow & & \downarrow & & \\
 & & \operatorname{coker} \partial^n & \longrightarrow & \operatorname{coker} \partial^n & \longrightarrow & \operatorname{coker} \partial^n & \longrightarrow & 0 \\
 & & \downarrow & & \downarrow & & \downarrow & & \\
 & & 0 & & 0 & & 0 & &
\end{array}
\tag{A.8}
$$

and

$$
\begin{array}{ccccccccc}
 & & 0 & & 0 & & 0 & & \\
 & & \downarrow & & \downarrow & & \downarrow & & \\
 & & H^{n-1}(G, A) & \xrightarrow{f^{n-1}} & H^{n-1}(G, B) & \xrightarrow{g^{n-1}} & H^{n-1}(G, C) & & \\
 & & \downarrow & & \downarrow & & \downarrow & & \\
 & & \operatorname{coker} \partial^{n-1} & \longrightarrow & \operatorname{coker} \partial^{n-1} & \longrightarrow & \operatorname{coker} \partial^{n-1} & \longrightarrow & 0 \\
 & & \downarrow{\scriptstyle \tilde{\partial}^n} & & \downarrow{\scriptstyle \tilde{\partial}^n} & & \downarrow{\scriptstyle \tilde{\partial}^n} & & \\
0 & \longrightarrow & \ker \partial^{n+1} & \longrightarrow & \ker \partial^{n+1} & \longrightarrow & \ker \partial^{n+1} & & \\
 & & \downarrow & & \downarrow & & \downarrow & & \\
 & & H^n(G, A) & \xrightarrow{f^n} & H^n(G, B) & \xrightarrow{g^n} & H^n(G, C) & & \\
 & & \downarrow & & \downarrow & & \downarrow & & \\
 & & 0 & & 0 & & 0 & &
\end{array}
\tag{A.9}
$$

By the snake lemma there exists $\delta_{n-1} : H^{n-1}(G, C) \longrightarrow H^n(G, A)$ such that the sequence

$$
H^{n-1}(G, A) \longrightarrow H^{n-1}(G, B) \longrightarrow H^{n-1}(G, C) \xrightarrow{\delta_{n-1}} H^n(G, A) \longrightarrow H^n(G, B) \longrightarrow H^n(G, C)
$$

is exact. □

A.2 Homology and Cohomology in Low Dimensions

Our goal in this section is to calculate homology groups H_n and cohomology groups H^n for $n = 0$, 1, or 2.

Let A be an arbitrary G-module. The homology sequence is

$$\cdots P_n \otimes A \stackrel{\partial_n \otimes 1_A}{\longrightarrow} P_{n-1} \otimes A \longrightarrow \cdots \longrightarrow P_1 \otimes A \stackrel{\partial_1 \otimes 1_A}{\longrightarrow} P_0 \otimes A \longrightarrow 0,$$

where $\{P_i\}_{i=0}^{\infty}$ is the canonical resolution given in Section A.1. In particular, $P_0 = \mathbb{Z}[G]$ and $P_0 \otimes A \cong A$. Then

$$H_0(G, A) = (P_0 \otimes A)/(\operatorname{im} \partial_1 \otimes 1).$$

Now, we have

$$(\partial_1 \otimes 1)((g_1, g_2) \otimes a) = g_1 a - g_2 a,$$

which implies that

$$\operatorname{im}(\partial_1 \otimes 1) = \langle a - ga \mid g \in G, a \in A \rangle = DA \subseteq A.$$

Thus $H_0(G, A) = A_G = A/DA$.

Here A_G is the maximal quotient module where G acts trivially.

Let $I_G = \left\{\sum_{\sigma \in G} a_\sigma \sigma \mid \sum_{\sigma \in G} a_\sigma = 0\right\}$. Here I_G is an ideal $\mathbb{Z}[G]$. Furthermore, $\mathbb{Z}[G]/I_G \cong \mathbb{Z}$. If $\sum_{\sigma \in G} a_\sigma \sigma \in I_G$ with $a_1 = -\sum_{\sigma \neq 1} a_\sigma$, we have

$$\begin{aligned}\sum_{\sigma \in G} a_\sigma \sigma &= a_1 1 + \sum_{\sigma \neq 1} a_\sigma \sigma = \left(-\sum_{\sigma \neq 1} a_\sigma\right) 1 + \sum_{\sigma \neq 1} a_\sigma \sigma \\ &= \sum_{\sigma \neq 1} a_\sigma (\sigma - 1) \in \langle \sigma - 1 \mid \sigma \in G \rangle .\end{aligned}$$

Conversely, we have $\sigma - 1 \in I_G$ for $\sigma \in G$. Thus

$$DA = \langle a - \sigma a \mid \sigma \in G, a \in A \rangle = I_G A.$$

Therefore

$$H_0(G, A) = A/I_G A.$$

Proposition A.2.1. *For any group G, we have $I_G/I_G^2 \cong G/G'$, where G' is the commutator subgroup of G.*

Proof. Let $f : G \longrightarrow I_G/I_G^2$ be the map defined by $f(\sigma) = (\sigma - 1) + I_G^2$. Now

$$\begin{aligned} f(\sigma\phi) &= (\sigma\phi - 1) + I_G^2 = (\sigma\phi - \sigma + \sigma - 1) + I_G^2 = \sigma(\phi - 1) + (\sigma - 1) + I_G^2 \\ &= (\sigma - 1)(\phi - 1) + (\phi - 1) + (\sigma - 1) + I_G^2, \end{aligned}$$

and since $(\sigma - 1)(\phi - 1) \in I_G^2$, we have $f(\sigma\phi) = f(\sigma) + f(\phi)$. Thus f is a homomorphism.

Since I_G/I_G^2 is abelian, $G/\ker f$ is abelian too. Therefore $[G, G] = G' \subseteq \ker f$. Consider the induced map $\tilde{f} : G/G' \longrightarrow I_G/I_G^2$ such that $\tilde{f}\left(\sigma G'\right) = (\sigma - 1) + I_G^2$.

Let $h : I_G \longrightarrow G/G'$ be defined by $h(\sigma - 1) = \sigma G'$. If $x \in I_G^2$, we have

$$\begin{aligned} x &= \left(\sum_{\sigma\in G} a_\sigma (\sigma - 1)\right)\left(\sum_{\sigma\in G} b_\sigma (\sigma - 1)\right) = \sum_{\sigma,\theta\in G} a_\sigma b_\theta (\sigma - 1)(\theta - 1) \\ &= \sum_{\sigma,\theta\in G} a_\sigma b_\theta \left[(\sigma\theta - 1) - (\sigma - 1) - (\theta - 1)\right]. \end{aligned}$$

Therefore

$$\begin{aligned} h(x) &= \prod_{\sigma,\theta\in G} \left[h(\sigma\theta - 1)\, h(\sigma - 1)^{-1}\, h(\theta - 1)^{-1}\right]^{a_\sigma b_\theta} \\ &= \prod_{\sigma,\theta\in G} \left(\sigma\theta\sigma^{-1}\theta^{-1}\right)^{a_\sigma b_\theta} G' = G'. \end{aligned}$$

It follows that $h(x) = 1$ and $I_G^2 \subseteq \ker h$. Thus h induces

$$\tilde{h} : I_G/I_G^2 \longrightarrow G/G' \quad \text{defined by} \quad \tilde{h}\left((\sigma - 1) + I_G^2\right) = \sigma G'.$$

Clearly $\tilde{f}$ and $\tilde{h}$ are inverse isomorphisms of groups. □

Definition A.2.2. Let X be an abelian group and let A be the G-module $\mathrm{Hom}(\mathbb{Z}[G], X)$, where the G-action on X is the trivial one. Any G-module of this kind is called *coinduced*. The action of G on A is defined explicitly as follows:

$$\text{For all} \quad \varphi \in A \quad \text{and} \quad g, g' \in G, \quad g \circ \varphi\left(g'\right) = g\varphi\left(g^{-1}g'\right) = \varphi\left(g^{-1}g'\right).$$

Definition A.2.3. Let X be an abelian group and let A be the G-module $\mathbb{Z}[G] \otimes_{\mathbb{Z}} X$. Any G-module of this type is called *induced* and G acts on A as follows:

$$\text{For all } g, g' \in G \quad \text{and} \quad x \in X, \quad g\left(g' \otimes x\right) = gg' \otimes x.$$

Proposition A.2.4. *Let* $A = \mathrm{Hom}(\mathbb{Z}[G], X)$. *Then for any* G*-module* B, *the groups* $\mathrm{Hom}_G(B, A)$ *and* $\mathrm{Hom}(B, X)$ *are isomorphic.*

Proof. Let $\varphi \in \mathrm{Hom}_G(B, A)$. Then $\varphi(b) \in \mathrm{Hom}(\mathbb{Z}[G], X)$ for all $b \in B$. Let $\theta_\varphi \in \mathrm{Hom}(B, X)$ be defined by $\theta_\varphi(b) = \varphi(b)(1)$. We have

$$\theta_\varphi\,(b+b_1)=\theta_\varphi\,(b)+\theta_\varphi\,(b_1)\,,$$

so $\theta_\varphi \in \operatorname{Hom}(B, X)$. We also have $\theta_{\varphi+\psi} = \theta_\varphi + \theta_\psi$, and θ is a group homomorphism from $\operatorname{Hom}_G(B, A)$ to $\operatorname{Hom}_{\mathbb{Z}}(B, X)$.

Now assume that $\theta_\varphi = 0$. Then

$$\theta_\varphi : B \longrightarrow X \quad \text{satisfies} \quad \theta_\varphi(b)=\varphi(b)(1)=0 \quad \text{for all} \quad b\in B.$$

Since $\varphi \in \operatorname{Hom}_G(B, A)$, we have $\varphi(g\,b) = g\,\varphi(b)$ for all $g \in G$ and $b \in B$. Now if $g' \in G \subseteq \mathbb{Z}[G]$, we have

$$(g\,\varphi(b))\,(g')=g\,\left[\varphi(b)\left(g^{-1}\,g'\right)\right]=\varphi(b)\left(g^{-1}\,g'\right).$$

In particular,

$$\varphi(g\,b)(1)=(g\,\varphi(b))\,(1)=\varphi(b)\left(g^{-1}\right),$$

or, equivalently,

$$\varphi\left(g^{-1}\,b\right)(1)=\varphi(b)(g).$$

Therefore

$$\theta_\varphi=0 \quad\Longrightarrow\quad \varphi(b)(g)=0 \quad \text{for all} \quad g\in G \quad \text{and} \quad b\in B.$$

It follows that $\varphi(b) = 0$ for all $b \in B$, that is, $\varphi = 0$. Therefore θ is injective.

Now let $\sigma \in \operatorname{Hom}_{\mathbb{Z}}(B, X)$. We wish to find $\varphi \in \operatorname{Hom}_G(B, A)$ such that $\sigma(b) = \varphi(b)(1)$ for all $b \in B$. Let $\varphi \in \operatorname{Hom}_G(B, A)$ be such that

$$\varphi(b):\mathbb{Z}[G]\longrightarrow X \quad \text{is defined by} \quad \varphi(b)(g)=\sigma\left(g^{-1}b\right) \quad \text{for all} \quad b\in B.$$

We have

$$\varphi(b+b')(g)=\sigma\left(g^{-1}(b+b')\right)=\sigma\left(g^{-1}\,b\right)+\sigma\left(g^{-1}\,b'\right)=\varphi(b)(g)+\varphi(b')(g),$$

so $\varphi \in \operatorname{Hom}(B, A)$. Now

$$[\varphi(g\,b)]\,(g')-\sigma\left((g')^{-1}\,g\,b\right),$$

$$(g\varphi(b))\,(g')=g\,\left(\varphi(b)\left(g^{-1}\,g'\right)\right)=\varphi(b)\left(g^{-1}\,g'\right)=\sigma\left((g')^{-1}\,g\,b\right)=\varphi(g\,b)(g').$$

Therefore $g\,\varphi(b) = \varphi(g\,b)$, i.e., $\varphi \in \operatorname{Hom}_G(B, A)$ and $\varphi(b)(1) = \sigma(1^{-1}b) = \sigma(b)$. □

Theorem A.2.5. *If $A = \operatorname{Hom}(\mathbb{Z}[G], X)$ is a coinduced module, then for all $n \geq 1$ we have $H^n(G, A) = 0$. (If A is an injective G-module, then $H^n(G, A) = 0$ for all $n \geq 1$.)*

Proof. The cohomology sequence is

$$0 \longrightarrow \operatorname{Hom}_G(P_0, A) \xrightarrow{\partial_1^*} \operatorname{Hom}_G(P_1, A) \xrightarrow{\partial_2^*} \cdots .$$

By Proposition A.2.4 this sequence is the same as

$$0 \longrightarrow \operatorname{Hom}(P_0, X) \xrightarrow{\partial_1^*} \operatorname{Hom}(P_1, X) \xrightarrow{\partial_2^*} \cdots .$$

Consider the sequence

$$\cdots \longrightarrow P_n \xrightarrow{\partial_n} P_{n-1} \longrightarrow \cdots \longrightarrow P_1 \xrightarrow{\partial_1} P_0 \xrightarrow{\partial_0} 0.$$

Since the groups P_i are free, it follows that the sequence is exact starting at P_1 and that the cohomology sequence is exact starting from the first index. Hence $H^n(G, A) = 0$ for all $n \geq 1$. □

In general, for any module A, it follows from the resolution

$$\cdots \longrightarrow P_n \xrightarrow{\partial_n} P_{n-1} \longrightarrow \cdots \longrightarrow P_1 \xrightarrow{\partial_1} P_0 \xrightarrow{\partial_0} \mathbb{Z} \longrightarrow 0$$

that

$$0 \longrightarrow \operatorname{Hom}_G(\mathbb{Z}, A) \xrightarrow{\partial_0^*} \operatorname{Hom}_G(P_0, A) \xrightarrow{\partial_1^*} \operatorname{Hom}_G(P_1, A)$$

is exact at $\operatorname{Hom}_G(\mathbb{Z}, A)$ and $\operatorname{Hom}_G(P_0, A)$. Therefore

$$H^0(G, A) = \ker \partial_1^* = \operatorname{im} \partial_0^* = \operatorname{Hom}_G(\mathbb{Z}, A) = (\operatorname{Hom}(\mathbb{Z}, A))^G \cong A^G.$$

Remark A.2.6. Note that the use of ∂_0 and ∂_0^* is not the same as that defined in Definition A.1.19, since here we have an extra term $\mathbb{Z}$ in the exact sequence.

In short, what we have obtained up to now for the 0-homology and cohomology groups, including the discussion previous to Proposition A.2.1, is the following theorem:

Theorem A.2.7. *For any G-module A, we have $H_0(G, A) = A/DA = A_G$, where $DA = \langle a - \sigma a \mid \sigma \in G \rangle$ and $H^0(G, A) = A^G$.* □

Corollary A.2.8. *If $0 \longrightarrow A \longrightarrow B \longrightarrow C \longrightarrow 0$ is an exact sequence of G-modules and B is a coinduced G-module, then $H^q(G, C) = H^{q+1}(G, A)$ for all $q \geq 1$.*

Proof. From Theorems A.1.24 and A.2.7 we obtain the exact sequence of groups

$$0 \longrightarrow A^G \longrightarrow B^G \longrightarrow C^G \longrightarrow H^1(G, A) \longrightarrow H^1(G, B) \longrightarrow H^1(G, C) \longrightarrow$$
$$\cdots \longrightarrow H^q(G, B) \longrightarrow H^q(G, C) \longrightarrow H^{q+1}(G, A) \longrightarrow H^{q+1}(G, B) \longrightarrow \cdots .$$

Since $H^q(G, B) = 0$ for $q \geq 1$, the result follows. □

Theorem A.2.9. *Let A be an induced G-module of the form $A = \mathbb{Z}[G] \otimes_{\mathbb{Z}} X$. Then $H_n(G, A) = 0$ for all $n \geq 1$. (If A is a projective G-module, then A is flat and $H_n(G, A) = 0$ for all $n \geq 1$.)*

Proof. We have

$$P_n \otimes_G A \cong P_n \otimes_G (\mathbb{Z}[G] \otimes_{\mathbb{Z}} X) \cong (P_n \otimes_G \mathbb{Z}[G]) \otimes_{\mathbb{Z}} X \cong P_n \otimes_{\mathbb{Z}} X.$$

Therefore, from the resolution

$$\cdots \longrightarrow P_n \longrightarrow \cdots \longrightarrow P_1 \longrightarrow P_0 \longrightarrow \mathbb{Z} \longrightarrow 0,$$

we obtain

$$\cdots \longrightarrow P_n \otimes_G A \longrightarrow \cdots \longrightarrow P_1 \otimes_G A \longrightarrow P_0 \otimes_G A \longrightarrow \mathbb{Z} \otimes_G A \longrightarrow 0,$$

which is equivalent to

$$\cdots \longrightarrow P_n \otimes_{\mathbb{Z}} X \longrightarrow \cdots \longrightarrow P_1 \otimes_{\mathbb{Z}} X \longrightarrow P_0 \otimes_{\mathbb{Z}} X \longrightarrow \mathbb{Z} \otimes_{\mathbb{Z}} X \cong X \longrightarrow 0.$$

Since P_i is a free abelian group, the sequence is exact. Therefore $H_n(G, A) = 0$ for $n \geq 1$.

For $n = 0$, we have $H_0(G, A) = (\mathbb{Z}[G] \otimes_{\mathbb{Z}} X) / \operatorname{im} \partial_1 \cong A / I_G A$. □

Corollary A.2.10. *If $0 \longrightarrow A \longrightarrow B \longrightarrow C \longrightarrow 0$ is an exact sequence of G-modules and B is induced, then $H_{q+1}(G, C) = H_q(G, A)$ for all $q \geq 1$.*

Proof. From Theorem A.1.24 we obtain the exact sequence

$$\cdots \longrightarrow H_{q+1}(G, B) \longrightarrow H_{q+1}(G, C) \longrightarrow H_q(G, A) \longrightarrow H_q(G, B) \longrightarrow \cdots .$$

Since $H_q(G, B) = 0$ for all $q \geq 1$, the result follows. □

Lemma A.2.11. *We have $\mathbb{Z}[G] \cong \operatorname{Hom}(\mathbb{Z}[G], \mathbb{Z})$ as G-modules. In particular, $\mathbb{Z}[G]$ is coinduced.*

Proof. Let $A = \operatorname{Hom}(\mathbb{Z}[G], \mathbb{Z})$. For $f \in A$, let $\varphi(f) = \sum_{\sigma \in G} f(\sigma)\sigma$. We have $\varphi \colon A \longrightarrow \mathbb{Z}[G]$. Then φ is a G-isomorphism. □

Proposition A.2.12. *We have* $H_1(G,\mathbb{Z}) \cong I_G/I_G^2 \cong G/G'$.

Proof. Let

$$0 \longrightarrow I_G \longrightarrow \mathbb{Z}[G] \xrightarrow{\pi} \mathbb{Z} \longrightarrow 0$$

be the exact sequence where

$$\pi\left(\sum_{\sigma\in G} a_\sigma \sigma\right) = \sum_{\sigma\in G} a_\sigma .$$

Now, since $\mathbb{Z}[G]$ is coinduced, we have the exact sequence:

$$\begin{aligned} 0 = H_1(G,\mathbb{Z}[G]) &\longrightarrow H_1(G,\mathbb{Z}) \longrightarrow H_0(G,I_G) \\ &\xrightarrow{f} H_0(G,\mathbb{Z}[G]) \xrightarrow{h} H_0(G,\mathbb{Z}) \longrightarrow 0. \end{aligned}$$

Therefore $H_1(G,\mathbb{Z}) = \ker\left(H_0(G,I_G) \xrightarrow{f} H_0(G,\mathbb{Z}[G])\right)$.

By Theorem A.2.7 and Proposition 2.5.1, we have

$$H_0(G,I_G) \cong I_G/I_G^2 \cong G/G' \quad \text{and} \quad H_0(G,\mathbb{Z}[G]) \cong \mathbb{Z}[G]/I_G \cong \mathbb{Z}.$$

From the exactness of the sequence we obtain that $\operatorname{im} f = \ker h$. On the other hand, we have

$$H_0(G,\mathbb{Z}) \cong \mathbb{Z}/I_G\mathbb{Z} \cong \mathbb{Z}$$

and since h is a surjective map from $\mathbb{Z}$ to $\mathbb{Z}$, we have $\ker h = 0 = \operatorname{im} f$. Thus

$$\ker f = H_0(G,I_G) \cong I_G/I_G^2 \cong G/G'.$$ □

Now we examine the cohomology. Consider the resolution

$$\cdots \longrightarrow P_n \xrightarrow{\partial_n} P_{n-1} \longrightarrow \cdots \longrightarrow P_1 \xrightarrow{\partial_1} P_0 \xrightarrow{\partial_0} \mathbb{Z} \longrightarrow 0,$$

where

$$P_n = \mathbb{Z}\left[G^{n+1}\right] \quad \text{and} \quad \partial_n(g_0,\ldots,g_n) = \sum_{i=0}^{n}(-1)^i\left(g_0,\ldots,\hat{g}_i,\ldots,g_n\right).$$

If $K_n = \operatorname{Hom}_G(P_n,A)$, we obtain that

$$0 \longrightarrow K_0 \xrightarrow{\partial_1^*} K_1 \longrightarrow \cdots \xrightarrow{\partial_n^*} K_n \xrightarrow{\partial_{n+1}^*} \cdots,$$

and $H^n(G,A) = \ker \partial_{n+1}^*/\operatorname{im}\partial_n^*$.

Observe that $f \in \operatorname{Hom}_G(P_n, A) = \operatorname{Hom}_G\left(\mathbb{Z}\left[G^{n+1}\right], A\right)$ is determined by its values on G^{n+1} and

$$f(g_0, \ldots, g_n) = g_0 f\left(1, g_0^{-1} g_1, \ldots, g_0^{-1} g_n\right).$$

Hence f is determined by the value it takes at elements of G^{n+1} of the from $(1, g_1, g_1 g_2, \ldots, g_1 g_2 \cdots g_n)$. We write

$$\varphi(g_1, \ldots, g_n) = f(1, g_1, g_1 g_2, \ldots, g_1 g_2 \cdots g_n).$$

Let $\left[g_1 \mid g_2 \mid \cdots \mid g_{n+1}\right] := (1, g_1, g_1 g_2, \ldots, g_1 g_2 \cdots g_{n+1})$. Then

$$\begin{aligned}
\partial_{n+1}\left(\left[g_1 \mid g_2 \mid \cdots \mid g_{n+1}\right]\right) &= (g_1, g_1 g_2, \ldots, g_1 g_2 \cdots g_{n+1}) \\
&\quad + \sum_{i=1}^{n+1} (-1)^i \left(1, g_1, \ldots, \widehat{g_1 \cdots g_i}, \ldots, g_1 g_2 \cdots g_{n+1}\right) \\
&= g_1 (1, g_2, \ldots, g_2 \cdots g_{n+1}) \\
&\quad + \sum_{i=1}^{n+1} (-1)^i (1, g_1, \ldots, g_1 \cdots g_{i-1}, g_1 \cdots g_i g_{i+1}, \ldots, g_1 g_2 \cdots g_{n+1}) \\
&= g_1 \left[g_2 \mid \cdots \mid g_{n+1}\right] + \sum_{i=1}^{n+1} (-1)^i \left[g_1 \mid \cdots \mid g_{i-1} \mid g_i g_{i+1} \mid \cdots \mid g_{n+1}\right].
\end{aligned}$$

Therefore

$$f \in \ker \partial_{n+1}^* \iff \partial_{n+1}^*(f) = f \circ \partial_{n+1} = 0 \iff \text{for } g_1, \ldots, g_{n+1} \in G,$$

$$\begin{aligned}
(f \circ \partial_{n+1})\left[g_1 \mid g_2 \mid \cdots \mid g_{n+1}\right] &= g_1 f\left(\left[g_2 \mid \cdots \mid g_{n+1}\right]\right) \\
&\quad + \sum_{i=1}^{n+1} (-1)^i f\left(\left[g_1 \mid \cdots \mid g_i \mid g_i g_{i+1} \mid \cdots \mid g_{n+1}\right]\right) = 0.
\end{aligned} \tag{A.10}$$

Since $\varphi(x_1, x_2, \ldots, x_{n+1}) = f\left(\left[x_1 \mid x_2 \mid \cdots \mid x_{n+1}\right]\right)$, formula (A.10) establishes that $\ker \partial_{n+1}^*$ consists of the functions $\varphi : G^n \longrightarrow A$ satisfying

$$\begin{aligned}
&g_1 \varphi(g_2, \ldots, g_{n+1}) \\
&\quad + \sum_{i=1}^{n+1} (-1)^i \varphi(g_1, \ldots, g_{i-1}, g_i g_{i+1}, g_{i+2}, \ldots, g_{n+1}) = 0.
\end{aligned} \tag{A.11}$$

Theorem A.2.13. *We have $H^1(G, A) \cong Z^1(G, A)/B^1(G, A)$, where*

$$Z^1(G, A) = \{f : G \to A \mid f(gh) = gf(h) + f(g) \text{ for all } g, h \in G\}$$

is the group of crossed homomorphisms from G to A and

$$B^1(G, A) = \{f : G \to A \mid \text{there exists } a \in A \text{ with } f(g) = ga - a \text{ for } g \in G\}.$$

In particular, if A is a trivial G-module, then $H^1(G, A) = \operatorname{Hom}(G, A)$.

Proof. We have $H^1(G, A) = \ker \partial_2^* / \operatorname{im} \partial_1^*$. By Equation (A.11) we have

$$\ker \partial_2^* = \{ f : G \longrightarrow A \mid gf(h) - f(gh) + f(g) = 0 \} = Z^1(G, A).$$

Now let $f \in \operatorname{im} \partial_1^*$. We can write

$$f = \partial_1^*(\varphi) = \varphi \circ \partial_1 \quad \text{with} \quad \varphi \in \operatorname{Hom}_G(P_0, A) \cong A.$$

Then $f(g) = \varphi \circ \partial_1([g])$. Let $a = \varphi(1) \in A$. We have

$$\begin{aligned} f(g) &= \varphi \circ \partial_1([g]) = \varphi(\partial_1(1, g)) = \varphi(g - 1) \\ &= \varphi(g) - \varphi(1) = g\varphi(1) - \varphi(1) = ga - a. \end{aligned}$$

Therefore $\operatorname{im} \partial_1^* = B^1(G, A)$.

In particular, if G is trivial then $ga - a = 0$ for all $g \in G$. Therefore $B^1(G, A) = \{0\}$ and

$$f \in Z^1(G, A) \quad \Longleftrightarrow \quad f(gh) = gf(h) + f(g) = f(g) + f(h)$$

for all $g, h \in G$, that is, $f \in \operatorname{Hom}(G, A)$. □

We also have $H^2(G, A) = Z^2(G, A)/B^2(G, A)$. By Equation (A.11) we have

$$Z^2(G, A) = \left\{ f : G^2 \longrightarrow A \mid gf(h, m) - f(gh, m) + f(g, hm) - f(g, h) = 0 \right\}.$$

An element $f \in Z^2(G, A)$ is called a *factor set*. These sets determine the groups E such that $A \lhd E$ and $E/A \cong G$, for some abelian group A. In other words, the factor sets determine the groups E given by an exact sequence

$$0 \to A \to E \xrightarrow{\pi} G \to 0$$

and such that $g \in G$ acts on A in the following way:

$$\text{If} \quad g = \pi(e) \quad \text{with} \quad e \in E \quad \text{then} \quad g \circ a = eae^{-1}.$$

Since A is abelian, the action of g does not depend on $e \in E$.

To see how E is determined, let $s : G \to E$ be a "section," that is, s satisfies $\pi \circ s = \operatorname{Id}_G$. We have

$$\pi(s(g)s(h)) = (\pi s)(g)(\pi s)(h) = gh = (\pi s)(gh).$$

Therefore

$$s(g)s(h)s(gh)^{-1} \in \ker \pi \cong A.$$

It follows that there exists an element $f(g, h) \in A$ such that

$$s(g)s(h) = f(g,h)s(gh) \quad \text{for any} \quad g,\ h \in G.$$

The knowledge of $f : G^2 \longrightarrow A$ allows us to know E. It can be verified that f is a factor set.

Two such extensions E and E' are called *equivalent* if there exists on isomorphism $\varphi : E \longrightarrow E'$ such that the diagram

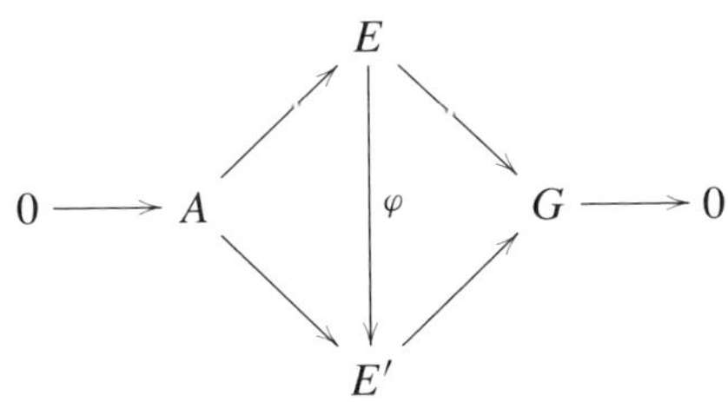

is commutative. This defines an equivalence relation whose classes are in bijective correspondence with $H^2(G, A)$. Note that if E and E' are equivalent, then they are isomorphic, but the converse does not hold (see Exercise A.5.13).

We end this section with some examples of "Galois cohomology." Consider a finite extension of fields L/K with Galois group $\mathrm{Gal}(L/K) = G$. Then L and L^* are G-modules in a natural way. Furthermore, L/K has a normal basis ([89, Theorem 13.1, p. 312]), that is, there exists $\alpha \in L$ such that $\{\sigma\alpha\}_{\sigma\in G}$ is a basis of L/K and the G-modules

$$L = \bigoplus_{\sigma\in G} K(\sigma\alpha) \quad \text{and} \quad K \otimes_{\mathbb{Z}} \mathbb{Z}[G]$$

are isomorphic. In particular, L is induced, and we obtain the folowing result:

Proposition A.2.14. *We have $H_n(G, L) = 0$ for all $n \geq 1$.*

Proof. The statement follows immediately from Theorem A.2.9. □

In fact, we have $\hat{H}^n(G, L) = 0$ for all $n \in \mathbb{Z}$, where $\hat{H}^n(G, L)$ denotes the nth Tate cohomology group (see Section A.3 below).

Proposition A.2.15. *Let F be any field. If S is any finite set of automorphisms of F, then S is linearly independent over F; in other words, if $S = \{\sigma_1, \dots, \sigma_n\}$ and $a_1, \dots, a_n \in F$ are such that*

$$a_1\sigma_1(x) + \cdots + a_n\sigma_n(x) = 0$$

for all $x \in F$, then $a_1 = \cdots = a_n = 0$.

Proof. Assume that

$$a_1\sigma_1(x) + \cdots + a_n\sigma_n(x) = 0 \tag{A.12}$$

for all $x \in F$ with some $a_i \neq 0$. Taking a minimal such relation, that is, having as few nonzero terms as possible, we may assume that n is minimal and $a_i \neq 0$ for $1 \leq i \leq n$. Note that $n > 1$. Since $\sigma_1 \neq \sigma_2$, we can choose $y \in F$ such that $\sigma_1(y) \neq \sigma_2(y)$.

Considering the element xy in (A.12) and multiplying by $\sigma_1(y)$ in (A.12), we obtain

$$\begin{aligned} &a_1\sigma_1(xy) + a_2\sigma_2(xy) + \cdots + a_n\sigma_n(xy) \\ &\qquad = a_1\sigma_1(x)\sigma_1(y) + a_2\sigma_2(x)\sigma_2(y) + \cdots + a_n\sigma_n(x)\sigma_n(y) = 0 \quad \text{(A.13)} \end{aligned}$$

and

$$a_1\sigma_1(x)\sigma_1(y) + a_2\sigma_2(x)\sigma_1(y) + \cdots + a_n\sigma_n(x)\sigma_1(y) = 0. \qquad \text{(A.14)}$$

Subtracting (A.14) from (A.13) we obtain

$$a_2\big(\sigma_2(y) - \sigma_1(y)\big)\sigma_2(x) + \cdots + a_n\big(\sigma_n(y) - \sigma_1(y)\big)\sigma_n(x) = 0$$

for all $x \in F$. Since $a_2\big(\sigma_2(y) - \sigma_1(y)\big) \neq 0$, this contradicts the minimality of n in (A.12) and proves the proposition. □

Finally, we have Hilbert's famous Theorem 90:

Theorem A.2.16 (Hilbert's Theorem 90). $H^1(G, L^*) = 0$.

Proof. Let $f \in Z^1(G, L^*)$. Then $f : G \longrightarrow L^*$ satisfies $f(\theta\sigma) = \theta(f(\sigma)) f(\theta)$ for any $\theta, \sigma \in G$. By the linear independence of automorphisms of L (Proposition A.2.15), there exists $x \in L^*$ such that $y = \sum_{\sigma \in G} f(\sigma)\sigma(x) \in L^*$.

We have, for $\theta \in G$,

$$\theta(y) = \sum_{\sigma \in G} (\theta f)(\sigma)(\theta\sigma)(x) = \sum_{\sigma \in G} f(\theta\sigma) f(\theta)^{-1} (\theta\sigma)(x) = f(\theta)^{-1} y.$$

Hence f satisfies

$$f(\theta) = \theta(y)^{-1} y \in B^1\left(G, L^*\right).$$

Therefore $H^1(G, L^*) = 0$. □

A.3 Tate Cohomology Groups

Definition A.3.1. Let G be a finite group. The element $N = \sum_{\sigma \in G} \sigma \in \mathbb{Z}[G]$ is called the *norm* of G.

For any G-module A, N defines an endomorphism of A given by $Na = \sum_{\sigma \in G} \sigma a \in A$. This endomorphism is also called the *norm* of A and in case of several G-modules A under discussion, we will use the symbol N_A in order to distinguish between the different norms.

Let $I_G = \langle \sigma - 1 \mid \sigma \in G \rangle \subseteq \mathbb{Z}[G]$. As we have seen before, I_G is the kernel of the map $\varepsilon : \mathbb{Z}[G] \longrightarrow \mathbb{Z}$ defined by $\varepsilon\left(\sum_{\sigma \in G} a_\sigma \sigma\right) = \sum_{\sigma \in G} a_\sigma$.

Now if $\sigma \in G$, we have

$$N\left((\sigma - 1)\, a\right) = \sum_{\theta \in G} \sigma\theta a - \sum_{\theta \in G} \theta a = Na - Na = 0,$$

so $I_G A \subseteq \ker N$. On the other hand, since $N\sigma a = \sigma N a = Na$, we have $NA = \operatorname{im} N \subseteq A^G$.

Recall that $H_0\,(G, A) = A/I_G A$, $H^0\,(G, A) = A^G$, so that at the quotient group level, N defines a homomorphism $N^* : H_0\,(G, A) \longrightarrow H^0\,(G, A)$.

Let $\hat{H}_0\,(G, A) = \ker N^* = \ker N / I_G A$ and $\hat{H}^0\,(G, A) = \operatorname{coker} N^* = A^G/NA$. We have the exact sequence

$$0 \longrightarrow \hat{H}_0\,(G, A) \longrightarrow H_0\,(G, A) \xrightarrow{N_A^*} H^0\,(G, A) \longrightarrow \hat{H}^0\,(G, A) \longrightarrow 0.$$

Theorem A.3.2. *Let G be a finite group and let $0 \longrightarrow A \longrightarrow B \xrightarrow{\pi} C \longrightarrow 0$ be an exact sequence of G-modules. Then the diagram*

$$\begin{array}{ccccccccc}
H_1(G, C) & \xrightarrow{\varepsilon_0} & H_0(G, A) & \longrightarrow & H_0(G, B) & \longrightarrow & H_0(G, C) & \longrightarrow & 0 \\
\downarrow & & \downarrow {\scriptstyle N_A^*} & & \downarrow {\scriptstyle N_B^*} & & \downarrow {\scriptstyle N_C^*} & & \downarrow \\
0 & \longrightarrow & H^0(G, A) & \longrightarrow & H^0(G, B) & \longrightarrow & H^0(G, C) & \xrightarrow{\delta_0} & H^1(G, A)
\end{array}$$

commutes and its rows are exact. Here ε_0 and δ_0 denote the connecting homomorphisms.

Proof. By Theorem A.1.24 the rows are exact. By definition, it is clear that the inside squares commute too. To see that the outside squares commute, we will use the explicit description of ε_0 and δ_0. We will just verify that the following square commutes, the proof for the other square being similar:

$$\delta_0 : H^0\,(G, C) = C^G \longrightarrow H^1\,(G, A) = Z^1\,(G, A)/B^1\,(G, A).$$

Let $c \in C^G$ and let $b \in B$ be such that $\pi(b) = c$. The function ∂b is defined by

$$(\partial b)(g) = gb - b \in A \quad \text{for all} \quad g \in G.$$

Now

$$\pi(gb - b) = g\pi(b) - \pi(b) = gc - c = c - c = 0$$

implies that $gb - b \in A$ and $\partial b \in Z^1\,(G, A)$.

Hence

$$\delta_0(c) = \partial\, b \bmod B^1\,(G, A) \text{ “ = ” } \partial\, \pi^{-1}(c) \bmod B^1\,(G, A)\,.$$

We want to show that $\delta_0 \circ N_C^* = 0$. Let

$$x \in H_0(G, C) = C/I_G C, \quad \text{say} \quad x = c + I_G C, N_C^* x = \sum_{\sigma \in G} \sigma c.$$

Then

$$\delta_0\left(N_C^* x\right) = \delta_0\left(\sum_{\sigma \in G} \sigma c\right) = \sum_{\sigma \in G} \delta_0(\sigma c) = \sum_{\sigma \in G} \partial \pi^{-1}(\sigma c) = \sum_{\sigma \in G} \partial(\sigma b),$$

where $\pi(b) = c$.

We have

$$\left(\sum_{\sigma \in G} \partial(\sigma b)\right)(g) = \sum_{\sigma \in G} (\partial(\sigma b))(g) = \sum_{\sigma \in G} (g\sigma b - \sigma b) = Nb - Nb = 0$$

for all $g \in G$. Therefore $\delta_0 \circ N_C^* = 0$. □

Corollary A.3.3. *There exists a canonical homomorphism*

$$\delta : \hat{H}_0(G, C) \longrightarrow \hat{H}^0(G, A)$$

that makes the group sequence

$$\hat{H}_0(G, A) \to \hat{H}_0(G, B) \to \hat{H}_0(G, C) \xrightarrow{\delta} \hat{H}^0(G, A) \to \hat{H}^0(G, B) \to \hat{H}^0(G, C)$$

exact.

Proof. This is just the snake lemma (Theorem A.1.16) applied to Theorem A.3.2. □

Theorem A.3.4. *δ gives an exact sequence:*

$$\begin{aligned} &\longrightarrow H_1(G, C) \xrightarrow{\varepsilon_0} \hat{H}_0(G, A) \longrightarrow \hat{H}_0(G, B) \longrightarrow \hat{H}_0(G, C) \\ &\xrightarrow{\delta} \hat{H}^0(G, A) \longrightarrow \hat{H}^0(G, B) \longrightarrow \hat{H}^0(G, C) \xrightarrow{\delta_0} H^1(G, A). \end{aligned}$$

Proof. We have

$$\begin{array}{ccc} \hat{H}_0(G, A) & \subseteq & H_0(G, A) \\ \| & & \| \\ \ker N_A / I_G A & \subseteq & A/I_G A \end{array} \quad ; \quad \hat{H}^0(G, C) = H^0(G, C)/\operatorname{im} N_C^*.$$

The connecting maps ε_0 and δ_0 given in Theorem A.3.2 satisfy $N_A^* \circ \varepsilon_0 = 0$ and $\delta_0 \circ N_C^* = 0$. Thus $\operatorname{im} \varepsilon_0 \subseteq \ker N_A^*$ and $\operatorname{im} N_C^* \subseteq \ker \delta_0$. The result follows immediately. □

Definition A.3.5. Let G be a finite group and let A be a G-module. We define the *Tate cohomology groups* with exponents in $\mathbb{Z}$ by

$$\begin{aligned}\hat{H}^n(G,A) &= H^n(G,A) \text{ for } n \geq 1,\\ \hat{H}^0(G,A) &= A^G/NA,\\ \hat{H}^{-1}(G,A) &= \ker N_A/I_G A,\\ \hat{H}^{-n}(G,A) &= H_{n-1}(G,A) \text{ for } n \geq 2.\end{aligned}$$

Theorem A.1.24 together with Theorem A.3.4 yields the following result:

Theorem A.3.6. *If*

$$0 \longrightarrow A \longrightarrow B \longrightarrow C \longrightarrow 0$$

is an exact sequence of G-modules, then

$$\begin{aligned}\cdots \longrightarrow \hat{H}^{n-1}(G,C) \longrightarrow \hat{H}^n(G,A) \longrightarrow \hat{H}^n(G,B)\\ \longrightarrow \hat{H}^n(G,C) \longrightarrow \hat{H}^{n+1}(G,A) \longrightarrow \cdots\end{aligned}$$

is exact for all $n \in \mathbb{Z}$. □

A.4 Cohomology of Cyclic Groups

Let G be a finite cyclic group of order n, say $G = \langle\sigma\rangle$. Let $N = \sum_{i=0}^{n-1}\sigma^i$ and $D = \sigma - 1$. Then

$$ND = DN = \left(\sum_{i=0}^{n-1}\sigma^i\right)(\sigma - 1) = \sum_{i=1}^{n}\sigma^i - \sum_{i=0}^{n-1}\sigma^i = \sigma^n - 1 = 0.$$

We have

$$\begin{aligned}I_G &= \langle g - 1 \mid g \in G\rangle\\ &= \left\langle\sigma^i - 1 = (\sigma - 1)\left(1 + \sigma + \cdots + \sigma^{i-1}\right) \mid i \in \mathbb{Z}\right\rangle\\ &= \langle\sigma - 1\rangle = D\,\mathbb{Z}[G].\end{aligned}$$

Thus N and D are maps from $\mathbb{Z}[G]$ to itself.

Proposition A.4.1. *We have* $\ker N = I_G = \operatorname{im} D$ *and* $\ker D = \mathbb{Z}[G]^G = \operatorname{im} N$.

Proof. Since $ND = 0$ and $DN = 0$, it follows that $\operatorname{im} D \subseteq \ker N$ and $\operatorname{im} N \subseteq \ker D$.

Conversely, if $s = \sum_{i=0}^{n-1} a_i\sigma^i \in \ker N$, we have

$$\begin{aligned}Ns &= \sum_{j=0}^{n-1} \sigma^j s = \sum_{j=0}^{n-1} \sigma^j \left(\sum_{i=0}^{n-1} a_i \sigma^i \right) = \sum_{i=0}^{n-1} a_i \left(\sum_{j=0}^{n-1} \sigma^{i+j} \right) \\ &= \sum_{i=0}^{n-1} a_i \left(\sum_{j=0}^{n-1} \sigma^j \right) = \sum_{j=0}^{n-1} \left(\sum_{i=0}^{n-1} a_i \right) \sigma^j = 0.\end{aligned}$$

This is equivalent to $\sum_{i=0}^{n-1} a_i = 0$, i.e., $s \in I_G = D\,\mathbb{Z}[G] = \operatorname{im} D$.

On the other hand,

$$s \in \ker D \iff (\sigma - 1)\, s = \sigma s - s = 0 \iff \sigma s = s \iff s \in \mathbb{Z}[G]^G.$$

Let $s = \sum_{i=0}^{n-1} a_i \sigma^i \in \mathbb{Z}[G]^G$. Then $\sigma s = \sum_{i=0}^{n-1} a_i \sigma^{i+1} = \sum_{i=0}^{n-1} a_{i-1} \sigma^i$ with $a_{-1} = a_{n-1}$. Therefore $\sigma s = s$ implies $a_i = a_{i-1}$, $i = 0, 1, \ldots, n-1$, and $a_i = a \in \mathbb{Z}$ for all i. We have

$$s = a \left(\sum_{i=0}^{n-1} \sigma^i \right) = N(a1) \in \operatorname{im} N. \qquad \square$$

Let $T_i = \mathbb{Z}[G]$ for $i = 0, 1, \ldots$, and define $\partial_i : T_i \longrightarrow T_{i-1}$ by

$$\partial_i = \begin{cases} D \text{ if } i \text{ is odd,} \\ N \text{ if } i \text{ is even,} \end{cases}$$

for $i = 1, 2, \ldots$. Let $\varepsilon : \mathbb{Z}[G] \longrightarrow \mathbb{Z}$ be the homomorphism defined by

$$\varepsilon \left(\sum_{i=0}^{n-1} a_i \sigma^i \right) = \sum_{i=0}^{n-1} a_i.$$

Proposition A.4.2. *The sequence of G-modules*

$$\cdots \longrightarrow T_i \xrightarrow{\partial_i} T_{i-1} \longrightarrow \cdots \longrightarrow T_1 \xrightarrow{\partial_1} T_0 \xrightarrow{\varepsilon} \mathbb{Z} \longrightarrow 0$$

is exact.

Proof. If i is even, then $\ker \partial_i = \ker N = \operatorname{im} D = \operatorname{im} \partial_{i+1}$. If i is odd, then $\ker \partial_i = \ker D = \operatorname{im} N = \operatorname{im} \partial_{i+1}$.

Finally, ε is surjective and $\ker \varepsilon = I_G = \operatorname{im} D = \operatorname{im} \partial_1$. $\square$

$\{T_i, \partial_{i+1}\}_{i=0}^{\infty}$ is a resolution of $\mathbb{Z}$ when G is a finite cyclic group. Therefore, for a G-module A, we obtain in cohomology:

$$0 \longrightarrow \operatorname{Hom}_G(T_0, A) \xrightarrow{D^*} \operatorname{Hom}_G(T_1, A) \xrightarrow{N^*} \cdots .$$

Now $\operatorname{Hom}_G(T_i, A) = \operatorname{Hom}_G(\mathbb{Z}[G], A) \cong A$. Thus we obtain:

$$0 \longrightarrow A \xrightarrow{D^*} A \xrightarrow{N^*} A \xrightarrow{D^*} \cdots ,$$

where

$$D^*a = Da = (\sigma - 1)(a) = \sigma a - a, N^*a = Na = \sum_{i=0}^{n-1} \sigma^i a.$$

We have

$$\hat{H}^{2n-1}(G, A) = H^{2n-1}(G, A) = \frac{\ker N^*}{\operatorname{im} D^*} = \frac{\ker N_A}{DA} = \hat{H}^{-1}(G, A),$$
$$\hat{H}^{2n}(G, A) = H^{2n}(G, A) = \frac{\ker D^*}{\operatorname{im} N^*} = \frac{\ker A^G}{NA} = \hat{H}^0(G, A)$$

for $n = 1, 2 \dots$.

Similarly, for homology we obtain $T_i \otimes_G A \cong \mathbb{Z}[G] \otimes_G A \cong A$ and

$$\cdots \xrightarrow{N^*} A \xrightarrow{D^*} A \longrightarrow 0.$$

Therefore, we obtain

$$\hat{H}^{-2n}(G, A) = H_{2n-1}(G, A) = \frac{\ker D^*}{\operatorname{im} N^*} = \frac{A^G}{NA} = \hat{H}^0(G, A),$$
$$\hat{H}^{-(2n+1)}(G, A) = H_{2n}(G, A) = \frac{\ker N^*}{\operatorname{im} D^*} = \frac{\ker N_A}{DA}$$
$$= \hat{H}^{-1}(G, A) = \hat{H}^1(G, A),$$

for $n = 1, 2 \dots$.

We have proved the following theorem:

Theorem A.4.3. *Let G be a finite cyclic group. Then for any G-module A, we have*

$$\hat{H}^{2n}(G, A) \cong \hat{H}^0(G, A) = \frac{A^G}{NA},$$
$$\hat{H}^{2n+1}(G, A) \cong \hat{H}^{-1}(G, A) = \frac{\ker N_A}{DA},$$

for all $n \in \mathbb{Z}$. □

Definition A.4.4. Let G be a finite cyclic group, and let A be a G-module such that $\hat{H}^0(G, A)$ and $\hat{H}^1(G, A)$ are finite of orders $h_0(A)$ and $h_1(A)$ respectively. We define the *Herbrand quotient of A* by $h(A) = \frac{h_0(A)}{h_1(A)}$.

Theorem A.4.5. *Let $0 \longrightarrow A \xrightarrow{f} B \xrightarrow{g} C \longrightarrow 0$ be an exact sequence of G-modules. We have the following exact hexagon,*

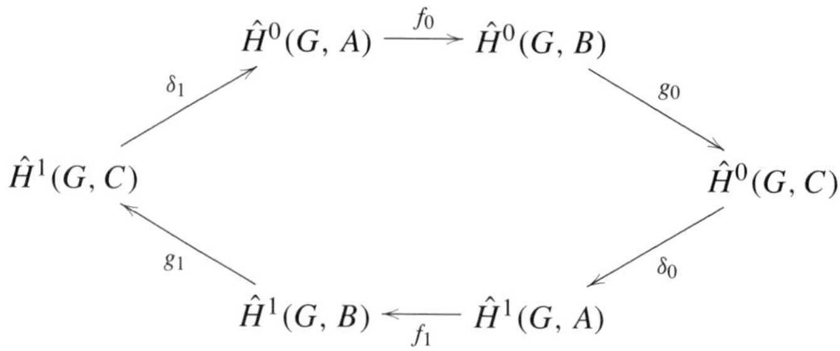

and if two of $h(A)$, $h(B)$, and $h(C)$ are defined, then the third one is defined and we have $h(B) = h(A)h(C)$.

Proof. The hexagon is simply the exact sequence given in Theorem A.3.6, and the result follows from the cyclicity of the Tate cohomology groups when G is a finite cyclic group (Theorem A.4.3).

Now say that $h(A)$ and $h(B)$ are defined. Then $h_0(C) \leq h_0(B)h_1(A) < \infty$ and $h_1(C) \leq h_1(B)h_0(A) < \infty$. Therefore $h(C)$ is defined.

Also, we have $h_0(B) = \left|\hat{H}^0(G, B)\right| = |\text{im}\, g_0|\, |\ker g_0|$, and similarly for A and C. We obtain

$$h(B) = \frac{h_0(B)}{h_1(B)} = \frac{|\text{im}\, g_0|\, |\ker g_0|}{|\text{im}\, g_1|\, |\ker g_1|},$$
$$h(A)h(C) = \frac{h_0(A)}{h_1(A)}\frac{h_0(C)}{h_1(C)} = \frac{|\text{im}\, f_0|\, |\ker f_0|}{|\text{im}\, f_1|\, |\ker f_1|}\frac{|\text{im}\, \delta_0|\, |\ker \delta_0|}{|\text{im}\, \delta_1|\, |\ker \delta_1|}.$$

From the exactness of the hexagon we obtain that $|\text{im}\, f_0| = |\ker g_0|$ and so on. The equality $h(B) = h(A)h(C)$ follows. □

Proposition A.4.6. *If A is a finite G-module, then $h(A) = 1$.*

Proof. The sequence

$$0 \longrightarrow A^G = \ker D_A \longrightarrow A \xrightarrow{D} A \longrightarrow A/DA = A_G \longrightarrow 0$$

is exact. Thus $|A_G| = \left|A^G\right|$.

Now

$$0 \longrightarrow \hat{H}^1(G, A) = \ker N^* \longrightarrow H_0(G, A) = A_G$$
$$\xrightarrow{N^*} H^0(G, A) = A^G \longrightarrow \hat{H}^0(G, A) \longrightarrow 0$$

is exact.

Therefore $h_1(A) = h_0(A)$. □

Corollary A.4.7. *If A and B are two G-modules and $f : A \longrightarrow B$ is a G-homomorphism such that* $\ker f$ *and* $\operatorname{coker} f$ *are finite, then $h(A)$ is defined if and only if $h(B)$ is defined, and in this case, we have $h(A) = h(B)$.*

Proof. The sequence

$$0 \longrightarrow \ker f \longrightarrow A \xrightarrow{f} \operatorname{im} f \longrightarrow 0$$

is exact. Thus $h(A)$ is defined if and only if $h(\operatorname{im} f)$ is defined. Now

$$0 \longrightarrow \operatorname{im} f \longrightarrow B \longrightarrow \operatorname{coker} f \longrightarrow 0$$

is exact. Therefore $h(B)$ is defined if and only if $h(\operatorname{im} f)$ is defined, if and only if $h(A)$ is defined.

In this case we have $h(A) = h(\ker f)h(\operatorname{im} f) = h(\operatorname{im} f) = \frac{h(B)}{h(\operatorname{coker} f)} = h(B)$. □

A.5 Exercises

Exercise A.5.1. Prove Proposition A.1.2.

Exercise A.5.2. Let $G = \langle\sigma\rangle$ be a finite cyclic group of order n. Prove that the G-modules $\mathbb{Z}[G]$ and $\mathbb{Z}[x]/(x^n - 1)$ are isomorphic, where $\sigma \mapsto x \bmod (x^n - 1)$, that is, the action of σ in $\mathbb{Z}[x]/(x^n - 1)$ is given by multiplication:

$$\sigma\big(f(x) \bmod (x^n - 1)\big) = xf(x) \bmod \big(x^n - 1\big).$$

Exercise A.5.3. Let G be a finite p-group and let A be a finite G-module whose order is a power of p. Prove that $A^G = \{0\}$ implies $A = \{0\}$.

Exercise A.5.4. Let G be any group, H a subgroup of G, and A a G-module. Consider the map

$$\operatorname{Res}\colon H^1(G, A) \longmapsto H^1(H, A)$$

defined as follows: if $f \in H^1(G, A)$ and $\chi \in Z^1(G, A)$ is such that $\chi \bmod B^1(G, A) = f$ with $\chi : G \to A$, then $\operatorname{Res} f = \chi|_H \bmod B^1(H, A)$.

Prove that Res is a group homomorphism. The homomorphism Res is called the *restriction homomorphism*.

Exercise A.5.5. With the notation of Exercise A.5.4 with H a normal subgroup of G, let

$$\operatorname{Inf}\colon H^1(G/H, A^H) \longmapsto H^1(G, A)$$

be defined as follows: for $f \in H^1(G/H, A^H)$ and $\chi \in Z^1(G/H, A)$ such that $\chi \bmod B^1(G/H, A^H) = f$ with $\chi : G/H \to A^H$, then $\operatorname{Inf}(f) = \chi \circ \pi \bmod B^1(G, A)$, where $\pi : G \to G/H$ is the natural projection.

Prove that Inf is a group homomorphism, called the *inflation homomorphism*.

Exercise A.5.6. With the notation of Exercises A.5.4 and A.5.5, prove that the sequence

$$0 \longrightarrow H^1(G/H, A^H) \xrightarrow{\text{Inf}} H^1(G, A) \xrightarrow{\text{Res}} H^1(H, A)$$

is exact.

Exercise A.5.7. Let G be any group and let H be a normal subgroup of G such that $[G : H] = n < \infty$. If $a \in A^H$ and $\sigma \in G$, prove that σa depends only on the left coset σ mod H. Let $N_{G/H}a := \sum_{\sigma \in G/H} \sigma a$. Prove that $N_{G/H}a \in A^G$ and that the map

$$N_{G/H} \colon \hat{H}^0(H, A) \longrightarrow \hat{H}^0(G, A)$$

is a well-defined group homomorphism; $N_{G/H}$ is called *corestriction* in dimension 0 and denoted by Cor.

Exercise A.5.8. Show that

$$(\text{Cor} \circ \text{Res})(z) = nz$$

for all $z \in \hat{H}^0(G, A)$, where $|G/H| = n$.

Exercise A.5.9. Let G be a cyclic group of order p, where p is a prime number. Define $A_1 := \mathbb{Z}_p$, $A_{p-1} := \mathbb{Z}_p[\zeta_p] = \mathbb{Z}_p[x]/(\Psi_p(x))$, and $A_p := \mathbb{Z}_p[G]$, where $\mathbb{Z}_p$ is the ring of p-adic integers, ζ_p is a primitive pth root of 1, and the action is as in Exercise A.5.2. Then A_1, A_{p-1}, A_p are G-modules. Prove that

$$\begin{aligned}
&\hat{H}^0(G, A_1) \cong C_p, && \hat{H}^{-1}(G, A_1) \cong 0,\\
&\hat{H}^0(G, A_{p-1}) \cong 0, && \hat{H}^{-1}(G, A_{p-1}) \cong C_p,\\
&\hat{H}^0(G, A_p) \cong 0, && \hat{H}^{-1}(G, A_p) \cong 0,
\end{aligned}$$

where C_p is the cyclic group of p elements.

Exercise A.5.10. Let G be a finite group and p^m the maximal power of p that divides $|G|$. Prove that $\hat{H}^1(G, \mathbb{Z}_p) \cong \hat{H}^{-1}(G, \mathbb{Z}_p) = \{0\}$ and $\hat{H}^0(G, \mathbb{Z}_p) \cong \mathbb{Z}_p/p^m\mathbb{Z}_p$. Also show that $H^i(G, \mathbb{Q}_p) = \{0\}$ for all i. You may use that $H^i(G, \mathbb{Q}_p)$ is a finite group.

Hint: Consider the isomorphism

$$\begin{aligned}
\mathbb{Q}_p &\xrightarrow{\ n\ } \mathbb{Q}_p\\
x &\longmapsto nx
\end{aligned}$$

for any $n \in \mathbb{Z} \setminus \{0\}$.

Exercise A.5.11. With the notation of Exercise A.5.10, prove that

$$\hat{H}^i(G, R) \cong \hat{H}^{i+1}(G, \mathbb{Z}_p)$$

for all i, where $R = \mathbb{Q}_p/\mathbb{Z}_p$.

Exercise A.5.12. Let G be a finite p-group and M a G-module. Assume that there exists $s \in \mathbb{N} \cup \{0\}$ such that the groups M and R^s are isomorphic, where R is as in Exercise A.5.11. Consider the exact sequence

$$0 \longrightarrow {}_pM \longrightarrow M \stackrel{p}{\longrightarrow} M \longrightarrow 0,$$

where the map denoted by p is multiplication by p and ${}_pM := \{m \in M \mid pm = 0\}$. Show that if

$$\alpha_i(M) = \dim_{\mathbb{F}_p} \frac{\hat{H}^i(G, M)}{p\hat{H}^i(G, M)} = \dim_{\mathbb{F}_p} {}_p\hat{H}^i(G, M),$$

then $\hat{H}^i(G, {}_pM) \cong C_p^{\alpha_{i-1}(M)+\alpha_i(M)}$.

Exercise A.5.13. Give an example of groups E and E' such that $0 \to A \to E \to G \to 0$ and $0 \to A \to E' \to G \to 0$ are exact sequences of groups, $A \lhd E$, $A \lhd E'$, A is abelian, $E \cong E'$, but E and E' are not equivalent.

Exercise A.5.14. Let K/k be a function field with k algebraically closed. Let L/K be a finite Galois extension with Galois group G. If D_L denotes the divisor group of L, prove that $\hat{H}^{-1}(G, D_L) = \{0\}$ and $\hat{H}^0(G, D_L) \cong \bigoplus_{i=1}^r C_{e_i}$, where $\mathfrak{p}_1, \ldots, \mathfrak{p}_r$ are the prime divisors in K ramified in L/K with ramification indices $e_1, \ldots, e_r$.

Notations

A^G = G-submodule of A where G acts trivially, 599.

A_G = maximal G-quotient module of A where G acts trivially, 615.

$A_\wp$ = localization of the commutative ring with unity A at the prime ideal $\wp$.

$|A|$ = cardinality of the set A.

$\operatorname{Aut}(L/K) = \operatorname{Aut}_K L$ = group of K-automorphisms of L, 118.

$\alpha\, d\beta$ = Hasse differential, 294.

$\mathbb{C}$ = field of complex numbers.

$\operatorname{char} K$ = characteristic of the field K.

C_K = divisor class group of the field K, 62.

$C_{K,0}$ = degree 0 divisor class group of classes of the field K, 65.

D_K = divisor group of the field K, 55.

$D_{K,0}$ = group of divisors of degree 0 of the field K, 64.

$d_K(\mathfrak{A})$ = degree of the divisor $\mathfrak{A}$, 65.

$\mathfrak{D}_{L/K}$ = different of the extension L/K, 149.

$\mathfrak{D}_{B/A}$ = different of the extension of Dedekind domains B/A, 154.

$D_{L/K}(\mathcal{P}|\wp)$ = decomposition group of the place $\mathcal{P}$ over the place $\wp$, 120.

$d_{L/K}(\mathcal{P}|\wp)$ = relative degree of the place $\mathcal{P}$ with respect to the place $\wp$, 43, 115.

dx = principal differential, 96.

$D(\mathfrak{A})$ = differentials divisible by the divisor $\mathfrak{A}$, 77.

Dif_K = differentials in the field K, 78.

$\partial_{L/K}$ = discriminant of the extension L/K, 149.

$\dim_k V$ = dimension of the k-vector space V.

$e_{L/K}(\mathcal{P}|\wp)$ = ramification index of the place $\mathcal{P}$ with respect to the place $\wp$, 114.

$f = o(g)$ means $|f(x)| \leq c|g(x)|$, for x large enough, 223.

$\mathbb{F}_q$ = finite field of q elements, 31.

g_K = genus of the field K, 69.

$\operatorname{Gal}(L/K)$ = Galois group of the extension L/K.

$\Gamma(\mathfrak{A}|S) = \{x \in K \mid v_{\mathcal{P}}(x) \geq v_{\mathcal{P}}(\mathfrak{A}) \text{ for all } \mathcal{P} \in S\}$, 56.

H_A = Hilbert class field of the Dedekind domain A, 517.

h_K = class number of the field K, 65.

$I_{L/K}(\mathcal{P}|\wp)$ = inertia group of the place $\mathcal{P}$ over the place $\wp$, 121.

$\operatorname{im}\varphi$ = image of the homomorphism φ.

$\operatorname{Irr}(\alpha, x, K)$ = irreducible polynomial in $K[x]$ of the element α.

$\ker\varphi$ = kernel of the homomorphism φ.

$K_{\mathcal{P}}$ = completion of the field K with respect to the valuation $v_{\mathcal{P}}$, 28, 29.

K_ρ = smallest field of definition of a Drinfeld module ρ, 508.

$k(\mathcal{P})$ = residue field with respect to the place $\mathcal{P}$, 29.

$K(x_1, x_2, \ldots, x_n)$ = rational function field in n variables with coefficients in K.

$K[x_1, x_2, \ldots, x_n]$ = ring of polynomials in n variables with coefficients in K.

$(K, \mathfrak{P}_\infty, \operatorname{sgn})$ = triple of a congruence function field K with a fixed prime divisor $\mathfrak{P}_\infty$ and a fixed sign function sgn, 510.

$\left[\frac{L/K}{\mathcal{P}}\right]$ = Frobenius symbol, 378.

$\left(\frac{L/K}{\wp}\right)$ = Artin symbol, 379.

$[L : K]$ = degree of the extension L/K.

$L_K(\mathfrak{A}) = \{x \in K \mid v_{\mathcal{P}}(x) \geq v_{\mathcal{P}}(\mathfrak{A}) \text{ for all } \mathcal{P} \in \mathbb{P}_K\}$, 58.

$\ell(\mathfrak{A})$ = dimension of the k-vector space $L(\mathfrak{A})$, 59.

$\varprojlim_{i \in I} A_i$ = inverse limit, 389.

$m \gg n$ means m larger enough than n.

$\mathbb{N}$ = set of natural numbers.

$\aleph_0 = |\mathbb{N}|$.

$\mathfrak{N}$ = unit divisor, 56.

$\mathfrak{N}_x$ = pole divisor of x, 62.

$N(C)$ = dimension of the class C, 69.

Pic A = class or Picard group of the Dedekind domain A, 505.

P_K = principal divisor group of the field K, 62.

$\mathfrak{P} \mid \mathfrak{p}$ = the prime divisor $\mathfrak{P}$ divides the prime divisor $\mathfrak{p}$, 114.

$\mathbb{P}_K$ = set of all places in the field K, 55.

quot A = field of quotients of the integral domain A, 19.

R^* = group of units of the commutative ring with unity R.

$\mathbb{Q}$ = field of rational numbers.

$\mathbb{Q}_p$ = field of p-adic numbers, 29.

$\mathbb{R}$ = field of real numbers.

tr L/K = transcendental degree of L over K, 3.

$\mathrm{Tr}_{L/K}\,\Omega$ = trace of a differential, 290.

$U_{\mathfrak{p}}$ = group of units of a local field, 471.

$\mathfrak{A} \mid \mathfrak{B}$ = the divisor $\mathfrak{A}$ divides the divisor $\mathfrak{B}$, 56.

$\mathfrak{A} \mid \xi$ = the divisor $\mathfrak{A}$ divides the repartition ξ, 71.

$\mathfrak{A} \mid \omega$ = the divisor $\mathfrak{A}$ divides the differential ω, 75.

$\mathfrak{A} \mid C$ = the divisor $\mathfrak{A}$ divides the class C, 85.

$v_{\mathcal{P}}$ = valuation with respect to the place $\mathcal{P}$, 43.

$(x)_K$ = principal divisor of the element $x \in K^*$, 62.

$\mathfrak{X}_K = \Lambda_K$ = repartitions or adeles of the field K, 70.

$\mathfrak{X}(\mathfrak{A}) = \Lambda(\mathfrak{A})$ = repartitions divisible by the divisor $\mathfrak{A}$, 71.

W_K = canonical class of the field K, 81.

$W_x(z_0, \dots, z_n)$ = Wronskian determinant with respect to D_x, 548.

$\mathfrak{H}$ = set of all Hayes–modules, 510.

$\mathbb{Z}$ = ring of integers.

$Z_K(u)$ = zeta function of the field K, 196.

$\mathbb{Z}_p$ = ring of p-adic integers, 29.

$\zeta_K(s)$ = zeta function of the field K, 195.

$\mathfrak{Z}_x$ = zero divisor of x, 62.

$\emptyset$ = empty set.

$\square$ = end of a proof.

References

1. Accola, Robert D. M., *Topics in the Theory of Riemann Surfaces*, Lecture Notes in Mathematics **1595**, Berlin: Springer-Verlag, (1994).
2. Adleman, Leonard M.; DeMarrais, Jonathan; Huang, Ming-Deh, A subexponential algorithm for discrete logarithms over the rational subgroup of the Jacobians of large genus hyperelliptic curves over finite fields, *Lecture Notes in Computer Science* **877**, Springer-Verlag, 28–40, (1994).
3. Artin, Emil, *Algebraic Numbers and Algebraic Functions*, New York–London–Paris: Gordon and Breach, Science Publishers (1967).
4. Atiyah, Michael Francis; Macdonald, Ian G., *Introduction to Commutative Algebra*, Reading, MA–London–Don Mills, Ont.: Addison–Wesley Publishing Co., (1969).
5. Bers, Lipman, *Riemann Surfaces*, New York University, (1957–1958).
6. Blackburn, Simon; Cid, Carlos; Galbraith, Steven, Cryptanalysis of a cryptosystem based on Drinfeld modules, http://eprint.iacr.org/2003/223.pdf, preprint.
7. Bombieri, Enrico, Counting points on curves over finite fields (d'après S. A. Stepanov), Sém. Bourbaki 1972/73, Exposé No.430, *Lecture Notes Math.* **383**, 234–241, (1974).
8. Boseck, Helmut, Zur Theorie der Weierstraßpunkte, *Math. Nachr.* **19**, 29–63, (1958).
9. Bourbaki, Nicolas, *Elements of Mathematics. Algebra I. Chapters 1–3*, Berlin–Heidelberg–New York: Springer-Verlag, (1989).
10. Bressoud, David M., *Factorization and Primality Testing*, Undergraduate Text in Mathematics, Springer-Verlag, (1989).
11. Brown, Kenneth S., *Cohomology of Groups*, Springer-Verlag, Graduate Texts in Mathematics **87**, (1982).
12. Buchmann, Johannes A., *Introduction to Cryptography*, New York: Springer-Verlag, (2001).
13. Carlitz, Leonard, On the representation of a polynomial in a Galois field as the sum of an even number of squares, *Trans. Amer. Math. Soc.* **35**, 397–410, (1933).
14. Carlitz, Leonard, On certain functions connected with polynomials in a Galois field, *Duke Math. J.* **1**, 137–168, (1935).
15. Carlitz, Leonard, A class of polynomials, *Trans. Amer. Math. Soc.* **43**, 137–168, (1938).
16. Cartier, Pierre, Une nouvelle opération sur les formes differentielles, *C. R. Acad. Sci., Paris* **244**, 426–428, (1957).
17. Cassels, J.W.S.; Fröhlich, Albrecht, editors, *Algebraic Number Theory*, Advanced Study Institute, London and New York: Academic Press (1967).

18. Cohen I. S., On the structure and ideal theory of complete local rings, *Trans. Amer. Math. Soc.* **59**, 54–106, (1946).
19. Cohn, Paul Moritz, *Algebraic Numbers and Algebraic Functions*, Chapman and Hall Mathematics Series, Chapman & Hall, London, (1991).
20. Cronheim, Arno, Ein Funktionenkörper von Primzahlcharakteristik ohne Automorphismen, *Math. Nachr.* **18**, 99–105 (1958).
21. Chevalley, Claude, On the composition of fields, *Bull. Am. Math. Soc.* **48**, 482–487, (1942).
22. Chevalley, Claude, *Introduction to the Theory of Algebraic Functions of One Variable*, Mathematical Surveys no. **6**, New York: American Mathematical Society XI, (1951).
23. Deligne, Pierre; Husemoller, Dale H., Survey of Drinfel'd modules, Current trends in arithmetical algebraic geometry, *Proc. Summer Res. Conf., Arcata/Calif. 1985, Contemp. Math.* **67**, 25–91 (1987).
24. Denef, Jan; Vercauteren, Frederik, An Extension of Kedlaya's Algorithm to Hyperelliptic Curves in Characteristic 2, *J. Cryptology online February 23*, (2005).
25. Deuring, Max, Zur arithmetischen Theorie der algebraischen Funktionen, *Math. Ann.* **106**, 77–102, (1932).
26. Deuring, Max, Invarianten und Normalformen elliptischer Funktionenkörper, *Math. Z.* **47**, 47–56, (1941).
27. Deuring, Max, Zur Theorie der elliptischen Funktionenkörpern, *Sem. Univ. Hamburg* **15**, 211–261, (1945).
28. Deuring, Max, *Lectures on the Theory of Algebraic Functions of One Variable*, Lecture Notes in Mathematics **314**, Berlin–Heidelberg–New York: Springer-Verlag, (1973).
29. Diffie, Whitfield; Hellman, Martin E., New directions in cryptography, *IEEE Trans. Inf. Theory* **22**, 644–654, (1976).
30. Drinfel'd, Vladimir G., Elliptic modules, *Math. USSR, Sb.* **23**, 561–592 (1974); translation from *Mat. Sb., n. Ser.* **94(136)**, 594–627 (1974).
31. Drinfel'd, Vladimir G., Elliptic modules. II, *Math. USSR, Sb.* **31**, 159–170 (1977).
32. Eichler, Martin, *Introduction to the Theory of Algebraic Numbers and Functions*, New York and London: Academic Press, (1966).
33. ElGamal, Taher, A public key cryptosystem and a signature scheme based on discrete logarithms, *IEEE Trans. Inf. Theory* **31**, 469–472, (1985).
34. Farkas, Hershel M.; Kra, Irwin, *Riemann Surfaces*, second edition, Graduate Texts in Mathematics **71**, New York: Springer-Verlag, (1992).
35. Fesenko Ivan B.; Vostokov Sergei V., *Local Fields and Their Extensions: A Constructive Approach*, Translations of Mathematical Monographs **121**, Providence, RI, American Mathematical Society, (1993).
36. Forster, Otto, *Lectures on Riemann Surfaces*, Graduate Texts in Mathematics, **81**, New York: Springer-Verlag, (1991).
37. Frey, Gerhard; Rück, Hans-Georg, A remark concerning m-divisibility and the discrete logarithm in the divisor class group of curves, *Math. Comput.* **62**, No. 206, 865–874, (1994).
38. Fried, Michael D.; Jarden, Moshe, *Field Arithmetic*, Ergebnisse der Mathematik und ihrer Grenzgebiete 3, Folge, Bd. **11**, Springer-Verlag, (1986).
39. Fulton, William, *Algebraic Curves*, Benjamin, (1969).
40. Galbraith, Steven D., Weil descent of Jacobians, *Discrete Appl. Math.* **128**, No.1, 165–180, (2003).
41. Galovich, Steven; Rosen, Michael, Distributions on rational function fields, *Math. Ann.* **256**, 549–560, (1981).

42. Galovich, Steven; Rosen, Michael, The class number of cyclotomic function fields, *Journal of Number Theory* **13**, 363–375, (1981).
43. Galovich, Steven; Rosen, Michael, Units and class groups in cyclotomic function fields, *Journal of Number Theory* **14**, 156–184, (1982).
44. Garcia, Arnaldo, On Weierstrass points on Artin–Schreier extensions of $k(x)$, *Math. Nachr.* **144**, 233–239, (1989).
45. Garcia, Arnaldo, On Weierstrass points on certain elementary abelian extensions of $k(x)$, *Commun. Algebra* **17**, No.12, 3025–3032, (1989).
46. Garcia, Arnaldo; Stichtenoth, Henning, Elementary abelian p-extensions of algebraic function fields, *Manuscr. Math.* **72**, No.1, 67–79, (1991).
47. Gaudry, Pierrick, An algorithm for solving the discrete log problem on hyperelliptic curves, *Advances in Cryptology—EUROCRYPT 2000* (Bruges), 19–34, Lecture Notes in Comput. Sci. **1807**, Berlin: Springer, (2000).
48. Gekeler, Ernst-Ulrich, *Drinfeld Modular Curves*, Lecture Notes in Mathematics **1231**, Springer-Verlag, (1986).
49. Gillard, Roland; Leprevost, Franck; Panchishkin, Alexi; Roblot, Xavier-François, Utilisation des modules de Drinfeld en cryptologie, *C. R. Acad. Sci. Paris, Ser I* **336**, 879–882, (2003).
50. Goldschmidt, David M., *Algebraic Functions and Projective Curves*, Graduate Text in Mathematics **215**, Springer-Verlag, (2002).
51. Goss, David, *Basic Structures of Function Field Arithmetic*, Berlin: Springer-Verlag, (1996).
52. Hasse, Helmut, Theorie der relativ–zyklischen algebraischen Funktionenkörper, insbesondere bei endlichen Konstantenkörper, *J. Reine Angew. Math.* **172**, 37–54, (1934).
53. Hasse, Helmut, Theorie der Differentiale in algebraischen Funktionenkörpern mit vollkommenem Konstantenkörper, *J. Reine Angew. Math.* **172**, 55–64, (1934).
54. Hasse, Helmut, Theorie der höheren Differentiale in einem algebraischen Funktionenkörper mit vollkommenem Konstantenkörper bei beliebiger Charakteristik, *J. Reine Angew. Math.* **175**, 50–54, (1936).
55. Hasse, Helmut, Zur Theorie der abstrakten elliptischen Funktionenkörpern I, *J. Reine Angew. Math.* **175**, 55–62, (1936).
56. Hasse, Helmut, Zur Theorie der abstrakten elliptischen Funktionenkörpern II, *J. Reine Angew. Math.* **175**, 69–88, (1936).
57. Hasse, Helmut, Zur Theorie der abstrakten elliptischen Funktionenkörpern III, *J. Reine Angew. Math.* **175**, 193–208, (1936).
58. Hasse, Helmut; Schmidt, Friedrich Karl, Noch eine Begründung der Theorie der höheren Differentialquotienten in einem algebraischen Funktionenkörper einer Unbestimmten, *J. Reine Angew. Math.* **177**, 215–237, (1937).
59. Hasse, Helmut; Witt, Ernst, Zyklische unverzweigte Erweiterungskörper vom Primzahlgrade p über einem algebraischen Funktionenkörper der Charakteristik p, *Monatsh. Math. Phys.* **43**, 477–492, (1936).
60. Hayes, David R., The Galois group of $x^n + x - t$, *Duke Math. J.* **40**, 459–461, (1973).
61. Hayes, David R., Explicit class field theory for rational function fields, *Trans. Amer. Math. Soc.* **189**, 77–91, (1974).
62. Hayes, David R., Explicit class field theory in global function fields, Studies in algebra and number theory, *Adv. in Math. Suppl. Stud.*, **6**, New York–London: Academic Press, 173–217, (1979).
63. Hayes, David R., A brief introduction to Drinfel'd modules, The arithmetic of function fields (Columbus, OH, 1991), *Ohio State Univ. Math. Res. Inst. Publ.*, **2**, de Gruyter, Berlin, 1–32, (1992).

64. hÉigeartaigh Ó Colm, A Comparison of Point Counting methods for Hyperelliptic Curves over Prime Fields and Fields of Characteristic 2, *Cryptology ePrint Archive*, Report 2004/241, http://eprint.iacr.org/2004/241, (2004).
65. Henn, Hans-Wolfgang, *Automorphismengruppe und Weierstraßpunkte von Funktionenkörpern*, Dissertation, Fakultät für Mathematik der Universität Karlsruhe, (1975).
66. Henn, Hans-Wolfgang, Funktionenkörper mit grosser Automorphismengruppe, *J. Reine Angew. Math.* **302**, 96–115, (1978).
67. Hess, Florian; Seroussi, Gadiel; Smart, Nigel P., Two topics in hyperelliptic cryptography, Vaudenay, Serge (ed.) et al., Selected areas in cryptography. 8th annual international workshop, SAC 2001, Toronto, Ontario, Canada, August 16–17, 2001. Revised papers. Berlin: Springer, *Lect. Notes Comput. Sci.* **2259**, 181–189 (2001).
68. Hilton, Peter John; Stammbach, Urs, *A Course in Homological Algebra*, Springer-Verlag, Graduate Texts in Mathematics **4**, (1970).
69. Hungerford, Thomas W., *Algebra*, Springer-Verlag, Graduate Texts in Mathematics **73**, (1974).
70. Hurwitz, Adolf, Über algebraische Gebilde mit eindeutigen Transformationen in sich, *Math. Ann.* **41**, 403–442, (1893).
71. Inaba, Eizi, Number of divisor classes in algebraic function fields, *Proc. Japan Academy* **26**, 1–4, (1950).
72. Iwasawa, Kenkichi, *Algebraic Functions*, Translations of Mathematical Monographs **118**, Providence: American Mathematical Society, (1993).
73. Iwasawa, Kenkichi; Tamagawa, Tsuneo, On the group of automorphisms of a function field, *J. Math. Soc. Japan* **3**, 137–147, (1951).
74. Iwasawa, Kenkichi; Tamagawa, Tsuneo, On the group of automorphisms of a function field. Correction, *J. Math. Soc. Japan* **4**, 100–101, (1952).
75. Iwasawa, Kenkichi; Tamagawa, Tsuneo, Correction: On the paper "On the group of automorphisms of a function field" (this journal, vol. **3** (1951), pp. 137–147), *J. Math. Soc. Japan* **4**, 203–204, (1952).
76. Iyanaga, Shôkichi, editor, *The Theory of Numbers*, North-Holland Mathematical Library, **8**, Amsterdam–Oxford: North-Holland Publishing Co., New York: American Elsevier Publishing Co., Inc., (1975).
77. Jacobson, Nathan, *Lectures in Abstract Algebra, Part III, Theory of Fields and Galois Theory*, Springer-Verlag, Graduate Texts in Mathematics **32**, (1964).
78. Janusz, Gerald J., *Algebraic Number Fields*, 2nd ed., Providence, RI: American Mathematical Society (AMS), (1996).
79. Kida, Masanari; Murabayashi Naoki, Cyclotomic function fields with divisor class number one, *Tokyo J. Math.* **14**, No. 1, 45–56, (1991).
80. Kitamura, Izuru; Katagi, Masanobu, Efficient Implementation of Genus Three Hyperelliptic Curve Cryptography over $\mathbb{F}_{2^n}$, *Cryptology ePrint Archive*, Report 2003/248, http://eprint.iacr.org/2003/248, (2003).
81. Koblitz, Neal, Elliptic curve cryptosystems, *Math. Comput.* **48**, 203–209, (1987).
82. Koblitz, Neal, *A Course in Number Theory and Cryptography*, Graduate Text in Mathematics **114**, Springer-Verlag, (1987).
83. Koblitz, Neal, Hyperelliptic cryptosystems, *J. Cryptology* **1**, 139–150, (1989).
84. Koblitz, Neal, *Algebraic Aspects of Cryptography*, Algorithms and Computation in Mathematics, Volume **3**, Springer-Verlag, (1998).
85. Koch, Helmut, *Algebraic Number Theory*, translated from the 1988 Russian edition, reprint of the 1992 translation, Berlin: Springer-Verlag, (1997).
86. Kontogeorgis, Aristides, The group of automorphisms of cyclic extensions of rational function fields, *Journal of Algebra* **216**, 665–706, (1999).

87. Lam-Estrada, Pablo, *Campos de Funciones Ciclotómicos y Extensiones Pseudo-Cogalois*, Ph.D. dissertation, CINVESTAV–IPN, México (1997).
88. Lam-Estrada, Pablo; Villa-Salvador, Gabriel, Some remarks on the theory of cyclotomic function fields, *Rocky Mountain Journal of Mathematics*, **31**, No. 2, 483–502, (2001).
89. Lang, Serge, *Algebra*, third edition, Reading, MA: Addison–Wesley Publishing Company, (1993).
90. Lang, Serge, *Algebraic Number Theory*, second edition, Graduate Texts in Mathematics **110**, New York: Springer-Verlag, (1994).
91. Lang, Serge; Tate, John, Principal homogeneous spaces over abelian varieties, *Am. J. Math.*, **80**, 659–684, (1958).
92. Lange Tanja, Koblitz curve cryptosystems, *Finite Fields and Their Applications* **11**, 200–229, (2005).
93. Le Brigand, Dominique, *Decoding of Codes on Hyperelliptic Curves*, Lecture Notes in Computer Science **514**, Springer-Verlag, 126–134, (1990).
94. Leitzel, James R.C.; Madan, Manohar L., Algebraic function fields with equal class number, *Acta Arith.* **30**, 169–177, (1976).
95. Leitzel, James R. C.; Madan, Manohar L.; Queen, Clifford S., Algebraic function fields with small class number, *Journal of Number Theory* **7**, 11–27, (1975).
96. Leopoldt, Heinrich-Wolfgang, Über die Automorphismengruppe des Fermatkörpers, *Journal of Number Theory* **56**, 256–282, (1996).
97. Leptin Horst, Ein Darstellungssatz für kompakte total unzusammenhängende Gruppen, *Arch. Math.* **6**, 371–373, (1955).
98. Lluis-Puebla, Emilio, *Álgebra Homológica, Cohomología de Grupos y K–Teoría Algebraica Clásica*, Addison–Wesley, (1990).
99. López-Bautista, Pedro Ricardo; Villa-Salvador, Gabriel, On the Galois module structure of semisimple holomorphic differentials, *Isr. J. Math.* **116**, 345–365, (2000).
100. MacRae, Robert E., On Unique factorization in certain rings of algebraic functions, *Journal of Algebra* **17**, 243–261, (1971).
101. Madan, Manohar L., On a theorem of M. Deuring and I. R. Šafarevič, *Manuscr. Math.* **23**, 91–102, (1977).
102. Madan, Manohar L.; Madden, Daniel J., Note on Class Groups of Algebraic Function Fields, *J. Riene Angew. Math.* **295**, 57–60, (1977).
103. Madan, Manohar L.; Madden, Daniel J., On the theory of congruence function fields, *Commun. Algebra* **8**, 1687–1697, (1980).
104. Madan, Manohar L.; Queen, Clifford S., Algebraic function fields of class number one, *Acta Arith.* **20**, 423–432, (1972).
105. Madan, Manohar; Rosen, Michael, The automorphism group of a function field, *Proc. Am. Math. Soc.* **115**, No.4, 923–929, (1992).
106. Madden, Daniel J.; Valentini, Robert C., The Group of Automorphisms of Algebraic Function Fields, *J. Angew. Math.* **343**, 162–168, (1983).
107. Matzat, B. Heinrich, *Über Weierstraßpunkte von Fermatkörpern*, Dissertation, Fakultät für Mathematik der Universität Karlsruhe, (1972).
108. McCarthy, Paul J., *Algebraic Extensions of Fields*, Baisdell Publishing Company, (1966).
109. Menezes Alfred J.; Okamoto Tatsuaki; Vanstone Scott A., Reducing elliptic curve logarithms to lagarithms in a finite field, *IEEE Trans. Inform. Theory*, **39** No. 5, 1639–1646, (1993).
110. Menezes Alfred J.; van Oorschot Paul C.; Vanstone Scott A., *Handbook of Applied Cryptography*, CRC Press, (1997).

111. Miller, Victor S., Use of elliptic curves in cryptography, *Advances in Cryptology* - CRYPTO '85, Proc. Conf., Santa Barbara/Calif. 1985, Lect. Notes Comput. Sci. **218**, 417–426, (1986).
112. Mollin, Richard A., *RSA and Public-Key Cryptography*, Chapman & Hall /CRC, (2003).
113. Moreno, Carlos J., *Algebraic Curves Over Finite Fields*, Cambridge Tracts in Mathematics, **97**, Cambridge University Press, (1991).
114. Mumford, David, *Tata Lectures on Theta. II: Jacobian Theta Functions and Differential Equations*. With the collaboration of C. Musili, M. Nori, E. Previato, M. Stillman, and H. Umemura, Progress in Mathematics, Vol. **43**, Boston–Basel–Stuttgart: Birkhäuser, (1984).
115. Neukirch, Jürgen, *Class Field Theory*, Springer-Verlag, Grundlehren der Mathematischen Wissenschaften **280**, (1986).
116. Pohlig, Stephen C.; Hellman, Martin E., An improved algorithm for computing logarithms over $GF(p)$ and its cryptographic significance, *IEEE Trans. Inf. Theory* **24**, 106–110, (1978).
117. Poincaré, Henri, Sur un théorème de M. Fuchs, *Acta Math.* **7**, 1–32, (1885).
118. Pollard, John H., Monte Carlo methods for index computation $\pmod p$, *Math. Comput.* **32**, 918–924, (1978).
119. Pollard, John H., Kangaroos, monopoly and discrete logarithms, *J. Cryptology* **13**, No.4, 437–447, (2000).
120. Pontriaguin, Lev Semióvich, *Grupos continuos*, Mir, Moscú, (1978).
121. Reichardt, Hans, Der Primdivisorsatz für algebraische Funktionenkörper über einem endlichen Konstantenkörper, *Math. Z.* **40**, 713–719, (1936).
122. Ribes, Luis; Zalesskii Pavel, *Profinite Groups*, Springer-Verlag, Ergebnisse der Mathematik und ihrer Ganzgebiete Folge 3, **40**, (2000).
123. Rivest, Ronald L.; Shamir, Adi; Adleman, Leonard, A method for obtaining digital signatures and public-key cryptosystems, *Commun. ACM* **21**, 120–126, (1978).
124. Roquette, Peter, Über die Automorphismengruppe eines algebraischen Funktionenkörpers, *Arch. Math.* **3**, 343–350, (1952).
125. Roquette, Peter, Abspaltung des Radikals in vollständigen lokalen Ringen, *Hbg. Math. Abh.* **23**, 75–113, (1959).
126. Roquette, Peter, Abschätzung der Automorphismenzahl von Funktionenkörpern bei Primzahlcharakteristik, *Math. Z.* **117**, 157–163, (1970).
127. Rosen, Michael, The Hilbert class field in function fields, *Expo. Math.* **5**, 365–378, (1987).
128. Rosen, Michael, *Number Theory in Function Fields*, Graduate Texts in Mathematics, **210**, New York: Springer, (2002).
129. Rosenlicht, Maxwell, Automorphisms of function fields, *Trans. Am. Math. Soc.* **79**, 1–11, (1955).
130. Rudin, Walter, *Real and Complex Analysis*, second edition, McGraw–Hill Series in Higher Mathematics, New York–Düsseldorf–Johannesburg, McGraw–Hill Book Co., (1974).
131. Rzedowski-Calderón, Martha; Villa-Salvador, Gabriel, Automorphisms of congruence function fields, *Pacific J. Math.* **150**, No.1, 167–178, (1991).
132. Scanlon, Thomas, Public key cryptosystems based on Drinfeld modules are insecure, *J. Cryptology* **14**, no. 4, 225–230, (2001).
133. Šafarevič, Igor R., Exponents of Elliptic Curves, *Dokl. Akad. Nauk SSSR.* **114**, 714–716, (1957).
134. Schmid, Hermann Ludwig, Über das Reziprozitätsgesetz in relativ zyklischen algebraischen Funktionenkörpern mit endlichem Konstantenkörper, *Math. Z.* **40**, 91–109, (1935).

135. Schmid, Hermann Ludwig, Zur Arithmetik der zyklischen p-Körper, *J. Reine Angew. Math.* **176**, 161–167, (1937).
136. Schmid, Hermann Ludwig, Über die Automorphismen eines algebraischen Funktionenkörpers von Primzahlcharakteristik, *J. Reine Angew. Math.* **179**, 5–15, (1938).
137. Schmidt, Friedrich Karl, Zur arithmetischen Theorie der algebraischen Funktionen. I. Beweis des Riemann–Rochschen Satzes für algebraische Funktionen mit beliebigem Konstantenkörper, *Math. Z.* **41**, 415–438, (1936).
138. Schmidt, Friedrich Karl, Die Wronskische Determinante in beliebigen differenzierbaren Funktionenkörpern, *Math. Z.* **45**, 62–74, (1939).
139. Schmidt, Friedrich Karl, Zur arithmetischen Theorie der algebraischen Funktionen. II: Allgemeine Theorie der Weierstraßpunkte, *Math. Z.* **45**, 75–96, (1939).
140. Serre, Jean–Pierre, *Local Fields*, Graduate Texts in Mathematics **67**, New York–Heidelberg–Berlin: Springer-Verlag, (1979).
141. Springer, George, *Introduction to Riemann Surfaces*, Reading, MA: Addison–Wesley Math. Series. , (1957).
142. Stichtenoth, Henning, Über das Geschlecht eines inseparablen Funktionenkörpers, *Manuscr. Math.* **14**, 173–182, (1974).
143. Stichtenoth, Henning, Algebraische Funktionenkörper einer Variablen mit Teilkörpern von beliebig hohem Geschlecht, *Arch. Math.* **25**, 379–384, (1974).
144. Stichtenoth, Henning, Algebraische Funktionenkörper einer Variablen, Vorlesungen aus dem Fachbereich Mathematik der Universität Essen. Heft 1. Essen: Universität, (1978).
145. Stichtenoth, Henning, Zur Realisierbarkeit endlicher Gruppen als Automorphismengruppen algebraischer Funktionenkörper, *Math. Z.* **187**, 221–225, (1984).
146. Stichtenoth, Henning, s–Erweiterungen algebraischer Funktionenkörper, *Arch. Math.* **43**, 27–31, (1984).
147. Stichtenoth, Hennning, Die Ungleichung von Castelnuovo, *J. Reine Angew. Math.* **348**, 197–202, (1984).
148. Stichtenoth, Henning, Algebraic Function Fields and Codes, Berlin–Heidelberg–New York: Springer-Verlag, Universitext, (1993).
149. Stöhr, Karl–Otto; Viana Paulo, A study of Hasse–Witt matrices by local methods, *Math. Z.* **200**, 397–407, (1989).
150. Subrao, Doré, The p–Rank of Artin–Schreier Curves, *Manuscr. Math.* **16**, 169–193, (1975).
151. Thakur, Dinesh S., *Function Field Arithmetic*, World Scientific, (2004).
152. Tate, John, Genus Change in Inseparable Extensions of Function Fields, *Proc. Am. Math. Soc.* **3**, 400–406, (1952).
153. Valentini, Robert C.; Madan, Manohar L., Weierstrass points in characteristic p, *Math. Ann.* **247**, 123–132, (1980).
154. Valentini, Robert C.; Madan, Manohar L., Automorphism groups of algebraic function fields, *Math. Z.* **176**, 39–52, (1981).
155. Villa-Salvador, Gabriel, *Introducción a la Teoría de las Funciones Algebraicas*, Fondo de Cultura Económica, México, (2003).
156. Washington, Lawrence C., *Introduction to Cyclotomic Fields*, 2nd. edition, Graduate Texts in Mathematics **83**, New York: Springer-Verlag, (1997).
157. Wahington, Lawrence C., *Elliptic Curves. Number Theory and Cryptography*, Chapman & Hall/CRC, (2003).
158. Weil, André, Sur les fonctions algébriques à corps de constantes fini, *C. R. Acad. Sci., Paris* **210**, 592–594 (1940).
159. Weil, André, On the Riemann hypothesis in function-fields, *Proc. Natl. Acad. Sci. USA* **27**, 345–347 (1941).

160. Weil, André, *Variétés abeliennes et courbes algébriques*, Paris: Hermann & Cie, (1948).
161. Weiss, Edwin, *Algebraic Number Theory*, McGraw–Hill, (1963).
162. Weierstrass, Karl Aus einem noch nicht veröffentlichten Briefe an Herrn Professor Schwarz, *Mathematische Werke*. II. Abhandlungen 2., New York, (1967).
163. Yan, Song Y., *Number Theory for Computing*, Springer-Verlag, (2000).
164. Zaldívar, Felipe, *Funciones Algebraicas de una Variable Compleja*, Universidad Autónoma Metropolitana, México, (1995).
165. Zaldívar, Felipe, *Cohomología de Galois de Campos Locales*, Aportaciones Matemáticas **17**, Textos, Sociedad Matemática Mexicana, México, (2001).
166. Zaldívar, Felipe, *Campos Locales*, Universidad Autónoma Metropolitana, México, (2001).
167. Zassenhaus, Hans, On the van der Waerden criterion for the group of an equation, Symbolic and algebraic computation, *EUROSAM '79, Int. Symp., Marseille 1979*, Lect. Notes Comput. Sci. **72**, 95–107, (1979).

Index